LISS2019

Juliang Zhang · Martin Dresner ·
Runtong Zhang · Guowei Hua ·
Xiaopu Shang
Editors

# LISS2019

## Proceedings of the 9th International Conference on Logistics, Informatics and Service Sciences

Springer

*Editors*
Juliang Zhang
School of Economics and Management
Beijing Jiaotong University
Beijing, Beijing, China

Martin Dresner
Logistics, Business and Public Policy
University of Maryland, Baltimore
Baltimore, MD, USA

Runtong Zhang
School of Economics and Management
Beijing Jiaotong University
Beijing, Beijing, China

Guowei Hua
School of Economics and Management
Beijing Jiaotong University
Beijing, Beijing, China

Xiaopu Shang
School of Economics and Management
Beijing Jiaotong University
Beijing, Beijing, China

ISBN 978-981-15-5681-4 ISBN 978-981-15-5682-1 (eBook)
https://doi.org/10.1007/978-981-15-5682-1



This Springer imprint is published by the registered company Springer Nature Singapore Pte Ltd.
The registered company address is: 152 Beach Road, #21-01/04 Gateway East, Singapore 189721, Singapore

# LISS 2019 Committees

## Editorial Committee

Juliang Zhang
Martin Dresner
Runtong Zhang
Guowei Hua
Xiaopu Shang
Anqiang Huang
Daqing Gong

## In Cooperation with

University of Maryland, USA
Informatics Research Centre, University of Reading, UK

## Hosted by

IEEE SMC Technical Committee on Logistics Informatics and Industrial Security Systems
The International Center for Informatics Research, Beijing Jiaotong University, China
National Academy of Economic Security, Beijing Jiaotong University, China
School of Economics and Management, Beijing Jiaotong University, China
Beijing Logistics Informatics Research Base, China

## Organizing Committee

### Honorary Chairmen

James M. Tien, Academician of America National Academy of Engineering, Past President of IEEE SMC, IEEE Fellow, University of Miami, USA
Shoubo Xu, Academician of China National Academy of Engineering, Beijing Jiaotong University, China
C. L. Philip Chen, Past President of IEEE SMC, IEEE Fellow, AAAS Fellow, University of Macau, China

### General Chairs

Runtong Zhang, Beijing Jiaotong University, China
Zhongliang Guan, Beijing Jiaotong University, China
C. L. Philip Chen, University of Macau, China
Kecheng Liu, University of Reading, UK
Ritu Agarwal, University of Maryland, USA
Gang-Len Chang, University of Maryland, USA

### Program Chairs

Zhenji Zhang, Beijing Jiaotong University, China
Martin Dresner, University of Maryland, USA
Juliang Zhang, Beijing Jiaotong University, China
Xianliang Shi, Beijing Jiaotong University, China

### Financial Chair

Shifeng Liu, Beijing Jiaotong University, China

### Secretariats

Dan Chang, Beijing Jiaotong University, China
Yisong Li, Beijing Jiaotong University, China

### Publication Chairs

Guowei Hua, Beijing Jiaotong University, China
Xiaopu Shang, Beijing Jiaotong University, China
Honglu Liu, Beijing Jiaotong University, China

### Organization Chairs

Xiaochun Lu, Beijing Jiaotong University, China
Chengri Ding, University of Maryland, USA

### Special Session/Workshop Chair

Hongjie Lan, Beijing Jiaotong University, China

### Publicity Chairs

Jie Gao, Beijing Jiaotong University, China

Xiaomin Zhu, Beijing Jiaotong University, China
Daqing Gong, Beijing Jiaotong University, China

### Web Master

Zikui Lin, Beijing Jiaotong University, China

## International Steering Committee

### Chairs

James M. Tien, University of Miami, USA
Shoubo Xu, Beijing Jiaotong University, China
C. L. Philip Chen, University of Macau, China

### Members

Francisco Vallverdu Bayes, Universitat Politècnica de Catalunya BarcelonaTech, Spain
Guoqing Chen, Tsinghua University, China
Jian Chen, Tsinghua University, China
Yu Chen, Renmin University of China, China
Zhixiong Chen, Mercy College, USA
T. C. Edwin Cheng, The Hong Kong Polytechnic University, China
Waiman Cheung, The Chinese University of Hong Kong, China
Jae-Sun Choi, Korea Maritime Institute, Korea
Jinhong Cui, University of International Business and Economics, China
Shicheng D, Royal Institute of Technology-KTH, Sweden
Dingyi Dai, China Federation of Logistics & Purchasing, China
Xuedong Gao, University of Science and Technology Beijing, China
David Gonzalez-Prieto, Universitat Politècnica de Catalunya BarcelonaTech, Spain
Liming He, China Federation of Logistics & Purchasing, China
Harald M. Hjelle, Molde University College, Norway
Joachim Juhn, Anglia Ruskin University, UK
Kim Kap-Hwan, Pusan National University, Korea
Harold Krikke, Tilburg University, The Netherlands
Erwin van der Laan, RSM Erasmus University, The Netherlands
Der-Horng Lee, The National University of Singapore, Singapore
Dong Li, University of Liverpool, UK
Menggang Li, China Center for Industrial Security Research, China
Cheng-chang Lin, National Cheng Kung University, Taipei
Kecheng Liu, University of Reading, UK
Oriol Lordan, Universitat Politècnica de Catalunya BarcelonaTech, Spain
Qining Mai, The University of Hong Kong, China
Theo Notteboom, University of Antwerp, Belgium
Yannis A. Phillis, The Technical University of Crete, Greece
Robin Qiu, Pennsylvania State University, USA

Jurgita Raudeliuniene, Vilnius Gediminas Technical University, Lithuania
Linda Rosenman, Victoria University, Australia
Ying Ren, Beijing Jiaotong University, China
Kurosu Seiji, Waseda University, Japan
Zuojun (Max) Shen, University of California, USA
Jiuh-Biing Sheu, National Chiao Tung University, Taipei
Carlos Sicilia, Universitat Politècnica de Catalunya BarcelonaTech, Spain
Pep Simo, Universitat Politècnica de Catalunya BarcelonaTech, Spain
Dong-Wook Song, Edinburgh Napier University, UK
Jelena Stankeviciene, Vilnius Gediminas Technical University, Lithuania
A Min Tjoa, Vienna University of Technology, Austria
Jaume Valls-Pasola, Universitat Politècnica de Catalunya BarcelonaTech, Spain
Lida Xu, Old Dominion University, USA
David C. Yen, Miami University, USA

## International Program Committee

Muna Alhammad, University of Reading, UK
Bernhard Bauer, University of Augsburg, Germany
Michael Bell, The University of Sydney, Australia
Ida Bifulco, University of Salerno, Italy
Christoph Bruns, Fraunhofer Institute for Material Flow and Logistics, Germany
Luis M. Camarinha-Matos, New University of Lisbon, Portugal
Mei Cao, University of Wisconsin – Superior, USA
Maiga Chang, Athabasca University, Canada
Sohail Chaudhry, Villanova University, USA
Gong Cheng, University of Reading, UK
Alessandro Creazza, University of Hull, UK
Lu Cui, Hohai University, China
Kamil Dimililer, Near East University, Cyprus
Chunxiao Fan, Beijing University of Posts and Telecommunications, China
Zhi Fang, Beijing Institute of Technology, China
Vicenc Fernandez, Universitat Politècnica de Catalunya BarcelonaTech, Spain
Juan Flores, Universidad Michoacana de San Nicolás de Hidalgo, Mexico
Alexander Gelbukh, National Polytechnic Institute, Mexico
Joseph Giampapa, Carnegie Mellon University, USA
Wladyslaw Homenda, Warsaw University of Technology, Poland
Wei-Chiang Hong, Oriental Institute of Technology, Taipei
Alexander Ivannikov, State Research Institute for Information Technologies and Telecommunications, Russian Federation
Taihoon Kim, Sungshin Women's University, Korea
Rob Kusters, Eindhoven University of Technology, The Netherlands
Der-Horng Lee, The National University of Singapore, Singapore
Kauko Leiviskä, University of Oulu, Finland
Zhi-Chun Li, Huazhong University of Science and Technology, China

# Contents

# The Risk Sharing of Inter-city Railway PPP Project in China

**Yuyu Zhang and Shengyue Hao**

**Abstract** In recent years, the PPP model has been widely used in China's inter-city railway construction, and the risk sharing of inter-city railway PPP projects has become the focus of research and practice. Based on stochastic cooperative game theory, this paper optimizes the Shapley value according to the characteristics of the construction pattern of the inter-city railway PPP projects in China, and constructs the risk sharing model applicable to the inter-city railway PPP projects in China, this model can determine the optimal risk allocation ratio of public services and private sector in order to achieve optimal revenue. In addition, relevant suggestions on risk management of intercity railway PPP projects are put forward from three dimensions: suggestions for national, suggestions for public sector level and suggestions for private sector level, which has certain reference significance for risk allocation management of inter-city railway PPP projects in China.

**Keywords** Risk sharing · Stochastic cooperative game · Intercity railway · Public-Private-Partnership (PPP)

---

Funding source of this paper-key technology research of Beijing-Shanghai high-speed railway standard demonstration line-key technology research of Beijing-Shanghai high-speed railway brand construction and operation management demonstration.

---

Y. Zhang (✉) · S. Hao
School of Economics and Management, Beijing Jiaotong University, Beijing, China
e-mail: 17120644@bjtu.edu.cn

S. Hao
e-mail: haoshyue@bjtu.edu.cn



J. Zhang et al. (eds.), *LISS2019*,
https://doi.org/10.1007/978-981-15-5682-1_1

# 1 Introduction

## *1.1 Research Background*

In recent years, in order to meet the needs of passengers in large and medium-sized cities, the government has made great efforts to promote the construction of intercity railway, hoping to meet the growing demand for short distance travel. Since the 18th national congress of the CPC, with the support of the government, the intercity railway construction market has been steadily promoted. According to the 13th five-year plan for railway development, the construction of intercity and municipal (suburban) railways should be promoted in an orderly manner. By the end of the 13th five-year plan, the scale of intercity and municipal (suburban) railways should reach about 2,000 km. The 13th five-year plan on the development of modern comprehensive transportation system once again emphasizes the promotion of intercity railway construction, pointing out that the investment of intercity railway construction will be further increased. In particular, the construction progress of intercity railway network will be promoted in Beijing-Tianjin-Hebei Region, Yangtze river delta, pearl river delta and other urban agglomerations. At present, China's "four vertical and four horizontal" high-speed railway network has been built and put into operation ahead of schedule, and the major heavy and large trunk lines have been basically completed. However, the "capillary lines" connecting these heavy and large trunk lines, namely the intercity high-speed railway network of local governments, still need to be improved. Therefore, the intercity railway market contains huge development opportunities.

Due to the increasing construction debt pressure of local governments, the government actively encourages private capital to participate in infrastructure construction. PPP mode has been rapidly developed in the field of infrastructure construction in China. In particular, the construction of intercity railway has the characteristics of huge investment and long payback period, which challenges the ability of local government and general railway to raise funds. In 2015, the government issued "on further encourage and expand the private sector invested in the construction of railway implementation opinion", it indicates that the government encourage and attract private capital into the inter-city railway. On the one hand, this can reduce the pressure of public sector fund, on the other hand, the reasonable interests of the private sector can also be through cooperation [1]. However, infrastructure projects financed by PPP often have the characteristics of huge investment, long contract period and imperfect contracts, which makes the participants of PPP projects face more risk [2]. Furthermore, an unreasonable risk sharing plan will lead to the increase of project cost, stagnation of cooperation and other adverse consequences [3]. Therefore, a reasonable risk sharing plan is of great importance to the public sector and private sector.

## 1.2 Literature Review

Through a review of a large number of literatures, it can be seen that from 1998 to now, research on PPP mode has been gradually heating up [4]. Risk sharing has always been one of the hot issues in PPP project research. A number of scholars at home and abroad have conducted relevant studies on risk allocation in the field of infrastructure construction. In recent years, PPP financing mode has increasingly become a hot topic in the academic circle, and some scholars have made further analysis on the risk sharing in infrastructure construction projects of PPP financing mode. According to the survey by Bing and Akintoye [5], the macro and micro risks of PPP projects in the UK are shared by the public sector or both, while the rest risks are mostly borne by the private sector. Jin [6] applied fuzzy neural network method to predict and evaluate the advantages and disadvantages of risk sharing strategies. Li [7] applied Bayesian posterior probability to deduce the qualitative risk sharing model, which is used to distribute the risk responsibility between the government and the private sector. Liu et al. [8] believed that the total project cost was closely related to the risks undertaken by the private sector, and the risk sharing ratio between the government and the private sector should be changed within a certain range, so as to achieve the highest efficiency and lowest cost of the project. Zhou [9] regarded the negotiation process of risk sharing between government and private sector in PPP projects as a bargaining game. He believed that the earlier the participants acted or the higher the discount rate, the more dominant they were in the game and the less risk they took. Liu et al. [10] believed that the status of the public sector and the private sector was unequal. They applied bargaining game to conclude that the risk sharing ratio of PPP projects was related to the negotiation loss coefficient of both parties. The more the participants learn about each other's information, the lower the negotiation loss. Chu [11] obtained the primary risk allocation of the public and private sectors according to the interval fuzzy shapely value, and then analyzed the impact of different factors on risk sharing by using ANP to obtain a more reasonable share ratio. He et al. [12] used the stochastic cooperative game model to deduce the best cooperative game model of risk sharing between the government and the private sector in PPP projects. Liu et al. [13] constructed the best risk sharing model between government departments and private enterprises based on the random cooperative game theory for PPP projects in the field of water conservancy construction in China. Li et al. [14] discussed the risk allocation of both parties involved in PPP projects under incomplete information symmetry. Zhao et al. [15] divided risk bearing subjects into single subjects and co-subjects for PPP mode intercity railway projects, and deduced a reward and punishment function to encourage the government and private enterprises to strengthen risk control.

### 1.3 Research Significance

Although domestic and foreign scholars have carried out in-depth and detailed studies on the risk sharing of PPP projects. There are relatively few research results on the risk sharing of China's intercity railway PPP projects, and most of the studies believe that the two parties involved are opposites, failing to consider that the two parties can reduce the project risk loss by transferring risks through cooperation. Based on the theory of stochastic cooperative game, this paper holds that risks can be transferred between the two parties involved, and improves the model by combining the risk utility function and the actual situation of the participants. Then, it establishes the risk sharing model for the PPP project of intercity railway, and determines the optimal risk sharing ratio between the public sector and private sector. In addition, according to the research conclusion and combined with the difficulties of risk sharing implementation in the field of intercity railway PPP projects in China, relevant Suggestions are put forward to provide certain theoretical basis and reference significance for the risk management of the participants of intercity railway PPP projects, which is conducive to the healthy development of intercity railway PPP projects in China.

## 2 Random Cooperative Game Model

### 2.1 Model Hypothesis

Cooperative game theory has been widely used in the distribution of costs and benefits of political and commercial cooperation. However, it can be affected by some uncertain factors in practical projects, costs and benefits are often not fixed. Hence, the traditional cooperative game theory is no longer applicable. Charnes and Granot [16] put forward the stochastic cooperative game theory with stochastic payment characteristics, and then Suijs and Borm [17] further improved the stochastic cooperative game model.

The stochastic cooperative game of intercity railway PPP project is represented by (1), where "$N$" represents the finite membership set of all participating members of the intercity railway PPP project, "$L$" represents a certain alliance, "θ" represents the alliance set composed of finite members, and "$R(L)$" represents the alliance $L$'s non-negative random income, "$i$" indicates a participating member, "$p_i \in L$" indicates that member $i$ allocates coefficients in alliance $L$, and distribution income is that $p_i * R(L)$ indicates the income earned by participating member $i$ in alliance $L$. "$A$" and "$B$" represent two different alliance structures. "$A \succ_i B$" means that $i$ members are more inclined to choose class $A$ alliances than class $B$ alliances; "A $\approx_i B$" indicates that $i$ members choose class $B$ alliances is tantamount to selecting class $A$ alliances. The PPP project of the intercity railway satisfies the premise of the stochastic cooperative game, as follows:

Hypothesis 1: all participants are rational people.

Hypothesis 2: income distribution of intercity railway PPP project alliance is effective distribution, $\sum_{i \in N} p_i = 1$.

Hypothesis 3: The public sector and private sector in the intercity railway PPP project are risk averse, and its expected utility is represented by $E(u_i(t))$ (the random benefit brought by the risk indicated by $W_i$, and $W_i$ is greater than $E(u_i(t))$, *ie* participant $i$ would rather give up random gain $W_i$, and get the expected utility $E(u_i(t))$. The risk expectation utility function is $u_i(t) = e^{-r_i t}$, the risk expectation utility function is a decreasing function, "$r_i$" represents the degree of risk aversion of participating members $i$, and the greater the $r_i$, the more disgusted the member is towards risk.

Hypothesis 4: Any participating member $i$ has a desired preference, that is, member $i$ tends to select a federation with a large expected return, where the expected return is expressed by the expected utility. That is, if and only if $E(u_i(A)) \geq E(u_i(B))$, then $A \succ_i B$.

$$\tau = (N, \{R(L)\}_{L \in \theta}, \{A \succ_i B\}_{i \in N} \tag{1}$$

## 2.2 *Shapley Value of Public Sector and Private Sector*

The Shapley value was proposed by Shapley in 1953 and was used to solve the distribution of income for political and business alliances. Under the premise that the alliance participants are rational people, Shapley believes that the income should match the contribution, so the Shapley value represents the expected value of the marginal contribution to the alliance. The Shapley value represents "Eq. 2", where $|L|$ represents the number of participants participating in Alliance $L$, and the coefficient $\omega(/L/)$ is the weighting factor, $R(L)$ is the overall benefit of Union $L$, and $R(L/i)$ is the Union $L$ exclusion participation. The total return after $i$. $R(L) - R(L/i)$ is the marginal contribution of participant i to cooperative alliance $L$, and the dictionary order of alliances including participant $i$ can be $n!$, and the order of $L/i$ and $\theta/L$ is $(n - |L|)!(|L| - 1)!$, the probability that the alliance $L$ appears is shown in "Eq. (3)".

$$\varphi_i = \sum_{L \in \theta} \omega(|L|)[R(L) - R(L/i)] \tag{2}$$

$$\omega(|L|) = \frac{(n - |L|)!(|L| - 1)!}{n!} \tag{3}$$

$$\varphi_i = \frac{1}{n!} \left[ \sum_{e \prod N} \varepsilon_i^e(N) \right] R(N) \tag{4}$$

The theorem proposed by Timmer [17]: there is a strictly monotonically increasing continuous function $f : X \to X^v$, and if and only if $x_1 \geq x_2$, $f(x_1) * R(A) \succ_i f(x_2) * R(B)$ is obtained (where $X \to X^v$ represents the mapping of $X$ to $X^v$, and $x_1, x_2 \in X, f(x_1), f(x_2) \in X^v$). At this time, the participating members $i$ tend to choose a higher-yielding alliance. There is a transfer function $(h(\alpha, \beta) : t \to T)$ between the two random federations ($A$ and $B$), such that $A \approx_i h(A, B)B$. That is, the existence of a conversion function can make the $A$ alliance the same as the $B$ alliance. Let $f(x_1 = x/E(u_i R(A))$, therefore, convert the coefficient $h_i(A, B) = f(x_2)/f(x_1) = E(u_i R(A)/E(u_i R(B)$, that is, by increasing the conversion coefficient, the expected utility of the two alliances $A$ and $B$ is the same. The major alliance of the public sector and private sector is denoted by $N$. By definition, the number of participants in the alliance is n, that is $|N| = n$, $\omega(|L|) = 1/n!$. And the Shapley value of the stochastic cooperative game of the intercity railway PPP project can be further derived:

Where $e \prod N$ represents a dictionary order of major league $N$, $\varepsilon_i^e(N) = h_i\left(\delta_i^e, R(N)\right)$, and $\delta_i^e$ represents the marginal contribution of participating member $i$ in the case of dictionary sorting $e$ (see Eq. (5) for detail).

$$\delta_i^e = \left[1 - \sum_{k=1}^{i-1} h_k\left(\delta_k^e, R(L^e)\right)R(L^e)\right] \tag{5}$$

$$L^e = (1, 2, 3 \cdots i)$$

The PPP project of intercity railway involves the participation of public sector and private sector. The grand alliance of public sector and private sector is $N$, and the number of participants is 2. According to (4), the Shapley values of public sector and private sector are (6) and (7):

$$x_1 = \frac{1}{2}[h_1(R(1), R(N)) + (1 - h_2(R(2), R(N)))]R(N) \tag{6}$$

$$x_2 = \frac{1}{2}[h_2(R(2), R(N)) + (1 - h_1(R(1), R(N)))]R(N) \tag{7}$$

# 3 Optimal Risk Sharing Model

## 3.1 Model Construction

Suji et al. [16] concluded that there is a deterministic equivalence benefit $(\pi_i)$ in the stochastic cooperative game model represented by (1). This deterministic equivalence benefit is no different from the expected utility of participant $i$. Here we think that $\pi_i$ can replace Expected utility $E(u(x_i))$, then $\pi_i = E(x_i) - \frac{1}{2} r_i Var(x_i)$; $E(x_i)$

represents the expectation of $x_i$, $Var(x_i)$ represents the variance of $x_i$, and $\frac{1}{2}r_i Var(x_i)$ represents the risk premium of participant *i*. In the intercity railway PPP project, the occurrence of risk will increase the project cost, and the corresponding project income will decrease, so the risk can be regarded as the negative influencing factor of the participant's income. Determine the conversion factor by determining the pricing function:

$$h_i(R(x_1), R(x_2)) = \frac{E(x_1) - \frac{1}{2}r_i Var(x_1)}{E(x_2) - \frac{1}{2}r_i Var(x_2)} \tag{8}$$

Because participants have different risk attitudes, participants will bear different proportions of risks according to their risk aversion, that is, each participant will have a risk transfer, so that the total project revenue will be maximized through cooperation. Therefore, the Shapley value of the intercity railway PPP project is further optimized here, the average weight $\left(\frac{1}{n!}\right)$ in (4) is replaced by the coefficient vector $p = \{p_1, p_2, \cdots, p_n\}$, and $\sum_{i \in N} p_i = 1$, the Shapley value of the optimized public sector and private sector:

$$x_1 = [p_1 h_1(R(1), R(N)) + p_2(1 - h_2(R(2), R(N)))]R(N) \tag{9}$$

$$x_2 = [p_2 h_2(R(2), R(N)) + p_1(1 - h_1(R(1), R(N)))]R(N) \tag{10}$$

(note: For the sake of convenience, use "$h_1$" instead of "$h_1(R(1), R(N))$", use "$h_2$" instead of "$h_2(R(2), R(N)))$".

Since Shapley value is a function of, it can be seen from the deterministic equivalent function that the risk premium is a function of, this paper maximizes the benefits by determining the optimal risk sharing proportion value of each participant, which can be expressed as:

$$p^* = arg\left[\sum_{i=1}^{n} \pi_i(x_i) \rightarrow max\right] \tag{11}$$

Equation (5) and (6) can be substituted into: $E(x_1) + E(x_2) = [p_1 h_1 + p_2(1 - h_2)]E(N) + [p_2 h_2 + p_1(1 - h_1)]E(N)) = E(N)$, since the sum of $E(x_1)$ and $E(x_2)$ is a fixed constant value, it can be converted into the solution of the overall deterministic equivalent optimization to the solution of the minimum overall risk premium. The optimization model is expressed as follows:

$$minf = \frac{1}{2}r_1 Var(x_1) + \frac{1}{2}r_2 Var(x_2)$$

$$p_1 + p_2 = 1 \tag{12}$$

## 3.2 Model Solution

Use Lagrange equation to solve $p_1^*$, $p_2^*$, the equation is $L = \frac{1}{2} r_1 Var(x_1) + \frac{1}{2} r_2 Var(x_2) + \lambda(1 - p_1 - p_2)$, and the variance relationship $D(aX) = a^2 D(X)$ and "(6) and (7)" can be obtained [18]:

$$L = \frac{1}{2} r_1 \left[ (p_1 h_1 + p_2(1 - h_2))^2 \sigma^2(N) \right] + \frac{1}{2} r_2 \left[ (p_2 h_2 + p_1(1 - h_1))^2 \sigma^2(N) \right] + \lambda(1 - p_1 - p_2) \tag{13}$$

(note: where $\sigma^2(N)$ denotes the variance of $N$ in major league).

For the Lagrangian function of the above construction, a first-order partial derivative is obtained for $p_1$, $p_2$, $\lambda$, respectively, and the first-order partial derivative is equal to zero, and the following three equations are obtained:

$$\frac{\partial L}{\partial p_1} = \left[ r_1 p_1 h_1^2 + r_2 p_1 (1 - h_1)^2 + r_1 p_2 h_1 (1 - h_2) + r_2 p_2 h_2 (1 - h_1) \right] \sigma^2(N) - \lambda = 0$$

$$\frac{\partial L}{\partial p_2} = \left[ r_2 p_2 h_2^2 + r_1 p_2 (1 - h_2)^2 + r_1 p_1 h_2 (1 - h_1) + r_2 p_1 h_2 (1 - h_1) \right] \sigma^2(N) - \lambda = 0$$

$$\frac{\partial L}{\partial \lambda} = 1 - p_1 - p_2 = 0 \tag{14}$$

The solution of the equation is $p_1^* = \frac{a_2 - a_3}{a_1 + a_2 - 2a_3}$, $p_2^* = \frac{a_1 - a_3}{a_1 + a_2 - 2a_3}$, $\lambda = \frac{a_1 a_2 - a_3^2}{a_1 + a_2 - 2a_3}$. ($a_1 = \left[ r_1 h_1^2 + r_1 (1 - h_1)^2 \right] \sigma^2(N)$, $a_2 = \left[ r_2 h_2^2 + r_1 (1 - h_2)^2 \right] \sigma^2(N)$, $a_3 = [r_1 h_1 (1 - h_2) + r_2 h_2 (1 - h_1)^2] \sigma^2(N)$, and, $a_1 a_2 \neq \frac{1}{2}$)

In the actual PPP infrastructure projects, the public sector and private sector are not equally risky. The two sides often coordinate and share the risks through negotiation. The final results of this model $p_1^*$ and $p_2^*$ determine the optimal sharing ratio of the two parties and realize the income. Maximize and minimize risk. The $(p_1*, p_2*)^T$ threshold is $(0, 1)^T$ or $(1, 0)^T$, that is the risk is borne by one party alone.

# 4 Case Analysis

In the *13th five-year plan of rail transit development of A province* released by A province in 2,016, A province proposed to build an intercity railway from a city to b city. Through calculation, it is known that the investment of this intercity railway is 12.6 billion yuan. If the traditional financing mode is adopted and the construction is based on the China Railway Crop and local government appropriation, it can be calculated by relevant departments that the payback period of the investment of this intercity railway is 30 years. By the end of the payback period, the income of the public sector follows the normal distribution $R(1) \sim N(12.6, 2.0)$. Because of public sector's high debt ratio, the project construction through bank loans as a way to raise funds need. However, this will further increase public sector debt. The public sector will need to bear all additional interest on loans. Therefore, the public sector want to take the PPP mode and introduce private sector (construction enterprise M) to share the construction costs of the inter-city railway. According to the calculation, when construction enterprise M does not participate in the investment, the income follows the normal distribution $R(2) \sim N(8.8, 1.2)$; However, the PPP financial commitment of the public sector will be the increasing income of construction enterprise M, so it is willing to choose cooperation and uses PPP mode to undertake the construction of this intercity railway. The calculation results of relevant departments show that the total project income is greater than the sum of the two incomes in the absence of cooperation when PPP mode is adopted, and the total cooperative income obey a normal distribution: $R(N) \sim N(25.2, 4.8)$. By comparison, the public sector is more risk resistant than the construction enterprise M due to its relatively abundant capital. After analysis and calculation, the risk aversion coefficients of the public sector is 1.2 ($r_1 = 1.2$), and the risk aversion coefficients of the private sector is 1.5 (($r_2 = 1.5$)). Both are risk averse.

Then, the above risk sharing model is applied for calculation and analysis, the optimal risk sharing ratio between the two parties can be obtained: $(p_1*, p_2*)^T = (0.60, 0.40)^T$. It can be shown that the share ratio of the public sector is higher than that of private sector. The risk premium is obtained under the optimal sharing ratio: $(0.90, 0.70)^T$, while another risk premium is obtained according to the not optimized Shapley value: $(1.00, 0.71)^T$. It can be shown that the risk premium of both sides decreases under the optimal sharing ratio, and the total risk premium also decreases. The deterministic equivalent is $(13.36, 10.13)^T$, when the risks are equally shared, while the deterministic equivalent is $(13.16, 10.43)^T$ by applying the optimal allocation proportion model. It can be seen that the total certainty equivalent increased by 10 million RMB. Through the risk transfer, the deterministic equivalent of public sector drops, but the deterministic equivalent of private sector increases. Private sector needs to pay the public sector a certain amount of risk sharing compensation cost, so that the public sector is willing to bear more risks. It can be seen that as long as the private sector is willing to pay the fee of 20 million, the public sector will bear the risk of 10%, while the private sector gains 30 million.

# 5 Suggestions

It can be seen from the above research results that the risk sharing ratio of intercity railway PPP project is optimal, which minimizes the overall risk cost and maximizes the total benefit value. However, due to the opportunistic behaviors of participants and other reasons, the two parties do not share risks in accordance with the optimal proportion, so the optimal effect of risk sharing cannot be achieved [19, 20]. Therefore, relevant suggestions are proposed from the perspectives of the country, private sector and public sector to urge participants to share risks in an optimal proportion and achieve the maximum overall benefits.

## 5.1 Suggestions for Country

The state should give full play to its macro-control role, and encourage the public sector and private sector involved in the PPP project of intercity railway to share risks reasonably through law. The state should pay attention to legislation and law enforcement to carry out macro-control on risk sharing of PPP project of inter-city railway through law. Legislation is the premise. After the improvement of relevant laws, laws should be taken as the starting point to regulate the risk management behaviors of both parties involved.

(1) The state should improve the legal system and implement special laws. The premise of promoting the public sector and private sector that participate in the PPP project of intercity railway to share risks reasonably through law is that there is a law to follow. In recent years, PPP mode has been developing rapidly in China, but the legislation development process of PPP project is relatively slow, which is also a weak link in the development process of PPP mode in China. Although, the state has been promoting the legislation of PPP projects in infrastructure and public service sectors, and has made some achievements. However, the law is still not unified, contradictory, incomplete, difficult to implement and other problems. For example, the definition of PPP project has not been unified so far; The definition of private sector is inconsistent with *the general guide to contracts for government-private partnership projects issued* by the national development and *Reform commission and the PPP project contract operation guide* issued by the ministry of finance. The latter explicitly excludes issues such as local government financing platforms and state-owned enterprises. Therefore, it is suggested to improve the legal framework, eliminate contradictions and strictly define important concepts. In addition, the state should further refine the PPP special law and promote the implementation of the special law, so that the development of PPP projects has laws to follow.

(2) The state should take law as the starting point to regulate the reasonable risks of participants. At present, China's PPP projects have a series of problems such as unreasonable risk sharing and government dishonesty. On the basis of establishing and improving the PPP project law, the state should take the law as the starting point, regulate the participation behavior of the government and private sector, reduce the abnormal termination of the project, and promote the healthy development of the PPP project. For example, on the one hand, legislation is passed to require the government to reasonably share risks and improve the government's credit. On the other hand, in the project in-depth verification of private sector's economic strength, according to the actual situation of private sector, reasonable allocation of risk.

## 5.2 *Suggestions for Private Sector*

Private sector should attach importance to risk management and establish a scientific and reasonable risk sharing mechanism. Risk sharing has always been a hot topic in academic circles, and many scholars agree that reasonable risk sharing is the key to the success of PPP projects [21]. Participants of intercity railway PPP project should attach importance to risk management, deeply analyze the risks existing in the forecast project, do the risk prediction work well in the early stage, and on this basis, formulate complete risk response strategies, including a series of measures such as risk retention, risk transfer, risk avoidance and risk reduction. In addition, private sector should communicate actively with the public sector on the risk sharing of the PPP project of intercity railway for many times, and formulated the risk sharing plan with the public sector scientifically, reasonably and fairly according to the construction characteristics of the project.

## 5.3 *Suggestions for Public Sector*

Public sector should bear risks reasonably and attach importance to government credit. For the problem of risk sharing, the government should reasonably bear some risks to ensure the fairness of the risk sharing mechanism. In PPP projects, government dishonesty is a common risk, which is one of the important reasons for the abnormal termination of PPP projects. There are many cases of project failure caused by government dishonesty at home and abroad. For example, as the earliest London subway in operation in the world, the old lines and equipment were updated by PPP mode, and the enterprise went bankrupt due to the government's withdrawal of 10% subsidy promise. The reason why the government breaks its promise is that the government takes too much risk because of the commitments it has made. In order to reduce the probability of abnormal termination of the project and promote the success of the project, the government should first ensure the rationality of taking risks and not increase risks by over-promising. Secondly, public sector should attach importance to credit and insist on fulfilling commitments.

## 6 Conclusion

It can be revealed from the relevant plans issued in recent years that the Chinese government will promote the construction of intercity railway in the next few years, and the government encourages private sector to participate in the construction of intercity railway. This article, the following results of inter-city railway PPP project risk allocation problem in China, provides a new way of thinking: Firstly, the optimal risk allocation model is got. The calculating formula of the optimal proportion is put forward, it helps to reduce the risk of loss of the inter-city railway PPP projects in China and improves the profit. Secondly, the optimized Shapely value is obtained through this model, it can be applied to formulate the benefit distribution scheme of in the PPP project of intercity railway in the future.

## References


1. Zhao, B. (2014). *Research on railway financing risk sharing based on PPP mode*. Ph.D. dissertation, Southwest Jiaotong University, China.
2. Woodward, D. G. (1995). Use of sensitivity analysis in build-own-operate-transfer project evaluation. *Journal of Project Management, 13*(4), 239–246.
3. Guo, Y. Y. (2017). *Risk mechanism analysis of PPP project based on random cooperative game*. Ph.D. dissertation, Qingdao University, China.
4. Zheng, C. B., Feng, J. C., Xue, S., & Lu, Q. Q. (2018). An empirical study on the impact of critical success factors on performance in PPP. *Journal of soft science, 31,* 124–129.
5. Bing, L., Akintoye, A., Edwards, P. J., et al. (2005). The allocation of Risk in PPP/PFI construction projects in the UK. *International Journal of Project Management, 23*(4), 25–35.
6. Jin, X. H. (2010). A neuro - fuzzy decision support system for efficient risk allocation in public-private partnership infrastructure projects. *Journal of Computing in Civil Engineering, 4,* 223–235.
7. Li, B. & Ren, Z. M. (2009). Bayesian technique framework for allocating demand risk between the public and private sector in PPP projects. In *6th International Conference on Service Systems and Service Management* (pp. 837–841).
8. Liu, S. P. & Wang, S. Q. (2009). Try to discuss the risk allocation principle and frame of PPP project. *Building Economy*, 837–841.
9. Zhou, X. (2009). Risk - sharing game analysis of PPP project financing model. *Cooperative Economy and Technology, 14,* 78–79.
10. Liu, S. K., & Li, G. (2018). Research on the optimal risk sharing mechanism of PPP projects. *Cooperative Economy and Technology, 28,* 48–51.
11. Chu, X. L. (2018). Research on risk sharing method of PPP mode based on interval fuzzy Shapley value. *Value Engineering, 36,* 106–107.

12. He, T., & Zhao, G. J. (2011). PPP project risk sharing based on stochastic cooperative game model. *Systems Engineering, 4,* 88–91.
13. Liu, B. B., Chen, Y. H., & Yi, J. C. (2016). Research on sharing of water conservancy PPP projects based on stochastic cooperative game. *Value Engineering, 33,* 26–28.
14. Li, H., & Wang, H. Q. (2016). A game model of public - private sharing risk allocation in PPP projects with incomplete information. *Value Engineering, 35*(14), 112–115.
15. Zhao, B., & Shuai, B. (2017). Risk sharing model of intercity railway project based on PPP model. *Journal of East China Jiaotong University, 5,* 65–71.
16. Granot, C. (1973). Prior solutions: extensions of convex nucleus solutions to chance constrained games. In *Proceedings of the Computer Science and Statistics Seventh Symposium at Iowa State University* (pp. 323–332).
17. Suijs, J. P. M., Borm, P. E. M., et al. (1995). Cooperative games with stochastic payoffs. *European Journal of Operational Research, 113*(1), 193–205.
18. Timmer, J., Borm, P., & Tijs, S. (2003). On three shapely-like solutions for cooperative games with random payoffs. *International Journal of Game Theory, 32,* 595–613.
19. He, J. (2018). Some thoughts on intercity railway investment. *Academic BBS, 13,* 158.
20. Zhou, Y. & Ronghua, H. (2018). Research on introducing private sector into intercity railway project based on PPP model. *Financial and Economic*, 158–170.
21. Zhang, W. K., Yang, Y. H., & Wang, Y. Q. (2010). Key influencing factors of public-private partnership performance - based on empirical studies in several transition countries. *Journal of public administration, 03,* 103–112.

# The Block-Wise Adaptive Modulation Technique for Frequency-Selective Fading Channels

**Yupeng Bao and Yanfei Lu**

**Abstract** Orthogonal frequency division multiplexing (OFDM) technology has the advantages of high frequency band utilization and resistance to frequency selective fading. However, when using high-order modulation for symbol transmission, interference and fading in the channel can significantly degrade system performance. This paper designs a subcarrier-block based adaptive modulation technique in OFDM Systems, and it uses the data structure of the 802.16d protocol to transmit the result of the link state analysis at the receiving end to the transmitting end through the feedback link for adjusting the subcarrier block used for the next transmission. The simulation results show that the adaptive OFDM system based on the subcarrier block has a lower bit error rate than the ordinary OFDM system in the frequency selective fading channel. At the same time, the system performance is better than that of the ordinary OFDM system when moving at high speed and increasing the Doppler shift.

**Keywords** Orthogonal frequency division multiplexing · Subcarrier block · Frequency selective fading · Adaptive modulation

## 1 Introduction

Adaptive technology was first proposed in the 1960s, but was ignored at the time due to backward hardware conditions. In the 1990s, communications systems and digital signal processing technologies developed rapidly, and the application of Orthogonal frequency division multiplexing (OFDM) technology became increasingly widespread. Adaptive algorithms were gradually gaining importance [1]. The fundamental idea of adaptive technology is to make the transmission scheme match

Y. Bao · Y. Lu (✉)
School of Communication and Information, Beijing Jiaotong University, Beijing, China
e-mail: yflu@bjtu.edu.cn

Y. Bao
e-mail: 17120040@bjtu.edu.cn



J. Zhang et al. (eds.), *LISS2019*,
https://doi.org/10.1007/978-981-15-5682-1_2

the channel as much as possible by properly adjusting and utilizing various parameters and resources when the transmitter knows some form of channel information [2].

In recent years, mobile communications have been continuously developed, while wireless services have been continuously increased, and high-speed and reliable data transmission has gradually become the goal and requirement of wireless communications [3]. The adaptive modulation technique dynamically allocates different transmission information bits and transmission power to subcarriers according to the instantaneous state information of the channel. When the channel state is good, the subcarrier adopts high-order modulation to increase the throughput of the system. Otherwise, the subcarrier will adopt low-order modulation to ensure system reliability [4]. In practice, the corresponding relationship between SNR and BER or SNR and system throughput will be established. According to the estimation of channel transmission quality, the optimal modulation and coding scheme will be selected [5]. Combining OFDM technology with adaptive modulation technology can significantly improve system performance. At present, typical adaptive modulation coding schemes are as follows. Water injection algorithm is not feasible in a practical system [6]. In reference [7], the proposed adaptive modulation algorithm makes it much easier to implement in hardware. Reference [8] makes full use of the channel capacity of the system under the condition of ensuring the BER performance of the system. The computational complexity of the algorithm is relatively low, and the information transmission rate of the system is improved at the same time. In reference [9], it reduces the computational complexity and enhances the real-time performance of the system. Reference [10] divides adjacent subcarriers into groups and uses the same modulation mode, which further reduces the computational complexity of the algorithm. However, these algorithms have the disadvantages of a large amount of calculation and a lot of feedback information [11].

Unlike the problems mentioned above, to make full use of the resources of the OFDM system to achieve the best match between the transmission rate and transmission reliability of the system, this paper introduces an adaptive transmission technology with low complexity, accurate estimation results and fewer iterations. Besides, a cross-layer sensing model is established through the joint Media Access Control layer (MAC), and channel state information is feedback through the cross-layer transmission. Based on this information, the transmitter can adjust the modulation mode used in each subcarrier block timely to minimize the bit error rate of the subcarrier block in the area of serious interference and deep fading, reduce the impact on data transmission, minimize the bit error rate and improve the transmission efficiency.

# 2 System Introduction and Algorithm Description

## *2.1 System Introduction*

Figure 1 is the structure of the subcarrier adaptive modulation OFDM system. As can be seen from the figure, after the data is encoded at the transmitting end, it enters the subcarrier adaptive modulation module to allocate the modulation mode. Here, the corresponding modulation mode is allocated for each subcarrier according to the channel estimation result at the time of the last data transmission. After that, the data enters the IFFT module and becomes the data in the time domain and adds the cyclic prefix (CP), and then enters the wireless channel for transmission. At the same time, the anti-Doppler shift interference performance of the system can be verified by changing the motion rate parameter in the channel. At the receiving end, after removing the CP, the data from the FFT are equalized by the preamble data, and then the data from the FFT will enter the Error Vector Magnitude (EVM) module for signal quality estimation. The result of channel estimation serves as the basis for allocating modulation modes for each subcarrier block in the next transmission of data. Next, the receiving end acquires the transmitting parameters of the transmitting end through the link state analysis (LQA) module, thereby determining the modulation mode used by each subcarrier block, so that the demodulation module can correctly demodulate the data of each subcarrier. Finally, restore the original data by decoding.

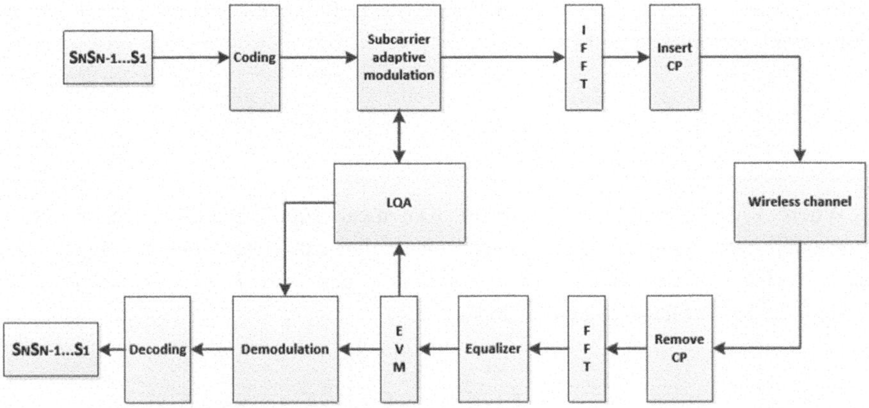


**Fig. 1** Adaptive OFDM system structure diagram

## 2.2 Algorithm Description

Assume that all subcarriers in the system are divided into N blocks. The system assumes that each subcarrier can use the following three modulation modes according to channel conditions for each transmission: QPSK, 16QAM, 64QAM. Then the use of the subcarrier block can be described by the quaternary vector $\lambda = [\lambda_1, \lambda_2, \ldots, \lambda_M]$, where $\lambda_1 \cdots \lambda_M$ can be 3, 2, 1, 0. Reference numeral 3 indicates that the modulation mode used by the subcarrier block is 64QAM, the same 2 represents 16QAM, 1 represents QPSK, and 0 represents no data transmission. The initial state can set all $\lambda_t$=1, $t \in [1, 2, \ldots, M]$, this does not affect the result. There are various parameters for characterizing the transmission quality of the channel. The system uses the EVM value of the subcarrier modulation signal [12].

For an ideal constellation, the number of levels along the in-phase or quadrature axis is expressed as: $n = \sqrt{N}$. For example, since N = 16 for 16QAM, there are four symbol levels (n = 4) for both the in-phase and quadrature axes. The integer coordinate of the ideal constellation point for each symbol can be expressed as:

$$\begin{aligned} C_{ideal,pq} &= C_{I,ideal,pq} + C_{Q,ideal,pq} \\ &= (2p - 1 - n) + j(2q - 1 - n) \end{aligned} \quad (1)$$

Where $1 \leq \mathrm{p} \leq n, 1 \leq \mathrm{q} \leq n$.

The EVM value is the proximity of the *I*, *Q* component of the received modulated signal to the ideal signal component. To calculate the EVM value, we must compare the symbol values in the ideal constellation with the actual measured values.

In the measured case, calculate the total power of all symbols in the constellation within a given frame length:

$$P_V = \sum_{r=1}^{T} \left[ (V_{I,meas,r})^2 + (V_{Q,meas,r})^2 \right] \quad (2)$$

Where $V_{I,meas,r}$ and $V_{Q,meas,r}$ are the root mean square values of the in-phase component and the quadrature component of the actual measured symbol value, respectively, T is the total number of symbols, generally $T \gg N$. Determine the normalized scale factor according to $P_V$:

$$|A_{meas}| = \sqrt{\frac{1}{P_V/T}} \quad (3)$$

Ideally, calculate the sum of the squares of all the symbols in the constellation:

$$
\begin{aligned}
P_C &= \sum_{p=1}^{n}\left[\sum_{q=1}^{n}\left(\left(C_{I,ideal,pq}\right)^2+\left(C_{Q,ideal,pq}\right)^2\right)\right] \\
&= \sum_{p=1}^{n}\left[\sum_{q=1}^{n}\left((2p-1-n)^2+(2q-1-n)^2\right)\right]
\end{aligned} \tag{4}
$$

Determine the normalized scale factor of the ideal symbol according to (4):

$$
|A_{ideal}| = \sqrt{\frac{1}{P_C/N}} \tag{5}
$$

The estimated EVM can be expressed as:

$$
EVM_{RMS} = \left[\frac{\frac{1}{T}\sum_{r=1}^{T}\left(|V_I \bullet |A_{meas}|| - C_I \bullet |A_{ideal}|^2\right)}{P_{S,avg}} + \frac{\frac{1}{T}\sum_{r=1}^{T}\left(|V_Q \bullet |A_{meas}|| - C_Q \bullet |A_{ideal}|^2\right)}{P_{S,avg}}\right]^{\frac{1}{2}} \tag{6}
$$

Where the $P_{S,avg}$ expression is:

$$
P_{S,avg} = \frac{1}{N}\sum_{p=1}^{n}\left[\sum_{q=1}^{n}\left((2p-1-n)^2|A_{ideal}|^2\right) + \sum_{q=1}^{n}\left((2q-1-n)^2|A_{ideal}|^2\right)\right] \tag{7}
$$

The algorithm uses the threshold setting method to evaluate the quality of the received data on the M-subcarriers. The smaller the value of the EVM, the closer the observation point is to the standard modulation point, and the smaller the deviation, the better the signal quality. The EVM vector is represented as $\rho = [\rho_1, \rho_2, \ldots, \rho_M]$, Threshold range set according to EVM value: $\rho_k \in [0, 1, 2, 3]$, $k \in [1, 2, \ldots, M]$. After obtaining the subscript K of each subcarrier, it is transmitted back to the transmitting end through the LQA, and the transmitting end allocates a corresponding modulation mode according to the condition of each subcarrier in the next transmission. As shown in Fig. 2, in the actual engineering application, the feedback link requires a MAC layer for coordination. The physical layer at the receiving end reports the channel information to the MAC layer, and the MAC layer aggregates the channel state information. Then send the physical layer to the transmitter. The physical layer of the receiving end reports the channel information to the MAC layer, and the MAC layer aggregates the channel state information and sends it to the physical layer of the transmitting end.

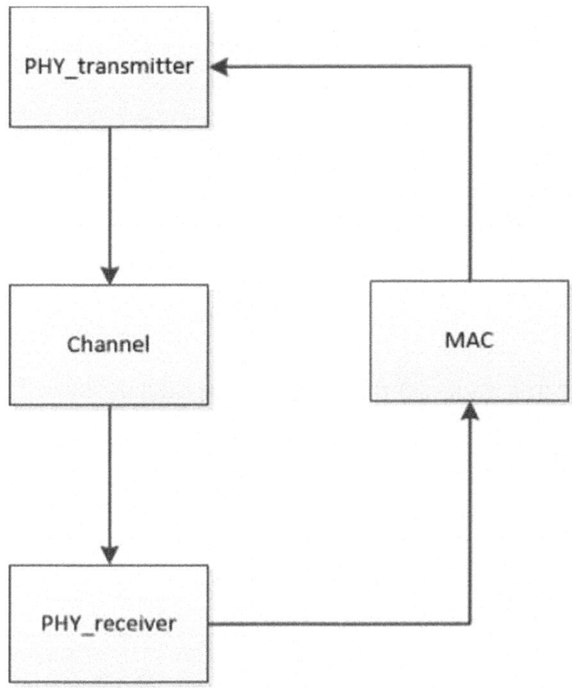


**Fig. 2** Cross-layer adaptive system structure diagram

# 3 Algorithm Performance Analysis

## *3.1 Complexity Analysis*

Low complexity algorithms are practical and easy to implement and can improve communication efficiency. The EVM calculation method mentioned in II is more complicated in practical applications. This value is a measure of how close the I and Q components of the received modulated signal are to the ideal signal component. The result can be reduced to the ratio of the root mean square of the average power of the error vector signal to the root mean square of the ideal signal:

$$EVM = \left[ \frac{\frac{1}{T}\sum_{r=1}^{T} \left| S_{ideal,r} - S_{meas,r} \right|^2}{\frac{1}{T}\sum_{r=1}^{T} \left| S_{ideal,r} \right|^2} \right]^2 \quad (8)$$

Where $S_{ideal,r}$ represents the ideal constellation point coordinates of the received symbol labeled $r$, and $S_{meas,r}$ represents the measured constellation point coordinates of the received symbol labeled r. For QPSK, 16QAM and 64QAM modulation modes, the power $\left|S_{ideal,r}\right|^2$ of the modulation signal corresponding to each constellation point is different. However, in this system, the number of received symbols T is large enough (each OFDM symbol contains 192 subcarrier symbols and at least 8 OFDM symbols per burst), and the signal is random. The number of modulations to the constellation points is similar, and the power of the three modulation methods at the transmitting end is normalized. Therefore, the average power of the ideal signals at the receiving end is the same. Therefore, only the average power of the error vector signal can be calculated instead of EVM, simplifying the calculation and reducing the implementation complexity. So only two additions (subtractions), one addition and two squares are needed in EVM calculation. The calculation results show that the algorithm has a lower complexity.

## 3.2 *Simulation Analysis*

According to the principle of block-wise adaptive OFDM (BA-OFDM) system, we simulated it in the software of MATLAB, analyzed the performance of frequency selective channel, and adjusted the motion rate to change the anti-jamming performance of Doppler frequency shift verification system. System simulation parameters are shown in the Table 1.

Figure 3 and Fig. 4 compare the performance curves of OFDM over AWGN channel and frequency selective channel using RS-CC coding. The modulation modes of the two graphs are 16QAM and 64QAM. As can be seen from Fig. 3, the performance of 16QAM modulation is better than that of frequency selective channel in AWGN channel. In AWGN channel, the BER is less than $10^{-3}$ after SNR is greater than 15 dB, while in frequency selective channel, the BER is less than $10^{-3}$ after SNR is greater than 18 dB, in Fig. 4, the performance of 64QAM modulation over AWGN channel is also better than that over frequency selective channel. When the bit error rate is $10^{-3}$, the SNR required for the AWGN channel is 14 dB less than that for frequency selective channel. According to the above results, the noise power of the two channels is the same, and the performance difference is

**Table 1** Parameters of simulation

| | |
|---|---|
| Number of subcarriers | 256 |
| Number of OFDM symbols | 8 |
| Coding type | RS+CC |
| Coding rate | 133/192 |
| Modulate type | QPSK 16QAM 64QAM |
| Carrier frequency | 600 MHz |
| Sample rate | 9 MHz |
| CP length | 32 |
| Channel type | AWGN and Frequency Selective Channel |
| Number of subcarriers | 256 |

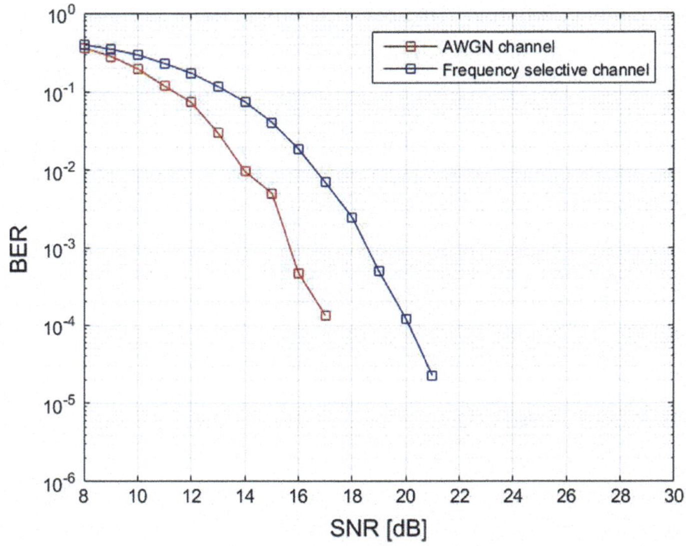


**Fig. 3** BER of two channel conditions under 16QAM mode

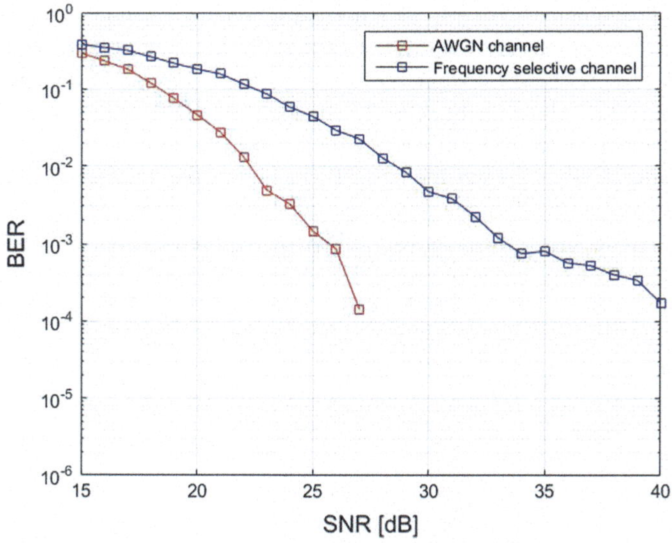


**Fig. 4** BER of two channel conditions under 64QAM mode

only caused by the frequency selective fading of the channel. Therefore, it is necessary to estimate the quality of each subcarrier, allocate appropriate modulation modes for each subcarrier adaptively, and reduce the impact of frequency selective fading on the signal.

Figure 5 , 6, 7 and 8 are performance comparison curves of BA-OFDM and OFDM in frequency selective channels. Figure 5 and Fig. 6 are the error performance comparison curves. As shown in Table 2, in practical applications, when the EVM and SNR values meet the requirements, the high-order modulation method is preferred. When SNR is less than 20 dB, OFDM uses 16QAM modulation mode, while BA-OFDM adaptively uses QPSK modulation mode on subcarrier blocks with poor quality according to EVM threshold. Therefore, the BER of OFDM is less than $10^{-3}$ after the SNR is greater than 16 dB, and OFDM can only be achieved after the SNR is greater than 19 dB, similarly, when SNR is greater than 20 dB, OFDM uses 64QAM modulation mode, while BA-OFDM adaptively uses 64QAM, 16QAM or QPSK on different subcarrier blocks according to EVM threshold. When the bit error rate is $10^{-3}$, the signal-to-noise ratio of BA-OFDM is 10 dB less than that of OFDM. Figure 7 and Fig. 8 show the performance comparison between BA-OFDM and OFDM throughput. Because of the large amount of data, the ordinate coordinates are expressed by the normalization method on the premise that the transmitted data can be decoded correctly: $K_{BA-OFDM} = B_{BA-OFDM}/B_{AWGN}$, $K_{Fading} = B_{Fading}/B_{AWGN}$, where $B_{\mathrm{BA-OFDM}}$ represents the amount of data correctly transmitted in frequency-selective channels using adaptive modulation based on subcarrier blocks. $B_{\mathrm{AWGN}}$ represents the amount of data correctly transmitted over an AWGN channel using a fixed modulation mode within

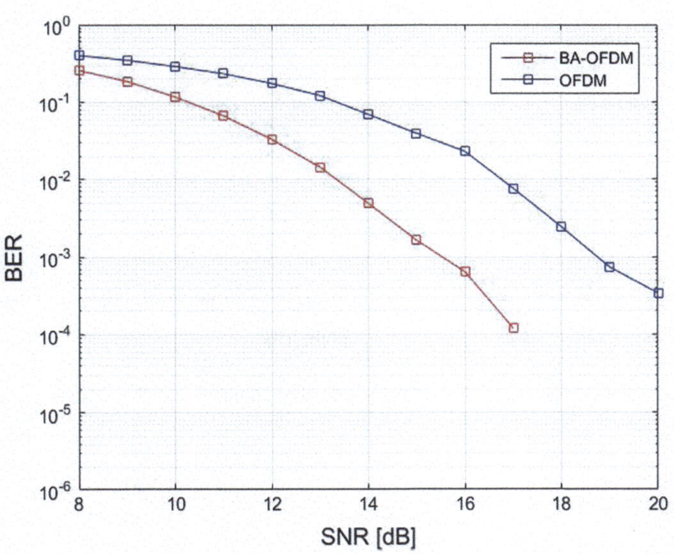


**Fig. 5** BER performance in frequency selective channels ($SNR \leq 20\,\mathrm{dB}$)

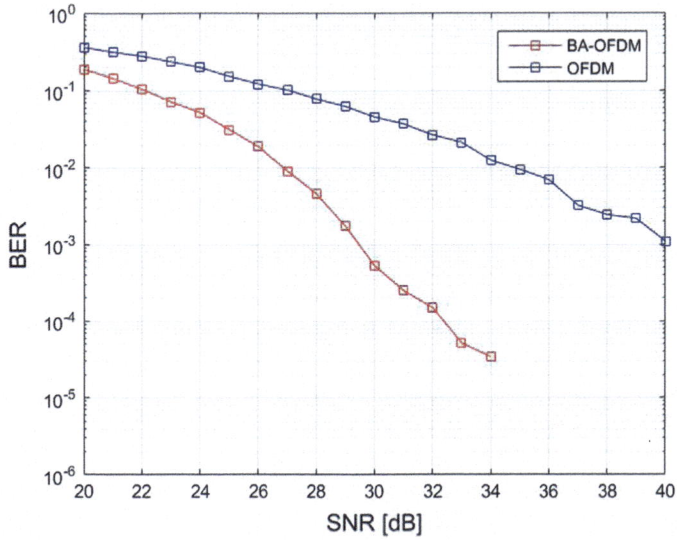


**Fig. 6** BER performance in frequency selective channels ($SNR \geq 20\,\text{dB}$)

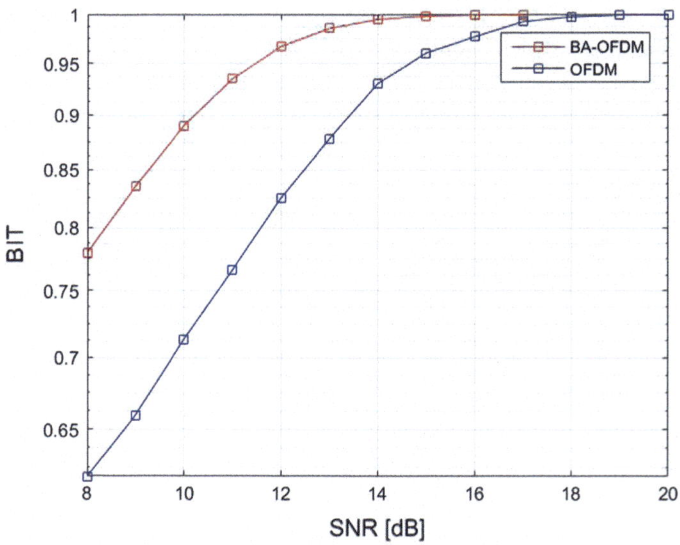


**Fig. 7** Throughput performance in frequency selective channels ($SNR \leq 20\,\text{dB}$)

**Table 2** Threshold parameters of BA-OFDM

| Modulation | EVM |
|---|---|
| 64QAM | $EVM \leq 2$ |
| 16QAM | $EVM \leq 5$ |
| QPSK | All values |

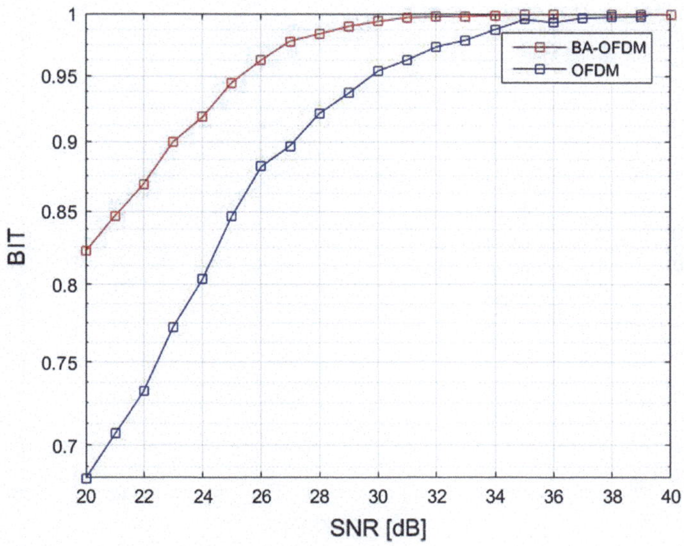


**Fig. 8** Throughput performance in frequency selective channels ($SNR \geq 20\,\mathrm{dB}$)

the SNR range. Similarly, $B_{\mathrm{Fading}}$ represents the amount of data correctly transmitted by a fixed modulation mode in the SNR range over frequency-selective channels. As can be seen from these figures that the performance of BA-OFDM is better than that of OFDM in a frequency selective channel. Therefore, BA-OFDM can estimate the quality of each subcarrier well in one frame, select the best quality subcarrier block to use high order modulation, and select the bad quality subcarrier block to use low order modulation. The above can conclude that BA-OFDM has a better performance against frequency selective fading.

Figure 9 and Fig. 10 are test results for increasing the Doppler shift. The method of calculating Doppler shift in this paper is as follows: $f_d = \frac{f}{c} \times v \times \cos\theta$, Where f is the carrier frequency, here set to 600 MHz, c is the electromagnetic wave propagation speed, here is $3 \times 10^8$ m/s, v represents the motion rate, set here as 0. It can be seen from the BER curve of the two figures: Under the Doppler shift interference with a motion rate of 300 km/h, the performance of BA-OFDM is the same as that of frequency selective channel OFDM. This phenomenon occurs because the motion rate is too fast and the channel time becomes strong, which causes the feedback channel state information to be unable to estimate the quality of the next channel. At this time, the adaptive modulation based on subcarrier is no longer applicable.

As can be seen from Fig. 9, the performance of BA-OFDM is better than that of ordinary OFDM when the SNR is less than 20 dB. At the rate of 200 km/h, the performance gain of at least 1 dB is obtained at the bit error rate of $10^{-3}$, and at the rate of 100 km/h, the performance gain of at least 2 dB is obtained at the bit error rate of $10^{-3}$. As can be seen from Fig. 10, when SNR is greater than 20 dB, the

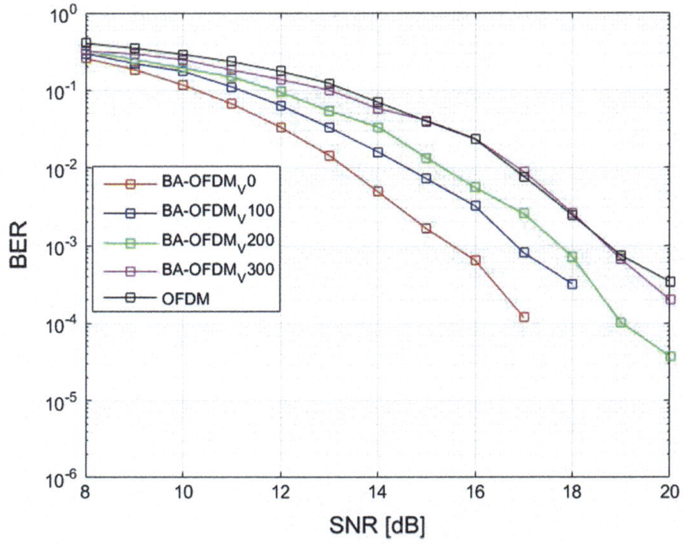


**Fig. 9** BA-OFDM test results at different motion rates ($SNR \leq 20\,\mathrm{dB}$)

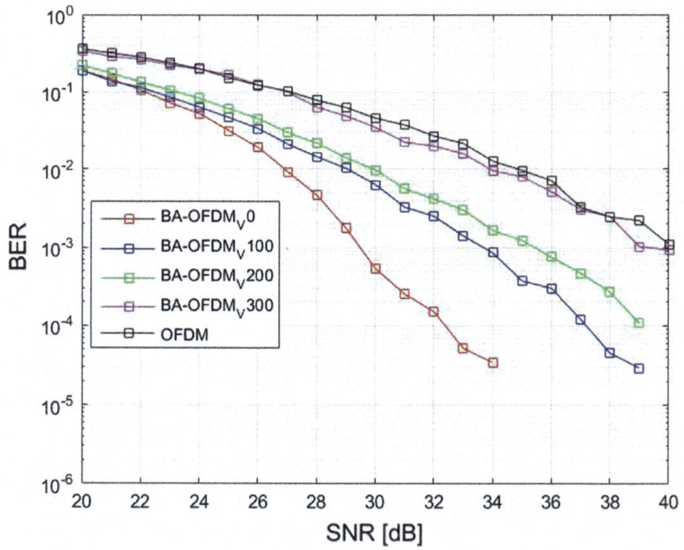


**Fig. 10** BA-OFDM test results at different motion rates ($SNR \geq 20\,\mathrm{dB}$)

performance of BA-OFDM is also better than that of ordinary OFDM. At the rate of 200 km/h, the performance gain of at least 4 dB is obtained at the bit error rate of $10^{-3}$, and at the rate of 100 km/h, the performance gain of at least 6 dB is obtained at the bit error rate of $10^{-3}$.

## 4 Conclusion

To effectively enhance the system's ability to resist frequency selective fading, this paper introduces the OFDM subcarrier block adaptive modulation technology into the physical layer of the system. By calculating the EVM value of each subcarrier block at the receiving end, it allocates appropriate modulation modes to each subcarrier block for minimizing the impact of interference and fading bands on it and improving the bit error rate and throughput of the system. In the case of keeping the original system complexity unchanged, it is only necessary to increase the EVM calculation module and the channel information feedback module to effectively function and obtain good performance gain. Compared with the common non-adaptive OFDM technology, it effectively perceives the channel quality and performs adaptive modulation on each subcarrier block, which improves the transmission efficiency to a great extent. The simulation results show that BA-OFDM technology possesses better error performance in the frequency selective fading channel than the OFDM system, and can improve system performance within a certain range of motion rates. In the future of mobile communications, it will become an effective means of adaptive transmission.

## References


1. Yu, Q., & Wang, Y. (2010). Improved chow algorithm used in adaptive OFDM system. In *2010 International Conference on Communications and Mobile Computing, Shenzhen* (430–432).
2. Pandit, S., & Singh, G. (2015). Channel capacity in fading environment with CSI and interference power constraints for cognitive radio communication system. *Wireless Networks, 21*(4), 1275–1288.
3. Hwang, Y. T., Tsai, C. Y. & Lin, C. C. (2005). Block-wise adaptive modulation for OFDM WLAN systems. In *2005 IEEE International Symposium on Circuits and Systems, Kobe* (Vol. 6, pp. 6098–6101).
4. Karaarslan, G., & Ertuğ, Ö. (2017). Adaptive modulation and coding technique under multipath fading and impulsive noise in broadband power-line communication. In *2017 10th International Conference on Electrical and Electronics Engineering, Bursa* (pp. 1430–1434).
5. Tato, A., Mosquera, C. & Gomez, I. (2016). Link adaptation in mobile satellite links: field trials results. In *2016 8th Advanced Satellite Multimedia Systems Conference and the 14th Signal Processing for Space Communications Workshop, Palma de Mallorca* (pp. 1–8).
6. Willems, F. M. J. (1993). Elements of Information Theory [Book Review]. *IEEE Transactions on Information Theory, 39*(1), 313–315.

7. Webb, W. T., & Steele, R. (1995). Variable rate QAM for mobile radio. *IEEE Transactions on Communications, 43*(7), 2223–2230.
8. Chow, P. S., Cioffi, J. M., & Bingham, J. A. C. (1995). A practical discrete multitone transceiver loading algorithm for data transmission over spectrally shaped channels. *IEEE Transactions on Communications, 43*(2/3/4), 773–775.
9. Fischer, R. F. H. & Huber, J. B. (1996). A new loading algorithm for discrete multitone transmission. In *Proceedings of GLOBECOM'96. 1996 IEEE Global Telecommunications Conference* (Vol. 1, pp. 724–728).
10. Grunheid, R., Bolinth, E., & Rohling, H. (2001). A blockwise loading algorithm for the adaptive modulation technique in OFDM systems. In *IEEE 54th Vehicular Technology Conference. VTC Fall 2001. Proceedings* (Cat. No.01CH37211) (Vol. 2, pp. 948–951).
11. Keller, T., & Hanzo, L. (2000). Adaptive multicarrier modulation: a convenient framework for time-frequency processing in wireless communications. *Proceedings of the IEEE, 88*(5), 611–640.
12. Qijun, Z., Qinghua, X. & Wei, Z. (2007). Notice of violation of IEEE publication principles a new EVM calculation method for broadband modulated signals and simulation. In *2007 8th International Conference on Electronic Measurement and Instruments* (pp. 2-661–2-665).

# Optimizing Retraining of Multiple Recommendation Models

**Michael Peran, Dan Augenstern, Josh Price, Rahul Nahar, and Pankaj Srivastava**

**Abstract** To increase recommendation accuracy we utilize multiple individual-item models instead of a single multi-classification model. Retraining all multiple models consumes significant time. To selectively retrain only models where new and historical consumers differ, we estimate the distance between historical and new consumers including both categorical and numerical consumer attributes to determine when to retrain each of the recommendation models.

**Keywords** Recommender system · Retraining statistical models · Model implementation

## 1 Introduction and Related Work

Our personalized recommendation engine is composed of multiple statistical models selecting items that are most likely to be of interest to a particular consumer by utilizing information known about this consumer. Such items can be travel packages [1], retail products [2, 3], entertainment [4, 5], services, offers of crowdsourcing jobs [6], learning activities like online or in-person courses [7], games, restaurants, promotions, etc. offered by an organization to its consumers.

M. Peran (✉) · D. Augenstern · J. Price · R. Nahar · P. Srivastava
IBM, Armonk, NY, USA
e-mail: mperan@us.ibm.com

D. Augenstern
e-mail: daugenst@us.ibm.com

J. Price
e-mail: joshprice@us.ibm.com

R. Nahar
e-mail: rnahar@us.ibm.com

P. Srivastava
e-mail: psrivast@us.ibm.com


J. Zhang et al. (eds.), *LISS2019*,
https://doi.org/10.1007/978-981-15-5682-1_3

The organization aspires to increase the number of consumed items by recommending a few most relevant items personally selected for each consumer.

In our case, items are learning activities offered to consumers in a large organization. Given large scale of the data, we need effective ways to overcome resource limitations. Some very similar items can be grouped together to be modelled as a single item, e.g., utilizing their attributes [8] or natural language processing of their descriptions [9]. Nevertheless, even after grouping items together we are dealing with thousands (or tens of thousands) of items offered to hundreds of thousands or millions of consumers.

Our approach is to train many individual-item models scoring consumer propensity to act on an item. This solution differs from item association-rules models and its enhancements like item-to-item collaborative filtering [2, 3, 5]. Our solution also differs from multi-classification models that take all items' and consumers' data as their input and from several other recommender systems, e.g. [7, 10].

To ensure our recommendations are the most accurate and relevant to each consumer, we do not want to utilize sampling like e.g. [8] to reduce training time. This means we utilize all available data within our time window. Since we utilize each consumer's attributes (and not only their consumption history) our solution addresses cold start problem for consumers (but not for items) differently from [11].

Our solution works the best when many consumer attributes, e.g. their education, years after graduation, business interests, etc. are either stored in an organization or are explicitly shared by consumers with an organization to improve their personalized recommendations. Such consumer data was not available in the most association rules applications including item-to-item collaborative filtering recommender systems [2–5]. We hope that in the future more consumers would be willing to share their data with organizations, so a combination of our solution with item-to-item collaborative filtering would further improve recommender systems.

In this paper we do not focus on what models or algorithms are optimal for individual-item models. We utilize standard models readily available "of-the-shelf". The focus of this paper is on efficient ways to incorporate new consumer data by retraining individual-item models.

The traditional recommendation models including our individual-item models need to be retrained from time to time to promptly include new information about consumers. The models we utilized do not allow uptraining, so retraining is our only feasible option. We found that learning recommender systems benefit from retraining and that retraining at fixed periods would not be enough unlike in other recommender applications, e.g. [11]. An alternative solution instead of retraining can be streaming clustering algorithms combining online-offline approaches [12, 13] that are trained practically constantly around the clock. Whereas such algorithms show promising potential, they are not readily available to calculate propensity to consume that is crucial in learning recommender systems and they require significant computational resources. We did not try streaming clustering partly to limit the load on our computing resources and partly to utilize only standard readily available packages for our individual-item models.

Due to limited resources our individual-item models are retrained only when distances between historical and new consumers for this item are significant. There are several algorithms, e.g. [14], to determine the distance between new consumers and historical consumers. However, those algorithms are not fast enough for our purpose of quickly estimating in real time whether to retrain a model. Moreover, those algorithms deal with numerical attributes and well-defined distances, whereas we need to consider both categorical and numerical consumer attributes.

This paper describes a fast method to determine whether the distance between new and historical consumers warrants retraining some of our individual-item models. This fast decision to retrain a model allows us to retrain each of our individual-item models in 'near' real-time and makes our set of multiple individual-item models superior to a single multiclassification model or association-rules model both in terms of accuracy and in terms of retraining efficiency.

## 2 Methods and Set-Up

We tried several approaches designing our recommendation engine that can be broken down into a single multiclassification model or an association-rules model vs. a set of multiple individual-item models. Both types of models were trained on the same historical data of items consumption. About 30% of historical data were set aside as test data and the remaining data were used for training.

A single multiclassification model as well as an association-rules model takes all items as its input. Its output ranks items for each of the consumers by predicted propensity to be consumed.

In contrast, a model set contains thousands of individual-item models. Each individual-item model is trained for only one item to predict propensity of this item to be consumed by each of the consumers. We score all consumers by each of individual-item models. We recommend a few items personalized to each of the consumers selecting items with the highest predicted propensity.

Multiclassification models we experimented with took about 3 to 5 h to train and less than an hour to score all the consumers. Relatively short initial training time is the advantage of those models vs. a set of several thousand individual-item models

It takes quite a long time to train a set of multiple individual-item models. Depending on the number of consumers, amount of historical data, and computer resources, it takes from 30 s to 300 s to train an individual-item model. Thus, it takes up to several days to train 3,000 to 5,000 models. Even with significant computer resources the training time might be quite long. If there are 120,000 items and training a model for each item takes 180 s (3 min), then training models for all items can last 2 full days even employing a feasible number of multi-threads and multi-instances. (10 instances with 12 threads each would require 3,000 min = 50 h to retrain all 120,000 models).

Out-of-sample validation revealed that the accuracy of a set of multiple individual-item models was superior to multiclassification models. The drawback of a long training time required for a set of multiple individual-item models is not a big impediment since we need to train all models only once initially, when we set up our recommendation models. After this long initial set-up, retraining of individual-item models can be done in ‘near’ real time, less than 300 s to retrain a model if, as shown in this paper, the models are retrained only when significant new consumer information becomes available. Additional individual-item models with large number of new consumers can be retrained from time to time even when the distance between historical and new consumers is small. They retrain during lower demand when less computer resources are utilized ensuring that all individual-item models are up-to-date.

## 3 Experimental Results

We have applied the approach outlined above to develop a personalized learning recommendation engine for a large organization.

Our set was composed of 2,500 individual-item models.

The initial training of the models took about 28 h. The individual-item models were subsequently retrained when new consumers took online learning activities. Retraining those individual-item models took about 2–4 h total over the course of 7 days.

Historical tests demonstrated that both a multi-classification model and an association-rules model were inferior in accuracy to a set of multiple individual-item models. Since initial training time was the only advantage of multi-classification models, we decided to implement multiple individual-item models and test them in real life vs. an existing (status quo, i.e. control) association-rules model.

Consumers (users) from both control group and our proposed models group were presented with personalized learning recommendations. When a consumer took any action on one of these recommendations it was recorded as a successful outcome.

A/B testing vs. control group during the first 3 months revealed that this set of multiple individual-item models was superior to a status quo association-rules model. A test example is shown in Fig. 1.

When a set of multiple individual-item models was implemented and demonstrated its advantages, we had to consider options to minimize retraining time by defining consumer features vector to estimate differences between new and historical consumers.

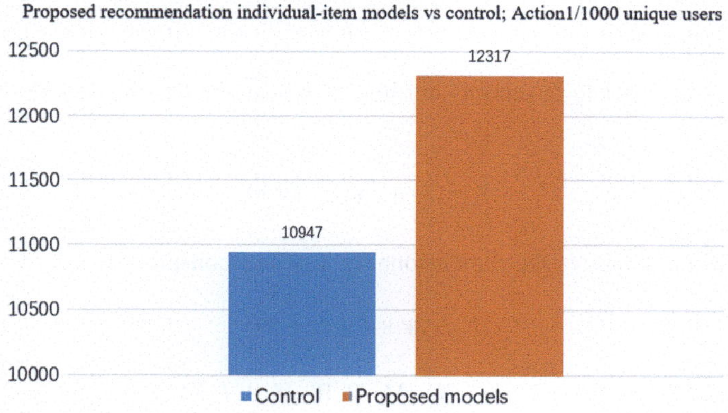


**Fig. 1** A/B testing of online user actions

# 4 Mathematical Background

Let us consider a model utilizing both categorical and numerical attributes of consumers as its inputs.

We define a numerical (i.e. quantitative) consumer feature vector as having each possible value of categorical attributes. Its value is 1 if a consumer has this attribute and 0 if a consumer does not have this attribute. (1 is if all attributes are equally weighted). If an attribute has a higher weight *w*, the possible values are between 0 and *w*. Weights can be useful when model weights of attributes are known, or weights can be defined based on business rules. Obviously, some categorical attributes with too many possible values are to be grouped together to reduce dimensionality. For example, instead of a consumer location name it is better to use a larger geographic or administrative region, like province or time zone.

In addition, such consumer feature vector employs normalized numerical attributes, i.e., a numerical attribute divided by maximum of this numerical attribute among all historical consumers.

For example, let us assume a model utilizes the following categorical and numerical attributes:

Consumer's geography defined as a time zone with 4 possible values (*tE, tC, tM, tP*).

Consumer's laptop, tablet, or cellphone operating system with 4 possible values ($W$ = Windows, $M$ = Mac, $A$ = Android, $I$ = iOS).

Consumer's years after high-school graduation $\tilde{Y}$ is a numerical (i.e. quantitative) attribute with continuous values.

Assuming all consumers are high school graduates, and the maximum value is 60 years, our feature vector component $Y$ is normalized as either (1) or (2).

If the "years after high school" attribute is equally weighted with other attributes

$$Y = \tilde{Y}/60 \tag{1}$$

If the "years after high school" attribute $\tilde{Y}$ is weighted 2 times more than other attributes

$$Y = 2 \cdot \tilde{Y}/60 = \tilde{Y}/30 \tag{2}$$

In the example above the dimensionality $J$ of our consumer feature vector $\vec{C}$ is

$$J = 4+4+1 = 9 \tag{3}$$

$$\vec{C} = (tE, tC, tM, tP, W, M, A, I, Y) \tag{4}$$

Let's continue our example giving a few more scenarios.

If all weights are equal and a consumer is in time zone E, her operating system is A = Android, and she is 12 years after high school graduation (12/60 = 0.2) then

$$\begin{gathered}\vec{C} = (tE = 1, tC = 0, tM = 0, tP = 0, W = 0, M = 0, \\ A = 1, I = 0, Y = 0.2)\end{gathered} \tag{5}$$

If consumer's operating system is weighted 3 times more than other attributes, then for the same consumer in time zone E with her operating system A = Android, and 12 years after high school graduation (12/60 = 0.2) we obtain

$$\begin{gathered}\vec{C} = (tE = 1, tC = 0, tM = 0, tP = 0, W = 0, M = 0, \\ A = 3, I = 0, Y = 0.2)\end{gathered} \tag{6}$$

If a new consumer has an "out-of-scope" categorical attribute value, the feature vector dimensionality is to be increased, e.g. a consumer has operating system $L$ = Linux and none of the historical consumers have Linux.

The dimensionality $J$ of our consumer feature vector $\vec{C}$ is then to increase by 1 compared to (3) as

$$J = 4+5+1 = 10 \tag{7}$$

$$\vec{C} = (tE, tC, tM, tP, W, M, A, I, L, Y) \tag{8}$$

All historical consumers get L = 0 in (4.8). The new consumer gets L = 1, e.g.

$$\begin{gathered}\vec{C} = (tE = 1, tC = 0, tM = 0, tP = 0, W = 0, M = 0, \\ A = 0, I = 0, L = 1, Y = 0.2)\end{gathered} \tag{9}$$

An “out-of-historical-scope” numerical attribute does not change the feature vector dimensionality $J$, e.g. if a new consumer in (4.6) is 72 years after high school graduation ($Y = 72/60 = 1.2$) and historical maximum is 60 years, then

$$\vec{C} = (tE = 1, tC = 0, tM = 0, tP = 0, W = 0, M = 0, A = 3, I = 0, Y = 1.2) \tag{10}$$

# 5 Method

We suggest several definitions of the distance $\gamma\{\tilde{G}, \tilde{F}\}$ between two consumer groups $\tilde{G}$ and $\tilde{F}$. The distance should be defined in such a way that it can be computed very fast.

Steps 1, 2 and 3 are the same for Sects. 5.1, 5.2 and 5.3.

Step 1: Define a numerical feature vector of each consumer utilizing both numerical (i.e. quantitative) and categorical consumer attributes utilized in a given individual-item model.

Step 2: Find the standard deviation $\sigma$ of lengths of feature vectors of historical consumers.

Step 3: Define a trigger threshold to retrain the individual-item model, e.g. 0.1% $\sigma$, 1% $\sigma$, 5% $\sigma$, or 200% $\sigma$.

## *5.1 Distance Between Centers of Gravity*

This distance is the fastest to compute but it is the least accurate. It means the new consumers can be different from the historical consumers even if the distance between their centers of gravity is small. However, if this distance is large then the new consumers are always different.

Step 4: Find the average feature vector of historical consumers (a historical center of gravity).

Step 5: Dynamically find the average feature vector of new consumers (their center of gravity) and compare it to historical consumers.

When the distance between new and existing consumers exceeds the trigger threshold, retrain.

Step 6: When retraining the model, calculate the average feature vector of historical consumers (the historical center of gravity) and historical $\sigma$ for the future, to compare to new consumers.

We define “center of gravity” – historical feature vector $\vec{F}$ as a simple average of $M$ historical consumer feature vectors $\vec{C}_m$

$$\vec{F} = \frac{\sum_{m=1}^{M} \vec{C}_m}{M} \tag{11}$$

The standard deviation of lengths σ is defined as

$$\sigma = \sqrt{var(\vec{C} - \vec{F})} \tag{12}$$

$$var(\vec{C} - \vec{F}) = \frac{\sum_{m=1}^{M} \left\| \vec{C}_m - \vec{F} \right\|^2}{M - 1} \tag{13}$$

where $\left\| \vec{C}_m - \vec{F} \right\|$ is the vector length as in linear algebra, (18).

Continuing our example (5): let us assume that 30% of consumers are in tE, 20% in tC, etc., 60% use $W$ = Windows, etc. and average is 6 years after high school graduation (6/60 = 0.1); then

$$\begin{aligned} \vec{F} = (tE = 0.3, tC = 0.2, tM = 0.1, tP = 0.4, W = 0.6, \\ M = 0.2, A = 0.1, I = 0.1, Y = 0.1) \end{aligned} \tag{14}$$

The new consumers' center of gravity vector $\vec{G}(N)$ is defined as

$$\vec{G}(N) = \frac{\sum_{n=1}^{N} \vec{C}_n}{N} \tag{15}$$

Note that if $\vec{G}(N)$ is stored for $N$ new consumers, it can be dynamically updated for $N + 1$ new consumers without recalculating the sum in (15).

Let us assume the model was not retrained after $N$ new consumers. New consumer $\vec{C}_{N+1}$ arrives and the number of new consumers becomes $N + 1$. Feature vector $\vec{G}(N + 1)$ is updated from $\vec{G}(N)$ without recalculating the sum for $N$ new consumers

$$\vec{G}(N + 1) = \frac{N * \vec{G}(N) + \vec{C}_{N+1}}{N + 1} \tag{16}$$

Then this updated feature vector $\vec{G}(N + 1)$ can be compared to historical feature vector $\vec{F}$ as described below to see whether the threshold for retraining the model is met.

Computational time of (16) and (17) is several orders of magnitude faster than retraining a model.

Distance $\gamma_1$ between historical and new consumers is defined as the distance between the historical "center of gravity" and a hypothetical "retrained" center of gravity if new consumers were added. It is proportional to the distance between feature vectors $\vec{F}$ and $\vec{G}(N)$:

$$\gamma_1 = \frac{N}{M+N} \left\| \vec{G}(N) - \vec{F} \right\| \tag{17}$$

Note that distance $\gamma_1$ in (17) increases with more $N$ new consumers; $\gamma_1$ in (17) triggers retraining when it exceeds a threshold in Step 3 expressed as units of $\sigma$. The lower is the threshold-the more frequent is the retraining and the higher is the cost of resources.

In addition, note that if the new consumers are exactly the same as the historical consumers, then $\left\|\vec{G}(N) - \vec{F}\right\| = 0$ and (17) yields $\gamma_1 = 0$ independent of how many $N$ new consumers take actions. Obviously, if many new consumers take actions (e.g., $N = M$) then the model is to be retrained out of caution even if $\gamma_1$ is below its threshold.

We define the vector length as:

$$\left\|\vec{X}\right\| = \sqrt{\sum\nolimits_{j=1}^{J} x_j^2} \tag{18}$$

Vector $\vec{X}$ in (18) is $J$ dimensional:

$$\vec{X} = \left(x_1, x_2, x_3, \cdots, x_j, \cdots, x_J\right) \tag{19}$$

## 5.2 *Nearest Neighbor Distance Without Replication*

Distance between a new consumer and its nearest neighbor among historical consumers without replication is more accurate than center of gravity distance in Sect. 5.1 but it takes longer to compute.

Step 4: Find distance $D_n = \left\|\vec{C}_n - \vec{C}_m\right\|$ between a new consumer $\vec{C}_n$ and its nearest neighbor $\vec{C}_m$ among historical consumers (without replication). For faster computation, a new consumer $\vec{C}_n$ finds its nearest neighbor $\vec{C}_m$ among "remaining" historical consumers that were not nearest neighbors of previous new consumers. There is no comparison of nearest neighbors across new consumers to minimize the sum of distances $D_n$.

Step 5: Measure the distance between new and historical consumers as the generalized average distance to the nearest neighbor.

$$\gamma_2 = \frac{\sum\nolimits_{n=1}^{N} D_n}{N^{\alpha}(N+M)^{(1-\alpha)}} \tag{20}$$

where $0 \le \alpha \le 1$. The greater $\alpha$ the more sensitive $\gamma_2$ is to the number of new consumers $N$. In practice $\alpha = 1/2$ is a "good enough" choice.

When the distance between new and existing consumers exceeds the trigger threshold, retrain.

### 5.3 Nearest Neighbor Distance with Replication

Distance between a new consumer and its nearest neighbor among historical consumers with replication is more accurate than without replication in Sect. 5.2 but it takes longer to compute.

Step 4: Find distance $D_n = \left\| \vec{C}_n - \vec{C}_m \right\|$ between a new consumer $\vec{C}_n$ and its nearest neighbor $\vec{C}_m$ among historical consumers (with replication). For more accurate computation, a new consumer $\vec{C}_n$ finds its nearest neighbor $\vec{C}_m$ comparing distances between all new and all historical consumers to minimize the sum of distances $D_n$.

Step 5: Measure the distance between new and historical consumers as the generalized average distance to the nearest neighbor as in (20).

When the distance between new and existing consumers exceeds the trigger threshold, retrain.

We did not yet test such distance-based retraining in our learning recommendation applications. We conject that increased accuracy of nearest neighbor distances vs. "center of gravity" distance may not justify longer computation time and a more complex logic, so our first choice would be "center of gravity" distance.

## 6 Conclusion

Our solution consisting of a set of many individual-item models is scalable and easily adjustable to market changes. When new items are added to the inventory or obsolete old items are removed, we can train and add corresponding individual-item models, usually in less than 300 s per new item.

In our learning applications, this solution produced recommendations superior to association rules models and to multi-classification models.

As demonstrated in this paper, our solution can be retrained in 'near' real time when the new consumers become sufficiently different from the historical consumers. Thus, we can retrain our individual-item models without waiting for model performance to deteriorate. Such retraining flexibility overcomes a drawback of a long training time required for our solution during an initial set-up, since a complete initial set-up is needed once only.

In the future our solution can be further improved by combining it with item-to-item collaborative filtering [2, 3, 5]. To develop such a combined recommender system, an organization should have the data on consumer historical consumptions and on their personal attributes.

**Acknowledgements** The authors thank Zhongzheng Zach Shu for his help with the model implementation.

## References

1. Huang, Y., & Bian, L. (2009). A Bayesian network and analytic hierarchy process based personalized recommendations for tourist attractions over the Internet. *Expert Systems with Applications, 36*(1), 933–943.
2. Smith, B., & Linden, G. (2017). Two decades of recommender systems at Amazon.com. *IEEE Internet Computing, 21*(3), 12–18.
3. Linden, G., Smith, B., & York, J. (2003). Amazon.com recommendations: item-to-item collaborative Filtering. *IEEE Internet Computing, 7*(1), 76–80.
4. Gomez-Uribe, C. A., & Hunt, N. (2016). The netflix recommender system: Algorithms, business value, and innovation. *ACM Transaction Management Information Systems, 6*(4), 1–19.
5. Davidson, J., Liebald, B., Liu, J., Nandy, P., Van Vleet, T., Gargi, U., et al. (2010) The YouTube video recommendation system. In *RecSys Proceedings of the fourth ACM conference on Recommender Systems* (pp. 293–296).
6. Jarrett, J., Blake, M. B. (2016). Using collaborative filtering to automate worker-job recommendations for crowdsourcing services. In *Proceedings IEEE International Conference* (pp. 641–645).
7. Tarus, J. K., Niu, Z., & Mustafa, G. (2018). Knowledge-based recommendation: a review of ontology-based recommender systems for e-learning. *Artificial Intelligence Review, 50*(1), 21–48.
8. Qu, Z., Yao, J., Wang, X., & Yin, S. (2018). Attribute weighting and samples sampling for collaborative filtering. In *IEEE International Conference on Big Data and Smart Computing (BigComp)* (pp. 235–241).
9. Vasudevan, S., Mondal, J., Zhou, R., Peran, M., Ticknor, M., Augenstern, D. (2018). Method of choosing exemplars from a large multi-class data-set given few exemplars of one or more of the classes. US Patent application 2018-15/866723.
10. Abbas, A., Zhang, L., & Khan, S. U. (2015). A survey on context-aware recommender systems based on computational intelligence techniques. *Computing, 97*(7), 667–690.
11. Wei, J., Hea, J., Chen, K., Zhou, Y., & Tang, Z. (2017). Collaborative filtering and deep learning based recommendation system for cold start items. *Expert Systems with Applications, 69,* 29–39.
12. Ding, S., Wu, F., Qian, J., Jia, H., & Jin, F. (2015). Research on data stream clustering algorithms. *Artificial Intelligence Review, 43*(4), 593–600.
13. Tidke, B. & Mehta, R. (2018). A comprehensive review and open challenges of stream big data. In Pant, M., Ray, K., Sharma, T., Rawat, S., Bandyopadhyay, A. (eds.) *Soft Computing: Theories and Applications, Advances in Intelligent Systems and Computing* (Vol. 584, pp. 89–99).
14. Everitt, B., Landau, S., Leese, M., & Stahl, D. (2011). *Cluster Analysis* (5th ed., pp. 43–69). West Sussex: Chichester.

# Task Assignment Optimization of Multi-logistics Robot Based on Improved Auction Algorithm

**Juntao Li, Kai Liu, and Huiling Wang**

**Abstract** At present, "parts-to-picker" picking mode is one of the hotspots in the field of logistics warehousing. In this paper, the problem of multi-logistics robot task assignment is studied. It is proposed that the load balancing of a single robot is based on minimizing the total cost of robot execution tasks, and an improved auction algorithm is set to solve the problem. Firstly, an optimization model that comprehensively considers the total cost of the robot to perform tasks and the load balancing of a single robot is established, so that the result of task assignment is more in line with the actual situation. Then consider the correlation between tasks, using the improved auction algorithm for the task solving of the robot. The experimental results show that compared with the traditional auction algorithm, under the condition of ensuring that the total cost of executing multi-logistics robot is minimized, the cost of each robot to perform tasks can be balanced, and the task assignment is more reasonable.

**Keywords** Multi-logistics robot · Task assignment · Auction algorithm · Load balancing · Intelligent warehouse

---

Young top-notch talent project of high level teacher team construction in Beijing municipal universities (CIT&TCD201704059); Funding Project for Beijing Intelligent Logistics System Collaborative Innovation Center; Funding Project for Beijing Key Laboratory of intelligent logistics systems; Beijing excellent talents support project (2017000020124G063); The logistics robot system scheduling research team of Beijing Wuzi University.

J. Li (✉) · K. Liu · H. Wang
School of Information, Beijing Wuzi University, Beijing, China
e-mail: ljtletter@126.com

K. Liu
e-mail: 18810250658@163.com

H. Wang
e-mail: 18810712038@163.com



J. Zhang et al. (eds.), *LISS2019*,
https://doi.org/10.1007/978-981-15-5682-1_4

## 1 Introduction

With the rapid development of e-commerce and the rising labor costs, traditional warehousing and logistics technology has become more and more difficult to adapt to the needs of e-commerce development. Mobile robot technology support-ed by artificial intelligence technology has begun to be applied to the field of warehousing and logistics. This sorting mode suitable for e-commerce logistics with multiple varieties, high-frequency and small batches has emerged as the times require, namely the multi-logistics intelligent storage picking system represented by KIVA System [1]. The intelligent storage system picking problem includes road network layout, storage allocation [2], order batching [3], task assignment, path optimization [4], etc. This paper focuses on Multi-robots Task Allocation (MRTA). That is, how to properly allocate the order task to the robot, so that the total cost of the entire system is the lowest.

The problem of multi-robot task assignment is an NP problem. Generally, only approximate solutions can be obtained. At present, there are many ways to solve such problems. The Multi-robot task assignment problem is similar to TSP [5], but the normal TSP cannot be applied to smart storage systems. There are many researches on robot task assignment problems at home and abroad. According to the dynamic characteristics of tasks [6], task assignment problems are divided into dynamic task assignment problems and static task assignment problems. Static task assignment does not take into account factors such as the complexity of the multi-robots system and the uncertainty of the operating environment. Dynamic task assignment is the focus and difficulty of multi-robot task assignment research. The dynamics of task assignment [7] includes many aspects, such as the generation of new tasks, delays in executing tasks, and system congestion, which need to be redistributed. Nan et al. [8] proposed a dynamic task redistribution using a combination of sequential single auction and re-auction for the task delay of multiple robots.

There are also many research methods for task assignment at home and abroad. Nediah et al. [9] used the improved particle swarm optimization algorithm to solve distributed task problems and solve multi-robot task assignment problems. Elango [10] classifies tasks by K-means, and then uses auction algorithms to assign tasks to robots at the lowest cost, achieving the lowest total system cost. Li [11] studied the task assignment problem of intelligent warehousing robots in the static task assignment environment, and used the improved ant colony algorithm to study the allocation of warehousing tasks. Guo [12] introduced the task correlation function and the task's own cost function, and used the auction algorithm to solve the task assignment problem. Qin et al. [13] proposed an improved ant colony algorithm for solving the problem of multi-robot task assignment convergence slow and easy to fall into local optimal problem.

At present, in the research of multi-logistics robot task assignment problems, most of the research focuses on the lowest total cost of robots to perform tasks, but always ignores the balance of robots performing tasks, that is, single robot load balancing. In this case, bias behavior occurs in a multi-robot system. For example, when a robot performs a task, it always completes the task with the lowest cost of execution, which causes some robots in the robot system to perform too many tasks. Some robots perform fewer tasks, which greatly reduces the overall efficiency of the robot system. Although Zhou et al. [14] considered the load balancing of robots, it did not consider the parallel operation of multiple picking stations. Therefore, this paper mainly focuses on the above problems. In the parallel operation mode of multiple picking workstations, an optimization model is proposed, which considers the minimum total cost of robot systems and robot load balancing, and designs an improved auction algorithm to solve the model. Robot task assignment is more reasonable.

## 2 Problem Description and Model Building

### *2.1 Description of the Problem*

The problem involved in this paper is the multi-logistics robot task assignment of the intelligent warehouse picking system. In order to describe the system reasonably, the following assumptions are made for the multi-robot system:

- Robot and the mission are all isomorphic, and the robots have the same speed and are uniform.
- Each robot can perform assigned tasks independently.
- Each robot can only carry one shelf at a time.
- The tasks on the shelves are stored randomly.
- The congestion problem of logistics robot handing process is not considered.

The intelligent warehouse picking system based on multi-logistics robot mainly includes: picking station, shelf, robot, charging station, temporary storage area, picking staff, etc., as shown in Fig. 1. One of the shelves contains multiple bins, each containing one type of cargo (task).

The specific process of the multi-logistics robot picking system is as follows: First, a series of orders are distributed to each picking station according to the corresponding order batching rules, and the picking station assigns the task to the robot according to a certain task allocation rule, and the robot reaches the designated shelf position from the current position, and transport the shelves to the designated picking station. The picking staff then picks up the tasks from the shelves and places them in the designated order box. Finally, the robot puts the shelf back to its original position and performs the next task, until all tasks are completed.

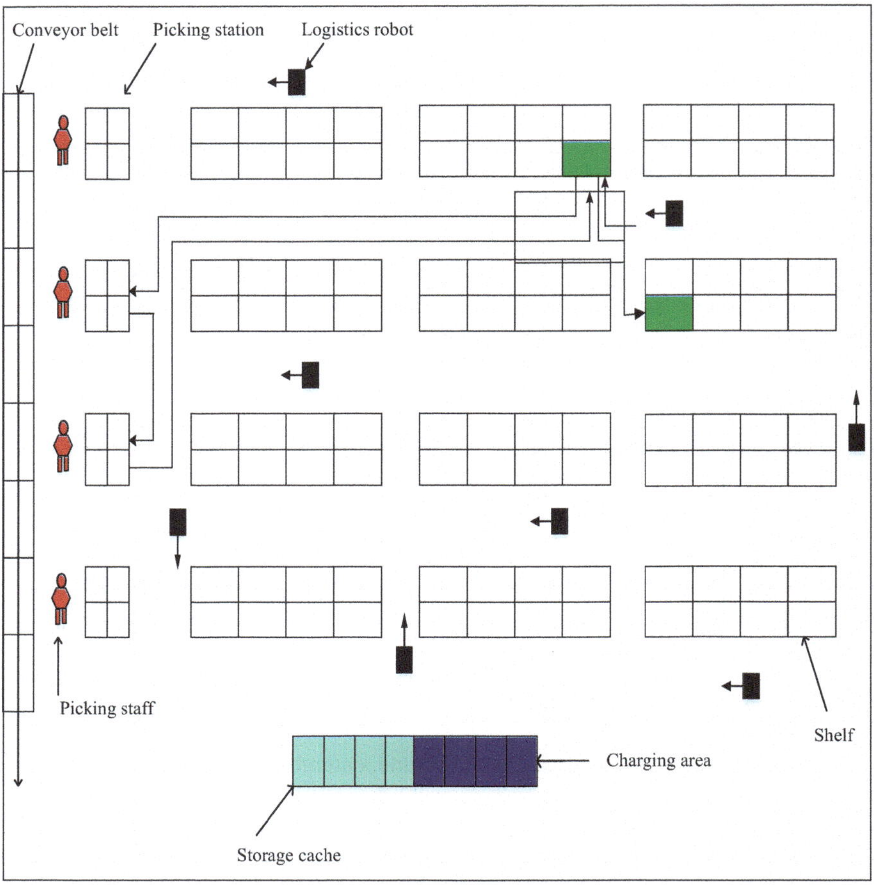


**Fig. 1** Multi-logistics robot picking system flow chart

In the problem studied in this paper, the warehouse picking system has already batched a series of orders to the designated picking station. The location of the task shelf and the location of the picking station are known, mainly considering how to assign tasks to each. The robot makes the robot system perform tasks with the best efficiency.

## *2.2 Model Construction*

In the multi-logistics robot task assignment problem model, the parameters and variables that need to be used are as follows:

$i$: robot, $i = 1, 2, \ldots, n$;
$j$: task, $j = 1, 2, \ldots, m$;
$s$: picking station, $s = 1, 2, \ldots, k$;
$c_{ij}$: Cost of Robot $i$ Completing Task $j$;
$T_i$: robot $i$ complete the collection of tasks;
$q$: shelf where the task is located, $q = 1, 2.., Q$;
$y_{qs}$: shelf $q$ serves the picking station s;

$$\eta_{ij} = \begin{cases} 1, & \textit{robot i performs task j} \\ 0, & \textit{otherwise} \end{cases}$$

$$x_{s\eta ij} = \begin{cases} 1, & \textit{robot i perform task j service picking station s} \\ 0, & \textit{otherwise} \end{cases}$$

$$\omega_{jq} = \begin{cases} 1, & \textit{task j on the shelf q} \\ 0, & \textit{otherwise} \end{cases}$$

According to the above variable description, assuming that the robot performs three adjacent tasks $j - 1$, $j$ and $j + 1$, considering the correlation between the two adjacent tasks, and the correlation between the two adjacent tasks $j - 1$ and $j$, the following definitions are made:

- If the two tasks are on the same shelf, there is a correlation between $j - 1$ and $j$, and $\gamma_{(j-1)j} = 1$; otherwise, when $\gamma_{(j-1)j} = 0$, the two tasks are on different shelves and are irrelevant.
- The two tasks are on the same shelf and serve the same picking station. There is a strong correlation between $j - 1$ and $j$, and $\varphi_{(j-1)j} = 1$; otherwise, $\varphi_{(j-1)j} = 0$, there is a weak correlation.

Considering that the cost of completing the task $j$ by the robot is $c_{ij}$, according to the operation mode of the intelligent storage system, it is divided into the following three cases. Among them, $d_{ij}$ represents the distance from robot $i$ to task $j$; $d_{js}$ represents the distance from task $j$ to picking workstation $s$; $d_{ss'}$ represents the distance between two picking workstations.

- If $\gamma_{(j-1)j} = 0$, unrelated between tasks, then $c_{ij} = d_{ij} + (1 + \left|\gamma_{j(j+1)} - 1\right|)d_{js} + d_{ss'}$;
- If $\gamma_{(j-1)j} = 1$ and $\varphi_{(j-1)j} = 1$,tasks are related and strongly correlated, then $c_{ij} = 0 + \left|\varphi_{j(j+1)} - 1\right| \times d_{ss'}$;
- If $\gamma_{(j-1)j} = 1$ and $\varphi_{(j-1)j} = 0$, tasks are related and weakly correlated, then $c_{ij} = d_{ss'} + \left|\varphi_{j(j+1)} - 1\right| \times d_{js}$.

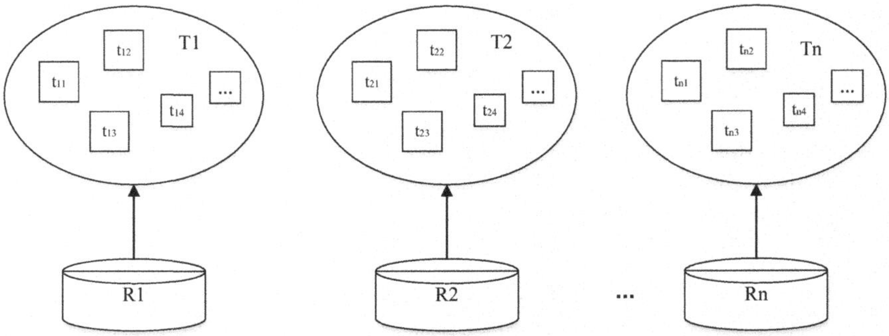


**Fig. 2** Each robot performs a task set

According to the workflow of the logistics robot and the above description, the cost of completing the assigned task $T_i$ by a single robot can be obtained, which is expressed as:

$$ITC = \sum_{s=1}^{k}\sum_{j=1}^{T_i} c_{ij} \times x_{s\eta ij} \tag{1}$$

Each logistics robot has a collection of execution tasks, as shown in Fig. 2. To ensure the lowest total operating cost of the system and to ensure that the logistics robot performs task load balancing, the multi-robot task assignment problem can be expressed as the following optimization model:

$$\underset{\Gamma}{Min}\,\underset{i}{Max}\ ITC(i, T_i) \tag{2}$$

Restrictions:

$$\sum_{s=1}^{k} x_{s\eta ij} = 1 \quad i = 1, 2, \ldots, n;\, j = 1, 2, \ldots, m \tag{3}$$

$$\sum_{i=1}^{n} \eta_{ij} = 1 \quad j = 1, 2, \ldots, m \tag{4}$$

$$\sum_{q=1}^{Q} y_{qs} \le \frac{\sum_{s=1}^{k}\sum_{q=1}^{Q} y_{qs}}{w} \quad s = 1, 2, \ldots, k;\ 1 \le w \le k \tag{5}$$

$$\sum_{q=1}^{l} \varpi_{jq} = 1 \quad j = 1, 2, \ldots, m \tag{6}$$

$$x_{s\eta ij} \in \{0, 1\} \;\; i = 1, \ldots, n; \; j = 1, \ldots, m; \; s = 1, \ldots, k \tag{7}$$

$$\eta_{ij} \in \{0, 1\} \qquad i = 1, 2, \ldots, n; \; j = 1, 2, \ldots, m \tag{8}$$

$$\varpi_{jq} \in \{0, 1\} \quad j = 1, 2, \ldots, m; \; q = 1, 2, \ldots, l \tag{9}$$

Objective is minimizing the total cost of robots performing tasks while balancing robot load. (3) means that a task is only assigned to one picking station. (4) a task is assigned to only one robot; (5) ensures that the picking station is load balanced, When $w = 1$, the equation is always true; when $1 < w < k$, the constraint becomes stronger as $w$ becomes larger; when $w = k$, the constraint is strongest. (6) means that a task is only on one shelf. (7)–(9) is binary variables.

## 3 Algorithm Design

As an improvement of the contract network, the auction algorithm is widely used in the task assignment of multi-logistics robots. There are many ways to auction algorithms, such as single auction, combined auction and so on. The auction algorithm is more and more widely used in multi-robot task assignment. The auction idea is simple and easy to understand. No central controller is needed. Each robot of hemostatic drugs receives all bids from all robots. Each robot can determine each auction. The winner, and assign the corresponding task to the winning robot to complete a round of auction.

The traditional auction algorithm, the robot does not consider the correlation between the tasks in the process of executing the task, which causes the cost of executing the task to be high. In addition, only the cost of executing the task is considered to be the lowest, and the balance of the task performed by the robot is not considered, so the robot is caused. The bias behavior occurs during the execution of the task, so this paper improves the traditional auction algorithm.

According to the consideration of minimizing the total cost of robots performing tasks and considering the cost of balancing each robot's execution tasks, an improved auction algorithm is proposed to solve the model. The basic steps are mainly divided into: First, determining the initial task; Second assigning the task according to the relevance of the task; Then considering the cost of the robot to complete the assigned task; and finally adjusting the allocation task. The algorithm flow chart is shown in Fig. 3. In order to better describe the multi-logistics robot task assignment problem, the Manhattan distance is used instead of the Euclidean distance. For example, two points $P_1(X_1, Y_1)$, $P_2(X_2, Y_2)$ are expressed as $d_{manhatton}(P_1, P_2) = |X_1 - X_2| + |Y_1 - Y_2|$.

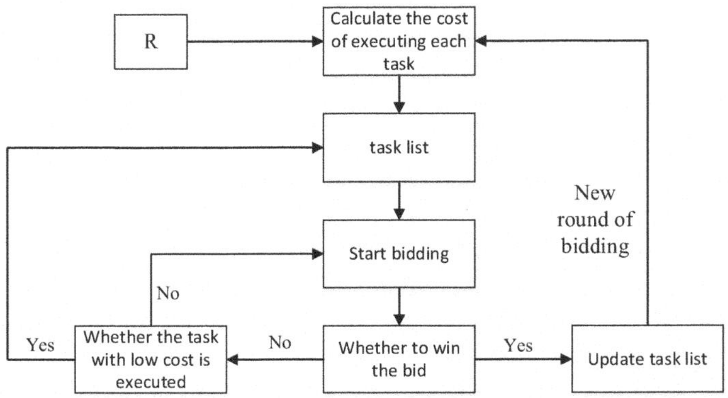


**Fig. 3** Traditional auction algorithm

Step 1: Initial task assignment

Each robot has a task set $T_i$, and the task set of the robot is just an empty set. Calculate the cost of each robot from the initial position to each shelf, and derive a cost matrix for moving the shelves. The shelf with the lowest robot handling cost is used as the robot to perform the initial task shelf until each robot has an initial task shelf, and goes to step 2.

Step 2: Correlation between tasks

According to the correlation between the tasks, the correlation matrix between the tasks is obtained, and the tasks with strong correlation and weak correlation on the shelf are added to the list of tasks performed by the robot until the tasks on the shelf are executed by the robot. The tasks are irrelevant, then put the shelves back, and go to step 3.

Step 3: Task assignment process

For the robot that completes the task, calculate the total cost of the robot to perform the assigned task, start a new round of auction process, re-bid the remaining tasks, and maximize the cost of the robot to perform the task. According to (2), while making all Robots have the lowest cost of performing tasks. Determine the robot that performs the least cost, judge the cost of executing the remaining tasks, and perform the task with the least cost as the next task of the robot, and then go to step 2 until all the tasks on the shelf are completed.

Step 4: Dynamic adjustment process

In the actual multi-logistics robot picking operation system, the robot may cause errors in the execution of tasks due to queue congestion or communication delay. Since the model does not take into account the delays or malfunctions of the robot performing tasks, consider the time window for such problems. Set the time node of each auction, return the remaining tasks executed by the robot back to the picking station, repeat step 2–step 3, perform the auction of the task, and finally realize the process of dynamic adjustment.

# 4 Experimental Verification

## *4.1 Experimental Description*

In order to verify the effectiveness of the task allocation model and algorithm of the intelligent warehouse multi-robot picking system, MATLAB is used for simulation and compared with the traditional auction algorithm. As shown in Fig. 4, there are 120 shelves in a 200-m$^2$ warehouse, and each shelf has 4–8 items. Note: Not every shelf has a task to perform. According to the relevant principle of order batching, a series of order assignment tasks are assigned to different picking stations for task assignment. Now assume that there are 40 orders, a total of 300 tasks need to be

**Fig. 4** Algorithm flow chart

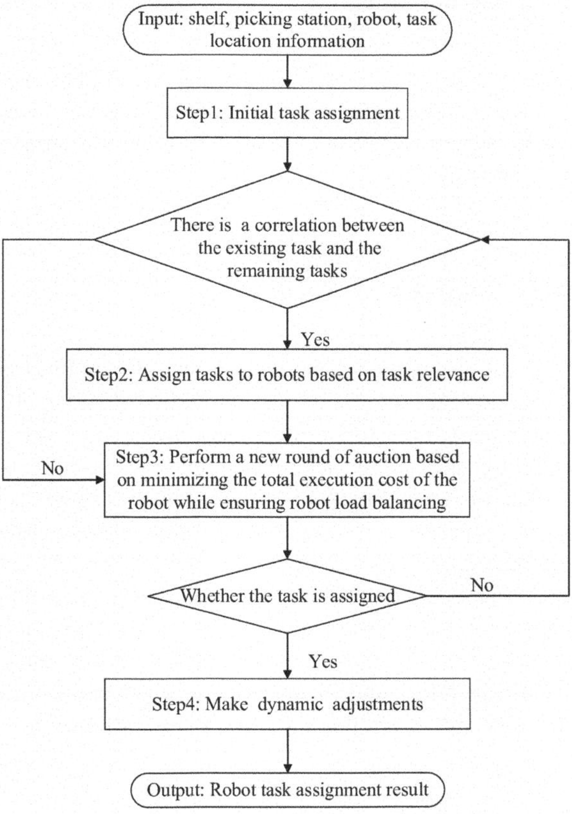

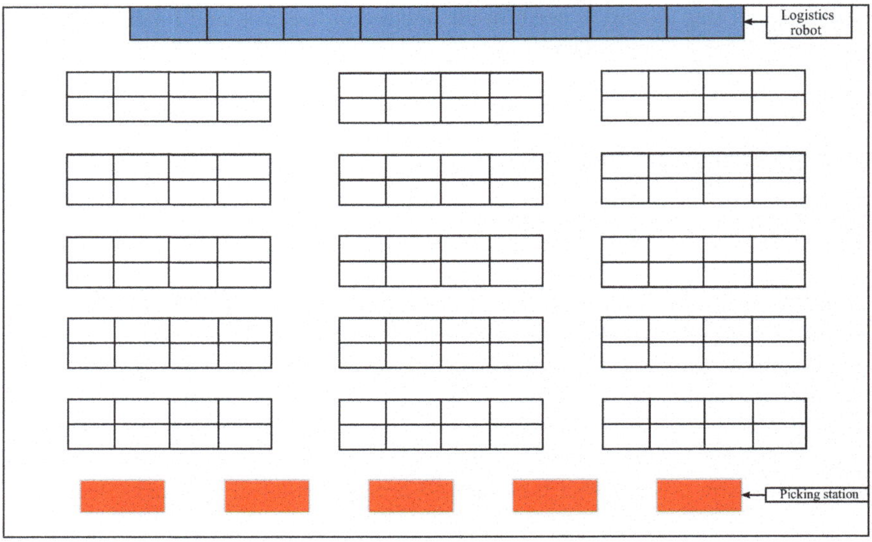


**Fig. 5** Lab environment

picked, 5 picking stations, and 8 logistics robots. Among them, the location of the task shelf and the picking station are known, the distance between any two points is calculated by Manhattan distance, and the time tempo for each task auction is set to 10 min (Fig. 5).

## *4.2 Analysis of Results*

The task assignment scheme is a record of multi-logistics robot task assignment. The downtime of the reduction task is the target of multi-logistics robot task assignment. For 100 orders, 300 tasks solve the task assignment problem of multi-logistics robot picking system.

The model and method of multi-logistics robot task assignment proposed in this paper considers the correlation of multi-logistics robots to perform tasks, and the ultimate goal is to minimize the total system cost and ensure the load balance of each robot. Through the improved auction algorithm, the optimization model is solved, and 300 tasks are assigned to 8 robots, and the assignment result of the robot execution task is obtained, including the total cost of each robot performing the assigned task and the sequence of executing the tasks, as shown in Table 1. And it can be concluded that the total cost of performing a multi-logistics robot is 1952.

**Table 1** The assignment result of the robot performing the task

| Number | Task allocation scheme | Execution cost |
|---|---|---|
| 1 | 44,48,14,19,20,21,1,2,3,4,9,13,16,29,34,12,18,15,17,24,5,6,7,16,29, 34,12,18,15,17,24,5,6,7,87,89,148,189,190,191,295,271,272,273 | 240 |
| 2 | 11,22,30,33,27,28,116,118,75,80,126,134,108,54,138,144,109,146,78, 128,49,116,118,75,80,156,157,280,169,170,186, 187,188,207,288,268,269,270 | 238 |
| 3 | 114,115,59,67,130,132,120,123,73,74, 64,65,66,72,122,124,127,143,107,111,92, 98,102,141,145,155,176,181,158,277,293 | 242 |
| 4 | 31,46,46,47,42,43,290,287,298,286,91, 137,105,112,113,8,10,83,99,25,32,36, 153, 154,164,165,199,200,289,193,194,285 | 251 |
| 5 | 119,121,37,60,84,49,52,79,78,282,222,223,224,110,149,86,90,100,103,69,71,77,159, 160,284,276,225,226,227,175,180, 185,198 | 231 |
| 6 | 50,51,61,68,125,88,95,58,62,93,94,96, 97,117,129,136,150,235,161,162,163, 240,243,246,228,229,230,201,206,202,253,274,275,279,283,299,292,211,219, 220,205,210,214 | 259 |
| 7 | 133,135,101,106,257,260,131,140,147, 76,241,244,254,234,56,57,40,41,53,217,218, 204,209,213,215,216,178,183,192,195,256,259,251,255,250,265,239,242, 245,291,262,263,264,281,296 | 246 |
| 8 | 23,26,35,38,139,142,63,70,82,85,81,297,203,208,212,266,267,171,172,173,174,166, 167,168,179,184,252,231,232,236, 237,247,233,258,261,248,249,221,238, 294,300 | 245 |

In order to verify the relevance of the logistics robots in this paper, the proposed task allocation method can effectively reduce the total cost of system operation and load balance the individual robots, and compare the traditional auction algorithms with simulation verification from the multi-logistics robot system. The total cost and the cost of a single logistics robot are analyzed.

(1) *Total cost analysis*

Select different orders n = 100, 200, 300 for simulation experiments. It can be seen from Fig. 6. that as the number of tasks increases, the total cost of the robot system increases, compared with the traditional auction algorithm. The proposed algorithm is obviously superior to the traditional auction algorithm. The proposed algorithm considers the relevance of logistics robots to perform tasks, reduces the number of shelves to be transported, and can effectively ensure the lowest total system cost.

(2) *Cost analysis of each robot*

The cost of each robot performing a task is shown in Fig. 7. It can be seen from the figure that during the execution of the eight logistics robots, the algorithm of this paper can ensure that the task cost of the logistics robots performing the tasks tends to be balanced; compared with the proposed auction algorithm in this paper, the traditional auction algorithm experiment There is a undulation in the execution of a single logistics robot, which causes the robot to have a biased behavior during the execution of the mission. It can be seen that the proposed algorithm is more reliable than traditional auction algorithms to ensure that each robot is load balanced and avoids bias behavior.

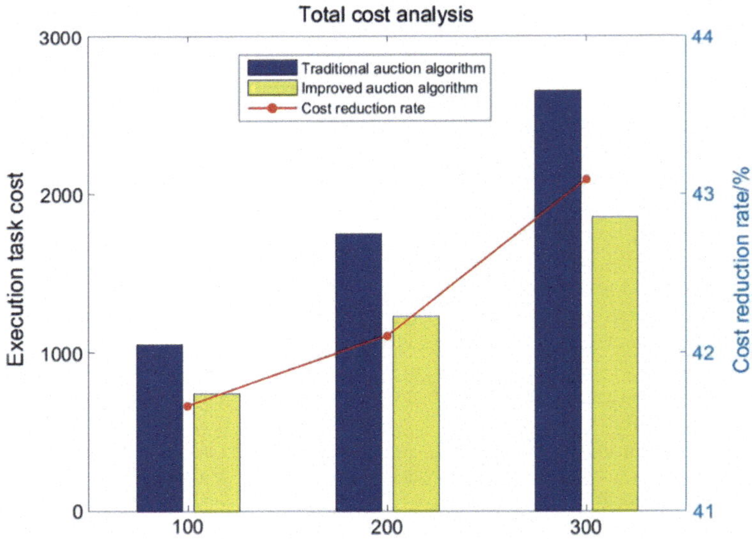


**Fig. 6** Comparison of total cost of traditional auction algorithm and improved auction algorithm

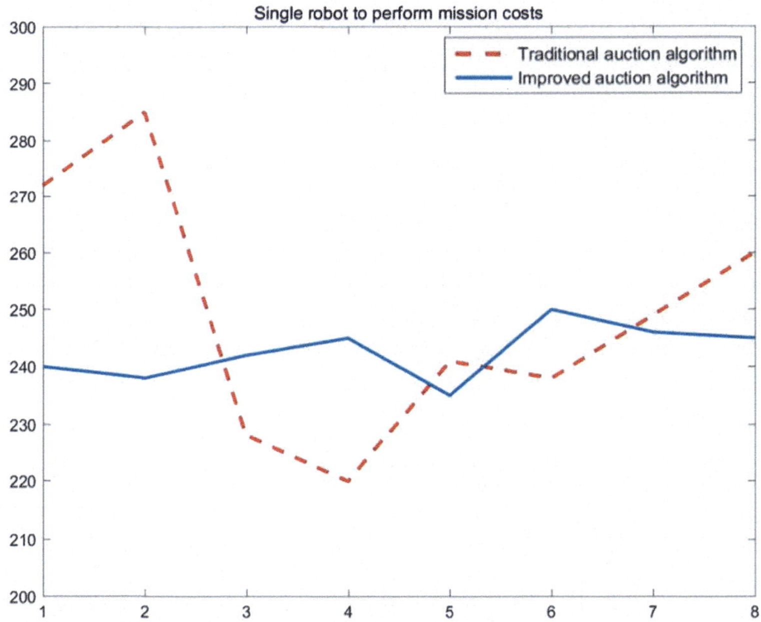


**Fig. 7** Cost analysis of single robot under traditional auction algorithm and improved auction algorithm

In summary, the optimization model and the improved auction algorithm constructed in this paper can obtain the optimization scheme of multi-logistics robot task allocation. On the one hand, as shown in Fig. 6, the total cost of the robot is the lowest, which improves the overall operation efficiency of the system. On the other hand, as can be seen from Fig. 7, the task solving algorithm in this paper can balance the load of a single logistics robot, which is conducive to the parallel operation of large-scale logistics robots.

## 5 Conclusion

In this study, the dynamic task assignment of multi-robot picking system under the "parts-to-picker" mode is studied. Considering the relevance of robots' execution tasks, an improved auction algorithm is proposed to optimize the task assignment problem of multi-robots. The research method can ensure the rationality of multi-robot task assignment, and can ensure the load balancing of a single robot and achieve reasonable task assignment under the condition that the total cost of the robot performing the task is the lowest. Therefore, the algorithm and optimization model in this paper are feasible for solving the task assignment problem of multi-robot. However, in the research of multi-robot task allocation, the congestion problem of picking workstation is not taken into account in this paper. In the parallel operation mode of multi-picking workstations, it is necessary to study the congestion problem of picking workstations when a robot carries a shelf to serve multiple picking workstations. In the next research, this problem will be studied, considering the load balancing of the picking table and the load balancing of the robot, to achieve the optimal result of the task allocation of the robot.

## References


1. Enright, J.J. & Wurman. P.R. (2011). Optimization and coordinated autonomy in mobile fulfillment systems. In *Conference on Automated Action Planning for Autonomous Mobile Robots, 11*(9), pp. 33–38. AAAI Press.
2. Boysen, N., Briskorn, D., & Emde, S. (2017). Parts-to-picker based order processing in a rack-moving mobile robots environment. *European Journal of Operational Research, 262*(2), 550–562.
3. Ma, H., Koenig, S. & Ayanian, N. (2017) Overview: Generalizations of multi-agent path finding to real-world scenarios. In *IJCAI-16 Workshop on Multi-Agent Path Finding.*
4. Yu, J. (2016). Intractability of Optimal Multirobot Path Planning on Planar Graphs. *IEEE Robotics & Automation Letters, 10*(1109), 33–40.
5. Koubaa, A., Bennaceur, H., & Chaari, I. (2018) *Robot Path Planning and Cooperation.* Cham: Springer.
6. Zhang, W., & Liu, S. (2008). Research and development of multi-robot task assignment. *Journal of Intelligent Systems, 3*(2), 115–120.

7. Cao, P., Hao, J., & Ding, Y. (2018). Event-driven immune network algorithm for multi-robot dynamic task assignment. *CAAI Transaction on Intelligent Systems, 13*(6), 952–958.
8. Nanjanath, M., & Gini, M. (2010). Repeated auctions for robust task execution by a robot team. *Robotics and Autonomous Systems, 58*(7), 900–909.
9. Nedjah, N. (2015). PSO-based distributed algorithm for dynamic task allocation in a robotic swarm. *Procedia Computer Science, 51*(05), 326–335.
10. Elango, M., Nachiappan, S., & Tiwari, M. K. (2011). Balancing task allocation in multi-robot systems using K-means clustering and auction based mechanisms. *Expert Systems with Applications, 38*(6), 6486–6491.
11. Li, G.J. (2013). *Research on task assignment of warehouse robot based on intelligent optimization,* Harbin Institute of Technology.
12. Guo, Y. (2010) *Research on multi-robot task assignment method based on auction in intelligent warehouse system*, Harbin Institute of Technology.
13. Qin, X., Zong, Q., Li, X., Zhang, B., & Zhang, X. (2018). Multi-robot task assignment based on improved ant colony algorithm. *Space Control Technology and Application, 44*(5), 55–59.
14. Zhou, L., Shi, Y., & Wang, J. (2014). A balanced heuristic mechanism for multirobot task allocation of intelligent warehouses. *Mathematical Problems in Engineering, 2014,* 380480.

# Analyzing on Inner-Cluster Hop Number of SPMA-Based Clustering Ad-Hoc Network

**Haizhao Liu, Wenjun Huang, and Wenqing Zhang**

**Abstract** Scalability and efficiency are two important indicators of a large-scale ad-hoc network. Multi-hop clustering is an effective mechanism to enhance the scalability by adjusting the inner-cluster hop number as a decisive factor. Compared to conventional schemes, Statistical Priority Multiple Access (SPMA) scheduling guaranties quality of service within high priority by its channel monitoring and back-off mechanism in large-scale networks. In this paper, a novel framework consist of SPMA multi-hop clustering is proposed where the mechanisms of cluster routing and SPMA scheduling are both considered. The relationship between inner-cluster hop number and network performance is investigated by deriving system throughput. The simulation results show the selection of inner-cluster hop number has a great impact on network performance.

**Keywords** Network structure · Cluster size · SPMA · System throughput

## 1 Introduction

Mobile Ad-hoc network (MANET) with multi-hop clustering is shown in Fig. 1, in which all nodes are divided into clusters. By constraining the routing range within a cluster, the influence of node mobility is limited and so the re-routing overhead due to link failure can be decreased. Thus, in network scenarios, such clustering networks can provide better performance than planar ones, as proposed in [1].

In recent researches, some progress of clustering network has been made, especially on selecting cluster head nodes and enhancing stability of cluster. But

H. Liu (✉) · W. Huang · W. Zhang
School of Communication and Information, Beijing Jiaotong University, Beijing, China
e-mail: 17120083@bjtu.edu.cn

W. Huang
e-mail: 16111027@bjtu.edu.cn

W. Zhang
e-mail: 16120167@bjtu.edu.cn



J. Zhang et al. (eds.), *LISS2019*,
https://doi.org/10.1007/978-981-15-5682-1_5

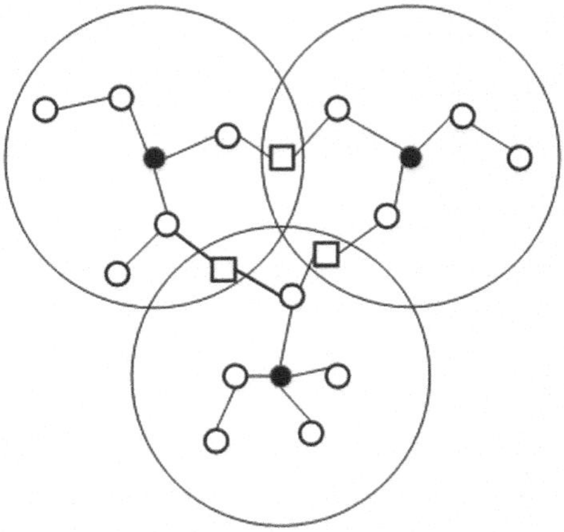

**Fig. 1** Multi-hop clustering MANET diagram

comprehensive research on cluster size in large-scale scenarios is inadequate. Max-min multi-hop clustering algorithm which is the first multi-hop clustering algorithm. Reference [2] proposes multi-hop clustering mechanism, but doesn't analyze the relationship between cluster size and system performance. An ant colony clustering algorithm was proposed in [3]. It elects the optimal cluster head by evaluating nodes' trust to maintain stability in clustering network, but analyzing on optimal cluster size isn't mentioned. Besides, in [4], Focusing on the effect of cluster size on cluster stability and routing overhead. But the influence of scheduling mechanism on system performance is not mentioned in that model. In [5], considering scheduling mechanism to analyze the cluster capacity in CDMA based wireless sensor clustering network which provides a thoughtful analysis of cluster size. But the model is not suitable for large-scale scenarios.

Besides, not every scheduling mechanism can perform well in clustering network. The scheduling schemes could be mainly classified into reservation and contention schemes. Time Division Multiple Access (TDMA) and Carrier Sense Multiple Access (CSMA) are the representative of these two access schemes, respectively. However, in large-scale scenarios, there are more collisions occurred under CSMA schemes, jeopardizing the QoS, as mentioned in [6]. TDMA schemes, on the contrary, can ensure collision-free transmissions through resources reservation was proposed in [7]. But the reservation procedure should occupy several slots before each data packet exchanging, and timely request traffic may be delayed.

Statistical Priority Multiple Access (SPMA) is a novel scheduling mechanism proposed by [8]. It has been regarded as an important part of next generation Data Link System. And, it supports different services with multiple priorities. Following traffic load increases, it guarantees reliability by putting back the packets with low priority. Besides, efficient Physical layer technique such as Turbo code and frequency hopping also been used in SPMA in [9]. Therefore, our research focuses on cluster size selection of SPMA multi-hop clustering network. We use inner-cluster hop number (ICHN) to represent cluster size in the following parts and analyze the impact of ICHN selection in clustering network which uses SPMA as scheduling mechanism in this paper. Deriving relationship between ICHN and system throughput by computing transmission probability per slot. And proving the importance of ICHN selection in clustering network.

The structure of the rest of this paper is summarized as follow. Section 2 introduces our system model including cluster structure analysis and throughput derivation. Section 3 exhibits the simulation results and verifies our model's correctness. Section 4 are the conclusion and acknowledgment of the paper.

# 2 System Introduction and Modeling

## 2.1 System Introduction

Our model is based on multi hop clustering network which scheduling by SPMA mechanism. As showed in Fig. 2, in SPMA scheduling process, each node has its own priority queue and get one packet from the queue after sending current packet.

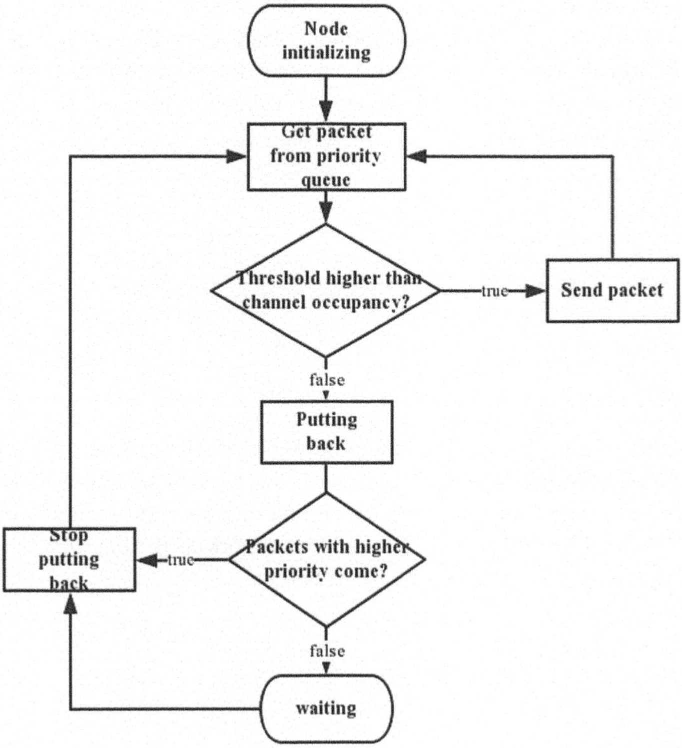


**Fig. 2** SPMA flow diagram

If channel occupancy higher than threshold of packet on service, SPMA will set current packet status to waiting. After its waiting period, SPMA measures the channel occupancy and make queueing or sending decision again.

A larger ICHN means one cluster will be consist of more nodes. The routing overhead of maintaining a cluster's structure increases, so as the burden of the cluster head, leading to a lower system throughput in [10]. On the contrary, choosing a smaller ICHN segments the network into too many clusters, which means the quality of service (QoS) between neighboring cluster heads cannot be guaranteed. An appropriate ICHN is decisive factor for the system performance. In clustering structure, each cluster head broadcast its own routing packets periodically to determine their cluster member. Every node use SPMA as their MAC layer protocol and each cluster has different transmission frequency points.

## 2.2 Cluster Structure Analysis

We assume N nodes in clustering network with Poisson Point Process, $\lambda_p$ is parameter of it. We use $h$ to represent ICHN, each cluster head determines their cluster members by sending and receiving routing packets to nodes in its h-hop range periodically. Routing table has been updated at the same time.

Each Node has its communication radius $r$, Hard-Core Point Processes (HCPP) can dilute notes distribution in network by specific rules, which is similar with cluster head selection in [11]. Therefore, we can get cluster heads' density and cluster numbers in network:

$$\lambda_h = \lambda_p \int_0^1 e^{-\lambda_p \pi h^2 r^2} = \frac{1 - e^{-\lambda_p \pi h^2 r^2}}{\pi r^2} \tag{1}$$

$$n_{clu} = \frac{N\lambda_h}{\lambda_p} = \frac{N\left(1 - e^{-\lambda_p \pi h^2 r^2}\right)}{\lambda_p \pi r^2} \tag{2}$$

Set of cluster heads make up a single network which delivers packets between different clusters. Following ICHN increases, the number of cluster heads in that network will decrease. And the average communication hop numbers depend on number of nodes in one-hop range and whole network as in [12]. Inner-cluster and Inter-cluster average transmission hop numbers is:

$$h_{\text{inn}} \approx \frac{\ln(n_n)}{\ln\left(\lambda_p \pi d_{\text{inn}}^2\right)} \frac{d_{\text{inn}}}{r} = \frac{\ln\left(\lambda_p \pi h^2 r^2\right)}{\ln\left(\lambda_p \pi r^2\right)} \tag{3}$$

$$h_{\text{int}} \approx \frac{\ln(N)}{\ln\left(\lambda_p \pi d_{\text{int}}^2\right)} \frac{d_{\text{int}}}{r} \tag{4}$$

$d_{\text{int}}$ is the distance of 1-hop transmission in cluster heads' network. We can analyze geometric situation to get it

$$d_{\text{int}} = \frac{\sum_{i=h+1}^{2h} (ir\lambda_{clu} s_i)}{\sum_{i=h+1}^{2h} (\lambda_{clu} s_i)} = \left(\frac{14h}{9} + \frac{1}{9h}\right) r \tag{5}$$

The routing overhead in network contains cluster members' and cluster heads' routing packets. And routing period is related to the time of 1-hop link keeping stable which depends on node's mobile speed.

In multi-hop network, cluster members not only send their own packets but also forward other members' packets. Cluster heads forward routing packets received from other cluster heads. We set routing maintain frequency of inner-cluster and inter-cluster network are $f_{inn}$ and $f_{\text{int}}$. The routing packet arrival rate of cluster member and head is:

$$\lambda_{mc} = f_{inn} \cdot h_{inn} \tag{6}$$

$$\lambda_{hc} = f_{inn} + f_{\text{int}} \cdot h_{\text{int}} \tag{7}$$

Data services can be divided into inner-cluster and inter -cluster service. We set $\gamma$ as inter-cluster service ratio, which describes demand of inter-cluster transmission. We can get $\gamma$ by assuming each cluster member has same service requirement:

$$\gamma = 1 - \frac{n_{inn}}{N} \tag{8}$$

## 2.3 System Throughput

Routing packets is regarded as the highest priority service in our model, because data transmissions are based on valid routes.

Assuming two data traffic flows with different priorities have the same arrival rate $\lambda$ following Poisson process. The actual arrival rate including forward service is $\lambda \cdot h_{inn}$. We assume three services with different priorities in network, two data services and one routing service. SPMA scheduling process is equivalent to M/G/1 queuing process within interruption mechanism. We can compute SPMA system throughput by inferring average slot transmission probability [13].

The cluster routing service will never be interrupted because it has the highest priority. And the probability that a data service is interrupted within a period of time

T is the arrival probability of any higher priority service in T. We mark the highest, medium and lowest priority level as 0, 1 and 2, respectively. According to routing service arriving rate which we derived in the previous subsection, the probability of a cluster member's packet of priority i will not be interrupted in a whole back-off period is:

$$b_i(T) = \prod_{j=0}^{i-1} e^{-\lambda_i \cdot h_{inn} \cdot T} \tag{9}$$

As $h_{inn}$ grows, the uninterrupted probability $b_{\mathrm{i}}(T)$ will decrease. It means the contention of inner-cluster channel resources become greater. Meanwhile the contention in the inter-cluster network become smaller.

Probability of packet within i-th priority could be sent in j-th back-off period is:

$$\mu_{i,j} = l_i^{j-1} \cdot (1 - l_i) \cdot \prod_{t=0}^{i-1} b_t(T_j) \tag{10}$$

$l$ is the probability of blocking as a result of high channel utilization. And each priority service has a different blocking probability determined by the numbers of sending frequency points in cluster and slots of detecting the channel occupancy. It's also related to slot transmission probability. Therefore, the packet deliver probability from a cluster member in allowed back-off times K is:

$$P_{ms,i} = \sum_{j=0}^{K} \mu_{i,j} \tag{11}$$

When the back-off time is greater than K, SPMA will stop back-off process and get a new packet from the queue. When channel quality is too poor to satisfy priority threshold. The packet within low priority cannot be sent in limit back-off times. In such situation, high priority packet will be served as first choice. This makes SPMA more stable than other contention schemes when channel quality is bad, which makes QoS of transmission can be guaranteed.

When channel occupancy gets better than previous detection and satisfies the threshold of packet on service, the packet could be sent. This mechanism depends on physical layer detection which guaranteed system reliability. It makes a full use of channel resource and avoiding invalid sending.

Next, the average service time of a packet in limit back-off times is related to the slot numbers of back-off window $W_i$ and the uninterrupted probability $b(T)$. In SPMA mechanism, $W_i$ increases by 1 slot in every back-off until back-off times reach K. Recall that the routing packets' uninterrupted probability is 1 because it has highest priority. Therefore, its average service time is:

$$\overline{X_0} = \sum_{i=0}^{K} W_i \cdot l_0^i \tag{12}$$

Data packets will be interrupted in back-off period when higher priority service arrive. And the average service time of data packets within 1 and 2 priority is:

$$\overline{X_1} = \sum_{i=0}^{K} W_i \cdot l_1^i \prod_{j=1}^{i} e^{-\lambda_0 \cdot h_{inn} \cdot W_i} \tag{13}$$

$$\overline{X_2} = \sum_{i=0}^{K} W_i \cdot l_2^i \prod_{j=1}^{i} e^{-\lambda_0 \cdot h_{inn} \cdot W_i} \cdot e^{-\lambda_1 \cdot h_{inn} \cdot W_i} \tag{14}$$

According to queuing theory, the average service probability in per slot is:

$$s_i = \frac{\rho_i}{\rho} = \frac{\lambda_i \cdot \overline{X_i}}{\lambda_0 \cdot \overline{X_0} + \lambda_1 \cdot \overline{X_1} + \lambda_2 \cdot \overline{X_2}} \tag{15}$$

After we derived key parameters of queuing process, we can get cluster member's data packet transmission probability per slot which is sum of service within priority 1 and 2's transmission probability per slot.

$$\overline{P_i} = \frac{s_i \cdot P_{ms,i}}{\overline{X_i}} \tag{16}$$

$$\overline{P} = \overline{P_1} + \overline{P_2} = \frac{s_1 \cdot P_{ms,1}}{\overline{X_1}} + \frac{s_2 \cdot P_{ms,2}}{\overline{X_2}} \tag{17}$$

Part of inner-cluster packets sends to other clusters under the help of cluster head. As mentioned in previous subsection, not all packets are inter-cluster service, it follows a certain proportion $\gamma$ which defines inter-cluster transmission demands. Then, the arrive rate of inter-cluster services should be:

$$\lambda_{hi} = \frac{n_{inn}\gamma \cdot \overline{P_i} \cdot h_{\text{int}} \cdot L_p}{\overline{\beta}} \tag{18}$$

$L_p$ is the length of data packet. $\beta$ is the timeslot length. It's also related to average ICHN and number of nodes in cluster. Following ICHN increases, number of cluster heads will decrease. It could reduce the competition between cluster heads which leads to better performance of cluster heads network.

Therefore, we can express system throughput as following equation:

$$S_i = n_{inn} \cdot n_{clu} \cdot L_p \cdot \left( \frac{\gamma \cdot \overline{P_{hs,i}} + (1+\gamma)}{\beta} \right) \quad (19)$$

$\overline{P_{hs,i}}$ is cluster head's transmission probability per slot, i is priority symbol. It can be derived as $\overline{P_i}$ which we have derived in this section. According to our derivation result, system throughput is relative to clustering network structure which affected by ICHN, length and arriving rate of data packets, and nodes number in network.

# 3 Simulation and Analysis

## 3.1 Simulation Introduction

In this section. We simulated the clustering network model on MATLAB. And analyze the simulation results which describe the relationship between ICHN and system performance.

N nodes in clustering network which use SPMA as scheduling mechanism. Each cluster use $N_f$ frequency points to send packets. Cluster heads' network use different frequency points from cluster members network. Either routing or data packet can be sent in one slot. Considering two data services arrive in same rate following poison process, just as mentioned in Sect. 2. And routing service arrive rate is related to maximum mobile speed. In cluster heads' network, service arrive rate decided by inner-cluster members' slot transmission probability. When ICHN grow, the back-off probability of inner-cluster members' services increase. It represents competition between cluster members become greater.

## 3.2 Simulation Result Analysis

The key parameters of system simulation are shown in Table 1.

**Table 1** Parameters of simulation

| Symbol | Meaning | Value |
|---|---|---|
| N | Number of nodes | 200 |
| V | Maximum mobile speed | 5 m/s |
| $L_P$ | Length of data packets | 1024 bit |
| $L_R$ | Length of route packets | 256 bit |
| $W_0$ | Minimum back-off window length | 16 slots |
| $W_K$ | Maximum back-off window length | 32 slots |
| h | Range of hop numbers | [1, 4] |

We simulated average transmission hop numbers at the beginning. According to conclusion in (1), (2) and (3), average transmission hop numbers is related to ICHN h and node's transmission radius *r*. Figures 3 and 4 describe trend of average

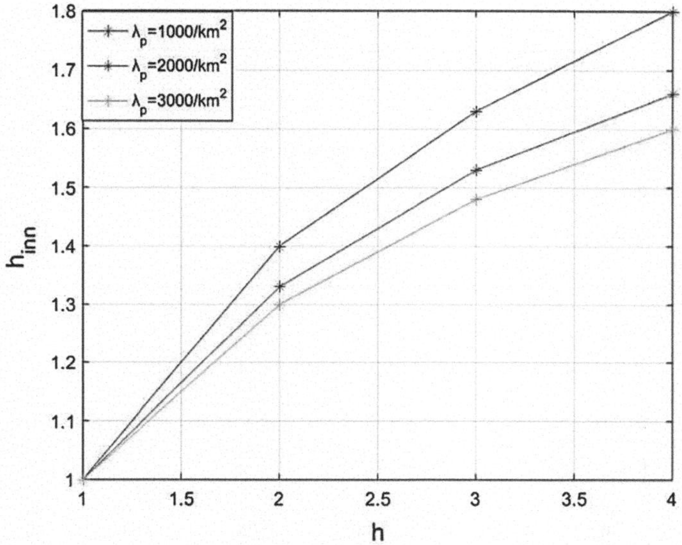


**Fig. 3** Relationship between average forward hop numbers inner-cluster and ICHN

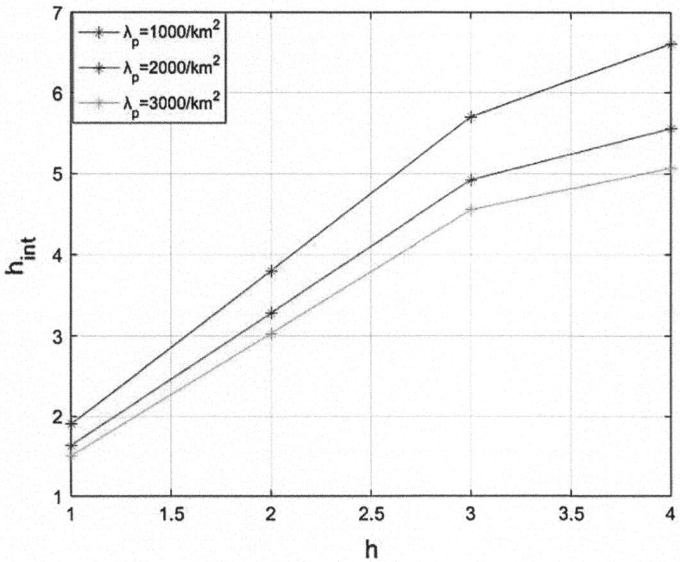


**Fig. 4** Relationship between average forward hop numbers inter-cluster and ICHN

transmission hop numbers when h increases, $h_{inn}$ and $h_{\mathrm{int}}$ both grow. It can lead to increment of forward tasks, which means more sending tasks.

Figures 5 and 6 describe trend of slot transmission probability when $h$ is 1 and 2. In (11), data transmission probability was divided into transmission probability of

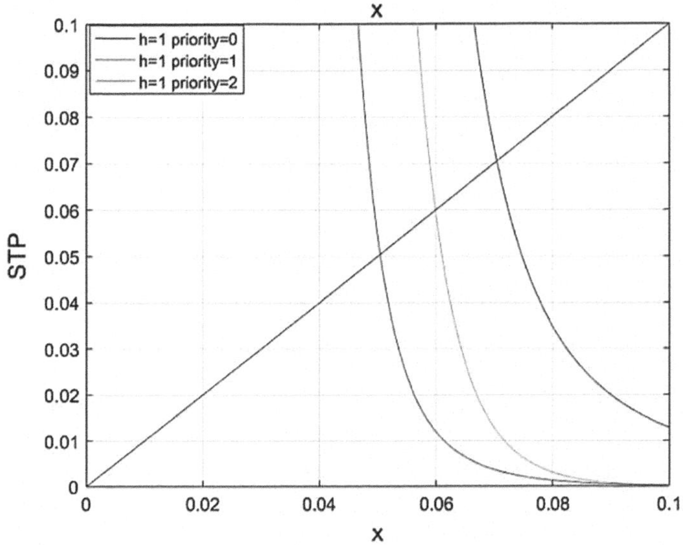


**Fig. 5** Slot transmission probability when h = 1

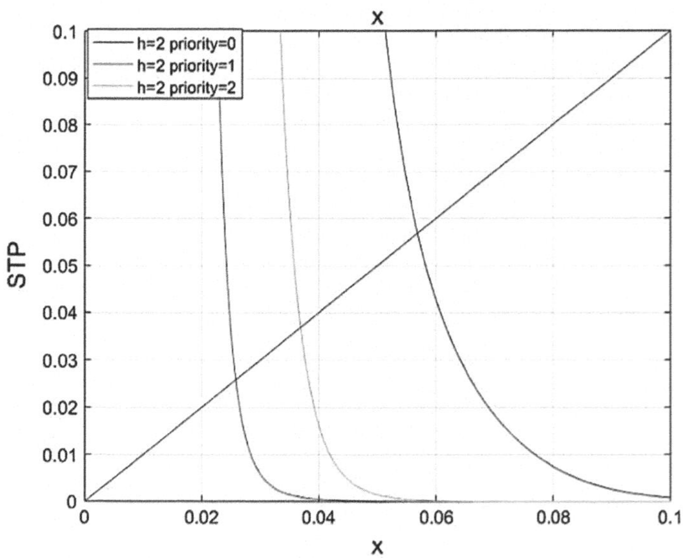


**Fig. 6** Slot transmission probability when h = 2

priority 1 and 2. Note that probability of blocking in (7) affects by transmission probability. Hence slot transmission probability is contained in right side of equal sign in (11). Therefore, the intersection of $y = \overline{P_i}$ and $\overline{P_i}$'s expression is slot transmission probability. The value is intersection's Y-axis coordinate. As shown in Figs. 5 and 6, service with priority 0 has higher transmission probability than priority 1 and 2 due to SPMA congestion control mechanism. Transmission probability per slot of packets with priority 0 reaches 0.07 when h = 1. Another two priorities' packets have less chance to be sent in transmission period.

Comparing two figure, it's obvious to know sending probability of inner-cluster services decrease when ICHN becomes bigger. And service within lowest priority's transmission probability loses nearly half. It seems smaller h is better for system performance. But smaller h means more cluster numbers in network. It will lead to resource shortage in cluster heads network. If every node has the same demands on communicate with other nodes. Inter-cluster transmission quality also needs to be guaranteed. Therefore, small ICHN can make inner-cluster service's quality to be guaranteed. But inter-cluster service will not that satisfied in that case. Conversely, too big ICHN could result in greater competition between cluster members which caused by increasing member numbers in cluster. Even if cluster heads network could have better performance than smaller ICHN. Cluster heads could not receive enough member's packets due to bad inner-cluster transmission performance in that case.

Figure 7 shows the relationship between ICHN and system throughput. Total throughput is sum of throughput of two data services within priority 1 and 2. And higher priority service corresponds more chances of sending packets. When h bigger than 3, throughput drops rapidly. According to our previous analyses, it caused by low transmission probability of inner-cluster service. And 1-hop clustering structure performs unsatisfied too. System throughput become best when hop numbers is 3.

In this section, we focus on how different ICHN affects system performance by computing throughput which is based on our model. First of all, we analyze the relationship between ICHN selection and network structure by deriving h under different ICHN.

After that, we derive transmission probability of each slot according to M/G/1 queuing model. And simulate the system throughput under different ICHN, The differences of system performance when ICHN is changing explains why inappropriate ICHN has negative effect on system performance. So, our simulation result shows the importance of ICHN selection in SPMA clustering network.

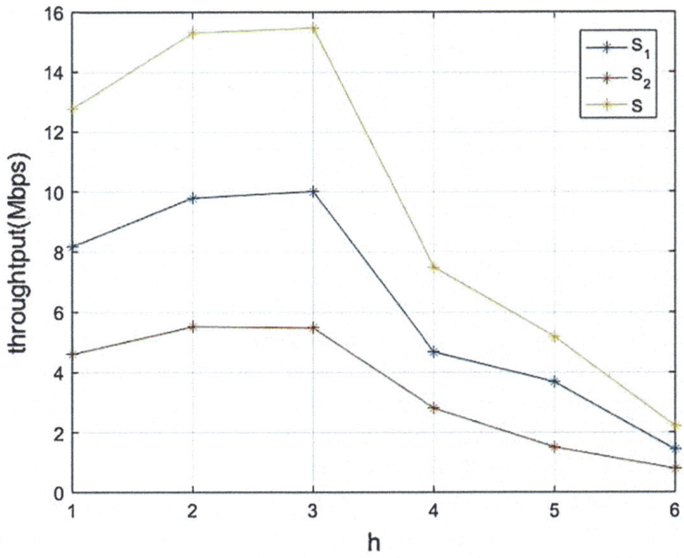


**Fig. 7** Relationship between system throughput and inner-cluster hop number

## 4 Conclusion

Clustering structure was applied to large-scale Ad-hoc network to solve planar structure's scalability problem. Cluster size selection under suitable scheduling mechanism is an unsolved problem. SPMA is a novel scheduling mechanism which has better performance than previous contention-based scheduling protocol in large-scale scenarios. In this paper, our research focuses on cluster size selection of SPMA clustering structure network. We use ICHN to represent cluster size and build a model to characterize SPMA clustering structure network. To explain the importance of ICHN selection and system performance, we derive throughput under different ICHN by considering routing and schedule mechanism. Simulation results show that ICHN should be adjust appropriately to achieve better system performance. Conversely, unsuitable ICHN can lead to bad system performance. So, ICHN selection has a great impact on network performance.

Other clustering network performance indicators such as queuing delay in scheduling process and maximum node number in a cluster when ICHN is different could be focused in future research.

## References

1. Wei, D., & Chan, H.A. (2006). Clustering ad hoc networks: Schemes and classifications. In *3rd Annual IEEE Communications Society on Sensor and Ad Hoc Communications and Networks, Reston, VA* (pp. 920–926).
2. Amis, A.D., Prakash, R., Vuong, T.H.P., & Huynh, D.T. (2000). Max-min d-cluster formation in wireless ad hoc networks. In *Proceedings IEEE INFOCOM 2000. Conference on Computer Communications. Nineteenth Annual Joint Conference of the IEEE Computer and Communications Societies (Cat. No.00CH37064), Tel Aviv, Israel* (Vol. 1, pp. 32–41).
3. Chen, Y., Xia, X. & Wang, R. (2013). An ad-hoc clustering algorithm based on ant colony algorithm. In *IEEE International Conference on Granular Computing (GrC), Beijing* (pp. 64–69).
4. Niu, X., Tao, Z., & Wu, G. (2006). Hybrid cluster routing: An efficient routing protocol for mobile ad hoc networks. In *IEEE International Conference on Communications.*
5. Kurniawan, D.W.H., Kurniawan, A., & Arifianto, M.S. (2017). An analysis of optimal capacity in cluster of CDMA wireless sensor network. In *International Conference on Applied Computer and Communication Technologies (ComCom), Jakarta* (pp. 1–6).
6. Ziouva, E., & Antonakopoulos, T. (2002). CSMA/CA performance under high traffic conditions: throughput and delay analysis. *Computer Communications, 25*(3), 313–321.
7. Ergen, S. C., & Varaiya, P. (2010). TDMA scheduling algorithms for wireless sensor networks. *Wireless Networks, 16*(4), 985–997.
8. Clark, S.M., Hoback, K.A. & Zogg, S.J.F. (2010, Mar 16). Statistical priority-based multiple access system and method. *U.S. Patent 7,680,077[P].*
9. Wang, L., Li, H. & Liu, Z. (2016). Research and pragmatic-improvement of statistical priority-based multiple access protocol. In *2nd IEEE International Conference on Computer and Communications (ICCC), Chengdu* (pp. 2057–2063).
10. Perevalov, E., Blum, R. S., & Safi, D. (2006). Capacity of clustered ad hoc networks: how large is "Large"? *IEEE Transactions on Communications, 54*(9), 1672–1681.
11. Haenggi, M. (2012). *Stochastic geometry for wireless networks: Description of point processes* (pp. 9–46). https://doi.org/10.1017/cbo9781139043816.2.
12. Pastor-Satorras, R., Rubi, M., & Diaz-Guilera, A. (2003). *Statistical mechanics of complex networks*. Heidelberg: Springer.
13. Liu, J., Peng, T., Quan, Q., & Cao, L. (2018). Performance analysis of the statistical priority-based multiple access. In *IEEE International Conference on Computer & Communications.*

# The Integration Degree of Logistics Industry and Regional Agglomeration Industry

**Jia Jiang and Tianyang Zhao**

**Abstract** On the basis of the research on the coordinated development of regional economy, the degree of agglomeration of related industries is obtained through the calculation and analysis of location quotient, and the integration measurement model of agglomeration industry and logistics industry is constructed to analyze the development level and integration degree of logistics industry and related agglomeration industry. According to the types of integration and coordination degree between logistics industry and related agglomeration industry, the main reasons that restrict the integration and development are found out. On this basis, the linkage mechanism between agglomeration industry and logistics industry based on PDCA cycle is innovatively established. Finally, taking Hebei Province as an example, this paper studies the common problems in logistics industry, and puts forward corresponding countermeasures and suggestions.

**Keywords** Logistics industry · Industrial agglomeration · Coordinated development · Coupling model

## 1 Introduction

The phenomenon of industrial agglomeration occurs when a certain industry is highly concentrated in the region and the elements of industrial capital are constantly converged. Nowadays, with the deepening of the process of coordinated

Project of Hebei Provincial Department of Science and Technology: Study on the integrated development strategy of logistics industry in Hebei province under the opportunity of Beijing-Tianjin-Hebei coordinated development (154576292).

J. Jiang (✉) · T. Zhao
School of Economics and Management, Hebei University of Science and Technology, Shijiazhuang, China
e-mail: jiang_jia67@163.com

T. Zhao
e-mail: 771834618@qq.com



J. Zhang et al. (eds.), *LISS2019*,
https://doi.org/10.1007/978-981-15-5682-1_6

development of regional economy, industrial agglomeration has taken shape in many developed regions of the world and is gradually evolving into a unique economic state. Taking Hebei Province as an example, as a key area of coordinated development between Beijing, Tianjin and Hebei, while exploring its own industrial transformation and upgrading, it makes every effort to promote the construction of industrial undertaking platform, actively undertake the transfer projects between Beijing and Tianjin, and simultaneously upgrade the industry and expand the development space. After the exploration and preliminary practice in recent years, Hebei Province has achieved the stage results of industrial transformation and upgrading and industrial transfer undertaking, basically forming a new pattern based on industrial agglomeration mechanism.

In the process of industrial restructuring and industrial agglomeration, logistics industry plays an important role in supporting and linking. If the development of logistics industry cannot well support the adjustment of industrial structure and the formation of cluster industry in the region, it will limit the economic development of the whole region. Based on this, this paper proposes a measurement model for studying the integration and evolution of regional logistics industry and related agglomeration industry, and explores the bottleneck of regional logistics industry agglomeration. It also puts forward specific countermeasures for the development of logistics industry and ways to optimize logistics industry, and puts forward relevant suggestions for the future development of logistics industry.

## 2 Research Summary of Industrial Agglomeration and Logistics Collaborative Development

In recent years, scholars have begun to pay attention to the issue of industrial agglomeration and the coordinated development of regional logistics, and many new research perspectives and methods have emerged in related literatures. However, the theoretical relevance between literatures is still lacking, and the research system framework has not been fully formed.

Cui and Song attempted to build an econometric model from the perspective of endogenous growth and new economic geography, to investigate the impact of spillover effect of logistics agglomeration and scale economy on the development of China's logistics industry. The results show that the spillover effect of logistics agglomeration plays an important role in the development of logistics in different provinces [1]. Wang studied the relevant mechanism of industrial agglomeration and coordinated development of regional logistics integration and linkage, established the model of regional logistics and other industries' linkage development by using gray correlation analysis, and proposed the coordinated development strategy and safeguard measures of regional logistics at the micro level of enterprises and the macro level of economic space [2]. Jia et al. described the agglomeration situation of logistics industry in Henan Province. Through stepwise regression, it was found that the spatial agglomeration degree of logistics industry was closely related to the upgrading of industrial structure. A systematic clustering was carried out for 15 cities in Henan Province, and the differences of agglomeration level of logistics industry in different cities were explained from the perspective of industrial structure [3]. Wang took automobile manufacturing industry cluster and regional logistics in Jiangxi Province as the research object, constructed the collaborative development evaluation model of automobile manufacturing industry and regional logistics through synergetic theory, and elaborated the cooperative evolution and development relationship between regional logistics and automobile manufacturing industry cluster in Jiangxi Province [4]. Guo and Qi constructed an evaluation index system for the coordinated development of regional logistics and regional economy. The coupling coordination between regional logistics and regional economic development in the Yangtze River Delta region between 2001 and 2016 was empirically analyzed by using the coupling degree mode [5]. Based on the analysis of the mechanism of the coupling and coordinated development of agriculture and logistics industry, Liang, Xu and Si constructed a coupling model to analyze the coupling and coordinated development of agriculture and logistics industry in the whole country and regions from 2004 to 2015 [6]. Hu established the coupling model between economy and logistics of urban agglomeration based on theoretical analysis and took the city group of Yangtze River Delta as an example to carry on the empirical research [7].

It can be seen from the above that relevant scholars have done a lot of research work on regional logistics and the development of regional industries. However, few literatures have systematically studied the integration degree of regional logistics and cluster industries. Therefore, how to improve the overall logistics industry to feed back the healthy development of relevant industry clusters has become an urgent problem for the logistics industry.

# 3 Establishment of Measurement Model for Integration of Logistics Industry and Related Agglomeration Industry

## *3.1 Calculation of Industrial Agglomeration Degree*

In order to study industrial agglomeration in the region and make better use of logistics industry to promote the development of other agglomeration industries, this paper adopts Location Quotient (*LQ*) as the main method to study industrial agglomeration after referring to relevant literatures and researches of scholars. The specific calculation equation is as follows:

$$LQ = \frac{P_i \Big/ \sum_{i=1}^{n} P_i}{\sum_{j=1}^{m} P_{ij} \Big/ \sum_{i=1}^{n}\sum_{j=1}^{m} P_{ij}} \tag{1}$$

In (1), $P_i$ represents the output value of $i$ industry in the region, $\sum_{i=1}^{n} P_i$ represents the total output value of all industries in the region, $\sum_{j=1}^{m} P_{ij}$ represents the output value of $i$ industry in the country, and $\sum_{i=1}^{n}\sum_{j=1}^{m} P_{ij}$ represents the total output value of all industries in the country. If $LQ > 1$, it indicates that $i$ industry has a higher degree of specialization and agglomeration in the region, and has certain development advantages. If $LQ < 1$, it indicates that the development of $i$ industry in the region is at a disadvantage with a low degree of agglomeration, lower than the average level of the industry development. If $LQ = 1$, it means that the development of $i$ industry in the region is in equilibrium.

## 3.2 Construction of Fusion Degree Model

Fusion refers to the integration of two or more different things, and the degree of fusion refers to the degree of correlation between different things. This paper studies the impact of integration elements between logistics industry and related agglomeration industry, constructs a measurement index system and evaluation criteria for quantitative calculation, and then provides decision-making basis for the integration of the two industries. This model is suitable for evaluating the integration degree of agriculture and tourism, manufacturing and service industries, urban development and economic environment, logistics and regional economy.

Hypothesis: $F(X^T)$ represents the development level of logistics industry in the region, and $X$ represents the main development index of logistics industry in the region. $G(Y^T)$ is the development level of relevant industries in the region, $Y$ is the main development index of relevant industries in the region, and $T$ represents the time point, usually referring to month, quarter and year, etc.

Thus, the comprehensive evaluation model of the development level of logistics industry and agglomeration industries in the region at time $T$ is established:

$$F\left(X^T\right) = \sum_{i}^{m} a_i x_i, \sum_{i}^{m} a_i = 1 \tag{2}$$

$$G\left(Y^T\right) = \sum_{j}^{n} b_j y_j, \sum_{j}^{n} b_j = 1 \tag{3}$$

In the equation, $m$ represents the number of evaluation indexes related to the logistics industry in the region, $a_i$ denotes the weight value of the logistics industry index $i$ in the region, $n$ represents the number of evaluation indexes involving the relevant industries in the region, and $b_j$ represents the weight value of the relevant industrial indicator $j$ in the region. The higher the calculated value of the evaluation function, the better the development of the system. $x_i$ and $y_j$ are normalized values in the range of (0, 1].

The quantitative model of integration degree of logistics industry and related industries in the region is established as follows:

$$D(T) = \sqrt{C(T) \times R(T)} \tag{4}$$

Among them:

$$C(T) = \sqrt{\frac{F(X^T)G(Y^T)}{[F(X^T) + G(Y^T)]^2}} \tag{5}$$

$$R(T) = \delta F(X^T) + \theta G(Y^T), \delta + \theta = 1 \tag{6}$$

In the equation, *R(T)* represents the comprehensive evaluation level of logistics industry and related agglomeration industries, and reflects their comprehensive benefits and development. δ and θ are undetermined parameters, and the general value range is (0,1), which is usually 0.5. *D(T)* indicates the degree of integration of logistics industry and related cluster industries. The higher the value is, the higher the degree of integration of logistics industry and related industries will be. Overall, the model has good stability and applicability.

## 3.3 Evaluation Criteria of Fusion Degree

Generally, the calculation results of $F(X^T)$ and $G(Y^T)$ can be divided into three situations: $F(X^T) > G(Y^T)$ indicates that the development speed of logistics industry in the region is faster than that of agglomeration industries, and the development speed of agglomeration industries is relatively slow. $F(X^T) < G(Y^T)$ indicates that the development speed of related agglomeration industries in the region is faster than that of logistics industry, and the development of logistics industry is slow. $F(X^T) = G(Y^T)$ indicates that the development speed of logistics industry in the region is synchronized with the development speed of related agglomeration industries. Specific evaluation criteria are shown in Table 1.

**Table 1** Evaluation criteria of fusion degree grade

| Fusion degree | Fusion level | Fusion degree | Fusion level |
|---|---|---|---|
| [0, 0.1) | Extremely incompatible | [0.5, 0.6) | General fusion |
| [0.1, 0.2) | Highly incompatible | [0.6, 0.7) | Mild fusion |
| [0.2, 0.3) | Moderate incompatibility | [0.7, 0.8) | Moderate fusion |
| [0.3, 0.4) | Mild incompatibility | [0.8, 0.9) | Highly integrated |
| [0.4, 0.5) | Barely fusion | [0.9, 1.0) | Extreme fusion |

# 4 Empirical Research

Taking Hebei Province as an example, this paper studies its industrial agglomeration under the background of the coordinated development of Beijing, Tianjin and Hebei. By calculating the degree of integration between logistics industry and agglomeration industry, it analyzes the coordinated development of logistics industry and related agglomeration industries in the new environment after undertaking the industrial transfer of Beijing, Tianjin and Hebei, so as to improve the development level of logistics industry, and at the same time rely on logistics industry to reverse the development level of related agglomeration industries, thus achieving the goal of mutual promotion and common development.

## *4.1 Analysis of Industrial Agglomeration*

With reference to relevant data in China statistical yearbook, this paper selects eight industries including agriculture, forestry, animal husbandry and fishery, industry, transportation, warehousing and post office in Hebei Province from 2010 to 2016 for cluster analysis and research, and summarizes transportation, warehousing and post office as logistics industry. Equation (1) is used to calculate the location quotient of relevant industries in Hebei Province. Meanwhile, in order to better highlight the advantageous industries of Hebei province in Beijing-Tianjin-Hebei region, the overall index of Beijing-Tianjin-Hebei region is selected to replace the national index as the denominator for calculation [8]. The calculation results and variation trend of specific location quotient are shown in Fig. 1.

In 1 *a* represents agriculture, forestry, animal husbandry and fishery, *b* represents industry, *c* represents construction industry, *d* represents wholesale and retail trade, *e* represents logistics industry, *f* represents accommodation and catering, *g* represents financial industry and *h* represents realty industry. It can be seen that under the background of the coordinated development of Beijing-Tianjin-Hebei, the location quotient of these eight industries in Hebei Province increased year by year to a certain extent from 2010 to 2016, and showed an upward trend. Among them, agriculture, forestry, animal husbandry and fishery showed obvious agglomeration advantages.

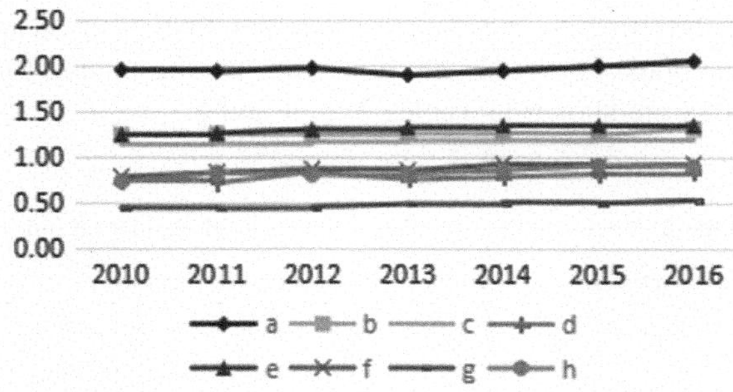


**Fig. 1** Development trend of location quotient of major industries in Hebei Province from 2010 to 2016

In 2016, its location quotient reached 2.15, and its development level has been significantly higher than the average level of Beijing-Tianjin-Hebei. In addition to agriculture, forestry, animal husbandry and fishery, the location quotient of industry, construction industry and logistics industry was also greater than 1. The location quotient of wholesale and retail, accommodation and catering industry, financial industry and real estate industry was all less than 1, which indicated that they did not show obvious agglomeration advantage in Hebei Province. From the above analysis, we can see that the agglomeration level of the primary industry and the secondary industry in Hebei Province was relatively high, while the tertiary industry did not show obvious agglomeration advantages except the logistics industry.

## 4.2 *Index Selection and Weight Calculation*

According to the principle of selecting indicators such as representativeness, systematicness, pertinence, operability and scientificity, this paper establishes the evaluation index system of the four agglomeration industries from three aspects of infrastructure, industrial scale and development capacity, on the basis of referring to relevant literature and the development status of Hebei industry, in order to achieve the degree of integration of logistics industry and related industries. In order to avoid the defects caused by subjective judgment, entropy weight method is selected in this paper to calculate and determine the weight of each index objectively, that is:

$$a_i = \frac{1 - E_i}{\sum (1 - E_i)}, \quad (i = 1, 2, \ldots, t) \tag{7}$$

$$b_j = \frac{1 - E_j}{\sum (1 - E_j)}, \quad (j = 1, 2, \ldots, t) \tag{8}$$

Equation $E_{\rm i} = -\ln(t)^{-1} \sum_{i=1} p_i \ln p_i$, $P_{\rm i} = \frac{x_i}{\sum_{i=1}^{t} x_i}$, If $p_i$= 0, define lim$p_i$ln$p_i$= 0.

Equation $E_j = -\ln(t)^{-1} \sum_{j=1} p_j \ln p_j$, $P_{\rm j} = \frac{y_j}{\sum_{i=1}^{t} y_j}$, If $p_j$= 0, define lim$p_{\rm j}$ln$p_j$= 0.

Referring to the data in National Statistical Yearbook and Hebei Economic Yearbook from 2011 to 2017, the specific index system and the weight of each index are shown in Tables 2, 3, 4 and 5.

**Table 2** Logistics industry development level evaluation index and weight

| Level indicators | Weight | Secondary indicators | Weight |
|---|---|---|---|
| Infrastructure | 0.358 | Per capita highway operating mileage | 0.150 |
| | | Per capita railway operating mileage | 0.125 |
| | | Cargo car ownership | 0.083 |
| Industrial scale | 0.416 | Number of people engaged | 0.097 |
| | | Per capita freight volume | 0.127 |
| | | Per capita postal and telecommunications business volume | 0.192 |
| Development ability | 0.226 | Industrial added value as a proportion of GDP | 0.078 |
| | | Fixed asset investment | 0.148 |

**Table 3** Agriculture, forestry, animal husbandry and fishery development level evaluation index and weight

| Level indicators | Weight | Secondary indicators | Weight |
|---|---|---|---|
| Infrastructure | 0.363 | Main machinery ownership | 0.083 |
| | | Effective irrigation rate | 0.096 |
| | | Cargo car ownership | 0.184 |
| Industrial scale | 0.294 | Number of people engaged | 0.196 |
| | | Unit area yield of agricultural products | 0.098 |
| Development ability | 0.343 | Product Value Added as a Proportion of GDP | 0.107 |
| | | Fixed asset investment | 0.236 |

**Table 4** Evaluation index and weight of industrial development level

| Level indicators | Weight | Secondary indicators | Weight |
|---|---|---|---|
| Infrastructure | 0.377 | Total assets | 0.177 |
| | | Contribution rate of total assets | 0.200 |
| Industrial scale | 0.203 | Total labor productivity | 0.099 |
| | | Main business income | 0.104 |
| Development ability | 0.420 | Industrial added value | 0.099 |
| | | Fixed asset investment | 0.199 |
| | | Total profit | 0.122 |

**Table 5** Construction industry development level evaluation index and weight

| Level indicators | Weight | Secondary indicators | Weight |
|---|---|---|---|
| Infrastructure | 0.274 | Technical equipment rate | 0.112 |
| | | Power equipment rate | 0.162 |
| Industrial scale | 0.234 | Number of people engaged | 0.126 |
| | | Floor space of buildings completed | 0.108 |
| Development ability | 0.492 | Industrial added value as a proportion of GDP | 0.183 |
| | | Fixed asset investment | 0.202 |
| | | Total profit | 0.107 |

## 4.3 Computation and Analysis of Fusion Degree

Based on the above evaluation index system and the coupling degree synergy model, the following calculation and analysis are made on the integration degree of logistics industry and related agglomeration industry in Hebei Province from 2010 to 2016.

As can be seen from Table 6, the integration level of logistics industry and agriculture, forestry, animal husbandry and fishery, industry as well as construction industry in Hebei Province has been increasing year by year from 2010 to 2016, with an overall upward trend. Among them, the average degree of integration between logistics industry and agriculture, forestry, animal husbandry and fishery are 0.492, and the highest degree of integration is 0.590 in 2016. The development of integration between logistics industry and agriculture, forestry, animal husbandry

**Table 6** The integration degree of logistics industry and related industries in 2010–2016

| Industrial integration degree | Logistics industry and agriculture, forestry, animal husbandry and fishery | | Logistics industry and industry | | Logistics industry and construction industry | |
|---|---|---|---|---|---|---|
| | Fusion degree | Fusion level | Fusion degree | Fusion level | Fusion degree | Fusion level |
| 2010 | 0.418 | Barely fusion | 0.336 | Mild incompatibility | 0.391 | Mild incompatibility |
| 2011 | 0.372 | Mild incompatibility | 0.375 | Mild incompatibility | 0.385 | Mild incompatibility |
| 2012 | 0.430 | Barely fusion | 0.467 | Barely fusion | 0.513 | General fusion |
| 2013 | 0.524 | General fusion | 0.564 | General fusion | 0.557 | General fusion |
| 2014 | 0.538 | General Fusion | 0.565 | General fusion | 0.568 | General fusion |
| 2015 | 0.570 | General fusion | 0.562 | General fusion | 0.569 | General fusion |
| 2016 | 0.590 | General fusion | 0.616 | Mild fusion | 0.557 | General fusion |

and fishery has gone through a stage from barely fusion, mild incompatibility to general fusion. Although the degree of integration between industries is gradually improving, the integration level is still at a medium level. The integration of logistics industry and industry is the best among the three. It has experienced four stages of development: mild incompatibility, barely fusion, general fusion and mild fusion. In 2016, it reached the highest integration level of 0.616, and entered a new stage of mild integration. Logistics and construction industry from the initial mild incompatibility to the general fusion stage, and the degree of integration is 0.557 in 2016. It can be seen that the fusion degree of the three is stable between 0.500 and 0.700. Although the fusion degree is constantly improving, it still does not reach a very ideal fusion state. In order to improve the overall level of coordinated development in Beijing, Tianjin and Hebei, it is necessary to further strengthen the integration of the three.

According to the model above, we can get the types of integration and coordination degree of logistics industry and related agglomeration industry, and make the following analysis for the reasons of imbalance of coordination between logistics industry and related industries.

From Table 7, we can see that the logistics industry in Hebei Province has developed slowly in the past few years, and has not kept pace with the development of agriculture, forestry, animal husbandry and fishery, industry and construction industry, which was at a disadvantage in the process of industrial integration. However, since 2013, due to the sudden outbreak of online shopping market, the construction of e-commerce logistics network system has been promoted. The development of China's logistics industry has attracted the comprehensive attention of the central and local governments. Hebei Province has also increased its attention to the logistics industry, resulting in the continuous improvement of the level of development of the logistics industry, and the development speed has exceeded the development speed of related agglomeration industries. In the following years, the logistics industry has maintained its development advantages, and in the process of integration with other industries, the development speed has been in the lead. At present, the degree of integration between logistics industry and other industries is not high, which is largely due to the fact that logistics industry and other agglomeration industries have not yet formed a cooperative development mechanism. Although the development speed of logistics industry is constantly improving, its development level still has no obvious advantages compared with other agglomeration industries. Therefore, in order to improve the degree of integration between logistics industry and related agglomeration industry, we need not only related agglomeration industry to promote the development of logistics industry, but also the logistics industry to rely on its own development advantages to promote the continuous progress of other industries, so as to achieve mutual coordination and linkage development and jointly improve the development level.

**Table 7** Types of integration and coordination degree between logistics industry and related industries in 2010–2016

| Fusion coordination type | Industry | | |
|---|---|---|---|
| | Logistics industry and agriculture, forestry, animal husbandry and fisheries | Logistics industry and industry | Logistics industry and construction industry |
| 2010 | Logistics industry develops slowly | Logistics industry develops slowly | Logistics industry develops slowly |
| 2011 | Logistics industry develops slowly | Logistics industry develops slowly | Logistics industry develops slowly |
| 2012 | Logistics industry develops slowly | Logistics industry develops slowly | Logistics industry develops slowly |
| 2013 | Agriculture, forestry, animal husbandry and fishery develops slowly | Industry develops slowly | Construction industry develops slowly |
| 2014 | Agriculture, forestry, animal husbandry and fishery develops slowly | Logistics industry develops slowly | Logistics industry develops slowly |
| 2015 | Agriculture, forestry, animal husbandry and fishery develops slowly | Industry develops slowly | Construction industry develops slowly |
| 2016 | Agriculture, forestry, animal husbandry and fishery develops slowly | Industry develops slowly | Construction industry develops slowly |

## *4.4 Establishing Linkage Mechanism between Logistics Industry and Agglomeration Industry*

In order to meet the new demands of the cluster industry on the logistics industry in Hebei Province, the logistics industry and other cluster industries can be developed in a coordinated way, so as to promote each other and make common progress. Therefore, this paper establishes the linkage mechanism of industrial agglomeration and logistics industry agglomeration based on PDCA cycle to coordinate the relationship between logistics industry and related industries. The linkage mechanism model is shown in Fig. 2.

PDCA cycle linkage mechanism is to divide the process of logistics industry agglomeration and related industry agglomeration into four stages, namely *plan*, *do*, *check* and *adjust*, and keep on circulation according to this step. First of all, in the *plan* stage, relevant government departments are required to formulate reasonable industrial development planning, and implement industrial collaborative planning according to the requirements. Secondly, in the *do* stage, the development of logistics industry agglomeration and other industrial agglomeration complement

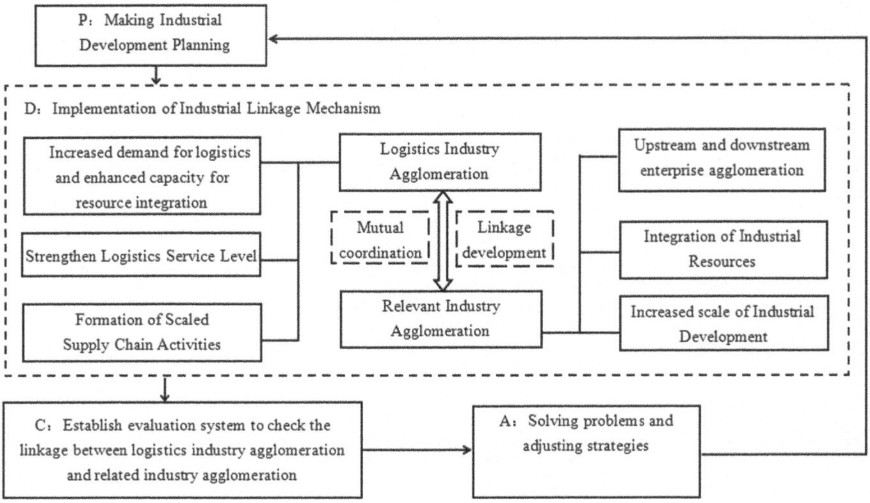


**Fig. 2** Linkage mechanism of industrial agglomeration and logistics industry agglomeration based on PDCA cycle

each other. On the one hand, related industrial agglomeration can promote the development of logistics industry agglomeration. On the other hand, the agglomeration of logistics industry can also drive the development of other industry agglomeration. For the *check* stage, it is necessary for the relevant government departments to formulate a reasonable evaluation system, regularly evaluate and analyze the current situation of logistics industry agglomeration and related industry agglomeration. Finally, in the *adjust* stage, according to the inspection results of the previous stage, we formulate adjustment strategies, solve the problems in the previous plan and implementation, and lay a good foundation for the next round of plan.

The linkage mechanism of industrial agglomeration and logistics industry agglomeration can meet the new demands of the agglomeration industry for logistics industry. At the same time, it not only provides a solution for the coordinated development of logistics industry and agglomeration industry, but also lays a foundation for their integration and development.

## 5 Conclusions and Recommendations

Based on the background of coordinated development of regional economy, this paper studies the integration of logistics industry and related agglomeration industry in this environment. Firstly, through the calculation and analysis of location quotient, the agglomeration degree of related industries is obtained. At the same time, in order to better grasp the degree of integration of logistics industry and agglomeration industry, so as to promote the future development of regional economy, this paper constructs a coupling degree coordination model, and on this basis, establishes an industrial agglomeration and logistics industry agglomeration mechanism based on PDCA cycle, hoping that this mechanism can provide a reference for the integration and development of logistics industry and related agglomeration industry. Taking Hebei Province as an example, this paper conducts an empirical study and finds that except for the logistics industry, agriculture, forestry, animal husbandry and fishery, industry and construction industry in Hebei Province have formed industrial agglomeration at the same time. This paper establishes the coupling degree collaborative model, and analyzes the data of Hebei Province from 2010 to 2016. Finally, it is found that the degree of integration of logistics industry and related industries in Hebei Province is constantly improving. However, the degree of industrial integration is still unsatisfactory.

In order to improve the integration of the logistics industry with other industries, and further improve the overall development level of the industry in Hebei Province, this paper proposes the following suggestions after synthesizing the above research results: (1) Break the industrial boundaries and form a situation of industrial convergence and development. After fully understanding the characteristics of each industry, the relevant departments in Beijing, Tianjin and Hebei should formulate development policies suitable for each industry, break the boundaries between different industries, avoid the isolated development of industries, take advantage of the advantages of integration and development among industries, and promote the coordination and common development of each industrial cluster. (2) Improve the degree of specialization of logistics industry and create advantages of industrial cluster. Because each region has its own characteristics, it will form corresponding specialized demand. Although many related industries in Beijing, Tianjin and Hebei have formed industrial agglomeration, their degree of specialization is not enough. Therefore, each agglomeration industry should widely tap the current demand situation in the region and improve its level of specialization. (3) Improve the level of logistics services and give full play to the mutual promotion between industries. In the process of coordinated development of Beijing, Tianjin and Hebei, logistics industry has been paid more and more attention, and its development speed is relatively fast, but the level of development still needs to be further improved. As the pillar of the service industry, while meeting the service needs of other industries, the logistics industry should take the initiative to attack the market and constantly cultivate its own new advantages to guide and support the development of other industries. Use services and potential services to leverage demand and potential demand, and form a continuous iteration of demand and service, so as to realize the coordinated development of Beijing, Tianjin and Hebei based on industrial integration mechanism.

## References

1. Cui, Y. Y., & Song, B. L. (2017). Logistics agglomeration and its impacts in China. *Transportation Research Procedia, 25,* 3875–3885.
2. Wang, Y. L. (2011). The regional logistics integration and the industrial agglomeration linkage development. *Economic Theory and Business Management, 11,* 78–87.
3. Jia, X. H., & Hai, F. (2015). Correlation analysis of logistics industry agglomeration and industrial structure——The case of 18 cities in Henan Province. *Lanzhou Academic Journal, 02,* 178–183.
4. Wang, B., & Wang, Y. Y. (2018). Evaluation of the synergy degree between automobile manufacturing industrial cluster and regional logistics of Jiangxi. *Science & Technology and Economy, 1,* 106–110.
5. Guo, H. B., & Qi, Y. (2018). Study on coordinative development between regional logistics and regional economy in Yangtze river delta based on coupling model. *Journal of Industrial Technological Economics, 37*(10), 51–58.
6. Liang, W., Xu, L. Y., & Si, J. F. (2018). Coupled coordination development between agriculture and logistics——An empirical analysis based on China's provincial panel data. *Review of Economy and Management, 34*(05), 150–161.
7. Hu, Y.Z. (2016). An analysis of dynamic coupling between logistics and economy of urban agglomeration based on Yangtze river delta urban agglomerationl. *Commercial Research 2016* (07), 180–186.
8. Liu, Y., Zhou, L.Y., Geng, C. (2017). Evaluation of coordinated industrial development in Beijing-Tianjin-Hebei metropolitan circle: Evidence from the grey relational analysis of 378 location quotient. *Journal of Central University of Finance & Economics 2017*(12), 119–129.

# Mortality Prediction of ICU Patient Based on Imbalanced Data Classification Model

**Xuedong Gao, Hailan Chen, and Yifan Guo**

**Abstract** The mortality prediction of Intensive Care Unit (ICU) patient has been an active topic in the medical filed. In recent years, data mining methods have been applied in this domain and have achieved good results. In this paper, we propose an imbalanced data classification algorithm to predict the mortality of ICU patient. Firstly, we introduce clustering algorithm to convert numeric data into categorical data. Then, imbalanced data processing method is used to obtain a balanced data. Finally, we apply some traditional classification algorithms to conduct experiment on a medical data set, and the experimental results show that the classification accuracy of balanced data set is more efficient than the imbalanced data set considering the minority death patients.

**Keywords** ICU · Mortality prediction · Imbalanced data · Classification

## 1 Introduction

Intensive Care Unit (ICU) is established to provide high quality medical treatment for critically ill patients. The patient mortality rate is one of the important factors to measure the ICU level. And effectively predicting the risk of death in ICU patients

This work is supported by national science fund of China (No. 71272161).

X. Gao · H. Chen (✉)
Donlinks School of Economics and Management, University of Science and Technology Beijing, Beijing, China
e-mail: chl_hld@163.com

X. Gao
e-mail: gaoxuedong@manage.ustb.edu.cn

Y. Guo
School of Management, China University of Mining and Technology, Beijing, China
e-mail: guo_onlyIdol@163.com



J. Zhang et al. (eds.), *LISS2019*,
https://doi.org/10.1007/978-981-15-5682-1_7

can reasonably allocate rescue resources and reduce the cost of first aid. The traditional prediction methods are medical scoring systems such as APACHE [1], SAPS [2], SOFA [3], etc., which are difficult to meet clinical practice and not effectively deal with case-mix [4] problems. Therefore, how to accurately predict the mortality of ICU patient according to the early admission records, and take appropriate measures to improve the patient survival rate has become a research hot spot and difficulty in the medical field.

In recent years, with the improvement of medical informatization, most medical information has been preserved in digital form, providing a reliable and rich source of data for developing new mortality prediction methods. On the other hand, the rapid development of data mining and machine learning technology provides a theoretical basis for the development of new mortality prediction methods. Many researchers [5, 6] introduce classification algorithms into medical domain and have good results. Lin et al. [7] applied SVM (Support Vector Machine) to predict in-hospital mortality risk of patients with acute kidney injury in ICU. Xie et al. [8] used random forest model to predict in-hospital mortality risk of ICU patients. Zhang et al. [9] applied SVM, ANN and C4.5 to verify the predictive performance based on a selected time interval after learning classification knowledge for the same data set. Davoodia et al. [10] proposed the deep rule-based fuzzy classifier to predict mortality of ICU patient.

Among these studies, most of them assume the data set to be well-balanced and focus on achieving overall classification accuracy, which may lead to classification errors in minority class samples. However, most medical data sets are usually imbalanced, with more cases of survival and fewer deaths. And we are more concerned with the prediction of death. So how to deal with imbalanced data to improve the classification performance is essential. There are two main methods. The first is to improve the data set by increasing the number of minority class samples by over-sampling and decreasing the number of majority class samples by under-sampling [11]. The other method is to improve the algorithm itself. There have been many studies [12] to improve the classification accuracy of minority class samples by improving the algorithm. But these methods are not universally suitable for all data sets. It is easier and more efficient to converting an imbalanced data set into a balanced data set. Wu et al. [11] proposed the cluster-based under-sampling method to establish a balanced data set in the data processing stage. In this paper, we apply this method to process the imbalanced data set.

The rest of this paper is organized as follows. In Sect. 2 we describe the data and summarize its properties. In Sect. 3 we propose the mortality prediction model of ICU Patient. In Sect. 4 we conduct experiment and analyze the experimental results. Finally in Sect. 5 we give the conclusion.

## 2 Data Description

In this section, we describe the experimental data in this paper including data sources and data properties.

### 2.1 Data Sources

In this paper, the dataset we used is downloaded from PhysioNet website [13], which is the PhysioNet/Computing in Cardiology Challenge provided by of the National Institutes of Health in 2012. It contains 4000 cases from four different ICUs (Coronary Care, Cardiac Surgery, Recovery Units, Medical and Surgical); all cases collected at most 36 physiological variables; all hospitalized for more than 48 h and all over 16 years of age. There were a total of 554 deaths and 3346 surviving cases, and the missing rate of the different physiological variables ranged from 1.6 to 95.3%. Among all variables, we select 16 physiological variables as shown in Table 1, which the missing data rate is less than 20%.

The data for each ICU patient is generally composed of three components, the basic information at admission, the sequence of physiological variables after admission, and the final survival or death status. A typical ICU data set is shown in Table 2.

**Table 1** Experimental physiological variables

| Abbreviation | Full name | Unit |
|---|---|---|
| BUN | Blood urea nitrogen | mg/dL |
| Creatinine | Creatinine | mg/dL |
| GCS | Glasgow coma scale | – |
| Glucose | Glucose | mg/dL |
| HCO3 | Serum bicarbonate | mmol/L |
| HCT | Red blood cell specific volume | % |
| HR | Heart rate | bpm |
| K | Kalium | mEq/L |
| Mg | Serum magnesium | mmol/L |
| Na | Sodium | mEq/L |
| NIDiasABP | Non-invasive diastolic arterial blood pressure | mmHg |
| NISysABP | Non-invasive systolic arterial blood pressure | mmHg |
| Platelets | Blood platelet count | cells/nL |
| Temp | Human body temperature | °C |
| Urine | Urine | mL |
| WBC | White blood cell | cells/nL |

**Table 2** Example of ICU data set

| ID$_n$ (demograph <vector>, physiological, outcome) | | | |
|---|---|---|---|
| HR | NIDiasABP | … | Temp |
| (00:00, 73) | (00:00, 88) | … | (00:00, 36.3) |
| (01:09, 67) | (00:09, 79) | … | (00:09, 35.8) |
| (01:24, 50) | (01:09, 80) | … | (05:09, 36.6) |
| (01:39, 63) | (01:39, 73) | … | (09:09, 36.8) |
| (01:54, 64) | (01:54, 76) | … | (13:09, 37.1) |
| … | … | … | … |

## 2.2 Properties of the Data

After preliminary exploration of the data, we summarize the properties as follows: high dimensionality, large amount of missing data, imbalance data, and different time intervals.

- High dimensionality: The ICU data set is highly dimensional in two respects. One performance is high in attributes. It has 36 physiological variables, such as temperature, blood pressure, and heart rate etc. Another one performance is high in time. For a physiological variable of one patient, there are multiple observations within 48 h. It leads to the high dimensionality of time.
- Large amount of missing data: The data set contains a large amount of missing data, and the missing rate ranged from 1.6 to 95.3% for different physiological variables. When the missing rate of the data set is not high, we can use the missing data filling methods to fill the data; while when the missing rate is high, the commonly missing data filling methods have no effect.
- Imbalance data: The data set contains more samples from one class (3446 survival cases) while the other is much smaller (only 554 dead cases). It is severely imbalanced. The phenomenon may lead to classification errors in minority class samples. Therefore, the classifier may behavior too bad to get satisfactory results. It is necessary to handle the imbalanced problem.
- Different time intervals: As the sensors and devices may break down, or medical staff may incorrectly operate these medical equipment, the variables of each record were sampled unevenly. For one variable of one patient, the time intervals of each two records may be different. It also can be regarded as a reflection of missing data.

Therefore, there is an urgent need to solve the above issues and develop an effective method to improve the mortality prediction accuracy of ICU patient.

# 3 Mortality Prediction Model of ICU Patient

In this section, we first propose the framework of ICU mortality prediction model. Then according to this framework, we introduce the method of data preprocessing, imbalanced data converting and classification. At last, we present evaluate indexes of prediction results.

## *3.1 The Framework of ICU Mortality Prediction Model*

How to predict the mortality of ICU patient is meet with many challenges and difficulties. So, in this subsection we propose the framework of ICU mortality prediction model as shown in Fig. 1.

According to the proposed framework, this prediction model mainly includes four steps: data preprocessing, imbalanced data processing, classification, and evaluation.

## *3.2 Data Preprocessing*

Since the ICU data have large amount of missing data and high dimensions, in this subsection, we preprocess the data to solve these challenges.

In order to avoid the instability caused by filling missing data, we select 16 physiological variables as shown in Table 1, which the missing data rate is less than 20%. These indexes are also considered to be closely related to death in medical. In this way, we reduce the attribute dimensions.

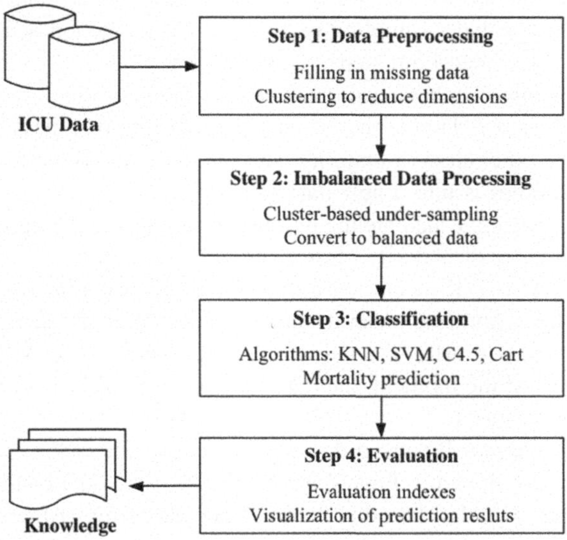


**Fig. 1** The framework of ICU mortality prediction model

**Table 3** Two-dimension information matrix

| ID | HR | NIDiasABP | … | Temp | Outcome |
|---|---|---|---|---|---|
| 132597 | 4 | 10 | … | 7 | 0 |
| 132598 | 1 | 1 | … | 9 | 1 |
| 132599 | 7 | 9 | … | 9 | 0 |
| 132601 | 8 | 5 | … | 2 | 0 |
| 132602 | 9 | 1 | … | 2 | 1 |
| 132605 | 9 | 8 | … | 1 | 1 |
| 132610 | 9 | 10 | … | 1 | 0 |

In Table 2, for a physiological variable of one patient, there are multiple observations within 48 h. It leads to the high dimensionality of time. So, we calculate the mean value of multiple observations for each attribute within 48 h to represent the attribute. And the missing value of each attribute is replaced by the mean of the attribute.

After the above operation, we get a matrix in which each patient has only one value per attribute. However, the variation range of each attribute is very different, we apply clustering algorithm to convert numeric data into categorical data. K-means clustering algorithm is used to do this operation, and we set k = 10. Completing clustering, the ICU data is converted into a matrix of two-dimension information (Table 3).

### *3.3 Imbalanced Data Processing Method Based on Cluster Under-Sampling*

In this subsection, we introduce an imbalanced data processing method based on cluster under-sampling to improve the prediction accuracy.

The algorithm steps are as follows.

*Input*: Imbalanced data set.

*Step 1*: Divide the original data set into one class of minority data samples and the other class of majority data samples. And directly put the class of minority data samples is into a new data set.

*Step 2*: Calculate the number of samples in the minority class, which is set to the number of majority clusters k.

*Step 3*: K-means algorithm is used to cluster the majority data samples of another class. And Euclidean distance is selected as the distance measurement.

*Step 4*: After clustering, the center points of k clusters are put into the new data set. At the moment, the data samples of two classes are equal, and the new data set is balanced.

*Output*: Balanced data set.

As shown in Figs. 2 and 3, Fig. 2 is the data distribution before clustering, and Fig. 3 is the data distribution after clustering. Although the amount of data after clustering is reduced, but the spatial distribution of the data set remains unchanged, and the distribution characteristics of the data set are preserved.

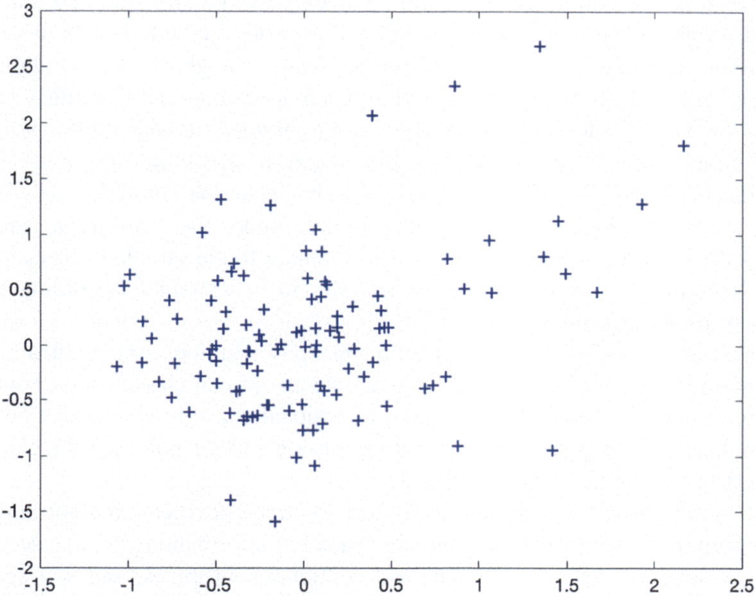


**Fig. 2** Data distribution before clustering

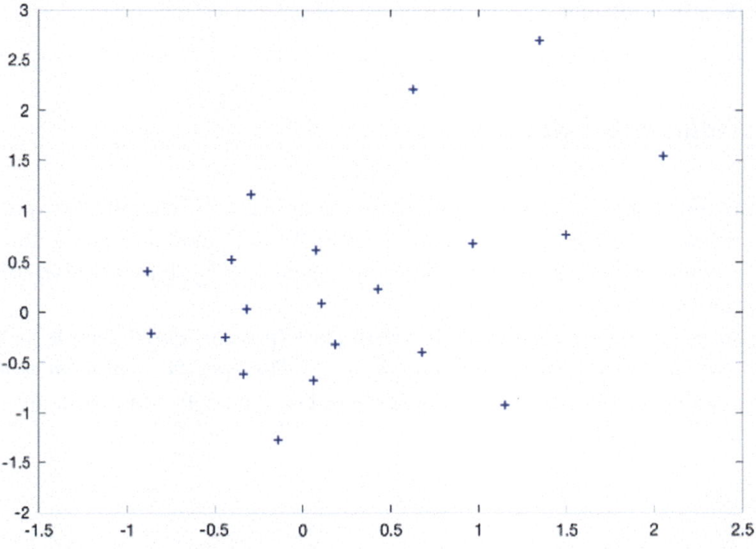


**Fig. 3** Data distribution after clustering

### 3.4 Classification Algorithms

Classification is an important data mining technique. Classification is usually used for prediction. The classification algorithms mainly include decision tree, statistical method, machine learning method, and neural network method [14]. In this paper, the classification algorithms we used are K-Nearest Neighbor (KNN) algorithm, Support Vector Machine (SVM) algorithm, C4.5 algorithm, and Classification and Regression Tree (CART) algorithm. They are traditional classification algorithms.

KNN algorithm is one of the best classification algorithms under the vector space model (VSM) [15]. The basic idea is to calculate the similarity between the sample to be classified and the training sample under the VSM representation. Finding the k nearest neighbors that are most similar to the sample to be classified, and determining the attribution of the samples to be classified according to the classes of the k neighbors.

As a classifier, SVM has proved itself being a powerful tool for classification and regression [7]. The goal of the SVM is to create an optimal classification boundary (maximum interval hyperplane) in high-dimensional space to distinguish between different classes of samples. The support vector refers to the point in each class that is closest to the largest interval hyperplane.

C4.5 is currently the most influential decision tree classification algorithm. The C4.5 algorithm uses information gain rate instead of information gain to select node attributes, supports discrete attributes and continuous attributes, and is a decision tree learning algorithm based on information entropy [16].

The CART algorithm uses the minimum Gini value as the demarcation point selection criterion, and is a binary tree. It has the advantages of small computation, fast speed and easy interpretation [17]. The algorithm works well for a large amount of real data in a relatively short time.

### 3.5 Evaluation Indexes

Currently, there are many evaluation indexes to measure the classification problem. In this work, we use confusion matrix to evaluate ICU mortality prediction. This matrix is composed of true positive (TP), true negative (TN), false positive (FP) and false negative (FN).

For the ICU data set, we define the death class (positive class) sample as P, and the survivor class (negative class) sample as N. The specific confusion matrix is shown in Table 4. The calculation indexes based on the confusion matrix are shown

**Table 4** Definition of confusion matrix

| Outcome | | Observed | |
|---|---|---|---|
| | | Death | Survivor |
| Predicted | Death | TP | FP |
| | Survivor | FN | TN |

in Eqs. (1) to (7), which are TPR (true rate) or Recall, TNR (true negative rate) or Sp (specificity), FPR (false positive rate), FNR (false negative rate), Precision, Accuracy, F-measure.

$$TPR(Recall) = TP/(TP + FN) \tag{1}$$

$$TNR(Sp) = TN/(TN + FP) \tag{2}$$

$$FPR = FP/(TN + FP) \tag{3}$$

$$FNR = FN/(TP + FN) \tag{4}$$

$$Precision = TP/(TP + FP) \tag{5}$$

$$Accuracy = (TN + TP)/(TN + TP + FP + FN) \tag{6}$$

$$F\text{-}measure = (2 * Precision * Recall)/Precision + Recall \tag{7}$$

Except for the above evaluation indexes, in this paper, we introduce min (Recall, Precision) [9] to measure the prediction results. Because it better reflects the classification accuracy of the class with few samples.

## 4 Experiments

In this section, we use MySQL to extract the data, data preprocessing and prediction model building is written by MATLAB.

In this experiment, we divide the ICU data set into training set and test set, and the ratio is 3:1. Then we calculate the performance indexes of each classification algorithm. In order to verify the effectiveness of the proposed method, we use the ICU data set without doing imbalanced data processing as comparison experiments to conduct classification. For imbalanced data set, there are two classes (554 dead cases and 3446 survival cases), and its corresponding ratio is 1:6.22. So it has 3000 cases in training set and 1000 cases in test set. For balanced data set, the number of two classes is equal (554 dead cases and 554 survival cases). So it has 831 cases in training set and 277 cases in test set. And the results is shown in Table 5.

**Table 5** Performance comparison of four classification algorithms on imbalanced data set and balanced data set

| Classification algorithms | | Evaluation indexes | | | | |
|---|---|---|---|---|---|---|
| | | Recall/% | Precision/% | min (Recall, Precision)/% | F-measure/% | Accuracy/% |
| KNN | Imbalanced data set | 10.48 | 18.33 | 10.48 | 13.33 | **85.70** |
| | Balanced data set | **42.45** | **54.63** | **42.45** | **47.77** | 53.43 |
| SVM | Imbalanced data set | 11.39 | 11.76 | 11.39 | 11.57 | **89.50** |
| | Balanced data set | **83.45** | **78.57** | **78.57** | **80.94** | 79.42 |
| C4.5 | Imbalanced data set | 35.24 | 24.50 | 24.5 | 28.91 | **81.80** |
| | Balanced data set | **72.66** | **74.81** | **72.66** | **73.72** | 74.01 |
| CART | Imbalanced data set | 24.39 | 23.26 | 23.26 | 23.81 | **85.35** |
| | Balanced data set | **81.13** | **78.18** | **78.18** | **79.63** | 78.74 |

In Table 5, five evaluation indexes (Recall, Precision, min (Recall, Precision), F-measure, Accuracy) are used to present the performance of four classification algorithms (KNN, SVM, C4.5, and CART) on imbalanced data set and balanced data set. It is obvious to see that balanced data set achieves better performance on four indexes: Recall, Precision, min (Recall, Precision), F-measure. We also see that the performance of imbalanced data set is not good on these four indexes. However, its accuracy is high. This phenomenon indicates that these classification algorithms are greatly affected by the imbalance of data, and tend to divide objects into classes with majority samples, ignoring the class with minority samples.

Although the accuracy of balanced data set is lower than using imbalanced data set, the difference of them are not obvious. When we classify the imbalanced data, we pay more attention to the accuracy of the class with majority samples, which is consistent with our real life.

The above experimental results demonstrate that the quality of the proposed framework of mortality prediction model is effective.

## 5 Conclusion

In this paper, we summarize the properties ICU data: high dimensionality, large amount of missing data, imbalance data, and different time intervals. In order to solve these challenges, we propose the framework of ICU mortality prediction model. In this framework, we first preprocess the data. We select some variables from many attributes, meanwhile we fill the missing data. And K-means clustering is applied to convert numeric data into categorical data. After completing the data preprocessing, we introduce an imbalanced data processing method based on cluster under-sampling to improve the prediction accuracy. Using this method, we can convert imbalanced data into balanced data. And it also reduces the quantity of data set. Then we present the tradition classification algorithms. Furthermore, we give the evaluation indexes of classification accuracy. Finally, experiments are carried out to evaluate the performance of the proposed framework of ICU mortality prediction model. And the experimental results verify that this model works effectively.

## References

1. Zimmerman, J. E., Kramer, A. A., & McNair, D. S. (2006). Acute physiology and chronic health evaluation IV: Hospital mortality assessment for today's critically ill patients. *Critical Care Medicine, 34,* 1297–1310.
2. Gall, J. R. L., Lemeshow, S., & Saulnier, F. (1993). A new simplified acute physiology score (SAPS II) based on a European/North American multicenter study. *Journal of the American Medical Association, 270,* 2957–2963.
3. Vincent, J. L., Moreno, R., & Takala, J. (1996). The SOFA (sepsis-related organ failure assessment) score to describe organ dysfunction/failure. *Intensive Care Medicine, 22,* 707–710.
4. Strand, K., & Flaatten, H. (2008). Severity scoring in the ICU: A review. *Acta Anaesthesiologica Scandinavica, 52,* 467–478.
5. Liu, S. J., Chen, X. X., Fang, L., Li, J. X., Yang, T., Zhan, Q., et al. (2018). Mortality prediction based on imbalanced high-dimensional ICU big data. *Computers in Industry, 98,* 218–225.
6. Xu, J., Zhang, Y., Zhang, P., Mahmood, A., Li, Y., & Khatoon, S. (2017). Data Mining on ICU Mortality prediction using early temporal data: A survey. *International Journal of Information Technology & Decision Making, 16,* 117–159.
7. Lin, K., Xie, J. Q., Hu, Y. H., & Kong, G. L. (2018). Application of support vector machine in predicting in-hospital mortality risk of patients with acute kidney injury in ICU. *Journal of Peking University (Health Sciences), 50,* 239–244.
8. Xie, J. Q., Lin, K., Li, C. X., & Kong, G. L. (2017). Application of random forest model in the prediction of in-hospital mortality risk in ICU Patients. *China Digital Medicine, 12,* 81–84.
9. Xu, J., Zhang, Y., Zhou, D., Li, D., & Li, Y. (2014). Uncertain multi-granulation time series modeling based on granular computing and the clustering practice. *Journal of Nanjing University (Natural Sciences), 50,* 86–94.
10. Davoodi, R., & Moradi, M. H. (2018). Mortality prediction in intensive care units (ICUs) using a deep rule-based fuzzy classifier. *Journal of Biomedical Informatics, 79,* 48–59.

11. Wu, S., Liu, L., & Lu, D. (2017). An Imbalanced data ensemble classification based on cluster-based under-sampling algorithm. *Chinese Journal of Engineering, 39,* 1244–1253.
12. Tao, X. M., Hao, S. Y., Zhang, D. X., & Xu, P. (2013). Overview of classification algorithms for unbalanced data. *Journal of Chongqing University of Posts and Telecommunications (Natural Science Edition), 25,* 101–111.
13. Predicting Mortality of ICU Patients: The PhysioNet/Computing in Cardiology Challenge 2012. National Institutes of Health NIH, 2012. http://physionet.org/challenge/2012/. Accessed 27 Aug 2012.
14. Luo, K., Lin, M. G., & Xi, D. M. (2005). Review of classification algorithms in data mining. *Computer Engineering, 31*(1), 3–11.
15. Geng, L. J., & Li, X. Y. (2014). Improvements of KNN algorithm for big data classification. *Application Research of Computers, 31,* 1342–1344.
16. Cherfi, A., Nouira, K., & Ferchichi, A. (2018). Very fast C4.5 decision tree algorithm. *Applied Artificial Intelligence, 32,* 119–137.
17. Xu, X. C., Saric, Z., & Kouhpanejade, A. (2014). Freeway incident frequency analysis based on cart method. *Promet - Traffic & Transportation, 26*(3), 191–199.

# High-Speed Railway in Yunnan Province: The Impacts of High-Speed Railway on Urban Economic Development

**Lili Hu, Ming Guo, Yunshuo Liu, and Long Ye**

**Abstract** In recent years, the rapid development of China's high-speed railway has brought great impact on regional economic development. This paper evaluates the impact of Shanghai-Kunming high-speed railway on the economic and social development of cities along Yunnan Province from two aspects: direct effect and indirect effect. The direct effects were evaluated by two indicators: railway transport capacity and regional accessibility. The indirect effects are evaluated by four indicators, namely regional economic development, regional industrial structure, tourism development and regional employment. The grey model prediction method is used to calculate the economic development index values of cities along the route of yunnan province in the case of "without" shanghai-kunming high-speed railway in 2017 and 2018, then compared with the actual values of that year. The results show that the Shanghai-Kunming high-speed railway enhances the transportation capacity of cities, improves regional accessibility, promotes economic development, promotes the optimization and upgrading of industrial structure, and promotes the increase of employment in cities along Yunnan Province. Based on this, the paper puts forward the countermeasures of developing Yunnan's high-speed economy from the aspects of giving priority to the development of tourism, optimizing the industrial layout and constructing the new town area, so as to maximize the positive effect of high-speed railway and avoid the appearance of "gray rhinoceros" of high-speed railway.

**Keywords** Shanghai-Kunming HSR way · Regional economy · Grey prediction model

L. Hu (✉) · M. Guo · Y. Liu · L. Ye
School of Economics and Management, Beijing Jiaotong University, Beijing, China
e-mail: 18113081@bjtu.edu.cn

M. Guo
e-mail: gming@bjtu.edu.cn

Y. Liu
e-mail: 18113085@bjtu.edu.cn

L. Ye
e-mail: yelong@bjtu.edu.cn



J. Zhang et al. (eds.), *LISS2019*,
https://doi.org/10.1007/978-981-15-5682-1_8

## 1 Introduction

By the end of 2018, China's high-speed railway (HSR) operation mileage reached 29,000 km, ranking first in the world. As a modern means of transportation with low social cost, high speed and convenience, the construction and operation of HSR will certainly have an impact on the economic and social development. With the formation of "eight vertical and eight horizontal" HSR in China, some scholars began to focus on the research concerned impact of HSR on regional accessibility, urban spatial structure, industrial layout (mostly tourism) and economic growth effect [1]. Scholars from different countries have different opinions on whether HSR can promote regional economic growth. One view is that the role of HSR on regional economic growth is not clear, even if it has a short-term impact, and in the long run, the economic growth rate of the marginal areas connected by HSR will decline [2]. Some scholars also believe that HSR has a significant role in promoting regional economic growth, such as Stokes'analysis of HS2, which was not started in Britain at that time, pointing out that the construction of HSR can significantly improve the economic benefits of traditional transportation [3]. After investigating the situation of HSR in Japan and Europe, Kim pointed out that by expanding regional accessibility, residents' living location choice and working methods can be gradually changed, thus stimulating regional economic growth [4]. Studies on China's wuhan-guangzhou HSR show that the opening of high-speed railway improves the accessibility level of cities along the route, and has a certain relationship with the negative economic growth of non-cities along the route [5]. Other scholars believe that HSR has a negative impact on regional economic development. Kim and Hyunwoo use the neoclassical economic growth model to conclude that large-scale land development projects such as HSR have a negative impact on local economic development, which is due to the mature development of western countries and the lack of space for regional development [6]. This is similar to the typical "siphon effect". HSR promotes the economic development of metropolitan cities along the line. Because HSR shortens the space-time distance between cities and regions, it also means that talents and capital will converge to more mature metropolitan cities. Small and medium-sized cities along the line will face the dilemma of brain drain and enterprise reduction, and the development of non-HSR lines and non-station surrounding areas. The exhibition will bring extremely adverse effects, break the original balance of regional development, and form a new regional advantage gap and development opportunity gap between regions and cities. At present, scholars' conclusions on the impact of HSR on regional economy are not uniform. Further research is needed to explore the relationship between the two, so as to better develop the positive effect of HSR and avoid the phenomenon of "gray rhino" of HSR.

The shanghai-kunming HSR (Shanghai-Kunming) has been in operation since 2010 and was connected to yunnan province in 2016. Kunming, qujing and other five cities are connected to the national HSR network. The opening and operation of Shanghai-Kunming HSR marked Yunnan's entry into a new era of HSR, realized the situation of full coverage of high-speed motor vehicles in the national railway bureau, greatly shortened the space-time distance between Southwest China and South, East and South China, and further revealed the network effect of HSR in China, which is of great significance to improving regional traffic conditions and promoting regional economic and social development.. However, there are relatively few studies on its impact on Yunnan Province. Tian and Xi believe that the operation of Shanghai-Kunming HSR makes the tourism agglomeration effect of Kunming, Qujing, Chuxiong and other cities along the HSR as well as close-distance cities larger, and makes the tourism agglomeration effect of Lijiang, Dali, Dezhou and other high-speed long-distance cities smaller [7]. Shanghai-Kunming HSR has played a role in promoting the economic development of Jiangxi Province, mainly in enhancing the transport capacity of Jiangxi Province, improving regional accessibility, promoting the growth of the total economic volume, promoting the optimization of industrial structure, and has a positive role in the increase of the number of employed people in Jiangxi Province [8]. This is because some studies only consider the direct impact of HSR on passenger transport, short-term impact, without considering its indirect and long-term impact on regional economy. HSR project has the characteristics of large investment, high technology level and long construction period. When evaluating the impact effect, we should not only pay attention to the impact on the national economy, but also consider the impact on transportation and social development. Therefore, considering the role of HSR in regional transport capacity, accessibility, economic development, industrial structure and employment, the mechanism of HSR is analyzed, and the positive effect of high-speed rail on traffic conditions and economic development along Yunnan high-speed rail is proved by empirical data.

## 2 Mechanism Analysis of the Economic Impact of Shanghai-Kunming HSR on Yunnan Province

### *2.1 Study and Analysis of the Direct Impact of HSR on Regional Economy*

The most direct manifestation of the impact of HSR on regional economy is in the aspect of transportation. The main direct impact analysis indicators are expected railway transport capacity and regional accessibility. Railway transport capacity is mainly measured by railway passenger volume, that is, the actual number of passengers transported by a certain regional transport mode in a certain period of time. Accessibility is mainly measured by weighted average travel times, economic potential and daily accessibility. On the one hand, the construction of HSR can significantly improve the economic benefits of traditional transportation and save time [2]. The selection scope of enterprise talent recruitment is also broader, creating more employment opportunities for the region [9]. On the other hand, HSR will expand the scope of isochronal circle, realize the overall optimization of urban daily accessibility, and form an unbalanced time convergence space, which will have a significant impact on the change of urban spatial pattern [10]. Further, Feng and other studies show that HSR construction will promote regional accessibility among provinces in China, and HSR construction will bring shorter and more accessible inter-provincial links. The "HSR effect" such as a large increase in the optimal area of accessibility, which makes the inter-provincial accessibility balanced by the operation of HSR, has a "bowl-shaped" characteristic of high accessibility in space around the middle concave. The provinces located near the center of the passenger railway network have a small change, while the provinces in the periphery, such as Yunnan and Fujian, have a large change in accessibility [11].

### 2.2 Research on Indirect Impact of HSR on Regional Economy

Scholars have done a lot of in-depth research on the indirect impact of high-speed rail on regional economic development. The indirect impact of high-speed railway on regional economic development mainly includes output value and population. Common indicators of output value include the indicator of regional GDP to promote regional economic development; The added value index of the regional primary industry, secondary industry and tertiary industry to optimize the industrial structure; An indicator of total tourism revenue that promotes the development of tourism. Population indicators include the total number of employees in the region, the number of employees in the region's primary industry, secondary industry and tertiary industry, which affect the number of employees.

Zhang believes that the high-speed rail network will intensify the competition between travel modes and different regions, expand the sources of tourists and change the spatial tourism pattern of China [12]. For example, the Zhengxi high-speed railway has played a significant role in promoting tourism along the line, which is reflected in the increase of short-distance and holiday tourists [13], Wang and Wen founds, it is considered that the impact of high-speed rail on tourism has a double-edged sword effect [14]. The positive effect is similar to that of previous scholars, but there are also negative effects of intensified competition in transport and inter-regional tourism. Secondly, scholars have done a lot of in-depth research on the regional spatial structure and economic development of high-speed rail. Fang and Sun used a comparative method to explore the impact of high-speed rail on the spatial economic connection strength and traffic accessibility of the cities along Beijing-Guangzhou line. It was found that the construction of high-speed rail significantly increased the attraction and traffic accessibility between cities, and improved the efficiency of provincial capital cities. The result is superior to that of small and medium-sized cities [15]. Qian et al. deepened the gravity model algorithm and analyzed the impact of high-speed rail on Wuhan urban agglomeration.

It was found that Wuhan urban agglomeration has three types of economic linkages: economic radiation type, economic dependence type and economic complex type [16]. In addition, Li, Liu and Cao constructed a quantitative analysis model of economic linkages intensity based on population human indicators (population mobility spatial linkages index, industrial population linkages index) [17].

Generally speaking, the current research on the impact of HSR on regional economy is divided into qualitative research and quantitative research. Qualitative research mainly discusses that the construction of HSR will bring the same city effect, agglomeration effect, siphon effect and so on, which has an important impact on the economic development, industrial structure, transportation and other cities along the line. Quantitative research mainly uses grey model prediction method and regression analysis method to select objective indicators to confirm the impact of HSR on regional economy. However, most of the studies focus on the economic impact of the whole HSR line, such as tourism, spatial structure, Urban Accessibility and so on. There is a lack of research on the impact of HSR construction on specific areas. Therefore, taking Shanghai-Kunming HSR as an example, this study uses grey prediction model to explore the impact of Shanghai-Kunming HSR on the contribution rate of various economic indicators in Yunnan Province before and after its opening (Table 1).

**Table 1** Impact index of HSR on regional economic development

| Target layer | Influence layer | Index level |
|---|---|---|
| Direct impact | Enhancing transport capacity | Railway passenger volume |
| | Provide regional accessibility | Weighted average travel time |
| Indirect impact | Promoting regional economic development | Gross regional product |
| | Optimizing industrial structure | Value added of regional primary, secondary and tertiary industries |
| | Promoting tourism development tourism income | Total tourism income |
| | Employment number | Total regional employment number, regional primary industry, secondary industry and tertiary industry employment number |

# 3 Research Design and Data Sources

## 3.1 Research Object

The data used in this paper are from China Statistical Yearbook [18]. In terms of time selection, the Shanghai-Kunming HSR was opened and compared in 2016. In fact, the Statistics Bureau data were used for the HSR, assuming that no HSR was opened by 2010–2016 Statistics Bureau data, and the situation in 2017 was predicted by using the comparison method.

## 3.2 Research Methods

"With and without comparison" means to predict the situation that may occur when the project is not completed, and to compare it with the actual situation of the project, so as to measure the specific impact of the project. In this study, we use the "whether or not" comparison method to analyze the differences of economic development indicators in Yunnan Province under the two conditions of "yes" and "no" HSR. These differences reflect the role of Shanghai-Kunming HSR in regional economic development. Among them, the index value of regional economic development in the case of "with" Shanghai-Kunming HSR is the actual statistical value, and in the case of "without" Shanghai-Kunming HSR, the index value of regional economic development in the case of "without" Shanghai-Kunming HSR in 2017 is predicted by using grey prediction (GM) model and multiple linear regression model, and based on the "with or without comparison method" and 2017. By comparing the actual indicators of social and economic development along the Shanghai-Kunming HSR in 2010, the contribution level of the Shanghai-Kunming HSR to the social and economic development along the Yunnan HSR was obtained. Grey prediction model has high prediction accuracy, which can maintain the original characteristics of the data system and reflect the changing trend of the system to a certain extent.

The forecasting principle of grey prediction (GM) model is that a set of new data series with obvious trend is generated by cumulative method for a certain data series, and a model is established to predict the growth trend of the new data series. The original data series are restored by inverse calculation with cumulative method, and the forecasting results are obtained. In theory, only more than four data can be used to build the model, and the prediction accuracy is high, which can better reflect the actual situation of the system. The GM modeling process is as follows [19]:

The original data of a predicted object are listed as follow:

$$X^{(0)} = \{X^{(0)}(1), X^{(0)}(2), X^{(0)}(3)\ldots X^{(0)}(i), i = 1, 2, \ldots, n\} \qquad (1)$$

In order to establish the grey prediction model, $X^{(0)}$ is accumulated once to generate an accumulated sequence.

$$X^{(1)} = \{X^{(1)}(1), X^{(1)}(2), X^{(1)}(3)\ldots X^{(1)}(i), i = 1, 2, \ldots, n\} \tag{2}$$

Superior 1 means one accumulation. Similarly, it can do accumulation.

$$X^{(m)}(k) = \sum_{\mathrm{i}=1}^{\mathrm{k}} X^{\mathrm{m}-1}(i) \tag{3}$$

Cumulative operation is the inverse operation of accumulation. Cumulative operation can restore the accumulated generated columns to non-generated columns and obtain incremental information in the model. The formula of a cumulative decrease is as follows:

$$X^{(1)}(k) = \sum_{i-1}^{k} X^{(0)}(i) = X^{(1)}(k-1) + X^{(0)}(k) \tag{4}$$

According to the grey theory, the first order differential equation about t can be established for $X^{(1)}$, GM (1,1):

$$\frac{dX^{(1)}}{dt} + \alpha X^{(1)} = \mu \tag{5}$$

Among them, $\alpha$ is the grey number of development and $\mu$ is the grey number of endogenous control.

Let's set $\hat{\alpha}$ as the parameter vector to be estimated,

$$\hat{\alpha} = \begin{pmatrix} a \\ - \\ \mu \end{pmatrix} \tag{6}$$

The solution can be obtained by using the least square method

$$\hat{\alpha} = \left(B^T B\right)^{-1} B^T Y_n \tag{7}$$

Based on the above analysis, construct data matrix A and data neighboring:

$$YA = \begin{bmatrix} -\frac{1}{2}[x^{(1)}(1) + x^{(1)}(2) & 1 \\ -\frac{1}{2}[x^{(1)}(2) + x^{(1)}(3) & 1 \\ \vdots & \vdots \\ -\frac{1}{2}[x^{(1)}(n-1) + x^{(1)}(n) & 1 \end{bmatrix} \tag{8}$$

$$Y = \{x^{(0)}(1), x^{(0)}(2), x^{(0)}(3), \cdots x^{(0)}(n)\}^T \tag{9}$$

Set $\hat{\alpha}$ as the parameter vector to be estimated and obtained by least square fitting m, n

$$\hat{\alpha} = \begin{bmatrix} m \\ n \end{bmatrix} = \left(A^T A\right)^{-1} A^T Y \quad (10)$$

By solving the differential equation according to GM (1,1), the prediction model can be obtained:

$$x^{(1)}(k) = \left[x^{(1)}(0) - \frac{n}{m}\right] e^{-mk} + \frac{n}{m}, k = 0, 1, 2, 3 \ldots n \quad (11)$$

# 4 Regression Model and Variable Selection

The impact of HSR on regional economic development is mainly manifested in two aspects: direct and indirect impact. Generally, the direct effect is measured by railway transport capacity and regional accessibility, and the indirect effect is measured by regional gross domestic product, industrial structure, tourism development and employment. Therefore, in the case of "yes" and "no" Shanghai-Kunming HSR from 2017 to 2018, the direct impact of Shanghai-Kunming HSR on the economic development of Yunnan Province is measured by analyzing the changes of Yunnan railway passenger volume and weighted average performance time of cities along the Yunnan section of Shanghai-Kunming HSR, and by analyzing the regional GDP, industrial structure (regional primary industry, secondary industry, and section). The indirect impact of Shanghai-Kunming HSR on the economic development of Yunnan Province is measured by the changes of the value-added of tertiary industry, the total income of tourism and the number of employment (the total number of regional employees, the number of employees in the primary industry, the secondary industry and the tertiary industry).

## *4.1 Analysis of Direct Impact*

(1) Railway transport capacity is generally measured by railway passenger volume, which is the actual number of passengers transported by means of transportation in a certain region in a certain period of time. The railway passenger volume of Yunnan Province from 2010 to 2018 is selected for modeling and forecasting. In the case of "no Shanghai-Kunming HSR" in 2017 and 2018, the forecast value of passenger volume is 118.074 and 132.783 million; in 2017 and 2018, the actual value of railway passenger volume of Yunnan Province is 131.896 and 150.989 million.

$$CR = \frac{X_{with} - X_{without}}{X_{with}} \quad (12)$$

CR means contribution rate, $X_{with}$ means actual value of railway passenger traffic in 2017, $X_{without}$ means forecast value of railway passenger traffic in 2017.

$$CR(2017) = \frac{X_{with} - X_{without}}{X_{with}} = \frac{131.896 - 118.074}{131.896} = 10.48\%$$

$$CR(2018) = \frac{X_{with} - X_{without}}{X_{with}} = \frac{150.989 - 132.783}{150.989} = 12.72\%$$

So the contribution rate of Shanghai-Kunming HSR to railway passenger volume of Yunnan Province in 2017 is10.48%, and in 2018 is about 12.72%. From this, we can see that the Shanghai-Kunming HSR is helpful to improve the passenger transport capacity of Yunnan Railway, and contribute to the promotion of Commerce and tourism in Yunnan and other provinces.

(2) Regional accessibility, weighted average travel time is chosen to measure regional accessibility. The running time of Shanghai-Kunming HSR is selected for the case of "you" Shanghai-Kunming HSR, and the running time of ordinary high-speed train before the completion of Shanghai-Kunming HSR is selected for the case of "no" Shanghai-Kunming HSR. The research cities are along Hunan, namely Changsha, Xiangtan, Loudi, Shaoyang, Huaihua and Zhuzhou. Shanghai, Hangzhou and Changsha are the three central cities of Shanghai-Kunming HSR Station. The economic links between these cities and Hunan Province are most obvious after the Shanghai-Kunming HSR is running through. The index values can also reflect the changes of accessibility level of each evaluation city to a large extent. The formula for calculating the weighted average travel time is as follows:

$$A_i = \frac{\sum_{j=1}^{n} T_{ij} M_j}{\sum_{j=1}^{n} M_j} \quad (13)$$

Among them, $A_i$ is the weighted average travel time value; $T_{ij}$ is the running time from the target city $I$ to the selected economic center $j$; $M_j$ is the social and economic factor flow of city $j$, which is calculated by using the formula of gross GDP and total sales of social commodities respectively, and then the average value of both is obtained.

The weighted average travel time index scores of cities along Guikun section of Shanghai-Kunming HSR in 2017 can be calculated as shown in Table 2. By comparing the opening and non-opening of Shanghai-Kunming HSR with the weighted average travel time as a comparative index, it is found that the indexes of Qujing and Kunming have decreased significantly, with a decrease of 71.16% in

Qujing, 75.81% in Kunming and 63.99% in Anshun. It can be seen that the operation of Shanghai-Kunming HSR has increased the weighted average travel time by more than 60% in Qujing, Kunming and Anshun. The Shanghai-Kunming HSR has shortened the travel time of urban residents along Guikun section and improved regional accessibility.

**Table 2** Impact of HSR on regional accessibility

| City | Without HSR | With HSR |
|---|---|---|
| Qujing | 1957.939 | 564.689 |
| Kunming | 2054.507 | 496.927 |
| Anshun | 1416.717 | 510.158 |

## *4.2 Analysis of Indirect Impact Effect*

(1) Regional economic development. The change of Yunnan's regional GDP under the "yes" and "no" Gui-Kun HSR in 2018 is compared by using the grey prediction (GM) model. The regional GDP of Yunnan Province from 2010 to 2018 is modeled and forecasted. In the case of "no Shanghai-Kunming HSR" in 2017 and 2018, the regional GDP is 16317.808 and 1779.748 billion yuan; in 2017 and 2018, the actual regional GDP of Yunnan Province is 1637.634 billion yuan and 1788.112 billion yuan. Therefore, the contribution rate of Shanghai-Kunming HSR to Yunnan's regional GDP in 2017 is 0.36%, and that of 2018 is about 0.47%. From this, we can see that the Shanghai-Kunming HSR is helpful to increase Yunnan's regional gross product, and to promote business and tourism between the province and other provinces.

(2) Regional industrial structure. By using grey prediction (GM) model, the contribution rates of Shanghai-Kunming HSR to industrial added value in 2017 are - 0.32, 0.99 and 2.23%, respectively. In 2018, the contribution rates of Shanghai-Kunming HSR to industrial added value in the first, second and third industries are 0.01, 4.07 and - 1.55%, respectively. Thus, the Shanghai-Kunming HSR has played a good role in promoting the development of Yunnan industry. Among them, the contribution rate of HSR to the primary industry is relatively low, and the secondary industry is greatly affected by HSR in 2018. This is due to the addition of new lines in Yunnan Province, such as Yukun HSR, Maile-Mengzi HSR, and the development of building materials, locomotive parts manufacturing, communication equipment and other related industries along the line will be due to the construction of HSR. The promotion of HSR to the tertiary industry was greatly affected by HSR in 2017 and decreased in 2018, which has a certain relationship with the strategy of industrial restructuring in Yunnan Province. In 2018, Yunnan Province built a modern industrial system of "two types and three modernizations" and promoted green development. The role of HSR in promoting the tertiary industry needs further observation, rational allocation of HSR

resources, and guard against the impact of HSR "gray rhinoceros" on the economic development of Yunnan Province, so as to achieve the goal of industrial structure optimization.

## 5 Conclusion and Discussion

### 5.1 Conclusion

The results show that the Shanghai-Kunming HSR enhances Yunnan's transportation capacity and regional accessibility, promotes economic development, promotes the optimization and upgrading of industrial structure, and promotes the increase of employment in Yunnan Province. Based on this, the paper puts forward three countermeasures to develop Yunnan's high-speed economy, namely, rational allocation of HSR resources, optimization of industrial layout and enhancement of tourism advantages, so as to maximize the positive effects of HSR and avoid the occurrence of high-speed "gray rhinoceros".

### 5.2 Discussion

The operation of Shanghai-Kunming HSR has opened the communication channels among the three economic circles of Central China, Yangtze River Delta and Changsha-Zhuzhou-Tan. At the same time, it has formed three fast direct channels in Central China, Pearl River Delta and Southwest China. These HSR lines constitute a huge HSR network, which plays an important role in promoting economic and social development along the line. First of all, the arrival of HSR has a great impact on the structural transformation and industrial upgrading of Kunming, the capital city of Yunnan Province, and the cities along it, thus promoting the diversified economic development of these cities. At the same time, the operation of HSR has improved the marginalization of Kunming to a certain extent, strengthened the connection between Kunming and the central and Eastern regions, better accepted the industrial transfer from the eastern region, and promoted the industrial restructuring of Yunnan Province. From the accessibility analysis, the Shanghai-Kunming HSR has significantly improved the accessibility with male students and the central and Eastern regions, and strengthened regional links. Secondly, Yunnan's tourism economy has great potential for development. The operation of Shanghai-Kunming HSR reduces the time for tourists to visit Yunnan, increases the mode of traveling, increases the number of tourists in Yunnan Province, and increases the value of tourism output. The research of Ahlfeldt and Feddersen (2010) points out that, because the huge temporary impact can not bring permanent changes to the regional layout of economic behavior, the short-term impact of HSR construction can not deny the contribution of HSR to regional economic development [20]. Therefore, in the long run, HSR will inject new vitality into Yunnan's economic development.

Based on the above research, the paper puts forward the countermeasures of developing yunnan high speed economy from the aspects of giving priority to the development of tourism, optimizing the industrial layout and building new cities and new districts. Yunnan province has obvious advantages in tourism development. By virtue of the convenience of HSR and the large increase of people flow, it can integrate the existing local resources, strengthen the unified planning and publicity of "HSR tour" brand routes in tourism cities along the HSR, and promote the transformation and upgrading of tourism industry. Before and after the completion of the HSR, it has a great impact on the industrial layout of cities along the HSR in yunnan province. With the help of passenger flow, capital flow and information flow brought by HSR passenger transport, it can improve the investment environment and undertake the transfer of industrial projects, so as to optimize the industrial layout. Finally, with the radiation and driving function of the HSR, the new city will be developed and built in line with the current HSR stations and become a sub-center or satellite city.

## References


1. Zeng, G. (2016). Basic connotations and development strategy of hsrway economic belt: A case of Shanghai-Kunming high-speed railway economic belt in Jiangxi. *Journal of Jinggangshan University (Social Sciences), 4,* 96–100.
2. Jia, S. M., & Lin, Q. C. (2014). Research trend of foreign high speed rail and regional economic development. *Human Grography, 2,* 7–12.
3. Stokes, C. (2012). *Does Britain need high speed rail?, Energy, transport, & the environment.* London: Springer.
4. Kim, K. S. (2000). High-speed railway developments and spatial restructuring: A case study of the Capital region in South Korea. *Cities, 17*(4), 251–262.
5. Lin, L., Zhang, J. R., Duan, Y. N., Xiao, Q., & Zhou, J. Y. (2011). Study on the impacts of accessibility along the Wuguang high speed railway in Hunan. *Economic Research Guide, 12,* 144–148.
6. Kim, H., Lee, D. H., Koo, J. D., Park, H. S., & Lee, J. G. (2012). The direct employment impact analysis of highway construction investments. *KSCE Journal of Civil Engineering, 16* (6), 958–966.
7. Tian, Y., & Xi, T. T. (2018). Study on the impact of Shanghai-Kunming high-speed railway on tourism spatial patterns in Yunnan Province. *Journal of Hebei Tourism Vocational College, 10,* 2943.
8. Guo, Z. H., & Wang, J. P. (2011). A study on the impact of Shanghai-Kunming high-speed railway on the economy of Jiangxi Province. *Railway Transport and Economy, 40*(11), 17–21.
9. Sun, S. F. (2012). Strategy research of high-speed railway promoting Liaoning regional economy development. In *Lecture notes in Electrical Engineering*, Vol. 148, pp. 93–105.
10. Jiang, H. B., Xu, J. G., & Qi, Y. (2010). The influence of Beijing-Shanghai high-speed railways on land accessibility of regional center cities. *Acta Geographica Sinica, 65*(10), 1287–1298.

11. Feng, C. C., Feng, X. B., & Liu, S. J. (2013). Effects of high speed railway network on the inter-provincial accessibilities in China. *Progress in Geography, 32*(8), 1187–1194.
12. Zhang, M. (2010). High-speed railway's prospective impact on Chinese tourism and consideration of countermeasures. *Value Engineering, 29*(11), 227–228.
13. Wang, J. X., Gao, Y., Li, W. B., & Pu, L. (2013). Study on the impact of high-speed railway construction on tourism development of cities along the line: A case study of Zheng Xi high-speed railway. *New West: Mid-Term Theory, 4,* 33–34.
14. Wang, F., & Wen, H. B. (2012). Analysis of the impact of the construction of Nan Guang high-speed railway on the tourism development of Guangxi and the countermeasure consideration. *Oriental Enterprise Culture, 10,* 228–229.
15. Fang, D. C., & Sun, M. Y. (2014). Study on the impact of high-speed railway construction on urban spatial structure in China–A case study of cities along Beijing-Guangzhou high-speed railway. *Regional Economic Review, 3,* 136–141.
16. Qian, C. L., Ye, J., & Lu, C. (2016). Gravity zoning in Wuhan metropolitan area based on an improved urban gravity model. *Progress in Geography, 34*(2), 237–245.
17. Li, X. M., Liu, Y. Z., & Cao, L. P. (2014). Research about the influence of population space of flow under the high-speed railway construction. *China Population, Resource and Environment, 24*(6), 140–147.
18. China, National Statistical Bureau of the People's Republic of China (2018). *China statistical yearboo*k. China Statistical Publishing House.
19. Deng, J. L. (1986). *The grey estimate and decision*. Wuhan: Huazhong University of Science and Technology Press.
20. Ahlfeldt, M. G., & Feddersen, A. (2010). *From periphery to core: Economic adjustments to high speed rail*. Working Papers Institut d'Economia de Barcelona (IEB).

# Improved Sweep Algorithm-Based Approach for Vehicle Routing Problems with Split Delivery

**Jianing Min, Cheng Jin, and Lijun Lu**

**Abstract** The vehicle routing problem with split delivery—an important branch of the classic vehicle routing problem—relaxes the constraint that each customer is visited only once. The objective is to minimize the transportation distance and the vehicles used. A two-stage algorithm based on the improved sweep heuristic approach is proposed for this problem. Customer points are clustered into the minimum number of groups via multi-restart iterations. The load demands and the split point in each group are determined by the load rate and the fine-tuned threshold coefficients. An optimal route is generated by a Tabu search algorithm to minimize the travel distance in each group. Numerical experiments are conducted on benchmark datasets to verify the feasibility and effectiveness of the proposed algorithm. The computational results show the near-optimal performance of the proposed algorithm with regard to the transportation distance and the computation time to the instances with a scattered distribution geographical characteristic.

**Keywords** Fine-tuning · Improved sweep algorithm · Split delivery · Tabu search · Vehicle routing problem

---

This work is sponsored by the National Natural Science Foundation of China (grant no. 61872077) and the Natural Science Fund of Jiangsu Province Education Commission (grant no. 17KJB520040).

---

J. Min
School of Business, Taihu University of Wuxi, Wuxi, China
e-mail: mjn3862@126.com

C. Jin (✉)
University Office, Taihu University of Wuxi, Wuxi, China
e-mail: jcm3988@126.com

L. Lu
School of Management, Nanjing University, Nanjing, China
e-mail: dg1702501@smail.nju.edu.cn



J. Zhang et al. (eds.), *LISS2019*,
https://doi.org/10.1007/978-981-15-5682-1_9

## 1 Introduction

The vehicle routing problem with split delivery (SDVRP) was first introduced by Dror and Trudeau [1] in 1989. In the SDVRP, a delivery demand to one point can be split into any number of vehicles [1]. Thus, the VRP can be optimized with regard to the vehicles used and transportation distance. Therefore, the SDVRP has rapidly become an important branch of the VRP and has received more and more attention [2].

Heuristics and exact solution approaches have been proposed for the SDVRP [2]. Dror and Trudeau [1] proposed a first heuristic algorithm for the SDVRP. There are two types of methods in the algorithm. The first one is the k-split interchange, i.e. a customer point with its demand is split into different routes in which the remaining capacity is sufficient for this split customer. The second one is route addition, i.e. a split customer point is removed from all the routes where the customer point lies and a new route is created, which is composed of that customer point only. The computational results showed that main savings are due to the high customer demands. Archetti et al. [3] proposed the first Tabu search algorithm for k-split named SPLITABU. There are three phases. In the first phase, an initial feasible solution is constructed. A traveling salesman problem (TSP) on all customers is generated and this large tour is cut into pieces to meet the vehicle capacity limitation. In the second phase, two procedures called order routes and best neighbor are performed in Tabu search algorithm to determine the best routes encountered so far. In the third phase, the further reducing the length of each route is done by using the GENIUS algorithm proposed by Gendreau, Hertz, and Laporte (1992) to further improve the solution. Jin et al. [4] proposed a two-stage algorithm with valid inequalities. In the first stage, the customers are clustered into groups. In the second stage, the TSP is solved for each group to determine the minimum transportation distance. The sum of the minimum distances transported over all groups is used to repeatedly update the objective function. Liu et al. [5] proposed a k-means clustering algorithm with strategy of "routing first grouping later". Wilck and Cavalier [6] developed a construction heuristic algorithm for solving the SDVRP. Wen [7] proposed a multi-restart iteration local search (MRSILS) algorithm. First, a "grouping after routing" strategy is adopted to partition all points into groups for satisfying the load demand limitations according to the vehicle capacity. Then, for each point, deletion and reinsertion operations are iterated until no more improvement occurs. Archetti et al. [8] implemented a branch–price–cut algorithm based on the principle of breaking up the problem into sub-problems. Each column was generated by the sub-problem (a route) with delivery quantities. These columns were used to find an optimal heuristic solution to the problem. The computational results showed that the algorithm could obtain the better solutions in most of the benchmark instances. Based on this algorithm, the commodity-constrained SDVRP on 40 customers with three commodities per customer was solved [9]. Xiang et al. [10] proposed an algorithm with the strategy of clustering first and routing later. The clustering is based on the "nearest" principle; in order to meet the capacity limitation, a split threshold is set to control the vehicle load within a certain range; the ant colony optimization algorithm is used to arrange the routes. Shi et al. [11] proposed a model for SDVRP with stochastic customers. Modified split insertion operators and an adaptive large neighborhood search heuristic are adopted. Han et al. [12] formulated the bike relocation problem (in bike-sharing systems) as an SDVRP. A constrained k-means algorithm is used to cluster stations, and a genetic algorithm is used to solve the TSP for the clustered stations. Shi et al. [13] proposed a local search-based particle swarm approach. Coding and decoding methods and best-position vectors are designed to optimize the SDVRP solutions. Ozbaygin et al. [14] proposed exact solution methods for SDVRP. First, a vehicle indexed flow-based formulation is developed. Then the size of this formulation is reduced by compositing the decision variables over all vehicles. The optimal solutions are obtained either by extending the formulation locally with vehicle-indexed variables or by node splitting.

The heuristic methods employed in the aforementioned studies are generally time-consuming, and most exact algorithms can solve the SDVRP only at a small scale. To resolve these issues, in the present study, an improved sweep algorithm (ISA) is proposed for partitioning all customer points into groups according to the maximum vehicle capacity fine-tuned by the appropriate load rate and the threshold coefficient (*TC*). A TSA is adopted to optimize the route in each group. Computational experiments on the benchmark datasets show that the proposed algorithm is feasible and effective, and can obtain near-optimal solutions in a few seconds to the instances with a scattered distribution geographical characteristic.

The remainder of this paper is organized as follows. In Sect. 2, the SDVRP is described. In Sect. 3, the proposed two-stage algorithm is introduced. In Sect. 4, the computational results are presented and discussed. Finally, the conclusions are presented in Sect. 5.

## 2 Problem Description

In the SDVRP, a vehicle departs from the depot, serves the customers with delivery demands, and finally returns to the depot. When vehicles serve customers, there is no limitation on the number of visits, and each customer can be visited by the same vehicle or different vehicles multiple times. The goal is to obtain a set of vehicle paths that minimizes the total traveling distance and the number of vehicles under the vehicle-capacity limitation.

The SDVRP is an un-digraph $G = (V, E)$, where $V = \{0, 1, 2, \ldots., m\}$ is the vertex set, and $E$ is the edge set. The other notations and parameters used in this paper are as follows.

$i, j$: Indices of the depot and customers; $i, j = (0, 1, 2, \ldots, n)$, where 0 represents the depot

$V$: Index of vehicles; $v = (1, 2, \ldots, m)$; vehicles in the fleet are homogenous
$Q$: Vehicle capacity
$d_i$: Delivery demand of customer $i$
$c_{ij}$: Distance between customer $i$ and point $j$; ($c_{ii} = 0$, $c_{ij} = c_{ji}$); it is non-negative and satisfies the triangle inequality
$d_{ij}$: Delivery demand moved from customer $i$ to customer $j$; $d_{ij} \geq 0$
$x_{ij}^{v}$: If vehicle $v$ travels from customer $i$ to customer $j$, $x_{ij}^{v} = 1$; otherwise, $x_{ij}^{v} = 0$; $i$, $j = (0, 1, 2, \ldots, n)$; $v = (1, 2, \ldots, m)$
$y_{iv}$: Demand of $i$ delivered by vehicle $v$; $i, j = (0, 1, 2, \ldots, n)$; $v = (1, 2, \ldots, m)$

Assume that there are $n$ customers and $m$ vehicles in the considered SDVRP. The minimum number of vehicles to be used is $\left\lceil \sum_{i=1}^{n} d_i/Q \right\rceil$ [2], where $\lceil x \rceil$ denotes the smallest integer that is not less than $x$. The customer requirements should be fully satisfied. The objective function is formulated as follows:

$$min \sum_{i=0}^{n} \sum_{j=0}^{n} \sum_{v=1}^{m} c_{ij} x_{ij},$$

subject to

$$\sum_{i=0}^{n} \sum_{v=1}^{m} x_{ij}^{v} \geq 1, j = 0, 1, \ldots, n; v = 1, 2, \ldots, m \tag{1}$$

$$\sum_{i=0}^{n} x_{ip}^{v} - \sum_{j=0}^{n} x_{pj}^{v} = 0, p = 0, 1, \ldots, n; v = 1, 2, \ldots, m \tag{2}$$

$$\sum_{i \in S} \sum_{j \in S} x_{ij}^{v} \leq |S| - 1, v = 1, 2, \ldots, m; S \subseteq V - \{0\} \tag{3}$$

$$y_{iv} \leq d_i \sum_{j=0}^{n} x_{ij}^{v}, i = 1, 2, \ldots, n; v = 1, 2, \ldots, m \tag{4}$$

$$\sum_{v=1}^{m} y_{iv} = d_i, i = 1, 2, \ldots, n \tag{5}$$

$$\sum_{i=1}^{n} y_{iv} \leq Q, v = 1, 2, \ldots, m. \tag{6}$$

Constraint (1) indicates that a customer point must be served at least once. Constraint (2) presents the flow-conservation restraint. Constraint (3) gives the sub-route elimination restraint. Constraint (4) indicates that customer point $i$ is served by vehicle $v$ only if vehicle $v$ visits customer point $i$. Constraint (5) ensures that all customer demands are satisfied. Constraint (6) guarantees that the load of each vehicle does not exceed its maximum load capacity.

## 3 Proposed Algorithm

The algorithm put forward in this paper is a two-stage algorithm based on the strategy of grouping followed by routing. In the first stage, an ISA is employed to divide the customers into clusters, and the split points in each partition are determined. The second stage involves route optimization; a TSA is used to determine the optimal route for each cluster. The details of the proposed algorithm are described in the following subsections.

### *3.1 Pre-process*

The deliveries of each customer point $d_i > Q$ should be considered at the very beginning. The weight $Q$ can be carried by one vehicle, and the remaining weight $d_i = (d_i - Q)$ of this point is dealt with using the following procedure.

### *3.2 ISA*

The sweep algorithm is a constructive heuristic approach proposed by Gillett and Miller [15] in the 1970s. It essentially partitions the nearest customers into a cluster. To satisfy the requirement of the SDVRP, the classic sweep algorithm is modified. $\lceil \sum_{i=1}^{n} d_i/Q \rceil$ is used to determine the number of groups. Multi-restart iteration is used to determine the last point and the split point for each group. The load rate and *TC* are used for fine-tuning the group partitions. The procedure of the ISA is explained in detail as follows.

*Step 1: Transform the coordinate systems*

(a) *Create a polar coordinate system:* Define the depot as the original point of this polar coordinate system.
(b) *Set angle* 0: Connect the origin and a customer point, and set it as angle 0.
(c) *Transform:* Convert all customer points to this polar coordinate system.
(d) *Sort:* Sort all angles of the customer points in ascending order.

*Step 2: Partition the customer domain*
*Step 2.1: Create the variables InitialValue and StoragePool. The former is used to store the initial value, and the latter is used to store the best value.*

*Step 2.2: First loop*

(a) *Set start point:* Set the point of angle 0 as the first starting point.
(b) *Sweep:* Sweep points one by one into this group in a clockwise (or counter-clockwise) manner until the cumulative value (*CV*) of $\sum_{i=1}^{n} d_i$ exceeds *Q*.
(c) *Create a group:* Split this last point into *lp1* and *lp2; lp1* makes that the load of the current group $d_{lp1}$ is equal to *Q*. This group ends with *lp1*.
(d) *Create the next group:* Start the next group from point *lp2* (with the demand $d_{lp2}$).
(e) *End the sweeping:* Repeat (b) to (d) until the last point is swept.
(f) *Compute the travel distance:* Calculate the total travel distance according to the spatial distribution of points in each cluster.
(g) *Store the travel distance:* the Put the total travel distance into *InitialValue* and *StoragePool*.

*Step 2.3: Restart iterations*
*Step 2.3.1: Sweep clockwise*

(a) *Iterate:* Set sequentially each point as the starting point, in a clockwise manner.
(b) *Continue:* Execute Step 2.2 (b–f).
(c) *Optimize:* Compare the current total travel distance with the value in *StoragePool*. If the value in *StoragePool* is higher, it is replaced with the current one.
(d) *Repeat:* Execute Step 2.3.1 in ascending order until the last point is reached.

*Step 2.3.2: Sweep counterclockwise*

(a) *Iterate:* Sequentially set each point as the starting point, in a counterclockwise manner.
(b) *Continue:* Execute Step 2.3.1 (b and c).
(c) *Repeat:* Execute Step 2.3.2 in descending order until the last point is reached.

*Step 2.3.3: Apply the load rate (ef) and TC*

(a) *Fine-turning:* Adjust the cluster partitions by fine-tuning the load rate (*ef*). The ef changes *Q* to *ef* **Q*.
(b) *Set control branches:* According to *ef* **Q*, refine Step 2.2 (b) into multiple control branches with the *TC*, described as follows.

- If $CV < ef * Q$, cumulating the next *i* value;
- If $CV = ef * Q$, ending the current group and starting the next group;
- If $CV > ef * Q$ and $CV < Q$, cumulating the next *i* value if $(d_i - (CV - ef * Q)) \leq TC * (1 - ef) * Q$; splitting the last point, ending the current group, and starting the next group if $(d_i - (CV - ef * Q)) > TC * (1 - ef) * Q$;
- If $CV = Q$, ending the current group and starting the next group;
- If $CV > Q$, splitting the last point, ending the current group and starting the next group;

(c) *Further fine-turning:* Set a *TC* (e.g., 2 or 4). The partitions can be further fine-tuned.

*Step 3: Output the results*

- Output the total travel distance in *InitialValue* and *StoragePool.*
- Output the points in each cluster.
- Output the split point and the corresponding split values in each cluster.

## 3.3 Route Optimization

After the two aforementioned *A* and *B*, the problem domain is divided into several smaller clusters [16]. There is one route in each group. A TSA is employed to optimize the route in each cluster.

The TSA is a meta-heuristic proposed by Glover in 1986. It is a dynamic neighborhood search algorithm. The detail procedure of the TSA is as follows.

*Step 1: Initialization*

(a) *Set the key variables:* the Tabu length (*TabuLength*), the terminal condition (*MaxIter*), and a Tabu list (*Tabu* [*TabuLength*]).
(b) *Generate the initial route:* Randomly generate the initial route solution of *SerialNum* [*CustomerNum*].

*Step 2: Compute the objective function value of SerialNum and assign it to the variable BestValue.*
*Step 3: If the number of the iteration is equal to MaxIter, stop the program and export the optimal results; otherwise, repeat continuously, and implement the following steps.*
*Step 4: Produce rr neighborhoods of the current solution by using the suitable selection functions (for example, 2-opts).*
*Step 5: Sort these rr neighborhoods in non-descending order and store this list in the variable TempDist [rr] as a candidate queue.*
*Step 6: Judge:*

If *TempDist* [0] < *BestValue*,

- Replace *BestValue* and *CurrentBestValue* with *TempDist* [0].
- Replace *BestQueue*, *CurrentBestQueue*, and *tabu* with the relevant objects of *TempDist* [0].
- Turn to Step 8.

Otherwise, execute Step 7.

*Step 7: Analyze the Tabu attributes of the relevant objects of TempDist [rr]*

- Replace *CurrentBestValue* with the best value of *TempDist* [*i*] (assume the $i^{th}$).

- Replace *CurrentBestQueue* with the relevant objects of *TempDist* [*i*].

*Step 8: Turn to Step 3.*

# 4 Case Studies

Computational experiments were performed to verify the feasibility and effectiveness of the proposed algorithm. Two case studies were adopted. Case study 1 comes from the Christofides and Eilon benchmark dataset and case study 2 comes from the Christofides, Mingozzi and Toth benchmark dataset of the Capacitated Vehicle Routing Problem (CVRPLIB) in the VRP Web [17]. The experiments are performed in *C* using a Windows 7 64-bit machine with an Intel (R) Core processor 2.50 GHz and 8 GB of memory.

## *4.1 Case Study 1*

There are 11 instances in the dataset of case study 1. The experimental results for instances are shown in Table 1.

For each instance, two operations are executed: one is clockwise, and the other is counterclockwise. In each operation for each instance, the results for three control branches are recorded: the load rate (*ef*) control and *TC* = 4 control. In each control branch column, three sub-columns indicate the rate load *ef*, the initial value generated in the $0^{th}$ loop, and the best value. Table 2 presents the experimental results for the 11 instances. The instance name "n21-k4" indicates "21points, 4 routes in Eil dataset". Opt. is the abbreviation for Operation. C-c is the abbreviation for Counterclockwise and C-w is the abbreviation for Clockwise.

As shown in Table 1, for each instance, (1) in the restart iteration (either counterclockwise or clockwise), the travel distances vary; (2) the best value is always lower than the initial value; (3) the best value in the *TC* = 4 control is lower than that in the *ef* control; (3) the best value in the *TC* = 4 control is slightly lower than that in the *TC* = 2 control for some instances.

Table 2 presents a comparison between the experimental results of the proposed algorithm and the reference values provided by the benchmark dataset. The column "*Time(s)*" indicates the computation time of the proposed algorithm. The relative deviation percentage (*RDP*) of the distance between the results of the proposed algorithm and the reference values is given as $RDP = ((Distance_{\text{proposed algorithm}} - Distance_{\text{reference}})/Distance_{\text{reference}}) \times 100\%$.

The comparisons in Table 2 show that the proposed algorithm can obtain near-optimal solutions. The *RDPs* of the best values are all less than 8%, and 7 out of the 11 are less than 5%. The largest amount of computation time is less than 3 s.

**Table 1** Experimental results for instances

| Instance | Opt. | Load-rate (ef) control | | | TC = 4 control | | |
|---|---|---|---|---|---|---|---|
| | | *ef* | *Initial value* | *Best value* | *ef* | *Initial value* | *Best value* |
| n21-k4 | C-c | 0.95 | 433 | 409 | 0.90 | 413 | 381 |
| | C-w | 0.98 | 432 | 409 | 0.90 | 432 | 413 |
| n22-k3 | C-c | 0.70 | 766 | 606 | 0.70 | 739 | 570 |
| | C-w | 0.70 | 768 | 587 | 0.65 | 764 | 587 |
| n29-k3 | C-c | 0.95 | 606 | 565 | 0.92 | 574 | 564 |
| | C-w | 1.0 | 705 | 571 | 0.94 | 567 | 518 |
| n32-k4 | C-c | 0.95 | 894 | 894 | 0.82 | 884 | 868 |
| | C-w | 0.95 | 911 | 911 | 0.83 | 876 | 876 |
| n50-k5 | C-c | 0.98 | 632 | 585 | 0.95 | 670 | 556 |
| | C-w | 0.96 | 632 | 585 | 0.96 | 610 | 570 |
| n75-k7 | C-c | 0.90 | 802 | 775 | 0.85 | 770 | 716 |
| | C-w | 0.90 | 826 | 774 | 0.85 | 780 | 747 |
| n75-k8 | C-c | 1.0 | 863 | 812 | 0.90 | 870 | 764 |
| | C-w | 0.95 | 867 | 820 | 0.92 | 815 | 798 |
| n75-k10 | C-c | 0.95 | 942 | 899 | 0.93 | 941 | 896 |
| | C-w | 0.95 | 931 | 900 | 0.93 | 905 | 882 |
| n75-k14 | C-c | 1.0 | 1189 | 1109 | 0.93 | 1169 | 1097 |
| | C-w | 1.0 | 1193 | 1108 | 0.93 | 1173 | 1095 |
| n100-k8 | C-c | 0.92 | 941 | 911 | 0.90 | 899 | 877 |
| | C-w | 0.92 | 960 | 911 | 0.90 | 891 | 857 |
| n100-k14 | C-c | 0.92 | 1231 | 1192 | 0.86 | 1217 | 1163 |
| | C-w | 0.95 | 1280 | 1173 | 0.88 | 1209 | 1180 |

**Table 2** Comparisons with the reference values

| Instance | Ref. value | Proposed algorithm | | | | *RDP%* |
|---|---|---|---|---|---|---|
| | | *ef* | *Initial value* | *Best value* | *Time (s)* | |
| n21-k4 | 375 | 0.90 | 413.2 | 381.1 | 0.53 | 1.64 |
| n22-k3 | 569 | 0.70 | 738.7 | 570.3 | 0.57 | 0.23 |
| n29-k3 | 534 | 0.94 | 566.8 | 517.6 | 0.69 | –3.07 |
| n32-k4 | 835 | 0.83 | 883.7 | 868.2 | 0.94 | 3.98 |
| n50-k5 | 521 | 0.95 | 609.6 | 555.8 | 1.28 | 6.68 |
| n75-k7 | 683 | 0.85 | 796.8 | 715.5 | 1.94 | 4.75 |
| n75-k8 | 735 | 0.90 | 809.6 | 764.4 | 2.07 | 4.0 |
| n75-k10 | 832 | 0.93 | 904.6 | 881.6 | 1.87 | 5.96 |
| n75-k14 | 1032 | 0.93 | 1173 | 1095 | 1.75 | 6.10 |
| n100-k8 | 817 | 0.90 | 891.2 | 857.3 | 2.848 | 4.93 |
| n100-k14 | 1077 | 0.86 | 1217.0 | 1162.8 | 2.502 | 7.96 |

The last best value can appear in either counterclockwise or clockwise operation and normally appears in the *TC* = 4 branch (of course, the best values in the branches of *TC* = 2 and *TC* = 4 are the same in some instances) in Table 1. This indicates that the ISA is necessary, feasible, and effective.

## 4.2 Case Study 2

There are 6 instances in the dataset of case study 2. The experimental results for instances are shown in Table 3.

Case study 2 is used to compare the performances of the proposed algorithm (ISA + TSA) with SPLITABU in [3] and the normal sweep algorithm + TSA (NSA + TSA).

Table 3 presents the executing results of SPLITABU, NSA + TSA and ISA + TSA on case study 2. In each algorithm column, the total travel distance (*Dis.*) and the computing time (*T*) are recorded, and the *RDP* of its travel distance against the ISA + TSA is calculated. The instance name "v1-50" indicates "vrpnc1 dataset, 50 points".

The experimental results in Table 3 show that (1) the *RDP*s of travel distances between SPLITABU and ISA + TSA except the instance "v11-120-7" are all less than 6%; (2) the *RDP*s of travel distances between NSA + TSA and ISA + TSA are all less than 0. The results indicate that ISA is necessary, feasible, and effective. The ISA + TSA can obtain near-optimal solutions to those instances with a scattered distribution geographical characteristic; (3) the computing time of ISA + TSA is much lower than SPLITABU. The largest amount of computation time is less than 15 s.

**Table 3** Experimental results on case study 2

| Instance | SPLITABU | | NSA + TSA | | | ISA + TSA | |
|---|---|---|---|---|---|---|---|
| | *Dis.* | *RDP %* | *Dis.* | *T* | *RDP %* | *Dis.* | *T* |
| v1-50 | 528 | 4.86 | 578 | 0.6 | –4.3 | 553 | 1.4 |
| v2-75 | 854 | 4.63 | 922 | 2.0 | –3.2 | 893 | 2.9 |
| v3-100 | 840 | 5.49 | 908 | 2.9 | –2.4 | 886 | 3.5 |
| v4-150 | 1055 | 4.17 | 1128 | 7.0 | –2.5 | 1099 | 7.3 |
| v5-199 | 1338 | 4.25 | 1444 | 12.6 | –3.4 | 1395 | 15 |
| v11-120 | 1057 | 20.9<br>6.9 | 1372 | 3.9 | –6.9<br>–17.6 | 1278<br>1130 | 6.6<br>7.8 |
| v12-100 | – | – | 1038 | 2.9 | –11.2<br>–20.1 | 922<br>829 | 3.89<br>4.36 |

But, Table 3 also presents that the *RDP*s of travel distances between SPLITABU and ISA + TSA of the instance "v11-120-7" are much larger. This states that ISA is not efficient on this instance.

The characteristic of geographical distribution of the instance "v11-120-7" is that all customer points are clustered distribution around the depot. To fit this feature, the strategy of "clustering first sweeping later" was adopted. First, Max-Min distance clustering method (briefly as Max-Min dis) was used to cluster the customer domain. Then, the ISA +TSA were used in each cluster. It is a three-stage approach of "Max-min dis" + ISA + TSA.

"Max-Min dis" is a type of pattern recognition method to cluster customer points according to their geographic positions into the groups. The number of groups is determined by the number of least used vehicles. At first, the principle of the maximum distance is used to look for the center point of each group; it means that groups are as far apart as possible. Later on, according to the principle of the minimum distance, the nearest points from a group center are collected into a group; it means that points in a group are as close as possible. The experimental results in [18] show that when the customer points are in a cluster distribution around the depot, the algorithm could achieve better performance.

The executing result is shown in the second line of the instance "v11-120-7" at the column ISA + TSA. The result presents that the travel distance of ISA + TSA is reduced. The *RDP*s are shown in the second line of this instance at the column of SPLITABU and NSA + TSA. The *RDP*s of SPLITABU and NSA + TSA are decreased about 14 and 10.8% respectively.

In order to further verify the execution performances on the "v11-120-7" by the three-stage approach of the "Max-min dis" + ISA + TSA, the instance "v12-100-10" with the similar characteristic to "v11-120-7" was conducted an experiment (No executing results could be found in SPLITABU [3]). The experimental results are shown in the second line of the instance "v12-100-10" at the column ISA + TSA. The executing results are shown in the second line of the instance "v11-120-7" at the column ISA + TSA. The results present that the travel distance of ISA + TSA is reduced. The *RDP* is shown in the second line of this instance. The *RDP* of NSA + TSA is decreased about 9%.

The experimental results indicate that the three-stage approach of the "Max-min dis" + ISA + TSA is much useful than the ISA + TSA to the instances with such a clustered distribution geographical characteristic.

## 5 Conclusion

A two-stage algorithm based on an ISA constructive heuristic is proposed for the SDVRP. In the first stage, a multi-restart iteration strategy is employed to sweep the customer domain in both the counterclockwise and clockwise directions and to start the operations at each point. The customer domain is divided into sub-domains according to the vehicle capacity. The load rate *ef* and the *TC* are used to fine-adjust the partitions, split points, and split load. In the second stage, a TSA is used to optimize a route in each sub-domain, for achieving the minimum total travel distance.

Numerical experiments using the benchmark datasets were conducted. Computational results show that the multi-restart iteration strategy is necessary in the ISA and that the proposed algorithm provides feasible and effective solutions to the SDVRP in most instances of the test datasets. The proposed algorithm obtained approximate optimal solutions ($RDP < 5\%$) in 7 out of 11 instances in dataset 1, and ($RDP < 6\%$) in 5 out of 6 instances in dataset 2. Additionally, the largest amount of computation time was less than 3 *s* in dataset 1 and 15 *s* in dataset 2, which is significantly shorter than those for the other algorithms evaluated. The further experiments showed that the proposed algorithm ISA + TSA was effective to the instances with a scattered distribution geographical characteristic, while the three-stage approach of the "Max-min dis" + ISA + TSA is much useful than the ISA + TSA to the instances with a clustered distribution geographical characteristic.

**Acknowledgement** The authors wish to express the gratitude to the Humanities and Social Sciences Research Base Fund of Jiangsu Province Education Commission (grant no. 2017ZSJD020) and the Jiangsu Key Construction Laboratory of IoT Application Technology, Taihu University of Wuxi for their financial support to this work.

The authors also would like to thank the reviewers and editors, who have carefully reviewed the manuscript and provided pertinent and useful comments and suggestions.

## References

1. Dror, M., & Trudeau, P. (1989). Savings by split delivery routing. *Transportation Science, 2,* 141–145.
2. Archettic, C., & Speranza, M. G. (2012). Vehicle routing problems with split deliveries. *International Transactions in Operational Research, 19,* 3–22.
3. Archettic, C., Hertz, A., & Speranza, M. G. (2006). A tabu search algorithm for the split delivery vehicle routing problem. *Transportation Science, 1,* 64–73.
4. Jin, M., Liu, K., & Bowden, R. O. (2007). A two-stage algorithm with valid inequalities for the split delivery vehicle routing problem. *International Journal of Production Economics, 1,* 228–242.
5. Liu, W., & Huang, J. (2011). Two-stage algorithm for split delivery vehicle routing problem. *Journal of Jimei University (Natural Science), 1,* 38–44.
6. Wilck, J. H., IV, & Cavalier, T. M. (2012). A construction heuristic for the split delivery vehicle routing problem. *American Journal of Operations Research, 2,* 153–162.
7. Wen, Z. Z. (2015) *Researches on iterated local search for the split delivery vehicle routing problem.* M.S. thesis, Beijing Transportation University, China.
8. Archettic, C., Bianchessi, N., & Speranza, M. G. (2011). A column generation approach for the split delivery vehicle routing problem. *Networks, 58*(4), 241–254.
9. Archetti, C., Bianchessi, N., & Speranza, M. G. (2015). A branch-price-and-cut algorithm for the commodity constrained split delivery vehicle routing problem. *Computer Operation Research., 64,* 1–10.

10. Xiang, T. T., & Pan, D. Z. (2016). Clustering algorithm for split delivery vehicle routing problem. *Journal of Computer Applications, 11,* 3141–3145.
11. Shi, J. L., & Zhang, J. (2017). Model and algorithm for split delivery vehicle routing problem with stochastic customers. *Kongzhi yu Juece/Control and Decision, 32*(2), 213–222.
12. Han, Z., Yang, Y., Jiang, Y., Liu, W. & Wang, E. (2018). SDVRP-based reposition routing in bike-sharing system. In *Lecture notes in computer science (including subseries lecture notes in artificial intelligence and lecture notes in bioinformatics), LNCS* (Vol. 11335, pp. 596–610).
13. Shi, J., Zhang, J., Wang, K., & Fang, X. (2018). Particle swarm optimization for split delivery vehicle routing problem. *Asia-Pacific Journal of Operational Research, 35*(2), 1840006.
14. Ozbaygin, G., Karasan, O., & Yaman, H. (2018). New exact solution approaches for the split delivery vehicle routing problem. *EURO Journal on Computational Optimization, 6*(1), 85–115.
15. Gillet, B. E., & Miller, L. R. (1974). A heuristic algorithm for the vehicle dispatch problem. *Operations Research, 22*(2), 340–349.
16. Min, J. N., Jin, C., & Lu, L. J. (2018) A three-stage approach for split delivery vehicle routing problem solving. In *Proceedings of 8th International Conference on Logistics, Informatics and Service Sciences (LISS), Toronto University, Toronto, Canada & Beijing Jiaotong University, Beijing, China* (pp. 96–101). IEEE Catalog Number: CFP18LIS-CDR.
17. VRP Web. http://www.bernabe.dorronsoro.es/vrp/.
18. Min, J. N., Jin, C., & Lu, L. J. (2019). Maximum-minimum distance clustering method for split delivery vehicle routing problem: Case studies and performance comparisons. *Advances in Production Engineering & Management, 14*(1), 125–135.

# The Probability Prediction of Mobile Coupons' Offline Use Based on Copula-MC Method

**Yue Chen and Yisong Li**

**Abstract** With the wide application of the O2O business model, merchants offer mobile coupons to attract customers to use offline. Offline use means that mobile coupons can be used when customers come to store to consume after receiving online. Using precision marketing methods to offer effective mobile coupons that meet consumer spending habits and preferences can increase receiving rate and offline use rate. In the background of O2O and precision marketing, this paper studies the behavior of mobile coupons' online receiving and offline use. Based on the real data of Alibaba's consumers, this paper uses Bayesian network, Copula function and Monte Carlo simulation to construct the consumption behavior model with receiving mobile coupons and the consumer behavior model without receiving mobile coupons. And then it analyzes two factors, which are discount and distance, how to impact on the offline use of mobile coupons. Finally, it predicts the offline use probability of mobile coupons under these two consumption behavior models.

**Keywords** Mobile coupons · Offline use · Copula · Monte carlo

## 1 Introduction

The O2O model makes it possible to interact between online and offline. It also makes transactions more convenient, providing consumers with a variety of choices and convenient services, as well as building a platform for merchants to communicate with customers. However, the massive amounts of information not only

This paper was financially supported by China Railway (Grant No. B19D00010: Optimization of railway material's management mode).

Y. Chen (✉) · Y. Li
School of Economics and Management, Beijing Jiaotong University, Beijing, China
e-mail: 18113039@bjtu.edu.cn

Y. Li
e-mail: ysli@bjtu.edu.cn


J. Zhang et al. (eds.), *LISS2019*,
https://doi.org/10.1007/978-981-15-5682-1_10

makes the choice of consumers difficult, but also makes it difficult for merchants to stand out among competitors that provide homogeneous service. Therefore, it is necessary to help consumers improve their selection efficiency and provide differentiated marketing strategies for merchants. Personalized recommendation for consumers is an effective way to solve this problem, that is, using precision marketing to identify target customers. Precision marketing is using big data mining methods to achieve accurate positioning, and then establish a personalized customer communication system. Precision marketing enhances the viscosity between company and consumer. On the one hand, consumers can experience a higher level of service because of accurate recommendation. On the other hand, merchants can also earn the profit growth caused by loyal customers.

With the wide application of the O2O business model and mobile Internet, merchants offer mobile coupons to attract customers to use offline. Offline use means that mobile coupons can be used when customers come to store to consume after receiving online. Consumers receive more and more mobile coupons online, at the same time, invalid mobile coupons that do not meet consumer spending habits and preferences have also been received. The invalid mobile coupons make the coupon receiving rate and the using rate low, which not only fails to achieve the purpose of attracting customers by the promotion, but also has negative effects to consumers such as message disturbance and information leakage. Therefore, it is of practical significance to use precision marketing methods to offer effective mobile coupons for consumers to increase the rate of receiving and offline use. Offering mobile coupons under O2O mode not only enables effective online and offline interactions, but also offers consumers benefits and a good experience. Mobile coupons as a medium, the consumer behavior can be predicted and the target customer group can be identified based on customers' historical track. Offering mobile coupons according to the probability prediction of consumption, can provide personalized service to the consumer and achieve the goal of low-cost promotions for merchants.

In the environment of mobile Internet, more and more scholars have researched on coupon marketing, including influencing factors of coupon redemption and sharing, coupon distribution considering customers' preferences, and coupon's discount design. Noble et al. studied the impact of family composition, family poverty, and basket size on coupon redemption rates [1]. Zhao et al. studied the factors affecting the willingness of consumers to share mobile coupons through social media [2]. Li et al. proposed a new social coupon endorsing mechanism to identify target customers with the tendency to receive, share, and accept recommended digital coupons [3]. Jiang et al. built a nonlinear mixed-integer programming model based on customer buying behavior and preferences for coupons marketing strategy design [4]. Greenstein-Messica et al. modeled the customer's personalized discount sensitivity and proposed a new CARS algorithm to conduct experiment on real data [5]. Khajehzadeh et al. studied on how the types of products, the convenience of visiting merchants, and consumers' shopping motives impact on mobile coupon redemption [6]. At present, most coupon marketing researches focus on the design of coupons, the analysis of customer's preferences and the influencing factors of coupon redemption. There is a lack of research on probability prediction of coupon using based on consumer's historical behavior trajectories.

Copula theory is used to describe the nonlinear correlation. Most scholars apply Copula theory to the financial field for investment risk analysis. A few scholars apply it to management optimization. Silbermayr et al. studied the economic and environmental performance of pooling strategy in inventory management, using the copula function to simulate the joint distribution of related requirements [7]. Kupka et al. studied the time evolution of the Copula family function and the foreign exchange market [8]. Nguyen et al. used the copula function to study the role of gold in the international stock market, and then analyze the complex dependence between the stock market and the gold price [9]. Mokni et al. studied the relationship between the international stock markets, using the copula function to analyze the dependency structure [10]. There is no research on the application of Copula theory to the correlation description of consumer behavior.

The Monte Carlo (MC) method is a statistical method used to predict the probability or expectation of a random variable. It is widely used in the numerical experiments. Janekova et al. used Monte Carlo simulation for optimization in production planning and project risk management [11]. Acebes et al. used extensive Monte Carlo simulation to obtain information about the expected behavior of the project and designed a method for project control under uncertain risk [12]. Few scholars have combined Copula theory with Monte Carlo simulation in marketing research.

In the background of O2O and precision marketing, this paper studies the consumption behavior of mobile coupons' online receiving and offline use. The remaining sections of this paper are arranged as follows. Section 2 constructs Bayesian network to determine the variables which will be used in the following models and abstract the mathematical probability expressions from the real situation. Then, Copula-MC method is introduced, which includes selecting the Copula function and designing the Monte Carlo simulation implementation process. Section 3 applies Copula-MC method to models construction, which are the consumption behavior model with receiving mobile coupons and the consumption behavior model without receiving mobile coupons, based on the real data of Alibaba's consumers. And then it analyzes two factors, which are discount and distance, how to impact on the offline use of mobile coupons. On the basis of that, it predicts the probability of using the mobile coupons offline under these two consumption behavior models. Section 4 derives the conclusions and provides advice on identifying target consumer groups and marketing strategies of designing mobile coupons for merchants.

# 2 Bayesian Network and Copula-MC Method Description

## 2.1 Bayesian Network Construction

Constructing Bayesian network can show the influencing factors of mobile coupons' offline use clearly and detailly, which is beneficial to derive the factors which have direct impact on mobile coupon's offline use. More importantly, it can derive the mathematical probability expressions which will be used in following models construction in Sect. 3. The offline use of mobile coupons is divided into two steps: one is to receive coupons online, and the other is to use coupons offline. According to the analysis of consumers' behavior of receiving coupons and using coupons, the factors affecting the two are summarized as the discount of mobile coupons and the distance from consumer to merchant. The discount influences whether consumers receive mobile coupons or not, and discount and distance influence whether consumers go to store to consume together.

In general, when the discount is attractive to customers, the probability of receiving mobile coupons is high. And it is a small probability event that consumers still receive mobile coupons when the discount is not attractive. In the same way, when the coupon is received and the distance from the customer to the merchant is convenient, the probability that the consumer uses the coupon offline is high. And it is a small probability event that consumers still go to the store to consume when the mobile coupon is not received and the distance is not convenient. In this paper, these two small probability events are not considered in the model building. The probability that can be obtained from the non-parametric estimation of historical data is regarded as known. And the other probabilities can be calculated in Bayesian formula derivation or model construction. The values of the random variables are shown in Table 1, and the Bayesian network with conditional probability distribution is constructed, as shown in Fig. 1.

From the perspective of the merchant, it focuses on the probability of consuming offline, including the consumption with receiving coupons and the consumption without receiving coupons. Therefore, these two scenario will be applied in the following models construction in Sect. 3. And assuming that consumers choose to consume under rational thinking, small-probability events are ignored in the modeling process. For the above two consumption scenarios, the mathematical probability expressions are as follows:

- Consumption with receiving mobile coupons: $P\left(d_1^1, r^1, d_2^1, c^1\right)$, build the model to obtain the probability values based on Copula-MC method. Section 3 will show the details.
- Consumption without receiving mobile coupons: $P\left(d_1^0, r^0, d_2^1, c^1\right)$, Bayesian network factorization is used to decompose it into several easy-to-obtain multiplicative conditional probability. Equation (1)–(7) show the decomposition process. Because of the assumption that small-probability events are ignored, (7)

**Table 1** The values of random variables

| Variable | Values | Definition | Probability calculation methods |
|---|---|---|---|
| Discount | $d_1^0$ | The discount isn't attractive to consumers | Non-parametric estimation |
| | $d_1^1$ | The discount is attractive to consumers | Non-parametric estimation |
| Receive | $r^0$ | Consumers didn't receive mobile coupon | Non-parametric estimation |
| | $r^1$ | Consumers received mobile coupons | Non-parametric estimation |
| Distance | $d_2^0$ | The distance between consumer and merchant is convenien | Non-parametric estimation |
| | $d_2^1$ | The distance between consumer and merchant isn't convenien | Non-parametric estimation |
| Consume | $c^0$ | Consumers go to the store to consume | Model construction |
| | $c^1$ | Consumers don't go to the store to consume | Model construction |

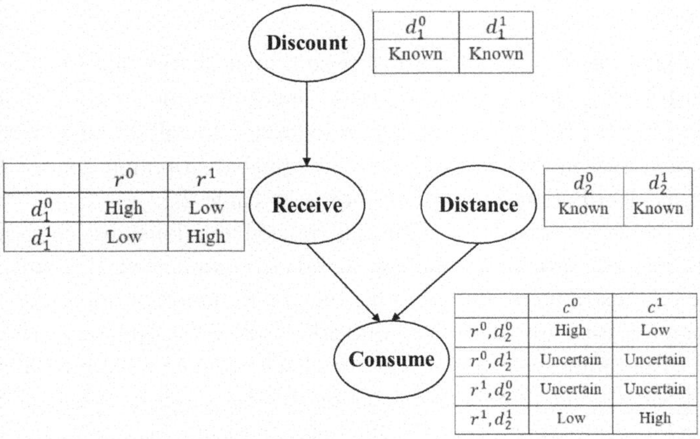


**Fig. 1** Bayesian network with conditional probability distribution

is derived. Use non-parametric estimation characterizing each decomposed factor in (7) to obtain the probability values. Section 3 will show the details.

$$P(d_1^0, r^0, d_2^1, c^1) = P(d_1^0) \bullet P(r^0|d_1^0) \bullet P(d_2^1|r^0) \bullet P(c^1|d_2^1) \quad (1)$$

$$P(d_1^0) \bullet P(r^0|d_1^0) \bullet P(d_2^1|r^0) \bullet P(c^1|d_2^1) = P(d_1^0) \bullet P(r^0|d_1^0) \bullet \frac{P(r^0, d_2^1, c^1)}{P(r^0)} \quad (2)$$

$$P(d_1^0) \bullet P(r^0|d_1^0) \bullet \frac{P(r^0, d_2^1, c^1)}{P(r^0)} = P(d_1^0) \bullet \frac{P(d_1^0|r^0)}{P(d_1)} \bullet P(r^0, d_2^1, c^1) \quad (3)$$

$$P(d_1^0) \bullet \frac{P(d_1^0|r^0)}{P(d_1)} \bullet P(r^0, d_2^1, c^1) = P(d_1^0|r^0) \bullet P(r^0, d_2^1, c^1) \quad (4)$$

$$P(d_1^0|r^0) \bullet P(r^0, d_2^1, c^1) = P(d_1^0|r^0) \bullet P(r^0) \bullet P(d_2^1, c^1|r^0) \quad (5)$$

$$P(d_1^0|r^0) \bullet P(r^0) \bullet P(d_2^1, c^1|r^0) = P(r^0, d_1^0) \bullet P(d_2^1, c^1|r^0) \quad (6)$$

$$P(r^0, d_1^0) \approx 1 - P(r^1, d_1^1) \quad (7)$$

## 2.2 Copula Function Selection

Copula is a function $[0, 1]^2 \rightarrow [0, 1]$. Let $H$ be a joint distribution function, and its marginal distribution functions are $F$ and $G$, then there must be a *Copula* $C$ that makes $H(x, y) = C(F(x), G(y))$ come into existence for all $x$ and $y$ in $\overline{R}$. Copula connects joint distribution functions and marginal distribution functions, and it contains many distribution families, as shown in Table 2.

The Copula theory excels in exploring related relationships. In general, the correlations between variables in the real world are complicated. It is difficult to get close to the real distribution with a simple linear correlation or normal distribution, and the complex joint distribution is difficult to obtain. Copula can simplify the description of the joint distribution function. Each variable's marginal distribution is easy to describe, and Copula functions are diverse. After comparison, it is possible to select the most appropriate Copula function to simulate the real distribution. Therefore, the joint distribution function of random variables can be obtained easily, and then the relationship between variables can be easily described and predicted.

**Table 2** Distribution families and Copula functions

| Distribution family | Copula functions |
|---|---|
| Normal distribution family | Normal Copula |
| t distribution family | t-Copula |
| Archimedean distribution family | Gumbel Copula |
| | Clayton Copula |
| | Frank Copula |

Monte Carlo simulation is a method of random variable generation. It inverts the random variables according to the dependence relationship between them, and obtains a large amount of simulation data. The simulation data provides a basis for solving practical problems. In general, the method of generating the random variable simulation data with the distribution function $F$ by inverting the distribution function is as follows:

- Generate variables u with uniform distribution across the interval $(0, 1)$ randomly.
- Let $x = F^{(-1)}(u)$, where $F^{(-1)}$ is the pseudo-inverse of F, then x is the value of the variable X of the distribution function F.

(1) *Marginal distribution construction:* The marginal distribution of the Copula function is the description of the random variables Discount and Distance. The empirical distribution of historical data may be consistent with some standard distributions. The empirical distribution function is normalized to the standard distribution function, which will eliminate the nonstandard data. It will make it possible to abstract the complex practical problem into mathematical problem that can be solved by some methods. The implementation steps for constructing the marginal distribution are shown in Table 3.

(2) *Copula function selection:* The Copula function can describe the correlation between random variables more accurately, when linear correlations can't describe well. Different Copula functions will produce different results in the application process. The key to choose the appropriate Copula function model contains two steps: one is to choose the form of Copula function, and the other is to select the unknown parameters of Copula function. Let Discount and Distance be $X$ and $Y$, respectively. The new sequence after the empirical

**Table 3** Methods of marginal distribution construction

| Marginal distribution construction | | Judgement standard |
|---|---|---|
| Skewness and kurtosis calculation | Skewness *bs* | If *bs* is close to 0, the distribution is symmetrical |
| | kurtosis *bk* | If *bk* is close to 3, the distribution is in accordance with normal |
| Normality test | Jarque-Bera test | If statistic is less than the quantile, the sample obeys normal distribution |
| | Kolmogorov-Smirnov test | If the statistic is within the limit, the sample obeys the given distribution |
| | Lilliefors test | If the mean and variance are within the limit, the sample obeys normal distribution |
| Non-parametric estimation | Empirical distribution | Transform discrete points into smooth curves using Spline interpolation |
| | Kernel density estimation | Smoothness and data authenticity of kernel density estimation curve |

distribution function is $(u_t, v_t)$, where $u_t = \hat{F}_x(x_t)$, $v_t = \hat{G}_y(y_t)$, $t = 1, 2, \cdots, T$ and $\hat{F}_x(x)$, $\hat{G}_y(y)$ are the empirical distribution functions of $X$ and $Y$, respectively. The selection steps are as follows:

- Make scatter plot for $(u_t, v_t)$. On the basis of that, judge the strength of the tail dependence of two random variables, and select three to four Copula functions as alternatives.
- The two-dimensional empirical distribution function is used as the estimator of the two-dimensional Copula function, and the most suitable Copula function is selected by comparing the squared Euclidean distance $d^2$ between them, see (8). The smaller the $d^2$ value is, the better the Copula function fits.

$$d^2 = \sum\nolimits_{i=1}^{n} \left| \hat{C}_n(u_i, v_i) - C(u_i, v_i) \right|^2 \tag{8}$$

- K-S test: Test whether the absolute difference D between the empirical distribution function and the Copula function is within the confidence interval. The common method is computing and making non-parameter estimation for $\tau$. For the Kendall's $\tau$ rank correlation coefficient of sample $(x, y)$, see (9). For the overall Kendall's $\tau$ rank correlation coefficient, see (10). Finally, the estimated values of parameters in Copula function can be calculated through (11).

$$\hat{\tau} = \frac{2}{n(n-1)} \sum\nolimits_{1 \le i \le j \le n} sign(x_i - x_j)(y_i - y_j) \tag{9}$$

$$\tau = 1 + 4 \int_0^1 \frac{\phi(t)}{\phi'(t)} dt \tag{10}$$

$$\hat{\tau} = \tau \tag{11}$$

## 2.3 Monte Carlo Simulation Implementation

The Monte Carlo simulation based on the Copula function is a process of generating multiple random variables according to the selected Copula function. The variable generation steps are as follows:

- Generate two independent random variables u and t that are uniformly distributed
- Let $v = C_u^{-1}(t)$, $C_U^{-1}$ denotes the pseudo-inverse of $C_u$

- $(u, v)$ is a random variable with Copula $C(u, v)$
- set $x = F^{(-1)}(u)$, $y = G^{(-1)}(v)$
- $(x, y)$ is a random variable that conforms to the joint distribution function $H(x, y)$.

# 3 Case Study

Based on the consumer's actual online and offline behavior data from Alibaba, Sect. 3 analyzes the consumer's historical consumption behavior and makes predictions.

## 3.1 Model Construction of "Consumption with Receiving Mobile Coupons" Based on Copula-MC Method

According to the Bayesian network, the probability that consumption with receiving mobile coupons is expressed as $P\left(d_1^1, r^1, d_2^1, c^1\right)$. The main variable is Discount and Distance. On the basis of that, the model is constructed with

**Table 4** Names and meanings of variables and parameters in 'consumption with receiving mobile Coupons' model

| $X$ | Discount | $Y$ | Distance |
|---|---|---|---|
| *xs* | Skewness of discount's sample | h | Normality test result |
| *ys* | Skewness of diatance's sample | $P$ | The probability value of accepting the null hypothesis |
| *kx* | Kurtosis of discount's sample | $U1$ | Set of empirical function values for discount sample |
| *ky* | Kurtosis of diatance's sample | $V1$ | Set of empirical function values for distance sample |
| $U$ | Set of discount kernel density estimates | *rho* | Linear correlation parameters of Copula function |
| $V$ | Set of distance kernel density estimates | *nuhat* | t-Copula degree of freedom parameter |
| *nuci* | t-Copula degree of freedom parameter | $d^2$ | Squared Euclidean distance |
| *UU* | Random variable with Copula function | *VV* | Random variable with Copula function |
| $p$ | Monte Carlo simulation data of discount | $q$ | Monte Carlo simulation data of distance |

Copula-MC method. The names and meanings of variables and parameters in the model are shown in the Table 4.

(1) *Marginal distribution model construction:* On the basis of methods of constructing marginal distribution in Table 3, use Matlab to characterize the marginal distribution of Discount and Distance.

   (a) Draw a histogram of the frequency distribution of $X$ and $Y$, as shown in Fig. 2. The horizontal axis represents values of two variables; the vertical axis represents the ratio of frequency to group distance. It can be seen from the figure that mobile coupons' receiving rate is higher when the discount is of 0.9–1, and the consumption rate is higher when the distance is 0–1 km away from merchants.
   (b) Use Matlab to calculate the skewness and kurtosis of $X$ and $Y$. the skewness is $xs = -1.4058$, $ys = 2.4167$, the kurtosis is $kx = 5.5015$, $ky = 7.6526$. It shows that $X$ is negative deviation, and $Y$ is positive deviation. Both of them have excessive kurtosis and do not meet the standard normal distribution.
   (c) Normality test: Use jbtest, kstest and lillietest function in Matlab to carry out Jarque-Bera test, Kolmogorov-Smirnov test and Lilliefors test for $X$ and $Y$ respectively. The test results show that $X$ and $Y$ do not obey the normal distribution.
   (d) Empirical distribution function: Use ecdf function in Matlab to derive the empirical distribution function of $X$ and $Y$. Use the spline interpolation method to find the value of the empirical distribution function, as shown in Fig. 3.
   (e) Kernel density estimation: Different kernel functions and window widths have different effects on curve smoothness and the data authenticity. The result shows that the default window width and Gaussian kernel function for kernel density estimation can better balance smoothness and data authenticity. The $X$ and $Y$ density estimates are shown in Fig. 3.

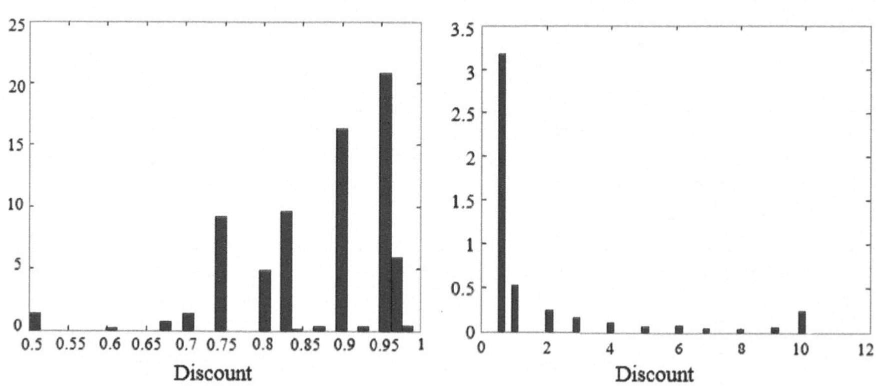


**Fig. 2** Frequency distribution histogram of Discount and Distance

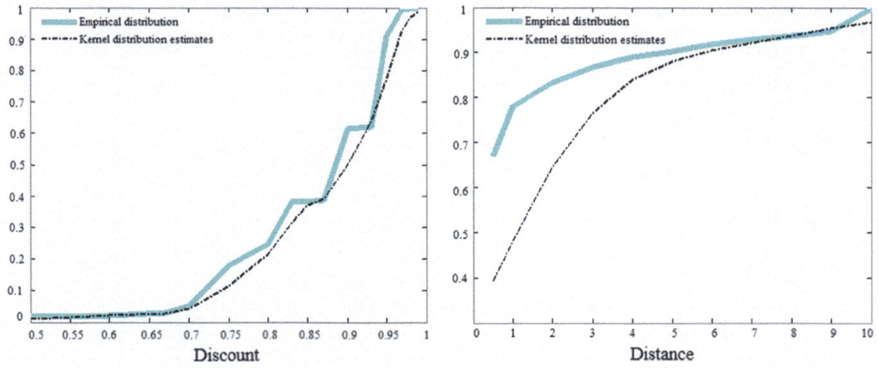


**Fig. 3** Empirical distribution and kernel density estimation

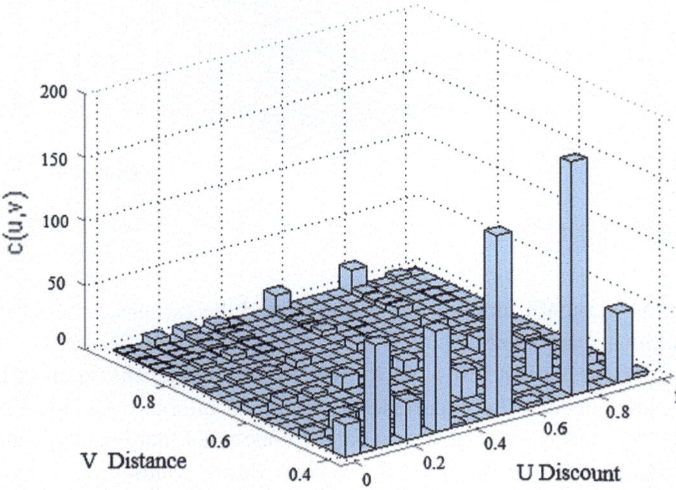


**Fig. 4** Binary frequency histogram of discount and distance

(f) According to kernel density estimation *U* and *V*, use Matlab to draw a binary frequency histogram shown in Fig. 4. As can be seen from the figure, the tail features of the joint density function are not obvious. Apparently, it is non-normal distribution, so Gumbel-Copula, t-copula, Clayton-Copula, Frank-Copula are used as alternative Copula functions.

(2) *Copula function selection:* According to the alternative Copula functions, after goodness-of-fit tests, the most appropriate Copula function for describing the relationship between variables Discount and Distance is determined. The steps are as follows:

**Table 5** Goodness-of-fit test results

| | Gumbel-Copula | t-Copula | Clayton-Copula | Frank-Copula |
|---|---|---|---|---|
| $d^2$ | 0.2997 | 0.0700 | 0.2997 | 0.0405 |

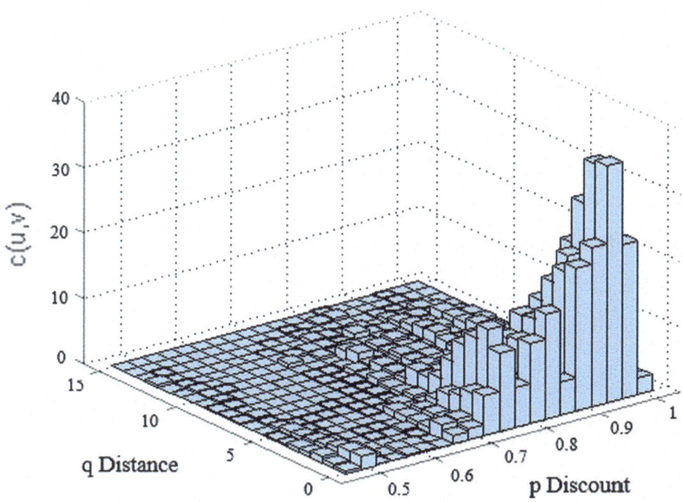


**Fig. 5** The probability prediction of consumption with receiving coupons

- Use the copulafit function in Matlab to find the estimated values of Copula functions' parameters.
- The goodness-of-fit test is performed by calculating the squared Euclidean distance between the experience Copula and alternative Copula functions, see (8). As shown in Table 5, it can be concluded that Frank-Copula has the smallest Euclidean distance and the best fitting condition.

(3) *Probability prediction:* Frank-Copula gives the joint distribution function of variables $X$ and $Y$, which provides a basis for predicting the probability of consumption with receiving coupons. Based on Monte Carlo method, using the copularnd function in Matlab generates 10000 pairs of random variables with the Frank-Copula function, and then removes data that does not satisfy the definition of the Discount and Distance. According to the marginal distribution function, using ksdensity function in Matlab inverts the $UU$ and $VV$ to obtain the simulation data of Discount and Distance, $p$ and $q$. The frequency distribution histogram of the simulation data $p$, $q$ is shown in Fig. 5.

So far, we have completed the probability prediction of consumption with receiving coupons, as shown in Fig. 5. It can be seen that there are more consumers going to the store to consume when the distance is between 0–5 km and the discount is 0.8–0.99. According to this forecast result, merchants can estimate the probability of consumers going to stores to consume based on the value of two variables. On the basis of predicted probability, merchants can design mobile coupons with differentiated discounts, and offer mobile coupons to targeted consumers. It can achieve targeted offering and reduce marketing costs for merchants.

## 3.2 *Model Construction of "Consumption Without Receiving Mobile Coupons" Based on Bayesian Network*

According to Bayesian decomposition in Sect. 2, we can see that the consumption probability without receiving mobile coupons is expressed as $P\left(d_1^0, r^0, d_2^1, c^1\right)$. According to (1)–(7), we only need to simulate the following two scenarios:

- Receiving coupons when the discount is attractive
- Consuming in the store without receiving mobile coupons, when the distance is convenient.

Figure 6 shows the frequency histogram of Discount and Distance. The data of Discount can be obtained from the behavior of receiving mobile coupons. The data of Distance can be obtained from the behavior that not receiving mobile coupons but consuming in the store. As can be seen from the figure, the lower the discount is, the higher the mobile coupons receiving rate is. And the closer the distance is, the higher the consumption rate is.

Based on the original data, the empirical distribution function values are used to estimate the kernel density. As shown in Fig. 7, the original data is modeled as a regular function. From the figure, it can be seen that when the discount is 0.9–1, the mobile coupons receiving rate rises more sharply, and when the distance is 0–3 km, the consumption rate increases greatly.

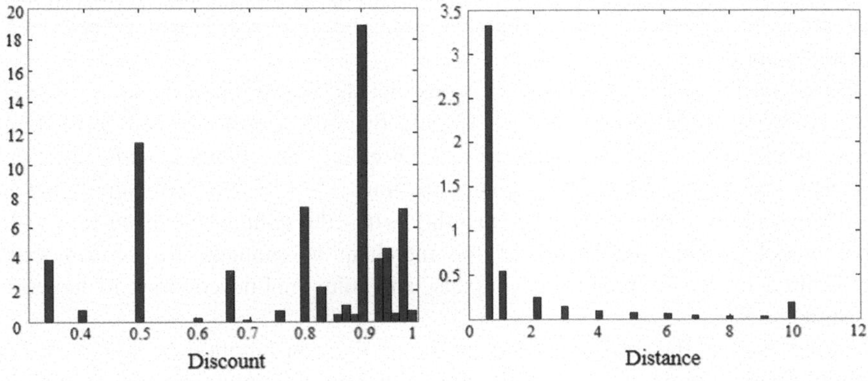


**Fig. 6** Frequency histogram of discount and distance

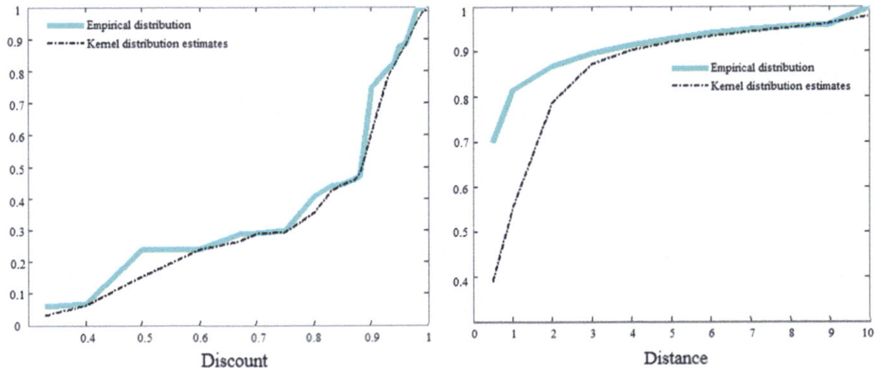


**Fig. 7** Empirical distribution and kernel density estimation

Above all, the probability prediction of consumption without receiving mobile coupons can be obtained from Figs. 6 and 7. The forecast results show that consumers who do not receive mobile coupons will have a high probability of going to the store to consume if the distance is convenient. It broadens the view of merchant, not only to focus on consumers who have received coupons, but also to focus on consumers' distance. Merchants can also estimate the probability of consumption without receiving coupons, and provide measures for promote sales and increasing consumer loyalty.

## 4 Conclusion

In the background of O2O and precision marketing, this paper studies the consumption behavior of mobile coupons' online receiving and offline use. Based on the real data of Alibaba's consumers, this paper use Bayesian network, Copula function and Monte Carlo simulation to construct the consumption behavior model with receiving mobile coupons and the consumer behavior model without receiving mobile coupons.

Based on the Copula-MC method, construct the consumption behavior model with receiving mobile coupons. Use Copula function to describe the correlation between the two variables, Distance and Discount. The Frank-Copula function performs best among alternative Copula functions. Then Monte Carlo simulation is used to predict the consumption probability when the coupon has been received. This model provides reference for the merchant to estimate the consumption probability, design differentiated coupons, and offer mobile coupons to targeted consumers.

Based on the Bayesian network, construct the consumption behavior model without receiving mobile coupons. A non-parametric estimation method is used in model building. Bayesian network is used for factorization to obtain factors that is easy to characterize in mathematical model. The two factors are Distance and Discount. The data of Discount is obtained from the behavior of receiving mobile coupons. The data of Distance is obtained from the behavior that not receiving mobile coupons but consuming in the store. The historical data is described by a regular function and then use non-parametric estimate to predict the consumption probability without receiving mobile coupons.

The probability prediction of these two models provides a reference for the merchant to identify the target customer group and offer mobile coupons accurately. On the one hand, the merchant can target customer group with high probability of using mobile coupons to offer mobile coupons, based on the discount of mobile coupons and the distance between the consumer and the merchant. On the other hand, the merchant can design a coupon that meets the consumer's preference according to the probability prediction.

## References

1. Noble, S. M., Lee, K. B., Zaretzki, R., & Autry, C. (2017). Coupon clipping by impoverished consumers: Linking demographics, basket size, and coupon redemption rates. *International Journal of Research in Marketing, 34*(2), 553–571.
2. Zhao, X., Tang, Q., Liu, S., & Liu, F. (2016). Social capital, motivations, and mobile coupon sharing. *Industrial Management & Data Systems, 116*(1), 188–206.
3. Li, Y. M., Liou, J. H., & Ni, C. Y. (2019). Diffusing mobile coupons with social endorsing mechanism. *Decision Support Systems, 117,* 87–99.
4. Jiang, Y., Liu, Y., Wang, H., Shang, J., & Ding, S. (2018). Online pricing with bundling and coupon discounts. *International Journal of Production Research, 56*(5), 1773–1788.
5. Greenstein-Messica, A., Rokach, L., & Shabtai, A. (2017). Personal-discount sensitivity prediction for mobile coupon conversion optimization. *Journal of the Association for Information Science and Technology, 68*(8), 1940–1952.
6. Khajehzadeh, S., Oppewal, H., & Tojib, D. (2015). Mobile coupons: What to offer, to whom, and where? *European Journal of Marketing, 49*(5/6), 851–873.
7. Silbermayr, L., Jammernegg, W., & Kischka, P. (2017). Inventory pooling with environmental constraints using copulas. *European Journal of Operational Research, 263*(2), 479–492.
8. Kupka, I., Kiseľák, J., Ishimura, N., Yoshizawa, Y., Salazar, L., & Stehlík, M. (2018). Time evolutions of copulas and foreign exchange markets. *Information Sciences, 467,* 163–178.
9. Nguyen, C., Bhatti, M. I., Komorníková, M., & Komorník, J. (2016). Gold price and stock markets nexus under mixed-copulas. *Economic Modelling, 58,* 283–292.
10. Mokni, K., & Mansouri, F. (2017). Conditional dependence between international stock markets: A long memory GARCH-copula model approach. *Journal of Multinational Financial Management, 42,* 116–131.
11. Janekova, J., Fabianova, J., Izarikova, G., Onofrejova, D., & Kovac, J. (2018). Product mix optimization based on Monte Carlo simulation: A case study. *International Journal of Simulation Modelling, 17*(2), 295–307.
12. Acebes, F., Pereda, M., Poza, D., Pajares, J., & Galán, J. M. (2015). Stochastic earned value analysis using Monte Carlo simulation and statistical learning techniques. *International Journal of Project Management, 33*(7), 1597–1609.

# Risk Assessment of Closed-Loop Supply Chain for Electronic Products Recycling and Reuse Based on Fuzzy BP Neural Network

**Wei Shao, Zuqing Huang, and Lianjie Jiang**

**Abstract** With the development of society and the progress of productivity, there is an increasing demand for electronic products. The number of waste electronic products is also growing. Thus, how to realize effective recycling and reasonable disposal of waste electronic products has become a problem for the participants in the supply chain. We built an index system of risk assessment for the manufactures and the retailers in this paper. And the risk assessment of closed-loop supply chain for the waste electronic products is discussed based on several improved fuzzy BP neural network models. Furthermore, a neural network model is designed and its algorithm is improved to be suitable to multi-dimension inputs by several optimization methods. Especially the LBFGS optimization algorithms with the Wolfe-type search are more accurate and practical by the experimental results. The results also show that the overall risk of the closed-loop supply chain for electronic products is bigger than general. The reason for the higher risk is mainly from the recycling process and the remanufacturing process. Finally, some suggestions for reducing risk are proposed for remanufactures and retailers.

**Keywords** Closed-loop supply chain · Risk assessment · BP neural network

The paper is funded by National Natural Science Foundation of China (NSFC) (No. 71871206).

W. Shao · L. Jiang
School of Economics and Management, China Jiliang University, Hangzhou, China
e-mail: 17826807857@163.com

L. Jiang
e-mail: 1768522162@qq.com

Z. Huang (✉)
School of Management, Guangzhou University, Guangzhou, China
e-mail: hzq1210@163.com


J. Zhang et al. (eds.), *LISS2019*,
https://doi.org/10.1007/978-981-15-5682-1_11

# 1 Introduction

Electronic products represented by smartphones had become necessities in life. Electronic products are very popular and renewed rapidly, so the number of used electronic products is bigger. In 2017, the total amount of the handling capacity of the used electronic products in China was 79 million units, but this has not yet reached the total output of electronic products in the first quarter of 2018. At present, the recycling system is still imperfect. Because of the complexity of the structure and the inadequacy of the recycling system, which bring great risk to the remanufacturers and the retailers. For example, according to Hey [1], the acquisition process of used products often faces high risk and results in both the uncertain quantity of the recycled products and market demand. For many remanufacturers such as IBM and Kodak, it is not easy to obtain the right quantity and the right quality of recycled products through their recycled channels. For these reasons, the closed-loop supply chain is not robust. Therefore, this paper can help remanufactures and retailers identify risk factors and evaluate risk.

The identification of risk factors is the foundation to implement the risk assessment for the closed-loop supply chain. The first step of the identification is to understand the structure of the closed-loop supply chain systematically. At present, Blome [2] divided the influence factors of the risk evaluation into internal risk and external risk. On the one hand, the internal risks consist of financial risk, market risk, etc. On the other hand, external risks include political risk, environmental risk, social risk, etc. Prakash [3] divided the risk factors into supply risk, process risk, financial risk, demand risk, etc. 30 risks across the three main pillars of sustainability (environmental, social, economic) are identified by the study of Mihalis [4]. Winter [5] divided the risk factors into performance related risk, control related risk, demand risk, and business related risk. These scholars identify risk factors only from the overall perspective of supply chain,which can hardly help the participants to identify the risks that need priority attention. Therefore, we attempt to identify the respective risks of each participant in the supply chain. It is helpful to control the risk for each participant.

To evaluate the risk factors and control the risk, there are many tools such as Delphi method, HAZOP-Based method [6] and so on. Recently, many methods are used for risk assessment such as Analytic Hierarchy Process (AHP) method, the fuzzy integrative assessment, gray systems analysis and so on. Mangla [7] used FAHP to prioritize risks in Green Supply Chain (GSC). Six kinds of risks and twenty-five specific risks, associated with GSC, were identified and prioritized. The results indicate that operational category risks are the most important risks in GSC. Reference [8] includes a model to evaluate risk based on the AHP and Fuzzy Analytic Hierarchy Process (FAHP). Further, the neural network and its application aroused wide concern in recent years, especially the BP neural network. For example, Tong [9] adopted the improved LSE method to optimize the parameters of BP neural network. But this method can easily lead to the problem of the slowness of convergence and the local minimum of the gradient search. In order to overcome these defects, we may use Adam or LBFGS methods. Adam and LBFGS method are well suited for problems with large parameters or data which show good empirical performance.

## 2 The Identification of Risk Factors

Most of the studies summary the risk factors based on economic, ecological and social factors. But it is not effective for participants to control risk in the closed-loop supply chain. It is necessary to identify the risk factors by analyzing the whole process of supply chain systematically. For closed-loop supply chain, the whole process consists of five parts [10]: the manufacturing process, the sale process, the recovery process, the remanufacturing process, and the resale process. To identify the risk factors and control the risk of the closed-loop supply chain, this paper attempts to identify the risk from two aspects: each process of the whole closed-loop supply chain and the external environmental factors.

In order to identify the main risks, we use the Delphi method by visiting 15 industry experts. At last, in this study mainly 20 factors (Table 1) are included.

The risk index system for the closed-loop supply chain includes 4 aspects, such as the recovery risk, the remanufacturing risk, the resale risk, and the external factors. Each aspect also includes 5 detailed risk factors.

**Table 1** Index system of risk assessment

| Target(A) | Risk(B) | Element(C) |
|---|---|---|
| Index system of risk assessment of closed-loop supply chain for waste electronic products | Recovery risk $u_1$ | The consumer's willingness to recycle $u_{11}$ |
| | | The integrity recovery system $u_{12}$ |
| | | The matching degree between recycled price and consumer's anticipation $u_{13}$ |
| | | Remanufacturer recovery cost $u_{14}$ |
| | | The recycling utilization rate of waste products $u_{15}$ |
| | Remanufacturing risk $u_2$ | Maintenance technology of recycling products $u_{21}$ |
| | | Dismantling technology of recycling products $u_{22}$ |
| | | The ability to dispose of the waste parts $u_{23}$ |
| | | Parts of shortage risk $u_{24}$ |
| | | The Technological level of the process of remanufacturing $u_{25}$ |

(continued)

**Table 1** (continued)

| Target(A) | Risk(B) | Element(C) |
|---|---|---|
| | Resale risk $u_3$ | The demand for remanufactured products $u_{31}$ |
| | | Quality of remanufactured products $u_{32}$ |
| | | Resale cost of remanufactured products $u_{33}$ |
| | | Management level for sales channel $u_{34}$ |
| | | The effectiveness of market information feedback $u_{35}$ |
| | External environment risk $u_4$ | The support of the government industry policy $u_{41}$ |
| | | Resource recovery rate $u_{42}$ |
| | | Enterprise operation competence $u_{43}$ |
| | | The environmental awareness $u_{44}$ |
| | | Corporate social responsibility $u_{45}$ |

# 3 Methodology

## 3.1 Standard BP Neural Network Model

BP neural network is a kind of error back-propagation algorithm based on the training of multilayer feedforward networks [11]. The network learns the relationship between the input and target data by adjusting the weights to minimize the error between the actual output and the desired output. The weight updates formula is

$$w_{ij}^{m}(t+1) = w_{ij}^{m}(t) - \eta_{ij}^{m}(t)$$

$$\eta_t = \alpha \cdot g_t$$

Where $\alpha$ learning rate, $g_t$ represents gradient direction, $t$ means the iteration. The error function is:

$$E_r = \frac{1}{2}\sum_{j=1}^{m}\left(y_{ij} - d_{ij}\right)^2$$

Where $y_{ij}$ and $d_{jk}$ denote the actual output and predicted output.

The weight $w$ needs to be found by reducing the error function $E_r$. Standard BP algorithm is a simple and steepest descent (SD) method. The gradient direction of the previous time cannot be taken into account in the standard BP algorithm, so it is liable to trap in local minimum value. In order to improve this defect, adaptive learning rate optimization method can be used to adjust different learning rates for different parameters, which can reduce the oscillation and improve the convergence speed.

## 3.2 *Fuzzy BP Neural Network Model*

Expert scoring is used in the process of risk evaluation, which has some subjective factors. Dealing with fuzzy information with the neural network can solve the problem of extracting fuzzy rules. Thus, fuzzy BP neural network model is designed in this paper. The fuzzed data is input to the BP neural network. Through the use of the optimization method to optimize the weights of the neural network, a model dealing with fuzzy information and accurate information can be obtained. Fuzzy rules will be acquired by training this model. The structure of the fuzzy BP neural network model is shown in the Fig. 1.

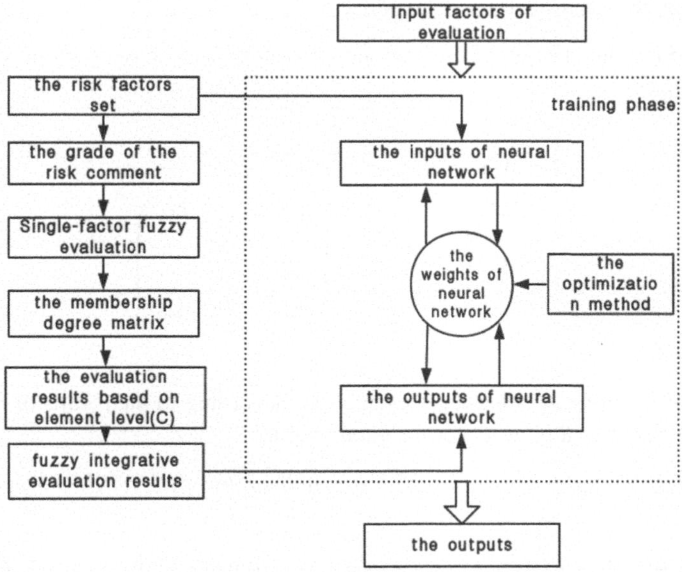


**Fig. 1** The structure of the fuzzy bp neural network model

(1) *Fuzzy integrative assessment*

The utilization of fuzzy integrative assessment includes 6 steps [12]:

Step 1: The first-class factor set $U = \{u_1, u_2, u_3, u_4\}$. And the second ones: $u_i = \{u_{i1}, u_{i2}, u_{i3}, u_{i4}, u_{i5}\}\ i = 1, 2, 3, 4$.
Step 2: As mentioned above, the reasons for closed-loop supply chain risk are complicated. We confirm the grade of the risk comment, that is $V = \{v_1, v_2, \cdots, v_5\}$, which means risk is $\{very\,small, smaller, general, bigger, very\,big\}$.
Step 3: Forming the weight set: By using the Analytic Hierarchy Process (AHP) method based on the expert's estimate, we set $A = \{a_1, a_2, \cdots a_n\}$.
Step 4: Based on single-factor, we evaluate the risk factors to determine the subordination of indicators, and form the fuzzy mapping $f : U \rightarrow F(V)$, suppose the membership vector of risk factors to the set of judgment $R_i = \{r_{i1}, r_{i2}, \cdots r_{im}\}$. We can get the membership degree matrix $R$ is represented by: $R = \begin{pmatrix} r_{11} & \cdots & r_{1n} \\ \vdots & \ddots & \vdots \\ r_{m1} & \cdots & r_{mn} \end{pmatrix}$.

$$r_{ij} = \frac{u_{ij}}{N} \tag{1}$$

Let $u_{ij}$ denote the score that the index $i$ given to the evaluation grade $j$. The $N$ represents the total quantity of experts.

Step 5: The first level of fuzzy evaluation: According to the basic factors, we set the results as $B_i$.

$$\begin{aligned} B &= A(\bullet,\oplus)R \\ &= (a_1, a_2, a_3)\,(\bullet,\oplus)\begin{pmatrix} r_{11} & \cdots & r_{1n} \\ \vdots & \ddots & \vdots \\ r_{m1} & \cdots & r_{mn} \end{pmatrix} \end{aligned} \tag{2}$$

Step 6: Multilevel fuzzy integrative evaluation: according to the results of level (B), we use the fuzzy mapping to get the final vector.

(2) *Adaptive Moment Estimation (Adam) method for Fuzzy BP neural network model*

Adam method, which is based on adaptive estimates of lower-order moments, shows good empirical performance and compares favorably to other SGD methods. It dynamically adjusts the learning rate for each parameter according to the first-order moment estimation and the second-order moment estimation. The sensitivity of the network to local details of the error function is reduced and the training time is shortened. So Adam method is well suited for problems with large parameters or data and it is also adapted to the processing of data with big noise.

Adam algorithm:

$E_r(w)$ represents the stochastic objective function with a parameter $w$,$\eta$ represents the step size, $\nabla w E_r(w)$ represents the gradients. Good default settings for the problems are $\eta = 0.001 \beta_1 = 0.9 \beta_2 = 0.999$ $\varepsilon = 10^{-8}$.

Require:
$m_0 \leftarrow 0$ (Initialize $1^{st}$ moment vector)
$v_0 \leftarrow 0$ (Initialize $2^{nd}$ moment vector)
$t \leftarrow 0$ (Initialize estimate)

While $\theta_t$ not converged do

$t \leftarrow t + 1$
$g_t \leftarrow \nabla w E_t(w_{t-1})$ (get gradients stochastic objective at estimate $t$)
$m_t \leftarrow \beta_1 \cdot m_{t-1} + (1 - \beta_1) \cdot g_t$ (update biased first moment estimate)
$v_t \leftarrow \beta_2 \cdot v_{t-1} + (1 - \beta_2) \cdot g_t^2$ (update biased second raw moment estimate)
$\hat{m}_t \leftarrow \frac{m_t}{1-\beta_1^t}$ (compute bias-corrected first moment estimate)
$\hat{v}_t \leftarrow \frac{v_t}{1-\beta_2^t}$ (compute bias-corrected second raw moment estimate)
$w_{t+1} = w_t - \frac{\eta}{\sqrt{\hat{v}_t} + \varepsilon} \hat{m}_t$ (update parameters)
End while
Return $w_t$ (resulting parameters)

(3) *Improved LBFGS optimization method for Fuzzy BP neural network model*

Newton's method in optimization explicitly calculates the Hessian matrix of the objective function and the inverse of the Hessian matrix [13]. It is better than the SD method in speed and accuracy because of the convergence of this method. Finding the inverse of the Hessian is difficult. Quasi-Newton methods thus appear to overcome this shortcoming. Currently, the most common quasi-Newton algorithm is the BFGS method [14] and its limited extension LBFGS [15]. We present an improved LBFGS method which presents the non-monotone line search technique for the Wolfe-type search. Its mathematical theory to support this algorithm can be seen from [16]. So the improved LBFGS method has the characteristics of high precision and fast convergence.

Algorithm LBFGS:
Step 0: Choose an initial point $x_0 \in R^n$, an initial positive definite matrix $H_0$, and choose a positive integer $m_1$. Let $t = 0$.
Step 1: If $\|g_t\| = 0$, then output $x_k$ and stop; otherwise, go to step 2.

Step 2: Solve the following linear equation to get $d_t$: $d_t = -H_t \nabla E_t$
Step 3: Find a step-size $\lambda_k > 0$ satisfying the Wolfe-type line search conditions:

$$E_r(x_t + \lambda_t d_t) \leq E(x_t) + \sigma_1 \lambda_t \nabla E_t^T d_t,$$

$$g(x_t + \lambda_t d_t)^T d_t \geq \sigma_2 \nabla E_t^T d_t.$$

Step 4: Let $x_{t+1} = x_t + \lambda_t d_{t+1}$ be the next iteration. Calculate $\nabla E_{t+1}$.
Step 5: Let $s_t = X_{t+1} - X_t$, $y_t = \nabla E_{t+1} - \nabla E_t$, $\gamma_t = \|g_t\|$, then $y_t^* = y_t + \gamma_t s_t$.
Step 6: Update $H_t$ following the formula

$$\begin{aligned} H_{t+1} = & \left(V_t^{*T} V_{t-1}^{*T} \cdots V_{t-\tilde{m}+1}^{*T}\right) H_{t-\tilde{m}+1}^0 \left(V_{t-\tilde{m}+1}^* \cdots V_{t-1}^* V_t^*\right) \\ & + \omega_{t-\tilde{m}+1} \left(V_{t-1}^{*T} \cdots V_{t-\tilde{m}+2}^{*T}\right) \left(S_{t-\tilde{m}+1} S_{t-\tilde{m}+1}^T\right) \left(V_{t-\tilde{m}+2}^* V_{t-\tilde{m}+3}^* \cdots V_t\right) \\ & + \cdots \omega_t^* S_t S_t^T \end{aligned}$$

Step 7: Let $t = t + 1$ and go to step 1.

In this algorithm, $\nabla E_t$ denotes the gradient of $E_r(x)$ at $x_t$.

# 4 Experiment Analysis

## 4.1 Data Collection and Preprocessing

An expert-team with 30 persons was organized to divide risky factors into five grades $\{very\, small, smaller, general, bigger, very\, big\}$. The expert team considers that the influence factor of the risk is mainly as above-mentioned Table 1, and they grade each factor respectively. The results of experts scoring are as shown in Table 2. According to the utilization of fuzzy integrative assessment, the index weights can be calculated. After a normalization processing, the analysis results can be obtained as shown in Table 3.

## 4.2 Establish the Comprehensive Evaluation Matrix

According to the (1) in the process of fuzzy integrative assessment, we can get the membership degree matrix.

**Table 2** Scoring table by experts

| Risk indicators | Risk degree | | | | | Total |
|---|---|---|---|---|---|---|
| | Very small | Smaller | General | Bigger | Very big | |
| $u_{11}$ | 2 | 4 | 5 | 17 | 2 | 30 |
| $u_{12}$ | 0 | 0 | 10 | 10 | 10 | 30 |
| $u_{13}$ | 0 | 3 | 7 | 18 | 2 | 30 |
| $u_{14}$ | 0 | 7 | 11 | 9 | 3 | 30 |
| $u_{15}$ | 3 | 5 | 10 | 8 | 4 | 30 |
| $u_{21}$ | 0 | 9 | 10 | 7 | 4 | 30 |
| $u_{22}$ | 2 | 3 | 9 | 12 | 4 | 30 |
| $u_{23}$ | 1 | 4 | 3 | 14 | 8 | 30 |
| $u_{24}$ | 2 | 5 | 8 | 12 | 3 | 30 |
| $u_{25}$ | 0 | 3 | 13 | 10 | 4 | 30 |
| $u_{31}$ | 0 | 7 | 14 | 8 | 1 | 30 |
| $u_{32}$ | 0 | 6 | 11 | 11 | 2 | 30 |
| $u_{33}$ | 1 | 7 | 10 | 11 | 1 | 30 |
| $u_{34}$ | 0 | 6 | 6 | 14 | 4 | 30 |
| $u_{35}$ | 0 | 9 | 5 | 12 | 4 | 30 |
| $u_{41}$ | 1 | 8 | 9 | 12 | 0 | 30 |
| $u_{42}$ | 0 | 7 | 10 | 8 | 5 | 30 |
| $u_{43}$ | 0 | 4 | 14 | 11 | 1 | 30 |
| $u_{44}$ | 1 | 3 | 11 | 11 | 4 | 30 |
| $u_{45}$ | 0 | 4 | 13 | 10 | 3 | 30 |

**Table 3** The weight of indicators in the evaluation system

| The weight of the risk level(B) to Target(A) | The weight of the risk level(C) to level(B) |
|---|---|
| $u_{1}(0.38)$ | $u_{11}(0.13)$ |
| | $u_{12}(0.26)$ |
| | $u_{13}(0.13)$ |
| | $u_{14}(0.26)$ |
| | $u_{15}(0.22)$ |
| $u_{2}(0.16)$ | $u_{21}(0.2)$ |
| | $u_{22}(0.31)$ |
| | $u_{23}(0.21)$ |
| | $u_{24}(0.12)$ |
| | $u_{25}(0.16)$ |
| $u_{3}(0.27)$ | $u_{31}(0.32)$ |
| | $u_{32}(0.21)$ |
| | $u_{33}(0.12)$ |
| | $u_{34}(0.14)$ |
| | $u_{35}(0.21)$ |
| $u_{4}(0.19)$ | $u_{41}(0.2)$ |
| | $u_{42}(0.16)$ |
| | $u_{43}(0.26)$ |
| | $u_{44}(0.23)$ |
| | $u_{45}(0.15)$ |

$$R_1 = \begin{bmatrix} 0.06 & 0.13 & 0.17 & 0.57 & 0.07 \\ 0 & 0 & 0.33 & 0.33 & 0.34 \\ 0 & 0.1 & 0.23 & 0.6 & 0.07 \\ 0 & 0.23 & 0.37 & 0.3 & 0.1 \\ 0.1 & 0.16 & 0.33 & 0.27 & 0.13 \end{bmatrix} \quad R_2 = \begin{bmatrix} 0 & 0.3 & 0.33 & 0.23 & 0.14 \\ 0.06 & 0.1 & 0.3 & 0.4 & 0.14 \\ 0.03 & 0.13 & 0.1 & 0.47 & 0.27 \\ 0.06 & 0.17 & 0.27 & 0.4 & 0.1 \\ 0 & 0.1 & 0.43 & 0.33 & 0.14 \end{bmatrix}$$

$$R_3 = \begin{bmatrix} 0 & 0.23 & 0.45 & 0.26 & 0.04 \\ 0 & 0.2 & 0.36 & 0.37 & 0.07 \\ 0.03 & 0.36 & 0.33 & 0.24 & 0.04 \\ 0 & 0.2 & 0.2 & 0.47 & 0.13 \\ 0 & 0.3 & 0.16 & 0.4 & 0.14 \end{bmatrix} \quad R_4 = \begin{bmatrix} 0.03 & 0.26 & 0.3 & 0.4 & 0 \\ 0 & 0.23 & 0.33 & 0.27 & 0.17 \\ 0 & 0.13 & 0.46 & 0.37 & 0.04 \\ 0.03 & 0.1 & 0.36 & 0.37 & 0.14 \\ 0 & 0.13 & 0.43 & 0.34 & 0.1 \end{bmatrix}$$

According to the (2) in the process of fuzzy integrative assessment, we can get the evaluation results based on an element level (C).

$$\begin{aligned} B_1 &= (0.03 \quad 0.12 \quad 0.31 \quad 0.38 \quad 0.16) \\ B_2 &= (0.03 \quad 0.15 \quad 0.28 \quad 0.37 \quad 0.17) \\ B_3 &= (0.01 \quad 0.25 \quad 0.34 \quad 0.32 \quad 0.08) \\ B_4 &= (0.01 \quad 0.16 \quad 0.38 \quad 0.35 \quad 0.1) \end{aligned}$$

Finally, fuzzy comprehensive evaluation results can be obtained by step 6 of the fuzzy integrative assessment.

$$B = (0.02 \quad 0.16 \quad 0.32 \quad 0.37 \quad 0.13)$$

According to the maximum subordination principle for the experimental results, the overall risk level of the closed-loop supply chain is bigger than general. Specifically, the risk of the recycling process is bigger than general, the risk of the remanufacturing process is also bigger than general. The risk of the resale process is general, the risk of external environment is also general.

## 4.3 The Results of The Fuzzy BP Neural Network Model

A BP neural network of the 3-layer structure is used to evaluate the risk. We use the membership degree matrix of the recovery risk, the remanufacturing risk and the external environment risk as the inputs of the training. The evaluation results $B_1$ $B_2$ $B_3$ as the outputs of the training sample. The membership degree matrix of the resale risk is used as the outputs of the testing sample. By training the several improved BP neural network models respectively, we compare the RMSE and the MAE. The results are as follows.

**Table 4** Standard BP neural network model

| Actual value | Predictive value | MAE | RMSE |
|---|---|---|---|
| 0.02 | 0.11 | 0.09 | 0.11 |
| 0.16 | 0.03 | | |
| 0.32 | 0.27 | | |
| 0.37 | 0.29 | | |
| 0.13 | 0.13 | | |

**Table 5** BP neural network optimized by adam method

| Actual value | Predictive value | MAE | RMSE |
|---|---|---|---|
| 0.02 | 0.13 | 0.12 | 0.15 |
| 0.16 | 0.06 | | |
| 0.32 | 0.28 | | |
| 0.37 | 0.25 | | |
| 0.13 | 0.12 | | |

**Table 6** BP neural network optimized by LBFGS method

| Actual value | Predictive value | MAE | RMSE |
|---|---|---|---|
| 0.02 | 0.02 | 0.05 | 0.06 |
| 0.16 | 0.14 | | |
| 0.32 | 0.32 | | |
| 0.37 | 0.37 | | |
| 0.13 | 0.15 | | |

From Tables 4, 5 and 6, the experimental results show that the model optimized by LBFGS has the minimum RMSE value and MAE value. The MAE value shows that the deviation between the predicted value and the true value. The lower RMSE value demonstrates fewer outliers. So, the BP neural network optimized by LBFGS method has better accuracy and performance than Adam method and standard BP method in this case.

Combining all of the experimental results, the overall risk level of the closed-loop supply chain is bigger than general. However, the value of the general risk is 0.32 by observing the results of the fuzzy comprehensive evaluation. The value is only lowered 0.05 than the value of risk level ‘bigger than the general’. It means that the overall risk level of the closed-loop supply chain is reduced to the general risk hopefully. Further, the results of the fuzzy comprehensive evaluation show that the risk level of the resale process and the external environment are general. These two aspects have lower risky level probably because of two reasons. On the one hand, the government support for recycling electronic products is increasing. On the other hand, with the acceptance of remanufactured products growing, the demand for remanufactured products is also increasing. So, the reason for the overall risky bigger than general is mainly from the recycling process and the remanufacturing process. At first, there is imperfect in the current recovery system, and the efficient system is urgent to make better. Then, awareness of recycling is not strong for consumers. For the remanufacturing process, the problem mainly comes from the technological level of remanufacturing. Furthermore, due to the limitation of recycling quantity, the cost of the remanufacturing is also quite high.

To solve these problems, the three following suggestions are proposed. Firstly, the construction of a recycling system is the foundation to control risk. So, it is necessary to establish a recycling system by expanding recycling channels and adding recycling stations for remanufactures. For example, O2O (online to offline mode) is a good way to expand recycling channels. And reasonable recycling station location is the key to expanding recycling channels. Secondly, for retailers, it is possible to increase the consumption willingness by implementing the points system of recycling, which means consumers recycle a single product to get the points to buy a new product. Finally, remanufacture industry is suitable for ecological and economical requirements. Of course, more government investment for the process of recycling and reuse is conducive to the sustainable development of global and regional economies.

## 5 Conclusion

The risk assessment of the closed-loop supply chain is an extraordinarily complicated evaluation system. The BP neural network model can effectively avoid the subjectivity of the process of the assessment. The experiments results show that the risk of the closed-loop supply chain for the electronic products is bigger than general, the BP neural network optimized by LBFGS algorithms has better performance than other improved BP neural network on this kind of risk assessment problem. According to the experimental results, higher risk factors are identified. By analyzing the reasons for the higher risk factors, some suggestions are proposed for remanufacturers and retailers respectively. The next step of the research is to study the optimal decision-making of each participant in the supply chain under the various risks. For example, how to add recycling stations reasonably for remanufacturers. Therefore, this study needs to be further improved and developed in the future.

## References

1. Hey, Y. (2017). Supply risk sharing in a closed-loop supply chain. *International Journal of Production Economics, 183,* 39–52.
2. Blome, C., & Schoenherr, T. (2011). Supply chain risk management in financial crises—A multiple case-study approach. *International Journal of Production Economics, 134*(1), 43–57.

3. Prakash, A., Agarwal, A., & Kumar, A. (2018). Risk assessment in automobile supply chain. *Materials Today: Proceedings, 5*(2), 3571–3580.
4. Mihalis, G., & Thanos, P. (2016). Supply chain sustainability: A risk management approach. *International Journal of Production Economics, 171*(4), 455–470.
5. Winter, A. (2018). A risk assessment model for supply chain design. In *International Conference on Logistics Operations Management.*
6. Adhitya, A., Srinivasan, R., & Karimi, I. A. (2010). Supply chain risk identification using a HAZOP-based approach. *AIChE Journal, 55*(6), 1447–1463.
7. Mangla, S. K., Kumar, P., & Barua, M. K. (2015). Risk analysis in green supply chain using fuzzy AHP approach: A case study. *Resources, Conservation and Recycling, 104,* 375–390.
8. Gordana, R., & Gajovic, V. (2013). Supply chain risk modeling by AHP and Fuzzy AHP methods. *Journal of Risk Research, 17*(3), 337–352.
9. Tong, L. (2017). Assessment of artillery fire application plan based on improved fuzzy BP neural network. *Command Control and Simulation.*
10. Nallusamy, S., Balakannan, K., & Chakrabortym, P. S. (2018). A mixed-integer linear programming model of closed loop supply chain network for manufacturing system. *International Journal of Engineering Research in Africa, 35,* 198–207.
11. Zhang, J., Lin, Y., & Yan, Y. (2009). Research of risk in supply chain based on fuzzy evaluation and BP neural network. In *International Conference on Computational Intelligence & Software Engineering.*
12. Hu, C., & Lv, C. (June 2010). Method of risk assessment based on classified security protection and fuzzy neural network. In *Asia-Pacific Conference on Wearable Computing Systems.*
13. Dennis, J. E., & Schnable, R. I. (2009). *Numerical methods for unconstrained optimization and nonlinear.* Englewood Cliffs: Prentice-Hall.
14. Dai, Y. (2013). A perfect example for the BFGS method. *Mathematical Programming, 13,* 501–530.
15. Hou, Y. T., & Wang, Y. T. (2013). *The modified limited memory BFGS method for large-scale optimization* (Vol. 24, pp. 15–19).
16. Zhang, J. P., Hou, Y. T., & Wang, Y. J. (2012). The LBFGS quasi-Newtonian method for molecular modeling prion AGAAAAGA amyloid fibrils. *Natural Science, 4,* 1097–1108.

# Government's Role Choice in Market Regulation from an Information-Disclosure Perspective

**Qinghua Li, Yisong Li, and Daqing Gong**

**Abstract** Government regulation on the market activities are vital to make sure social welfare is enhanced and not impaired. There are multiple perspectives of role strategy analysis, one of which is to identify government-related projects and appropriate handling strategy. In order to explore the relationship of information disclosure and governments' role-playing, this paper presents three-scenario analysis of government regulation roles from an information disclosure perspective and proposes three models: infomediary government model, efficiency-oriented government model and supportive government model. Some insights and discussion thereof are put forward to strengthen the government's supervision and guidance of market environmental information disclosure. Additionally, possibility of combinations of different scenario models are discussed and future research on measurement of each type of combination of scenario models, more quantified research on various involved factors are expected.

**Keywords** Information disclosure · Government regulation · Game theory

This work was financially supported by China Railway (Grant No. B19D00010: Optimization of railway material's management mode).

Q. Li (✉) · Y. Li · D. Gong
School of Economics and Management, Beijing Jiaotong University, Beijing, China
e-mail: 16113137@bjtu.edu.cn

Y. Li
e-mail: ysli@bjtu.edu.cn

D. Gong
e-mail: gongtuipigua@163.com



J. Zhang et al. (eds.), *LISS2019*,
https://doi.org/10.1007/978-981-15-5682-1_12

## 1 Introduction and Literature Review

Generally, governments' regulation of market activities remains important though market self-regulating dominates internally. Though there are various ways to perceive and analyze the mechanism of government and companies' motivation and output, it's seen that increasing attention and importance has been casted to government regulation of market by game theory after major global economic crisis. Additionally, it is not rare that profit-driven market fails to push business into certain new areas that favor social welfare or certain development expectations, electrical vehicles promotion for pollution reduction purpose being an example. Therefore, it is vital to research on each participant' s motivation in market activities. The paper explores information disclosure and governments' role-playing upon three different information disclosure scenarios. It is not only a theoretical discussion of energy management, but also the concrete practice to analyze decision behaviors of government and enterprises through the basic idea of cooperative game theory of equilibrium. By looking into those equilibrium scenarios, a government can choose their best strategy for positioning itself in relation to information disclosure tactics. The enterprise's decision-making behavior is subject to government's regulation, while the government seeks maximum utility upon enterprise's response once its role and policy is set. After looking into the features of each type of information disclosure strategies, possibilities of combining strategies are explored and possible practical actions are inspired.

The information disclosure literature has been on a rise over the recent decades. Verrecchia et al. respectably provide extended overviews over the analytical and empirical literature [1]. The major analytical conclusion is all information will be voluntarily disclosed by self-interested agents due to concerns of negative interpretation by potential recipients of the information [2]. Subsequent analytical literature can be classified according to which assumption of the unraveling results is violated, leading to partial non-disclosure [2]. Haejun concludes that the more valuable technology is and the less bargaining power the innovator has, the more information the innovator discloses under asymmetric information [9]. There are three streams of literature of disclosure by non-profit organizations. First stream is of formal-model studies on effects of disclosure in other sectors such as in government, health care, and the financial markets. Second stream is of fundraising campaigns related studies, which is economics-based. Third stream is of empirical literature from accounting and nonprofit studies in the market for charitable contributions [3].

Though there are abundance of published articles on disclosure, and some of them are on government applications, there still lacks of quantified research works on how information disclosure can service government for its interaction with enterprises to seek maximum social welfare in certain circumstances. It is this paper's value to contribute to this to-be-further-explored area.

Government regulation to the market is to make sure it is on the right path for booming more and better social welfare. However, it is often faced with complex and various situations. To better identify and cope with specific regulating tasks are vital and valuable to all parties involved [6]. One of important regulation methods is to promote/supervise/support certain enterprise involving projects. Considering the profit-driven market characteristic, the government should deliberate what to do to reduce the market participant into actions towards the desired direction [7]. There are mainly three types of role models, which the government can play by, namely infomediary model, efficiency-oriented model and supportive model. Infomediary model refers to scenario that the government provide full information to the market. This applies when the government supervises mature projects, in which the participants are all veteran. Efficiency-oriented model refers to scenarios that the government for some reason need to lean forward to efficiency over other considerations somehow [4]. Lastly the supporting model applies to scenarios when prospect uncertainty is the main concern that hold back the participants of potential enterprises [8].

The remainder of this paper is organized as follows. We describe the overview of the model in Sect. 2. Sections 3, 4 and 5 discuss three types of models, respectively. Section 6 provides a brief summary of the paper.

## 2 Models Analysis

Upon consideration of governments' roles in relation to market information disclosure, three scenarios are presented and discussed separately: the first scenarios is infomediary model in which the government solely provide full information to enterprises who will by their own discretion make free decisions on whether and how to participate in market activities. The second scenarios is efficiency-oriented model in which the government prefer efficiency and decide to disclose more optimistic information than pessimistic information with subsequence of enterprise investing more into market. The third one is supporting model in which the government provide clear positive assurance signal by financially aiding the enterprises who participate in market activates [5]. Three separate models are formed in an attempt to find out what is the best theoretical response the government should choose.

Normally there are three typical scenarios of information disclosure when governments are to decide how information should be disclosed to the participating enterprise. In scenario 1, governments disclose information to the participating enterprises fully as it is. It's titled as infomediary model in the following model analysis. In this scenario, the enterprises receives full transparent information and solely decide their action in participating in the activities the governments refer to. This is the typical scenario for those existing mature projects in which governments hold no tendency and merely provide fully information with normal regulations. In scenario 2, governments disclose more optimistic other than pessimistic information to the participating enterprises. It's titled as efficiency-oriented model in the following model analysis. In this scenario, the enterprises receive more optimistic information and upon which decide their action in participating in the activities the governments refer to. This is the typical scenario for those emerging pre-mature projects in which governments hold certain pro tendency and therefore provide more optimistic information with tendency regulations. In scenario 3, governments not only disclose more optimistic other than pessimistic information to the participating enterprises, but also provide clear positive assurance signal by financially aiding the enterprises who participate in market activates It's titled as supporting model in the following model analysis. In this scenario, the enterprises receives more optimistic information and additionally, they see the government financially support the enterprise involving in the projects. Enterprises are usually stimulated significantly by this direct and explicit tendency and to participate the advocated activities with much less concern of risks thereof. This is the typical scenario for those government-dominated and in-sketch projects which governments are to promote for economic or other social warfare purpose. While these types of projects are still in their sketch phase in which enterprise are usually not willing to participate actively consider economic and policy risks. To alleviate those potential enterprises' concern over these risks, the government usually take more drastic measures to signal assurance to the enterprises, which are the actions such like bearing the risk partially or fully by financially supporting those participating enterprises.

We can see from Fig. 1 that there is a continuous increase in its policy tendency intensiveness in term of how strongly the government is willing to promote the activities withScenario1 weakest and Scenario 3 the strongest. It's the purpose of the paper to introduce 3 usual signal policies and analyze how the mechanism works by looking into its models. It natural that when trimming the real situation into these modeled scenarios, we made assumptions. Additionally, there are other ways to analyze the interactions between the dominating government and the enterprises who solely receive information from the government and make their independent decisions as of how they should react to the government's signals. In the real world the government seldom take only one scenario policy, rather they can combine and tailor the policy taken. The combinations can be chosen according to the real need. Some example are: the government classify the enterprises into 3 different grades by their related feathers, three different types of scenario policies can apply to the different grades of enterprises separately; or along with the progress of the project implementation, the government may apply different types of scenario policies to the enterprises involved according the need of promotion phases. In a summary, upon understanding the mechanism of information disclosure scenarios model, the government can freely play combination of the models and apply the model policy with other regulating strategies.

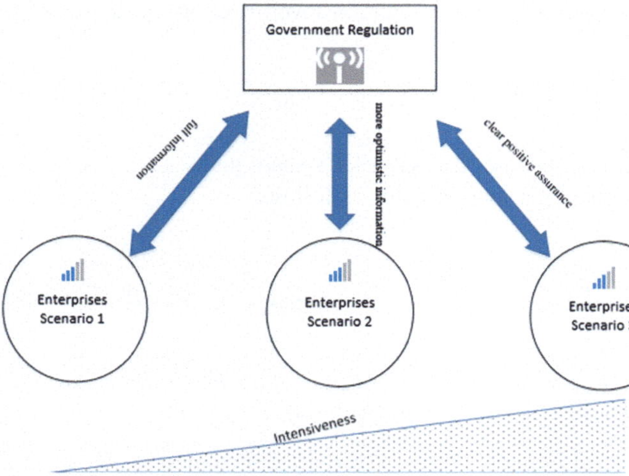


**Fig. 1** Three information disclosure scenarios

## 3 Infomediary Government Model

The utilities $U$ (that yield $Q$) are the market gains when the firm provides products or services to the users with a certain level of effort, and $U$ is subject to the firm's ability, market prosperity and market randomness. Thus, the utilities $U$ is described as follows:

$$U = Af(e) + B + \theta \tag{1}$$

where $A$ is the firm's ability ($A > 0$), $e$ is the effort, $f(e)$ is the function, $f'(e) > 0$ denotes that marginal yields are positive and that yields are positively correlated with firm ability, $f''(e) < 0$ denotes that the rate of yields is decreasing, $B$ is market prosperity, $\theta$ is market randomness, and $\theta \sim N(0, \sigma^2)$ ($\sigma^2$ is the output variance).

**Corollary 1** *The linear relation of the model is still reasonable when* $f(e) = e$.

***proof*** We set $f(e) = e^{r1}$, where $0 < r1 < 1$, so equation $f(e) = e^{r1}$ is consistent with the restrictions above. Supposing $E = e^{r1}$, so $U = AE + B + \theta$. Additionally, we can see the full linear relations in expression $e^{r1} \to E \to U$. When $f(e) = e$, Eq. (1) is simplified as follow:

$$U = Ae + B + \theta \tag{2}$$

The firm generates sales with a certain level of effort, so the firm can get pay from their work:

$$s(U) = \alpha + \beta U \tag{3}$$

where α is the fixed income, β is the user's share gains.

Therefore, we can get market gains as follow:

$$\begin{aligned} E(U - s(U)) &= -\alpha + E(1 - \beta) \\ &= -\alpha + (1 - \beta)(Ae + B) \end{aligned} \tag{4}$$

**Corollary 2** *$C(e)$ can be simplified as $be^2/2$, if the firm's direct cost is $C(e)$ when providing products or services, and $C'(e) > 0$, $C''(e) \geq 0$.*

***proof*** We set $C(e) = me^{r2}$, where $m > 0, r2 > 1$, and $E = e^{r1}$, so $C(e) = mE^{r3}$ $(r3 = r2/r1 > 1)$.

$C(e) = mE^{r2}$ is consistent with the restrictions above, so $C(e) = mE^{r3}$ can be simplified as $C(e) = be^2/2$ when $m = b/2, r3 = 2$, where $b$ is the cost coefficient.

θ is market randomness and $\theta \sim N(0, \sigma^2)$, so random variable can be represented as follows:

$$I_1 = \alpha + \beta(Ae + B + \theta) - be^2/2 \tag{5}$$

Equation (5) obeys a normal distribution, and the firm's expected profits are

$$EI_1 = \alpha + \beta(Ae + B) - be^2/2 \tag{6}$$

In addition, the firm must pay the opportunity cost when participating in collaborative incentive contracts, and the subsidiary firm's certainty equivalent profits ($I$) can reflect its actual income.

**Theorem 1** *If firm's utilities u is exponential distribution, and certainty equivalent profits I is normal distribution $N(m, n^2)$, so $I = m - \frac{rn^2}{2}$, where r is the absolute risk aversion factor.*

***proof*** Set the utilities $u = -e^{rI}$ (exponential distribution) for the firm, and $r > 0$, $I \sim N(m, n^2)$. The firm's expected utilities are

$$\begin{aligned} E(u) &= \int_{-\infty}^{+\infty} -e^{-rI} \frac{1}{\sqrt{2\pi}n} e^{-\frac{(z-m)^2}{2n^2}} dI \\ &= -e^{-r\left(m - \frac{rn^2}{2}\right)} \end{aligned} \tag{7}$$

so $E(u) = u(I)$, where

$$-e^{-r\left(m-\frac{rn^2}{2}\right)} = -e^{-rI} \tag{8}$$

$$I = m - \frac{rn^2}{2} \tag{9}$$

Donate the transformation of some contents in (7), respectively.

In (9), $m$ is the mean of the firm's profit, $n$ is the variance of the firm's profit. And the firm's expected profits are $EI_1 = \alpha + \beta(Ae + B) - be^2/2$, the mean of the firm's profit are $EI_1 = \alpha + \beta(Ae + B) - be^2/2$, The variance of the firm's profit $DI_1 = \beta^2\sigma^2$.

Therefore, the firm's certainty equivalent profits are

$$I = \alpha + \beta(Ae + B) - \frac{be^2}{2} - \frac{r\beta^2\sigma^2}{2}. \tag{10}$$

where $r\beta^2\sigma^2/2$ is the opportunity cost, $s$ is the lowest profit that the firm requires, and the firm will not participate in an incentive contract when the equivalent profit is less than $s$. Therefore, the prerequisite that the firm participates in the incentive contract is

$$\alpha + \beta(Ae + B) - \frac{be^2}{2} - \frac{r\beta^2\sigma^2}{2} \geq s \tag{11}$$

The model based on principal-agent theory [5] is

$$E(U - s(U)) = max\{-\alpha + (1 - \beta)(Ae + B)\} \tag{12}$$

$$s.t.\begin{cases} argmax\left\{\alpha + \beta(Ae + B) - \frac{be^2}{2} - \frac{r\beta^2\sigma^2}{2}\right\} \\ \alpha + \beta(Ae + B) - \frac{be^2}{2} - \frac{r\beta^2\sigma^2}{2} \geq s \end{cases} \tag{13}$$

## 4 Efficiency-Oriented Government Model

In this model, due to preference to efficiency, the government prefer to disclose more optimistic information than pessimistic information to enterprises by which the enterprises is induced into more enthusiastic engagement into the market.

If there are no income, the loss of the enterprise is $\varepsilon \in [-be^2/2, 0]$, which confirms to normal distribution.

In case the government accurately disclose its supportive policy information and the enterprise is aware of $\varepsilon$'s probability density function $f(\varepsilon)$, the enterprise expects is

$$E[\varepsilon] = \int_{-be^2/2}^{0} \varepsilon f(\varepsilon) d\varepsilon \tag{14}$$

In case the government vaguely disclose its supportive policy information, the enterprise will estimate a $\hat{f}(\varepsilon)$, then the expectation is:

$$E[\varepsilon]' = \int_{-be^2/2}^{0} \varepsilon \hat{f}(\varepsilon) d\varepsilon \tag{15}$$

so

$$E[\varepsilon]' - E[\varepsilon] = \int_{-be^2/2}^{0} \varepsilon \hat{f}(\varepsilon) d\varepsilon - \int_{-be^2/2}^{0} \varepsilon f(\varepsilon) d\varepsilon \tag{16}$$

$E[\varepsilon]' - E[\varepsilon]$ is the enterprise' understanding deviation due to vague disclosure. When $E[\varepsilon]' - E[\varepsilon] > 0$, the enterprise overestimate risks that will depress market activeness (H type); when $E[\varepsilon]' - E[\varepsilon] < 0$, market risks is underestimated (L type).

The probability of no income is $p_i (i = H, L)$, $0 < p_L < p_H < 0.3$. Information disclosure Utility is $\pi_i U$, from which the enterprise share is $\gamma_i^C U$, $\gamma_i^C < \pi_i$. The government's cost of information disclosure is $C_i^C$, $U > C_i^C$. The information disclosed is of private type, therefore the enterprise can only judge its risk level by information collected. $G(\gamma_i^C)$ represents the government's expected utility, and $V(\gamma_i^C)$ represents the firm's expected utilities.

Government's expected utilities are

$$G_H(\gamma_H^C) = (1 - p_H)(\pi_H - \gamma_H^C)(Ae + B) - C_H^C \tag{17}$$

$$G_L(\gamma_L^C) = (1 - p_L)(\pi_L - \gamma_L^C)(Ae + B) - C_L^C \tag{18}$$

The firm's expected utilities are

$$V(\gamma_H^C) = (1 - p_H)\gamma_H^C(Ae + B) + p_H \int_{-\frac{be^2}{2}}^{0} \varepsilon \hat{f}(\varepsilon) d\varepsilon \tag{19}$$

$$V(\gamma_L^C) = (1 - p_L)\gamma_L^C(Ae + B) + p_L \int_{-\frac{be^2}{2}}^{0} \varepsilon f(\varepsilon) d\varepsilon \tag{20}$$

The model based on principal-agent theory is

$$E\left(G_H\left(\gamma_H^C\right)\right) = max\{(1-p_H)\left(\pi_H-\gamma_H^C\right)(Ae+B)-C_H^C \tag{21}$$

s.t.

$$argmax\{(1-p_H)\gamma_H^C(Ae+B)+p_H\int_{-\frac{be^2}{2}}^{0}\hat{\varepsilon f}(\varepsilon)d\varepsilon\} \tag{22}$$

or

$$E\left(G_L\left(\gamma_L^C\right)\right) = max\{(1-p_L)\left(\pi_L-\gamma_L^C\right)(Ae+B)-C_L^C\} \tag{23}$$

s.t.

$$argmax\{(1-p_L)\gamma_L^C(Ae+B)+p_L\int_{-\frac{be^2}{2}}^{0}\varepsilon f(\varepsilon)d\varepsilon\} \tag{24}$$

**Lemma 1** *In the case of complete information market and accurate disclosure of information, the market equilibrium of the contract will make the share of the company depend mainly on their own efforts.*

***proof*** In a full competitive market, the enterprise's gain exactly equals to cost of information collection ($V\left(\gamma_L^C\right) = C_L^C$), therefore:

$$(1-p_L)\gamma_L^C(Ae+B)+p_L\int_{-\frac{be^2}{2}}^{0}\varepsilon f(\varepsilon)d\varepsilon = C_L^C \tag{25}$$

$$\gamma_L^C = \frac{C_L^C - p_L\int_{-\frac{be^2}{2}}^{0}\varepsilon f(\varepsilon)d\varepsilon}{(Ae+B)(1-p_L)} \tag{26}$$

While $-p_L\int_{-\frac{be^2}{2}}^{0}\varepsilon f(\varepsilon)d\varepsilon > 0$, is function of $e$ (open information), obviously, the enterprise return rate depends on its efforts and information collection cost.

## 5 Supportive Government Model

Due to market prospect uncertainty, the government input investment of $d_i$, $d_i \leq \left(1+\gamma_i^S\right)U$ and its expected utility is:

$$G_H\left(\gamma_H^S\right) = (1-p_H)\left(\pi_H-\gamma_H^S\right)(Ae+B)-p_Hd_H-C_H^S \tag{27}$$

$$G_L(\gamma_L^S) = (1 - p_L)(\pi_L - \gamma_L^S)(Ae + B) - p_L d_L - C_L^S \tag{28}$$

When $G_H(\gamma_i^S) < 0$, the government rather exits, therefore its constraint is:

$$(1 - p_i)(\pi_i - \gamma_i^S)U - p_i d_i - C_i^S \geq 0 \tag{29}$$

i.e.

$$\gamma_i^S \leq \pi_i - \frac{C_i^S + p_i d_i}{(1 - p_i)(Ae + B)} \tag{30}$$

The firm's expected utilities are:

$$V(\gamma_H^S) = \gamma_H^S U - f_H U + \int_{-\frac{be^2}{2}}^{0} \varepsilon \hat{f}(\varepsilon) d\varepsilon \tag{31}$$

$$V(\gamma_L^S) = \gamma_L^S U - f_L U + \int_{-\frac{be^2}{2}}^{0} \varepsilon f(\varepsilon) d\varepsilon \tag{32}$$

$f_i U$ is the fund the government grants to the enterprise, $f_H > f_L$, therefore no constraint consideration is needed for the enterprise.

The model based on principal-agent theory is:

$$E(G_H(\gamma_H^S)) = max\{(1 - p_H)(\pi_H - \gamma_H^S) * (Ae + B) - p_H d_H - C_H^S\} \tag{33}$$

$$\text{s.t. } argmax\{\gamma_H^S U - f_H U + \int_{-\frac{be^2}{2}}^{0} \varepsilon \hat{f}(\varepsilon) d\varepsilon\} \tag{34}$$

or

$$E(G_L(\gamma_L^C)) = max\{(1 - p_L)(\pi_L - \gamma_L^S) * (Ae + B) - p_L d_L - C_L^S\} \tag{35}$$

$$\text{s.t. } argmax\{\gamma_L^S U - f_L U + \int_{-\frac{be^2}{2}}^{0} \varepsilon f(\varepsilon) d\varepsilon\} \tag{36}$$

In the case of complete information market and accurate disclosure of information, the market equilibrium contract will make the share of the company not fully depend on the level of their efforts.

***proof*** Similarly, the enterprise's gain exactly equals to cost of information collection, therefore:

$$V(\gamma_L^S) = \gamma_L^S U - f_L U + \int_{-\frac{be^2}{2}}^{0} \varepsilon f(\varepsilon) d\varepsilon = C_L^S \tag{37}$$

The equilibrium is:

$$\gamma_L^{S^*} = \frac{C_L^S - \int_{-\frac{be^2}{2}}^{0} \varepsilon f(\varepsilon) d\varepsilon + f_L(Ae + B)}{(Ae + B)} \tag{38}$$

While $-\int_{-\frac{be^2}{2}}^{0} \varepsilon f(\varepsilon) d\varepsilon > 0$, is $e$'s function (open information). Obviously, the enterprise return rate depends on its efforts, information collection cost, and the granted fund by the government.

**Lemma 2** *In a non-full-competitive market, $\gamma_i^S$ and $d_i$ are alternatives to each other. In the market of incomplete information, $\gamma_i^S$ can be substituted by $d_i$.*

***proof*** It can be concluded from (27) and (28) that:

$$\frac{\partial G_i(\gamma_i^S)}{\partial \gamma_i^S} = -(1 - p_i)U \tag{39}$$

$$\frac{\partial G_i(\gamma_i^S)}{\partial d_i} = -p_i \tag{40}$$

$$MRS_{\gamma_i^S d} = -\frac{\frac{\partial G_i(\gamma_i^S)}{\partial \gamma_i^S}}{\frac{\partial G_i(\gamma_i^S)}{\partial d_i}} = -\frac{-1(1 - p_i)U}{-p_i} = -\frac{(1 - p_i)b}{p_i} \tag{41}$$

This means $\gamma_i^S$ and $d_i$ can alternate. It can be concluded that: the greater loss the market failure can cause to the government, the higher rate of return it will set. Since:

$$\frac{\partial MRS_{\gamma_i^S d_i}}{\partial p_i} = \frac{b}{p_i^2} > 0 \tag{42}$$

When faced with the same potential loss, low risk government has to set a lower return compared with higher risk government.

## 6 Conclusion

In consideration of its complexity and vitality characteristics, this paper contributes to variety of ways of governments regulating over the market. Generally, when handling mature projects where all market participants are veteran and willing to invest, the government usually can choose the role of infomediary model, to do nothing but provide full accurate information. When faced with efficiency-oriented projects, the government rather disclose more positive prospects information, thus the risk thereof may be under-estimated by the enterprise. However, the enterprise perceiving that more effort will yield more profit, will engage in the projects more enthusiastically, consequentially enhancing social welfare. Lastly, for those projects with uncertain market prospect, thus, external support is needed; the government may choose to signal a strong assurance information to offset potential enterprise participants' concern over market uncertainty. However, in this circumstance, the enterprises understand that the yields are not only dependent on their effort, and may divert much attention to seek more government funding, which is not rare in practice.

Additionally, this paper only initially study the pure model directly related information factors upon assumption that these factors are the only considerations when the player of government and enterprise make decisions as to whether or not they should take part in the activities it refers to. combination of the different scenarios models are more close to the real world. The government may play freely the combinations with enterprise grading and classification as needed by the real situations.

Finally, upon understanding the information disclosure scenarios, the future research can explore looking into the measurement of different combination of information disclosure models. Thus we can reveal in more details the intensiveness of each model in applications and design in-depth models will can adapt better into the real work. Also more quantified analysis can be designed into each models to make them more precise and practical.

## References

1. Araujo, A., Gottlieb, D., & Moreira, H. (2007). A model of mixed signals with applications to countersignaling an the GED. *Ensaios Economicos da EPGE, 38*(4), 1020–1043.
2. Arya, A., & Glover, J. (2003). Abandonment option and information system design. *Review of Accounting Studies, 8*(1), 4–29.
3. Seitanidi, M., & Austin, J. (2012). Collaborative Value Creation: A Review of Partnering Between Nonprofits and Businesses. *Nonprofit and Voluntary Sector Quarterly, 41,* 723–755.
4. Clements, M. (2011). Low quality as a signal of high quality. *Economics, 5,* 1–22.
5. Guo, B., Zhang, R., & Yuan, C. (2012). A study on government regulation mechanism of promoting enterprises in China in energy conservation. *Kybernetes, 41,* 874–885.
6. Ioannis, A. (2010). The principal-agent problem in economics and in politics. *Humanomics, 26* (4), 259–263.

7. Verhoest, K., Petersen, H., Scherrer, W., & Murwantara Soecipto, R. (2015). How do governments support the development of public private partnerships? Measuring and comparing PPP governmental support in 20 European countries? *Urban Transport of China, 35,* 118–139.
8. Rayo, L., & Segal, I. (2010). Optimal Information Disclosure. *Journal of Political Economy, 118*(5), 949–987.
9. Jeon, H. (2019). Licensing and information disclosure under asymmetric information. *European Journal of Operational Research, 276,* 314–330.

# Context-Aware Aviation Auxiliary Services Recommendation Based on Clustering

**Yingmin Zhang, Wenquan Luo, Dong Wang, Tingting Chen, and Jiewen Zhang**

**Abstract** Personalized auxiliary service is an important profit source for airlines. Considering massive passengers' online and offline travelling context data, the traditional collaborative filtering (CF) algorithm has low accuracy in auxiliary service recommending. This paper adds real-time travel context factors to passenger auxiliary service preference modelling, using five-dimensional data sets to construct a context-aware aviation auxiliary service recommendation model. Aiming to solve the Cold Start and data sparsity, the paper proposes a context-aware recommendation method which calculating similarity between the current context and historical context other than passenger's reviews on services. And recommends the target passenger top-N auxiliary service items that under historical similar travel context. Finally, the experimental simulation method is used to evaluate the recommended effect and accuracy of the recommendation system. The result shows that the context-aware recommendation method is higher accurate in personalized aviation auxiliary services recommending.

**Keywords** Aviation auxiliary service · Context-aware recommendation · Preferences clustering · Collaborative filtering · Data sparsity

---

Y. Zhang · T. Chen
School of Civil Aviation Management, Guangzhou Civil Aviation College, Guangzhou, China
e-mail: zhangyingmin@caac.net

T. Chen
e-mail: chentingting@caac.net

W. Luo
Office of Educational Administration, Guangzhou Civil Aviation College, Guangzhou, China
e-mail: luowenquan@caac.net

D. Wang
School of Business Administration, Guangzhou University, Guangzhou, China
e-mail: wangdong@gzhu.edu.cn

J. Zhang (✉)
Department of Finance, Guangzhou University, Guangzhou, China
e-mail: zhangjiewen@gzhu.edu.cn



J. Zhang et al. (eds.), *LISS2019*,
https://doi.org/10.1007/978-981-15-5682-1_13

# 1 Introduction

Civil aviation auxiliary service refers to additional services provided by airlines other than traditional passenger transportation. It makes income from direct sales or indirectly generated as a travel experience, mainly divided into pre-departure reservation, baggage, airport service, on-board service, destination service, etc. [1]. At present, most of China's domestic airlines account for less than 1% [2] of additional service revenue, mainly for onboard services, with a single service content. It is urgent to explore the personalized service demand of passengers.

With the rapid development and popularization of the mobile Internet, passengers' online and offline travel behaviors and context data are recorded, analyzed and utilized. But huge information overload [3] makes it difficult to quickly find the needed services from a large amount of information. Therefore, it is urgent to study timely and accurate information recommendation methods to provide a better travel experience for passengers. At present, personalized active recommendation [4] mainly includes Association Rules (AR), Content-Based (CB) and Collaborative Filtering (CF). The CF method does not need to consider the content of the recommended items [5], it has diversified recommendation forms, potential interest discovery and recommendation, and has been widely studied and applied [6, 7]. However, the CF method relies on the user's historical rating of items to evaluate the similarity of the recommended resources to predict and recommend the interested items. For new users or new items, the recommendation may be invalid due to data sparsity. In China, aviation services have gradually extended from the high-end market to the popularized market. The demand for new personalized auxiliary services has surged, which makes the cold start problem worse. In addition, as the number of passengers and service items increases, the sparsity of the rating matrix in collaborative filtering also seriously affects the accuracy of recommendation systems.

This paper improves the similarity matrix in the traditional collaborative filtering algorithm for the data sparsity of information recommendation and the cold start problem of collaborative filtering recommendation in massive context data environment. The passengers and auxiliary services data are clustered, and the similarity of passengers' evaluation of service items is replaced by context similarity. When a new passenger enters the service system, the interested service items are predicted by calculating the similarity between the current situation of that passenger and the historical situations, and the nearest top-N neighbor services is recommended to the new passenger. Hence to ease the Cold start problem when missing service evaluation information.

# 2 Literatures Review

Collaborative filtering has obtained a lot of research results in personalized recommendation systems, especially the information recommendation based on user browsing behavior has been mature, and is widely used in e-commerce shopping [8, 9], news pushing [10], movie program recommending [11] and other fields. With the rapid development of mobile Internet, massive online and offline user travel situation data has been added to the recommendation system. Context-sensitive mobile recommendation has become the focus of current research. Researchers have proposed various methods and strategies for the compatibility of context information [12, 13], including adding context information in preference modelling, contextualization of item similarity calculation, matrix decomposition, nearest neighbor clustering, etc. The following is a summary of the existing researches from three aspects.

## *2.1 User Preferences Based Contextual Recommendation*

User preferences based contextual recommendation refers to adding a situation dimension to the user and recommendation item preferences, and constructing a contextual preference model of "users-context-items". Yu et al. [14], Peng et al. [15] used topic model (LDA) to increase the contextual perception of user preference model. Domain ontology is widely used in knowledge modeling. Considering the domain characteristics of context perception, researchers [16–18] proposed to establish context ontology to solve the context sensitivity and scalability of the recommendation system. Ai et al. [16] constructed a two-layer context ontology model for O2O mobile recommendation system based on CONON model, describing service context from three core concepts: user, object and environment, established inference rules for mobile business catering for mobile catering recommendations. Liu et al. [19] proposed a tourism location situational recommendation based on demographic persona. Shin [20], Liu [21] from the point of "users are interested in similar items which in similar situations", represented context as a vector, combined the similarity of context and the similarity of interested item in a particular context, calculated the preference new items and recommended to users. These methods increase the context sensitivity and improve the recommendation accuracy, but exacerbate data sparsity for additional context information and greatly reduce the recommending effectiveness.

## *2.2 Matrix Decomposition*

The tensor method integrates situations information with users and items into n-dimensional vectors [22], but the combination of information will bring the algorithm complexity too high and reduce the efficiency of recommendation. In order to solve this problem, the researchers proposed matrix decomposition, such as the decomposition machine [23], and Ye [24] proposed a probability matrix decomposition model which has higher efficiency in the field of time context recommendation. However, when considering complex situations, such as location, weather and so on, there are still problems such as excessively high decomposition matrix dimensions and low efficiency.

## 2.3 Clustering-Based Contextual Recommendation

This method aims to improve the efficiency of high-dimensional tensor recommendation through clustering methods. The neighborhood clustering is commonly used. Chen [25] calculated the situation similarity based on the user's rating, evaluated the neighboring users with the situation similarity, and predicted the similar behavior of target user with the behavior of the nearest neighbor in a specific situation. Zhang et al. [26] studied an improved information core extraction method based on frequency and ranking, and proposed optimized set clustering to find the most similar neighbors for target user. The above researches did not consider the impact of new users and their features on the recommendation system, and when the description of context attribute is lacking, the calculated nearest neighbor users will be greatly inaccurate.

Overall, the above researches show that traditional recommendation method relies on historical feature extraction and evaluation of interested contents, or limits to single situational data processing and calculating, has low computational efficiency and recommendation accuracy when facing massive real-time situation data.

This paper introduces the preference clustering mechanism to optimize the tensor decomposition. Considering the situational characteristics of aviation auxiliary services, the context-aware auxiliary service preference ontology is constructed, and the neighboring passengers are presented according to clustering optimized travel context similarity, so as to achieve higher accuracy.

# 3 Aviation Auxiliary Service Recommendation Modelling

The aviation auxiliary service preference is affected by the service content and the travel and service situations in which the passengers are located, such as weather conditions and air traffic control.

This paper adds situation and auxiliary service preference dimensions to the "user-item", and builds a five-tuple auxiliary service preference model $M_A$ $\{U, C, S, P, I\}$. Where, $U$, $C$, and $S$ respectively represent passengers, service situations, and aviation auxiliary service sets; $P$ represents a set of preferences for auxiliary services for passengers in a specific context; and $I$ represent a set of interactions between passengers. The auxiliary service preference instance

$\{u_i, c_j, s_k, p_{i,j,k}\}$ is expressed as the user $u_i$'s preference for the auxiliary service $s_k$ in the context $c_j$ is $p_{i,j,k}$. The context-aware auxiliary service recommendation framework is shown in Fig. 1.

As shown in Fig. 1, the implementation process of context-aware aviation auxiliary service recommendation can be divided into data resource layer, data mining layer and application layer. Firstly, the data resource layer is the basis for obtaining data. It collects passengers' basic attributes, auxiliary service preferences, service context and interaction data from airlines' CRM systems, websites, airline check-in systems, flight information system, and WeChat, mobile APP platforms. And the obtained data is preprocessed by data cleaning, regulating and conversing. Secondly, the data mining layer is the core component of the whole recommendation system. It extracts key feature information from the massive passengers' travelling data through cluster analysis, association mining and deep learning, and constructs a context-aware aviation auxiliary service ontology combined with air passengers, additional services and service context. Finally, the application layer

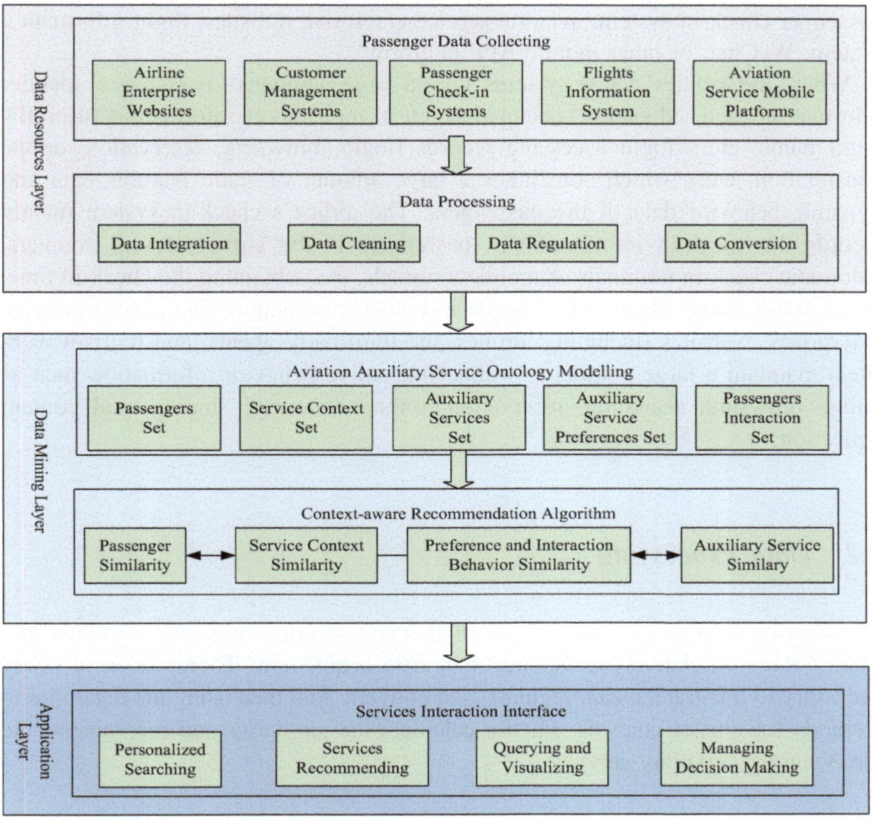


**Fig. 1** Context-aware aviation auxiliary service recommendation framework

calculates the context similarity for the target passenger, compound the passenger's current service situation (time, location, weather conditions, air traffic control, etc.) and the corresponding preference feature tags to push target passenger with the top-N matching personalized services.

## 4 Passenger Auxiliary Service Data Collecting

### *4.1 Data Collecting*

The acquisition of passengers' auxiliary service preferences is inseparable from passenger data. The more data, the more specific the passenger characteristics reflected, and higher accurate services recommended. Context-aware service recommendation can break through the data sparsity problem of traditional content-based collaborative filtering. The passengers' contextual auxiliary service preference data can generally be obtained through the airlines' CRM systems, passenger check-in system, aviation service enterprise websites, flight information system, WeChat, or other mobile APP platforms.

Where the airlines' CRM systems store a large number of passengers' identity information (age, education, occupation, etc.), registration information (user ID, login name, etc.), login accessing records (login, browsing, reservation, online consultation, etc.), which constitutes a large amount of static features data and dynamic behavior data of the passengers. The airline's check-in system mainly records the check-in information of passengers at the airport service counters, automatic check-in terminals or mobile terminals, etc., obtaining the check-in time, place, status (travel alone or in groups) and other information. The aviation service enterprises' websites (including airlines, the third-party agents, and tourism websites) maintain a large number of online interactive behavior information such as online browsing, searching, reservation, online comment, sharing, and content collection.

### *4.2 Data Processing*

The aviation passengers' data is distributed among multiple air service systems, and each system is relatively independent in data acquisition. Therefore, it is firstly necessary to integrate, clean, regulate, and convert. And then using the data mining methods for cluster analysis. Finally calculate the similarity and recommend the top-N similar auxiliary services.

# 5 Similarity Calculating

Similarity calculation is crucial to service recommendation. According to consumer behaviors and the recommendation principles of recommendation system, this paper establishes the following assumptions: (1) Passengers tend to choose auxiliary services that they have used and their similar ones; (2) Similar types of passengers have the same interested auxiliary services; (3) Passengers have similarities in preferences for auxiliary services in the same aviation service scenarios. Therefore, this paper aims to realize service preference prediction and recommendation through three aspects of similarities in passengers, service context and passengers' behaviors.

## *5.1 Passenger's Similarity*

The passenger's similarity is generally calculated by different values between passengers' attributes (the information filled in when the passenger registers or purchases the airline ticket). For passenger U, the attribute vector is $A(a_1, a_2, \cdots, a_n)$, where $a_i$ represents the value of the i-th attribute. The mean absolute difference of attributes is generally used to measure the similarity.

The dimensional difference of attributes leads to different data granularity. The similarity calculation method is different. In this paper, passenger's attribute information is divided into numerical attributes (such as age, income, etc.) and nominal attributes (such as gender, place of origin, occupation, etc.), and their similarities are calculated separately.

- Similarity calculation of numerical attributes

Due to different measurement units the comparability (similarity) of mean absolute difference of attributes is greatly different. Standardization is required to eliminate influence of dimension. This paper uses the difference between each attribute's value and the mean value to map between [0, 1] intervals, which shown in (1)

$$S_{ai} = \frac{a_i - \frac{1}{n}\sum_{i=1}^{n} a_i}{\frac{1}{n}\sum_{i=1}^{n} |a_i - M_A|} \tag{1}$$

Where, $S_{ai}$ represents the normalization value, $a_i$ represents the i-th attribute, $M_A$ is the mean value of attributes. The Manhattan distance [27] is used to calculate the similarity of the passenger's numerical attributes, shown in (2).

$$Sim_{num}(U_i, U_j) = \sum_{k=1}^{n} |S_{aik} - S_{ajk}| \tag{2}$$

Where, the numerical property similarity of passengers $U_i$ and $U_j$ is the sum of the differences of the two passengers' attributes after standardization.

- Similarity calculation of nominal attributes

The similarity of the nominal properties of two passengers is generally represented by the number of matching attributes. Assuming that passengers $U_i$ and $U_j$ have m nominal attributes, the similarity of their nominal attributes $Sim_{nom}\left(U_i, U_j\right)$ can be represented by the matching numbers of nominal attributes.

Therefore, the similarity of the passengers proposed as in (3).

$$Sim_{U_i,U_j} = Sim_{num}\left(U_i, U_j\right) + Sim_{nom}\left(U_i, U_j\right) \tag{3}$$

## 5.2 Service Context Similarity

The aviation auxiliary service context C represented as $C(c_1, c_2, \cdots, c_n)$. Where, $c_i$ is the value of the i-th context attribute. The standardized method is similar to the normalization shown in passenger's similarity. In addition, considering the different impacts of travel situation factors. This paper uses different coefficient $p_k$ to represent similarity weighting. Where, $k$ represents the k-th contextual factor. The value of $k$ is among *1, 2, ⋯, n.*

Hence, the similarity of auxiliary service context $c_x$ and $c_y$ is in (4).

$$Sim_{c_x,c_y} = \sum\nolimits_{k=1}^{n} p_k Sim\left(c_x, c_y\right) \tag{4}$$

## 5.3 Passenger's Preference Behavior's Similarity

For passenger U, the selected auxiliary service set is $S_U$. Where, $S_U = \left\{s_{1,}s_{2,\cdots,}s_n\right\}$.

**Definition 1** If passenger $U_i$ and $U_j$ have interactive behavior ($x_j$) such as selecting, evaluating, praising, etc. under the same context set C. Where, $x_j \in X, X = \{x_1, x_2, \cdots, x_n\}$ is a set of interactions of the passengers with the auxiliary services. It is said that passengers $U_i$ and $U_j$ have similar preferences for auxiliary services $s_i$. And the similarity between the two passengers is the interact numbers (*Num*) of service $s_i$. The mathematical expression is: $U_i \cong U_j, \exists\left(U_i \rightarrow s_k \cap U_j \rightarrow s_k\right)$.

Then, the passenger auxiliary service preference behavior similarity calculation equation is as in (5).

$$Sim_{pre}(U_i, U_j) = \sum\nolimits_{j=1}^{n} x_j Num \tag{5}$$

Where, *Num* is the number of interactions of the same additional service. The larger of *Num*, the higher the similarity.

Considering that passengers' preference of auxiliary services in diverse context is differently affected by passenger features, context, and preference interaction behaviors, the weight coefficient α, β, γ is added to the overall similarity (6).

$$Sim = \alpha Sim(U_i, U_j) + \beta Sim(C_x, C_y) + \gamma Sim_{pre}(U_i, U_j) \tag{6}$$

Where, $\alpha + \beta + \gamma = 1$.

Similar passengers usually have similarities in auxiliary services selecting, but this similarity is relatively fixed. The similarity of passengers is calculated from the static attributes of passengers, so the sensitivity to service preferences of dynamic situations is poor. In contrast, the travel and service scenarios of aviation passengers have a more significant impact on the choice of auxiliary services, the weight is higher than that of passengers. The interaction of passengers with auxiliary services is more complicated, and needs to be expressed through praising, comment, sharing, etc., and can express the feelings or thoughts of passengers, so it has the highest weight.

In view of the above analysis, this paper assigns weights of 0.2, 0.3, and 0.5 to the three similarities of passengers, service context, and passenger's preference behavior. That is, α, β, and γ are respectively 0.2, 0.3, and 0.5. (The weight distribution can be determined separately according to actual situation. For example, when β = 0, the equation degenerates into the traditional rating-based service recommendation.)

## 6 Context-Aware Aviation Auxiliary Service Recommendation

The interest of passengers on auxiliary service items is greatly related on their travel context. In a similar travel scenario, the same type of passenger has similarities in the choice of auxiliary services. Therefore, in the context-aware aviation auxiliary service recommendation system, firstly, the similarity between the current and historical situation is calculated, and the most similar historical situation is proposed. Secondly, the interest degree of the passenger to auxiliary service items is ranked in that historical situation. Finally, the ranked top N additional services are recommended to the target passenger.

Specifically, different types of context such as service time, location, weather and air traffic control, is marked as $C = \{C_1, C_2, \ldots, C_n\}$. Where, $C_i$ is a vector representing the attributes of a certain context. It reflects the current travel situation of passengers, such as time information, location, weather, air traffic control, service status, etc. Passenger's current travel context is marked as $C_{now}$, while $C_{past}$ is historical context. The similarity between the two context is recorded as $Sim(C_{now}, C_{past})$. If the situation similarity $Sim(C_{now}, C_{past}) \geq \theta$ (a threshold) then refers that $C_{now}$ is highly similar with $C_{past}$. Therefore, the predicted preference of the auxiliary services in context $C_{now}$ is ranked by $Pre_{U \times S}^{c=C(Top_i)=C_{now}}(u, s)$.

Through above, getting the interested auxiliary service items under historical context $C_{past}$. Then match these services with the items to be recommended, and finally recommend the top-N similar items to the target passenger.

## 7 Experiment and Result Analysis

Aiming to verify the feasibility of the above contextual recommendation method, this paper takes an airline's passengers for experimental verification. Collects, classifies and processes data including personal information (name, age), auxiliary service information (name, price), and passengers' purchase behaviors (searching, reservation, payment, collection, sharing). Where passenger's travel or check-in time would reflect the context, attributes associated with the service: time of day (working day, weekend, winter or summer vacation, etc.), weather, air traffic control, etc. The above data set is respectively substituted into the traditional CF recommendation system and the supposed context-aware one, the corresponding recommended services list A and list B is sent to the tested passengers. According to the feedback satisfaction to compare the accuracy of the two recommendation methods.

In the experiment, 50 tested passengers were randomly selected. Firstly, the passengers' information was identified and processed. For example, a male passenger P's age was 55. By querying the website log, it was learned that A had frequent logins to airline website, reservations, and online consultations, so system inferred P was an active passenger. The passenger's check-in time stamp recorded part of the context information. For P, he checked-in about 3:00 pm in March 9, 2019. The weather condition was moderate to heavy rain that time, and his check-in location was the T2 terminal of Guangzhou Baiyun Airport. The status was travel alone. According to the above data, the top-5 recommendation is performed to P, where auxiliary service list A was recommended based on the traditional CF method and list B was recommended based on the context-aware preference cluster recommendation. According to passenger P's feedback, he had a strong interest in three recommend auxiliary services in list B, and only interested in one of list A. The result indicated that the proposed contextual recommendation had achieved a higher accuracy.

In the same way, the experiment provided recommendation for the 50 passengers, and investigated the satisfaction on the two recommendation results. It was measured with the five-point Likert scale, where quite dissatisfied (1 point), dissatisfied

**Table 1** Passenger satisfaction rating for auxiliary service recommendations

| Recommendation methods | Quite satisfied | Satisfied | Acceptable | Dissatisfied | Quite dissatisfied |
|---|---|---|---|---|---|
| | 5 | 4 | 3 | 2 | 1 |
| Traditional CF method | 8 | 12 | 18 | 9 | 3 |
| The paper's method | 12 | 20 | 14 | 3 | 1 |

(2 points), acceptable (3 points), satisfied (4 points), and quite satisfied (5 points). The tested result shown in Table 1.

From Table 1, it can be calculated that the average satisfaction in traditional CF recommendation is:

$$SR_{CF} = (8 \times 5 + 12 \times 4 + 18 \times 3 + 9 \times 2 + 3 \times 1) \div 50 = 3.26$$

While the proposed context-aware recommending average rating is:

$$SR_{cox-aware} = (12 \times 5 + 20 \times 4 + 14 \times 3 + 3 \times 2 + 1 \times 1) \div 50 = 3.78$$

The experimental result showed that the paper proposed context-aware recommendation achieved higher satisfaction. The main reason is that in the massive data environment, there is great sparsity of passengers' interactive data on auxiliary services. And the traditional CF methods are difficult to effectively and accurately personalized recommend with sparse item or evaluation data. The proposed context-aware recommendation using context data to calculate the similarity of passengers' preference other than using historical evaluation data. Thereby effectively alleviating influence of sparsity in passenger interaction data and improving the accuracy of recommendation.

## 8 Conclusions

Based on the five-dimensional characteristic information of air passengers' basic information, auxiliary service information, service context, auxiliary service preferences and passenger interaction behaviors, this paper constructs a context-aware air passenger auxiliary service ontology model. It further introduces collaborative filtering recommendation ideas and calculates the auxiliary service preference similarity through experimental verification. The result showed that the context-aware recommendation alleviates data sparsity problem caused by user rating in traditional CF, improves the recommendation accuracy. However, it should be pointed out that in the recommendation process of aviation auxiliary services, the passenger's service demand is dynamically changed. Only by dynamically tracking the passenger's interest changes and timely adjusting the content and methods recommended by the service resources can be perfectly matched. The dynamic recommendation system is intended to be further explored in future research. And in addition, the number of samples surveyed in this experiment is relatively small, which may influence the results. It is proposed to increase the experimental sample size in the follow-up study.

**Acknowledgements** This work was supported by the National Natural Science Foundation of China under Grant No. 71801059, Research Platform and Projects of Guangdong Education Department Grant No. 2018GWTSCX053, and Research Project of the Guangzhou Civil Aviation College under Grant No. 18X0332 and No.17X0304.

## References

1. Liu, H. Y., Li, R. M., & Qiu, Y. J., et al. (2015). Research on passenger interest model for civil aviation ancillary services. *Journal of Beijing University of Posts and Telecommunications, 38*(6), 82–86.
2. Lin, L. (2016). The airlines are shouting for transformation, and the spring of auxiliary service business is really coming. China Civil Aviation Resource Network, from http://news.carnoc.com/list/341/341632.html.
3. Borchers, A., Herlocker, J., & Konstan, J., et al. (1998). Ganging up on information overload. *Computer, 31*(4), 106–108.
4. Li, C., & Ma, L. (2016). *Research on bottleneck problem of e-commerce recommendation system* (p. 9). Beijing: Science Press of China.
5. Glodberg, D., Nichols, D., & OkiB, M., et al. (1992). Using collaborative filtering to weave an information tapestry. *Communications of the ACM, 35*(12), 61–70.
6. Ansari, A., Essegaier, S., & Kohli, R. (2000). Internet recommender systems. *Journal of Marketing Research, 37*(3), 363–375.
7. Koren, Y. (2010). Collaborative filtering with temporal dynamics. *Communications of the ACM, 53*(4), 89–97.
8. Sun, T., & Trudel, A. (2002). An implemented e-commerce shopping which makes personal recommendations. In *Sixth IASTED International Conference on Internet and Multimedia Systems and Applications (IMSA 2002)* (pp. 58–62). Hawaii: IASTED/ACTA Press.
9. Schafer, J. B., Konstan, J. A., & Riedl, J. (2001). E-commerce recommendation applications. *Data Mining and Knowledge Discovery, 5*(1–2), 115–153.
10. Billsus, D., & Pazzani, M. (2007). Adaptive news access. In *The Adaptive Web, LNCS 4321* (pp. 550–570). Heidelberg: Spinger.
11. ColomboMendoza, L. O., Valencia-García, R., & Rodríguez-González, A., et al. (2015). RecomMetz: A context -aware knowledge-based mobile recommender system for movie showtimes. *Expert Systems with Applications, 42*(3), 1202–1222.
12. Adomavicius, G., Mobasher, B., & Ricci, F., et al. (2008). Context-Aware recommender systems. *International Journal of Information Technology and Web Engineering, 32*(3), 335–336.
13. Ai, S. T. (2012). Research on a hybrid personalized recommendation platform based on Service-Oriented Architecture (SOA). *Information Studies: Theory & Application, 35*(5), 107–111.
14. Yu, K. F., Zhang, B. X., & Zhu, H. S., et al. (2012). Towards personalized context-aware recommendation by mining context logs through topic models. In *Proceedings of the 16th Pacific-Asia conference on Advances in Knowledge Discovery and Data Mining* (pp. 431–443).

15. Peng, M., Xi, J. J., & Dai, X. Y., et al. (2017). Collaborative filtering recommendation based on sentiment analysis and LDA topic model. *Journal of Chinese Information Processing, 31* (2), 194–203.
16. Ai, D. X., Zhang, Y. F., & Zuo, H., et al. (2015). Study on O2O mobile recommendation system based on contextual semantics inference. *Journal of Intelligence, 34*(8), 183–189.
17. Ai, D. X., Zhang, Y. F., & Liu, G. Y., et al. (2016). Contextual semantics modeling and rule-based reasoning for mobile business catering recommendation. *Information Studies: Theory & Application, 39*(2), 82–88.
18. Li, H. J., & Zhang, F. (2018). Research on context-aware information recommendation service on mobile device from the perspective of activity theory——based on the contextual ontology model and rule-based reasoning. *Journal of Intelligence, 37*(3), 187–192.
19. Liu, H. O., Sun, J. J., & Su, Y. Y., et al. (2018). Research on the tourism situational recommendation service based on persona. *Information Studies: Theory & Application, 41* (10), 87–92.
20. Shin, D., Lee, J. W., & Yeon, J., et al. (2009). Context-aware recommendation by aggregating user context. In *IEEE Conference on Commerce and Enterprise Computing, Washington, DC, USA* (pp. 423–430).
21. Liu, L., Lecue, F., & Mehandjiev, N., et al. (2010). Using context similarity for service recommendation. In *IEEE Fourth International Conference on Semantic Computing* (pp. 277–284).
22. Shi, H. Y., & Han, X. J. (2018). Review on context-aware recommender system. *Information Science, 36*(7), 163–169.
23. Rendle, S., Gantner, Z., & Freudenthaler, C., et al. (2011). Fast context-aware recommendations with factorization machines. In *Proceeding of the International Acm Sigir Conference on Research and Development in Information Retrieval, Beijing* (pp. 635–644).
24. Ye, X. N., & Wang, M. (2018). The research about TFPMF of the personalized music recommendation algorithm. *Journal of System Simulation*, from http://kns.cnki.net/kcms/detail/11.3092.V.20180322.1721.078.html.
25. Chen, A. (2005). Context-aware collaborative filtering system: predicting the user's preference in the ubiquitous computing environment. In *International Conference on Location-& Context-awareness* (pp. 244–253).
26. Zhang, W. J., Li, J. P., & Yang, J. (2018). Improved extraction method of information core in collaborative filtering recommendation. *Application Research of Computers,* from https://doi.org/10.19734/j.issn.1001-3695.2018.05.0450.
27. Shang, Y. F., Chen, D. Y., & Yang, H. L. (2018). A personalized recommendation algorithm for mobile application. *Journal of Harbin University of Science and Technology, 23*(5), 116–123.

# Emergency Evacuation Model for Large Multistory Buildings with Usual Entrances and Emergency Exits

**Rong Liu, Xingyi Chen, Yuxuan Tian, Sha Wang, and Zhenping Li**

**Abstract** When emergencies such as fires, earthquakes or terrorist attacks occur, the effective evacuation of all personnel from buildings is the primary objective. Sometimes, in order to further reduce the losses, we should also consider how to make firefighters, medical staffs and other emergency rescue personnel more convenient to access to the interior of the building. Especially when there are valuable property, dangerous goods or casualties inside the building. For buildings with multiple usual entrances and emergency exits, this paper studies how to plan evacuation routes making all people complete evacuation within a specified time and use emergency exits as few as possible so that some emergency exits can be left for firefighters and other rescue workers. In order to achieve this, an integer programming model with three steps is established in this paper. Finally, we designed an emergency evacuation scheme for the Louvre in Paris using our model to show how to use our model in practice.

**Keywords** Emergency evacuation · Emergency exits · Usual entrances · Integer programming

R. Liu (✉) · X. Chen · Y. Tian · S. Wang · Z. Li
School of Information, Beijing Wuzi University, Beijing, China
e-mail: buptrong@163.com

X. Chen
e-mail: 461982286@qq.com

Y. Tian
e-mail: 18813016775@163.com

S. Wang
e-mail: saya1016@163.com

Z. Li
e-mail: lizhenping@bwu.edu.cn


J. Zhang et al. (eds.), *LISS2019*,
https://doi.org/10.1007/978-981-15-5682-1_14

# 1 Introduction

In recent years, with the accelerating urbanization process, the construction industry has been hot and a large number of large-scale buildings have emerged. Due to the high floor, large floor space and high crowded, large-scale buildings are prone to major casualties in the event of fires and other accidents. Therefore, it has great practical significance to study the emergency evacuation of large multistory buildings. Liu proposed a model combining of heuristic algorithm and network flow control under the limiting condition of fire smoke and routes capacity [1], which aims to minimize the total evacuation time for all people. Krasko presented a two-stage stochastic mixed-integer nonlinear programming model for post-wildfire debris flow hazard management which minimizes expected damages taking several storm scenarios into account and considered loss of life [2]. Rozo proposed an agent-based simulation model incorporating the reaction of people in an emergency and their route choice behavior in a building [3]. Sheeba presented an analytical model of the evacuation scenario in buildings on accidental fire using stochastic petri nets [4], which took into account human behavior parameters. Cheng established and evaluated an Emergency Evacuation Capacity model for key evacuation facilities in subway stations through analyzing key factors [5]. Li constructed an optimization model for evacuation routes planning with optimizing strategies that minimizes the total evacuation distance in the dangerous area under fire [6]. Renne examined large-scale and multimodal emergency evacuation planning for carless and vulnerable populations in the US and UK [7], and suggested four key recommendations to improve it. Chen established a model based on network distribution which aimed the shortest emergency time, and the optimal solution is acquired using the Pontryagin minimum principle [8]. Yang developed a multimodal evacuation model that considers multiple transportation modes and their interactions, and designed a method of successive average based sequential optimization algorithm for large-scale evacuations [9]. Singhal proposed an altered ant colony optimization algorithm which minimized the entire rescue time of all evacuees [10].

It can be seen from the above researches that many scholars have adopted lots of different models for the emergency evacuation. Unfortunately, the entrances for the emergency personnel entering into the building are almost ignored, which is very important for rapid and safe evacuation sometimes. In some special cases, such as fires and terrorist attacks, we should reserve some emergency exits for the firefighters, police and medical staffs to come into the buildings. Therefore, the emergency evacuation of the large multistory buildings with several usual entrances and emergency exits are investigated in this paper. And we try to solve the problem: besides the usual entrances, how to evacuate all of the people in the building through the fewest emergency exits within the security time bound, so that the additional spare emergency exits could be used for the emergency personnel.

## 2 Emergency Evacuation Model

### 2.1 Assumption of the Model

- The management system of the building can get the amount of people in every area of the building in time. Yoshimura [11] and others used the method of installing sensors at the entrance of corridors and exhibition areas to record the number of visitors in some areas of the Louvre.
- All evacuees can receive the evacuation route instructions issued by the building managers in time, and all of the evacuees leave the building in an orderly manner according to the routes specified by the management system.
- After issuing the evacuation notice, all of the opened usual entrances and emergency exits can be used at their maximum throughput capacity.
- The evacuees are orderly evacuated under the instructions of the system and do not form artificial blockage at the passage and exports. A queue may be formed at the exits, which is a state of orderly passage. And there is no disorderly blocking in the system.

### 2.2 Parameter Descriptions

(See Table 1).

**Table 1** Description of parameters

| Parameter | Description |
|---|---|
| $a_i$ | Number of visitors in area $i$ |
| $b_j$ | Number of visitors passing usual entrance $j$ per unit time |
| $c_j$ | Number of visitors passing emergency exit $j$ per unit time |
| $d_{ij}$ | Time used to evacuate from area $i$ to usual entrance $j$ |
| $e_{ij}$ | Time used to evacuate from area $i$ to emergency exit $j$ |
| $T_{\max}$ | Maximum time allowed for safe evacuation |
| $V$ | The set of all areas |
| $EX$ | The set of all usual entrances |
| $EM$ | The set of all emergency exits |
| $E$ | The set of all edges in topological graph |
| $arc_{ij}$ | Edge connecting node $i$ and node $j$ |
| $Q_{ij}$ | Flow on $arc_{ij}$ |
| $t_{ij}$ | Time to use from node $i$ to node $j$ |

## 2.3 Variable Descriptions

For a specific problem, we should divide the space inside the building into multiple areas. Then we could use Dijkstra algorithm or Bellman-Ford algorithm to calculate the shortest route from each area to each exit. So, we obtain $d_{ij}$ and $e_{ij}$. We can get $a_i$ from the building management system. $b_j$ and $c_j$ can be set based on existing researches or experiments (Tables 2).

## 2.4 The Integer Programming Model with Three Steps

- Step 1: generate initial routes

In this step, we establish an integer programming model to generate the initial routes. In the initial routes, once a person in an area chooses to leave from an exit, they will move along the shortest path from the area to the selected exit. At the same time, we should ensure that these routes are not too long to exceed the security time limit, and that the total flow of each exit is not greater than the maximum capacity of the exit within the security time limit. The definitions of the model are as follows.

$$\min \sum_{j \in EM} z_j \tag{1}$$

$$s.t.$$

$$\sum_{j \in EX} x_{ij} + \sum_{j \in EM} s_{ij} = a_i \quad \forall i \in V \tag{2}$$

$$\sum_{i \in V} x_{ij} \leq T_{\max} b_j \quad \forall j \in EX \tag{3}$$

**Table 2** Description of variables

| Variable | Description |
|---|---|
| $z_j$ | Indicate whether the emergency exit $j$ will be used |
| $x_{ij}$ | Number of visitors evacuated from area $i$ to usual entrance $j$ |
| $s_{ij}$ | Number of visitors evacuated from area $i$ to emergency exit $j$ |
| $p_{ij}$ | Flow from node $i$ to node $j$ in topological graph |
| $q_{ij}^m$ | Flow on $arc_{ij}$ which come from area $m$ |
| $r_{ij}^m$ | Indicate whether there is flow from area $m$ on $arc_{ij}$ |

$$\sum_{i \in V} s_{ij} \leq T_{\max} c_j z_j \quad \forall j \in EM \tag{4}$$

$$x_{ij} \leq a_i y_{ij} \quad \forall i \in V; \forall j \in EX \tag{5}$$

$$s_{ij} \leq a_i r_{ij} \quad \forall i \in V; \forall j \in EM \tag{6}$$

$$d_{ij} y_{ij} \leq T_{\max} \quad \forall i \in V; \forall j \in EX \tag{7}$$

$$e_{ij} r_{ij} \leq T_{\max} \quad \forall i \in V; \forall j \in EM \tag{8}$$

$$x_{ij} \geq 0, \quad y_{ij} \in \{0, 1\} \quad \forall i \in V; \forall j \in EX \tag{9}$$

$$s_{ij} \geq 0, \quad r_{ij} \in \{0, 1\} \quad \forall i \in V; \forall j \in EM \tag{10}$$

$$z_j \in \{0, 1\} \quad \forall j \in EM \tag{11}$$

Equation (1) is the objective function, which minimizes the number of emergency exits used in the evacuation. Equation (2) means the total number of people through the usual entrances and the emergency exits is equal to the total number of people in the building. Equation (3) ensures every usual entrance to satisfy the security time limit. Equation (4) ensures all the emergency exits we use satisfy the security time limit. Moreover, Eq. (5) indicates the logistic relationship between $x_{ij}$ and $y_{ij}$. Equation (6) indicates the logistic relationship between $s_{ij}$ and $r_{ij}$. Equations (7) and (8) means that every route we chose satisfy the security time limit. Equations (9)-(11) define some variables should be nonnegative or binary.

- Step 2: satisfy flow capacity constraints

In practical applications, during a certain period of time, the flow passing through a certain route is limited. It is meaningful to add flow capacity constraints on the evacuation routes obtained in Step 1.

It can be treated as a problem of finding feasible flows. In this problem we have already known the starting points, terminal points, flow capacity constraints on edges and flow from each starting point. We still establish an integer programming model to solve this problem.

$$\min \sum_{i \in V} \sum_{j \in V \cup EX \cup EM} p_{ij} \tag{12}$$

$$s.t.$$

$$p_{ij} = 0 \qquad arc_{ij} \not\subset E \tag{13}$$

$$\sum_{j \in V} p_{ij} = \sum_{j \in V} p_{ji} + a_i \qquad \forall i \in V \tag{14}$$

$$\sum_{i \in V} p_{ij} \le b_j \qquad \forall j \in EX \tag{15}$$

$$\sum_{i \in V} p_{ij} \le c_j \qquad \forall j \in EM \tag{16}$$

$$\sum_{j \in EX \cup EM} \sum_{i \in V} p_{ij} = \sum_{i \in V} a_i \tag{17}$$

$$p_{ij} \in Z^+ \qquad \forall i \in V, j \in V \cup EX \cup EM \tag{18}$$

The aim is minimizing the sum of flow on all edges. This objective function ensures that there is no cycle in the paths we found. Equation (13) means if there is no edge connecting node $i$ and node $j$ directly, then $p_{ij}$ is set as 0. Equation (14) ensures the flow balance on nodes. Equations (15) and (16) mean that the number of people through each usual entrance and emergency exit cannot exceed exit's flow limit individually. Equations (17) ensures that all the people in every area are evacuated.

- Step 3: Redistribute routes for each area

Now we have the routes satisfy the flow capacity constraints, but it is not what we want. In practical, we hope to get the evacuation routes of each area individually, not the routes contain the flows from all the areas. We can solve the problem by calculating the composition of flow on each edge. In other words, we need to figure out the flow from each area individually on each edge. We design an integer programming model to solve this problem.

$$\min \sum_{m \in V} \sum_{i \in V} \sum_{j \in V \cup EX \cup EM} r_{ij}^m \tag{19}$$

$$s.t.$$

$$\sum_{m \in V} q_{ij}^m = Q_{ij} \qquad \forall i \in V, j \in V \cup EX \cup EM \tag{20}$$

$$\sum_{j \in V} q_{ij}^m = \sum_{j \in V} q_{ji}^m \qquad \forall m \in V, i \in V, m \neq i \tag{21}$$

$$\sum_{j \in V} q_{ij}^m = a_i \qquad \forall m \in V, i \in V, m = i \tag{22}$$

$$q_{ij}^{m} \leq M \cdot r_{ij}^{m} \qquad \forall i \in V, j \in V \cup EX \cup EM, m \in V \tag{23}$$

$$q_{ij}^{m} = 0 \qquad \forall arc_{ij} \not\subset E, m \in V \tag{24}$$

$$\sum_{i}\sum_{j} t_{ij} \cdot r_{ij}^{m} \leq T_{\max} \quad \forall m \in V,\ where\ if\ arc_{ij}\ belongs\ to\ a\ possible\ path\ that\ cantains\ flow\ from\ area\ i \tag{25}$$

$$q_{ij}^{m} \in Z^{+}, r_{ij}^{m} \in \{0, 1\} \tag{26}$$

The aim is to minimize the number of routes we used, because we always want to evacuate one area trough just one route, that will be more feasible and convenient in practice. Equations (20)-(22) ensures the flow balance on nodes. Equation (23) indicates the logistic relationship between $q_{ij}^{m}$ and $r_{ij}^{m}$. Equation (24) is similar to (13) in Step 2. Equation (25) indicates that every route we used cannot exceed the time limit. Equation (25) also means we should figure out the possible routes of each area ahead of time. Since the topographic is sparse, it will not be difficult. If the model is infeasible, we should consider open more emergency exits, we will should this in next section.

## 3 Applications

The Louvre in Paris is one of the most famous and most visited museums all over the world. It is really an important and challenging problem to design an emergency evacuation scheme for the Louvre. In this section, we apply the above proposed model to the Louvre.

### 3.1 Data Processing

According to the layout of the Louvre, each floor of the museum is divided into several small visiting areas. The principle of division is to make the exhibition halls closed to each other as far as possible in the same area, and try to ensure that each divided area has staircases leading to the upper and lower floors. All of the divided visiting areas are numbered as area0 $\sim$ area18, which are shown in Figs. 1, 2, 3 and 4. The Louvre has four usual entrances for visitors, namely Galerie Du Carrousel entrance (Carrousel), Passage Richelieu entrance (Richelieu), Porte des Lions entrance (Lions) and Pyramid main entrance (Pyramid). The four usual entrances for visitors are numbered as exit0 to exit3, which are shown in Fig. 5. Whereas the emergency exits of the Louvre has not been announced. Without loss of generality, we assume that an area shares an emergency passage with the areas

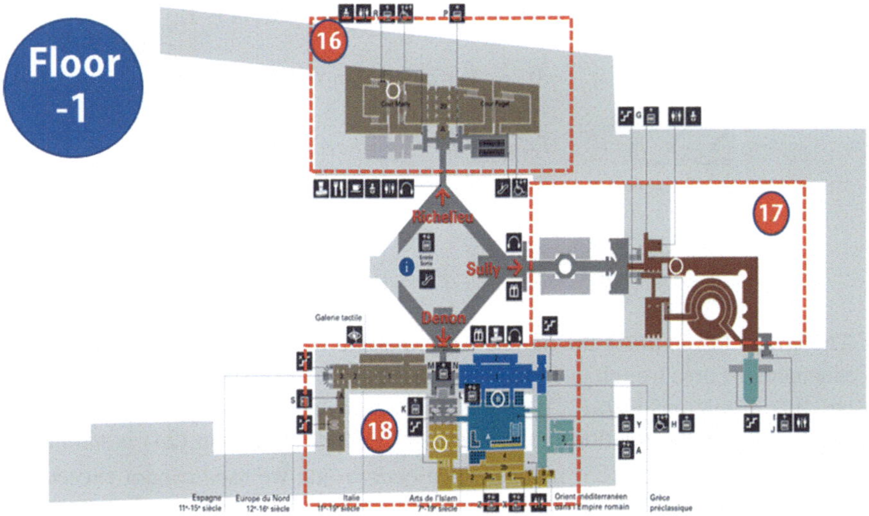


Fig. 1 Visiting areas in floor - 1

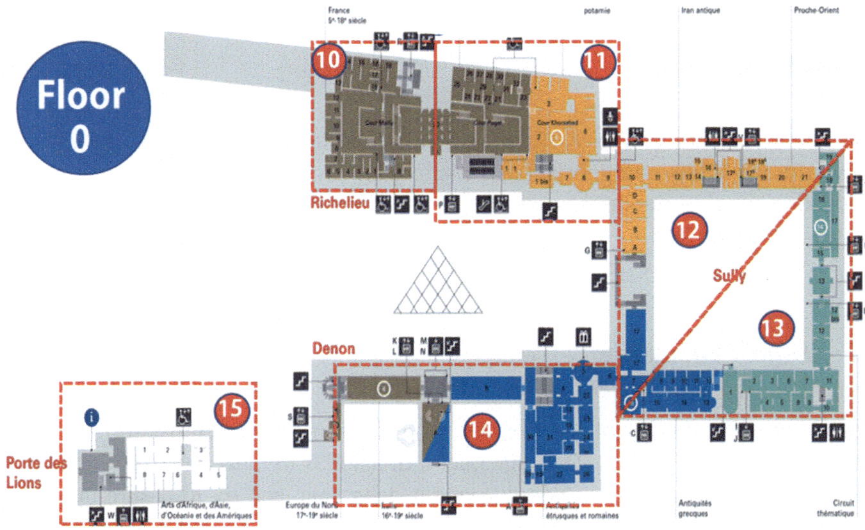


Fig. 2 Visiting areas in floor 0

directly above and below it. Then, regarding each divided visiting area as a node and each passage between the areas as an edge, a topological graph is obtained in Fig. 6. For convenience, we set up corresponding virtual nodes for the 16 visiting areas above the ground, which are numbered as em0 ~ em15. This is to display

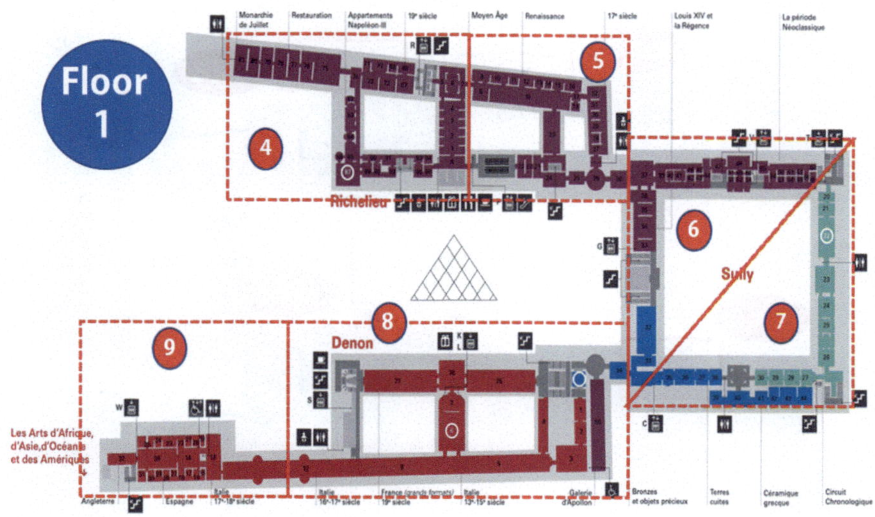


**Fig. 3** Visiting area in floor 1

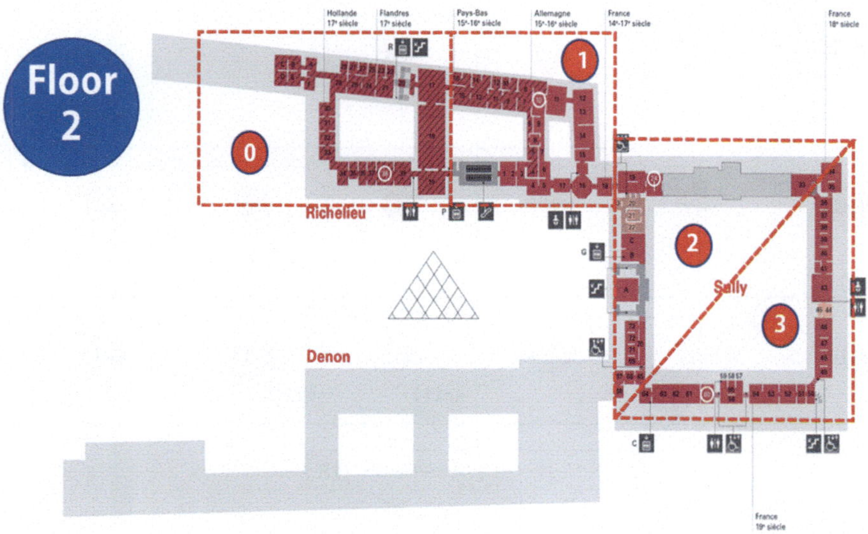


**Fig. 4** Visiting area in floor 2

usual passage and emergency passage individually. Let's give an example, as shown in Fig. 1, the edge between em0 and em4 represents the emergency passage between area0 and area4. Em10 $\sim$ em15 are emergency exits because they are on the ground. Overall, the building is divided into nineteen areas, with four usual entrances and six emergency exits.

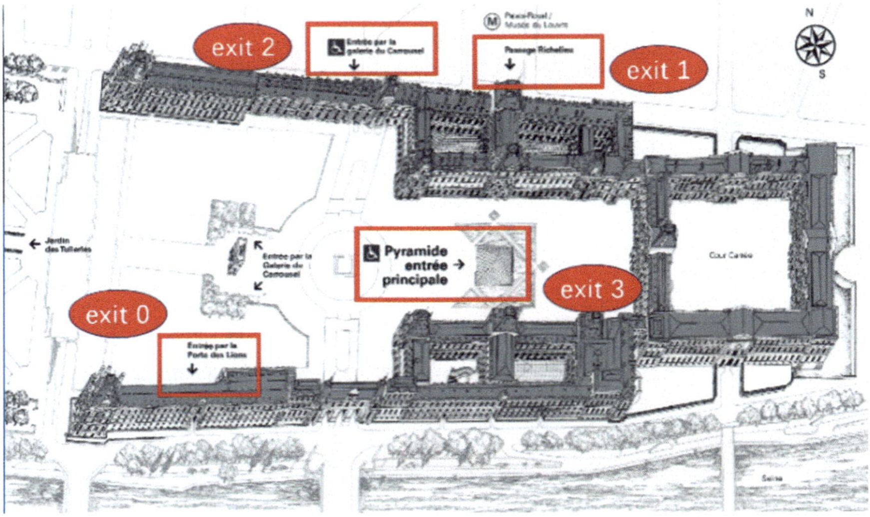


**Fig. 5** Four usual entrances

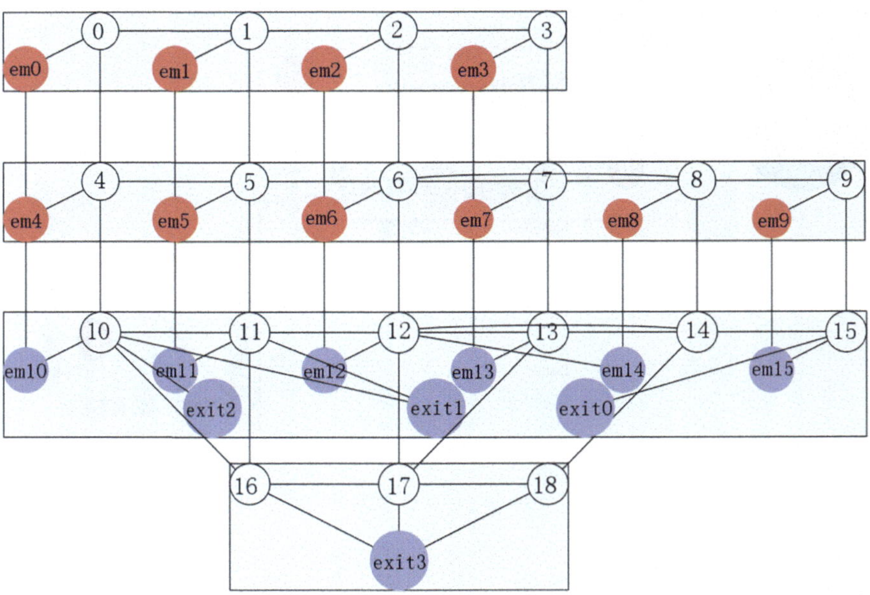


**Fig. 6** Topological structure of visiting areas

Then we need to estimate the shortest distance from each area to each usual entrance and emergency exit. The distance between interconnected areas on the same layer is obtained by multiplying the Manhattan distance between the

**Table 3** The distance between each visiting area to each usual entrance and each emergency exit

| | exit0 | exit1 | exit2 | exit3 | em10 | em11 | em12 | em13 | em14 | em15 |
|---|---|---|---|---|---|---|---|---|---|---|
| area0 | 971.2 | 296 | 339.2 | 396 | 200 | 308 | 497.6 | 600.8 | 616 | 913.6 |
| area1 | 971.2 | 308 | 447.2 | 396 | 308 | 200 | 389.6 | 492.8 | 616 | 913.6 |
| area2 | 807.2 | 497.6 | 639.2 | 468 | 500 | 389.6 | 200 | 303.2 | 452 | 749.6 |
| area3 | 819.2 | 612.8 | 764 | 468 | 624.8 | 504.8 | 303.2 | 200 | 464 | 761.6 |
| area4 | 871.2 | 196 | 239.2 | 296 | 100 | 258.4 | 421.6 | 524.8 | 516 | 813.6 |
| area5 | 871.2 | 208 | 359.2 | 296 | 220 | 100 | 289.6 | 392.8 | 516 | 813.6 |
| area6 | 707.2 | 397.6 | 548.8 | 368 | 409.6 | 289.6 | 100 | 203.2 | 352 | 649.6 |
| area7 | 719.2 | 512.8 | 664 | 368 | 524.8 | 404.8 | 203.2 | 100 | 364 | 661.6 |
| area8 | 402.4 | 612 | 655.2 | 320 | 516 | 606.4 | 352 | 364 | 100 | 344.8 |
| area9 | 157.6 | 856.8 | 900 | 564.8 | 760.8 | 851.2 | 596.8 | 608.8 | 344.8 | 100 |
| area10 | 771.2 | 96 | 139.2 | 196 | 0 | 220 | 321.6 | 424.8 | 416 | 713.6 |
| area11 | 771.2 | 108 | 259.2 | 196 | 120 | 100 | 201.6 | 304.8 | 416 | 713.6 |
| area12 | 607.2 | 309.6 | 460.8 | 268 | 321.6 | 301.6 | 0 | 103.2 | 252 | 549.6 |
| area13 | 619.2 | 412.8 | 564 | 268 | 424.8 | 404.8 | 103.2 | 0 | 264 | 561.6 |
| area14 | 355.2 | 512 | 555.2 | 220 | 416 | 516 | 252 | 264 | 0 | 297.6 |
| area15 | 57.6 | 809.6 | 852.8 | 517.6 | 713.6 | 813.6 | 549.6 | 561.6 | 297.6 | 0 |
| area16 | 671.2 | 196 | 239.2 | 96 | 100 | 200 | 301.6 | 356.8 | 316 | 613.6 |
| area17 | 692.8 | 409.6 | 496 | 168 | 356.8 | 401.6 | 100 | 100 | 337.6 | 635.2 |
| area18 | 455.2 | 412 | 455.2 | 120 | 316 | 416 | 337.6 | 337.6 | 100 | 397.6 |

geometric centers of two areas on the plane by the corresponding scale. The distance between two connected areas of adjacent floors is set as 100 m and the distance between one area and it's corresponding virtual node is 0. Dijkstra algorithm is used to calculate the shortest distance between the nineteen areas and four usual entrances and six emergency exits. The result is shown in Table 3.

According to the annual visitor data published on the official website of the Louvre Museum over the past ten years (2008-2018), the total number of visitors in each year is selected. The average number of visitors per day during the opening period of the Louvre Museum in each year is calculated, with the minimum average number of visitors per day $g_{\min}$ and the maximum average number of visitors per day $g_{\max}$ as the intervals, and the distribution of the time spent by visitors in the Louvre Museum [11]. The average stay time of visitors is 3.21 h, and the interval of the total number of visitors in the Louvre Museum is estimated. The estimation formula is shown as Eq. (27). Among them, 9.92 h is the average daily opening time calculated by the daily opening time published by the official website of the Louvre Museum.

$$[l_{\min}, l_{\max}] = \left[\frac{3.21 \times g_{\min}}{9.92}, \frac{3.21 \times g_{\max}}{9.92}\right] \tag{27}$$

By gridding the layout of the Louvre Museum, we calculate the area of each region approximately. The number of visitors is distributed in a manner proportional to the area of the region, that means larger the region is, more visitors there will be. However, because there are Mona Lisa and the Goddess of Victory in area8 and Venus in area13, the amount of visitors in these two areas will be significantly larger than in other areas according to experience. The initial population data constructed in this paper are shown in Table 4.

It is assumed that the throughput capacity is proportional to the width of the exit. The width of the four entrances of the Louvre is measured by the distance measuring tool of Google Map. The width of Lion Gate entrance, Richelieu entrance, Carrousel entrance and Pyramid entrance are 3 m, 2 m, 3 m and 7 m respectively. The evacuation experiment organized by Yue et al. obtained the relationship between exit width and throughput capacity during emergency evacuation [12]. Based on this, the throughput capacity of four usual entrances are estimated as in Table 5. The throughput capacity of each emergency exit is 1 person per second.

**Table 4** The initial population in different visiting areas

| Area | Population | Area | Population |
|---|---|---|---|
| area 0 | 485 | area 10 | 406 |
| area 1 | 505 | area 11 | 518 |
| area 2 | 330 | area 12 | 337 |
| area 3 | 228 | area 13 | 872 |
| area 4 | 452 | area 14 | 525 |
| area 5 | 333 | area 15 | 238 |
| area 6 | 373 | area 16 | 409 |
| area 7 | 363 | area 17 | 353 |
| area 8 | 1498 | area 18 | 584 |
| area 9 | 261 | | |

**Table 5** The throughput capacity of each usual entrance (Unit: person/second)

| Usual entrances | Lions | Richelieu | Carrousel | Pyramid |
|---|---|---|---|---|
| Throughput capacity | 6 | 4 | 6 | 14 |

## 3.2 Model Calculation

We use Gurobi 8.1.0 (Python API) to solve the model of each step on a Core i7 (2.6 GHz) processor. The security evacuation time is set as 310 s and the flow capacity of each passage in 310 s is set as 1500. Each step takes less than one second for calculation.

Solving the model of Step 1, we can get the result that em12 and em13 must be opened. All the passages used and the corresponding flow on them are shown in Fig. 7. The passages that exceed the flow capacity was marked with red boxes in Fig. 7.

To satisfy the flow capacity constraints, we solve the model of Step 2. In Fig. 7, we can see some cycles in the graph, like area14 and area18. We can fix this issue in Step 2 too. Since the passages that exceed the flow capacity are all in the lowest second floors, we can simplify the calculation by just using the lowest two floors in our model of Step 2. The result of this step is shown in Fig. 8.

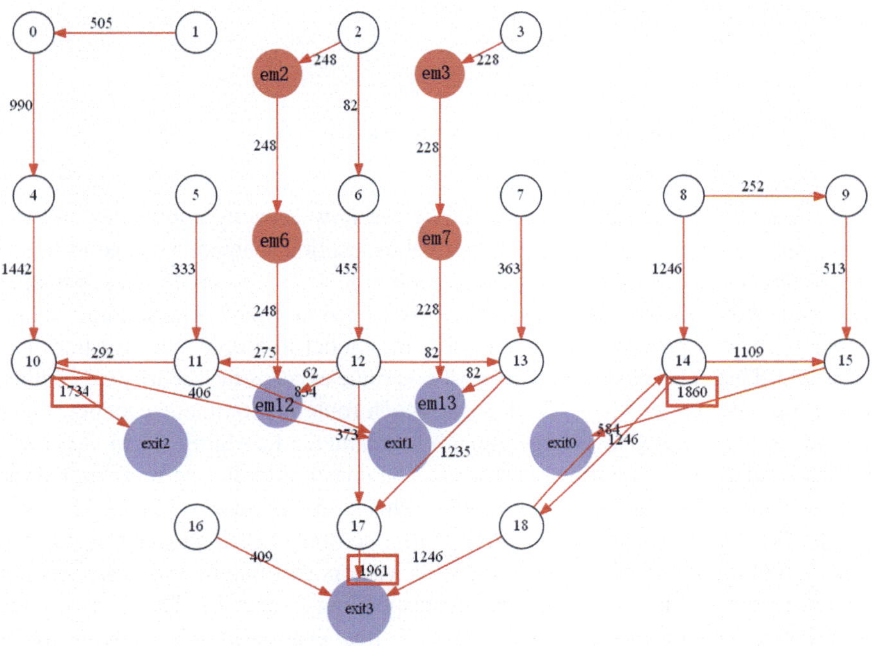


**Fig. 7** Initial routes obtained by Step 1

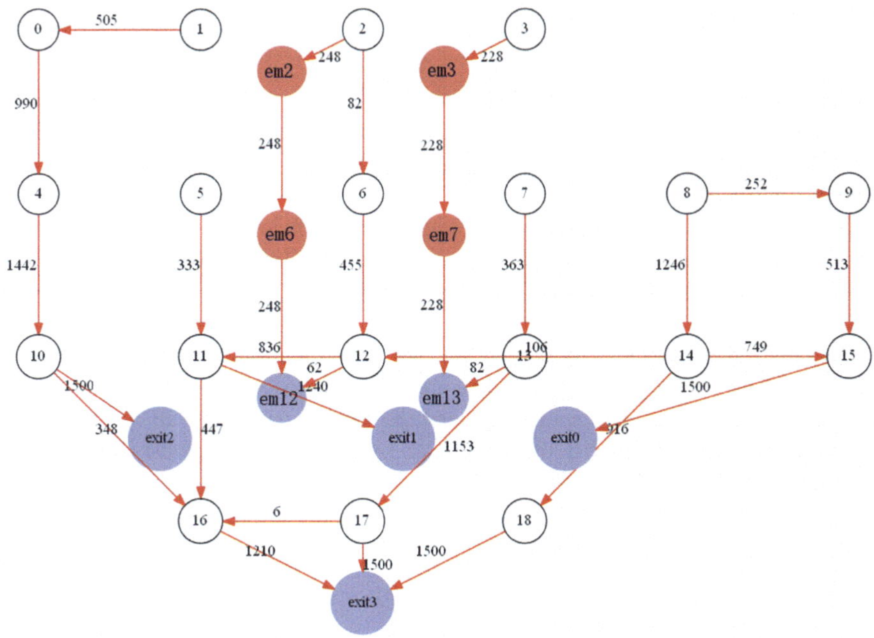


**Fig. 8** Routes satisfy the flow capacity constraints obtained by Step 2

We can see that all the passages satisfy the flow capacity constraints and the cycles are eliminated. At last, to get the final evacuation scheme, we need to figure out the flow components on each passage in Fig. 8. That is to say for each passage, we should find out what area the flow comes from and the corresponding value. Step 3 is for this purpose. When we solve the model in Step 3, we will find it is infeasible. That is because there is at least one route exceeds the security evacuating time which was set as 310 s. To find out which route is the time-exceeding route, we can set the security time large enough to guarantee feasibility. Then we check the slack variables of the time constraints. The constraint with the minimum slack variable shows the time-exceeding route. In this case, the route is [14-12-11-exit1] and the flow is 44. Fortunately, there is only one time-exceeding route, if not, we need to find out all of them. To evacuate all visitors still within 310 s, an obvious method is to open the emergency exit corresponding to area 14. That is to say we should open three emergency exits at last. The final evacuating scheme is shown in Table 6.

**Table 6** Final evacuating scheme

| Area | Route | Flow |
|---|---|---|
| 0 | 0-4-10-exit2 | 485 |
| 1 | 1-0-4-10-exit2 | 505 |
| 2 | 2-29-30-em12 | 248 |
| 2 | 2-6-12-11-exit1 | 82 |
| 3 | 3-32-33-em13 | 228 |
| 4 | 4-10-exit2 | 452 |
| 5 | 5-11-exit1 | 333 |
| 6 | 6-12-11-exit1 | 373 |
| 7 | 7-13-17-exit3 | 363 |
| 8 | 8-14-18-exit3 | 497 |
| 8 | 8-14-15-exit0 | 749 |
| 8 | 8-9-15-exit0 | 252 |
| 9 | 9-15-exit0 | 261 |
| 10 | 10-exit2 | 58 |
| 10 | 10-16-exit3 | 348 |
| 11 | 11-16-exit3 | 447 |
| 11 | 11-exit1 | 71 |
| 12 | 12-11-exit1 | 337 |
| 13 | 13-17-exit3 | 790 |
| 13 | 13-em13 | 82 |
| 14 | 14-em14 | 44 |
| 14 | 14-12-em12 | 62 |
| 14 | 14-18-exit3 | 419 |
| 15 | 15-exit0 | 238 |
| 16 | 16-exit3 | 409 |
| 17 | 17-exit3 | 347 |
| 17 | 17-16-exit3 | 6 |
| 18 | 18-exit3 | 584 |

## 4 Conclusion

In the case of the Louvre, we can find that not all exits must be opened to meet certain security evacuation time limit, which makes it possible for us to optimize the number of exits to be opened. In addition to reducing the computational scale, the three-steps model designed in this paper has other benefits. The first two steps make the routes obtained naturally have the following characteristics. In the first step, the routes of the higher floors will be concentrated, and in the second step, the routes of the lower floors will be dispersed. This is in line with the fact that the lower floors will have more people and be more crowded. Finally, we get the evacuation routes of each area. For the convenience in practice, the model tries to allocate an unique path for each area as far as possible. However, some areas have a large number of people, so it is inevitable to allocate multiple routes for them.

The main deficiencies of this paper is that our model assumes that the exits will be used from the very begin to the very end, which is an ideal situation. We will try to consider the evacuees arrive in batches or subjecting to a certain probability distribution in future research, which may lead to a stochastic programming model.

## References

1. Liu, C., Mao, Z., & Fu, Z. (2016). Emergency evacuation model and algorithm in the building with several exits. *Procedia Engineering, 135,* 12–18.
2. Krasko, V., & Rebennack, S. (2017). Two-stage stochastic mixed-integer nonlinear programming model for post-wildfire debris flow hazard management: Mitigation and emergency evacuation. *European Journal of Operational Research, 263*(1), 265–282.
3. Rozo, K. R., et al. (2019). Modelling building emergency evacuation plans considering the dynamic behaviour of pedestrians using agent-based simulation. *Safety Science, 113,* 276–284.
4. Sheeba, A. A., & Jayaparvathy, R. (2019). Performance modeling of an intelligent emergency evacuation system in buildings on accidental fire occurrence. *Safety Science, 112,* 196–205.
5. Cheng, H., & Yang, X. (2012). Emergency evacuation capacity of subway stations. *Procedia-Social and Behavioral Sciences, 43,* 339–348.
6. Li, J., & Zhu, H. (2018). A risk-based model of evacuation route optimization under fire. *Procedia Engineering, 211,* 365–371.
7. Renne, J. L. (2018). Emergency evacuation planning policy for carless and vulnerable populations in the United States and United Kingdom. *International Journal of Disaster Risk Reduction, 31,* 1254–1261.
8. Chen, Y., & Xiao, D. (2008). Emergency evacuation model and algorithms. *Journal of Transportation Systems Engineering and Information Technology, 8*(6), 96–100.
9. Yang, X., Ban, X., & Mitchell, J. (2017). Modeling multimodal transportation network emergency evacuation considering evacuees' cooperative behavior. *Transportation Research Procedia, 23,* 1038–1058.
10. Singhal, K., & Sahu, S. (2016). Fire evacuation using ant colony optimization algorithm. *International Journal of Computer Applications, 139*(8), 0975–8887.
11. Yoshimura, Y., et al. (2014). An analysis of visitors' behavior in the Louvre Museum: a study using Bluetooth data. *Environment and Planning B: Planning and Design, 41*(6), 1113–1131.
12. Yue, H., et al. (2014). Pedestrian pedestrian facilities emergency evacuation exit setting method. *Journal of Beijing Jiaotong University* 38(6).

# Crude Steel Production and Procurement Joint Decision Modeling Under the National Regulation of Production and Emissions Limitation

**Jingyu Huang, Chang Lin, and Yuehui Wu**

**Abstract** Considering the steel industry rectification goal of improving the converter scrap ratio proposed by Chinese authority recently, a production procurement model is proposed, which integrates procurement, transport, inventory and production cost. The model is aimed at proposing the production scrap ratio decision on different production stages under the requirements of the domestically restricted production and emission limit. Besides, the procurement strategy of key raw materials is determined according to the variable demand speed of raw material, thereby increasing the total profit of the mill's crude steel production. The model is the mixed integer nonlinear programming model (MINLP) which is divided into two stages for solving in the solution stage, the scrap ratio decision stage and the procurement decision stage. The scrap ratio decision stage is solved by Cplex, and the procurement decision stage is solved by the artificial bee colony algorithm (ABC algorithm). The final solution result is reduced by 21.92% compared with the total cost calculated by the classic EOQ model. And based on the results of the ABC algorithm, the rationality and feasibility of the authority to promote the target ratio of converter scrap are verified.

**Keywords** Scrap ratio · Long process steelmaking · Production and procurement decision model · Artificial bee algorithm

---

This paper is sponsored by Natural Science Fund of Shanghai (18ZR1441600).

---

J. Huang · C. Lin (✉) · Y. Wu
Key Laboratory of Road and Traffic Engineering of the Ministry of Education, Tongji University, Shanghai, China
e-mail: linch@tongji.edu.cn

J. Huang
e-mail: huang_jingyu@126.com

Y. Wu
e-mail: wuyuehui88@gmail.com



J. Zhang et al. (eds.), *LISS2019*,
https://doi.org/10.1007/978-981-15-5682-1_15

## 1 Introduction

In the steel manufacturing process, it is divided into two categories: "long process steelmaking" and "short process steelmaking": the long one (converter steelmaking) use iron ore as the main raw material and scrap as the auxiliary material. In 2016, China's long-process refining crude steel was about 760 million tons, accounting for 93.6% of China's total crude steel output in 2016. The main charge material of electric steelmaking is scrap steel, while converter steelmaking is based on economic alternatives to choose scrap or iron ore as the main charge. In order to reflect the using condition of scrap steel in the whole crude steel industry, the scrap ratio is introduced to describe the amount of scrap used in the production of crude steel:

$$scrap\ ratio = \left(\frac{scrap}{scrap + steel\ ore}\right) \times 100\% \quad (1)$$

In the past decade, the global scrap ratio has kept on the level at 35–40%. Among developed countries, the US has the highest scrap ratio, which is around 75%. The ratio in the EU is also high, generally at the level of 55–60%. At present, the scrap ratio in China is still at a low level, which is roughly 11–17% between 2016 and 2018. This is mainly because a complete scrap steel recycling system has not been established within the whole industry (Fig. 1).

As a recycled resource, scrap steel has high economic, environmental and social benefits. China's scrap steel industry has huge development space and a considerable development trend. In recent years, the state's policy of banning "substandard steel" has led to an increase in scrap resources, creating conditions for continuously increasing the scrap ratio produced in China. The quantity of scrap consume should be increased in crude steel manufacture. Factors that affect the scrap ratio in the 13th Five-Year Plan include: (1) Converter scrap ratio: 20% or more in 2020 and more than 30% in 2025. According to the relevant literature, theoretically, the ratio of converter scrap can reach 30–40% [1], so it is considered that there are no technical obstacles; (2) the proportion distribution of electric steelmaking and converter steelmaking: The proportion of electric steelmaking should be increased to the best level in history at the final stage of the 13th Five-Year Plan: The highest record is in 2003, when the output of crude steel is

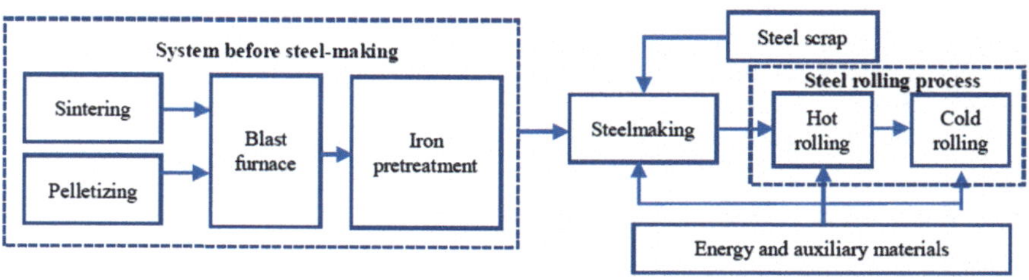


**Fig. 1** Flow chart of long process steelmaking

223.34 million tons, and the output of electric furnace steel is 39.06 million tons, accounting for 17.6%.

Since 2016, limited production and restriction of emissions have been imposed on steel mills with the serious pollution problem, which has driven steel mills to increase the scrap ratio or to select more environmentally-friendly electric steel-making. In February 2017, five red lines have been set up for steel companies by the State Council. Any steel production capacity that does not meet the standards must be withdrawn. The steel production capacity, of which pollutant discharge does not meet the requirement will be punished. In addition, in heating season in some cities, steel production capacity is limited to 50%, which limits the amount of molten iron output of the system before steel-making, prompting many companies to improve scrap ratio which will increase production in another way to compensate for the decline in profits caused by production limit.

This paper focuses on the converter steelmaking process as the background to explore the impact of rising scrap ratio on the upstream cost of the supply chain, including raw material procurement, transportation, inventory and production cost. A mathematical model will be established to derive the appropriate production and procurement decision under the dual objectives of the government's environmental protection requirements and industrial upgrading requirements.

## 2 Literature Review

There are plenty of literature on the strategy of combining inventory and procurement. After Harris proposed the classic economic order quantity (EOQ) model in 1913 [2], many scholars have studied the variants of economic order quantity models [3–4], thus forming a relatively complete inventory system. Kingsman [5] considered the impact of price fluctuations on procurement decisions in raw material procurement and proposed linear programming to solve raw material procurement decisions. Melis [6] studied the price-related EOQ model, which determined the functional relationship between price and production input, and finally determined the economic order quantity. Back to the research on the raw materials procurement decision in the steel industry: due to the large variety of raw materials, large quantity and high cost of steel production, Gao et al. [7] have established the minimization model of procurement cost of raw materials. The model is used to make the raw material procurement plan to determine the variety and quantity of the raw materials. Arnold [8] considered the purchase price, inventory cost and demand change over time in raw material procurement, and believed that the purchase price has a functional relationship with time. Miletic [9] considered the relationship between the production strategy of scrap steel input and market price, purchase and inventory, and established mixed integer programming to solve the problem of electric steelmaking.

The innovation of this paper is to integrate raw material procurement, transportation, inventory, and production costs in the steel industry to propose combination procurement decisions. The literature on procurement decision in the steel industry rarely involves this aspect, and thus draws on relevant papers in other fields. Karimi et al. [10] argued that most mass production models considered dividing a limited planning range into a discrete set of time periods. Multi-Level Capacitated Lot Sizing Problem (MLCLSP) arises. Lee [11] proposed a comprehensive inventory control model that established the joint economic batches ordered by the buyer. Velez and Maravelias [12] review the progress of chemical batch production scheduling solutions technology over the past three decades based on the integration of production planning and procurement strategies. Choudhary [13] solved the problem of purchasing a single product from multiple suppliers in a multi-stage situation with limited production capacity by considering the economies of scale in procurement and transportation costs, and proposed a multi-objective integer linear programming model to solve the problem of a single product in multi-stage inventory lot, supplier selection and shipping tools selection. Cunha [14] considered integrating raw material procurement and production planning, and aimed to reduce procurement and operating costs.

It can be seen that although the integration of production planning and the inventory-purchasing decision is a relatively concentrated and important issue in industrial production, it has not been comprehensively solved in this part. On the one hand, most of the inventory-purchasing model is based on the fact that raw material demand is known (demand is constant or known demand fluctuation). On the other hand, in the production batch model, only some production characteristic parameters are considered, but, the change in production cost caused by different production decisions, the consumption speed of raw materials and fluctuating market price are not taken into consideration. In order to meet the national requirements to increase the scrap ratio of crude steel production, this paper focus on the problem of raw material procurement strategy under variable production scrap ratio.

# 3 Analysis of the Impact of Increasing Scrap Ratio

## *3.1 Impact on Steel Mill Level*

As for the entire steel supply chain, the increase in scrap ratio has less impact on the quality of crude steel. Therefore, the increase of scrap ratio mainly affects the raw material logistics and production process in the upstream of the supply chain. The raw material logistics include procurement, transportation and inventory.

The impacts are mainly reflected in the changes in raw material procurement, transportation, inventory and production costs caused by the change of the raw materials demand, and the changes in production processes to increase scrap steel.

At present, it is considered to add scrap steel in the blast furnace refining process and the converter steelmaking process. Among them, the way adding scrap steel in the blast furnace is to use scrap steel instead of iron ore to produce molten iron, which is divided into two types: pre-iron addition and post-iron addition. The pre-iron addition includes two methods of adding under the blast furnace tank and sintering adding, post-iron addition also including molten iron ladle addition and iron gutter addition two methods. Steelmaking characteristics analysis of each process is shown in Table 1. In addition, it is also possible to directly add scrap steel in the converter. This method can increase the scrap ratio to 16% on the basis of no major changes to the existing steelmaking process and cost.

The scrap ratio makes the cost change. For the single analysis on the influence on raw material logistics and production cost due to the scrap ratio, the control variable method is adopted. When the raw material market price is constant, the transportation mode is fixed, the transportation rate is constant and the production capacity is 10 million tons for a year, the cost changes of various parts caused by the change of scrap ratio are shown in Fig. 2. In general, the overall change in total cost above has increased with the scrap ratio increasing, and the growth rate has gradually increased.

However, considering the demand for limited production of molten iron, it is an inevitable trend for steel mills to increase the scrap ratio for production increasing to make up the decline in profits caused by limited production.

**Table 1** Process comparison list

| Period of adding scrap | Method | Feeding requirement | Workload transform | The upper limit of scrap ratio | Economy |
|---|---|---|---|---|---|
| Before molten iron manufacture | Adding scrap under the blast furnace trough | Unlimited | No additional workload | 20% | No increased cost with increased labor cost and decreased fuel cost |
| | Adding scrap by sintering | The sintering process is the primary process for controlling dust. It isn't recommended in the context of national restrictions | | | |
| After molten iron manufacture | Adding scrap in molten iron ladle | Lightweight scrap | Manual checking | 4% | Increased labor cost |
| | Adding scrap in molten iron gutter | Small size steel grit | Adding scrap is inefficient. So it's used less due to increased labor intensity | | |

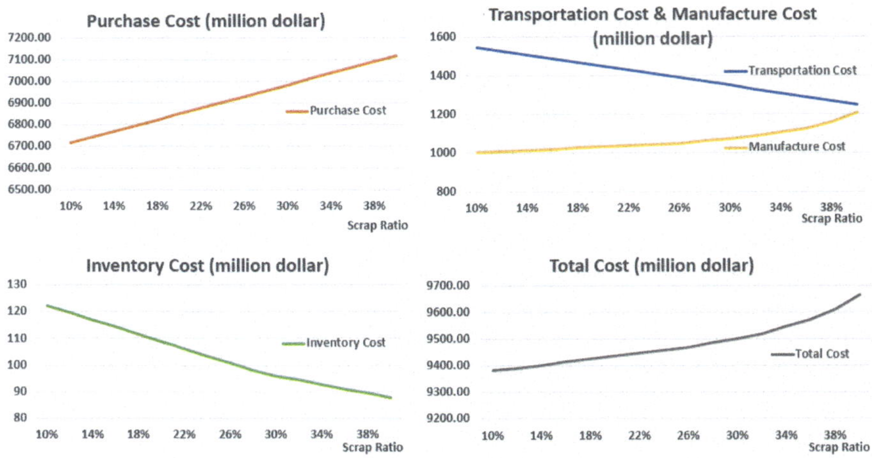


**Fig. 2** Cost change chart of steel mill

## *3.2 Impact on Society Level*

From a social perspective, improving the scrap ratio has high economic, environmental and social benefits, not only energy saving and emission reduction, but also effectively solving the problem of excessive scrap steel resource. According to the calculation data from China Association of Metal Scrap Utilization, compared with iron ore steelmaking, the use of 1 ton of scrap steel can reduce the emission of 1.6 tons of carbon dioxide, reduce 3 tons of solid emissions, and save 1 ton of raw coal in refining molten iron. To sum up, the improvement of the scrap ratio has profound environmental, economic and social benefits on the society, which is also the fundamental reason for the country to increase the scrap ratio. For steel mills, it is necessary to increase the scrap ratio to seek profit maximization under the pressure of the iron-making limit.

# 4 Production and Procurement Decision Model

## *4.1 Model Assumptions and Symbol Description*

(1) The price per ton of crude steel is assumed to be a constant P to ignore the impact of the retail on total profit.
(2) Raw material purchase orders in the spot market are random, and the order arrival time is subject to a normal distribution.
(3) Iron ore quarter pricing purchase contract and monthly pricing purchase contract pricing model is determined, that is, the quarterly price is determined by the average price of China's iron ore market in the previous three months.

Monthly pricing is determined by the average price of China's iron ore market in the previous month. The quarterly contract and monthly contract are signed for one year, and the number of transactions is determined at the time of signing.

(4) When the iron ore arrives at the port through maritime transportation, it is sent to the steel mill only by rail transport and road transport. Since water transportation is relatively small, this mode of transportation is not considered in this model.

(5) Scrap purchase only considers the domestic purchase.

(6) The daily manufacturing output of crude iron is constant, determined by the scrap ratio of the production decision periods, which means the demand speed of each raw material in the same production decision period is fixed.

## 4.2 Symbol Description

(See Table 2).

**Table 2** Symbol description of model

| Symbol | Definition |
|---|---|
| Decision variables | |
| $\gamma_{im}$ | Increased scrap ratio by the method $m$ in the production decision period $i$ |
| $Q_{rkj}$ | The purchase amount of raw material $r$ in the time $j$ by the procurement method $k$ |
| $a_{rkj}$ | The procurement point of raw material $r$ in the time $j$ by the procurement method $k$ |
| Parameters | |
| $i$ | Production decision period, $i = 1, 2, \ldots I$ |
| $r$ | Raw material number, $r = 1, 2, \ldots R, r = 1$ is scrap steel, $r = 2$ is iron ore, $r = 3$ is coke |
| $k$ | Purchasing method of raw material $r$ , $k = 1, 2, \ldots K_r$, Iron ore: $k = 1$ is quarterly pricing contract, $k = 2$ is monthly pricing contract, $k = 3$ is spot market purchase. Scrap:$k = 1$ is Primary crusher, $k = 2$ is secondary crusher |
| $j$ | The procurement of raw material $r$ by the procurement method $k$ , $j = 1, 2, \ldots J_{rk}$ |
| $m$ | Method to increase the scrap ratio, $m = 1, 2, \ldots M, m = 1$ is adding under the blast furnace tank, $m = 2$ is iron ladle addition, $m = 3$ is converter addition |
| $w$ | Emission type, $w = 1, 2, \ldots W$, $w = 1$ is particulates, $w = 2$ is $SO_2$ , $w = 3$ is $NO_x$ |
| $n$ | Number of arrival points of raw material $r$ $n = 0, 1, 2, \ldots N_r$ |
| $T$ | Length of the decision period |
| $\Delta t_i$ | Duration of the production decision period $i$ |
| $C$ | Total cost, $C_P$ is purchase cost, $C_T$ is transportation cost, $C_W$ is inventory cost, $C_M$ is manufacture cost |

(continued)

**Table 2** (continued)

| Symbol | Definition |
|---|---|
| $\alpha$ | The crude steel conversion rate of iron ore |
| $\beta_m$ | The crude steel conversion rate of scrap in the method $m$ |
| $\eta$ | Coke ratio for making molten iron |
| $d_{ri}$ | Demand speed of raw material $r$ in the production decision period $i$ |
| $P$ | Crude steel market price |
| $D_i$ | Production capacity in the production decision period $i$ |
| $D_r$ | The total demand for raw material $r$ |
| $\Phi_m$ | Scrap ratio upper limit in the method $m$ |
| $e_{iw}$ | Unit emissions of emission $w$ in the production decision period $i$ |
| $E_{iw}$ | Emissions upper limit of emission $w$ in the production decision period $i$ |
| $p_{rkj}$ | The unit price of raw material $r$ in the time $j$ by the procurement method $k$ |
| $\Omega_r$ | The fixed purchase cost of raw material $r$ |
| $q_r$ | Emergency order quantity of raw material $r$ |
| $p_r^{'}$ | Emergency order price of raw material $r$ |
| $\varepsilon_{rk}$ | The unit shipping rate of raw material $r$ by the procurement method $k$ |
| $\Omega_r^{''}$ | The transportation batch cost of raw material $r$ |
| $I_r$ | The maximum storage capacity of raw material $r$ in the steel mill's own warehouse |
| $h_1$ | The unit storage fee for steel mill's own warehouse |
| $h_2$ | The unit storage fee for the steel mill's renting a warehouse |
| $I_{max}$ | The maximum storage of raw material $r$ |
| $l_{rik}$ | Lead time for purchasing raw material $r$ in the time $j$ by the procurement method $k$ |
| $s_{rik}$ | The order point of raw material $r$ by the procurement method $k$ in the production decision period $i$ |
| $ll_r$ | The arrival time for an emergency order of raw material $r$ |
| $ss_{ri}$ | Safety stock of raw material $r$ in the production decision period $i$ |
| $b_r$ | $b_r = \{b_{0r}, b_{1r}, \ldots, b_{Nr}, b_{N+1,r}\}$, $N = \sum_{k=1}^{K} J_{rk}, b_{nr}$ is the arrival point of raw material $r$ in the time $k, b_{0r} = 0$ |
| $g_r$ | $g_r = \{g_{0r}, g_{1r}, \ldots, g_{Nr}, g_{N+1,r}\}$, $N = \sum_{k=1}^{K} J_{rk}, g_{nr}$ is the inventory level of raw material $r$ at the point $b_{nr}$ , $g_{0r}$ is the initial stock |

## 4.3 Production-Procurement Decision Model

The objective function of the model is to determine the scrap ratio and the procurement strategy to maximize the total profit of the mill for the production of crude steel:

$$\max F = PD - C \tag{2}$$

(1) Total cost:

$$C = C_P + C_T + C_W + C_M \tag{3}$$

(2) Purchase cost:

$$C_p = \sum_{r=1}^{R} \sum_{k=1}^{K_r} \sum_{j=1}^{J_{rk}} Q_{rkj} p_{rkj} + \sum_{r=1}^{R} N_r \Omega_r + \sum_{r=1}^{R} \sum_{n=1}^{N_r} q_{nr} p_r^{'} \tag{4}$$

(3) Transportation cost:

$$C_T = \sum_{r=1}^{R} \sum_{k=1}^{K_r} \sum_{j=1}^{J_{rk}} Q_{rkj} \varepsilon_r + \sum_{r=1}^{R} N_r \Omega_r^{'} \tag{5}$$

(4) *Inventory cost:* the raw materials for steelmaking are mainly stored in the form of ground stacking. Generally, the mill has a certain area of storage area. When the inventory level exceeds the maximum storage capacity of the own warehouse, the mill needs to pay a certain rent to rent an external warehouse for storage. The inventory inspection strategy adopts a continuous inspection strategy. According to historical data, the lead time is normally distributed $l_{rik} \sim N(\mu_{rik}, \ \sigma_{rik}^2)$. Therefore, the inventory level of the steel company should change as shown in Fig. 3.

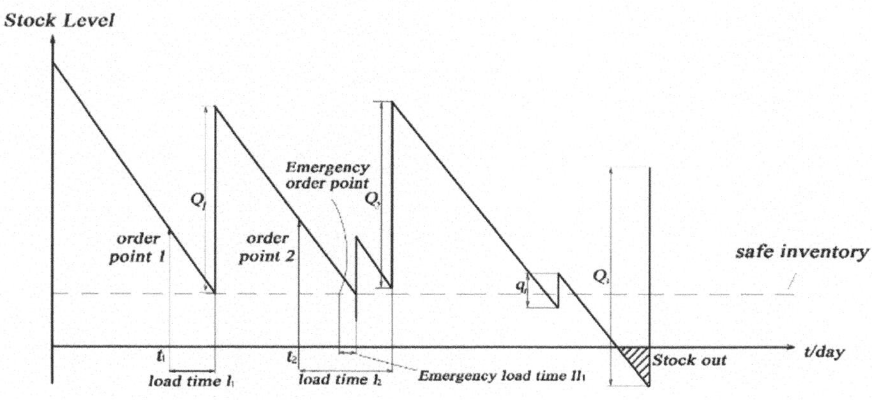


**Fig. 3** Inventory level change chart

$$C_W = \sum_{r=1}^{R} \sum_{n=1}^{N_r} C_{W,nr} \tag{6}$$

$$C_{W,nr} = \begin{cases} \int_{b_{n-1,r}}^{b_{nr}} h_1(g_{n-1,r} - d_{ri}t)dt + h_1 \frac{q_{nr}^2}{2d_{ri}}, g_{n-1,r} \le I_r \\ \int_{b_{n-1,r}}^{\frac{g_{n-1,r}-I_r}{d_{ri}}} h_2(g_{n-1,r} - d_{ri}t)dt + \int_{\frac{g_{n-1,r}-I_r}{d_{ri}}}^{b_{nr}} h_2(I_r - d_{ri}t)dt + h_1 \frac{q_{nr}^2}{2d_{ri}}, g_{n-1,r} > I_r \end{cases} \tag{7}$$

(5) *Manufacture cost:* the labor cost rises due to the increase in the scrap ratio. The fuel cost increases in order to ensure that the scrap quickly reaches the melting point. Through production cost accounting and function fitting, the production cost has the following functional relationship with the scrap ratio:

$$C_M = \sum_{i=1}^{I} (a\gamma_i^3 + b\gamma_i^2 + c\gamma_i + d) \tag{8}$$

$$s.t. \qquad \gamma_i = \sum_{m=1}^{M} \gamma_{im}, \ \forall i \tag{9}$$

$$\gamma_{im} = \frac{D_{1im}}{D_{1i} + D_{2i}}, \forall i, m \tag{10}$$

$$\gamma_{im} \le \Phi_m, \quad \forall i, m \tag{11}$$

$$D_r = \sum_{i=1}^{I} \sum_{m=1}^{M} D_{rim}, \ \forall r \tag{12}$$

$$D = \sum_{i=1}^{I} (\alpha D_{2i} + \sum_{m=1}^{M} \beta_m D_{1im}) \tag{13}$$

$$e_w D_{2i} \le E_{iw}, \forall i, w \tag{14}$$

$$\sum_{k=1}^{K_r} \sum_{j=1}^{J_{rk}} Q_{rkj} \ge \sum_{i=1}^{I} D_{ri}, \ \forall r \tag{15}$$

$$Q_{rkj} = \frac{Q_{rk}}{J_{rk}}, \ r = 2, k = 1, 2, \forall j \tag{16}$$

$$d_{ri} = \frac{D_{ri}}{\Delta t_i}, \ \forall r \tag{17}$$

$$g_{0r} = \int_0^{\mu_{ri}} d_{ri}dt + ss_{ri}, i = 1, \forall r \tag{18}$$

**Table 3** The corresponding value of service level and $z$

| The service level | 90% | 91% | 92% | 93% | 94% |
|---|---|---|---|---|---|
| $z$ | 1.29 | 1.34 | 1.41 | 1.48 | 1.56 |

$$g_{nr} = g_{n-1,r} + Q_{nr} + q_{nr} - \int_{b_{n-1,r}}^{b_{nr}} d_{ri}dt,\ n = 1, 2, \ldots, N_r, N_r + 1, \forall r, I \tag{19}$$

$$b_r = \underset{k \in K_r, j \in J_{rk}}{R} \{a_{rkj} + l_{rik}\},\ \forall r \tag{20}$$

$$g_{nr} \le I_{max},\ \forall r, n \tag{21}$$

$$q_{nr} = \begin{cases} 0, & g_{n-1,r} - \int_{b_{n-1,r}}^{b_{nr}} d_{ri}dt > ss_{ri} \\ t_q d_{ri}, & g_{n-1,r} - \int_{b_{n-1,r}}^{b_{nr}} d_{ri}dt \le ss_{ri} \end{cases}, \forall r, n, I \tag{22}$$

$$ss_{ri} = \min_k z\sqrt{d_{ri}\sigma_{rik}^2}, \forall r, i, k \tag{23}$$

$$s_{rik} = \int_0^{\mu_{rik}} d_{ri}dt + ss_{ri}, \forall r, i, k \tag{24}$$

Equations (9)–(11) define the calculation of the scrap ratio. (12) and (13) define the relationship between raw material demand and capacity during each production decision period. (14) constrains the upper limit of ironmaking emissions during each production decision period. (15) constraints that the total purchase amount of each raw material must meet the production demand. (16) constrains the iron ore quarterly pricing contract and monthly pricing contract purchasing the same quality each time. (17) defines the demand speed of each raw material during the production decision period. (18)–(21) constrain the update mode of inventory level. (22) defines the situation of emergency ordering and the amount of emergency order before each arrival point. (23) defines the safety stock of various types of raw materials stored, and $z$ is the corresponding safety factor under the determined service level. As shown in the following table, the service level of the model is 93%, so the value of $z$ is 1.48 (Table 3).

## 5 Model Solving

Considering that scrap ratio mainly affects the actual production capacity of crude steel to increase the total profit and need to be determined in advance, the model is split into two stages to solve: the scrap ratio decision stage and the procurement decision stage.

**Table 4** Parameter description

| Symbol | Value | Symbol | Value |
|---|---|---|---|
| $P$ | 487 | $E_{iw}$ | $E_{i1} = \{30.39, 30.39, 42.56, 54.73, 54.73, 51.68, 51.68, 54.73, 54.73, 30.39, 30.39, 30.39\}$<br>$E_{i2} = \{49.09, 49.09, 68.76, 88.4, 88.4, 83.49, 83.49, 88.4, 88.4, 49.09, 49.09, 49.09\}$<br>$E_{i3} = \{81.81, 81.81, 114.6, 147.34, 147.34, 139.15, 139.15, 147.34, 147.34, 81.81, 81.81, 81.81\}$ |
| $\gamma_{om}$ | 10% | | |
| $\alpha$ | 0.567 | | |
| $h_1$ | 0.1 | | |
| $\bar{t}$ | 12 | | |
| $\beta_m$ | $\beta_m = \{0.531, 0.531, 0.91\}$ | | |
| $\Phi_m$ | $\Phi_m = \{20\%, 4\%, 16\%\}$ | | |
| $e_w$ | $e_w = \{0.16, 0.26, 0.44\}$ | | |
| $\varepsilon_r$ | $\varepsilon_r = \{33, 36, 47\}$ | | |

**Table 5** The solution results

| Symbol | Value |
|---|---|
| $\gamma_i$ | $\gamma_i = \{30.24\%, 36\%, 16\%, 16.24\%, 16.24\%, 21.12\%, 21.12\%, 16.24\%, 16\%, 16\%, 16\%, 16\%\}$ |
| $\gamma_{im}$ | $\gamma_{i1} = \{14.23\%, 20\%, 0, 0.24\%, 0.24\%, 5.12\%, 5.12\%, 0.24\%, 0, 0, 0, 0\}$<br>$\gamma_{i2} = \{0, 0, 0, 0, 0, 0, 0, 0, 0, 0, 0, 0\}$<br>$\gamma_{i3} = \{16\%, 16\%, 16\%, 16\%, 16\%, 16\%, 16\%, 16\%, 16\%, 16\%, 16\%, 16\%\}$ |
| $D_{ri}$ | $D_{1i} = \{81, 105, 50, 64, 64, 65, 65, 64, 64, 36, 36, 36\}$<br>$D_{2i} = \{187, 187, 262, 337, 337, 318, 318, 337, 337, 187, 187, 187\}$<br>$D_{3i} = \{52, 52, 73, 94, 94, 89, 89, 94, 94, 52, 52, 52\}$ |

## 5.1 The Scrap Ratio Decision Stage

In the case of the same procurement strategy, the cost and profit changes caused by the change of scrap ratio are considered in a single way:

$$max\,\Delta F = P\Delta D - \Delta C \tag{25}$$

Assume that the steel enterprises with a capacity of 30 million tons will make a decision on the ratio of production scraps once a month. The relevant parameters are as follows (Table 4):

The improved scrap ratio decision model is a simple nonlinear plan, which can be directly solved by Cplex software. The solution results are as follows (Table 5).

## 5.2 The Procurement Decision Stage

After determining the ratio of production scrap, the demand speed of each raw material is determined accordingly. The objective function of making purchasing decisions is to minimize the total cost of raw material procurement logistics while meeting production requirements:

$$minC = C_P + C_T + C_I \tag{26}$$

The artificial bee colony algorithm (ABC algorithm) divides the artificial bee colony into three categories by simulating the nectar collecting mechanism of the actual bees: collecting bees, observing bees and scouting bees. The goal of the entire colony is to find the source of the most nectar. The position of each nectar source represents a feasible solution to the problem. The feasible solution is a two-dimensional matrix of n × 3. The first column is used to store the raw material procurement method. The second column of the matrix is used to store the procurement time point of the raw material procurement batch, and column 3 is used to store the raw material lot purchase quantity. The amount of nectar of the nectar source corresponds to the fitness of the corresponding solution. The higher the fitness of the nectar source, the higher the nectar content. The smaller the total cost, the greater the fitness value should be assigned, so the fitness value is solved by the following formula:

$$fitness_{sn} = \frac{1}{C} \tag{27}$$

The collecting bees search for other values of a certain range of in each fitness_ {{sn}} of the n-dimensional feasible solution to obtain a better solution and use greedy selection strategy to retain better solution (Fig. 4). Each observing bee uses roulette to select a nectar source based on probability:

$$p_i = \frac{fitness_{sn}}{\sum_{sn=1}^{SN} fitness_{sn}} \tag{28}$$

$fitness_{sn}$is the fitness value of the feasible solution. For the selected nectar source, the observation bees searche for a new possible solution based on the above probability formula and its update way is shown in Fig. 5.

**Algorithm steps:**

**STEP 1:** Form an initialization nectar source;

**STEP 2:** Update the honey source information according to the above first method, and determine the amount of nectar of the nectar source;

**STEP 3:** Observation bees to select the nectar source according to the information provided by collecting bee, and update the honey source information according to the second method above, and determining the amount of nectar of the honey source;

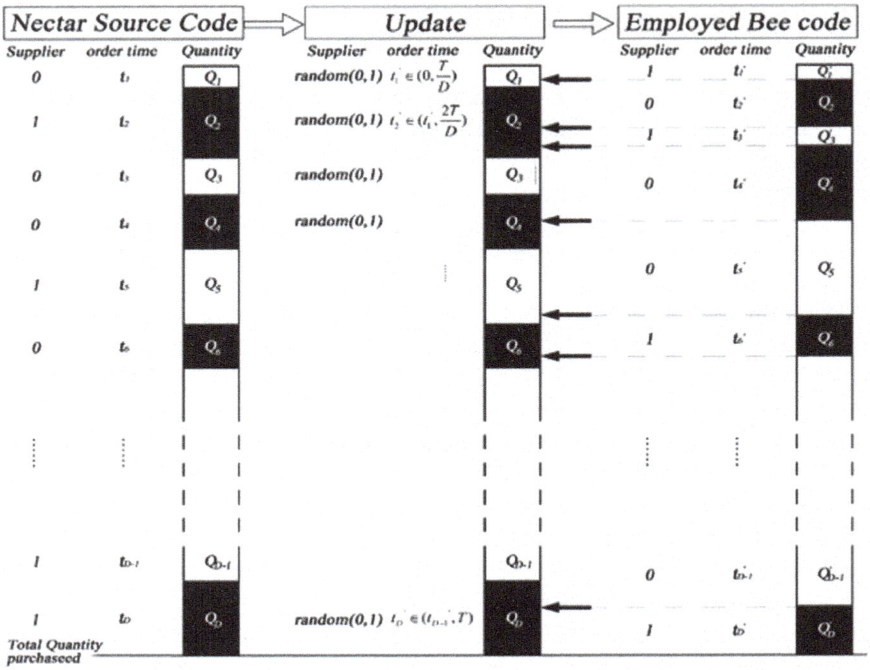


**Fig. 4** The collecting bees update chart

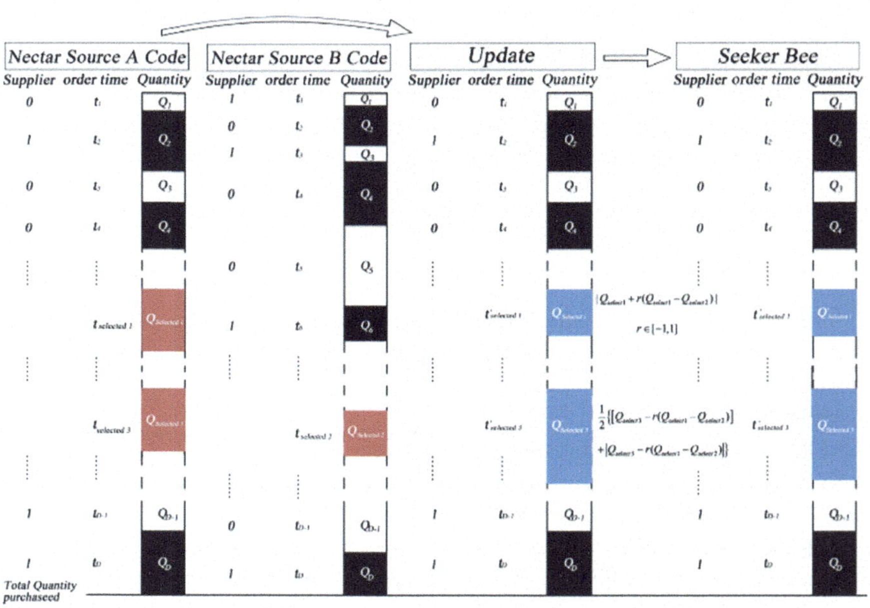


**Fig. 5** The observation bees update chart

| | | | | | | | | | | | | | | | | | | | |
|---|---|---|---|---|---|---|---|---|---|---|---|---|---|---|---|---|---|---|---|
| **Iron Ore Purchase: (Supplier no.1: Quarterly Pricing, Supplier no.2: Monthly Pricing, Supplier no.3:Spot Market)** | | | | | | | | | | | | | | | | | | | |
| Order Point | 4 | 9 | 20 | 31 | 33 | 40 | 52 | 57 | 63 | 68 | 80 | 84 | 88 | 100 | 112 | 116 | 122 | | |
| Supplier No. | 1 | 3 | 2 | 3 | 2 | 2 | 2 | 1 | 3 | 1 | 2 | 3 | 1 | 3 | 3 | 1 | 2 | | |
| Quantity Purchased | 137 | 84 | 137 | 47 | 135 | 156 | 98 | 87 | 59 | 100 | 117 | 53 | 96 | 77 | 67 | 85 | 109 | | |
| Order Point | 125 | 136 | 141 | 145 | 153 | 157 | 162 | 168 | 174 | 184 | 188 | 196 | 203 | 213 | 214 | 241 | | | |
| Supplier No. | 3 | 1 | 3 | 1 | 3 | 2 | 1 | 2 | 1 | 3 | 1 | 2 | 1 | 1 | 3 | 3 | | | |
| Quantity Purchased | 52 | 137 | 53 | 149 | 63 | 79 | 191 | 93 | 103 | 57 | 154 | 77 | 113 | 120 | 56 | 40 | | | |
| **Scrap Purchase** | | | | | | | | | | | | | | | | | | | |
| Order Point | 5 | 15 | 29 | 36 | 53 | 67 | 73 | 85 | 103 | 113 | 128 | 132 | 155 | 158 | 179 | 189 | 213 | | |
| Quantity Purchased | 65 | 62 | 57 | 66 | 26 | 23 | 35 | 46 | 43 | 56 | 28 | 77 | 0 | 59 | 38 | 42 | 52 | | |
| **Coke Purchase** | | | | | | | | | | | | | | | | | | | |
| Purchase Period | 10 | 18 | 33 | 41 | 51 | 68 | 76 | 91 | 104 | 114 | 125 | 135 | 146 | 159 | 172 | 190 | 197 | 207 | 221 |
| Quantity Purchased | 42 | 38 | 47 | 37 | 68 | 53 | 66 | 49 | 58 | 52 | 42 | 49 | 52 | 45 | 56 | 38 | 35 | 21 | 39 |

| | | | |
|---|---|---|---|
| Total cost | 1.57E+07 | Quarterly Pricing Rate | 46.27% |
| Purchase cost | 1.53E+07 | Monthly Pricing Rate | 31.47% |
| Transportation cost | 3.11E+05 | Spot Market Rate | 22.26% |
| Inventory cost | 8.60E+04 | | |

**Fig. 6** Procurement decision result

**STEP 4:** Determine if a detection bee is generated. If it isn't, skip to **STEP 6**; if it is, go to **STEP 5**

**STEP 5:** Find a new source of nectar according to the first method by using the best nectar source has been found;

**STEP 6:** Remember the best nectar source information;

**STEP 7:** Determine whether the termination condition is true. If not, return to **STEP 2**; if it is, output the solution of the optimal nectar source.

After the scrap ratio at each production decision stage known, the bee colony algorithm is used to solve the raw material procurement decision. The results are shown in Fig. 6.

The solution results show that iron ore procurement adopts quarterly, monthly pricing contracts and market spot procurement in parallel. Quarterly pricing contract accounts for 46%, monthly pricing contract 31%, and market spot purchase 22%. Compared with the single-mode procurement, the hybrid procurement method has the advantages of flexibility and high reliability. By analyzing the relationship between the historical market price trend and the proportion of each procurement method, it is possible to determine the future procurement method based on the forecast of future raw material price trends to reduce the cost and risk of raw material procurement. Besides, the total cost of the procurement decision proposed in this paper is 21.92% lower than the total cost calculated by the classic EOQ model.

## 6 Conclusion

In summary, the model uses the production background of 2017 as an example to calculate the best production scrap ratio that maximizes profits in the context of national production limits. The results show that the heating season with the highest limit, the scrap ratio of production is up to 36%. The annual scrap ratio is nearly

20% which fully demonstrates that the goal of the National 13th Five-Year Plan about converter scrap ratio to 20% is feasible and in accordance with steel enterprises' actual pursuit of capital reduction. In addition, the proposed raw material procurement decision can further reduce the logistics cost of raw material, thus achieving a further increase in steel enterprise profits.

**Acknowledgment** The authors confirm contribution to the paper as follows: Modeling and algorithm design: Jingyu Huang, Chang Lin; data collection: Jingyu Huang, Chang Lin; analysis and interpretation of results: Jingyu Huang, Yuehui Wu; draft manuscript preparation: Jingyu Huang. All authors reviewed the results and approved the final version of the manuscript.

## References

1. Dai, Y., & Wang, W. (1994). Converter's energy consumption and raising of scrap return ratio. *Journal of Northeastern University, 04,* 384–389.
2. Harris, W. (1913). How many parts to make at once. *The Magazine of Management, 10*(2), 135–136.
3. Hariga, M., & Haouari, M. (1999). An EOQ lot sizing model with random supplier capacity. *International Journal of Production Economics, 58*(1), 39–47.
4. Bertazzi, L. (2003). Rounding off the optimal solution of the economic lot size problem. *International Journal of Production Economics, 81*(82), 385–392.
5. Kingsman, B. G. (1986). Purchasing raw materials with uncertain fluctuating prices. *European Journal of Operational Research, 25*(3), 358–372.
6. Melis Teksan, Z., & Geunes, J. (2016). An EOQ model with price-dependent supply and demand. *International Journal of Production Economics, 178,* 22–33.
7. Gao, Z., & Tang, L. X. (2003). A multi-objective model for purchasing of bulk raw materials of a large-scale integrated steel plant. *International Journal of Production Economics, 83*(3), 325–334.
8. Arnold, J., Minner, S., & Eidam, B. (2009). Raw material procurement with fluctuating price. *International Journal of Production Economics, 121*(2), 353–364.
9. Miletic, I. (2007). Model-based optimization of scrap steel purchasing. *IFAC Proceedings Volumes, 40*(11), 263–266.
10. Karimi, B., Ghomi, S. M. T. F., & Wilson, J. M. (2003). The capacitated lot sizing problem: a review of models and algorithms. *Omega, 31*(5), 365–378.
11. Lee, W. (2005). A joint economic lot size model for raw material ordering, manufacturing setup, and finished goods delivering. *Omega, 33*(2), 163–174.
12. Velez, S., & Maravelias, C. T. (2014). Advances in mixed-integer programming methods for chemical production scheduling. *Annual Review of Chemical and Biomolecular Engineering, 5,* 97–121.
13. Choudhary, D., & Shankar, R. (2014). A goal programming model for joint decision making of inventory lot-size, supplier selection and carrier selection. *Computers & Industrial Engineering, 71,* 1–9.
14. Cunha, A. L., et al. (2018). An integrated approach for production lot sizing and raw material purchasing. *European Journal of Operational Research, 269*(3), 923–929.

# Trends and Challenges of Data Management in Industry 4.0

**Eduardo A. Hinojosa-Palafox, Oscar M. Rodríguez-Elías, José A. Hoyo-Montaño, and Jesús H. Pacheco-Ramírez**

**Abstract** Trends and challenges of data management are presented in the context of Industry 4.0 to know the impact that is being generated by the development of new models and architectures that consider the Internet of Things, Cloud Computing and Big Data in its different levels of integration to allow intelligent analytics. To achieve this purpose, we developed a research protocol that follows the guide of systematic literature mapping. With this base, we elaborated an industry 4.0 classification that considers the life cycle of the data. The results show that Big Data in Industry 4.0 is in its infancy, so few proposals for prescriptive analytics have been developed. Based on the evidence found, we believe that it is necessary to align technology, modeling, and optimization under a methodology focused on data management.

**Keywords** Industry 4.0 · Internet of Things · Cloud computing · Big data

## 1 Introduction

New digital technologies in industries such as agri-food, logistics, and manufacturing are allowing humans, machines, products, and resources to exchange information among themselves. The industry is migrating from a traditional

E. A. Hinojosa-Palafox (✉) · O. M. Rodríguez-Elías ·
J. A. Hoyo-Montaño
DEPI/I. T. de Hermosillo Tecnológico Nacional de México, Hermosillo, Mexico
e-mail: ehinojosa@ith.mx

O. M. Rodríguez-Elías
e-mail: omrodriguez@ith.mx

J. A. Hoyo-Montaño
e-mail: jhoyo@ith.mx

J. H. Pacheco-Ramírez
Depto. de Ingeniería Industrial, Universidad de Sonora, Hermosillo, Mexico
e-mail: jesus.pacheco@unison.mx



J. Zhang et al. (eds.), *LISS2019*,
https://doi.org/10.1007/978-981-15-5682-1_16

approach to one wherein a machine is not only limited to produce but has to do so in an intelligent and energy efficient manner, it must also be able to provide information on the process to various ranges of the hierarchy of the organization [1]. This new approach known as Industry 4.0 marks an important milestone in industrial development and expresses the idea that we are at the beginning of a fourth industrial revolution [2]. Its foundation is that with the connection of machines, systems and assets organizations can create intelligent networks along the value chain to control production processes autonomously. In this new scenario, the focus is not only on new technologies but also on how they combine considering three levels of integration from the perspective of the data [3].

The Internet of Things (IoT): Consists in creating networks of physical objects, environments, vehicles, and machines through integrated electronic devices that allow the collection and exchange of data. The systems that operate in the IoT are equipped with sensors and actuators, cyber-physical systems, and are the basis of Industry 4.0 [4].

Industrial Cloud Computing: we can understand cloud computing in a general way [5] as a focus on the use of computing resources (hardware and/or software) accessed at will by contracting services to third parties. The main focus of industrial cloud computing is vertical integration and vertical solutions instead of horizontal ones, which is the focus of general cloud computing, this means that the industrial cloud solutions focus on creating more value within industry boundaries rather than expanding its breadth.

Big data and analytics: In Industry 4.0 [6], data contexts are generated by various sources, such as machine controllers, sensors, manufacturing systems, among others. All this volume of data, which arrive at high speed and different formats is called big data. Then, analytics is the processing of big data to identify useful ideas, patterns or models; is the key to sustainable innovation in Industry 4.0. There are four types of analytics that can be applied to the industry 4.0: Descriptive, Diagnostic, Predictive and Prescriptive.

Descriptive analytic. Production processes are often complex and involve thousands of components from a large number of suppliers. The descriptive analysis is the most basic or preliminary and through reports on historical data seeks to inform what is happening.

Diagnostic analytics. Big data where house are used in business intelligence for modeling data multidimensionally to do OLAP analysis and take to the correct level of aggregation to inform and observe, and shows how big or small the problem is, so we can understand what happened.

The predictive analysis. Big data keeps the data of the origins of IoT events and other sources and makes them available for further processing, using it for predictive modeling based on big data mining methods to provide forecasts to answer the question: what could happen?

The prescriptive analytic. It uses the relevant data coming from IoT devices by allowing the analysis of information in real time while offering means to execute previously trained optimization models, using machine learning algorithms and data science. Therefore, we can understand how to be prepared and how to handle it.

Due to all the above, the scientific community has awakened a significant and unstoppable activity to develop new architectures and models for data management in Industry 4.0. The map systematic literature review allows discovering intersections in the current trends that can be used to know the possibilities in the potential development and future tendency. The purpose of this paper is to find out what proposals exist for data management in Industry 4.0 and what their trends and challenges are. This paper is structured as follows: Sect. 2 describes the protocol used in this research work. In Sect. 3 we present the results that respond to the objective and research question. In Sect. 4, we show the conclusion of this article, also future work based on the opportunities found to fill gaps in knowledge.

## 2 Methodology

The review protocol was developed based on the guide for a map systematic literature review as proposed in [7], and describes the research questions (and objectives), inclusion/exclusion criteria, databases and search engines, search terms, extraction of content and relevant data, evaluation of the quality of these results, and gathering of the most outstanding results for analysis.

### *2.1 Related Work*

We developed this research to provide a report of the data management in the different contexts of the technologies of the industry 4.0. Also, to try to extract suggestions on what type of data model helps to build architectures whose nuclear component is the data management for the creation of value.

However, we have been able to find forty literature reviews that address different research topics related to big data, the internet of things, cloud computing and machine learning. In Table 1 we can find a recount by the methodology used for the literature review. The main topics addressed are trends in research, security, tools, related technologies and different domains of application.

Nevertheless, only four of these papers are related to this research: The [1]'s literature review is concerning to the areas in manufacturing where big data research is focused and the outputs from these research efforts. The purpose of [8] is

**Table 1** Review articles related to big data topics

| Types of review papers | Papers |
|---|---|
| Map and systematic literature review | 15 |
| Literature review | 12 |
| Survey | 13 |
| Total | 40 |

to find out how open source is used in the Industrial Internet of Things and how the usage of open source has evolved throughout the years. [9] focuses on a review of recent fault diagnosis applications in manufacturing that is based on several prominent machine learning algorithms, [10] reviews the current research of major IoT applications in industries and identifies research trends and challenges.

## 2.2 Objective and Research Question

The purpose of this paper is to identify elements of data management in industry 4.0. The research aims to answer the research question: What proposals exist for data management in the context of Industry 4.0?

## 2.3 Search Strategy

cWe search for scientific articles in electronic databases that were accessible online. The type of document was limited to conference publications, research journals, and papers in digital format. To find keywords and their synonyms more appropriate to the goals of a map systematic literature, a preliminary search based on a literature sample was made. We used different combinations of keywords in the search string as shown in Table 2. The total number of scientific papers found was 610.

**Table 2** Search string and articles by source of consultation

| Source | Search string | Papers |
|---|---|---|
| ACM | recordAbstract:("Data model " AND ("Big Data" OR "machine learning" OR "Cloud Computing" OR "Internet of Things")) | 77 |
| Google Scholar | "Data management model " and ("Big Data" or "Cloud Computing" or "Internet of Things" | 335 |
| IEEE | (("Data model ") AND ("Big Data" OR "machine learning" OR "Cloud Computing" OR "Internet of Things"))<br>Filters Applied: Journals & Magazines | 26 |
| ISI Web of Science | (TS = (("Data model ") AND ("Big Data" OR "machine learning" OR "Cloud Computing" OR "Internet of Things")))<br>AND Document type: (Article) | 51 |
| Science@Direct | ("Data model ") AND ("Big Data" OR "machine learning" OR "Cloud Computing" OR "Internet of Things") | 89 |
| CONRICyT | (("Data management model ") AND ("Big Data" OR "machine learning" OR "Cloud Computing" OR "Internet of Things")) | 32 |

**Table 3** Accepted and rejected items by inclusion and exclusion criteria

| Inclusion criteria | | Exclusion criteria | |
|---|---|---|---|
| It clearly shows a contribution to data management in Industry 4.0 | 24 | It is not related to the research question | 358 |
| It presents the challenges or trends in the cloud, Big Data or Internet of Things in the context of Industry 4.0 | 24 | It only refers to "cloud computing", Internet of Things ", or " big data "as a reference | 17 |
| It focuses on models in the cloud, big data or the internet of thing | 79 | It is not completely available | 51 |
| It includes a real case or a case study | 47 | It is a research in progress, or conclusions are not available in the document | 6 |

## 2.4 *Inclusion and Exclusion Criteria*

The objective of the definition of selection criteria was to find all the relevant published literature. We applied the inclusion criteria for each search that was carried out in the query source (see Table 3). In each result includes title, author, summary, year of publication and keywords and we read the summary for each one. If the paper was relevant for the study, then we did an additional evaluation to do a complete reading. When we applied the inclusion and exclusion criteria to the 610 selected papers, we accepted 174 and rejected 436. Most papers accepted are from specialized journals (124), and 50 are from conferences.

## 2.5 *Quality Evaluation*

Quality was evaluated to verify the relevant aspects in the map systematic literature according to the research objectives through a checklist for each document (174).

A quality assessment checklist consists of three parts (see Table 4): a list of questions, a list of predefined answers, a cutoff score.

The maximum score is 100 and the minimum evaluation to accept an article was 90. Forty-four articles meet the minimum quality assessment.

## 2.6 *Data Extraction*

Once completed screening items, we extracted the relevant information. The data extraction forms are useful for extract data from the selected papers and analyze them later (see Table 5). Additionally, after this level of detailed analysis, some studies considered irrelevant were excluded, so that thirteen main papers for this research were finally accepted [3, 4, 6, 11–20].

**Table 4** Quality criteria form

| Question | Answer | | |
|---|---|---|---|
| Is the topic covered in the research relevant to the systematic review? | High<br>20 | Medium<br>15 | Low<br>10 |
| Are the objectives of the research specified clearly? | High<br>20 | Medium<br>15 | Low<br>10 |
| Does the data model bring novelty in the context of industry 4.0? | High<br>20 | Medium<br>15 | Low<br>10 |
| Is there a description of the characteristics of the data model or the implementation of the architecture in the research work? | High<br>20 | Medium<br>15 | Low<br>10 |
| Has the data model been validated on a reliable scale (either in the academy or in the industry)? | High<br>20 | Medium<br>15 | Low<br>10 |

**Table 5** Data extraction form

| Description | Values |
|---|---|
| Country | Undefined |
| Industry Classification | Agriculture, Smart City, Construction, Health Care, Energy, General, Geospace, Industry 4.0, Smart grid, Transportation, Not applicable |
| The scope of the data model | Big Data, Combined, Cloud Computing, Industry 4.0, Internet of things, None |
| Contribution | Architecture, Framework, Tool, Methodology, Model, Platform, Processes, Theory |
| Big Data | Analytics, Computing Infrastructure, Data, Not Applicable, Security and Privacy, Storage Infrastructure, Visualization |
| IoT | Architecture, General Challenges, Hardware, Healthcare, Not applicable, Security and privacy challenges, Intelligent infrastructure, Social applications, Software, Supply/logistics chains |
| Industry 4.0 | Cyber-Physical Systems (CPS), Interoperability, Machine to Machine (M2M), Intelligent Factory (Product/Service), Not applicable |
| Synthesis | Undefined |
| Paradigm | Undefined |

## *2.7 Resume of the Paper Selection Process*

Figure 1 shows the search and selection process, which includes multiple steps and was carried out according to the protocol described above.

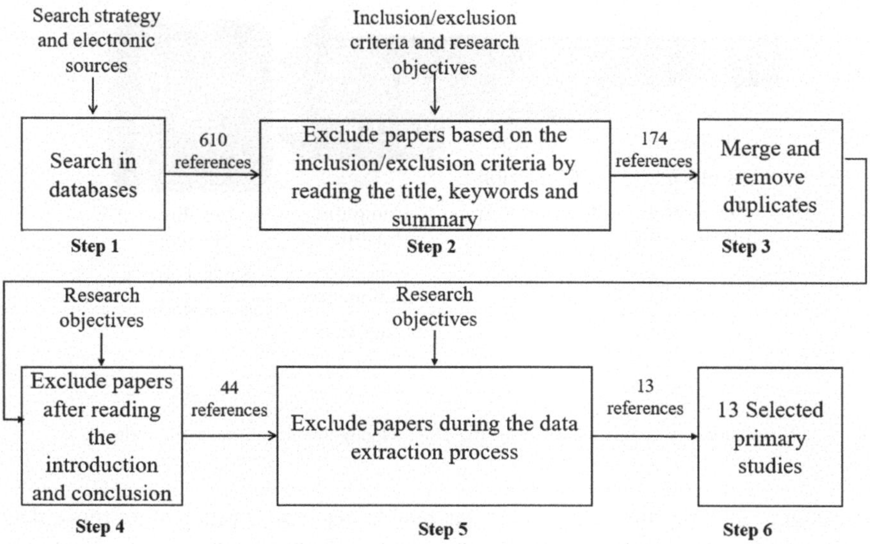


**Fig. 1** The process of paper selection

# 3 Results

## *3.1 Convergence of Technologies in Data Management for Industry 4.0*

To understand more clearly the life cycle of the data in Industry 4.0, based on the literature reviewed, we elaborated the classification shown in Fig. 2. We classified The IoT category according to [21], and we divided into the architecture subcategory which, in turn, can be hardware/networks, software, and processes. The hardware subcategory includes RFID, NFC and sensor network technologies; the subcategory of software includes middleware (information exchange logic) and device search/navigation. Finally, intelligent infrastructure integrates smart objects into a physical infrastructure that can provide flexibility, reliability, and efficiency in the operational infrastructure.

We developed a classification regarding the cloud, based on [22]. In storage, the cloud is used to store big data from the IoT, in the connector's category, cloud services are used to integrate different data sources. Finally, apps for analytics use big data analysis as a cloud service (Analytics as a Service, AaaS).

Big data category was developed based on [5]. In the big data subcategory, there are papers focused on the process ETL (extraction, transformation, and load) from IoT (sensors, RFID). In the subcategory of computing is found batch processing, data transmission processing, and real-time data processing. In the storage architecture subcategory, the studies can address proposals to store data in structured (SQL) or unstructured (NoSQL) databases and big data warehouse. The analytical

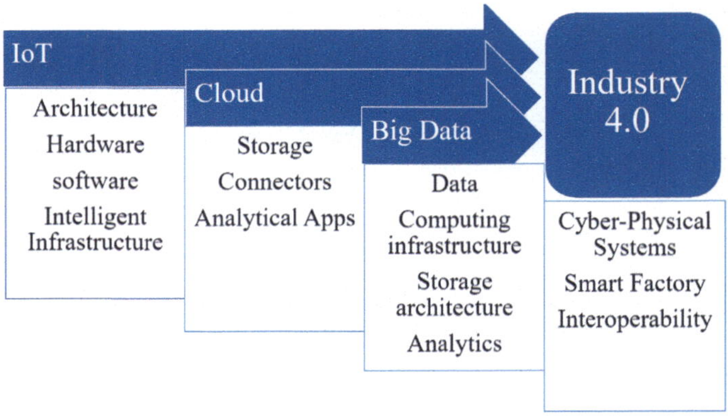


**Fig. 2** Classification of technologies from the Data Management approach for Industry 4.0

subcategory includes the components that enable machine learning algorithms, data mining, and advanced data visualization.

We classified Industry 4.0 according to [23]. The cyber-physical systems integrate computation, networks, and physical processes. The smart factory subcategory seeks to make manufacturing processes more flexible, and the subcategory interoperability is the transparent intercommunication between systems, people and information in cyber-physical systems allowing exchange information between machines, processes, interfaces, and people.

## *3.2 The Scope of Data Management in Industry 4.0*

To gain insight into the current state of management and organization of data on industry 4.0 convergence, we analyzed the proposals that exist for industry 4.0 and in particular, emerging technologies that comprise the IoT, Big Data and cloud, as well as their possible combinations (see Fig. 3).

With the development of growing IoT technologies, different management approaches have emerged, which is why different architectures for sensor clouds have been proposed in recent years.

For [11] cloud sensor architecture must implement in data: security, speed, and reliability by providing adequate data processing and management services. For [12] a software-based model for managing data in the IoT is based on an architecture for controlling the internet of things in three layers: the sensor cluster layer, the middleware layer that manages the data, and the application layer that considers data as services.

Regarding the Cloud and industry 4.0, the authors in [13] propose a service as a platform (SAAP) called Hana to capture manufacturing information from IoT with

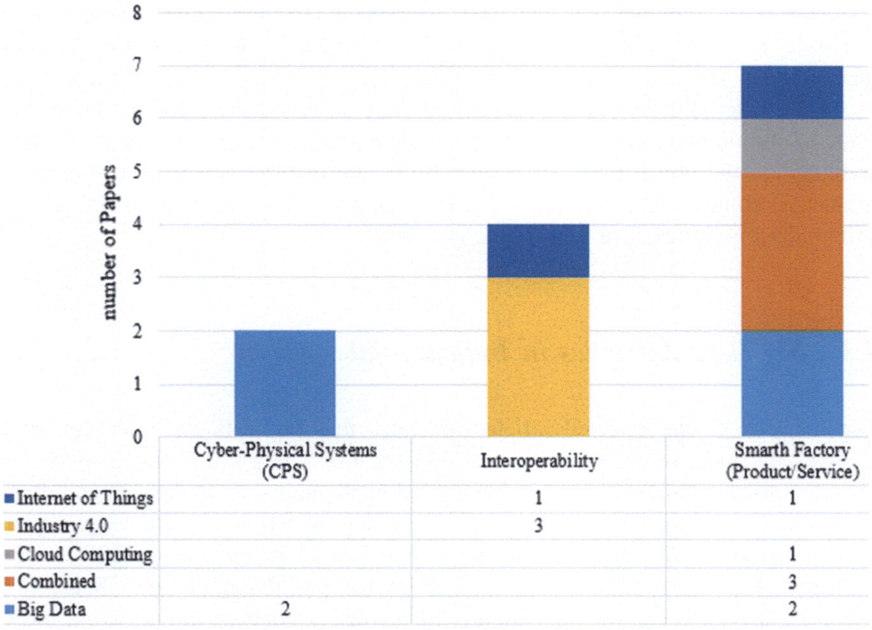


**Fig. 3** Data Management Classification papers in Industry 4.0

technologies for real-time analytics that can be integrated with enterprise resource planning systems (ERP).

Currently, there is a tendency to use Big Data in Industry 4.0, in this sense in [14] it provides a big data engine defined for big data industrial analysis models, which processes data in parallel to improve performance, using in one implementation of the map-reduce computation based on a hierarchical model. In [3] they present the multidimensional model for data analysis in a big data warehouse implemented in Hive, an open source Apache project, used as a tool to create warehouses for structured datasets and support decision making. In [6] they publish an architecture for big data analytics developed for industry 4.0 that supports the collection, integration, storage, and distribution of data designed to consider the volume, variety, and speed of the data that may come from different processing needs, different end users and their roles in the decision-making process. For [15] the development of a big data system is different from the development of a small data system, and they propose a first attempt for a methodology that combines architectural design with data modeling.

We refer to a combination when the IoT, big data and the cloud are integrated in any of its possible combinations to respond to a problem. In this sense, in [3] they propose a data architecture based on five layers to integrate sensors, actuators, networks, cloud computing and IoT technologies for the generation of applications for industry 4.0. To maintain persistence between layers, they consider a data response layer that manages the data. Reference [16] the COIB framework is proposed to integrate big data with IoT to implement an architecture to data management.

For Industry 4.0 the authors in [17] propose an architecture for the entire life cycle of aluminum industry 4.0. This theoretical proposal integrates into a six-layer architecture, the physical sensor system (IoT), the platform for data management (industrial cloud), and a model for big data analytics that allows decision making and enables monitoring of real-time applications.

## *3.3 Big Data Analytics in Industry 4.0*

The analysis of the selected articles indicates that big data analytics currently receives the most research interest in industry 4.0. The process and planning in manufacturing with diverse interdepartmental applications, maintenance, and diagnosis present the challenge of prediction accuracy, which is a desirable quality in decision-making. We can attribute the importance of predictive analytics to the presence of theories and methods related to the prediction of other fields (for example, statistics) and the applicability of predictive analysis to real-world problems. On the other hand, the lack of prescriptive analytics implementations is evident from the results. Therefore, we can associate this by the challenge of developing applications of prescriptive analytics; they are inherently complex compared to descriptive and predictive analytics, given the need to align technology, modeling, forecasting, optimization, and experience in the field. Therefore, given that the big data area in the industry is still in its infancy, it is not surprising that there are only a few applications of prescriptive analytics that have been developed.

## *3.4 Analytics Privacy*

Machine learning for big data is the typical services that companies tend to outsource in the cloud, due to their nature of intensive data use and the complexity of the algorithm [4]. What is attractive in the industry is relying on external experience and infrastructure to calculate the analytical results and models that data analysts require to understand the processes under observation. Although it is advantageous to achieve a sophisticated analysis, there are several privacy problems in this paradigm [18], one of them is that the cloud servers have access to valuable data from the industry and they can potentially obtain confidential information from it [19]. This relevant information security problem is known as corporate privacy [5]. Unlike personal privacy, which only considers the protection of personal information recorded about individuals, corporate privacy [20] requires that both the individual elements and the data collection pattern, like corporate assets, should be protected.

## 4 Discussion

We have not found a contribution that considers data management as a core component in the design of a methodology to allow integrating the whole life cycle of data in Industry 4.0 into a management model and an architecture that allows creating applications, which opens the possibility of developing future work that considers the elements described here.

We found the following significant future research aspects:

The integration of the demanding IoT environment with cloud computing including the challenges of velocity, volume, variety, and veracity, challenges the design of big data analytics systems. That is, it is necessary to describe a new design process that reflects the change required for the development of big data analytics systems that adapt to the paradigm shift of industry 4.0. The determination of architectural styles or patterns, or instead, reference architectures, facilitate the design of solutions to problems that have common aspects or characteristics.

We consider that a novel model is necessary that takes into account the paradigm changes in the convergence of technologies into industrial cyber physical systems and that, unlike the presented proposals, underlines the importance of a focused design approach in the data model that allows creating architectures that are the basis for the development of big data analytics architectures systems applied in industrial contexts.

Recently anomaly detection in real-time analysis has enabled a novel way to optimize systems [24, 25], not just machines to unknown anomalies, helping industrial analysts and operators in the resolution of possible invisible problems.

Finally, we want to acknowledge that a limitation of our current research could be the absence of a mathematical model to describe, for instance, the trends in data management in the context of industry 4.0, or for proposing a specific approach for data management in such a context. Although the field of each of the technologies that have been integrated into the so called industry 4.0 have much work as separated fields, their integration is an emerging area in research and practice, therefore, it is still difficult to analyze current work statistically. As we have shown, our initial search found 610 works, but after a systematic and carefully analysis, only 13 fulfilled the criteria for the research protocol, so we focused on providing a descriptive analysis.

Nevertheless, we firmly believe that the results presented in this paper could have significant practical application either for research or practice. For instance, as a result of our work, we have provided a framework for classifying data extracted from research papers in the field (see for instance Sect. 2.6), that can be used for performing future statistical analysis of a wider review of works related to the filed. As well, we have identified and discussed some important areas that require more work, so it could guide future developments and research.

## 5 Conclusions

The study presented in this paper provides a map systematic literature review as proposed in [7] to identify the elements that structure data management in industry 4.0. We developed a classification of the technologies that converge in Industry 4.0 based on the paper reviewed. For IoT are considered architectural elements, hardware, software, and intelligent infrastructure. For cloud computing, we include storage, connectors, and apps for analytics. For Big Data, the elements considered are data, computing infrastructure, storage architecture, and analytics. In Industry 4.0 we consider cyber-physical systems, interoperability and smart factory.

The analysis of the selected papers indicates that big data analytics currently receives the most research interest in the context of data management in Industry 4.0. The process and planning in manufacturing with diverse interdepartmental applications, maintenance, and diagnosis present the challenge of prediction accuracy, which is a desirable quality in decision making.

This work yields the foundations to develop a future methodology to allow integrating the entire life cycle of data within an Industry 4.0 data management model.

**Acknowledgment** Partially supported by "CONACYT with grant No. 890778".

## References

1. Ochs, T., & Riemann, U. (2017). Smart manufacturing in the internet of things era. In *Internet of Things and Big Data Analytics Toward Next-Generation Intelligence* (pp. 199–217). Cham: Springer.
2. Skilton, M., & Hovsepian, F. (2017). The 4th industrial revolution : responding to the impact of artificial intelligence on business. Springer.
3. Santos, M. Y., Martinho, B., & Costa, C. (2017). Modelling and implementing big data warehouses for decision support. *Journal of Management Analytics, 4*(2), 111–129.
4. Madakam, S., Ramaswamy, S., & Tripathi, R. (2015). Internet of Things (IoT): a literature review. *Journal of Computer and Communications* 3, 164–173.
5. Vora, R., Garala, K., & Raval, P. (2016). An era of big data on cloud computing services as utility: 360 of review, challenges and unsolved exploration problems. In *Smart Innovation, Systems and Technologies. Proceedings of First International Conference on Information and Communication Technology for Intelligent Systems, Ahmedabad, India* (vol. 2, pp. 575–583).
6. Santos, M. Y. et al. (2017). A big data analytics architecture for industry 4.0. In *Advances in Intelligent Systems and Computing, Madeira, Portugal* (vol. 570, pp. 175–184).
7. Kitchenham, B., & Charters, S. (2007). Guidelines for performing Systematic Literature reviews in Software Engineering.
8. Helmiö, P. (2017). Open source in industrial Internet of Things: A systematic literature review.
9. Ademujimi, T. T., Brundage, M. P. & Prabhu, V. V. (2017). A review of current machine learning techniques used in manufacturing diagnosis. In *IFIP Advances in Information and Communication Technology, Hamburg, Germany* (vol. 513, pp. 407–415).

10. Da Xu, L., He, W., & Li, S. (2014). Internet of things in industries: A survey. *IEEE Transactions on Industrial Informatics, 10*(4), 2233–2243.
11. Atif, M. A., & Shah, M. U. (2017). OptiSEC: In search of an optimal sensor cloud architecture. In *2017 23rd International Conference on Automation and Computing (ICAC)* (pp. 1–6).
12. Jararweh, A., Al-Ayyoub, Y., Benkhelifa, M., Vouk, E., & Rindos, M. (2015). SDIoT: a software defined based internet of things framework. *Journal of Ambient Intelligence and Humanized Computing, 6*(4), 453–461.
13. Stankevichus, I. (2016). Data Acquisition as Industrial Cloud service. In Jamk.
14. Basanta-Val, P. (2018). An efficient industrial big-data engine. *IEEE Transactions on Industrial Informatics, 14*(4), 1361–1369.
15. Chen, K., Li, X., & Wang, H. (2015). On the model design of integrated intelligent big data analytics systems. *Industrial Management & Data Systems, 115*(9), 1666–1682.
16. Mishra, N., Lin, C. C., & Chang, H. T. (2015). A cognitive adopted framework for IoT big-data management and knowledge discovery prospective. *International Journal of Distributed Sensor Networks, 11*(10), 718390.
17. Cao, B., Wang, Z., Shi, H., & Yin,Y. (2016). Research and practice on Aluminum Industry 4.0. In *Proceedings of 6th International Conference on Intelligent Control and Information Processing, ICICIP 2015, Wuhan, China* (pp. 517–521).
18. Borhade, M. S. S., & Gumaste, S. V. (2015). Defining privacy for data mining- an overview. *International Journal of Science, Engineering and Computer Technology, 5*(6), 182–184.
19. Zissis, D., & Lekkas, D. (2012). Addressing cloud computing security issues. *Future Generation computer systems, 28*(3), 583–592.
20. Dev Mishra, A., Beer Singh, Y. (2016). Big data analytics for security and privacy challenges. In *2016 International Conference on Computing, Communication and Automation (ICCCA), Noida, India* (pp. 50–53).
21. Whitmore, A., Agarwal, A., & Da Xu, L. (2015). The Internet of Things—A survey of topics and trends. *Information Systems Frontiers, 17*(2), 261–274.
22. Sharma, S. (2016). Expanded cloud plumes hiding big data ecosystem. *Future Generation Computer Systems, 59,* 63–92.
23. Pertel, V. M., Saturno, M., Deschamps, F., Loures, E. D. R. (2017). Analysis of it standards and protocols for industry 4.0. *DEStech Transactions on Engineering and Technology Research*, 622–628.
24. Bagozi, A., Bianchini, D., De Antonellis, V., Marini, A., & Ragazzi, D. (2017). *Big data summarisation and relevance evaluation for anomaly detection in cyber physical systems* (pp. 429–447)., Lecture Notes in Computer Science (including subseries Lecture Notes in Artificial Intelligence and Lecture Notes in Bioinformatics) Rhodes: Greece.
25. López-Estrada, F. R., Theilliol, D., Astorga-Zaragoza, C. M., Ponsart, J. C., Valencia-Palomo, G., & Camas-Anzueto, J. (2019). Fault diagnosis observer for descriptor Takagi-Sugeno systems. *Neurocomputing, 331,* 10–17.

# The Influence of Government Subsidies on the Development of New Energy Vehicle Industry

**Bu Erma and Jinjing Li**

**Abstract** Accelerating the cultivation and development of new energy vehicles will not only effectively alleviate the energy and environmental pressures, but also foster new economic growth points and international competitive advantages of China's automobile industry. This paper makes a quantitative analysis of the economic performance and growth ability of new energy vehicle listed companies, and draws a conclusion. The government subsidy has a significant negative correlation with enterprise economic performance and enterprise growth ability, indicating that there are problems in China's subsidy system. Government in support of listed companies, at the same time, also should help them to improve the core competitiveness. Improving the regulatory mechanism, transforming the mode of subsidies and promoting the ability of the independent research and development of enterprises can also help enterprises to achieve economies of scale.

**Keywords** Subsidy · New energy vehicle · Enterprise · Economic performance · Enterprise growth ability

## 1 Introduction

Since the 21st century, the security of oil and gas resources has become a major hidden danger for China's economic development. In 2015, China's oil consumption external dependence reached 60.6%. The development of new energy has a strong strategic significance for China. New energy vehicles use unconventional vehicle fuel as their power source, or use conventional vehicle fuel, but the use of new vehicle power device, advanced technical principle, with new structure, new technology of the car.

B. Erma (✉) · J. Li
School of Economics and Management, Beijing Jiaotong University, Beijing, China
e-mail: 18120508@bjtu.edu.cn

J. Li
e-mail: lijj@bjtu.edu.cn



J. Zhang et al. (eds.), *LISS2019*,
https://doi.org/10.1007/978-981-15-5682-1_17

Most consumers hold a wait-and-see attitude towards new energy vehicles. There are three reasons. First, the price of new energy vehicles after subsidies will still be more than 50,000 yuan higher than that of traditional cars with the same configuration. Second, the charging infrastructure is not perfect, and incomplete charging facilities or long charging time will also affect consumers' purchasing behavior. Third, the current battery has a shorter range. New energy vehicles are a rare field in which Chinese cars can surpass developed countries, and China may make up for or even surpass the gap with powerful countries. Government subsidy support is of great significance to new energy vehicles, but the government's choice of subsidy intensity and form remains to be explored.

China's new energy vehicle subsidy policy can be divided into two stages. From 2009 to 2012, it is a heavily subsidized stage to implement demand-based policies; after 2012, a supply policy has been implemented to gradually reduce subsidies. Through quantitative analysis, this paper empirically analyzes the data of 19 new energy vehicle listed companies from 2009 to 2016, and discusses the impact of government subsidies on the two aspects of enterprise economic performance and enterprise growth ability. It provides reference for the government to adjust subsidy intensity, improve subsidy forms and allocate resources efficiently. It also avoids the unilateral conclusion that the subsidy policy of new energy vehicles is "effective" or "invalid".

## 2 Literature Review

There is still disagreement over whether the government should subsidies the new energy and car industry, and scholars have not agreed on the choice of subsidies, and the review of the following literature and abroad is mainly discussed in terms of the necessity and subsidy of subsidies.

### *2.1 The Need for Government Subsidies for the New Energy Vehicle Industry*

Through qualitative interview and quantitative research, McKinsey & Company believes that subsidy policy can promote the growth of new energy vehicle sales. Li and Wang [1] believe that scientific and reasonable subsidy policies can effectively stimulate enterprises, enable the government and enterprises to complete resource integration, and enhance the core competitiveness of China's new energy automobile industry. Gu [2] believes that the research and development and promotion of new energy vehicles are of great significance for promoting economic development and protecting the ecological environment, and government subsidies are very necessary. Peneder [3] pointed out that subsidies would effectively stimulate enterprise R&D investment, thus promoting the growth of economic benefits of enterprises. Liang [4] believes that the concept of Chinese consumers is far behind the development speed of the industry in the field of new energy vehicles. Government subsidies can promote the purchase of consumers and realize the transition from the new energy vehicle industry to the mature stage.

## *2.2 Subsidy Method for New Energy Vehicles*

The innovation ability of Chinese new energy enterprises is generally low, and the role of government subsidies in promoting the innovation activities of new energy enterprises has not been effectively played. According to Wang [5], there is a problem that the subsidy targets are not comprehensive. There is no capital subsidy or allocation for the enterprises producing spare parts, and there are few subsidies for subsequent corresponding links. Gillian [6] found that short-term government subsidy policies are able to promote the improvement of new energy vehicle technology, and the promotion effect of subsidy policies on vehicle technology gradually weakens with time. Liao [7] proposed that the threshold of subsidies should be raised to strengthen the management and supervision of the mode and actual utilization of subsidies. Zhang and other scholars [8] believe that the dynamic subsidy mechanism will be better than the static subsidy mechanism, and the government should start from the field of market demand and attach importance to enterprises with excellent performance.

To sum up, it is still controversial whether government subsidies can promote the development of new energy automobile enterprises. Current research mainly focuses on the theoretical aspects, lacking of in-depth study on the new energy vehicle subsidy policy, and has not yet formed a complete theoretical system.

# 3 Theoretical Analysis and Research Hypothesis

## *3.1 Theoretical Analysis*

The new energy automobile industry is a strategic emerging industry, the core of development is technological innovation, and the new technology benefits the society far more than the individual benefits, but the enterprise will have a technical spilt in the production and development, and the positive externalities will be generated, causing market failures. The effective measures to solve externalities are government subsidies, which make up for the marginal private income less than the marginal social income, and realize the maximization and scale effect of social welfare. This is based on the theory of financial subsidies for new energy cars, but government subsidies can lead to crowding out and discourage companies from technological innovation.

## 3.2 Research Hypothesis

Under the background of China's special economic system, wind power industry, photovoltaic industry and other new energy industries have seen explosive growth in recent years, and new energy vehicle industry is no exception. From 2010 to 2015, the total amount of government subsidies received by listed new energy vehicle enterprises in China reached 51.568 billion yuan. As a result of the government's strong subsidies, enterprises rushed to obtain subsidies regardless of the actual effect of subsidies. Some enterprises even made use of loopholes in the current policies to make false declarations and other "cheat subsidies". Therefore, the government's heavy subsidies are likely to lead to enterprises' pursuit of short-term interests and blind production without technological innovation, resulting in a large number of homogenization overcapacity in the new energy vehicle industry. From the original data of the sample enterprises, the economic performance of each enterprise is not ideal, and it is preliminarily judged that the huge subsidy has not played its due role. This paper attempts to deeply explore whether the government's active intervention in the new energy vehicle industry promotes the economic performance and scale expansion of enterprises, so the following hypotheses are proposed:

$\mathbf{H_{1a}}$: Government subsidies are significantly negatively correlated with the economic performance of new energy vehicle enterprises.
$\mathbf{H_{1b}}$: Government subsidies are significantly negatively correlated with the growth capacity of new energy vehicle enterprises.

According to the externality theory, government subsidies can solve the externality problem of enterprises, so enterprises will have enough enthusiasm to carry out scientific and technological innovation, and technological innovation can help enterprises reduce costs, thus achieving scale effect. As a technology-intensive industry, new energy vehicle industry will lose its core competitiveness if it fails to take the lead in the technology field. By comparing and analyzing whether government subsidies have a more significant promoting effect on the return on assets and scale expansion ability of enterprises with strong R&D ability, this paper further explores the influence of government subsidies on the economic performance and growth ability of enterprises, and puts forward the following hypotheses:

$\mathbf{H_{2a}}$: Government subsidies are significantly positively correlated with the economic performance of enterprises with strong R&D capabilities.
$\mathbf{H_{2b}}$: There is a significant positive correlation between government subsidies and the growth capacity of enterprises with strong R&D capacity.

The research framework of the paper is shown in Fig. 1.

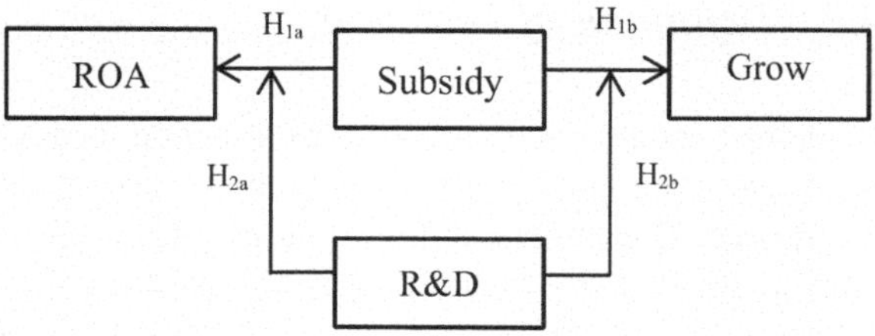


**Fig. 1** The research framework

# 4 Empirical Analysis

## *4.1 Sample Selection and Data Sources*

The sample enterprises in this paper are selected from the list of complete vehicle manufacturing enterprises of new energy vehicles in stock markets of Shanghai and Shenzhen. Based on the main business income of each enterprise and the new energy vehicle model catalog issued by the state, 19 sample enterprises are selected according to the availability of data to analyze the relevant data from 2009 to 2016. The situation of sample enterprises is shown in Table 1.

The selection of listed enterprises as research objects in this paper depends on their financial situation, the openness of enterprise information, and the accuracy of data. Data Total assets, total liabilities, development expenditure, operating income and net profit are all selected from the annual reports of the listed companies of the Securities Star. Some enterprises are listed late, and the main financial data are compiled according to the annual reports and financial statements of the official website of the company.

**Table 1** Basic information of sample enterprises

| CODE | Operating income in 2016 | CODE | Operating income in 2016 |
|---|---|---|---|
| 000559 | 10,785,821,704.5 | 600066 | 35,850,442,042.7 |
| 000625 | 78,542,441,757.2 | 600104 | 746,236,741,228.6 |
| 000800 | 22,709,984,165.5 | 600166 | 46,532,069,535.5 |
| 000868 | 4,757,326,623.7 | 600213 | 3,395,743,970.3 |
| 000967 | 9,257,190,233.1 | 600303 | 3,736,692,124.8 |
| 002594 | 103,469,997,000.0 | 600418 | 52,490,556,761.3 |
| 600686 | 21,827,961,681.6 | 601238 | 49,417,676,151.0 |
| 000550 | 26,633,948,551.0 | 000572 | 13,890,070,950.7 |
| 000980 | 1,693,500,373.8 | 600006 | 16,018,020,957.6 |
| 601633 | 98,443,665,116.0 | | |

[a] The data source: The database of CSMAR

## 4.2 Definition of Indicator

Explained variable: refer to the index selection method of Liu Jibing [9], This paper uses return on total assets (ROA) to represent the economic performance of enterprises. The growth ability of an enterprise (Grow) refers to the future development trend and development speed of the enterprise, and the ability of the enterprise to expand its operation. This article uses the expansion of the enterprise scale to express it.

Explanatory variable: subsidy, including subsidies directly allocated by the central and local governments. Research and Development Costs (R&D) is the part that can be capitalized in the development of intangible assets.

Control variable: This paper mainly controls the characteristic variables at the enterprise level, Including the age of the business, the scale of the enterprise, based on the enterprise division method issued by the Ministry of Industry and Information Technology, measures the scale of the enterprise by operating income (Size); financial leverage, this paper takes the ratio of total annual liabilities to total assets (Lev) as a measure (Table 2).

## 4.3 Model Building

According to the above research assumptions and variable Settings, the following regression model is established in this paper:

$$\begin{aligned} ROA_{it} &= \alpha_0 + \beta_1 Subsidy_{it} + \lambda_1 Age_{it} \\ &\quad + \lambda_2 Size_{it} + \lambda_3 Lev_{it} + \varepsilon_{it} \end{aligned} \tag{1}$$

**Table 2** Variable declaration

| Variable | Index definition |
|---|---|
| *Explained Variable:* | |
| ROA | Corporate return on total assets |
| Grow | (Total current operating income-Total revenue of the previous period) /Total revenue of the previous period |
| *Explaining variable:* | |
| Subsidy | Central and local subsidies |
| R&D | The research and development cost |
| *Control variable:* | |
| Age | The establishment period of the enterprise |
| Size | Operating income |
| Lev | Asset-liability ratio |

$$ROA_{it} = \alpha_0 + \beta_1 Subsidy_{it} + \beta_2 R\&D_{it} + \lambda_1 Age_{it} + \lambda_2 Size_{it} + \lambda_3 Lev_{it} + \varepsilon_{it} \quad (2)$$

$$Grow_{it} = \alpha_0 + \beta_1 Subsidy_{it} + \beta_2 R\&D_{it} + \lambda_1 Age_{it} + \lambda_2 Size_{it} + \lambda_3 Lev_{it} + \varepsilon_{it} \quad (3)$$

$$Grow_{it} = \alpha_0 + \beta_1 Subsidy_{it} + \beta_2 R\&D_{it} + \lambda_1 Age_{it} + \lambda_2 Size_{it} + \lambda_3 Lev_{it} + \varepsilon_{it} \quad (4)$$

In (1) to (4), I represented enterprise, t represented time. It is the random disturbance term. Considering the problem that some index values are too large or the units are not uniform, this paper takes the logarithm of the data corresponding to explanatory variables and control variables, so the above four models are all semi-logarithmic models. Model 1 is used to test $H_{1a}$; Model 3 is used to test $H_{1a}$; Model 2 was used to test $H_{2a}$; Model 4 is used to test $H_{2b}$.

## *4.4 Data Description*

The minimum value of ROA is −0.13, the maximum is 0.18, and the 75th quantile is 0.08, so there is a big gap in the economic performance of each company. The standard deviation of grow is 0.25, which is relatively stable. The 75-digit quantile is 0.32, and the maximum value is 1.25, indicating that the growth capacity of new energy auto companies is generally low, and the gap between strong companies and other companies is too large. The profit of some companies is negative, which may be caused by poor sales in the current period. In terms of control variables, there is a large difference between the minimum and maximum values of Age and Lev, indicating that the control variables have large changes in time or space, but the fluctuations are not significant. The standard deviation of size data is large, indicating that the scale difference between new energy auto companies is obvious.

The mean value of ROA of enterprises with development expenditure is the same as the general situation, and the standard deviation is 0.04. Compared with the general situation, the fluctuation is smaller, but its minimum value is larger. The average grow value is 0.19, which is larger than the average grow value (0.17), indicating that the new energy automobile enterprises attach importance to R&D and innovation ability have stronger overall economic performance and growth ability. The standard deviation of enterprise development expenditure data is 1.61, with the maximum value of 21.86 and the minimum value of 14.69, indicating that there is a huge gap between enterprises in development expenditure (Fig. 2).

It is basically possible to exclude outliers and calculation errors, and these data can be used for subsequent analysis. In addition, this study further gives a scatter plot of the correlation of these variables. We can only see that the size of the firm has a positive correlation with the age of the enterprise and the subsidies. There is

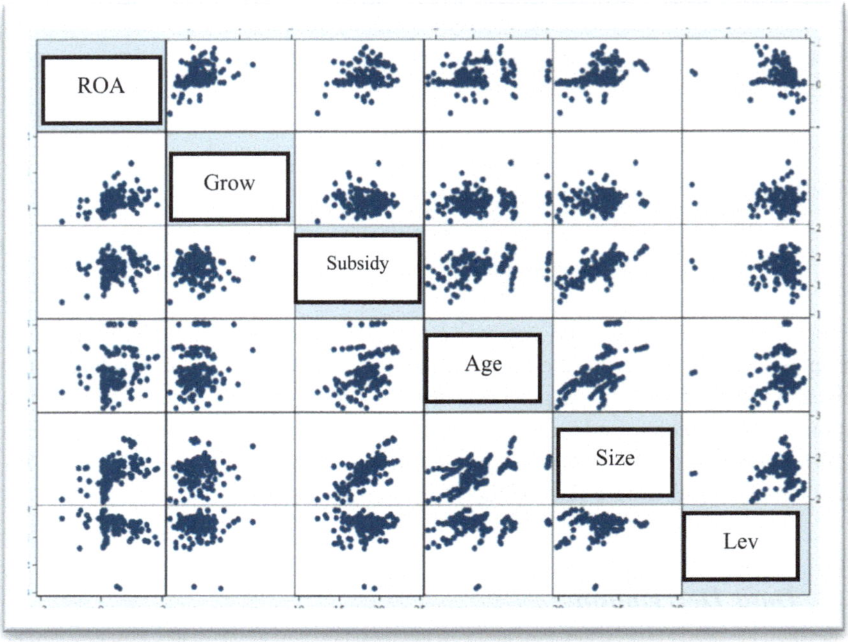


**Fig. 2** Scatter plots of all variable correlations

not a clear correlation between subsidies and the economic performance of enterprises or the ability of enterprises to grow, which indicates that government subsidies have not significantly promoted the economic performance of enterprises and the Growth of enterprises. The preliminary judgment assumes that hypothesis $H_{1a}$ and $H_{1b}$ are established.

# 5 Analysis Method and Empirical Test

## *5.1 Analytical Method*

In order to test the influence of government subsidies on the economic performance and growth capacity of enterprises, this paper selected relevant panel data from 2009 to 2016 for analysis. Panel data have two dimensions of cross section and time, which can solve the problem of missing variables. The missing variables are often caused by unobservable individual differences or "heterogeneity". The regression analysis of panel data is mainly divided into fixed effect model and random effect model. If the intercept term representing individual heterogeneity is related to an explanatory variable, it is called a "fixed effect model"; if it is not related to all explanatory variables, it is called a "random effect model". In this paper, Hausman test is used to select the appropriate model. The null hypothesis of Hausman test is that there is no fixed effect, and the estimation result of random effect should be used. The fixed-effect model is consistent whether the null hypothesis is true or not. If the null hypothesis is not true, the random-effect model is inconsistent. If the null hypothesis is true, the random-effect model is more effective than the fixed-effect model. In addition, in this paper, the method of IPS is used to conduct the unit root test on the data. The hypothesis of IPS test is that the unit root exists in $H_0$. If the p value is greater than 0.05, the null hypothesis cannot be rejected and the data is non-stationary.

## 5.2 *Test Ideas*

In Table 3, model 1 does not include any control variables, enterprise age is added to model 2, and enterprise size is added to model 3. Model 4 further introduces financial leverage as a control variable. Model 5 introduces the explanatory variables R&D based on the relevant data of some enterprises with development expenditures, thus testing the impact of government subsidies on the economic performance of enterprises with strong R&D capabilities, and also analyzing the effect of development expenditures on economic performance of enterprises. The analysis of the growth ability of enterprises in Table 4 is the same.

**Table 3** Empirical results of enterprise economic performance

| Explained variable | ROA | | | | |
|---|---|---|---|---|---|
| Variable | (1) | (2) | (3) | (4) | (5) |
| Subsidy | −0.00435* | −0.0027* | −0.00636*** | −0.00809*** | −0.00808** |
| | (0.00236) | (0.00255) | (0.00237) | (0.00227) | (0.00368) |
| Age | | −0.035 | −0.100*** | −0.0231* | −0.249*** |
| | | (0.0214) | (0.0222) | (0.0127) | (0.0489) |
| Size | | | 0.0437*** | 0.0320*** | 0.0672*** |
| | | | (0.00756) | (0.00559) | (0.0121) |
| Lev | | | | −0.0395*** | −0.0277** |
| | | | | (0.0104) | (0.0123) |
| R&D | | | | | 0.00866*** |
| | | | | | (0.00309) |
| Constant | 0.126*** | 0.207*** | −0.542*** | −0.506*** | −0.752*** |
| | (0.043) | (0.0655) | (0.142) | (0.106) | (0.235) |
| IPS | −1.16365 | −15.7726*** | −13.6555*** | −12.2191*** | −14.1582*** |
| Hausman | 7.38** | 10.13** | 14.22*** | 10.52* | 22.62**** |
| Remark | FE | FE | FE | RE | FE |

b. *Note:* (1) ***, **, *are significant at the level of 1, 5 and 10%, respectively. (2) Hausman represents the statistic of Hausman test, FE and RE represents the fixed effect and the random effect. (3) IPS test showed that the data was stable

**Table 4** Empirical results of enterprise growth capability

| Explained variable | Grow | | | | |
|---|---|---|---|---|---|
| Variable | (1) | (2) | (3) | (4) | (5) |
| Subsidy | −0.0516*** | −0.0232* | −0.0596*** | −0.0609*** | −0.0704** |
| | 0.0177 | 0.0127 | 0.0161 | (0.0162) | (0.0284) |
| Age | | 0.0466 | −0.0174 | −0.0150 | −0.0128 |
| | | 0.0367 | 0.0421 | (0.0432) | (0.0567) |
| Size | | | 0.0867*** | 0.0888*** | 0.139*** |
| | | | 0.0253 | (0.0257) | (0.0515) |
| Lev | | | | −0.0468 | −0.147 |
| | | | | (0.0607) | (0.0934) |
| R&D | | | | | 0.0400* |
| | | | | | (0.0238) |
| Constant | 1.12*** | 0.45** | −0.715* | −0.777* | −2.70** |
| | 0.324 | 0.224 | 0.415 | (0.427) | (1.24) |
| IPS | −4.8287*** | −18,778*** | −16.2536*** | −14.5445*** | −15.4200*** |
| Hausman | 6.39** | 5.66 | 8.94* | 8.79 | 10.65* |
| Remark | FE | RE | RE | RE | RE |

c. *Note:* (1) ***, **, *are significant at the level of 1, 5 and 10%, respectively (2) Hausman represents the statistic of Hausman test, FE and RE represents the fixed effect and the random effect (3) IPS test showed that the data was stable

## *5.3 Empirical Test*

In Table 3, the results of Models (1)–(5) show that the coefficient of government subsidies are all negative, and both are significant at different levels of 1%–10%, indicating that the government has made a large amount of financial subsidies for listed companies of new energy vehicles. However, it did not achieve good results. Taking Model 4 as an example, government subsidies are significantly negatively correlated with the economic performance of new energy auto companies at the level of 1%, with a coefficient of −0.00809, because the model is a semi-logarithmic model, and the explanatory variable changes when the relative variable is interpreted. Absolute quantity, so for every 1% increase in government subsidies, the economic performance of enterprises will fall by 0.0000809 units. Explain that the government subsidies obtained by new energy auto companies have reduced corporate performance and confirmed the hypothesis $H_{1a}$.

In terms of control variables, the enterprise age coefficient is significantly negatively correlated with the enterprise economic performance at the level of 10%, while the enterprise size is significantly positively correlated with the enterprise economic performance at the level of 1%. In line with the theory of scale economy, the improvement of production efficiency brought by the expansion of production scale can achieve the purpose of reducing average cost, increasing profit space and improving the economic performance of enterprises. In terms of financial leverage, the asset-liability ratio is significantly negatively correlated with the economic performance of new energy automobile enterprises at the level of 1%, with a coefficient of −0.0395.

In model 5, the government subsidy coefficient is negative and significant at 5%, indicating that there is a significant negative correlation between government subsidies and economic performance of companies with strong R&D capabilities. However, government subsidies have not produced positive effects, but have adversely affected the technological innovation of enterprises.

In Table 4, the coefficient of government subsidies in Models 1–5 are all negative, both of which are significant at different levels of 1%–10%, further indicating that the government's huge subsidies for listed companies of new energy vehicles have not achieved the desired results. Similarly, in the model 4 cases, the coefficient of government subsidy (−0.0609) is significantly negatively correlated with the growth capacity of enterprises at 1%, indicating that for every 1% increase in government subsidies, the growth capacity of enterprises will decrease by 0.000609 units, indicating that new energy vehicles The government subsidies obtained by the enterprises hindered the expansion of the scale of production and confirmed the hypothesis $H_{1b}$.

The samples in model 5 are enterprises with development expenditure, and government subsidies are significantly negatively correlated with the growth ability of enterprises. However, the coefficient of government subsidies is very small different from that in model 2, which further indicates that government subsidies do not play their due role in the growth ability of new energy automobile enterprises with strong development ability. Development expenditure plays a significant role in promoting the growth ability of enterprises, with a coefficient of 0.04, indicating that enterprises can enhance their competitiveness through increasing R&D and technological innovation. There is a significant positive correlation between the size of an enterprise and its growth capacity at the level of 1%, and the coefficient is 0.0888, indicating that larger enterprises have stronger growth capacity, and the improvement of production efficiency brought by the expansion of production scale can achieve the purpose of reducing the average cost, thus improving the profit level and helping enterprises to achieve the scale effect.

## 5.4 Empirical Result

Based on the empirical analysis of relevant data of listed new energy enterprises, this paper studies the impact of government subsidies on the economic performance and growth capacity of enterprises. The empirical results are summarized as follows:

Firstly, government subsidies do not contribute to the development of new energy vehicle enterprises, but have an inhibitory effect, which confirms the hypothesis $H_{1a}$ and $H_{1b}$ and exposes the drawbacks and deficiencies of China's new energy vehicle subsidy system.

Secondly, the age of the company is significantly negatively correlating with economic performance. The reason may be that the long-established enterprise management model is backward and cannot adapt to market changes quickly, resulting in poor management. There is no significant correlation between the age of the enterprise and the growth ability of the enterprise. It may be that the development of the enterprise tends to be stable, and the influence of the age of the enterprise on the expansion of the scale of the enterprise gradually disappears. The size of the firm has a positive effect on economic performance and growth ability, in line with the theory of scale effect. The lower the asset-liability ratio of a company, the smaller the financial risk is, which is more conducive to the growth of the company's economic performance, and the company can use government subsidies more effectively.

Thirdly, R&D can promote the improvement of enterprise economic performance and the expansion of enterprise scale. Government subsidies are also significantly negatively correlated with the economic performance and growth capacity of enterprises with strong R&D capacity, which is inconsistent with hypothesis $H_{2a}$ and hypothesis $H_{2b}$. The inconsistency between the effect of government subsidies and the expected results is probably due to the fact that subsidies crowd out the investment of private capital in R&D activities, which verifies the results obtained from the theory of crowding out effect in the above theoretical analysis, and indicates that there are problems in the selection of objects of government subsidies.

## 6 Conclusions and Policy Recommendations

Based on the current research results, this paper analyzes the impact of government subsidies on the economic performance and growth capacity of enterprises, and conducts an empirical test, and draws the following conclusions and Suggestions:

Firstly, although government subsidies can stimulate to some extent, "all-inclusive" subsidy system [4] and imperfect regulatory mechanism will cause enterprises to slack in operation and management and trigger a large number of "cheat subsidy" behaviors [10]. China's current subsidy system is not reasonable, and enterprises often adopt low-price and homogeneous competition to obtain subsidies [11], which eventually leads to enterprises' neglect of long-term interests and loss of sustainable development ability. Government subsidies are prone to low efficiency and other problems. Lax supervision will lead to dependence of some enterprises and insufficient motivation for subsequent development.

Secondly, the new energy vehicle industry is a technology-intensive industry, and the enterprise's early-stage research and development investment cost is very large. Government subsidies have no positive impact on the operation of enterprises with strong research and development ability, indicating that the subsidies do not provide targeted effective incentives to enterprises. The ways and targets of subsidies still need to be improved, and the maintenance of intellectual property rights should be paid attention to, and special subsidy funds should be given to key technology research and development areas.

Thirdly, the empirical results show that the enterprise scale effect has played. The significant negative correlation between the age of enterprises and the economic performance of enterprises indicates that some enterprises with long years of establishment are likely to have problems such as backward management modes. Enterprises also need to carry out management mode reform to adapt to the current market environment, strengthen the research and development of cutting-edge technologies, and improve their core competitiveness.

## References

1. Quan, Q. L., & Wang, X. X. (2012). Fiscal and tax policy study on promoting the development of new energy vehicles – a case study of listed automobile enterprises. *Economics and Management, 26,* 37–43.
2. Gu, R. L. (2013). *Study on fiscal and tax policies to promote the development of China's new energy vehicle industry*, Institute of Fiscal Science, Ministry of Finance.
3. Peneder, M. (2008). The problem of private under-investment in innovation: A policy mind map. *Technovation, 28,* 518–530.
4. Liang, X. S. (2016). On the current situation and prospect of China's new energy vehicles. *Science and Technology Outlook, 26,* 294.
5. Wang, W., Li, H. Z., Qiao, P. H., & Gui, J. W. (2017). Study on the impact of government subsidies on the performance of new energy vehicle enterprises – based on in-depth analysis of enterprise growth. *Science and Technology Progress and Policy, 34,* 114–120.
6. Harrison, G., & Thiel, C. (2017). An exploratory policy analysis of electric vehicle sales competition and sensitivity to infrastructure in Europe. *Technological Forecasting and Social Change, 114,* 165–178.
7. Liao, J. Q., & Sun, X. S. (2017). Study on fiscal and tax policy effects of new energy vehicles. *Tax and Economy, 31,* 86–93.
8. Zhang, H. B., Sheng, Z. H., & Meng, Q. F. (2015). Study on government subsidy mechanism of new energy vehicle market development. *Management Science, 28,* 122–132.
9. Liu, J. B., Wang, D. C., & Xia, L. (2014). The study of the efficiency of the innovation efficiency of the government subsidies. *Scientific and Technological Advances and Countermeasures, 31,* 56–61.
10. Pang, L. (2017). *An analysis of government support and subsidy policies for new energy vehicles in China*, Beijing Jiaotong University.
11. Ma, L., Zhong, W. J., & Mei, S. (2017). Study on subsidy policy innovation for new energy vehicle industry under the background of supply side reform. *Systems Engineering Theory and Practice, 37,* 2279–2288.

# A Data-Driven Methodology for Operational Risk Analytics Using Bayesian Network

**Weiping Cui, Rongjia Song, Suxiu Li, and Lei Huang**

**Abstract** Traditional research in port risk analytics is expert- and survey-based, relying on domain knowledge and data collected from questionnaires, which is not optimal for data-driven proactive prediction. Therefore, we propose a novel data-driven approach to formulate predictive models for bulk cargo theft in port. We use event data extracted from inspection records of delivery trucks and operational records of port logistics information systems to construct and evaluate predictive models in a real-world scenario. More specifically, we apply various feature ranking methods and classification algorithms to select an effective feature set of relevant risk elements. Then implicit, Bayesian networks are derived with the features to graphically present the relationship with the risk elements of the bulk cargo theft. The resulting Bayesian networks are then comparatively analyzed based on outcomes of model validation and testing, as well as essential domain knowledge. The experimental results show that predictive models are effective with both

Science and Technology Project of State Grid Corporation of China, entitled "Implementation Roadmap and Application Research on Data Fusion bases on Operation and Monitoring Business".

W. Cui (✉) · S. Li
State Grid Energy Research Institute, State Grid Corporation of China, Beijing, China
e-mail: cuiweiping@sgeri.sgcc.com.cn

S. Li
e-mail: lisuxiu@sgeri.sgcc.com.cn

R. Song
Department of Information Management, Beijing Jiaotong University & KU Leuven, Beijing, China
e-mail: rjsong@bjtu.edu.cn; rongjia.song@kuleuven.be

L. Huang
School of Economics and Management, Beijing Jiaotong University, Beijing, China
e-mail: lhuang@bjtu.edu.cn



J. Zhang et al. (eds.), *LISS2019*,
https://doi.org/10.1007/978-981-15-5682-1_18

accuracy and recall greater than 0.8. These predictive models are not only useful for understanding the dependency between relevant risk elements, but for supporting strategy optimization of risk management.

**Keywords** Risk prediction · Predictive modeling · Bayesian network · Sea port · Bulk cargo theft

## 1 Introduction

Many specific methods including qualitative methods and quantitative methods have been developed to assess risk and safety in the port area or operation. Data that is used in risk assessment and risk detection in the literature, usually comes from surveys [1] or manual classification [2]. However, research has demonstrated that the amount of data produced and communicated in logistics and supply chain is significantly increasing, thereby creating challenges for the organizations that would like to reap the benefits from analyzing real operational data [3]. On the other hand, there is no specific data-driven method or framework to cope with the risk analytics and risk detection of bulk port operations.

We focus on cargo theft risk of bulk port in this paper. More specifically, this means that the drivers deliver more bulk cargo during transportation than the amount in the bill of lading, e.g., by cheating or other illegal means. These risk events involve much uncertainty due to the effects of various interrelated elements including the port company, cargo interest, cargo type, carrier, operational environment, etc. The main aim of this paper is to use a proposed method to develop risk detection models to identify bulk cargo theft and further analyze the dependency between risk elements and the risk event. Furthermore, we present a real-world case study for a Chinese bulk port to evaluate the proposed data-driven method.

This paper is organized as follows. Section 2 reviews the related work. Section 3 describes the proposed methodology. Section 4 provides a case study to demonstrate the use of the proposed methodology and models. Section 4.5 presents the results of our experiments. Conclusions and future work are discussed in Sect. 5.

## 2 Related Work

In ports, a high quality risk management is absolutely necessary for their sustainable development [4]. The extant focus of risk management in port are rather high level, mostly on development, management, organization and commercial issues of ports and terminals. Nevertheless, a few studies have tried to solve risk problems in practical operation of cargo delivery. Nowadays a variety of techniques of risk analysis have been developed in industrial settings, such as physical inspections, flow charts, safety review, checklist analysis, relative ranking, cause-consequence analysis, 'what-if' analysis, failure modes and so forth, all excessively depend on expert knowledge and manual preprocessing. As a result, in complex operational environment, experts may not be able to be fully aware of all situations, which may have negative influence on the result of risk detection and analysis.

The widespread use of digital technologies has led to the intelligent port operation paradigm, the amount of data collected during operational processes is growing at a rapid pace. Data-driven risk management emerges to provide companies with better means to obtain early warning from an increasingly massive amount of data and gain a powerful competitive advantage, especially in the operational context [3–5]. More specifically, data from surveys, interviews and expert classification plays a significant role in extant research on port risk analysis. For instance, Constantinos [2] qualitatively assessed the potential risks for ports using data from interviews with the administrations of a container terminal in an empirical study. Furthermore, 19 port-centric disruptive events were identified based on the literature and used data from surveys to generate risk metrics for analyzing port-centric supply chain disruption threats in [1]. Mokhtari [4] used fault tree analysis and event tree analysis to assess the risk factors associated within sea ports and offshore terminals operations and management based on experts' judgement.

Cargo theft is referred to as a silent crime as it accounts for huge losses that frequently go unreported, and can even occur without the product owner being aware until days or weeks after the theft occurred [6]. Regarding current research on cargo theft, most studies focus on the problem of theft prevention by introducing software systems, hardware facilities or a set of managerial measures [7, 8]. In addition to academic papers, there are also some regulations aiming at theft prevention, for instance the Freight Security Requirement and the Trucking Security Requirement issued by Technology Assets Protection Association. However, the programs designed are not effective enough for preventing theft. To our knowledge, there are no data-driven methods to cope with cargo theft risk analytics in the real-world operational scenario in general and bulk cargo port in particular.

Bayesian networks are acyclic directed graphs that represent the conditional dependencies, which are also considered as a graphical inference technique used to express the causal relationship among the variables. Hence, Bayesian networks are able to perform both predictive analysis and diagnostic analysis [9]. Hence, it is widely used in modeling of complex systems and risk analysis of a wide variety of accidents based on probabilistic and uncertain knowledge thanks to its flexible structure and reasoning engine. Bayesian networks are applied as the modeling technique and the model parameters are based on expert elicitation and learning from historical data. It is powerful for analyzing relations between risk components, visualizing specific risk structures and reasoning risk potential as well.

This paper aim at proposing a data-driven methodology based on Bayesian networks for analyzing the relations between risk components of cargo theft and early detection of theft risk, which is also applied it to a real-world bulk cargo port in the south of China.

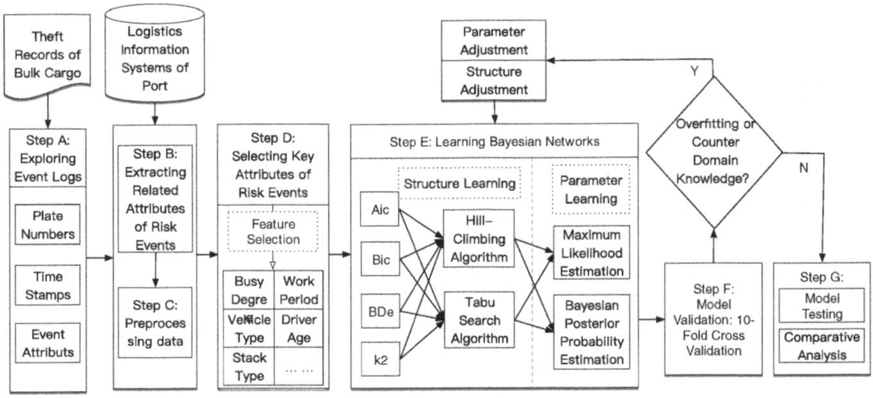


**Fig. 1** The data-driven methodology for risk analysis based on Bayesian Networks

# 3 Methodology

In this study, we consider the cargo theft detection as a two-class prediction problem. As a supervised learning problem, the performance of a classifier relies on how effectively it can learn from the training set. More specifically, one aspect that affects learning quality is the predictive power of the selected features. Another aspect is the proper adjustment of the learning process. The methodology is depicted in Fig. 1, which consists of three main parts: (a) feature selection; (b) Bayesian network learning; and, (c) model validation, testing and comparative analysis.

## *3.1 Step A. Exploring Event Logs*

The historical records are usually in the form of unstructured data, such as text. Hence, word segmentation techniques are needed to extract meaningful features from the theft reports in order to obtain structured data, such as license plate numbers, timestamps and other related features.

## *3.2 Step B. Extracting Related Features of Risk Events from Databases*

Related operational logs in the logistics information system need to be extracted in terms of license plate numbers and timestamps. All irrelevant features need to be further filtered out from the extracted operational logs. As a result, the obtained initial dataset consists of historical records of the bulk cargo theft with structured storage and related operational logs in the logistics information system including all relevant features.

### 3.3 Step C. Preprocessing Data

After data exploration and data collection, we need to preprocess the initial dataset, which mainly consists of two tasks: data balancing and data discretization.

From a model-refinement perspective, the fact that bulk cargo thefts are the focus of the prediction but have a much smaller sample size than normal operations provide the opportunity of utilizing imbalanced classification techniques to improve the model sensitivity on the minority class. Hence, we use the Synthetic Minority Oversampling Technique (SMOTE) [10] method to balance the initial dataset. On the other hand, we keep binary variables and categorical variables in the dataset, but use an entropy based technique: the Minimum Description Length Principle (MDLP) [11] to discretize continuous variables. More specifically, we set bulk cargo theft risk as the target variable for the supervised data discretization.

### 3.4 Step D. Selecting Key Features of Risk Events

After data preprocessing, feature selection (Fig. 2) is the next step to reduce the dimensionality of the feature space by removing redundant, irrelevant, or noisy data. A Bayesian network that is constructed directly from data related to the cargo theft in the historical theft records and bulk ports' database can be rather complicated and thus lack readability because too much unimportant data is involved. Feature selection is therefore crucial to obtain the most effective feature set for the model construction. It supports a faster learning process, improves the data quality and thereby the performance of prediction, as well as increasing the comprehensibility of the analytical results [12, 13]. More specifically, we use feature ranking methods and feature selection methods with two basic steps of subset generation and subset evaluation for the ranking of each feature in every sub-dataset. Ranking methods with different classification algorithms might give different results for the classification accuracy. Comparative analysis is then performed based on the result of cross validation. Finally, we choose the most effective composition of ranking method and classification technique to select and ensure a subset of features giving the highest effectiveness.

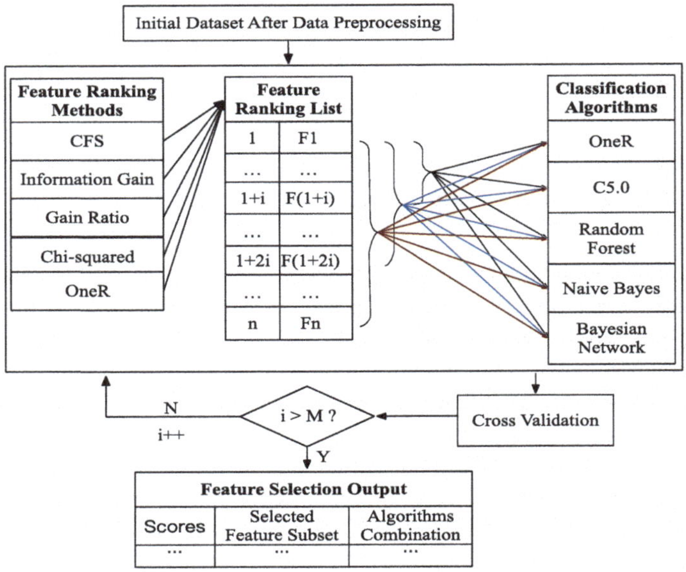


**Fig. 2** The methodology of feature selection

## 3.5 Step E. Learning Bayesian Networks

After feature selection, learning the Bayesian networks is the next step. This includes two major tasks: learning the graphical structure, and learning the parameters for that structure. We design comparative experiments for the structure learning by utilizing combinations of four score functions and heuristic structure search algorithms including Hill Climbing and Tabu Search. More specifically, score functions include Akaike information criterion (Aic), Bayesian information criterion (Bic), Bayesian Dirichlet equivalent score (BDe) and k2 algorithm. In addition, the goal of parameter learning is to obtain the optimal parameter values for a given structure with a given corpus of complete data. Thus, we could simply use the empirical conditional frequencies from the data. As parameter learning is straightforward, we will mainly focus on learning the Bayesian network structure.

## 3.6 Step F. Model Validation

After the Bayesian network learning, we will use k-fold cross validation to evaluate the performance of the classifiers. We choose the classification error as the loss function to estimate how accurately the model will perform in practice, in order to limit problems like overfitting and give insights on how the model will generalize to an independent dataset.

### 3.7 Step G. Model Testing and Comparative Analysis

The last step is the model testing and comparative analysis. We test all the optimal models of different combinations of score functions and heuristic structure search algorithms by applying them to the separate test dataset. The output is a confusion matrix, which can be used for analyzing the accuracy, the sensitivity and the specificity in order to choose the optimum predictive model for the bulk cargo theft risk.

## 4 Empirical Study

In this section, we apply the proposed methodology in the previous section to a real-word cargo theft dataset, and we present the empirical study which consists of the feature selection, the structure learning, the parameter learning, stratified cross validation and model testing. Because of the limited space we present the whole process of structure learning using only the combination of the score function with k2 algorithm and the structure search algorithm with Hill Climbing. In addition, in the following figures, we use the abbreviations HC for Hill Climbing; HC. Aic for the combination of Hill Climbing and Aic and TABU. Aic for Tabu Search combined with Aic. For convenience, all combinations adhere to this naming format.

### 4.1 Data Description and Preprocessing

The data used in this study originates from Port Corporation and Port Public Security Bureau in the south of China. It consists of inspection records of delivery trucks at the port from September 2013 until May 2014. In total, 5,320 records are used in this case study, of which 4898 were reports were of normal deliveries and 422 were records of delivery with theft, which we refer to as delivery risk. Table 1 depicts a report example of risky delivery from sampling inspection.

**Table 1** A report example of risky delivery

| ID | Event description |
|---|---|
| 1 | On the night of March, 14, 2014, driver Zhou drove the truck with plate number XiangX. He stole 3.88 tons of brown coal (Indonesia coal, Lower Heating Value: 4,400 kcal/Kg) from terminal A of Guangzhou Port by reforming or retrofitting the water tank of the truck. The truck came into the terminal A with water in the tank and released the water before loading. Then he secretly transported the stolen cargo, identified with a value of 455 yuan per ton and 1,765.4 yuan in total, to the factory B in DongGuan. |

First, we segment the text of inspection records using Rapidminer to obtain features with structured storage. Then we extract related operational records from the logistics information system in terms of license plate numbers and timestamps. Based on these data, the domain experts of the practitioner and the police help us to identify 38 relevant risk elements of the bulk cargo theft as the initial feature set. After data preprocessing which includes data cleaning, data balancing using SMOTE method and data discretization using MDLP method, we were left with 9,796 records in structure storage with 38 features in addition to the response feature, i.e., normal of risk delivery.

## *4.2 Feature Selection*

We use CFS (Correlation-based Feature Selection), GI (Information Gain), GR (Gain Ratio), Chi (Chi-squared) and OneR to rank the features, in other words, features in the initial feature set and output different ranking results. In Table 2, a lower number means a more relevant feature for theft risk.

Based on the feature rankings in Table 2, we find that: (a) Different ranking results are obtained by using various feature ranking methods. Some of these outcomes are similar, especially using GI, GR and Chi. (b) The features 'Cargo Weight Deviation', 'Reformed or Not', 'Arrival time' always rank the highest independent of the method used. This means that these three features are closely relevant to the theft risk of bulk cargo. (c) The features 'Education Background', 'Weighbridge of Empty Cargo Truck', 'Weighbridge of Heavy Cargo Truck' and 'Operation Period Deviation' always rank the lowest. It means that these four features are most irrelevant features to the theft risk of bulk cargo. We use OneR, Decision Tree (DT), Random Forest (RF), Naive Bayesian (NB) and Bayesian Network (BN) as classifiers to evaluate the subsets, and classification accuracy is estimated using ten-fold cross validation. In this practical problem, we test a given classifier on a number of feature subsets, obtained from different ranking indices, using classic evaluation indexes based on the classification confusion matric, i.e., accuracy, precision and recall. In particular, recall is the key evaluation index because it measures what percentage of all risk events are correctly predicted. Moreover, variance is used for evaluating the reliability of classification results. We process 100 rounds of ten-fold cross validation in total. In each round, a given classifier is applied to a feature set of one of the ranking methods ($5 \times 38 \times 5$ times cross validation in one round).

We decide the final feature set must fulfill the following three criteria: (a) the accuracy must be above 70%; (b) the precision and recall must be above the zeroR baseline (0.5); (c) the variance of the accuracy, precision and recall must be between −20% and 20%. In addition, in every round, one combination of classifier, selected feature set and ranking method must have the most effective performance in terms of the recall. More specifically, as the number of selected features is

**Table 2** Feature ranking result

| Feature | CFS | GI | GR | Chi | OneR |
|---|---|---|---|---|---|
| 1- Age | 8 | 7 | 9 | 8 | 10 |
| 2- Education Background | 34 | 34 | 34 | 34 | 34 |
| 3- Credit in Port | 14 | 16 | 15 | 15 | 12 |
| 4- Region | 32 | 33 | 31 | 27 | 29 |
| 5- Cargo Type | 18 | 17 | 16 | 16 | 19 |
| 6- Trade Type | 30 | 29 | 28 | 33 | 3 |
| 7- Import or Export | 38 | 38 | 38 | 38 | 35 |
| 8- Customs Supervision | 7 | 5 | 7 | 5 | 9 |
| 9- Cargo Price | 33 | 32 | 33 | 32 | 28 |
| 10- Truck Manufacturer | 19 | 19 | 21 | 21 | 20 |
| 11- Year of Manufacture | 9 | 10 | 10 | 11 | 13 |
| 12- Registration Region | 27 | 31 | 32 | 30 | 33 |
| 13- Truck Category | 12 | 13 | 13 | 14 | 14 |
| 14- Reformed or Not | 1 | 3 | 3 | 3 | 5 |
| 15- Ownership | 26 | 28 | 29 | 29 | 31 |
| 16- Times of Filing | 15 | 18 | 17 | 18 | 18 |
| 17- Blacklist | 11 | 12 | 11 | 9 | 8 |
| 18- Yard Category | 6 | 6 | 6 | 7 | 4 |
| 19- Specialized Yard | 21 | 25 | 24 | 25 | 24 |
| 20- Cargo Volume | 29 | 26 | 25 | 24 | 27 |
| 21- Consignor Level | 13 | 11 | 12 | 13 | 17 |
| 22- Contract Category | 20 | 22 | 19 | 19 | 26 |
| 23- Means of Payments | 5 | 8 | 4 | 4 | 6 |
| 24- Arrival time | 2 | 2 | 2 | 6 | 1 |
| 25- Leave time | 22 | 20 | 20 | 22 | 21 |
| 26- Operation Period | 4 | 4 | 5 | 2 | 7 |
| 27- Operation Period Deviation | 37 | 36 | 36 | 36 | 36 |
| 28- Operation Period Fluctuation | 28 | 30 | 26 | 28 | 30 |
| 29- Cargo Weight Deviation | 3 | 1 | 1 | 1 | 2 |
| 30- Cargo Weight Fluctuation | 31 | 27 | 30 | 31 | 32 |
| 31-Weighbridge of Empty Cargo | 35 | 35 | 35 | 35 | 36 |
| 32- Weighbridge of Heavy Cargo | 36 | 37 | 37 | 37 | 38 |
| 33- Arrival Date | 24 | 24 | 22 | 20 | 22 |
| 34- Shift Category | 25 | 23 | 27 | 26 | 23 |
| 35- Number of Workers | 17 | 15 | 18 | 17 | 16 |
| 36- Number of Machineries | 23 | 21 | 23 | 23 | 25 |
| 37- Weather | 10 | 9 | 8 | 10 | 15 |
| 38- Busy Degree | 16 | 14 | 14 | 12 | 11 |

**Table 3** Compositional classification results of all rounds

| Rounds | *Number of selected features* | *Ranking method* | *Classifier* |
|---|---|---|---|
| 35 | 17 | GR | Bayesian Network |
| 21 | 19 | OneR | Random Forest |
| 21 | 17 | GI | Naïve Bayesian |
| 12 | 15 | GR | Bayesian Network |
| 6 | 16 | GR | Random Forest |
| 5 | 17 | CFS | OneR |

increasing in a given composition of classifier and ranking method, the recall of the classification increases until it reaches a peak and then decreases again.

Table 3 presents the compositional classification results of all round 10-fold cross validation. The classification quality for this dataset is obviously influenced by the choice of ranking indices. When 17 features are selected, most rounds obtain the best performance of cross validation with 61 of 100 rounds. Moreover, 47% rounds gain the best performance by using the combination of GR and Bayesian network. In this section, we focused on selecting the most effective feature set for classification. Classification recall with GR ranking method is the highest for Bayesian network, and also quite high for other classification algorithms as presented in Fig. 3. Therefore, we select these 17 features using the GR ranking method for constructing the predictive model of cargo theft risk (features in blue of Table 2).

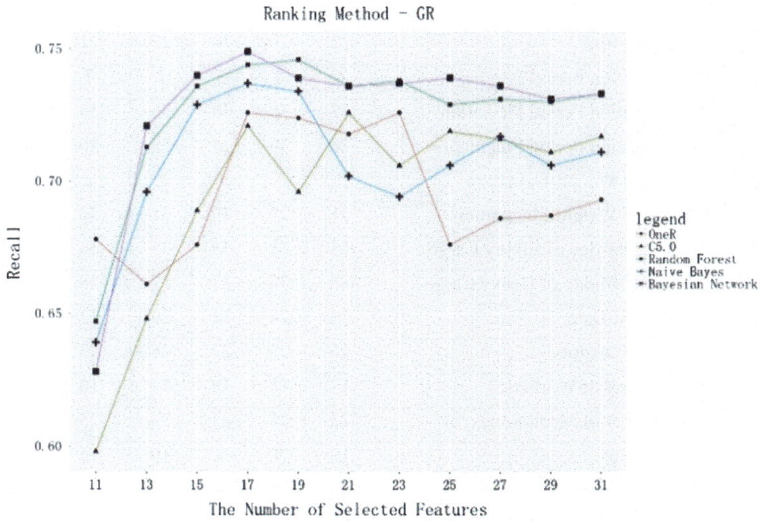


**Fig. 3** The classification results using GR ranking method

## 4.3 Structure Learning and Analysis

We first equally divide the whole dataset into three subsets, keeping the same proportion of risk delivery in each subset as in the original dataset. Then, two subsets are used as the training set for learning and validating the model. The last subset is used as the test set for testing the model performance. Bayesian networks are able to graphically present the dependency between features through its structure topology. However, it could be difficult to understand or explain in case of overfitting, which means a very complicated network structure. Hence, in this study, we try to choose an optimal value for the maximum of father nodes while learning the network structure, in order to insure that the structure learned is both explainable and effective. As the parameter 'mapx' increases, errors of training and cross validation decline sharply and the level of using combinations of structure search algorithms, score functions and parameter learning algorithms as well. Moreover, in the case of having maximum five father nodes, errors are at their minimum independent of the combination. Thus we determine the maximum number of father nodes as five.

Eight Bayesian network structures were built using the structure search algorithm with HC and TABU, as well as various score functions. Even when applying the same structure search algorithms, the learnt Bayesian network structures are different due to different score functions. Nevertheless, all network structures derived are rather similar, also in terms of arc strength, especially when HC is applied. Moreover, meaningful insights are graphically provided in these network structures. Especially coupling these results from quantitative information with qualitative knowledge offered by domain experts is useful. The network structures need to be further refined based on the domain knowledge and expert expertise, as well as adding counter-measures for risk management. A network structures learnt with HC.k2 is presented as an example in Fig. 4.

## 4.4 Parameter Learning and Analysis

In order to learn from discrete data, Bnlearn provides two methods for the parameter learning: Maximum Likelihood estimation and Bayesian posterior probability estimation [14]. In case that the training dataset is very large, the results of the parameter learning are the same using these two methods. However, in reality and especially in our case, the training set is rather small. Hence, we should consider the choice of these two methods from two perspectives of computational complexity and computational veracity. On the one hand, Maximum Likelihood estimation with differential calculus has lower computational complexity than Bayesian posterior probability estimation with multiple integrals. On the other hand, Bayesian posterior probability estimation normally perform better computational veracity than Maximum Likelihood estimation.

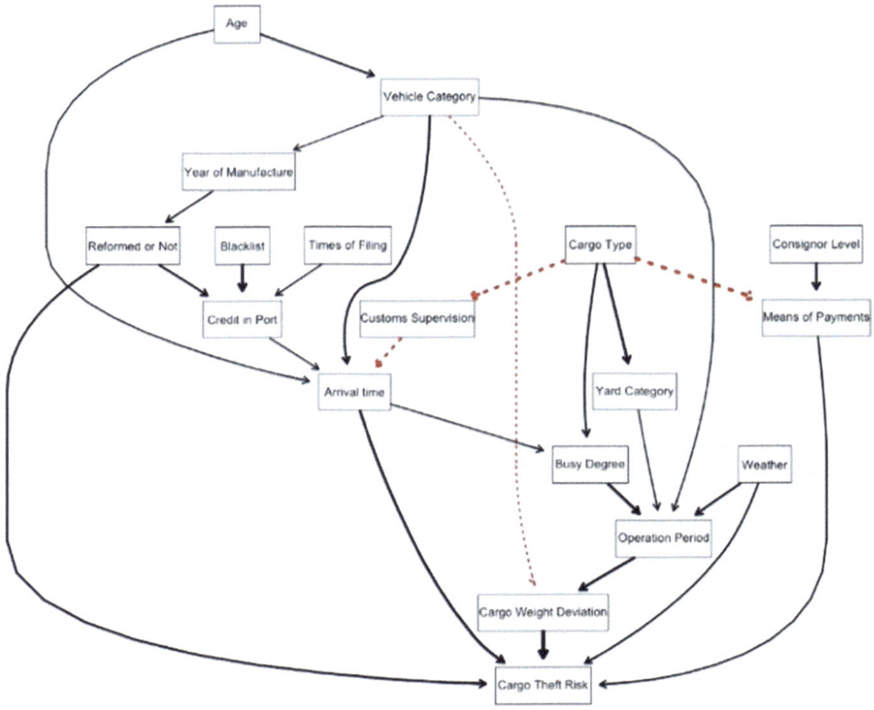


**Fig. 4** The Bayesian network structure derived using HC and k2

Since the training set is not big, Bayesian posterior probability estimation performs better in this case. For instance, Fig. 5 shows the conditional probability distribution of the 'Cargo Theft Risk' node using the structure learning algorithm with HC, the score function with k2, the parameter learning algorithm with Bayesian posterior probability estimation and five as the maximum number of father nodes.

The subfigure in the bottom right corner (highlight in green rectangle) of Fig. 5 indicates that the cargo theft risk is highest when 'Cargo Weight Deviation' equals 2, 'Means of Payments' equals 2, 'Reformed or Not' equals 1, 'Arrival time' equals 2 and 'Weather' equals 1. In other words, this means that when a reformed truck arrives at the port between 10 pm and 6 am to pick up a delivery of prepaid cargo and the cargo weight deviates greatly from the normal value, then there is a significant risk of the bulk cargo theft.

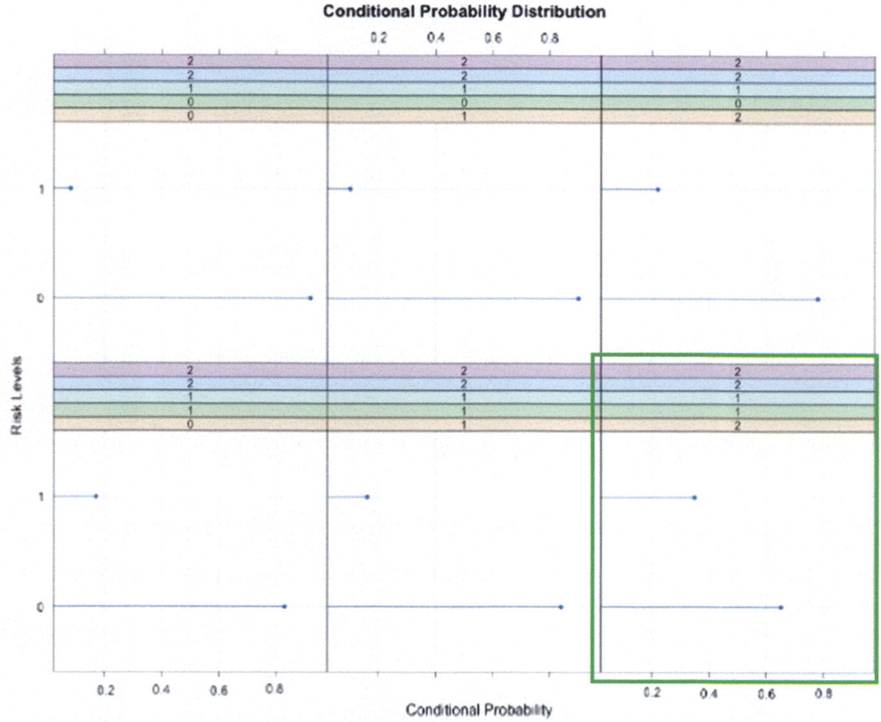


**Fig. 5** The conditional probability distribution of 'Theft Risk' node

## 4.5 Model Validation and Model Testing

We performed cross validation using Bayesian posterior probability estimation and Maximum Likelihood estimation. For instance, Fig. 6 presents the predictive error of Bayesian posterior probability estimation.

In the case of structure learning algorithm with HC, the medians of the predictive errors with Aic, Bic, BDe and k2 are similar, and the range is relatively small. Nevertheless, there is an outlier in the predictive errors of HC.Aic, which indicates the volatility. Moreover, the predictive error of HC.k2 has the smallest range while also being most stable. By comparing the medians of the predictive errors, Bayesian posterior probability estimation performs better than Maximum Likelihood estimation as a whole. Moreover, HC performs better than TABU. Furthermore, the HC.k2 with Bayesian posterior probability estimation performs the best.

Figure 7 presents the results of the model testing with various combinations of structure learning algorithms and score functions. All models except for TABU. Aic, have accuracy above 80%, which indicates that the risk prediction model based on Bayesian network has high reliability and accuracy. The predictive model of HC.

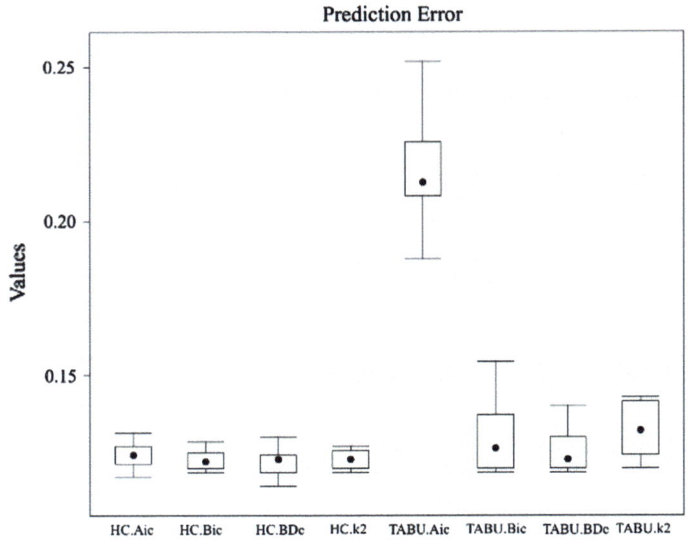


**Fig. 6** The predictive error of cross validation using Bayesian posterior probability estimation

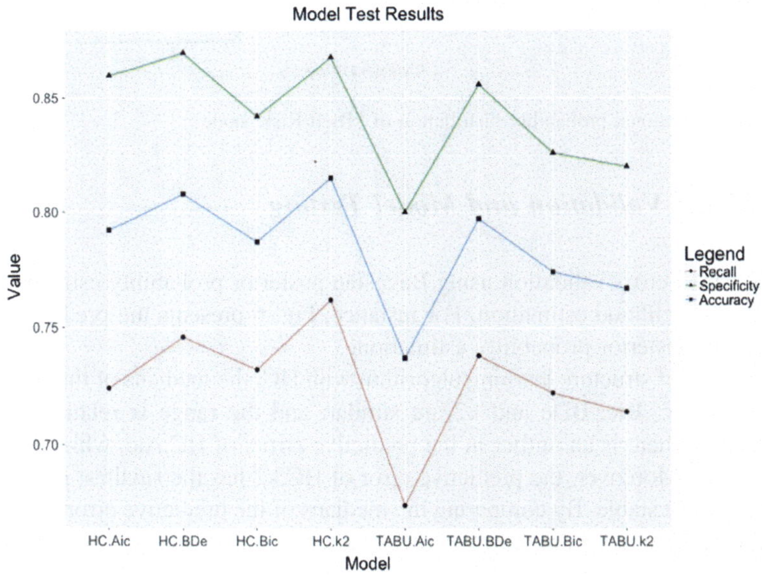


**Fig. 7** The results of the model testing

k2 has the highest accuracy and recall, and the predictive model of HC. BDe has the highest specificity. Considering three indexes of accuracy, recall, specificity, the predictive models using HC performs better than models using TABU.

## 5 Conclusion and Future Work

In this paper, we presented a novel data-driven methodology to predict the theft risk of bulk cargo in port. First, key features of the theft risk were selected using combinations of feature ranking methods and classification algorithms. Then Bayesian networks were constructed from the event data of the bulk cargo theft extracted from sampling inspection records of delivery trucks and the operational records in logistics information systems of port. Furthermore, we evaluated and comparatively analyzed these predictive models derived using various structure learning algorithms and parameter learning algorithms considering the real world scenario. Finally, domain knowledge and expert expertise were explored to refine the model and add counter-measures.

The validation and testing results with indexes of accuracy, recall, specificity all above 80% indicate that the predictive models are capable to effectively predict the theft risk of bulk cargo in port. The optimal predictive model is selected using a comparative analysis, which can be utilized for organizations of interest to pre-control the theft risk and narrow down the list of suspect trucks. Moreover, the predictive model of Bayesian network provides a visually understandable representation of the cargo theft risk, with the relations between relevant elements involved. Other than the 'black box' characteristic of some classification methods, Bayesian network provide better interpretability, which is essential for management optimization.

Potential future work directions are: a) Refine the predictive model via expert interview and practical application. b) Include additional features in the model to improve the accuracy and the effectiveness, e.g., currently we have not considered the communication of driver of delivery trucks, while these theft risk actually spread through the social contact of truck drivers.

## References

1. Hui, S. L., Thai, V. V., Wong, Y. D., Yuen, K. F., & Zhou, Q. (2013). Portfolio of port-centric supply chain disruption threats. *International Journal of Logistics Management, 28*(6), 1368–1386.
2. Constantinos, I. C., Petros, L. P., & Emestos, S. T. (2016). A risk assessment methodology in container terminals: The case study of the port container terminal of Thessalonica, Greece. *Journal of Traffic and Transportation Engineering, 4*(5), 251–258.
3. Wang, G., Gunasekaran, A., Ngai, E., & Papadopoulos, T. (2016). Big data analytics in logistics and supply chain management: Certain Investigations for research and applications. *International Journal of Production Economics, 176,* 98–110.
4. Mokhtari, K., Ren, J., Roberts, C., & Wang, J. (2011). Application of a generic bow-tie based risk analysis framework on risk management of sea ports and offshore terminals. *Journal of Hazardous Materials, 192*(2), 465–475.
5. Chen, H., Chiang, R., & Storey, V. C. (2012). Business intelligence and analytics: from big data to big impact. *MIS Quarterly, 36*(4), 1165–1188.

6. Miorandi, D., Sicari, S., Pellegrini, F. D., & Chlamtac, I. (2012). Internet of Things: vision, applications and research challenges. *Ad Hoc Networks, 10*(7), 1497–1516.
7. Toth, G. E., Depersia, A. T., & Pennella, J. J. (1998). CargoTIPS: An innovative approach to combating cargo theft. *The International Society for Optical Engineering, 3375,* 315–319.
8. Pallis, L. P. (2017). Port risk management in container terminals. *Transportation Research Procedia, 25,* 4411–4421.
9. Khakzad, N., Khan, F., & Amyotte, P. (2011). Safety analysis in process facilities: Comparison of fault tree and Bayesian network approaches. *Reliability Engineering and System Safety, 96*(8), 925–932.
10. Chawla, N. V., Bowyer, K. W., Hall, L. O., & Philip Kegelmeyer, W. (2012). SMOTE: Synthetic Minority Over-sampling Technique. *Journal of Artificial Intelligence Research, 16*(4), 321–357.
11. de Sá, C. R., Soares, C., & Knobbe, A. (2016). Entropy-based discretization methods for ranking data. *Information Sciences, 329,* 921–936.
12. Chandrashekar, G., & Sahin, F. (2014). A survey on feature selection methods. *Computers & Electrical Engineering, 40*(1), 16–28.
13. Li, J., Cheng, K., Wang, S., Morstatter, F., Trevino, R. P., Tang, J., & Liu, H. (2016) Feature selection: A data perspective. *ACM Computing Surveys, 50*(6), 1–45.
14. Scutari, M. (2009). Learning Bayesian networks with the bnlearn R Package. *Journal of Statistical Software, 35*(3), 1–22.

# Evolutionary Game Analysis of Service Outsourcing Strategy for Manufacturing Enterprises Based on Servitization

**Ruize Ma, Changli Feng, and Lin Jiang**

**Abstract** The transformation of manufacturing enterprises from traditional manufacturing to service-oriented manufacturing is one of the most important ways to upgrade and enhance competitiveness. Whether the manufacturing enterprise outsources the service business to the producer service providers or internalize the service business is an important problem in the transformation process. Based on the profit function of manufacturing enterprises and producer service providers, this paper establishes a two-party asymmetric evolution game model, explores the rules of manufacturing enterprises and producer service providers to implement servitization strategies, and uses MATLAB simulation to deeply study the influence of relevant factors on the implementation of servitization strategy in cooperation between manufacturing enterprises and producer service enterprises. The results show that the successful implementation of service outsourcing strategies in manufacturing enterprises depends on manufacturers' own conditions and the market environment. The elasticity of demand for service product quality in external markets, and the cost coefficient of enterprises in providing service products are both positively related to the willingness of manufacturing enterprises and producer service providers to cooperate in implementing servitization strategies. The loss caused by unsuccessful cooperation is negatively related to the willingness of cooperation between the two parties.

**Keywords** Manufacturing enterprises · Servitization · Evolutionary game · Service outsource

---

Fund Project: National Social Science Fund Project (16BGL08).

---

R. Ma (✉) · C. Feng · L. Jiang
School of Economics and Management, Dalian University of Technology, Dalian, China
e-mail: rzma@mail.dlut.edu.cn

C. Feng
e-mail: fengcl@dlut.edu.cn

L. Jiang
e-mail: 1956861381@qq.com



J. Zhang et al. (eds.), *LISS2019*,
https://doi.org/10.1007/978-981-15-5682-1_19

# 1 Introduction

Since Vandermerwe and Rada state that the servitization of manufacturing enterprises is a concept that the manufacturing enterprise has changed from providing only physical products, or the combination of physical products and simple services to providing "good+service package" [1], there have been growing studies in this field. By analyzing North American automotive manufacturing industry data, Vickery found that integrating service elements into manufacturing products can increase corporate profits [2], Ambroise analyzes the financial data of 184 manufacturing enterprises and finds that after the implementation of servitization, manufacturing enterprises can interact with clients, so that they can discover more potential profit opportunities [3]; Based on the data from Swedish manufacturing enterprise, Visnjic and Van claim that after adding services to products, the manufacturers' profits increased exponentially with the increase of the service scales [4]. Shanghai Electric Group expands its market share by implementing a servitization strategy, increasing 10 billion yuan of sales value every year by services such as design and installation [5]. The Chinese government also attaches great importance to the servitizaion work of manufacturing enterprises, in the promulgation of "Made in China 2025", service-oriented manufacturing is listed as an important strategic task [6]. Therefore, the research on promoting the implementation of the servitization strategies is an important theoretical and practical work.

Manufacturing enterprises are constrained by their own resources in the process of service transformation, and often need to outsource service business to professional producer service providers to complete servitization strategy [4]. For example, Shenyang Machine Tool outsources the financial services to Hua Xia Bank, and meets the individual needs of customers through financial instruments [7]. Haier Group cooperates with New Zealand FPA and GE's R&D department to design fine decoration and intelligent systems for customers [8], Shaanxi blower group outsources the equipment maintenance business to Beijing Zhongru Technology company to provide after-sales maintenance services [9]. However, IBM has gradually withdrawn from the low value-added manufacturing sector through its own service business, and has completely transformed into an information system solution provider [10], GE also achieves higher returns than its competitors through self-financial and medical services, using the "goods+service package" service model [8]. Therefore, in the process of transformation, whether enterprises outsource service business to producer service providers, or internalize service business is a very valuable problem [8]. However, in the existing research, the manufacturing enterprises mainly put eyes on their own servitization strategies, while less attention is paid to the research on the servitization of manufacturing enterprises with the help of producer service enterprises. Focus on the lack of existing research, this paper analyzes how manufacturing enterprises can conduct servitization strategies with producer service providers.

The process of jointly implementing servitization strategies by manufacturing enterprise joint producer service providers is accompanied by the problems of income distribution and cost sharing, and is the game process of related enterprises [8]. This process is affected by many factors, including: the elasticity of demand for the quality of service products in the external market of enterprises [11]; the cost coefficient of providing service products by enterprises [12]; the cooperation costs of the other parties due to the unwillingness of one party to cooperate loss [13]. After studying the above influencing factors, this paper combines the cost-benefit theory in economics to construct the profit function of each stakeholder, and establishes an asymmetric evolutionary game model to solve the evolutionary stability strategy (ESS) of both sides in the game. Evolutionary game theory is based on bounded rationality, combined with game theory analysis and dynamic evolutionary process analysis. It is an effective method to analyze such problems. Finally, exploring the mechanism of the various influencing factors on the implementation of cooperation between manufacturing enterprises and producer service providers through MATLAB simulation. The results can not only provide theoretical guidance for enterprises to implement servitization strategies, but also fill the research gap of the implementation of servitization between manufacturing enterprises and producer service providers, enrich and develop the servitization theory.

## 2 Evolutionary Game Model Construction

### *2.1 Problem Description*

With the advent of the service economy era, the personalized needs of customers have driven the increasing demands of manufacturing enterprises for service products, and manufacturers have transformed from traditional product manufacturers to product service integrators. Manufacturing enterprises can directly carry out services, and can also outsource services to producer service providers, so that service producer service providers can replace manufacturers in the market. The mode of joint service between manufacturing enterprises and producer service providers can increase the income source of producer service providers and enhance the competitiveness of manufacturing enterprises [3], making manufacturing enterprises in an advantageous position in market competition [11]. However, in the process of outsourcing, manufacturing enterprises and producer service providers will be affected by many factors, which may cause loss of profits for both sides of the game and lead to the failure of service outsourcing. What's more, the strategy space of manufacturing enterprises is (outsourcing, self-operated), and that of producer service providers is (cooperation, non-cooperation). The game model is built based on the strategy and revenue relationship of both sides. The behavior evolution of both sides is studied according to the replicated dynamic equation.

## 2.2 Model Assumption

**Assumption 1.** In the initial stage of the game, the probability of manufacturing enterprises choosing "outsourcing" strategy is $x(0 \leq x \leq 1)$, the probability of selecting "self-operated" strategy is $1-x$; the probability that a producer service provider chooses to "cooperate" with a manufacturing enterprises is $y\ (0 \leq y \leq 1)$, and the probability of "uncooperative" is $1-y$. Moreover, both sides of the game are considered with bounded rationality, and the strategy of evolutionary stability (ESS) is achieved after learning, imitating and trial-and-error in the repeated games.

**Assumption 2.** From the perspective of the external environment of the manufacturing enterprise, with increased customer demand for product services, the enterprises are unable to respond quickly to customer needs due to limited resources, thus the enterprises' willingness to outsource the service business to a professional producer service provider is more intense [14]. We assume that $\alpha$ is the demand quality elasticity, which means the degree of customer demand for product service quality ($0 \leq \alpha \leq 1$). When $\alpha$ tends to 1, the customer's requirements for product service quality are extremely high, which is conducive to the cooperation between manufacturing enterprises and producer service providers. Conversely, when $\alpha$ tends to 0, the customer's demand for product service quality is extremely low, which is not conducive to the cooperation between manufacturing enterprises and producer service providers.

**Assumption 3.** From the perspective of the internal environment of manufacturing enterprises, enterprises are constrained by their own resources, the cost of entering the service field is higher, which makes enterprises tend to outsource service business that they are not good at to producer service providers [13]. Therefore, we assume that $\beta$ is the quality cost coefficient, which represents the degree of difficulty in providing services ($0 \leq \beta \leq 1$). When $\beta$ tends to 1, it means that the cost of the service business provided by the enterprise is extremely high. Conversely, when $\beta$ tends to 0, the cost of providing services is extremely low.

**Assumption 4.** In order to highlight the leading role of services in the "good +service package", we assume that the quality of the service-oriented product mainly depends on the quality of the service, regardless of the influence of other factors, so the demand of manufacturing enterprises is a function of the price and quality of service products they provide.

**Assumption 5.** Similar to the practice of many scholars (such as Bernstein et al. [12]), the demand function is a linear function of price and quality, demand price elasticity is 1.

**Assumption 6.** According to Sunghee's [8] latest research and related literature on quality cost [12], it is assumed that the technical conditions of self-service provision by manufacturing enterprises are the same as those of productive service enterprises, and the cost of service is a quadratic index (concave) function of quality.

**Assumption 7.** When a manufacturing enterprise implements a service outsourcing strategy but a producer service provider adopts a non-cooperative strategy, the manufacturing enterprise needs to pay extra cost $c_G$ (such as opportunity cost, time cost et al.). Conversely, when a producer service provider adopts a cooperative strategy and the intention to manufacture enterprise service outsourcing is not strong, the service provider needs to pay an additional cost $c_S$. When a manufacturing enterprise outsources the service business to a producer service provider, the income generated by producer service providers is $u$.

**Assumption 8.** When a manufacturing enterprise purchases a service product from a producer service provider, it must pay a related fee at a time, such as purchasing a technology, and the cost does not change with the demand of the product in a short period of time.

**Assumption 9.** The service capability of the manufacturing enterprise is lower than the producer service provider($q_S > q_M$). After manufacturing enterprise successfully implements the service outsourcing strategy, the service cost is reduced with the help of the producer service provider. The model parameters and meanings are shown in Table 1.

## 2.3 *Model Building*

Evolutionary game theory is that in the initial state of a given group, each of the bounded rationality individuals learns from each other and makes decisions over time, finally reaches a stable state. Relatively independent manufacturing enterprises and producer service providers coexist in the market to form a group. After reading researches from Bernstein [12] and Huang [13], this paper combines the cost-benefit theory in economics to propose the market profit function of manufacturing enterprises ($i = M$) and producer service providers ($i = S$):

**Table 1** Model parameter summary table

| Parameter | Meaning |
|---|---|
| $\alpha$ | Demand quality elasticity ($0 \leq \alpha \leq 1$) |
| $\beta$ | Quality cost coefficient ($0 \leq \beta \leq 1$) |
| $p_i$ | Product price (i = *M*, *S*) |
| $q_i$ | Product quality (i = *M*, *S*) |
| $D_i$ | Demand function (i = *M*, *S*) |
| $C_i$ | Cost function (i = *M*, *S*) |
| $\pi_i$ | Profit function (i = *M*, *S*) |
| $c_i$ | Unilateral service cost (i = *M*, *S*) |
| $u$ | Income increase |
| $u_i$ | Market foundation (i = *M*, *S*) |

[a] *Note* Subscript M stands for manufacturing products and S stands for service

$$D_i = u_i + \alpha q_i - p_i \tag{1}$$

$$C_i = \frac{1}{2}\beta q_i^2 \tag{2}$$

$$\pi_i = u_i p_i + \alpha q_i p_i - p_i^2 - \frac{1}{2}\beta q_i^2 \tag{3}$$

Equation (1) represents the demand function of the manufacturing enterprise or the producer service providers. The demand of the enterprise product equal to the enterprise market basic demand plus the demand increase caused by the service quality, minus the demand reduction caused by the price. Cause the cost function ($C_i$) is a quadratic exponential function of the quality of service ($q_i$). (2) represents the enterprise cost function, from which the profit function ($\pi_i$) of the enterprise of (3) is obtained, which is equal to the demand function multiply by the price product minus the cost.

According to the definition of the assumptive partial variables and the above definition of the profit function, the game income matrix of the manufacturing enterprise and the producer service providers is constructed, as shown in Table 2.

When a manufacturing enterprise games with a producer service provider, the benefits of the two sides of the game are different under different strategic combinations. If the manufacturing enterprise and the producer service provider adopt an outsourcing and cooperation strategy respectively, in (1), the quality of products put into the market by manufacturing enterprises is $q_S$. According to Assumption 8, the manufacturing enterprise must pay a one-time fee $p_S$ for the service, and the producer service provider obtains the profit($u$) due to the service outsourcing. Therefore, the profit of the manufacturing enterprise is $u_M p_M + \alpha q_S p_M - p_M^2 - p_S$, and the profit of the producer service provider is $\pi_S + u$.

If manufacturing enterprises adopt service self-operated strategy and producer service providers adopt cooperative strategy, service enterprises will pay extra cost $c_S$. At this time, the income of producer service providers is $\pi_S - c_S$, and that of manufacturing enterprises is $\pi_M$.

**Table 2** Game income matrix

| Expected benefits under different strategies | | Manufacturing enterprises | |
|---|---|---|---|
| | | *Outsourcing x* | *Self-operated 1 − x* |
| Producer service providers | *Cooperate y* | $\pi_S + u$, $u_M p_M + \alpha q_S p_M - p_M^2 - p_S$ | $\pi_S - c_S$, $\pi_M$ |
| | *Uncooperative1 − y* | $\pi_S$, $\pi_M - c_M$ | $\pi_S$, $\pi_M$ |

If producer service providers adopt non-cooperative strategy and manufacturing enterprises adopt service outsourcing strategy, manufacturing enterprises need to pay extra cost $c_M$. At this time, the income of manufacturing enterprises is $\pi_M$-$c_M$, while that of producer service providers is $\pi_S$.

If manufacturing enterprises and producer service providers adopt self-operated and non-cooperative strategies respectively, the profits of manufacturing enterprises and producer service providers are $\pi_M$ and $\pi_S$.

We calculate the expected revenue and average revenue of both players. As shown in (4)–(6), $f_1$ is the expected revenue when manufacturing enterprises choose service outsourcing strategy, $f_2$ is the expected revenue when manufacturing enterprises choose service self-operated strategy, and $\overline{f_{12}}$ is the average revenue of manufacturing enterprises.

$$f_1 = y(u_M p_M + \alpha q_S p_M - p_M^2 - p_S) + (1-y)(\pi_M - c_M) \tag{4}$$

$$f_2 = y\pi_M + (1-y)\pi_M = \pi_M \tag{5}$$

$$\begin{aligned}\overline{f_{12}} = xf_1 + (1-x)f_2 = x[y(u_M p_M + \alpha q_S p_M \\ - p_M^2 - p_S) + (1-y)(\pi_M - c_M)] + (1-x)\pi_M\end{aligned} \tag{6}$$

Similarly, as shown in (7)–(9), $f_3$ is the expected revenue when producer service providers choose cooperation strategy, $f_4$ is the expected revenue when producer service providers choose non-cooperation strategy, and $\overline{f_{34}}$ is the average revenue of producer service providers.

$$f_3 = x(\pi_S + u) + (1-x)(\pi_S - c_S) \tag{7}$$

$$f_4 = x\pi_S + (1-x)\pi_S = \pi_S \tag{8}$$

$$\begin{aligned}\overline{f_{34}} = yf_3 + (1-y)f_4 = y[x(\pi_S + u) \\ + (1-x)(\pi_S - c_S)] + (1-y)\pi_S\end{aligned} \tag{9}$$

The strategy growth rates *F(x, y)* and *G(x, y)* of both sides of the game are not only proportional to the proportion of individuals who choose this strategy at a certain time(*t*), but also proportional to the difference between the revenue and the average revenue. The dynamic replication equation between manufacturing enterprises and producer service providers is obtained, as shown in (10) and (11).

$$\begin{aligned} F(x,y) &= \tfrac{d_x}{d_t} = x\left(f_1 - \overline{f_{12}}\right) \\ &= x(1-x)(f_1 - f_2) \\ &= x(1-x)\left[y(\alpha q_S p_M - \alpha q_M p_M \right. \\ &\quad \left. - p_S + \tfrac{1}{2}\beta q_M^2 + c_M) - c_M\right] \end{aligned} \tag{10}$$

$$\begin{aligned} G(x,y) &= \tfrac{d_y}{d_t} = y\left(f_3 - \overline{f_{34}}\right) \\ &= y(1-y)(f_3 - f_4) \\ &= y(1-y)[x(u+c_S) - c_S] \end{aligned} \tag{11}$$

# 3 Evolutionary Game Model Solution

## 3.1 Local Stability Analysis of Model

Let $F(x, y) = 0$ and $G(x, y) = 0$, we can get 5 local equilibrium points, which are (0, 0), (0, 1), (1, 0), (1, 1), ($x^*$, $y^*$), where $x^* = \frac{c_S}{u+c_S}$, $y^* = \frac{c_M}{\alpha q_S p_M - \alpha q_M p_M - p_S + \frac{1}{2}\beta q_M^2 + c_M}$. The Jacobian matrix (12) is derived from the replicated dynamic (10) and (11), and the local stability of the equilibrium points of the system is analyzed.

$$J = \begin{bmatrix} \frac{dF}{d_x} & \frac{dF}{d_y} \\ \frac{dG}{d_x} & \frac{dG}{d_y} \end{bmatrix} = \begin{bmatrix} e_{11}, & e_{12} \\ e_{21}, & e_{22} \end{bmatrix} \tag{12}$$

$$\begin{aligned} e_{11} &= (1-2x)\left[y\left(\alpha q_S p_M - \alpha q_M p_M - p_S + \tfrac{1}{2}\beta q_M^2 + c_M\right) - c_M\right] \\ e_{12} &= x(1-x)\left(\alpha q_S p_M - \alpha q_M p_M - p_S + \tfrac{1}{2}\beta q_M^2 + c_M\right) \\ e_{21} &= y(1-y)(u+c_S) \\ e_{22} &= (1-2y)[x(u+c_S) - c_S] \end{aligned}$$

The determinant (13) and trajectory (14) of Jacobian matrix are calculated according to 5 equilibrium points.

$$\mathrm{Det}J = \frac{dF}{d_x}\Big|_{(x_i,y_i)} \times \frac{dG}{d_y}\Big|_{(x_i,y_i)} - \frac{dF}{d_y}\Big|_{(x_i,y_i)} \times \frac{dG}{d_x}\Big|_{(x_i,y_i)} \tag{13}$$

$$\mathrm{Tr}J = \frac{dF}{d_x}\Big|_{(x_i,y_i)} + \frac{dG}{d_y}\Big|_{(x_i,y_i)} \tag{14}$$

Based on the above assumptions and using the Jacobian matrix local stability analysis method, 5 local equilibrium points are brought into (13) and (14) respectively for local stability analysis. The results are shown in Table 3.

From the analysis of Table 3, we can see that there are (1,1) and (0,0) two evolutionary stabilization strategies (ESS) in the 5 equilibrium points due to the

**Table 3** Local stability analysis

| Equilibrium point | DetJ | Symbol | TrJ | Symbol | Local stability |
|---|---|---|---|---|---|
| $x = 0, y = 0$ | $c_M c_S$ | + | $-(c_M + c_S)$ | – | ESS |
| $x = 1, y = 0$ | $c_M u$ | + | $c_M + u$ | + | Instable |
| $x = 0, y = 1$ | $(\frac{1}{2}\beta q_M^2 - p_S)u$ | + | $p_S - \frac{1}{2}\beta q_M^2 + u$ | + | Instable |
| $x = 1, y = 1$ | $-c_S(p_S - \frac{1}{2}\beta q_M^2)$ | + | $\frac{1}{2}\beta q_M^2 - p_S + c_S$ | – | ESS |
| $x = x^*$, $y = y^*$ | $\frac{-c_M c_S u(\frac{1}{2}\beta q_M^2 - p_S)}{(u + c_S)(\alpha q_S p_M - \alpha q_M p_M - p_S + \frac{1}{2}\beta q_M^2 + c_M)}$ | – | 0 | 0 | Saddle point |

limited rationality of both sides of the game. They are two extreme situations in the game process of manufacturing enterprises and producer service providers, namely (outsourcing, cooperation) and (self-operated, non-cooperation) strategies. Among them, (0, 1) and (1, 0) are unbalanced points, ($x^*$, $y^*$) are saddle points, the saddle points are unstable, and the broken line composed of unbalanced points is the critical line convergent to the above two extreme cases. The dynamic process of the game between manufacturing enterprises and producer service providers is analyzed on the equilibrium solution domain $R = \{(x,y)|0 < x = 1, 0 < y = 1\}$. As shown in Fig. 2, when the initial probability $(x,y)$ of the strategies chosen by both sides falls in the right area of the broken line BOD, the strategies converge to the equilibrium point C(outsourcing, cooperation), that is, the strategy of service outsourcing is successfully implemented. Conversely, when the initial strategy probability falls on the left side of the broken line, it converges to equilibrium point A (self-operated, non-cooperation), that is, the implementation of service outsourcing strategy fails.

Figure 1 When $S_{ABOD} > S_{BCDO}$, the probability of failure to implement a service outsourcing strategy for manufacturing enterprises is large. When $S_{ABOD} = S_{BCDO}$, the probability of the system evolving to two stable strategies is equal. When $S_{ABOD} < S_{BCDO}$, the probability of service outsourcing success is greater than that of failure. In this paper, the area of $S_{ABOD}$ is denoted as $S_1$, the area of $S_{BCDO}$ is denoted as $S_2$, and $S_1 + S_2 = 1$. The formula of $S_1$ is (15). Therefore, the parameters affecting the area of $S_1$ are the parameters affecting the evolution strategies of both players in the game. Therefore, through the parameter analysis of $S_1$ formula, it can be transformed into the analysis of factors influencing the successful implementation of service outsourcing strategy.

$$S_1 = \frac{c_S(\alpha q_S p_M - \alpha q_M p_M - p_S + \frac{1}{2}\beta q_M^2) + c_M u + 2c_M c_S}{2(u + c_S)(\alpha q_S p_M - \alpha q_M p_M - p_S + \frac{1}{2}\beta q_M^2 + c_M)} \tag{15}$$

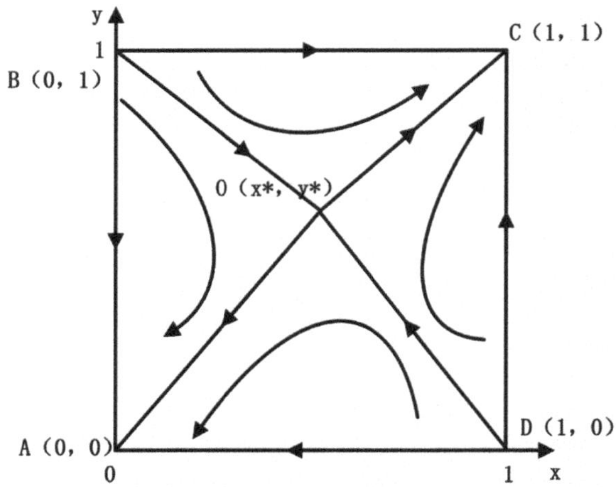


**Fig. 1** Phase diagram of strategic evolution

## 3.2 Analysis of the Results of Evolutionary Game Models

The following uses MATLAB to find the first-order partial derivative of the demand quality elasticity α, the quality cost coefficient β of the area (11) and the unilateral service outsourcing cost coefficient $c_i$ of the two sides of the game, and analyze the impact of the demand quality elasticity, the quality cost coefficient and the unilateral service outsourcing costs on the outcome of the game.

(1) *Demand quality elasticity*

For the equation $S_1$, the first-order partial derivative of the demand quality elasticity (α) is obtained $\frac{dS_1}{d_\alpha} = -0.5p_M(c_S+u)(q_S-q_M)(3c_Sc_M-2\alpha c_Sq_Mp_M-2c_Sp_S+\beta c_Sq_M^2+c_Mu+2\alpha c_Sq_Sp_M) < 0$, so $S_1$ is the monotonic decreasing function of α, that is, as the customer's demand for product service quality increases, the value of $S_1$ decreases, and the value of $S_2$ increases. Increased probability of game outcomes evolving into service outsourcing strategies, and the easier it is for manufacturing enterprises to successfully implement service outsourcing strategies.

(2) *Quality Cost Coefficient*

For the equation $S_1$, the first-order partial derivative of the quality cost coefficient (β) is obtained $\frac{dS_1}{d\beta} = -0.25q_M^2(c_S+u)(\beta c_Sq_M^2-2\alpha c_Sp_Mq_M+3c_Mc_S-2c_Sp_S+c_Mu+2\alpha c_Sp_Mq_S)<0$, so $S_1$ is the monotonical decreasing function of β, that is, as the degree of service difficulty of the enterprise increases, the value of $S_1$ decreases, the value of $S_2$ increases, the probability of both sides of the game evolving to service outsourcing strategy increases, making it easier for manufacturing enterprises successfully implement service outsourcing strategies.

(3) *Unilateral Service Outsourcing Costs*

For the equation $S_1$, the first-order partial derivative of the unilateral service outsourcing cost $c_G$ of the manufacturing enterprise and the unilateral service outsourcing cost $c_S$ of the producer service providers are obtained $\frac{dS_1}{dc_G} = (c_S + u)(c_M c_S + 0.5 c_M u + 0.5 c_S(\alpha q_S p_M - \alpha\ q_M p_M - p_S + \frac{1}{2}\beta q_M^2)) + (c_S + u)(c_S + 0.5u)(\alpha q_S p_M - \alpha\ q_M p_M - p_S + \frac{1}{2}\beta q_M^2 + c_M) > 0$, $\frac{dS_1}{dC_s} = 0.125(2\alpha q_s p_M - 2\alpha q_M p_M - 2p_S + \beta q_M^2 + 2c_M)(8c_M c_S - 4c_S p_S + 6c_M u - 2p_S u + 2\beta c_S q_M^2 + \beta q_M^2 u - 2\alpha q_M p_M u + 2\alpha q_S p_M - 4\alpha c_S q_M p_M + 4\alpha c_S q_S p_M) > 0$ respectively, so that $S_1$ is a monotonically increasing function of $c_G$ and $c_S$, that is, with the increase of $c_G$ and $c_S$, the value of $S_1$ increases, the value of $S_2$ decreases, the probability of both sides of the game evolving to service outsourcing strategy is reduced, making it difficult for manufacturing enterprises to implement service outsourcing strategies.

# 4 Numerical Simulation

## 4.1 Parameter Setting

In order to intuitively analyze the influence of different parameters on the evolution of outsourcing strategy in manufacturing enterprises, the parameters of this paper are set according to replication dynamic equation and assumptions conditions. The dynamic evolution process of the strategy is simulated by using MATLAB, and the influence of the factors on the evolutionary stability strategy (ESS) trajectory in the outsourcing process of manufacturing enterprises is further analyzed. The numerical examples are $\alpha = 0.7$, $\beta = 0.88$, $p_M = 0.25$, $p_S = 0.175$, $q_M = 0.2$, $q_S = 0.45$, $c_M = 0.7$, $c_S = 0.5868$, $u = 0.2$.

## 4.2 Simulation Analysis

Based on the above experimental data, numerical simulation is carried out to more intuitively analyze the effects of demand quality elasticity $\alpha$, quality cost coefficient $\beta$ and unilateral outsourcing cost $c_i$ on the evolution of game strategy.

(1) *The Impact of Demand Quality Elasticity on Strategy Evolution*

The numerical simulation results of the effect of demand quality elasticity $\alpha$ on $S_1$ are shown in Fig. 2 (a), with the increase of $\alpha$ from 0 to 1, the value of $S_1$ decreases linearly in the interval [0.52, 0.64], which makes the game strategy

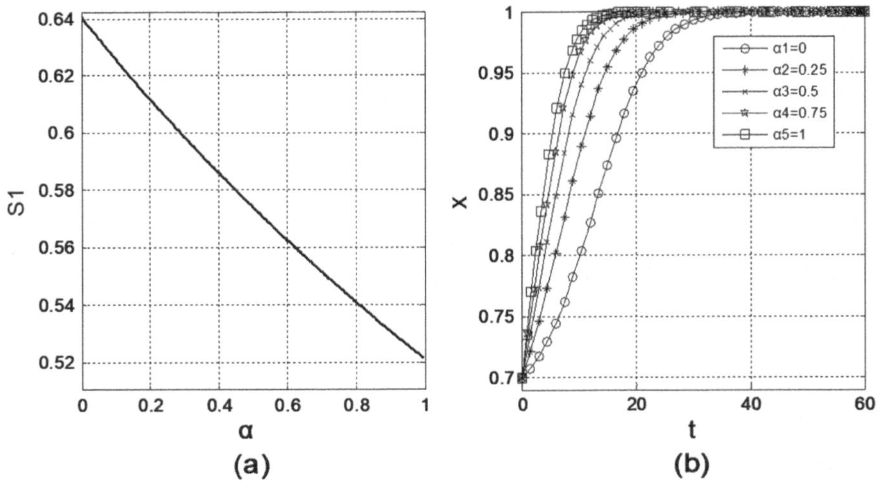


**Fig. 2** The effect of demand quality elasticity on strategy evolution

converge to (outsourcing, cooperation). This shows that demand quality elasticity is one of the factors to promote the service outsourcing strategy.

The abscissa *t* of Fig. 2 (b) is the dynamic evolution time, and the ordinate *x* is the probability of choosing service outsourcing strategy of manufacturing enterprises. With the gradual increase of α, that is, the higher the customer's demand for service quality of service products, the greater the slope of the curve, the faster the evolution speed, and the stronger the willingness of manufacturing enterprises to cooperate with producer service providers, which can give full play to the professional service capabilities of producer service providers, so the manufacturing enterprises can provide integrated solutions to customer needs, improve the level of end-customer service and increase manufacturing enterprises' income, enhance the competitiveness of manufacturing enterprises, and promote manufacturing enterprises and producer service providers to reach a servitization strategy alliance.

(2) *The Effect of Quality Cost Coefficient on Strategy Evolution*

The simulation results of the effect of quality cost coefficient β on the value of $S_1$ are shown in Fig. 3 (a). As β increases from 0 to 1, the value of $S_1$ decreases approximately linearly in the interval [0.47, 0.74], which makes the game strategy converge to (outsourcing, cooperation). This shows that the increase of quality cost coefficient is one of the factors to promote the success of service outsourcing strategy. Comparing the slope of Fig. 2 (a) and Fig. 3 (a), we can see that β has a stronger influence on the evolution of service outsourcing strategy than α, that is,

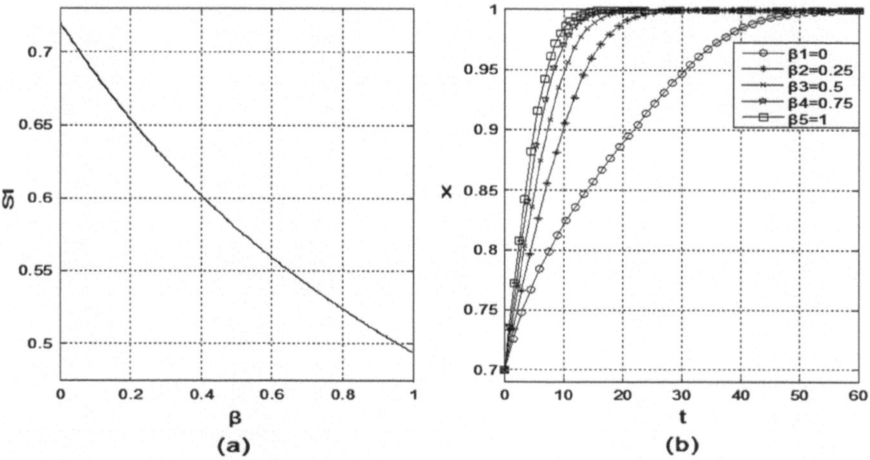


**Fig. 3** Impact of quality cost coefficient on strategy evolution

demand quality elasticity has a more significant effect on the promotion of service outsourcing strategy than quality cost coefficient.

Figure 3 (b) shows that with the increase of β value, manufacturing enterprises and producer service providers evolve into service outsourcing strategy. And the higher the quality cost coefficient, the faster the evolution rate of service outsourcing strategy. It can be explained that, with the increasing difficulty of providing services, manufacturing enterprises will outsource services that they are not good at to producer service providers. Compared with the "good+service package" of service providers after establishing strategic alliance relationship, manufacturing products can better meet the individual needs of customers, enhance customer loyalty and bring competitive advantages to service enterprises alliance. and the profits of manufacturing enterprises and service providers are increased. Thus, it promotes the implementation of service outsourcing strategy in manufacturing enterprises.

(3) *The Impact of Unilateral Service Outsourcing Cost on Strategic Evolution*

As shown in Fig. 4 (a), $S_1$ increases with the increase of manufacturing enterprises' unilateral outsourcing cost $c_G$ and producer service providers' unilateral servicing cost $c_S$. In the later period, $S_1$ grows slowly, and the slope of $c_S$ is larger. This shows that with the increase of the cost of unilateral service outsourcing, the game strategy converges to (self-operated, non-cooperation). Compared with manufacturing enterprises, producer service providers are more vulnerable to the impact of unilateral service outsourcing costs and shake the willingness of service outsourcing, and the impact of unilateral service outsourcing costs on service strategy also has the law of diminishing marginal benefits.

According to Fig. 4 (b) and (c), with the gradual reduction of the cost of unilateral service outsourcing and the acceleration of its evolution, manufacturing

**Fig. 4** The impact of unilateral service outsourcing cost on strategic evolution

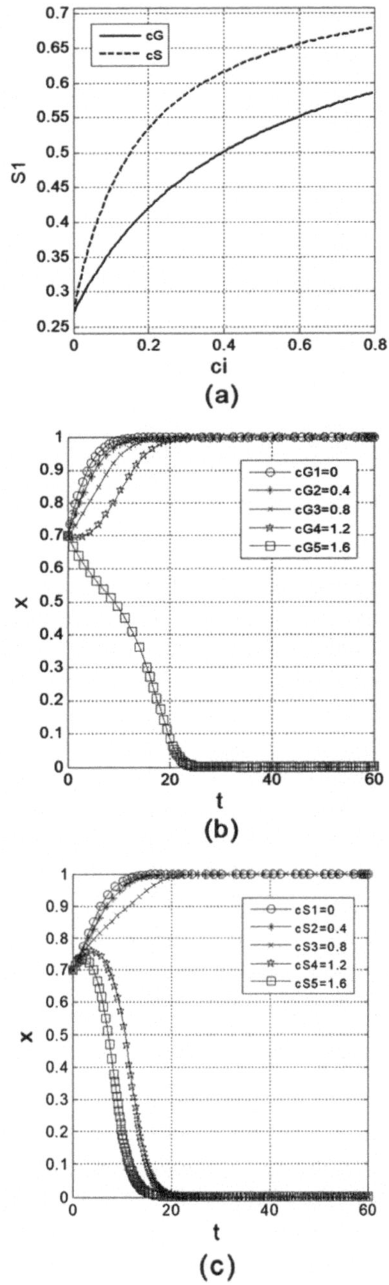


enterprises and producer service providers can choose service outsourcing strategies. When the unilateral service cost of manufacturing enterprises is too high ($c_G = 1.6$) and the unilateral service cost of producer service providers is too high

($c_S = 1.2$ or $c_S = 1.6$), the anticipation of service input cost of enterprises is increased, which leads to the evolution of game strategy towards (self-operated, non-cooperation).

## 5 Conclusion

The servitization strategy is an important path for the transformation and upgrading of manufacturing enterprises, and how to carry out servitization is an important issue. In order to explore whether the manufacturing enterprise should self-operate the service business, or outsource the service to the producer service providers, this paper uses the cost-benefit theory to construct the main stakeholder profit function in the market, establish the evolutionary game model of the manufacturing enterprise and the producer service providers, analysis the factors that affect the implementation of service outsourcing strategies between manufacturing enterprise and producer service providers through the MATLAB Simulation. According to the conclusions of this paper, the decision makers of manufacturing enterprises are provided with suggestions on the implementation of servitization strategy under the cooperation of producer service providers.

The higher the customer's requirements for service quality of service products, the stronger the willingness of manufacturing enterprises to outsource services, and the more favorable to the cooperation between manufacturing enterprises and producer service providers. Manufacturing enterprise decision-makers should pay attention to the customers' demand for service quality, focus on establishing cooperation relationships with their product-related service departments, establish a good interaction mechanism, create a good communication environment, actively communicate and coordinate with partners, and develop effective solutions for customers, making the manufacturing enterprise and the producer service providers take service as a link of mutual benefit.

The greater the difficulty of providing high-quality services, the more incentives for manufacturing enterprises to actively cooperate with producer service providers to implement servitization strategies. Decision makers of manufacturing enterprises should give full play to the professional service capabilities of producer service providers, recognize their own shortcomings, actively cooperate with producer service providers from the perspective of cost reduction, integrate service elements into products, improve product quality, and meet the personalized needs of customers.

Reducing the cost of unilateral service outsourcing can promote the successful implementation of service outsourcing strategies, and producer service providers are more susceptible to unilateral service costs and shake the willingness of service outsourcing. Therefore, decision-makers in manufacturing enterprises should pay attention to the control of unilateral service costs of producer service providers, reduce the cost expectations of producer service providers, and enhance the willingness of cooperation. Because the impact of unilateral service outsourcing cost on service outsourcing strategy has a law of diminishing marginal benefits, enterprise decision makers should pay attention to the control of initial cost of service.

## References


1. Sandra, V., & Juan, R. (1988). Servitization of business: Adding value by adding services. *European Management Journal, 6*(4), 314–324.
2. Vickery, S. K., Jayaram, J., & Droge, C. (2003). The effects of an integrative supply chain strategy on customer service and financial performance: An analysis of direct versus indirect relationships. *Journal of Operations Management, 21*(5), 523–539.
3. Laure, A., Isabelle, P., & Christine, T. (2018). Financial performance of servitized manufacturing firms: A configuration issue between servitization strategies and customer-oriented organizational design. *Industrial Marketing Management, 71*(4), 54–68.
4. Kastalli, I. V., & Looy, B. V. (2013). Servitization: Disentangling the impact of service business model innovation on manufacturing firm performance. *Journal of Operations Management, 31*(4), 169–180.
5. Jgartner, D. (1999). Go downstream: The new imperative in manufacturing. *Harvard Business Review, 77*(5), 133–141.
6. Davies, A. (2003). *Are firms moving 'downstream' into high-value service* (Vol. 71, no. 4, pp. 314–324). London: Imperial College Press.
7. Guerrieri, P., & Meliciani, V. (2005). Technology and international competitiveness: The interdependence between manufacturing and producer services. *Structural Change and Economic Dynamics, 16*(4), 489–502.
8. Lee, S., Yoo, S., & Kim, D. (2016). When is servitization a profitable competitive strategy? *International Journal of Production Economics, 173*(4), 43–53.
9. Eades, K. M. (2014). *The new solution selling* (3rd ed.) (Vol. 2, pp. 68–73). New York: McGraw-Hill.
10. Wang, W. J., Lai, K. H., & Shou, Y. Y. (2018). The impact of servitization on firm performance: A meta-analysis. *International Journal of Operations and Production Management, 38*(7), 1562–1588.
11. Maciosek, M. V. (1995). Behind the growth of services. *Journal of Illinois Business Review, 52*(3), 3–6.
12. Bernstein, F., & Federgruen, A. (2004). A general equilibrium model for industries with price and service competition. *Operations Research, 52*(6), 868–886.
13. Huang, J., Leng, M., & Parlar, M. (2013). Demand functions in decision modeling: A comprehensive survey and research directions. *Decision Sciences, 44*(3), 557–609.
14. Gebauer, H. (2007). Identifying services strategies in product manufacturing companies by exploring environment strategy configurations. *Industrial Marketing Management, 37*(3), 278–291.

# Variable-Scale Clustering Based on the Numerical Concept Space

**Ai Wang, Xuedong Gao, and Minghan Yang**

**Abstract** Traditional data mining application is an iterative feedback process which suffers from over-depending on both business and data specialists' decision ability. This paper studies the variable-scale decision making problem based on the scale transformation theory. We propose the numerical concept space to model significant information and knowledge after business and data understanding. An algorithm of variable-scale clustering is also put forth. A case study on TYL product management demonstrates that our method is able to achieve accessible and available performance in practice.

**Keywords** Variable-scale clustering · Concept space · Product management · Decision making

## 1 Introduction

Traditional data mining application (see Fig. 1) is an iterative feedback process accomplished by both business and data specialists [1]. These highly people-dependent process affects not only the algorithm performance but also results accuracy. However, seldom studies focus on how to automatically determine whether next iteration should be launched, rather than human decision making.

Research in decision theory finds that people make decisions by means of multiple perspectives, hierarchies, and dimensions [2]. Moreover, the problem-solving

---

A. Wang (✉) · X. Gao
Donlinks School of Economics and Management, University of Science and Technology Beijing, Beijing, China
e-mail: wangai22222@126.com

X. Gao
e-mail: gaoxuedong@manage.ustb.edu.cn

M. Yang
Guanghua School of Management, Peking University, Beijing, China
e-mail: hankmyang@icloud.com


J. Zhang et al. (eds.), *LISS2019*,
https://doi.org/10.1007/978-981-15-5682-1_20

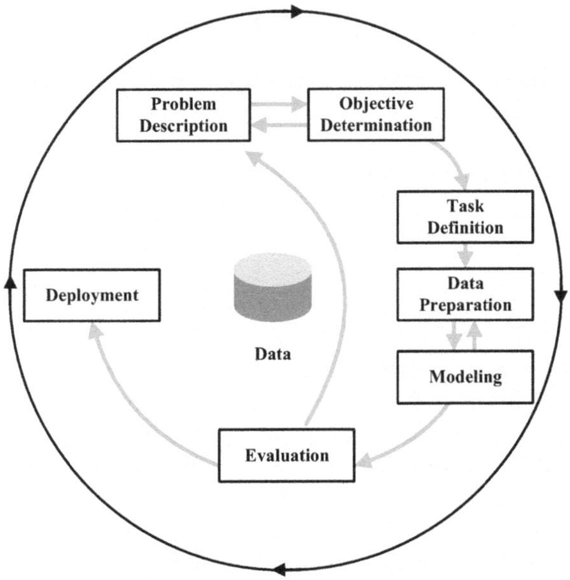


**Fig. 1** The application process of data mining project [1]

model in cognitive psychology proves that people solving problems through transforming different analysis dimensions in the problem space [3]. It can be seen that scale transformation contributes to the intelligent decision making.

Several researches have made some breakthrough in scale transformation analysis [4, 5]. Figure 2 describes the mechanism of variable-scale decision making. There are three main phases: (1) Multi-scale dataset establishment. The purpose of this stage is to model business and data understanding results through the concept space. (2) Scale transformation process control. This stage aims to obtain better analysis results through automatically transforming analysis scale (based on the scale transformation rate). (3) Result satisfaction degree evaluation. Since decision making prefer to get an acceptable solution for management activities instead of an unexplainable optimal result, this stage is to discover satisfy results by granularity deviation.

The current concept space model is built for character type data, which is not able to solve variable-scale decision making tasks in numerical dataset [6, 7]. Therefore, this paper studies the numerical variable-scale clustering problem for intelligent decision making.

The main contributions are as follows: First, we establish the numerical concept space ($CS^n$) to describe numerical dataset. Second, we propose a variable-scale clustering algorithm in the perspective of scale down transformation. Related measurements, such as the scale down transformation rate (SDTR) and granular deviation ($\widetilde{GrD}$), are also improved. Third, a difference management solution of TYL products is put forth. Experiment shows that our proposed algorithm is able to obtain accessible and available results in practice.

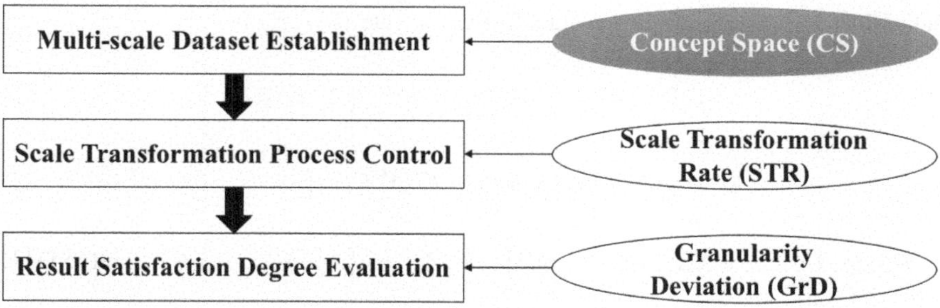


**Fig. 2** The mechanism of variable scale decision making

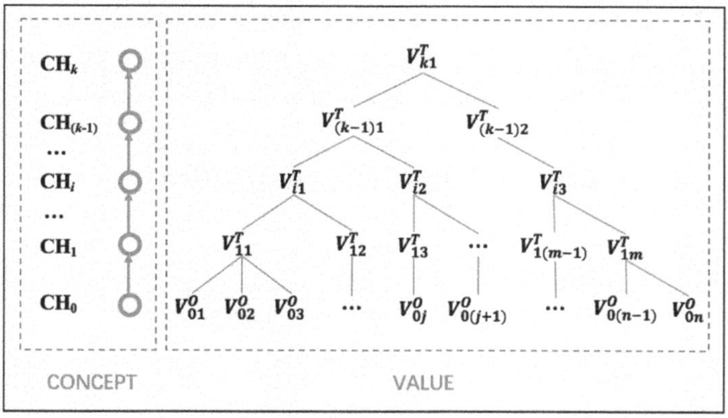


**Fig. 3** The numerical concept space

## 2 Research Method

### *2.1 Numerical Concept Space*

Concept space consists of two parts, i.e., concept chain and value space (see Fig. 3). In terms of a data mining project, concept chain is utilized to model the knowledge of business understanding, while value space aims to describe the information of data understanding [5].

According to the feature of numerical data, the hierarchy structure of numerical concept space ($CS^n$) has several principles: (1) there is the partial order relation between concepts on the lower and higher hierarchy, i.e. $CH_i \preccurlyeq CH_{(i+1)}$, (2) concept determines the value on the same hierarchy, i.e., $CH_i \Leftrightarrow \cup V_{ij}(V_{ij} \in N^+)$, (3) value obeys the relation of the concept on the same hierarchy, i.e., $CH_i \preccurlyeq CH_k \rightarrow (V_{ij} \not\preceq V_{kj}) \wedge (V_{ij} > V_{kj})$.

**Definition 1:** The scale down transformation rate (SDTR) is:

$$\mathrm{SDTR}\left(A^O, A^T\right) = \sum\nolimits_{j=1}^{m} \left|A^O\_\left(A_j^T\right)\right| / |U| \tag{1}$$

$$A^O\_\left(A_j^T\right) = \cup\{A_i^O | A_i^O \subseteq A_j^T\} \tag{2}$$

Where $U/A^O = \{A_1^O, A_2^O, \cdots, A_n^O\}$, $U/A^T = \{A_1^T, A_2^T, \cdots, A_m^T\}$, and SDTR $(A^O, A^T) \in [0, 1]$.

It can be seen that original scale $A^O$ is more similar to target scale $A^T$ when $\mathrm{SDTR}(A^O, A^T)$ grows larger.

**Definition 2:** Let $C_I$ represents a cluster with $n$ instances and $m$ attributes $\left(x_{ij} \in C_I\right)$, $a$ is an attribute of $C_I$, $C_P$ represents the nearest cluster of $C_I$ with $l$ instances $\left(x_{pj} \in C_P\right)$, the granular deviation of scale-down mechanism $(\widetilde{GrD})$ is:

$$\widetilde{\mathrm{GrD}}(C_I) = \sqrt{\sum\nolimits_{j=1}^{m} \bar{d}(x_{ij}, x_{Pj}) / G(U/a)} \tag{3}$$

$$\bar{d}\left(x_{ij}, x_{Pj}\right) = \sum_{i=1}^{n} \left|x_{ij} - x_{Pj}\right| / n \tag{4}$$

$$G(U/a) = \sum_{l=1}^{k} |a_i|^2 / n^2 \tag{5}$$

$$x_{Pj} = \sum_{p=1}^{l} x_{pj} / l \tag{6}$$

Where $x_{Pj}$ is the cluster center of $C_P$, $\bar{d}\left(x_{ij}, x_{Pj}\right)$ is the mean deviation towards $x_{Pj}$, $U$ is the universe of $C_I$, $G(U/a)$ is the knowledge granularity of $U/a$ $(U/a = \{a_1, a_2, \cdots, a_k\})$.

It can be seen that the cluster $C_I$ is more satisfy for mangers when $\widetilde{\mathrm{GrD}}(C_I)$ grows larger.

## 2.2 Variable-Scale Clustering Method

In this section, we improve the original variable-scale clustering algorithm in [5] from the perspective of scale-down transformation analysis via the numerical concept space ($CS^n$), scale down transformation rate (SDTR) and granular deviation $(\widetilde{GrD})$. The pseudo code is shown in Algorithm 1.

**Algorithm 1** *ScaleDownClustering* $(D, CS^n, S_0, k)$

```
C=D.normalized.initialCluster (k)
C_P = C_i.nearestCluster;
R_0 = min(GrD(C_i.qualified, C_P))
D.delete (C_i.qualified)
for D ≠ ∅ do
    for all A_j ∈ D do
        if SDTR(A_j, CH(A_j)) > S_0 then
            D.update (A_j, CH(A_j))
        break
        end if
    end for
    R_0 = min(R_0, GrD(C_i.closestQualified))
    C=D.normalized.Cluster (k-count (C_i.qualified))
    C_P = C_i.nearestCluster;
    for all C_i ∈ C do
        if GrD(C_i, C_P) > R_0 then
            D.delete (C_i)
        end if
    end for
end for
```

The time complexity of scale-down clustering algorithm is $O(nkt)$, where $n$ is the number of instances, $k$ is the number of clusters, and $t$ is the number of iterations.

## 2.3 Scale Transformation Theorem

**Theorem 1.** (Scale Down Transformation Theorem) A cluster that has already been qualified on a certain scale, is also qualified on the lower scale.

**Proof 1.** Given a cluster $C_I$, that has already been qualified on the scale $A^O$, a is the $k$ th attribute of $C_I$ ($a \in A^O$), $C_P$ represents the nearest cluster of $C_I$, $R_0$ is the threshold of $A^O$, consequently, $\widetilde{\mathrm{GrD}}(C_I) \geq R_0$. According to the hypothesis, there is only one attribute a that transforms to ȧ. Let $R_0$ represents the threshold of the target scale $A^T$ ($\dot{a} \in A^T$), $A^O \succeq A^T \rightarrow a < \dot{a}$.

$$\widetilde{\mathrm{GrD}}(C_I) = \sqrt{\frac{\sum_{j=1 \wedge j \neq k}^{m} \bar{d}(x_{ij}, x_{Pj}) + \bar{d}(x_{ik}, x_{Pk})}{G(U/a)}},$$

$$\widetilde{\mathrm{GrD}}(\dot{C}_I) = \sqrt{\frac{\sum_{j=1 \wedge j \neq k}^{m} \bar{d}(x_{ij}, x_{Pj}) + \bar{d}(\dot{x}_{ik}, \dot{x}_{Pk})}{G(U/\dot{a})}},$$

Let $\sum_{j=1 \wedge j \neq k}^{m} \bar{d}(x_{ij}, x_{Pj}) = b(b \in N)$.

$$\widetilde{\text{GrD}}(C_I) = \sqrt{\frac{\bar{d}(x_{ik}, x_{Pk}) + b}{G(U/a)}},\ \widetilde{\text{GrD}}(\dot{C}_I) = \sqrt{\frac{\bar{d}(\dot{x_{ik}}, \dot{x_{Pk}}) + b}{G(U/\dot{a})}}.$$

(1) Since $U/a = \{a_1, a_2, \cdots, a_l\}$, we assume that only the equivalence class $a_l$ is divided into the ST of a, i.e., $\dot{a}_l$ and $\dot{a}_{(l+1)}$.

$$U/\dot{a} = \{\dot{a}_1, \dot{a}_2, \cdots \dot{a}_l, \dot{a}_{l+1}\} = \{a_1, a_2, \cdots, \dot{a}_l, \dot{a}_{l-1}\}$$
$$a_l = \dot{a}_l, \dot{a}_{l-1}.$$

$$G(U/\dot{a}) = \frac{\sum_{r=1}^{l-1} |a_r|^2 + |\dot{a}_l|^2 + |\dot{a}_{(l+1)}|^2}{n^2},$$

$$G(U/a) = \frac{\sum_{r=1}^{l-1} |a_r|^2 + |a_l|^2}{n^2} = \frac{\sum_{r=1}^{l-1} |a_r|^2 + (|\dot{a}_l| + |\dot{a}_{(l+1)}|)^2}{n^2}.$$

Because $|\dot{a}_l|^2 + |\dot{a}_{(l+1)}|^2 < (|\dot{a}_l| + |\dot{a}_{(l+1)}|)^2$,

$$G(U/\dot{a}) < G(U/a).$$

(2) According to the feature of numerical concept space, $\dot{x}_{ik} \preccurlyeq x_{ik} \rightarrow \dot{x}_{ik} > x_{ik}$, and $\dot{x}_{Pk} \preccurlyeq x_{Pk} \rightarrow \dot{x}_{Pk} > x_{Pk}$.
Since $<\dot{x}_{ik}, \dot{x}_{Pk}>$ and $<x_{ik}, x_{Pk}>$ are both stays on the same hierarchy, let $\dot{x}_{ik} = \partial x_{ik} - \Delta_i$ and $\dot{x}_{Pk} = \partial x_{Pk} - \Delta_P$, where $\partial > 1$ and $\Delta \in [1, \partial]$.

$$\bar{d}(x_{ik}, x_{Pk}) = \frac{\sum_{i=1}^{n} |x_{ik} - x_{Pk}|}{n},$$

$$\bar{d}(\dot{x}_{ik}, \dot{x}_{Pk}) = \frac{\sum_{i=1}^{n} |\dot{x}_{ik} - \dot{x}_{Pk}|}{n},$$

$$\bar{d}(\dot{x}_{ik}, \dot{x}_{Pk}) - \bar{d}(x_{ik}, x_{Pk}) = \frac{\sum_{i=1}^{n} (|\dot{x}_{ik} - \dot{x}_{Pk}| - |x_{ik} - x_{Pk}|)}{n}.$$

(a) *If* $x_{ik} = x_{Pk}$, $|x_{ik} - x_{Pk}| = 0$.
Because $|\dot{x}_{ik} - \dot{x}_{Pk}| \geq 0$, $|\dot{x}_{ik} - \dot{x}_{Pk}| \geq |x_{ik} - x_{Pk}|$.

(b) *If* $x_{ik} > x_{Pk}$, $|x_{ik} - x_{Pk}| = x_{ik} - x_{Pk}$ *and* $|\dot{x}_{ik} - \dot{x}_{Pk}| = \dot{x}_{ik} - \dot{x}_{Pk}$.
Because $|\dot{x}_{ik} - \dot{x}_{Pk}| - |x_{ik} - x_{Pk}|$

$$= (\partial - 1)(x_{ik} - x_{Pk}) + (\Delta_P - \Delta_i)$$

$$> (\partial - 1) + (\Delta_P - \Delta_i)$$

$$\geq (\partial - 1) + (1 - \partial) = 0,$$

$$|\dot{x}_{ik} - \dot{x}_{Pk}| > |x_{ik} - x_{Pk}|.$$

(c) *If* $x_{ik} < x_{Pk}$, $|x_{ik} - x_{Pk}| = x_{Pk} - x_{ik}$ *and* $|\dot{x}_{ik} - \dot{x}_{Pk}| = \dot{x}_{Pk} - \dot{x}_{ik}$.

Similarly, $|\dot{x}_{ik} - \dot{x}_{Pk}| > |x_{ik} - x_{Pk}|$.
Because $\forall i(i \in [1, n])$, $|\dot{x}_{ik} - \dot{x}_{Pk}| \geq |x_{ik} - x_{Pk}| \geq 0$,
$\sum_{i=1}^{n} |\dot{x}_{ik} - \dot{x}_{Pk}| \geq \sum_{i=1}^{n} |x_{ik} - x_{Pk}|$.
Consequently, $\overline{d}(\dot{x}_{ik}, \dot{x}_{Ik}) \geq \overline{d}(x_{ik}, x_{Ik})$.

(3) Because $\overline{d}(\dot{x}_{ik}, \dot{x}_{Pk}) \geq \overline{d}(x_{ik}, x_{Ik})$ and $G(U/\dot{a}) < G(U/a)$, $\widetilde{\mathrm{GrD}}(\dot{C}_I) > \widetilde{\mathrm{GrD}}(C_I)$.
According to Algorithm 1, $\dot{R}_0 = \min(R_0, \widetilde{\mathrm{GrD}}(C_K))$,

$$R_0 \geq \dot{R}_0.$$

Besides, $\widetilde{\mathrm{GrD}}(C_I) \geq R_0$,
Thus, $\widetilde{\mathrm{GrD}}(\dot{C}_I) > \dot{R}_0$,

## 3 Experimental Analysis

In this section, we utilize our proposed method VSC to solve the product management problem in practice. A large machinery and equipment R&D group CALVT in China planned to establish a difference management system on TYL products, which refer to kinds of metal materials that could be applied universally to various engineering projects.

Figure 4 shows the management framework of TYL products in three dimensions, i.e., volume, variety and value.

According to business and data understanding, we establish the multi-scale dataset (see Table 1), where inventory represents volume, similarly, average usage means variety, and price expresses value. Every original scale has three meaningful

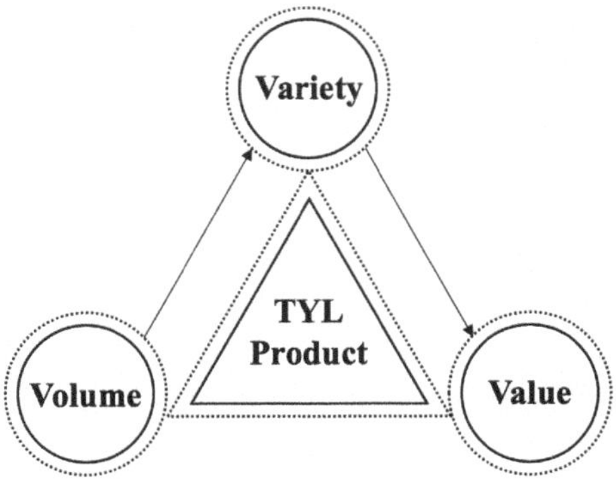


**Fig. 4** The TYL product management framework

target scales (shown in dark background) following the principles of numerical concept space.

According to the steps in Algorithm 1, we get the variable-scale decision making results (see Tables 2, 3 and 4), which includes the center, instance count and granular deviation ($\widetilde{\boldsymbol{GrD}}$) of every cluster. At the beginning, we select all the highest level scale, that is inventory (t), average usage (102 kg) and price (¥/g), in order to recognize satisfy product clusters efficiently. We discover three qualified clusters, which account for 38% of the total TYL products. During the second iteration, we lower the value dimension to price (102 ¥/kg) and also obtain three qualified TYL clusters. Finally, the remains six instances are divided into two qualified clusters in the third iteration under lower inventory (102 kg). It can be seen that none of the products is considered as an isolated outlier, which decreases the resource consumption in management process.

After considering the characteristics of these eight TYL product groups, different operation strategies and regulations are able to be designed directly. Cluster 2, 5, 7 have either minimum or maximum volume, compared to other TYL classes. Hence, fine inventory policies should be proposed towards products. Cluster 4 owns the highest average usage, while Cluster 3 the lowest. Consequently, termly procurement plan needs to be optimized. Last but not least, Cluster 1, 6, 8 have different price, especially Cluster 1 (over 8 ¥/g). Relevant TYL product value evaluation system should be established.

**Table 1** The multi-scale dataset

| TYL | Inventory (10^0 kg) | Inventory (10^1 kg) | Inventory (10^2 kg) | Inventory (10^0 t) | AvgUsage (10^2 g) | AvgUsage (10^0 kg) | AvgUsage (10^1 kg) | AvgUsage (10^2 kg) | Price (10^0 ¥/kg) | Price (10^1 ¥/kg) | Price (10^2 ¥/kg) | Price (10^0 ¥/g) |
|---|---|---|---|---|---|---|---|---|---|---|---|---|
| 1 | 1024 | 103 | 11 | 2 | 5612 | 562 | 57 | 6 | 1525 | 153 | 16 | 2 |
| 2 | 2417 | 242 | 25 | 3 | 4523 | 453 | 46 | 5 | 1231 | 124 | 13 | 2 |
| 3 | 2458 | 246 | 25 | 3 | 7824 | 783 | 79 | 8 | 2534 | 254 | 26 | 3 |
| 4 | 9232 | 924 | 93 | 10 | 6888 | 689 | 69 | 7 | 3512 | 352 | 36 | 4 |
| 5 | 3413 | 342 | 35 | 4 | 9034 | 904 | 91 | 10 | 3179 | 318 | 32 | 4 |
| 6 | 2367 | 237 | 24 | 3 | 5613 | 562 | 57 | 6 | 5136 | 514 | 52 | 6 |
| 7 | 4734 | 474 | 48 | 5 | 5842 | 585 | 59 | 6 | 2673 | 268 | 27 | 3 |
| 8 | 4794 | 480 | 49 | 5 | 2463 | 247 | 25 | 3 | 7434 | 744 | 75 | 8 |
| 9 | 5758 | 576 | 58 | 6 | 4567 | 457 | 46 | 5 | 6713 | 672 | 68 | 7 |
| 10 | 6831 | 684 | 69 | 7 | 1222 | 123 | 13 | 2 | 3677 | 368 | 37 | 4 |
| 11 | 3585 | 359 | 36 | 4 | 1356 | 136 | 14 | 2 | 2315 | 232 | 24 | 3 |
| 12 | 3546 | 355 | 36 | 4 | 7023 | 703 | 71 | 8 | 3678 | 368 | 37 | 4 |
| 13 | 4379 | 438 | 44 | 5 | 9054 | 906 | 91 | 10 | 6234 | 624 | 63 | 7 |
| 14 | 7989 | 799 | 80 | 9 | 8956 | 896 | 90 | 10 | 3623 | 363 | 37 | 4 |
| 15 | 1112 | 112 | 12 | 2 | 2893 | 290 | 30 | 4 | 4433 | 444 | 45 | 5 |
| 16 | 5612 | 562 | 57 | 6 | 6048 | 605 | 61 | 7 | 3242 | 325 | 33 | 4 |
| 17 | 8980 | 899 | 90 | 10 | 6724 | 673 | 68 | 7 | 5156 | 516 | 52 | 6 |
| 18 | 9041 | 905 | 91 | 10 | 5136 | 514 | 52 | 6 | 7234 | 724 | 73 | 8 |
| 19 | 6963 | 697 | 70 | 8 | 8045 | 805 | 81 | 9 | 3896 | 390 | 40 | 5 |
| 20 | 2635 | 264 | 27 | 3 | 5834 | 584 | 59 | 6 | 1262 | 127 | 13 | 2 |
| 21 | 4835 | 484 | 49 | 5 | 3658 | 366 | 37 | 4 | 1724 | 173 | 18 | 2 |
| 22 | 5628 | 563 | 57 | 6 | 4787 | 479 | 48 | 5 | 1794 | 180 | 19 | 2 |
| 23 | 3512 | 352 | 36 | 4 | 4923 | 493 | 50 | 6 | 2785 | 279 | 28 | 3 |
| 24 | 3588 | 359 | 36 | 4 | 5723 | 573 | 58 | 6 | 5788 | 579 | 58 | 6 |
| 25 | 6947 | 695 | 70 | 8 | 8234 | 824 | 83 | 9 | 3567 | 357 | 36 | 4 |
| 26 | 6934 | 694 | 70 | 8 | 6168 | 617 | 62 | 7 | 3623 | 363 | 37 | 4 |
| 27 | 3446 | 345 | 35 | 4 | 7824 | 783 | 79 | 8 | 4253 | 426 | 43 | 5 |
| 28 | 5678 | 568 | 57 | 6 | 9045 | 905 | 91 | 10 | 5232 | 524 | 53 | 6 |
| 29 | 7451 | 746 | 75 | 8 | 6924 | 693 | 70 | 8 | 4555 | 456 | 46 | 5 |
| 30 | 5680 | 569 | 57 | 6 | 5890 | 590 | 60 | 7 | 3527 | 353 | 36 | 4 |
| 31 | 9096 | 910 | 92 | 10 | 9392 | 940 | 95 | 10 | 4723 | 473 | 48 | 5 |
| 32 | 3527 | 353 | 36 | 4 | 8934 | 894 | 90 | 10 | 4235 | 424 | 43 | 5 |
| 33 | 8123 | 813 | 82 | 9 | 7845 | 785 | 79 | 8 | 4768 | 477 | 48 | 5 |
| 34 | 3461 | 347 | 35 | 4 | 7977 | 798 | 80 | 9 | 4723 | 473 | 48 | 5 |
| 35 | 4456 | 446 | 45 | 5 | 2571 | 258 | 26 | 3 | 2241 | 225 | 23 | 3 |
| 36 | 4662 | 467 | 47 | 5 | 1424 | 143 | 15 | 2 | 1423 | 143 | 15 | 2 |
| 37 | 4712 | 472 | 48 | 5 | 1572 | 158 | 16 | 2 | 3623 | 363 | 37 | 4 |
| 38 | 8236 | 824 | 83 | 9 | 5362 | 537 | 54 | 6 | 3613 | 362 | 37 | 4 |
| 39 | 7812 | 782 | 79 | 8 | 2571 | 258 | 26 | 3 | 3627 | 363 | 37 | 4 |
| 40 | 9234 | 924 | 93 | 10 | 2789 | 279 | 28 | 3 | 4778 | 478 | 48 | 5 |
| 41 | 5812 | 582 | 59 | 6 | 6946 | 695 | 70 | 8 | 6834 | 684 | 69 | 7 |
| 42 | 5671 | 568 | 57 | 6 | 6803 | 681 | 69 | 7 | 5673 | 568 | 57 | 6 |
| 43 | 4513 | 452 | 46 | 5 | 6535 | 654 | 66 | 7 | 7272 | 728 | 73 | 8 |
| 44 | 5624 | 563 | 57 | 6 | 5123 | 513 | 52 | 6 | 4723 | 473 | 48 | 5 |
| 45 | 5442 | 545 | 55 | 6 | 1525 | 153 | 16 | 2 | 3758 | 376 | 38 | 4 |
| 46 | 2312 | 232 | 24 | 3 | 1678 | 168 | 17 | 2 | 4623 | 463 | 47 | 5 |
| 47 | 2468 | 247 | 25 | 3 | 2683 | 269 | 27 | 3 | 3778 | 378 | 38 | 4 |
| 48 | 3521 | 353 | 36 | 4 | 8724 | 873 | 88 | 9 | 3623 | 363 | 37 | 4 |
| 49 | 2487 | 249 | 25 | 3 | 9036 | 904 | 91 | 10 | 6123 | 613 | 62 | 7 |
| 50 | 2111 | 212 | 22 | 3 | 9999 | 1000 | 11 | 2 | 2225 | 223 | 23 | 3 |
| 51 | 4713 | 472 | 48 | 5 | 9245 | 925 | 93 | 10 | 3712 | 372 | 38 | 4 |
| 52 | 4903 | 491 | 50 | 6 | 6346 | 635 | 64 | 7 | 1262 | 127 | 13 | 2 |
| 53 | 3478 | 348 | 35 | 4 | 7245 | 725 | 73 | 8 | 3674 | 368 | 37 | 4 |
| 54 | 4522 | 453 | 46 | 5 | 3333 | 334 | 34 | 4 | 7247 | 725 | 73 | 8 |
| 55 | 6724 | 673 | 68 | 7 | 4513 | 452 | 46 | 5 | 8588 | 859 | 86 | 9 |
| 56 | 6823 | 683 | 69 | 7 | 3425 | 343 | 35 | 4 | 8235 | 824 | 83 | 9 |
| 57 | 1124 | 113 | 12 | 2 | 6235 | 624 | 63 | 7 | 3758 | 376 | 38 | 4 |
| 58 | 1804 | 181 | 19 | 2 | 4892 | 490 | 50 | 6 | 8245 | 825 | 83 | 9 |
| 59 | 8934 | 894 | 90 | 10 | 4899 | 490 | 50 | 6 | 3656 | 366 | 37 | 4 |
| 60 | 9235 | 924 | 93 | 10 | 5855 | 586 | 59 | 6 | 6346 | 635 | 64 | 7 |
| 61 | 7846 | 785 | 79 | 8 | 5734 | 574 | 58 | 6 | 4666 | 467 | 47 | 5 |
| 62 | 7978 | 798 | 80 | 9 | 2573 | 258 | 26 | 3 | 3623 | 363 | 37 | 4 |
| 63 | 4523 | 453 | 46 | 5 | 7795 | 780 | 79 | 8 | 2451 | 246 | 25 | 3 |
| 64 | 5612 | 562 | 57 | 6 | 2790 | 280 | 29 | 3 | 2555 | 256 | 26 | 3 |
| 65 | 3684 | 369 | 37 | 4 | 9045 | 905 | 91 | 10 | 3612 | 362 | 37 | 4 |
| 66 | 5723 | 573 | 58 | 6 | 5645 | 565 | 57 | 6 | 1256 | 126 | 13 | 2 |
| 67 | 6892 | 690 | 70 | 8 | 4778 | 478 | 48 | 5 | 9999 | 1000 | 11 | 2 |
| 68 | 5892 | 590 | 60 | 7 | 7234 | 724 | 73 | 8 | 4742 | 475 | 48 | 5 |

**Table 2** First iteration of variable-scale decision making

| No | Inventory (10^0 t) | AvgUsage (10^2 kg) | Price (10^0 ¥/g) | Count | $\widetilde{GrD}$ |
|---|---|---|---|---|---|
| **1** | 5.29 | 4.86 | **8.29** | 7 | **1.2520** |
| **2** | **2.75** | 6.00 | 3.38 | 8 | **0.7084** |
| **3** | 4.64 | **2.73** | 3.18 | 11 | **0.6972** |
| 4 | 8.40 | 3.20 | 3.80 | 5 | 0.4828 |
| 5 | 5.73 | 6.91 | 4.27 | 11 | 0.3668 |
| 6 | 4.25 | 9.33 | 4.92 | 12 | 0.3288 |
| 7 | 8.67 | 9.00 | 4.67 | 6 | 0.2753 |
| 8 | 9.38 | 6.38 | 5.25 | 8 | 0.2178 |

**Table 3** Second iteration of variable-scale decision making

| No | Inventory (10^0 t) | AvgUsage (10^2 kg) | Price (10^2 ¥/kg) | Count | $\widetilde{GrD}$ |
|---|---|---|---|---|---|
| **4** | 6.80 | **8.00** | 66.20 | 5 | **1.4440** |
| **5** | **7.33** | 7.33 | 50.08 | 12 | **1.2736** |
| **6** | 6.68 | 7.32 | **37.47** | 19 | **0.6595** |
| 7 | 6.67 | 6.00 | 12.33 | 3 | 0.5118 |
| 8 | 4.67 | 8.00 | 28.00 | 3 | 0.3175 |

**Table 4** Third iteration of variable-scale decision making

| No | Inventory (10^2 kg) | AvgUsage (10^2 kg) | C2 (10^2 ¥ /kg) | Count | $\widetilde{GrD}$ |
|---|---|---|---|---|---|
| **7** | **55.33** | 6.00 | 12.33 | 3 | 1.5400 |
| **8** | 43.00 | 8.00 | **28.00** | 3 | 1.3367 |

# 4 Conclusions

Variable-scale clustering is an emerging research area for intelligent decision making, which successfully transform the single scale analysis of tradition data mining application to multiple scale analysis.

The main contributions of our research are as follows. First, we establish the numerical concept space to describe numerical dataset, where concept chain is utilized to model the knowledge of business understanding, while value space aims to describe the information of data understanding. Second, we propose a variable-scale clustering algorithm in the perspective of scale down transformation. Related measurements, such as the scale down transformation rate and granular deviation, are also improved. Third, a difference management solution of TYL products is put forth. Experiment shows that our proposed algorithm is able to obtain accessible and available results in practice.

## References


1. Wu, S., Gao, X., & Bastien, M. (2003). *Data warehousing and data mining* (9th ed.) (Vol. 1, pp. 148–155). China: Metallurgical Industry Press.
2. Zhang, Q., Wang, G., & Hu, J. (2013). *Multi-granularity knowledge acquisition and uncertainty measurement* (1st ed.) (Vol. 1, pp. 16–25). Beijing: Science Press.
3. Wang, A., & Gao, X. (in press). Automatic data analysis technique: Data mining tasks discovery based on the concept network. *International Journal of Information Technology and Management*.
4. Wang, A., & Gao, X. (2017). Technique of data mining task discovery for data mining. In *Proceeding of the 7th International Conference on Logistics, Informatics and Service Sciences*, Kyoto, Japan.
5. Gao, X., Wang, A. (2018). Variable-scale clustering. In *Proceeding of the 8th International Conference on Logistics, Informatics and Service Sciences*, Toronto. Canada.
6. Wang, A., & Gao, X. (2019). Multifunctional product marketing using social media based on the variable-scale clustering. *Technical Gazette, 26*(1), 193–200.
7. Wang, A., & Gao, X. (2019). Hybrid variable-scale clustering method for social media marketing on user generated instant music video. *Technical Gazette, 26*(3), 771–777.

# An Empirical Test of Beijing Industry Transfer Policy

**Lu Feng, Kun Zhang, Yingmin Yu, and Jia Zuo**

**Abstract** The rapid development of economic globalization increases the process of urbanization, which leads to high resource consumption and labor intensity. Relieving of the non-capital function is a policy guideline proposed by the Chinese Government to promote the production upgrade and the transformation of economic development mode of Beijing. As the foundation of Beijing economy, manufacturing industry transfer is given increasingly attention by academic circles. By using District X of Beijing as example, this paper tests current policies with empirical methods. According to the analysis results, strategies for improving transfer policies are offered. As a new research perspective, it can be a good solution.

**Keywords** Industry transfer policy · Public policy analysis · PCA

## 1 Introduction

As a national political center, economic center and cultural center, Beijing has achieved remarkable economic development and is a major economic center in the northern region. In the years of development, a large number of industries have

L. Feng
Economic and Social Survey Office of Beijing Economic-Technological Development Area, Beijing Municipal Bureau of Statistics, Beijing, China
e-mail: uibefl@126.com

K. Zhang (✉) · Y. Yu
Technology Strategy Department, China National Software & Service Co., Ltd., Beijing, China
e-mail: zhangkun@css.com.cn

Y. Yu
e-mail: yuym@css.com.cn

J. Zuo
Investment Bank Department, Guotai Junan Securities, Beijing, China
e-mail: zuojia@gtja.com



J. Zhang et al. (eds.), *LISS2019*,
https://doi.org/10.1007/978-981-15-5682-1_21

accumulated, including some low-end industries with high resource consumption and high labor intensity. As the positioning of the "four centers" becomes clearer, it's imperative to remove the low-end industries, develop high-grade, precision and sophisticated industries for building a science and technology innovation center [1].

To achieve this goal, many public policies have been published at the municipal and even national levels to restrict or remove these industries which are not suitable for Beijing. Especially the Catalogue of Prohibited and Restricted New Additional Industries (hereafter referred as the Catalogue), which is firstly launched on 2014 and most recently revised on 2018, applies to any new enterprise, in order to improve the city's quality, to guarantee the city's operation and to better serve the people. More than half of the industry subcategories are covered by the Catalogue, including 28 manufacturing industry categories out of 31. As an outcome of the Catalogue, the number of newly-launched enterprises in 2018 in Mining, Manufacturing, Wholesale and Retail Trades, Agriculture Industry, is less than half of that in 2013, before the Catalogue was firstly implemented, demonstrating that the Catalogue effectively controls the non-capital function growth. By the end of August 2018, a total of 20300 applications for business registration were rejected.

District X is one of the key new regions for the development of Beijing's manufacturing industry. In 2015, the industrial added value was 74.73 billion yuan, accounting for 61.4% of the regional GDP, and accounting for 19.5% of Beijing's whole industrial added value. In 2015, there were 274 industrial enterprises above designated size in District X, with a total industrial output value of 255.55 billion yuan [2]. Compared with other regions of Beijing, the manufacturing industry in District X is characterized by high efficiency, excellent structure, green development and clustering advantages.

As a manufacturing region, in order to continuously improve the quality of the regional economy and maintain a high level of development, District X continues to explore effective ways for the transformation and upgrading of the manufacturing industry to "change the cage for birds" for successfully completing industrial settlement. The current industry-removing policy is based on the industry divided according to the final product or service of the enterprise, and fails to directly reflect the impact of factors such as enterprise production efficiency, resource consumption, and labor intensity. Therefore, this paper draws on the idea of evaluating the competitiveness of the industry, and through analyzing the empirical data of manufacturing enterprises in a certain area of Beijing (replaced by District X) as an indicator, verifies the feasibility of the current manufacturing industry-removing policy in the area.

## 2 Literature Review

Industrial restructuring usually comes with the adjustment and change of the city positioning. Therefore, it has great significance to systematically research the city positioning for the understanding of industrial transfer policy. In this area, Su adopted affluent and authentic historical literatures and gave a detailed city positioning history of Beijing. She divided this history from the founding of People's Republic of China to nowadays into four phases: City layout and positioning before 1949; Exploration of the capital's positioning during 1949–1979; Definite of capital's positioning during 1980–2000 and deepen capital's positioning after 2000 [3].

The so-called "big city syndromes" is not originated from Beijing, so there are lots of experiences and bitter lessons to learn. Yang et al. introduced the evaluation method of population transfer, industrial transfer, land use changes, and researched the experiences of urban transfer in foreign countries from a contrastive perspective. For example, London adopts the "Central City-Green Belt-Satellite town" pattern to accomplish population suburbanization. Paris, Seoul and some other cities all make sure that their new towns are not too close to the central city and that there's convenient traffic between the new town and the central city. As to the method, some similar policies were implemented, such as infrastructure construction, differentiated industry structure, economic methods [4]. Zhang et al. studied Seoul's transfer policies, which are institution relocation, masterplan, new town construction, the practice of moving the capital, transportation policies, and then gave six suggestions for Beijing: taking regional coordination development into consideration while function transfer, making the city masterplan under integrate leadership and making sure that it's operational, controlling the population with business, improving multicore structure, taking a forward-looking perspective and establishing transportation network of Beijing-Tianjin-Hebei [5]. Yang summarized transfer approach from the perspective of population transfer in international cities, such as Tokyo, Seoul, Hongkong, New York and Paris, made a contrast of Beijing's transfer policies to these cities from four aspects, and gave suggestions based on the contrast [6].

As to non-capital function transfer, the latest researches are mostly done on background of Beijing-Tianjin-Hebei coordinate development. Zhu stated that the goal of Beijing-Tianjin-Hebei coordinate development was to build a world-known megalopolis. The Beijing-Tianjin-Hebei megalopolis should participate global competition as the highest level of China's Researching and high-end services, and should become the growth pole of China in the future [7]. Gu mentioned the scientific classification of Beijing's functions is the first step of function transfer. The step should be taken on three levels: country-level, region-level and city-level. The problem-orientation principle should be used during this step [8]. Wei also stated that defining city function should come first during function transfer [9]. Li put forward the point of view that comparative advantage theory could be used in function definition [10]. Lu, Yang and Sun had a similar conclusion, and further stated that industry transfer should achieve a mutually beneficial win-win corporation through complementarities. So, the major task for Beijing was to transfer its non-capital functions, and that for Tianjin was to develop high-edge manufacturing industries with geographic advantages, and for Hebei was to construct infraction and accept the relocated industries [11]. Song from the perspective of government responsibility, draw the conclusion with empirical approach that the government's guiding responsibility included incentive and inhibitive responsibilities. Power division during transfer, financial corporation, reversed transmission are new for the government today [12].

The non-capital function transfer policies are the object that this paper studies. In this aspect, Ma proposed three issues influencing Beijing's industry relocation, and came up with the policy suggestions: structuring collaborative development framework, promoting industrial planning in capital region, working on the mechanism of responsibility division on industry relocation and public service providing, as well as cultivating human resource in industry recipient cities [13]. Ye studied industry transfer of Beijing and listed the industries that needs to be transferred in the secondary and third industry, as well as ways to transfer the industry [14]. Ding measured the effects of population transfer, industrial transfer as well as land limit on Beijing's economy, and the study showed that they have remarkable effect on the economy, and so Ding suggest that appropriate measures should be taken to reduce these negative effects during the industry transfer [15].

## 3 Objectives of Verification

According to the literature, we know that current policies use industry as standard when it comes to decide the scope to be transferred. However, industry division is based on the final product or service of an enterprise and usually fails to directly show the level of production efficiency, resource consuming, labor density etc. of an enterprise. Using the method of industrial competitiveness evaluation, this paper will examine current policies with enterprise microdata instead of industry data.

Based on the policy problems constructed in the analysis stage, the quantitative model of enterprise data in the District X is selected to verify the problem of the method for dividing the objects' scope of the removing policy, in order to obtain more information on the rationality of the problem.

## 4 Construction of Indicator System

There are three main references for the selection of model indicators. The first are the statements of industrial characteristics which are not suitable for Beijing. These characteristics appear in large numbers and are very similar in the policy, indicating that the value orientation of the policy is very clear and it is very referential. For example, the new general regulation proposes that enterprises with "high pollution and high energy consumption and water consumption" should be eliminated on the spot. The strategic goals for manufacturing proposed in the Beijing Action Plan for "Made in China 2025" are divided into four categories: innovation ability, quality efficiency, dual integration, and green development. The second is to refer to the relevant results in existing studies. In the 2016 study, Zhang and others selected indicators such as the increase rate, operating income profit margin, per capita tax, energy consumption, water consumption, employment, R&D input strength, and invention patents of the major and middle categories of the Beijing manufacturing industry to calculate industry scores and research the removing objects' scope [16]. Li et al. divided removing objects through policy, economy, and resources [17]. Li uses indicators such as added value, income, profit, employment, energy consumption, and land use to obtain a list of suggestions for Beijing's industry-removing policy through the method of principal component analysis [18]. The third is based on the policy issues constructed at the policy analysis stage. According to the previous analysis, the industrial remove mainly aims to promote the manufacturing industry from large-scale growth to high efficiency growth, and further guide the manufacturing economy to a high-grade, precision and sophisticated structure, through selective remove of manufacturing stocks and strictly control of the increment. Therefore, when constructing the indicator system, we should fully consider the factors such as technology, efficiency, and energy consumption etc.

Based on the analysis above, we decide to construct an indicator system which is consistent with Beijing's rule as capital. To reveal degree of conformity between the corporates and the capital role, this indicator system can be divided into 4 first index signs, namely technical indicator, energy consumption indicator, performance indicator and output indicator. As shown in Table 1, 8 s index signs are included.

It is critical to set weight of the indicators in a multi-index comprehensive evaluation method. Be different with other traditional procedures which are subjective, PCA (Principal Component Analysis) is a statistical procedure that uses an orthogonal transformation to convert a set of observations of possibly correlated variables. In this paper, we use PCA to set weight.

**Table 1** The index of removing urgency degree

| | |
|---|---|
| Technical indicator | Per capita R&D expenditure (ten thousand yuan)<br>Per capita invention patent (piece) |
| Energy consumption indicator | Energy consumption of output value per ten thousand yuan (tons of standard coal)<br>Water consumption of output value per ten thousand yuan (cubic meters) |
| Performance indicator | Per capita corporate tax (ten thousand yuan)<br>Per capita operating profit (ten thousand yuan) |
| Output indicator | Per capita gross output (ten thousand yuan)<br>Per capita added value (ten thousand yuan) |

## 5 Model and Data

The data source is the corporate annual reports of the enterprises in District X. After sampling out, data of 250 enterprises are available. Considering the disunity of units and different orders of magnitude, standardized treatment or dimensionless treatment must be used. The standardized values of technical indicator, output indicator and performance indicator can be computed, which is mathematically expressed as:

$$z_{ij} = \frac{a_{ij} - \min a_{ij}}{\max a_{ij} - \min a_{ij}} \tag{1}$$

$z_{ij}$ is standardized value, $a_{ij}$ is original value, *i* refers to Distinct and *j* refers to the number of indicator. The standardized value of energy consumption indicator is different from other indicators, so it can be mathematically expressed as $1 - z_{ij}$.

After getting the standardized value, we use SPSS 19.0 to analyze the KMO and Bartlett's test. The check value of KMO test is 0.616 and P value is 0.000. According to the judgment standard, PCA can be used in the model [19].

The factor analysis shows that there are 3 factors' eigenvalue are greater than 1, the Cumulative Rate Contribute to total variance is 79.33%. Based on the methods of determining weight, we can get the Removing Urgency Degree indexes' score. $\omega 1 = 0.1167$, $\omega 2 = 0.1147$, $\omega 3 = 0.1512$, $\omega 4 = 0.1507$, $\omega 5 = 0.1325$, $\omega 6 = 0.1302$, $\omega 7 = 0.0566$, $\omega 8 = 0.1474$ (Table 2).

**Table 2** Eigenvalue and variance contribution

| | The initial eigenvalue | | | Extraction sums of squared loadings | | |
|---|---|---|---|---|---|---|
| | Total | % of variance | Cumulative % of variance | Total | % of variance | Cumulative % of variance |
| 1 | 3.117 | 38.959 | 38.959 | 3.117 | 38.959 | 38.959 |
| 2 | 1.853 | 23.157 | 62.116 | 1.853 | 23.157 | 62.116 |
| 3 | 1.377 | 17.214 | 79.330 | 1.377 | 17.214 | 79.330 |
| 4 | 0.736 | 9.195 | 88.524 | | | |
| 5 | 0.548 | 6.845 | 95.370 | | | |
| 6 | 0.238 | 2.977 | 98.347 | | | |
| 7 | 0.074 | 0.921 | 99.2607 | | | |
| 8 | 0.059 | 0.733 | 100.000 | | | |

## 6 Results and Analysis

According to the results, we can get the median and the mean of the 250 samples' scores. The median is 0.3878 and the mean is 0.403. As shown in Table 1, 155 enterprises' scores are between 0.3 to 0.4, 77 enterprises' scores are between 0.4 to 0.5, and 22 enterprises' scores are in others' interval. Then the industries' weighted mean scores can be got based on the enterprises' scores (Fig. 1).

According to the results shown in Table 3, 6 industries' scores are higher than the mean score and 6 industries' scores are lower than the mean score in the main industries. The chemical materials and products manufacturing's score is the highest because of the cosmetics and inert gas manufacturing's good performance in Technology, energy consumption, efficiency. Automobile manufacturing and medicine manufacturing which have high value-added rate also rank the top in District X. At the same time, score of the emerging area such as electronic communication equipment and computer manufacturing is lower than mean score because those enterprises in District X are almost foundries with low per capita output.

Based on the analysis above, most of the low-scoring industries, which should be shut down or relocated, are already covered by current policy. That shows the validity of current policies. Meanwhile, it's also demonstrated that score of an industry isn't determined by its product. For instance, some highest-scoring industries are prohibited to launch or expand according to the Catalogue of Prohibited and Restricted New Additional Industries, such as food manufacturing, printing and recording media reproducing, and chemical materials and products manufacturing. At the same time, some low-scoring industries are even encouraged by the government. For example, instrument meter stationery and office machine manufacturing are one of the high-grade, precision and sophisticated industries according to relevant policies. Some sub-categories of electronic communication equipment and computer manufacturing, although low-scoring, are excluded from the Catalogue of Prohibited and Restricted New Additional Industries. What else needs to be explained is that all above is about district X and the situation may differ from other districts of Beijing because of the difference in industry structure.

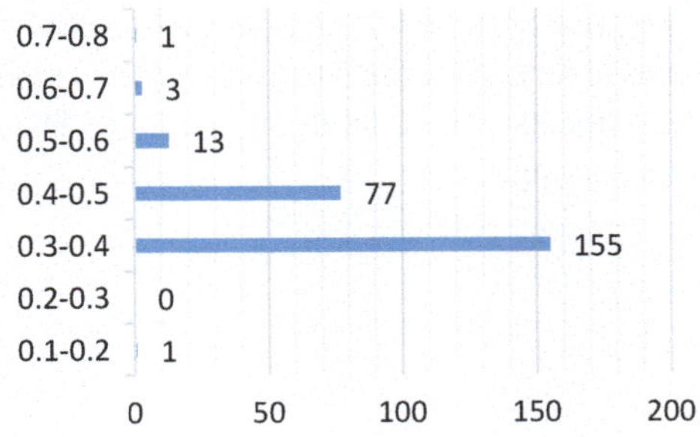


**Fig. 1** The interval distribution of the enterprises' scores in district X

**Table 3** Scores of industries in district X

| Code | Name | Mean scores |
|---|---|---|
| 14 | Food manufacturing | 0.4485 |
| 23 | Printing and recording media reproducing | 0.3860 |
| 26 | Chemical materials and products manufacturing | 0.4630 |
| 27 | Medicine manufacturing | 0.4205 |
| 33 | Metalwork manufacturing | 0.3887 |
| 34 | Universal equipment manufacturing | 0.3974 |
| 35 | Special equipment manufacturing | 0.4038 |
| 36 | Automobile manufacturing | 0.4269 |
| 38 | Electrical machinery and equipment manufacturing | 0.4149 |
| 39 | Electronic communication equipment and computer manufacturing | 0.3832 |
| 40 | Instrument meter stationery and office machine manufacturing | 0.3887 |
| 41 | Others | 0.3793 |

# 7 Policy Recommendation on Manufacturing Industry Removing in Beijing

This paper analyzes the Beijing's industrial policy history and lists current policies, and tries to summarize industry relocation experiences with data modeling and case study. In this part, the paper gives four suggestions on manufacturing industry relocation policy of Beijing.

## 7.1 *Specify the Policy Issues*

"General manufacturing industry" is currently used in the policies to describe a low-end industry that needs to be shut down or relocated. However, this term is too vague and general to be put into action, probably leading to failure in the upcoming policy. To solve this problem, this paper suggests to model with direct indexes such as labor intensity, resource consumption, R&D and pollution index, etc. [20, 21]. The model could be used to determine whether or not an existing manufacturing enterprise or industry ought to be removed. The indexes above come from different government departments, so in the modeling process, cooperation of these departments will certainly be required. The cooperation includes but not limited to data sharing.

### *7.2 Improve Facilities at the Undertake Places of the Relocated Industries, Especially Taking the Establish of Xiongan New Area as Opportunity*

The Central Committee of the Communist Party of China and the State Council decided to establish Xiongan New Area in 2017. Under the background of coordinated development of Beijing-Tianjin-Hebei, the establish of Xiong'an New Area is essential to urban agglomeration in Beijing-Tianjin-Hebei area. Also, one of the most important functions of Xiong'an New Area is to accept the relieved non-capital functions of Beijing. Nowadays, high land and resource price is a heavy burden to many manufacturing enterprises, some of which are high-edge and should be encouraged. If these companies move to Xiong'an New Area, not only will the goal of relieving the non-capital function be achieved, these relocated companies can also develop much better, at least cost-wise.

### *7.3 Promote Industrialization of Science and Technology, and Cultivate High-Quality and Cutting-Edge Industries to Reduce Negative Effects of the Policies*

Transformation and industrialization of science and technology is essential to technical content of economy and to readjusting industrial structure and mode of economic development [22–24]. The paper suggests that government should learn from developed countries and introduce policies to promote industrialization of science and technology. All possible kinds of support should be taken into action, such as direct financing, credit policy, intellectual property protection, tax cuts, etc. Business environment is also important for the goal of attracting companies with high content of science and technology.

### *7.4 Further Improve Public Policy and Build the Participatory System of Beijing-Tianjin-Hebei to Achieve "All-Win"*

Instead of "competition over cooperation", understanding and full cooperation is crucial for Beijing-Tianjin-Hebei coordinate development and non-capital function relocation. To achieve this goal, central government order and policy is not enough. Consensus of "all-participation, all-win system" needs to be reached. This consensus is also foundation of cooperation and guarantee of the Megalopolis in the long run. Beijing, Tianjin and Hebei should all focus more on their common interests and maximize these interests in order to get an all-win. There are some public policies on tax sharing along the relocation process, but not enough. There's still room to implement or improve public policies in the all-win environment, especially systematic and long-term policies. Besides, negotiation system should be built to encourage local governments to demand their interests and to negotiate with each other to maximize and balance the interest. This system has already been used between different departments within one government and should be used among different local governments. Hopefully, a competition-cooperation balanced system will be reached among Beijing-Tianjin-Hebei and the cities within this area will form a Megalopolis that is competitive world-wise and could participate in international division in the process of internationalization.

## References


1. Zhang, K. (2017). *Research on promotion path of manufacturing enterprises' integration of informatization and industrialization in Beijing*. Dissertation.
2. Beijing Municipal Bureau of Statistics (2017). Beijing statistical yearbook 2017. China Statistics Press.
3. Su, F. (2014). Several evolutions of Beijing's urban orientation. *Beijing DangShi, 4,* 35–38.
4. Yang, C. F., Han, H. R., Zhang, X. B., & Song, J. P. (2016). Progress of urban function decentralization research. *Human Geography, 1,* 8–15.
5. Zhang, K. Y., & Dong, J. J. (2015). Seoul case and its implication to Beijing in relieving non-capital function. *China Business and Market, 11,* 64–71.
6. Yang, G. (2013). Review and reference on the population spatial layout policy of international metropolis and Beijing. *Northwest Population Journal, 3,* 43–48.
7. Zhu, E. J. (2014). To promote the thinking and emphases of the coordinated development of Beijing-Tianjin-Hebei Region. *Economy and Management, 28,* 10–12.
8. Gu, G. Z. (2015). Relieving of the non-capital function and big city syndromes.
9. Wei, H. K. (2014). Polarization in China's urbanization: tendency and outlook. *China Economist, 9,* 12.
10. Li, G. P. (2014). Research on development policies of science and technology innovation integration in Beijing-Tianjin-Hebei Region. *Economy and Management, 28,* 13–18.
11. Lu, J. P., Yang, Z. W., & Sun, J. W. (2014). Research on economic ties dynamic in Beijing-Tianjin-Hebei Region-from the perspective of urban flow intensity. *Inquiry into Economic Issues, 12*, 99–104.
12. Song, B. (2016). The governmental responsibilities for industrial shift program of the capital. *Leaders' Companion, 19,* 63–68.
13. Ma, S. P. (2016). Suggestions on Beijing industry transfer policy. *Beijing Observation, 9,* 42–43.
14. Ye, Z. Y., & Ye, S. Y. (2015). Thinking of Beijing industry transfer policy. *City, 1,* 20–25.
15. Ding, C. R., Shi, X. D., Li, Z., & Niu, Y. (2016). Applying an extended input-output model to investigate impacts of growth control policies of population and industries in Beijing. *Urban Insight, 6,* 52–66.
16. Zhang, J., Xu, Y. Y., & Zhang, Y. C. (2016). Promoting and easing manufacturing industry in megacity – observation from Beijing. *Research on Economics and Management, 7,* 103–111.
17. Li, T. K., Liu, Y., Wang, C. R., & Wang, Z. S. (2017). Environmental impact of Beijing's industrial decentralization based on a discrete choice modeling approach. *Journal of Tsinghua University (Science and Technology), 11,* 46–52.
18. Li, J. (2015). *Research on industrial adjustment and organization in Beijing*. Dissertation.
19. Chen, H. C. (2010). Empirical study on evaluation of competitiveness of high-tech industry. *Science & Technology and Economy, 24,* 21–23.

20. Ma, S. P. (2016). Study on improving the policy of Beijing industry transfer policy. *Beijing Observation, 9,* 42–43.
21. Du, L. Q., & He, C. D. (2015). Research on transfer of the non-capital function in Beijing. *Shanghai Urban Planning Review, 6,* 17–20.
22. Sheng, L., & Hong, N. (2014). The progress and trend of US's revitalizing manufacturing strategy and it's influences on industry upgrading of China. *Word Economy Studies, 7,* 80–86.
23. Tu, P., & Ke, Y. (2011). Legislative issues on the transformation of Beijing's scientific and technological achievement. *Think Tank of Science & Technology, 2,* 78–83.
24. Zhang, K., & Zhang, Z. J. (2016). An overview on the integration of informatization and industrialization (IOII). *International Journal of Hybrid Information Technology, 9,* 1–12.

# Optimization of Automated Warehouse Storage Location Assignment Problem Based on Improved Genetic Algorithm

**Wenyi Zhang, Jie Zhu, and Ruiping Yuan**

**Abstract** Due to increasing in logistics speed, the efficiency requirements of warehousing are getting higher. Therefore, a storage location assignment model with the main goal of reducing the time traveling between the I/O point and the slot, decreasing the distance between correlation of items and ensuring the stability of shelves is proposed for the optimization of storage location in automated warehouses. To solve the model better, it is necessary to change the disadvantage of traditional genetic algorithm (GA) which is weak local search ability, thus adding an inversion operation to solve this problem. Two kinds of algorithms were used to simulate the experiment with the experimental data. After comparing the experimental results, the effectiveness of the improved GA was verified and the operational efficiency and shelf stability were increased.

**Keywords** Storage optimization · Inversion · Improved GA

This research is sponsored by Young Top-notch Talent Project of High Level Teacher Team Construction in Beijing Municipal Universities (grant number: CIT&TCD201704059) and Beijing Excellent Talents Support Project (grant number: 2017000020124G063); This research is supported by Funding Project for Beijing Intelligent Logistics System Collaborative Innovation Center, Beijing Key Laboratory of intelligent logistics systems and Logistics Robot System Scheduling Research Team of Beijing Wuzi University.

W. Zhang (✉) · J. Zhu · R. Yuan
School of Information, Beijing Wuzi University, Beijing, China
e-mail: victoria428@126.com

J. Zhu
e-mail: zhujie_bwu@163.com

R. Yuan
e-mail: angelholyping@163.com



J. Zhang et al. (eds.), *LISS2019*,
https://doi.org/10.1007/978-981-15-5682-1_22

## 1 Introduction

The automated warehouse is mainly used in logistics warehousing, which make full use of the advantages of height and can realize multi-layer storage of items. As the speed of logistics increasing and the number of express deliveries increasing, the requirements for warehousing capacity are also greatly improved, so it is necessary to optimize the storage arrangement of automated warehouses. That is rescheduling the slot of items in order to achieve the purpose of improving warehouse efficiency and reducing costs.

Research on the optimization of storage location assignment in automated warehouses has already achieved some results. Deng et al. [1] mainly consider the location optimization from the items turnover efficiency with time as a reference, using GA to solve. Yang et al. [2] constructed a mathematical model with the goal of minimizing travel time, using a two-layer genetic algorithm to solve. Li et al. [3] establish a mathematical model from the perspective of items turnover and shelf stability, and solved by the virus cooperative genetic algorithm. According to the characteristics of automobile storage, [4] decided to establish a model from the aspects of storage efficiency and shelf stability, and solved it by using hybrid genetic algorithm. Wang et al. [5] mainly studied the influence of turnover rate rules, items weight characteristics rules and items correlation rules to the slot, using the virus cooperative genetic algorithm to solve. Bie [6] using reducing the traveling distance of the storage, increasing the effective utilization of the shelf to establish a model, and integrate the Pareto principle into the GA. Jiao et al. [7] establish a mathematical model from the perspective of reducing the time of traveling, increasing the stability of the shelf and storing the related items, and using multi-population genetic algorithm to solve. Kim [8] proposed a model of MIP problem to minimize the picking period as the objective function, and then use the improved algorithm based on simulated annealing to solve. Park [9] proposed a heuristic algorithm based on GA with improved crossover operation to solve the model of minimizing the moving distance. Pan [10] proposed to establish a mathematical model by minimizing the difference between the expected quantity and the actual inventory and the deviation between the workload and the average workload, using a GA to solve and using the heuristic correction mechanism to achieve the results.

From the above references, it can be seen that most of the previous studies have established objective functions to improve the efficiency of warehousing operations and shelf stability, lack the objective function for the association rules of items. The algorithms solved basically use traditional GA and are easy to fall into immature convergence. This paper introduces the minimizing slot distance of the relevant items as the objective function and adds an inversion operation in the algorithm to improve the local search ability to prevent into immature convergence.

# 2 Storage Allocation Mathematical Model

## 2.1 Automated Warehouse Storage

The main body of the automated warehouse is composed of shelves, roadway stacking cranes, loading and unloading workbench, automatic conveyor belt and operation control system. The shelf is a standard size storage space, mainly used for the storage of items; the roadway stacking crane runs through the roadway between the shelves to complete the work of depositing and picking up the goods; the work of the automated conveyor belt is mainly to transfer the items from I/O point to the tunnel. As a new technology that combines material handling and warehousing management, the automated warehouse can not only effectively reduce labor input, achieve precise operation, but also help to improve warehouse operation efficiency and reduce logistics costs. This is definitely a core element that stimulates the competitiveness of the company and fulfills a range of customer needs (Fig. 1).

As the logistics speed is increasing now, the requirements for the storage chain are increasing, and the overall operational efficiency of the automatic warehouse must be improved to meet the demand for logistics and storage. Therefore, it is necessary to establish a mathematical model to optimize the storage allocation in the automated warehouse, while achieving the two objectives of ensuring shelf stability and improving operational efficiency.

The allocation of storage refers to the rational planning and use of the storage space in the warehouse by rationally utilizing the allocation strategy and distribution principle of the storage space to ensure that the goods are dispatched to the storage location that is most favorable for picking.

At the same time, as far as the development of logistics technology itself is concerned, when the degree of automation, information and intelligence of the

**Fig. 1** Schematic diagram of warehouse storage

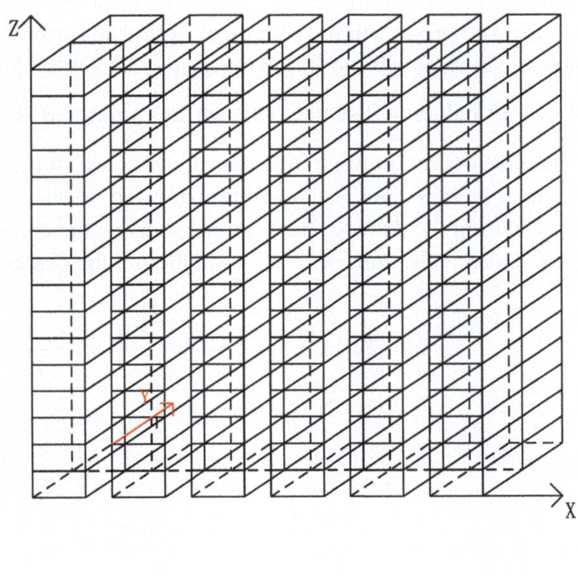

storage system is continuously improved, the work intensity of manual work is largely replaced, and the accuracy and efficiency of manual work are also better. Higher, so its advantages continue to highlight. At the same time, the replacement of labor by equipment has greatly reduced the cost of logistics operations. The use of automated warehouses in the warehousing segment and the rational allocation of storage can effectively reduce their costs and increase the return on investment.

The operation of the automatic warehouse is only by the stacker and the conveyor belt. Therefore, the storage method in this kind of work environment is mainly to place the items with large turnover rate in the slot closer to the I/O point and make the correlation items get closer in order to reduce costs during operation. To ensure shelf stability, high quality items need to be placed on lower layer. It can be assumed that the items in the warehouse have a unique item number and will not change with the slots.

## 2.2 *Storage Optimization Model*

In order to simplify the problem, the following assumptions were made:

1) Each slot is a square with the same length, width, height and all the items are adapted to size of slot;
2) Ignore the time when the stacker accesses the items, starts and brakes;
3) Only one item can be stored in one slot;
4) The conveyor moves only in the direction of the $x$ axis with the speed $v_x$; the stacker moves along $y$ and $z$ axes and the speeds are $v_y$ and $v_z$.

In the process of building the storage model, decision variables $x_{ij}$ indicates the line number of $i$ group $j$ item. $y_{ij}$ indicates the column number, $z_{ij}$ indicates the layer number. Other parameters are shown in Table 1.

The storage location assignment optimization problem of the automated warehouse can establish the following mathematical model:

- F1 to reducing distance between I/O point and slots

The most important thing in storage location assignment is to improve the efficiency of the warehousing. That is minimizing the distance between I/O point and slots. Considering the distance between the shelves, the unit distance the conveyor travels on the axis is the sum of the slot length and the distance between the shelves $h + l$.

$$\min F_1 = \sum_{i=1}^{n} \sum_{j=1}^{k_i} p_{ij} [\frac{x_{ij}(l+h)}{v_x} + \frac{y_{ij} \times h}{v_y} + \frac{(z_{ij}-1)h}{v_z}] \tag{1}$$

**Table 1** Mathematical model parameter definition

| Parameter | Description | Parameter | Description |
|---|---|---|---|
| $A$ | The number of shelves | $h$ | Length width and height of slot |
| $B$ | Number of slot on one floor of shelf | $l$ | The distance between shelves |
| $C$ | Layer of shelves | $Q$ | Mean of $q_i$ |
| $v_x$ | Conveyor speed x-axis | $k_i$ | The quantity of items in $i$ group |
| $v_y$ | Stacker speed on y-axis | $q_i$ | Average coordinate of all items in $i$ group |
| $v_z$ | Stacker speed on z-axis | $n$ | Total number of items group |
| $p_{ij}$ | Turnover of the $i$ group $j$ item | $M_{ij}$ | The weight of $j$ items in group $i$ |
| $d$ | The sum of distances of $i$ group $j$ item items to the average coordinates | $G$ | The distance between the I/O point and mean of $q_i$ |
| $D$ | The sum of distance from average distance to total average coordinates in $n$ groups | | |

- F2 to reducing the distance between correlated items

Reducing the distance between slots of highly correlated items can reduce the cost of items traveling. The association rule mining algorithm can solve the items correlated in the sales order, that is, the items that often appear in the same order and define the items with high correlation as the same group of items. First, the average coordinate $q_i$ in a group is defined.

$$q_i = \frac{1}{k_i}\sum_{j=1}^{k_i}[x_{ij}, y_{ij}, z_{ij}] \tag{2}$$

Define an expression to calculate the sum of the distances between the items and the average coordinates in the corresponding group.

$$d = \sum_{i=1}^{n}\sum_{j=1}^{k_i}\sqrt{[x_{ij} - q_i(x)]^2 + [y_{ij} - q_i(y)]^2 + [z_{ij} - q_i(z)]^2} \tag{3}$$

Define $Q$ as the total average coordinate of the items for all groups.

$$Q = \frac{1}{n}\sum_{i=1}^{n} q_i \tag{4}$$

In order to ensure a certain dispersion between groups, (5) is defined to calculate the sum of distances from average distance to total average coordinates in n groups.

$$D = \sum_{i=1}^{n} \sqrt{[q_i(x) - Q(x)]^2 + [q_i(y) - Q(y)]^2 + [q_i(z) - Q(z)]^2} \tag{5}$$

Then calculate the distance between the I/O point and the mean coordinates.

$$G = \sqrt{Q(x)^2 + Q(y)^2 + Q(z)^2} \tag{6}$$

In order to ensure that the items can be distributed as close as possible to the I/O point, and the same group of items are placed on close slots, F2 can be obtained by combining (3) (4) and (5).

$$\min F_2 = \frac{d + G}{D} \tag{7}$$

- F3 to ensure shelf stability

In order to ensure the stability of the shelf, it is necessary to place the larger quality items on the lower layer, and the lighter items on the upper.

$$\min F_3 = \sum_{i=1}^{n} \sum_{j=1}^{k_i} M_{ij} \times z_{ij} \times h \tag{8}$$

The final objective function of the storage optimization can be derived from (1), (7) and (8) as follows.

$$\begin{aligned} \min F_1 &= \sum_{i=1}^{n} \sum_{j=1}^{k_i} p_{ij} [\tfrac{x_{ij}(l+h)}{v_x} + \tfrac{y_{ij} \times h}{v_y} + \tfrac{(z_{ij}-1)h}{v_z}] \\ \min F_2 &= \tfrac{d+G}{D} \\ \min F_3 &= \sum_{i=1}^{n} \sum_{j=1}^{k_i} M_{ij} \times z_{ij} \times h \end{aligned}$$

$$0 < x_{ij} \le A(x_{ij} \in N) \tag{9}$$

$$0 < y_{ij} \le B(y_{ij} \in N) \tag{10}$$

$$0 < z_{ij} \le C(z_{ij} \in N) \tag{11}$$

Constraints (9), (10), and (11) represent the range of the slot. $x_{ij}$, $y_{ij}$, $z_{ij}$ are all integers.

In order to prevent the multi-objective mathematical model from interacting with each other in the solution, it is necessary to add a corresponding coefficient $\omega_s \in [0, 1]$ to the objective function in order to get the linear sum. The mathematical model is expressed as follows:

$$\min F = \sum_{s=1}^{3} \omega_s \cdot F_s \tag{12}$$

# 3 Improved Genetic Algorithm

The crossover and mutation of GA can make chromosomes more diverse. The GA has good global search ability and higher robustness, but the local search ability is weak. In order to make up for the defects of GA, the inversion operation is added to improve the local search ability and achieve better solution (Fig. 2).

## *3.1 Coding*

Coding adopts the decimal non-negative integer method, and slot coordinate consists of three genes, that is, the slot code is prepared according to the x-axis, the y-axis and z-axis of the slot and each slot has a unique number.

## *3.2 Population Initialization*

The initial population is a two-dimensional matrix composed of NIND chromosomes. One chromosome represents the location number of N kinds of items. A chromosome is composed of randomly generated 3 × N numbers. Where N represents the type of items and NIND represents the size of the population.

## *3.3 Fitness*

The fitness function of GA is an important condition that can help assess the complexity of an algorithm and whether a chromosome will be selected in the next iteration. The model is to solve the minimum value, but the fitness function needs

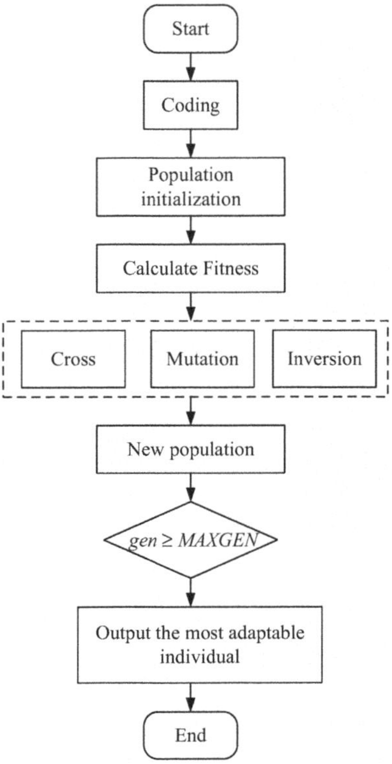


**Fig. 2** Improved genetic algorithm

the maximum value of the objective function. So it needs to be reciprocal to obtain the corresponding fitness function.

$$fit = \frac{\omega_1}{F1} + \frac{\omega_2}{F2} + \frac{\omega_3}{F3} \tag{13}$$

## 3.4 Select

The selection is to better preserve the excellent individuals in the population, improve the efficiency of the algorithm, that is, select the individual from the old population with a certain probability to form a new one to carry out the subsequent operation of the algorithm, wherein the individual with higher fitness. The higher the probability of being selected, thus ensuring item convergence of the algorithm.

## 3.5 *Cross*

According to the characteristics of the coordinates, a two-point crossover method is selected for crossover operation, that is, generates two numbers between 1 and 3 × N are randomly as the cross-cut point of the chromosome, and then swap the part located within the tangent point. Its specific operation is shown in Fig. 3.

## 3.6 *Mutation*

Mutation operation is exchanging the selected position, randomly generating two numbers between1 and 3 × N, then exchanging the codes on the corresponding positions. Its specific operation is shown in Fig. 4.

**Fig. 3** Cross operation

```
5 3 1 4 ┆ 2 2 1 1 ┆ 3 5 1 2
4 1 5 1 ┆ 3 1 1 4 ┆ 5 1 3 5
            ↓
5 3 1 4 ┆ 3 1 1 4 ┆ 3 5 1 2
4 1 5 1 ┆ 2 2 1 1 ┆ 5 1 3 5
```

**Fig. 4** Mutation operation

```
5 3 1 4 ┆ 2 ┆ 2 1 1 ┆ 3 ┆ 5 1 2
            ↓
5 3 1 4 ┆ 3 ┆ 2 1 1 ┆ 2 ┆ 5 1 2
```

**Fig. 5** Inversion operation

| | | | | | | | | | | | |
|---|---|---|---|---|---|---|---|---|---|---|---|
| 5 | 3 | 1 | 4 | 2 | 2 | 1 | 1 | 3 | 5 | 1 | 2 |
| | | | | | ↓ | | | | | | |
| 5 | 3 | 1 | 4 | 2 | 5 | 3 | 1 | 1 | 2 | 1 | 2 |

### *3.7 Inversion*

The inversion operation can solve the problem that the local search ability of the GA is poor and the traditional crossover is difficult to make the offspring inherit the superior gene from parents. The inversion operation is irreversible. Only the individual with increased fitness is accepted, that is, randomly generate two numbers r1 and r2 within the [1, 3 × N] interval, and then flip the chromosome segments in the middle of r1 and r2. The operation is shown in Fig. 5.

## 4 Simulation Results and Analysis

The simulation experiment was carried out according to the improved GA flow shown in Fig. 3. The program design of the improved GA was adopted by MATLAB 2014a, and the basic optimization parameters were designed according to Table 2.

This paper selects 30 items in the warehouse, and they belong to 5 different groups, there are 6 items in one group. The turnover rate and weight information of the items are shown in Table 3.

In order to compare the traditional and improve GA's the computational result, MATLAB is used to record the iterative process of the two algorithms. The comparison graph is shown in Fig. 6.

It can be seen from the Fig. 6 that the improved GA has a significant downward trend in the iterative process of the simulation experiment, but the decline trend of the traditional GA in the iterative process is not obvious, and the computational process once fell into local optimal solution. It can be known from the data in Table 4 that the objective function value obtained by the improved GA is more optimized than the traditional GA, and the storage location assignment can be better accomplished.

From the comparison of the optimized storage allocation and the initial storage allocation in Table 5, it can be seen that the storage coordinates after the optimization of the improved GA obviously store the storage allocation and the tendency to approach the I/O point, which can prove the improvement. The improved GA effectively optimizes the storage allocation of automated warehouses.

**Table 2** Parameters of storage optimization

| Parameter | Value | Parameter | Value |
|---|---|---|---|
| X-axis speed | 1.5 m/s | Line | 9 |
| Y-axis speed | 1.5 m/s | Layer | 9 |
| Z-axis speed | 1 m/s | Iteration number | 500 |
| Slot length | 1 m | Initial population size | 200 |
| Slot width | 1 m | Cross probability | 0.9 |
| Slot height | 1 m | Mutation probability | 0.1 |
| Row | 9 | Generation gap | 0.9 |

**Table 3** Item information

| Group | Turnover | Quality | Group | Turnover | Quality |
|---|---|---|---|---|---|
| *1* | 0.25 | 670 | *3* | 0.15 | 430 |
| | 0.2 | 450 | | 0.3 | 480 |
| | 0.2 | 900 | | 0.15 | 500 |
| | 0.15 | 430 | *4* | 0.2 | 400 |
| | 0.36 | 780 | | 0.65 | 600 |
| | 0.35 | 800 | | 0.25 | 430 |
| *2* | 0.35 | 350 | | 0.1 | 360 |
| | 0.54 | 580 | | 0.34 | 560 |
| | 0.14 | 450 | | 0.15 | 780 |
| | 0.17 | 400 | *5* | 0.35 | 780 |
| | 0.15 | 570 | | 0.25 | 560 |
| | 0.14 | 670 | | 0.45 | 760 |
| *3* | 0.73 | 350 | | 0.2 | 560 |
| | 0.25 | 430 | | 0.16 | 230 |
| | 0.32 | 420 | | 0.47 | 250 |

Figure 7a shows the initial storage distribution graph. The items are placed in a disorderly manner. The distance from I/O point to slots is far and the distance between the items in the same group is large. Such storage allocation will increase the time of items traveling the warehouse, resulting in an increase in storage costs, which is not conducive to the operation of automated warehouses. Fig. 7b shows the storage distribution graph after optimization using the improved GA with coefficient of (0.4, 0.4, 0.2). It can be seen that the placement of the items is more standardized and convenient than before. Moreover, the distance between the items and I/O point is smaller. And the distance between the items in the same group is also smaller. Meanwhile the storage allocation of the items is lower, which can better ensure the stability of the shelves. In this way, the overall operation safety can be ensured, and the cost incurred when the items traveling the warehouse can be reduced. It indicates that the improved GA is effective solved the problem.

In summary, improving GA has advantages over other algorithms in reducing the time of items traveling the warehouse, increasing the aggregation of items in the same group, and reducing the center of gravity of the shelf. From the effect of simulation experiment, improving GA has certain effect on solving the optimization problem of automated warehouse storage allocation. This verifies the effectiveness of improving GA to solve the storage optimization problem.

**Fig. 6** Algorithm iteration comparison graph

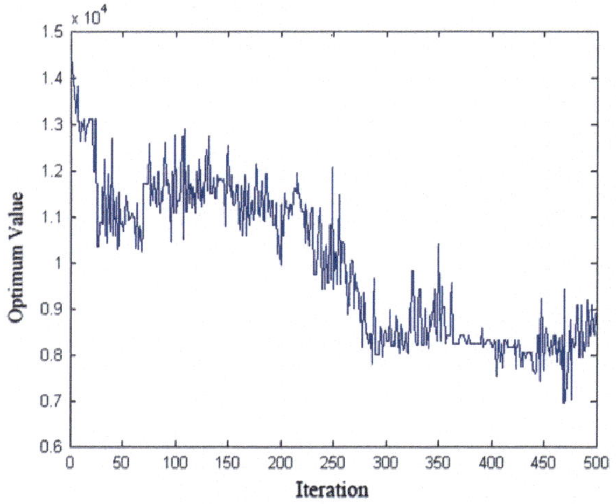


(a) Genetic algorithm iterative graph

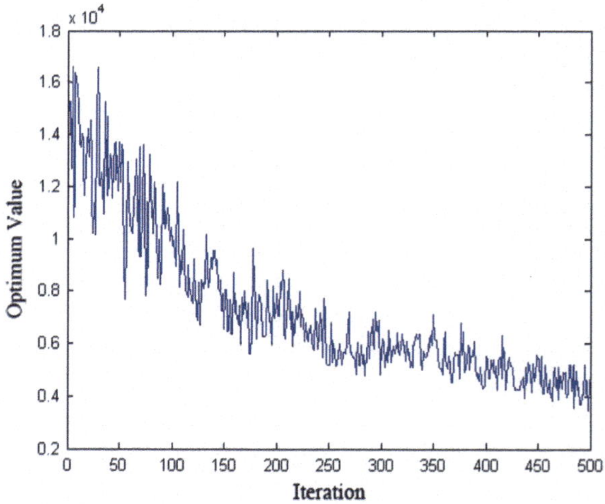


(b) Improved genetic algorithm iterative graph

Fig. 7 Comparison of storage distribution before and after optimization

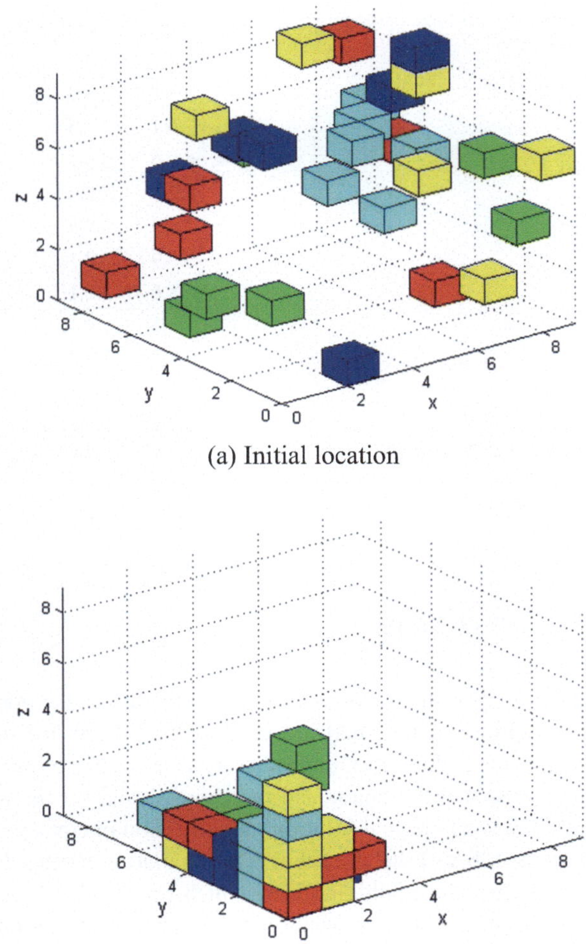


(b) Optimized location

**Table 4** Comparison of GA and improved GA objective function values

| Before optimization | GA | Decrease percentage | Improved GA | Decrease percentage |
|---|---|---|---|---|
| 1.40802e + 04 | 8.7282e + 03 | 38.01% | 4.8036e + 03 | 65.88% |

**Table 5** Optimized storage allocation

| Group | Initial | Optimized | Group | Initial | Optimized |
|---|---|---|---|---|---|
| *1* | 3,4,2 | 2,4,1 | *3* | 7,8,9 | 2,1,3 |
| | 9,2,4 | 2,4,2 | | 9,1,7 | 1,5,1 |
| | 2,6,1 | 2,2,5 | | 8,2,2 | 1,1,2 |
| | 6,9,5 | 2,4,1 | *4* | 9,7,5 | 2,2,2 |
| | 8,2,7 | 2,2,6 | | 2,6,6 | 3,1,2 |
| | 1,4,3 | 2,5,5 | | 9,9,8 | 2,1,2 |
| *2* | 9,6,9 | 2,2,1, | | 1,8,2 | 1,5,2 |
| | 5,8,6 | 3,2,1, | | 8,4,1 | 1,1,1 |
| | 3,8,5 | 1,4,1, | | 4,9,2 | 1,4,2 |
| | 9,7,7 | 1,3,2, | *5* | 9,6,5 | 1,2,2 |
| | 3,1,1 | 1,3,1, | | 9,8,4 | 1,2,5 |
| | 2,2,9 | 1,1,2, | | 8,5,4 | 1,2,1 |
| *3* | 3,7,8 | 1,1,5 | | 7,7,4 | 1,2,3 |
| | 9,6,8 | 1,1,3 | | 8,6,3 | 1,6,1 |
| | 6,2,7 | 2,1,1 | | 7,6,6 | 1,1,4 |

## 5 Conclusions

In this paper, the traveling distance of the items, the position of the items group and the stability of the shelf are used as the criteria for judging the automated warehouse, and then the improved GA is used to optimize the storage location assignment of the automated warehouse. And for the traditional GA is easy to fall into the local optimal solution, the inversion operation is added to improve the local search ability of the algorithm. The simulation results show that the improved GA is better than the traditional GA, which proves that it has certain value and can be applied to the storage location assignment of the automated warehouse.

Optimization of location allocation needs to consider more optimization principles. This paper only considers three optimization principles, but in the reality, factors such as balancing the workload of each region have not been taken into account. In the future research work, it is necessary to carry out research on the target enterprises, take into account more specific and more realistic factors, and propose more reasonable optimization plans and will continue to consider the impact of stacker access time and braking time on operational efficiency.

The research of the model solving algorithm needs further study. Multi-objective optimization problems can also be solved by other algorithms, such as particle swarm optimization, simulated annealing algorithms and nondominated sorting genetic algorithms, or by designing a combination of multiple algorithms to solve the storage allocation problem. In the future research, more attention is paid to the research on the timeliness of the algorithm, and the running time of the algorithm is shortened. This aspect can be further studied.

**Acknowledgment** First of all, I would like to thank Professor Jie Zhu and Professor Ruiping Yuan for his guidance on this paper, and we acknowledge the suggestions from editor and referees. This paper was supported by the logistics robot system scheduling research team of Beijing Wuzi University support.

## References

1. Deng, A. M., Cai, J., & Mao, L. (2013). The optimization model of automated warehouse slotting allocation based on time. *China Scientific Management, 21*(6), 107–112.
2. Yang, W., Li, J., & Yue, T. (2019). Integrated optimization of location assignment and job scheduling in multi-carrier automated storage and retrieval. *Computer Integrated Manufacturing Systems, 25*(1), 247–255.
3. Li, P. F., & Ma, H. (2017). Virus based on cooperative genetic algorithm automated warehouse space optimization model. *China Scientific Management, 25*(5), 70–77.
4. Xiao-Zheng, E., Zu, Q. H., & Cao, M. M. (2013). Slotting optimization of AS/RS for automotive parts based on genetic algorithm. *Journal of System Simulation, 25*(3), 430–435.
5. Wang, T. C., Qiu, J. D., & Shang, Q. J. (2014). Research on application of automated storage and retrieval system slotting optimization based on virus co-evolutionary genetic algorithm. *Computer Science, 41*(b11), 35–38.
6. Bie, W. Q., & Li, Y. J. (2009). Dynamic location assignment of AS/RS based on genetic algorithm. *Computer Engineering and Applications, 4*(29), 211–213.
7. Jiao, Y. L., Zhang, P., & Tian, G. D. (2018). Slotting optimization of automated warehouse based on multi-population GA. *Journal of Jilin University (Engineering and Technology Edition), 48*(5), 1398–1404.
8. Kim, B. S., & Smith, J. S. (2012). Slotting methodology using correlated improvement for a zone-based carton picking distribution system. *Computers & Industrial Engineering, 62*(1), 286–295.
9. Park, C., & Seo, J. (2010). Comparing heuristic algorithms of the planar storage location assignment problem. *Transportation Research Part E Logistics & Transportation Review, 46* (1), 171–185.
10. Pan, C. H., & Shih, P. H. (2015). A storage assignment heuristic method based on genetic algorithm for a pick-and-pass warehousing system. *Computers & Industrial Engineering, 81* (C), 1–13.

# Optimization of Transport Vehicle Path Based on Quantum Evolution Algorithm

**Xiao Zhang, Jilu Li, and Jie Zhu**

**Abstract** The path optimization problem with capacity constraints has always been a combinatorial optimization problem, and the quantum evolution algorithm has excellent performance in solving combinatorial optimization problems. Firstly, after analyzing the traditional path optimization problem and the path optimization problem with capacity constraints, the distribution center is numbered 0 based on the customer-based decimal coding method, and the quantum evolution algorithm is improved according to the capacity constraint method. Fixed quantum rotation angle to dynamic rotation angle. The improved algorithm is then applied to vehicle path optimization problems with capacity constraints. Finally, through the MATLAB and LINGO methods to verify the comparison, it can be concluded that the improved quantum evolution algorithm can accurately solve the vehicle path optimization problem, and the improved quantum evolution algorithm has a good influence on the path optimization problem.

**Keywords** Quantum evolutionary algorithm · VRP · CVRP

## 1 Introduction

The study of quantum evolutionary algorithms began in the 1950s. At that time, scientists in several computer fields independently began to study evolutionary systems. The idea was to introduce evolutionary processes in nature into the field of engineering research to solve optimization problems in engineering [1]. After Benioff and eFynman proposed the concept of quantum computing in the early

X. Zhang (✉) · J. Li · J. Zhu
School of Information, Beijing Wuzi University, Beijing, China
e-mail: bjwzxyzhangxiao@163.com

J. Li
e-mail: jilu_li320@163.com

J. Zhu
e-mail: 1774070756@qq.com



J. Zhang et al. (eds.), *LISS2019*,
https://doi.org/10.1007/978-981-15-5682-1_23

1980s, and demonstrated the high efficiency of quantum computing, quantum computing has the characteristics of exponential storage capacity, parallelism, and exponential acceleration. The entanglement, superposition, and interference of the quantum states in its theory are helpful to solve the problem in intelligent algorithm calculation [2]. In 1994, Shor proposed a quantum algorithm for the decomposition of large integer prime factors. In 1996, Grover proposed a quantum search algorithm for a disordered database; Tayaran Shanghai design of a positive West number adaptive population size adjustment, effectively increase the diversity of QEA algorithm population [3]. On the basis of the Q-bit probability model, Shen proposed the encoding method of multiple probability angles. Through the probability angle encoding, the speed of QEA algorithm is improved, which effectively reduces the computational complexity of the computer [4]. Ma was adaptive quantum genetic algorithm using real numbers and Q-bits shows good advantages in terms of convergence speed and convergence precision [5]. Mou and others used an improved quantum evolution algorithm in the balance of supply and demand of vehicles and goods, and proposed an adaptive attenuation method with restraint punishment, which solved the problem of selecting the optimal quantum individual when the quantum group had no strong and feasible solution. Improving the exit mechanism of quantum evolution algorithm by introducing quantum group maturity [6], quantum evolution algorithm uses quantum rotation gate as an evolutionary strategy, so that quantum individuals gradually approach the optimal solution, and the required results are increased in the form of probability, and the unwanted results are reduced in the form of probability.

Based on the quantum evolution algorithm, this paper seeks the shortest path and can solve the path optimization problem quickly and effectively. On the basis of the original, we can improve the quantum rotation angle, change from the original fixed quantum rotation angle to the dynamic rotation angle, and can more accurately solve the path optimization problem.

## 2 QEA Model

In general, a Q-bit can be expressed as follows:

$$|\varphi> = \alpha|0> + \beta|1> \tag{1}$$

Where $\alpha$ and $\beta$ are complex numbers, representing the state Probability range of $|0>$ and $|1>$ . Thus, the $|\alpha|^2$and $|\beta|^2$ represents the probability size of the Q-bit in state 0 and state 1, and must satisfy the normalization condition, that is, $|\alpha|^2 + |\beta|^2= 1$ and the chromosome population in the $t$ generation is:

$$Q(t) = \{q_1^t, q_2^t, \ldots, q_n^t\} \tag{2}$$

$n$ is the population size; t is an evolutionary algebra; $q_j^t$ is chromosomes, m represents chromosome length, namely,

$$q_j^t = \begin{bmatrix} \alpha_{j1} & \alpha_{j2} & \ldots & \alpha_{jm} \\ \beta_{j1} & \beta_{j2} & \ldots & \beta_{jm} \end{bmatrix}, j = 1, 2, \ldots, n. \tag{3}$$

Quantum rotation doors can be described as:

$$\begin{bmatrix} \alpha^{t+1}(m,n) \\ \beta^{t+1}(m,n) \end{bmatrix} = \begin{bmatrix} \cos(\mu\theta) & -\sin(\mu\theta) \\ \sin(\mu\theta) & \cos(\mu\theta) \end{bmatrix} \times \begin{bmatrix} \alpha^t(m,n) \\ \beta^t(m,n) \end{bmatrix} \tag{4}$$

$\mu\theta = s(\alpha, \beta) * \theta, s(\alpha, \beta)$ represents the direction of rotation to ensure the convergence of the algorithm; $\theta$ is the rotation angle, control the convergence speed of the algorithm, available from Table 1.

*x(m, n)* represents a bit in the current 0–1 matrix chromosome, *b(m, n)* represents the corresponding bit in the current optimal chromosome, and *f(x)* is the objective function.

Table 2 is a transformation on the initial probability matrix. After the above changes, the probability matrix is updated so that the population continues to evolve in the direction of the optimal solution.

**Table 1** Quantum rotation angles

| x(m, n) | b(m, n) | f(x) < f(b) | θ | s(α, β) | | | |
|---|---|---|---|---|---|---|---|
| | | | | $\alpha\beta > 0$ | $\alpha\beta < 0$ | $\alpha = 0$ | $\beta = 0$ |
| 0 | 0 | F | 0 | 0 | 0 | 0 | 0 |
| 0 | 0 | T | 0 | 0 | 0 | 0 | 0 |
| 0 | 1 | F | 0 | 0 | 0 | 0 | 0 |
| 0 | 1 | T | 0.05π | −1 | +1 | ±1 | 0 |
| 1 | 0 | F | 0.01π | −1 | +1 | ±1 | 0 |
| 1 | 0 | T | 0.025π | +1 | −1 | 0 | ±1 |
| 1 | 1 | F | 0.05π | +1 | −1 | 0 | ±1 |
| 1 | 1 | T | 0.025π | +1 | −1 | 0 | ±1 |

**Table 2** Transform matrix table

| x(m, n) | b(m, n) | f(x) < f(b) | θ | $\mu\theta$ | αt | βt | αt + 1 | βt + 1 |
|---|---|---|---|---|---|---|---|---|
| 0 | 0 | F | 0 | 0 | $1/\sqrt{2}$ | $1/\sqrt{2}$ | $1/\sqrt{2}$ | $1/\sqrt{2}$ |
| 0 | 0 | T | 0 | 0 | $1/\sqrt{2}$ | $1/\sqrt{2}$ | $1/\sqrt{2}$ | $1/\sqrt{2}$ |
| 0 | 1 | F | 0 | 0 | $1/\sqrt{2}$ | $1/\sqrt{2}$ | $1/\sqrt{2}$ | $1/\sqrt{2}$ |
| 0 | 1 | T | 0.05π | −0.05π | $1/\sqrt{2}$ | $1/\sqrt{2}$ | 0.7090 | 0.7052 |
| 1 | 0 | F | 0.01π | −0.01π | $1/\sqrt{2}$ | $1/\sqrt{2}$ | 0.7075 | 0.7067 |
| 1 | 0 | T | 0.025π | 0.025π | $1/\sqrt{2}$ | $1/\sqrt{2}$ | 0.7061 | 0.7081 |
| 1 | 1 | F | 0.05π | 0.05π | $1/\sqrt{2}$ | $1/\sqrt{2}$ | 0.7052 | 0.7090 |
| 1 | 1 | T | 0.025π | 0.025π | $1/\sqrt{2}$ | $1/\sqrt{2}$ | 0.7061 | 0.7081 |

# 3 CVRP and QEA

## 3.1 VRP and CVRP

Vehicle Routing Problem (VRP) since 1959 by Danting and Ramser in "The Track Dispatch Problem" [7], since the article was put forward, it has been the focus of many scholars and experts. Capacity constrained Vehicle Path Optimization (CVRP) is a problem in which a known capacity vehicle at a distribution center delivers services to multiple service demand points [8].

## 3.2 Evolutionary QEA to Solve CVRP Problems

CVRP can be described as: Distribution services from a distribution center to $S$ customers with $N$ vehicles. The maximum load of each vehicle is $Q_n$ = (1, 2, …., n), The demand for each customer is $R_{0j}$ = *(1, 2, …, S)*, The distance from customer $i$ to $j$ is $d_{ij}$, and the distance from the parking lot to each customer is $L_{0j}$ = (*i*, *j* = 1, 2, .., S), Additional Ln is the customers for the n vehicle distribution ($L_n$ = 0 means that the $n$ vehicle is not used), adopt set $C$ to represent the n path, $r_{ni}$ is the order of the client $r_{ni}$ in the path n is $i$ (excluding distribution center), set $r_{n0} = 0$ indicate distribution center, the goal of the solution is to make the total mileage of the vehicle distribution the shortest under the condition of meeting the constraints. The constraint conditions are as follows: (1) The total customer demand on each distribution route does not exceed the rated load of the service vehicle; (2) Each customer's needs must be met, and each customer can only be visited once.

The model of CVRP as follows:

$$\min Z = \sum_{n=1}^{N} \left[ \sum_{i=1}^{L_n} d_{r_{n(i-1)} r_{ni}} + d_{r_{nL_n} r_{n0}} \times sign(L_n) \right] \tag{5}$$

$$\sum_{i=1}^{L_n} q r_{ni} \le Q_n \tag{6}$$

$$0 \le L_n \le S \tag{7}$$

$$\sum_{n=1}^{N} L_n = S \tag{8}$$

$$C_n = \{ r_{ni} | r_{ni} \in \{1, 2, \ldots, S\}, i = 1, 2, \ldots, L_n \} \tag{9}$$

$$C_{n_1} \cap C_{n_2} = \phi, \forall n_1 \neq n_2 \tag{10}$$

$$sign(L_n) = \begin{cases} 1, L_n \geq 1 \\ 0, else \end{cases} \tag{11}$$

In the model, (5) solves the optimization goal for the algorithm, so that the total length of the vehicle service path or delivery mileage is the shortest; (6) guarantees that the total amount of goods required by each customer on each route does not exceed the load limit of the distribution vehicle; (7) guarantees that the number of customers on each path does not exceed the total number of customers; (8) means that each customer is provided with distribution services; (9) represents the customer composition of each path; (10) restricts the distribution of each customer only by one vehicle; (11) indicates that when the number of customers served by the n car is greater than or equal to 1, that means the car was in the distribution, then *sign* $(L_n)$ = 1; when the number of customers served by the n car is less than 1, it means that the car is not used, then $sign(L_n)$ = 0.

The Q-bit observation model is defined as a two-dimensional Q-bit probability matrix β of $L \times L$, as follows:

$$\beta = \begin{bmatrix} |\beta_{11}|^2 & |\beta_{12}|^2 & \cdots & |\beta_{1L}|^2 \\ |\beta_{21}|^2 & |\beta_{22}|^2 & \cdots & |\beta_{2L}|^2 \\ \vdots & \vdots & \ddots & \vdots \\ |\beta_{L1}|^2 & |\beta_{L2}|^2 & \cdots & |\beta_{LL}|^2 \end{bmatrix} \tag{12}$$

$|\beta_{ij}|^2(1 \leq i,j \leq L, 0 \leq |\beta_{ij}|^2 \leq 1)$ represents the probability that the column J element in line I has a value of 1. $\beta'_{ij}s$ initial value is taken $\sqrt{2}/2$. By observing $\beta$, the 0–1 observation matrix $G$ can be obtained as follows:

$$G = \begin{bmatrix} G_1 \\ G_2 \\ \vdots \\ G_L \end{bmatrix} = \begin{bmatrix} g_{11} & g_{12} & \cdots & g_{1L} \\ g_{21} & g_{22} & \cdots & g_{2L} \\ \vdots & \vdots & \ddots & \vdots \\ g_{L1} & g_{L2} & \cdots & g_{LL} \end{bmatrix} \tag{13}$$

$G_i = [g_{i1}, g_{i2}, \ldots, g_{iL}]$ the value of $g_{ij}(1 \leq i,j \leq L)$ is determined by the Q-bit $|\varphi>$ observation corresponding to $\beta_{ij}$. That is, by producing $a[0,1]$ The random number $r$ that is evenly distributed between them, when $r \leq |\beta_{ij}|^2$, then $g_{ij}$ = 1, or $g_{ij}$ = 0.

Set $\pi = [\pi_{[1]}, \pi_{[2]}, \ldots, \pi_{[s]}](S \geq 2+L)$ be the delivery path or solution for all participating vehicles, $N_1$ is the number of participating vehicles $(1 \leq N_1 \leq N)$, $S$ is the length of $\pi$, $C = [c_{[1]}, c_{[2]}, \ldots, c_{[L]}]$ is arranged for customers. In this paper, the decimal code based on customer arrangement $C$ is used, and the distribution center number is 0. The vehicle travels from the distribution center to the customer, serves a certain number of customers, and then returns to the distribution center. Since each row in the observation matrix $G$ may contain a plurality of elements of 1, it is not possible to determine the customer arrangement $C$, so the $G$ needs to be transformed. Set $X$ be a 0–1 matrix of $L \times L$, each row of X and only one element in each column is 1. $x_{ij}$ $(1 \le i, j \le L)$ is the $i$ row and $j$ column element in matrix $X = \{clc = 1, \ldots, L \text{ and } gic = 1\}$ is a set of 1 corresponding column numbers in $G_i = \{clc = 1, \ldots, L \text{ and } gic = 0\}$ is a set of 0 corresponding column numbers in $G_i$, $\eta_{pre\_i,j} = 1/d_{pre\_i,j}$ is the reciprocal of the distance between customer $pre\_i$ and customer j. ($\eta_{pre\_i,j}$ is defined as visibility); = $\max\{|\beta_{ij}|^2 \times \eta_{pre\_i,j}\}$ is a set of j in $|\beta_{ij}|^2 \times \eta_{pre\_i,j}$ that maximizes the value of $j$, = $\max\{|\beta_{ij}|^2 \times \eta_{pre\_i,j}\}$ is a set of $j \in M_i^0$ in $|\beta_{ij}|^2 \times \eta_{pre\_i,j}$ that maximizes the value of $j$. The symbol " $\leftarrow$ " randomly selects an element in the right set of the symbol and assigns it to the variable to the left of the symbol. In order to correctly produce the customer arrangement $C$, you first need to convert $G$ to $X$ and then generate $C$ from $X$. The following steps can be used to convert $G$ to $X$:

Step 1: set $i = 1$, $X = G$;
Step 2: When there is an element of 1 in $G_i$ $(1 \le i \le L)$, and $m_i^1 = \arg_{j \in M_i^1} \max\{|\beta_{ij}|^2 \times \eta_{pre\_i,j}\}$, $m_i^0 = \arg_{j \in M_i^0}, \max\{|\beta_{ij}|^2 \times \eta_{pre\_i,j}\}$, $pre\_i = m_{i-1}^1 (m_0^1 = 0)$, then $j' \leftarrow m_i^1$, $x_{ij}' = 1$, $x_{ij} = 0$ $(j \in M_i^1 - \{j'\})$, $x_{i'j'} = -1$ $(i' \in \{1, 2, \ldots L\} - \{i\})$.
Step 3: When there isn't an element of 1 in $G_i$ $(1 \le i \le L)$, and $m_i^0 = \arg_{j \in M_i^0} \max\{|\beta_{ij}|^2 \times \eta_{pre\_i,j}\}$, $m_i^0 = \min\{m_i^0\}$, $pre\_i = m_{i-1}^0 (m_0^0 = 0)$, then $j' \leftarrow m_i^0$, $x_{ij'} = 1$, $x_{i'j'} = -1$, $(i' \in \{1, 2, \ldots L\} - \{i\})$;
Step 4: set $i = i + 1$, if $i \le L$ turn to step 2, otherwise set 0 for all elements in G that are $-1$ and output $X$.

Set *fun(X, i)* be a value function for extracting the column number of elements 1 in line $i$ of matrix $X$. The following steps can be used to change $X$ to $C$:

Step 1: set $i = 1$;
Step 2: $c[i] = fun(X, i)$, $i = i + 1$;
Step 3: if $i \le L$, then turn to step 2, otherwise output $C$;

In summary, quantum evolution algorithm solves the process of CVRP [9]:

(1) $T = 0$, randomly initializing the Q-bit population.
(2) Through the generation, random numbers between the observation populations are generated and decoded to generate lines.

(3) For the lines that have been generated, the order within the line is repeatedly optimized by the closest method.
(4) Individual evaluation, find the current optimal individual.
(5) Determine whether the termination conditions are met. If satisfied, if the output optimal solution is not satisfied, jump to 6.
(6) Judge whether the disaster conditions are met. If satisfied, save the current optimal individual, and re-initialize the Q-bit population if it is not satisfied, jump directly to the next step.
(7) Rotating doors update the Q-bit population, jump to 2 and continue.

The detailed algorithm flow chart is shown in Fig. 1.

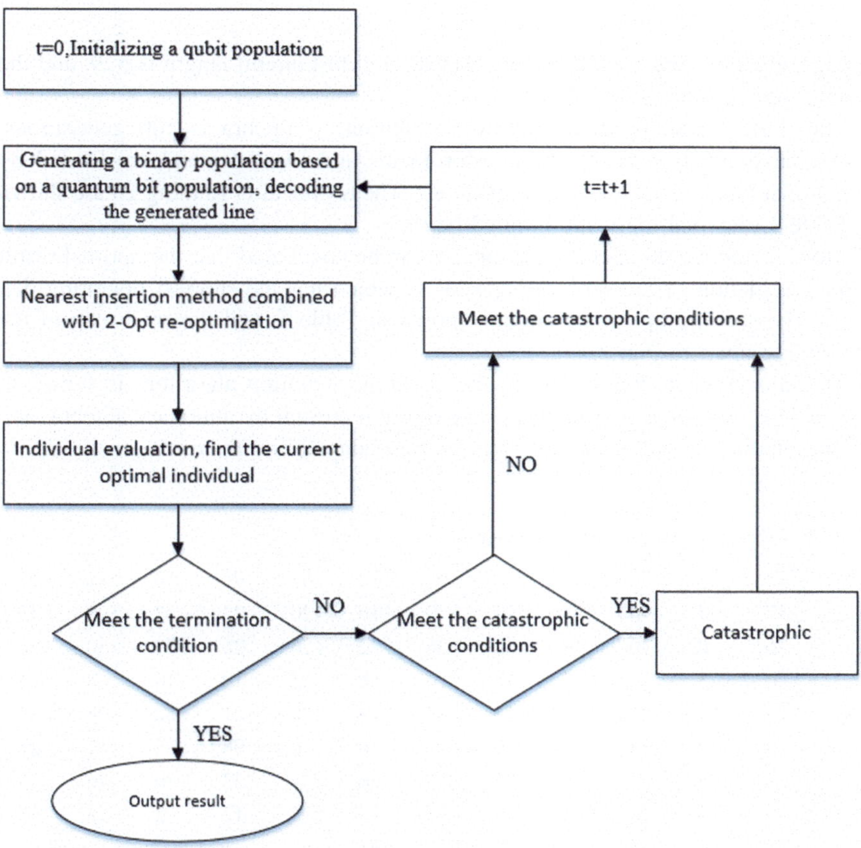


**Fig. 1** Algorithm flow chart

# 4 Experiment and Analysis

## 4.1 Problem Description

Distribution of goods from distribution center 1 to 30 other cities requires meeting the needs of each city. The vehicle's load capacity is limited to 100, and the optimal route is found so that the final distribution route is the shortest. Distribution is 1, n31-5 vehicles.

## 4.2 Problem Solving

The calculation using LINGO shows that the shortest circuit length is 529, and the optimal path is shown in Table 3.

The initial value is set as follows: evolutionary algebra is 500 generations, individual population is 50, urban coordinates and local demand are loaded by data_b.mat file, and vehicle load is 100 tons. The results after running 20 are shown in Table 4 (decimal digits are omitted here).

From the above 20 operating results, it can be concluded that the optimal result is the calculation of the serial number of 1, which has the shortest operating distance. The specific operating route is shown in Table 5, and the road map of the vehicle is shown in Fig. 2.

The convergence diagram using the quantum evolution algorithm is shown in Fig. 3. The transverse coordinates in the figure represent evolutionary algebra, and the longitudinal coordinates represent the best adaptability for each generation.

**Table 3** Urban coordinates and requirements

| City | 1 | 2 | 3 | 4 | 5 | 6 | 7 | 8 | 9 | 10 | 11 | 12 |
|---|---|---|---|---|---|---|---|---|---|---|---|---|
| x | 17 | 24 | 96 | 14 | 14 | 0 | 16 | 20 | 22 | 17 | 98 | 30 |
| y | 76 | 6 | 29 | 19 | 32 | 34 | 22 | 26 | 28 | 23 | 30 | 8 |
| D | 0 | 25 | 3 | 13 | 17 | 16 | 9 | 22 | 10 | 16 | 8 | 3 |
| City | 13 | 14 | 15 | 16 | 17 | 18 | 19 | 20 | 21 | 22 | 23 | 24 |
| x | 23 | 19 | 34 | 31 | 0 | 19 | 0 | 26 | 98 | 5 | 17 | 21 |
| y | 27 | 23 | 7 | 7 | 37 | 23 | 36 | 7 | 32 | 40 | 26 | 26 |
| D | 16 | 16 | 10 | 24 | 16 | 15 | 14 | 5 | 12 | 2 | 18 | 20 |
| City | 25 | 26 | 27 | 28 | 29 | 30 | 31 | | | | | |
| x | 28 | 1 | 27 | 99 | 26 | 17 | 20 | | | | | |
| y | 8 | 35 | 28 | 30 | 28 | 29 | 26 | | | | | |
| D | 15 | 8 | 22 | 15 | 10 | 13 | 19 | | | | | |

**Table 4** Optimal path table

| Vehicle | Path |
|---|---|
| 1 | 1-35-5-26-6-19-17-22-1 |
| 2 | 1-23-7-10-14-18-8-1 |
| 3 | 1-9-31-24-13-29-27-1 |
| 4 | 1-4-2-20-25-12-16-15-1 |
| 5 | 1-21-28-11-3-1 |

**Table 5** Table of operating results

| No. | 1 | 2 | 3 | 4 | 5 | 6 | 7 | 8 |
|---|---|---|---|---|---|---|---|---|
| Optimal value | 676 | 688 | 676 | 690 | 679 | 680 | 699 | 681 |
| Vehicles | 5 | 5 | 5 | 5 | 5 | 5 | 5 | 5 |
| No. | 9 | 10 | 11 | 12 | 13 | 14 | 15 | 16 |
| Optimal value | 682 | 692 | 685 | 688 | 689 | 676 | 690 | 691 |
| Vehicles | 5 | 5 | 5 | 5 | 5 | 5 | 5 | 5 |
| No. | 17 | 18 | 19 | 20 | | | | |
| Optimal value | 697 | 693 | 684 | 680 | | | | |
| Vehicles | 5 | 5 | 5 | 5 | | | | |

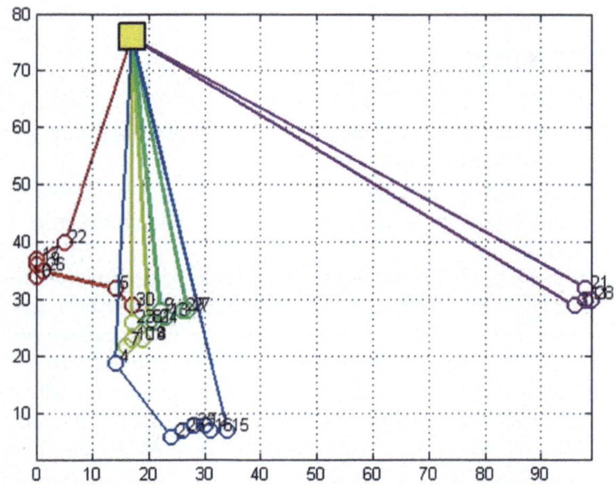


**Fig. 2** Vehicle drive map

According to the results of the solution, the optimal path uses five vehicles, and the total distance of the optimal path is 676. The LINGO knows that the exact optimal distance of the case is 674, so the error is 0.3% < 3%. It can accept the result. And through observation, only one of the 20 results is 699, the error is more than 3%. The accuracy rate is higher, which shows that the algorithm has achieved a better result in dealing with this problem at this time.

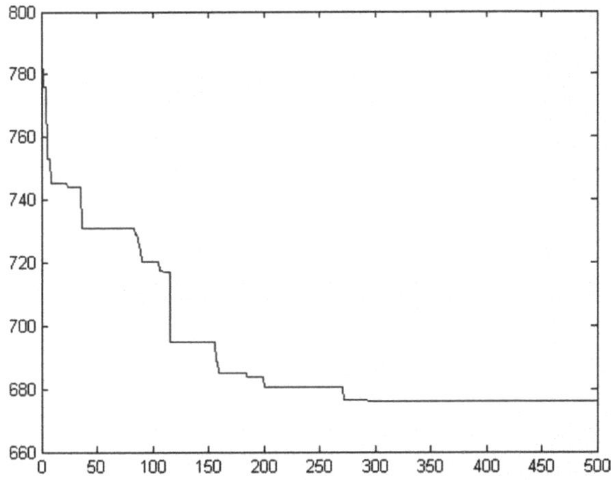


**Fig. 3** Algorithm convergence diagram

Through analysis, it can be seen that when the number of cities is small, the quantum evolution algorithm has a strong solving ability, converges faster, and cooperates with the disaster change function. It is easy to obtain better results, but as the number of cities increases, its operating speed will become relatively slow. The algorithm needs to be improved in a new step to make its application more efficient.

## References

1. Wang, Q. X. (2007). *Improvement and application of quantum evolution algorithm*. M. S. thesis, Jilin University, Jilin.
2. Zhao, Y. W., Peng, D. J., Zhang, J. L., & Wu, B. (2009). Quantum evolutionary algorithm for capacitated vehicle routing problem. *Systems Engineering Theory and Practice, A29*(2), 159–166.
3. Cao, G. L. (2015). *Research on hybrid intelligent optimization algorithm based on vehicle path problem*. M. S. thesis, Kunming University of Science and Technology, Kunming.
4. Ning, T. (2013). *Study of application of hybrid quantum algorithm in vehicle routing problem*. Ph. D. dissertation, Dalian Maritime University, Dalian.
5. Ma, Y., Wang, H. X., & Liu, H. (2018). Research on self-adaptive quantum genetic algorithm. *Computer Engineering and Applications, A54*(20), 99–103.
6. Mou, X. W., Chen, Y., & Gao, S. J. (2016). Research on the matching method of vehicle supply and demand based on improved quantum evolution algorithm. *Chinese Journal of Management Science, 24*(12), 166–176.
7. Dantzig, G. B., & Ramser, J. H. (1959). The truck dispatching problem. *Management Science, A6*(1), 80–91.
8. Wang, P. D. (2009). *Application research on improved ant colony algorithms for CVRP and mobile robot path planning*. M. S. thesis, North MinZu University, Yinchuan.
9. Zhao, Y. W., Zhang, J. L., & Wang, W. L. (2014). *Vehicle path optimization method for logistics distribution*. Beijing: Beijing Science Press.

# Order Batching of Intelligent Warehouse Order Picking System Based on Logistics Robots

**Ruiping Yuan, Juntao Li, and Huiling Wang**

**Abstract** Through the analysis of operation process of intelligent warehouse order picking system based on logistics robots, it is found that reducing the times of shelf moves and balancing the picking time among different picking stations are key factors to improve the efficiency of order picking. Based on the correlation of different orders demand on the same shelf, the order batch model of parallel picking mode is established. The "order correlation factor" is proposed to describe the number of items located on the same shelf between two orders, and the longest picking time of picking stations is minimized to achieve time balance among different stations operated in parallel. An improved dynamic clustering algorithm is proposed to solve the model, wherein the orders are first clustered according to their correlation factor, and this is followed by dynamic balancing of the picking time among the stations. Simulation results show that, compared with order batching method which only considers reducing the times of shelf moves, the proposed method is more effective for improving the picking efficiency of one wave.

**Keywords** Order batching · Logistics robots · Intelligent warehouse system · Improved dynamic clustering algorithm

---

National Natural Science Foundation of China (71831001); Young Top-notch Talent Project of Beijing Municipal Education Commission (CIT&TCD201704059); Beijing Excellent Talents Support Project (2017000020124G063); Funding Project for Beijing Intelligent Logistics System Collaborative Innovation Center; Funding Project for Beijing Key Laboratory of Intelligent Logistics Systems.

---

R. Yuan (✉) · J. Li · H. Wang
School of Information, Beijing Wuzi University, Beijing, China
e-mail: angelholyping@163.com

J. Li
e-mail: ljtletter@126.com

H. Wang
e-mail: 18810712038@163.com



J. Zhang et al. (eds.), *LISS2019*,
https://doi.org/10.1007/978-981-15-5682-1_24

# 1 Introduction

The traditional picker-to-parts picking mode in distribution centers is labor-intensive and characterized by low picking efficiency and high cost. According to a survey, traditional manual picking accounts for about 55–75% of the operating costs of a distribution center [1]. In order to reduce labor costs and improve picking efficiency, parts-to-picker picking system based on logistics robots has gradually gained attention and application in e-commerce distribution centers. Amazon's Kiva system is a successful application case of this new picking mode [2].

In the parts-to-picker picking system based on logistics robots, the items are stored on movable storage shelves, also known as inventory pods, and brought to the order pick stations by logistics robot. Due to the large number of robots which can transport many inventory pods simultaneously, an order picker can complete the order in a shorter time and can complete more orders compared to traditional picker-to-parts systems [3]. The parts-to-picker picking system also provides added benefits of flexibility and scalability in addition to the associated advantages of automation. By adding more robots, pods and/or work stations, the throughput capacity of handling additional orders can be addressed economically and in a relatively short time span. So, the new picking mode is particularly suitable for e-commerce distribution centers that handle strong demand fluctuations, large assortment, and small orders with tight delivery schedules.

The parts-to-picker picking system based on logistics robots heavily influences all traditional planning problems in warehouses and numerous operational decisions problems under the new picking mode are yet to be examined in depth, for example storage allocation, the assignment of orders to picking stations, the allocation of tasks to robots and the path planning of multi robots. Previous studies [4–7] mainly focused on storage allocation, task assignment and path planning. In this research, we specifically investigate the assignment of orders to picking stations in the picking system based on logistics robots when orders are picking in batches, namely order batching, in which multiple orders are unified to a batch of orders jointly assembled in a picking station to improve picking efficiency.

Most studies on order batching are based on the traditional picker-to-parts picking mode. There are basically two criteria for batching in manual picking: the proximity of pick locations and time windows. Reference [8, 9] gave a very detailed review of solution methods to the order batching in manual picking. The order batching problem is proven to be NP-hard. For larger problem instances the use of heuristics becomes unavoidable. Heuristics for order batching problem can be distinguished into five groups: priority rule-based algorithms, seed algorithms, savings algorithms, data mining approaches, and metaheuristics [10]. Hwang and Kim [11] develop a cluster analysis based batching algorithm, Chen et al. [12] proposed a data mining method based on association rules for order batching. Henn and Wäscher [13] built a mathematical model based on the relative distances among the goods to achieve the shortest walking distance for the picker, and the solution was obtained by two modified tabu search algorithms. Menéndez et al. [14]

proposes several strategies based on the variable neighborhood search methodology to tackle the order batching problem. Jiang et al. [15] solved the order batching and sequencing problem with limited buffers with a modified seed algorithm.

Due to the different working ways between the parts-to-picker and picker-to-parts picking modes, namely, that pickers in the former walk to pick goods one by one, whereas the logistics robots used in the latter carry shelves to stationary pickers, the relative distances among goods cannot be used as the primary consideration during the order batching for the parts-to-picker mode. Consequently, traditional order batching models in manual picking cannot be directly applied to the parts-to-picker mode. Compared to the large variety of research on order batching in manual picking system, the research on order assignment to picking stations in parts-to-picker mode is very limited. Boysen et al. [16] constructed a mathematical model of the order picking sequence and shelf-moving sequence to a station, with the purpose of reducing rack movements and speeding up order delivery. Wu et al. [17] further tackled order sequencing by building a mathematical model for decreasing the frequency of bins being moved around. A modified K-Means clustering algorithm was used to obtain the desired solution. However, these previous studies only considered the number of shelves moves in the developed models and algorithms, but did not take the time balance among picking stations into consideration. In intelligent warehouse system based on logistics robots, the picking time for stations operated in parallel is ultimately determined by the station with the longest picking time, and this necessitates time balance among the stations.

Based on the order batching models and algorithms developed in previous studies, this paper develops a mathematical model for balancing the picking time among multiple stations and reducing the number of shelves moves to make order batching more time-efficient. A modified clustering algorithm, which considers the order correlation factor and balanced station picking time, is proposed to solve the model. Simulation experiments are then performed to verify the picking efficiency.

The rest of this paper is organized as follows. Section 2 gives a detailed description of the order batching problem of robots based order picking system. The "order correlation factor" is introduced in Sect. 3 and used to describe shelf sharing by goods of two orders. A mathematical model for more time-efficient order batching is then developed. Section 4 presents a modified dynamic clustering algorithm for solving the developed model, while Sect. 5 compares the picking time, efficiency, and the times of shelf moves using the proposed algorithm with those using the traditional algorithm. Finally, Sect. 6 presents the conclusions drawn from the present study and the prospects.

## 2 Problem Description

The parts-to-picker order picking system based on logistics robots is mainly composed of pickers, picking stations, shelves, aisles, cache regions, and logistics robots, as shown in Fig. 1.

The order picking process based on logistics robots is as follows. Firstly, orders are batched and merged into pick lists, which are assigned to picking stations. Then the tasks on pick lists are allocated to logistics robots, which move the corresponding shelves to the designated picking stations, as shown in ① of Fig. 1. After pickers at the station pick the target tasks on the shelves, the logistics robots transport the shelves back to their original locations, as in ② of Fig. 1. The logistics robots go on to execute next task, as in ③ of Fig. 1. This is continued until all the assigned tasks are performed.

E-commerce orders are usually of large-variety, small-batch and high-frequency, so wave-picking is usually adopted with the purpose of improving picking efficiency. Wave-picking is to gather orders together within a certain period for picking [18]. That is, orders arrive at the order pool continually, and then a certain quantity of subsequent orders is removed as one wave. In this method, the number of orders in a wave is first determined, and the orders are then divided into batches based on certain rules, and one batch is distributed to one picking station. The orders in the same batch are merged into a pick list with many tasks, which are the items to be picked. After order batching, the tasks are subsequently assigned to robots that perform the picking. The order batching process is illustrated in Fig. 2.

In application, e-commerce orders are numerous, and the relationship among the orders, picking stations, and shelves is complex. A reasonable distribution of arriving orders to stations is thus important to achieve high picking efficiency. A robot needs to move the entire shelf to picking stations, so it is better to pick as

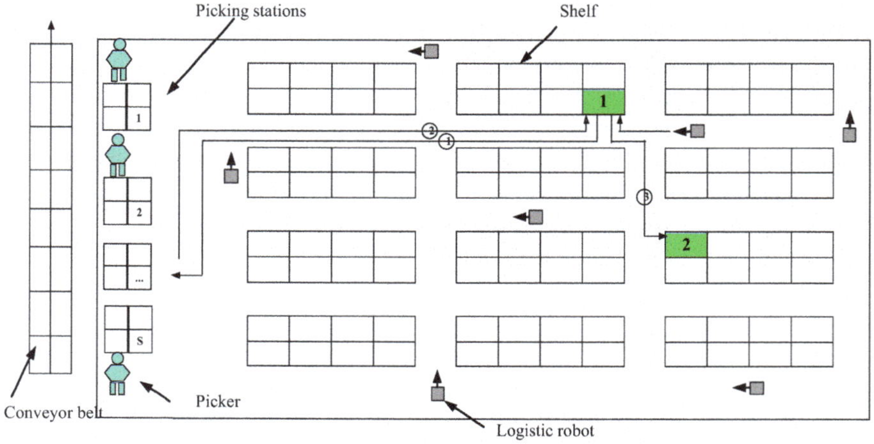


**Fig. 1** Schematic diagram of intelligent warehouse picking system

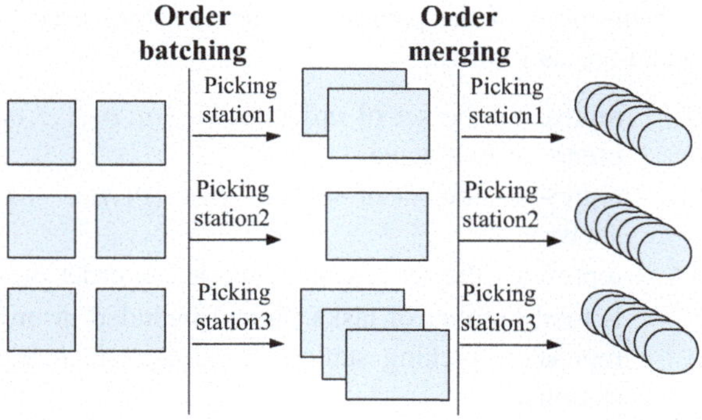


**Fig. 2** Order batching process

many as items for a single shelf move. Therefore, the orders which consist of items on the same shelf should be assigned to one station, which obviously can reduce the times of shelve moves. So, minimization of the times of shelf moves can improve the picking efficiency. In addition, the parallel operation of multiple stations is usually adopted in intelligent warehouses. So, the time required to complete the picking for a wave of orders is determined by the station with the longest picking time, inappropriate order batching may cause unevenly distributed busy or idle times of the pickers. Balancing the picking time among different stations is thus also important to improve the order picking efficiency.

# 3 Model Formulation

## *3.1 Model Assumptions and Parameter Definitions*

Based on an analysis of the work flow in an intelligent warehouse system based on logistics robots, the following assumptions are made in the modeling:

- Relationship among orders, tasks (items), and pick lists: an order may entail multiple tasks, but a task can only be included in one order. A task is an indivisible unit. A pick list contains at least one order and multiple tasks; a pick list may thus be considered as a large order.
- Relationship between a pick list and picking station: In wave-picking, there is only one pick list for each picking station.
- Relationship among shelves, tasks, and picking stations: A shelf may store items for multiple tasks, but can serve only one picking station at a time. The location of the shelf for a picking task is known. A picker works at only one station. The picking time for each task is the same for all pickers.
- All the logistics robots run at the same uniform speed.

Definitions of the parameters and variables used to develop the order batching model are as follows:

$O$: represents the set of orders, $O = \{o_1, o_2, \ldots, o_m\}$. $m$ is the total number of orders of one wave.

$P$: represents the set of shelves, $P = \{p_1, p_2, \ldots, p_n\}$. $n$ is the total number of shelves.

$A_i$: represents the set of tasks included in order $o_i$, $A_i = \{a_{i1}, a_{i2}, \ldots, a_{i|A_i|}\}$. $|A_i|$ is total number of tasks (items) included in order $o_i$.

$s$: represents picking station, $s = 1, 2, \ldots h$, $h$ is the total number of picking stations.

$O_s$: represents the set of orders allocated to station $s$, $O_s = \{o_{s1}, o_{s2}, \ldots, o_{s|O_s|}\}$.$|O_s|$ is the total number of orders in $O_s$.

$d_{a_{ij}s}$: distance between the shelf of task $a_{ij}$ and picking station $s$.

$v'$: running speed of logistics robots.

$t'$: picking time of a task by a picker.

$q'_s$: maximum number of orders received by station $s$

$$\gamma_{a_{ij}p_k} = \begin{cases} 1, & \textit{if task } a_{ij} \textit{ is located on shelf } p_k \\ 0, & \textit{otherwise} \end{cases}$$

$$x_{is} = \begin{cases} 1, & \textit{if order } o_i \textit{ is assigned to pickingstation } s \\ 0, & \textit{otherwise} \end{cases}$$

## 3.2 Mathematical Model of Order Batching

Based on the common demand for a shelf by different orders, an "order correlation factor" is proposed in this sector to describe the occupation of the same shelf by goods from different orders. The "order correlation factor" is then used to develop a mathematical model of order batching which can reduce the times of shelf moves and balance the picking time among stations.

The correlation factor of each two orders is denoted by $Q_{o_io_l}$, which indicates the number of any two items from order $O_i$ and order $O_l$ that are located on the same shelf. $a_{i\partial}$ and $a_{l\beta}$ represents a task from order $o_i$ and order $o_l$ respectively. When the items of these two tasks are both located on shelf $p_k$, we say these two tasks are correlative tasks. $Q_{o_io_l}$ is defined as the total number of correlative tasks of these two orders.

$$Q_{o_io_l} = \sum_{k=1}^{n}\left(\sum_{\partial=1}^{|A_i|}\sum_{\beta=1}^{|A_l|}\gamma_{a_{i\partial}p_k}\gamma_{a_{l\beta}p_k}\right), \quad i \neq l \tag{1}$$

The correlation factor between order $o_i$ and picking station $s$ is denoted by $Q_{is}$, which is the sum of the correlation factor between order $o_i$ (unassigned order) and the orders in $O_s$ (the set of orders assigned to station $s$). $Q_{is}$ is defined as follows:

$$Q_{is} = \sum_{w=1}^{|O_s|} Q_{o_i o_{sw}} \tag{2}$$

The picking time of a task mainly consists of two parts: (1) the logistics robots travel time. $t_1$ is the time required to transport the shelf to the appropriate picking station, and $t_2$ is the time required to return the shelf back to its original location after completion of the picking. Here, the distances covered in the transportation of the shelf in both ways are assumed to be equal, and referred to as the Manhattan distance. So, in the assumption, it is obvious that $t_1 = t_2$. (2) The picking time $t'$, which is the time required to pick the items by the picker. The time cost $t_{a_{ij}}$ of task $a_{ij}$ can be expressed as:

$$t_{a_{ij}} = t_1 + t_2 + t' = 2 \times \frac{d_{a_{ij}s}}{v'} + t' \tag{3}$$

The time cost $t_{is}$ of order $o_i$ when it is allocated to station $s$ is:

$$t_{is} = \left(\sum_{j=1}^{|A_i|} t_{a_{ij}}\right) \times (|A_i| - Q_{is})/|A_i| \tag{4}$$

The picking time at station $s$ is given by:

$$t_s = \sum_{i=1}^{m} t_{is} \times x_{is} \tag{5}$$

The objective is to minimize the picking time at the station with the longest picking time:

$$f = \min \max_s t_s = \min \max_s \left(\sum_{i=1}^{m} t_{is} \times x_{is}\right) \tag{6}$$

*s.t.*

$$\sum_{s=1}^{h} x_{is} = 1, \forall i \in \{1, 2, \ldots, m\} \tag{7}$$

$$1 \le \sum_{i=1}^{m} x_{is} \le q_s', \forall s \in \{1, 2, \ldots, h\} \tag{8}$$

Equation (7) imply that an order can only be assigned to one station. Equation (8) shows that at least one order is allocated to each station, with the maximum number of allocated orders not exceeding the capacity of the cache region. The proposed order batching model for logistics robot based order picking system uses the time cost as the order allocation criterion, with comprehensive consideration of reducing the times of shelf moves and balancing the picking time among stations.

## 4 Improved Dynamic Clustering Algorithm for Solving Order Batching Problem

The order batching problem is a NP-hard problem of combinatorial optimization, which is difficult to optimally solve using an exact algorithm. This combinatorial optimization problem can be converted into a clustering problem. There are many different solving methods for clustering problems. Among them, dynamic clustering algorithm can modify and optimize clustering results on the basis of classification [19, 20]. So, the dynamic clustering algorithm is improved and used in this study to solve order batching model. The advantages of this algorithm lie in: (1) It ensures that the number of orders in a wave will not be too large due to the limit of the cache region, and the clustering algorithm is particularly good for medium-scale problems. (2) While the clustering algorithm does not guarantee optimal final results, it offers fast computation. (3) The present study considers both the order correlation factor and balance of the picking time among stations, and thus requires adjustment of the results, which necessitates solving the model by dynamic cluster analysis.

In our method, firstly the generation of the initial cluster centers and the clustering rules is improved according to the batching problem, and then the solutions are dynamically adjusted by balancing the picking time among stations to get better order batching solutions. The specific steps of the improved clustering algorithm are as follows:

### *4.1 Initial Clustering Based on Order Correlation Factor*

Step 1: Generation of the initial cluster center for each picking station.

(1) The order correlation factors between each two orders are calculated by (1). The order with the largest sum of order correlation factor is taken as the initial cluster center $a$, and is arbitrarily assigned to a station.

(2) Using the initial cluster center $a$ as the starting point, the order with the least order correlation factor with $a$ is set as the initial cluster center $b$, and is randomly assigned to any available station.
(3) $a$ and $b$ are combined and the order that has the least order correlation factor with them is set as the initial cluster center $c$, and is randomly assigned to any available station.
(4) Repeat (3) until all the picking stations are assigned an initial cluster center. The algorithm then proceeds to Step 2.

Step 2: Clustering rules for orders

(1) Calculate the number of orders of each cluster center, if it has not reached the maximum capacity of the corresponding order cache region, the cluster center is available and can accept new orders. Else, the cluster center is unavailable.
(2) The correlation factors between a remaining order and available cluster centers are calculated by (2). An order is assigned to and merged with the cluster center with which it has the largest correlation factor. If the correlation factors are the same, the order is assigned to the cluster center with the shortest picking time to balance the picking time among stations.
(3) Repeat (1) and (2) until all the orders are distributed. The algorithm then proceeds to next step for time balance adjustment.

## 4.2 *Dynamic Clustering Algorithm for Balancing the Picking Time among Stations*

Step 3: Time balance between stations

(1) Based on the clustering results and the picking times of the different stations calculated by (5), the station with the longest picking time $t_1$ is set as station $s_1$.
(2) The correlation factors between each two order at station $s_1$ are calculated by (1). The order with the smallest sum of correlation factors is identified as $o_l$.
(3) The correlation factors between order $o_l$ and other available cluster centers are calculated by (2).
(4) Order $o_l$ is merged with the available cluster center with the largest correlation factor. Recalculate the longest picking time $t_2$. If $t_2 < t_1$, Order $o_l$ is reassigned to the available the station with the largest correlation factor and return to (1). Otherwise, try to assign order $o_l$ *to* the available station with the second largest correlation factor, and so on. If all failed, the order batching terminates.

## 5 Simulation Analyses

An e-commerce company A is used to assess the validity of the proposed order batching model and algorithm. The orders received by company A are highly varied, with small-batch and high frequency. The company utilizes logistics robots based intelligent warehouse picking system with a total storage area of 6000 $m^2$ divided into four identical zones. There are 800 shelves and 2000 categories of goods per zone, with each shelf holding 10–15 items. The warehouse is equipped with three picking stations, each having a cache capacity of 12 orders per wave. The number of tasks for each order varies from 1 to 8. The warehouse is equipped with logistics robots that move at a speed of 2 m/s. The path between each shelf and station is known and fixed. The picking efficiency of the pickers is 4 s/item. MATLAB 7.11 is used in this study to perform multiple simulation iterations on orders of different quantities (100, 200, …, 1000).

The proposed order batching optimization algorithm (here referred to as Algorithm 2) comprehensively considers the reduction of the times of shelf moves and balancing of the picking time among the stations to minimize the time cost. Conversely, the algorithm (referred to as Algorithm 1) which comes from [10] only considers the reduction of the times of shelf moves. In Algorithm 1, the correlation is calculated for pairs of orders to identify and merge the pair with the largest correlation, with the merged order assigned to the first picking station. The correlation is then computed again for the merged order and the other orders, and the orders with the largest correlation factor are merged and assigned to the first picking station. The procedure is repeated until the orders of first station reach its limit, and the remaining orders are distributed to other stations in a similar way.

The results of the two methods were compared to assess the validity of the proposed model and algorithm. In each case, the picking time of the station with the longest picking time was adopted as the total picking time.

The two algorithms were respectively used to perform multiple simulations for different quantity of orders to determine the picking time and the times of shelf moves. The results are shown in Fig. 3 and Fig. 4 respectively. Time difference and order quantities assigned to different stations for the two algorithms in the case of 600 orders is presented in Table 1. Time difference is the largest picking time minus the picking time of each picking station. The following can be observed from Fig. 3, Fig. 4 and Table 1:

(1) Figure 3 and Fig. 4 show that, with increasing order quantity, there is a decrease in the increase rate of both the times of shelf moves and the order picking time for both algorithms. This implies reasonable order batching can improve picking efficiency.
(2) Though the times of shelf moves is almost the same for both algorithms as shown in Fig. 3, the picking time for Algorithm 2 is less than that for Algorithm 1 in a given scenario, as shown in Fig. 4. And with the increase of order quantity, the efficiency of algorithm 2 is more obvious.

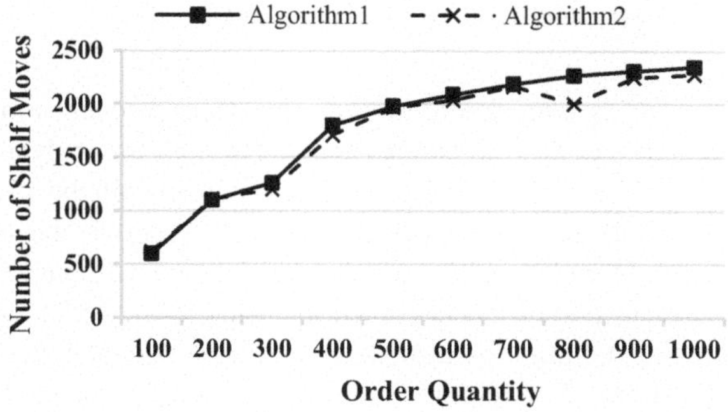


**Fig. 3** Comparison of the times of shelf moves for the two order batching algorithms

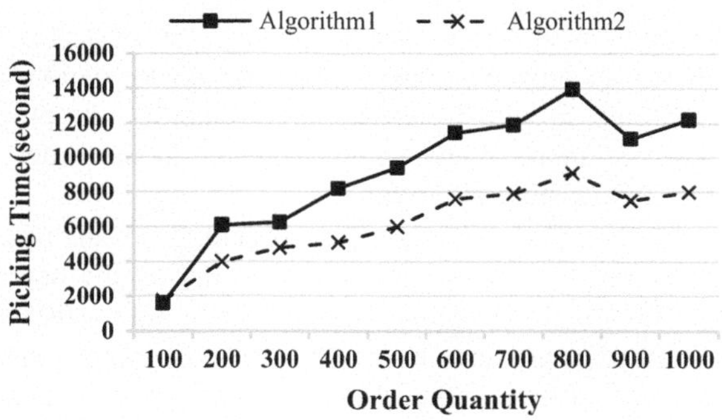


**Fig. 4** Comparison of the picking time for the two order batching algorithms

**Table 1** Picking time differences (in seconds) and order quantities assigned to different stations for the two order batching algorithms

| Order batching algorithm | Picking station 1 | | Picking station 2 | | Picking station 3 | |
|---|---|---|---|---|---|---|
| | Time diff. | Order qty. | Time diff. | Order qty. | Time diff. | Order qty. |
| Algorithm 1 | 1792 | 200 | 0 | 200 | 783 | 200 |
| Algorithm 2 | 169 | 195 | 0 | 207 | 57.92 | 202 |

(3) It can be observed from Table 1 that the picking time differences between stations of Algorithm 1 are greater than those of Algorithm 2. That is because Algorithm 1 only considers the number of shelves moves, but as noted earlier, the picking time is determined by the station with the longest picking time. So, while Algorithm 1 affords savings in the number of shelves moves, it offers no advantage with regard to the picking time. And it also could cause uneven distributions of the busy and idle time of the pickers.

## 6 Conclusions and Prospects

This study considered the order batching problem in intelligent warehouse picking system based on logistics robots. Through an analysis of the work flow, reduction of the times of shelf moves and balancing of the pick time among stations were identified as two important factors to improve the order batching efficiency. An "order correlation factor" was introduced to characterize the situation in which goods of two orders occupied the same shelf. A detailed mathematical model of the order batching was also developed and solved by a modified dynamic clustering algorithm. Simulation experiments were performed to compare the developed model and algorithm with an existing order batching algorithm. Based on the observations, the following conclusions were drawn: (1) an appropriate order batching method improves the picking efficiency; (2) reduction of the times of shelf moves and balancing the picking time among stations are equally important to order batching; (3) with increasing order quantity, the proposed order batching method produces more obvious improvements in the order picking efficiency compared with the existing algorithm.

Further study is, however, required to investigate the effects of dynamic uncertainty factors on the time cost during the operation of logistics robots, i.e., considering the time cost as a random variable rather than a fixed value. This would facilitate improvement of the dynamic adaptability in the implementation of logistics robot based picking systems. Item (3) in the preceding paragraph also requires deeper examination, taking into consideration dynamic uncertainties of picking tasks.

**Acknowledgements** The authors gratefully acknowledge the helpful comments and suggestions of the reviewers, which have improved the presentation.

## References

1. Chiang, M. H., Lin, C. P., & Chen, M. C. (2011). The adaptive approach for storage assignment by mining data of warehouse management system for distribution centres. *Enterprise Information Systems, 5*(2), 219–234.
2. Wurman, P. R., & Mountz, M. (2007). Coordinating hundreds of cooperative, autonomous vehicles in warehouses. In *National Conference on Innovative Applications of Artificial Intelligence* (pp. 1752–1759). AAAI Press.
3. Enright, J. J., & Wurman, P. R. (2011). Optimization and coordinated autonomy in mobile fulfillment systems. In *AAAI Conference on Automated Action Planning for Autonomous Mobile Robots* (pp. 33–38). AAAI Press.
4. Lamballais, T., Roy, D., & de Koster, R. (unpublished). Inventory allocation in robotic mobile fulfillment systems.
5. Zou, B., Gong, Y., Xu, X., et al. (2017). Assignment rules in robotic mobile fulfillment systems for online retailers. *International Journal of Production Research, 55*(20), 6157–6192.
6. Zhou, L., Shi, Y., Wang, J., et al. (2014). A balanced heuristic mechanism for multirobot task allocation of intelligent warehouses. *Mathematical Problems in Engineering, 2014,* 1–10.

7. Ma, H., Koenig, S., & Ayanian, N., et al. (2016). Overview: generalizations of multi-agent path finding to real-world scenarios. In *IJCAI 2016 Workshop on Multi-agent Path Finding* (pp. 1–4).
8. de Koster, R., Le-Duc, T., & Roodbergen, K. J. (2007). Design and control of warehouse order picking: A literature review. *European Journal of Operational Research, 182*(2), 481–501.
9. Cergibozan, Ç., & Tasan, A. S. (2019). Order batching operations: an overview of classification, solution techniques, and future research. *Journal of Intelligent Manufacturing, 30*(1), 335–349.
10. Henn, S. (2012). Algorithms for on-line order batching in an order picking warehouse. *Computers & Operations Research, 39*(11), 2549–2563.
11. Hwang, H., & Kim, D. G. (2005). Order-batching heuristics based on cluster analysis in a low-level picker-to-part warehousing system. *International Journal of Production Research, 43*(17), 3657–3670.
12. Chen, M. C., Huang, C. L., Chen, K. Y., et al. (2005). Aggregation of orders in distribution centers using data mining. *Expert Systems with Applications, 28*(3), 453–460.
13. Henn, S., & Wäscher, G. (2012). Tabu search heuristics for the order batching problem in manual order picking systems. *European Journal of Operational Research, 222*(3), 484–494.
14. Menéndez, B., Pardo, E. G., Alonso-Ayuso, A., et al. (2017). Variable neighborhood search strategies for the order batching problem. *Computers & Operations Research, 78,* 500–512.
15. Jiang, X. W., Zhou, Y. X., Zhang, Y. K., et al. (2018). Order batching and sequencing problem under the pick-and-sort strategy in online supermarkets. *Procedia Computer Science, 126,* 1985–1993.
16. Boysen, N., Briskorn, D., & Emde, S. (2017). Parts-to-picker based order processing in a rack-moving mobile robots environment. *European Journal of Operational Research, 262,* 550–562.
17. Wu, Y. Y., Meng, X., Wang, Y. Y., et al. (2016). Order sequence optimization for part-to-picker order picking system. *Journal of Mechanical Engineering, 52*(4), 206–212.
18. Li, Y. D. (2013). Model and algorithm for cartonization and slotting optimization simultaneously in wave-picking zone-based system. *Systems Engineering—Theory & Practice, 33*(5), 1269–1276.
19. Liu, J., & Wang, D. (2009). An improved dynamic clustering algorithm for multi-user distributed antenna system. In *International Conference on Wireless Communications & Signal Processing* (pp. 1–5). IEEE.
20. Bukchin, Y., Khmelnitsky, E., & Yakuel, P. (2012). Optimizing a dynamic order-picking process. *European Journal of Operational Research, 219*(2), 335–346.

# Resource Allocation with Carrier Aggregation in Wireless Ad Hoc Networks

**Tingyu Yang, Yanfei Liu, and Wentao Lu**

**Abstract** Carrier aggregation (CA) has become one of the promising technologies in long term evolution-advanced (LTE-A) due to its ability to provide higher data rates. All previous research is based on cellular systems. But in this paper, resource allocation with carrier aggregation in wireless ad hoc networks is studied. Wireless ad hoc networks are multi-hop networks compared to one-hop cellular networks. Therefore, the innovation of this paper is that we established a new model in which the hop count and the frequency reuse which aims at maximizing system throughput are considered. At the same time, the model also incorporates adaptive modulation coding and guarantees the minimum transmission rate requirement of each node. Then we formulate the resource allocation problem as a 0–1 integer linear programming problem and solved by the branch implicit enumeration algorithm (BIEA). Simulation results show that resource allocation with carrier aggregation can achieve higher system throughput in wireless ad hoc networks, and the proposed algorithm can also meet the quality of service (QoS) requirements of each node.

**Keywords** Carrier aggregation · Resource allocation · Hop count · Frequency reuse · 0–1 integer linear programming

---

This work was supported by the basic research business of the Ministry of Education. The project number is 2018JBZ102.

---

T. Yang (✉) · Y. Liu · W. Lu
School of Communication and Information, Beijing Jiaotong University, Beijing, China
e-mail: 17120154@bjtu.edu.cn

Y. Liu
e-mail: 17125037@bjtu.edu.cn

W. Lu
e-mail: 17111026@bjtu.edu.cn



J. Zhang et al. (eds.), *LISS2019*,
https://doi.org/10.1007/978-981-15-5682-1_25

# 1 Introduction

Wireless ad hoc networks have the characteristics of fast and flexible networking, dynamic topology changes, multi-hop, and strong resistance to damage. They have appeared in a wide variety of applications such as the military, disaster relief, sensing, monitoring and etc. [1]. With the continuous expansion of ad hoc networks and the increasing demand for node services, mobile data traffic has grown exponentially [2]. In order to improve system capacity and the transmission rate of each node, a large amount of spectrum resources is needed to solve the problem. However, it is difficult to find a whole continuous large bandwidth spectrum with severe electromagnetic interference. Carrier aggregation (CA) proposed by 3GPP can increase system bandwidth by aggregating multiple continuous or discrete frequency bands, thereby achieving faster data transmission rate and higher spectrum utilization [3].

A lot of research on resource allocation with CA has been carried out in LTE-A. Initially, the authors considered component carrier (CC) selection and resource block (RB) allocation separately, resulting in lower network performance [4]. Liao et al. [5], proposed the greedy algorithm that takes into account joint allocation of CCs and RBs and modulation and coding schemes (MCS). An efficient RB allocation algorithm was proposed, which allocates CC, RB and MCS according to the users' CA capability [6]. A large-scale CA scenario was considered in [7] and a continuous geometric programming approximation method was used to solve the resource allocation problem. The authors considered the quality of service (QoS) mechanism of different traffic types and used different utility functions for heterogeneous traffic [8]. They proposed a two-step resource allocation algorithm including resource grouping. The utility proportional fair method [9] was used to guarantee the minimum QoS of each user according to the utility percentage of the percentage of the application run by the user and the priority criterion of the application type.

So far, the above research has been directed to single-hop cellular networks. In these scenarios, any RB can be assigned exclusively to maximum one user equipment. In contrast, wireless ad hoc networks are multi-hop systems, so the contribution of this paper is to consider its multi-hop characteristics. At the same time, in order to maximize system throughput, we also introduced frequency reuse factors into the model. That is, one resource block can be simultaneously allocated to multiple nodes. CCs, RBs and MCS are jointly allocated and the minimum transmission rate requirement of each node is guaranteed. We transform the resource allocation problem into a 0–1 integer linear programming problem and use the branch implicit enumeration algorithm (BIEA) to solve it.

# 2 System Model and Problem Formulation

## 2.1 Three-Hop Multiplexing Mechanism

The three-hop multiplexing mechanism in wireless ad hoc networks is specifically analyzed in [10]. And the authors use channel utilization as an indicator of performance analysis.

The data transmission rate of control slots $R_C$ and data slots $R_D$ can be calculated according to Shannon's theorem.

$$R_C = B_C \log_2(1 + SIR_c)$$

$$R_D = B_D \log_2(1 + SIR_D)$$

$B_C$ and $B_D$ represent the bandwidth of control slots and data slots respectively. The signal-to-interference ratios of control slots and data slots are $SIR_C$ and $SIR_D$.

The amount of data transmitted in each control and data slot is denoted by $L_C$ and $L_D$. Then the interval $t_C$ of each control slot and the $t_D$ of each data slot are expressed as follows

$$t_C = \frac{L_C}{R_C}$$

$$t_D = \frac{L_D}{R_D}$$

In the frame structure, each frame contains C control slots and D data slots. The expression of channel utilization in a network that does not involve frequency reuse is given by

$$\eta = \frac{Dt_D}{Ct_C + Dt_D}$$

Under the multi-hop external frequency reuse mechanism, the channel utilization with the reuse coefficient $\xi$ is

$$\eta = \frac{\xi}{\frac{C}{D}\frac{L_C}{L_D}\log\frac{1+SIR_D}{1+SIR_C} + 1}$$

It can be seen that the channel utilization is related to the ratio of the number of control slots to the number of data slots (C-D ratio), the length of control message and data message, the signal-to-interference ratios of control slots and data slots and multiplexing coefficient. Among these the C-D ratio is related to the number of neighbor nodes maintained [11].

By simulating and analyzing the relationship between channel utilization and the number of neighbor nodes maintained [10], it can be concluded that the three-hop multiplexing mechanism can obtain the highest channel utilization. Therefore, we choose the three-hop frequency reuse mechanism to build our system model.

## 2.2 System Model

We consider the ad hoc network based on the OFDM system consisting of $K$ nodes as $\mathrm{K} = \{1, 2, \ldots, \mathrm{K}\}$. Also the network has $M$ orthogonal CCs as $M = \{1, 2, \ldots, \mathrm{M}\}$. Each CC can be consisted of a different number of RBs. The set of RBs is modelled as $B = \{1, 2, \ldots, \mathrm{B}_m\}$ and the frequency outside $j$ hop can be reused. $\mu_{k,m,b} = 1$ takes a value of 0 or 1. $\mu_{k,m,b} = 1$ indicates that the $b^{th}$ RB on the $m^{th}$ CC is assigned to the $k^{th}$ node, and $\mu_{k,m,b} = 0$ indicates the opposite. The MCS is shown in Table 1, where $i = \{0, 1, \ldots, \mathrm{I}\}$ represents the index of the modulation and coding scheme. We propose six schemes. The first scheme means that the RB has a low signal-to-noise ratio and is no longer able to transmit information, so it is not assigned to any nodes.

We consider the chain network topology. The model after applying the three-hop multiplexing mechanism is shown in Fig. 1. F(1), F(2), and F(3) are frequency sets composed of RBs that have different center frequency points, that is, any two of the three sets intersect as an empty set. The total frequency set can be expressed as

$$F = \begin{cases} \{F(1), F(2), \ldots, F(\mathrm{K})\}(1 \leq \mathrm{K} \leq j) \\ \{F(1), F(2), \ldots, F(j)\}(\mathrm{K} > j) \end{cases}$$

and any two sets in F intersect as an empty set. In Fig. 1, node 1, 4, and 7 share the same frequency set, node 2 and 5 share the same frequency set, and nodes 3 and 6 share the same frequency set. Starting from node 1, each j node is a group, and the remaining less than j nodes are a group. For example, nodes 1, 2 and 3 are a group, node 4, 5, and 6 are a group. The frequency set allocation of the remaining nodes is analogized according to this rule.

**Table 1** Modulation and coding scheme

| $i$ | Modulation | Code rate |
|---|---|---|
| 0 | Out of range | |
| 1 | QPSK | 1/2 |
| 2 | QPSK | 3/4 |
| 3 | 16QAM | 1/2 |
| 4 | 16QAM | 3/4 |
| 5 | 64QAM | 3/4 |

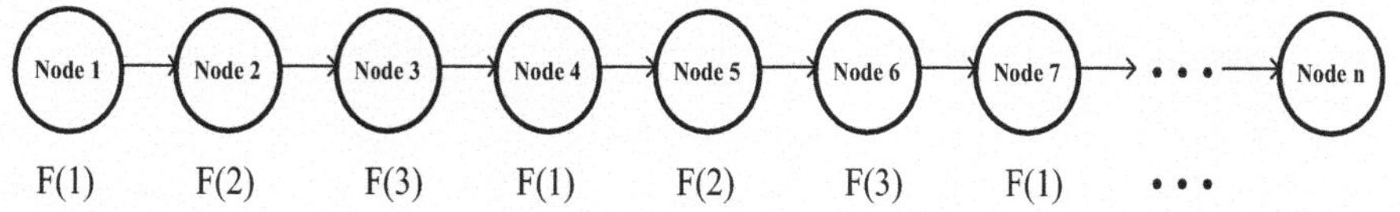


**Fig. 1** System topology

## 2.3 Problem Formulation

We set the system throughput as the objective function, and the system throughput can be defined as:

$$T = \sum_{k=1}^{K} \sum_{m=1}^{M} \sum_{b=1}^{B_m} \sum_{i=1}^{I} \mu_{k,m,b} r_b^{(i)} \tag{1}$$

$r_b^{(i)}$ is the transmission rate of the $b^{th}$ RB on the $m^{th}$ CC when using the $i^{th}$ modulation and coding scheme, which can be expressed as:

$$r_b^{(i)} = \frac{1}{t_s} R_i \log_2(M_i) \cdot N_{sc} \cdot N_o \tag{2}$$

$i = 0$ means that the RB does not carry data, and the RB transmission rate is zero at this time. $t_s$ represents the time of one slot. $R_i$ is the coding rate. $M_i$ refers to the modulation order. The meaning of $N_{sc}$ is the number of subcarriers included in one RB. And $N_o$ represents the number of OFDM symbols contained in one RB in time domain. The constrained joint optimization problem with the goal of maximizing the system objective function $T$ can be defined as follows:

$$\max T = \max \sum_{k=1}^{K} \sum_{m=1}^{M} \sum_{b=1}^{B_m} \sum_{i=1}^{I} \mu_{k,m,b} r_b^{(i)} \tag{3}$$

subject to the following constraints:

$$\sum_{k=1+jz}^{(z+1)j} \mu_{k,m,b} \le 1 \text{ for } z \in N, K \ge (z+1)j \tag{4}$$

$$\sum_{k=1+jz}^{K} \mu_{k,m,b} \le 1 \text{ for } z \in N, K < (z+1)j \tag{5}$$

$$\mu_{k,m,b} = \mu_{(k+j),m,b} = \mu_{(k+2j),m,b} = \ldots = \mu_{(k+vj),m.b}, v \in N \tag{6}$$

$$\mu_{k,m,n} \in \{0, 1\} \tag{7}$$

$$\sum_{m=1}^{M} \sum_{n=1}^{N_m} \mu_{k,m,n} r_n^{(i)} \geq R_k \tag{8}$$

(4) and (5) indicate that the $b^{th}$ RB on the $m^{th}$ CC is allocated to at most one node in each group of nodes. (6) ensures that the $b^{th}$ RB on the $m^{th}$ CC is assigned to the corresponding node in each group of nodes. (7) illustrates that $\mu_{k,m,b}$ is a binary variable whose value can only take 0 or 1. (8) guarantees the minimum rate requirement of each node.

## 3 Branch Implicit Enumeration Algorithm

The simplest and easiest way to solve 0–1 integer linear programming is exhaustive algorithm. It checks each combination of decision variables that are 0 or 1 in order to obtain the optimal solution of the objective function value. However, this method has a large amount of computation and needs to check $2^n$ combinations of variable values. The branch implicit enumeration algorithm greatly reduces the amount of computation by hiding some cases that do not require enumeration. The specific implementation steps of this algorithm are as follows:

### 3.1 Model Standardization

Convert (3), (4), and (5) into

$$\min(-T) = \min(-\sum_{k=1}^{K} \sum_{m=1}^{M} \sum_{b=1}^{B_m} \sum_{i=1}^{I} \mu_{k,m,b} r_b^{(i)}) \tag{9}$$

$$-\sum_{k=1+jz}^{(z+1)j} \mu_{k,m,b} \geq -1 \tag{10}$$

$$-\sum_{k=1+jz}^{K} \mu_{k,m,b} \geq -1 \tag{11}$$

Let the number of total RBs be B. And the set of decision variables is labelled as $U = \{\mu_1, \mu_2, \ldots, \mu_{K\times B}\}$. For each element in the set, let $\mu_1 = 1 - \mu'_1$, $\mu_2 = 1 - \mu'_2$, …, $\mu_{K\times B} = 1 - \mu'_{K\times B}$ and substitute them into (9), (10) and (11). Then, the variables in the objective function are arranged according to the coefficients from large to small. Accordingly, the order of the variables in the constraint condition is consistent with the order of the variables in the objective function.

## 3.2 Determine the Basic Scheme

Let $\mu'_1 = \mu'_2 = \mu'_3 = \ldots = \mu'_{K\times B} = 0$ as the benchmark scheme, and check whether the scheme satisfies all the constraints. If it is satisfied, the scheme is tentatively determined as the optimal solution of the objective function; if not, the next step is performed.

## 3.3 Branch and Select the Best

Specify a variable as a fixed variable and the remaining variables as free variables. The problem is divided into two sub-problems. The fixed variable in one sub-question takes 1 and the fixed variable in the other sub-question takes 0, and the free variables take 0. Find the target function value and check whether the value of the variable can satisfy the constraint, and then decide whether to continue the branch according to the following principles:

(1) When a sub problem of a branch is a feasible solution, the branch stops continuing to branch. The branch with the smallest value of the objective function in all feasible solutions is retained, and the branch with large boundary value in the feasible solution is removed.
(2) Regardless of whether it is a feasible solution, the branch is stopped as long as the boundary value of the branch is greater than the boundary value of the remaining feasible solution.
(3) Some variables in the branch have been determined, and the branch is stopped when at least one of the constraints is not satisfied, regardless of the value of the other variables.

In addition to the above three cases, continue to branch until all the branches except the reserved branch have been removed. The feasible solution that is retained at this time is the optimal solution of the objective function.

# 4 Simulation Results

The simulation parameters are shown in Table 2. Each node is evenly distributed in the system and has the same transmission power. Rayleigh fading channel is selected as the channel model. First, we consider the spectrum utilization in both single-band and multi-band scenarios and the system throughput under different number of nodes and number of CCs. The single band consists of four consecutive CCs in the 2.4 GHz band. The bandwidth of the four CCs is 3 M. The multi-band is composed of two consecutive CCs in the 2.4 GHz band and two consecutive CCs in the 900 MHz band. The bandwidth of the four CCs is also 3 M. Meanwhile, the system throughput of the ad hoc network and the cellular network which use the same algorithm is compared. Secondly, the throughput under the node index is given to verify whether the minimum transmission rate requirement of each node is guaranteed. Finally, the effect of different frequency reuse coefficients on system throughput is provided.

Figure 2 shows that the system throughput with different aggregation bands and number of nodes. The reason why the multi-band has higher throughput than the single band in the same network environment is that the low frequency band of 900 MHz has better channel conditions, so that the nodes have good signal-noise ratio. Therefore, some nodes can choose high order modulation and higher coding rate. When the number of CCs is constant, as the number of nodes increases, the system throughput increases. This is because a larger number of nodes will enhance the diversity effect of multiple nodes.

We also analyze the system throughput with different aggregation bands and number of CCs. As shown in Fig. 3, in the case of a certain number of nodes, the more CCs, the greater the system throughput. This is because the number of CCs increases, that is, the number of RBs increases, which enhances the selectivity of the nodes for RBs with better channel conditions.

**Table 2** Simulation parameters

| Parameter | Numerical value |
|---|---|
| K | 10 |
| M | 4 |
| B | 15 |
| Carrier frequency | 900 MHz, 2.4 GHz |
| Maximum transmission power | 46 dBm |
| TTI length | 1 ms |
| Multiplexing mechanism | three-hop |
| Channel model | Rayleigh fading channel |

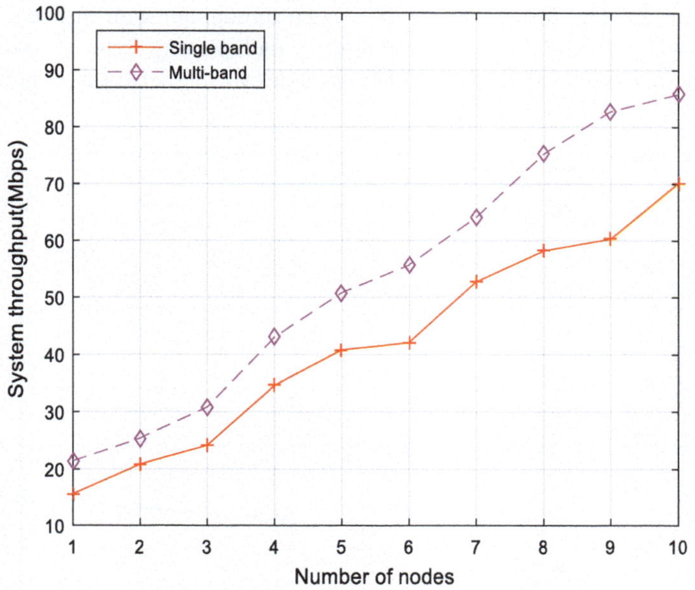


**Fig. 2** System throughput with different aggregation bands and number of nodes

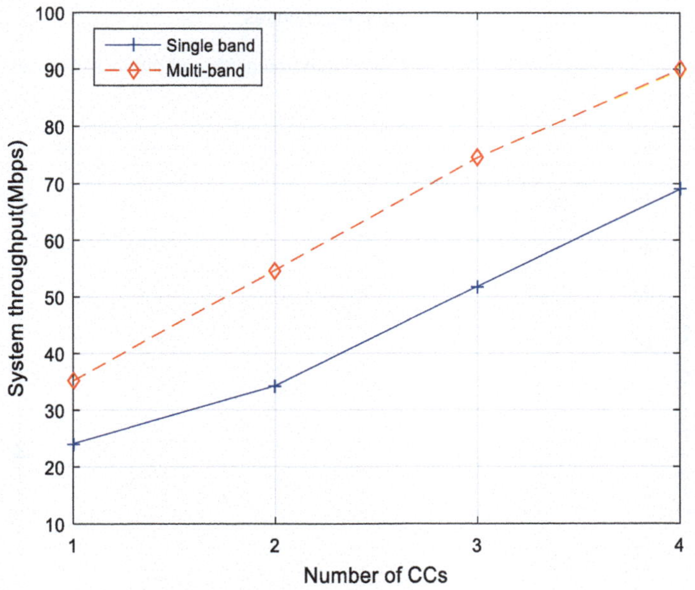


**Fig. 3** System throughput with different aggregation bands and number of CCs

Figure 4 and Fig. 5 mainly analyze system throughput with different network environments. In Fig. 4, since the co-frequency multiplexing factor is considered in the ad hoc networks, when the number of nodes in the networks is greater than

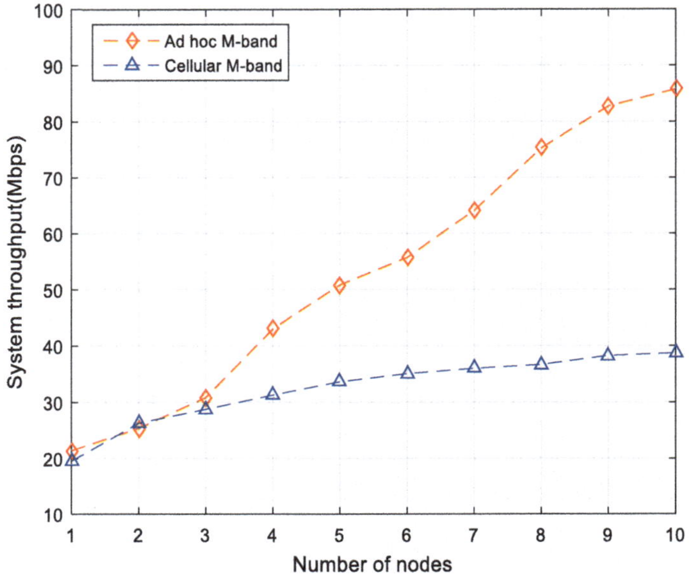


**Fig. 4** System throughput with different network environments and number of nodes

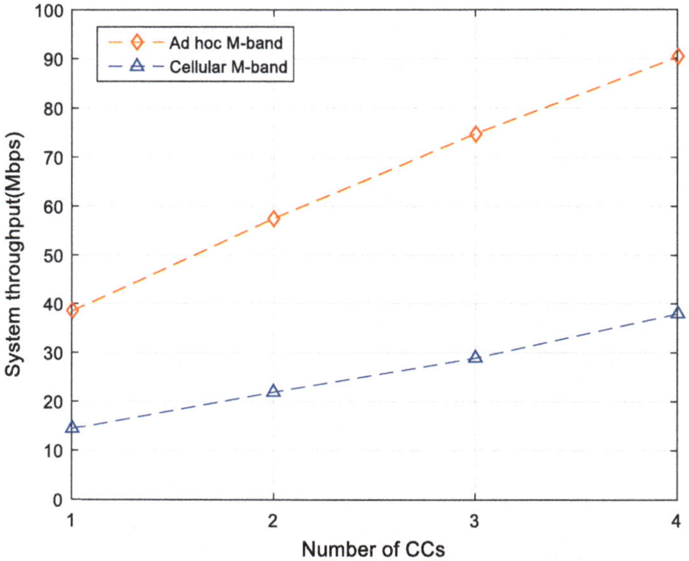


**Fig. 5** System throughput with different network environments and number of CCs

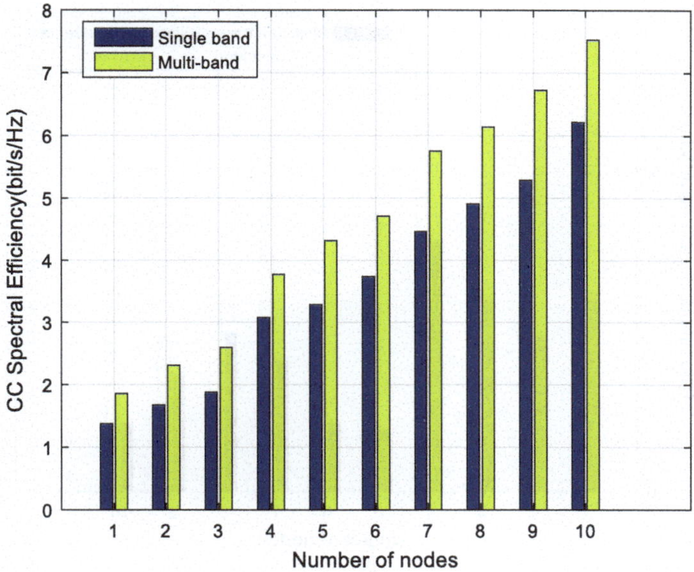


**Fig. 6** Comparison of spectral efficiency between multi-band and single band

three, the ad hoc networks are equivalent to indirectly having more available RBs than the cellular networks, thus having higher throughput. It can also be seen from Fig. 5 that the system throughput is higher due to the co-frequency multiplexing factor of the ad hoc networks. It is proved once again that the more nodes and CCs, the higher the system throughput.

As shown in Fig. 6, comparison of spectral efficiency between multi-band and single band is given. When the number of nodes is fixed, multi-band CA achieves higher total spectral efficiency than single-band CA.

Considering that there are 10 nodes in the system, the minimum transmission rate requirement for each node in Fig. 7 is compared with the actual transmission rate of each node after applying the BIEA. The comparison results illustrate that the designed algorithm can meet the minimum transmission rate requirement of each node.

Figure 8 shows the effect of different hop frequency reuse factor on system throughput. The result shows that the smaller the reuse hop count, the higher the system throughput. This is because the system with lower frequency reuse coefficients indirectly have more RBs than the systems with higher frequency reuse coefficients.

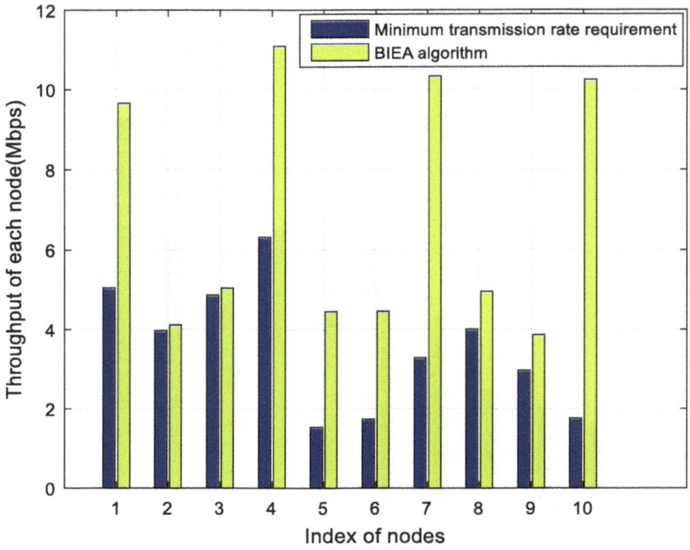


**Fig. 7** Comparison of the minimum transmission rate of each node with the actual transmission rate

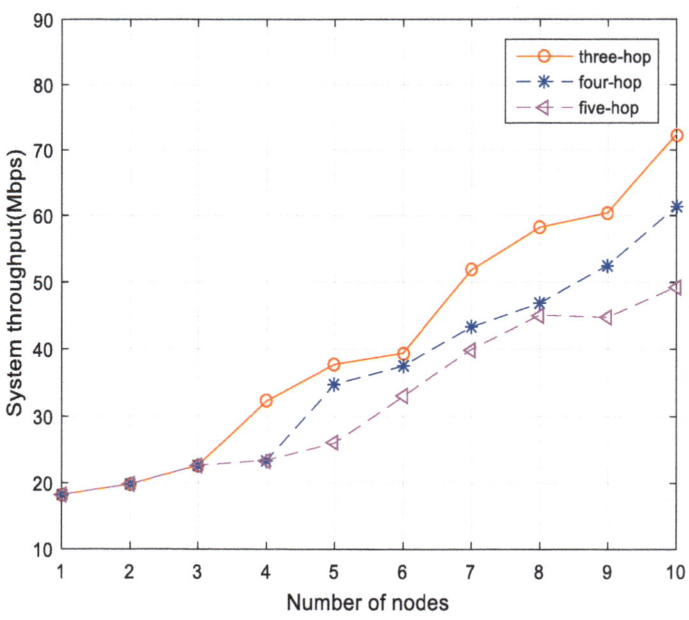


**Fig. 8** Effect of different hop frequency reuse factor on system throughput

## 5 Conclusion

In this paper, we establish a frequency reuse model for resource allocation with CA in ad hoc networks. The model is based on the multi-hop frequency reuse mechanism and takes into account the minimum transmission rate requirement of each node. The model is transformed into a 0–1 integer linear programming model. It is solved by the BIEA and measured by system capacity and spectral efficiency. The results show that CA technology can improve system throughput and spectrum efficiency. When the number of CCs is constant, the more nodes, the greater the system throughput. When the number of nodes is constant, the more CCs, the greater the system throughput. Compared with the resource allocation with CA in cellular networks, wireless ad hoc networks obtain higher system throughput by considering multi-hop frequency reuse factor. The BIEA proposed in this paper solves the 0–1 integer linear programming problem well and can also satisfy the QoS requirements of each node.

## References


1. Chakradhar, P., & Yogesh, P. (2015). An optimal CDS construction algorithm with activity scheduling in ad hoc networks. *The Scientific World Journal, 9*(5), 1–12.
2. Wang, Y., Pedersen, K. I., Sorensen, T. B., & Mogensen, P. E. (2011). Utility maximization in LTE-advanced systems with carrier aggregation. In *Proceedings of the VTC Spring, Yokohama, Japan*, (pp. 1–5).
3. Lim, C., Yoo, T., Clerckx, B., Lee, B., & Shim, B. (2013). Recent trend of multiuser MIMO in LTE-advanced. *IEEE Communications Magazine, 51*(3), 127–135.
4. Lee, H., Vahid, S., & Moessner, K. (2014). A survey of radio resource management for spectrum aggregation in LTE-advanced. *IEEE Communications Surveys and Tutorials, 16*(2), 745–760.
5. Liao, H. S., Chen, P. Y., & Chen, W. T. (2014). An efficient downlink radio resource allocation with carrier aggregation in LTE-advanced networks. *IEEE Transactions on Mobile Computing, 13*(10), 2229–2239.
6. Jiang, L. B. & Liew, S. C. (2005). Proportional fairness in wireless LANs and ad hoc networks. In *IEEE Wireless Communications and Networking Conference* (vol. 3, pp. 1551–1556).
7. Stefanatos, S., Foukalas, F., & Tsiftsis, T. (2017). Low complexity resource allocation for massive carrier aggregation. *IEEE Transactions on Vehicular Technology, 66*(10), 9614–9619.
8. Lee, H., Vahid, S., & Moessner, K. (2017). Traffic-aware carrier allocation with aggregation for load balancing. In *European Conference on Networks and Communications* (pp. 1–6).
9. Shajaiah, H., Abdelhadi, A., & Clancy, T. C. (2016). Towards an application aware resource scheduling with carrier aggregation in cellular systems. *IEEE Communications Letters, 20*(1), 129–132.
10. Kang, L., Li, X., Huo, J., & Huang, W. (2018). Inter-area interference modeling in distributed wireless mobile ad hoc networks. *IEEE Access, 6,* 58510–58520.
11. Li, X., Gao, H., Liang, Y., Xiong, K., & Liu, Y. (2016). Performance modeling and analysis of distributed multi-hop wireless ad hoc networks. In *IEEE International Conference on Communications* (pp. 1–6).

# Secure Routing Based on Trust Management in Ad-Hoc Networks

**Jia Ni, Wenjun Huang, and Wenqing Zhang**

**Abstract** Malicious nodes attack brings efficiency decline of routing protocols in multi-hop ad-hoc networks, especially when the network resource is limited. To deal with this challenge, a secure routing mechanism T-AODV with low routing overhead is proposed, in which trust management is deployed. The impact of time attenuation factor and trust-threshold over the performance of T-AODV is analyzed under two attack mode: black hole attack (BLA) and on-off attack. Simulation results show it can resist internal attacks when malicious nodes work and improve security of network. T-AODV with appropriate protocol parameters can significantly improve the ability against internal attacks, especially BLA and on-off attack.

**Keywords** Trust management · Black hole attack · Internal attacks · On-off attack · Securing routing

## 1 Introduction

Ad-Hoc network, it is a typical decentralized self-organizing network, rely on collaboration between nodes. There may be selfish and malicious behavior in the limited resources network. BLA is a typical internal attacks [1], where malicious nodes will discard all packets routed by themselves. Bad mouthing attack [2] is also a common internal attack as malicious nodes can pollute ordinary nodes and hinder data transmission. Other internal attacks like on-off attack, conflicting behavior attack, collusion attack, etc., affect the network throughput, packet loss rate, delay etc.

J. Ni (✉) · W. Huang · W. Zhang
School of Electronic and Information Engineering, Beijing Jiaotong Univesity, Beijing, China
e-mail: 17120097@bjtu.edu.cn

W. Huang
e-mail: 16111027@bjtu.edu.cn

W. Zhang
e-mail: 16120167@bjtu.edu.cn



J. Zhang et al. (eds.), *LISS2019*,
https://doi.org/10.1007/978-981-15-5682-1_26

Trust management [3] is usually used to handle internal attacks in ad-hoc network [4], point-to-point network and WSN network [5]. Trust management identifies malicious nodes by evaluating the trust of positive behaviors (forwarding data) and negative behaviors (dropping data packets). In reference [1], Active-Trust scheme based on active detection of trust security routing mechanism was proposed, which can actively detect the security and reliability of the path by consuming the energy of the monitoring node, so as to create as many secure routes as possible. This mechanism can detect the BLA caused by malicious nodes in advance. However, on-off attack that nodes discarding data after gaining high trust through good behavior in the early stage of data transmission cant be recognized and processed. In reference [2], a recommendation based trust management mechanism is proposed to filter misbehaving nodes, and the robustness and stability of this model in MANET are verified in the topological changes frequently scenarios.

In reference [6], authors proposed the multi-objective optimization algorithm considering the link factors, distance, energy, link lifetime, which could optimize data transmission delay, delivery rate and other indicators on the premise of ensuring security. In reference [7], the sensitivity of trusted routing schemes based on attack pattern discovery is analyzed by setting different parameters in different attack modes. It is concluded that trust update interval and trust threshold will affect network performance, and the parameters of different networks should be adjusted in consideration of attack modes.

The selection of protocol parameters is very important for trust management mechanism. If the trust threshold is wet too high, the non-malicious packet loss behavior of non-malicious nodes can easily be misjudged as malicious nodes, and the re-routing will cause unnecessary additional overhead. If the threshold of trust is too low to identify malicious nodes in time, the network performance will be degraded. However, few papers have analyzed the selection of protocol parameters. Aiming at malicious node attacks and combining AODV with trust management, a secure routing mechanism T-AODV that can resist black hole attacks and on-off attacks with low routing overhead is proposed. The influence of trust threshold and time attenuation factor on network performance under different scenarios is analyzed. The main contributions are as follows:

(1) A secure routing mechanism T-AODV that guarantees the overall performance of the network with low routing overhead under internal attacks is proposed.
(2) Analyses the influence of T-AODV trust threshold and time attenuation factor protocol parameters on the overall performance of the network in different network scenarios in this paper.

## 2 Trust Management Mechanism

To identify malicious nodes, usually establish a trust model, evaluate the positive and negative behaviors of nodes and calculate node's trust degree. Trust model mainly consists of direct and indirect trust value. Through promiscuous mode, upstream node can directly monitor whether the downstream node has completed the data forwarding, then calculation the direct trust value. Data upstream nodes can directly monitor the downstream nodes to evaluate whether the packets forwarded by the nodes are valid packets in the network, so as to avoid malicious nodes forwarding useless packets to deceive trust. Indirect trust is a trust table that interacts with neighbor nodes. In sparse data situation, nodes not on data transmission link can quickly obtain trust information through interacting with neighbor nodes.

$T_{ij}$ is node j's trust value in node i's record, can be computed by (1):

$$T_{ij} = \delta \cdot T_{ij}^{D} + (1 - \delta) \cdot T_{ij}^{I} \tag{1}$$

where $T_{ij}^{D}$ represents node j's direct trust value in node i's record, $T_{ij}^{I}$ is the node j's indirect trust value of node j in node i's record, $\delta \in (0, 1)$ is proportion of trust.

## 2.1 Direct Trust Mode

Beta reputation model [8], which has the property of conjugate priori can be used to predict the probability of node forwarding data, can be used to calculate direct trust. It can be considered that the behavior of node forwarding data packet obeys beta distribution Beta $(a = f, b + d)$, in which $a$, $b$ represents initial trust, $f$ represents number of node forwards, $d$ represents number of packet losses. Based on this, the trust degree of neighbor nodes is constructed. Probability density function of beta distribution:

$$\mathrm{p}(x) = \frac{\Gamma(a+b)}{\Gamma(a)\Gamma(b)} x^{a-1}(1-x)^{b-1},\ 0 < x < 1 \tag{2}$$

$f_{ij}(t_c)$ is the sum of forwards data by node j in node i's record at $t_c$. $d_{ij}(t_c)$ represents the sum of losses data by node j in node i's record at $t_c$, $c$ represents number of records. To resist on-off attack [9], we use time attenuation factor $e^{\alpha t}$ which can timely identify on-off attack and timely update data to avoid misjudgment of behaviors of nodes based on outdated historical data:

$$f_{ij}(t_c) = e^{-\alpha(t_c - t_{c-1})} f_{ij}(t_{c-1}) + \Delta f_{ij} \tag{3}$$

$$d_{ij}(t_c) = e^{-\beta(t_c - t_{c-1})} d_{ij}(t_{c-1}) + \Delta d_{ij} \tag{4}$$

where $\alpha$ is positive behaviors time attenuation factor, $\beta$ is negative behaviors time attenuation factor. In order to punish negative behaviors of nodes, the forgetting speed of negative behaviors of nodes should be lower than that of positive behaviors of nodes. So $\alpha$, $\beta$ should be satisfied $\alpha > \beta > 0$. $\Delta f_{ij}$, $\Delta d_{ij}$ represent whether the node forwards the packet. If node i detects node j forwarding the packet, then $\Delta d_{ij} = 0$, $\Delta f_{ij} = 1$. Otherwise $\Delta d_{ij} = 1$, $\Delta f_{ij} = 0$. In order to avoid misjudging packet loss caused by channel quality, detection of channel quality is introduced. When channel quality SNR < $C_{th}$, considering listening result is invalid. In this case $\Delta d_{ij} = \Delta f_{ij} = 0$.

According to the beta trust model, the mathematical expectation of beta distribution represents the direct trust value:

$$T_{ij}^{D} = \frac{a + f_{ij}}{a + b + d_{ij} + f_{ij}} \tag{5}$$

## 2.2 Indirect Trust Model

Indirect trust $T_{ij}^{I}$ is node j's trust value in other nodes' record that in node i's communication range $D_i$. In order to avoid defamation and collusion attacks, reference [2] proposed filter out the untrustworthy by using confidence degree and deviation degree.

$V_{ik}$ is the confidence degree of k for i, can be computed by (6):

$$V_{ik} = 1 - \sqrt{12\sigma_{ik}} = 1 - \sqrt{\frac{12 f_{ik} d_{ik}}{(f_{ik} + d_{ik})^2 (f_{ik} + d_{ik} + 1)}} \tag{6}$$

$Q_{ik}$ is the deviation degree of k for i, can be computed by (7):

$$Q_{ik} = \left| T_{ij}^{D} - T_{kj}^{I} \right| \tag{7}$$

When $V_{ik}$ is within the confidence interval $[d_{min}\ , d_{max}]$ and $Q_{ik}$ is less than the deviation threshold $d_{th}$, $T_{kj}^{I}$ is the effective value, recommended coefficient $\eta_{ik} = 1$. Otherwise, $\eta_{ik} = 0$.

$T_{ij}^{I}$ can be computed by (8):

$$T_{ij}^{I} = \frac{\sum_{k \in D_i \setminus \{j\}} \left( \eta_{ik} T_{kj}^{I} \right)}{\sum_{k \in D_i \setminus \{j\}} \eta_{ik}} \tag{8}$$

## 2.3 Routing Mechanism Based on Trust Management

The routing mechanism mainly improves AODV by combining trust management, and chooses the route according to the principle of maximum path trust and minimum hops. Assuming there are six randomly distributed nodes in network, through the interaction of HELLO messages, the nodes in the network establish the initial trust list of neighbors and initialize the trust degree to 0.5, as shown in Fig. 1.

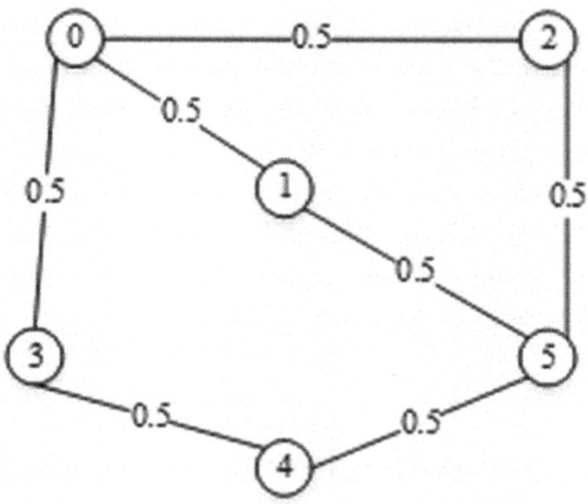


**Fig. 1** Network initialization diagram.

When node 0 has data sent to node 5, node 0 queries whether there is a path to the node 5. If no node 0 broadcast RREQ. Through RREQ message it is found that the path of 0–5 are 0,2,5 which trust is 0.25 with 2 hops, 0,1,5 which trust is 0.25 with 2 hops, 0,3,4,5 which trust is 0.125 with 3 hops. The route finding process is completed by selecting path 0,2,5 based on the principle of maximum link trust and minimum hops.

After the completion of road construction, node 0 selects node 2 to forward packets to node 5. After forwarding data packet to node 2, it starts the promiscuous mode to monitor node2 and calculate node 2 trust value according to Algorithm 1.

**Algorithm 1** Direct trust update

Query trust list, if no establish it
**if** Node k is not the destination node **then**
  **if** ks trust more than trust-threshold **then**
    The state of detection is true
  **else**
    Send RERR message
  **end if**
**end if**
Forward data()
Open timer of detection
**while** Timer of detection **do**
  **if** Data forward by k **then**
    $f_{ik}(t_c) = e^{-\alpha(t_c - t_{c-1})} f_{ik}(t_{c-1}) + 1$
    $d_{ik}(t_c) = e^{-\beta(t_c - t_{c-1})} d_{ik}(t_{c-1})$
  **else**
    $f_{ik}(t_c) = e^{-\alpha(t_c - t_{c-1})} f_{ik}(t_{c-1})$
    $d_{ik}(t_c) = e^{-\beta(t_c - t_{c-1})} d_{ik}(t_{c-1}) + 1$
  **end if**
  **if** ks trust less than trust-threshold **then**
    send RERR message
  **end if**
**end while**

If the trustworthiness of node 2 is normal forwarding, the trustworthiness will accumulate while the trustworthiness decrease if node 2 drops packets. When the trustworthiness of node 2 is lower than the threshold of trustworthiness, node 0 will redirect and choose the path 0,1,5.

When node 0 monitors node 2, value of $T_{02}^{D}$ updates. To reduce routing overhead, node 0 adds the changed trust list entries to the HELLO message payload and completes the trust information interaction with the neighbor node through the HELLO message, as Algorithm 2 shows.

## 3 Simulation Results and Analysis

NS2 is used for analyzing the performance of T-AODV in identifying and dealing with internal attacks, and we analyze the impact of protocol parameters on network performance in varies network scenarios by changing the threshold and other parameters. In an area of 1000 m x 1000 m, 36 nodes are randomly distributed which execute T-AODV or AODV. In part A and part B, all the malicious nodes start dropping packets at the same time, but in part C, malicious nodes is set as normal participation in data transmission for a period of time to simulate the on-off attack in the network, and then the malicious packet loss behavior can be started after obtaining a high degree of trust, malicious nodes start working in different time. Table 1 are experiment parameters.

**Algorithm 2** Indirect trust update

```
1: Query whether the trust table of node k, if no establish it
2: if V_ik ∈ [d_min, d_max] then
3:    for each k ∈ D_i \ {j} do
4:       Gets the list of interactions carried by the HELLO message
5:       if Q_ik = |T^D_ij − T^I_kj| ≤ d then
6:          η_ik = 1
7:       else
8:          η_ik = 0
9:       end if
10:   end for
11: else
12:   η_ik = 0
13: end if
```

**Table 1** Experiment parameters

| Parameter | Value |
|---|---|
| Coverage area | 1000m × 1000m |
| MAC layer protocol | IEEE802.11 |
| Communication range | 400m |
| Traffic type | CBR-UDP |
| Packet size | 500–800bits |
| Number of node | 36 |
| Packet interval | 0.2s |
| Number of traffic | 4 |
| Hops of traffic | 1–3hops |
| Simulation time | 200s |
| Initial trust | 0.5 |
| Confidence interval | [0.5,0.9] |
| Deviation degree | 0.4 |

### *3.1 Misjudgment Performance Analysis Without Malicous Nodes*

This part analyzes the performance of T-AODV and AODV by varying values of trust-threshold in the case of non-malicious and $\alpha = 0.2$, $\beta = 0.1$.

In Fig. 2, blue line represents performance of AODV and green line represents performance of T-AODV. It shows that the performance of T-AODV is slightly lower

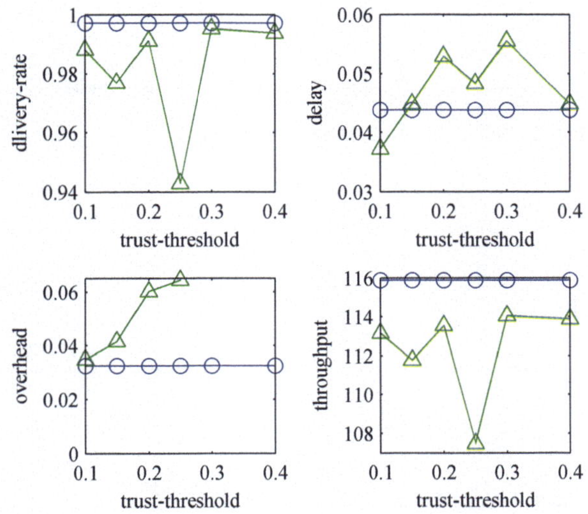


**Fig. 2** Network performance with non-malicious.

than that of the AODV in the case of no malicious nodes. Some nodes may drop packets because of limited network resources and link conflicts. It increases the probability that a node will be misjudged as malicious, resulting in additional overhead. When the threshold is set to 0.3, T-AODV is close to the performance of AODV in the network with non-malicious.

## 3.2 Varying Trust Threshold

This part analyzes the performance of T-AODV and AODV by varying values of trust-threshold and the proportion of malicious. Malicious nodes start to drop packets at the same time, $\alpha = 0.2$, $\beta = 0.1$ in the network. Figures 3, 4, 5, 6 are simulation results. As simulation results shows, when network is under internal attack caused by malicious nodes, AODV protocol is unable to identify the malicious nodes and restore data transmission, while T-AODV protocol can identify malicious nodes and restore data transmission. Compared with AODV, T-AODV increases routing overhead and delay, but can significantly improve network delivery rate and throughput.

With proportion of malicious nodes increasing, the overall performance of T-AODV shows a trend of decline as more network resources are needed to deal with

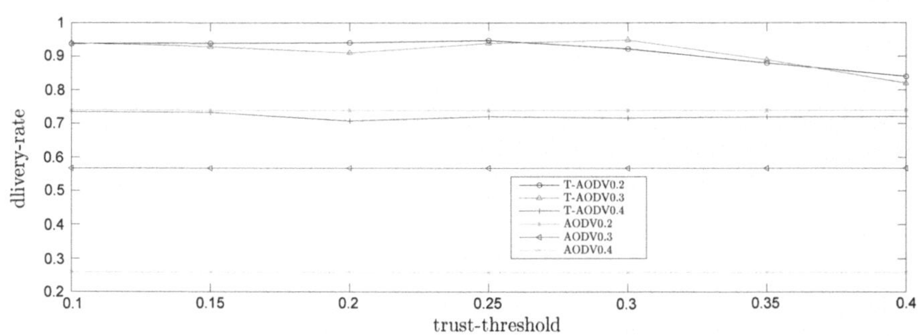


**Fig. 3** Delivery rates of AODV and T-AODV.

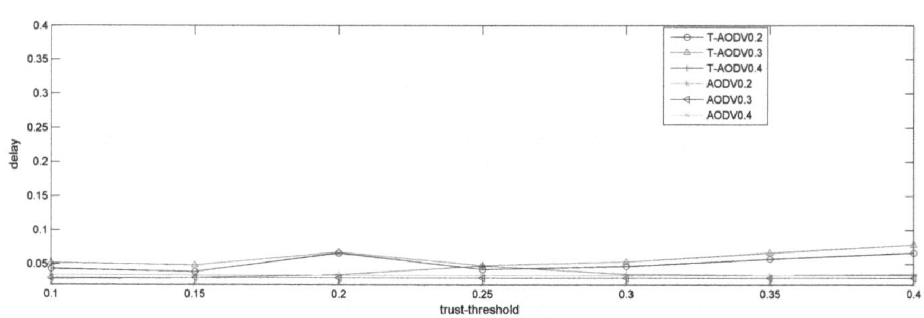


**Fig. 4** Delay rates of AODV and T-AODV.

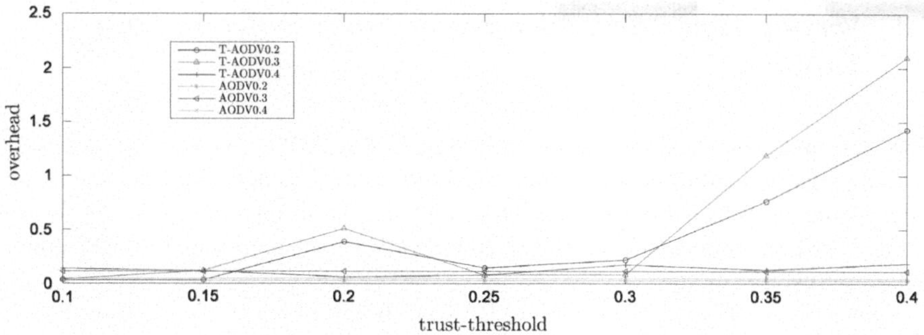


**Fig. 5** Overhead of AODV and T-AODV.

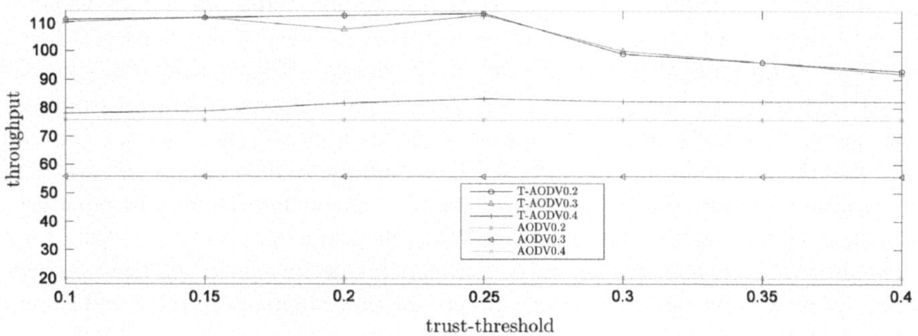


**Fig. 6** Throughput of AODV and T-AODV.

the adverse effects caused by malicious nodes. When malicious nodes in all available paths between source and destination nodes, the data transmission cant be repaired.

It also shows that in different network scenarios, the selection of trust threshold will affect the network performance. When malicious nodes' proportion is 0.2, 0.3 and 0.4, the threshold of trust is 0.25, 0.3 and 0.4, and the overall performance of the network is better.

The selection of the optimal trust threshold increases with malicious nodes' proportion increases, because it is necessary to set higher trust-threshold to identify malicious nodes as early as possible in a scenario with a high malicious nodes' proportion. But when malicious nodes' proportion is low, we should set a moderate trust-threshold to avoid the situation where a node is misjudged as malicious node due to an excessively high trust-threshold. The trust threshold is closely tied to network state. To ensure network performance, different network scenarios should be evaluated first and then appropriate threshold is selected.

### 3.3 *Varying Time Attenuation Factor Under Different Trust Threshold*

This part analyzes the performance of T-AODV and AODV by varying values of time attenuation factor under different threshold. Malicious nodes' proportion in the network increases by 0.1 per 10s, until reaches 0.4 . The simulation results are shown in Fig. 7, red line represents the AODV and the other lines represent T-AODV under different trust threshold.

Values of [$\alpha$, $\beta$] are [(0.98,0.96) (2.0,1.0) (3.0,1.0) (4.0,1.0) (5.0,1.0)] in Fig. 7. When value of [$\alpha$, $\beta$] is [0,0] and threshold is 0.25, value of delivery rate is 0.7497, value of delay is 0.031239 s, value of overhead is 0.1192, and value of throughput is 83.1583kb/s. When value of [$\alpha$, $\beta$] is [0.98,0.96] and threshold is 0.25, value of delivery rate is 0.7770, value of delay is 0.031648 s, value of overhead is 0.1707, and value of throughput is 85.5864kb/s. As it shown in Fig. 6, when the value of [$\alpha$, $\beta$] is [0.98,0.96], the overall performance of the network reaches the optimal level and its higher than that without time attenuation factor. The simulation results show that the $\alpha/\beta$ ratio should not be too large, otherwise the network performance will decline. The numerical selection of the time attenuation factor will obviously affect the overall network performance. When the numerical selection of the time attenuation factor is appropriate, the overall network performance will be improved; on the contrary, the inappropriate selection of the time attenuation factor will lead to the decline of the overall network performance.

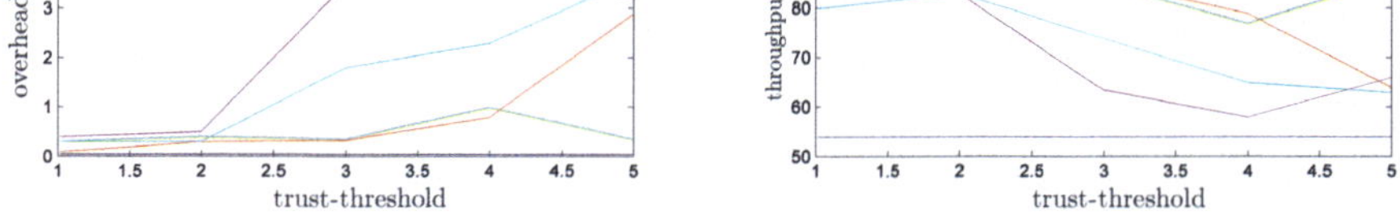


**Fig. 7** Vary time attenuation factor.

## 4 Conculsion

This paper mainly focuses on the malicious node attack in multi-hop ad-hoc network, proposes a secure routing mechanism T- AODV based on AODV and trust management, and analyzes the performance of T-AODV and AODV by varying values of time attenuation factor and trust-threshold under BLA and on-off attack. Simulation results show it can resist internal attacks caused by malicious nodes and improve network's security. Protocol parameters as trust threshold and time attenuation factor will have a marked impact on the network, which lies a solid foundation for future researches on adaptive parameter selection under dynamic network situation.

## References


1. Liu, Y., Dong, M., Ota, K., & Liu, A. (2016). ActiveTrust: Secure and trustable routing in wireless sensor networks. *IEEE Transactions on Information Forensics and Security*, *11*(9), 2013–2027.
2. Shabut, A. M., Dahal, K. P., Bista, S. K., & Awan, I. U. (2015). Recommendation based trust model with an effective defence scheme for MANETs. *IEEE Transactions on Mobile Computing*, *14*(10), 2101–2115.
3. Yu, Y., Li, K., Zhou, W., & Li, P. (2012). Trust mechanisms in wireless sensor networks: attack analysis and countermeasures. *Journal of Network and Computer Applications*, *35*(3), 867–880.
4. He, Q., Wu, D., & Khosla, P. (2004). SORI: A secure and objective reputation-based incentive scheme for ad-hoc networks. In IEEE Wireless Communications and Networking Conference, USA, (pp. 825-830).
5. Leligou, H. C., Trakadas, P., Maniatis, S., Karkazis, P., & Zahariadis, T. (2012). Combining trust with location information for routing in wirelesssensor networks. *Wireless Communications and Mobile Computing*, *12*(12), 1091–1103.
6. Chintalapalli, R. M., & Ananthula, V. R. (2018). M-LionWhale: Multi-objective optimisation model for secure routing in mobilead-hocnetwork. *IET Communications*, *12*(12), 1406–1415.
7. Jhaveri, R. H., Patel, N. M., Zhong, Y., & Sangaiah, A. K. (2018). Sensitivity analysis of an attack-pattern discovery based trusted routing scheme for mobile ad-hoc networks in industrial IoT. *IEEE Access*, *6*, 20085–20103.
8. Josang, A., Ismail, R., & Boyd, C. (2007). A survey of trust and reputation systems for online service provision. *Decision Support Systems*, *43*(2), 618–644.
9. Qin, D., Yang, S., Jia, S., Zhang, Y., Ma, J., & Ding, Q. (2017). Research on trust sensing based secure routing mechanism for wireless sensor network. *IEEE Access*, *5*, 9599–9609.

# Inventory Optimization of General-Purpose Metal Materials Based on Inventory Time Series Data

**Tingting Zhou and Guiying Wei**

**Abstract** In this paper, a time series-based inventory optimization management model is established based on the purchase data of general-purpose metal materials in the supply chain system of a certain aerospace enterprise logistics center, then the rules of inventory management for general-purpose metal materials are determined. In order to establish new ordering rules, firstly, an ordering strategy diagram is sketched to help analyze the actual operation mechanism of the general-purpose metal material inventory. Secondly, an inventory optimization management model is constructed by using the inventory data for nearly three years of the supply chain system in the logistics center as a dynamic driver. Then, a grid search algorithm is used to solve the problem, and new reorder points and replenishment ordering quantitives of the general-purpose metal materials are obtained. These new results can fill the gaps in management rules for some general-purpose metal materials and further correct the management rules previously formulated by managers through empirical rules. The experimental results show that the proposed method greatly improves the inventory turnover of general-purpose metal materials and makes the ordering strategy more scientific.

**Keywords** Supply chain system · Time series · Ordering strategy · Grid search algorithm · Inventory optimization

---

T. Zhou · G. Wei (✉)
Donlinks School of Economics and Management, University of Science and Technology Beijing, Beijing, China
e-mail: weigy@manage.ustb.edu.cn

T. Zhou
e-mail: 1105220939@qq.com



J. Zhang et al. (eds.), *LISS2019*,
https://doi.org/10.1007/978-981-15-5682-1_27

# 1 Introduction

Metal materials, as an important part in the procurement process of model used materials in a certain aerospace enterprise, have a great impact on the supply of model used materials. At present, metal materials management is divided into the management of general-purpose metal materials and non-general-purpose metal materials. General-purpose metal materials refer to metal materials for multi-model use, which with stable quality, controllable delivery cycle, low risk of inventory loss and high frequency of use [1, 2].

At present, in the management of materials, there is a lack of rules to accurately judge the general-purpose metal materials, which leads to the subjective impression of managers easily mixed in the process of identifying general-purpose metal materials and affects the accuracy of identification. Therefore, in the inventory management of general-purpose metal materials, it is necessary to establish an ordering strategy and inventory optimization model that can optimize the inventory level of general-purpose metal materials, which has certain theoretical significance and practical application value for improving inventory turnover.

Researches on inventory management started early in Western countries. Since the 1950s, foreign scholars have studied inventory management and formed the theory and implementation method of inventory optimization. In 1951, Dickie applied Activity Based Classification method to management, which was called ABC method. In 1963, Peter Drucker extended ABC method to all social phenomena, which made ABC method widely used in enterprise management. As an important method for inventory range division and reasonable classification, ABC method classifies a wide variety of inventory according to the proportion of the number of categories and amounts, and improves the efficiency of inventory management with different management methods. With the application of computers in inventory management in the 1960s, Joseph and Orlicky divided materials into independent and related requirements according to the nature of demand, and formed a kind of material demand plan based on computer to compile production plan and implement production control. Cetinkaya and Lee aimd at two-echelon supply chain, assuming that the retailer's demand obeys Poisson distribution, while the supplier makes joint replenishment with retailers whose geographical location is close to and whose demand nature is similar. Cachon discussed the use of VMI to coordinate the inventory management of a single supplier and multiple retailers in the supply chain. The results showed that VMI alone can not guarantee the integration of all links in the supply chain unless all members of the supply chain are willing to pay a fixed transportation cost [3–5].

China's inventory management is still in accordance with the traditional way. Small companies have a weak management force and a low starting point. Their internal management mode was loose and extensive, and there was no awareness of the necessity of enterprise inventory control in the process of operation; Domestic scholars began to study storage theory in the 1980s. Through the analysis of EOQ, MRP, MRPII and JIT inventory control strategy, Zhao proposed that the inventory management model must be based on the actual situation of the enterprise and adopt the most suitable enterprise itself, only in this way can the inventory management level be truly improved [6]. Xie put forward that the inevitable choice for enterprises to reduce logistics costs is to implement supply chain inventory management. Then he analyzed the existing problems of supply chain inventory and put forward solutions to these problems. i.e., supply chain inventory management implements VMI, JMI, 3PL and other modes, under which, enterprises can achieve the goal of reducing logistics costs [7]. Da et al. studied a special EOQ problem. Assuming that the purchase price decreases with time exponentially and the demand increases with time, an algorithm for finding the optimal service level and the number of purchases was proposed [8]. Through studying bullwhip effect and optimal batch size, Liu summarized and analyzed the factors affecting the inventory cost of supply chain, and put forward some measures such as choosing suitable suppliers and implementing JIT management mode, which improved the efficiency of supply chain and reduced the inventory cost [9]. On the basis of full analysis of the use frequency and consumption of general-purpose materials, Quan put forward a rolling stock preparation method. By establishing reasonable safety stock, reorder point and order batch, the method makes the safety stock act as buffer stock of urgently needed materials, and effectively improves the timeliness of supply of general-purpose materials [10].

In recent years, the inventory level control of space manned rocket materials has attracted extensive attention from academia and business managers. However, the current situation shows that the inventory management method of such heavy industry enterprise still relies on the management of empirical rules to a large extent, and this ambiguous management method lead the inventory to deviate from the normal level, occupying a large amount of fluid capital. Therefore, it is necessary to establish a scientific inventory management rule which is in line with the actual operation mechanism of the logistics center of aerospace enterprises, and to control the inventory of metal materials in a high degree.

The paper is organized as follows: In Sect. 2, a material ordering strategy diagram is introduced, and the data preprocessing is carried out, then basing on what, the Inventory optimization model of general-purpose metal materials is established. In Sect. 3, The Grid Search Algorithm is chose to stimulate the model and analyze the results of experiment. In Sect. 4, as a conclusion, this paper discusses the significance of establishing scientific inventory management rules of general-purpose metal materials and the shortcomings of the experiment.

# 2 Establishment of an Optimized Inventory Management Model for General-Purpose Metal Materials

## 2.1 Inventory Management Ordering Strategy

The inventory management indicators include: inventory cost, inventory turnover rate, customer service level and on-time delivery rate. Because of the situation that the logistics center of the aerospace enterprise must meet the supply of materials required by the downstream departments in a timely and sufficient manner, this study intends to establish an inventory management optimization model from the perspective of improving inventory turnover, and then control the inventory. Firstly, the following five hypotheses are made to facilitate the development of research activities:

(1) Materials arrive in bulk, not in succession;
(2) Carry out ordering and stock-taking when demand occurs;
(3) Enterprise capital is sufficient, there will be no shortage of funds affecting the situation of ordering;
(4) Inventory cost is calculated from the average of the initial inventory and the end inventory;
(5) At the beginning of the model, there is no undelivered materials (i.e. the quantity of goods in transit is 0).

Based on the above five assumptions and the actual operation mechanism of logistics center inventory, the following ordering strategy chart is established, as shown in Fig. 1.

In Fig. 1, $t_0$ is the initial state and the corresponding inventory level is $I_0$; horizontal line $S$ represents the reorder point of materials, i.e., when the inventory level is lower than S, the ordering will be arranged; the replenishment ordering quantity of each ordering is constant variable $\bar{Q}$, which is shown in the figure as a

**Fig. 1** Ordering strategy diagram

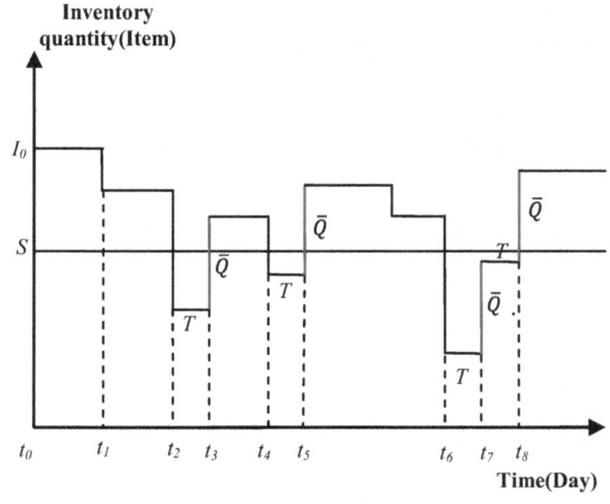

vertical red line segment; the arrival cycle of each ordering is constant variable $T$; when the demand occurs, the quantity of goods is needed to be counted at the same time as the allocation of goods. When the inventory level is lower than the reorder point $S$, the ordering will be made. Among them, $t_2$ is the time point when the materials are out of stock from the warehouse and the inventory level decrease; $t_2$, $t_4$, $t_6$, $t_7$ are all the ordering time points, while $t_3$, $t_5$, $t_7$, $t_8$ are all the arrival time points of materials.

As a means of describing the inventory situation of logistics center, general-purpose metal material ordering strategy diagram clearly illustrates the importance of reorder point and replenishment ordering quantity for inventory control. And this discrete inventory level curve is more in line with the actual inventory status, besides, the inventory management optimization model established on the basis of this theory will be more scientific and realistic.

## 2.2 Data Preprocessing

In this study, 13 business data tables of ERP management system of logistics center of aerospace enterprise were selected. The original business data from January 1, 2015 to May 1, 2018 covered about 205 attributes related to the whole process of metal materials from ordering to warehousing, totaling more than 310,000 data. These raw data basically contain all the information used for analysis, but there are still some problems as follows:

(1) Data volume is large, data redundancy and duplication are serious, and the correlation between data tables is poor.
(2) The original data contains 205 attributes, which makes it impossible to identify the attributes of inventory turnover rate quickly and accurately.
(3) Each data serves the actual business. Generally, one data can not be uniquely identified by one or two certain attributes, which is inconvenient to use.

So is necessary to process the original data. In order to facilitate the solution pf monthly average inventory turnover rate of general-purpose metal materials, 12 attributes need to be obtained: materials number $i$, total number of months $M$, month serial number $j$, days of month $N_j$, date sorting $k$, whether demand occurs $T_k^i$, time when requirements occur $t_k^i$, outbound unit price $p_k^i$, actual outbound quantity $Q_k^i$, average arrival cycle $\bar{T}^i$, average warehousing unit price $\bar{P}^i$ and initial inventory $I_0^i$ et al., where $I, M, j, N_j, k, \bar{T}^i, \bar{P}^i, I_0^i$ are constant values, $T_k^i$ is (0–1) matrix, $t_k^i$, $p_k^i$, $Q_k^i$ are all numerical matrixes. The following is the processing description of each attribute value:

(1) $i$: Number the 422 general-purpose metal materials of the original data file.
(2) $M, j, N_j, k$: $M$ means the total number of months in the original data; $j$ means the serial number of months, $j \in [1, M]$; $N_j$ means days of the $j$-th month; And $k$ is the order of historical time, counted by days.

(3) $\bar{T}^i$: The arrival cycle is the difference between the arrival date and the reviewing date of order, and the average of the arrival cycle of each material in the historical time series is represented by $\bar{T}^i$.
(4) $\bar{P}^i$: The average warehousing unit price is the average of all warehousing unit prices of each material in the historical time series.
(5) $I_0^i$: Screen out the demand occurrence date of each material when the demand arises for the first time, and filter out the inventory at the beginning of the month corresponding to the date as the initial inventory of the *i*-th material.
(6) $T_k^i$: It indicates whether demand of the *i*-th material occurs on the *k*-th day of the historical time series. $T_k^i = 0$, indicates that no demand has occurred, while $T_k^i = 1$, indicates that demand has occurred, so whether the demand has occurred is represented by a (0–1) time series matrix.
(7) $t_k^i$: It means the time when all the requirements of each material occur in the historical time, accurated to the *k*-th day of the historical time series;
(8) $p_k^i$, $Q_k^i$: They are both time-series matrices, respectively representing the outbound unit price and actual outbound quantity of the *i*-th material on the *k*-th day of the historical time series.

After data preprocessing, the total amount of data and the number of attributes are obviously much less than the original data, as shown in Table 1. A large amount of redundant information is removed, and the processed data features are obvious and can be accurately used to solve the inventory turnover rate. Table 2 is a summary of the attributes, used to solve the inventory turnover rate.

**Table 1** Data prepeocessing results

| | Raw data | Processed data |
|---|---|---|
| Data size | Over 31000 | 1768 |
| Total number of attributes | 205 | 12 |

**Table 2** Inventory turnover solution attributes

| Attribute symbol | Attribute name |
|---|---|
| $i$ | Materials number |
| $M$ | Total number of months |
| $j$ | Month serial number |
| $N_j$ | Days of month |
| $k$ | Date sorting |
| $\bar{T}^i$ | Average arrival period |
| $\bar{P}^i$ | Average warehousing unit price |
| $I_0^i$ | Initial inventory |
| $T_k^i$ | Whether demand occurs |
| $t_k^i$ | Time when requirements occur |
| $Q_k^i$ | Actual outbound quantity |
| $p_k^i$ | Outbound unit price |

## 2.3 Establishment of Inventory Management Optimization Model

According to the result of data preprocessing and the material ordering strategy diagram, first of all need to determine the initial inventory $I_0^i$ of each material, then introduce two control variables: reorder point $S^i$ and replenishment ordering quantity $\bar{Q}^i$. Assume that each inventory liquidation time point is the time point when demand occurs. Compare the inventory $I_k^i$ with the next actual demand $Q_{k+1}^i$ in the historical time series, and constantly adjust $S^i$ and $\bar{Q}^i$ to make sure $I_k^i > Q_{k+1}^i$ be satisfied, until the whole process of historical inventory level is simulated successfully and the demand of each allocation is met finally. All above is defined as a constraint.

Secondly, the maximum monthly average inventory turnover rate $(ITO_A^i)$ is calculated as the optimal objective function for each general-purpose metal material. The solution equation of inventory turnover rate is as follows: inventory turnover rate = cost of materials sold/average inventory cost, the monthly cost of materials sold is calculated by multiplying the outbound unit price of several times a month with the actual outbound quantity. The average monthly inventory cost is equal to each warehousing unit price times the average monthly inventory quantity, averaging the inventory quantity at the beginning of each month and the inventory quantity at the end of each month. Then average the inventory turnover rate of $M$ months in the historical time to obtain the monthly average inventory turnover rate. The following is the dynamic programming equation for solving the reorder point and the replenishment ordering quantity of general-purpose metal materials. In the equation, $a_j$ is the beginning date of the $j$-th month and $b_j$ is the end date of the $j$-th month.

*Objective Function:*

$$Max\left(ITO_A^i\right) = \frac{\sum_{j=1}^{M} \frac{\sum_{k=a_j}^{b_j} P_k^i \cdot Q_k^i}{P_j^i\left(I_{j1}^i + I_{jN_j}^i\right)/2}}{M} \tag{1}$$

*Constraints:*

$$I_k^i \geq Q_{k+1}^i \tag{2}$$

In this constraint, there are three points about the updating of inventory $I_k^i$ as follows:

(1) As for the inventory of the $i$-th material after being shipped out on the $k$-th day, if $I_k^i$ is greater than $S^i$, then no replenishment is needed, $\beta = 0$ ($\beta = 0$, no order is made; $\beta = 1$, order), the order time point $t_\beta^i$ = null; while if $I_k^i < S^i$ ,replenishment is needed, then $\beta = 1$, $t_\beta^i = t_k^i$, $t_\beta^i$ and $\beta$ are process variables. Among them, $t_k^i$ is the demand occurrence date on the $k$-th day of the $i$-th material.

(2) If the $i$-th material has been ordered before the $k$-th day, i.e. $\beta = 1$. If $t_k^i \geq t_\beta^i + \bar{T}^i$, then update $I_k^i = I_{k-1}^i + \bar{Q}^i - Q_k^i$ in time, and update $\beta = 0$, $t_\beta^i = null$; while if the $i$-th material has been ordered before this date and the order has not yet arrived, i.e., $t_k^i < t^\beta + \bar{T}^i$ then update $I_k^i$ to be: $I_k^i = I_{k-1}^i - Q_k^i$.
(3) The month beginning inventory of the $i$-th material in month $j$ is equal to the month end inventory of the $i$-th material in month $j - 1$.

To sum up, the updated equation for inventory is shown as follows:

$$I_k^i = \begin{cases} I_{k-1}^i + \bar{Q}^i - Q_k^i & \left(\beta = 1, T_k^i = 1, t_k^i \geq t_\beta^i + \bar{T}^i\right) \\ I_{k-1}^i + \bar{Q}^i & \left(\beta = 1, t_k^i \geq t_\beta^i + \bar{T}^i, T_k^i = 0\right) \\ I_{k-1}^i & (\beta = 1, t_k^i < t_\beta^i + \bar{T}^i, T_k^i = 0; \text{ or } \beta = 0, T_k^i = 0) \\ I_{k-1}^i - Q_k^i & (\beta = 1, t_k^i < t_\beta^i + \bar{T}^i, T_k^i = 1; \text{ or } \beta = 0, T_k^i = 1) \end{cases} \tag{3}$$

$$I_{j0}^i = I_{(j-1)N_j}^i \tag{4}$$

## 3 Simulation and Result Analysis

Nowadays, many algorithms can be used to solve the optimal value of the model. According to the development history of the algorithms, they can be divided into traditional algorithms and modern optimization algorithms. According to the analytic expression of the objective function, the traditional optimal algorithm could obtain the global minimum of the objective function, which mainly includes simplex method, steepest descent method, Newton method, common roll gradient method and so on. Modern optimization algorithms are mainly designed to solve the problems that traditional optimization algorithms are difficult to solve. Modern optimization algorithms are mainly some direct methods, such as pattern search, Rosenbrock algorithm, Powell method, especially stochastic intelligent optimization algorithms, such as evolutionary algorithm, neural network, swarm intelligence algorithm, simulated annealing algorithm, particle swarm optimization algorithm, etc. Particle swarm optimization algorithm is more prominent in solving the optimal value of the model, and its calculation speed is very fast, it can converge to the optimal solution in a relatively short time with a relatively high efficiency. However, due to the influence of its own computational characteristics, the result of particle swarm optimization is likely to be a local optimal value, and cannot get the best global optimal value. In the traditional algorithm, the grid search algorithm has greater advantages in solving the optimal solution of the model. It much suits the situation in this study, that the solution of model must be global optimal solution and high accuracy requirements.

The grid search method is an exhaustive search method for specified parameter values. It optimizes the parameters of the estimated function through cross-validation to obtain the optimal learning algorithm. Nowadays, many heuristic algorithms are prevalent. A lot of experiments prove that the optimal solution determined by grid search algorithm is generally global optimal solution. Because of its comprehensive search performance and independent of each group of parameters, it often does not fall into local extreme value. Therefore, this study chooses to use grid search algorithm to simulate inventory optimization model.

First of all, the traversal space $p^i$ and $q^i$ are set for the two parameters of reorder point $S^i$ and replenishment ordering quantity $\bar{Q}^i$, which are used as the input values of the optimal inventory management model. Secondly, select the initial inventory $I_0$ as the initial inventory level of the simulation, Thirdly, take the known parameters $Q_k^i$, $p_k^i$, $\bar{P}^i$, $\bar{T}^i$, $t_k^i$, $N_j$, $M$, $T_k^i$ as input variables and $t_\beta^i$, $\beta$ as process variables, all of what are simultaneously put into the grid search algorithm to traverse the optimal solution to obtain $S^i$ and $\bar{Q}^i$, and then achieving the maximum value of $ITO_A^i$. The steps for solving the inventory optimization model using the grid search algorithm are presented in Fig. 2 below.

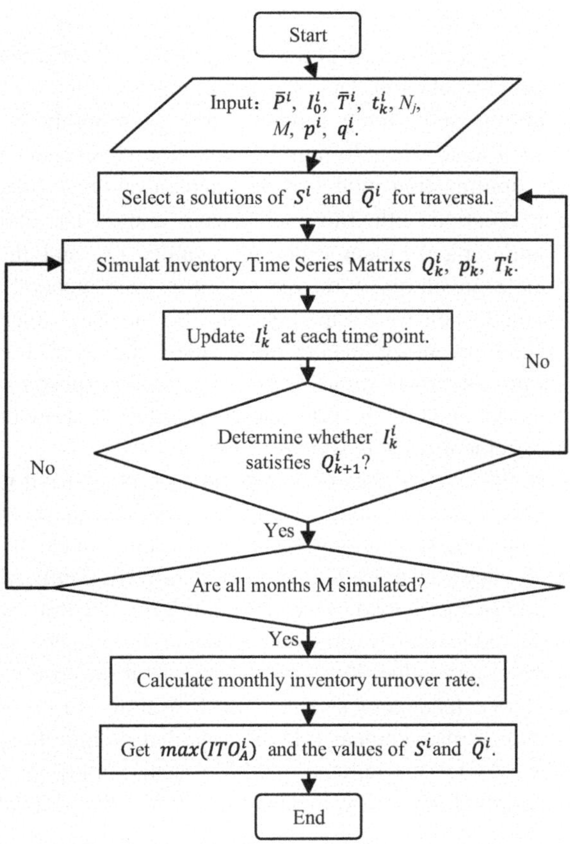


**Fig. 2** Input and output data solving model based on grid search algorithms

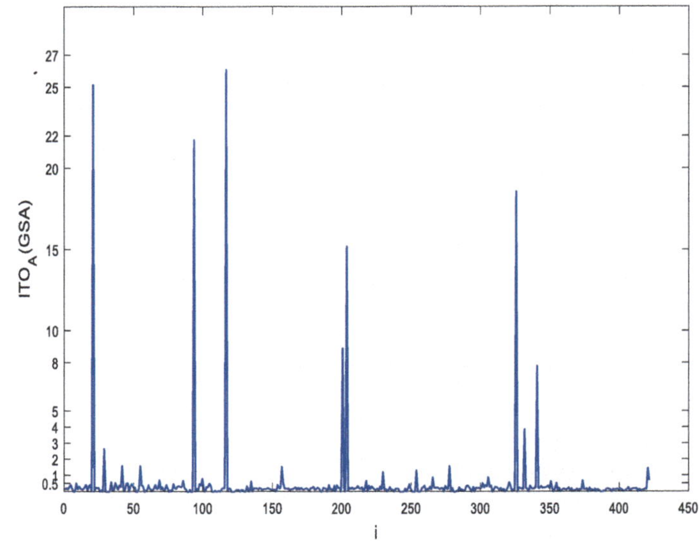


**Fig. 3** Monthly average inventory turnover rate solved by grid search algorithms

This paper uses MATLAB language to compile simulation experiments. Driven by all attributs and data after data preprocessing, with $I_k^i \geq Q_{k+1}^i$ as the constraint condition, and aiming at maximizing the monthly average inventory turnover rate of each material, a 40-month simulation experiment was conducted on the inventory optimization model of 422 general-purpose metal materials by adjusting the input solution of the traversal space $p^i$ and $q^i$. The optimal solution of reorder point $S^i$ and replenishment ordering quantity $\bar{Q}^i$, and the optimal value of monthly average inventory turnover rate for each general-purpose metal material are obtained, then by comparing with the monthly inventory turnover rate under the original inventory management rules, the overall increase rate of the average monthly inventory turnover rate of general-purpose metal materials is obtained. The result obtained by the grid search algorithm is shown in Fig. 3 and the result of old management rules is shown in Fig. 4.

In the original data of 422 general-purpose metal material, 280 of which had no inventory management rules in the past, i.e., there was no fixed ordering strategy and the ordering arrangement was usually made by the managers according to experience, causing the inventory data were confused, and the inventory recorded could not correspond to the actural ordering quantity and actural allocation quantity, so the old inventory turnover rate could not be calculated scientifically and regarded as the empty value. Here, for these general-purpose metal materials without the old inventory management rules, the inventory turnover rate obtained by the new inventory management rules is regarded as infinite and far superior to the empty value. As for the remaining 142 of general-purpose metal materials, according to the original inventory management rules, the simulation results show that 88 of

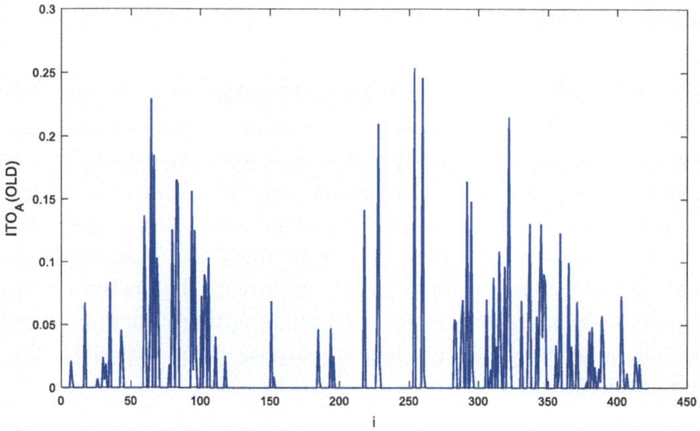


**Fig. 4** Monthly average inventory turnover rate solved by old inventory management rules

**Table 3** The overall improvement of $ITO_A$

| Indicator name | $OITO_A$(OLD) | $OITO_A$(GSA) | Overall increase rate |
|---|---|---|---|
| Indicator value | 0.08430 | 0.59765 | 7.08956 |

them cannot meet the actual demand, and their inventory turnover rate cannot be solved, which is regarded as empty value. Then corresponding to the new inventory management rules, the increase rate of the inventory turnover rate is also regarded as infinite. Therefore, under the old inventory management rules, only 54 of general-purpose metal materials can be solved the average monthly inventory turnover rate, while results of the other 368 materials is regarded as 0.

Under the old inventory management rules, the average monthly inventory turnover of 54 general-purpose metal materials is concentrated between (0, 0.3) and the overall average monthly inventory turnover rate $OITO_A$ (OLD) is 0.08430. While under the simulation experiment of grid search algorithm, 54 of general-purpose metal materials have not found new inventory management rules and monthly average inventory turnover rate. The $ITO_A$ of the remaining 368 general-purpose metal materials has been greatly improved, basically maintained in the range of (0, 1). The monthly average inventory turnover rate of nearly 10 of materials has exceeded 10, the maximum value has reached 26.12458, the overall monthly average inventory turnover rate $OITO_A$ (GSA) is 0.59765. The overall increase rate between $OITO_A$ (OLD) and $OITO_A$ (GSA) is 7.08956 times. The effect is very ideal, as shown in Table 3.

## 4 Conclusion and Prospect

Under the background of scale adjustment and production mode transformation of aerospace enterprises, this paper studies the material inventory management model of a aerospace enterprise, and designs the inventory management strategy and optimization model of general-purpose metal materials. Under the environment of MATLAB, driven by the time series data of inventory, this paper explores the inventory management rules which can fully meet the customer's demand of general-purpose metal materials, and greatly improving the inventory turnover rate of general-purpose metal materials, then determining more scientific reorder points and replenishment orderings for all general-purpose metal materials. The achievements of this research are as follows:

(1) As an important part of material purchasing, general-purpose metal materials have a great influence on supply for model used materials. By putting forward new inventory management rules for each general-purpose metal material, the inventory level of general-purpose metal materials can be controlled more scientifically, and it can be more adapted to the situation of high-density launching mission of aerospace model, and enhance the active supply capacity further.
(2) Through the analysis and feature identification of the inventory data of general-purpose metal materials, the supply and demand situation of general-purpose metal materials can be more clearly defined, which can lay a foundation for the differentiated management of general-purpose metal materials in combination with the supply chain system. And it can play an important role in helping enterprises carry out inventory management reform and save the cost of inventory occupation in the future.

In this paper, the ordering rules of each general-purpose metal materials have be obtained successfully, while the association of combined ordering between general-purpose metal materials has not be considered. Therefore, we can carry out a deeper study on the basis of this study in the future to seek the correlation of ordering rules between materials. In addition, due to the limited research data in this experiment, it is necessary to seek more related attributes of ordering and logistics to improve the model and conduct further simulation. Inventory management optimization has always been a topic of concern for enterprises. In the future, I believe more methods will be put forward and improved to solve practice business problems.

## References

1. Wang, F., Wang, C., Duan, L., et al. (2017). Solutions to improve inventory turnover of general purpose metal bars for aerospace models. *Aerospace Industry Management, 9,* 33–35.

2. Yuan, W., Fan, J., Wang, H., et al. (2017). Promote the improvement of space materials management mode with the concept of zero inventory. *Space Industry Management, 3,* 7–13.
3. Cui, W. (2015). ABC classification and Kraljic matrix analysis in inventory management. *Logistics Engineering and Management, 12,* 35–36.
4. Büyükkaramikli, N. C., Gürler, Ü., & Alp, O. (2014). Coordinated logistics: Joint replenishment with capacitated transportation for a supply chain. *Production and Operations Management, 23*(1), 110–126.
5. Fang, Y., & Shou, B. (2015). Managing supply uncertainty under supply chain Cournot competition. *European Journal of Operational Research, 243*(1), 156–176.
6. Xing T. B. (2014). *Study on optimization of raw fuel inventory management in. S ironmaking plant.* M.S. thesis, Fuzhou University, Fuzhou, Fujian, China.
7. Li, L. (2014). An analysis of enterprise inventory management in supply chain. *Market Modernization, 28,* 132.
8. Dai, X., Fan, Z., & Liu, Y. (2015). Rolling purchasing strategy model for raw materials under uncertain prices. *Industrial Engineering and Management, 20*(3), 60–65.
9. Zhang, C. (2015). *Study on the optimized inventory management scheme of buy Baole commercial company.* Changsha: Hunan University.
10. Quan, L. (2018). Study on general material management in space model material supply. *Technology and Innovation, 116*(20), 118–119.

# Passenger Travel Mode Decision-Making Research in Transport Corridor Based on the Prospect Theory

**Yunqing Feng, Xirui Cao, and Xuemei Li**

**Abstract** The passengers' travel decision-making process can be regarded as limited rational decision-making behavior under the certain and uncertain comprehensive condition. In this study, in consideration of the different types of passengers' heterogeneity characteristics, the passengers are divided into 5 heterogeneous categories according to the travel purpose. The heterogeneous passengers' prospected comprehensive evaluation to the high-speed railway, common railway, highway and aviation are used as the reference point. The frequency distributions of these heterogeneous passengers' actual evaluation score to the 4 travel modes, which are obtained through questionnaire investigation, are used as the probability distribution. The prospect theory is used to establish a passenger travel mode decision-making model to calculate the 5 kinds of heterogeneous passengers' comprehensive prospect value to the 4 travel modes. Besides, combined with the Logit model, the 5 kinds of heterogeneous passengers' fuzzy choice probability to the 4 modes are calculated and a comparative analysis has been made to reflect the heterogeneous passengers' fuzzy travel choice preference.

**Keywords** Integrated transportation · Prospect theory · Travel decision-making · Passenger corridor · Reference point

This paper is supported by the National Natural Science Fund Project "Passenger Share Research of different travel modes based on the Passengers' Travel Preferences in the Passenger Corridor (51778047)".

Y. Feng (✉) · X. Cao
Planning Institute of CCID Thinktank, China Center for Information Industry Development, Beijing, China
e-mail: 09120785@bjtu.edu.cn

X. Cao
e-mail: caoxirui@ccidthinktank.com

X. Li
School of Economics and Management, Beijing Jiaotong University, Beijing, China
e-mail: xmli@bjtu.edu.cn



J. Zhang et al. (eds.), *LISS2019*,
https://doi.org/10.1007/978-981-15-5682-1_28

# 1 Introduction

In recent years, the high-speed railway's booming development provides a new travel choice for the people and changes the passengers' travel behavior, which has broken the equilibrium of the original passenger transport marketing and changed the passenger share. Besides, with the continuous improvement of income level, the ticket price is no longer the only decisive factor for passengers' travel choice. The indexes such as the passengers' value of time, the comfort of transportation products, Convenience and safety have also become important factors affecting the passengers' choice of different travel modes. Therefore, to research the passengers' travel decision-making behavior under the comprehensive effect of multiple factors has an important significance to construct a comprehensive transportation corridor with reasonable resource allocation and equilibrium passenger share.

# 2 Research Status of the Passengers' Travel Choice and Decision-Making Behavior

The study of passenger travel decision-making behavior can be divided into 2 aspects: the passenger share research based on the expected utility theory, the passengers' travel decision-making behavior research based on the limited rationality.

## *2.1 The Passenger Share Research Based on the Expected Utility Theory*

The expected utility theory (EU) was proposed by Morgenstern and von Neumann in 1944, which has put forward the hypothesis that the decision makers are always rational and describes the decision-making behavior of "rational people" under risk conditions [1]. In the 1970s, econometric economist McFadden introduced the EU into the study of travel mode partition and put forward the Logit model based on the stochastic utility theory [2]. Since then, the passenger share and travel behavior research have been carried out one after another [3]. However, the premise of the expected utility theory supposes that the decision maker (passenger) is an individual with complete rationality under the condition of certainty, who has complete information in the decision-making process, can exhaust all alternatives, and select the best one based on the accurate evaluation of all alternatives. In reality, the external information is usually inadequate and the passengers are not pure rational people. Their decision-making is also affected by complex psychological mechanism. The emotion, intuition, comprehension ability, attitude factors often influence and even dominate the passengers' personal decision. Therefore, the expected utility theory can't reflect the decision makers' attitude to subjective factors, such as risk, and does not consider the ambiguity of individual utility and subjective probability in real life.

## 2.2 The Bounded Rational Behavior Research

The bounded rationality research was first put forward by Simon in 1947. He believed that it was impossible for people to obtain all the necessary information to make rational decisions. In actual decision-making, people only have limited rationality. The result of decision-making was a "satisfactory" scheme chosen under limited conditions rather than the optimal scheme. Based on Simon's research, Kahneman and Tversky introduced the decision maker's preferences into decision-making process and put forward the prospect theory (PT) [4] from the bounded rationality prospect in 1979. This theory transformed the decision-maker's preference into the choice of reference point and setting of weighting function and value function parameters, which researched the decision maker's irrational decision behavior under uncertainty conditions starting from the people's actual decision behavior. Compared with the EU theory, the PT was put forward according to a large number of actual investigation and experimental analysis, which more accurately describes the decision makers' judgment and decision-making behavior under uncertainty. At present, it has been widely used in the research of travel route, travel mode choice and departure time. In the travel route choice research, Katsikopoulos et al. [5] first used the prospect theory in the traffic domain and made a stated preference experiment investigation for the traveler's route choice behavior. Avineri et al. [6] studied the effect of the loss and gain perception on the route choice decision-making process and test the possibility of using the cumulative prospect theory to express the stochastic user equilibrium in the framework of PT. Xu [7] combined the passenger's route choice and network randomness and established an equivalent variational inequality model under network stochastic user equilibrium based on cumulative PT. Connors et al. [8] used the cumulative PT to set up a general network equilibrium model and researched the route choice behavior based on risk perception. Based on the cumulative PT, Liu et al. [9] established a traveler perceive utility model under the condition of continuous random distribution for the route travel utility. In the departure time research, Jou et al. [10] have used the PT to analyze the departure time choice problem under the uncertain traffic network environment. Xia et al. [11] have used the arrival time as the reference point for route choice and discussed the setting method of different reference point for commuters. In the travel mode choice research, Luo et al. [12] have analyzed the decision process of resident travel mode choice and established the prospect function travel mode choice model according to the travel characteristics of the residents.

The EU is mainly used to describe people's rational behavior, and can make correct analysis of some simple decision-making problems. The PT is a description of actual behavior, which can describe complex human decision-making behavior under uncertain circumstances. Therefore, the behavioral decision-making model based on the PT has important guiding significance for the study of human decision-making behavior in reality.

## 3 Passenger Travel Decision-Making Factor Analysis

In recent years, the passengers' requirements for travel quality are no longer simply to accomplish the purpose of travel, but to pay more attention to the various services enjoyed during travel. To analyze the influencing factors of passenger travel decision-making and establish the index system of influencing factors of passenger travel decision-making can help transport departments to find out what can be improved, which is of great significance to improve the passenger's transport service quality and enhance the attractiveness of passenger flow.

Passenger travel decision-making process is a comprehensive decision affected by multiple factors. Most of the existing research analyzes the passenger travel decision-making factor from the passenger satisfaction prospective. "Passenger satisfaction" is the result of comparison for the railway passenger service performance and passenger expectations. Whether the railway passengers are satisfied with the railway transport service lies in two aspects. The first is that whether the passenger is satisfied with the transportation service purchased. The second is whether the passengers' paying worth the money. Li [13] studied the passengers' basic requirements on the railway transport service quality from safety, function, service and time and considerate. Shi et al. [14] put forward that the influence factors of passenger travel satisfaction mainly include three aspects: Travel time, cost and comfort. He et al. [15, 16] selected the economy, rapidity, comfort, convenience, safety as the factors affecting the passenger travel demand. Based on this, Deng et al. [17] added the service, hardware facilities, staff quality and service supervision to research the impact on the passenger travel choice. Jiang et al. [18] selected the speed, comfort, ticket price, convenience and security as the main factors that affect the passengers' travel decision-making and calculated the weight of each factor. This paper quotes Jiang's index system and summarize the factors that influence the passenger travel decision-making into 5 aspects: Ticket price (PRI), comfort (Com), speed (SPE), convenience (Con) and Security (Sec). The comfort factor includes 4 indexes: Service Quality of the Train (Com1), Waiting Room Environment and Congestion Degree (Com2), Train Stationary (Com3) and Carriage Environment and Congestion Degree (Com4). The convenience factor includes 4 factors, Convenience of Buying the Tickets (Con1), Convenience of Departure to Station (Con2), Convenience of In and Out the Station (Con3) and Convenience of Leaving the Station (Con4). The security factor includes 3 indexes, Security of the Carry-on Items (Sec1), Personal Safety (Sec2) and Security of the Train (Sec3).

# 4 The Passenger Perception Difference Decision-Making Model Based on the PT

## 4.1 Overview of the PT

The PT selects the human in reality as the subjects and units the rational profit and value perception into an individual, which reveals the individual's decision-making mechanism under the risk and uncertainty conditions and well describes the people's bounded rationality characteristics.

This theory divides the risk decision-making into editing and evaluation process. In the editing phase, the individual collects and processes the evaluation framework and reference information. In the evaluation stage, the individual makes a judgment for the information depends on the value function and decision weights function. The value function has three characteristics. (1) People are risk averse when confronted with gains. (2) People are risk seeking when confronted with losses. (3) People are more sensitive to losses than gains. Therefore, people are reluctant to take risks and be more cautious when faced with gains. They will be more inclined to take risks when faced with losses. People's sensitivity on the loss and gain are different. Pain from the loss greatly exceed to the happiness from the gain. The formula and diagram of the value function is shown as follow in Fig. 1.

$\Delta x > 0$ means the decision-maker is faced with gains. $\Delta x < 0$ means the decision-maker is faced with losses. $\alpha, \beta$ represents the risk preference and risk aversion coefficient respectively. Higher value of these parameters means people are more prone to take risks. The curve is convex curve and the decision-maker is risk seeking when facing with loss. The slope in loss in bigger than in gain and the curve is steeper, which means that decision-maker is more sensitive to losses than gain.

$$v(\Delta x) = \left\{ \begin{array}{ll} x^{\alpha} & \text{if } \Delta x \geq 0 \\ -\lambda(-x)^{\beta} & \text{if } \Delta x < 0 \end{array} \right\} (0 < \alpha, \beta < 1; \lambda > 1) \quad (1)$$

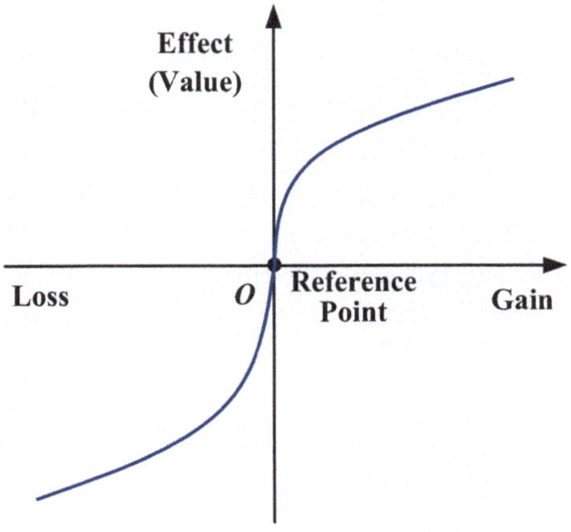


**Fig. 1** Diagram of the value function

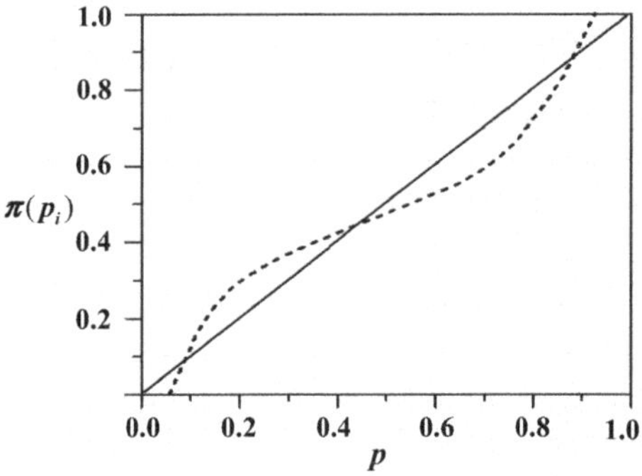


**Fig. 2** Diagram of the decision weighting function

The equation and diagram of the decision weights function is shown as follow in Fig. 2.

$$\pi(p) = \begin{cases} \pi^+(p) = p^\gamma/[p^\gamma + (1-p)^\gamma]^{\frac{1}{\gamma}} & Gain \\ \pi^-(p) = p^\delta/[p^\delta + (1-p)^\delta]^{\frac{1}{\delta}} & Loss \end{cases} (\gamma > 0, \delta > 0) \quad (2)$$

The decision weighting function is not a probability but a nonlinear function of probability $p$. In the prospect theory, for an uncertain event $(x_1, p_1; x_2, p_2; \ldots x_n, p_n)$, the probability of result $x_n$ is $p_n \cdot p_1 + p_2 + \ldots + p_n = 1$. The prospect effect of the event can be expressed as follow.

$$f(x_1, p_1;\ x_2, p_2;\ \cdots;\ x_n, p_n) = \sum_{i=1}^{n} \pi(p_i)v(x_i) \quad (3)$$

## 4.2 *Reference Point Setting of the Heterogeneous Passengers*

(1) *Classification of the heterogeneous passengers*

The passengers can be divided into different types according to different travel demands. This paper reference Aike Jiang's research and divide the passenger into 5 types by travel purpose: Business, Traveller, Commute, Student and Migrant Worker. According to Aike Jiang's r research, the weights of each factor under different types of passengers are shown in Table 1 as follow.

**Table 1** Weights of different factors for the passengers with different travel purpose

| Factors | Weights | | | | |
|---|---|---|---|---|---|
| | Business | Traveller | Commute | Student | Migrant worker |
| Ticket price | 0.175 | 0.179 | 0.183 | 0.204 | 0.239 |
| Speed | 0.197 | 0.159 | 0.229 | 0.172 | 0.174 |
| Comfort | 0.203 | 0.22 | 0.144 | 0.187 | 0.187 |
| Com1 | 0.06 | 0.058 | 0.043 | 0.046 | 0.039 |
| Com2 | 0.048 | 0.052 | 0.025 | 0.051 | 0.05 |
| Com3 | 0.054 | 0.048 | 0.041 | 0.04 | 0.042 |
| Com4 | 0.041 | 0.062 | 0.036 | 0.049 | 0.057 |
| Convenience | 0.193 | 0.182 | 0.183 | 0.2 | 0.201 |
| Con1 | 0.039 | 0.051 | 0.039 | 0.063 | 0.061 |
| Con2 | 0.041 | 0.042 | 0.06 | 0.046 | 0.046 |
| Con3 | 0.052 | 0.04 | 0.05 | 0.052 | 0.048 |
| Con4 | 0.061 | 0.049 | 0.033 | 0.038 | 0.047 |
| Security | 0.232 | 0.259 | 0.261 | 0.238 | 0.199 |
| Sec1 | 0.08 | 0.087 | 0.086 | 0.085 | 0.074 |
| Sec2 | 0.083 | 0.09 | 0.103 | 0.082 | 0.078 |
| Sec3 | 0.069 | 0.083 | 0.072 | 0.071 | 0.046 |

(2) *Passenger travel expected perceptual evaluation questionnaire survey*

In some extent, the passengers' travel choice is dependent on the expected perceptual evaluation. This study selects the satisfaction as the passenger perception evaluation factor and establishes a questionnaire for the passenger expected perceptual evaluation. We can get the passengers' expected perceptual evaluation data from the passengers' satisfaction evaluation on the rapidity, comfort, economy, convenience, safety, service and punctuality factor.

The questionnaire mainly includes two parts. The first part is the basic personal information, including gender, age, occupation, income level, travel purpose, travel expenses, travel mode, seat type, etc. The passenger travel type can be divided reasonable according to the basic personal information survey. The second part is the passengers' expected perceptual evaluation on economic, rapidity, comfort, convenient, safety, service, punctuality factors and the whole travel satisfaction evaluation. The measurement of each index all adopts the Likert 5 scale design. The respondents (passengers) should choose the appropriate perceptual evaluation according to their own travel experience. Different values reflect the different satisfaction degree. The higher the value is, the more satisfied the passenger is. The specific rating scale is as follow.

Very dissatisfied-1 points. A little dissatisfied-2 points. Generally satisfied-3 points. A little satisfied-4 points. Very satisfied-5 points.

(3) *Reference point setting based on the expected perception*

The reference point is a key factor to measure the individual's gain or loss in the editing stage of prospect theory and play a key role in the value function. Different reference point will affect the individual's judgment to gain or loss. The traditional reference point selection method assumes that all of the traveler's reference point is the same. In reality, the heterogeneous passengers' expected perceptual evaluations on travel modes evaluation is different because of the differences in income levels, travel preferences, etc. therefore, the reference point is also different. The passengers can be divided into 5 categories according to the travel purpose: traveller, business, commute, migrant workers, students. They all have different weight and expected perceptual evaluation on each travel decision factor. Therefore, the heterogeneous reference point can better describe the diversity and individuality of the passenger travel decision behavior in the real environment.

Suppose $c = \{c_{PRI}, c_{Spe}, c_{Com}, c_{Con}, c_{Sec}\}$ is the set of factors that influence the passenger travel decision-making. $\boldsymbol{P_{ij}}$ means that the passengers' travel type is $\boldsymbol{i}$ and the travel mode is $\boldsymbol{j}$. $v_{ijk}^0$ stands for the mean expected perceptual evaluation of passenger $\boldsymbol{P_{ij}}$ for the travel decision influencing factor $k$. $v_{i,j}^{RP}$ stands for the expected comprehensive perception evaluation reference point of passenger $\boldsymbol{P_{ij}}$.

$$v_{i,j}^{RP} = \sum_{k \in c} v_{i,j,k}^0 \cdot w_{i,j,k} \tag{4}$$

## 4.3 The Passengers' Travel Decision-Making Model Based on the Perception Difference

In the evaluation stage, the decision maker calculates the comprehensive prospect value under the condition of gains or losses and makes a decision by establishing the value function and decision weights function.

(1) *Travel decision value function-v(Δx)*

Make a questionnaire survey for the passenger actual perception and use the survey data as the passengers' actual perception evaluation. Use the heterogeneous passengers' perception difference ($\Delta x$) between the actual perception and comprehensive expected perception evaluation to measure the passenger gain or loss and establish the travel decision value function.

$$v(\Delta x_{ijm}) = \left\{ \begin{array}{ll} (\Delta x_{ijm})^{\alpha} & \text{if } \Delta x_{ijm} \geq 0 \\ -\lambda(-\Delta x_{ijm})^{\beta} & \text{if } \Delta x_{ijm} < 0 \end{array} \right\} (0 < \alpha, \beta < 1; \lambda > 1) \tag{5}$$

$$\Delta x_{ijm} = v_{i,j,k=m} - v_{i,j}^{RP} = m - \sum_{k \in c} v_{i,j,k}^{0} \cdot w_{i,j,k} \tag{6}$$

$\Delta x_{ijm}$ means that the perception difference of passenger $\boldsymbol{P_{ij}}$ compared with the reference point $v_{i,j}^{RP}$ when the actual evaluation is $m$ ($m$ = 1, 2, 3, 4, 5). $\Delta x_{ijm} > 0$ means that the passenger $i$ is perception gained for travel mode $j$. $\Delta x_{ijm} < 0$ means that the passenger $i$ is perception lossed for travel mode $j$. According to the research [19], $\alpha = 0.37$, $\beta = 0.59$, $\lambda = 1.51$.

(2) *Travel decision weighting function-*$\pi(p)$

Use the various types of passengers' actual evaluation score frequency distribution for different travel modes as the probability distribution of the actual evaluation. Suppose $p_{ijm}$ is the probability of getting the actual evaluation score $m$ ($m$ = 1, 2, 3, 4, 5) for the passenger $j$ to travel mode $i$. The probability distribution of getting the actual evaluation score is $p_{(v_{ij}=m)} = p_{ijm}$. Based on the passenger perception evaluation probability distribution, the passenger travel decision weight function is shown as follow.

$$\pi(p_{ijm}) = \begin{cases} \pi^{+}(p_{ijm}) = p_{ijm}^{\gamma} / [p_{ijm}^{\gamma} + (1 - p_{ijm})^{\gamma}]^{\frac{1}{\gamma}} \\ \pi^{-}(p_{ijm}) = p_{ijm}^{\delta} / [p_{ijm}^{\delta} + (1 - p_{ijm})^{\delta}]^{\frac{1}{\delta}} \end{cases} (\gamma > 0, \delta > 0) \tag{7}$$

(3) *Comprehensive prospect value evaluation for the heterogeneous passengers*

Based on the establishment of heterogeneous passengers' perception evaluation reference point, travel decision value function and decision weights function, the PT is used to calculate the comprehensive prospect value of the passenger with travel type $i$ for travel mode $j$.

$$f_{ij} = \sum_{m=1}^{5} \pi(p_{ijm}) v(\Delta x_{ijm}) \tag{8}$$

The technology roadmap of the passengers' travel mode choice based on the PT is shown in Fig. 3.

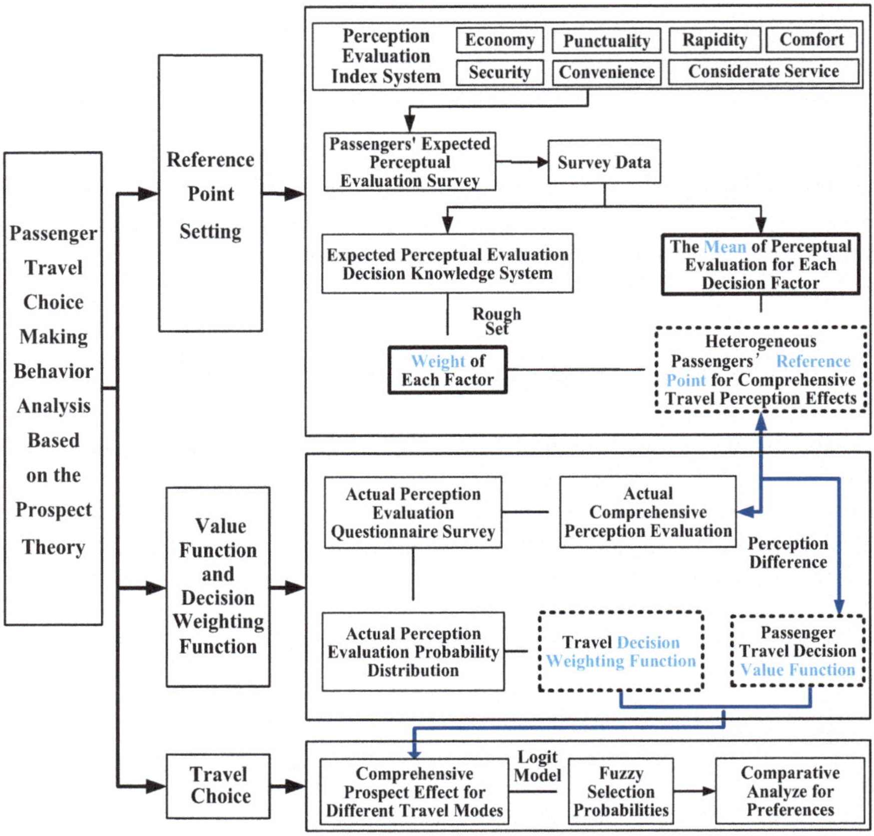


**Fig. 3** Technology roadmap

## 5 Case Analysis of the Passengers' Travel Mode Choice - Empirical Research of the Beijing-Shanghai Line

The Beijing-Shanghai passenger corridor has the most development potential in economic and has total four travel mode, HSR, traditional rail, road and air. Therefore, selecting the Beijing-Shanghai Line as the research object to research the change of the passengers' travel preference and the passenger share before and after the operation of the HSR is representative to study the effect of HSR to other travel modes.

### 5.1 *Statistical Analysis of the Questionnaire for the Passengers' Travel Perceptual Evaluation in the Beijing-Shanghai Line*

The investigation totally sent 1200 questionnaires and 1020 valid questionnaires were collected. The *Chronbach's Alpha* value of each factor is shown in Table 2.

From Table 2 we can see that the coefficient of all the measuring factors is higher than 0.75, which means that each factor has a high level of reliability and the measured results are stable and reliable. Use the KMO and Bartlet test to make the validity test. The results are shown in Table 3. From Table 3, the KMO test value is 0.847 and the Bartlett test significant probability is all less than 0.05, which explain that the research data is effective and suitable for further analysis.

### 5.2 *The Passenger Travel Mode Choice Behavior Analysis in the Beijing-Shanghai Line*

Use the heterogeneous passengers' weight on different factors in Table 1 and combine the formula (4) to calculate the heterogeneous passengers' reference point $v_{ij}^{RP}$ and probability distribution of actual comprehensive evaluation. The result is shown in Table 4.

**Table 2** Reliability analysis test value of each travel influence factor

| Factors | Cronbach's alpha | Factors | Cronbach's alpha |
|---|---|---|---|
| Ticket price | **0.784** | Convenience | **0.814** |
| Speed | **0.847** | Safety | **0.866** |
| Comfort | **0.836** | Overall evaluation | **0.812** |

**Table 3** Validity test value of each factor

| Kaiser-Meyer-Olkin measurement | The sphericity test of Bartlett | | |
|---|---|---|---|
| | Approximate chi square | df | Sig. |
| 0.847 | 1438.15 | 176.00 | 0.00 |

**Table 4** Heterogeneous passengers' reference point and probability distribution of actual comprehensive evaluation

| Type | Mode | RP($v_{ij}^{RP}$) | Probability distribution $v_{ij}$ | | | | |
|---|---|---|---|---|---|---|---|
| | | | $p(v_{ij}=1)$ | $p(v_{ij}=2)$ | $p(v_{ij}=3)$ | $p(v_{ij}=4)$ | $p(v_{ij}=5)$ |
| Business | HSR | 3.34 | 2.98% | 9.26% | 34.10% | 41.77% | 11.89% |
| | Rail | 3.2 | 11.64% | 24.18% | 34.92% | 21.48% | 7.78% |
| | Road | 3.02 | 18.93% | 41.63% | 29.46% | 7.54% | 2.44% |
| | Avia | 3.11 | 4.31% | 10.59% | 31.23% | 33.26% | 20.61% |
| Traveller | HSR | 3.21 | 3.28% | 10.13% | 35.74% | 33.69% | 17.16% |
| | Rail | 3.07 | 5.25% | 15.16% | 39.34% | 30.41% | 9.84% |
| | Road | 3.18 | 13.33% | 25.57% | 37.87% | 20.33% | 2.90% |
| | Avia | 3.36 | 18.20% | 25.98% | 31.80% | 16.23% | 7.79% |
| Commute | HSR | 3.28 | 0.98% | 26.23% | 32.95% | 25.90% | 13.94% |
| | Rail | 2.98 | 11.31% | 17.54% | 33.70% | 23.11% | 14.34% |
| | Road | 3.16 | 7.62% | 12.62% | 22.62% | 42.38% | 14.76% |
| | Avia | 3.7 | 20.16% | 15.25% | 23.11% | 23.61% | 17.87% |
| Migrant workers | HSR | 3.48 | 4.50% | 18.23% | 52.95% | 20.90% | 3.42% |
| | Rail | 3.08 | 3.31% | 12.54% | 32.31% | 33.11% | 18.73% |
| | Road | 3.12 | 7.62% | 17.62% | 27.63% | 37.38% | 9.75% |
| | Avia | 3.79 | 23.16% | 45.25% | 23.11% | 5.61% | 2.87% |
| Student | HSR | 3.32 | 2.98% | 11.23% | 22.95% | 43.90% | 18.94% |
| | Rail | 3.1 | 1.31% | 7.54% | 28.61% | 58.11% | 4.43% |
| | Road | 3.16 | 2.62% | 2.62% | 22.62% | 57.38% | 14.75% |
| | Avia | 3.64 | 15.17% | 40.25% | 28.11% | 13.61% | 2.87% |

For the business passengers, use the comprehensive prospect value formula to calculate the heterogeneous passengers' comprehensive prospect effects as follows.

$$f_{Business}^{HSR} = \sum_{i=1}^{5} \pi(p_{Business}^{v_{HSR}=i}) v(\Delta x_{Business}^{v_{HSR}=i}) = -0.2717$$

$$f_{Business}^{Rail} = \sum_{i=1}^{5} \pi(p_{Business}^{v_{Rail}=i}) v(\Delta x_{Business}^{v_{Rail}=i}) = -0.7352$$

$$f_{Business}^{Road} = \sum_{i=1}^{5} \pi(p_{Business}^{v_{Road}=i}) v(\Delta x_{Business}^{v_{Road}=i}) = -0.9845$$

$$f_{Business}^{Aviation} = \sum_{i=1}^{5} \pi(p_{Business}^{v_{Aviation}=i}) v(\Delta x_{Business}^{v_{Aviztion}=i}) = -0.1271$$

combine the prospect value and the Logit model to calculate the business passengers' fuzzy choice probability on three travel modes (railway, road, aviation) $\overline{p_{Business}^{Rail}}$, $\overline{p_{Business}^{Road}}$, $\overline{p_{Business}^{Aviation}}$ before the operation of the HSR.

$$\overline{p_{Business}^{Rail}} = e^{f_{Business}^{Rail}} / (e^{f_{Business}^{Rail}} + e^{f_{Business}^{Road}} + e^{f_{Business}^{Aviation}}) = 0.2765$$

$$\overline{p_{Business}^{Road}} = e^{f_{Business}^{Road}} / (e^{f_{Business}^{Rail}} + e^{f_{Business}^{Road}} + e^{f_{Business}^{Aviation}}) = 0.2155$$

$$\overline{p_{Business}^{Aviation}} = e^{f_{Business}^{Aviation}} / (e^{f_{Business}^{Rail}} + e^{f_{Business}^{Road}} + e^{f_{Business}^{Aviation}}) = 0.508$$

The business passengers' fuzzy choice probability on the four travel modes (HSR, railway, road, aviation) $\overline{p_{Business}^{HSR}}'$, $\overline{p_{Business}^{Rail}}'$, $\overline{p_{Business}^{Road}}'$, $\overline{p_{Business}^{Aviation}}'$ after the operation of HSR are as follow.

$$\overline{p_{Business}^{HSR}}' = e^{f_{Business}^{HSR}} / (e^{f_{Business}^{HSR}} + e^{f_{Business}^{Rail}} + e^{f_{Business}^{Road}} + e^{f_{Business}^{Aviation}}) = 0.3053$$

$$\overline{p_{Business}^{Rail}}' = e^{f_{Business}^{Rail}} / (e^{f_{Business}^{HSR}} + e^{f_{Business}^{Rail}} + e^{f_{Business}^{Road}} + e^{f_{Business}^{Aviation}}) = 0.1921$$

$$\overline{p_{Business}^{Road}}' = e^{f_{Business}^{Road}} / (e^{f_{Business}^{HSR}} + e^{f_{Business}^{Rail}} + e^{f_{Business}^{Road}} + e^{f_{Business}^{Aviation}}) = 0.1497$$

$$\overline{p_{Business}^{Aviation}}' = e^{f_{Business}^{Aviation}} / (e^{f_{Business}^{HSR}} + e^{f_{Business}^{Rail}} + e^{f_{Business}^{Road}} + e^{f_{Business}^{Aviation}}) = 0.3529$$

Similarly, use the prospect theory to calculate all the passengers' comprehensive prospect value and the fuzzy choice probability. The result of before and after the HSR operation is shown in Tables 5 and 6 correspondingly.

According to Tables 5 and 6, to draw the heterogeneous passengers' travel preference comparison chart shown in Fig. 4. From Fig. 4, we can see that the business passengers mainly prefer the aviation travel before the operation of HSR. They have a low preference on railway and road. The passenger of Traveller and Commute mainly prefer the railway travel. They have a low preference on road and aviation. Both of the migrant workers and students prefer railway and road. They have a low preference on aviation. From the horizontal contrast of the travel mode preference, the passenger who prefers the railway most is the migrant worker. The business has the least preference. The passenger who prefers the road most is the student. The business passenger has the least preference. The passenger who prefers the aviation most is the business people. The migrant worker has the least preference. After the operation of HSR, the business passengers mainly prefer the aviation and HSR. They have a low preference on railway and road. The passenger of Traveller and Commute mainly prefer the railway travel. They have a low preference on road and aviation. Both of the migrant workers and students prefer railway and road. The migrant workers mainly prefer the railway. The students mainly prefer the road.

**Table 5** The passengers' comprehensive prospect value and the fuzzy choice probability on three travel modes before HSR operation

| Type | Comprehensive prospect value | | | Fuzzy choice probability | | |
|---|---|---|---|---|---|---|
| | Railway | Road | Aviation | Railway | Road | Aviation |
| Business | −0.7354 | −0.9845 | −0.1271 | 0.2765 | 0.2155 | 0.508 |
| Traveller | −0.2728 | −0.8466 | −1.0528 | 0.4946 | 0.2787 | 0.2267 |
| Commute | −0.015 | −0.2836 | −1.0098 | 0.4685 | 0.3582 | 0.1733 |
| Migrant worker | −0.1022 | −0.382 | −1.7382 | 0.5127 | 0.3875 | 0.0998 |
| Students | −0.006 | 0.107 | −1.4421 | 0.4242 | 0.4749 | 0.1009 |

**Table 6** The passengers' comprehensive prospect value and the fuzzy choice probability on four travel modes after HSR operation

| Type | Comprehensive prospect value | | | | Fuzzy choice probability | | | |
|---|---|---|---|---|---|---|---|---|
| | HSR | Rail | Road | Aviation | HSR | Rail | Road | Aviation |
| Business | −0.271 | −0.735 | −0.984 | −0.127 | 0.305 | 0.192 | 0.149 | 0.353 |
| Traveller | −0.199 | −0.272 | −0.846 | −1.052 | 0.347 | 0.322 | 0.181 | 0.148 |
| Commute | −0.399 | −0.015 | −0.283 | −1.009 | 0.241 | 0.355 | 0.271 | 0.131 |
| Migrant worker | −0.844 | −0.102 | −0.382 | −1.738 | 0.196 | 0.412 | 0.312 | 0.081 |
| Student | −0.189 | −0.006 | 0.107 | −1.442 | 0.260 | 0.313 | 0.351 | 0.074 |

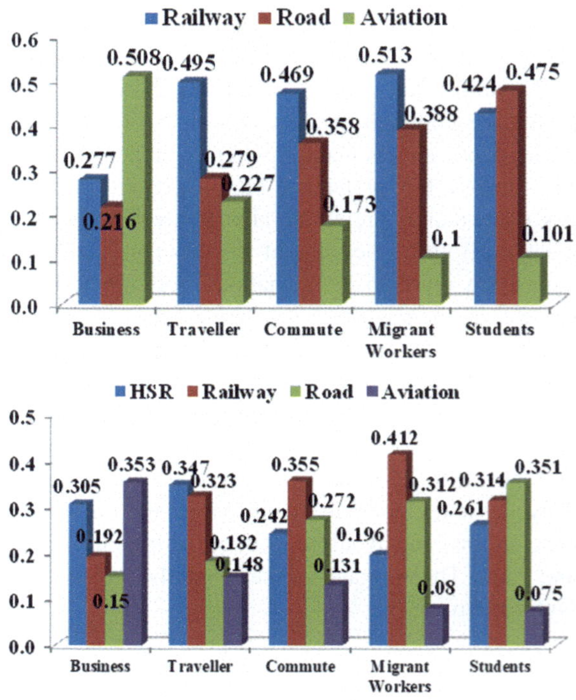


**Fig. 4** Heterogeneous passengers' travel preference comparison chart before and after the HSR operation

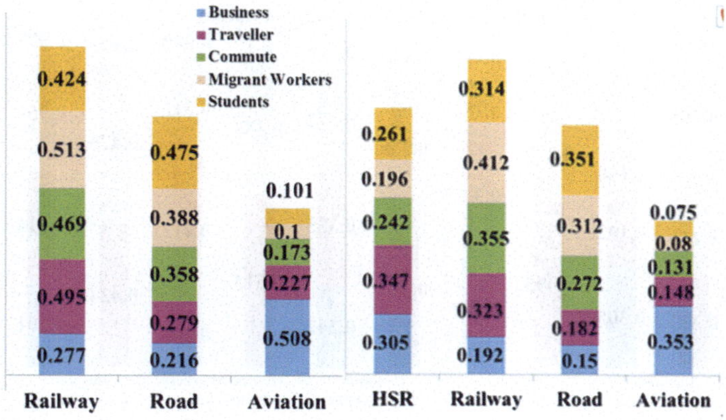


**Fig. 5** Passenger's overall preference comparison chart before and after the HSR operation

From the horizontal contrast of the travel mode preference, the passenger who prefers the HSR most is the passenger of commuting. The migrant workers have the least preference. The passenger who prefers the railway most is the migrant workers. The business passengers have the least preference. The passenger who preference the road most is the student. The business passenger has the least preference. The passenger who preference the aviation most is the business people. The students have the least preference.

The heterogeneous passenger's overall preference comparison chart before and after the operation of HSR is shown in Fig. 5.

On the whole, before the operation of the HSR, the descending order of the passengers' travel preference on the 3 travel modes is railway > road > aviation. After the operation of the HSR, the descending order of the passengers' travel preference on the 4 travel modes is railway > HSR > road > aviation. Therefore, the operation of the HSR has an impact on the passengers' travel preference. Each type of passenger' preference comparison chart is shown in Fig. 6.

From Fig. 6 it can conclude that the operation of the HSR attracts a large number of business passengers to change their travel preference from aviation to HSR, which reduces the passengers' travel preference on aviation. The traveller passengers' travel preference has been changed from railway to HSR. A part of passengers for commuting have turned to HSR, but the railway is still their most preferred travel mode. Some of the migrant worker and student passengers changed their travel preference from railway and road to HSR. But the migrant workers still preference the railway and road. The students' preference changed from road and railway to HSR, railway and road. The migrant worker and student passengers' preference on aviation changed a little. Therefore, the operation of the HSR has influenced all the passengers' preference. The most influenced passengers are the business and traveller passengers. The most influenced travel mode is the aviation. The least influenced passengers are the migrant workers. The other 3 types of passengers have been influenced a little by the high-speed railway's operation.

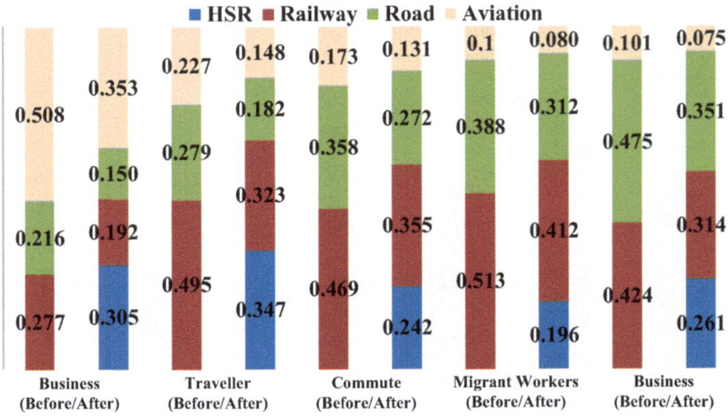


**Fig. 6** Each type of passenger' preference comparison chart

## 6 Conclusion

The operation of the HSR influenced the heterogeneous passengers' travel preference. From the overall perspective, before the operation of the HSR, the descending order of the passengers' travel preference on the 3 travel modes is railway > road > aviation. After the operation of the HSR, the descending order of the passengers' travel preference on the 4 travel modes is railway > HSR > road > aviation. Therefore, the operation of the HSR generates a great impact on the travel mode of aviation and makes a large number of passengers change their travel preferences from the road and aviation to HSR.

From each passenger' preference prospective, the operation of the HSR has the most impact on the business and traveller passengers' travel preference and changes their travel preference from aviation to HSR. A part of passenger of commuting and students also choose HSR as their travel mode, but not the main travel mode. The HSR and their original preference undertake the main passenger flow together. The HSR has little impact on the migrant workers' travel preference. Most of them still choose their original travel preference (railway).

From the impact on the travel mode prospective, for the business passengers, the HSR mainly influences the travel mode of aviation. For the traveller passengers, the HSR mainly influences the travel mode of railway. For the students and the passenger of commuting, the HSR mainly influences the travel mode of railway and road and has little impact on the aviation.

## References


1. Neuman, J. V., & Morgenstern, O. (1944). *Theory of games and economic behavior*. Upper Saddle River: Princeton University Press.
2. McFadden, D. (1974). Conditional logit analysis of qualitative choice behavior. In P. Zarembka (Ed.), *Frontiers in econometrics* (pp. 105–142). New York: Academic Press.
3. Zhang, J. N., & Zhao, P. (2012). Research on passenger choice behavior of trip mode in comprehensive transportation corridor. *China Railway Science, 33*(3), 123–131.
4. Kahneman, D., & Tversky, A. (1979). Prospect theory: an analysis of decision under risk. *Econometrica, 47*(2), 263–291.
5. Katsikopoulos, K. V., Duse-Anthony, Y., Fisher, D. L., et al. (2000). The framing of drivers' route choices when travel time information is provided under varying degrees of cognitive load. *Human Factors: The Journal of the Human Factors and Ergonomics Society, 42*(3), 470–481.
6. Avineri, E. (2006). The effect of reference point on stochastic network equilibrium. *Transportation Science, 40*(4), 409–420.
7. Xu, H. L., Zhou, J., & Xu, W. (2011). Cumulative prospect theory-based user equilibrium model for stochastic network. *Journal of Management Sciences in China, 14*(7), 1–7.
8. Connors, R. D., & Sumalee, A. (2009). A network equilibrium model with travellers' perception of stochastic travel times. *Transportation Research Part B: Methodological, 43*(6), 614–624.
9. Liu, Y. Y., Liu, W. M., & Wu, J. W. (2010). Route choice model of traveler based on cumulative prospect theory. *Journal of South China University of Technology (Natural Science Edition), 38*(7), 84–90.
10. Jou, R. C., Kitamura, R., Weng, M. C., et al. (2008). Dynamic commuter departure time choice under uncertainty. *Transportation Research Part A: Policy and Practice, 42*(5), 774–783.
11. Xia, J. J., Juan, Z. C., & Gao, J. X. (2012). Travel routing behaviors based on prospect theory. *Journal of Highway and Transportation Research and Development, 42*(5), 126–131.
12. Luo, Q. Y., Wu, W. J., Jia, H. F., & Hu, P. F. (2012). Analysis of residents travel mode choice based on prospect theory. *Journal of Transport Information and Safety, 42*(5), 37–40.
13. Li, Y. W. (2002). Railway passenger satisfaction survey analysis. *Railway Economics Research*, 33–35.
14. Shi, F., Deng, L. B., & Huo, L. (2007). Boarding choice behavior and its utility of railway passengers. *China Railway Science, 28*(6), 117–121.
15. He, Y. Q., Mao, B. H., Chen, T. S., & Yang, J. (2006). The mode share model of the high-speed passenger railway line and its application. *Journal of the China Railway Society, 28*(3), 18–21.
16. Feng, Y. Q., Li, X. W., & Li, X. M. (2011). Comprehensive evaluation of the railway passenger's satisfaction based on rough set and entropy. In *International Conference on Enterprise Information Systems,* SAIC, HCL, (pp. 629–635).
17. Deng, J. B., & Li, X. Y. (2012). Research on the railway passenger satisfaction based on neural network. *Journal of Qiqihar University (Natural Science Edition), 28*(2), 73–75.
18. Jiang, A. K., Feng, Y. Q., Li, X. M., & Li, X. W. (2013). Railway passenger's satisfaction evaluation based on entropy method and fuzzy theory. In *2013 International Conference on Logistics, Informatics and Services Sciences*, no. 1, (pp. 578–583).
19. Xu, H. L., Zhou, J., & Xu, W. (2011). A decision-making rule for modeling travelers' route choice behavior based on cumulative prospect theory. *Transportation Research Part C, 19*(2), 218–228.

# Price Risk Measurement Model of Pledge Financing of Lending Institution in Natural Rubber Supply Chain Based on VaR-GARCH Method

**Xuezhong Chen, Yang Liu, and Anran Chen**

**Abstract** Enterprises that carry out inventory pledge business will face a variety of risks. Among them, the price risk of inventory is a risk that is difficult to prevent and control, because it is affected by many factors such as macro environment and market price fluctuations and supply and demand balance. Based on the analysis of the statistical characteristics of inventory yield, the risk of natural rubber is studied, and the VaR-GARCH(1,1)-GED model is constructed to measure the yield risk of the inventory. The following conclusions are drawn: (1) The higher the confidence level, the higher the risk control requirements and the smaller the business risk; (2) The proposed method improves the calculation of VaR and its accuracy and provides a more comprehensive estimate of the characteristics of spikes that occur in extreme fluctuations of the inventory. The more extreme the reality, the more serious the error will occur, which will lead to unnecessary losses for financing companies.

**Keywords** Supply chain finance · Pledge financing · Price risk measurement · VaR-GARCH(1,1)-GED

---

This paper was financially supported by the National Natural Science Foundation of China (Grant No. 71371085: Evolution of Decision-Making Governance and Growth Path of China Multinational Corporations Based on Computational Experiment).

---

X. Chen (✉) · Y. Liu
Business School, University of Jinan, Jinan, China
e-mail: sm_chenxz@ujn.edu.cn

Y. Liu
e-mail: m15335319076@163.com

A. Chen
College of Science and Engineering, University of Minnesota Twin Cities, Minneapolis, USA
e-mail: Chen6265@umn.edu



J. Zhang et al. (eds.), *LISS2019*,
https://doi.org/10.1007/978-981-15-5682-1_29

## 1 Introduction

As a new type of financing, supply chain finance effectively solves the problem of financing difficulties for small and medium-sized enterprises in the upstream and downstream of the supply chain. In recent years, the inventory pledge financing business integrating logistics and financial services has developed rapidly in China and the development mode has gradually diversified. With the rapid growth of China's third-party logistics enterprises and the continuous improvement of business functions, logistics enterprises play an increasingly important role in the inventory pledge financing business, and a new model of inventory pledge financing with logistics enterprises as the main body of lending is gradually emerging. In the inventory pledge financing model, the fluctuation of the inventory price during the pledge period will affect the balance between supply and demand, and affect the income of the loan enterprise, and generate default risk, and bring losses to the logistics enterprises. This paper uses the VaR indicator of risk management in the economic and financial systems to measure the price risk of pledge financing of lending institution in natural rubber supply chain.

## 2 Literature Review

The concept of risk value was proposed in 1993 and has been widely used as a tool for risk analysis and measurement [1]. VaR (Value at Risk), translated as "risk value" or "insurance value", refers to the highest loss of financial assets or portfolios at a specific level of confidence and target period, or refers to the potential loss of financial assets combining the probability distribution of random variables with the monetary unit of measurement.

Since the return rate of assets does not obey the independent normal distribution under the assumption of effective financial market, VaR has many characteristics including a certain volatility agglomeration and the feature of sharp peak and fat tail. The generalized autoregressive conditional heteroscedasticity model (GARCH) can effectively describe the feature of the agglomeration of the return rate of assets [2]. Ricardo (2006) uses the GARCH family model to predict the VaR of the asset's return thick-tailed distribution and measure the maximum loss that may be encountered in the future [3]. Chen and Panjer believe that the market credit spread data and the simple model can be used to obtain the information on the default strength of the trusted company, and the structural model can obtain the complete information of the value of the company's assets and the obstacles of default, so as to evaluate the credit status of the trusted company. This method uses the JLT (Jarrow-Lando-Turnbull) model to describe the change in credit rating and simulates the VaR risk value of the portfolio investment to manage the transmission of financial risk [4]. Gu by comparing the theory and technology of financial risk management in different periods, proposes that the VaR method not only has a good effect in managing financial market risks, but also can be further extended to manage financial credit risk, financial liquidity risk and financial operation risk [5]. Guo also studies the application of VaR risk management model in financial risk management. The most typical feature of VaR is that it can calculate not only the risk of a single financial instrument, but also the risk of multiple financial instrument portfolios. However, for low-probability financial risks, the VaR method has the drawback that the VaR method cannot fully use the information in the data, so it is necessary to use the method of extreme event processing to give a relatively high confidence VaR estimate [6]. Rosenberg and Schuermann give a comprehensive VaR metric "H-VaR" for credit risk, market risk and operational risk from the overall perspective of financial risk management [7]. Wei and Zheng give the main management techniques of financial risk from the financial engineering perspective, including the combination of risk diversification, the hedging method of transfer risk and the insurance law to avoid risks, and they also applies financial risks management techniques to the supply chain management [8]. To keep the conditional variances generated by the GARCH($p$,$q$) model nonnegative, Bollerslev imposed nonnegativity constraints on the parameters of the process. Nelson, Daniel and Cao show that these constraints can be substantially weakened and so should not be imposed in estimation. They also provide empirical examples illustrating the importance of relaxing these constraints [9]. Österholm investigates the relation between treasury yields and corporate bond yield spreads. This is done by estimating VaR models on monthly Australian data from January 2005 to March 2017. The results suggest, in line with mainstream theoretical models, that a higher risk free rate compresses the corporate bond yield spread and a higher corporate bond yield spread lowers the three-month treasury bill rate [10]. Ayturk investigates the relationship between government borrowing and corporate financing decisions in 15 developed European countries for the period of 1989–2014. A robust negative relationship between government borrowing and corporate debt in developed European countries is found. However, any significant relation between government debt and equity is not identified [11].

## 3 Model Algorithm

### *3.1 Calculation Method of Yield*

Assume that the inventory pledge $t$-day transaction price is $sp_t$, and the market price on $t-1$ is $sp_{t-1}$, logarithmic yield sequence $\{r_t\}$ can be obtained from (1).

$$r_t = lnsp_t - lnsp_{t-1} \tag{1}$$

## 3.2 Calculation of the Volatility of the Pledge Yield Rate

The volatility of the yield pledge rate of the inventory, i.e., the standard deviation ($\sigma$) of the rate of return on the continuous compound interest rate per unit time. The source of volatility estimates is historical data or option prices. This paper intends to use the GARCH family model to predict the volatility of the rate of return. The GARCH model has the following characteristics.

(1) The GARCH model can describe the time-varying characteristics of the volatility of financial assets.
(2) The GARCH model has the feature of mean regression.
(3) The GARCH model effectively avoids a large number of estimations for high-order terms in the ARCH model.
(4) GARCH is a symmetric model. When $p = 1$, $q = 1$, GARCH(1,1) can effectively capture the agglomeration of the return on assets.

The GARCH model is built according to the following mean and conditional equations.

$$r_t = \mu_t + \varepsilon_t \tag{2}$$

$$\sigma_t^2 = \omega + \alpha\varepsilon_{t-1}^2 + \beta\sigma_{t-1}^2 \tag{3}$$

Where, $\mu_t = \rho r_{t-1}$ is conditional mean of yield, object to AR(1) process, $\varepsilon_t = \sigma_t z_t$ is random disturbance, $\sigma_t^2$ is conditional variance, $\sigma_{t-1}^2$ is forecast variance of the previous period, $\omega$ is a constant term, $\alpha$ is the ARCH coefficient estimate, $\beta$ is the GARCH coefficient estimate, so $\varphi = \alpha + \beta$ is the durability, then the long-term unconditional variance of GARCH(1,1) is as follows.

$$h = \frac{\omega}{1 - (\alpha + \beta)} \tag{4}$$

When modeling with the GARCH model, it is usually assumed on the premise of normal distribution, but the financial time series usually has the characteristics of sharp peaks and fat tails. The generalized error distribution (GED) can better fit the fat tail problem of the yield series, and can adjust the parameters to reflect the fat tail characteristics of the financial assets yield.

In order to make comparison choices in the GARCH family model, we try to fit the volatility with EGARCH (1,1) and TGARCH (1,1) at the same time.

The EGARCH model, known as the exponential GARCH model, can better reflect the asymmetric characteristics of volatility fluctuations. It was proposed by Nelson to adjust the response of heteroscedasticity to the shocks of positive and negative information [12].

The EGARCH(1,1) formula is as follows.

$$r_t = \mu_t + \varepsilon_t \tag{5}$$

$$\ln\left(\sigma_t^2\right) = \omega + \alpha\left(\left|\frac{\varepsilon_{t-1}}{\sigma_{t-1}}\right| - \sqrt{\frac{2}{\pi}}\right) + \beta\ln\left(\sigma_{t-1}^2\right) + \gamma\frac{\varepsilon_{t-1}}{\sigma_{t-1}} \tag{6}$$

Where, $\alpha$, $\beta$ has no restricted condition, $\sigma_t^2$ is a positive number, the model can be applied in a wider range.

The GARCH model cannot reflect the defect of the leverage effect of financial sequence. Glosten et al. improved the GARCH model and proposed the TGARCH model [13]. The formula for TGARCH(1,1) is as follows.

$$r_t = \mu_t + \varepsilon_t \tag{7}$$

$$\sigma_t^2 = \omega + \alpha\varepsilon_{t-1}^2 + \gamma\varepsilon_{t-1}^2 I_{t-1} + \beta\sigma_{t-1}^2 \tag{8}$$

Where, $I_k = 1, \varepsilon_k < 0; I_k = 0, \text{otherwise}$, $\varepsilon_k < 0$ means negative information, $\varepsilon_k \geq 0$ means positive information.

The Mathematical definition of VaR is as follows.

$$p(X > VaR(c)) = 1 - c \tag{9}$$

Here, $p$ denotes the probability of occurrence of some event, $X$ is defined as the loss of the pledge during the target period, $c$ is the confidence level, and VaR($c$) is the value at risk in the confidence level $c$. The calculation formula for VaR is as follows.

$$\text{VaR} = p\,\sigma\,z_c\sqrt{T} \tag{10}$$

Where, $p$ is the value at the beginning of an asset, $\sigma$ is the variance, $z_c$ is the lower quantile, and $T$ is the holding period.

Measurement of VaR is usually characterized by a probability distribution and a monetary unit of measure. When VaR is measured, the agglomeration and spike-tail characteristics of the asset return and the leverage effect should be considered, so the student distribution and the generalized error distribution are used for fitting.

# 4 Empirical Analysis

## 4.1 Statistical Analysis of Sample Data

The data as shown in Fig. 1 in this article is taken from the trading price of natural rubber daily from January 17, 2017 to December 29, 2017, except for the rest day of the exchange. There are 234 observation points in total, all of which are from the website of Shanghai Futures Exchange. Data processing and analysis software is Excel and Eviews.

As can be seen from Fig. 1, the fluctuation rate of natural rubber has a certain correlation, and the yield has a certain agglomeration effect, that is, large fluctuations will be followed by large fluctuations, and small fluctuations will be accompanied by small fluctuations. The phenomena of results of application is in accordance with the theoretical analysis.

As shown in Fig. 2, natural rubber fluidity index is

$$p_s = \frac{sp_{max} - sp_{min}}{sp} \tag{11}$$

While mean is 0.0298, standard deviation is 0.01369. The outliers are small, indicating that the market liquidity is relatively stable and the VaR model can be used for further analysis.

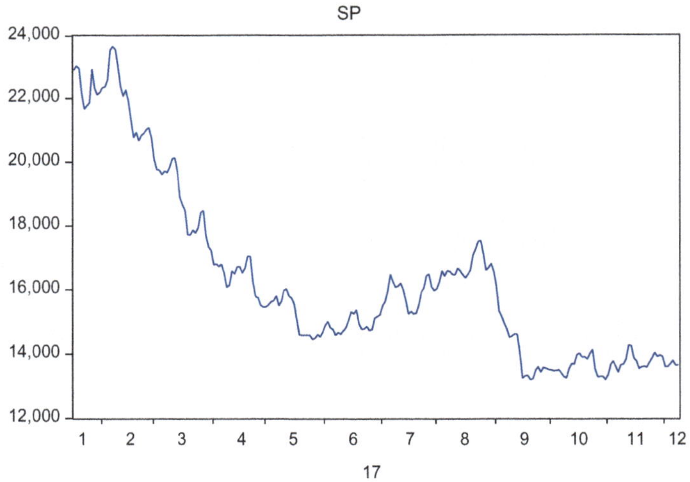


**Fig. 1** Natural rubber trading price time series

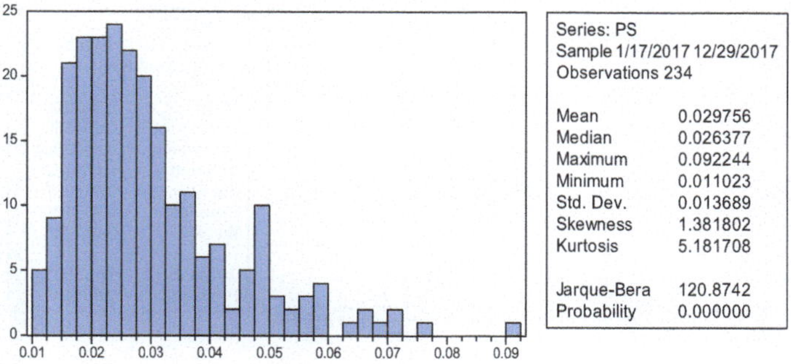


**Fig. 2** Natural rubber yield liquidity index statistics

**Table 1** Natural rubber yield sequence ADF single root test results

| | | *t*-Statistic | Prob.* |
|---|---|---|---|
| Augmented Dickey-Fuller test statistic | | −10.29793 | 0.0000 |
| Test critical values | 1% level | −2.574968 | |
| | 5% level | −1.942199 | |
| | 10% level | −1.615787 | |

## 4.2 *Natural Rubber Yield Sequence Volatility Analysis*

(1) *Natural rubber yield stability test*

In this paper, the unit root ADF test method is selected to perform the time series stability analysis. The single root test results in Table 1 show that the *t* statistic is −10.29793, which is less than the critical value at each significant level, and the *P* value is <0.05. The natural rubber yield sequence can be considered as stable, so the GARCH model can be fitted.

(2) *Natural rubber yield normal distribution test*

In Fig. 3, during the sample period, the average yield of natural rubber was −2.21E−03 with a standard deviation of 0.016. The skewness is −0.477, which is less than zero and exhibits a left-biased feature. The kurtosis is 3.957, which is greater than the normal distribution kurtosis $K = 3$, showing the characteristics of a sharp peak and fat tail. The Jarque-Bera statistic is 17.84 and the *P* value is 0.000134, rejecting the assumption that the yield rate series follows a normal distribution.

(3) *Natural rubber yield heteroscedasticity test*

From the time series of the mean residual of the yield as shown in Fig. 4, the fluctuation of the residual sequence of the natural rubber yield mean equation has

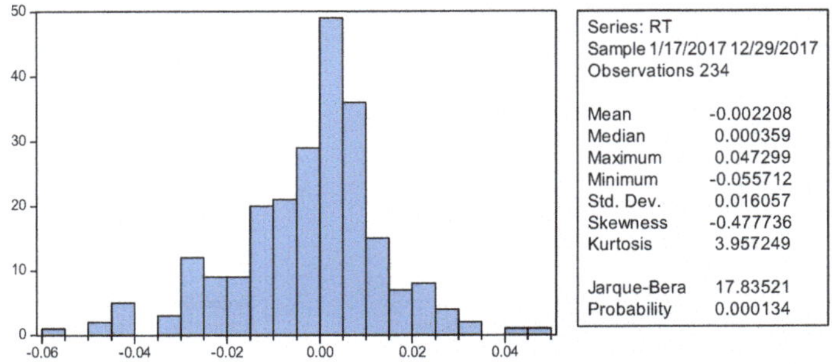


**Fig. 3** Natural rubber yield normality test and basic descriptive statistical results

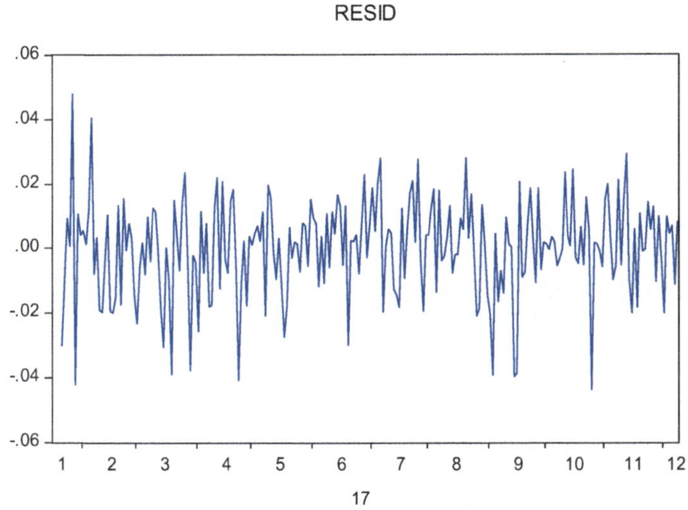


**Fig. 4** Residual timing chart of the mean value equation of natural rubber yield

obvious agglomeration effect, and the volatility of the yield has the conditional heteroskedasticity.

The residual squared autocorrelation test is performed on the mean equation and the Grange multiplier test (ARCH-LM test) is performed on the mean equation to test the heteroscedasticity.

As shown in Fig. 5, the 3rd-order autocorrelation and partial autocorrelation coefficients are not significantly zero, and the $Q$ statistics are large. It can be considered that the residual square has autocorrelation, i.e., the mean equation has heteroscedasticity.

The $F$ statistic is 1.391916, the accompanying probability is 0.2460, the $T \times R^2$ value is 4.17256, and the ARCH effect exists in the residual of the yield series as shown in Table 2. The volatility can be modeled by using GARCH.

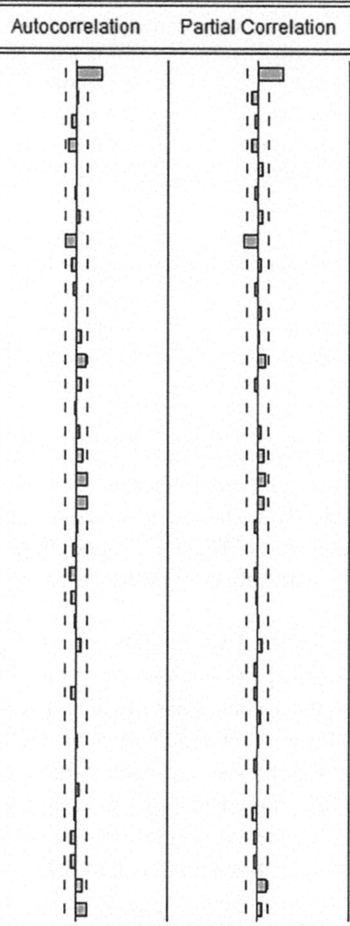


| | AC | PAC | Q-Stat | Prob |
|---|---|---|---|---|
| 1 | 0.289 | 0.289 | 19.667 | 0.000 |
| 2 | 0.014 | -0.076 | 19.712 | 0.000 |
| 3 | -0.053 | -0.039 | 20.385 | 0.000 |
| 4 | -0.095 | -0.073 | 22.537 | 0.000 |
| 5 | -0.004 | 0.047 | 22.540 | 0.000 |
| 6 | -0.018 | -0.040 | 22.615 | 0.001 |
| 7 | 0.032 | 0.046 | 22.857 | 0.002 |
| 8 | -0.115 | -0.160 | 26.081 | 0.001 |
| 9 | -0.057 | 0.033 | 26.874 | 0.001 |
| 10 | -0.039 | -0.046 | 27.248 | 0.002 |
| 11 | 0.002 | 0.031 | 27.249 | 0.004 |
| 12 | 0.062 | 0.023 | 28.194 | 0.005 |
| 13 | 0.102 | 0.094 | 30.771 | 0.004 |
| 14 | 0.043 | -0.034 | 31.240 | 0.005 |
| 15 | -0.024 | -0.001 | 31.389 | 0.008 |
| 16 | 0.029 | 0.035 | 31.595 | 0.011 |
| 17 | 0.067 | 0.073 | 32.726 | 0.012 |
| 18 | 0.126 | 0.086 | 36.783 | 0.006 |
| 19 | 0.126 | 0.079 | 40.865 | 0.003 |
| 20 | 0.022 | -0.026 | 40.992 | 0.004 |
| 21 | -0.039 | 0.006 | 41.390 | 0.005 |
| 22 | -0.070 | -0.032 | 42.678 | 0.005 |
| 23 | -0.059 | -0.021 | 43.584 | 0.006 |
| 24 | -0.011 | 0.010 | 43.615 | 0.008 |
| 25 | 0.046 | 0.050 | 44.180 | 0.010 |
| 26 | 0.004 | -0.031 | 44.185 | 0.014 |
| 27 | -0.053 | -0.026 | 44.940 | 0.017 |
| 28 | -0.001 | 0.034 | 44.940 | 0.022 |
| 29 | 0.024 | 0.014 | 45.097 | 0.029 |
| 30 | 0.001 | -0.048 | 45.098 | 0.038 |
| 31 | 0.027 | 0.007 | 45.295 | 0.047 |
| 32 | -0.015 | -0.055 | 45.354 | 0.059 |
| 33 | -0.054 | -0.028 | 46.149 | 0.064 |
| 34 | -0.050 | -0.033 | 46.843 | 0.070 |
| 35 | 0.079 | 0.105 | 48.555 | 0.064 |
| 36 | 0.118 | 0.047 | 52.410 | 0.038 |

**Fig. 5** Residual squared autocorrelation test

**Table 2** ARCH-LM test

| | | | |
|---|---|---|---|
| F-statistic | 1.391916 | Prob. F(3,226) | 0.2460 |
| Obs*R-squared | 4.172560 | Prob. ChiSquare (3) | 0.2434 |

## *4.3 GARCH Model and Parameters Determination*

After the above analysis, the natural rubber yield rate is in line with the GARCH model. The results of fitting three GARCH models by Eviews6.0 are shown in Table 3.

**Table 3** Estimation results of three garch model

| Parameters | GARCH(1,1) | TGARCH(1,1) | EGARCH(1,1) |
|---|---|---|---|
| ρ | 0.0016 | 0.1082 | 0.0016 |
| ω | 0.0038<br>(0.0000) | 0.00037<br>(0.0002) | 12.7675<br>(0.0000) |
| α | 0.1197<br>(0.0424) | −0.5515<br>(0.0984) | 0.1673<br>(0.1055) |
| β | 0.5755<br>(0.0567) | 0.1179<br>(0.0480) | 0.2415<br>(0.0170) |
| γ | | 0.2773<br>(0.0064) | 0.5221<br>(0.0702) |
| AIC | −5.41 | −5.41 | −5.42 |
| SC | −5.35 | −5.35 | −5.35 |

It can be seen from Table 3 that the values of parameter γ are 0.0064 and 0.0702 respectively in EGARCH(1,1) and TGARCH(1,1), which are not significant and have certain influence on the model, therefore GARCH(1,1) modeling is selected. The results of GARCH(1,1) under three distribution hypotheses are estimated by Eviews in Table 4.

As shown in Table 4 the parameters in the two models are significant. The coefficient α reflects the influence of historical fluctuations on future fluctuations. The larger the value of β indicates that the initial volatility has a longer-lasting effect and long-term memory. From the fitting results, the values of AIC and SC are the smallest under the GED distribution hypothesis, and each parameter is significant, indicating that the GARCH(1,1)-GED model is more suitable for fitting the characteristics of natural rubber yield rate sequence. Therefore, the mean equation and the conditional variance equation are as follows under the GED distribution.

$$r_t = 3.75E{-}004 r_{t-1} + \varepsilon_t \qquad (12)$$

**Table 4** Estimation results of ARCH(1)-GARCH(1,1) model parameters under three distribution hypotheses

| Parameters | Normal distribution | Student's distribution | GED distribution |
|---|---|---|---|
| ρ | 0.0016 | 0.0006 | 0.000375 |
| ω | 0.0038<br>(0.0000) | 0.0004<br>(0.0069) | 0.0004<br>(0.0066) |
| α | 0.1197<br>(0.0424) | 0.1431<br>(0.1240) | 0.1505<br>(0.1146) |
| β | 0.5755<br>(0.0567) | 0.5594<br>(0.1297) | 0.5217<br>(0.2086) |
| AIC | −5.41 | −5.44 | −5.46 |
| SC | −5.35 | −5.37 | −5.39 |

**Table 5** AR(1)-GARCH (1,1)-GED fitting ARCH LM test results

| | | | |
|---|---|---|---|
| F-statistic | 0.31851 | Prob. F(3,226) | 0.8120 |
| Obs*R-squared | 0.96835 | Prob. ChiSquare(3) | 0.8089 |

$$\delta_t^2 = 0.0004 + 0.1505\varepsilon_{t-1}^2 + 0.5217\delta_{t-1}^2 \tag{13}$$

$\varphi = \alpha + \beta = 0.1509 < 1$, satisfying the stationary condition of the GARCH model. The unconditional variance of the random error term $\varepsilon_t$ converges to $h$.

$$h = \frac{\omega}{1 - (\alpha + \beta)} = 0.0047 \tag{14}$$

The autocorrelation test was performed on the residual square of the model after fitting. Table 5 shows the ARCH LM test results of the residual of the AR(1)-GARCH(1,1)-GED model after fitting. The LM statistic is less than the critical value. There is no autoregressive conditional heteroscedasticity in residuals, i.e., the original hypothesis is accepted, and the existence of homovariance is considered, and the model is predictable.

## 4.4 VaR Calculation Results

The results obtained in Table 6 are consistent with the hypothesis according to the theory.

When the confidence is improved, the calculated VaR result becomes larger, indicating that the improvement of the confidence level requires an increase in the risk control level of the inventory.

**Table 6** Results of VaR calculation for AR(1)-GARCH(1,1)-GED model under different confidence levels

| VaR | 95% | 99% |
|---|---|---|
| Maximum | 0.093944 | 0.132505 |
| Minimum | 0.033906 | 0.047823 |
| Means | 0.061192 | 0.088852 |

## 5 Conclusion

This paper calculates the daily return rate VaR value of the inventory pledge combination based on the AR(1)-GARCH(1,1)-GED model. Finally, by comparing the experimental results, the following conclusions are drawn:

The GARCH(1,1) model of GED distribution can better fit the selected data compared with the GARCH model of the student distribution, the GARCH model of the normal distribution, the EGARCH model of the normal distribution, and the TGARCH model of the normal distribution. The higher the confidence level, the higher the risk control requirements and the smaller the business risk.

Compared with the traditional VaR calculation method, the method that this paper proposed improves the calculation of VaR and improves its accuracy. This method provides a more comprehensive estimate of the characteristics of spikes that occur in extreme fluctuations of the inventory. In contrast, the traditional VaR calculation based on the time series normal distribution hypothesis and the linear correlation hypothesis of the portfolio will underestimate these problems and thus the accurate judgments cannot be made. The more extreme the reality, the more serious the error will occur, which will lead to unnecessary losses for financing companies.

## References


1. Jorion, P. (2007). *Value at Risk* (3rd ed., pp. 139–230). New York: Mc Graw-Hill.
2. Bollerslev, T. (1986). Generalized autoregressive conditional heteroskedasticity. *Journal of Econometrics, 31,* 307–327.
3. Ricardo, A. (2006). *The estimation of market VaR using Garch models and a heavy tail distributions*. Working Paper Series.
4. Chen, C., & Panjer, H. (2003). Unifying discrete structural models and reduced-form models in credit risk using jump-diffusion process. *Insurance: Mathematics and Economics, 33,* 357–380.
5. Gu, X. J. (2007). Financial risk management: the variance and development of theory and technology. *Economic Survey, 1,* 140–143.
6. Guo, G. (2010). Analysis on the application and challenge of VAR method of financial risk management. *Commercial Accounting, 15,* 28–29.
7. Rosenberg, J. V., & Schuermann, T. (2006). A general approach to integrated risk management with skewed, fat tailed risks. *Journal of Financial Economics, 79*(3), 569–614.
8. Wei, Y. F., & Zheng, S. L. (2007). On the application of financial risk management in supply chain management. *Market Weekly (Theoretical Research), 6,* 104–105.
9. Nelson, D. B., Daniel, B., & Cao, C. (1992). Inequality constraints in the univariate GARCH model. *Journal of Business & Economic Statistics, 10,* 229–235.
10. Österholm, P. (2018). The relation between treasury yields and corporate bond yield spreads in Australia: evidence from VARs. *Finance Research Letters, 24,* 186–192.
11. Ayturk, Y. (2017). The effects of government borrowing on corporate financing: evidence from Europe. *Finance Research Letters, 20,* 96–103.

12. Nelson, D. B. (1991). Conditional heteroskedasticity in asset returns: a new approach. *Econometrica, 59*(2), 347–370.
13. Glosten, L. R., Jagannathan, R., & Runkle, D. E. (1996). On the relation between the expected value and the volatility of the nominal excess return on stocks. *Journal of Finace, 48*(5), 1779–1801.

# Modeling and Analysis on the Capacity of Centralized Directional Multi-hop Ad Hoc Networks

**Ying Liu and Hao Xing**

**Abstract** Directional antenna technique enables multi-hop ad hoc networks to achieve a better anti-interference ability with higher signal intensity. In the mean time, it is safer and more efficient to communicate by directional antennas. However, there are few performance models and analyses for centralized directional multi-hop networks in existing research, and there is no theoretical basis for the design of protocol parameters. This paper targets the theoretical framework of multi-hop ad hoc networks with directional antennas, and analyses the effect of protocol parameters on the average network interference based on the hard core point process (HCPP). The throughput capacity of the network is further derived. The influence of the density of nodes, the number of antenna beams and the maintained neighbor hops on system performance is analyzed by simulation. The optimal design of protocol parameters is given.

**Keywords** Multi-hop ad hoc · Centralized · Directional antenna · Interference · Capacity

## 1 Introduction

Wireless multi-hop ad hoc network is a reliable wireless network architecture which has the advantage of self-organization, self-recovery and self-configuration. So it has been widely used in military, vehicular and emergency communications. In recent years, centralized scheduling mechanisms are more used in wireless multi-hop networks. Because this kind of network has a base station which can manage and dispatch the whole network resources [1, 2].

In traditional centralized multi-hop ad hoc networks, the deterioration of channel quality and the interference between links will affect the sending and receiving of

Y. Liu · H. Xing (✉)
School of Electronic and Information Engineering, Beijing Jiaotong University, Beijing, China
e-mail: 16120149@bjtu.edu.cn

Y. Liu
e-mail: liuying@bjtu.edu.cn



J. Zhang et al. (eds.), *LISS2019*,
https://doi.org/10.1007/978-981-15-5682-1_30

messages, which seriously degrades the performance of the network. Directional antennas are proposed for centralized ad hoc networks in order to meet the demands of reducing interference and improving throughput capacity [3–5]. Compared with omnidirectional antennas, they have higher sending and receiving gains in some specified directions. In other directions, it can be reduced to almost zero under ideal conditions. The directional antennas have stronger signal intensity and anti-interference ability than omnidirectional antennas, which makes the communicate process safer. At the same time, directional antennas can also realize spatial multiplexing, thus increasing system capacity in limited space [5]. However, little theoretical analysis supporting for the protocol design of centralized directional ad hoc networks has been done so far. Performance modeling and analyzing of the centralized directional ad hoc networks for rational protocol parameters design has become a problem to be studied.

Capacity is one of the important indicators to measure the performance of wireless networks. On the research of wireless multi-hop ad hoc network capacity, Gupta [6] proposed the concepts of throughput capacity and transmission capacity and pointed out that when capacity analysis is promoted to the network level, people should focus on exploring the interference caused by spatial multiplexing. The theory laid the foundation for later research. Motivated by this, in this paper, a novel model of the throughput capacity is proposed based on centralized directional scheduling scheme, which considers the average network interference.

The structure of the rest of this paper is organized as follows. Section 2 reviews the related work. Section 3 describes the centralized directional multi-hop network protocol. Section 4 is devoted to the modeling and analysis of average interference and net throughput with the simulation results. Section 5 concludes the paper.

## 2 Related Work

At present, the throughput model of directional scheduling mechanism is relatively simple. Yi [7] studies the capacity of Ad Hoc wireless networks when the beam width of directional antenna is approximately reduced to 0, which means wireless network can be equivalent to wired network. The article considered an over-ideal situation. Grossglauser [8] proposed the capacity of ad hoc networks with mobile nodes under the limits imposed by the interference model. This paper focuses on the mobility of nodes, and obtains an asymptotic throughput capacity under ideal conditions. Reference [9] studies a hypothetical interference cancellation technology, which proves that the improvement of transmission capacity of wireless ad hoc network is proportional to the size of interference cancellation area. This interference cancellation mechanism has certain reference significance for the design of protocol parameters in centralized multi-hop ad hoc networks, but the influence of antenna beam numbers should also be considered.

Stochastic geometry theory, especially the point process theory, provides an effective tool to model the randomness of spatial distribution of nodes in a wireless communication networks. The hard core point process (HCPP), introduced by Matrn in [10], is one of the most popular model. Different from the traditional Poisson point process (PPP), HCPP conditions on having a minimum distance to separate the points of the process [11].

In this paper, we adopt the HCPP model of stochastic point process to analyze the interference. A throughput model based on average interference is further derived. We analyze the relationship among network node density, maintained neighbor range and the numbers of antenna beam through simulation results, and present a best design of protocol parameters in a certain network node density.

## 3 Overview of Centralized Directional Multi-hop Networking Protocol

We consider a multi-hop network in which nodes are presented in tree structure. As shown in Fig. 1, centralized directional multi-hop network is composed of a base station and user stations. Base station is the original node and responsible for establishing the network. It helps for the resource scheduling of all nodes in the network. At the same time, as a root node, it forms a scheduling tree with user stations.

In Fig. 2, under TDMA-based centralized scheduling, each MAC frame is divided into control slots and data slots. Control slots are used to transmit control signals including network entry (NENT) message, network configuration (NCFG) message, position probe (PROB) message, centralized schedule configuration (CSCF) message, centralized schedule request (CSCR) message and centralized schedule grant (CSCG) message, while data slots are used to transmit users' traffic. Whether control signals or data messages, their transmission time is allocated in slots. Each node in the network needs to maintain its neighbor nodes within several hops from itself. When the node sends messages, the neighbor nodes within its maintenance range can not send messages at the same time.

Centralized directional multi-hop protocol specifies that each node is equipped with an omnidirectional antenna and a directional antenna. The omnidirectional antenna is used for general control signals, and the directional antenna is used to locate nodes and transmit data. The first node establishes the network, and then it periodically broadcasts NCFG messages. New nodes synchronize with the network by scanning NCFG messages and request to join the network by sending NENT messages. To communicate with a directional antenna, the location of the target node must be determined first. PROB messages are sent by directional antenna in turn. The target node judges the beam number by receiving the PROB message, and informs the sending node in the next omnidirectional control message. Packet transmission in the network is based on the process of "configuration-request-confirmation". The base station sends CSCF messages, and other nodes forward CSCF with scheduling tree information to the whole network. Then the nodes in the scheduling tree send

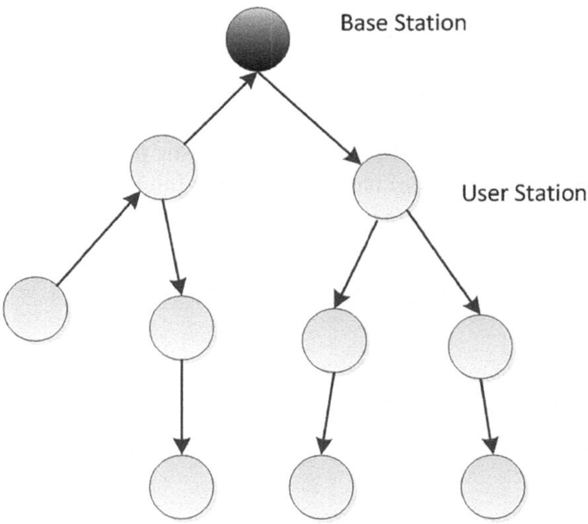


**Fig. 1** Centralized scheduling tree architecture

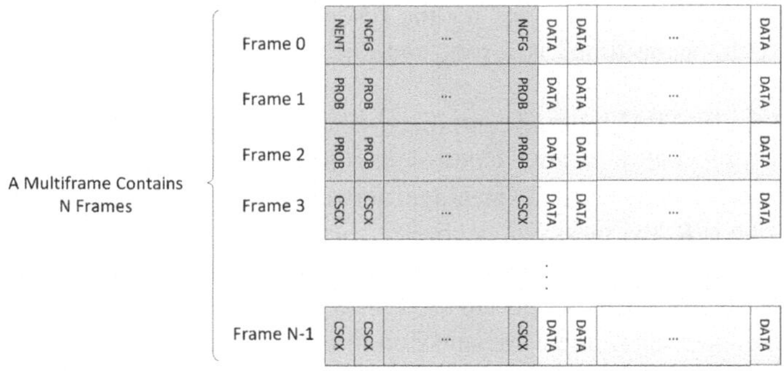


**Fig. 2** Centralized directional MAC layer frame structure

CSCR messages to the parent nodes, aggregate the CSCR with the request information of each node until the base station receives. Finally, the base station sends CSCG messages, and other nodes forward the CSCG with the result of resource allocation in turn until the whole network receives it.

# 4 Performance Modeling and Analysis

Motivated by the aforementioned protocol, we utilize Hard Core Point Process (HCPP) theory to quantify the link interference intensity in different maintained neighbor range and directional beam numbers. In addition, we propose an effective throughput model based on interference.

## *4.1 Density of Interference Nodes*

Paper [12] establishes a distributed omnidirectional multi-hop ad hoc network interference model based on stochastic point process theory. In this paper, the influence of directional antenna beam on interference is considered, and the centralized multi-hop ad hoc network protocol is introduced to analyze the capacity of centralized directional multi-hop network.

On the basis of PPP process, HCPP process can be obtained by diluting the original node density according to certain rules. The main steps of the algorithm include:

1) Firstly, a Poisson point process $\Phi_P$ with a density of $\lambda_P$ is generated.
2) Attach a random tag $m_x \in U[0, 1]$ to each point $x \in \Phi_p$.
3) For each point in circle $B(x, \delta)$, if the tag of point $x \in \Phi_p$ is the smallest, it is retained and the rest of the nodes are deleted.

$U[0, 1]$ indicates a uniform distribution with the range of 0 to 1. $\delta$ is the smallest distance between two sending nodes namely the dilution radius. In Fig. 3, the black nodes are retained because they have the minimum values in the range they maintain. While those white nodes whose values are not the minimum are deleted by HCPP.

The model formed by all the retained points is the HCPP process. The density of remained points can be regarded as the density of the interference nodes after interference cancellation, which can be concluded as followed [10]:

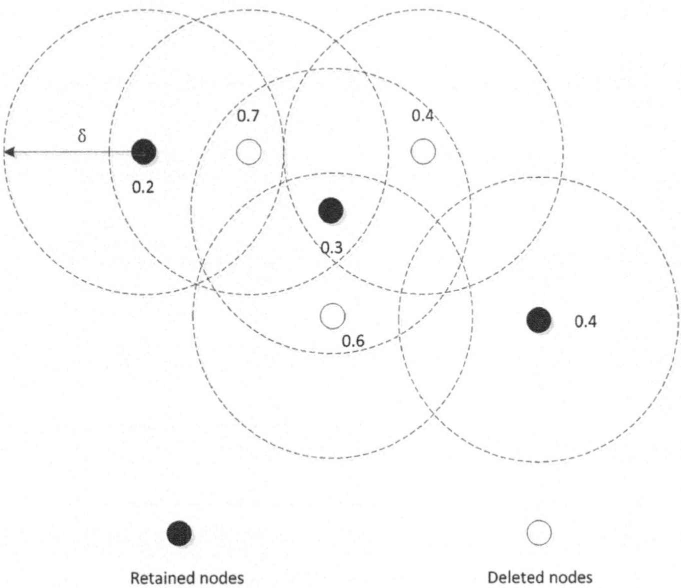


**Fig. 3** Schematic diagram of HCPP

$$\lambda_{HS} = \frac{1 - \exp(-\lambda_p \pi \delta^2)}{\pi \delta^2} \tag{1}$$

$\delta$ directly determines the "dilution" of HPPP and has a great influence on the distribution of interference nodes. The larger the value of $\delta$, the smaller the probability of each node being retained, the lower the density of interference nodes. However, if the value of $\delta$ is too large, the number of nodes that can send data at the same time in the network will be greatly reduced, which will lead to the decline of spatial reuse and reduce the network capacity. Therefore, it is of great significance to find the most suitable value of $\delta$.

In centralized scheduling mechanism, the node maintains the neighbor nodes within the H-hop range, and the dilution radius can be expressed as:

$$\delta = Hr \tag{2}$$

$r$ is the transmission range of the node.

The transmitting direction of nodes equipped with a directional antenna has an antenna beam width, and only the nodes covered by the antenna beam will be interfered. We assume that the antenna direction of nodes in space obeys random distribution. Some nodes are selected from HPPP with density of $\lambda$ by probability $k$ and constitute a new HPPP with the density of $\lambda_k$. If the number of beam directions of each node is $N$, the node density in one direction becomes 1/N of the original HPPP process. So the density of interference nodes in the directional ad hoc network is:

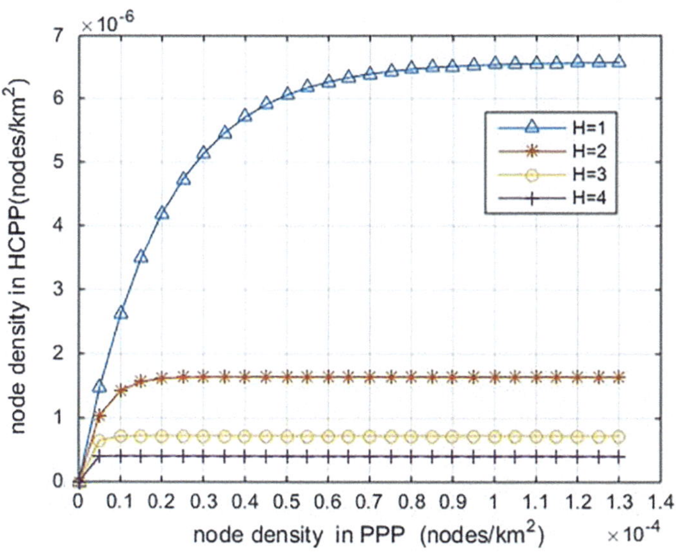


**Fig. 4** The interference nodes density versus network node density

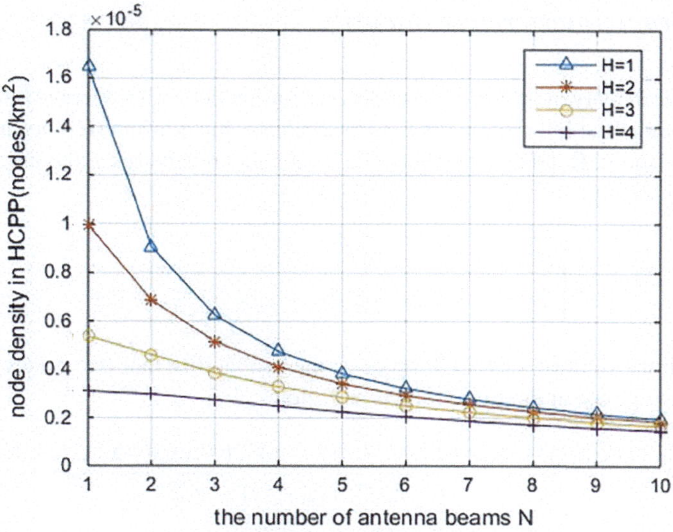


**Fig. 5** The interference nodes density versus antenna beams

$$\lambda_{HM} = \frac{1 - \exp(\frac{-\lambda_p \pi H^2 r^2}{N})}{\pi H^2 r^2} \tag{3}$$

The number of beams, the number of neighbor maintenance hops and the initial node density all affect the distribution of interference nodes after dilution. In Fig. 4, when the node density exceeds a certain threshold, the interference node density tends to be stable. Because most of the nodes are in the range node maintenance, so they are diluted out.

Based on the centralized directional multi-hop network protocol, we exclude the initial node density and analyze the influence of the number of antenna beams and the number of neighbor maintenance hops on the interference node density in Fig. 5.

With the increase of neighbor maintenance hop $H$, the interference avoidance mechanism in the protocol plays a more and more important role in the interference node density. As the number of antenna beams increases, the density of interference nodes decreases gradually, because narrow beams reduce the effect of omnidirectional interference. However, with the increase of maintenance hops, the influence of antenna beams becomes weaker, because the larger the maintenance area is, the fewer interference nodes remain in the network.

## 4.2 Average Interference Intensity

Based on the density of interference nodes, we further derive the average interference intensity. Under the interfere avoidance mechanism, the non-interference area is not circular any more. If the distance between sender and receiver is $s$, the union area of the two circles is [13]:

$$V_\delta(s) = \begin{cases} 2\pi\delta^2, & s \geq 2\delta \\ 2\pi\delta^2 - 2\delta^2 \arccos(\frac{s}{2\delta}) + s\sqrt{\delta^2 - \frac{s^2}{4}}, & \delta \leq s < 2\delta \end{cases} \tag{4}$$

Through the dilution of HCCP process, the probability that two nodes with distance of can coexist at the same time is as follows:

$$k(s) = \frac{V_\delta(s)(1 - \exp(-\lambda_p\pi\delta^2/N)) - \pi\delta^2(1 - \exp(-\lambda_p V_\delta(s)/N))}{\frac{1}{2}\lambda_p\pi\delta^2 V_\delta(s)[V_\delta(s) - \pi\delta^2]} \tag{5}$$

In order to show the relationship between the two nodes, paper [14] mentions the spatial correlation function between any two nodes is:

$$g(s) = \frac{{\lambda_p}^2 k(s)}{{\lambda_{HM}}^2} \tag{6}$$

$\lambda_{HM}$ is the density of interference nodes after HCPP process. If the distance between sender and receiver $s \geq 2\delta$, they are out of each other's clearance area. The two nodes are independent in space, $g(s) = 1$. If $\delta \leq s \leq 2\delta$, there is a positive correlation between nodes, $g(s) > 1$.

Set a wireless network $\Psi$, the sending node is the origin $o \in \Psi$, and the receiving node is located in $z_o = \{dcos\theta_o, dsin\theta_o\} \notin \Psi$. $d$ denotes the distance between the receiving node and the sending node, and $\theta_o$ denotes the direction of the receiving node. The expected interference intensity of the receiving node is expressed as follows:

$$E(I) = E_o[\sum_{x_i \in \Psi} h_{x_i z_o} l(x_i - z_o)] \tag{7}$$

$h_{x_i z_o}$ is the average power attenuation coefficient per unit distance, $l(\cdot)$ is the path loss model, $l(x_i) = \|x_i\|^{-\alpha}$, the path loss index $\alpha > 2$. The interference value can be expressed by the correlation function $g(s)$:

$$E_o(I) = 2\pi\lambda_{HM} \int_0^\infty l(s)g(s)sds \tag{8}$$

Nodes are divided into two parts according to spatial correlation:

$$
\begin{aligned}
E_o(I) &= E_o(I_{\delta \le s < 2\delta}) + E_o(I_{s \ge 2\delta}) \\
&= 2\pi\lambda_{HM} \int_{\delta}^{2\delta} l(s)g(s)sds + 2\pi\lambda_{HM} \int_{2\delta}^{\infty} l(s)g(s)sds
\end{aligned} \tag{9}
$$

In interval $[2\delta, \infty)$, $g(s) = 1$, the above formula can be further calculated:

$$
E_o(I) = \begin{cases} 2\pi\lambda_{HM} \int_{\delta}^{2\delta} l(s)g(s)sds + \frac{2\pi\lambda_{HM}}{\alpha-2}(2\delta)^{2-\alpha} & , \alpha > 2 \\ \infty & , \alpha = 2 \end{cases} \tag{10}
$$

It is very complicated to calculate the interference. We introduce the approximate upper bound of the correlation function to simplify the calculation.

$$
\begin{aligned}
g(s) &< \tfrac{g(2\delta)-g(\delta)}{\delta} s + 2g(\delta) - g(2\delta) \\
&= \tfrac{1-g(\delta)}{\delta} s + 2g(\delta) - 1
\end{aligned} \tag{11}
$$

Average interference intensity can be obtained from (3) (4) (10) (11):

$$
\begin{aligned}
E_o(I) = 2\pi\lambda_{HM} &\int_{\delta}^{2\delta} l(s)(\tfrac{1-g(\delta)}{\delta} s + 2g(\delta) - 1)sds \\
+ 2\pi\lambda_{HM} &\int_{2\delta}^{\infty} l(s)sds
\end{aligned} \tag{12}
$$

$E_o(I)$ depends on node density $\lambda_{HM}$ and dilution radius $\delta$. In the protocol, these two parameters are closely related to node maintenance neighbor hop $H$, antenna beam number $N$ and node density $\lambda_P$.

We set the transmission range of the node $r = 100$ and the path loss index $\alpha = 4$. The average interference of the node is obtained by numerical simulation as shown in Fig. 6. The influence of neighbor maintenance hops and antenna beam numbers on the average interference is the same as that of the interference node density in Fig. 5, because the average interference is directly related to the interference node density.

## 4.3 Net Throughput Capacity

The average throughput capacity of the network refers to the average maximum transmission rate of each node in the network. The nodes in the wireless multi-hop network are equal and independent of each other. Therefore, we use the HCPP process to describe the average interference intensity of each node in the interference cancellation area.

Although Shannon information theory is based on the point-to-point communication system where noise is the main interference in the network. For wireless multi-hop networks, each independent node is regarded as one end of the point-to-point system, while other interference nodes in the network are regarded as the other end of the system. The average interference of the interference nodes is regarded as noise. We can still analyze the throughput capacity with the help of the classical Shannon information theory.

On the premise that the same transmission power is used in the whole network node, the signal-to-interference-plus-noise ratio(SINR) can be simplified as follows:

$$SINR = \frac{S}{I+N} \approx \frac{S}{I} = \frac{d^{-\alpha}}{E_o(I)} \tag{13}$$

According to Shannon's theorem, the maximum information rate which can be transmitted in channel is determined by the following formula:

$$R = B\log_2(1 + \frac{d^{-\alpha}}{E_o(I)}) \tag{14}$$

We set the bandwidth to $B = 10$, and get the relationship between the node throughput capacity and the number of antenna beams under different neighbor maintenance hops as shown in Fig. 7. It can be seen that the increase of maintenance

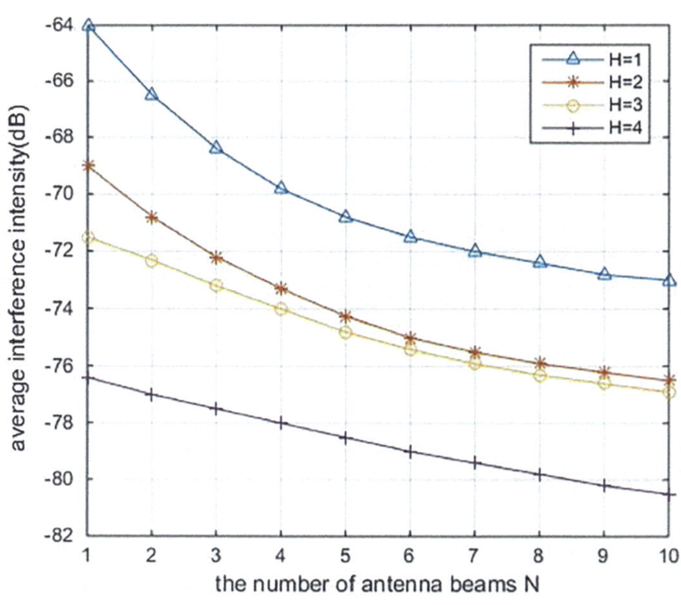


**Fig. 6** The average interference versus antenna beams

**Table 1** The structure of control message

| Struture | FRH | CMD | INT | GMH | CMH | MSG | CRC | FRI |
|---|---|---|---|---|---|---|---|---|
| Length/byte | 1 | 1 | 14 | 4 | 1 | Data | 4 | 1 |

hops and antenna beams leads to the decrease of interference in the network, which leads to the optimization of capacity.

Messages in the network include control messages and data messages. Net throughput of nodes refers to the maximum sending rate of data messages, that is, the average throughput minus the overhead of control messages. According to the centralized directional wireless multi-hop ad hoc network protocol, control messages include NENT, NCFG, PROB, CSCF, CSCR and CSCG. The control message structure is shown in the following Table 1.

Each part represents Frame Header, Control Header, Interface Message, Common MAC Header, Control Sub-header, Message Body, Cyclic Redundancy Check and Frame Tail respectively. The length of message body needs to be designed according to specific network requirements. The length of various control messages is related to the state of the network and protocol parameters.

The protocol specifies that the length of NENT is fixed at 29 Byte.

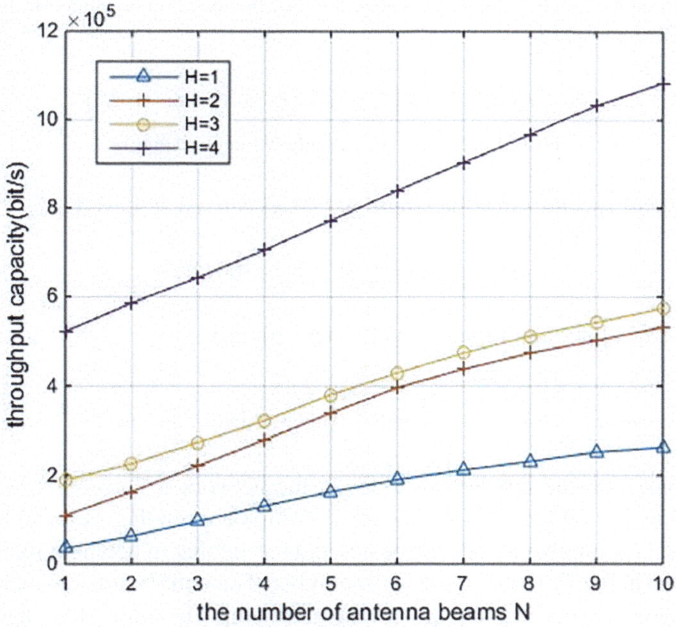


**Fig. 7** The average throughput capacity versus antenna beams

$$L_{nent} = (3 + 26) \cdot 8 = 232\,\text{bits} \tag{15}$$

NCFG messages contain neighbor node information within the maintenance range. The number of neighbor nodes can be expressed as node density $(\pi \lambda_p H^2 r^2)$.

$$\begin{aligned} L_{ncfg} &= \left[17 + \left(\pi \lambda_p H^2 r^2 - 1\right) + 26\right] \cdot 8 \\ &= 8\pi \lambda_p H^2 r^2 + 336\,\text{bits} \end{aligned} \tag{16}$$

The total length of PROB messages is affected by both the number of neighbor nodes and the antenna beams $N$.

$$\begin{aligned} L_{prob} &= \left[6 + N \cdot \left(\pi \lambda_p H^2 r^2 - 1\right) + 26\right] \cdot 8 \\ &= 8N \cdot \left(\pi \lambda_p H^2 r^2 - 1\right) + 256\,\text{bits} \end{aligned} \tag{17}$$

The number of nodes in the network is expressed by $uNodeNum$. The length of CSCF messages is related to the number of nodes working in the network.

$$\begin{aligned} L_{\mathrm{csc}\,f} &= (4 \cdot uNodeNum + 26) \cdot 8 \\ &= 32uNodeNum + 208\,\text{bits} \end{aligned} \tag{18}$$

The length of the CSCR message is related to the number of services in the network. Assuming that there is only one service request in the network, the length of the CSCR message is:

$$\begin{aligned} L_{\mathrm{csc}\,r} &= \left\{uNodeNum\left[5 \times (\lambda_p \pi r^2 - 1) + 2\right] + 26\right\} \cdot 8 \\ &= 8 \cdot (5\lambda_p \pi r^2 - 3) \cdot uNodeNum + 208\,\text{bits} \end{aligned} \tag{19}$$

The protocol specifies that the length of CSCG is fixed at 45 Byte.

$$L_{\mathrm{csc}\,g} = (19 + 26) \cdot 8 = 360\,\text{bits} \tag{20}$$

Total overhead per unit time can be further deduced.

$$R_{overhead} = \frac{1}{T}\left(L_{nent} + L_{ncfg} + L_{prob} + L_{\mathrm{csc}\,f} + L_{\mathrm{csc}\,r} + L_{\mathrm{csc}\,g}\right) \tag{21}$$

We assume that the number of nodes in the network $uNodeNum$ is 100 and the scheduling period length is $T$. The relationship between the overhead of control messages and the number of antenna beams and the number of neighbor maintenance hops is shown in Fig. 8. The increase of the number of antenna beams and maintenance hops will lead to the corresponding increase of the length of control messages, which will give rise to the increase of control overhead.

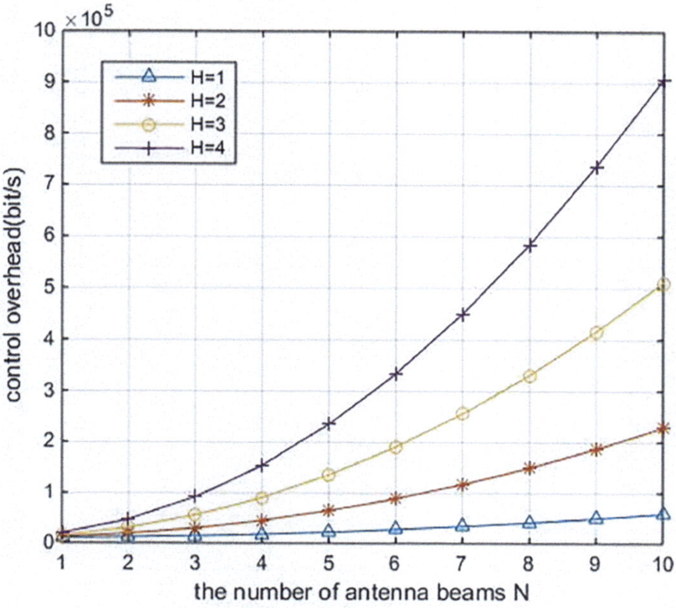


**Fig. 8** The control overhead versus antenna beams

The less signaling overhead, the better for wireless networks. Therefore, the determination of the number of maintenance hops and antenna beams in the protocol should be considered from the perspective of the net throughput capacity.

Net throughput of nodes refers to the average throughput minus the overhead of control messages.

$$S = \frac{R}{\pi \delta^2 \lambda_p - 1} - R_{overhead} \tag{22}$$

The relationship between the net throughput capacity and the number of antenna beams under different maintenance hops is shown in Fig. 9. With the increase of antenna beams, the net throughput first reaches its peak value and then decreases. When the number of antenna beams is 3 and the neighbors are maintained for 4 hops, the net throughput reaches the maximum. We can reasonably design protocol parameters based on the simulation results.

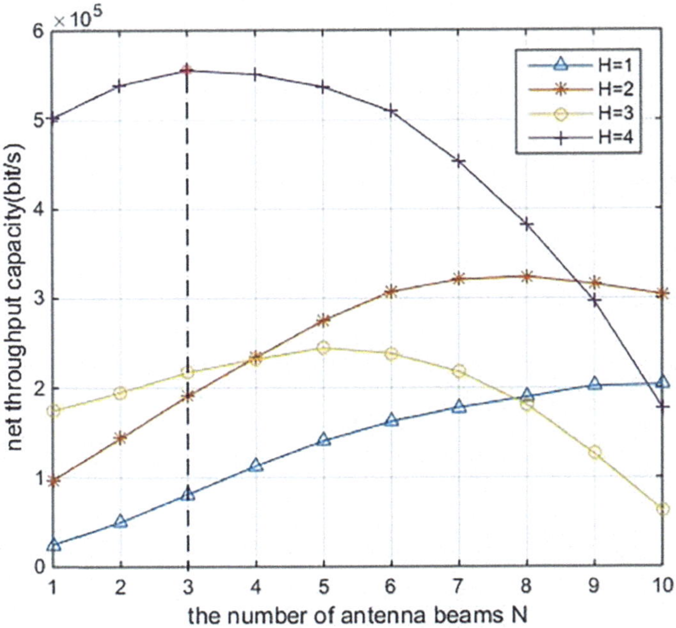


**Fig. 9** The net throughput capacity versus antenna beams

## 5 Conclusion

In this paper, we address the interference in centralized directional multi-hop ad hoc networks. Starting from the HCPP model, we derive the analytical expression for the interference as a function of the node density, maintained neighbor hops and antenna beams. The analysis results show that expanding the maintained area of neighborhood and adding antenna beams are the inevitable choices to resist interference. On this basis, we put forward the net throughput of nodes considering the impact of control overhead on the system. The simulation results show that nodes maintain 4 hops of neighbors and use 3 antenna beams in the design of the centralized directional wireless multi-hop ad hoc network protocol, which not only does not occupy too much spectrum resources, but also maximizes the net throughput. This paper provides theoretical guidance for the design of centralized directional wireless multi-hop ad hoc network protocol.

## References

1. Oikonomou, K., Vaios, A., Simoens, S., Pellati, P., & Stavrakakis, I. (2003). A centralized ad-hoc network architecture (CANA) based on enhanced HiperLAN/2. In *14th IEEE proceedings on personal, indoor and mobile radio communications* (Vol. 2, pp. 1336–1340).
2. Cheng, S. M., Lin, P., Huang, D. W., & Yang, S. R. (2006). A study on distributed/centralized scheduling for wireless mesh network. In *Proceedings of the 2006 international conference on Wireless communications and mobile computing* (pp. 599–604).
3. Choudhury, R. R., Yang, X., Ramanathan, R., & Vaidya, N. H., (2002). Using directional antennas for medium access control in ad hoc networks. In *Proceedings of ACM international conference on mobile computing and networking (MobiCom)* (pp. 59–70).
4. Ramanathan, R., Redi, J., Santivanez, C., Wiggins, D., & Polit, S. (2005). Ad hoc networking with directional antennas: a complete system solution. *IEEE Journal on Selected Areas in Communications*, *23*(3), 496–506.
5. Ko, Y. B., Shankarkumar, V., & Vaidya, N. H., (2000). Medium access control protocols using directional antennas in ad hoc networks. In *Proceedings IEEE INFOCOM 2000. Conference on computer communications. Nineteenth annual joint conference of the IEEE computer and communications societies* (Vol. 1, pp. 13–21).
6. Gupta, P., & Kumar, P. R. (2000). The capacity of wireless networks. *IEEE Transactions on Information Theory*, *46*(2), 388–404.
7. Yi, S., Pei, Y., Kalyanaraman, S., & Azimi-Sadjadi, B. (2007). How is the capacity of ad hoc networks improved with directional antennas? *Wireless Networks*, *13*(5), 635–648.
8. Grossglauser, M., & Tse, D. (2002). Mobility increases the capacity of ad hoc wireless networks. *IEEE/ACM Transactions Networking*, *10*(4), 447–486.
9. Weber, S. P., Andrews, J. G., Yang, X., & De Veciana, G. (2007). Transmission capacity of wireless ad hoc networks with successive interference cancellation. *IEEE Transactions on Information Theory*, *53*(8), 2799–2814.
10. Matrn, B. (1986). *Spatial variation.*, Lecture Notes in Statistics Heidelberg: Springer.
11. Chiu, S. N., Stoyan, D., Kendall, W. S., & Mecke, J. (2013). Stochastic geometry and its applications (3rd ed.). *Advances in Mathematics 40*(3), 257–269.
12. Gao, H. (2017). *Performance modeling and design of MAC protocol in distributed self-organizing networks based on link quality*. Master thesis, Beijing Jiaotong University
13. Chiu, S. N., Stoyan, D., Kendall, W. S., & Mecke, J. (2013). *Stochastic geometry and its applications*. New York: Wiley.
14. Xu, K., Gerla, M., & Bae, S. (2002). How effective is the IEEE 802.11 RTS/CTS handshake in ad hoc networks? In *IEEE Global Telecommunications Conference* (Vol. 2, pp. 72–76).

# Vehicle Scheduling Problem of Logistics Companies Under Genetic Tabu Hybrid Algorithm

**Wei Xu, Wenli Liang, and Qinqin Yu**

**Abstract** Vehicle Scheduling Problem (VSP) plays a very important role in the logistics distribution process. This paper proposes a two-stage algorithm. In the first stage, the fuzzy clustering algorithm is used to divide the customer group into multiple small-scale groups according to different clustering index factors. In the second stage, a new genetic tabu hybrid algorithm was used to solve the vehicle scheduling scheme in each customer class. In addition, a new heuristic crossover operator and inversion mutation operator are used in the genetic tabu hybrid algorithm, which can not only preserve the excellent genes of the parent chromosome, but also generate new individuals quickly. Finally, H Logistics Company is used as an analysis case to solve the optimal distribution plan for 120 customer points. Comparing the results obtained by the algorithm described in this paper with the traditional genetic algorithm, it is concluded that the genetic tabu hybrid algorithm is better than the traditional genetic algorithm.

**Keywords** Vehicle scheduling problem · Genetic tabu hybrid algorithm · Fuzzy clustering algorithm

---

This work was supported by Shandong social science planning research project (18CCXJ25) and Qingdao social science planning project (QDSKL1801134).

---

W. Xu · W. Liang · Q. Yu (✉)
College of Transportation, Shandong University of Science and Technology, Qingdao, China
e-mail: 571209360@qq.com

W. Xu
e-mail: xuwei972@163.com

W. Liang
e-mail: lwl0301@126.com



J. Zhang et al. (eds.), *LISS2019*,
https://doi.org/10.1007/978-981-15-5682-1_31

## 1 Introduction

At present, the logistics industry, as an emerging industry, is developing rapidly. The logistics and transportation costs of enterprises take up a large proportion of the total costs. Reducing the transportation costs will promote the growth of the benefits of enterprises. Nowadays, logistics integrates energy saving, high efficiency and economy as the overall goal of development. In order to achieve the optimal distribution, it is necessary to optimize between the distribution vehicles and customers. At present, the country proposes sustainable economic development, and logistics enterprises should also respond to the proposed environmental protection laws and regulations, and make changes to the development model. Logistics enterprises should achieve energy saving and emission reduction, and should design the optimal transportation route in the distribution area.

Vehicular scheduling problem (VSP) was first proposed by Dantzig and Ramser in 1959, providing thoughts and methods for the future improvement of algorithm in [1]. The current scholars' research results are as follows: Desrochers and Verhoog research on VSP based on various vehicle types. This model uses a variety of vehicles with different loadings for transportation, so as to reduce vehicle waste and to meet the maximization of benefits in [2]. Moghaddam et al. established a mathematical model to meet random needs, which can satisfy diversified customer demands and increase customer satisfaction in [3]. Yang established a vehicle scheduling model with the objective function of the least total postal cost and the least penalty cost, taking into account the regional division, the route of postal vehicles and the number of postal vehicles in [4]. Ren et al. emphasizes the factors in the cost and the time, and puts forward a vehicle scheduling model, which meets the emergency time constraint and costs minimum in [5]. Balázs and Miklós present heuristic methods for the vehicle scheduling problem that solve it by reducing the problem size using different variable fixing approaches [6]. Li in [7] proposed a multi-objective function, in which the three-layer targets respectively are the lowest transportation cost, the largest vehicle load and the shortest transportation distance. Because the path planning problem is NP problem and the calculation amount is huge, most scholars inclined to use heuristic algorithm to solve this kind of problem. Heuristic algorithms mainly include simulated annealing algorithm, distribution estimation algorithm, tabu algorithm, ant colony algorithm, genetic algorithm, et al. [8]. In this paper, a two-stage algorithm is proposed based on the above research status. In the first stage, the fuzzy clustering algorithm is used to divide the customer group into several small groups according to different clustering index factors. In the second stage, a new genetic tabu hybrid algorithm is applied to solve each customer class's vehicle scheduling scheme.

# 2 Soft Time Window Vehicle Scheduling Problem and Mathematical Model Establishment

Vehicle distribution is a typical optimization problem in the operation of third-party logistics, its delivery shall be made according to the customer requirements of the delivery time, shipping mode and demand. Vehicle routing problem can be divided into vehicle scheduling problem with time window and vehicle scheduling problem without time window according to whether the delivery time is required. For the vehicle scheduling problem with time windows, customers have clear regulations on the delivery time [9]. According to the requirements of time regulations, they can be divided into vehicle scheduling problem with hard time window (VSPHTW) and vehicle scheduling problem with soft time window (VSPSTW) [10, 11]. The VSPHTW is very strict with the time requirement. If the delivery is made outside the required delivery time, the goods will be refused to be received, which will cause huge economic losses to the logistics company [12]. VSPSTW also has clear provisions for the delivery time, the goods delivered outside the specified delivery time can be accepted, but will be to the delivery vehicle for some economic compensation [13]. At present, with the rise of logistics companies, the competition between logistics companies is getting stronger, in order to improve their competitiveness, they are increasingly demanding delivery times.

## 2.1 Vehicle Scheduling Problem with Soft Time Window

The VSPHTW has strict requirements on the concept of arrival time of delivery, which weakens the possibility of dealing with emergencies. While the VSPSTW is relatively more reasonable, the delivery car can deliver goods outside the window of the specified delivery time, the penalty cost is much less than the hard time window, relatively more relaxed. Figure 1 shows the penalty cost of the soft time window.

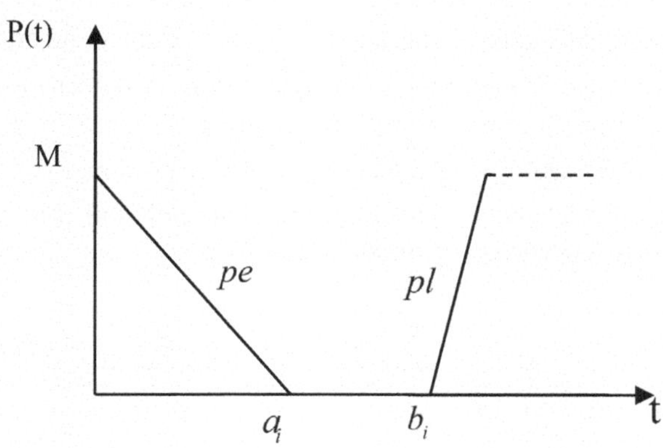


**Fig. 1** Soft time window penalty cost diagram

As shown in Fig. 1, the vehicle delivery time period $[a_i, b_i]$ of the soft time window represents the optimal delivery interval required by the customer, where $a_i$ represents the optimal earliest delivery time and $b_i$ represents the optimal latest delivery time. This period of time is the most appropriate delivery time for during this time goods delivery deserves no penalty cost. If the delivery time is earlier than the optimal earliest time $a_i$ or later than the optimal latest time $b_i$, the customer can reluctantly accept the delivery of goods, but there will be a corresponding penalty cost. If the delivery vehicle delivers the goods before $a_i$, the customer will accept the goods, and the corresponding penalty coefficient is $pe$. If the delivery vehicle delivers the goods after $b_i$, great economic loss will be generated due to lateness, so the corresponding penalty coefficient is $pl$, and the penalty coefficient is $pe < pl$.

## 2.2 Establishment of Mathematical Model

The VSPSTW is generally expressed as: logistics distribution center has $k$ distribution cars with the same load capacity. The maximum loading capacity for each car is $q$, there are totally $n$ customer points for delivery, set $V = \{1, 2, \cdots, m\}$ for collection vehicles, $N = \{0, 1, 2, \cdots, n\}$ represent the points of collection and distribution centers for all customers, 0 represents distribution center, $1, 2, \ldots,$ n represent customer points, the quantity of delivery required by the $i^{th}$ customer point is $d_i$ $(i = 1, 2, \cdots, n)$, max $d_i \leq q$, $d_0 = 0$. Each customer can only be delivered by one vehicle, each vehicle starts to distribute goods from the logistics company goods centralized point of delivery, when finish to deliver the customer's order the vehicle will return to the starting point. $l_{ij}$ represents the distance between any two customers in set $N$, $C_v$ represents the costs of the unit distribution distance of the delivery vehicle, and $C_f$ represents the starting cost of the delivery vehicle. As long as one more vehicle is used, there will be an additional starting cost. $C_{ij}$ means the delivery fees of vehicles driving from customer $i$ to customer $j$, $C_{ij} = C_v l_{ij}$. $T_i^k$ represents the starting departure time of delivery vehicle $k$, and need to start within the time window $[a_i, b_i]$. If the delivery vehicle arrives at customer $i$ before $a_i$, it needs to wait at $i$, if it is later than $b_i$, the service will be delayed, and both early arrival and late arrival will result in a certain amount of fine. $T_{ij}^k$ represents the driving time of vehicle $k$ from customer $i$ to customer $j$. $s_i^k$ represents the service time of vehicle $k$ for customer $i$. Set $s_0^k = 0$, variable $x_{ij}^k = 0$ or 1. If the vehicle arrive at customer $j$ after service customer $i$, $x_{ij}^k = 1$, otherwise $x_{ij}^k = 0$. Then the vehicle route scheduling problem with soft time window can be expressed by mixed integer programming as follows:

$$Z = \sum_{i=1}^{n} \sum_{k=1}^{m} C_f x_{0i}^k + \sum_{i=0}^{n} \sum_{j=0}^{n} \sum_{k=1}^{m} C_{ij} x_{ij}^k + \sum_{i=0}^{n} \sum_{k=1}^{m} PT_i^k$$

s. t.

$$\sum_{i=1}^{n}\sum_{k=1}^{m} x_{ij}^{k} = 1, \forall j \in C, \forall k \in V \tag{1}$$

$$\sum_{i=1}^{n} d_i \sum_{j=0}^{n} x_{ij}^{k} \leq q, \forall k \in V \tag{2}$$

$$\sum_{j=1}^{n} x_{0j}^{k} = 1, \forall k \in V \tag{3}$$

$$\sum_{i=0}^{n} x_{ih}^{k} - \sum_{j=0}^{n} x_{hj}^{k} = 0, \forall k \in V, \forall h \in C \tag{4}$$

$$\sum_{i=1}^{n} x_{i0}^{k} = 1, \forall k \in V \tag{5}$$

$$T_j^k = T_i^k + T_{ij}^k + s_i^k, \forall ij \in N, \forall k \in V \tag{6}$$

$$PT_i^k = \max\{pe(a_i - T_i^k), 0, pl(T_i^k - b_i)\}, \forall k \in V, \forall i, \in N \tag{7}$$

$$C_{ij} = C_v l_{ij}, \forall k \in V, \forall k \in V \tag{8}$$

$$x_{ij}^k \in \{0, 1\}, \forall k \in V, \forall k \in V \tag{9}$$

The objective function is divided into two parts: minimum shipping cost and minimum penalty cost. The former including the fixed starting cost of delivery vehicles and expenses incurred in transit. The fixed cost per vehicle is the same, so more delivery vehicles are used, the total fixed cost is higher. The expenses incurred in the course of transportation are related to the transportation distance and the volume of vehicles; The later including the soft time window has clear requirements on the delivery time, and it is best to deliver within the service time window, but it also allows delivery vehicles to deliver earlier than the best earliest time $a_i$ and later than the best latest time $b_i$. If the delivery is carried out outside the delivery time window required by the customer, the delivery vehicle shall be subject to a fee penalty.

In the above model, (1) to (8) are constraints. Formula (1) indicates that each customer can only be accessed once. Formula (2) indicates that the delivery vehicle cannot be overloaded, and the delivery quantity of each vehicle to the customer shall not be greater than the loading capacity. Formula (3) indicates that delivery vehicles are delivered from the centralized point of logistics. Formula (4) is that the cars arriving and departing from each customer are the same. Formula (5) refers to the centralized return to the origin after each vehicle delivers the goods. Formula (6) represents the time when vehicle $k$ arrives at customer $j$. Formula (7) represents the penalized of delivery vehicle $k$'s early or late arrival during its delivering time; $pe$, $pl$ are punish coefficients, set the coefficient as positive number. In general, it is better to make delivery before the service time than after the end time. Therefore, the penalty for early arrival is less than that for late arrival, as the $pe < pl$. Formula (8) represents the costing incurred by the delivery vehicle driving on the way, which is related to the driving distance. Formula (9) is an integer constraint.

## 3 Fuzzy Clustering Analysis

Clustering analysis is to classify elements with the same attributes according to the characteristics of some elements. In this way, the elements of the same category are be grouped together, while the elements of different categories are be separated. Different clustering methods will generate different clustering schemes. The same clustering method can generate different clustering schemes based on different parameters. The calculation steps are as follows:

### *3.1 Get the Information of Delivery Customer Points*

According to the problems, the data to be classified generally include the location coordinates of customer points, the location coordinates of distribution centers and the demand of each customer.

### *3.2 Selection of Evaluation Indicators*

Through analyzing the object of study, the element indexes that can determine the similar characteristics are selected. Here, the selection of indicators is very important and plays a decisive role in the subsequent classification. Therefore, it is necessary to select reasonable classification indicators through comprehensive analysis.

### *3.3 Conduct Standardized Processing of Data*

The reason for the standardization of data is that different data have different units and cannot be unified or calculated. The data are standardized in a certain way to let them all belong to the range of [0, 1]. The general data standardization can be carried out in the following ways:

$$\hat{x}_i^p = \frac{x_i^p - \overline{x_i^p}}{s_i}, \overline{x_i^p} = \frac{1}{n}\sum_{i=1}^{n} x_i^p, s_i = \sqrt{\frac{1}{n}\sum_{i=1}^{n} (x_i^p - \overline{x_i^p})^2}, \quad (10)$$

Where $x_i^p$ represents the $p^{th}$ attribute of the $i^{th}$ data.

$$\hat{x}_i^p = \frac{x_i^p - m_p}{M_p - m_p}, i = 1, 2, 3, \cdots, 120;\ p = 1, 2, \cdots, 5 \quad (11)$$

Where $x_i^p$ represents the $p^{th}$ attribute of the $i^{th}$ data, $M_p$ and $m_p$ are respectively the maximum and minimum values of the $p^{th}$ attribute of each data $x_i$.

## 3.4 Establishment of Fuzzy Similarity Matrix

The construction of fuzzy similarity matrix is generally processed by the numeral product method, the angle cosine value method, the correlation coefficient method, the distance method, etc. [14]. If the similarity coefficient between two data is very large, it can indicate that the two data have a high degree of similarity. The similarity coefficient is solved for $r_{ij} = 1 - cd(x_i, x_j)$, and an appropriate value is set for $c$ to make $r_{ij} \in [0, 1]$. Distance can be solved in many ways: the commonly used distance solving formula is the absolute value distance $d(x_i, x_j) = \sum_{k=1}^{m} |x_{ik} - x_{jk}|$, chebyshev distance $d(x_i, x_j) = V_{K-1}^{m} |x_{ik} - x_{jk}|$ and euclidean distance $d(x_i, x_j) = \sqrt{\sum_{k=1}^{m} (x_{ik} - x_{jk})^2}$.

## 3.5 Establishment of Fuzzy Equivalent Matrix

The establishment of fuzzy equivalence matrix is based on the calculation of fuzzy similarity matrix ($R \to R^2 \to R^3 \to \ldots R^n$). After a finite number of times, $R^n \times R = R^n$ is made to obtain the fuzzy equivalent matrix. $R^n$ can also be called transitive closure matrix, it has certain characteristics, can avoid the reflexivity, symmetry and transitivity of the matrix.

## 3.6 Cluster Analysis

Set multiple different confidence levels $\lambda$ for the fuzzy equivalence matrix and calculate the $R_\lambda$ intercept. Dividing the numerical data of 1 into category, if the value of $\lambda$ becomes smaller and smaller, the classification results gradually become thicker from fine. Select the appropriate value and find the best clustering result.

When the total demand of the customer point in the designated class is greater than the loading capacity of the distribution vehicle, the customer point is classified into another most similar class.

# 4 Solution of Path Planning Problem Based on Genetic Tabu Hybrid Algorithm

Genetic algorithm (GA) is proposed to search for the optimal solution based on the evolution principle in nature. GA adopts the principle of "survival of the fittest" to randomly search for the optimal solution. This method is global and can search for the optimal solution quickly. However, GA also has certain limitations, as it is prone to "precocity" or falling into a locally optimal cycle [14]. Tabu search algorithm uses tabu table to record the searched solutions and tabu them to prevent them from double counting, so as to improve the searching ability of solutions. In the search process, it can accept bad solutions, so it has a good "climbing" ability, which can avoid the solution falling into the local optimal solution. Tabu search algorithm can make up for the shortcomings of GA, but tabu search algorithm is mostly used for local search, while GA is mostly used for global search, which can make up for each other. Therefore, this paper proposes to combine tabu algorithm with genetic algorithm, and the resulting hybrid algorithm can quickly and efficiently obtain the optimal solution.

## *4.1 Solution Flow of Genetic Tabu Hybrid Search Algorithm*

(1) Firstly, the initial values of related parameters were set, such as population size *nPop*, maximum number of genetic iterations *MaxIt*, tabu list length *TL*, and maximum number of tabu iterations *MaxTs*.
(2) Classify the original data with fuzzy clustering algorithm, and get several independent systems, one vehicle delivery one small system, so as to get the initial solution, produce $N$ initial populations and $gen = 1$.
(3) Calculate the fitness value for all chromosome individuals in the population.
(4) Judge whether the operation of population is over, if it meets the termination criterion $gen > MaxIt$, operate step (11), otherwise operate step (5).
(5) Select some individuals with the best individual retention method, and use the roulette method select the rest. This selection strategy can select excellent population individuals well.
(6) Select individuals according to the adaptive crossover probability $Pc$ to perform the improved crossover operation.

(7) Select individuals according to the adaptive mutation probability *Pm* to perform the improved mutation operation.
(8) Set empty tabu table, make $K = 1$.
(9) If $K \leq MaxTs$, execution (a), or execute (d).

   (a) Generate neighborhood solutions and determine candidate solutions;
   (b) Evaluate whether the candidate solution conforms to the flout criteria, and if so, replace the current optimal solution with the candidate solution, and then execute (c); if not, judge whether the candidate solution is taboo, and then execute (c).
   (c) Make $K = K + 1$, and returns (9).
   (d) Form new individuals from the subpaths improved by tabu search algorithm, $gen = gen + 1$.

(10) Return to (3) for the operation of the new generation $GA - TS$.
(11) End of operation, output the optimal distribution plan.

Each customer group obtained by clustering was calculated with the genetic tabu hybrid algorithm, and the operation process of the genetic tabu hybrid algorithm was expressed by the flow chart, as shown in Fig. 2.

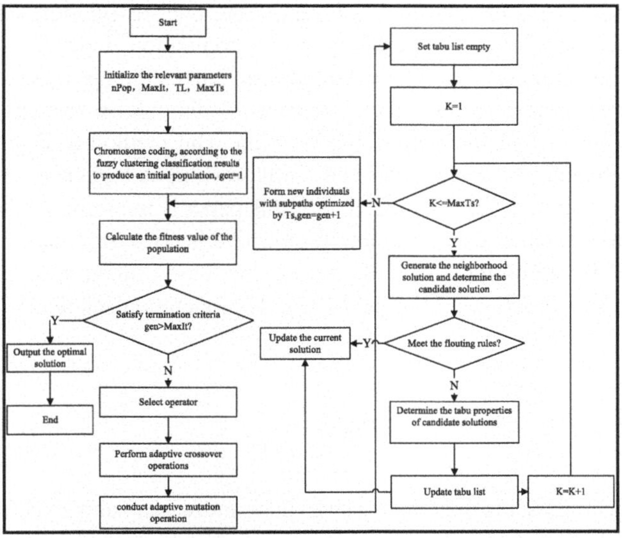


**Fig. 2** Flow chart of genetic tabu hybrid algorithm

## 4.2 Algorithm Design

(1) *Coding rules*: Genetic algorithm has many kinds of coding methods and every method has its own requirements and characteristics. Real number coding is the most simple and clear coding method. Use real number represent every element's gene value has the advantages of high precision and clear expression. In this paper, we use the real number for coding, which is directly represented by the customer code number. For example, path 0-1-2-4-5-0-8-6-9-0 is directly represented as (0124508690), number 0 represents the distribution center.

(2) *Initial population.* Genetic algorithm is population oriented, all operations are carried out in the population, so the selection of the initial population is very important for the solution and it's also a key step in the operation process. For that in the initial group design, the final result of the problem cannot be obtained in advance, and the number of final results and the dispersion of all feasible solutions cannot be determined, the initial solution is generally formed randomly. This article has used the fuzzy clustering algorithm to classify customers, one vehicle delivers one kind of customer groups, generate initial solution within each partitions randomly, remove the individuals that do not meet the constraint conditions and retain the individuals that meet the rules based on the constraint in the model. When the required number of chromosomes is obtained, the initial chromosome population is obtained.

(3) *Fitness function.* One chromosome represents one distribution scheme, olves the fitness value of each chromatin in the population under the condition of meeting constraints. When the GA algorithm solves the optimal target value, it relies on the value of the fitness function to distinguish the advantages and disadvantages of each chromosome solution. The method to evaluate the distribution scheme is: if the individual's fitness value is bigger, it means that the solution is better. Since the objective of this paper is to minimize the cost of the distribution scheme, so the minimum objective is converted into the maximum objective. Based on the above discussion, the fitness function is designed $F(x)$ and objective function is designed $Z(x)$. The operation relation is as follows:

$$F(x) = \frac{1}{Z(x)} \tag{12}$$

Where $F(x)$ represents the fitness value of the individual, and $Z(x)$ represents the target value of the individual.

(4) *Select operator.* Select operator as the evaluation method of chromosomes in the objective function, represents the evolution phenomenon of "fitness survival" of GA algorithm iteration, represents the characteristics of chromosomes in the population, and can determine the rate of GA algorithm to obtain the optimal solution and the final result. In order to achieve the optimal

preservation method, this paper combining the advantages of optimal individual preservation method and roulette selection method. Firstly, the optimal chromatids in the current population were copied to the next generation to participate in the rest of the process (crossover and mutation). Other individuals use a roulette selection strategy to determine the probability of survival based on the fitness ratio of each individual. The likelihood that a chromosome individual will be selected is related to its own fitness value and is in direct proportion. If the fitness value of the chromosome individual is big, the probability of the chromosome individual being selected is bigger. Conversely, if the fitness value of an individual chromosome is small, the individual is less likely to be selected.

(5) *Heuristic crossover operator.* In general, methods such as cyclic crossover (CX), partial matching crossover (PMX) and sequential crossover (OX) are mostly used to conduct crossover processing for operators, but there are some defects, such as gene duplication and the inability to retain excellent gene fragments of the parent generation. The main purpose of crossover operation in genetic algorithm is to form more and better offspring, and to be able to inherit the excellent fragments in the parent chromosome, without destroying the good fragments in the parent. To meet these requirements, heuristic hybrid crossover operators are used in this paper.

In this study, a new heuristic crossover operation is adopted, and the operation processes of searching for offspring 1 are as follows:

(a) Firstly judge the first gene location in offspring 1, delivery vehicles are start from the distribution center, finally returned to the starting point. Starting from the first gene of the two parents, calculate the distance from distribution center 0 to the first customer $a_1$ and $a_2$ of the two parents' chromosomes as $d_1$ and $d_2$, compare the size of the two distances.
(b) If $d_1 \leq d_2$, insert the first parent individual genes point $a_1$ to offspring 1, and delete the point $a_1$ in parent individuals. If $d_2 < d_1$, insert the second parent individual genes point $a_2$ to offspring 1, and delete the point $a_2$ in parent individual.
(c) Judge the next gene point of offspring 1, find the next customer point $c_1$ and $c_2$ of the current customer point in the two chromosomes, calculate the distance between the current customer point and the next customer point, compare the size of the two distances, and return (b) for cycle.
(d) If the next gene point in one parent chromosome is 0, the next gene point in the other parent individual is taken as the gene point in the offspring 1. If there is only one customer point in the two parent chromosomes, the operation is completed and the customer point is inserted into the offspring 1.

Offspring 2 uses the distance of the previous customer point to find the gene point to determine and starts from the last customer gene point. The method to find the gene point is same as offspring 1.

(6) *Greedy inversion mutation operator.* Inversion mutation operator refers to the random selection of two gene points, from the first selected gene point to the second selected gene point to form a gene fragment, the gene fragment will be inverted order to generate a new chromosome. For example, the parent chromosome is P (1625473), the randomly selected gene customer points are 6 and 7, then the new individual generated after the operation of inversion mutation is P′ (1674523).

In order to be able to produce more excellent individuals, this study use a mutation operation based on inversion greedy, the operation process is: Firstly, randomly select a genetic customer point $a$. Secondly, Find the closest customer point $a'$ to customer point $a$ in addition to the adjacent genetic point around customer point $a$. Finally, reverse the sequence of the gene fragments between the customer's right gene point $a$ and the gene point $a'$. For example, the parent chromosome is P (0451230), randomly select the customer's point 5. After judgment, the closest customer's point to customer's point 5 except customer's point 4 and 1 is point 3, so the gene fragment between 1 and 3 is used for example sequencing, and the new individual is P′ (0453210). In this way, the fitness of the new individuals will be larger, better than the parent individuals, and can improve the convergence speed of the algorithm better.

(7) *Adaptive adjustment of crossover and mutation probabilities.* In the traditional GA algorithm, crossover probability and mutation probability are usually given a value in advance whatever the end result, these probability values are certain, so local convergence or slow convergence will be likely to occur. In order to make the species of the population diverse, strengthen the global search convergence of the algorithm, and prevent the algorithm from falling into the local optimum, this study used the multi-variation crossover mutation probability.

$$p_c = \begin{cases} \frac{k_1(f_{\max} - f')}{f_{\max} - \bar{f}} & f' \geq \bar{f} \\ k_2 & f < \bar{f} \end{cases} \tag{13}$$

$$p_m = \begin{cases} \frac{k_3(f_{\max} - f')}{f_{\max} - \bar{f}} & f' \geq \bar{f} \\ k_4 & f < \bar{f} \end{cases} \tag{14}$$

In the above formula, $k_1$, $k_2$, $k_3$, $k_4$ are constants in [0, 1].

(8) *Generate neighborhood solutions.* Neighborhood structure refers to the process in which the current feasible solution changes to another feasible solution through certain operations. In this study, three basic neighborhood search algorithms are used in tabu algorithm, which is a common strategy for solving path planning problems. Generally, there are three kinds of neighborhood rules: reverse operation, insert operation and exchange operation [15].

(a) *Reverse operation.* Reverse operation refers to the random selection of two customer points *a* and *b* in an arranged distribution path, and then carry out the reverse order of the paths between the two customer points. The operations are as follows:
Distribution path 1: 3-2-1-4-6-8-5-7-9

Randomly selected customer points 1 and 5, after the reverse operation, become as follows:
Distribution path 2: 3-2-5-8-6-4-1-7-9

(b) *Insertion operation.* The insertion operation refers to the random selection of a customer point *a* in the arranged distribution path and insert point *a* into another point in the path. The operation is as follows:
Distribution path 1: 3-2-1-4-6-8-5-7-9

Select customer point 1 and insert point 1 into position 8. After the insertion operation, it will become as follows:
Distribution path 2: 3-2-4-6-8-5-7-1-9

(c) *Exchange operation.* The exchange operation represents the selection of two customer points *a* and b on an aligned distribution path to exchange the two customer points. The operation process is as follows:
Distribution path 1: 3-2-1-4-6-8-5-7-9

Randomly select customer points 1 and 5, and exchange the positions of the two customer points, then the exchange operation turns into the following:
Distribution path 2: 3-2-5-4-6-8-1-7-9

(9) *Tabu list and amnesty criteria.* Tabu table is used to mark tabu objects. Tabu objects refer to the neighborhood operation. Generally, there are three kinds of operations in neighborhood structure movement, namely exchange, insertion and inversion. Each neighborhood operation corresponds to a tabu record.

The amnesty principle in this paper is: if running a neighborhood structure can produce the best solution so far, the neighborhood structure, whether not tabu, will replace the current optimal solution and be adopted. If all neighborhood structures are recorded in the tabu list and do not meet amnesty criteria, one of these neighborhood structures will be chosen.

(10) *Termination criteria.* GA algorithm is a random parallel search method, which cannot judge whether to end the operation in advance. Therefore, it is necessary to make conditions for the end of the algorithm in advance, which are called termination criteria. General GA algorithm termination criteria usually make the following three settings:

(a) Set the target function value in advance, if the individual fitness value and the target value are same, then end the operation.

(b) If the population of chromosome individuals haven't obtained the better value after a number of successive generations, end the operation.
(c) Set the number of iterations in advance, when the number of operations reach the seted iterations number, end the operation.

In this paper, the third termination criterion is selected, and the substitution number is set in advance. When the loop reaches the maximum iteration number, the operation ends and the optimal solution is output.

## 5 Case Analysis and Evaluation

This paper takes a private logistics distribution enterprise as an example to process case analysis. The enterprise is mainly engaged in road transport of goods at LTL, and H logistic company can provide logistics and supply chain integration services. 120 delivery customers are selected in this paper. H logistics has a maximum load of 12 tons of distributed vehicles. Through the field investigation of the company, the daily demand of 120 delivery customers, the delivery time window of customers and the position coordinates of each customer were obtained, the number 0 represents the logistics goods distribution center. Partial basic information of the customers is shown in Table 1.

**Table 1** Partial customer information table

| Customer number | Customer location | | Service window | | Customer demand |
|---|---|---|---|---|---|
| | X coordinate | Y coordinate | Start time | End time | |
| 0 | 100 | 90 | 6 | 20 | 0 |
| 1 | 15 | 20 | 8 | 18 | 0 |
| 2 | 20 | 21 | 8 | 20 | 0.4 |
| 3 | 16 | 19 | 6 | 18 | 0.8 |
| 4 | 18 | 20 | 8 | 20 | 1 |
| 5 | 5 | 18 | 7 | 20 | 0.6 |
| 6 | 12 | 17 | 6 | 20 | 0.4 |
| 7 | 14 | 22 | 6 | 18 | 0.8 |
| … | … | … | … | … | … |
| 118 | 170 | 159 | 8 | 18 | 0.4 |
| 119 | 168 | 170 | 6 | 19 | 0.6 |
| 120 | 40 | 45 | 7 | 19 | 0.6 |

## 5.1 Fuzzy Clustering Results

In this paper, the five clustering indicators, namely, abscissa $x_1$ of customer point, ordinate $x_2$ of customer point, starting time $x_3$ of customer acceptance and distribution of goods, ending time $x_4$ of customer acceptance and distribution of goods, and daily demand data $x_5$, are used for clustering Taking the customer information of H Logistics Company as an example [16], after the fuzzy clustering in Sect. 3, the obtained classification results is shown in Table 2.

## 5.2 Results of Genetic Tabu Hybrid Algorithm

The maximum load capacity of each delivery vehicle is 12 tons, according to the classification results of fuzzy clustering, it can be divided into nine categories, so we use 9 vehicles to do the distribution. The fixed cost of each vehicle is 200 yuan per vehicle. The unit running cost of the vehicle is 3 yuan per kilometer. The average speed of the vehicle is 50 km per hour. The penalty cost coefficient of early arrival of distribution vehicles is 30 yuan per hour and the penalty cost coefficient of late arrival of distribution vehicles is 60 yuan per hour. The unit demand service time of each customer is 0.05 h per ton. For each customer group, the genetic tabu hybrid algorithm proposed in this paper was applied to calculate the optimal route planning for vehicles. Using Matlab software to calculate the data and the parameters of genetic tabu hybrid algorithm are as follows: The population size of the genetic algorithm is $nPop = 50$, the maximum number of iterations is $MaxIt = 1200$, the crossover probability is $k_1 = k_2 = 0.9$, the mutation probability is $k_3 = k_4 = 0.2$, the roulette selection parameter is $beat = 8$, the length of tabu list in the tabu search algorithm is $TL = 20$, the maximum number of iterations is 1200 and the number of neighborhood candidate solutions is 100 (Fig. 3).

**Table 2** Cluster analysis of customer distribution classification results

| Sequence number | Demand | Customer point |
|---|---|---|
| 1 | 56 | (1, 2, 3, 4, 5, 6, 7, 8, 10, 18, 21, 27, 35, 120, 56) |
| 2 | 59 | (20, 28, 29, 32, 36, 37, 42, 51, 58, 62, 64, 70, 71, 110) |
| 3 | 56 | (19, 53, 54, 60, 69, 73, 76, 79, 80, 82, 84, 86, 90, 93) |
| 4 | 59 | (12, 16, 17, 25, 26, 33, 34, 41, 46, 59, 63, 68, 72, 74, 94) |
| 5 | 50 | (14, 15, 23, 24, 31, 38, 50, 78, 109) |
| 6 | 53 | (13, 83, 85, 88, 89, 91, 92, 95, 98, 99, 100, 101) |
| 7 | 56 | (39, 40, 44, 45, 48, 49, 52, 57, 65, 66, 67, 75, 77, 81, 87) |
| 8 | 49 | (22, 30, 43, 102, 103, 104, 105, 106, 107, 108, 111, 112, 61) |
| 9 | 56 | (9, 11, 47, 55, 96, 97, 113, 114, 115, 116, 117, 118, 119) |

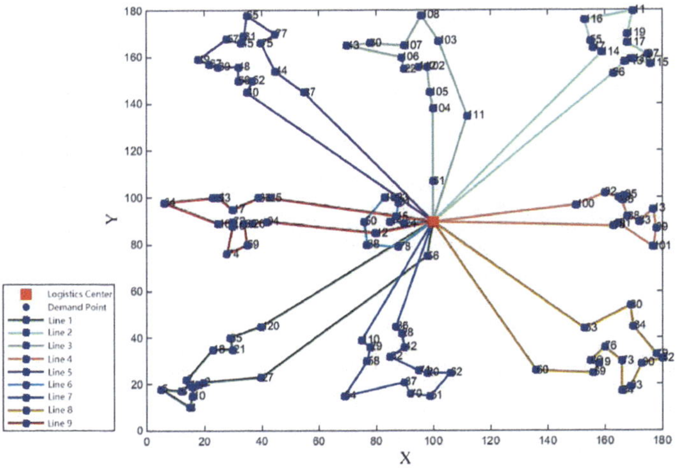


**Fig. 3** Genetic-tabu hybrid algorithm circuit diagram

Matlab software is used to run the genetic tabu hybrid algorithm for 20 times, and the final optimal solution is shown in Table 3.

From the calculation results, it can be concluded that the optimal distribution plan requires a total of 9 delivery vehicles for delivery, and the optimal total transportation cost is 7770.9 yuan, and the total transportation mileage is 1990.3 km. It can be seen from the results that the loading requirements required by the customer were fulfilled without overload, and each vehicle was loaded to the maximum extent, so that there was no empty or wasted vehicles. Customers of vehicles on each distribution route meet the requirements of time window without penalty cost, indicating that the genetic tabu hybrid algorithm proposed in this paper can obtain a better solution and is applicable to the vehicle scheduling problem in this paper. The results from 20 times of operation are shown in Table 4.

We can see from Table 4 that in the 20 times of operation, the optimal solution is obtained for 4 times, and the operation efficiency is quite high. The average distribution cost is 7802.7 yuan and the average distribution mileage is 2000.9 km.

Figure 4 shows the convergence process curve of the genetic tabu hybrid algorithm. It can be seen from the curve that the genetic tabu mixture algorithm is able to find the optimal solution quickly and has a fast convergence speed. The algorithm has better performance and higher solving efficiency.

## 5.3 *Other Heuristic Algorithm Solutions*

This paper uses traditional genetic algorithm to solve the above problems, the initial parameters setting are consistent with that of genetic tabu hybrid algorithm. The population size of the genetic algorithm is $nPop = 50$, the maximum number of

**Table 3** Distribution scheme of genetic-tabu hybrid algorithm

| Vehicle | Distribution path | Load (ton) | Distribution mileage (km) |
|---|---|---|---|
| 1 | 120→35→21→18→7→1→6→5→8→10→3→4→2→27→56 | 56 | 268.1 |
| 2 | 96→113→9→118→115→97→117→119→11→116→55→47→114 | 58 | 258.5 |
| 3 | 111→103→108→107→30→43→106→22→112→102→105→104→61 | 53 | 226.0 |
| 4 | 100→92→89→95→85→88→83→13→99→101→91→98 | 53 | 187.4 |
| 5 | 40→52→66→48→39→67→49→57→45→81→65→77→75→44→87 | 56 | 263.9 |
| 6 | 78→38→50→109→23→31→15→14→24 | 50 | 79.9 |
| 7 | 110→29→58→64→37→70→51→62→20→71→32→42→28→36 | 59 | 216.4 |
| 8 | 53→80→84→79→82→90→93→54→73→76→86→19→69→60 | 58 | 267.4 |
| 9 | 25→33→17→63→41→34→16→72→46→74→59→68→26→94→12 | 59 | 222.8 |

**Table 4** Calculation results of genetic tabu hybrid algorithm

| Number of operations | Distribution cost (yuan) | Distribution range (km) | Number of operations | Distribution cost (yuan) | Distribution range (km) |
|---|---|---|---|---|---|
| 1 | 43925.9 | 2012.5 | 11 | 43482.5 | 1990.3 |
| 2 | 43500.2 | 1990.9 | 12 | 43665.7 | 1999.0 |
| 3 | 43536.5 | 1993.5 | 13 | 43845.1 | 2009.1 |
| 4 | 43876.4 | 2012.5 | 14 | 43738.1 | 2001.1 |
| 5 | 43751.5 | 2004.5 | 15 | 43662.1 | 2001.1 |
| 6 | 43482.5 | 1990.3 | 16 | 44025.7 | 2017.6 |
| 7 | 43482.5 | 1990.3 | 17 | 43706.8 | 2002.4 |
| 8 | 43791.4 | 2005.1 | 18 | 43652.3 | 2000.1 |
| 9 | 43482.5 | 1990.3 | 19 | 43728.7 | 2004.0 |
| 10 | 43542.4 | 1993.2 | 20 | 43867.1 | 2010.5 |

iterations is $MaxIt = 1200$, the crossover probability is $k_1 = k_2 = 0.9$, the mutation probability is $k_3 = k_4 = 0.2$, the roulette selection parameter is $beat = 8$. Initial population is obtained randomly without fuzzy clustering classification. In this study, tabu search algorithm is used again for calculation and the parameters are seted as follows: the length of tabu list in the tabu search algorithm is $TL = 20$, the maximum number of iterations is 1200 and the number of neighborhood candidate solutions is 100. Using Matlab software, run each algorithm 20 times, the optimal distribution results of different algorithms are compared as shown in Table 5.

**Fig. 4** Genetic tabu hybrid algorithm convergence figure

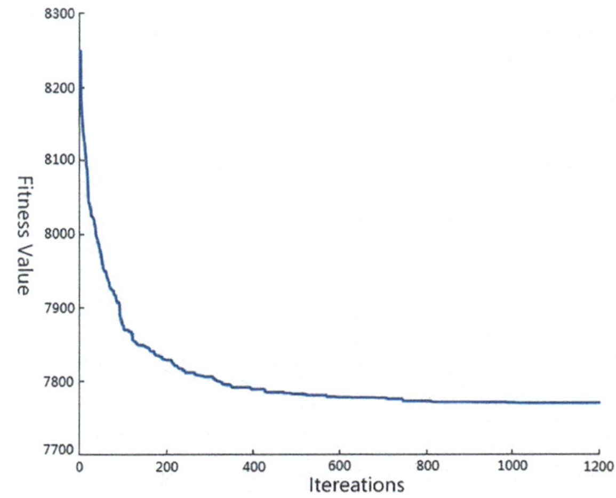


**Table 5** Comparison between genetic tabu algorithm and traditional genetic algorithm for optimal distribution scheme

| Algorithm | Distribution cost (yuan) | Distribution range (km) |
|---|---|---|
| The optimal solution of genetic tabu mixing algorithm | 7770.9 | 1990.3 |
| Average value of genetic tabu hybrid algorithm (20 times) | 7802.7 | 2000.9 |
| The optimal solution of traditional genetic algorithm | 10452.6 | 2884.2 |
| Average value of traditional genetic algorithm (20 times) | 10506.9 | 2902.3 |
| Tabu search algorithm optimal solution | 9795.9 | 2665.3 |
| Average value of tabu search algorithm (20 times) | 9867.6 | 2689.2 |

From Table 5, the genetic tabu hybrid algorithm proposed in this paper is compared with the optimal distribution scheme solved by other heuristic algorithms. We can analyze that under the same parameter setting, the solution efficiency and quality of the genetic tabu hybrid algorithm are higher than the traditional genetic algorithm and tabu search algorithm. Compared with the genetic algorithm, the genetic tabu hybrid algorithm can save 25.66% of the total distribution cost and 893.9 km of driving mileage. Compared with tabu search algorithm, the total delivery cost is saved by 20.67% and the driving mileage is saved by 675 km. Experimental results show that the proposed algorithm can achieve better results and higher efficiency when solving the distribution vehicle route scheduling model.

## 6 Conclusion

In this study, a two-stage algorithm is proposed. In the first stage, a fuzzy clustering algorithm is used to classify large-scale customer groups, and customer points are divided into different categories according to clustering indexes. Each customer point in the class is delivered by one vehicle and the total customer point demand in the class needs to meet the loading needs of the vehicle. In the second stage, an improved genetic tabu hybrid algorithm is used to solve the problem of the vehicle routing in the customer groups for customer points in each assigned classification region. The initial solution of the genetic algorithm is generated randomly by customer points in each region.

In this paper, an improved heuristic crossover operator and inversion inversion mutation operator are proposed, which can better inherit the excellent genes of the parent generation and generate new operators efficiently, so as to avoid the solution falling into local optimum and improve the efficiency and quality of solution. Finally, taking H logistics company's delivery customers as the research object, using the genetic taboo mixed algorithm proposed in this paper to solve the vehicle distribution path. The research results show that the improved algorithm proposed in this paper has a better calculation effect on the vehicle scheduling problem, can reduce the distribution cost of logistics enterprises, and has a great guiding effect on the efficiency and emission reduction of H logistics company.

## References

1. Dantzig, G. B., & Ramser, J. H. (1959). The truck dispatching problem. *Management Science, 6,* 80–91.
2. Desrochers, M., & Verhoog, T. W. (1991). A new heuristic for the fleet size and mix vehicle routing problem. *Computers & Operations Research, 18*(3), 263–274.
3. Moghaddam, B. F., Ruiz, R., & Sadjadi, S. J. (2011). Vehicle routing problem with uncertain demands: an advanced particle swarm algorithm. *Computers & Industrial Engineering, 62*(1), 306–317.
4. Yang, X. J. (2014). *Research on postal transport vehicle scheduling based on tabu genetic algorithm*. Chongqing University of Posts and Telecommunications.
5. Ren, J., Li, D. F., & Qin, Z. (2012). *Genetic algorithm for the vehicle scheduling problem of emergency logistics distribution in City*. CA: World Automation Congress Proceedings.
6. Balázs, D., & Miklós, K. (2013). Application oriented variable fixing methods for the multiple depot vehicle scheduling problem. *Acta Cybernetica, 21*(1), 53–73.
7. Li, S. B. (2010). *Research on open vehicle routing problem based on tabu search algorithm*. Zhengzhou University.
8. Li, Z. Y., Ma, L., & Zhang, H. Z. (2014). Adaptive cellular particle swarm optimization for 0/1 knapsack problem. *Computer Engineering, 40*(10), 198–203.
9. Solomon, M. M., & Desrosiers, J. (1988). Survey paper-time window constrained routing and routing problems. *Transportation Science, 22*(1), 1–13.
10. Chen, J. F., & Wu, T. H. (2005). Vehicle routing problem with simultaneous deliveries and pickups. *Journal of the Operational Research Society, 57*(5), 579–587.

11. Ceschia, S., Gaspero, L. D., & Schaerf, A. (2011). Tabu search techniques for the heterogeneous vehicle routing problem with time windows and carrier-dependent costs. *Journal of Scheduling, 14*(6), 601–615.
12. Bent, R., & Hentenryck, P. V. (2004). A two-stage hybrid local search for the vehicle routing problem. *Transportation Science, 38*(4), 515–530.
13. Repoussis, P. P., Tarantilis, C. D., & Ioannou, G. (2007). The open vehicle routing problem. *Journal of the Operational Research Society, 58,* 355–367.
14. Li, S. L. (2010). *Research on regional planning of cigarette sales and distribution based on fuzzy theory*. Shanghai Jiaotong University.
15. Lv, Y. H. (2016). *Research on vehicle routing problem based on mixed tabu distribution estimation algorithm*. Liaoning University of Science and Technology.
16. Zhao, M., & Wu, L. (2009). Study on improved text clustering algorithm. *Journal of Yangtze university, 6*(2), 73–75.

# Doppler Diversity for OFDM System in High-Speed Maglev Train Mobile Communication

**Hancheng Ma, Ying Liu, and Xu Li**

**Abstract** According to the characteristics of maglev train application, the traditional Doppler diversity receiver structure is optimized in this paper. In this paper, channel estimation, maximum ratio merging, equalization and selection of diversity branch parameters in the process of Doppler diversity reception are described in detail, and a time-frequency channel estimation algorithm is designed in the Doppler diversity system in which the maximum ratio merging is combined with equalization. Simulation results show that when the normalized maximum Doppler broadening is greater than 0.5/T, the bit error rate of the system is one order of magnitude lower than that of the traditional OFDM system. When the SNR is less than 16 dB, the SNR required to achieve the same bit error rate is reduced by 0.8 dB.

**Keywords** High-speed · Maglev train · OFDM · Doppler diversity

## 1 Introduction

In the high-speed railway environment, the Doppler effect is very obvious when the train is moving faster than 250 km/h [1]. When the maximum speed of the high-speed maglev is 600 km/h, the Doppler frequency offset will seriously damage the orthogonality of OFDM subcarriers, which will result in inter-carrier interference and thus reduce the OFDM system performance. The Doppler diversity can use Doppler frequency offset as a source of diversity gain, which will improve the OFDM system performance in high-speed mobile environment. There is a strong magnetic field and an alternating magnetic field that are suspending and propelling

H. Ma · Y. Liu (✉) · X. Li
School of Communication and Information, Beijing Jiaotong University, Beijing, China
e-mail: 17120094@bjtu.edu.cn

H. Ma
e-mail: 13211063@bjtu.edu.cn

X. Li
e-mail: xli@bjtu.edu.cn


J. Zhang et al. (eds.), *LISS2019*,
https://doi.org/10.1007/978-981-15-5682-1_32

the train respectively in the high-speed magnetic levitation environment. The environment is complicated and varied, which is difficult to ensure the receiver to obtain a high signal to noise ratio (SNR). The error caused by traditional LS channel estimation in such environment have an adverse effect on the Doppler diversity performance. Therefore, it is necessary to study and optimize the channel estimation algorithm of low to medium SNR by using the Doppler diversity.

Several previous researches have studied the Doppler diversity techniques of single carrier systems [2, 3]. Sayeed proposed the Doppler diversity technology of CDMA system, which reduced the influence of multipath and Doppler effect on system performance by designing a time-frequency related receiver [2]. Boudreau studied the Doppler diversity technique in fast time-varying flat-fading channels, and partly adopted the matched filtering in the positive and negative frequency offsets of the received signal, and then merged to achieve Doppler diversity [3]. Classical Doppler diversity of OFDM system was to make frequency shifts of the received signal, producing multiple diversity branches and finally performing weighted addition [4]. The simplified Doppler diversity technique changed the weighting of each branch from the frequency domain to the time domain, which reduced the complexity of implementation without reducing the Doppler diversity performance [5]. Considering about the channel time variation, a new time-frequency processing scheme was proposed to reduce the bit error rate [6]. The LS channel estimation method in Doppler diversity technology was described in detail, and the equalization technique was added in the Doppler diversity [6].

In this study, we focused on the characteristics of high-speed mobile and low-to-medium SNR in the application scenarios of maglev, and optimized the channel estimation in the traditional Doppler diversity technique, which improved the channel estimation accuracy of the receiver, and verified the advantages of this program by using simulation analysis.

## 2 Formulation

### 2.1 Channel and Signal Models

The time-variant impulse response of a wireless channel can be represented by

$$h(t,\tau) = \frac{1}{\sqrt{N_p}} \sum_{p=1}^{N_p} \alpha_p e^{j[2\pi f_{D_p} t + \phi_p(t)]} \delta(\tau - \tau_p) \tag{1}$$

where $N_p$ is the number of paths, $\alpha_p$, $f_{Dp}$, $\phi_p$, and $\tau_p$ are the fading amplitude, Doppler frequency shift, phase shift and delay of signal of the $p$th multipath arrival. We assume that complex path gain $\{\alpha_p\}$ are i.i.d. Gaussian random variables with zero mean and variance $\sigma^2$, and $\{\tau_p;\ 0 < \tau_p < \tau_{max}\}$, $\{\phi_p(t);\ 0 < \phi_p(t) < 2\pi\}$ and $\{f_{Dp};\ -F_D < f_{Dp} < F_D\}$ are i.i.d random variables with uniform distribution. $\tau_{max}$ is the maximal multipath delay of the channel and $F_D$ is the maximal Doppler shift.

Consider an OFDM signal containing N sub-carriers

$$s(t) = \frac{1}{\sqrt{N}} \sum_{i=0}^{N-1} d_i e^{j\frac{2\pi}{N}it} g(t) \tag{2}$$

where $d_i$ is the transmitted symbol on the *i*th sub-carrier and $N$ is the number of sub-carriers. *g(t)* is a symbol window such that

$$g(t) = \begin{cases} 1, -T_g \leq t < T \\ 0, otherwise \end{cases} \tag{3}$$

where T is the duration of an OFDM symbol and $T_g$ is the duration of cyclic-prefix (CP) which is decided by the maximum delay spread of channel. So $T_g + T$ is the duration of a complete OFDM symbol.

## 2.2 System Model

The structure of the designed OFDM Doppler diversity receiver with $Q$ branches is shown in Fig. 1. The receiver removing the *CP* of the received time-domain signal, and then performs frequency shift processing on the received signal to obtain a plurality of diversity branches. Then each branch signal is transformed into frequency domain through FFT algorithm. The results of the channel estimation are used in the maximal-ratio combining (*MRC*) of subcarriers on each branch, and then the signal is balanced and demodulated. Finally, the bit stream is output.

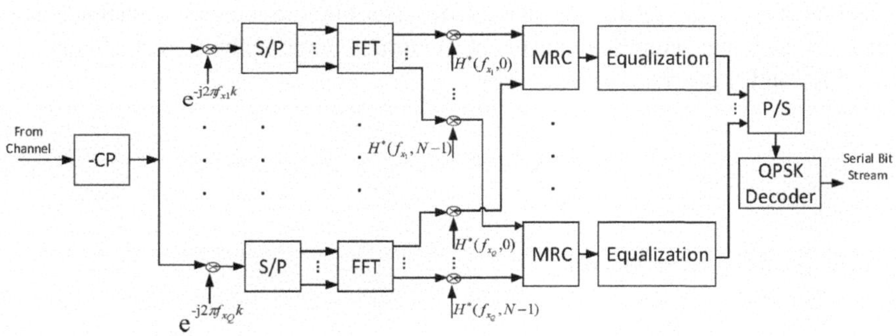


**Fig. 1** OFDM Doppler diversity receiver

The received sample sequence for a symbol after removing *CP* can then be represented as

$$x(nT_s) = \frac{1}{\sqrt{N_p N}} \sum_{p=1}^{N_p} \alpha_p e^{j[2\pi f_{D_p} nT_s + \varphi_P(nT_s)]} \times \sum_{i=0}^{N-1} d_i e^{j\frac{2\pi}{N} i(nT_s - \tau_p)} + v(nT_s) \tag{4}$$

where $T_s$ is the sampling interval and $v(nT_S)$ denotes the additive white Gaussian noise. In the paper, for convenience and without loss of generality, $T_s$ will set to be one, so the symbol duration can be calculated by $T = NT_s = N$. After removing the cyclic-prefix, DFT matching filtering is performed on the frequency offset signal whose frequency offset is $f_x$. The frequency domain signal on the *i*th subcarrier of the branch with frequency offset of $f_x$ can be express as

$$\begin{aligned} X(f_x, i) &= \frac{1}{\sqrt{N}} \sum_{l=0}^{N-1} x(l) e^{-j2\pi f_x l} e^{-j2\pi \frac{i}{N} l} \\ &= X_s(f_x, i) + X_i(f_x, i) + N(f_x, i) \end{aligned} \tag{5}$$

where $X_s(f_x, i)$ is the effective signal term, $X_i(f_x, i)$ is the interference signal term and $N(f_x, i)$ is the noise term. $X_s(f_x, i)$ can be represented by

$$X_s(f_x, i) = \frac{1}{N\sqrt{N_p}} \sum_{p=1}^{N_p} \alpha_p e^{-j\frac{2\pi}{N} i \tau_p} \sum_{l=0}^{N-1} e^{-j[2\pi l (f_x - f_{D_p}) + \phi_p(l)]} d_i \tag{6}$$

Define $X_s(f_x, i)$ as $d_i H(f_x, i)$. Here, $H(f_x, i)$ is the corresponding channel gain which is a linear combination of $N_p$ complex exponentials with frequency $\{f_{Dp} - f_x\}$ and it can be represented by

$$H(f_x, i) = \frac{1}{N\sqrt{N_p}} \sum_{p=1}^{N_p} \alpha_p e^{-j\frac{2\pi}{N} i \tau_p} \sum_{l=0}^{N-1} e^{-j[2\pi l (f_x - f_{D_p}) + \phi_p(l)]} \tag{7}$$

$X_i(f_x, i)$ can be represented by

$$X_i(f_x, i) = \frac{1}{N\sqrt{N_p}} \sum_{p=1}^{N_p} \alpha_p \sum_{\substack{k=0 \\ k \neq i}}^{N-1} d_i e^{-j\frac{2\pi}{N}\tau_p} e^{j\phi_p(k)} e^{j\pi(N-1)(\frac{k-i}{N} + f_{D_p} - f_x)} \times \tag{8}$$

and $N(f_x, i)$ can be represented by

$$N(f_x, i) = \frac{1}{\sqrt{N}} \sum_{l=0}^{N-1} v(k) e^{-j2\pi f_x l} e^{-j\frac{2\pi}{N} il} \tag{9}$$

# 3 Design of the Doppler Diversity Receiver

## 3.1 Channel Estimation

According to the above analysis, multipath time-varying channel gain $H(f_x, i)$ should be estimated in the process of Doppler diversity reception. Channel estimation methods used in OFDM system include channel estimation based on known data, semi-blind estimation and blind estimation. Channel estimation based on known data has a good balance between complexity and performance.

The frame structure of channel estimation adopted in this paper is shown in Fig. 2. A known sequence is added before multiple OFDM symbols whose number can be determined according to the channel coherence time in the application scenario of high-speed maglev train. Combining with the Doppler diversity scheme, the paper adopts the frequency-domain channel estimation algorithm based on training sequence. In this scheme, an OFDM symbol can be regarded as $N$ parallel single-carrier systems, and each subsystem is affected by multiplicative interference and additive Gaussian noise.

The frequency domain of each OFDM symbol received is denoted as

$$Y(k) = H(k)X(k) + N(k) \quad k = 0, 1 \cdots, N-1 \tag{10}$$

where $H(k)$ is the complex fading coefficient of the $k$th subcarrier, $X(k)$ is training sequence in channel estimation and OFDM symbol in equalization and $N(k)$ is the Gaussian white noise on the $k$th subcarrier. The real part and imaginary part of $N(k)$ are independent of each other, and both obey the Gaussian distribution with the mean value of 0. Define the frequency domain channel response estimation obtained by traditional *LS* channel estimation as

$$H_{LS}(k) = \frac{Y(k)}{X(k)} = H(k) + \frac{N(k)}{X(k)} \tag{11}$$

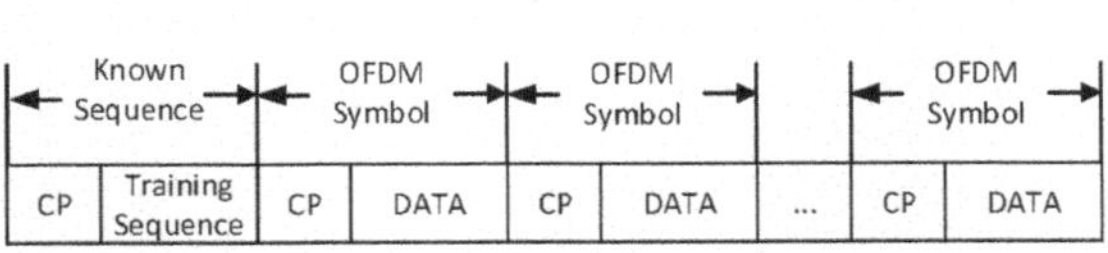


**Fig. 2** OFDM frame structure design based on training sequence channel estimation algorithm

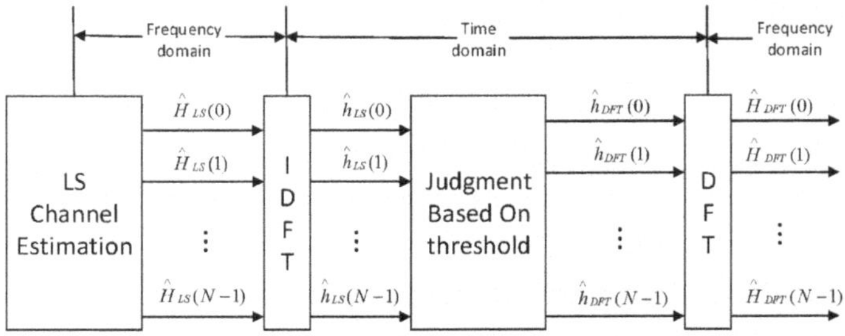


**Fig. 3** LS-noise-reduction algorithm implementation process

This method can estimate the channel gain on each subcarrier. When the SNR is relatively high, the performance of this algorithm is ideal, but when the SNR is low than 15 dB, the deviation generated by traditional LS estimation will affect the performance of MRC and equalization. In the application scenario of maglev train, it is difficult to ensure the SNR above 15 dB, so it is necessary to optimize the channel estimation algorithm.

Figure 3 show the channel estimation algorithm this paper uses. The channel response obtained from the LS channel estimation in the frequency domain is converted to the time domain through IDFT conversion, and the point with gain less than the threshold is set to zero. Then the channel estimation result in the time domain is transformed to the frequency domain by DFT matching filtering, which not only reduces the impact of Gaussian noise, but also does not increase the complexity of MRC and equalization.

After IDFT transformation, the time-domain response is

$$\hat{h}_{LS}(n) = \frac{1}{N}\sum_{k=0}^{N-1} \hat{H}_{LS}(k)e^{j2\pi \frac{n}{N}k} \tag{12}$$

After threshold decision, the time-domain channel response is expressed as

$$\hat{h}_{DFT}(m) = \begin{cases} \hat{h}_{LS}(m), & \left|\hat{h}_{LS}(m)\right| > \mathrm{A} \\ 0, & other \end{cases} \tag{13}$$

where A is the decision threshold. Finally, after DFT transform, the final channel gain is

$$\hat{H}(k) = \hat{H}_{DFT}(k) = DFT(\hat{h}_{DFT}(m)) \tag{14}$$

## 3.2 MRC and Equalization

The traditional Doppler diversity scheme does not include the equalization process and only uses the estimation channel gain to do the MRC. However, in the application scenarios of high-speed maglev trains, intercarrier interference caused by Doppler expansion has a great impact on the OFDM system. Therefore, in order to reduce ICI and enhance the effective signal, it is necessary to add equalization into the Doppler diversity system.

The signal on the *i*th subcarrier after the MRC can be expressed as

$$\begin{aligned} Y_{MRC}(i) &= \hat{H}_1^*(i)Y_1(i) + \hat{H}_2^*(i)Y_2(i) + \cdots + \hat{H}_Q^*(i)Y_Q(i) \\ &= \hat{H}_1^*(i)\hat{H}_1(i)X_1(i) + \hat{H}_2^*(i)\hat{H}_2(i)X_2(i) + \cdots \\ &+ \hat{H}_q^*(i)\hat{H}_q(i)X_q(i) + I(i) \end{aligned} \tag{15}$$

where $\hat{H}_q(i)$ and $\hat{Y}_q(i)$ are the channel estimation and the received signal on the *i*th subcarrier of the *q*th deviation branch. *I*(*i*) is the interference on the *i*th subcarrier, and its power includes Gaussian white noise power and intercarrier interference power.

When the interference power is small, the interference term *I*(*i*) can be ignored and forced zero (ZF) equalization can be used, then the *i*th subcarrier frequency-domain equalization coefficient is

$$W(i, f_{x_1}, f_{x_2}, \cdots, f_{x_Q}) = \frac{1}{\hat{H}(i, f_{x_1}, f_{x_1}, \cdots, f_{x_Q})} \tag{16}$$

In (16), $\hat{H}(i, f_{x_1}, f_{x_2}, \cdots, f_{x_Q})$ can be expressed as

$$\hat{H}(i, f_{x_1}, f_{x_2}, \cdots, f_{x_Q}) = \hat{H}_1^*\hat{H}_1 + \hat{H}_2^*\hat{H}_2 + \cdots + \hat{H}_Q^*\hat{H}_Q \tag{17}$$

After frequency domain equalization, data on the ith subcarrier can be expressed as

$$X(i, f_{x_1}, f_{x_2}, \cdots, f_{x_Q}) = Y_{MRC} W(i, f_{x_1}, f_{x_2}, \cdots, f_{x_Q}) \tag{18}$$

When the interference power is large, the interference term needs to be taken into account. MMSE method can be adopted for equalization, and the equalization coefficient is

$$W(i, f_{x_1}, f_{x_2}, \cdots, f_{x_Q}) = \frac{\hat{H}^*(i, f_{x_1}, f_{x_2}, \cdots, f_{x_Q})}{\left|\hat{H}(i, f_{x_1}, f_{x_2}, \cdots, f_{x_Q})\right|^2 + \frac{P_I}{P_S}} \tag{19}$$

where $P_I/P_S$ is the ratio of the interference power to the signal power, and its calculation method is shown in Eqs. (20–21).

Considering that the introduction of CP can reduce ISI and ICI can be reduced through the parameter design of OFDM system, so the ZF equilibrium method is adopted.

## 3.3 Doppler Branches

It needs to be considered that the effective signal power decreases and the interference power increased as the frequency offset increase gradually. But if the shift is small, the correlation of multiple Doppler branches will be so high that the system can not obtain diversity gain. Both will adversely affect the final result of Doppler diversity. Therefore, it is important to select the appropriate Doppler frequency offset.

According to Eqs. (6) and (7), the effective signal power $P_S(f_x, i)$ on the $i$th subcarrier of the Doppler branch whose frequency offset is $f_x$ can be expressed as

$$\begin{aligned} P_S(f_x, i) &= R_S^i(f_x, f_x) = E\{X_S(f_x, i)X_S(f_x, i)\} \\ &= \sigma_d^2 E\{H(f_x, i)H(f_x, i)\} = \frac{\sigma_\alpha^2 \sigma_d^2}{N^2 N_p} \sum_{p=1}^{N_p} E\left\{ \frac{\sin^2 N\pi(f_x - f_{D_p})}{\sin^2 \pi(f_x - f_{D_p})} \right\} \end{aligned} \tag{20}$$

According to Eq. (9), the interference power $P_I(f_x, i)$ on the $i$th subcarrier of the Doppler branch whose frequency offset is $f_x$ can be expressed as

$$\begin{aligned} P_I(f_x, i) &= R_I^i(f_x, f_x) = E\{X_I(f_x, i)X_I(f_x, i)\} \\ &= \frac{\sigma_\alpha^2 \sigma_d^2}{N^2 N_p} \times \sum_{\substack{l=1 \\ l \neq i}}^{N_p} E\left\{ \frac{\sin^2 N\pi(\frac{l-i}{N} + f_{x_1} - f_{D_p})}{\sin^2 \pi(\frac{l-i}{N} + f_{x_1} - f_{D_p})} \right\} \end{aligned} \tag{21}$$

The relationship between the signal interference ratio (SIR) and $F_D$ is shown in Fig. 4. Although with $f_x$ increases, the effective signal power decreases and the interference power increased gradually, it can be seen from Fig. 4 that SIR will increase when $f_x$ increases if $f_xT$ is equal to 0.75 and 0.8. It shows that this branch performance gradually get better when the Doppler shift increases.

The absolute value of the power correlation coefficient of the two branch interference signals of $f_{x1}$ and $f_{x2}$ can be expressed as

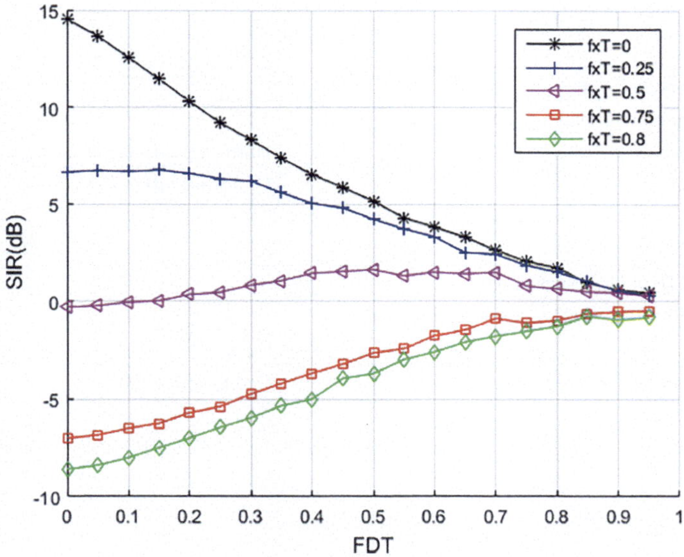


**Fig. 4** SIR vs $F_DT$ for 5 different $f_x$

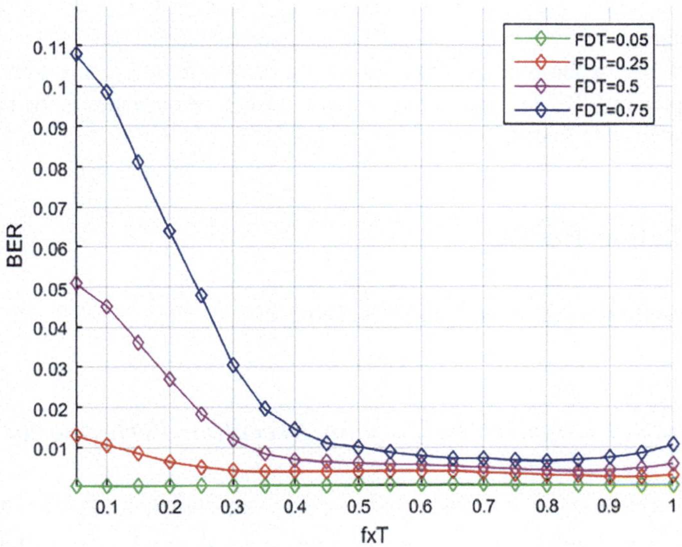


**Fig. 5** BER vs $f_xT$ for 3 branch Doppler diversity

**Table 1** Simulation parameters

| Parameter | Value |
|---|---|
| Number of paths ($N_p$) | 6 |
| Maximal multipath delay ($\tau_{max}$) | 32 |
| Delay of paths ($\tau_p$) | Uniform(i.i.d), $0 < \tau_p < \tau_{max}$ |
| Doppler shift ($f_{Dp}$) | Uniform(i.i.d), $-F_D < f_{Dp} < F_D$ |
| Number of subcarriers per symbol ($N$) | 256 |
| Length of CP ($N_k$) | 32 |
| Number of Doppler branches (Q) | 3 |
| Decision threshold of channel estimation (A) | $\frac{1}{N}\sum_{n=0}^{N-1}\left\|\hat{h}_{LS}(n)\right\|$ |
| Frequency shifts for Doppler branches ($f_x$) | {−0.8/T, 0, 0.8/T} |

$$\begin{aligned}|\rho_{II}(f_{x_1}, f_{x_2})| &= \frac{|R_{II}(f_{x_1}, f_{x_2})|}{\sqrt{R_{II}(f_{x_1}, f_{x_1})R_{II}(f_{x_2}, f_{x_2})}} \\ &= \frac{|E\{X_i(f_{x_1}, i)X_i^*(f_{x_2}, i)\}|}{\sqrt{E\{X_i(f_{x_1}, i)X_i^*(f_{x_1}, i)\} \times E\{X_i(f_{x_2}, i)X_i^*(f_{x_2}, i)\}}}\end{aligned} \tag{22}$$

According to Eq. (22), it is found that when $f_xT$ is close to 0.8, the correlation is the minimum.

Figure 5 shows the relationship between bit error rate and $f_x$ at different maximum Doppler shift ($F_DT$). It can be seen that the BER performance is the best when $f_xT$ is between 0.75 and 0.85.

## 4 Simulation Results

The simulation parameters are shown in Table 1.

### 4.1 *Effect of Doppler Diversity on Reception Performance*

For *SNR* = 15 dB case, the *BER* performances are shown in Fig. 6. When the normalized maximum Doppler frequency shift is small ($F_DT < 0.25$), the performance improvement of Doppler diversity is not obvious, but with the increase of the maximum Doppler frequency shift, the performance improvement of Doppler diversity is more and more significant. When $F_DT$ is greater than 0.5, the bit error rate of Doppler diversity is reduced by an order of magnitude compared with that of the traditional OFDM system without Doppler diversity. This means that the faster the train moves, the higher diversity gain the Doppler diversity will get. So Doppler

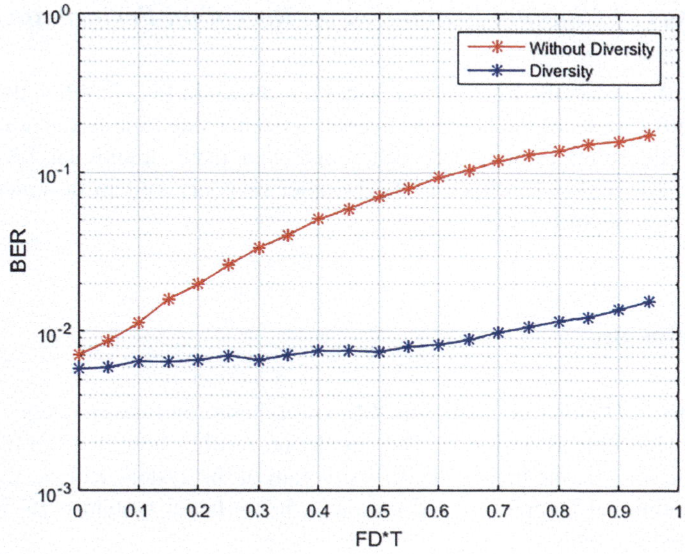


**Fig. 6** BER vs $F_DT$ for 15 dB *SNR*

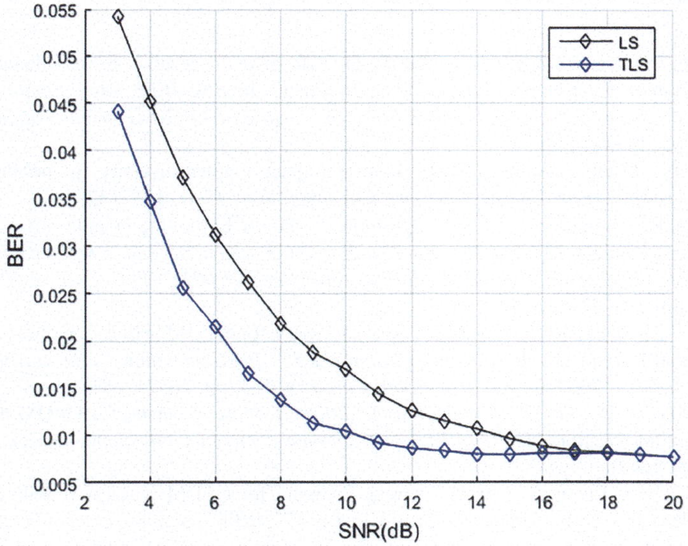


**Fig. 7** BER vs SNR for $F_DT = 0.75$ case

diversity can significantly improve the performance of OFDM system in the communication environment of high-speed maglev trains with a maximum moving speed of 600 km/h.

### 4.2 Effect of Channel Estimation on Reception Performance

For $F_D T = 0.75$ case, the *BER* performances are shown in Fig. 7. When the SNR is low, the performance of the Doppler diversity system using time-frequency channel estimation algorithm is obviously better than that using traditional LS channel estimation. For SNR <16 dB case, the Doppler diversity system designed in this paper can gain about 0.8 dB.

## 5 Conclusion

In this paper, a Doppler diversity OFDM system based on time-frequency channel estimation algorithm is designed. The simulation results show that the higher the speed of the train is, the higher the diversity gain of the system will be. Compared with the traditional Doppler diversity system, it has better system performance at low and medium SNR.

## References

1. Wang, X., Tan, Z. -H., & Chen, X. (2005). Doppler diversity receiver for broadband wireless OFDM system under high-speed mobile environments. In *2005 IEEE international system on microwave, antenna, propagation and EMC technologies for wireless communications, Beijing* (Vol. 2, pp. 1444–1447).
2. Sayeed, A., & Aazhang, B. (1999). Joint multipath-doppler diversity in mobile wireless communications. *IEEE Transactions on Communications, 47*(1), 123–132.
3. Boudreau, R., Chouinard, J. Y., & Yongacoglu, A. (2000). Exploiting Doppler-diversity in flat, fast fading channels. In *2000 Canadian conference on electrical and computer engineering. Conference proceedings. Navigating to a new era (Cat. No. 00TH8492), Halifax, NS, Canada* (Vol. 1, pp. 270–274).
4. Kim, B. -C., & Lu, I. -T. (2003). Doppler diversity for OFDM wireless mobile communications. Part I: frequency domain approach. In *The 57$^{th}$ IEEE semiannual vehicular technology conference, VTC 2003-Spring, Jeju, South Korea* (Vol. 4, pp. 2677–2681).
5. Xin, W., Gang, Z., Xia, C., & Zhen-hui, T. (2006). Doppler diversity for OFDM high-speed mobile communicaitons. In *2006 IEEE international conference on communications, Istanbul* (pp. 4665–4669).
6. Kim, B. -C., & Lu, I. -T. (2003). Doppler diversity for OFDM wireless mobile communications. Part II: time-frequency processing. In *The 57$^{th}$ IEEE semiannual vehicular technology conference, VTC 2003-Spring, Jeju, South Korea* (Vol. 4, pp. 2682–2685).

# Site Selection and Optimization for B2C E-Commerce Logistics Center

**Shuai Wang**

**Abstract** With the development of China's Internet and consumer habits, the B2C e-commerce website development by leaps and bounds. We should note that although a large number of B2C companies, however, firm size is difficult to expand. The construction of distribution centers has been the bottleneck of the development of B2C, delay distribution, distribution costs are too high has hampered the development of B2C e-commerce site. In this paper, we study distribution costs and reduce delivery time, B2C e-commerce sites to the distribution center site selection for research. By the method of combining genetic algorithms and MATLAB software, we can solve the model problems. The model of the distribution center based on distribution of the lowest cost and shortest delivery time, taking into account the service radius of the cost implications, by influence the service radius the operation cost to join model, the model with close to reality. Through research, from the delivery time and distribution costs to achieve the best balance of perspective B2C e-commerce businesses of the distribution center site, to determine the more suitable for the distribution center location of the B2C e-commerce logistics center, for the type of enterprise distribution center location offers some helpful suggestions.

**Keywords** B2C · Logistics center · Genetic algorithm · Location

## 1 Introduction

In recent years, along with the rapid development of China's Internet, various types of shopping sites have sprung up in the eyes of people. With the change of consumption pattern of Chinese residents, the scale of e-commerce enterprises is expanding rapidly, and the volume of business is growing exponentially. Especially

S. Wang (✉)
School of Economics and Management, Beijing Jiaotong University, Beijing, China
e-mail: 16113139@bjtu.edu.cn


J. Zhang et al. (eds.), *LISS2019*,
https://doi.org/10.1007/978-981-15-5682-1_33

the scale of the shopping website which is directly to the general consumer B2C (Business-to-Customer) can make the fastest development.

The main three sides of China's B2C e-commerce business distribution model to third party distribution, postal delivery and independent distribution. With the growth of business volume and the gradual improvement of customer demand, the third-party distribution and postal delivery cannot keep up with the pace of development of B2C e-commerce enterprises. In the case of Jingdong, Jingdong is the typical representative of our independent sales B2C, it started to sell electronic products, sales of products will gradually expand to the type of food, daily necessities and so on, with the increase of the sales type, Jingdong customer volume will continue to increase, the average annual growth rate of turnover has reached a staggering 300%. Due to the shortcomings of low efficiency and poor quality of service in the third-party distribution companies in China, the distribution of the third-party enterprises cannot meet the requirements of the Jingdong. It is in the face of the high rate of complaints about the customer distribution business logistics cost remains high under the situation, according to the characteristics of the enterprise, in order to maintain a 300% average annual growth rate of business, reduce logistics costs, and reduce the rate of customer complaints, the Jingdong has its own distribution system, to meet the needs of the growth of electronic commerce; but with the continuous expansion of business, the distribution center location caused by unreasonable problems appears: one is the customer complaint is not timely delivery due to the rising proportion of two is large and that there is a big pressure distribution; so how to choose the reasonable logistics center became the primary problem faced in Jingdong the construction of the distribution center [1]. The logistics cost increases in certain supply chain nodes and the risk of backlog or running out of stock is high because of the different income and risk expectations of entities, as well as the varying degrees of information sharing between them. Clearly, a one-stop delivery mode crossing numerous transaction links in online shopping can effectively solve these problems and save social resources [2]. Ghobakhloo had research that study is to provide a better and clearer understanding of business-to-business (B2B) EC value and success within SMEs, particularly in emerging economies [3].

At present, the B2C e-commerce enterprises advanced modern logistics distribution center is an indispensable infrastructure of distribution system was successfully constructed, the distribution center should be a collection of information flow, logistics and business flow in one, a modern logistics distribution system. Whether the location of the distribution center is the key factor of whether the goods can be delivered in a timely manner, so the scientific and reasonable distribution center location has become an indispensable step in building a successful distribution system. Due to the completion of the construction of distribution center has the address cannot be changed, but in the process of the construction of large investment, long construction period and cost recovery slow, so the location of the distribution center planning is not only the construction of a new distribution center, the first step is the most critical step one. Scientific and reasonable distribution center location will not only save the cost of the enterprise, but also promote the rapid development of the enterprise. Otherwise it will result in the waste of manpower, material resources and capital.

The location of logistics distribution center is not only conducive to reduce the cost of distribution, but also conducive to improve the distribution characteristics (including the distribution of convenience, timely delivery, etc.). Not only reasonable distribution center location can increase the service outsourcing, but also reduce operating costs, delivery time, and customer complaints about the quality of distribution. The timeliness and regional distribution are very important for a B2C e-commerce enterprise, under the environment of e-commerce, with the continuous improvement of logistics distribution environment, influences of distribution center location factors are more and more. The influence factors of the traditional logistics distribution center location used in the lack of understanding of e-commerce environment model of distribution center, lack of understanding of information, automation and network, so the traditional distribution center location theory is not entirely applicable to B2C e-commerce enterprise requirements. We need to establish a new theory in line with the requirements of the new distribution center location, to provide theoretical guidance for the construction of scientific and reasonable distribution center of B2C enterprises, and to provide scientific suggestions for the development of enterprises. Therefore, it is possible to provide a theoretical support for the development of China's B2C e-commerce enterprise distribution center by the further research on the location of B2C e-commerce enterprise distribution center.

Based on the special requirements of the distribution of the current B2C e-commerce enterprises, in-depth study of the distribution center location problem of this type of enterprise, according to the characteristics of influence factors of location selection of distribution center of B2C e-commerce enterprises, set the appropriate constraints, finally establish the location model of distribution center, distribution center location theory through the model of the process B2C e-commerce enterprises to enrich and supplement, and provide reference for the distribution center of B2C e-commerce enterprise location.

## 2 Literature Review

Su and Hu is put forward and applied to the electronic commerce enterprise self-distribution center location model, the model with the minimum cost as the target, using simulated annealing algorithm for model analysis, and use computer programming method to solve the model, the final model is verified by an example, prove that the model is practical and feasible [4]. Hao, with the lowest cost and the best service for the location of the target, considering the influence of distribution distance distribution cost, establish the location model, and the method of using genetic algorithm and ILOG software to solve the model, finally obtained the best plan to meet the goals of the site [5]. Lin and Li, based on the analysis of the traditional logistics distribution center location algorithm, the establishment of logistics distribution center location model, first determine the objectives and influence factors of site selection of distribution center location, and then using GIS spatial analysis technology, quantitative analysis of the factors affecting the performance of the logistics distribution center location in the electronic map, draw a series of candidate locations, finally using genetic algorithm, obtain the best location selection, and through the case study shows that the feasibility of algorithm [6]. Yao analyses optimization of the contents and objectives of one-stop delivery scheduling to construct a multi-objective optimization of the mathematical model and propose a solving algorithm. Finally, this study uses an application case to verify the feasibility and effectiveness the mechanism and algorithm [2]. S. Moons, K. Ramaekers, it is the first time that an order picking problem and a vehicle routing problem are integrated [7]. Applications of Site Selection and Optimization for B2C E-commerce Logistics center are proposed by many researchers [8–10].

## 3 Problem Description and Hypothesis

### 3.1 Problem Description

Based on the characteristics of logistics under the electronic commerce has carried on the simple analysis, analyzed and summarized the characteristics of B2C e-commerce site, based in the analysis of influencing factor analysis and distribution center of distribution requirements in B2C e-commerce, B2C e-commerce enterprises established distribution center location model. The main contents of this paper are as follows:

- Through the analysis of the characteristics of the current B2C e-commerce enterprise distribution center, this paper studies the location problem of distribution center, and discusses the establishment of the B2C e-commerce enterprise distribution center location model. In theory, the theory of distribution center location of B2C e-commerce enterprises is supplemented. The difference between the traditional model and location model of distribution center is: first, the model considering the influence of service radius of distribution center operating costs, according to the service radius increasing operation cost more rules and model into variable costs by the service radius, make the model closer to reality. Two, according to the distribution center location is more likely to be affected by the distribution time; the distribution time is introduced into the model, the establishment of the shortest delivery time as the objective function.
- In the solution of the model, this method using genetic algorithm combined with MTLAB, first introduced the related knowledge of genetic algorithm, then according to the characteristics of the model design a reasonable algorithm flow, design reasonable encoding process and fitness function, selection of binary coding method of genetic algorithm in parent population the choice of the roulette selection, single point crossover in the crossover operation, the ultimate

combination of genetic algorithm toolbox in MATLAB software programming, finally realize the solution of the model.

**B2C e-commerce business logistics in the development of the difficulties encountered in the following:**

1) *Long delivery time, slow response*

Consumers shopping online is to seek fast and convenient, at present most of the B2C logistics cannot reach the expected level of consumers, consumers in the online shopping process is complete, enter the long waiting period, the actual arrival time from the shopping time longer, showing no advantage or the goods to spend energy to consumers without hearing a word about. Pay attention to the flow of goods, this slogan B2C e-commerce site hit the save shopping time without any appeal, appeal to consumers online shopping will gradually disappear. The main reasons for the above problems are:

The website operator's information processing is slow, the processing flow is long, the website backstage support system is not perfect, causes to the customer order processing response delay, cannot complete the operation effectively.

Due to the small number of B2C orders, shopping varieties scattered, B2C operators is difficult to find a stable cooperation with upstream suppliers, supply cannot be met at any time, resulting in consumers waiting too long.

When it comes to Payment system, online payment cost a longer time because the banking system has not yet achieved a good docking.

2) *High delivery costs*

Due to the small number of goods, the distribution of the client is more dispersed characteristics of B2C distribution and is difficult to form economies of scale. Coupled with the logistics and distribution system is not developed, a lot of B2C sites and did not get the support of professional logistics companies, so the cost of distribution must be high. High distribution costs may be added to the commodity price in disguise, thereby weakening the advantages of B2C low price strategy.

3) *Limited distribution area*

In our country, domestic economic development level is not balanced, the number of different parts of the network there is a great difference between the existing distribution systems is limited in large and medium-sized city, small city and the vast rural areas is insufficient. Some e-commerce companies and distribution companies in the process of cooperation in the delivery order and the B2C operator, and will eventually be delivered scorched by the flames, so that consumers will all return to the sins of B2C operator, has a bad influence on the credibility of the site B2C.

These are some of the difficulties in the development of B2C e-commerce because of the backward logistics cannot match the fast electronic means. B2C e-commerce website if you want to be able to continue to develop, only to establish efficient, low-cost, fast response, low error rate B2C logistics distribution system.

## 3.2 Characteristics Analysis

In the era of e-commerce, distribution centers can be summarized as the following characteristics:

Fast response: under the environment of e-commerce, distribution center customer requirements on the reaction rate increases gradually, and the information processing time decreases, because the distribution process planning more scientific, reasonable, delivery time is getting shorter, the speed of delivery increased.

Function: a new integrated logistics distribution center functions to constantly improve and perfect, the new distribution center in the logistics function integration, for example, the integration between logistics and manufacturing.

Service serialization: in the e-commerce environment, the distribution center in addition to the traditional logistics services, but also to the upper and lower reaches of the extension. To provide accurate market research report for suppliers; provide scientific and reasonable suggestions to the supplier's inventory control; provide consulting services for customers, including the consultation of the distribution plan, etc. Serialization of services provides a profit growth point for the distribution center.

The Target Systematization means distribution center develops the overall objectives from the perspective of the system. In view of the relationship between the processing system of logistics, business flow, good balance between the various distribution activities, the pursuit of the overall benefit maximization.

Organization network: e-commerce logistics distribution center, with more and more perfect logistics distribution network, through the network distribution center to provide customers with fast, accurate and comprehensive service. Through the overall control of the network, the distribution center can monitor all the logistics activities in the network.

Process automation: a new type of distribution center of goods sorting, warehousing, handling and other activities are automated or semi-automated, through the process of automation, logistics efficiency of the distribution center to improve. In the management system, the distribution center has strict rules and regulations; each process has its corresponding management approach to achieve the management.

## 3.3 Hypothesis

Liu in his master's thesis in the distribution center location model with the lowest shipping cost as the goal, according to the characteristics of e-commerce distribution center location model, considering the effect of delivery time on the distribution center location, the delivery delay coefficient and delivery early coefficient distribution time into specific model of distribution cost.

$$\min E = \sum_{i \in I} G_i y_i + r_1 \sum_{k=1}^{K} \sum_{i=1}^{N} \sum_{j=1}^{N} D_{ij} x_{ijk} + \sum_{k \in K} \frac{C}{r_1} + r_2 \sum_{i=1}^{I} Max(E_i - T_i, 0) + r_3 \sum_{i=1}^{I} (T_i - L_i, 0) \quad (1)$$

$$p_1 = r_1 \sum_{k=1}^{K} \sum_{i=1}^{N} \sum_{j=1}^{N} D_{ij} x_{ijk} + \sum_{k \in K} \frac{C}{r_1} \quad (2)$$

$$P_2 = r_2 \sum_{i=1}^{I} Max(E_i - T_i, 0) + r_3 \sum_{i=1}^{I} (T_i - E_i, 0) \quad (3)$$

$P_1$ is the distribution cost. The distribution cost consists of service costs and the costs of using vehicle.

$P_2$ represents the cost of delivery time. With the soft time window of the delivery time, the delay penalty coefficient and the penalty coefficient are specified, and the delivery time is transformed into the distribution cost, and finally the objective function of the minimum distribution cost is solved.

From the point of view of supply chain, the paper puts forward the distribution center location model which is suitable for the supply chain environment, and establishes the distribution center location model with the minimum cost as the goal. Considering the different operation cost of different service radius, the variable cost is introduced into the model, and r is the service radius. The introduction of variable costs in the model is closer to the actual operation of the model, is the location of the distribution center more three-dimensional. Variable cost is the operating cost with the change of service radius.

This paper is established in accordance with the B2C e-commerce distribution center location model based on the two models, the introduction of changes in the cost of distribution based on the minimum cost, the cost in the model is closer to the actual operating conditions of enterprises. Setting a constant average cost will be converted into the distribution of the cost of delivery, and ultimately the establishment of the distribution of the minimum cost of the distribution center location model. Specific model sees below.

According to the characteristics of B2C e-commerce logistics distribution, the following assumptions are made before the model is proposed:

- The location and number of alternative distribution centers are known.

In this paper, the optimal distribution center is selected by using genetic algorithm and MATLAB in a given distribution center.

- The number of customers in a region and the demand for goods is known.
- The number of goods loaded per vehicle is the same.
- The time constant for the distribution of vehicles per kilometer.

- The cost of each distribution center is not the same, but it is fixed and known.
- Transportation costs are proportional to the transport distance.
- The distance between the distribution center and the customer is represented by the shortest straight line distance between the two.
- The vehicles of the distribution vehicles are in the legal norms, not overloaded.
- Each customer can only be served once.
- The variable cost of the distribution center is a continuous function of its coverage radius.
- The distance between the distribution center and the customer is expressed as the coverage radius of the distribution center.
- Other assumptions.

If the goods are not damaged in transit, do not take into account the restrictions on working hours, do not take into account the passage of the road, do not consider the transport rules and regulations etc.

## 3.4 Symbol Definition

$I$ represents the distribution center location set;
$J$ represents a potential distribution center location set;
$B$ represents a collection of distribution vehicles;
$K$ said the customer point set; $F$ delivery time;
$D_{ij}$ represents the distance between $i$ and $j$;
$W_i$ represents $i$ distribution center construction cost;
$a_j$ represents the demand for customer $j$;
$S$ represents the maximum load of the vehicle;
$C$ represents the unit's distance from the vehicle when the task is carried out;
$r_{ij}$ represents the service radius of the distribution center $i$ to the customer $j$;
$t_{ij}$ represents the time spent per kilometer of the vehicle;
$Q$ represents the cost per hour of the distribution center in the delivery of goods.

$$x_{ijb} = \begin{cases} 1 & \textit{Distribution center is selected} \\ 0 & \textit{Other} \end{cases}$$

$$f_i = \begin{cases} 1 & \text{Distribution center is selected} \\ 0 & \text{Other} \end{cases}$$

# 4 Model Establishment and Analysis

## 4.1 Model Establishment

According to the characteristics of B2C e-commerce, this paper establishes a multi-objective function which includes the operating cost (service radius), the minimum delivery cost and the shortest delivery time:

$$\min R = \sum_{i \in I} W_i f_i + \sum_{i \in I} \varphi(r_{ij}) f_i + C \sum_{b=1}^{B} \sum_{i=1}^{K} \sum_{j=1}^{K} D_{ij} x_{ijb} + \sum_{i \in I} \max\left(t_{ij} D_{ij} x_{ijb}\right) \tag{4}$$

Subject to:

$$\varphi(r_{ij}) = Tr_{ij}^2 \tag{5}$$

$$\sum_{b \in B} \sum_{i \in k \cup \{0\}} x_{ijb} = 1, \forall_j = 1, 2, 3 \ldots K \tag{6}$$

$$\sum_{b \in B} \sum_{i \in k \cup \{0\}} x_{ijb} - \sum_{b \in B} \sum_{i \in k \cup \{0\}} x_{jib} = 0 \tag{7}$$

$$0 \leq \sum_{j \in J} a_j \sum_{i \in I \cup J} x_{ijb} \leq S \tag{8}$$

$$\sum_{i \in I} \sum_{j \in J} x_{ijb} \leq 1 \tag{9}$$

Equation (4) is the objective function of the model, which is composed of the fixed cost of the distribution center, the cost of the distribution center and the operation cost of the distribution center. Equation (5) express distribution center to change j the cost of customers within the service area required for the type. Equation (6) said the customer is only a service type; Eq. (7) said the customer is a conservation type. Equation (8) any delivery vehicles are not overloaded type. Equation (9) said a customer only corresponds to a distribution center, it only accepts the distribution center service provided.

Genetic algorithm is an algorithm simulation of the natural evolution of the algorithm, first select the initial population, calculate the fitness of individuals (fitness must be non-negative), then the initial population according to the calculation of fitness, the excellent individual genetic to the next generation, the crossover operation can make excellent genetic properties of from generation to generation, through three kinds of genetic operators are more adapt to the environment of the individual, this cycle continue to evolve, Until the problem is optimal solution.

## 4.2 Numerical Example Analysis

This paper assumes that there are 5 candidate distribution centers, and 20 potential customers, each of the fixed cost of the distribution center is known.

The relevant data of the candidate distribution centers are shown in Table 1. The potential customer data are shown in Table 2.

Through the calculation of the 5 distribution centers in the election of the two responsible for the distribution of the task of the 20 customers, we achieve our goal of optimizing the distribution costs and delivery time. The following parameters were calculated: unit distance distribution cost per kilometer $C$ is 5; time spent in $t_{ij}$ was 30; Q distribution operation cost per hour is 10; the evolution algebra is 20, the population size is 10, the mutation probability is 0.75.

Table 3 gives the initial population and the objective function.

The evolution of the population and the objective function values are calculated in Table 4.

Through the calculation of the first and the third distribution center for goods distribution, the ultimate goal of the program function value is 2613.8, consistent with the traditional location method, the results prove the reliability and practicability of the algorithm.

**Table 1** Customer data

| Customer number | Ordinate | Abscissa | Requirement | Customer number | Ordinate | Abscissa | Requirement |
|---|---|---|---|---|---|---|---|
| 1 | 43 | 71 | 300 | 11 | 25.5 | 33 | 100 |
| 2 | 45 | 63 | 100 | 12 | 53.3 | 10 | 400 |
| 3 | 23 | 84 | 200 | 13 | 41 | 16 | 100 |
| 4 | 63 | 13 | 100 | 14 | 66 | 23.5 | 100 |
| 5 | 30 | 55 | 200 | 15 | 5 | 33.3 | 100 |
| 6 | 66.2 | 37.9 | 100 | 16 | 33 | 81 | 100 |
| 7 | 12.1 | 49 | 100 | 17 | 42 | 30 | 100 |
| 8 | 10 | 17 | 200 | 18 | 65 | 27.5 | 200 |
| 9 | 50 | 40 | 500 | 18 | 25 | 55 | 100 |
| 10 | 60 | 86 | 600 | 20 | 14 | 75 | 200 |

**Table 2** Data of distribution center

| Distribution center number | Ordinate | Abscissa | Fixed cost |
|---|---|---|---|
| 1 | 50 | 43 | 190 |
| 2 | 36 | 45 | 220 |
| 3 | 40 | 23 | 200 |
| 4 | 36 | 63 | 210 |
| 5 | 63 | 30 | 220 |

**Table 3** Gnitial population and objective function

| Initial population | | | | | Objective function value |
|---|---|---|---|---|---|
| 1 | 1 | 0 | 0 | 0 | 3081.5 |
| 1 | 0 | 1 | 0 | 0 | 2958.7 |
| 1 | 0 | 0 | 1 | 0 | 2851.1 |
| 1 | 0 | 0 | 0 | 1 | 2798.2 |
| 0 | 1 | 1 | 0 | 0 | 2989.9 |
| 0 | 0 | 1 | 1 | 0 | 2687.4 |
| 0 | 0 | 0 | 1 | 1 | 2876.3 |
| 0 | 1 | 0 | 1 | 0 | 2986.1 |
| 0 | 0 | 1 | 0 | 1 | 2721.7 |
| 0 | 1 | 0 | 0 | 1 | 2802.5 |
| Initial population objective function mean value | | | | | 2875.3 |

**Table 4** Evolutionary population and objective function values

| Post evolution population | | | | | Objective function value |
|---|---|---|---|---|---|
| 0 | 1 | 1 | 0 | 0 | 2970.5 |
| 1 | 1 | 0 | 0 | 0 | 2682.6 |
| 1 | 0 | 1 | 0 | 0 | 2613.8 |
| 0 | 0 | 1 | 0 | 1 | 2735.2 |
| 1 | 0 | 0 | 1 | 0 | 2851.5 |
| 0 | 0 | 1 | 1 | 0 | 2650.8 |
| 0 | 0 | 0 | 1 | 1 | 2661.2 |
| 0 | 1 | 0 | 1 | 0 | 2850.4 |
| 0 | 1 | 0 | 0 | 1 | 2901.8 |
| 1 | 0 | 0 | 0 | 1 | 2741.2 |
| Mean value of population objective function after evolution | | | | | 2767.2 |

## 5 Conclusion

Based on the characteristics of logistics under e-commerce environment is studied, analyzed and summarized the distribution center of B2C e-commerce enterprise features, based on B2C e-commerce enterprise distribution center location factors on the analysis of B2C e-commerce enterprise established distribution center location model. The main results of this paper are as follows:

1) *The distribution center location model of B2C e-commerce enterprises with the lowest cost and shortest delivery time is established*

In this paper, through the analysis of the characteristics of the current B2C e-commerce enterprise distribution center, and then in-depth study of the distribution center location problem, discussed the B2C e-commerce enterprise distribution center location model. In theory, it is of distribution center location of B2C e-commerce enterprises is supplemented. The difference between the traditional model and location model of distribution center is: first, the model considering the influence of service radius of distribution center operating costs, according to the service radius increasing operation cost more rules and model into variable costs by the service radius, make the model more closely to reality. Two, according to the distribution center location is more likely to be affected by the distribution time, the distribution time is introduced into the model, the establishment of the shortest delivery time as the objective function.

2) *Using genetic algorithm and MATLAB to solve the model, the results show that the model is feasible*

In the model, this method uses genetic algorithm combined with MTLAB, first introduced the related knowledge of genetic algorithm, then according to the characteristics of the model design a reasonable algorithm flow, design reasonable encoding process and fitness function, selection method of binary encoding, genetic algorithm in parent population selection roulette selection, single point crossover in the crossover operation, the ultimate combination of genetic algorithm toolbox in MATLAB software programming, finally realize the solution of the model.

## References

1. Chen, Y., & Wang, G. W. (2006). Study on B2C logistics delivery model under electronic commerce. *Journal of Liaoning Provincial College of Communications, 8*(3), 63–64.
2. Yao, J. (2016). Optimization of one-stop delivery scheduling in online shopping based on the physical Internet. *International Journal of Production Research, 55*(2), 1–19.
3. Ghobakhloo, M., Hong, T. S., & Standing, C. (2015). B2B e-commerce success among small and medium-sized enterprises. *Journal of Organizational & End User Computing, 27*(1), 1–32.
4. Su, S. X., & Hu, Y. (2011). Simulated annealing on B2C e-commerce companies' distribution center selection. *Logistics engineering and management, 33*(205), 75–77.

5. Hao, D. (2011). Distribution network optimization at CRE Beijing branch. *Logistics Technology, 30*(4), 75–77.
6. Lin, N., & Li, Z. (2010). Based on GIS and genetic algorithm. *Remote Sensing Information, 5,* 110–114.
7. Moons, S., Ramaekers, K., & An, C. (2017). Integration of order picking and vehicle routing in a B2C e-commerce context. *Flexible Services & Manufacturing Journal, 30,* 1–31.
8. Bai, J. F., Wei, X. Y. & Yan, J. C. (2017). Research on the selection of business-to-customer e-commerce logistics model based on analytic hierarchy process method. In *23rd International Conference on Industrial Engineering and Engineering Management 2016* (pp. 11–15).
9. Deng, X. P., Zhou, X. M., & Tian, S. H. (2016). Research on integrated optimization of location and routing for B2C E-commerce logistics center. *Journal of Chongqing University of Posts & Telecommunications, 28*(4), 593–600.
10. Gao, B. X., & Wang, D. (2016). Logistics network optimization of B2C e-commerce based on constrained clustering algorithm. *Journal of Harbin University of Commerce, 32*(1), 71–76.

# Inventory Strategy of Supply Chain Under Delay in Payments

**Shanshan Gao and Peng Meng**

**Abstract** In the field of supply chain, pricing and inventory strategies have always been important issues for scholars to study. This paper uses Stackelberg game theory as a basic analysis tool to study the effect of deferred payment period and interest rate on the profit of each component of the supply chain, as well as the profit and inventory cost of the whole supply chain. From the angle of maximum profit of manufacturer, retailer and the whole supply chain, the manufacturer determines the optimal deferred payment period, and the retailer determines the optimal retail price and order quantity. The model provides decision reference for each member of the secondary supply chain. The conclusion is that: After the supply chain implements the deferred payment strategy, the profits of manufacturers, retailers, and the entire supply chain are increased, and the inventory cost of the entire supply chain is reduced; the lowering of the manufacturer's interest payment rate will lead to optimal extension of deferred payment period. The extension of its term will increase the profits of each subject in the supply chain, and the inventory costs of the supply chain will decrease, ultimately realizes optimal inventory management.

**Keywords** Supply chain · Inventory strategy · Game theory · Delay in payments

S. Gao (✉) · P. Meng
School of Economics and Management, Qingdao University of Science and Technology, Qingdao, China
e-mail: gaoshan2929@163.com

P. Meng
e-mail: roc998@163.com

P. Meng
Business Management Department, Qingdao Vocational and Technical College of Hotel Management, Qingdao, China


J. Zhang et al. (eds.), *LISS2019*,
https://doi.org/10.1007/978-981-15-5682-1_34

# 1 Introduction

At the beginning of the 20th century, supply chain inventory management came into being. At that time, Harris (1915) put forward the view of inventory model for the first time, but this view did not arouse much reaction. Wilson (1934) got Harris's conclusion again. Many scholars began to understand the theory of inventory and apply it in practice. Attention. Wang [1] and others studied the problem of transportation under certain constraints and the distribution of goods with waiting time game. Similarly, some scholars studied the effect of supply and demand on inventory under the condition of asymmetric information. For example, Thangam [2] explored the optimal pricing and ordering model under demand structure constraints of two-tier deferred payment strategy. Ozen and Sisco [3] discussed that the redistribution of inventory and the effect of demand on inventory allocation when the relevant demand information updated. Overview of foreign research, we can find that there are two aspects of understanding of the theory of inventory research, one is to optimize the inventory management of individual enterprises, the other is to optimize the inventory management in the context of supply chain.

Compared with foreign countries, it is relatively late to study the inventory problem using the theory of game. Shao [4] studied the optimal optimization strategy mainly for the hypothesis of the secondary supply chain, and designed the inventory contract on this basis. Yang [5] studied the game model of single manufacturer and multiple retailers under the condition of deferred payment, and obtained the optimal deferred payment strategy. Wang [6] established a Steinberg model led by supplier under the strategy of deferred payment and price discount, and studied the setting of conditional deferred payment strategy based on price discount from the supplier's point of view. Ran [7] studied the inventory problem of a class of shelf and warehouse goods which jointly affect demand, and considered the two-tier deferred payment strategy. Cui [8] studied the problem of two-stage deferred payment inventory strategy, in which the demand for perishable goods depends on both inventory and deferred payment period, considering that the market demand is not only affected by the retailer's inventory, but also often depends on the retailer's deferred payment strategy. Kang [9] constructed a newsboy model and formulated an ordering strategy with consideration of product defects and allowable late payment.

The problem of supply chain inventory model under uncertain information and delayed payment has aroused widespread concern of scholars both at home and abroad. However, most of the previous studies are based on the game model of transportation and retailer waiting time, or on the study of supply chain inventory separately. But few studies have considered Selling costs of retailers. Therefore, on the basis of the mature secondary supply chain already completed by the former, this paper puts forward the conditions of deferred payment to obtain the best deferred payment period of the supply chain, and adds the sales cost in the expression of retailer's profit, and compares the different of retailer's sales price before implementing deferred payment strategy and after implementing deferred payment strategy. The profit of each party and the profit and cost of the whole supply chain are changed. By comparing, the profit of all parties and the profit of the whole supply chain are improved, and the inventory cost of the whole supply chain is reduced. This paper also studies the relationship between the optimal deferred payment period and the manufacturer's interest rate, and studies the influence of the change of the manufacturer's interest rate on the retailer's optimal selling price, the retailer's order quantity, the profit of all parties and the profit and inventory cost of the whole supply chain.

## 2 Mathematical Model

### 2.1 *Model Assumptions*

In order to simplify the model, the supply chain system considered in this paper is a two-level supply chain, a level has only one member. So it involves only one manufacturer and one retailer. In order to streamline the model, it involves only one product. Because the time limit for this delay is determined by the manufacturer, the manufacturer is in the leading position. The specific assumptions are as follows:

(1) This paper studies the two level supply chain system with only one manufacturer and one retailer. In contrast, manufacturers are more active.
(2) The product is a normal commodity, and its demand will be reduced by the price rise. In order to facilitate discussion, it is positioned as a linear relationship.
(3) The retailer's inventory will decrease as the sales time increases.
(4) The retailer's order period is fixed and no shortage is allowed.
(5) The manufacturer's pricing is fixed.

### 2.2 *Model Notations*

$p$: unit price of retailer
$c_1$: unit price of manufacturer
$D$: unit time requirement $D = a - bp$
$a, b$: parameter
$A$: ordering cost for retailers
$s$: unit transportation cost
$B$: selling expenses
$f$: unit sales cost
$c_2$: unit cost of products for manufacturer
$e$: unit time output of manufacturer

$T$: ordering cycle
$H$: unit inventory cost for manufacturer
$M$: delay in payment
$Q$: order quantity for unit period
$\pi$: profit of retailer
$\Pi$: profit of manufacturer
$\Gamma$: profit of the supply chain
$C$: inventory cost of the supply chain

## 2.3 Model Description

The relationship between manufacturer, retailer and consumer is shown in the Fig. 1.

According to the model, the profits of the manufacturer, the retailer and the whole supply chain are shown below.

Manufacturer's profit:

$$\begin{aligned} &\Pi(p) \\ &= (c_1 - c_2)Q - H(eT - DT) \\ &= T(a - bp)(c_1 - c_2 + HT) - HeT \end{aligned} \tag{1}$$

Retailer's profit:

$$\begin{aligned} &\Pi(p) \\ &= (p - c_1)Q - (B + fQ) - \frac{DT^2h}{2} - (A + sQ) \end{aligned} \tag{2}$$

The supply chain's profit:

$$\begin{aligned} &\Gamma(p) = \pi(p) + \Pi(p) \\ &= (p - c_1)Q - (B + fQ) - \frac{DT^2h}{2} - (A + sQ) \\ &= T(a - bp)(c_1 - c_2 + HT) - HeT \end{aligned} \tag{3}$$

The inventory of the whole supply chain:

$$\begin{aligned} C &= \frac{DT^2h}{2} + H(eT - DT) \\ &= \frac{(a-bp)T^{2h}}{2} + H(e - a + bp)T \end{aligned} \tag{4}$$

**Fig. 1** Model without consideration of deferred payment

### 2.4 Model Solution

The retailer's pricing will affect the market demand, and then affect the profit. Retailers can maximize their profits by considering the price of goods.

When $\pi'(p) = 0, p^*$ can be obtained,

$$p^* = \frac{c_1 + s + f + \frac{Th}{2}}{2} + \frac{a}{2b} \tag{5}$$

Therefore, the optimal order quantity, the optimal profit of the retailer, the optimal profit of the manufacturer, the optimal profit and inventory of the supply chain can be obtained, and individually expressed by $Q(p^*)$, $\pi(p^*)$, $\Pi(p^*), \Gamma(p^*)$, $C(p^*)$.

## 3 Strategy Model Under Delay Payment

### 3.1 Model Assumptions

On the basis of the assumptions in the mathematical model, the manufacturer gives the retailer appropriate delay in payment. During the period of delay payment, the retailer's sales revenue will bring some interest income to the retailer. After the deadline, the retailer pays all the payments to the manufacturer.

### 3.2 Model Notations

Based on the parameter definition in the second section, two new parameters definitions are added in this section:

$I_c$: interest payment
$I_d$: interest income

### 3.3 Model Description

According to the model, The assumptions of the deferred payment model are the same as those in the previous section. Interest on deferred payments is added to this. The profits of the manufacturer, the retailer and the supply chain are shown below (Fig. 2).

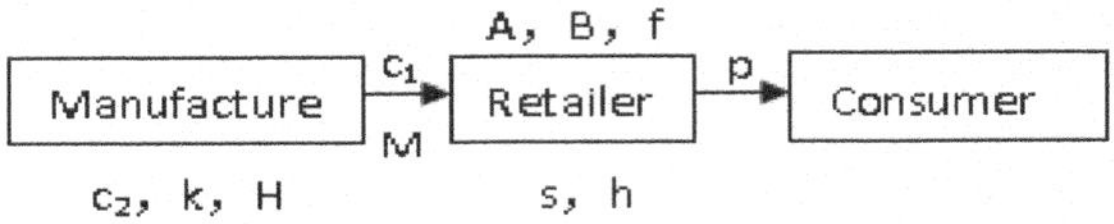


**Fig. 2** Model considering delay payment

*(1) Retailer's profit*

Whether or not the delay payment strategy is implemented, the expressions of the retailer's profit, inventory cost and order cost will not change, but the interest income caused by the implementation of deferred payment strategy will change.

*(a)* $M \geq T$

The retailer has sold out all the orders at the time of *T*. At the time of *T*, the retailer has sales revenue *pDT*, which is not delivered to the manufacturer until the time of *M*. Interest is not paid during the *T-M* period. The image relationship between retailer's sales income *y* and time *t* is shown in Fig. 3.

So the interest income from sales revenue is:

$$\begin{aligned} & pI_d\left(\int_0^T Dtdt + DT(M-T)\right) \\ & = \frac{pI_dDT^2}{2} + pI_dDT(M-T) \\ & = pI_dDT(T-M) \end{aligned}$$

When the period of deferred payment is longer than the order cycle, the retailer's profit is obtained by subtracting inventory cost, order cost and sales cost from the sales profit, plus the interest income from deferred payment.

In this case, retailer's profit is:

$$\begin{aligned} & \pi_1(p, M) \\ & = (p-c_1)Q - (B+fQ) - \frac{DT^2h}{2} \\ & -(A+sQ) + pI_dDT\left(M - \frac{T}{2}\right) \end{aligned} \tag{6}$$

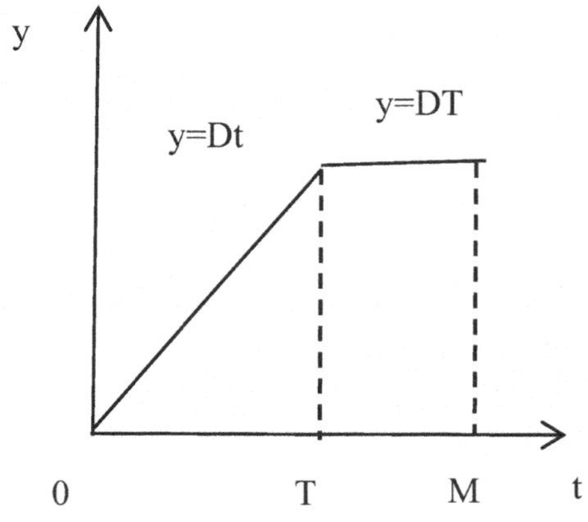


**Fig. 3** Functional image between retailer sales revenue and time

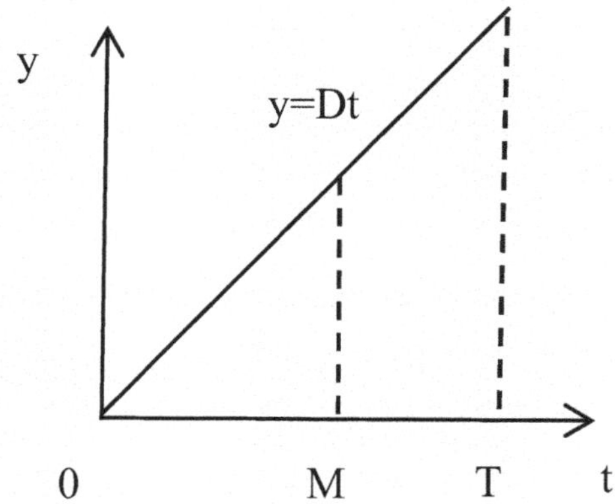


**Fig. 4** Functional image between retailer sales revenue and time

In this case, retailer's profit is:

*(b) M < T*

The retailer does not sell all the goods, but only part of the goods. The retailer's sales revenue *y* and time *t* are graphically related, as shown in Fig. 4.

The interest income generated by this part of the income is:

$$pI_d \int_0^M Dtdt = \frac{pI_dDM^2}{2}$$

In this case, retailer's profit is:

$$\begin{aligned}&\pi_2(p, M) \\ &= (p - c_1)Q - (B + fQ) - \tfrac{DT^2h}{2} - (A + sQ) + \left(\tfrac{pI_dDM^2}{2}\right)\end{aligned} \tag{7}$$

From (6) and (7) we can find that (6) and (7) are both incremental functions, so for retailers, the longer the deferred payment period, the higher the retailer's profit.

*(2) Manufacturer's profit*

The sales profit and inventory cost of the manufacturer are not related to the delay in payment, but only to the interest loss. Due to the implementation of the delayed payment strategy, the manufacturer's interest loss is $c_1QI_cM$.

Thus, we obtain the profit of manufacturer.

$$\begin{aligned}\Pi &= (c_1 - c_2)Q - H(eT - DT) - c_1QI_cM \\ &= T(a - bp)(c_1 - c_2 - Mc_1I_c + HT) - HeT\end{aligned} \tag{8}$$

(8) is a monotone decreasing function for M. For the manufacturer, the shorter the deferred payment period, the higher the profit of the manufacturer when the retailer's order quantity remains unchanged.

*(3) Supply chain's profit*

The profit of the entire supply chain includes the profit of the retailer and the profit of the manufacturer. There are two possible scenarios for the profits of the entire supply chain. We discuss them separately.

*(a)* $M \geq T$

The profit equation of the supply chain can be written as:

$$\begin{aligned} &\Gamma_1(p,M) = \pi_1(p,M) + \Pi_1(p,M) \\ &= \left[(p-c_1)Q - \tfrac{DT^2h}{2} - (A+sQ) - (B+fQ)\right] + pI_dDT \\ &\times \left(M - \tfrac{T}{2}\right) + [T(a-bp)(c_1 - c_2 - Mc_1I_c + HT) - HeT] \end{aligned} \tag{9}$$

*(b)* $M < T$

The profit equation of the supply chain can be written as:

$$\begin{aligned} &\Gamma_2(p,M) = \pi_2(p,M) + \Pi_2(p,M) \\ &= \left[(p-c_1)Q - \tfrac{DT^2h}{2} - (A+sQ) - (B+fQ)\right] + \tfrac{pI_dDM^2}{2} \\ &+ [T(a-bp)(c_1 - c_2 - Mq_1I_c + HT) - HeT] \end{aligned} \tag{10}$$

*(4) The inventory cost of the supply chain*

The inventory cost of supply chain includes two parts: retailer's inventory cost and manufacturer's inventory cost.

$$\begin{aligned} C &= \tfrac{DT^2h}{2} + H(eT - DT) \\ &= \tfrac{(a-bp)T^2h}{2} + H(e - a + bp)T \end{aligned} \tag{11}$$

## 3.4 Model Solution

First, the manufacturer decides to the payment period, according to the manufacturer's biggest profit. Then, the retailer determines the optimal retail price according to the manufacturer's deferred payment period, and then determines the order quantity according to the optimal retail price and the order period.

*(1) Optimal delay payment period*

When $\pi_1(P)' = 0$, $\pi_2(P)' = 0$, we can obtain:

$$P^{**} = \begin{cases} P_1 = \frac{c_1 + \frac{Th}{2} + s + f}{2\left(I + MI_d - \frac{I_d T}{2}\right)} + \frac{a}{2b}, M \geq T \\ P_2 = \frac{c_2 + \frac{Th}{2} + s + f}{2\left(I + \frac{I_d M^2}{2T}\right)} + \frac{a}{2b}, M < T \end{cases} \tag{12}$$

It is not difficult to find that after the implementation of deferred payment strategy, the retailer's sales price has decreased. From (12) we can know that the price decreases gradually with the increasing of $M$, Because the demand function is $D = a - bp$, so with the decreasing of price, the demand will increase, so the order quantity will also increase. That is to say, if the manufacturer extends the deferred payment period, the retailer's order volume will increase. However, increasing the time limit of deferred payment may also lead to a decrease in manufacturer' profits, because the manufacturers will have a certain loss of interest if the deferred payment policy is implemented.

Then we put $p_I$ into (8), and we can obtain:

$$\begin{aligned} &\Pi_1(p, M) \\ &= T\left[\frac{a}{2} - \frac{b\left(c_1 + \frac{Th}{2} + s + f\right)}{2\left(I + MI_d - \frac{I_d T}{2}\right)}\right] \\ &\times (c_1 - c_2 - Mc_1 I_c + HT) - HeT \end{aligned} \tag{13}$$

When

$$2 - I_d T = R_1,$$

$$b\left(c_1 - \frac{Th}{2} + s + f\right) = R_2,$$

$$c_1 - c_2 - HT = R_3$$

(13) is changed to:

$$\begin{aligned} &\Pi_1(p_1, M) \\ &= 2\left(\frac{a}{2} - \frac{R_2}{R_1 + 2MI_d}\right)(R_3 - Mc_1 I_c) - HeT \end{aligned} \tag{14}$$

$\frac{\partial \Pi_1(p_1, M)}{\partial M} = 0$, Thus, the optimal delay payment period is:

$$M_1^* = \frac{\sqrt{\frac{4R_2R_3I_d + 2R_1R_2c_1I_c}{ac_2I_c}} - R_1}{2I_d} \tag{15}$$

So there is a maximum value $\Pi_1^*$ in the profit function of the manufacturer, and the optimal deferred payment period is $M^*$.

In the same way, we can find out:

$$\begin{aligned}&\Pi_2(p_1,M)\\&=2\left(\frac{a}{2}-\frac{R_2}{2T+2DM^2}\right)(R_3-Mc_1I_c)-HeT\end{aligned} \tag{16}$$

Then

$$\begin{aligned}&\frac{\partial\Pi_2(p_1,M)}{\partial M}\\&=T\left[\frac{2R_2R_3I_dM+k_2c_1I_cI_dM^2+2R_2c_1I_cY}{(2T+I_dM^2)^2}-\frac{ac_1I_c}{2}\right]\end{aligned} \tag{17}$$

$$\begin{aligned}&\frac{\partial^2\Pi_2(p_1,M)}{\partial M}\\&=\frac{(2R_2R_3I_d+R_2c_1I_cI_dM^2)(I-4I_dM^2)-8I_dMR_2c_1I_cT}{(2T+I_dM^2)^2}\end{aligned} \tag{18}$$

From (9) we can know that $\Pi_2(p,M)$ does not always have a maximum value. If it has the maximum value, the manufacturer will get the maximum profit. When $M^*=M_2^*$,compare $\Pi_1^*$ and $\Pi_2^*$, if $\Pi_1^*>\Pi_2^*$, $M^*=M_1^*$, otherwise $M^*=M_2^*$.

*(2) Optimal retail price and order quantity*

According to the deferred payment period $M^*$ given by the manufacturer, the retailer will take it into the retailer's optimal retail price function to get the optimal retail price.

$$P^{**}(M^*)=\begin{cases}P_1=\dfrac{c_1+\frac{Th}{2}+s+f}{2\left(I+MI_d-\frac{I_dT}{2}\right)}+\frac{a}{2b},M\geq T\\[2ex] P_2=\dfrac{c_2+\frac{Th}{2}+s+f}{2\left(I+\frac{I_dM^2}{2T}\right)}+\frac{a}{2b},M<T\end{cases} \tag{19}$$

So the optimal order quantity is:

$$Q(p^{**}(M^*))=[a-bp^{**}(M^*)]T$$

*(3) The inventory cost of the supply chain*

Combining the optimal delay time determined by the manufacturer and the optimal retail price and order quantity determined by the retailer, the inventory cost of the whole supply chain can be obtained.

$$\begin{aligned}&C(p)\\&=\frac{DT^2h}{2}+H(eT+DT)\\&=\frac{(a-bp)T^2h}{2}+H(e-a+bp)T\end{aligned} \tag{20}$$

So

$$C'(p) = bT\left(H - \frac{T}{2}\right) \tag{21}$$

Because the delay payment strategy is mainly applied to products with higher inventory costs, so, as the price increases, inventory costs will increase. By comparing (5) with (12), it is found that the optimal selling price of retailers decreases after the implementation of delay payment strategy, and the inventory cost decreases.

## 4 Analysis

*(1) The retailer's retail price is reduced after the manufacturer implements the strategy of delay payment to the retailer.*

By comparing (5) with (12), it is found that $p^* > p^{**}$. The retailer's retail price decreases when the manufacturer allows the retailer to delay payment.

Manufacturers must always update information and keep the timeliness of information, especially retailers' information such as the retailer's inventory costs, transportation costs, and so on, because these factors have a great impact on manufacturers to determine the deadline for delay payment.

*(2) After implementing the strategy of delay payment, the inventory cost of the entire supply chain is reduced.*

Because $C'(p) > 0$,and the retail price will be reduced after the manufacturer implements the delay payment strategy to the retailer, we can know that the inventory cost of the whole supply chain will be reduced.

From the above analysis, the delay payment strategy helps to reduce the inventory cost of the entire supply chain. Therefore, from the perspective of the entire supply chain, delay payment should be implemented.

*(3) With the decrease in interest rate paid by manufacturers, the optimal delay payment period is extended.*

When $I_c$ decreases, $M^*$ increases. The optimal delay payment period given by the manufacturer to the retailer increases with the decrease of the interest rate paid by the manufacturer.

Those enterprises that do not implement deferred payment strategy, can consider implementing deferred payment strategy, because the inventory cost of the whole supply chain is lower after the implementation of deferred payment strategy than before.

After implementing the delay payment strategy, the manufacturer should adjust the delay payment period properly according to the change of the influencing factors of the delay payment period, so that the delay payment period time fluctuates near the optimal value. When the interest rate is lower, the manufacturer can extend the period of delay payment appropriately; when the interest rate is higher, the manufacturer can shorten the period of delay payment appropriately.

The manufacturer must update the information at all times to keep the timeliness of the information, especially the information of the retailer. For example, the retailer's inventory costs, transportation costs and so on, because these factors have a great impact on the manufacturers to determine the deferred payment period.

## 5 Conclusions and Discussion

This paper establishes two inventory models with deferred payment policy and without deferred payment policy. Through comparative analysis, it is found that the delay payment strategy plays a positive role in reducing inventory cost. The conclusion is as follows:

Firstly, the profit of manufacturer, retailer and the whole supply chain is increased and the inventory cost of the whole supply chain is reduced after the enterprises implement the deferred payment strategy. The delay payment strategy has a positive effect on the optimization of inventory cost.

Secondly, the interest rate paid by the manufacturer has an effect on the deferred payment period. When the interest rate paid by the manufacturer decreases, the optimal deferred payment period will increase. When the interest rate changes, the manufacturer should adjust the deferred payment period appropriately.

Thirdly, the order quantity of the retailer, the profit of each member of the supply chain and the whole supply chain, and the inventory cost of the whole supply chain all change with the delay of payment.

Based on Stackelberg's game theory, this paper further deepens the study of deferred payment strategy, and analyses the influence of manufacturer's interest rate on deferred payment period. The sale cost is added to the calculation of retailer's profit, and the inventory cost of the manufacturer is introduced into the Stackelberg game model.

In the case of implementing deferred payment strategy in supply chain and not implementing deferred payment strategy, the changes of profits of all parties and the whole supply chain are compared. It is concluded that the implementation of deferred payment strategy can reduce the inventory cost of the whole supply chain and improve the profits of the members of the supply chain as well as the profits of the whole supply chain. Moreover, the decrease of interest rate of manufacturer will lead to the extension of the optimal deferred payment period and the increase of profits of manufacturer, retailer and the whole supply chain, which will reduce the inventory cost of supply chain.

There are also deficiencies in this study.

Firstly, the research model of this paper contains only one manufacturer and one retailer. But in the actual enterprise management, the complexity of the supply chain is far beyond the model in this paper. The expansion of this model and the introduction of multistage supply chain model will be the next research direction.

Secondly, this paper only studies the effect of manufacturer's interest rate on the deferred payment period, but does not study the effect of retailer's interest income rate on the deferred payment period. In reality, retailer's interest income rate certainly has influence on the deferred payment period, so the influence of retailer's interest income rate on deferred payment is the next research direction.

In summary, the study of deferred payment is of great significance. In the inventory optimization problem, deferred payment is undoubtedly an incentive for dealers to purchase in large quantities. After the retailer determines its own sales cycle, it requests the manufacturer to deliver a batch of goods at a certain time to reduce its cost, and the manufacturer can also increase its control over the retailer, even if the default occurs, whether the manufacturer or the retailer, can handle it calmly.

The advantage of deferred payment is that the manufacturer can not only reduce inventory, but also reduce inventory, and to a certain extent stimulate the retailer's large-scale order behavior. Thus, the manufacturer can get long-term and effective orders without interruption of orders, but also reduce its inventory cost. At the same time, retailers can also increase orders in batches and get discounts. Although this makes the retailer's inventory increase, retailers can get interest and order discounts in the case of deferred payment. Based on the effect of deferred payment, retailers can also reduce the price of products appropriately, in order to seize market share. Although the retailer's profit of per unit product has decreased, the total profit has increased due to the increase of sales volume. And it enlarges the market share and gains more customers.

Considering the deferred payment strategy in supply chain, we analyze the profit changes of suppliers, retailers and the whole supply chain. We find that when the supply chain adopts deferred payment strategy, the total profit of the supply chain increases. Moreover, the inventory cost of suppliers has been reduced, and the market share of retailers has increased.

The influence of the time limit and interest of deferred payment on supply chain is also discussed. Through discussion, the corresponding measures that suppliers and retailers should take to cope with the adverse effects are put forward.

Of course, there are still some shortcomings in the research. In the follow-up research, we will devote ourselves to solving these shortcomings, so that our research will have more practical value. The study of supply chain is of great practical significance. So there is still a lot of research space in the future.

**Acknowledgment** We would like to thank the anonymous research participants for donating their valuable time and knowledge to our study.

## References

1. Wang, H. (2003). Supply chain operation performance roles. *Integrated manufacturing System, 8*(2), 70–78.
2. Thangam, A., & Uthanyakunmar, R. (2010). Optimal pricing and lot-sizing policy for a two-warehouse supply chain system with perishable items under partial trade credit financing. *Operational Reswarch, 13*(10), 289–295.
3. Ozen, U., Erkip, N., & Slikker, M. (2012). Stability and monotonicity in newsvendor situations. *European Journal of Operational Research, 22*(6), 416–425.
4. Shao, X. F., & Ji, J. H. (2004). Research on impact of postponement strategies on inventory cost in mass customization. *Systems Engineering Theory Methodology Applications, 21*(5), 385–389.
5. Yang, S., Liang, L., & Dong, J. F. (2007). An inventory model for single manufacture and multiple retailers. *Systems Engineering, 15*(4), 9–14.
6. Wang, Y. J., & Meng, F. X. (2014). Determination of conditional delay in payment policy based on pricediscount. *Control and Decision., 36*(8), 1413–1418.
7. Ran, C. L., He, W., & Xu, F. Y. (2015). Supply chain inventory model with two levels of trade credit and stock dependent demand. *Mathematics in Practice and Theory, 11*(22), 9–13.
8. Cui, L., Feng, Y. C., & Yao, L. G. (2016). Survey of supply chain game theory. *Systems Engineering Theory Methodology Applications, 7*(5), 385–389.
9. Kang, W. L., Wang, L., & Hang, B. (2017). Busy newsboy model and ordering strategy selection considering product defect rate and delay payment. *Journal of Shanghai Jiaotong University, 51*(3), 379–384.

# Optimization of Urban Emergency Refuge Location Based on the Needs of Disaster Victims

**Shaoqing Geng, Hanping Hou, and Jianliang Yang**

**Abstract** After large-scale disasters, the earthquake refuge site-allocation is the key content of emergency management to quickly resettle the victims and launch emergency rescue. From the perspective of the needs of victims' evacuation, considering various factors such as traffic convenience, environmental safety, facility completeness and service efficiency, this paper constructs the site-distribution optimization model based on the maximization of earthquake refuge weights. The case verifies the validity of the model. Finally, the sensitivity analysis provides inspiration to management decision makers: when the refuge carries more than a certain range, the overall optimality of the refuge becomes worse.

**Keywords** Evacuation demand · Earthquake refuge · Site selection-distribution · Sensitivity analysis

## 1 Introduction

In recent years, various natural disasters have occurred frequently, and the scope of their influence has been expanding, bringing huge loss of life and property to the people in the world. According to relevant data, the economic losses caused by global disasters in the last 10 years of the 20th century were more than five times that

Fund Project: the Fundamental Research Funds for the Central Universities 2019YJS061.

S. Geng (✉) · H. Hou
School of Economics and Management, Beijing Jiaotong University, Beijing, China
e-mail: 794670702@qq.com

H. Hou
e-mail: 18113040@bjtu.edu.cn

J. Yang
College of Economics and Management, Beijing University of Chemical Technology, Beijing, China
e-mail: 17120573@bjtu.edu.cn



J. Zhang et al. (eds.), *LISS2019*,
https://doi.org/10.1007/978-981-15-5682-1_35

of the 1960s, and nearly 80% of all disaster losses occurred in cities and communities [1]. With the rapid development of the economy and the increase in urban population density, disaster prevention and mitigation has become an important part of urban safety research. As a "life shelter", the emergency shelter is an important infrastructure to protect public safety. At present, all countries attach great importance to the planning and construction of evacuation sites [2, 3], and have successively issued relevant management regulations [4, 5], and improving the existing emergency evacuation system will improve urban disaster management capabilities.

The planning of evacuation site selection belongs to the category of emergency facility location, which has attracted wide attention from scholars. Toregas et al. officially proposed the location of emergency facilities, selected the least number of locations in the candidate points that met the time and distance restrictions, set up emergency facilities, and solved the set coverage problem through linear programming [6]. Lin et al. solved the problem of site selection of temporary warehouses around the disaster site after the earthquake and assigned the disaster sites to the corresponding warehouse [7]. Kilci et al. aimed at maximizing the minimum weight of the refuge, and by combining the integer linear programming model, determined the location of the refuge and specified the refuge to which the victims were assigned [8]. Sabouhi et al. established a logistics system for integrated evacuation and material distribution. The victims evacuated from the affected areas to the shelters and provided them with necessary relief materials to minimize the sum time of arrival of vehicles in the affected areas, shelters and distribution centers [9]. Barzinpour and Esmaeili built a multi-objective model to improve decision-making efficiency by constructing virtual partitions, solving evacuation sites and disaster resettlement models [10].

Chen and Ren analyzed the characteristics of the fire station layout planning problem, optimized the average fire-fighting distance, simplified the covered area as the "node" to realize the site selection of fire-fighting facilities and to divide the responsibility area of the fire station [11]. Chen et al. considered the capacity limitation of the shelter, combined with the uneven spatial distribution of urban population and targeted the evacuation of all evacuated personnel to distribute the victims to different emergency shelters [12]. Ma et al. constructed a multi-objective model for disaster evacuation sites, minimized the total evacuation distance of the evacuated population to the shelter and the total area of the shelter under the conditions of safety constraints, distance constraints and capacity constraints [13]. Castro et al. mainly aimed at the densely populated towns, considered the distribution characteristics of the residential area, terrain and other influencing factors, and constructed the earthquake evacuation site model [14].

The above researches mostly base on the facility capacity limitation, and optimize the time, distance or coverage rate. Finally, the evacuation site selection or distribution optimization scheme is given. However, those literatures less considered to use the resources around the facility to quickly rescue the victims of the shelters, select the location of the shelters according to local conditions, and effectively use the evacuation resources to distribute the victims with the need for evacuation.

From the perspective of covering all disaster victims and meeting the needs of asylum, based on various factors such as traffic convenience, environmental safety, facility completeness and service balance. Therefore, this paper optimizes the overall weight of earthquake evacuation site, and constructs emergency shelters selection-distribution model. Finally, an example verifies the validity and feasibility of the model, and the sensitivity analysis of important parameters is carried out.

## 2 Problem Description

In the urban earthquake evacuation selection problem, it is assumed that there are $I(I = 1, 2, \ldots, i)$ residential area, and the number of residents in each residential area is $a_i(i = 1, 2, 3\ldots)$. And there are $J(J = 1, 2, \ldots, j)$ alternative refuge points, from which to choose the facilities that need to be opened, each candidate point accommodates $\beta_j$ $(j = 1, 2, 3\ldots)$ disaster victims, and the corresponding serviceable radius is $d_1$. The open refuges can accommodate all people with refuge needs.

This paper constructs an emergency evacuation site selection-distribution simulation map as shown in Fig. 1. Once an earthquake disaster occurs, the residents of the residential area quickly evacuate to the assigned earthquake shelter to avoid danger. Relevant departments quickly rescue with the surrounding facilities, and cooperate with external rescue forces, which reduce disaster losses and improve comprehensive urban support capabilities.

**Fig. 1** Simulation map of emergency evacuation site selection-distribution after earthquake

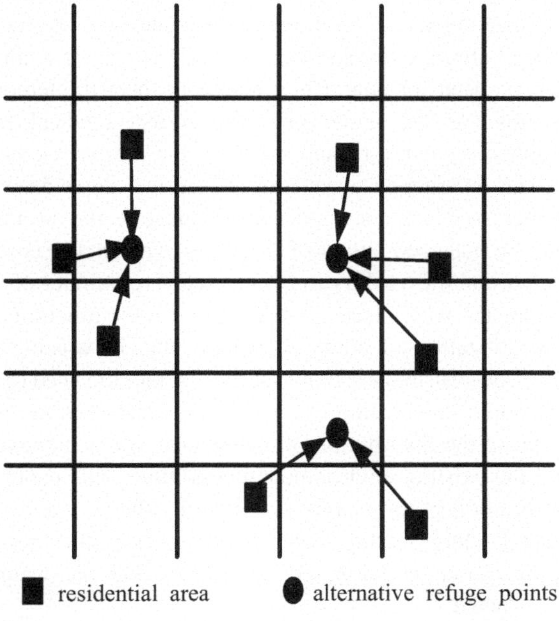

In order to determine the factors affecting the earthquake evacuation site selection and disaster victims allocation, through the analysis of relevant literature and field research, the following four influencing factors are obtained: traffic convenience, environmental safety, facility completeness and service efficiency.

The site of the earthquake refuge should achieve the transition between normal and emergency, that is, to provide space for people's education, fitness, entertainment and other activities, and to provide emergency refuge after the disaster to provide reliable security for the victims. Therefore, alternative refuge points mainly include existing parks, schools and squares. After the earthquake, it is necessary to meet the needs of the victims to quickly walk to the nearby shelters to avoid danger. The service radius of the general emergency shelters is 1000 m [15]. At the same time, in order to transfer victims after a period of time, distribution of materials and rescue vehicles pass smoothly, it is necessary to consider the distribution of traffic networks around the shelters. The closer to the main roads, the more favorable the follow-up disaster relief activities.

The selection of the earthquake refuge needs to consider the natural conditions and surrounding environment of the area where it is located. Surrounding environment includes whether an alternative refuge is in a geological disaster-prone area and near to danger sources. Secondary disasters often occur after the earthquake. Therefore, the site selection must avoid dangerous areas such as landslides, floods and mudslides. In addition, consideration should be given to sources of danger in towns, such as hazardous chemical plants.

The refuge is an important place for emergency evacuation and concentrated rescue of the victims. It should try to ensure that the surrounding material storage warehouses or large-scale commercial super-warehouses and medical institutions are fully equipped and functions are perfect. The material storage warehouse stores the drinking water, food, tents, medicines and lighting equipment that are urgently needed after the disaster occurs. The large-scale commercial super-warehouses can also be used for emergency rescue to meet the temporary needs of the evacuated victims. At the same time, the perfect medical facilities guarantee the basic healthcare conditions and order for the injured victims.

The emergency evacuation effect also depends on the service capacity of the shelter, so it is necessary to consider the effective utilization of the emergency shelter and the balanced layout of resources. The shelter ensures that the victims maintain the normal life, and try to avoid excessive number of victims exceeds the burden of the shelter, which leads to refuges losing the function of emergency evacuation. The low utilization of some shelters results in wasting resources. Therefore, refuge selection-distribution should fully consider the need of the affected people, rationally determine the location of the earthquake shelter and its number.

From the distribution of settlements, the actual evacuation needs of the victims and the existing environmental conditions, this paper plans the refuge site and the distribution of the victims, considering the convenience of transportation, environmental safety, facility integrity and service efficiency, etc. A selection-distribution optimization model based on maximizing the weight of earthquake shelters is constructed.

# 3 Earthquake Refuge Selection-Distribution Optimization Model

## 3.1 Model Construction

$K$: Collection of material reserve warehouses or large-scale commercial super-warehouses $k \in K$

$T$: The number of influencing factors in the objective function $t \in T$

$w_t$: Important coefficient of the $t^{th}$ influencing factor in the objective function

$d_{ij}$: Distance between residential point $i$ and alternative earthquake shelter $j$

$d_{jk}$: Distance between earthquake shelter $j$ and adjacent warehouse $k$

$d_j^{road}$: Distance between earthquake shelter $j$ and adjacent roads

$d_j^{health}$: Distance between earthquake shelter $j$ and nearby medical institutions

$d_j^{factory}$: Distance between earthquake shelter $j$ and adjacent dangerous sources

$d^{road}$: Distance constraint between earthquake shelter and road

$d^{health}$: Distance constraint between earthquake shelter and medical institutions

$d^{factory}$: Distance constraint between earthquake shelter and dangerous sources

$\rho_j$: Utilization of earthquake shelter $j$

$\rho$: Earthquake shelter minimum utilization limit

M: Infinite number

$s_j = 1$: Indicates that the earthquake shelter $j$ is in a safe area, otherwise 0

$x_{ij} = 1$: Indicates that some residents of the residential area $i$ are assigned to the earthquake shelter $j$, otherwise 0

$u_j = 1$: Indicates that the earthquake shelter $j$ is open, otherwise 0

$z_{ij}$: Indicates the number of victims assigned to the earthquake shelter $j$ at the residential area $i$

In order to construct the earthquake refuge selection-allocation model, this paper assumes that the total number of disaster victims in the earthquake shelter does not exceed the maximum capacity of each shelter, and the maximum evacuation distance is met during the evacuation process. To simplify the analysis, this paper calculates the population size of victims who need to evacuate according to 30% of the local resident population [16]. After the disaster, as victims and the familiar people together will increase the sense of security, we treat each family in the residential area as an overall to the nearest earthquake shelter, which is very beneficial to relieve the panic.

Objective function

$$\max F_1 = w_1 \sum\nolimits_{i \in I} \sum\nolimits_{j \in J} \frac{d_{ij} z_{ij}}{0.3 a_i} + w_2 \sum\nolimits_{j \in J} d_j^{road} u_j + w_3 \sum\nolimits_{j \in J} \sum\nolimits_{k \in K} d_{jk} u_j + w_4 \sum\nolimits_{j \in J} d_j^{health} u_j \quad (1)$$

$$\min F_2 = \sum\nolimits_{j' > j \in J} \left(\rho_j - \rho_{j'}\right)^2 \quad (2)$$

Constraint

$$\sum_{i \in I} x_{ij} z_{ij} \leq \beta_j u_j \ \forall j \in J \tag{3}$$

$$\sum_{j \in J} z_{ij} \geq 0.3 a_i \ \forall i \in I \tag{4}$$

$$x_{ij} d_{ij} \leq u_j d_1 \ \forall i \in I, j \in J \tag{5}$$

$$z_{ij} \leq M x_{ij} \ \forall i \in I, j \in J \tag{6}$$

$$\rho_j = \frac{\sum_{i \in I} z_{ij}}{\beta_j} \geq \rho u_j \ \forall j \in J \tag{7}$$

$$u_j \leq s_j \ \forall j \in J \tag{8}$$

$$d_j^{factory} u_j \geq d^{factory} \ \forall j \in J \tag{9}$$

$$d_j^{road} u_j \geq d^{road} \ \forall j \in J \tag{10}$$

$$\sum_{t \in T} w_t = 1 \tag{11}$$

$$x_{ij}, u_j, s_j \in \{0, 1\} \ \forall i \in I, j \in J \tag{12}$$

$$z_{ij} \geq 0, \ integer \tag{13}$$

The first two of the objective (1) optimize traffic convenience, including the average walking distance of the victims and the distance between the shelter and the road. The latter two optimize facilities completeness, namely material reserves and medical facilities. (2) tries to use the shelter resources as balanced as possible. (3) is the capacity constraint of the earthquake shelter, and the total number of victims in the shelter does not exceed its maximum capacity. (4) ensures that victims with asylum needs can be resettled. (5) indicates the distance constraint between the victims and the shelters. (6) indicates victims can only take refuge in the assigned refuge. (7) is the minimum utilization constraint of the shelter. (8) indicates that only open shelters that are safe in natural geological conditions. (9) is the distance constraint between the shelter and the dangerous sources. (10) is the distance constraint between the shelter and the main road. (11) represents the sum of the important coefficients of each influencing factor is 1 in (1). (12) is the 0–1 variable constraint. (13) is the integer constraint.

### 3.2 Model Solution

Due to the different range of various distances in the objective function (1), this paper standardizes variables $d_{ij}$, $d_j^{road}$, $d_{jk}$ and $d_j^{health}$, replaces them with $d'_{ij}, d_j^{road\prime}$, $d'_{jk}$ and $d_j^{health\prime}$, $d'_{ij} = \frac{d_{ij}^{max} - d_{ij}}{d_{ij}^{max} - d_{ij}^{min}}$. $d'_{ij} = 1$ indicates the best results, and $d'_{ij} = 0$ the worst effect. Similarly, $d_j^{road\prime} = \frac{d_j^{roadmax} - d_j^{road}}{d_j^{roadmax} - d_j^{roadmin}}$, $d'_{jk} = \frac{d_{jk}^{max} - d_{jk}}{d_{jk}^{max} - d_{jk}^{min}}$, $d_j^{health\prime} = \frac{d_j^{healthmax} - d_j^{health}}{d_j^{healthmax} - d_j^{healthmin}}$.

Combine (1) and (2) to construct a single objective function as follows:

$$\max F = F_1 - F_2 \tag{14}$$

The (14) replaces (1) and (2), and the multi-objective optimization model is transformed into a single-objective optimization model. The constraints are linear, so the model is solved by LINGO11.0.

## 4 Numerical Simulation and Analysis

This article takes a block that currently needs to build refuges as a research object.

The block has built a shelter. The terrain is relatively flat. The natural address conditions of the area are safe. There are 14 residential areas with evacuation requirements. The emergency relief supplies reserve mainly depends on the surrounding supermarkets and there are 2 medical facilities. The source of danger is a gas station, as shown in Fig. 2 (No. 22 is a shelter that has been built).

### 4.1 Solution of the Example

First, the number of households in each residential area in the study area is estimated by the population density of the block (the average household size is composed of 3 people). The number of households with different evacuation needs is shown in Table 1.

According to the calculation of the per capita area of 1.5 m$^2$ after the disaster [17], the relevant data of the alternative earthquake shelters are shown in Table 2.

The distance between the residential area and the alternative earthquake shelter is shown in Table 3, where the distance is the European distance measured by Baidu map. The source of danger is mainly the gas station, according to the standard distance shelter 50 m [17].

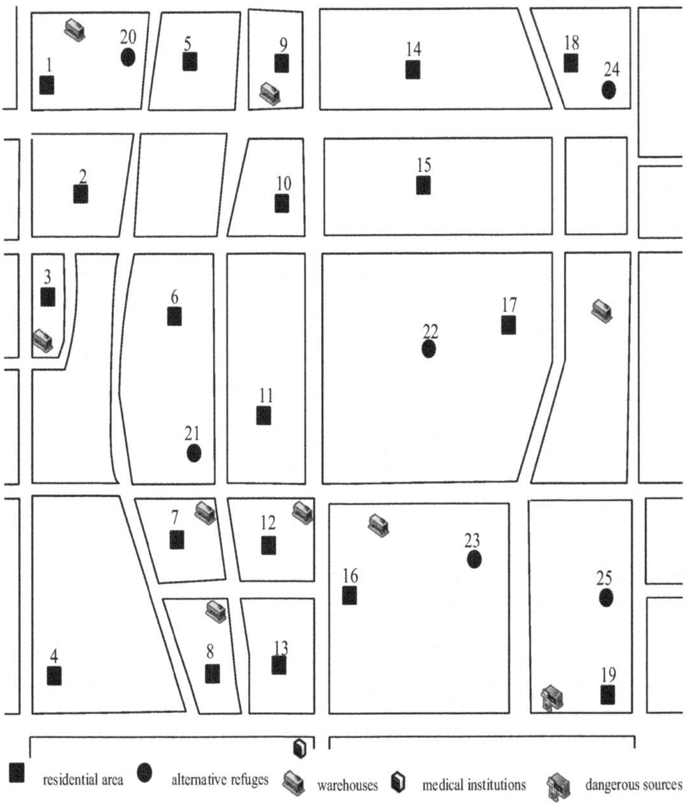


**Fig. 2** Block map

**Table 1** Number of households with refuge needs in different settlements

| Numbering | Number of families with asylum needs | Numbering | Number of families with asylum needs |
|---|---|---|---|
| 1 | 740 | 11 | 1500 |
| 2 | 2120 | 12 | 1290 |
| 3 | 800 | 13 | 420 |
| 4 | 990 | 14 | 4020 |
| 5 | 1620 | 15 | 4640 |
| 6 | 970 | 16 | 2490 |
| 7 | 1340 | 17 | 1210 |
| 8 | 1070 | 18 | 970 |
| 9 | 1100 | 19 | 560 |
| 10 | 1310 | – | – |

**Table 2** Alternative earthquake shelter related data

| Numbering | 20 | 21 | 22 | 23 | 24 | 25 |
|---|---|---|---|---|---|---|
| Capacity (number of households) | 790 | 10640 | 5360 | 27710 | 1500 | 2430 |
| Distance from adjacent roads (m) | 110 | 130 | 240 | 110 | 50 | 90 |
| Distance to neighboring supermarkets (m) | 180 | 280 | 470 | 300 | 440 | 680 |
| Distance from neighboring medical institutions (m) | 1500 | 810 | 920 | 730 | 1730 | 1000 |
| Distance from dangerous source (m) | 1920 | 1270 | 790 | 380 | 1340 | 290 |

**Table 3** Distance between residential area and alternative earthquake shelter (unit: meter)

| Numbering | 20 | 21 | 22 | 23 | 24 | 25 |
|---|---|---|---|---|---|---|
| 1 | 160 | 870 | 1340 | 1680 | 1750 | 2090 |
| 2 | 250 | 600 | 1200 | 1520 | 1720 | 1940 |
| 3 | 550 | 430 | 1190 | 1440 | 1820 | 1830 |
| 4 | 1240 | 590 | 1280 | 1370 | 2120 | 1710 |
| 5 | 290 | 790 | 1010 | 1380 | 1320 | 1770 |
| 6 | 580 | 240 | 810 | 1110 | 1480 | 1520 |
| 7 | 1020 | 240 | 850 | 950 | 1660 | 1320 |
| 8 | 1240 | 490 | 890 | 880 | 1740 | 1220 |
| 9 | 540 | 830 | 800 | 1210 | 1060 | 1590 |
| 10 | 560 | 640 | 650 | 1050 | 1090 | 1440 |
| 11 | 860 | 250 | 560 | 760 | 1330 | 1190 |
| 12 | 1090 | 350 | 640 | 710 | 1460 | 1090 |
| 13 | 1410 | 470 | 910 | 890 | 1750 | 1230 |
| 14 | 1060 | 1130 | 670 | 1050 | 580 | 1330 |
| 15 | 1070 | 990 | 410 | 810 | 570 | 1100 |
| 16 | 1350 | 650 | 510 | 410 | 1330 | 790 |
| 17 | 1420 | 1120 | 310 | 500 | 570 | 700 |
| 18 | 1440 | 1460 | 800 | 1080 | 140 | 1230 |
| 19 | 2000 | 1430 | 820 | 450 | 1240 | 100 |

After the earthquake disaster, the victims quickly and safely arrive at the earthquake shelter and relevant departments quickly launched rescue. Then we should consider the completeness and perfection of the supporting facilities around the earthquake shelter, and try to ensure that the shelter resources are balanced. Therefore the importance coefficients in (1) are set to $w_1 = 0.3$, $w_2 = 0.3$, $w_3 = 0.2$, and $w_4 = 0.2$, respectively. The remaining parameters are shown in Table 4, and the distance unit is meters. The problem was solved with LINGO11.0. All the candidate points were selected as the address of the earthquake shelter. The utilization rates of the shelters were 59.87%, 59.88%, 59.89%, 59.89%, 59.93% and 59.88%, respectively. The objective function value is 2.57, where the value of (2) is 0,

**Table 4** Parameter settings

| Parameter | $\rho$ | $d_1$ | $d^{road}$ | $d^{factory}$ |
|---|---|---|---|---|
| Value | 0.5 | 1000 | 500 | 50 |

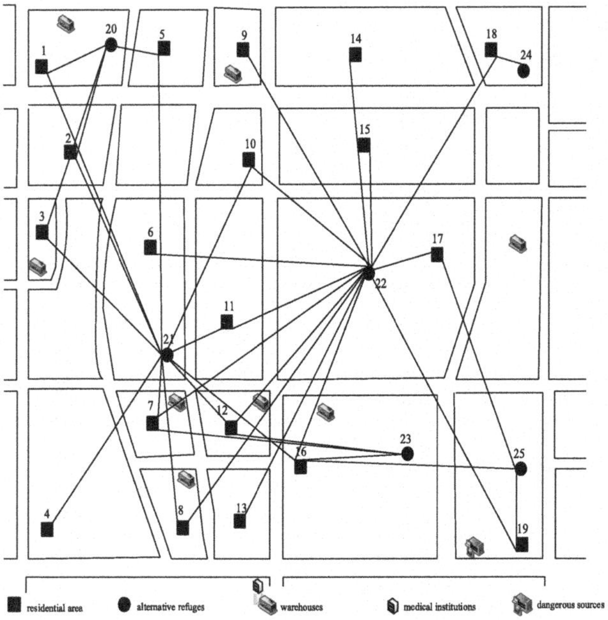


**Fig. 3** Earthquake refuge selection-allocation scheme

indicating that the evacuation resources are used balanced, and finally the earthquake refuges selection-allocation scheme as shown in Fig. 3.

There are many residential areas around the shelters 21 and 22, which locate at the center of the research object. The larger capacity accommodates more victims. The surrounding traffic is convenient and large-scale commercial super-warehouses distribute densely. Shelters 21 and 22 can meet the temporary needs of the victims and help the relevant departments implement rescue operations, so they provided shelters to victims from 10 and 14 residential areas. The distances the victims with evacuation needs have to walk to the assigned shelters after the disaster are shown in Fig. 4. When the service radius of the earthquake shelter was set at 1000 m, most of the victims were assigned to shelters that were 400–900 m away, and the victims were evacuated in short distances.

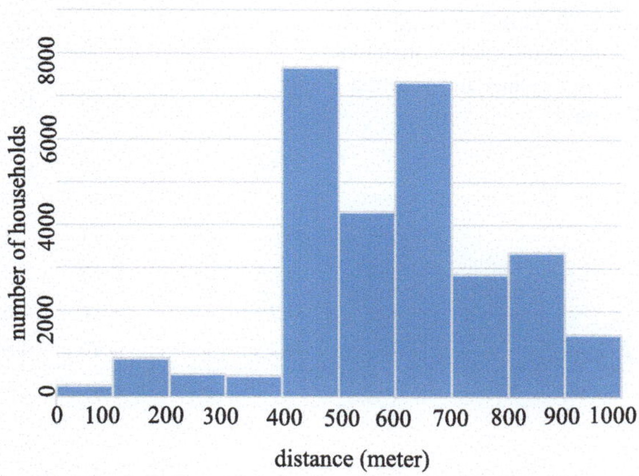


**Fig. 4** Distribution scheme walking distance histogram

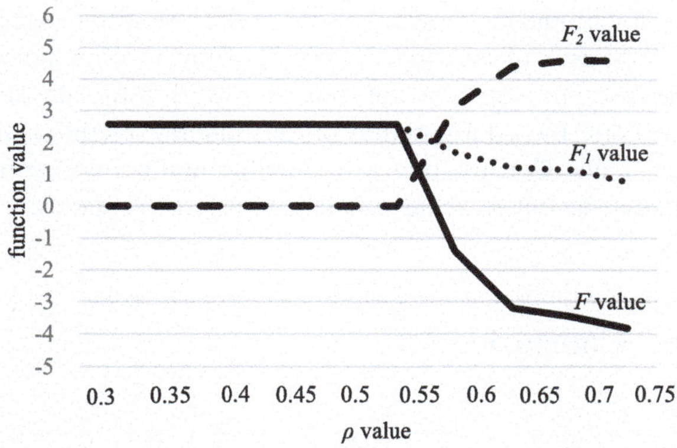


**Fig. 5** The effect of the lowest utilization ρ on the function value

## 4.2 *Sensitivity Analysis*

When the parameters were set, it is found that the value of the lowest utilization rate ρ of the earthquake shelter affected the selection-allocation scheme. For a clearer observation, the lowest utilization ρ was increased from 0.3 to 0.75 in the case where other parameter values of the example were unchanged in the above study area, and the final result is shown in Fig. 5. As the parameter value increases, the objective function value $F$ mutated at $\rho = 0.55$, as shown in Table 5. Then the $F$ value decreased continuously, indicating that the ability of surrounding facilities to meet the needs of the victims was worse and worse. The function value $F_2$ increased constantly, indicating that the refuge resources were difficult to use in a balanced manner. At the same time, when the minimum utilization rate of the set shelter varied from 0.3 to 0.75, the candidate points 21 and 22 were selected as open refuge, indicating that the above two places better satisfied the emergency evacuation needs of the victims in service area compared with other candidate points. Therefore, the area can first build shelters here.

**Table 5** Different parameter values correspond to objective function values and location schemes

| Parameter value | $F_1$ | $F_2$ | $F$ | Open refuge |
|---|---|---|---|---|
| 0.3 | 2.57 | 0 | 2.57 | 20, 21, 22, 23, 24, 25 |
| 0.55 | 2.57 | 0 | 2.57 | 20, 21, 22, 23, 24, 25 |
| 0.6 | 1.71 | 3.11 | −1.40 | 21, 22, 23, 25 |
| 0.65 | 1.22 | 4.42 | −3.20 | 20, 21, 22 |
| 0.7 | 1.16 | 4.61 | −3.46 | 21, 22, 24 |
| 0.75 | 0.76 | 4.59 | −3.83 | 21, 22 |

From above analysis, it is known that the decision-makers and managers of relevant departments should consider the victims successively avoid the post-earthquake danger, and make full use of the capacity of the shelters. From the perspective of meeting the needs of the victims, managers also think out the distribution of materials, the smooth flow of traffic, the actual conditions for the use of shelters, the configuration of surrounding facilities and the distribution of urban residents. Constructing a reasonable number of earthquake shelters optimizes its spatial layout to achieve balanced use of resources.

## 5 Conclusion

The location of the earthquake shelter and the distribution of evacuation to the people after the disaster are conducive to improving the security capabilities. This paper constructs a selection-distribution optimization model based on the maximum weight of the shelters from the perspective of meeting the needs of the victims. Under the premise of meeting the capacity of the shelter and the minimum utilization limit, we consider the various factors such as traffic convenience, environmental safety, facility completeness and service efficiency. The earthquake shelters are optimally selected and the victims are allocated nearby. Finally, an example illustrated the validity of the model, and the management suggestions for the construction of urban earthquake shelters are proposed through the sensitivity analysis of key parameters.

## References


1. Mirza, M. M. Q. (2003). Climate change and extreme weather events: Can developing countries adapt? *Climate Policy, 3*(3), 233–248.
2. Mcpeak, B. G., & Ertas, A. (2012). The good, the bad, and the ugly facts of tornado survival. *Natural Hazards, 60*(3), 915–935.

3. Jiang, W., Zhang, J., Lu, X. C., & Lu, J. (2011). Crustal structure beneath Beijing and its surrounding regions derived from gravity data. *Earthquake Science, 24*(3), 299–310.
4. Renne, J. L. (2018). Emergency evacuation planning policy for carless and vulnerable populations in the United States and United Kingdom. *International Journal of Disaster Risk Reduction, 31,* 1254–1261.
5. Hino, K., Tanaka, Y., & Schneider, R. H. (2018). Characteristics of fear of crime in evacuation shelters after the Great East Japan Earthquake. *Environmental Hazards, 17*(5), 1–14.
6. Toregas, C., Swain, R., Revelle, C., & Bergman, L. (1971). The location of emergency service facilities. *Operations Research, 19*(6), 1363–1373.
7. Lin, Y. H., Batta, R., Rogerson, P. A., Blatt, A., & Flanigan, M. (2012). Location of temporary depots to facilitate relief operations after an earthquake. *Socio-Economic Planning Sciences, 46*(2), 112–123.
8. Kilci, F., Kara, B. Y., & Bozkaya, B. (2015). Locating temporary shelter areas after an earthquake: A case for Turkey. *European Journal of Operational Research, 243*(1), 323–332.
9. Sabouhi, F., Bozorgi-Amiri, A., Moshref-Javadi, M., & Heydari, M. (2018). An integrated routing and scheduling model for evacuation and commodity distribution in large-scale disaster relief operations: A case study. *Annals of Operations Research, 1,* 1–35.
10. Barzinpour, F., & Esmaeili, V. (2014). A multi-objective relief chain location distribution model for urban disaster management. *International Journal of Advanced Manufacturing Technology, 70*(5–8), 1291–1302.
11. Chen, C., & Ren, A. Z. (2003). Computer method for layout optimization of fire stations. *Journal of Tsinghua University (Science and Technology), 43*(10), 1390–1393.
12. Chen, X., Kwan, M. P., Li, Q., & Chen, J. (2012). A model for evacuation risk assessment with consideration of pre-and post-disaster factors. *Computers, Environment and Urban Systems, 36*(3), 207–217.
13. Ma, Y. J., Zhao, X. J., Qin, L. J., Liang, P. J., Zhou, H. J., Yuan, Y., et al. (2018). Optimization of multi-constrained and multi-objective disaster evacuation site selection—A case study of Wenchang City, Hainan Province. *Journal of Catastrophology, 1,* 218–224.
14. Castro, C. P., Ibarra, I., Lukas, M., Ortiza, J., & Sarmiento, J. P. (2015). Disaster risk construction in the progressive consolidation of informal settlements: Iquique and Puerto Montt (Chile) case studies. *International Journal of Disaster Risk Reduction, 13,* 109–127.
15. Li, H., Zhao, L., Huang, R., Huang, R. B., & Hu, Q. M. (2017). Hierarchical earthquake shelter planning in urban areas: A case for Shanghai in China. *International journal of disaster risk reduction, 22,* 431–446.
16. Bian, R., & Wilmot, C. G. (2017). Measuring the vulnerability of disadvantaged populations during hurricane evacuation. *Natural Hazards, 85*(2), 691–707.
17. Li, Q. G., Song, W. H., Xie, F., & Chen, Z. (2011). Mathematical simulation of damage range of fire and explosion accidents in oil storage tank area of gas station. *Journal of Nankai University (Natural Science), 5,* 7–13.

# Adaptive Routing Based on Wireless Ad Hoc Networks

**Xinxiang Yin, Wenjun Huang, and Wenqing Zhang**

**Abstract** In order to adapt to various scenarios of different network congestion, topology switching and service arrival rate, an adaptive routing method combining multiple network state information is proposed. Considering signal-to-interference ratio, connectivity probability and resource utilization of the link, the adaptive routing protocol establishes a link quality model and calculates link quality values for each link in different network scenarios, and further obtains the optimal path from source node to destination node based on link quality values. The simulation results show that the model reflects the same trend of link quality under different network state information, and can adapt to various scenarios of different network congestion, topology switching and service arrival rate.

**Keywords** Network scenario · Adaptive routing · Link quality model · Network state information

## 1 Introduction

Routing protocol is one of the key technologies to realize efficient transmission, flexible survivability and fast self-healing of networks. However, in the face of various network scenarios with drastic network topology fluctuations and inconsistent service arrival rates, routing protocols that use single network state information as the basis for route discrimination often fail to adapt to scene changes. How to combine various network state information effectively to overcome the

X. Yin (✉) · W. Huang · W. Zhang
School of Electronic and Information Engineering, Beijing Jiaotong University, Beijing, China
e-mail: 17120157@bjtu.edu.cn

W. Huang
e-mail: 16111027@bjtu.edu.cn

W. Zhang
e-mail: 16120167@bjtu.edu.cn


J. Zhang et al. (eds.), *LISS2019*,
https://doi.org/10.1007/978-981-15-5682-1_36

influence of the complex transient environment on the optimal path selection, and put forward the adaptive routing protocol for different network scenarios has become one of the research hot spots in the field of ad hoc networks.

The research and application of adaptive routing protocol are very extensive, Because of the particularity of large-scale network, frequent disconnection of links and rapid change of topology, vehicle-to-vehicle network needs to combine the environmental information of different roads to select the route to the destination adaptively [1]. A new cross-link network routing algorithm is proposed [2], which overcomes the limitations of traditional fully adaptive routing and achieves good results in reducing latency, increasing throughput and increasing the number of idle channels, thus reducing the number of congested routes. A fully adaptive routing algorithm for partially interconnected cross-link grid topology is proposed [3], which connects four additional diagonal ports in a partially interconnected cross-link network topology to optimize the performance of overhead, network delay and so on. In view of the characteristics of vehicle-to-vehicle network and cross-link network, the above-mentioned references proposed adaptive algorithms for static analysis and optimization of scenarios, but fail to demonstrate the effectiveness of the algorithm in dynamic scenarios or scenarios with different network size and channel quality.

An adaptive routing algorithm based on Hamiltonian routing model is introduced [4]. The algorithm achieves high adaptability by finding the minimum number of paths between each pair of source node and destination node, and through better distribution of service in the whole network to reduce the number of hot spots to improve the overall performance of the network. The classical method of data routing in multi-hop networks is introduced [5], which emphasizes the disadvantage of traditional multi-hop routing algorithm which uses hop number as the basis of route measurement. Then the author proposes an adaptive routing algorithm based on reinforcement learning theory, which relies on feedback information from neighbor nodes to select more effective route. The above two articles analyze the performance of the adaptive routing protocol based on service flow and maintenance information of neighbor nodes, but fail to consider whether the algorithm can continue to achieve self-adaptation when service flow and neighbor nodes maintenance information are not the main factors affecting routing choice of the network.

Although the above-mentioned references can adaptively select the path to transmit service for the proposed scenario, the adaptive routing algorithm will not continue to be applicable when the scenario changes dynamically or the network state information are no longer the main factors affecting the routing choice of the network. In view of various scenarios of different network congestion, topology switching and service arrival rate, this paper synthetically considers the signal-to-interference ratio, connectivity probability and resource utilization of the link and calculates link quality values for each link in different network scenarios, and further obtains the best path from source node to destination node based on link quality values. Finally, through simulation, the adaptability of routing algorithm to scenarios under different network state information is researched.

## 2 System Model

The goal of ARNS (Adaptive Routing Protocol Based on Network State Information) routing protocol is to model and analyze network performance parameters, such as SIR (Signal-to-Interference Ratio), connectivity probability and resource utilization, based on a variety of network state information (including node distribution density, channel loss coefficient, service arrival rate, etc.), and ultimately find the best path for service transmission performance.

First of all, we assume that the source node to the destination node has a service need to be transmitted, and the transmission of a service may need to be forwarded by multiple nodes, so that *h* is the node hop from the source node to the destination node, and the link quality of the whole link is decided by *h* hops links. $CP(e_i)$ is the connected probability of the *i*th link. We consider that when a link is in the process of service transmission, its connectivity probability should be judged first. If the connectivity probability is high, then the link is considered to be better connectivity. Then the communication quality of the link should be judged, SIR is the network performance parameter representing the communication quality of the link. The signal-to-interference ratio of the *i*th link is $S(e_i)$, and the $S_{th}$ is the threshold of SIR. If the SIR value is higher than the threshold value, the link transmission quality is considered to be better, and the more SIR value is higher than the threshold value, the better the link transmission quality is. If the source node to the destination node needs to pass through the multi-hop node for forwarding, $Sv(y)$ is the variance of the signal-to-interference ratio. We think that the smaller the change of the signal-to-interference ratio and the smaller the value of $Sv(y)$, the more stable the link will be and the more suitable for the transmission of service. Finally, the Resource utilization ratio of the link is determined. $RS(e_i)$ is the resource utilization rate of the *i*th link. For a connected link with good communication quality, the higher the proportion of the data slot in the total slot, the larger the value of $RS(e_i)$, the higher the efficiency of service transmission. In conclusion, if the link quality value based on connectivity probability, SIR and resource utilization is the largest, then the link is the best route for service transmission from source node to destination node in the current network scenario.

Based on the above considerations, the ARNS routing problem can be defined as a maximum value problem. The objective function is as follows:

$$Max\,F(\mathrm{h}) = CP(\mathrm{h}) \bullet \frac{S(\mathrm{h}) - \mathrm{S_{th}}}{S(\mathrm{h})} \frac{1}{1 + S_v(\mathrm{h})} \bullet RS(\mathrm{h}) \tag{1}$$

$$\begin{cases} CP(\mathrm{h}) = \prod_{i=1}^{h} CP(\mathrm{e_i}) \\ S(\mathrm{h}) = \frac{\sum_{i=1}^{h} S(\mathrm{e_i})}{h} \\ S_v(\mathrm{h}) = \frac{1}{h}\sum_{i=1}^{h} [\mathrm{S}(\mathrm{e_i}) - S(\mathrm{h})]^2 \\ \mathrm{RS}(\mathrm{h}) = \prod_{i=1}^{h} \mathrm{RS}(\mathrm{e_i}) \end{cases} \quad (2)$$

*F(h)* is the link quality value of the target link based on the ARNS routing protocol. The larger the value of *F(h)* of the link, the more effective the link can be in the current network scenario.

# 3 Network Performance Analysis

Network scenarios are determined by network information parameters. Switching between different network scenarios is essentially the change of the values of network information parameters. ARNS routing protocol aims to perceive different network information parameters to different network scenarios, and reflect the performance of link service transmission through the link quality value of the objective function. Therefore, according to the objective function, this chapter will analyze the signal-to-interference ratio, connectivity probability and resource utilization of links separately, and find out which network information parameters determine the performance parameters of network. Then we can get which network information parameters determine link quality value.

## *3.1 Analysis of SIR*

For a wireless network, the node density obeys the poisson point process with the mean value of $\lambda_{\mathrm{p}}$. Communication nodes can eliminate interference nodes in a certain range by using a certain interference avoidance mechanism.

In such a network, the probability that two nodes with a distance of *s* can exist simultaneously is *k(s)*. In order to reflect the interaction between two nodes, the spatial correlation function between two nodes is introduced, which is defined as:

$$g(s) = \frac{\lambda_P^2 k(s)}{\lambda_H^2} \quad (3)$$

Set the sending node as the origin $o$, $d$ is the distance between the receiving node and the sending node, and the direction of the receiving node is $\varphi$. The intensity of interference experienced by the receiving node is as follows:

$$E_o^!(I) = \lambda_H \int_0^{\infty} \int_0^{2\pi} l(s)g(s)sd\varphi ds \tag{4}$$

In the equation, $l(s)$ is the path model, if only considers the channel loss, then

$$l(s) = s^{-\alpha} \tag{5}$$

$\alpha$ is the channel loss coefficient. The key to calculate the average interference intensity is to analyze the correlation function $g(s)$ with the interference avoidance mechanism.

Nodes are set to maintain the scheduling information of H hops neighbors [4], send nodes compete with other nodes within the $H$ hops range. This will form a circular interference removal area with $HR$ radius around the sending node, where $R$ is the effective transmission radius of DSCH messages. It can be obtained that the density of interference nodes under the electoral mechanism is as follows:

$$\lambda_H = \lambda_P \int_0^1 e^{-\lambda_P x \pi H^2 R^2} dx = \frac{1 - e^{-\lambda_P \pi H^2 R^2}}{\pi H^2 R^2} \tag{6}$$

From the above equation, it can be seen that the interference density is not only affected by the original node density, but also has an important relationship with the number of hop of the node. When the node density is small, the interference node density increases with the increase of the original node density, but then it tends to a stable value ($\frac{1}{\pi H^2 R^2}$), which is the function of resource scheduling and conflict avoidance mechanism. The larger the $H$ is, the lower the density of the interference nodes in the network, which means the stronger the role of interference management, and the higher the reliability of transmission.

$s$ represents the distance between two nodes in the network with density $\lambda_H$. Two circles with centers $s$ apart, each of which has a radius of $R$, have a joint area of:

$$V(s) = \begin{cases} 2\pi H^2 R^2, \; s \geq 2HR \\ 2\pi H^2 R^2 - 2H^2 R^2 \arccos\left(\frac{s}{2HR}\right) + s\sqrt{H^2 R^2 - \frac{s^2}{4}}, \; HR \leq s < 2HR \end{cases} \tag{7}$$

After the density of the original network node is diluted by the election algorithm, the probability that the node $s$ away from the receiver can be retained is as follows:

$$k(s) = \frac{2V(s)\left(1 - e^{-\lambda_P \pi H^2 R^2}\right) - 2\pi H^2 R^2 \left(1 - e^{-\lambda_P V(s)}\right)}{\lambda_P \pi H^2 R^2 V(s)[V(s) - \pi H^2 R^2]} \quad (8)$$

Divide the nodes into two parts according to their spatial correlation:

$$E_o^!(I) = E_o^!(I_{\delta \leq s < 2\delta}) + E_o^!(I_{s \geq 2\delta}) \quad (9)$$

In the interval $[2\delta, \infty)$, the nodes are outside of each other's exclusion zone, so the two nodes are independent in space. Further calculation of the expected receiver-to-receiver interference ratio:

$$E_o^!(SIR) = 2\pi\lambda_H d^{-\alpha} \int_{\delta}^{2\delta} l(s)g(s)sds + \frac{2\pi\lambda_H d^{-\alpha}}{\alpha - 2}(2\delta)^{2-\alpha}, \alpha > 2 \quad (10)$$

From the results, we can see the signal-to-interference ratio of the link depends on the node density, the path loss coefficient, and the number of neighbor nodes maintenance hops. In network scenarios, the number of neighbor nodes maintenance hops is given by the designed routing protocol. With the change of network size and link topology switching, node density and channel loss coefficient become the key information parameters affecting network performance.

### 3.2 Analysis of Connected Probability

In order to analyze which network information parameters determine the connected probability, this paper uses the statistical method to calculate the probability of successful transmission by changing node density $\lambda_{\mathrm{p}}$, channel loss factor $\alpha$ and neighbor maintenance hop $h$ under the influence of link interference.

First of all, the interference interval *Am* is introduced to better quantify the additive interference from different transmission distances: Node *a* sends a message to node *b*, if *m* or more nodes send a message at the same time with *a* in the *Am* interval, node *b* will fail to accept them.

The transmission power of the node is *P*, and $\gamma$ represents the threshold of SIR of the received message, and $d_k$ is the distance between the interference node *k* and the receiving node *b*. The condition for node *b* to receive successfully is:

$$\frac{P_{\mathrm{a}} d^{-\alpha}}{\sum_{k=1}^{m} P_k {d_k}^{-\alpha}} \geq \gamma \quad (11)$$

Assuming that all nodes in the network transmit the same amount of power, the following result is obtained:

**Fig. 1** Schematic diagram of connectivity probability model

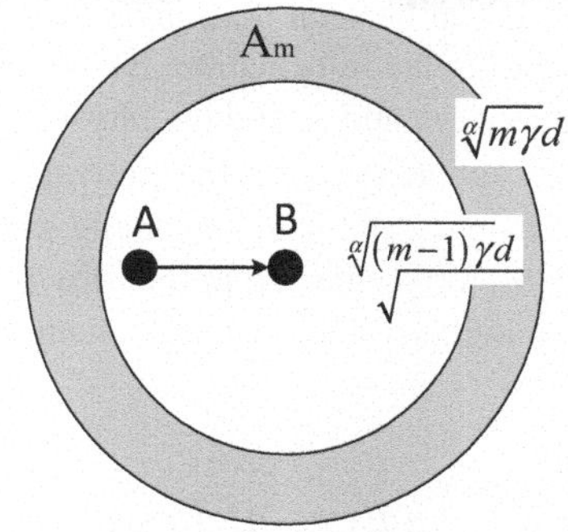


$$\frac{1}{d^{\alpha}} \geq \gamma \sum_{k=1}^{m} \frac{1}{\min\{d_k^{\alpha}\}} \tag{12}$$

If the distance $d_k$ from node $k$ to the receiving node $b$ satisfies the following equation, then the node $k$ is located in the $Am$ interval (Fig. 1).

$$\sqrt[\alpha]{m\gamma d} \leq d_k < \sqrt[\alpha]{(m+1)\gamma d} \tag{13}$$

The value of $m$ is selected according to the location of the nearest interference node, and the node in the farther interval is equivalent to the node in the $Am$ interval according to the proportional interference intensity: only consider the channel loss, the interference intensity is inversely proportional to the interference distance. The equivalent ratio is calculated as follows:

$$\mu_m^k = \left( \frac{(m-1)^{\frac{2}{\alpha}} + m^{\frac{2}{\alpha}}}{(m+k-1)^{\frac{2}{\alpha}} + (m+k)^{\frac{2}{\alpha}}} \right)^{\frac{\alpha}{2}}, \ (k \geq m+1) \tag{14}$$

$N_{Am}$ indicates the number of summary points in the Am interval. $N'_{A_m}$ is the total number of nodes in the Am interval plus equivalent nodes.

$$N'_{A_m} = N_{A_m} + \sum_{k=m+1}^{M_{\max}} \mu_m^k N'_{A_k} \tag{15}$$

$\lambda$ is the probability that the node sends the message, and the probability that node $b$ receives successfully is:

$$P_{RCV} = \sum_{m=M_{\min}}^{M_{\max}} \left\{ P_{\left(\sum_{k=1}^{m-1} I_{A_k}=0\right)} P_{(I_{Am} \neq 0)} \left[ 1 - \sum_{j=m}^{M_{\max}} \binom{j}{N'_{A_m}} \lambda^j (1-\lambda)^{N'_{Am}-j} \right] \right\} \tag{16}$$

It can be seen that, given the probability (service transmission rate) of the node sending the information, the connectivity probability of a link is related to the value of $N'_{A_m}$, $m$ and $\lambda$, and the physical meaning of $N'_{A_m}$ is the total number of nodes after the $Am$ interval plus the equivalent nodes. This value varies with the change of node density $\lambda_p$ and path loss coefficient $\alpha$, while $m$ depends on neighbor maintenance hop $h$. So the connected probability is determined by the following network information parameters: node density, channel loss coefficient and service arrival rate.

### 3.3 Analysis of Resource Utilization

In the coordinated distributed scheduling model, a frame is divided into control sub-frame and data sub-frame, and sub-frame is divided into multiple time slots. Each node maintains the $h$ hops neighbor information by sending and receiving scheduling information.

Remember the number of control slots is $C$, the number of data slots is $D$, so the resource utilization of the service transmission is:

$$\eta = \frac{D}{C+D} \tag{17}$$

In the model, we assume that the resource utilization of each link is the same, that is, the resource utilization of the whole network is equal to the resource utilization of the link. The next step is to find the average number of forwarding hops between nodes. According to the conclusion, in a network with $N$ nodes [6, 8], if the number of nodes in the coverage range of a single node is $n$, the average forwarding hop between nodes is:

$$E(h_f) \approx \frac{\ln(N)}{\ln(n)} \tag{18}$$

So in a network with a maximum communication hop of $H_{max}$, the average node forwarding hop is:

$$h_f = \frac{\ln(\lambda_p \pi H_{max}^2 R^2)}{\ln(\lambda_p \pi R^2)} \tag{19}$$

Since data services need multi-hop forwarding, we assume that the nodes send services at the service transmission rate $\lambda_a$ and each data packet occupies a data time slot in the service transmission, then the number of data slots required by all services in a scheduling cycle of the whole network:

$$D = N\lambda_a h \tag{20}$$

The number of control time slots depends on the maintenance overhead of the network. Maintenance overhead refers to the overhead of maintaining table-driven routing between single-hop neighbors of each node, which is related to the following two factors: broadcast frequency $f$ of maintenance Hello messages and average forwarding hops $h_f$. The broadcasting frequency $f$ of Hello messages is the rate at which nodes maintain routing with neighbors and periodically update the signal-to-noise ratio, connectivity probability and resource utilization of neighbor links, which should be set as the reciprocal of link maintenance time:

$$f = 1/T_{link} \tag{21}$$

The one-hop average link holding time as follows [7, 9]:

$$T_{link} = \frac{r}{2\pi v_{max}^2} \int_0^{2\pi} \int_0^{v_{\max}} \int_0^{v_{\max}} \frac{1}{G_{v_1,v_2}^{\theta}} dv_1 dv_2 d\theta \tag{22}$$

Therefore, the number of control slots for the whole network is [10, 11]:

$$C = fN(\lambda_p \pi R^2) \tag{23}$$

The resource utilization of the whole network link is as follows [12, 13]:

$$\eta = \frac{D}{C + D} = \frac{N\lambda_a h_f}{fN(\lambda_p \pi R^2) + N\lambda_a h_f} \tag{24}$$

From the result of resource utilization, we can see that resource utilization is different from signal-to-interference ratio and connectivity probability. Resource utilization does not depend on channel loss coefficient and neighbor nodes maintenance hops, but on service arrival rate and node density [14].

## 4 Simulation

When we propose the system model, we define the routing problem of ARNS as a maximum problem. In the last chapter, we analyze the different network performance parameters in the objective function, and get the network information parameters that affect the network performance in different scenarios, including node density, channel attenuation coefficient and service arrival rate. Next, we bring the results of each part of the network performance parameters into the objective function, and simulate the performance of ARNS routing protocol with node density, channel loss coefficient and service arrival rate as independent variables.

**Table 1** Simulation parameters

| Parameters | Values |
|---|---|
| Scenario setting ($m^2$) | 1000 × 1000 |
| Scale of nodes (nodes/$km^2$) | 50,100,150,200,250 |
| Simulation time (s) | 1000 |
| Maximum Speed (m/s) | 15 |
| Communication range (m) | 200 |
| Channel loss coefficient | 2,3 |
| Service arrival rate (packages/s) | 50,100 |

This paper simulates ARNS routing protocol based on NS2 platform. According to the network information parameters which influence the network performance obtained from the above analysis, the network scene is adjusted by changing the parameters. The specific simulation scene parameters set in this paper are shown in Table 1:

Through the simulation results, this paper mainly compares ARNS and AODV from two aspects of delay and delivery rate. Firstly, as shown in Fig. 2, with the increase of node density, the number of messages for communication and maintenance of nodes increases significantly during the process of finding routing, and the decrease of average distance between nodes leads to the increase of interference. The end-to-end communication delay of ARNS and AODV routing protocols increases continuously. However, under the same channel loss coefficient and service arrival rate, AODV finds routing based on the number of hops, while ARNS calculates network performance parameters based on network information parameters to find the best service transmission performance routing in the current scenario. The communication quality and resource utilization of routing based on ARNS are better than those based on retransmit probability caused by poor routing quality in the process of sending packets, and thus reduces the overall end-to-end delay.

As shown in Fig. 3, ARNS routing protocol has a slightly higher delivery rate than AODV in the same network scenario represented by the same network information parameters. This is because ARNS improves the effectiveness of high-quality routing selection in routing selection process. Higher and more stable routing with higher connectivity probability and truncation ratio can effectively reduce the probability of loss caused by routing failure or interruption during packet transmission.

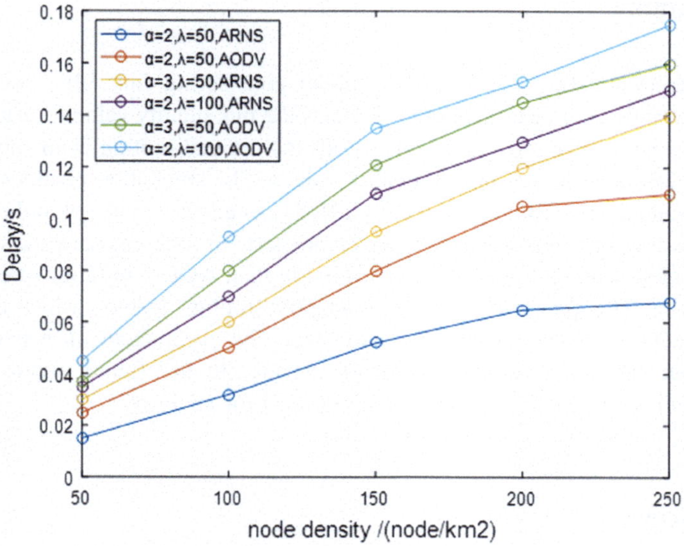


**Fig. 2** Average end-to-end delay comparison graph of ARNS and AODV

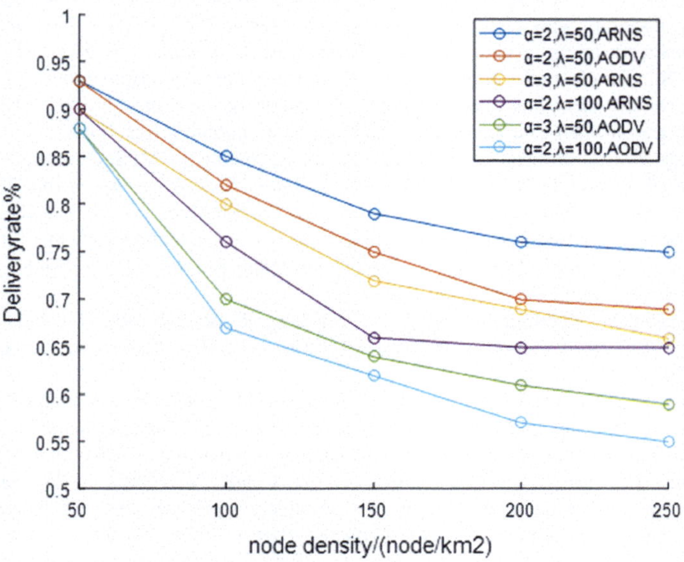


**Fig. 3** Delivery ratio comparison graph of ARNS and AODV

## 5 Summary

This paper establishes a link quality model considering link SIR, connectivity probability and resource utilization, calculates the link quality value for each link in different network scenarios, and further finds the optimal routing from source node to destination node based on link quality values. The simulation results show that the proposed adaptive routing protocol (ARNS) can effectively adapt to the changes of network topology and other network parameters. In different network scenarios, ARNS is more suitable for service transmission than AODV in terms of delay and delivery rate. This paper focuses on an adaptive routing method which combines multiple network state information. However, the parameters of network state information are not considered comprehensively. In the future, more complex channel and scene parameters will be introduced for analysis.

## References


1. Li, G., Boukhatem, L., & Wu, J. (2017). Adaptive quality-of-service-based routing for vehicular ad hoc networks with ant colony optimization. *IEEE Transactions on Vehicular Technology, 66*(4), 3249–3264.
2. Mahar, R., & Choudhary, S. (2017). Performance evaluation of cross link fully adaptive routing algorithm with cross link architecture for Network on Chip. In *2017 International Conference on Inventive Computing and Informatics, Coimbatore* (pp. 576–583).
3. Mahar, R., Choudhary, S., & Khichar, J. (2017). Design of fully adaptive routing for partially interconnected cross-link mesh topology for Network on Chip. In *2017 International Conference on Intelligent Computing and Control, Coimbatore* (pp. 1–6).
4. Bahrebar, P., & Stroobandt, D. (2013). Adaptive routing in MPSoCs using an efficient path-based method. In *2013 International SoC Design Conference, Busan* (pp. 31–34).
5. Dugaev, D. A., Matveev, I. G., Siemens, E., & Shuvalov, V. P. (2018). Adaptive reinforcement learning-based routing protocol for wireless multihop networks. In *2018 XIV International Scientific-Technical Conference on Actual Problems of Electronics Instrument Engineering, Novosibirsk* (pp. 209–218).
6. Daneshtalab, M., Ebrahimi, M., Xu, T. C., Liljeberg, P., & Tenhunen, H. (2011). A generic adaptive path-based routing method for MPSoCs. *Journal of Systems Architecture, 57*(1), 109–120.
7. Kurniawan, D. W. H., Kurniawan, A., & Arifianto, M. S. (2017). An analysis of optimal capacity in cluster of CDMA wireless sensor network. In *2017 International Conference on Applied Computer and Communication Technologies, Jakarta* (pp. 1–6).
8. Dowling, J., Curran, E., Cunningham, R., & Cahill, V. (2005). Using feedback in collaborative reinforcement learning to adaptively optimize MANET routing. *IEEE Transactions on Systems, Man and Cybernetics, Part A, 35*(3), 360–372.
9. Ascia, G., Catania, V., Palesi, M., & Patti, D. (2008). Implementation and analysis of a new selection strategy for adaptive routing in networks-on-chip. *IEEE Transactions on Computers, 57*(6), 809–820.
10. Jeang, Y.-L., Wey, T.-S., Wang, H.-Y., Hung, C.-W., & Liu, J.-H. (2008). An adaptive routing algorithm for mesh-tree architecture in network-on-chip designs. In *3rd International Conference Innovative Computing Information on Control, ICICIC 2008* (p. 182).

11. Sun, Y., Luo, S., Dai, Q., & Ji, Y. (2015). An adaptive routing protocol based on QoS and vehicular density in urban VANETs. *International Journal of Distributed Sensor Networks, 1* (1), 1–14.
12. Li, G., & Boukhatem, L. (2013). Adaptive vehicular routing protocol based on ant colony optimization. In *Proceedings of the IEEE VANET, Taipei, China* (pp. 95–98).
13. Caro, G., Ducatelle, F., & Gambardella, L. (2005). AntHocNet: An adaptive nature-inspired algorithm for routing in mobile ad hoc networks. *European Transactions on Telecommunications, 16*(5), 443–455.
14. Wu, Y., Min, G., Ould-Khaoua, M., & Yin, H. (2009). An analytical model for torus networks in the presence of batch message arrivals with hot-spot destinations. *International Journal of Automation and Computing, 6*(1), 38–47.

# Path Network Arrangement of "Goods-to-Man" Picking System Based on Single-Guided Path Network

**Li Li, Ke Qin, Kai Liu, Zhixin Chen, and JunTao Li**

**Abstract** Aiming at the single-guided path network in e-commerce storage system, we proposed an improved genetic algorithm. The model considered loaded travel distance searched by Dijkstra and A* algorithm, and the guide-path network was preprocessed to reduce the complexity of the algorithm. The genetic algorithm used binary coding, which chromosome represented the direction of the path and adopted the Hamming distance to maintain population diversity when forming a new generation of populations. In order to improve the convergence speed of genetic algorithm, neighborhood search operations were added after selection, crossover, and mutation. Experimental results showed that the improved algorithm had better overall performance compared with the traditional genetic algorithm.

**Keywords** Single-guided · AGV · Path planning · Genetic algorithm · Goods-to-Man

---

Young top-notch talent project of high level teacher team construction in Beijing municipal universities (CIT&TCD201704059); Funding Project for Beijing Intelligent Logistics System Collaborative Innovation Center; Funding Project for Beijing Key Laboratory of intelligent logistics systems; Beijing excellent talents support project (2017000020124G063); The logistics robot system scheduling research team of Beijing Wuzi University; Ningxia Science and Technology Key Research and Development Project (2018BEG03003).

---

L. Li (✉)
Maritime Supply Department, Naval Logistics Academy, Tianjin, China
e-mail: 8235727@qq.com

K. Qin · K. Liu · J. Li
School of Information, Beijing Wuzi University, Beijing, China
e-mail: 1483701476@qq.com

K. Liu
e-mail: 18810250658@163.com

J. Li
e-mail: ljtletter@126.com

Z. Chen
School of Business, Ningxia Polytechnic, Yinchuan, Ningxia, China
e-mail: ithink@sina.com



J. Zhang et al. (eds.), *LISS2019*,
https://doi.org/10.1007/978-981-15-5682-1_37

# 1 Introduction

With the development of Artificial Intelligence Technology and Robot Technology, more and more Automated Guided Vehicles (AGVs) are widely applied in the warehousing system. Amazon acquired Kiva Systems company in 2012, and applied Kiva system to warehouse logistics system, which not only increased the flexible automation and the rapid response of warehousing system, but also enhanced the overall working efficiency [1]. However, at present, there are also many problems in the designing process of AGV system, such as the design of path network layout, the optimization of slotting, and the dispatching of AGV system [2–4], particularly the design of path network layout whose generation algorithm is the bottleneck that restricts the efficiency of the entire system operation process.

In the field of manufacturing, Guan [5] Xiao [6] study the design of the AGV path network and present an improved genetic algorithm that based on topological map modeling, which significantly improves the operating efficiency of the AGV system.

However, in the field of E-commerce warehousing, there are few studies on the AGV path network planning, which mainly focus on the AGV path planning. Zhang [7] proposes a path planning method for AGV picking system that combined different task assignment methods with traffic rules and improved A* algorithm, but the traffic rules are artificially defined, whose rationality is not scientifically verified. In practical application, with the increase of orders and task shelf, multiple AGVs which carry shelf in storage area face a series of problems, if there are no unified rules of path network to guide the operation of AGVs, the whole system will be deadlock or even paralysis. Consequently, we proposed an improved genetic algorithm based on output rate and aimed at minimizing the total distance traveled by AGVs.

# 2 Problem Description and Model Establishment

## *2.1 Problem Description*

In the process of designing path network based "goods to man", the chief requisite was to create a suitable environment model. On the basis of Kiva system, we built models in two dimensions by combining grid map [8] with topological graph, as shown in Fig. 1. In this environment model, each shelf unit area had four shelves length and two shelves width. The grid map, described different attributes in warehouse environment, where gray represented the shelf, which was not targeted would be considered obstacle. The grid of path channel represented by white and each grid had four directions of "East, South, West, North" which of them can be searched. Topological graph was mainly to determine the direction of each path and correlated the path direction to the gird map, then judged whether the network was strongly connected by its adjacency matrix. In the graph, the topology nodes corresponded to the intersection points of the grid graph channels, the edges between the nodes represented the channel path and all the channels were unidirectional.

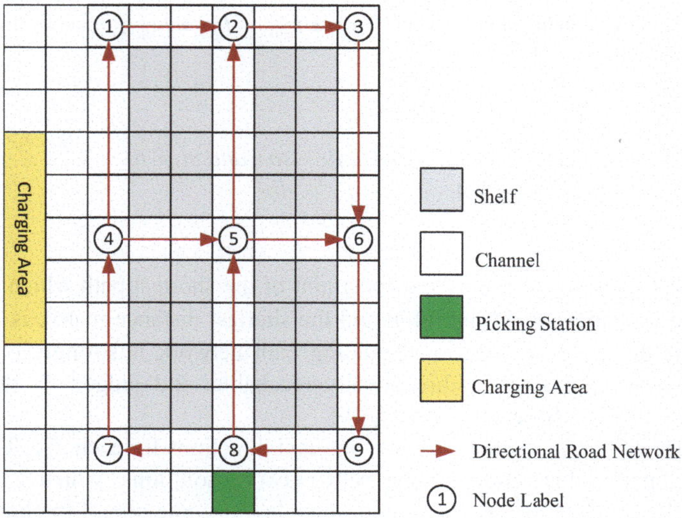


**Fig. 1** Simulation warehouse partial schematic

## 2.2 *Model Hypothesis and Establishment*

Model hypothesis:

- AGV can be steered 90° in the channel.
- AGV is carried by the longitudinal channel.
- Only one picking station is served by one shelf per transfer.
- Do not consider the AGV empty moving distance.
- AGV starts from the charging area and returns to the charging area after completing tasks.

Modeling:
Abstract the topology layer path network of k nodes to the graph G = (V, E), and the Q matrix of r order is its adjacency matrix. V is vertex set, E is edge set, $e_{mn}$ is the edge from node m to node n, whose direction is $Z_{mn}$, and the initial edge is undefined. Suppose that the picking station set is P, which picking station number is p, and the shelf area set is S which quality is s. The flow of the picking station to the shelf is expressed as $f_{pisj}$, the flow from the j shelf to the *i* picking station is expressed as $f_{sjpi} \bullet L_{sjpi}$ represents the shortest path from the *i* picking station to the *j* shelf, and represents the shortest path from the *j* shelf to the *i* picking station.

*J*, the goal of the path network design, represents the minimization of the total distance of the load.

$$Min\ J = \sum_{\substack{1 \le i \le p \\ 1 \le j \le s}} (f_{pisj} \bullet L_{pisj} + f_{sjpi} \bullet L_{sjpi}) \tag{1}$$

$$Z_{mn} = \begin{cases} 1, & from\,node\,m\,to\,n\,and\,m \neq n \\ 0, & else \end{cases} \tag{2}$$

$L_{pisj}$ and $L_{sjpi}$ needs to satisfy the constraint of the shortest path which searched by Dijkstra algorithm or A*algorithm and the shortest distance matrix is $D_{ps}$, $D_{sp}$. Further, the designed unidirectional guiding path network is strongly connected, that is, any two points are bidirectionally reachable, and judged by the strong connectivity of the adjacency matrix Q [9].

In the design of a path, there are two optional directions for each age. Therefore, there are 2^n possible unidirectional path network solutions, which is NP-hard problem [10, 11]. Intelligent search algorithms had become one of the hot spots in current research with its excellent capability and performance, such as neighborhood search, genetic algorithm and so on. But they also have their own limitations, for example, neighborhood search has strong local search ability but poor global search ability. Although genetic algorithm has strong global search ability, genetic algorithm has slow convergence speed. Therefore, in this study, a genetic algorithm with neighborhood search is used to solve the problem, which not only can make the improved genetic algorithm having strong local search, but also having strong global search.

The chromosome coding is used to represent the direction of the unidirectional navigation network in the improved genetic algorithm. Dijkstra and A* algorithms are used to search according to the direction of each unidirectional navigation network to obtain the corresponding total distance and calculate individual fitness degree. Through the neighborhood search and genetic iteration, until the optimal solution or suboptimal solution is searched.

## 3 Proved Genetic Algorithm for Unidirectional Path Network Design

In this paper, the genetic algorithm is mainly analyzed as following: coding and decoding, generating initial population, the selecting and calculating of individual fitness degree, selecting operation, intercrossing and mutation operation. In order to improve the convergence speed of the algorithm, a neighborhood search operation is introduced, the improved genetic algorithm processing flow chart is shown in Fig. 2.

**Fig. 2** The genetic algorithm processing flow chart

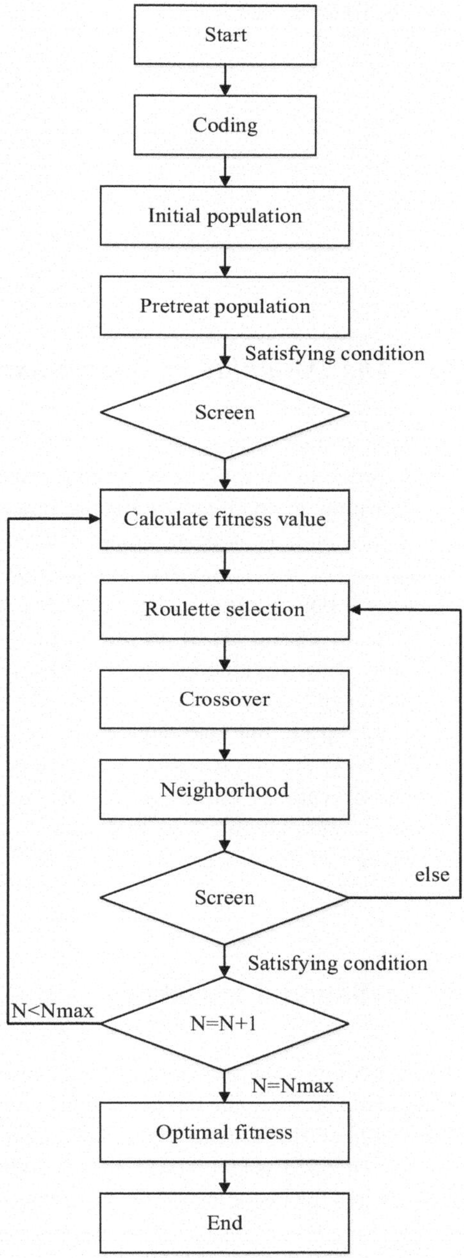

**Table 1** Synthetic edges and their combinations

| No. | Synthetic edge | Combination of edge |
|---|---|---|
| 1 | (4, 2) | (4, 1) (1, 2) |
| 2 | (2, 6) | (2, 3) (3, 6) |
| 3 | (4, 8) | (4, 7) (7, 8) |
| 4 | (8, 6) | (8, 9) (9, 6) |
| 5 | (2, 5) | (2, 5) |
| 6 | (4, 5) | (4, 5) |
| 7 | (5, 6) | (5, 6) |
| 8 | (5, 8) | (5, 8) |

### *3.1 The Designing of Coding and Decoding*

(a) *Path network preprocessing*
In order to simplify the coding and reduce the complexity of the genetic algorithm, the path network is pre-processed before coding, in which the degree of each node is at least 2, that is, at least one ingress and one outbound, which can make the composed unidirectional guided-path network strongly connected. For instance, in Fig. 1, node 1 can satisfy this condition only when the edge (4, 1) and (1, 2) are the same direction. Consequently, when designing unidirectional guided-path networks, edges (4, 1) and (1, 2) can be combined into a synthetic edge (4, 2) for encoding, which not only effectively reduces the search space, but also reduces the computation and increases search speed. In this way, the preprocessed guidance path network diagram is shown in Fig. 1, whose synthetic edges and corresponding edges are shown in Table 1. There are 4 combination edges, that is, $N_c = 4$, and simple edges $N_s = 4$, so the total amount of the processed edges is 8.

## 4 Coding and Decoding

Binary coding is the direction of the channel in the grid map which means that the relationship between the topological map nodes. The length of the chromosome code is the number of synthetic edges. The directed grid topology path network shown in Fig. 1 is encoded as [1 1 0 0 0 1 1 0]. A code uniquely corresponds to a directed path network and the decoding process is to map the code to a directed path network. For instance, the binary encoding of the first synthetic edge (4, 2) is 1, which indicates that the direction on the gird topological map is from node 4 to node 2, at the same time, the corresponding grid direction (4, 1) is north, (1, 2) is east. On the contrary, when the corresponding code is 0, the direction is from node 2 to node 4, of course, the corresponding grid direction (4, 1) is south, and (1, 2) is west. And the decoding process is to map the code to an unidirectional guided-path network.

### 4.1 Fitness Function Calculation

Determine whether the coding is feasible by judging whether the corresponding directional path network is strongly connected. If strong connectivity is available, the coding is feasible; else, it is not feasible. In this paper, the strong connectedness of the graph is determined by the strong connected discriminant method of the adjacency matrix. For a code, if it is feasible, the total distance J is calculated, and the fitness value of the code is F = 1/J.

### 4.2 The Method of Generating the Initial Population

Randomly generate an initial population with $N_p$ population and generate a random vector of $1 \times N_B$. If the random number in an individual is less than 0.5, the position is coded as 0, otherwise it is 1. The resulting binary code needs to be checked for its feasibility. If possible, keep the code, otherwise regenerate the code. To ensure the differences in genotypes between individuals, we introduce the Hamming distance to measure, which calculation formula is:

$$H = (X^i, X^j) = \sum_{k-1}^{NB} \left|x_k^i - x_k^j\right| \tag{3}$$

It is required that the Hamming distance between any two bodies in the initial population is greater than a certain lower limit $H_{min}$, $H_{min} = NB/6$. Repeat the above process until $N_P = N_B/2$ feasible codes are obtained. These feasible codes constitute the initial population.

### 4.3 Selection

Choose the roulette method. First, calculate the fitness of each individual. All individuals in the population will be sorted according to their fitness, and the probability that each individual is selected is directly proportional to its fitness, the greater the fitness, the greater the probability of being selected. Second, the two best individuals of each generation go directly to the next generation to retain the best individuals, and the other individuals is selected by the probability of fitness.

## 4.4 Crossover

This paper uses two-point crossover with a crossover probability of $P_c = 0.8$. Two sites $N_1$ and $N_2$ are randomly selected, and two selected parent individuals exchanged genes between $N_1$ and $N_2$ to obtain two new individuals. For instance, two parental chromosomes $\alpha_1 = [11111111]$ and $\alpha_2 = [00000000]$ are selected, and if the cross positions are selected to be 2 and 7, the new individuals are $\alpha_1' = [10000001]$ and $\alpha_2' = [01111110]$. And then, randomly select one of the new individuals generated to enter the mutation operation.

## 4.5 Mutation

Adopt two points mutation operation. Two positions are randomly selected, and the genes in these two positions are inverted, that is, 1 becomes 0 and 0 becomes 1. For $\alpha_1' = [10000001]$ in Sect. 4.4, if the position of 5 and 8 are mutated, then $\beta_1 = [10001000]$ is obtained. The same applies to $\alpha_2'$. The mutation operation is determined by the mutation probability $P_m$, which is generally taken as 0.01 to 0.10. This paper takes 0.1. After mutation operation, it is necessary to verify the feasibility of the coding. If the coding is feasible, it will be saved, otherwise, the selection, crossover and mutation operations will be repeated until $N_p$ new individuals without restrings are formed, which will form the transitional populations.

## 4.6 Neighborhood Search

Neighborhood search is a simple and effective local search algorithm, only for a particular area in the solution space [12]. The purpose of using the neighborhood search operation in this paper is to improve the local search ability of the genetic algorithm.

Neighborhood search algorithm principle: Given an initial solution $x$, a neighborhood solution generator is used in the neighborhood of the initial solution x to find a solution that is better than the initial solution $x$. If the solution found is superior to the initial solution $x$, the initial solution $x$ is replaced by this solution, and continue the above process until no more optimal solution is found. Generally, the algorithm can only find the local optimal solution, and the choice of the initial solution will affect the quality of the solution to a large extent. Thus, the algorithm is executed repeatedly with different initial solutions, which can effectively improve the quality of the final solution.

In this study, the coding obtained by selection, crossover, and mutation was used for the neighborhood search operation. Selecting an individual's chromosome code such as $\beta_1 = [10001000]$, reverse the first digit and leave the other digits unchanged, get $\beta_1' = [00001000]$, and so on, each individual has a total of $N_B$ neighborhood solutions. Then verify whether the unidirectional guided-path networks corresponding to the chromosomes of the solution path networks in each neighborhood are strongly connected. If they are connected, they will be saved and calculated the fitness, else, they will be eliminated. If the solution is superior to the individual, the original individual will be replaced.

### *4.7 Termination Conditions (Parameter Settings)*

In this paper, the evolutionary generation continues $N$ (the maximum number of evolutions is $N_{MAX}$). If the optimal individuals in the population do not change, stop the operation, and $N = N_{MAX}$.

## 5 Calculation Examples and Analysis

In order to verify the effectiveness of the improved genetic algorithm, firstly we study the model examples in Sect. 5.1 and make the following four comparisons.

- Standard Genetic Algorithm and Dijkstra, namely, SGA&D
- Standard Genetic Algorithm and A*, namely, SGA& A*
- Improved Genetic Algorithm and Dijkstra, namely, IGA& D
- Improved Genetic Algorithm and Dijkstra, namely IGA& A*

So far, we had not found direct design of literature based on the "goods to man" model of e-commerce warehousing path network layout, so we cannot be directly compared with other literature.

### *5.1 Example 1*

The layout of the experience path network corresponding to Example 1 is shown in Fig. 1, and its load flow is shown in Table 2. In view of this situation, we will directly compare the total distance under the empirical path network layout with the total distance under the path network layout generated by this paper. The program was written in MATLAB 2016. And the operating environment is Windows 10, Intel(R) Core (TM) i5-7200 CPU @ 2.50 GHz 2.70 GHz dual-core CPU, 4 GB of memory. The simulation experiment was carried out under the above environment, and the results are as follows:

**Table 2** Load flow table for Example 1

| No. | Shelf coordinates | Assign | No. | Shelf coordinates | Assign |
|---|---|---|---|---|---|
| 1 | (2, 2) | 9 | 17 | (2, 5) | 6 |
| 2 | (3, 2) | 5 | 18 | (3, 5) | 8 |
| 3 | (4, 2) | 2 | 19 | (4, 5) | 8 |
| 4 | (5, 2) | 0 | 20 | (5, 5) | 0 |
| 5 | (7, 2) | 7 | 21 | (7, 5) | 4 |
| 6 | (8, 2) | 5 | 22 | (8, 5) | 4 |
| 7 | (9, 2) | 9 | 23 | (9, 5) | 6 |
| 8 | (10, 2) | 8 | 24 | (10, 5) | 2 |
| 9 | (2, 3) | 6 | 25 | (2, 6) | 9 |
| 10 | (3, 3) | 7 | 26 | (3, 6) | 1 |
| 11 | (4, 3) | 8 | 27 | (4, 6) | 5 |
| 12 | (5, 3) | 0 | 28 | (5, 6) | 0 |
| 13 | (7, 3) | 4 | 29 | (7, 6) | 1 |
| 14 | (8, 3) | 3 | 30 | (8, 6) | 6 |
| 15 | (9, 3) | 10 | 31 | (9, 6) | 2 |
| 16 | (10, 3) | 9 | 32 | (10, 6) | 5 |

**Table 3** Experiment results of each algorithm in Example 1

| Algorithm | Average total distance | Algorithm run time (unit: second) | The coding of better total distance | Better solution times |
|---|---|---|---|---|
| Experience path network layout | 4190 | / | / | / |
| SGA&D | 4144.6 | 17.8734592 | 0 1 1 0 0 0 1 0 | 2 |
| | | | 1 0 0 1 1 1 0 1 | |
| SGA& A* | 4031.8 | 4.1435035 | 0 1 1 0 0 0 1 0 | 1 |
| IGA& D | 3929.6 | 146.5095046 | 0 1 1 0 0 0 1 0 | 9 |
| | | | 1 0 0 1 1 1 0 1 | |
| IGA& A* | 3978.4 | 32.9255581 | 1 0 0 1 1 1 0 1 | 7 |
| | | | 0 1 1 0 0 0 1 0 | |

In this simple example, the grid topology has one picking station and 32 shelves, and 8 synthetic edges (i.e. $N_B = 8$). Experimental algorithm parameters: Population size: $N_p = N_B/2 = 4$, Hamming distance: $H_{min} = N_B/6 = 1.3$, take 1. Crossover probability: $P_c = 0.8$, Variation probability: $P_m = 0.1$, Number of iterations: $N_{MAX} = 20$. Each algorithm will be run 10 times. The experimental results are shown in Table 3.

From Table 3, it can be seen that the average total distance of the path network generated in this paper is less than the total distance of the empirical path network.

The average total distance traveled by SGA&A* is less than SGA&D, and its search time is short, but only one of the 10 runs has one better solution and the corresponding code of the better solution is only one. And the convergence of the standard genetic algorithm is not good. Although the genetic algorithm with neighborhood search increases the search time, the improved algorithm greatly improves the quality of solutions. More than 7 times of the 10 operations obtained better solutions, and the average total distance is close to the better solution. Therefore, it can be seen that the neighborhood search increases the local search ability of the genetic algorithm.

## 5.2 Example 2

For example 2, its environment model is abstracted into a grid topology path network graph and its load flow is shown in Table 4. The total distance under the experience path network layout is: 10818, and the path network layout is shown in Fig. 3. In this path the network has one picking station, 64 shelves, 18 synthetic edges ($N_B = 18$). Experimental algorithm parameters: Population size: $N_p = N_B/2 = 9$, Hamming distance: $H_{min} = N_B/6 = 3$, Crossover probability: $P_c = 0.8$, Variation probability: $P_m = 0.1$, Number of iterations: $N_{MAX} = 20$. The experimental results are shown in Table 5.

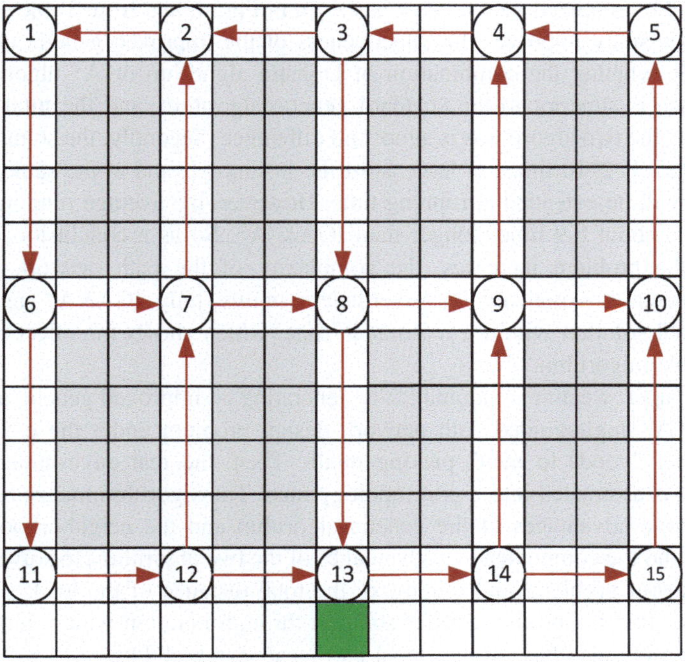


**Fig. 3** Example 2 experience path network layout

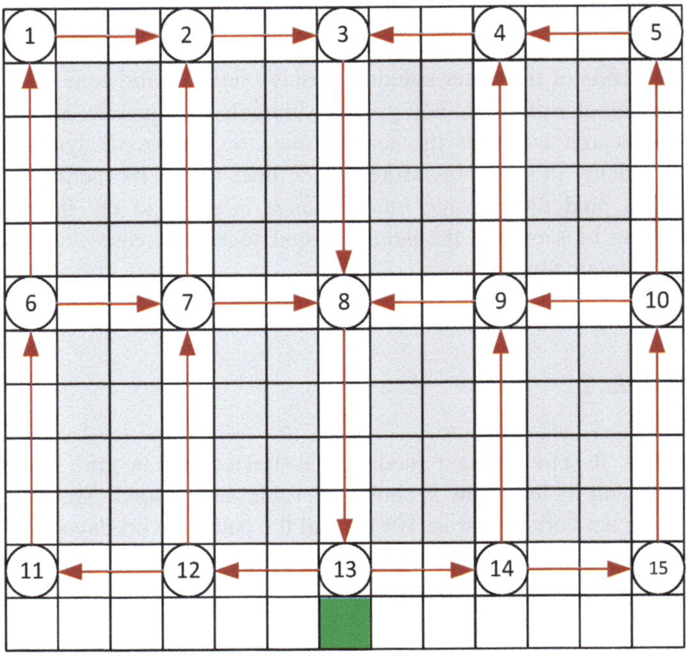


**Fig. 4** Automatically generated path network layout

One of the optimized path layouts is shown in Fig. 4, and from Table 5, we can see that this study improves the effectiveness of the improved genetic algorithm. First of all, whether the combination of Dijkstra algorithm or A* algorithm, the convergence is superior to the standard genetic algorithm, and the total distance obtained by the two algorithms is almost no difference. Secondly, the solution space of example 2 is 2^10 times as large as of the instance 1, and at the same time, its algorithm will be extended at running time. However, the average running time of IGA& D is about 8.9 times longer than IGA& A*. So as a conclusion, with the scale of the problem increases, the complexity of the path network planning problem increases exponentially. Nevertheless, in this study, IGA&A* can obtain a near-optimal solution within a reasonable time, which shows the effectiveness of the IGA&A* algorithm.

In this paper, we aim at automatically generating an improved genetic algorithm for the AGV single-guided path network design problem under the e-commerce warehousing "goods to man" picking mode. Then, the real environment of the warehouse is abstracted into a grid topology map. The algorithm makes full use of the respective advantages of the genetic algorithm and the neighborhood search algorithm and the comprehensive advantages of the two algorithms to solve the path network layout problem, and minimizes the total distance of the load, which not only proves that it can find a better solution through iteration with a certain population size, but also has strong global and local search ability.

**Table 4** Load flow table for Example 2

| No | Shelf coordinates | Assign | No | Shelf coordinates | Assign |
|---|---|---|---|---|---|
| 1 | (2, 2) | 7 | 33 | (2, 8) | 6 |
| 2 | (3, 2) | 8 | 34 | (3, 8) | 8 |
| 3 | (4, 2) | 9 | 35 | (4, 8) | 8 |
| 4 | (5, 2) | 0 | 36 | (5, 8) | 0 |
| 5 | (7, 2) | 6 | 37 | (7, 8) | 7 |
| 6 | (8, 2) | 7 | 38 | (8, 8) | 6 |
| 7 | (9, 2) | 3 | 39 | (9, 8) | 7 |
| 8 | (10, 2) | 9 | 40 | (10, 8) | 6 |
| 9 | (2, 3) | 7 | 41 | (2, 9) | 4 |
| 10 | (3, 3) | 1 | 42 | (3, 9) | 5 |
| 11 | (4, 3) | 5 | 43 | (4, 9) | 7 |
| 12 | (5, 3) | 0 | 44 | (5, 9) | 0 |
| 13 | (7, 3) | 9 | 45 | (7, 9) | 5 |
| 14 | (8, 3) | 2 | 46 | (8, 9) | 7 |
| 15 | (9, 3) | 8 | 47 | (9, 9) | 8 |
| 16 | (10, 3) | 4 | 48 | (10, 9) | 3 |
| 17 | (2, 5) | 7 | 49 | (2, 11) | 2 |
| 18 | (3, 5) | 1 | 50 | (3, 11) | 7 |
| 19 | (4, 5) | 6 | 51 | (4, 11) | 1 |
| 20 | (5, 5) | 0 | 52 | (5, 11) | 0 |
| 21 | (7, 5) | 1 | 53 | (7, 11) | 4 |
| 22 | (8, 5) | 0 | 54 | (8, 11) | 6 |
| 23 | (9, 5) | 3 | 55 | (9, 11) | 5 |
| 24 | (10, 5) | 0 | 56 | (10, 11) | 9 |
| 25 | (2, 6) | 1 | 57 | (2, 12) | 0 |
| 26 | (3, 6) | 6 | 58 | (3, 12) | 1 |
| 27 | (4, 6) | 0 | 59 | (4, 12) | 5 |
| 28 | (5, 6) | 0 | 60 | (5, 12) | 0 |
| 29 | (7, 6) | 9 | 61 | (7, 12) | 2 |
| 30 | (8, 6) | 7 | 62 | (8, 12) | 4 |
| 31 | (9, 6) | 8 | 63 | (9, 12) | 6 |
| 32 | (10, 6) | 8 | 64 | (10, 12) | 2 |

The algorithm does not consider AGV no-load conditions, and the problem of this paper is smaller. Issues to be addressed in future work include considering the no-load problem, increasing the problem size, and further optimizing the algorithm.

**Table 5** Experiment result of each algorithm in Example 2

| Algorithm | Average total distance | Algorithm run time (unit: second) | The coding of better total distance | Better solution times |
|---|---|---|---|---|
| Experience path network layout | 10818 | / | / | / |
| SGA&D | 7647.4 | 87.6673186 | 0 0 0 0 1 0 0 1 0 1 1 0 0 0 1 0 0 1 | 2 |
| | | | 1 1 0 0 1 0 0 1 0 1 1 0 0 0 1 0 0 1 | |
| SGA& A* | 7567.2 | 10.7650058 | 0 0 0 0 1 0 0 1 0 1 1 0 0 0 1 0 0 1 | 3 |
| | | | 1 0 0 0 1 0 0 1 0 1 1 0 0 0 1 0 0 1 | |
| | | | 1 1 0 0 1 0 0 1 0 1 1 0 0 0 1 0 0 1 | |
| IGA& D | 7295 | 7783.250805 | 1 0 0 0 1 0 0 1 0 1 1 0 0 0 1 0 0 1 | 7 |
| | | | 1 1 0 0 1 0 0 1 0 1 1 0 0 0 1 0 0 1 | |
| | | | 0 0 0 0 1 0 0 1 0 1 1 0 0 0 1 0 0 1 | |
| IGA& A* | 7297.6 | 871.8217027 | 1 0 0 0 1 0 0 1 0 1 1 0 0 0 1 0 0 1 | 7 |
| | | | 1 1 0 0 1 0 0 1 0 1 1 0 0 0 1 0 0 1 | |
| | | | 0 0 0 0 1 0 0 1 0 1 1 0 0 0 1 0 0 1 | |

# References


1. Wu, J. (2015). Application analysis and prospect of Amazon warehouse Kiva robot. *Logistics Technology and Application, 10,* 159–164.
2. Wang, J. (2014). *Research on single and bidirectional mixed path planning and traffic management technology for automatic guided vehicle system.* Nanjing University of Aeronautics and Astronautics.
3. Pan, C. (2017). *Simulation research on routing planning of storage and logistics robot.* North University of China.
4. He, C. (2017). *Study on the optimization of goods' place under the mode of delivery.* Zhejiang Sci-Tech University.
5. Guan, X., Dai, X., & Li, J. (2009). Design of AGV path network based on variable neighborhood niche genetic algorithm. *China Mechanical Engineering, 21,* 2581–2586.
6. Xiao, H., Lou, P., & Qian, X. (2013). Flexible guided job-shop environment under single guided path network design method. *Journal of Mechanical Engineering, 3,* 122–129.
7. Zhang, X. (2015). *Research on picking operation optimization and algorithm based on Kiva system.* Beijing University of Posts and Telecommunication.

8. Yu, C., & Qiu, Q. (2013). Hierarchical robot path planning algorithm based on grid map. *Journal of University of Chinese Academy of Sciences, 4,* 528–538.
9. Liu, X., & Qin, F. (2005). Strong connectivity analysis and discriminant algorithm for digraph. *Computer Applications and Software,* (4), 138–139.
10. Guan, X., Dai, X., & Li, J. (2009). Variable neighborhood niche genetic algorithm based agv flow path design method. *China Mechanical Engineering, 21*, 2581–2586.
11. Guan, X., Dai, X., & Li, J. (2011). Revised electromagnetism-like mechanism for flow path design of unidirectional AGV systems. *International Journal of Production Research, 49*(2), 401–429.
12. Lu, Z., Guo, G., & Jiang, J. (2005). Hybrid genetic algorithm based on neighborhood search and its application in symmetric TSP. *Computer Engineering and Applications,* (7) 79–81.

# The Influence of Personality Traits on Organizational Identity

**Hongyu Li, Long Ye, Zheng Yang, and Ming Guo**

**Abstract** This paper deeply studies how the personality of employees affect organizational identity. Organizational identity is a special form of social identity, focusing on the consistency of employee thinking and corporate values. The construction of organizational identity has gradually become a key task of organizational development. There are still a lot of research space on the antecedent variables that lead to organizational identity, and few scholars study the impact of individual employee differences on organizational identity. Therefore, this paper takes the front-line staff of a large enterprise as the object, studies the influence of personality on organizational identity, compares and analyzes the influence path of job satisfaction and engagement as the mediator, and analyzes the internal mechanism of personality on organizational identity. Finally, according to the conclusion, we propose the management suggestions for employees with different personality.

**Keywords** Organizational identity · Big-five personality · Job satisfaction · Engagement

H. Li (✉) · L. Ye · Z. Yang · M. Guo
School of Economics and Management, Beijing Jiaotong University, Beijing, China
e-mail: caulhy2017@163.com

L. Ye
e-mail: yelong@bjtu.edu.cn

Z. Yang
e-mail: 16113169@bjtu.edu.cn

M. Guo
e-mail: gming@bjtu.edu.cn



J. Zhang et al. (eds.), *LISS2019*,
https://doi.org/10.1007/978-981-15-5682-1_38

## 1 Introduction

In the 21st century, with the implementation of "The Belt and Road Initiative" and "Go globally" strategy, the transnational operation of various enterprises has become the new normal, which brings new opportunities as well as new challenges. These enterprises have to face the employees with different cultural backgrounds and different social conditions. The differences in the degree of thinking and value identification of these employees bring great challenges to enterprise management. Nowadays, our attitude and values about jobs have changed dramatically. The work motivation has changed from the initial material needs to more self-realization and sense of value. At the same time, the stable development of the organization requires employees to highly identify with the values and strategic goals of the enterprise, so as to maximize the benefits of the enterprise and individuals. Excellent enterprises will regard employees as the most important wealth of the enterprise and will focus on the consistency of employees' thinking and corporate values.

Multinational corporations should apply different management methods to improve employees recognition and internalization of organizational values according to the cultural characteristics of different countries and regions. For example, there is a survey about one of Top 100 multinational corporations' employees, which collects data every two years (see Table 1). It mainly investigates employees' identification with strategies and management institutions, as well as their engagement and career development. Around 2012, the enterprise completed a merger and acquisition events, After the merger, the employee's work pressure increased significantly, and the balance of work and life of the employees was affected, but the organizational recognition and engagement of the employees were significantly improved, and then leading to the rapid and stable development after the merger. On the one hand, organizational identity can improve employees' understanding and recognition of the strategic objectives, work environment and work pressure of the organization; on the other hand, it can enhance employees' trust in the organization. Employees believe that the organization can meet their spiritual and material needs, and they can realize self-improvement and effective growth in the organization. Therefore, employees' recognition of organizational values will promote organizational performance and promote the healthy and stable development of enterprises.

In the organizational behavior theories, organizational identity has always been one of the hot issues in research. Organizational identity refers to the individual's sense of we-ness in the organization, and expressed as the individual's choice, acquisition, and retention behavior for a particular role in the organization [1]. Organizational identity originated from the concept of social psychology, the classic social identity to distinguish the two kinds of identity, social identity and personal identity. Personal identity is mainly refers to the individual in comparison with other individuals associated with self-concept, and social identity is refers to the individual self-concept in social category or group relations [2, 3]. As a special

**Table 1** Company's workers work conditions investigation result

| Variable | Description | 2010 | | | 2012 | | | 2014 | | | 2016 | | |
|---|---|---|---|---|---|---|---|---|---|---|---|---|---|
| | | High | Mid | Low | High | Mid | Low | High | Mid | Low | High | Mid | Low |
| Strategy | I clearly know the company's strategy and positioning | 83 | 13 | 4 | 91 | 8 | 1 | 93 | 4 | 2 | 92 | 5 | 2 |
| | I understand the relationship between my job and company's goals | 87 | 9 | 4 | 91 | 9 | 0 | 92 | 5 | 3 | 92 | 5 | 4 |
| Personal development | I am well aware of the company's performance requirements | 78 | 16 | 6 | 86 | 10 | 4 | 89 | 1 | 1 | 98 | 1 | 1 |
| | I have a big goal and I can achieve it | 75 | 16 | 8 | 86 | 10 | 4 | 88 | 6 | 6 | 89 | 5 | 6 |
| | The company gives a fair appraisal of its performance | 48 | 23 | 30 | 64 | 23 | 13 | 73 | 8 | 20 | 74 | 8 | 19 |
| | I can get a very firm appraisal of my performance | 58 | 23 | 19 | 64 | 23 | 13 | 72 | 11 | 16 | 74 | 10 | 16 |
| Work environment | The company provides a very satisfactory working environment | 56 | 33 | 12 | 66 | 27 | 6 | 88 | 6 | 6 | 88 | 7 | 5 |
| | The company provides a stable working environment | 71 | 23 | 6 | 64 | 26 | 10 | 90 | 2 | 8 | 91 | 3 | 6 |
| Work pressure | There are no major obstacles to my work | 67 | 21 | 12 | 55 | 36 | 9 | 86 | 5 | 9 | 86 | 6 | 8 |
| Organizational identity | I was valued and respected in the company | 67 | 19 | 14 | 71 | 22 | 8 | 88 | 5 | 8 | 87 | 5 | 8 |
| | I am free to express my opinions without worry or fear | 55 | 24 | 21 | 68 | 24 | 8 | 69 | 10 | 22 | 71 | 9 | 20 |
| | The company can help me achieve work-life balance | 55 | 25 | 21 | 42 | 38 | 19 | 78 | 7 | 15 | 81 | 7 | 12 |
| Engagement | I'm very proud to be a part of the company | 72 | 20 | 8 | 79 | 18 | 3 | 87 | 6 | 6 | 88 | 6 | 6 |
| | I would recommend my company to my family or friends | 54 | 26 | 21 | 56 | 29 | 15 | 80 | 7 | 13 | 80 | 8 | 12 |
| | I feel motivated to perform duties outside of work | 50 | 28 | 22 | 49 | 30 | 21 | 78 | 6 | 16 | 78 | 7 | 15 |
| | If given a choice, you would like to work in company for a long time | 83 | 10 | 7 | 80 | 16 | 4 | 82 | 10 | 9 | 78 | 15 | 6 |

*Note* The data in the table are all percentage (%) and the total number is 401

form of social identity, organizational identity refers to the individual's perception that an individual defines himself as an organizational member and thus belongs to the organization [3]. The lack of organizational identity may cause serious problems for the organization, such as anti-social behaviors and other extreme behaviors [4]. Today, the construction of organizational identity has become a key task in the development of organizations.

Throughout the studies on organizational identity, the existing studies still have the following problems: firstly, the study on the antecedent variables of organizational identity is not sufficient, and there is still a lot to study; Secondly, there are few studies on the impact of individual differences of employees on organizational identity, and the internal mechanism of individual differences and organizational identity also needs to be further revealed. Therefore, this paper focuses on the impact of individual differences of organizational employees on organizational identity, and carries out the study from the following three aspects: firstly, exploring whether there are differences in job satisfaction among individual employees with different personality; secondly, exploring whether job satisfaction and engagement affect organizational identity, and whether personality affect organizational identity through job satisfaction; finally, on the basis of empirical research, establish the impact model of job satisfaction on organizational identity through job satisfaction, and propose management suggestions for individuals with different personality characteristics.

## 2 Theory and Hypotheses

Organizational identity originates from social psychology. There have been a lot of studies on its manifestations, concepts, but the antecedent variables of organizational identity is still not enough. Pratt believes that there are two basic motivations for organizational identity: One is the need for self-classification, so as to define the "status of individuals in society" [2]; Second is the need for self-improvement, hoping to be rewarded as an organization member [5, 6]. Male believes that studies on organizational identity should explore the propensity to identify [7], so there may be some internal connection between organizational identity and individual personality. A recent study shows that the different personalities have a significant impact on organizational identity. For example, individuals with higher individualism and individuals with higher collectivism have obvious differences in their perceived corporate social responsibility, thus significantly affecting organizational identity [8]. Some scholars have found that extroversion positively affects organizational identity and neuroticism negatively affects organizational identity [9], but the internal mechanism of these effects has not been thoroughly discussed.

Personality is stable and unchangeable [9]. In the study of personality, different scholars have developed a number of measurement dimensions, but most scholars have an amazing and fairly consistent view: the basic structure of personality is composed of five factors, known as the "big five personality" [9–13]. And relatively consistent into five factors: extroversion, agreeableness, conscientiousness, neuroticism, openness to experience [14]. Extroversion refers to the individual's ability to interact with each other. High scores represent courageous, strong social skills. Agreeableness represents affinity, and individuals with higher scores tend to have more intimate, satisfactory and safe interpersonal relationships [15]. Individuals with high scores for conscientiousness often have strong personal achievement motivations and a sense of responsibility, working seriously and devotedly. They hope to continuously improve their own quality and work performance through their own efforts and obtain the support and recognition of the organization [9]. Neuroticism is a kind of negative emotional state, which refers to the instability of individual emotions. Individuals with openness to experience generally have highly imagination and creativity, and are willing to change and seek new ideas [16]. Based on the big five model, Digman proposed a high-order big two model, which mainly includes two dimensions of stability factor and elasticity factor [17]. The stability factor is the common variance that can be explained by agreeableness, neuroticism and conscientiousness, which corresponds to concepts such as self-control and social adaptation. The elasticity factor is the common variance that can be explained by extraversion and openness to experience, which corresponds to concepts such as self-resilience and personal growth [18, 19]. The study of the influence path of personality on organizational identity can reveal the internal mechanism of personality affecting organizational identity.

Job satisfaction is an evaluation of job characteristics and emotional experience of employees. According to encouragement theory, individual's sense of responsibility, achievement and growth can lead to job satisfaction. The higher the individual's job satisfaction is, the more he can experience joyful experience and emotion, and the more likely he tend to generate emotional commitment and identity to work and organization [20]. On the contrary, employees with low job satisfaction are likely to have negative emotions such as job burnout and turnover tendency. Employees' awareness and attitude towards the organization can be enhanced by improving job satisfaction, so as to improve work performance and reduce turnover. In recent years, many scholars have studied the mediating effect of job satisfaction on organizational identity and related variables [20–23]. For example, Roger et al. studied the mediating effect of teachers' job satisfaction on occupational identity and emotional commitment [20], while Tidwell studied the mediating effect between job satisfaction and organizational commitment and organizational identity.

In Kahn's conceptual model, Kahn believes that satisfaction of basic needs of employees will lead to job satisfaction, and when employees have emotional connection with other individuals in the organization on the premise of cognitive vigilance, job satisfaction will lead to engagement [24]. Since Kahn first proposed engagement, a large number of studies have emerged. However, the antecedents of engagement are mostly at the organizational level. The researches focus on external environmental factors, but pay little attention to internal factors of employees. To sum up, current researches on engagement mainly focus on job characteristics, salary and benefits, career development, organizational atmosphere and personality.

In terms of organizational characteristics, many scholars believe that a satisfactory job can attract the interest of employees, so that they can devote themselves to the work and make their own efforts to help the organization complete its mission [25]. In terms of salary and welfare, some scholars believe that satisfactory salary and welfare can significantly affect engagement. In terms of career development, Buckingham and Coffman believe that opportunities for learning and growth are one of the most important factors affecting engagement, while Macisaac believes that a good organizational atmosphere is very important for improving engagement. In terms of personality, from the perspective of demographics, Schaufeli shows that employee engagement is related to gender, and female engagement is significantly higher than that of male [26]. Robinson et al. have found that engagement changes with age [27].

Based on previous studies, this paper explores the mediating effect of job satisfaction and engagement on personality and organizational identity. To sum up, the model established in this paper is shown in Fig. 1, and the following research hypotheses are proposed:

**Hypothesis 1a&1d.** Neuroticism negatively affects engagement and job satisfaction.
**Hypothesis 1b&1c&1e&1f.** Conscientiousness and agreeableness positively affect engagement and job satisfaction.
**Hypothesis 2a&2b.** Extroversion and openness to experience positively affect engagement.
**Hypothesis 2c&2d.** Extroversion and openness to experience positively affect job satisfaction.
**Hypothesis 3a.** Engagement positively affects organizational identity.
**Hypothesis 3b.** Job satisfaction positively affects organizational identity.
**Hypothesis 3c.** Engagement has a greater impact on organizational identity than job satisfaction.

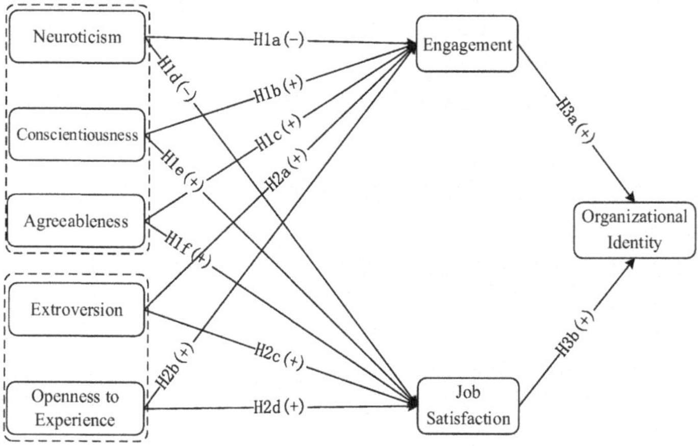


**Fig. 1** Research model

# 3 Methods

## 3.1 Research Setting, Sample and Procedures

The research was carried out among the front-line employees of a large enterprise. The survey used a combination of online and offline methods. A total of 500 questionnaires were distributed and 477 were effectively recovered with a recovery rate of 95.4%. There were 423 questionnaires in line with this study, and the effective rate was 84.6%. According to the survey results, 89.1% of the respondents are male employees, and 85.1% are under 45 years old. They are the backbone of the company's business.

## 3.2 Measures

The 5-point Likert scale was used for all items in the questionnaire, with "1" indicating strong disagreement, "5" indicating strong agreement, and 1 to 5 increasing successively.

- *Personality Characteristics.* The personality mainly adopt the NEO questionnaire of 60 questions. The scale includes five dimensions of extraversion, agreeableness, conscientiousness, neuroticism and openness to experience. The internal consistency coefficients of the above five dimensions is 0.733, 0.579, 0.774, 0.845 and 0.668 respectively. The overall internal consistency coefficient of the table is 0.607.
- *Job Satisfaction.* The scale of job satisfaction adopts the 12-question scale compiled by Gallup [23], which is divided into two dimensions: job satisfaction and engagement. The internal consistency coefficient of the scale is 0.868.
- *Organization Identity.* The organization identity tends to adopt the 6-question scale compiled by Mael, which has been widely used and has high reliability and validity. The internal consistency coefficient is 0.891.
- *Control Variables.* Age and gender were taken as control variables in this study.

## 3.3 Descriptive Statistical Analysis

The results of descriptive statistical analysis of each research variable are shown in Table 2. Neuroticism was negatively correlated with organizational identity, job satisfaction and engagement; Extroversion was positively correlated with organizational identity, job satisfaction and engagement; Openness to experience is positively correlated with job satisfaction and engagement; Conscientiousness is positively correlated with job satisfaction; Agreeableness is positively correlated

**Table 2** Variable's mean, standard deviation and correlation coefficient

| Variable | Mean | SD | 1 | 2 | 3 | 4 | 5 | 6 | 7 | 8 | 9 |
|---|---|---|---|---|---|---|---|---|---|---|---|
| 1. Sex | – | – | | | | | | | | | |
| 2. Age | 2.42 | 1.07 | −0.157$^{**}$ | | | | | | | | |
| 3. Neuroticism | 2.77 | 0.85 | 0.100$^{*}$ | −0.01 | | | | | | | |
| 4. Extroversion | 3.56 | 0.78 | −0.02 | 0.02 | −0.296$^{**}$ | | | | | | |
| 5. Openness to experience | 3.25 | 0.69 | −0.08 | −0.06 | −0.107$^{*}$ | 0.226$^{**}$ | | | | | |
| 6. Conscientiousness | 3.44 | 0.49 | 0.04 | −0.01 | −0.114$^{*}$ | 0.07 | −0.02 | | | | |
| 7. Agreeableness | 3.91 | 0.62 | −0.05 | 0.02 | −0.301$^{**}$ | 0.420$^{**}$ | 0.202$^{**}$ | 0.189$^{**}$ | | | |
| 8. Organizational identity | 3.92 | 0.73 | 0.01 | −0.07 | −0.230$^{**}$ | 0.202$^{**}$ | 0.08 | 0.07 | 0.222$^{**}$ | | |
| 9. Job satisfaction | 3.75 | 0.62 | −0.05 | −0.02 | −0.309$^{**}$ | 0.234$^{**}$ | 0.131$^{**}$ | 0.122$^{*}$ | 0.351$^{**}$ | 0.447$^{**}$ | |
| 10. Engagement | 3.4 | 0.70 | −0.07 | −0.04 | −0.353$^{**}$ | 0.219$^{**}$ | 0.111$^{*}$ | 0.04 | 0.230$^{**}$ | 0.569$^{**}$ | 0.555$^{**}$ |

*Note* *. $p < 0.05$; **. $p < 0.01$

with organizational identity, job satisfaction and engagement; Organizational identity is positively correlated with job satisfaction and engagement. These provide preconditions for studying the relationship between variables.

## 3.4 Structural Equation Models

Confirmatory factor analysis (CFA) was used to test the convergent validity and discriminative validity of structural variables such as personality, organizational identity and engagement in this study [28]. The result of confirmatory factor analysis were obtained using Lisrel8.5 (see Table 3). A total of 7 models were analyzed, and it reported $\chi^2$/df, NNFI, CFI, SRMR and RMSEA.

$$\chi^2 = (N-1)F \tag{1}$$

$$NNFI = \frac{\frac{\chi_b^2}{df_b} - \frac{\chi^2}{df}}{\frac{\chi_b^2}{df_b} - 1} \tag{2}$$

$$CFI = 1 - \frac{max(\chi^2 - df, 0)}{max\left(\chi_b^2 - df_b, 0\right)} \tag{3}$$

$$SRMR = \sqrt{\frac{2\sum\sum\left(\frac{s_{ij}-\widehat{\sigma_{ij}}}{s_{ij}s_{jj}}\right)^2}{p(p+1)}} \tag{4}$$

**Table 3** Confirmatory factor analysis results

| Model | $\chi^2$ | $\chi^2$/df | NNFI | CFI | SRMR | RMSEA |
|---|---|---|---|---|---|---|
| Eight-factor | 1681.16 | 2.22 | 0.93 | 0.94 | 0.069 | 0.054 |
| Seven-factor | 1775.03 | 2.33 | 0.93 | 0.94 | 0.062 | 0.056 |
| Four-factor | 2695.64 | 3.49 | 0.89 | 0.90 | 0.076 | 0.077 |
| Three-factor | 2830.82 | 3.84 | 0.89 | 0.89 | 0.080 | 0.082 |
| Two-factor | 3752.83 | 5.07 | 0.85 | 0.86 | 0.086 | 0.098 |
| Single factor | 5149.11 | 6.95 | 0.80 | 0.81 | 0.099 | 0.119 |

*Note* The single factor model is big five personality + engagement + organizational identity; The two-factor model is big five personality, engagement + organizational identity. The three-factor model is the big five personality, engagement and organizational identity. The four-factor model is composed of stability factor, elasticity factor, engagement factor and organizational identity. The seven-factor models were extraversion, agreeableness, conscientiousness, neuroticism, openness, engagement and organizational identity. The eight-factor model was extroversion, agreeableness, conscientiousness, neuroticism, openness, job satisfaction, engagement, and organizational identity

$$RMSEA = \sqrt{\frac{\chi^2}{(N-1)-(df-1)}} \tag{5}$$

From the fitting parameters, the seven-factor model and the eight-factor model are optimally fitted. According to the theory of engagement, this study chooses the eight-factor model to analyze the structural equation. The model fitting indicators are: Chi-Square(423) = 1681.16, NNFI = 0.93, CFI = 0.94, SRMR = 0.069, RMSEA = 0.054. According to the research recommendations of relevant scholars, it is estimated that the model of this study is acceptable. There is the path standardization coefficient of the structural equation model (see Fig. 2). By comparison, it can be found that H3c is supported. All hypothesis test results are shown in Table 4. Seven of the 13 hypotheses proposed in this study are supported.

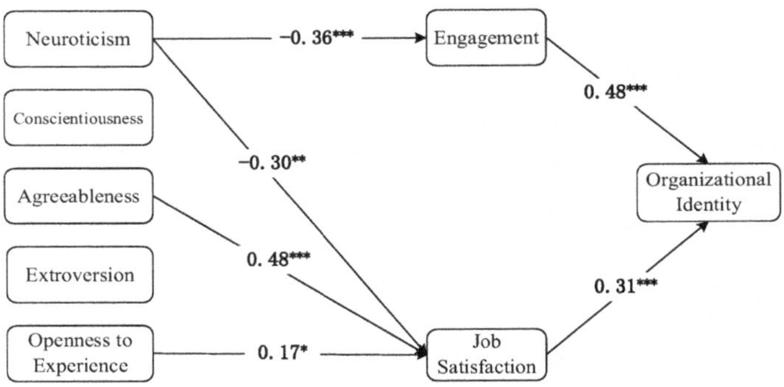


**Fig. 2** Structural equation model

**Table 4** Hypothesis test results

| Hypothesis | Conclusion |
|---|---|
| H1a: Neuroticism negatively affects engagement | **Support** |
| H1b: Conscientiousness positively affects engagement | Refuse |
| H1c: Agreeableness positively affects engagement | Refuse |
| H1d: Neuroticism negatively affects job satisfaction | **Support** |
| H1e: Conscientiousness positively affects job satisfaction | Refuse |
| H1f: Agreeableness positively affects job satisfaction | **Support** |
| H2a: Extroversion positively affects engagement | Refuse |
| H2b: Openness to experience positively affects engagement | Refuse |
| H2c: Extroversion negatively affects job satisfaction | Refuse |
| H2d: Openness to experience positively affects job satisfaction | **Support** |
| H3a: Engagement positively affects organizational identity | **Support** |
| H3b: Job satisfaction positively affects organizational identity | **Support** |
| H3c: Engagement has a greater impact on organizational identity than job satisfaction | **Support** |

# 4 Discussion

## 4.1 Conclusion

This research takes the front-line employees of a large enterprise as the object, which studies the internal influence mechanism of personality on organizational identity, and contrastively analyzes the influence path with job satisfaction and engagement. It is found that neuroticism of the big five negatively affects engagement and job satisfaction, agreeableness positively affects job satisfaction, and openness to experience positively affects job satisfaction. The study found that only three dimensions of neuroticism, agreeableness and openness to experience in the big five model were related to job satisfaction, which was consistent with the previous conclusions of scholars on career satisfaction and proved the effectiveness of this study. Engagement and job satisfaction positively affect organizational identity, among which engagement has a greater impact on organizational identity than job satisfaction. According to the analysis, neuroticism negatively affects engagement, and then negatively affects organizational identity. Agreeableness positively affects job satisfaction and then organizational identity.

## 4.2 Practical Value

Studies have shown that neuroticism has a negative predictive effect on engagement and organizational identity. Neurotic employees, in most cases, find it difficult to cope with stress at work, and they are prone to a various negative emotions, which can reduce work efficiency and damage interpersonal relationships with colleagues and leaders. Therefore, it is difficult for employees with high neuroticism to obtain job satisfaction and the sense of identity. At the same time, employees with high neuroticism tend to sink themselves into negative emotions, thus reducing engagement and organizational identity. Therefore, for employees with high neurotic personality, the most important thing is to change their mindset to improve their engagement, and the second thing is to increase their job satisfaction through benefits, so as to improve their identification with the organization more effectively.

Agreeableness positively affects job satisfaction and then organizational identity. Employees with high agreeableness are good at interpersonal communication and are more popular among people. They are amiable and easy to integrate into the collective with high agreeableness. However, in some organizations with a strong competitive atmosphere, high agreeableness does not affect engagement, and their habit of being "nice" will make them unwilling to compete with others. Therefore, under these circumstances, the performance of employees with agreeableness may suffer setbacks. Such employees' recognition of the organization can be guaranteed by other means (such as humanistic care, birthday gifts and family member care, etc.) to improve their job satisfaction. In addition, for employees with high openness to experience, enterprises should try to give full play to their strengths and arrange them in positions requiring innovation or flexibility. If the enterprise is limited by the real situation, it is better to implement the post rotation system to keep them fresh about their work, which can also improve their job satisfaction and organizational identity.

Job satisfaction is one of the basic conditions of engagement behavior. When the basic needs of employees are satisfied, job satisfaction will come into being. However, only when employees have an emotional connection with the organization cognitively can they identify with the organization. Personality can significantly affect individual job satisfaction, thus affecting individual organizational identity. Engagement and job satisfaction are two sub-dimensions of the Gallup engagement scale, both of which can positively influence employees' organizational identity. The difference is that the influence of personality on organizational identity through engagement is greater than that of job satisfaction. Therefore, it is not accurate for some enterprises to simply believe that improving welfare treatment can retain talents, because when people's basic material life is satisfied, they will pursue spiritual satisfaction and pay more attention to the realization of self-value. So enterprises should focus on how to improve engagement, rather than blindly improve welfare treatment, so as to make employees engaged and engaged in work ideologically and generate positive emotional connection with the organization.

## *4.3 Theoretical Value and Prospect*

(1) This study innovatively takes personality as the antecedent variables of organizational identity and is supported. It further enriches the research on the antecedent variables of organizational identity in terms of theoretical research and empirical research, and enriches the research on the relationship between individual differences and organizational identity;
(2) In this study, engagement and job satisfaction are selected as the internal mechanism of personality affecting organizational identity. And it reveals the internal mechanism of individual differences affecting organizational identity and enriching the current researches on big five personality and organizational identity;
(3) Taking individual differences and personality as the entry point, this paper proves the influence of personality on organizational identity, and further demonstrates the influence mechanism of individual characteristics on organizational development from the perspective of management practice.

This study uses front-line employees as the research object to explore the influence of openness to experience on organizational identity. It may be because the employees' low innovation requirements make their relationship not proven. It is hoped that the subsequent research can choose the research subjects who need innovative behavior to continue to explore its impact mechanism and make up for the shortcomings of this research. Meanwhile, the research in the field of organizational behavior has paid more and more attention to individual characteristics and the influence of individuals on organizational development. It has become a hotspot on the current management research. Next research can continue to explore the mechanism of personality (such as proactive personality) or other individual factors affecting organizational identity, further enrich the research on organizational identity antecedent.

## References


1. Mael, F. A., & Tetrick, L. E. (1992). Identifying organizational identification. *Educational and Psychological Measurement, 52,* 813–824.
2. Turner, H., & John, C. (1986). The social identity theory of intergroup behavior. *Psychology of Intergroup Relations, 13,* 7–24.
3. Wu, Z., Liu, J., & Hui, C. (2010). Workplace ostracism and organizational citizenship behavior: The roles of organizational identification and collectivism. *Nankai Business Review, 13*(3), 36–44.
4. Li, Y., & Wei, F. (2011). Do practices in human resources with high performance contribute to the organizational identification? *Management World,* 109–117.
5. Smidts, A., Pruyn, A. T. H., & Van Riel, C. B. M. (2001). The impact of employee communication and perceived external prestige on organizational identification. *Academy of Management Journal, 44,* 1051–1062.
6. Pratt, M. G. (1998). To be or not to be: Central questions in organizational identification. In D. A. Whetten & P. C. Godfrey (Eds.), *Identity in Organizations: Building Theory Through Conversation* (pp. 171–207). Thousand Oaks: Sage Publications, Inc.
7. Mael, F., & Ashforth, B. E. (1992). Alumni and their alma mater: A partial test of the reformulated model of organizational identification. *Journal of Organizational Behavior, 13,* 103–123.
8. Farooq, O., Rupp, D., & Farooq, M. (2016). The multiple pathways through which internal and external corporate social responsibility influence organizational identification and multifoci outcomes: The moderating role of cultural and social orientations. *Academy of Management Journal, 60*(3), 954–985.
9. Weng, Q. X., Peng, C. H., Cao, W. L., & Xi, Y. M. (2016). The relationship between big-five personality and subjective career success: A meta-analysis of the prior 15-year researches. *Management Review, 28,* 83–95.
10. Judge, T. A., Heller, D., & Mount, M. K. (2002). Five-factor model of personality and job satisfaction: A meta-analysis. *Journal of Applied Psychology, 87,* 530.
11. Schinka, J. A., Kinder, B. N., & Kremer, T. (1997). Research validity scales for the NEO-PI-R: Development and initial validation. *Journal of Personality Assessment, 68,* 127.
12. Bai, X. W., Wang, E. P., & Li, Y. J. (2006). Five-factor model of personality and performance: Review of Team-level research. *Advances in Psychological Science, 14,* 120–125.
13. Meng, H., & Li, Y. X. (2004). A research on the relationship of the big-five personality and leadership effectiveness. *Psychological Science, 27,* 611–614.
14. Mccrae, R. R., & Costa, P. T., Jr. (1997). Personality trait structure as a human universal. *American Psychologist, 52,* 509.
15. Mansuralves, M., Floresmendoza, C., & Abad, F. J. (2010). Multi-source assessment of neuroticism trait in school children. *Estudos De Psicologia, 27,* 315–327.

16. Costa, P. T., & Mccrae, R. R. (1992). Revised NEO Personality Inventory (NEO-PI-R) and NEO Five-Factor Inventory (NEO-FFI) professional manual.
17. Digman, J. M. (1997). Higher-order factors of the Big Five. *Journal of Personality and Social Psychology, 73,* 1246.
18. Deyoung, C. G. (2006). Higher-order factors of the Big Five in a multi-informant sample. *Journal of Personality and Social Psychology, 91,* 1138.
19. Liu, J. P., Song, X. X., Fang, Q. H., & Zhu, L. (2017). How do personality traits affect innovative employees' performance: Taking employee engagement as mediator variables. *Science and Technology Management Research, 37,* 149–154.
20. Luo, R., Zhou, Y., Chen, W., Pan, Y., & Zhao, S. Y. (2014). Teachers professional identity and affective commitment: Mediating role of job satisfaction. *Psychological Development and Education, 30,* 322–328.
21. Dick, R. V., Hirst, G., Grojean, M. W., & Wieseke, J. (2007). Relationships between leader and follower organizational identification and implications for follower attitudes and behaviour. *Journal of Occupational & Organizational Psychology, 80,* 133–150.
22. Tidwell, M. V. (2005). A social identity model of prosocial behaviors within nonprofit organizations. *Nonprofit Management & Leadership, 15,* 449–467.
23. Harter, J. K., Schmidt, F. L., & Hayes, T. L. (2002). Business-unit-level relationship between employee satisfaction, employee engagement, and business outcomes: A meta-analysis. *Journal of Applied Psychology, 87,* 268.
24. Kahn, W. A. (1990). Psychological conditions of personal engagement and disengagement at work. *Academy of Management Journal, 33,* 692–724.
25. May, D. R., Gilson, R. L., & Harter, L. M. (2004). The psychological conditions of meaningfulness, safety and availability and the engagement of the human spirit at work. *Journal of Occupational & Organizational Psychology, 77,* 11–37.
26. Schaufeli, W. B., Salanova, M., Vicente, G., & Bakker, A. B. (2002). The measurement of engagement and burnout: A two sample confirmatory factor analytic approach. *Journal of Happiness Studies, 3,* 71–92.
27. Robinson, D., Perryman, S., & Hayday, S. (2004). The drivers of employee engagement. *Institute for Employment Studies.*
28. O'Leary-Kelly, S. W., & Vokurka, R. J. (1998). The empirical assessment of construct validity. *Journal of Operations Management, 16,* 387–405.

# Benefit Sharing in the Context of Default Risk Cooperation

**Haoxiong Yang, Ding Zhang, and Hao Wang**

**Abstract** Contract farming, as a representative model of current agricultural industrialization operation, has the greatest impact on compliance rate caused by market risk. Currently, the agricultural product industry is increasingly susceptible to market price risk, and the paper mainly studies the benefit sharing in the context of default risk cooperation. It analyzes the five modes of non-cooperation and cooperation. The conclusions of the study will help to design effective contracts in connection with the contract farming supply chain featured by "processing enterprises + farmers".

**Keywords** Contract farming · Default risk · Benefit sharing · Supply chain

## 1 Introduction

Due to a variety of uncertainties, the prices of agricultural products fluctuate frequently. In some countries, agricultural production is dominated by smallholders, which often leads to the contradiction between "farmer" and "market". Contract farming is an agricultural business model that in the process of agricultural production and management, the farmers and the enterprise sign a legally valid production and sales contract to specify the rights and obligations of both parties, according to which the farmers organize production, and the enterprise purchases products thus produced [1]. With the rapid development of contract farming, the

H. Yang
Business School, Beijing Technology and Business University, Beijing, China
e-mail: yanghaoxiong@126.com

D. Zhang (✉) · H. Wang
School of Computer and Information Engineering, Beijing Technology and Business University, Beijing, China
e-mail: justdzhang@163.com

H. Wang
e-mail: W18810255886@163.com



J. Zhang et al. (eds.), *LISS2019*,
https://doi.org/10.1007/978-981-15-5682-1_39

issue of breach of contract is increasingly awkward, which not only damages the interests of both parties to the contract, but also affects normal development of contract farming, thus hindering the development of agricultural industrialization [2].

In reality, due to information asymmetry, incomplete information, limited rationality, and opportunistic behavior, the contract cannot be a complete contract, and it can not always be fulfilled [3]. Contract production is a form of long-term trading in nature. For farmers, as long as the benefits in not performing the contract are greater than those otherwise, the signed contract will be at the risk of being broken. In order to avoid default, both parties to the contract may set forth warranties in the contract [4]. A one-time game is the primary cause for a breach of contract. In a static game, as long as one party has a strategy to get more benefits, it will choose it unconditionally if the loss of the other party will bring more benefits to the defaulter while there is no effective external constraint [5]. Default in contract farming is not surprising, and it is of great practical significance to promote the development of modern agriculture by thoroughly analyzing the behaviors of default in contract farming and study the fulfillment of agricultural orders with the fluctuation of agricultural products prices.

## 2 Literature Review

Our effort draws upon three streams of literature: contract farming, default risk and benefit sharing in the agricultural supply chain.

Literature on contract farming is extensive. Contract farming refers to a vertical cooperation model between agricultural enterprises and rural households. After the contract is signed, the farmer conducts agricultural production according to the enterprise's requirements, and delivers the agricultural products to the enterprise according to the contract [6]. Contract farming is to adapt to changes in consumer demand. Moreover, the special production and consumption characteristics of agricultural products are also intrinsic reasons for the development of contract farming [7]. Farmers tend to produce on a contract basis to obtain new technical and material inputs, scarce production funds, and relatively stable purchase prices. Contract farming model can well reflect the advantages of vertical integration of agriculture and is an agricultural industrialization operation model that is easy to popularize and promote [8].

In addition, the analysis of default risk may be made from different perspectives. The use of contract farming will benefit farmers and agricultural enterprises. However, there are still many problems in practice that lead to defaults in contract farming. Farmers will choose default to ensure that their own interests are maximized by increasing market purchase prices [9]. For agricultural products priced by the buyer, a fixed-price purchase contract is advisable, while for those subject to changes in market price, it is suggested to sign a purchase contract with a guaranteed purchase and proper market price terms [10]. If the enterprise does not have a stable supply of raw materials, the special investment already paid can hardly be used for other purposes, so the opportunity cost will be very great, in which case the enterprise would like to establish a long-term contractual relationship with the relevant raw material suppliers [11].

Many scholars believe that the supply chain coordination contract can help eliminate the double marginalization, and improve the performance level of the supply chain to perfect coordination. The revenue sharing contract has been considered as a major form for such purpose. With stochastic market demand, when a single demander has access to fully competitive retailers in the downstream market, it can adopt the revenue sharing contract to realize the coordination of the supply chain [12]. The revenue sharing contract is a demand risk sharing mechanism that coordinates the supply chain channels in the case of fixed purchase price and unfixable retail prices. If differential pricing strategy by the benefit sharing coordination is used, the efficiency of the scattered closed-loop supply chain will level up with that of the integrated closed-loop supply chain, realizing the perfect coordination of the supply chain [13].

In addition, some scholars have studied the three-level supply chain system composed of farmers, manufacturers and distributors, and the products circulating in the system are perishable agricultural products. For the three-level agricultural product supply chain coordination under the benefit sharing contract, the benefit sharing contract can effectively coordinate the supply chain and get win-win under certain conditions [14]. In the case of uncertain market demand, the supply chain can be coordinated through the benefit sharing cooperation contract mechanism to increase profits. Through research, it is found that in the face of market demand uncertainty, the lead time of the supply chain will become longer and the wholesale price will be decreased.

## 3 Establishment of Model

### *3.1 Model Building*

In the production and sales process of contract farming, farmers are responsible for the production of agricultural products, and processing enterprises are responsible for processing and sales. The decision-making process can be described as regarding two situations. First, farmers and processing enterprises do not cooperate. Second, farmers and processing enterprises cooperate on the basis of profit benefit sharing. Farmers and processing enterprises will choose to default or cooperate according to the change of market prices (Fig. 1).

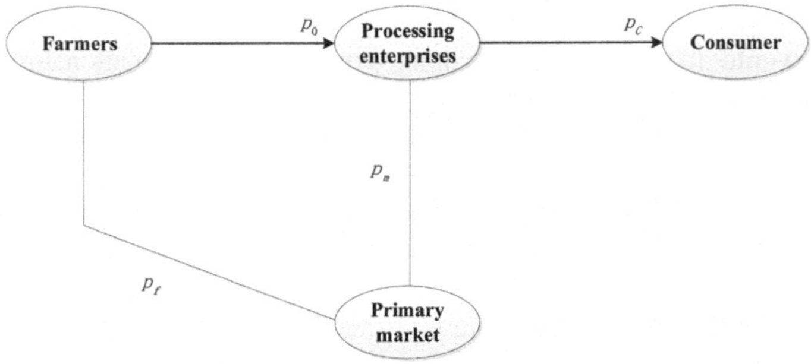


**Fig. 1** Relationship between farmers, primary markets and processing enterprises

## 3.2 Basic Assumptions

Before building the model, the following hypotheses are proposed:

- Consider that only one processing enterprise, one primary market and one farmer are included in the supply chain.
- The production cost of farmers is related to the yield of agricultural product $d$ put into production and the production cost of agricultural product is assumed as $c_1(d) = c_1 d$, where $c_1$ is the unit production cost of farmers. The processing cost of the processing enterprises is related to the purchase volume $d$, and the processing cost is assumed as $c_2(d) = c_2 d$, where $c_2$ is the unit processing cost.
- Market demand $D$ satisfies the equation $D = a - bp_c$, where $a$ is market capacity, $b$ is a constant representing the sensitivity of consumers to price, and $p_c$ is the sales price of the processing enterprises.
- Do not take into account the loss for shortage of agricultural products and the ending residual value gains.
- Take into account the penalty for breach of contract by both parties, $\beta d$, where $\beta$ is the penalty coefficient.
- $\pi_F$ and $\pi_P$ represent the profits of farmers and processing enterprises respectively (Table 1).

## 3.3 Decision-Making Model if the Two Parties Do not Cooperate

The market price of agricultural products fluctuates frequently, so if the farmers didn't follow the profit sharing cooperation with the processing enterprises, they may sell the agricultural products to the primary market at the market price rather than to the processing enterprises at the agreed purchase price. Specifically, when the market price $p_f$ is lower than the purchase price $p_0$ of the processing enterprise,

**Table 1** Notations

| Notation | Table column head |
|---|---|
| a | Market capacity |
| $d_2$ | Demand for processing enterprises in the absence of cooperation |
| $d_1$ | Demand for processing enterprises in the case of cooperation |
| D | Market demand |
| b | Price sensitivity coefficient |
| $p_0$ | The sales price of the farmers |
| $p_c$ | The sales price of the processing enterprises |
| $p_f$ | The purchase price in the primary market |
| $p_m$ | The sales price in the primary market |
| $\lambda$ | The benefit sharing coefficient, and $\lambda \in [0, 1]$ |
| $\beta$ | The penalty coefficient |
| $\pi_F^i$ | The profits of farmers in case $i$ |
| $\pi_P^i$ | The profits of processing enterprises in case $i$ |
| $c_1$ | Market capacity |
| $c_2$ | Demand for processing enterprises in the absence of cooperation |
| $\alpha$ | Demand for processing enterprises in the case of cooperation |
| $p_{0i}^*$ | Market demand |
| $p_{ci}^*$ | Price sensitivity coefficient |

i.e., $p_f < p_0$, farmers will sell their products to the processing enterprise. Otherwise, the farmers will sell products to the primary market, and the processing enterprises will purchase products from the primary market at price $p_m$ to meet the needs of processing and sales.

When $p_0$ is greater than $p_f$, the farmers will sell products to the processing enterprises which will, to maximize the expected income from their perspective, determine the retail price $p_c$ and contract price $p_0$. In such case, the expected profit of the processing enterprises is given by:

$$\pi_P = p_c D - P_0 d_2 - c_2 d_2 \tag{1}$$

where,

$$d_2 = (1 + \alpha_2) D \tag{2}$$

Therefore, the profit of processing enterprises is given by:

$$\pi_P = (p_c - (P_0 + c_2)(1 + \alpha_2))(a - bp_c) \tag{3}$$

If $p_0$ is greater than $p_f$, the expected income of farmers is given by:

$$\pi_F = P_0 d_2 - c_1 d_2 \tag{4}$$

Substitute (2) into (4):

$$\pi_F = (P_0 - c_1)(1 + \alpha_2)(a - bp_c) \tag{5}$$

**Theorem 1** *There is a unique optimal retail price for any given agricultural product market demand D and agricultural product conversion factor* $\alpha$:

$$p_{c1}^* = \frac{b(c_1 + c_2)(1 + \alpha_2) + 3a}{4b} \tag{6}$$

*It can be seen that* $p_{c1}^*$ *increases along with the increase of* $\alpha$*, namely, the greater* $\alpha$ *is, the lower the conversion rate of agricultural products.* (6) *indicates that the lower the conversion rate, the higher the retail price of the processing enterprises, but the demand will be reduced accordingly.*

**Theorem 2** *There is a unique optimal wholesale price for any given agricultural product market demand D and conversion coefficient* $\alpha$:

$$p_{01}^* = \frac{a + b(c_1 - c_2)(1 + \alpha_2)}{2b(1 + \alpha_2)} \tag{7}$$

*It can be seen that the price* $p_{01}^*$ *decreases along with the increase of* $\alpha$*; and the greater* $\alpha$ *is, the lower the conversion rate of agricultural products. Therefore, with the increase of* $\alpha$*,* $p_{01}^*$ *will decrease and* $p_{c1}^*$ *will increase. When* (6) *and* (7) *are substituted into* (3) *and* (5) *respectively, the following equations can be obtained:*

$$\pi_F^1 = \frac{[a - b(c_1 + c_2)(1 + \alpha_2)]^2}{8b} \tag{8}$$

$$\pi_P^1 = \frac{[a - b(c_1 + c_2)(1 + \alpha_2)]^2}{16b} \tag{9}$$

**Corollary 1** *If* $0 < \alpha < \frac{a}{b(c_1 + c_2)} - 1$*,* $\pi_F^1$ *and* $\pi_P^1$ *decrease monotonously, and if* $\frac{a}{b(c_1 + c_2)} - 1 < \alpha < +\infty$*,* $\pi_F^1$ *and* $\pi_P^1$ *increase monotonously.*

*If* $p_0$ *is smaller than* $p_f$*, farmers sell their products to the primary market, and processing enterprises purchase products from the primary market. In such case, the expected profit of processing enterprises is given by:*

$$\pi_P = p_c D - P_m d_2 - c_2 d_2 \tag{10}$$

*Substitute* (2) *into* (10), *and the profit of the processing enterprises can be expressed as:*

$$\pi_P = (p_c - (P_m + c_2)(1 + \alpha_2))(a - bp_c) \tag{11}$$

*If* $p_0$ *is smaller than* $p_f$, *the expected income of farmers is:*

$$\pi_F = P_f d_2 - c_1 d_2 \tag{12}$$

*Substitute* (2) *into* (12):

$$\pi_F = (P_f - c_1)(1 + \alpha_2)(a - bp_c) \tag{13}$$

**Theorem 3** *There is a unique optimal retail price for any given agricultural product market demand D and agricultural product conversion factor* $\alpha$:

$$p_{c2}^* = \frac{b(p_m + c_2)(1 + \alpha_2) + a}{2b} \tag{14}$$

*It can be seen that the price* $p_{c2}^*$ *increases with the increase of* $\alpha$; *and the greater the price, the lower the conversion rate of agricultural products* $\alpha$. (14) *indicates that the higher the conversion rate, the higher the retail price of processing enterprises.*

*When* (14) *is substituted into* (11) *and* (13) *respectively, the following equations can be obtained:*

$$\pi_F^2 = \frac{(a - b(p_m + c_2)(1 + \alpha_2))(p_f - c_1)(1 + \alpha_2)}{2} \tag{15}$$

$$\pi_P^2 = \frac{[a - b(p_m + c_2)(1 + \alpha_2)]^2}{4b} \tag{16}$$

**Corollary 2** *If* $-\infty < \alpha < \frac{a}{2b(p_m + c_2)} - 1$, $\pi_F^2$ *increases monotonously, and if* $\frac{a}{2b(p_m + c_2)} - 1 < \alpha < +\infty$, $\pi_F^2$ *is monotonically decreasing.*

*If* $-\infty < \alpha < \frac{a}{b(p_m + c_2)} - 1$, $\pi_P^2$ *will decreases monotonously, and if* $\frac{a}{b(p_m + c_2)} - 1 < \alpha < +\infty$, $\pi_P^2$ *increases monotonously.*

## *3.4 Decision-Making Model if Both Parties Cooperate*

(1) *Decision-making model based on benefit sharing cooperation model*
When farmers improve product quality, in order to guarantee a steady supply of raw materials and to improve supply chain performance level, the processing

enterprises will adopt revenue sharing contract mechanism to encourage farmers when the conversion rate will be increased. After that, the decision-making process of the processing enterprises and the farmers can be described as: first, the processing enterprises and the farmers will sign a benefit-sharing based contract agreement, with benefit sharing coefficient and the output of agricultural products to be determined by negotiation of both parties, after which the farmers organize production accordingly; after the production, the processing enterprises purchase all agricultural products at the agreed price, then process and deliver them to the retail market to sell at a certain price, after which, the processing enterprise and the farmer will redistribute the actual sales profits according to the contract.

Based on the above assumptions, the profit of processing enterprises is given by:

$$\pi_P = \lambda(p_c D - p_0 d_1 - c_2 d_1) \tag{17}$$

In such case, the income of farmers is given by:

$$\pi_P = p_0 d_1 - c_1 d_1 + (1 - \lambda)(p_c D - p_0 d_1 - c_2 d_1) \tag{18}$$

**Theorem 4** *Similar to the case that both sides don't cooperate, if processing enterprises use benefit sharing mechanism to coordinate the supply chain, for any given market demand for agricultural products D and conversion factor* $\alpha$*, the only definite optimal retail price is given by:*

$$p_{c3}^{*} = \frac{a(2\lambda + 1) + b(c_1 + c_2)(1 + \alpha_1)}{2b(1 + \lambda)} \tag{19}$$

*The optimal wholesale price is given by:*

$$p_{03}^{*} = \frac{\lambda a + b(c_1 - \lambda c_2)(1 + \alpha_1)}{b(1 + \lambda)(1 + \alpha_1)} \tag{20}$$

*After variable substitution calculation, the expected return function of the processing enterprise is given by:*

$$\pi_F^3 = \frac{[a - b(c_1 + c_2)(1 + \alpha_1)]^2}{4b(1 + \lambda)} \tag{21}$$

*The expected return of farmers is given by:*

$$\pi_P^3 = \frac{\lambda[a - b(c_1 + c_2)(1 + \alpha_1)]^2}{4b(1 + \lambda)^2} \tag{22}$$

**Corollary 3** *If* $0 < \alpha < \frac{a}{b(c_1 + c_2)} - 1$, $\pi_F^3$ *and* $\pi_P^3$ *decrease monotonously, and if* $\frac{a}{b(c_1 + c_2)} - 1 < \alpha < +\infty$, $\pi_F^3$ *and* $\pi_P^3$ *increase monotonously.*

(2) *Decision-making model based on default risk cooperation model*
If the market prices $p_f$ reaches a certain value, farmers will only consider the short-term interests for their own sake, ignore the future long-term interests of profit sharing and choose default. For the products sold to the primary market, the farmers shall pay the processing enterprise a certain amount, $\beta d$, as penalty due to breach of contract; after losing the farmers' supply of products, the processing enterprises will choose to purchase products from the primary market to complete the production plan, and get the farmers' penalty $\beta d$.

In the case of the farmers' default, based on the above hypothesis, the income of processing enterprises is given by:

$$\pi_P = p_c D - p_m d_1 - c_2 d_1 + \beta d_1 \tag{23}$$

In such case, the income of farmers is given by:

$$\pi_F = p_f d_1 - c_1 d_1 - \beta d_1 \tag{24}$$

**Theorem 5** *It is similar to both sides if they don't cooperate. If the farmers broke the contract, for any given market demand for agricultural products and agricultural products conversion coefficient* $\alpha$*, the only definite optimal retail price is given by:*

$$p_{c4}^* = \frac{b(p_m + c_2 - \beta)(1 + \alpha_1) + a}{2b} \tag{25}$$

*After variable substitution calculation, the expected profit of farmers and processing enterprises are given by:*

$$\pi_F^4 = \frac{(a - b(p_m + c_2 - \beta)(1 + \alpha_1))(p_f - c_1 - \beta)(1 + \alpha_1)}{2} \tag{26}$$

$$\pi_P^4 = \frac{[a - b(p_m + c_2 - \beta)(1 + \alpha_1)]^2}{4b} \tag{27}$$

**Corollary 4** *If* $-\infty < \alpha < \frac{a}{2b(p_m + c_2 - \beta)} - 1$, $\pi_P^4$ *increases monotonously, and if* $\frac{a}{b(p_m + c_2 - \beta)} - 1 < \alpha < +\infty$, $\pi_P^4$ *decreases monotonously.*

*When market price* $p_f$ *drops, the processing enterprises will choose to break the contract for their own sake, purchase all products from the primary market, and pay a certain amount of liquidated damages to the farmers,* $\beta d$*. After losing the* *processing enterprise to purchase the products, the farmers will choose to sell the products to the primary market to complete the sales plan and get the liquidated damages paid by the processing enterprise,* $\beta d$.

*In the case of the default of the processing enterprise, based on the above hypothesis, the profit of the processing enterprise is given by:*

$$\pi_P = p_c D - p_m d_1 - c_2 d_1 - \beta d_1 \tag{28}$$

*In such case, the income of farmers is given by:*

$$\pi_F = p_f d_1 - c_1 d_1 + \beta d_1 \tag{29}$$

**Theorem 6** *Similar to the case that both sides don't cooperate, if the processing enterprises broke the contract, for any given market demand for agricultural products and agricultural products conversion factor, the only definite optimal retail price is given by:*

$$p_{c5}^* = \frac{b(p_m + c_2 + \beta)(1 + \alpha_1) + a}{2b} \tag{30}$$

*After variable substitution calculation, the expected profits of farmers and processing enterprises are given by:*

$$\pi_F^5 = \frac{(a - b(p_m + c_2 + \beta)(1 + \alpha_1))(p_f - c_1 + \beta)(1 + \alpha_1)}{2} \tag{31}$$

$$\pi_P^5 = \frac{[a - b(p_m + c_2 + \beta)(1 + \alpha_1)]^2}{4b} \tag{32}$$

**Corollary 5** *If* $-\infty < \alpha < \frac{a}{2b(p_m + c_2 + \beta)} - 1$, $\pi_P^5$ *increases monotonously, and if* $\frac{a}{b(p_m + c_2 + \beta)} - 1 < \alpha < +\infty$, $\pi_P^5$ *decreases monotonously.*

## 4 Numerical Analysis

Assuming that the agricultural products market demand capacity $a = 100$, retail market demand $D$ satisfies the equation $D = a - bp_c$, where, the price elasticity $b = 5$, the production cost of farmers $c_1 = 5$, the production cost of processing enterprises $c_2 = 5$, the initial primary market purchase price $p_f = 6$, sales price $p_m = 8$, the purchase price after the market price drops $p_f^1 = 5.5$, the sales price $p_m^1 = 7$, the purchase price after market price rises $p_f^2 = 7$, and the sales price $p_m^2 = 8.5$. In addition, the benefit sharing coefficient $\lambda = 0.8$, and the default penalty coefficient $\beta = 1.4$.

According to Fig. 2, within the range from 0 to 1, with $\alpha$ increasing gradually, namely agricultural conversion rate decreases, the farmer's profit is gradually reduced, which suggests that a higher conversion rate is beneficial to increase farmers' income. $\pi_F^1$ and $\pi_F^3$ shows obvious advantages over those in the other three cases, which implies that with the same conversion rate, the profits will be the highest if the farmers cooperate with the processing enterprises based on benefit sharing and sell the products to the processing enterprise in the case of cooperation failure, with the profit in the former case greater than that in the latter. In addition, both the default of processing enterprises and that of farmers themselves will bring down the farmers' income, so the cooperation between the two parties will increase farmers' income.

According to Fig. 3, within the range from 0 to 1, with $\alpha$ increasing gradually, namely agricultural conversion rate decreases, the change of the processing enterprises' profit is divided into two cases: for the first case, $\pi_P^1$ and $\pi_P^3$ are reduced gradually, which shows that greater conversion rate in the two models will increase the processing enterprises' income. For the other case, $\pi_P^2$, $\pi_P^4$ and $\pi_P^5$ decrease first and then increase, which is an uncommon phenomenon.

The simulation results show that to guarantee a higher income, the farmers should cooperate with the processing enterprise and ensure the latter's compliance to the contract. For such purpose, the farmers should control the conversion rate $\alpha$ to between 0.36 and 0.67, in which case the processing enterprises will choose cooperation to maximize the profit, and it will continue to cooperate and ensure sound cooperation.

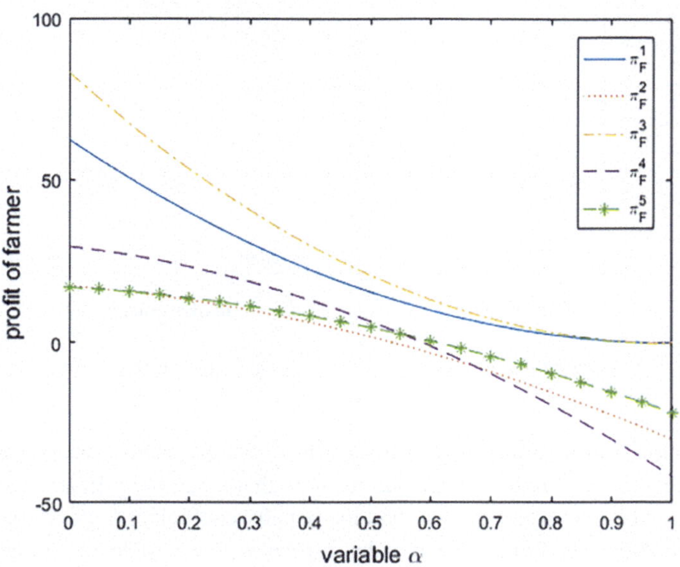


**Fig. 2** Impact of agricultural product conversion coefficient on profit of farmers

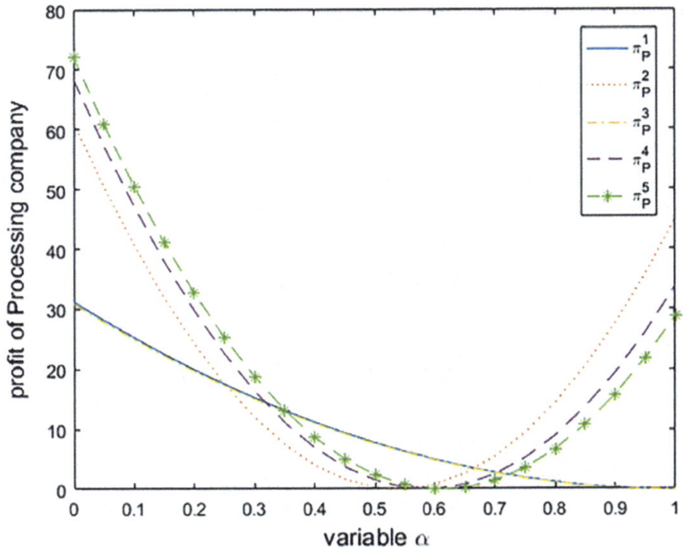


**Fig. 3** Impact of agricultural product conversion coefficient on profit of processing enterprises

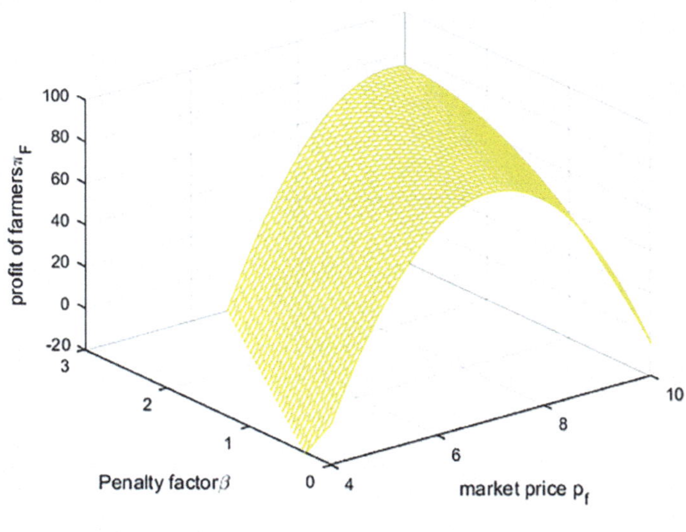


**Fig. 4** Impact of market price fluctuations and changes in penalty factor on profit of farmers

In addition, Fig. 4 shows that as the market price fluctuates, the farmer's profit will either increase or reduce. Calculation show that when the primary market price $p_f$ is increased by an extent greater than the penalty coefficient $\beta$, the farmers' profit will exceed that by compliance, when the farmers will choose to default and sell the products to the primary market. In order to prevent farmers from breach of contract, the processing enterprises can determine a proper penalty coefficient when concluding the contract with the farmers based on past market price change range, so as to ensure the maximum benefits of cooperation.

## 5 Conclusion

In this paper, we have built the "benefit sharing decision-making model in the context of default risk cooperation" based on the analysis of contract farming characteristics, and conducted a comparative study on the optimal decision-making behaviors of the processing enterprises and farmers with five cases involving non-cooperation and cooperation. The main conclusions are as follows: (1) The optimal retail price of agricultural products when the two parties default is higher than that in the case of profit sharing, followed by that in the case of non-cooperation. (2) The models ensure the equivalence of status between the processing enterprise and the farmer in the contract farming business model, and both can choose transaction according to the market price. (3) To prevent the processing enterprises from defaulting when the market price is unfavorable, the farmers can control the conversion rate of the products to ensure the profit-sharing based contract cooperation between the two parties. This mechanism can not only realize the coordination of contract farming supply chain featured by "processing enterprises + farmers", but also lead to an increase in social welfare. (4) Market price fluctuations will bring about changes in farmers' profits. When the primary market price $p_f$ is increased by an extent greater than the penalty factor $\beta$, the farmers' profit will exceed that in the case of cooperation and thus the farmers will choose to default and sell all the products to the primary market. In order to prevent farmers from breach of contract, the processing enterprises can determine a proper penalty coefficient when concluding the contract with the farmers based on past market price change range, so as to ensure the maximum benefits of cooperation. (5) Farmers are economic individuals in small scales. The majority of farmers make a living by the production of agricultural products, which determines that farmers are typical risk averter who not only determine the yield based on the expected profit but also the risk for the profit. Therefore, it is closer to reality to incorporate the risk aversion characteristics of farmers into the decision-making model, which is also the focus of future studies. If the farmers act as risk-averting decision makers, they may control the conversion rate of agricultural products, and design an effective contract to prevent the processing enterprises from defaulting so as to coordinate the contract farming supply chain featured by "processing enterprises + farmers".

# References

1. Kalamkar, S. S. (2012). Inputs and services delivery system under contract farming: A case of broiler farming. *Agricultural Economics Research Review, 25*(2012), 515–521.
2. Hardaker, J. B., Lien, G., Anderson, J. R., & Huirne, R. B. M. (2015). *Coping with risk in agriculture: Applied decision analysis*. Oxfordshire: CABI Publishing.
3. Nagaraj, N., Chandrakanth, M. G., Chengappa, P. G., Roopa, H. S., & Chandakavate, P. M. (2008). Contract farming and its implications for input-supply, linkages between markets and farmers in Karnataka. *Agricultural Economics Research Review, 21,* 1–10.
4. Ragasa, C., Lambrecht, I., & Kufoalor, D. S. (2018). *Limitations of contract farming as a pro-poor strategy the case of maize outgrower schemes in Upper West Ghana*. Rochester: Social Science Electronic Publishing.
5. Goodhue, R., & Simon, L. (2016). Agricultural contracts, adverse selection, and multiple inputs. *Agricultural & Food Economics, 4*(1), 19.
6. Raynolds, L. T. (2000). Negotiating contract farming in the Dominican Republic. *Human Organization, 59*(4), 441–451.
7. Shankar, B., Posri, W., & Srivong, T. (2010). A case study of a contract farming chain involving supermarkets and smallholders in Thailand. *Canadian Journal of Development Studies, 31*(1–2), 137–153.
8. Barrett, C. B., Bachke, M. E., Bellemare, M. F., Michelson, H. C., Narayanan, S., & Walker, T. F. (2012). Smallholder participation in contract farming: Comparative evidence from five countries. *World Development, 40*(4), 715–730.
9. Upton, J., & Lentz, E. (2017). Finding default? Understanding the drivers of default on contracts with farmers' organizations under the World Food Programme Purchase for progress pilot. *Agricultural Economics, 48*(S1), 43–57.
10. Burer, S., Jones, P. C., & Lowe, T. J. (2008). Coordinating the supply chain in the agricultural seed industry. *European Journal of Operational Research, 185*(1), 354–377.
11. Mertens, A., Meensel, J. V., Willem, L., Lauwers, L., & Buysse, J. (2018). Ensuring continuous feedstock supply in agricultural residue value chains: A complex interplay of five influencing factors. *Biomass and Bioenergy, 109,* 209–220.
12. Gerchak, Y., & Wang, Y. (2010). Revenue-sharing vs. wholesale-price contracts in assembly systems with random demand. *Production & Operations Management, 13*(1), 23–33.
13. Jayne, T., & Rashid, S. (2013). Input subsidy programs in sub-Saharan Africa: A synthesis of recent evidence. *Agricultural Economics, 44*(6), 547–562.
14. Sarkar, S. (2014). Contract farming and McKinsey's plan for transforming agriculture into agribusiness in West Bengal. *Journal of South Asian Development, 9*(3), 235–251.

# Analysis and Forecast of Employee Turnover Based on Decision Tree Method

**Haining Yang and Xuedong Gao**

**Abstract** The decision tree is established by the employee turnover as a target variable, and gender, marital status, household registration, registered permanent residence, age, driving duration, education background, post, and category of employees as input variables using C5.0 decision tree algorithm, and recommendations about human resources management are proposed based on the mined rules. The research results show that this approach has more accurate forecast of employee turnover, and the knowledge is found easily to be understood. It can help companies to master the law of employee turnover and reduce the brain drain level.

**Keywords** Employee turnover · Decision tree · Data mining

## 1 Presentation of the Problem and Review of Literature

In recent years, the economic development of our nation has gradually move from the industrial age to the information age, and human resources, as the medium of knowledge and creators, have become the core driving force behind the development of enterprises. At the same time, the increasingly open nature of employment information and increasing specialization of recruitment services have led to more frequent personnel turnover, with the turnover rate of mid-level to senior workers becoming as high as 50–60% in some domestic enterprises, with average employment periods of less than 3 years [1]. Mass departures not only brings about loss in enterprise recruitment and training costs, but also has negative effects on the operation of the enterprise such as loss of important client, leakage of key technology, decrease the enterprise competitive power, even cause the big business

H. Yang (✉) · X. Gao
Donlinks School of Economics and Management, University of Science and Technology Beijing, Beijing, China
e-mail: yanghaining@apiins.com

X. Gao
e-mail: gaoxuedong@manage.ustb.edu.cn



J. Zhang et al. (eds.), *LISS2019*,
https://doi.org/10.1007/978-981-15-5682-1_40

failure. Therefore, effectively predicting employee turnover patterns and reducing losses incurred in these turnovers has become an important question in the practice and theoretical research of human resource management.

So far, research on employee turnover mainly focuses in the fields of economics and organizational behavior, with research in economics utilizing methods of econometric analysis and game theory analysis to analyze supply and demand patterns in the labour market behind employee turnover. Xu and Zhao used the Probit model to study employee turnover, demonstrating that regional economical development level will affect employee turnover [2]. Based on data from 676 listed companies from 1997 to 2007, Yang's research on CEO turnover demonstrated that CEO turnover in listed companies of our nation is not only an financial process, but also a social-political process [3]. Zou and Dong constructed game models from the perspective of a library and its employees, and used evolutional game theory to analyze stability strategies from both sides [4].

Research in organizational behavior mainly employs questionnaires and statistical analysis methods to analyze influencing factors and its mechanisms behind employee turnover. Wang and others conducted a statistical analysis on 932 questionnaires on the basis of stress interaction theory and job requirement resource models, and demonstrated that employees' job requirement had a positive impact on intention of leaving employment, impulsive personality has a mediative effect on the impact of job requirement on intention of leaving employment, and social support has an intermediary effect for the relationship between impulsive personality, and job requirements and intention of leaving employment [5]. Through statistical analysis of 433 questionnaires, Dan and others found that a negative correlation exists between employees' perception of status and employees' perception of union status, with intention of leaving employment [6]. After analyzing 267 employees and 92 supervisors' paired samples, Ye and others demonstrated that employees' sense of workplace exclusion and marginalization have a positive impact on turnover intention [7].

Research results in economics demonstrated market trends of employee turnover from a macro perspective, but these research abstracts their research subject, ignoring the influence of gender, age, needs, perception, emotion, and other individual factors in a worker's decision to leave a job. Research in organizational behavior reveals the intrinsic motivation of employee turnover from the aspects of employees' psychological factors, organizational factors and incentive factors, but the research data used in these studies are mostly based on the "intention of leaving employment" questionnaire, rather than the actual turnover behavior of employees, and are therefore unable to make direct predictions on turnover behavior. And so, this essay uses actual turnover data from enterprises along with decision tree algorithms to analyze and predict staff turnover, and to provide relevant strategical suggestions.

Decision tree is a data mining method that uses inductive learning. It is a tree-like decision graph with additional probability results and an intuitive graphical method using statistical probability analysis. In machine learning, decision tree is a prediction model. It represents a mapping between object attributes and object values. Each node in the tree represents the judgment conditions of object attributes, and its branches represent the objects that meet the node conditions. The leaf nodes of the tree represent the predicted results of the object's ownership. The reasons why decision tree classification model is widely used are as follows:

(1) *Compared with neural network or Bayesian classification, the classification principle of decision tree is simple and easy to understand and accept by users.*
(2) *In the process of decision tree classification, there is no need for human activities. Setting any parameters is more suitable for the requirement of knowledge discovery.*
(3) *Decision Tree Classification does not require any data sets other than training data sets. And test additional information beyond the data set to ensure that the decision tree and others compared with other classification methods, this method has higher classification speed.*
(4) *Compared with other classification models, the decision tree classification method has the following advantages.*

It has very good classification accuracy.

To sum up, this method summarizes corresponding classification rules from a set of unordered and irregular cases, and can also test the classification rules using actual classification results. Widely used in a variety of research fields [8]. Gu and Xu used decision tree classification method to model the user loyalty of professional virtual communities [9]. Peng first used a metrology method to analyze user characteristics, and then used a decision tree to classify the user characteristics [10]. Sun and other, using 16009 traffic accident data to analyze the factors affecting traffic accidents, using a method based on the C5.0 decision tree [11]. C5.0 is an algorithm in decision tree model. It was developed by J. R. Quinlan in 79 years and ID3 algorithm was proposed. It mainly aims at discrete attribute data, and then continuously improves to form C4.5. It adds the discretization of contiguous attributes on the basis of ID3. C5.0 is a classification algorithm of C4.5 applied to large data sets. It mainly improves the execution efficiency and memory usage. C4.5 algorithm is a revised version of ID3 algorithm. Gain Ratio is used to improve the method. The segmentation variable with the largest Gain Ratio is selected as the criterion to avoid the problem of over-matching of ID3 algorithm. C5.0 algorithm is a revised version of C4.5 algorithm, which is suitable for processing large data sets. Boosting method is used to improve the accuracy of the model, also known as Boosting Trees. It is faster in software calculation and occupies less memory resources. Application results in various fields show that the data mining method using decision tree method has better capacity for interference and interpretability, and that this method can process multiple types of data, and has low requirements for the data used.

## 2 Indicator Selection

In the research of factors influencing employee intention in leaving employment, some scholars use statistical analysis on questionnaires to discover the impact of factors such as employee gender, age, education, marital status, registered permanent residence, job position, continuous service age, number of resignations, probability of promotion, number of promotions, etc. on employee-organizational commitment and intention of leaving employment [12, 13]. Based on research results and data of existing literature, this paper chose 9 indicators: gender, age, time in current company, education, marriage, household registration, registered permanent residence, job category, and job level, to analyze the status of employee turnover for an electronic material manufacturing enterprise. Mobley believes that leaving a job means that the employee has worked in a position in the organization for a period of time [14]. After consideration, the individual is deliberately (Deliberate Willfulness) to leave the original job, thus losing his position and related interests, and the original company/Organization is no longer relevant. Based on Mobley's previous separation factors, Rodger, Peter & Stefan also performed post-analysis to update the effects of various factors on turnover behavior in a number of studies, in terms of demographic variables: on-the-job cognitive ability [15]. There is a low correlation between the test and the resignation. The high-education, family responsibilities (marital status, number of children, minimum child age, age) are low, women, young people, and younger people tend to leave, and the recruitment test has helping to stay in the job; in terms of job satisfaction, organizational and work environment factors: the distribution of performance compensation justice helps to improve organizational commitment, low overall satisfaction, does not meet employee expectations, salary dissatisfaction expectant, dislikes supervisor, no satisfied colleagues, poor cooperation with colleagues, lower social status, lower understanding of company organization and procedures, lower self-awareness of work-related abilities, increased conflicts and burdens at work, less opportunities for promotion. The low autonomy of work tends to leave the company, and the salary increase and the positive effect of Role-orientation can help employees to stay; At the level of work content compared with the external environment: the scope of work is positively related to the high-growth needs of employees and job satisfaction, high work repetition, high unemployment rate, high job opportunities, high intentions to find work, tend to leave, Compared with the outside world, the job satisfaction is high, and those who have high investment in work are helpful to staying in the job; in other projects: absentee, latecomer and poor performers tend to leave; in the sense of organizational retreat and behavior: high organizational commitment. Those who help to retain their jobs, have high intentions to leave their jobs, those who have low willingness to work, and the effectiveness of the evaluation work are more likely to leave than the current high.

Xu was using the "China Employer-Employee Matching Data Tracking Survey" data, using the Probit model system to analyze me Factors influencing the departure of employees of state-owned enterprises [16]. Assuming that there is no off-the-job search, employees leave the company generally through three decision-making processes: The first is to measure the cost of job conversion, the second is to get a new job, and finally the desire and ability to accept a new job. We found the special characteristics of the departure of Chinese employees, for example, the turnover rate of agricultural permanent residence employees is relatively high; the factors affecting compensation are more complicated, and the benefits are due to the influence is weak; the opportunity to obtain new jobs depends on the macro environment, so the turnover rate is also affected by the regional economic level and wage level. In addition, there is a clear gender difference between men and women and a lower turnover rate for female employees.

Feng summarizes and analyzes the research situation of foreign work family conflict [17]. It is found that in recent years the number of papers and quotations in working family conflict show exponential growth. From the published journals and disciplinary attributes, mainly concentrated in high-level journals of the psychology and organizational behavior field, and disciplines of cross-fusion properties become increasingly evident, showing a diversified development trend. From the issue of countries and research institutions, mainly concentrated in the United States, Europe and other developed countries and regions, the influence of Chinese scholars in this academic field continues to rise, the number of published in the world ranks second. From the high citation and co-citation literature, the classical literature is still the main source of knowledge in this field, the literature review, the construction of different theories and variables, and the empirical research on the conflict of work family from different levels The literature constitutes the main content of the classical literature research, and the theoretical and empirical research system of work family conflict has been formed. From the core author group, the number of papers published in the field become more, but has not yet formed a stable core of the author group. From the keywords collinear and strategic coordinates, pressure, satisfaction, gender, social support, resources, role conflict, health are high-frequency keywords, representing hot spots from 30 years. Strategic coordinates show that the role of conflict and social support research despite the frequency of the composition of the hot spots, but in the study of the external links and internal development has not yet formed a high degree of recognition of the research field, "work family conflict type", "stress conflict and control", "work life satisfaction" form the hotspot of work-family conflict research, "gender and emotional exhaustion", "job performance", "work family promotion", "work family balance", stress and "workplace needs", "psychological distress", "mental health and quality of work", "working time flexibility and health" constitute the work of the family conflict in the hot areas, "character and emotions", "shift work", "social support and self-assessment", "role conflict", "time-based conflict and balance", "organizational support and work attitude", "workplace flexibility", "female research" constitute a work family conflict relative Popular areas.

Ma adopting questionnaire survey to 29 manufacturing enterprises and 772 industrial workers as the research object, this study investigates the influencing mechanism of employee-oriented human resource practices and career growth on turnover intention by using multi-level linear model [18]. The empirical study results show that the employee-oriented human resource practices can significantly improve employee career growth and reduce turnover intention, and employee career growth is the cross-level mediating variable between employee-oriented human resource practice and turnover intention. The employee-oriented human resources practice has no significant moderating effect on the relationship between employee's career growth and employee turnover intention. But there are significant differences in the influence of employee career growth on turnover intention among different enterprises.

## 3 Data Processing and Descriptive Statistics

To analyze the characters of turnover staff more precisely, this paper classifies and discretizes some indicators. The details are as follows:

### *3.1 Age*

Continuous age data was discretized, with the 25–39-year-old employees were converted into "25–29 years old", "30–34 years old" and "35–39 years old". Due to the small number of employees under the age of 24 and over 40, they were not classified in detail, and are collectively referred to as "under 24 years old" and "40 years old or older".

### *3.2 Job Category*

Job categories are divided into five categories based on the work content of employees: "management", "production", "technical", "distribution" and "support". "Management" refers to the general manager and deputy general manager of the enterprise and the strategic management positions responsible for production planning and market development; "production" refers to the operators and on-site management personnel and equipment maintenance personnel of the production workshop of the enterprise; "technical" refers to technical positions such as quality inspection, technology development, and information system maintenance; "distribution" includes procurement positions and sales positions of the enterprise; "support" includes auxiliary positions such as administrative, financial, and personnel affairs of the enterprise.

## 3.3 Job Level

Job level is divided based on employee's responsibility and power within their position: "senior manager", "mid-level manager", "junior manager" and "general employee", among which "senior manager" refers to the general manager of the company and deputy general manager; "mid-level manager" refers to the shop supervisor, department manager, etc.; "junior-level manager" refers to team leaders in the company's workshop.

Using 237 employee data of the company from 2015, and using SPSS 20. 0 software to produce descriptive statistics of each indicator, the statistical results are shown in Table 1.

The results of descriptive statistical analysis show that the company's turnover rate in 2015 was 32.50%, which is relatively high. The employees of the company are mainly male, aged between 25–34 years old; non-local employees and agricultural household registry holders are relatively higher in percentage, and the education level overall is low, with the number of employees with vocational college education or above being less than 40%. Job positions are concentrated in quality inspection and technology research and development, and the job level distribution system present typical linear hierarchy characteristics.

## 3.4 Decision Tree Construction and Results Analysis

70% of the 237 data were randomly selected as training samples. The SPSS modeler software was used to model the decision tree using C5.0 algorithm. The 8 variables of employee gender, marital status, household registration, registered permanent residence, age, time in current company, education, and job category were entered (only one of all departing employees belongs to the level of middle manager, with the rest being general employees, which holds no analytical significance, and so post level was not used as an input variable), a decision tree is constructed for the classification of employee turnover (on-the-job/departure) status as shown in Fig. 1.

Using the remaining 30% of the data as test samples, the prediction results of the decision tree are verified as shown in Table 2. The verification results show that the prediction accuracy rate is 88.16%, and the prediction result has relatively high accuracy.

**Table 1** Descriptive statistics of each indicator

| Turnover Status | *Departed* | | *Currently employed* | | |
|---|---|---|---|---|---|
| | **32. 50%** | | **67. 50%** | | |
| Gender | *Male* | | *Female* | | |
| | 86. 10% | | 13. 90% | | |
| Marital Status | *Unmarried* | | *Married* | | |
| | 51. 50% | | 48. 50% | | |
| Household Registration | *Local* | | *Non-Local* | | |
| | 6. 30% | | 93. 70% | | |
| Registered permanent residence | *Urban* | | *Rural* | | |
| | 20. 30% | | 79. 70% | | |
| Age | *24 & Below* | *25-29* | *30-34* | *35-39* | *40 & Above* |
| | 13. 10% | 44. 70% | 26. 60% | 8. 00% | 7. 60% |
| Time in Current Company | *1 Year* | | *2 Years* | | |
| | 20. 30% | | 11. 80% | | |
| | *3 Years* | | *4 Years* | | |
| | 12. 70% | | 20. 30% | | |
| | *5 Years* | | *6 Years* | | |
| | 15. 60% | | 3. 00% | | |
| | *Less than a year* | | | | |
| | 16. 50% | | | | |
| Education | *Elementary School* | | *Middle School* | | |
| | 0. 42% | | 6. 75% | | |
| | *Vocational High School* | | *High School* | | |
| | 21. 94% | | 36. 71% | | |
| | *Vocational College* | | *Undergraduate* | | |
| | 20. 68% | | 11. 39% | | |
| | *Masters* | | *Doctorate* | | |
| | 1. 27% | | 0. 84% | | |
| Job Type | *Management* | | *Technical* | | |
| | 9. 30% | | 28. 30% | | |
| | *Distribution* | | *Production* | *Support* | |
| | 5. 50% | | 44. 70% | 12. 20% | |
| Job Level | *Senior Manager* | | *Junior manager* | | |
| | 1. 27% | | 7. 17% | | |
| | *Mid-Level Manager* | | *Regular Worker* | | |
| | 5. 91% | | 85. 65% | | |

**Table 2** Test results

| Partition | Training sample | | Testing sample | |
|---|---|---|---|---|
| | Quantity | Proportion (%) | Quantity | Proportion (%) |
| Accurate | 150 | 93.17 | 67 | 88.16 |
| | 11 | 6.83 | 9 | 11.84 |
| Accurate | 161 | 100 | 76 | 100 |

Using rules set of the decision tree mining and the actual situation of the enterprise, meaningful rules found are as shown in Table 3.

Using the above rules, it is found that the employees leaving employment have the following three characteristics, and that the enterprise should adjust its human resource management method accordingly.

(1) *Time in company is an important factor in staff turnover.* Employees employed for more than 3 years have a turnover rate higher than 60%. The reason behind this phenomenon might be due to problems in this company's salary adjustment model, which may not be giving continuous stimulation to employees. The company should adjust their compensation policy in a timely fashion, to reflect the effect of work experience on salary level, and lower loss rate of older employees.

(2) *Marital status has a relatively large influence on staff turnover, with unmarried staff having a significantly higher departure rate than married staff.* Most of the employees in this enterprise are composed of non-local males from rural areas. Unmarried males have no familial burdens, and therefore change their employment locations frequently. Whereas married male workers often come to cities to work as a family unit, with their spouse and children working and living in the same city, therefore having lower mobility rates. When employing workers, the enterprise should notice the influence of marital status, and increase the hiring proportion of married personnel.

**Table 3** Meaningful rules

| No. | Condition IF | Result THEN |
|---|---|---|
| 1 | Time in Company in [1 year/2 years/Less than a Year] | Currently employed |
| 2 | Time in Company in [3 years/4 years/5 years/6 years] + Household Registration = Local | Currently employed |
| 3 | Time in Company in [3 years/4 years/5 years/6 years] + Household Registration = Non-Local + Marital Status = Married + Job Type in [Technical/Distribution/Production/Management] | Currently employed |
| 4 | Time in Company in [3 years/4 years/5 years/6 years] + Household Registration = Non-Local + Marital Status = Unmarried | Departed |
| 5 | Time in Company in [3 years/4 years/5 years/6 years] + Household Registration = Non-Local + Marital Status = Married + Job Type in [Support] | Departed |

(3) *Household registration is an important factor affecting staff turnover.* Staff with local or foreign household registration, married and highly skilled jobs is relatively stable, while those with non-local household registration, unmarried and engaged in simple labor, have a higher turnover rate. This shows that staff with non-local household registration is highly mobile. Staff with local household registration should be recruited as much as possible during the recruitment process. It also shows that household registration plays an important role in the human resource management.

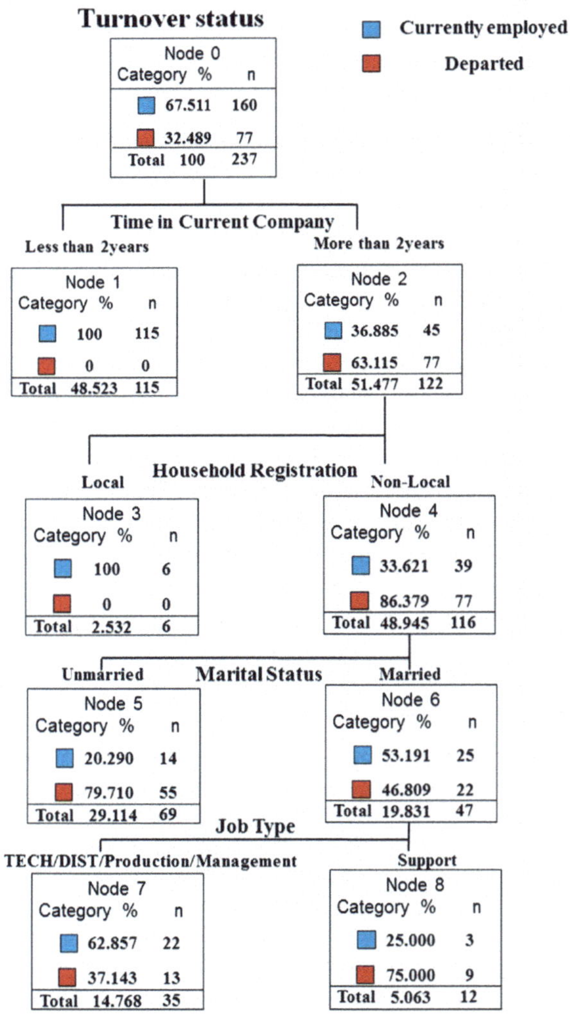


**Fig. 1** The structure of the decision tree

## 4 Conclusion

This paper uses employee's gender, age, time in current company, education, marriage, household registration, registered permanent residence, and job category as input variables, constructs a decision tree with employee turnover status as the target variable, mining information related to employee turnover, and uses the results of decision tree analysis to make recommendations for the enterprise's human resource management. The research results show that when using the decision tree method to predict and analyze the actual turnover of employees, difficulty of data collection is relatively low, and the results have clear testable standards. Information produced by this method is more easily accepted and adopted by enterprises, and using this information, enterprises can grasp the rules of employee turnover more fully, and adjust their human resources management activities accordingly, moreover, the data-driven employee turnover prediction method is mainly based on objective experiments and is not affected by subjective factors. Therefore, the proposed method can be integrated into the support decision-making system to help improve the employee turnover behavior prediction ability of human resources decision makers; furthermore, the analysis of the main factors affecting employee turnover can help enterprise decision makers to target employee turnover. They are inclined to adopt corresponding coping plans, or formulate policies to try to retain excellent employees, or take measures to avoid the enterprise losses caused by staff turnover to the greatest extent.

## References

1. Yang, C., & Dong, X. (2016). Research on warning and management of enterprise employee turnover crisis—based on a non-language communication perspective. *Economic Problems,* (8), 83–86.
2. Xu, F., & Zhao, Z. (2015). Analysis of enterprise employee turnover behavior—an economic review of China's experience. *Economic Perspectives,* (11), 55–67.
3. Yang, D. (2012). Efficiency logic or power logic—corporate politics and compulsory resignation of CEOs of listed companies. *Chinese Journal of Sociology, 32*(5), 151–178.
4. Zou, P., & Dong, Y. (2015). Analysis on the countermeasures of employee turnover crisis in university libraries based on game theory. *Information Science,* (I) 43–48.
5. Wang, X., Li, A., Xiong, G., & Sun, H. (2016). Research on the impact of job requirements of frontline employees on turnover intention and chain adjustment mechanism. *Journal of Management and Business Research, 13*(8), 1191–1198.
6. Shan, H., Hu, E., Bao, J., & Zhang, M. (2015). Research on the influence of level of employee status perception on intention of leaving employment in non-state-owned enterprises. *Journal of Management and Business Research, 12*(8), 1144–1153.
7. Ye, R., Ni, C., & Huang, S. (2015). The impact of workplace exclusion and workplace marginalization on employee intention of leaving employment: the adjustment of employee performance. *Management Review, 27*(8), 127–140.
8. Zhang, Y., & Zhang, Z. (2015). A customer loss prediction model based on C5.0 decision tree. *Statistics & Information Forum,* (1), 89–94.

9. Gu, B., & Xu, J. (2015). User loyalty mining based on knowledge sharing in professional virtual communities. *Information Science,* (1), 105–110.
10. Peng, X., Zhu, Q., & Liu, W. (2015). Analysis of microblog user features and classification research—using Sina Weibo as a case study. *Information Science,* (1), 69–75.
11. Sun, Y., Shao, C., Zhao, D., & Ouyang, S. (2014). C5.0 decision tree prediction model of traffic accident severity. *Journal of Chang'an University (Natural Science Edition), 34*(5), 111–116.
12. Cui, X. (2003). Research on the influence of personal characteristics of employees on organizational commitment and intention of leaving employment. *Nankai Business Review, 6*(4), 4–11.
13. Wang, G. (2015). Comparative study on commitment differences of enterprise information workers organizations from the perspective of individual characteristic variables. *Journal of Guangzhou University (Social Science Edition), 14*(5), 47–53.
14. Mobley, W. H., Griffeth, R. W., Hand, H. H., & Meglino, B. M. (1979). Review and conceptual analysis of the employee turnover process. *Psychological Bulletin, 86*(3), 493–522.
15. Rodger, W. G., Peter, W. H., & Stefan, G. (2000). A meta-analysis of antecedents and correlates of employee turnover: Update, moderator tests, and research implications for the next millennium. *Journal of Management, 26*(3), 463–488.
16. Xu, F., & Zhao, Z. (2015). Analysis of the turnover behavior of enterprise employees—economic review of chinese experience. *Economic Perspectives,* 55–66.
17. Feng, L. (2018). Research situation of work family conflict in foreign countries: based on bibliometric analysis. *Business Management Journal, 4,* 187–208.
18. Ma, W., Yan, K., Chen, X., & Weng, Y. (2019). The influencing mechanism employee-oriented human resource practice and career growth on turnover intention: a multi-level model based on industrial workers in the new era. *Soft Science,* 1–11.

# Key Technologies of High Speed Maglev Soft Handoff for MAC Layer

**Yi Zhang, Xu Li, and Xinxiang Yin**

**Abstract** As a high speed vehicle, maglev train requires a safe and reliable train-ground communication. Firstly, this paper analyses the high-speed maglev vehicle-to-ground communication system. In order to meet the requirements of low-delay, high-reliability maglev train transport management service transmission and ensure the probability of successful handoff, this paper designs a wireless network soft handoff mechanism based on MAC layer control of TDD system for handoff of high-speed maglev train. The relationship between the packet loss rate and the number of retransmissions is analyzed, and a reasonable frame structure is designed. The retransmit mechanism and signaling protection mechanism are adopted to improve the probability of successful soft handoff. Finally, through simulation, the improved algorithm's probability of success achieves 99.6% and the handoff delay is 176 ms when the train speed is 600 km/h. Compared with the A3 algorithm, the probability of successful handoff is significantly improved, and the time delay of train handoff is guaranteed.

**Keywords** Maglev · Frame structure · Soft handoff · The probability of successful handoff · Time delay

Y. Zhang (✉) · X. Li · X. Yin
School of Electronic and Information Engineering, Beijing Jiaotong University, Beijing, China
e-mail: 17120173@bjtu.edu.cn

X. Li
e-mail: xli@bjtu.edu.cn

X. Yin
e-mail: 17120157@bjtu.edu.cn


J. Zhang et al. (eds.), *LISS2019*,
https://doi.org/10.1007/978-981-15-5682-1_41

# 1 Introduction

Maglev train originated in the 20th century. It is a kind of modern high-tech rail transit that overcomes gravity by using magnetic force, so that the train has no mechanical contact with the track, thereby reducing friction and improving running speed. There are four subsystems in magnetic levitation system, which are line turnout, vehicle, traction power supply and communication system. As an important subsystem of Maglev system, communication system needs to ensure the efficient, safe, real-time and controllable operation of Maglev train, which has a decisive impact on the drive and operation of the whole system. Because of the high-speed operation of magnetic levitation, the communication system must ensure the reliability of data transmission and low delay of service transmission.

For the high-speed maglev communication business, it can be divided into train transport management business and passenger information service business according to the application category [1]. Train transport management business carries train control safety data transmission and dispatch communication business, while passenger information service business is to satisfy the passenger's high quality and comfort experience, and provides services to meet the passenger's mobile communication needs. At present, most of the research on railway handoff mainly focuses on passenger information service.

When the train enters the new base station coverage area from the current base station (old base station), the link between the old base station and the mobile station will be disconnected, and the new base station and the mobile station will establish a new communication link. At this time, the handoff between the old base station and the new base station is required. Handoff process is one of the most important processes in mobile communication system, which affects the reliability and communication quality of train service transmission at the boundary.

In the traditional mobile communication system, the most typical handoff technology is hard handoff. Hard handoff has the characteristics of "first disconnect, then switch". At any time, the mobile station only connects a cell. GSM-R and LTE-R used in high-speed railway scenarios are hard handoff technology based on the received pilot signal strength [2–4]. When the pilot signal strength is higher than the handoff threshold, the system will trigger handover. However, if the pilot signal intensity of two base stations varies dramatically in a certain area, the "ping-pong effect" will occur, and the train will switch back and forth between the two base stations in the area. In order to improve the probability of successful handoff, research on hard handoff mainly focuses on handoff algorithm [5, 6], and handoff interrupt time [7, 8]. Although these studies increase the probability of successful handoff and reduce the time of handoff interruption to a certain extent, the reliability of high-speed maglev train has a very high requirement for the operation of train transport management, and the probability of successful handoff is more stringent.

Therefore, the high-speed maglev train can only use the soft handoff technology with the process of "establishment, comparison and release" and the characteristics of "first switching, then disconnecting". At any time of the train switching state, the mobile station can link multiple cells at the same time. In order to improve the probability of successful handoff, people sometimes choose to depend on the selection of optimal handoff threshold [9–11], adopt higher arithmetic average window or set different handoff threshold values. The higher the arithmetic average window, the more difficult it will be to trigger, and the lower settings, the more switching errors will occur. Reference [12] proposes to reduce the number of soft handoffs by adjusting the azimuth angle of the whole day line, but the adjustment cycle is long and the optimization effect takes a long time to reflect. For maglev train in high-speed mobile environment, reasonable handoff threshold and adjusting the azimuth angle of antenna cannot fully guarantee the reliability of message transmission.

Because most of the existing research is aimed at the passenger information service business, in the limited network resources to pursue more reliable business communications. Train transportation management business is of great significance to railway transportation command, safety management, equipment monitoring, emergency management and rescue. It is an important guarantee for the safe operation of trains. Compared with passenger information service business, it has higher reliability requirements. Therefore, for the high-speed maglev vehicle-to-ground communication system, it is necessary to increase the consumption of system resources appropriately to ensure the safety and reliability of train transport management business transmission.

This paper studies the soft handoff technology of wireless network controlled by MAC layer in high-speed maglev scenario based on TDD system, introduces the high-speed maglev vehicle-ground communication system, designs a reasonable frame structure to meet the transmission requirements of high-reliability maglev train transport management business, establishes a soft handoff model based on train speed and position, avoids the ping-pong effect, improves the probability of successful handoff of train in high-speed operation environment, and reduces the consumption of system resources as much as possible to ensure the handoff delay.

## 2 High Speed Maglev Vehicle-Ground Communication System

In the high-speed mobile scenario, within the overlapping range of the adjacent base stations, the signal intensity of the current base stations is gradually weakening, and the signal intensity of the adjacent base stations is gradually increasing. When the difference between the two satisfies certain conditions, the train will cut off the communication link, and then try to establish a communication connection with the adjacent base stations.

The maximum speed of magnetic levitation can reach 600 km/h, while the speed of ordinary high-speed railway can only reach 350 km/h. Because of high-speed operation, the channel fades rapidly, so compared with high-speed railway, the error rate of magnetic levitation message transmission is higher. In order to reduce the loss of signal in the transmission process and reduce the bit error rate, leaky cable will be used to replace the traditional tower antenna to communicate with the train.

The design of high-speed maglev ground communication system mainly consists of two parts, one is ground equipment, the other is vehicle equipment. Ground equipment mainly includes core network equipment, zone controller, ground communication equipment (ground base station, leakage cable). The schematic diagram of the system structure is shown in the Fig. 1.

The core network equipment corresponds to multiple dispatching partitions, and one partition control equipment manages a communication partition, that is, to manage multiple ground communication equipment, which is connected to the ring network and connected to the partition control equipment through switches. The partition control equipment is connected to the core network equipment through ethernet.

The time division duplex (TDD) communication mode is adopted in the train-ground communication system of high-speed maglev train, so the whole system must ensure strict time synchronization. The MAC layer is divided by time and the frame is used as the periodic unit. The mechanism design of MAC layer is based on frame structure. The interaction between train and base station is accomplished by signaling. The frame structure designed in this paper not only achieves high handoff reliability, but also considers whether the system can bear the dual backup of resources in the overlapping area of two base stations. Therefore, the mechanism designed in this paper should not only shorten the handoff time and improve the probability of successful handoff, but also reduce the waste of resources as much as possible.

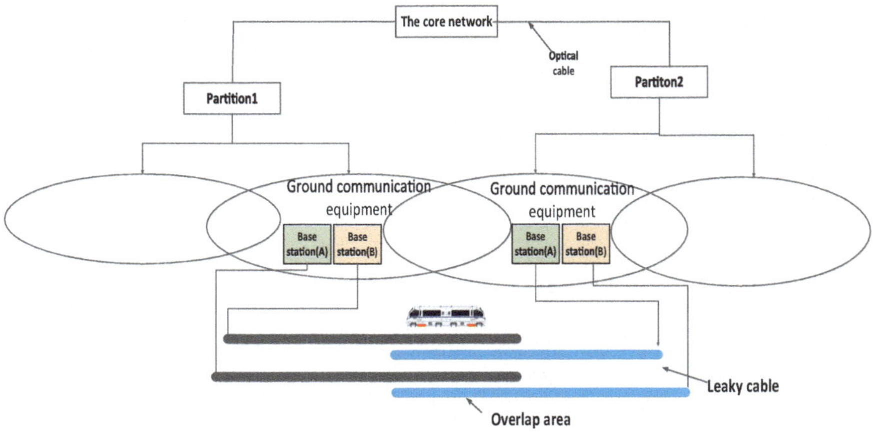


**Fig. 1** The schematic diagram of the system structure

# 3 Improved Handoff Mechanism for High Speed Mobile Scene

LTE protocol stipulates the triggering mechanism of measurement report, including A1, A2, A3, A4, A5, B1 and B2. Reference [13] proposes soft handoff based on the event judgment of A3 to measure transmission power. When the current service area of the train detects that the current service signal intensity of the train is less than that of the adjacent area, that is to say, the current signal intensity is less than the threshold value of the adjacent area signal, the switch trigger is started. So, we can get the trigger probability

$$\mathrm{P}_x(i,j) = P[R(j,x) - R(i,x) > \mathrm{T}_h] \tag{1}$$

$R(j,x)$ is the signal strength of adjacent cells, $R(i,x)$ is the signal strength of current cells.

Usually, we define the uninterrupted train signal in the process of handoff as the successful handoff, then the traditional probability of successful handoff is:

$$\mathrm{P}_{sucess} = \mathrm{P}_x(i,j) \cdot (1 - \mathrm{P}_{out}(i,j)) \tag{2}$$

$P_{out}(i,j)$ is signal interruption probability.

This soft handoff method does not consider whether the resources are enough when the overlap area length and resources are backed up, nor how to further improve the probability of successful handoff when the signal intensity changes dramatically in the overlap area covered by the two base stations.

Therefore, from the point of view of MAC layer control, this paper first designs a reasonable frame structure which meets the requirements of soft handoff mechanism with high the probability of successful handoff and high reliability of service transmission, and then chooses the optimal handoff point according to the current speed and location of the train. In order to prevent the actual distance of train running from exceeding the threshold distance and miss the handoff time, a protection mechanism based on the interaction of control signals is designed, which requires the signaling interaction between the train and the base station to complete the handoff and improve the probability of successful handoff.

## 3.1 Design of Frame Structure

The maglev physical layer adopts OFDM technology, and its frame structure adopts the communication mode of time division duplex (TDD). Each frame is composed of time slots. The time slots in the physical layer represents a burst. The burst is composed of preamble, OFDM symbol and guard. Each OFDM symbol can carry 210 bits data. A frame is composed of control sub-frame and data sub-frame, in which control sub-frame is used to realize Network maintenance, information synchronization and resource scheduling between train and base station, including NCFG (Network Configuration) and NENT (Network Entry) messages, where NCFG messages provide a basic level of communication between different adjacent network nodes from the same or different device vendors; NENT messages are network access messages that provide Network access for new nodes. The data sub-frames are allocated with short or long data slots according to the data volume, data transfer rate and transmission cycle of each service.

Because the digital signal in the wireless channel transmission process will inevitably have the interference, which produces the error. In order to improve the reliability of communication, the data packet retransmission mechanism is adopted, and the number of retransmissions is *N*. After retransmission *N* Times, the failure probability of data service delivery is less than *P*. BER is the error rate of wireless transmission of data between maglev vehicle-mounted equipment and leaky cables. The data volume of a service package is *N*, and the loss rate of the data service without retransmit is *p*.

$$p = 1 - (1 - BER)^N \tag{3}$$

The relationship between the probability of data service delivery failure *P* and the number of retransmissions *n* is as follows.

$$1 - [p^0(1 - p) + \ldots + p^n(1 - p)] < P \tag{4}$$

The relationship between the probability of data service delivery failure *P* and the volume *N* of service data is shown in Fig. 2.

It can be seen from the graph that the failure probability of data service delivery is high when *n* is 0, that is, data service is not retransmitted. When *n* is 1, the data service is retransmitted once, the failure probability of the data service delivery decreases dramatically. With the increase of the number of retransmissions, the reduction of the delivery failure probability of the data service decreases obviously, and with the increase of the number of retransmissions, the time slot occupied by the same data service increases, which will increase the transmission delay of the data service. Therefore, the retransmission number of 1 is the most appropriate, which not only reduces the packet loss rate, but also ensures the low delay of the service transmission.

Soft handoff is implemented in the handoff to minimize the delay caused by resource scheduling, and the frame structure is designed to pre-allocate the time slots of transmitting, receiving and retransmitting data between two ground communication devices. If there is a spare time slot, it can be used for redundant transmission to improve the reliability and real-time performance or to schedule other business requirements. So, the frame structure is designed as follows (Fig. 3):

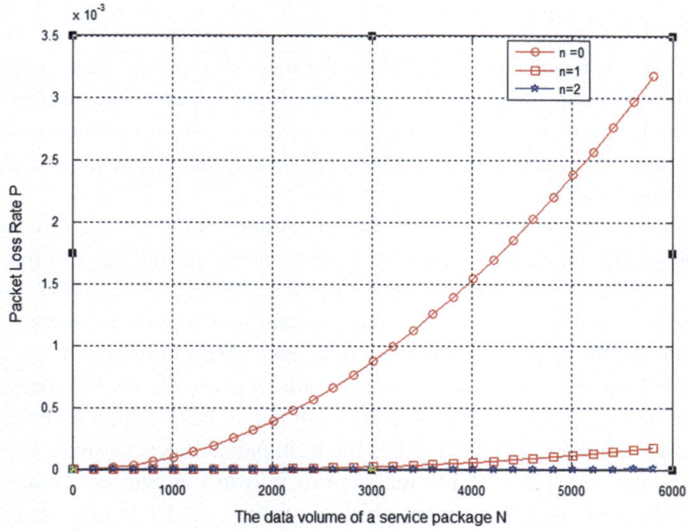


**Fig. 2** Schematic diagram of the relation between packet loss rate and data volume of service package

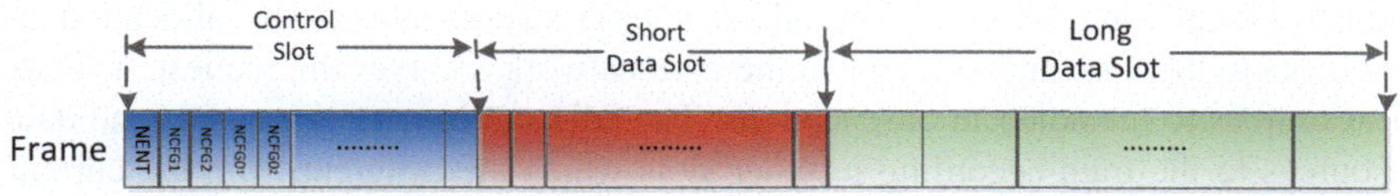


**Fig. 3** The diagram of designed frame structure

The first control slot of each frame is NENT message, the second and third control slots are NCFG message sent by base station, the fourth control slot is NCFG message sent by train, and the fifth control slot is NCFG message retransmitted by train.

## 3.2 Improved Handoff Algorithm

This algorithm is different from the traditional A3 event decision algorithm. Instead of relying on signal strength, it provides the location information, speed and direction of the train by transmitting the train location service (PRW) every 5 ms during the operation of the maglev train, so as to avoid the ping-pong effect. In order to prevent the train from missing the preset handoff point, a control signaling protection mechanism is designed to improve the probability of successful handoff by signaling interaction between the train and the base station.

The algorithm is as follows:

Step 1. Judging Distance Threshold. When the magnetic levitated train is running, it will periodically judge whether the running distance meets the threshold of the handoff condition distance, so as to judge whether the train is running within the coverage range of adjacent base stations. If the distance reaches the threshold, proceed to step 5, and if not, proceed to step 2.
Step 2. Scanning Channel. When the handoff distance is not satisfied threshold, the magnetic levitated train and the adjacent base stations continue to scan the channel. When the train receives NCFG message from the adjacent base station, proceed to step 3, otherwise return step 1. When the adjacent base station receives the NENT message sent by the train, proceed to step 4, otherwise return step 1.
Step 3. Train Executes Handoff. When the train receives the NCFG message from the adjacent base station, it judges that the train parent base station is different from the base station number, then it modifies the train parent base station as the adjacent base station, and sends the NCFG message to inform the adjacent base station to perform handoff, and the adjacent base station receives the NCFG message from the train. Then proceed to step 5.
Step 4. The Base Station Executes Handoff. When the adjacent base station receives the NCFG message sent by the train and knows that the train has entered the coverage area of the base station, it immediately executes the step 5.
Step 5. Completing Handoff. The adjacent base station request the allocation of resources to the core network. When the core network receives the request, it sends all resources to the adjacent base stations. The adjacent base stations will send data resources to the train according to other slots which are different from the current base stations. At this time, the train will set the adjacent base station as the parent base station, the current base station as the former parent base station, receiving all data messages sent by the parent base station and the former parent base station. Train and adjacent base stations establish good communication connection and handoff have be completed.

## 4 Performance Analysis

The communication between the train and the leaky cable is shown in Fig. 4.

The transmission power of base station is $P_t$, h is the vertical distance between the train and the leaky cable, and the free space loss of the *i*th slot of the train is $L_{fi}$.

When the train runs to the *i*th slot, the received signal power *R(i, x)* is

$$R(i,x) = P_t - L_i - L_{fi} \tag{5}$$

When the signal propagates in free space, the noise obeys the gauss distribution of $N(0, \sigma^2)$, and the average power of the noise is $E(n_0) = \sigma^2$.

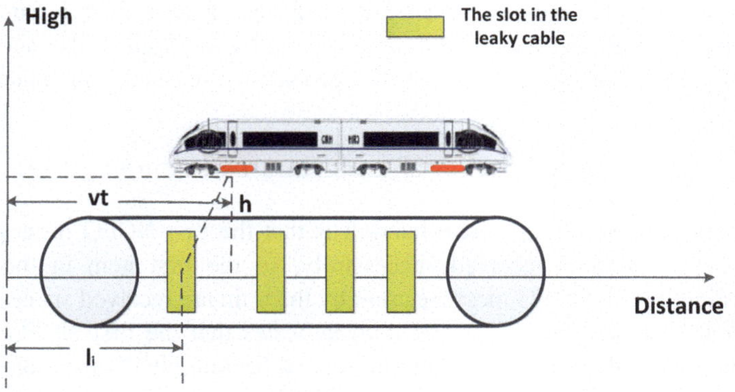


**Fig. 4** Schematic diagram of communication between train and leaky cable

Signal-to-noise ratio of received signals is

$$SNR = \frac{R(i,x)}{E(n_0)} \tag{6}$$

Bit error rate of received signals is

$$BER = \frac{1}{2} erfc\sqrt{\frac{SNR}{2}} \tag{7}$$

## 4.1 Probability of Successful Handoff

The train receives NCFG messages from adjacent base stations and needs to send NCFG messages to inform the base stations of the need to perform handoff. Assuming that the length of overlap area of adjacent base stations is $\Delta L$, the train passes through the overlap area for n scheduling cycles.

$$n = \frac{\Delta L - \int vdT_{rep}}{\int vdT_p} \tag{8}$$

$T_p$ is the time of a scheduling cycle.

Assuming that the bit error rate is *BER* when the train speed is *v* km/h, for the control message of *N* bits length, the packet loss rate *P* is

$$p = 1 - (1 - BER)^N \tag{9}$$

If the first NCFG message sent by the adjacent base station is successfully received by the train and the NCFG message sent by the train is also successfully received by the adjacent base station, the probability of successful handoff is as follows:

$$P_{s1} = (1-p)(1-p)^2 + (1-p)p(1-p) + (1-p)(1-p)p \tag{10}$$

The main premise of successful handoff is that the first NCFG message of the adjacent base station is received successfully, so the first item in the formula indicates that the two NCFG messages sent by the train are received successfully by the adjacent base station; the second item indicates that the first NCFG message sent by the train failed to receive, and the second backup NCFG message received successfully; the third item is that the first NCFG message sent by the train was received successfully, and the second item is that the first NCFG message sent by the train was received The backup NCFG message failed to receive.

The above formula can be transformed into

$$P_{s1} = (1+p)(1-p)^2 \tag{11}$$

If the *n*th NCFG message train sent by the adjacent base station receives successfully and the NCFG message sent by the train is also received successfully by the adjacent base station, the probability of successful handoff is as follows

$$P_{sn} = [1 - \sum_{i=1}^{n-1} P_{si}] \cdot P_{s1} \tag{12}$$

The total probability of successful handoff is

$$P_s = \sum_{i=0}^{n} P_{si} \tag{13}$$

The improved handoff algorithm is simulated by MATLAB with the handoff algorithm proposed in [5] and [13], and the relationship between the probability of successful handoff and train running speed is obtained, as shown in Fig. 5.

As can be seen from the graph above, when the train speed is more than 350 km/h, the probability of the successful handoff in [5] based on the arithmetic of A3 and [13] based on the average arithmetic value drops sharply. When the train speed is 600 km/h, the probability of the successful handoff is 99.6%, which is much higher than that in [5] and [13].

## 4.2 Time Delay of Handoff

The total delay of soft handoff is composed of signaling interaction time and signaling reporting time. Signaling interaction delay refers to the time when the train and the base station interact to switch control messages. Signaling reporting

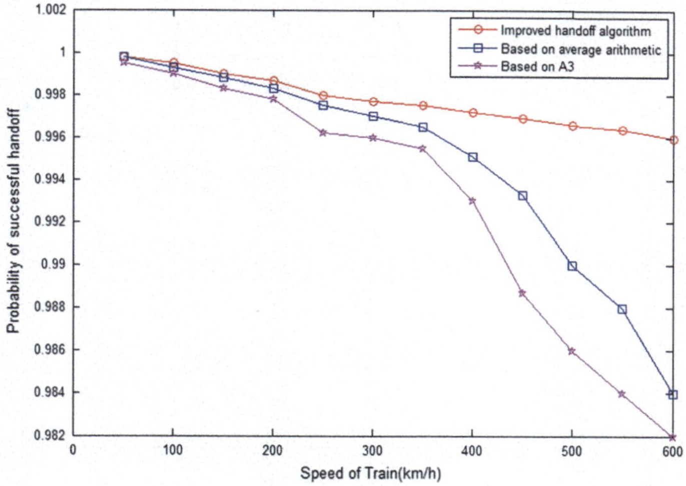


**Fig. 5** Schematic diagram of the relation between probability of successful handoff and speed of train

delay refers to the time when the ground base station requests resources from the partition controller and the core network.

$$T_t = T_{exc} + T_{rep} \tag{14}$$

$T_t$ is the total delay of soft handover, $T_{exc}$ is the signaling interaction delay, $T_{rep}$ is the signaling reporting delay.

The communication between ground base station equipment and partition controller, partition controller and core network adopt the way of token ring network. The signal propagation rate of token ring network is $v_{ring}$, and the length of ring network is $S_H$, so the signaling reporting time is $T_{rep} = \frac{S_{ring}}{v_{ring}}$.

The total delay of soft handoff is as follows:

$$T_t = T_{exc} + T_{rep} = \sum\nolimits_{i=0}^{n} P_{si} \cdot i \cdot T_p + \frac{S_{ring}}{v_{ring}} \tag{15}$$

Figure 6 shows the relationship between the total delay of soft handoff and train speed.

As can be seen from the figure, the handoff delay increases with the increase of train speed. When the train speed is 600 km/h, the channel condition is bad and the error rate increases to 10–3, the average handoff delay increases to 176 ms. Due to the double backup of resources in overlapping area, the system resources are

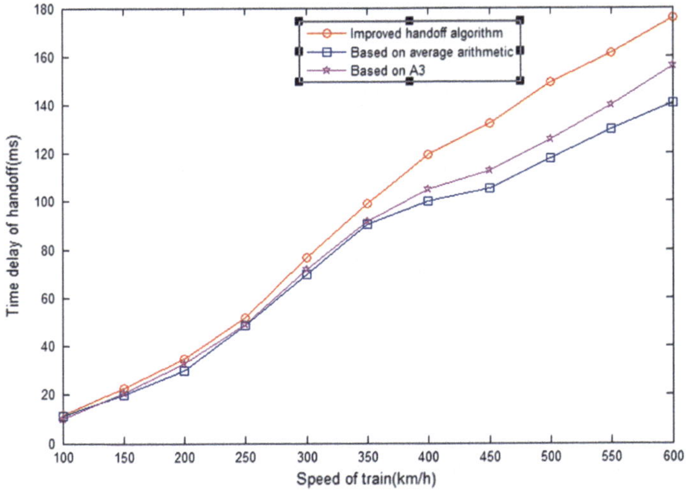


**Fig. 6** Schematic diagram of the relation between time delay of handoff and speed of train

properly sacrificed. Compared with the time delay based on A3 algorithm given in [5] and the handoff time based on arithmetic average given in [13], the total time delay increases, but it ensures the safe and reliable transmission of train transportation management business.

## 5 Conclusion

The handoff is one of the important functions of the high-speed maglev vehicle-ground communication system. The reliability of the handoff is of great significance to the safe and efficient operation of the Maglev Operation Control system. This paper studies the soft handoff technology of wireless network controlled by MAC layer in high-speed maglev scenario. Firstly, the high-speed maglev vehicle-ground communication system is introduced. According to the requirement of packet loss rate, the number of retransmissions is obtained, and the required frame structure is designed to ensure the safe and reliable transmission of maglev train transport management business. Soft handoff mechanism based on location information is adopted, and retransmit mechanism and signaling protection mechanism are designed to improve the success rate of soft handoff and prevent ping-pong handover. Finally, the performance analysis is carried out. When the train speed is higher than 350 km/h, the probability of successful handoff based on the arithmetic of A3 and the average arithmetic value is less than 99.5%, and the average handoff delay is 95–150 ms. The improved handoff mechanism has a 99.6% probability of handoff success when the train speed is up to 600 km/h and the handoff delay is only increased by 26 ms. In the coverage area, the train can communicate with the current base station and the neighboring base station simultaneously, which reduces the blocking rate of the vehicle-ground communication system and ensures the stable and reliable transmission of the service of the maglev vehicle-ground communication system.

**Acknowledgements** This work was supported in part by the Fundamental Research Funds for the Central Universities under Grant 2018JBZ102, the National Key R&D Program of China 2017YFF0206202.

## References

1. Li, G., Boukhatem, L., & Wu, J. (2017). Adaptive quality-of-service-based routing for vehicular ad hoc networks with ant colony optimization. *IEEE Transactions on Vehicular Technology, 66*(4), 3249–3264.
2. Mahar, R., & Choudhary, S. (2017). Performance evaluation of cross link fully adaptive routing algorithm with cross link architecture for Network on Chip. In *2017 International Conference on Inventive Computing and Informatics in Coimbatore* (pp. 576–583).
3. Mahar, R., Choudhary, S., & Khichar, J. (2017). Design of fully adaptive routing for partially interconnected cross-link mesh topology for network on chip. In *2017 International Conference on Intelligent Computing and Control in Coimbatore* (pp. 1–6).
4. Bahrebar, P., & Stroobandt, D. (2013). Adaptive routing in MPSoCs using an efficient path-based method. In *2013 International SoC Design Conference, Busan* (pp. 31–34).
5. Dugaev, D. A., Matveev, I. G., Siemens, E., & Shuvalov, V. P. (2018). Adaptive reinforcement learning-based routing protocol for wireless multihop networks. In *2018 XIV International Scientific-Technical Conference on Actual Problems of Electronics Instrument Engineering in Novosibirsk* (pp. 209–218).
6. Wang, J., Zhu, X., & Luo, Y. (2009). Link availability for any time in mobile ad hoc networks. *Journal of Circuits Systems, 14*(1), 1–7.
7. Dowling, J., Curran, E., Cunningham, R., & Cahill, V. (2005). Using feedback in collaborative reinforcement learning to adaptively optimize MANET routing. *IEEE Transactions on Systems, Man and Cybernetics, Part A, 35*(3), 360–372.
8. Ascia, G., Catania, V., Palesi, M., & Patti, D. (2008). Implementation and analysis of a new selection strategy for adaptive routing in networks-on-chip. *IEEE Transactions on Computers, 57*(6), 809–820.
9. Jeang, Y. L., Wey, T. S., Wang, H. Y., Hung, C. W., & Liu, J. H. (2008). An adaptive routing algorithm for mesh-tree architecture in network-on-chip designs. In *ICICIC 2008, 3rd International Conference on Innovative Computting Information on Control* (p. 182).
10. Sun, Y., Luo, S., Dai, Q., & Ji, Y. (2015). An adaptive routing protocol based on QoS and vehicular density in urban VANETs. *International Journal of Distributed Sensor Networks, 1* (1), 1–14.
11. Li, G., & Boukhatem, L. (2013). Adaptive vehicular routing protocol based on ant colony optimization. In *Proceedings of IEEE VANET in Taipei, China* (pp. 95–98).
12. Caro, G., Ducatelle, F., & Gambardella, L. (2005). AntHocNet: an adaptive nature-inspired algorithm for routing in mobile ad hoc networks. *European Transactions on Telecommunications, 16*(5), 443–455.
13. Daneshtalab, M., Ebrahimi, M., Xu, T. C., Liljeberg, P., & Tenhunen, H. (2011). A generic adaptive path-based routing method for MPSoCs. *Journal of Systems Architecture, 57*(1), 109–120.

# Grid Path Planning for Mobile Robots with Improved Q-learning Algorithm

**Lingling Peng and Juntao Li**

**Abstract** According to the problem of obstacles avoidance path planning in mobile robot system, this article uses the Q-learning algorithm in reinforcement learning to solve it. The original Q-learning algorithm has the problem of low learning efficiency, then an improved algorithm is proposed that adds a layer of learning process based on it. So that the mobile robot can find obstacles and target positions as soon as possible, which accelerates the efficiency of path planning and improves the efficiency. Finally, the path planning is established by a grid method on Python, and the comparison between the original algorithm, the improved algorithm proves that the learning efficiency of the algorithm is improved.

**Keywords** Mobile robot · Path planning · Q-learning algorithm · Grid method · Learning efficiency

## 1 Introduction

Path planning is an important part of mobile robot research. Its goal is to find a collision-free path from the starting point to the end point in a complex environment. The core of path planning is the design of the algorithm. The machine learning that comes with it is a discipline that studies how to use machines to simulate human learning activities, including supervised learning, unsupervised learning, and

---

Young top-notch talent project of high level teacher team construction in Beijing municipal universities (CIT&TCD201704059); Funding Project for Beijing Intelligent Logistics System Collaborative Innovation Center; Funding Project for Beijing Key Laboratory of intelligent logistics systems; Beijing excellent talents support project (2017000020124G063); The logistics robot system scheduling research team of Beijing Wuzi University.

---

L. Peng (✉) · J. Li
School of Information, Beijing Wuzi University, Beijing, China
e-mail: 15612819110@163.com

J. Li
e-mail: ljtletter@126.com



J. Zhang et al. (eds.), *LISS2019*,
https://doi.org/10.1007/978-981-15-5682-1_42

reinforcement learning. In recent years, with the development of the machine learning, the application of various types of mobile robots has entered a period of rapid development. They can accomplish tasks that humans can't complete or finish which are inefficient and time-consuming. In the field of the mobile robots, effective obstacle avoidance and path planning [1] is a critical issue. The mobile robots are expected to detect obstacles and deal with them in the shortest time. For now, there are many ways to learn to solve this issue at domestic and foreign. Commonly used path planning methods for mobile robots are divided into two categories, the global environment and the local environmental planning, The global path planning is to search for a collision-free path according to a specific algorithm under static environmental conditions when the global static environment map is known, and the local path planning mainly considers that the robot only knows part of the environment information or completely does not understand the environmental information. Update the path planning scheme based on the acquired environmental information. Including: artificial potential field method [2], a genetic algorithm based on path coding [3], ant colony algorithm based on random search [4], neural network based on learning training [5–13] and reinforcement learning method [14–24], grid map method [25], particle group [26], A* algorithm [27], and some mixed processing models based on various basic methods. When mobile robots can't complete tasks based on prior knowledge or by imitating others, they need to try and learn. The success or failure of the test should help to change the behavior in the right direction, and reinforcement learning is the study of this behavior. One of the scientific methods. Many scholars have made a lot of research and achieved certain results for the path planning of mobile robots in complex environments. However, Q-learning [25, 28–30] proposed by Watikins was widely used in the path planning application of mobile robots, it is characterized by the absence of prior knowledge of the environment. Mobile robots establish an interaction with complex dynamic environments, the environments will return the robots a reward, at this point the robot will evaluate the action based on the current reward. The iteration of this algorithm is a process of continuous trial and error and exploration. The condition for its convergence is to make multiple attempts for each possible state and action, and the robot learns the optimal strategy. In other words, according to some criteria, an optimal or sub-optimal path from the start point to the target point that can successfully avoid obstacles is found in the unknown environment.

In this paper, some shortcomings of Q-learning are improved, for example, it has the problem of low learning efficiency and slow convergence, it also does not have good adaptability to complex environments. The improved Q-learning algorithm is applied to the path planning for mobile robots. Most of the related studies in this area remain at the theoretical level, lacking of solutions to practical problems and background. Firstly, this article introduces the principle of reinforcement learning and the basic principle of Q-learning algorithm. Secondly, an abstract model is established for the path planning problem of mobile robots and solved by the improved Q-learning algorithm, so that the optimal path can be planned in a relatively short time. Finally, making relevant analysis of the results to verify the efficiency of the algorithm.

# 2 Reinforcement Learning

## 2.1 Theory of Reinforcement Learning

Compared with supervised learning and unsupervised learning, according to whether the training samples are labeled to give a certain orientation to the training process and objectives, reinforcement learning has a more independent theoretical system. The classical reinforcement learning is based on the Markov decision process (MDP), reinforcement learning refers to a kind of learning which from environmental state to action mapping, it maximizes the rewards that actions receive from the environment, thus the objective function is maximized. Learning framework is shown in Fig. 1.

Reinforcement learning algorithm is different from supervised learning and unsupervised learning, its learning process is dynamic. It needs to interact with the environment for learning. The data generated by the interaction will be passed directly to the agent to carry out the next step of learning. However, in the supervised and unsupervised learning, the learning process is static, including the data in learning generate without interaction with the environment. It can directly input the sample data into the deep network of the neural network for training. The reinforcement learning method obtains the feedback signal through the interaction between the robot and the environment, and judges the value of the previous action according to the quality of the feedback signal. The goal is to update the action selection strategy by using the feedback signal in the process of continuous iteration. Complete the optimal action selection strategy for the task. Therefore, it can be seen that there are many objects involved in reinforcement learning, including actions, state transition probabilities, non-essential environment, reward and punishment functions, and value functions.

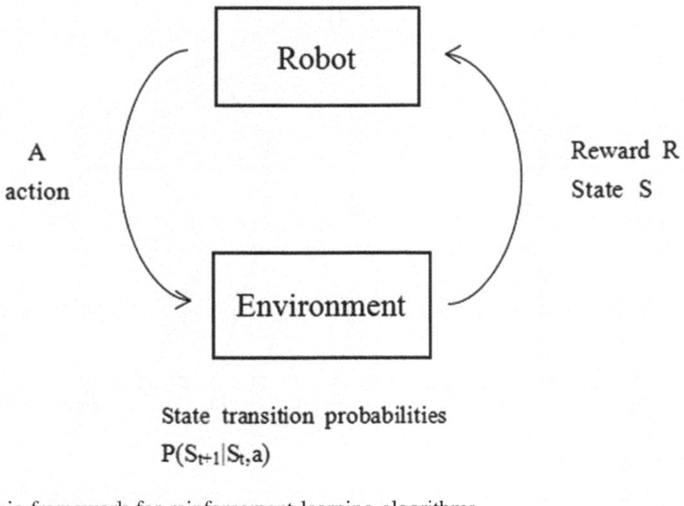


**Fig. 1** Basic framework for reinforcement learning algorithms

Reinforcement learning algorithm reported include Dynamic Programming method (DP), Monte Carlo (MC), TD and other methods. The Dynamic Programming method requires two conditions, one can be decomposed into sub-problems, and the other is that sub-problems can be stored and utilized. In the case of consistent models, the unknowns only have a state value function, which solves the state value function, that is, constantly finds the optimal value function to determine the optimal strategy. However, if there is no model, the experience can be used to estimate the expectation, that is, using the MC method to repeat the experiment, obtain the experimental data, calculate the empirical average, and evaluate and improve the strategy based on the empirical average. From the above, DP method is a learning method based on deep search, and MC is a learning method based on breadth search. However, there is a problem in the MC method that the average experience is not until the end of an experiment. The exploration process of DP and MC algorithm is shown in Figs. 2 and 3.

However, the learning speed is slow and the learning efficiency is not high. Therefore, the most studied reinforcement learning method is the time difference algorithm. This algorithm does not need to be known. For the model, the experimental cumulative experience average uses the state value function at the next moment. Although the expected estimation is biased, the variance is small, and the predicted value for the next moment action is continuously updated. TD method also has a SARSA algorithm based on the same policy and Q-Learning algorithm based on the different strategy. This paper uses Q-Learning algorithm based on different strategies. The exploration process of TD algorithm is shown in Fig. 4.

Since the dynamic programming algorithm estimates the state by calculating the mathematical expectation, it contains all the next steps, the breadth is best, there is no depth level of exploration, but the precise transition probability requires accurate environmental model, and the calculation MC method uses random sampling to estimate the state and simulate the stochastic characteristics of the system. Each

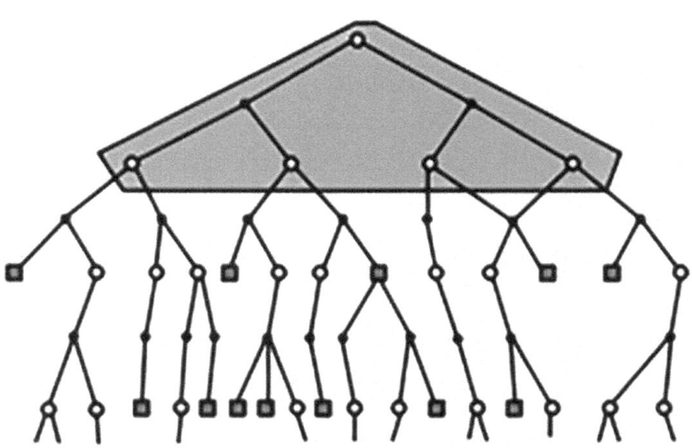

**Fig. 2** The exploration process of DP

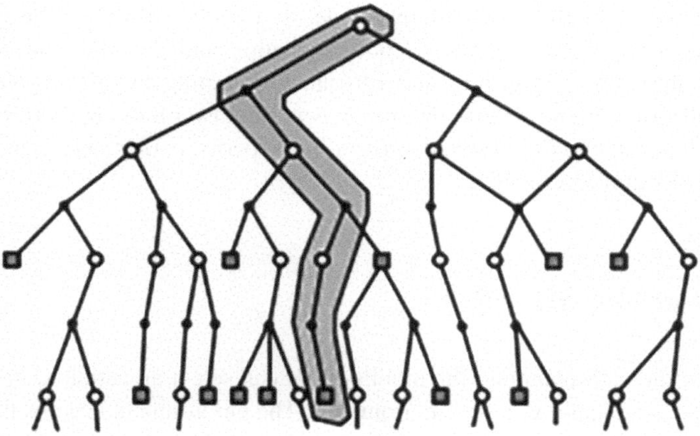

**Fig. 3** The exploration process of MC

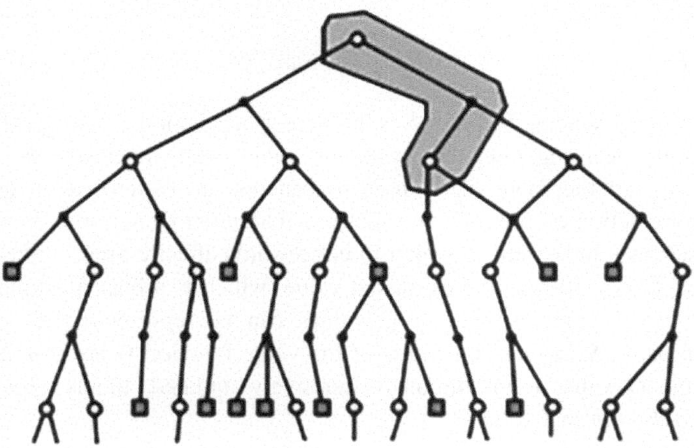

**Fig. 4** The exploration process of TD

sampling is considered to be a complete episode, that is, it is executed from the initial state to the termination state, and the depth is optimal. There is almost no breadth level of exploration. If the Monte Carlo method wants to require more accurate estimates, it needs as many samples as possible. The time difference method combines the characteristics of the above two algorithms, taking into account the depth and breadth direction of the search. By iteratively solving the problem of time reliability allocation, the rewards in the reinforcement learning problem can be transmitted in the opposite direction to the state transition; the control algorithm of the time difference idea is divided into the SARSA in the strategy according to whether the action prediction and the selection action are performed the same. Algorithm and off-rule Q-Learning algorithm, along with Learning theories sound system, the researchers but also on the basis of the above algorithm was improved, improved version has superior performance, but the core idea is discussed above content.

## 2.2 Q-learning Algorithm

At time $t$ of the path planning, the mobile robot can select an action $a_t$ in a limited set of actions and apply it to the environment. The environment accepts that action and performs the transfer of the state $S_t$ to $S_{t+1}$, and the robot receives the bonus value $R$ at the same time. And $R$ is the evaluation of the state $S_{t+1}$. Q-learning uses state-action pairs $Q(s, a)$, which uses this iterative way to get the optimal strategy. The equation for updating the value function of the algorithm is:

$$Q(s_t, a_t) \leftarrow Q(s_t, a_t) + \alpha(R_{t+1} + \gamma \max{}_a Q(s_{t+1}, a) - Q(s_t, a_t)) \tag{1}$$

In the formula, state $S_t$, $S_{t+1} \in S$, $S$ is the state space; action $a \in A$, $A$ is the action place; $\alpha$ is the learning rate, the larger the value of $\alpha$, the faster the value of $Q$ converges, but the more likely it is to generate oscillation; $\max{}_a Q(s_{t+1}, a)$ shows that the action choice $a$ from $A$ makes a maximum $Q(s_{t+1}, a)$; $\gamma$ is a discount factor, it indicates the extent to which future rewards affect current movements. In the episode, $Q(s_{t+1}, a)$ uses the estimated value, which is a biased estimate. From these state-action pairs, a table of $Q$ values can be represented, in which the state-action pairs $(S_t, a_t)$ and the value of the value function $Q$ are stored.

When the $Q$ value is continuously iteratively updated, finally $Q(s_t, a_t)$ will converge to the optimal value:

$$Q(\mathrm{s_t}, a_\mathrm{t}) \leftarrow R_{t+1} + \gamma \max{}_a Q(s_{t+1}, a) \tag{2}$$

Therefore, the convergence of the former state of the algorithm depends on the latter state, it is independent of the initial value, and the result can be guaranteed to converge without environment model.

However, Q-learning has the disadvantages of low learning efficiency and slow convergence, it will be improved. A new learning process is added to the original algorithm, which enables the mobile robot to find obstacles and target positions one step ahead and make decisions in advance.

# 3 Mobile Robot Path Planning Based on Improved Q-learning Algorithm

## 3.1 Environmental Model

This article will use the grid map method [31, 32] based on Python to build the environment space, and will define a 20 × 20 grid environment. 10 means passable points, 0 means obstacles, 7 means the start point, 5 means the end point. So, the white parts are the robot's passable points, the black parts are the obstacles, the yellow part is the start point, and the red part is the end point. As shown in Fig. 5.

## 3.2 Improvements to the Q-learning Algorithm

Since the Q-learning algorithm only makes the mobile robot explore one step, the search scope is limited. According to the acquired environmental information, the mobile robot can perform deeper search than before. The ultimate goal of mobile robot path planning is to find a path with the shortest time and the shortest distance and reach the end point. Therefore, this article will improve the update function of the original Q-learning algorithm value function, adding the deep learning factor $\omega \in (0, 5, 1)$. $Q(s_{t+2}, a)$ is the $Q$ value of the next two steps. It prompts the mobile

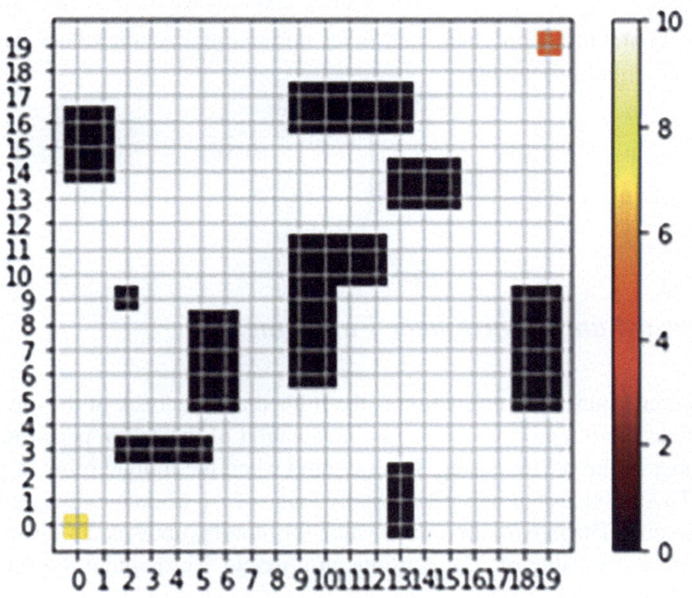


**Fig. 5** A grid map obtained by abstracting obstacles

robot to anticipate obstacles or the target point earlier and updates $Q$ as early as possible. The improved value function's update rule is:

$$\begin{aligned} Q(s_t, a_t) \leftarrow & Q(s_t, a_t) + \alpha(R_{t+1} + \gamma(\omega \max{}_a Q(s_{t+1}, a) \\ & + (1 - \omega)\max{}_a Q(s_{t+2}, a)) - Q(s_t, a_t)) \end{aligned} \tag{3}$$

In order to ensure the convergence of the $Q$ value, $\omega$ is introduced, and the updated learning rule uses the deep learning factor $\omega$ to weigh the return obtained in the first step and the return obtained in the second step. Since the action of the mobile robot is determined by the surrounding environment, it is stipulated that $\omega > 0.5$, which can guarantee the return weight of the larger first step. There is no case that the first step obstacle is ignored due to the second step. When $\omega = 1$, the value rule of $Q$ is determined by the first step, which is the original Q-learning algorithm update.

## 3.3 Representation of Action Space

The simulation experiment in this article will abstract the mobile robot into a mass point, centering on the mobile robot, and define the four action spaces that the mobile robot can execute: $A = \{E, S, W, N\}$. Action strategy chooses ε-greedy strategy, its meaning is to choose the action with the largest action value, that is, the probability of greedy action is $1 - \varepsilon + \frac{\varepsilon}{|A(s)|}$, the probability of other non-greedy actions is equal probability $\frac{\varepsilon}{|A(s)|}$. This strategy can balance the use and exploration, using the maximum action value, and other non-optimal action values continue to explore with equal probability.

$$\pi(s, a) = \begin{cases} 1 - \varepsilon + \frac{\varepsilon}{|A(S)|} \\ \frac{\varepsilon}{|A(S)|} \end{cases} \tag{4}$$

## 3.4 Reward and Punishment Functions

The reward and punishment function is the immediate feedback of the environment to the mobile robot, and it is also a good evaluation of the action performed by the mobile robot in the previous step. It is oriented when the mobile robot is optimally planned. The mobile robot will get a penalty value of −1 while moving a grid until it reaches the end. The maximum reward value for mobile robots is 200. It will get the maximum penalty value of −50 if it hits an obstacle. Throughout the process, the mobile robot will select actions with higher reward values, which will cause the robot to reach the end point faster and eventually get the maximum total reward value.

$$R = \begin{cases} -1 \\ 200 \\ -50 \end{cases} \tag{5}$$

### 3.5 *Design of Overall Algorithm*

See Fig 6.

## 4 Results and Analysis of Experiments

This article will carry out the optimal path planning of mobile robots in the same environment with obstacles. The grid environment is 20 × 20. Then, the widely used Q-learning algorithm is used to verify and compare with the improved Q-learning algorithm. This article lists the number of learning times before and after the improved algorithm under different parameters. The improved algorithm has universal versatility, as shown in Table 1.

In the experiment, the discount factor is assumed to be $\gamma = 0.2$, the deep learning factors are $\omega = 0.6, 0.7, 0.8, 0.9$.

Under the above conditions, with the increase of the deep learning factor, the number of learning times of the improved algorithm is more and more, but the convergence speed and learning efficiency of the algorithm are improved. It can be

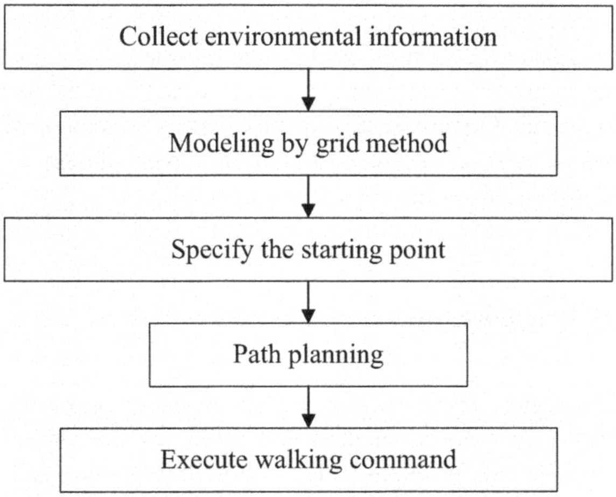


**Fig. 6** Design of algorithm

**Table 1** Comparison of learning times and efficiency under different parameters

| Discount factor $\gamma$ | 0.2 | | | |
|---|---|---|---|---|
| Deep learning factor $\omega$ | 0.6 | 0.7 | 0.8 | 0.9 |
| The number of improved algorithm before | 50 | 50 | 50 | 50 |
| The number of after improving the algorithm | 40 | 43 | 44 | 46 |
| Improved efficiency after improved the algorithm/% | 20 | 14 | 12 | 8 |

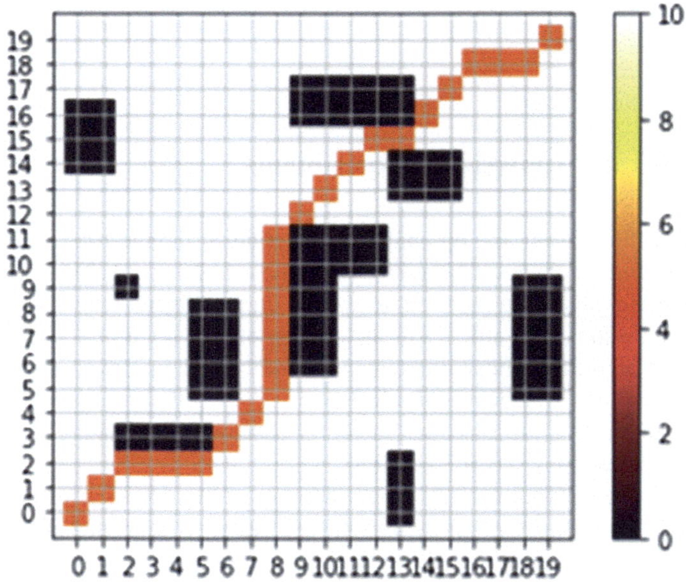


**Fig. 7** Final learning result

seen from the experiment that the mobile robot uses the Q-learning algorithm and the improved Q-learning algorithm to achieve the same path planning results. The improved Q-learning algorithm is more adaptable to complex environments than before, and can perform the shortest path planning more accurately and quickly. Figure 7 shows the final learning result.

## 5 Conclusion

This article proposes a new method, based on the map environment model established by the grid method and improves the Q-learning algorithm. This method improves the problem of low learning efficiency and slow convergence. Adding a deep learning factor when updating the $Q$ value enables the mobile robot to find the target point faster and earlier in an environment with irregular complex obstacles, and to explore farther with all the information of the environment. Finally, the original Q-learning algorithm and the improved algorithm are used to optimize the path planning of the mobile robot in the same static obstacle environment. The results are consistent. The comparison of the learning times proves that the improved algorithm learning efficiency will improve, and the convergence speed will increase. Successfully realized the path planning of the mobile robot, which has certain versatility. However, this article only solves the complex path planning of mobile robots in the environment with static obstacles. Further research is needed on the path planning of mobile robots in more complex environments with dynamic obstacles than before.

**Acknowledgements** First of all, I would like to thank Teacher JunTao Li for his guidance on this article. And then thank the Young top-notch talent project of high level teacher team construction in Beijing municipal universities (CIT&TCD201704059); Funding Project for Beijing Intelligent Logistics System Collaborative Innovation Center; Funding Project for Beijing Key Laboratory of intelligent logistics systems; Beijing excellent talents support project (2017000020124G063); The logistics robot system scheduling research team of Beijing Wuzi University support.

## References

1. Polydorosa, A. S., & Nalpantidis, L. (2017). Survey of model-based reinforcement learning: applications on robotics. *Journal of Intelligent and Robotic Systems, 86*(2), 1–21.
2. Huang, L. X., & Geng, Y. C. (2017). Research on path planning of mobile robot based on dynamic artificial potential field method. *Computer Measurement and Control, 25*(2), 164–166.
3. Yu, H. C., & Lu, F. (2014). Multi-mode and multi-standard path planning method based on genetic algorithm. *Journal of Surveying and Mapping, 01,* 89–96.
4. You, X. M., Liu, S., & Lv, J. Q. (2017). Ant colony algorithm based on dynamic search strategy and its application in robot path planning. *Control and Decision, 32*(3), 552–556.
5. Lillicrap, T. P., Hunt, J. J., Pritzel, A., et al. (2015). Continuous control with deep reinforcement learning. *Computer Science, 8*(6), 187–195.
6. Sun, Z. J., Xue, L., Xu, Y. M., et al. (2012). Overview of deep learning. *Application Research of Computers, 29*(8), 2806–2810.
7. Volodymyr, M., Koray, K., & David, S. (2015). Human-level control through deep reinforcement learning. *Nature, 518*(7540), 529–536.
8. Mnih, V., Kavukcuoglu, K., Silver, D., et al. (2013). Playing Atari with deep reinforcement learning. *Computer Science, 56*(1), 201–220.
9. Schulman, J., Levine, S., Moritz, P., et al. (2015). Trust region policy optimization. *Computer Science, 24*(1), 1889–1897.
10. Van, H. V., Guez, A., & Silver, D. (2015). Deep reinforcement learning with double Q-learning. *Computer Science, 34*(2), 2094–2100.
11. Glascher, J., Daw, N., Dayan, P., et al. (2010). States versus rewards: dissociable neural prediction error signals underlying model-based and model-free reinforcement learning. *Neuron, 66*(4), 585–595.
12. Sylvain, G., & Silver, D. (2011). Monte-Carlo search and rapid action value estimation in computer go. *Artificial Intelligence, 175*(11), 1856–1875.

13. Mnih, V., Puigdomenech, A., & Mehdi, M. (2016). Asynchronous methods for deep reinforcement learning. *Journal of Machine Learning Research, 33*(6), 1928–1937.
14. Mo, F. F. (2016). *Research on Path Planning Method of Mobile Robot Based on Q Learning Algorithm*. Beijing University of Technology.
15. Sutton, R., & Barto, A. G. (2018). *Reinforcement learning an introduction* (pp. 1–10). Massachusetts: The MIT Press Cambridge.
16. Barto, A. G. (1998). Reinforcement learning. In A *bradford book* (Vol. 15, no. 7, pp. 665–685).
17. Leong, Y. C., Radulescu, A., Daniel, R., et al. (2017). Dynamic interaction between reinforcement learning and attention in multidimensional environments. *Neuron, 93*(2), 451–463.
18. Yahya, A., Li, A., & Kalakrishnan, M., et al. (2017). Collective robot reinforcement learning with distributed asynchronous guided policy search. In *International Conference on Intelligent Robots and Systems* (pp. 79–86).
19. Williams, J. D., Asadi, K., Zweig, G. (2017). Hybrid Code Networks: practical and efficient end-to-end dialog control with supervised and reinforcement learning. arXiv preprint arXiv pp. 665–677.
20. Lewis, F. L., & Vrabie, D. (2015). Reinforcement learning and adaptive dynamic programming for feedback control. *IEEE Circuits and Systems Magazine, 9*(3), 32–50.
21. Asadi, M., & Huber, M. (2005). Autonomous subgoal discovery and hierarchical abstraction for reinforcement learning using Monte Carlo method. In *The Twentieth National Conference on Artificial Intelligence and the Seventeenth Innovative Applications of Artificial Intelligence Conference in Pittsburgh* (pp. 1588–1589).
22. Silver, D., & Sutton, R. (2012). Temporal-difference search in computer Go. *Machine Learing, 97*(2), 183–219.
23. Go, C. K., Lao, B., & Yoshimoto, J., et al. (2016). A reinforcement learning approach to the shepherding task using SARSA. In *IEEE International Joint Conference on Neural Networks in Killarney* (pp. 3833–3836).
24. Watkins, C. J. C. H., & Dayan, P. (1992). Q-learning. *Machine Learning*, pp. 279–292.
25. Yu, H. B., & Li, X. A. (2005). Fast path planning of robot based on grid method. *Microelectronics and Computer, 22*(6), 98–100.
26. Zhang, W. X., Zhang, X. L., & Li, Y. (2014). Intelligent robot path planning based on improved particle swarm optimization. *Journal of Computer Applications, 34*(2), 510–513.
27. Sun, W., Lv, Y. F., Tang, H. W., & Xue, M. (2017). Mobile robot path planning based on an improved A* algorithm. *Journal of Hunan University (Natural Science Edition), 44*(4), 94–101.
28. Qiao, L. (2012). *Research on Q-learning algorithm in multi-agent system*. Nanjing: College of Automation, Nanjing University of Posts And Telecommunications.
29. Wang, Z. Q., & Wu, J. G. (2014). Mobile robot path planning based on RDC-Q learning algorithm. *Journal of Computer Engineering, 40*(6), 211–214.
30. Wei, R. M. (2015). *Research and Implementation of Path Planning for Mobile Robot Based on Reinforcement Learning*. South China University of Technology.
31. David, L. C., & Wen, Y. (2017). Path planning of multi-agent systems in unknown environment with neural kernel smoothing and reinforcement learning. *Neurocomputing, 233*, 34–42.
32. Aleksandr, I. P., Konstantin, S. Y., & Roman, S. (2018). Grid path planning with deep reinforcement learning: preliminary results. *Procedia Computer Science, 123*, 347–353.

# Two-Sided Matching Venture Capital Model Based on Synthetic Effect

**Xiaoxia Zhu**

**Abstract** In the paper, a venture capital model is considered with fuzzy information, which main be related to evaluation indexes data between venture capitalists and venture enterprises. So, the expected level value is selected as a reference point, the profit and loss values are calculated by using prospect theory, and the psychological characteristics of decision makers' risk perception are considered, and the overall perception value of decision makers is calculated, then Two-sided Matching Venture Capital Model is established. An example analysis shows that the model is feasible, effective and reasonable, and can be providing method support for venture capital decision-making.

**Keywords** Venture capital · Two-sided Matching · Fuzzy information

## 1 Introduction

Venture capital has played an important role in the process of economic development, and is often called the engine of economic growth [1]. In venture capital activities, there are three main bodies, such as Venture capitalists (VCs), Venture Enterprises (VEs) and Investment Intermediaries, and the matching of venture capital can be realized through intermediaries. The best match can make both parties achieve satisfactory results as far as possible, which are conducive to improving the success rate of venture capital activities, reducing risks and achieving win-win results. The best matching process is actually a two-side selection process for venture capitalists and venture enterprises, which can be summed up as a bilateral matching decision-making problem [2].

Project of Hebei Provincial Department of Science and Technology Project number: 154576292.

X. Zhu (✉)
School of Sciences, Hebei University of Science and Technology, Shijiazhuang, Hebei, China
e-mail: zhuxiaoxia66@126.com



J. Zhang et al. (eds.), *LISS2019*,
https://doi.org/10.1007/978-981-15-5682-1_43

The Two-sided Matching problem originates from the Two-sided Matching theory. The theory was proposed by Gale & Shapley in 1962, and the existence of stable matching, the optimal stable matching and the linear programming algorithm of matching problem were creatively studied [3]. Roth [4] proposed that there is a match between bilateral agents in market economic activities, and analyzed practical cases of bilateral matching. After decades of development, various fields of bilateral matching decision-making have a wide range of practical backgrounds, especially the problem of venture capital decision-making in financial markets. Sorensen [5] established a bilateral matching model between venture capital and enterprises according to the specific conditions of the US venture capital market, and believed that bilateral matching had positive effects on both sides. Cao et al. applied Gale and Shapley model to construct a stable bilateral matching model between venture capitalists and entrepreneurs from the perspective of game theory [6]. Wu equality aiming at the choice of bilateral matching for venture capital under the Internet financial environment, taking full account of both parties' psychological expectations and risk perception, a bilateral matching decision model based on prospect theory is established under the linguistic information index system [7]. Li et al. used disappointment theory to describe the psychological perception of the subjects in venture capital, and then constructed a real information multi-index bilateral matching optimization model for the comprehensive perceived utility of both subjects [8].

Existing research results show that this kind of venture capital is a multi-index multi-objective bilateral matching decision-making process that fully takes into account the psychological and behavioral characteristics of both parties. The characteristics of bilateral matching decision of venture capital studied in this paper are summarized as follows: (1) Fuzzy information index and processing method. In the bilateral matching of venture capital, the evaluation indexes of different subjects may be expressed in one or more of real number, language, triangular fuzzy number and interval number. In this paper, the method of document [9] is used to deal with fuzzy index information. (2) Psychological behavior and expected utility of both parties.

Based on the above analysis, aiming at the characteristics of the psychological behavior and multi-index for both parties of venture capital, this paper mainly does the following work: (1) The fuzzy index information aimed at the two-side matching model of venture capital should be used the comprehensive effect function to be applied to make it clear; (2) The model is took into account the investment intermediary objectives, and a multi-objective two-side matching model is established based on comprehensive perceived utility. An example analysis shows that the model is feasible, effective and reasonable; it can be providing method support for venture capital decision-making.

## 2 Problem Description

In the venture capital activities, VCs expect to obtain high profits, VEs expect to obtain better development, and Investment intermediaries aggregate the expected level information and the actual evaluation information, and forming a reasonable matching pair, so as to achieve satisfaction for both parties.

### 2.1 *Matching Description of VCI to VE*

VCs put their capital into VEs in order to obtain high investment returns. In the investment process, it is hoped to find VEs with advanced technology, short payback period, high return rate, low risk, good reputation, good investment environment, favorable tax revenue and high management level to invest.

The sets of VCI is given by $A = \{A_1, A_2, \cdots, A_i, \cdots, A_n\}$, here, $A_i$ is denoted as $i^{th}$ VCI $i = (1, 2, \cdots, n)$. The satisfaction evaluation index (attribute) set of venture investors to venture enterprises is $X = \{X_1, X_2, \cdots, X_h, \cdots, X_k\}$, here $X_h$ is the $h^{th}$ venture enterprises, $h = (1, 2, \cdots, k)$. Indicator assignment includes both quantitative and qualitative. For quantitative indicators, real numbers or interval numbers or triangular fuzzy numbers can be used, while for qualitative indicators, linguistic information can be used.

The weight of indicator sets $X$ is expressed as $w = \{w_1, w_2, \cdots, w_h, \cdots, w_k\}$, here, $w_h$ is the weight of $X_h$ index, and they are satisfied with $\sum_{h=1}^{k} w_h = 1$, $0 \leq w_h \leq 1$.

Let $P = [p_i^h]_{n \times k}$ be the expectation level matrix of VCI $A_i$ to VE $B_j$ on indicator sets $X$, here $p_i^h$ is the expectation level value of VCI $A_i$ to VE $B_j$ on indicator $X_h$.

Let $Q = [q_{ij}^h]_{n \times k \times m}$ be the Actual Evaluation Matrix of Investment Intermediaries to VE $B_j$ on indicator sets $X$, here $q_{ij}^h$ is the Actual Evaluation Value of investment intermediaries to VE $B_j$ on indicator $X_h$.

### 2.2 *Matching Description of VE to VCI*

In the development process, VEs will encounter many problems, such as innovation, transformation and expansion production. In order to get rid of the financial pressure, VEs expect to provide stable and continuous financial support through VCs' investment. During this period, VEs can choose venture investors with fast lending speed, high quota, low rate of return and good credit standing to finance through investment intermediaries.

The sets of VE is given by $B = \{B_1, B_2, \cdots, B_j, \cdots, B_m\}$, here $B_j$ is denoted as $j^{th}$ VE, $j = (1, 2, \cdots, m)$. Let $Y = \{Y_1, Y_2, \cdots, Y_f, \cdots, Y_l\}$ be the satisfaction evaluation indicators (attributes) sets of VE to VCI, here $Y_f$ is denoted as $f^{th}$ evaluation indicator, $f = (1, 2, \cdots, l)$.

Similarly, the indicators values can be expressed by quantitative and qualitative data. For quantitative indicators, they can be given by real numbers or interval numbers or triangular fuzzy numbers. For qualitative indicators, they can be given by linguistic information.

The weight of indicator sets $Y$ is expressed as $u = \{u_1, u_2, \cdots, u_f, \cdots, u_l\}$, here, $u_f$ is the weight of the index $Y_f$, and they are satisfied with $\sum_{f=1}^{l} u_f = 1, 0 \leq u_f \leq 1$.

Let $E = [e_j^f]_{m \times l}$ be the expectation level matrix of VE $B_j$ to VCI $A_i$ on indicator sets $Y$, here $e_j^f$ is the expectation level value of VE $B_j$ to VCI $A_i$ on indicator $Y_f$.

Let $R = [r_{ji}^f]_{m \times l \times n}$ be the actual evaluation matrix of Investment Intermediaries to VCI $A_i$ on indicator sets $Y$, here $r_{ji}^f$ is the actual evaluation value of Investment Intermediaries to VCI $A_i$ on indicator $Y_f$.

# 3 Two-Sided Matching Model

## *3.1 Fuzzification Rules*

In this paper, the evaluation index of VCs and VEs are given different data types, such as real number, interval number, triangular fuzzy number and language information. In order to obtain more scientific and reasonable matching results and achieve comparability of different data types, these data need to be mapped to real number space [9, 10]. So, language information, interval numbers and triangular fuzzy numbers are all converted into real numbers by the comprehensive effect functions.

Assuming that the evaluation index values of VCs and VEs are given triangular fuzzy numbers, that is $A = (a,\ b,\ c)$. We use the comprehensive effect method to convert it into real numbers.

For Triangular fuzzy number $A = (a,\ b,\ c)$, $a \in [0,\ \infty)$, when $L(\lambda) = \lambda^a$, we have

$$I_L(A) = \frac{a + 2ab + c}{2(a + 2)},\ U_L(A) = \frac{c - a}{(a + 1)(a + 2)} \tag{1}$$

Here, $I_L(A)$ is called the centralized quantification principle value of A, and $U_L(A)$ is the dispersion of A [11, 12]. In order to transform triangular fuzzy numbers into real numbers, a comprehensive effect function is defined.

If we regard $x$ and $y$ as $I_L(A)$ and $U_L(A)$ of $A$ respectively, then

$$S_L(A) = S\ (I_L(A), U_L(A)) \tag{2}$$

A continuous function $S(x, y)$: $(-\infty, +\infty) \times [0, +\infty) \to (-\infty, +\infty)$ is called a maximum synthesizing effect function, if it satisfies: (1) $S(x, y)$ is monotone non-decreasing on $x$ for any $y \geq 0$; (2) $S(x, y)$ is monotone non-increasing on $y$ for any $x \in (-\infty, \infty)$; (3) $S(x, y) \leq x$, $S(x, 0) = x$. Then, $S(x, y)$ is called synthesizing effect function $S(x, y)$ is a compound quantification method of fuzzy information considering both $I_L$-metric and $U_L$-dispersion. And this method not only contains $I_L$-metric method, but it has better interpretability.

In order to eliminate the differences between each attribute to the influence of the evaluation results, the index weights of VCs and VEs need to standardize the original assessed value. In Literature (Fang, 2014, Zhu, 2014) the standardization equation is given.

## 3.2 Two-Sided Matching Model Be Constructed

In the venture capital investment process, both VCs and VEs have their own expectations when they can make two-sided choices, and they can avoid potential risks. Under the new era financial market, there is not much information symmetry between the matching parties, and the best matching can be successful by the aid of investment intermediaries. Investment intermediaries assist the matching parties in choosing suitable partners through decision analysis. Firstly, the mutual expectation of the matching parties is determined as the reference point. Secondly, the profit and loss decision of the actual level of the matching parties relative to the reference point is calculated. Thirdly, we calculate the perceived value of the profit and loss value of each matching party, and then we get the comprehensive perceived value of each matching party. Finally, taking the comprehensive perceived value of the matching parties as the maximum objective function, a multi-objective function programming model is established.

The Profit-Loss Matrix and Perceived Value of VCs

Assuming that the expectation level value $e_{ih}$ of the venture capitalist $A_i$ on the index $X_h$ is considered to be a reference point, which is compared with the actual evaluation value $r_{jh}$ of the venture enterprise $B_j$ on the index $X_h$, and we may establish the profit and loss value matrix of the venture investor, as following:

$$D\left(d_{ij}^{h}\right) = \begin{cases} d\left(r_{jh}, e_{ih}\right), & r_{jh} \geq e_{ih} \\ -d\left(r_{jh}, e_{ih}\right), & r_{jh} < e_{ih} \end{cases} \tag{3}$$

$$i = 1, 2, \cdots, n; j = 1, 2, \cdots, m, h = 1, 2, \cdots, k$$

Here, $d\left(r_{jh}, e_{ih}\right)$ is denoted as the distance between $r_{jh}$ and $e_{ih}$. When $r_{jh} \geq e_{ih}$ , $d\left(r_{jh}, e_{ih}\right)$ is considered to be the profit value, otherwise it is the loss value. Thus, $D = [d_{ij}^{h}]_{n \times k \times m}$ is regarded as the profit-loss matching decision matrix of VCs.

According to the prospect theory, we may calculate the perceived value of the venture capitalist $A_i$ by the (3). As following:

$$v\left(d_{ij}^{h}\right)=\begin{cases} D(d_{ij}^{h})^{\alpha_1}, & r_{jh}\geq e_{ih} \\ -\theta_1\left(-D\left(d_{ij}^{h}\right)^{\beta_1}\right), & r_{jh}<e_{ih} \end{cases} \tag{4}$$
$$i=1,2,\cdots,n;j=1,2,\cdots,m,h=1,2,\cdots,k$$

Here, $v\left(d_{ij}^{h}\right)$ is denoted as the perceived value of the venture capitalist $A_i$. Thus, $V=[v\left(d_{ij}^{h}\right)]_{n\times k\times m}$ is denoted as the risk perception matrix of the different risk attitudes of VCs towards loss and income. $\alpha_1\in(0,1)$, $\beta_1\in(0,1)$ are denoted as the convex-concave degree of the perceived value $v\left(d_{ij}^{h}\right)$. $\theta_1>1$ is indicated as the degree of loss aversion of venture capitalist $A_i$, and its value is the greater that it is regarded as the higher degree of loss aversion of venture capitalist, otherwise it is regarded as the lower. If $\alpha_1$, $\beta_1$,$\theta_1$ are assigned value, and can be obtained by the results in the literature [7]. Usually, let $\alpha_1=\beta_1=0.88$, $\theta_1=2.25$.

According to the (4), we can calculate the overall risk perceived value which is regarded as VC $A_i$ to VE $B_j$. As following:

$$U_{ij}=\sum_{h=1}^{k}v(d_{ij}^{h}) \tag{5}$$
$$i=1,2,\cdots,n;j=1,2,\cdots,m,h=1,2,\cdots,k$$

In the (5), $U_{ij}$ is the greater that it is the higher degree of coincidence between the expected level and the actual level of VCs, and be the greater propensity to invest in VEs.

The Profit-Loss Matrix and Perceived Value of VEs

Assuming that the expectation level value $e_{jf}$ of VE $B_j$ on the index $Y_f$ is considered to be a reference point, which is compared with the actual evaluation value $r_{if}$ of VC $A_i$ on the index $Y_f$, and we may establish the profit and loss value matrix of VEs, as following:

$$D\left(d_{ji}^{f}\right)=\begin{cases} d\left(r_{if},e_{jf}\right), & r_{if}\geq e_{jf} \\ -d\left(r_{if},e_{jf}\right), & r_{if}<e_{jf} \end{cases} \tag{6}$$
$$i=1,2,\cdots,n;j=1,2,\cdots,m,f=1,2,\cdots,l$$

Here, $d\left(r_{if},e_{jf}\right)$ is denoted as the distance between $r_{if}$ and $e_{jf}$. When $r_{if}\geq e_{jf}$, $d\left(r_{if},e_{jf}\right)$ is considered to be the profit value, otherwise it is the loss value. Thus, $D=[D\left(d_{ji}^{f}\right)]_{n\times l\times m}$ is regarded as the profit-loss matching decision matrix of VEs.

According to the prospect theory, we may calculate the perceived value of VE $B_j$ by the (6), as following:

$$v\left(d_{ji}^{f}\right)=\begin{cases} D(d_{ji}^{f})^{\alpha_2}, & r_{if}\geq e_{jf} \\ -\theta_2(-D\left(d_{ji}^{f})^{\beta_1}\right), & r_{if}<e_{jf} \end{cases} \tag{7}$$
$$i=1,2,\cdots,n;j=1,2,\cdots,m,f=1,2,\cdots,l$$

Here, $v\left(d_{ji}^{f}\right)$ is denoted as the perceived value of VE $B_j$. Thus, $V=[v\left(d_{ji}^{f}\right)]_{n\times l\times m}$ is denoted as the risk perception matrix of the different risk attitudes of VEs towards loss and income. $\alpha_2\in(0,1)$, $\beta_2\in(0,1)$ are denoted as the convex-concave degree of the perceived value $v\left(d_{ij}^{h}\right)$. $\theta_2>1$ is indicated as the degree of loss aversion of VE $B_j$, and its value is the greater that it is regarded as the higher degree of loss aversion of VE, it is regarded as the lower. If $\alpha_2$, $\beta_2$, $\theta_2$ are assigned value, and similarly, let $\alpha_1=0.88$, $\beta_1=0.88$, $\theta_1=2.25$.

According to the (7), we can calculate the overall risk perceived value which is regarded as VE $B_j$ to VC $A_i$. As following:

$$U_{ji}=\sum_{f=1}^{l}v(d_{ji}^{f}) \tag{8}$$
$$i=1,2,\cdots,n;j=1,2,\cdots,m,f=1,2,\cdots,l$$

In the (8), $U_{ji}$ is the greater that it is the higher degree of coincidence between the expected level and the actual level of VEs, and be the greater propensity to invest in VCs.

- The Multi-objective Two-Sided Matching Model

By the above calculation process, we can construct the Multi-objective Two-sided Matching Model (MTMM). In the model, it takes into account the matching objectives and the psychological and behavioral characteristics of both parties, it is the practical significance to solve the problem of investment and financing under the current situation of highly asymmetric information in the financial market.

In the investment process, both parties pursue the maximization of their overall perceived value. So, a target planning model for the maximization of overall perceived value is established in this paper.

Let $T=[\tau_{ij}]_{n\times m}$, here, when $\tau_{ij}=0$, it is denoted that VC $A_i$ and VE $B_j$ can be not matched, $\tau_{ij}=1$ when $\tau_{ij}=0$, it is denoted that VC $A_i$ and VE $B_j$ can be matched. The multi-objective matching model for both parties is as follows:

$$\max z_1 = \sum\nolimits_{i=1}^{n} \sum\nolimits_{j=1}^{m} U_{ij}\tau_{ij} \tag{9a}$$

$$\max z_2 = \sum\nolimits_{i=1}^{n} \sum\nolimits_{j=1}^{m} U_{ij}\tau_{ij} \tag{9b}$$

$$\max z_3 = \sum\nolimits_{i=1}^{n} \sum\nolimits_{j=1}^{m} \rho\sigma_{ij}\tau_{ij} \tag{9c}$$

$$s.t\begin{cases} \sum\nolimits_{i}^{n} \tau_{ij} \le \varsigma_j \quad (j = 1, 2, \cdots, m) \\ \sum\nolimits_{j}^{m} \tau_{ij} \le \xi_i \quad (i = 1, 2, \cdots, n) \\ \tau_{ij} = 0, 1 \quad (i = 1, 2, \cdots, n; j = 1, 2, \cdots, m) \end{cases} \tag{9d}$$

Here, the (9a) is indicated that all VCs have the largest overall induction value to all VEs, the (9b) is indicated that all VEs have the largest overall response value to all VCs, in the (9c), $z_3 = \sum\nolimits_{i=1}^{n} \sum\nolimits_{j=1}^{m} \sigma_{ij}\tau_{ij}$ is indicated the number of successful matches between VCs and VEs, If the investment intermediary collects commissions according to the matches number, the (9c) is indicated as the maximum income of the investment intermediary, $\rho$ is indicated as the income ratio of financial consulting services after the investment intermediary receives the successful matches. The (9d) is indicated that VCs can choose at most $\varsigma_j$ VEs to invest, while VEs can choose at most $\xi_i$ VCs to finance. In order to solve the multi-objective matching model, in the paper, we will transform the multi-objective optimization model into a single-objective optimization model through the linear weighting method [6–8].

In order to solve the multi-objective matching model, in the paper, we will transform the multi-objective optimization model into a single-objective optimization model through the linear weighting method [6–8]. Let $\sum\nolimits_{i=1}^{3} \vartheta_i = 1, \vartheta_i \in [0, 1]$, $i = 1, 2, 3$, so the 9a–9b model is can be expressed as the following.

$$\begin{aligned} \max Z = {} & \vartheta_1 \sum_{i=1}^{n} \sum_{j=1}^{m} U_{ij}\tau_{ij} \\ & + \vartheta_2 \sum\nolimits_{i=1}^{n} \sum\nolimits_{j=1}^{m} U_{ij}\tau_{ij} + \vartheta_3 \sum\nolimits_{i=1}^{n} \sum\nolimits_{j=1}^{m} \rho\sigma_{ij}\tau_{ij} \end{aligned} \tag{10a}$$

$$s.t\begin{cases} \sum\limits_{i}^{n} \tau_{ij} \le \varsigma_j \quad (j = 1, 2, \cdots, m) \\ \sum\limits_{j}^{m} \tau_{ij} \le \xi_i \quad (i = 1, 2, \cdots, n) \\ \tau_{ij} = 0, 1 \quad (i = 1, 2, \cdots, n; j = 1, 2, \cdots, m) \\ \sum\limits_{i=1}^{3} \vartheta_i = 1, \vartheta_i \in [0, 1], i = 1, 2, 3 \end{cases} \tag{10b}$$

## 4 Example Calculation and Analysis

The investment intermediary has received the formation base on 4 venture investors (A1, A2, A3, A4) and 6 venture enterprises (B1, B2, B3, B4, B5, B6). VCs choose five indexes as their own index system to determine their investment, such as the return period of investment, technical level, market accessibility, quality of enterprise managers and investment environment. VEs choose four indexes as their own index system to determine who will invest, such as investment strength, investment success rate, reputation and quality of VCs. According to experience and market survey, VCs give their index expectation level values to VEs as shown in Table 1, and VEs give their index expectation level values to VCs as shown in Table 2.

The investment intermediary may actually evaluate the index set (x1, x2, x3, x4, x5) of the 4 VCs to VEs, such as Table 3. Let $u$ = (0.25, 0.25, 0.15, 0.20, 0.15) be the corresponding index weight value. The investment intermediary may actually evaluate the index set (y1, y2, y3, y4), of 6VEs to VCs, such as Table 4. Let $u'$ = (0.40, 0.25, 0.20, 0.15) be the corresponding index weight value.

Assuming $S(x, y) = x(1 + ay)^{-b}$, a = 0.2, b = 0.5 is the comprehensive effect function, The model is solved by LINGO; the result is as follows: $\tau_{11} = 1$, $\tau_{16} = 1, \tau_{25} = 1, \tau_{34} = 1, \tau_{42} = 1, \tau_{43} = 1$, the rest $\tau_{ij} = 0$. From this, we can see that VC $A_1$ may be matched VE $B_1$, VC $A_2$ may be matched VE $B_5$, VC $A_3$ may be matched VE $B_4$, VC $A_4$ may be matched VE $B_2$ and VE $B_3$. The matching results recommended by the intermediary will provide decision support for the venture capital entities.

**Table 1** Expected level value of VCs to VEs

| | $x_1$ | $x_2$ | $x_3$ | $x_4$ | $x_5$ |
|---|---|---|---|---|---|
| $A_1$ | 3 | [7%, 10%] | (0.5, 0.75, 1) | (5, 7.5, 10) | G |
| $A_2$ | 4 | [10%, 12%] | (0.75, 0.75, 1) | (6, 8, 10) | VG |
| $A_3$ | 3.5 | [9%, 15%] | (0.5, 0.75, 1) | (1, 3, 5) | VG |
| $A_4$ | 5 | [10%, 15%] | (0.75, 0.75, 1) | (3, 6, 9) | G |

**Table 2** Expected level value of VEs to VCs

| | $Y_1$ | $Y_2$ | $Y_3$ | $Y_4$ |
|---|---|---|---|---|
| $B_1$ | 900 | [8%, 12%] | (0.5, 0.65, 0.7) | G |
| $B_2$ | 800 | [10%, 15%] | (0.3, 0.35, 0.4) | G |
| $B_3$ | 500 | [3%, 5%] | (0.1, 0.2, 0.25) | VG |
| $B_4$ | 400 | [1%, 3%] | (0.25, 0.3, 0.4) | G |
| $B_5$ | 700 | [5%, 7%] | (0.4, 0.65, 0.8) | VG |
| $B_6$ | 600 | [12%, 16%] | (0.2, 0.41, 0.5) | VG |

**Table 3** VEs' actual evaluation value

| | $B_1$ | | | | | $B_2$ | | | | |
|---|---|---|---|---|---|---|---|---|---|---|
| | $x_1$ | $x_2$ | $x_3$ | $x_4$ | $x_5$ | $x_1$ | $x_2$ | $x_3$ | $x_4$ | $x_5$ |
| $A_1$ | 3.5 | [8%, 10%] | (0.5, 0.75, 1) | (3, 7, 9) | G | 2.5 | [9%, 11%] | (0.25, 0.5, 0.75) | (3, 5, 7) | VG |
| $A_2$ | 4.5 | [10%, 11%] | (0.25, 0.5, 0.75) | (1, 4, 6) | G | 6.5 | [8%, 12%] | (0.5, 0.75, 1) | (1, 6, 9) | G |
| $A_3$ | 3.5 | [8%, 10%] | (0, 0.25, 0.5) | (2, 3, 5) | VG | 3.5 | [10%, 12%] | (0.5, 0.75, 1) | (2, 5, 9) | G |
| $A_4$ | 4 | [10%, 11%] | (0.5, 0.75, 1) | (4, 6, 9) | G | 4.5 | [7%, 9%] | (0.5, 0.75, 1) | (3, 7, 9) | G |
| | $B_3$ | | | | | $B_4$ | | | | |
| | $x_1$ | $x_2$ | $x_3$ | $x_4$ | $x_5$ | $x_1$ | $x_2$ | $x_3$ | $x_4$ | $x_4$ |
| $A_1$ | 3.5 | [8%, 10%] | (0.25, 0.5, 0.75) | (4, 5, 7) | G | 3.5 | [8%, 10%] | (0.5, 0.75, 1) | (4, 6, 9) | G |
| $A_2$ | 4.5 | [6%, 8%] | (0, 0, 0.25) | (5, 7, 10) | P | 4.5 | [10%, 11%] | (0.25, 0.5, 0.75) | (1, 5, 7) | VG |
| $A_3$ | 3.5 | [11%, 12%] | (0, 0.5, 0.75) | (1, 4, 6) | G | 3.5 | [8%, 10%] | (0.25, 0.5, 0.75) | (2, 4, 6) | G |
| $A_4$ | 2.5 | [8%, 11%] | (0.5, 0.75, 1) | (2, 3, 5) | M | 3 | [9%, 11%] | (0.5, 0.75, 1) | (1, 3, 5) | G |
| | $B_5$ | | | | | $B_6$ | | | | |
| | $x_1$ | $x_2$ | $x_3$ | $x_4$ | $x_5$ | $x_1$ | $x_2$ | $x_3$ | $x_4$ | $x_4$ |
| $A_1$ | 2.5 | [6%, 8%] | (0.4, 0.65, 0.8) | (3, 5, 7) | G | 3 | [6%, 8%] | (0.4, 0.65, 0.8) | (3, 7, 9) | VG |
| $A_2$ | 4.5 | [10%, 15%] | (0.25, 0.5, 0.75) | (5, 7, 10) | VG | 4.5 | [10%, 11%] | (0.25, 0.5, 0.75) | (1, 5, 7) | G |
| $A_3$ | 3.5 | [8%, 10%] | (0.5, 0.75, 1) | (1, 6, 9) | G | 3.5 | [5%, 7%] | (0.5, 0.75, 1) | (2, 4, 8) | G |
| $A_4$ | 6 | [12%, 16%] | (0.5, 0.75, 1) | (4, 6, 9) | P | 4 | [10%, 11%] | (0.25, 0.5, 0.75) | (3, 6, 10) | VG |

**Table 4** VCs' actual evaluation value

| | A1 | | | | A1 | | | |
|---|---|---|---|---|---|---|---|---|
| | $Y_1$ | $Y_2$ | $Y_3$ | $Y_4$ | $Y_1$ | $Y_2$ | $Y_3$ | $Y_4$ |
| B1 | 900 | [8%, 12%] | (0.5, 0.65, 0.7) | G | 600 | [9%, 12%] | (0.5, 0.75, 1) | G |
| $B_2$ | 800 | [10%, 15%] | (0.3, 0.35, 0.4) | G | 200 | [1%, 3%] | (0.25, 0.3, 0.4) | P |
| $B_3$ | 500 | [3%, 5%] | (0.1, 0.2, 0.25) | VG | 300 | [5%, 7%] | (0.4, 0.65, 0.8) | VG |
| $B_4$ | 300 | [6%, 8.5%] | (0.1, 0.2, 0.3) | G | 400 | [6%, 7.5%] | (0.5, 0.6, 0.7) | M |
| $B_5$ | 700 | [6%, 8%] | (0.2, 0.3, 0.34) | M | 600 | [12%, 16%] | (0.2, 0.41, 0.5) | VG |
| $B_6$ | 900 | [16%, 20%] | (0.2, 0.4, 0.5) | G | 600 | [4%, 7%] | (0.5, 0.75, 1) | G |
| | A1 | | | | A1 | | | |
| | $Y_1$ | $Y_2$ | $Y_3$ | $Y_4$ | $Y_1$ | $Y_2$ | $Y_3$ | $Y_4$ |
| B1 | 600 | [9%, 12%] | (0.5, 0.75, 1) | G | 500 | [3%, 5%] | (0.1, 0.2, 0.25) | P |
| $B_2$ | 300 | [6%, 8.5%] | (0.1, 0.2, 0.3) | G | 700 | [6%, 8%] | (0.25, 0.5, 0.75) | G |
| $B_3$ | 700 | [6%, 8%] | (0.2, 0.3, 0.34) | VG | 800 | [10%, 15%] | (0.5, 0.75, 1) | VG |
| $B_4$ | 900 | [16%, 20%] | (0.2, 0.4, 0.5) | G | 600 | [4%, 7%] | (0.1, 0.2, 0.3) | G |
| $B_5$ | 400 | [3%, 9%] | (0.25, 0.3, 0.4) | G | 600 | [12%, 16%] | (0.4, 0.65, 0.8) | M |
| $B_6$ | 300 | [5%, 7%] | (0.4, 0.65, 0.8) | P | 900 | [8%, 12%] | (0.3, 0.35, 0.4) | G |

## 5 Conclusion

In this paper, the fuzzy two-side matching problem is researched that it is complicated and fuzzy among the subjects involved in venture capital activities and in the external environment. In the matching model, the expected level value is selected as a reference point, the profit and loss values of each index are calculated by using prospect theory, and the psychological characteristics of decision makers' risk perception are considered, and then the overall perception value of decision makers is calculated. An example analysis shows that the model is feasible, effective and reasonable; it can be providing method support for venture capital decision-making.

**Acknowledgements** Project of Hebei Provincial Department of Science and Technology—Study on the integrated development strategy of logistics industry in Hebei province under the opportunity of Beijing, Tianjin and Hebei coordinated development. Project number: 154576292.

## References

1. Chen, X., & Fan, Z. P. (2010). Problem of two-sided matching between venture capitalists and venture enterprises based on axiomatic design. *Systems Engineering, 28*(6), 70–75.
2. Fang, S. P., & Li, D. F. (2014). Decision Making Method for Multi-attribute Two-sided Matching Problem between venture capitalists and venture enterprises with deferent Kinds of information. *Chinese Journal of Management Science, 22*(2), 40–47.
3. Gale, D., & Shapley, L. (1962). College admissions and the stability of marriage. *American Mathematical Monthly, 69*(1), 9–15.
4. Roth, A. E. (1985). Common and Conflicting Interests in Two-sided Matching Markets. *European Economic Revie, 27*(1), 425–427.
5. Sorensen, M. (2007). How smart is smart money? A two sided matching model of venture capital. *Journal of Finace, 62*(6), 2725–2762.
6. Cao, G. H., & Hu, Y. (2009). Study on the two-sided matching model between venture capitalists and entrepreneurs. *Science & Technology Progress and Policy, 26*(5), 31–38.
7. Wu, F. P., Zhu, W., & Cheng, T. J. (2016). Study on the venture capital two-sided matching decision-making in internet finance. *Science & Technology Progress and Policy, 33*(4), 25–30.
8. Li, M. Y., Li, B., Cao, P., & Huo, C. H. (2018). A method of Two-Sided Matching base on multiple criteria considering disappointment and elation of both sides. *Journal of Zhejiang University (Science Edition), 45*(2), 169–179.
9. Wang, L. L., & Li, D. F. (2018). Research on two-sided match methods on venture capitalists and entrepreneurs with heterogeneous information. *Mathematics in Practice and Theory, 48* (7), 43–55.
10. Chen, P., Egesdal, M., Pycia, M., et al. (2016). Medianstable matching in two-sided markets. *Games and Economic Behavior, 97*(1), 64–69.
11. Li, F. C., Li, L., Jin, C. X., Wang, R. J., & Hong, W. (2012). Research on the 3PL supplier selection based on set-valued satisfaction. *Computers & Operations Research, 39*(8), 1879–1884.
12. Wang, R. J., Zhu, X. X., & Li, F. C. (2014). Supply chain risks evaluation model in fuzzy environment. *Journal of System and Management Sciences, 2*, 37–43.

# Revenue Sharing Contracts of Risk-Averse Retailers Under Different Modes

**Xiaojing Liu, Wenyi Du, and Xigang Yuan**

**Abstract** The paper studies the revenue sharing contract with risk factors in different dominant modes and establishes two game models under the risk aversion environment. In supplier and retailer different modes, it is found that risk aversion coefficient and benefit sharing ratio can affect the decision-making strategy and performance of supply chain members. In both models, suppliers will set lower wholesale prices for retailers with higher risk aversion coefficient; The larger the retailer's risk aversion coefficient is, the smaller the retailer's order quantity is in supplier mode; The profit of the retailer increases with the increase of risk aversion, and the profit of the suppliers increases with the increase of revenue sharing ratio.

**Keywords** Supplier dominant · Retailer dominant · Risk aversion · Revenue-sharing contract

## 1 Introduction

In the rapid development of global economy, countries and countries, enterprises and enterprises are facing both competition and cooperation. In many markets, [1] thought that the new products will enter the market, when they are produced. What kind of retailers or distributors can new products choose to enter the market? This

This research is supported by the Youth Project of Natural Science Foundation of China (Grant number 71702067) and PhD Research Project of Jiangsu Normal University (Grant number 16XWR011).

X. Liu · W. Du (✉) · X. Yuan
Business School, Jiangsu Normal University, Xuzhou, China
e-mail: wydu@jsnu.edu.cn

X. Liu
e-mail: xjliu@jsnu.edu.cn

X. Yuan
e-mail: yxg200811606@126.com



J. Zhang et al. (eds.), *LISS2019*,
https://doi.org/10.1007/978-981-15-5682-1_44

mainly depends on the dominant mode of the market. Sometimes the market is dominated by retailers. They choose retailers to sell new products. Sometimes the market is dominated by distributors. They choose distributors to sell new products. We can also understand that a retailer or distributor may not be the only sole agent to provide product sales to upstream suppliers. Many retail industries have a variety of sales channels, such as KFC and McDonald's, Petro China and Sinopec, which are all sold in China. Reference [2] considers that this kind of marketing channel seems very common in China. In the supply market, [3] and [4] discussed how do risk-averse retailers adopt ordering strategies. Literature [5] studied water resources supply chain under different dominant modes. Reference [6] showed that enterprise resources are relatively limited, usually not easy to reuse. It necessary for every enterprise and every country to change to the behavior strategy in today's world. How to find cooperation within the supply chain becomes particularly important. The strategic behavior of revenue sharing just solves this problem.

Revenue-sharing contract is a kind of contractual commitment between members of supply chain and a strategic behavior for their common interests. This strategy may contribute to the growth of one side's interests, while harming the other side's interests, but for the benefit of the whole supply chain is growing. It is also possible that such strategic actions can benefit both sides, and thus contribute to the increase of the interests of the whole supply chain. As early as the video rental market, there has been a shadow of revenue sharing. Based on the famous Blockbuster Video Rental Company, a two-stage game model of suppliers and leasers is established. Reference [7] found that the revenue-sharing contract helps to improve the efficiency of the whole market. Considering the three-stage supply chain with demand affecting price, [8] found that profit allocation among members can be achieved by means of revenue sharing agreement. Reference [9] used differential game theory to study information sharing in food supply chain, and designed a coordination mechanism. Reference [10] analyzed the operation decision of two-stage supply chain under retailer-led situation. Based on the elastic demand market, [11] studied the revenue-sharing contract of retailer-led model. Reference [12] studied the profit-sharing contract of retailer's dominant revenue-sharing coefficient under the cost of capital. It was found that under decentralized and centralized decision-making, the revenue-sharing contract could increase the order quantity and profit of the supply chain system. Reference [13] analyzed the revenue sharing contract mechanism under uncertain production and demand, and found that the sharing mechanism reduced the impact of uncertain production and demand on supply chain. Reference [14] considered the pricing and coordination of commodities under different dominant modes. In the revenue-sharing contracts mentioned above, [15] and [16] showed that retailers are assumed to be risk-neutral and risk aversion has no effect on their profits. Reference [17] considered the issue of revenue-sharing contracts between traditional and electronic channels in a price-sensitive environment. Reference [18] assumed the supply chain coordination problem of revenue sharing contract with risk preference under carbon trading environment. When studying the competition of chain enterprises under different dominant modes, [19] estimated the form of competition under two modes, designed two different supply chain contracts, namely revenue sharing contract and quantity discount contract, and compared the system performance under two supply chain contracts. Quantity discount contract improves the performance of supply chain system. In practice, whether retailers or suppliers, they are more or less faced with the impact of risk factors. How risk aversion coefficient affects the decision making of retailers and even their profits, it is a problem worth studying in real life. From the perspective of CVAR, [20] constructed Stackelberg two-stage game model, and found that brand enterprises could share part of the profits with suppliers and increase the quantity of orders. However, the above literature only depends on the risk-neutral side to determine the profit-sharing coefficient. When the other side is no longer risk-neutral, how do the risk aversion coefficient and the profit-sharing coefficient dominated by different members affect the decision-making of the members of the supply chain and even the overall profit of the supply chain? When considering risk aversion, how does the dominant revenue sharing coefficient of different members affect the operational decision-making of supply chain members? These will be the practical problems to be solved in this paper.

The rest of this paper is organized as follows. In Sect. 2, we briefly introduce the description of practical problems and some basic assumptions. In Sect. 3, we establish the decision-making model of supply chain revenue sharing under supplier leading and retailer leading modes, and give the optimal decision-making strategy of each supply chain member by means of reverse induction. By means of numerical analysis, Sect. 4 give the relationship between optimal decision-making, performance and parameter variables. In Sect. 5, we find the shortcomings of the paper and possible future research directions.

## 2 Problem Description

This paper considers a supply chain structure model, a single retailer and a single supplier. In the structure, the retailer is risk-averse, and the manufacturer is risk-neutral. The production cost of the product is $c$, the wholesale price is $w$, the order quantity of the retailer is q, and $p$ is the retail price. The market demand is expressed by random variables $x$. Its probability density function and probability distribution function are $f(x)$ and $F(x)$, they are continuous and differentiable. $F^{-1}(x)$ is an inverse function, and the $f(x)x/1 - F(x)$ is an incremental function, i.e. the demand distribution IGFR (increasing generalized failure), and the general distribution function satisfies, such as uniform distribution is constructed in [21].

For risk-averse retailers, this paper uses loss-averse model which used in [22] in prospect theory to characterize them. Retailers have the following piecewise linear utility functions:

$$U(\pi_R) = \begin{cases} \pi_R, & \pi_R \geq 0 \\ \lambda\pi_R, & \pi_R < 0 \end{cases} \quad (1)$$

The parameter $\lambda$ in (1) can represent the retailer's loss aversion coefficient, whose value is greater than or equal to 1. $\lambda > 1$, the retailer's loss aversion to risk is lower with $\lambda$ smaller loss aversion. When $\lambda = 1$, it means the retailer is neutral to risk.

# 3 Model Solution

Different members of the supply chain can dominate the profit sharing coefficient as different subjects, which will change the order of game in each case, and then affect the decision-making behavior of the members of the supply chain.

## *3.1 Supplier Leading Revenue Sharing (SR)*

Suppliers determine wholesale prices and require retailers to share their $1 - \phi$ revenue. Then, the retailer determines the order quantity according to the supplier's decision-making situation and realizes the demand. In such a supplier leading supply chain, the supplier decides the proportion of revenue sharing and is responsible for signing supply chain contracts. Therefore, there are two decision variables for suppliers, namely wholesale price and revenue sharing ratio. There is only one decision variable for retailers, which is only the number of orders. Therefore, the Stackelberg game process of them can be simply described: step one, the supplier requests the retailer to give the share ratio and the wholesale price of the product from its own interests; step two, the retailer decides the order quantity that can maximize its own profit, where $\phi p - w > 0$ is establish.

At this point, the expected profit of the retailer is

$$\pi_R(q) = \phi p \min(q, x) - wq = \begin{cases} \phi px - wq & x < q \\ \phi pq - wq & x \geq q \end{cases} \quad (2)$$

The first item in (2) denotes the retailer's share of market sales revenue, and the second item denotes the retailer's order cost.

Note $q$ is the break-even requirement (when the demand is at that point, the retailer's profit is zero). Let $\pi_R = 0$, and we can get $q_0 = wq/(\varphi p)$. Once the order quantity is less than $q_0$, the retailer's order can't meet the market demand, so the retailer can't make profits, and its value is negative. Only when the retailer orders more than $q_0$, the market demand is satisfied. When $x < q_0$, then there is $\pi_R < 0$;

When $q_0 \le x \le q$, then there is $\pi_R \ge 0$; When $x > q$, then there is $\pi_R > 0$. Thus, the expected loss, expected profit and expected utility of retailers are respectively,

$$\begin{aligned} L_R &= \int_0^{q_0} \pi_R f(x)dx \\ &= \int_0^{\frac{wq}{\phi p}} (\phi px - wq)f(x)dx = -\phi p \int_0^{\frac{wq}{\phi p}} F(x)dx \end{aligned} \tag{3}$$

$$\begin{aligned} E(\pi_R) &= \int_0^{+\infty} \pi_R f(x)dx \\ &= \int_0^{q_0} (\phi px - wq)f(x)dx + \int_{q_0}^{+\infty} (\phi pq - wq)f(x)dx \\ &= \phi pq - wq - \phi p \int_0^{q_0} F(x)dx \end{aligned} \tag{4}$$

$$\begin{aligned} E(U(\pi_R)) &= \lambda \int_0^{q_0} \pi_R f(x)dx + \int_{q_0}^{+\infty} \pi_R f(x)dx \\ &= (\lambda - 1) \int_0^{q_0} \pi_R f(x)dx + \int_0^{+\infty} \pi_R f(x)dx \\ &= E(\pi_R) + (\lambda - 1)L_R \end{aligned} \tag{5}$$

It is known in (5) that the expected utility of a retailer consists of two parts: expected profit and expected loss. When $\lambda = 1$, Retailers are risk-neutral, and their expected utility is equal to the expected profit, $E(U(\pi_R)) = E(\pi_R)$; When $\lambda > 1$, Retailers are loss averse, and their expected utility is always less than expected profit, $E(U(\pi_R)) < E(\pi_R)$, and when given expected profits and expected losses are larger, the greater the loss aversion coefficient is, the smaller the expected utility of retailers is.

The goal of retailers is to determine the quantity of orders based on the wholesale price and revenue sharing ratio of suppliers in order to maximize their expected utility. (5) calculates the first and second derivatives of the order quantity for the expected profit.

$$\frac{dE(U(\pi_R))}{dq} = \varphi p - w\left[1 + \lambda F\left(\frac{wq}{\varphi p}\right)\right] \tag{6}$$

$$\frac{d^2E(U(\pi_R))}{dq^2} = -\frac{w^2\lambda}{\varphi p} f\left(\frac{wq}{\varphi p}\right) < 0 \tag{7}$$

**Proposition 1:** *When* $\lambda > 0$, *for the order quantity,* $E(U(\pi_R))$ *is a concave function, and the optimal order quantity of the retailer satisfies* $F\left(\frac{wq}{\phi p}\right) = \frac{\phi p - w}{w\lambda}$, *i.e.* $q^* = \frac{\phi p}{w} F^{-1}\left(\frac{\phi p - w}{w\lambda}\right)$.

**Corollary 1:** *Under SR-dominated model, the order quantity of the retailer is a decreasing function of the wholesale price. The order quantity of retailers is an incremental function of the proportion of revenue sharing. And the order quantity of retailers is a subtraction function of loss avoidance.*

Prove: First-order derivatives of wholesale price, profit-sharing ratio and loss aversion coefficient for both sides of equation in proposition 1, then $\frac{dq}{dw} = -\frac{\lambda wq + \phi p(1+F(q_0))}{\lambda w^2 f(q_0)} < 0$, $\frac{dq}{d\phi} = \frac{\lambda w^2 qf(q_0) + \phi^2 p^2}{\lambda \phi w^2 f(q_0)} > 0$, $\frac{dq}{d\lambda} = -\frac{\phi q(\phi p - w)}{\lambda^2 w^2 f(q_0)} < 0$.

Corollary 1 is proved.

Corollary 1 shows that suppliers can change the order quantity of retailers by adjusting the profit sharing ratio coefficient and setting the wholesale price. For retailers with different risk aversion, the order quantity will also be different. For the same supplier, retailers with higher risk aversion coefficient will be cautious in ordering behavior.

The expected profit is in the following expressions

$$\begin{aligned} E\pi_S(\phi, w) &= wq - cq + (1-\phi)p\min(q, x) \\ &= \begin{cases} (w-c)q + (1-\phi)px & x < q \\ (w-c)q + (1-\phi)pq & x \geq q \end{cases} \\ &= (w-c)q^* + (1-\phi)p\left[q^* - \int_0^{q^*} F(x)dx\right] \end{aligned} \tag{8}$$

The supplier's profit is a function of the revenue sharing ratio and the wholesale price, so the first step is to determine whether the supplier has the optimal profit sharing ratio and the optimal wholesale price. According to Eq. (8), the second derivative of profit-sharing ratio and wholesale price can be obtained. Then we can get the Heisen matrix of the supplier about the revenue sharing ratio and wholesale price. The value of Heisen matrix is less than zero, and it has the largest expected profit.

The expected profit function expression about the profit-sharing ratio and the wholesale price is zero, so that the supplier's optimal profit-sharing ratio and the optimal wholesale price can be obtained. According to the reverse induction method, the optimal order quantity of retailers can be obtained by substituting the share ratio and wholesale price into the function of order quantity.

The expected profits of retailers, suppliers and supply chains can be expressed in the following expressions, respectively,

$$E(U(\pi_R)) = (\phi^* p - w^*)q^* - \lambda\phi^* p\int_0^{q_0} F(x)dx \tag{9}$$

$$E\pi_S(\phi, w) = (w^* - c)q^* + (1-\phi^*)p\left(q^* - \int_0^{q^*} F(x)dx\right) \tag{10}$$

$$
\begin{aligned}
E\pi_C &= EU(\pi_R) + E\pi_S \\
&= (p-c)q^* - \lambda\phi^* p\int_0^{q_0} F(x)dx \\
&\quad - (1-\phi^*)p\int_0^{q^*} F(x)dx
\end{aligned} \tag{11}
$$

## 3.2 Retailer Leading Revenue Sharing (RR)

In this situation, the retailer first determines the quantity of products ordered according to the market demand and promises to give the supplier $\varphi$ share of revenue. Subsequently, the supplier determines the wholesale price according to the retailer's decision-making situation to maximize its own profits, and meets the retailer's orders and needs. The supplier's decision variable is only the wholesale price $w_1$. Although in theory retailers make decisions in favor of themselves before the supplier announces the wholesale price, the actual decision-making (ordering) behavior of retailers take place after the supplier announces the wholesale price. Therefore, we can think that retailer's decision-making behavior channel is a contingent strategic behavior, that is to say, the share ratio and order quantity published by retailer can be the function of wholesale price.

In this way, we assume that after the retailer draws up a contract and publishes the revenue-sharing ratio, the supplier first decides the wholesale price, and the retailer then decides the order quantity.

The retailer's revenue is,

$$
\begin{aligned}
\pi_R^1(q_1,\varphi) &= (1-\varphi)\,p\min(q,x) - w_1 q_1 \\
&= \begin{cases} (1-\varphi)\,px - w_1 q_1 & x<q_1 \\ (1-\varphi)\,pq_1 - w_1 q_1 & x\geq q_1 \end{cases}
\end{aligned} \tag{12}
$$

The first item in (12) denotes the retailer's share of market sales revenue, and the second item denotes the retailer's order cost.

Note $q_{01}$ is the break-even requirement (when the demand is at that point, the retailer's profit is zero). Let $\pi_R^1 = 0$, and we can get $q_{01} = w_1 q_1/(1-\phi)p$. Once the order quantity is less than $q_{01}$, the retailer's order can't meet the market demand, so the retailer can't make profits, and its value is negative. Only when the retailer orders more than $q_{01}$, the market demand is satisfied. When $x < q_{01}$, then there is $\pi_R^1 < 0$; When $q_{01} \leq x \leq q$, then there is $\pi_R^1 \geq 0$; When $x > q$, then there is $\pi_R^1 > 0$. Thus, the expected loss, expected profit and expected utility of retailers are respectively,

$$
\begin{aligned}
L_R^1 &= \int_0^{q_{01}} \pi_R^1 f(x)dx = \int_0^{\frac{w_1 q_1}{(1-\varphi)p}} ((1-\varphi)px - w_1 q_1) f(x)dx \\
&= -(1-\varphi)p \int_0^{\frac{w_1 q_1}{(1-\varphi)p}} F(x)dx
\end{aligned} \tag{13}
$$

$$
\begin{aligned}
E(\pi_R^1) &= \int_0^{+\infty} \pi_R^1 f(x)dx \\
&= \int_0^{q_{01}} ((1-\varphi)px - w_1 q_1) f(x)dx \\
&+ \int_{q_{01}}^{+\infty} ((1-\varphi)pq_1 - w_1 q_1) f(x)dx \\
&= (1-\varphi)pq_1 - w_1 q_1 - (1-\varphi)p \int_0^{q_{01}} F(x)dx
\end{aligned} \tag{14}
$$

$$
\begin{aligned}
E(U(\pi_R^1)) &= \lambda \int_0^{q_{01}} \pi_R^1 f(x)dx + \int_{q_{01}}^{+\infty} \pi_R^1 f(x)dx \\
&= (\lambda - 1) \int_0^{q_{01}} \pi_R^1 f(x)dx + \int_0^{+\infty} \pi_R^1 f(x)dx \\
&= E(\pi_R^1) + (\lambda - 1) L_R^1
\end{aligned} \tag{15}
$$

It is known in (15) that the expected utility of a retailer consists of two parts: expected profit and expected loss. When $\lambda = 1$, retailers are risk-neutral, and their expected utility is equal to the expected profit, $E(U(\pi_R^1)) = E(\pi_R^1)$; When $\lambda > 1$, Retailers are loss averse, and their expected utility is always less than expected profit, $E(U(\pi_R^1)) < E(\pi_R^1)$, and when given expected profits and expected losses are larger, the greater the loss aversion coefficient is, the smaller the expected utility of retailers is.

The goal of retailers is a joint function of revenue sharing ratio and order quantity, which can be further regarded as a single function of wholesale price. According to Eq. (15), the second derivative of profit-sharing ratio and order quantity can be obtained. Then we can get the Heisen matrix of the supplier about the proportion of revenue sharing and order quantity. The value of Heisen matrix is less than zero and the retailer's expected profit function is a joint concave function about the proportion of revenue sharing and order quantity, which has the largest expected profit.

The expected profit function expression is zero with respect to the proportion of revenue sharing and the order quantity, so that the retailer's optimal proportion of revenue sharing and the optimal wholesale price can be obtained respectively.

$$
\varphi^* = 1 - \frac{w_1}{p}(1 + F(q_{01})) \tag{16}
$$

$$q_1^* = \frac{\lambda(1+\lambda F(q_{01}))(p - w_1 - \lambda w_1 F(q_{01})) \int_0^{q_{01}} F(x)dx}{p^2(1+F(q_{01}))} \quad (17)$$

Substitute (16) (17) into (18), simplify the supplier's function about wholesale price, and make its first derivative about wholesale price zero. According to the reverse induction method, the optimal profit-sharing ratio and the optimal order quantity of retailers can be obtained by substituting wholesale price into (16) and (17) respectively.

The expected profit of the supplier can be expressed as,

$$\begin{aligned} E\pi_S^1(w_1) &= w_1 q_1 - c q_1 + \varphi p \min(q_1, x) \\ &= \begin{cases} (w_1 - c)q_1 + \varphi p x & x < q_1 \\ (w_1 - c)q_1 + \varphi p q_1 & x \geq q_1 \end{cases} \\ &= (w_1 - c)q_1 + \varphi p \left[ q_1 - \int_0^{q_1} F(x)dx \right] \end{aligned} \quad (18)$$

So, expected Profit of supply chain members are respectively,

$$\begin{aligned} E(U(\pi_R^1)) &= (1 - \varphi^*(w_1^*))p q_1^*(w_1^*) - w_1^* q_1^*(w_1^*) \\ &\quad - \lambda(1 - \varphi^*(w_1^*))p \int_0^{q_{01}} F(x)dx \end{aligned} \quad (19)$$

$$\begin{aligned} E\pi_S^1(w_1^*) &= (w_1^* - c)q_1^*(w_1^*) \\ &\quad + \varphi^*(w_1^*)p \left[ q_1^*(w_1^*) - \int_0^{q_1^*(w_1^*)} F(x)dx \right] \end{aligned} \quad (20)$$

$$\begin{aligned} E\pi_C^1 &= (p - c)q_1^*(w_1^*) - \lambda(1 - \varphi^*(w_1^*))p \int_0^{q_{01}} F(x)dx \\ &\quad - p\varphi^*(w_1^*) \int_0^{q_1^*(w_1^*)} F(x)dx \end{aligned} \quad (21)$$

## 4 Examples Analysis

In order to better observe the influence of risk aversion coefficient and profit sharing ratio on the operation decision and the member' expected profit, we assume that in a certain product market in a certain region, the cost of a product is 2 yuan per unit, the market price is 10 yuan per unit, and the product's regional distribution function is $F(x) = x/100$. Under SR and RR modes, the operation decisions and the relationship between profit size and change are shown in Tables 1 and 2.

**Table 1** Variation of parameters in supply chain under SR mode

| $\lambda$ | 1.5 | | | | | 2 | | | | | 3 | | | | |
|---|---|---|---|---|---|---|---|---|---|---|---|---|---|---|---|
| $\phi$ | 0.1 | 0.3 | 0.5 | 0.7 | 0.9 | 0.1 | 0.3 | 0.5 | 0.7 | 0.9 | 0.1 | 0.3 | 0.5 | 0.7 | 0.9 |
| $w$ | 0.769 | 2.142 | 3.228 | 3.699 | 16.257 | 0.712 | 1.989 | 3.009 | 3.466 | 16.25 | 0.636 | 1.787 | 2.719 | 3.157 | 16.243 |
| $q$ | 13.434 | 15.347 | 18.318 | 25.785 | 1.48 | 13.357 | 15.145 | 17.91 | 24.896 | 1.285 | 13.294 | 14.882 | 17.374 | 23.725 | 1.052 |
| $E(\pi_S)$ | 68.002 | 67.437 | 62.863 | 35.446 | 22.575 | 67.264 | 65.321 | 59.747 | 34.032 | 19.593 | 66.271 | 62.551 | 55.658 | 31.854 | 16.034 |
| $E(U(\pi_R))$ | 0.347 | 3.3 | 11.785 | 40.847 | −11.17 | 0.98 | 5.188 | 14.79 | 44.275 | −9.691 | 1.817 | 7.603 | 18.546 | 48.299 | −7.928 |
| $E(\pi_C)$ | 68.349 | 70.737 | 74.648 | 76.293 | 11.405 | 68.244 | 70.509 | 74.537 | 78.307 | 9.902 | 68.088 | 70.154 | 74.204 | 80.153 | 8.106 |

**Table 2** Variation of parameters in supply chain under RR mode

| $\lambda$ | 1.5 | | | | | 2 | | | | | 3 | | | | |
|---|---|---|---|---|---|---|---|---|---|---|---|---|---|---|---|
| $\varphi$ | 0.1 | 0.3 | 0.5 | 0.7 | 0.9 | 0.1 | 0.3 | 0.5 | 0.7 | 0.9 | 0.1 | 0.3 | 0.5 | 0.7 | 0.9 |
| $w_1$ | 10.837 | 10.587 | 10.321 | 10.024 | 9.621 | 10.757 | 10.501 | 10.23 | 9.928 | 9.531 | 10.616 | 10.358 | 10.087 | 9.789 | 9.418 |
| $q_1$ | 0.905 | 0.89 | 0.876 | 0.863 | 0.857 | 0.912 | 0.897 | 0.883 | 0.871 | 0.865 | 0.924 | 0.909 | 0.914 | 0.882 | 0.874 |
| $E(\pi_S^1)$ | 8.901 | 10.309 | 11.664 | 12.958 | 14.234 | 8.897 | 10.313 | 11.676 | 12.995 | 14.29 | 8.884 | 10.321 | 11.955 | 13.036 | 14.34 |
| $E(U(\pi_R^1))$ | −1.695 | −3.253 | −4.809 | −6.661 | −21.4 | −1.646 | −3.222 | −4.815 | −6.8326 | −26.06 | −1.558 | −3.174 | −4.963 | −7.18 | −35.243 |
| $E(\pi_C^1)$ | 7.206 | 7.0556 | 6.855 | 6.297 | −7.168 | 7.251 | 7.091 | 6.861 | 6.1624 | −11.77 | 7.326 | 7.147 | 6.992 | 5.856 | −20.903 |

Under SR mode, the supplier decides the proportion of revenue sharing and wholesale price, and the retailer decides the order quantity. As can be seen from Table 1, suppliers can adjust their decision-making behavior at any time in the face of retailers with different risk aversion. Once the supplier determines the profit-sharing ratio, the supplier will give a discount to the wholesale price of the retailer with high risk aversion coefficient, which means the higher risk aversion coefficient, the lower wholesale price set by the supplier. Although the revenue of suppliers has decreased, the sales volume of products is increasing, the revenue of the whole supply chain is also increasing, and the performance of the whole product market is flourishing. For retailers, the order quantity increases first and then decreases with the increase of wholesale price and revenue sharing ratio. The optimal order quantity is between 0.7 and 0.9. At this time, retailers can obtain the maximum expected profit, and the supply chain system can also obtain higher overall profit. Under SR mode, risk aversion is beneficial to the profits of the retailers, but not to suppliers and the whole supply chain. The profits of the suppliers increase with the increase of revenue sharing ratio, while the profits of retailers and supply chain system decrease with the increase of revenue sharing ratio.

Under RR mode, retailers decide the proportion of revenue sharing and suppliers decide the wholesale price. From Table 2, it can be seen that retailers can adjust their decision-making behavior at any time in the face of suppliers with different wholesale prices, that is, the proportion of revenue sharing and the quantity of orders. Once the supplier determines the wholesale price, retailers with high risk aversion coefficient can order more and sell more to increase their sales and thus increase their sales profits. Although the retailer's revenue has decreased, the sales volume of products is increasing, but the overall supply chain revenue is decreasing. For suppliers, the wholesale price increases with the increase of revenue-sharing ratio, but the profits of retailers and retailers decrease rapidly, which affects the profits of the whole supply chain. Finally, cooperative retailers may choose to withdraw from the market eventually, which requires suppliers to find new partners to enter the market. Under SR mode, risk aversion is beneficial to the profit of retailers and the whole supply chain, but not to the profit of suppliers. Revenue sharing ratio is not conducive to the profit of retailers and the whole supply chain, but is conducive to the profit of suppliers.

## 5 Conclusions

Under the revenue sharing contract, paper studies the issue of t with risk factors in different dominant modes. The supply chain has a single retailer and a single supplier. The retailer is risk-averse, and the manufacturer is risk-neutral. Under different mode structures with a single retailer or a single supplier is the leader, this study built different decision models of different mode structures supply chain. Under two different dominant models, we give the optimal decision-making strategies of each member in the supply chain, and further analyze how risk aversion affects the optimal decision-making strategies of each member.

It is found that under SR and RR modes, the risk aversion coefficient has different effects on the operation decisions and profits of supply chain members. Under SR mode, the higher risk aversion coefficient, the lower wholesale price set by suppliers, while the order quantity of retailers increases first and then decreases with the increase of wholesale price and revenue sharing ratio. In RR mode, once the retailer is more afraid of risk, that is, the retailer's risk aversion coefficient is larger, the supplier will set a lower wholesale price. However, the order quantity of retailers will decrease with the increase of wholesale price, and increase with the increase of revenue sharing ratio and risk aversion coefficient. For the members of supply chain and system profits, profits of the retailer increase with the increase of risk aversion, and the supplier's profits increase with the increase of revenue sharing ratio.

The limitation of our study is to consider only a supply chain structure model, a single retailer and a single supplier. In the supply chain structure, the retailer is risk-averse, and the manufacturer is risk-neutral. One important extension of this work in the near future is to consisting two competitive mode structures. We can further analyze how competition intensity and risk aversion coefficient affect the optimal decision-making and system performance of supply chain members. Another future research can consider the optimal decision-making problem of risk-averse supply chain members participating in competitive supply chain under the environment of chain-to-chain competition.

## References

1. Barnes, D. (2006). Competing supply chains are the future. *Financial Times, 11*(8).
2. Ha, A. Y., & Tong, S. L. (2008). Contracting and information sharing under supply chain competition. *Management Science, 54*(4), 701–715.
3. Liu, L., & Li, F. T. (2018). Optimal decision of deferred payment supply chain considering bilateral risk-aversion degree. *Mathematical Problems in Engineering, 5,* 1–11.
4. Li, B. X., Zhou, Y. W., & Niu, B. Z. (2013). Contract strategies in competing supply chains with risk-averse suppliers. *Mathematical Problems in Engineering, 3,* 1–12.
5. Du, W. Y., Fan, Y. B., & Yan, L. N. (2018). Pricing strategies for competitive water supply chains under different power structures: an application to the south-to-north water diversion project in China. *Sustainability, 10*(8), 1–13.
6. Peteraf, M. A., & Barney, J. B. (2003). Unraveling the resource-based tangle. *Managerial and Decision Economics, 24*(4), 309–323.
7. Cachon, G. P., & Lariviere, A. M. (1999). Capacity choice and allocation: Strategy behavior and supply chain performance. *Management Science, 45*(8), 1091–1100.
8. Hao, S. F., Liu, M. J., Ding, W., & Huang, X. Y. (2008). Research on revenue sharing contract coordination based on three-level supply chain. *Journal of Northeast University (Natural Science Edition), 29*(11), 1652–1656.
9. Song, H., Wang, R. M., & Ma, W. (2018). Research on coordination mechanism of source information sharing behavior in food supply chain based on differential game. *Journal of Huazhong Agricultural University (Social Science Edition), 3,* 144–151.

10. Nie, J. J., & Xiong, Z. K. (2006). Research on two-cycle supply chain dominated by retailers. *Statistics and Decision-Making, 13,* 156–157.
11. Zhang, S. H., Du, X. J., & Zhou, B. G. (2019). Design and optimization of competitive product channel structure based on revenue sharing contract. *Journal of Northeast University (Natural Science), 40*(3), 441–446.
12. Du, W. Y., & Fan, L. B. (2017). Revenue-sharing contract with deferred payment period under capital cost. *Mathematics Practice and Cognition, 47*(11), 9–15.
13. Liang, Y. (2016). Revenue sharing contract mechanism of supply chain with uncertain production and demand at two levels. *Practice and Recognition of Mathematics, 46*(12), 1–10.
14. Huang, Y. S., Lin, S. H., & Fang, C. C. (2017). Pricing and coordination with consideration of piracy for digital goods in supply chains. *Journal of Business Research, 77,* 30–40.
15. Zhao, Z. G., Li, X. Y., Liu, X. H., & Zhou, Y. C. (2007). Research on profit-sharing contract and its response method for supply chain coordination. *China Management Science, 15*(6), 78–85.
16. Hong, D. J., Ma, Y. K., & Tang, X. W. (2017). Supply chain coordination with revenue sharing contracts. *Operations Research and Management Science, 26*(6), 70–80.
17. Xu, G. Y., Dan, B., & Xiao, J. (2010). Study on dual channel supply chain coordination with new revenue-sharing contract. *Chinese Journal of Management Science, 18*(6), 59–64.
18. Sheng, L. L., & Wang, C. X. (2014). Revenue sharing coordination of loss-averse supply chain under cardon trading environment. *Mathematics Practice and Cognition, 44*(3), 94–102.
19. Zhao, H. X., Ai, X. Z., & Tang, X. W. (2013). Vertical alliance and profit sharing contract based on diseconomies of scale under chain-to-chain competition. *Journal of Management Sciences in China, 17*(1), 48–56.
20. Chen, Z. M., & Chen, Z. X. (2015). Coordination decision of OEM supply chain with stochastic supply and demand under risk avoidance. *Systems Engineering Theory and Practice, 35*(5), 1123–1132.
21. Lariviere, M. A., & Porteus, E. L. (2001). Selling to the newsvendor: an analysis of price-only contracts. *Manufacturing & Service Operations Management, 3*(4), 293–305.
22. Liu, X., Pan, J. M., & Tang, X. W. (2011). Perishable supply chain markdown money contract with a loss-averse retailer and a loss-averse supplier. *Journal of Industrial Engineering/ Engineering Management, 25*(3), 24–30.

# The Application of Taxis Time-Sharing Pricing Under the Influence of Sharing Economy

**Fei Wang and Jing Li**

**Abstract** The popularity of DiDi, Uber and other online car-hailing service apps has had an impact on the modes of residents' travel and has a strong influence on the taxi market. This paper builds the Agent modeling based on time-sharing pricing and studies the choice of travel decisions for complex individuals and their changes in the share of different modes of travel in transportation systems. It takes Beijing actual data as an example to analyze the sharing rate of peak and off-peak periods under the shared economic environment. The simulation results show that the taxi time-sharing pricing strategy uses price levers to achieve peak shaving and valley filling, which is conducive to improving the imbalance between supply and demand of taxis and the problem of high no-load rates during off-peak hours. In addition, it improves the utilization rate of resources and balance the sharing rate of different modes of travel. Finally, it realizes the optimal configuration of public transportation system.

**Keywords** Agent model · Sharing economy · Sharing rate · Time-sharing pricing

## 1 Introduction

As an important component of urban public transport, taxis have experienced violent periods, lucrative periods, and profit-making periods since their rise in the late 1980s. Taxi pricing has become a key issue for the urban transportation sector and passengers. With the rise of big data, the Internet and mobile payments, the information asymmetry between traditional taxi operators and taxi drivers has been exposed and taxi travel difficulties occur frequently in peak hours. In recent years, the emergence of Internet-based vehicles in the shared economy is the result of the

F. Wang (✉) · J. Li
School of Economics and Management, Beijing Jiaotong University, Beijing, China
e-mail: 18125492@bjtu.edu.cn

J. Li
e-mail: jingli@bjtu.edu.cn



J. Zhang et al. (eds.), *LISS2019*,
https://doi.org/10.1007/978-981-15-5682-1_45

combination of Internet technology and traditional transportation industries. It provides users with a variety of travel options, breaks the monopoly of the taxi market, has the advantages of affordable, comfortable and fast. Consequently, it has a huge impact and threat on the taxi industry. Therefore, optimizing the taxi pricing method from the actual supply-demand relationship is conducive to the rational allocation of resources [1], improving operational efficiency and balancing the sharing rate of the transportation system. In addition, it is of great practical significance to provide affordable and convenient travel for passengers and promote the development of the shared economy [2, 3]. This article takes the taxi price adjustment policy as the mainstay and take the car taxi software business as a supplement. Based on the analysis of the influencing factors of residents' travel modes, this paper evaluates the choice of residents' travel modes before and after the introduction of taxi price adjustment policy and taxi software business by using the method of agent multi-agent modeling and simulation [4]. We analyze the changes of the share rate of different travel modes under different policies and realistic conditions, and puts forward corresponding suggestions on taxi price adjustment policy in the process of traffic demand management to assist the government in making taxi management and operation decisions. At the same time, it can ensure the effective revenue and stable and orderly operation of the taxi market, optimize the traffic structure and maintain the healthy and sustainable development of taxi industry.

## 2 Literature Review

At present, there are many research results on taxis, However, relatively few research results on taxi pricing and travel mode selection, and most of them are analyzed from the macro level. As a typical complex adaptive system, the traditional mathematical analysis method has some limitations on the selection, prediction and sharing rate calculation of the travel mode [5–10]. The current research on the application of the multi-agent system modeling and simulation method based on complex adaptive system (CAS) theory at home and abroad shows that this method has many applications in the transportation field [11, 12], it is also applied to the analysis of residents' taxi travel options. But few people conduct research for optimizing taxi prices.

American economist Milton has widely attracted economists' attention by adjusting the price of urban taxi and controlling market failure in the theory of price [13]. Urban taxi prices are always the life issues concerned by the government and residents. Price regulation is a necessary condition for ensuring a balanced transportation system and plays an important role in the efficient operation of taxis. Taxi prices vary in different cities, and there are differences in the prices of different models. Some scholars have established a mathematical model based on the gradual policy of discounting taxi fares and maximizing the average profit of taxis in order to determine the best taxi discount rate and price breakpoints. Chauhan Li focus on government agencies, drivers and passenger tripartite stakeholders and analyzes the current situation of taxi fare system and establishes a taxi pricing model based on the incentive and restraint mechanism to improve and guide the taxi practice [14]. Chang Ying considered the quality of service on the basis of analyzing taxi operating costs and established a dynamic adjustment mechanism of freight tariff linked with cost and price, which helps to shorten the period and efficiency of price adjustment and provide the transparency of decision making [15]. With the advent of network-based vehicles, the number of scholars who study the impact on taxi industry based on the characteristics of network-based vehicles has been increased with the aim of reducing the taxi-free rate. However, there are few studies on taxi price optimization. Taking the taxi software platform as an example, Lin analyzed the feasibility of taxi dynamic pricing by using a variety of methods such as Internet of Things technology, game theory, bilateral market and queuing theory and so on [16, 17]. Xue argues that the relationship between taxi and online booking vehicles is the relationship between differentiation and integration. Both should play their respective advantages and functions to achieve coordinated distribution, positive interaction and reasonable sharing of market share, together to form the urban transport system for residents to provide convenient services [18]. In addition, some scholars established a dynamic pricing model for multi-service desk queuing system using the queuing theory to set the waiting time as a dynamic pricing variable and based on the analysis of taxi pricing methods and models, in order to obtained an optimal pricing strategy through comparison.

Online car-hailing as a new mode of transport appeared, the taxi market operations have been greatly affected [19]. In this paper, an Agent simulation model is established to analyze the sharing rates of different modes of travel, such as taxi and online booking vehicles in the traffic system. Based on the tendency of taxi sharing rate during peak hours and off-peak hours, we put forward different time-sharing pricing tactics and analyzes its influence on the taxi sharing rate so as to discover the optimal time-sharing pricing method.

## 3 Time Differential Pricing Theory

### *3.1 Taxi Time-Sharing Pricing Theory*

(1) Goal management theory. The ideal management objective of taxi market is the management subject, consumers, the harmony of market, it not only conducive to business, but also facilitate consumers and achieve an orderly win-win. Currently, the taxi market adopts the administrative fixed price management mechanism. The administrative department will make corresponding adjustments according to the changes in the market. However, it is stable and rigid within a certain period of time and cannot response to the market supply and demand in real time. Time-sharing pricing strategy can adjust the rationality and effectiveness of market operation according to supply and demand relationship.

(2) *Characteristic price theory.* Time–sharing pricing is based on different periods of time, using price leverage to alleviate crowding in peak hours [20]. The price of urban taxi is made up of starting price, basic unit price, fuel surcharge and night surcharge. On the basis of analyzing the factors that affect the price of taxis, we can adjust the pricing methods of taxis at peak and off-peak hours at the macrograph and microcosmic levels according to the actual conditions of the cities. This can play a role in cutting peaks and valleys and improve the congestion during peak hours, increase the off-peak hour taxi sharing rate and adjust the flow of different periods of time to optimize the allocation of resources.
(3) *Price sensitivity theory [21].* The passenger's expectation of the taxi price is a range. Price sensitivity is a low-cost, simple and easy pricing method. The optimal pricing threshold can be determined through investigation and analysis. The price sensitivity test model (PSM) determines the optimal price level within a certain tolerance range from the perspective of the passenger as shown in Fig. 1.

## *3.2 Taxi Time-Sharing Pricing Feasibility Analysis*

Taxi is convenient, flexible, comfortable and has the feature of point to point service. The characteristics of individual travel have consistency and otherness. The passenger flow is large during rush hour, the resident is less sensitive to price and the time value is paid more attention to [25]. In non-peak period, residents travel has a great control over time, the price has become relatively flexible demand, residents tend to choose cost-effective way to travel. Taking time-sharing pricing strategy in non-peak hours, as the preferential prices can attract more passengers to choose a taxi as a way of travel and prompting some of the peak travel passengers to non-peak hours. It's beneficial to relieve peak travel pressure, but also help to

**Fig. 1** Price sensitivity test model

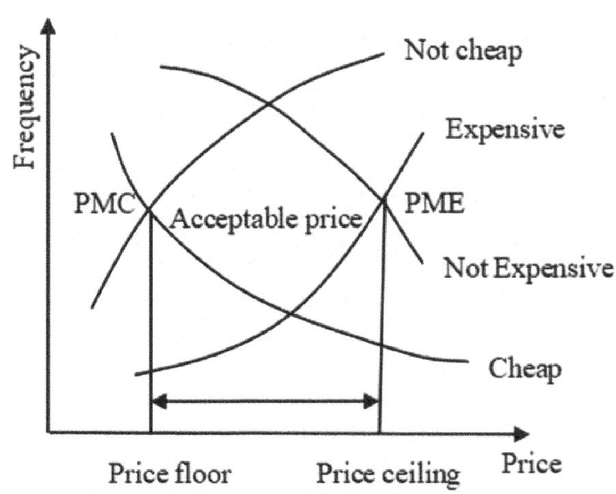

improve sharing rate of taxi. Therefore, adjust the price of taxis according to the peak hours and off-peak hours of residents, the implementation of time-sharing pricing is conducive to the rational allocation of taxis to reduce the empty rate and improve operational efficiency [26].

### *3.3 Taxi Time-Sharing Pricing Factors*

The factors that influence the taxi price in the city can be divided into two aspects: macrograph and microcosmic as well as other special factors [22–24], as shown in Fig. 2:

## 4 Construction of Multi-body Simulation Model

### *4.1 Model Parameters*

(1) *Vehicle agent generation model.* Vehicle Agent generation model is the most basic model of micro-traffic system simulation. This model generates vehicle Agent that meets the given probability distribution according to the requirements of simulation experiment, which the input of simulation traffic flow. It is assumed that the vehicle Agent is randomly generated according to the Poisson distribution in the initial location of the road network and the initial appearance time. Vehicle Agent generation model is as follows:

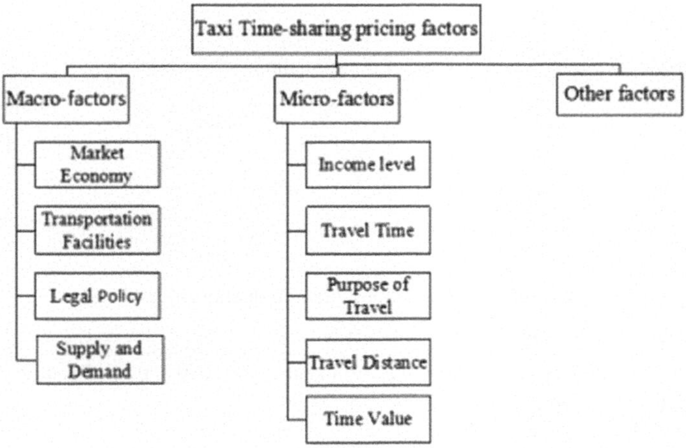


**Fig. 2** Taxi Time-sharing pricing factors

$$t_{n+1} = t_n - \frac{\ln(r_n)}{a} \tag{1}$$

In (1), $t_n$ is the time when the nth vehicle agent enters the simulation road network; a is the arrival rate of vehicle agent;$r_n$ is a random number obeying uniform distribution.

(2) *Vehicle Agent driving model.* In the microscopic traffic system simulation, the vehicle driving model can be divided into the vehicle free driving model, following the car model, overtaking model, lane change model and parking model. As we all known, subway doesn't occupy the road network resources, while other buses, taxis and private cars share road network resources, so this paper set the two-way two-lane road network. Metro uses a one-way road network alone, while others share another one-way road network in the two-way road network. In this paper, only the free driving, following and parking models of vehicles are involved in the simulation, there is no overtaking and changing behavior in the simulation. In the process of vehicle driving simulation, if the stopping vehicle is scanned within stopping line of sight (set up to two cells and one cell set), then the vehicle agent will stop at a certain deceleration speed to avoid the risk of collision.

As a result of the diversity in the speed of traffic mode will affect the travel mode decision, it is necessary to set the travel speed of different modes in the simulation system. The operating speed of various mode of traffic in the traffic system can be seen in Table 1

Subway travel time is not with the previous agent for speed following but has a specific standard; Taxis, buses, and private cars are calculated by moving the vehicle's Agent at a rate of 1 per Starlogo time (The refresh rate is average every 0.01 s in Starlogo) at Starlogo and setting the speed of each mode of transportation in Starlogo as Table 2 [27].

**Table 1** Operating various speed modes of transportation

| Transportation | Private car | Taxi | Bus | Subway |
|---|---|---|---|---|
| Speed | 25–50 | 20–50 | 16–25 | 30–40 |

**Table 2** Moving various speed modes of transportation in simulation

| Transportation | Private car | Taxi | Bus |
|---|---|---|---|
| Speed | Random 1000/1000 + 0.2 | Random 1000/1000 + 0.1 | Random 500/1000 |
| Expect speed | 1 | 1 | 0.5 |

## 4.2 Agent Traffic Mode Decision-Making Model

Passengers from the place of departure to the destination have a variety of transportation options. There are various service characteristics of different modes and the degree of satisfaction is somewhat distinct.

As in the Didi information platform, the vehicles and drivers are from legally qualified car rental companies and service companies through strict screening and training. Each special car is high-grade car with the price of more than 200,000 yuan and mainly engaged in the premium service market. It is essentially different from the traditional taxi, so the special cars are more comfortable and faster than the express cars. Given all this, in addition to billing rules, we use the characteristics of private cars to simulate the rest attributes of special cars. Meanwhile, we use characteristics of taxi to simulate the express cars so as to distinguish special and express cars. There are no travel difficulties at night, so we only studies daytime travel conditions.

Didi special cars' travel expenses:

$$z(d) = \begin{cases} 12 + 2.9 * d + 0.6 * t, d <= 15 \& T \in non - peakhours \\ 12 + 2.9 * d + 0.6 * t + 1.5 * (d - 12), 15 < d \& T \in non - peakhours \\ 12 + 2.9 * d + 0.6 * t + random(5, 20), d <= 15 \& T \in peakhours \\ 12 + 2.9 * d + 0.6 * t + random(5, 20), 15 < d \& T \in peakhours \end{cases} \tag{2}$$

Didi express cars' travel expenses:

$$\mathrm{k}(d) = \begin{cases} [10, 15 * d + 0.35 * t], d <= 12 \& T \in non - peakhours \\ [10, 15 * d + 0.35 * t + 0.8 * (d - 12)], 12 < d \& T \in non - peakhours \\ [10, 15 * d + 0.35 * t + random(5, 20)], d <= 12 \& T \in peakhours \\ [10, 15 * d + 0.35 * t + 0.8 * (d - 12) + random(5, 20)] \\ , 12 < d \& T \in peakhours \end{cases} \tag{3}$$

Taxi travel expenses:

$$\mathrm{t}(d) = \begin{cases} 13 + 1 + 2 * 2.3 * \frac{t}{5}, d <= 3 \\ 13 + 2.3 * (d - 3) + 1 + 2 * 2.3 * \frac{t}{5}, 3 < \mathrm{d} \end{cases} \tag{4}$$

Private car owners consider travel expenses not only the fuel costs incurred while driving, but also the cost of each purchase which including car purchase, vehicle insurance, and vehicle maintenance. At present, there are many studies on the travel cost of automobiles, so this paper chooses the personal payment cost of automobiles as the travel cost. Cars travel expenses [28]:

$$C(d) = 0.9 * d \tag{5}$$

Bus travel cost is less than 2 yuan within 10 km (including) and 1 yuan per 5 km increase. Bus travel expenses:

$$\mathrm{b}(d) = \begin{cases} 2, d < \; = 10 \\ 2 + \mathrm{roundup(d\ -\ 10)/5}, 10 < \mathrm{d} \end{cases} \tag{6}$$

The standard of Metro travel cost is 3 yuan within 6 km, 4 yuan for 6–12 km, 5 yuan for 12–22 km, 6 yuan for 22–32 km, and 1 yuan for every 20 km after 32 km. Subways travel express:

$$\mathrm{S}(d) = \begin{cases} 3, d < \; = 6 \\ 4, 6 < \mathrm{d} < \; = 12 \\ 5, 12 < \mathrm{d} < \; = 22 \\ 6, 22 < \mathrm{d} < \; = 32 \\ 6 + \mathrm{roundup(d} - 32)/20, 32 < \mathrm{d} \end{cases} \tag{7}$$

## *4.3 Simulation Results Analysis*

The simulation model is introduced into the Didi special car and express business to simulate the distribution rates of different types of vehicles under peak and non-peak times, now we analyze the simulation results.

(1) *Analysis of travel sharing rate in non-peak time.* With the development of service of Didi express cars and special cars, express cars are comfortable and rapid as well as taxi while travel expenses are cheaper than taxi. According to their different attributes, travel Agent prefer Didi to subway and taxi. Although special cars charge more than taxis, they offer better services than taxis. Some high-income travelers who focus on comfort and quality service will choose special cars to travel according to their needs.

Figure 3 shows the change of travel pattern distribution in different time by introducing the simulation model of special vehicle and fast travel mode. It can be seen from Fig. 3 that the change of the car travel mode before the introduction of the DIDI travel mode is not obvious. The average sharing rate of buses, subways, and taxis was 25, 18, and 3%, respectively, which was significantly lower by 5, 4, and 5% before the introduction of the express train business. Most of these passengers are transferred because the express car is cheaper than that of the taxi. Of course, some business travelers pay more attention to the comfort and special

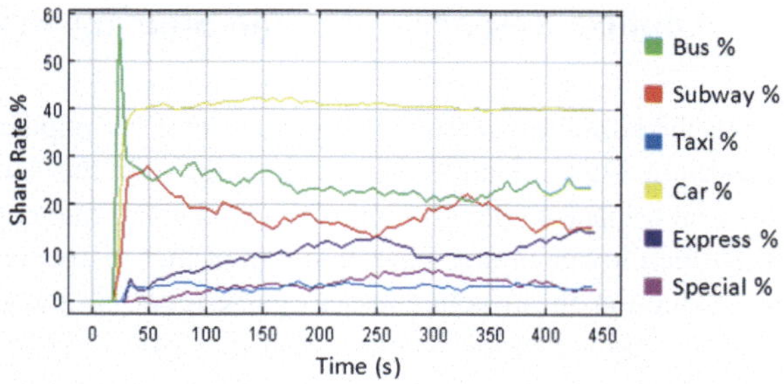


**Fig. 3** Trip sharing rate in different time with taxi-hailing apps in non-peak

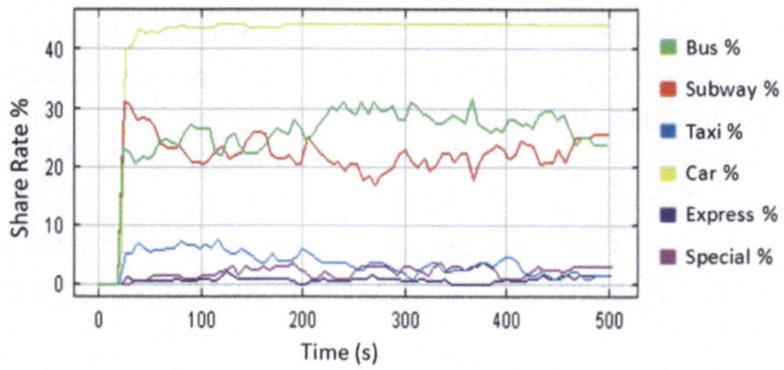


**Fig. 4** Trip sharing rate in different time with taxi-hailing apps in peak

services provided by the express car. It can be seen that the dedicated car and express service provided by the software meets the high quality, diversification and different needs of the transportation market.

(2) *Analysis of travel sharing rate in peak time.* According to the data of traffic statistics of different periods in the Beijing Statistical Yearbook, the traffic flow during peak hours is about three times of that during non-peak hours. Therefore, this paper sets the vehicle arrival rate at rush hour to 60%. In addition, the fees charged by the special cars and express cars during peak hours are also different from those charged during non-peak periods. Figure 4 shows the change of travel mode sharing rate in different periods of peak traffic period after introducing special car and express train mode into the simulation model. Figure 4 shows the change of mode sharing rate at different times on the road network during peak period after introducing special and express train modes into the simulation model. Compared with the non-peak period, the average sharing rate of bus and subway is 27 and 21%, respectively, which is about 2% higher than that of non-peak period. This phenomenon is consistent with the change of car and subway in different travel periods. Because of the traffic congestion in peak time, passengers can not reduce the efficiency of vehicle agents, so they want to choose other modes of travel. It can be seen from this that taxis account for the largest proportion among the three modes of travel in peak and non-peak periods, while express trains account for the lowest proportion in non-peak periods. Indicating that in order to alleviate difficulty of Taking taxis, express car's charging rules are more effective than taxi's charging rules.

# 5 Applied Research of Time-Sharing Pricing

From the simulation results, we can see that the popularity of taxi software business providing diversification of travel options, at the same time, it will undoubtedly have an impact on the taxi industry. Taxi share rate is the highest in peak hours, but it is much lower in non-peak hours than that of special and express cars because of the reduction in attractiveness to passengers, so the charging rules of special and express cars are more effective than taxis. In order to cope with the impact of taxi software business on the taxi industry, taxis can adopt time-sharing pricing to improve the competitiveness of taxi tourism.

According to the latest taxi price adjustment plan of the Beijing Municipal Development and Reform Commission, the starting price has been adjusted to 13 yuan for 3 km and the standard for taxi has increased to 2.3 yuan per kilometer, and the average increase of passengers was 13% each time after the rental price adjustment, with an increase of approximately 3.3 yuan. The adjustment of taxi time-sharing strategy mainly according to the starting price and unit price, the change of time-sharing rates of taxis, special cars and express cars in different periods is discussed so that we can give policy suggestions based on the simulation results.

## 5.1 *Simulation Results Analysis*

Taxi travel cost in non-peak period is higher than that of special cars and express cars, which makes the attraction of taxi agents decrease, resulting in low taxi travel share rate in non-peak period, which easily makes the taxi industry weak in non-peak period. Therefore, considering the reduction of the taxi starting price and the unit price during non-peak hours, this paper simulates the distribution ratio of residents' travel modes and analyzes the changes of the sharing rates of taxi trips (taxis, special cars and express cars) to get the related pricing strategy. Table 3 shows the results of different modes of sharing rates and specific pricing strategies.

**Table 3** Share rate under different price strategy in non-peak time

| Starting Price (Yuan) | Unit Price (Yuan/km) | Taxi | Special car | Express car | Sum |
|---|---|---|---|---|---|
| 13 | 2.3 | 3.40% | 3.60% | 9.30% | 16.30% |
| 12 | 2.2 | 4.10% | 3.10% | 9.10% | 16.30% |
| | 2.1 | 4.35% | 3.00% | 8.00% | 15.35% |
| | 2.0 | 4.60% | 2.87% | 7.80% | 15.27% |
| 11 | 2.2 | 5.84% | 2.50% | 5.80% | 14.14% |
| | 2.1 | 6.20% | 2.38% | 5.20% | 13.78% |
| | 2.0 | 6.68% | 2.20% | 4.60% | 13.48% |

From the adjustment of the starting price, when the basic unit price is 2.2 yuan/km and the starting price changes from 12 yuan to 11 yuan, the taxi share rate increases by 1.6%, while when the starting price is 12 yuan and the basic unit price changes from 2.2 to 2.1, the increase of the taxi share rate is less than 1%, which shows that the adjustment of the starting price is sensitive to the taxi share rate.

From the perspective of the three modes of travel, with the reduction of taxi starting price and basic unit price, the taxi sharing rate has an obvious upward trend in non-peak period, while the share rate of special and express cars has a downward trend. When the starting price is 11 yuan and the basic unit price is 2.2 yuan per kilometer, the taxi share rate is greater than the express share rate for the first time, but at this time the taxi share rate is still less than the sum of the taxi software car and the express car travel share rate. When the starting price is 11 yuan and the basic unit price is set at 2 yuan per kilometer, although the taxi share rate is close to the total share rate of the taxi software, the total share rate of the taxi share is lower than that of the basic unit price of 2.1 yuan per kilometer.

In terms of taxis, in order to make taxis more attractive for traveler than special car and express and prevent the capacity of taxis from becoming faintest in the non-peak period, the starting price is 11 yuan and the unit price between 2.2 to 2.1 are more appropriate which taking into consideration the operational benefits of taxis. However, at the government level, taxi attractiveness should also balance the sharing rates of various modes of travel, this means it is necessary to choose a pricing strategy that the taxi sharing rate is greater than that of the special car and express and the total sharing rate is relatively high. Therefore, it is suggested that the taxi pricing strategy with a starting price of 11 yuan and unit price of 2.2 yuan is suitable during the off-peak hours.

## 5.2 *Pricing Strategies in Peak Time*

During peak hours, the dynamic pricing strategy of the special cars and express cars makes the attractiveness of the taxi to the vehicle agent relatively higher than non-peak time during the same period of time. From the foregoing simulation results, it can be seen that the taxi share rate has indeed declined since the taxi price adjustment, which has relatively alleviated the difficulty of taxi travel. But with the rise of taxi software business, it is found that taxi pricing strategy is far less effective than that of special cars and express cars in alleviating the difficulty of taxi travel in peak time.

We analyze the sharing rate by simulating the distribution rate of residents' travel modes, considering reducing the starting price and the unit price of taxi during peak hours. Table 4 shows the results of different modes of sharing rates and specific pricing strategies.

As can be seen from Table 4, with the increase of taxi starting price and unit price, the attractiveness of taxis to travelers decreases during peak time and the taxi sharing rate gradually declines while the share of special cars and express cars still

**Table 4** Share rate under different price strategy in peak time

| Starting Price (Yuan) | Unit Price (Yuan/km) | Taxi | Special car | Express car | Sum |
|---|---|---|---|---|---|
| 12 | 2.3 | 4.80% | 1.60% | 0.80% | 7.20% |
| | 2.4 | 4.76% | 1.52% | 0.83% | 7.11% |
| | 2.5 | 4.65% | 1.55% | 0.79% | 6.99% |
| 11 | 2.3 | 4.58% | 1.53% | 0.80% | 6.91% |
| | 2.4 | 4.54% | 1.56% | 0.83% | 6.93% |
| | 2.5 | 4.48% | 1.49% | 0.76% | 6.73% |

remains basically unchanged. From the trend of simulation results of taxi price increase, when the price reaches a certain value, the taxi share ratio will be equal to the total share ratio of the special car and the express car, but the total share ratio of the taxi travel mode (taxi, special car, express car) will decrease, thus increasing the travel pressure of other modes (bus, subway and car). Therefore, from the perspective of the government and the taxi industry, in order to balance the sharing rate of different modes during the peak period and prevent the travel pressure of other modes, it is more appropriate for the taxi to maintain the current pricing strategy.

## 6 Conclusion

Because of the different arrival of vehicles, the taxi time-sharing price should be based on comprehensive analysis of various factors, and then choose the best solution. In this paper, agent simulation modeling method is used to compare the travel rate of different travel modes in different time periods and compare the change of sharing rate under different pricing conditions. After analyzing the simulation results, the following conclusions are drawn: In order to balance the sharing rate of different trip model, it is suggested that taxis maintain the pricing strategy of starting price of 13 yuan and basic unit price of 2.3 yuan/km. so as to maximize the sharing rate of different modes of travel and balance the sharing rate during peak hours. At the same time, in order to make taxis more attractive than special cars and express cars, it is suggested to set a starting price of 11 yuan and a basic unit price of 2.2 yuan in off-peak periods.

In summary, this paper makes a profound analysis on the choice of residents' travel modes. Using multi-agent modeling and simulation method, we compare the change of sharing rate by simulating the choices of residents' travel modes and analyses the effect of taxi-related traffic demand management policy to provide scientific support for improving the level of urban taxi management. At the same time, the multi-agent modeling and simulation method is established to analyze the model of residents' travel choices and it is conducive to improving the analysis method of residents' travel mode choice and providing a feasible method support for residents' travel mode choice analysis. According to the model and calculation of taxi time-sharing pricing in this paper, we can also optimize the current taxi pricing mode in different time periods, adjust the distribution of traffic flow in different time periods, adapt to the shared economic environment under the condition of the relative balance of the sharing rate of different modes of travel, and ensure the effective revenue and stable and orderly operation of the taxi market.

## References


1. Biao, L. X. (2013). Analysis of the price elasticity of taxi demand in Beijing—based on the annual data from 2001 to 2011. *Transport Accounting, 48,* 69–72.
2. Liu, T.T. (2016). Research on forecast of city taxi demand under intelligent travel platform. Jilin University.
3. Xing, D.C. (2016). China's smart travel 2015 big data report (vol. 1, no. 1, p. 1). Beijing: Drip travel.
4. Ping, P. X., & Ming, T. Y. (2007). Significance of complex system simulation for the research and innovation of complex systems. *China Association of Science and Technology, 9,* 2607–2624.
5. Doglasg, W. (1972). Price regulation and optimal service Standards: the taxicab industry. *Journal of Transport Economics and Policy, 6*(2), 116–127.
6. Schaller, B. (2007). Entry controls in taxi regulation: Implications of US and Canadian experience for taxi regulation and deregulation. *Transport Policy, 14*(6), 490–506.
7. Joo, K. Y., & Hark, H. W. (2008). Incremental discount policy for taxi fare with price-sensitive demand. *International Journal of Production Economics, 112*(2), 895–902.
8. Hai, Y., Fung, C., & Wong, K. (2010). Nonlinear pricing of taxi services. *Transportation Research Part A-Policy and Practice, 44*(5), 337–348.
9. Hai, Y., & Cowina, L. (2010). Equilibria of bilateral taxi-customer searching and meeting on networks. *Transportation Research PART B-Methodological, 44*(8), 1067–1083.
10. Na, L. L., Yan, C. Y., & Ge, Z. W. (2010). Research on forecasting model of taxi passenger demand in Beijing. *Communications Standardization, 13,* 89–92.
11. Ping, L. X., Ming, T. Y., & Ping, Z. L. (2008). A review of the simulation of complex systems. *Journal of System Simulation, 20*(23), 6303–6315.
12. Jun, N. J., & Zhong, X. L. (2006). Research progress of multi-agent modeling and simulation method based on CAS theory. *Computer Engineering and Science, 28*(5), 83–97.
13. Friedman, M. (1976). Price theory. Price theory
14. Heng, L. C. (2015). Study on taxi pricing in Guangzhou from the perspective of stakeholder theory. *Journal of Guangdong Economics, 7,* 72–75.
15. Ying, C. (2016). Study on pricing mechanism of taxis in taxi network. *Price Theory and Practice, 10,* 98–101.
16. Fei, L.Y.: Research on the dynamic pricing of internet of things. Tianjin University, vol. 1, pp. 1 (2016).
17. Xiong, Y. H., & Wei, B. (2016). Research on the competition game between internet vehicles and taxi: Taking the platform subsidy as the background. *Beijing Social Sciences, 5,* 68–76.
18. Long, X. W. (2017). Law problems and solutions to the development of network vehicles - from the perspective of shunt and aggregate development. *Journalism, 5,* 111–112.
19. Tong, J.: 2016 Beijing Transportation Development Annual Report, vol. 1, no. 1, pp. 1. Beijing Transportation Development Research Institute, Beijing (2016).
20. Ting, H. Z. (2016). Research on time-share pricing of Guangzhou Metro. *Juan Zong, 11,* 279–280.

21. Joaquin, J. D., & Enrique, M. J. (2006). A diagrammatic analysis of the market for cruising taxis. *Transportation Research Part E, 42*(6), 498–526.
22. Yi, W., Xi, C. J., & Wu, C. (2013). Research on urban taxi reasonable capacity scale. *Applied Mechanics and Materials, 29*(2), 194–198.
23. Sen, H. S., Rui, S., & Tao, Y. (2018). A study on selecting behavior and influencing factors of residents' travel modes in large cities——taking Beijing as an example. *Communications Standardization, 35*(1), 142–148.
24. Min, L. S., & Zhuang, A. S. (2011). Analysis of influencing factors of travel time value based on disaggregate mode. *Traffic Standardization, 14,* 136–138.
25. Guri, D. F. (2003). An economic analysis of regulated taxicab markets. *Review of Industrial Organization, 23*(3), 255–266.
26. Chang, T. (2004). Optimal taxi market control operated with a flexible initial fare policy, *2*(2), 1335–1340.
27. Xia, Y. W., & Jia, W. X. (2006). Application of starlogo in complex system modeling and simulation based on agent. *Journal of Wuhan University, 14,* 136–138.
28. Guo, M. (2007). Evidential reasoning-based preference programming for multiple attribute decision analysis under uncertainty. *European Journal of Operational Research, 182*(3), 1294–1312.

# The Effect of Cross-sales and Retail Competition in Omni-Channel Environment

**Jian Liu, Junxia He, and Shulei Sun**

**Abstract** This paper studies the equilibrium price and service level of both the e-retailer (she) and omni-channel retailer (he) in a competitive market. We find that the optimal price and service level in Bertrand-Nash game is always higher than that in Stackelberg game and both the acceptance of online channel product and cross-sales effect are key factors that influence the equilibrium results. The total profit of the e-retailer always precedes the omni-channel retailer's, cross-sales effect exists between online-offline channels of the omni-channel retailer helps improve his profit.

**Keywords** Competitive retailers · BOPS strategy · Cross-sales effect · Game theory

## 1 Introduction

The last few decades have seen a rapid growth of e-commerce, resulting in a huge revolution of consumer habits due to the convenience of online shopping pattern. As consumers get accustomed to online shopping, more and more brick-and-motor retailers attempted to develop their own online-channels as a supplementary to

This paper is supported by the National Natural Science Foundation of China (Nos. 71501060, 71502049), and Natural Science Foundation of the Higher Education Institutions of Jiangsu Province (No. 17KJB120006) and the Postgraduate Research & Practice Innovation Program of Jiangsu Provience (No. KYCX19_1432).

J. Liu (✉) · J. He · S. Sun
School of Management Science and Engineering, Nanjing University of Finance and Economics, Nanjing, China
e-mail: liujane1124@126.com

J. He
e-mail: sunshine_he1228@163.com

S. Sun
e-mail: sunshulei@njue.edu.cn



J. Zhang et al. (eds.), *LISS2019*,
https://doi.org/10.1007/978-981-15-5682-1_46

enhance their competitiveness, and thus have been called as dual-channel retailers. Dual-channel retailers can enhance customer satisfaction and ultimately consumer loyalty [1]. A large proportion of brick-and motor retailers in China have established online sales channels, such as Suning, Gome, Metro, etc. However, the dual-channel retailers are still in a weak position in the e-commerce market [2]. Huge competitive pressure from pure e-retailers prompts dual-channel retailers to adopt different integration strategies to win as much as market share by providing seamless experience to consumers, which makes them have been called as omni-channel retailers. At the same time, consumers are no longer satisfied with single channel to meet their demands, and this induces a wide variety of channel synergy strategy alternatives, and retailers are continuously trying kinds of integration strategies to keep pace. As a result, many channel integration strategies emerge, such as buy-online pickup-at-store (BOPS), buy-online ship-to-store (BOSS), and buy-online ship from-store (BOFS) [3], among which BOPS is the most important and widely adopted omni-channel strategy [4]. Large retailers, such as Walmart, Best Buy, and Gap all adopted BOPS strategy [5]. Reasonably, BOPS implementation becomes our specific study with respect to the synergy strategy of the competitive retailers.

BOPS implementation benefits omni-channel retailers in different aspects. On the one hand, it reduces the delivery cost of online product due to consumers' offline picking-up by themselves; on the other hand, it brings additional customer foot traffic as well, which has been mentioned in Gallino and Moreno [6]. However, BOPS implementation is not always an optimal strategy for omni-channel retailers because of high operation cost and consumers' established shopping habit, resulting in an unexpected contribute margin. Therefore, it is necessary for retailers to trade-off between a relatively higher cooperation cost and overall sales. Actually, while online and offline channels compete with each other, there also exists cross-sales or additional profits along with both online and offline demands. What is interesting is that such profits all come from each channel's competitor. This is because behaviors such as "research online purchase offline" or "research offline purchase online" are very common. With the consumers' increasing requirement for products, competitors also pay attention to providing various services to earn competitive advantage. For example, in 2018, Walmart launched personal shopping guidance service to compete with Amazon.com [7]. In real practice, some omni-channel retailers are still in a relative weak position, either in terms of market share or consumer familiarity. On the contrary, e-retailers, as the leader of the market, need also take measures to cater to the market in order to maintain their competitive position. Therefore, the optimal operation strategy for both e-retailers and omni-channel retailers under different market power structure becomes continuous research and of great practical significance.

Motivated by the above observation, we set out to study the effect of cross-sales and retail competition on the optimal pricing and service strategies of both the e-retailer and the omni-channel retailer. Specifically, we are endeavored to answer the following questions:

(1) What are the optimal pricing and service level for each retailer under different game scenarios?
(2) What is equilibrium profit brought by optimal pricing and service level strategies due to different game sequences?

To this end, we consider a market consisting of an omni-channel retailer and an e-retailer who sell substitute product at different price with different types of service. In accordance with the real practice, the omni-channel retailer adopts "same product same price" strategy. To study the specific competition case between the two retailers, we propose two different game scenarios: Bertrand-Nash game and Stackelberg game. In Bertrand-Nash game, the e-retailer and the omni-channel retailer decide their price and service level simultaneously. In Stackelberg game, the e-retailer, as the leader of the game, decides price and service level firstly, followed by the price and service decision of the omni-channel retailer. By formulating different game models and changing the two retailers' decision sequence, we draw different results which can well elaborate how different game sequences influence the optimal pricing and service level strategies. We find that both the acceptance of online channel product and cross-sales effect affect the retailers' equilibrium pricing and service level decisions. Profits of the e-retailer always precede the omni-channel retailer either in Bertrand-Nash game or in Stackelberg game. For each retailer, the equilibrium pricing and service level decisions in Bertrand-Nash game are always higher than that in Stackelberg game due to different game sequences.

Major contributions of this paper are the consideration of cross-sales effect due to BOPS functionality adopted by the omni-channel retailer in competitive market and examining the effect of different game sequences on the equilibrium decisions. To the best of our knowledge, this is the first study considering both cross-sales effect and interactive game behavior simultaneously in competitive environment where pricing and service level are two crucial decision variables.

## 2 Literature Review

Related research of this paper mainly focuses on the BOPS implementation and some other studies. [6] analyzes the impact of BOPS project by performing an empirical research. They show that the BOPS project is associated with a reduction in online sales and an increase in store sales and traffic, which can be explained as the joint result of cross-selling effect and channel-shift effect. By building a stylized model, [8] also study the impact of the BOPS initiative on store operations and consumer choices. They find that BOPS strategy attracts consumers by providing inventory availability information and reducing the hassle cost of shopping process. However, in their model they don't consider different channel preferences of consumers and heterogeneous shopping costs due to distance limits, and this is at odds with reality [9]. From the perspective of inventory, [3] introduces the differences among several omni-channel strategies and develop a stylized model to investigate the impact of such omni-channel strategies on store operations. They divide consumers into two categories and indicate that BOPS is a double-edged sword for the retailer, depending on the retailer's operational capabilities. [10] considers three information mechanisms: physical showrooms, virtual showrooms, and availability information, and find that such information mechanisms are generally profitable to retailers by reducing consumer uncertainty about product value and inventory availability. What needs to be pointed out is that their model is based on a predetermined price and their focus is retailer's inventory decision. [2] studies the BOPS strategy for a dual-channel retailer under both a monopoly case and a competitive case with an e-retailer. They show that BOPS is not always an optimal strategy, especially under a monopoly case, where BOPS strategy decreases the retailer's market share as well as his profit. They also indicate that only when the operational cost is low and the degree of customer acceptance of the online channel is low should the dual-channel retailer adopt the BOPS strategy, even though it is in a competitive case. The above literature investigates the BOPS functionality and its impacts from different aspects, and further study the optimal conditions for BOPS implementation. But they all don't consider the impact of BOPS implementation under competitive situation, much less the case where the relative competition position of both retailers is different. And those will be discussed in our paper. [11] compares the price and service level of dual-channel retailers before and after BOPS implementation from the perspective of price competition and consumer purchase behavior between online and offline channels. They consider additional consuming and indicate that the increase of additional sales is always profitable for dual-channel retailers and such benefit is more obvious in the presence of BOPS implementation. They show that within limited additional consumption, channel competition lead to a better performance under dual-price strategy than uniform pricing strategy of BOPS setting. Especially when more customers are inclined to BOPS mode, such strategy will be more profitable for all supply chain members. The above literatures also study the optimal conditions for BOPS adopting, different from the previous research, such conditions are concluded based on consumer's channel preference and service requirements, as well as the service capability of retailers themselves. Competition case involved in the above literature only refers to the selling channels between manufacturer and retailer. Different from their work, we study the competition between different retailers. [12] sets up a profit function for the retailer who is a follower in the Stackelberg game dominated by the advantaged manufacturer and show that only when the offline operation cost is within certain range and online sales reach certain scale will it be beneficial for the traditional retailer to build an additional online channel. [13] compares a BOPS model with traditional dual-channel model in both centralized and decentralized contexts. Their work focuses on analyzing whether it is always beneficial for companies to implement BOPS strategy on the basis of dual-channel, and their conclusion is that the size of BOPS-consumer and consumers' service sensitivity degree are influential factors to some extent. Similarly, the competition case mentioned above is also confined to retailers and manufacturers and their concentration is on the supply chain coordination. Therefore, from the perspective of competitive game, our work is specifically different and of good practical significance due to the consideration of cross-sales effect which has also been discussed in many related literatures.

One work involves the factors that influence consumers' channel choice. Reference [14] examines the influential factors of consumer channel selection in an omni-channel retail setting via empirical analysis. They show that channel transparency, channel convenience, and channel uniformity positively influence customer perceived behavioral control, further, customer perceived behavioral control and channel price advantage positively affect customer channel selection intention in the omni-channel retail environment. From the perspective of information, inventory availability, viewed as some information signal that can reduce product uncertainty, has been discussed in [10]. While in our paper, consumers make purchase decision by comparing different price and service level, they also consider travel cost to offline stores for picking up their ordered items.

## 3 Problem Statement and Model

We consider a market consisting of an omni-channel retailer with BOPS strategy and an e-retailer who sell substitute product at different price with different types of service. We assume that the dual-channel retailer adopt the "same product same price" strategy at price $p_B$, and the e-retailer sets the price $p_0$. We use $e_0$ and $e_B$ to represent the service level of the e-retailer and the omni-channel retailer, respectively. And $c_0$ and $c_B$ to represent their respective cost. According to Li et al. [15], we assume that $c_o = \frac{1}{2}k_1 e_0^2$ and $c_B = \frac{1}{2}k_2 e_B^2$, where $k_1$ and $k_2$ are positive service cost coefficients and refers to the cost differences in service effort. This is also adopted in Tsay and Agrawal [16]. Meanwhile, we assume that consumers are heterogeneous in product valuation $v$, and $v$ follows uniform distribution in $[0, 1]$. At the same time, when browsing goods online, consumers are more inclined to assign $\theta v$ for the same product. As is explained by Chiang et al. [17], $\theta$ represents the degree of consumer acceptance for the online channel. Due to product uncertainty along with a virtual experience, $0 < \theta < 1$ is assumed. Therefore, the consumer's utilities of buying product from the e-retailer and from the omni-channel retailer with BOPS are $u_0 = \theta v - p_0 + e_0$ and $u_B = v - p_B + e_B - t$, respectively. Where $t$ is consumer's travel cost when fetching the product in brick-and-motor store. When $u_0 = u_B$, i.e. $\theta v - p_0 + e_0 = v - p_B + e_B - t$, there is no difference for consumers to choose the product between the two retailers, so we have $\overline{v} = \frac{p_B - e_B + t - p_0 + e_0}{1-\theta}$ as the threshold of valuation. Accordingly, the demand for the e-retailer is:

$$D_o = \frac{p_B - e_B + t - p_o + e_o}{1-\theta} - \frac{1}{\theta}(p_o - e_o) = \frac{1}{\theta(1-\theta)}[\theta(p_B - e_B + t) - (p_o - e_o)] \tag{1}$$

The demand for the dual-channel retailer is:

$$D_B = 1 - \frac{p_B - e_B + t - p_o + e_o}{1-\theta} \tag{2}$$

In addition, $r$ is the additional unit profit coming along with the cross-sales. To be simplified, we assume that cross-sales only happens within the omni-channel retailing.

Therefore, the profit for the e-retailer and the omni-channel retailer is:

$$\Pi_o = p_o D_o - \frac{1}{2}k_1 e_o^2 \tag{3}$$

$$\Pi_B = (p_B + r)D_B - \frac{1}{2}k_2 e_B^2 \tag{4}$$

To study the specific competition case between the two retailers, we propose different game scenarios: Bertrand-Nash game and Stackelberg game.

## 4 Equilibrium Results Under Different Game Sequences

In this section, we analyze the equilibrium results of the Bertrand-Nash game and Stackelberg game respectively.

### 4.1 The Equilibrium Results in Nash Game

In Bertrand-Nash game, the e-retailer and the omni-channel retailer decide their price and service level simultaneously to maximize their profits. We assume that service decisions of both retailers are made on the basis of pricing. That is to say, for each single retailer, his (her) service level decision follows the corresponding pricing strategy. Therefore, we can obtain the optimal service level with backward induction:

$$e_0^{N*} = \frac{p_0^N}{k_1\theta - k_1\theta^2} \tag{5}$$

$$e_B^{N*} = \frac{p_B^N + r}{k_2 - k_2\theta} \tag{6}$$

Substituting (5) and (6) into (3) and (4) respectively, we have:

$$\Pi_0 = -\frac{k_1 p_0^2}{2\left(k_1\theta - k_1\theta^2\right)^2} - \frac{p_0\left(\frac{p_0}{k_1\theta - k_1\theta^2} - p_0 + \theta(p_0 - p_B + t)\right)}{\theta(\theta - 1)} \tag{7}$$

$$\Pi_B = (p_B + r)\left(\frac{p_B - p_0 + t - \frac{p_B + r}{k_2 - k_2\theta} + \frac{p_0}{k_1\theta - k_1\theta^2}}{\theta - 1} + 1\right) - \frac{k_2(p_B + r)^2}{2(k_2 - k_2\theta)^2} \tag{8}$$

From (7) and (8), we can get $\frac{\partial^2 \Pi_0}{\partial p_0^2} < 0$ and $\frac{\partial^2 \Pi_B}{\partial p_B^2} < 0$, so it can be proved that there exists the optimal retail price $p_0^{B*}$ and $p_B^{B*}$ by setting $\frac{\partial \Pi_0}{\partial p_o} = 0$ and $\frac{\partial \Pi_B}{\partial p_B} = 0$.

Finally, we can obtain the unique optimal price as follows:

$$p_0^{B*} = \frac{k_1\theta^2(\theta - 1)a}{b} \tag{9}$$

$$p_B^{B*} = \frac{c + d + l}{b} \tag{10}$$

Wherein

$$a = k_2 + r + t - k_2 r - 3k_2 t - 2k_2\theta + k_2\theta^2 + k_2 r\theta + 3k_2 t\theta$$

$$\begin{aligned} b = {} & k_2\theta - 2k_1\theta - 2k_2 + 4k_1\theta^2 - 2k_1\theta^3 + k_2\theta^2 - 11k_1k_2\theta^2 + 10k_1k_2\theta^3 \\ & - 3k_1k_2\theta^4 + 4k_1k_2\theta + 1 \end{aligned}$$

$$c = k_2 r - r - k_2 + k_2 t + 2k_2\theta - k_2\theta^2 - 8k_1k_2\theta^2 + 12k_1k_2\theta^3 - 8k_1k_2\theta^4 + 2k_1k_2\theta^5$$

$$d = -4k_1 r\theta^2 + 2k_1 r\theta^3 + k_2 t\theta^2 + 2k_1k_2\theta + 2k_1 r\theta - k_2 r\theta - 2k_2 t\theta$$

$$\begin{aligned} l = {} & -2k_1k_2 r\theta - 2k_1k_2 t\theta + 6k_1k_2 r\theta^2 - 6k_1k_2 r\theta^3 + 2k_1k_2 r\theta^4 + 7k_1k_2 t\theta^2 \\ & - 8k_1k_2 t\theta^3 + 3k_1k_2 t\theta^4 \end{aligned}$$

Substituting (9) and (10) into (5) and (6) respectively, we have:

$$e_0^{B*} = \frac{k_1\theta^2(\theta - 1)a}{\left(k_1\theta - k_1\theta^2\right)b} \tag{11}$$

$$e_B^{B*} = \frac{rb + c + d + l}{bk_2(1 - \theta)} \tag{12}$$

## 4.2 The Equilibrium Results in Stackelberg Game

In Stackelberg game, without loss of generality, we assume that the e-retailer is the leader of the game, who decides pricing and service level firstly, followed by the pricing and service decision of the omni-channel retailer. Similarly, we have:

$$e_0^{S*} = \frac{p_0^S}{k_1\theta - k_1\theta^2} \tag{13}$$

$$e_B^{S*} = \frac{p_B^S + r}{k_2 - k_2\theta} \tag{14}$$

Under this game sequence, we first set $\frac{\partial \Pi_B}{\partial p_B} = 0$ with backward reduction and get:

$$p_B^S = \frac{k_2^2(1-\theta)[k_1\theta(1-\theta)-1]p_0^S - tk_1k_2^2\theta(1-\theta)^2 - k_1k_2^2\theta^2(1-\theta) + 3k_1k_2r\theta - k_1k_2r\theta^2}{k_1k_2\theta[3-2k_2(1-\theta)]} \tag{15}$$

Substituting (15) into (7) and set $\frac{\partial \Pi_0}{\partial p_o} = 0$, we get:

$$p_0^{S*} = \frac{A^2\theta\varsigma}{2A^2 - A + 2A^2\alpha\theta} \tag{16}$$

Thus:

$$e_0^{S*} = \frac{A\theta\varsigma}{2A^2 - 2A + 2A^2\alpha\theta + k_1\theta - k_1\theta^2} \tag{17}$$

And:

$$p_B^{S*} = \frac{AB^2t + AB^2\theta + AB^2\omega + AB^2r - AB^2 - 2ABr - B^2\omega - Ak_2r\theta + Ak_2r}{AB - 2AB^2} \tag{18}$$

$$e_B^{S*} = \frac{AB^2t + AB^2\theta + AB^2\omega + AB^2r - AB^2 - 2ABr - B^2\omega - Ak_2r\theta + Ak_2r + r}{(AB - 2AB^2)B} \tag{19}$$

Wherein

$$B = k_2 - k_2\theta$$

$$\alpha = \frac{(1-A)B^2}{AB - 2AB^2} - 1$$

$$\varsigma = t - \frac{(t+r+\theta-1)B^2 - 2rB - k_2 r\theta + k_2 r}{B - 2B^2}$$

$$\omega = \frac{A^2\theta\varsigma}{k_1\theta(\theta-1) - 2A^2\theta\alpha + 2A - 2A^3}$$

## 4.3 Analysis of the Equilibrium Results

From the above results, we can preliminarily get the following observations:

- The service level is influenced by the corresponding cost efficient $k_i$ and a larger $k_i$ means lower service level. This is intuitively reasonable because larger $k_i$ implies heavy operation cost that retailers must trade off prudently. For the omni-channel retailer, cross-sales can also contribute to a positive effect for higher service level.
- In Bertrand-Nash game and Stackelberg game, both the optimal retail price and service level of retailers are influenced by $r$ and $\theta$.
- From the equilibrium pricing and service level of the two game we can easily conclude that $\frac{\partial p}{\partial t} > 0$ and $\frac{\partial e}{\partial t} > 0$ can always be true for both retailers, which means that as consumers' travel cost increases retailers would also increase their pricing as well as service level.

## 4.4 Numerical Studies

Since the analytical solutions are too complicated to get visualized observations, we conduct numerical examples to further illustrate our analytic findings when parameters' range making sense. Within different analysis, we set $k_1 = k_2 = 100$, $t = 0.4$, $r = 0.1$ and $\theta = 0.8$ respectively (see [5, 18, 19]).

(1) *The impacts of $\theta$ on the equilibrium service level and price:* From Fig. 1 we can find that for both of the retailers, the equilibrium service level in Bertrand-Nash game is always higher than that in Stackelberg game. And when $\theta$ is small, the service level of the dual-channel retailer is better than the e-retailer. Once $\theta$ increases such situation is just the opposite. This implies that when the acceptance for online channel is low, the dual-channel retailer is more willing to improve his service. This is because extra cost along with higher service level can well offset by more demands coming from offline channel. However, when consumer preference changes and the acceptance of online channel get promoted, the service provided by the e-retailer also will substantially get improved and precede the dual-channel retailer with larger margin. From Fig. 2, we observe as $\theta$ increases, the e-retailer keeps improving her pricing

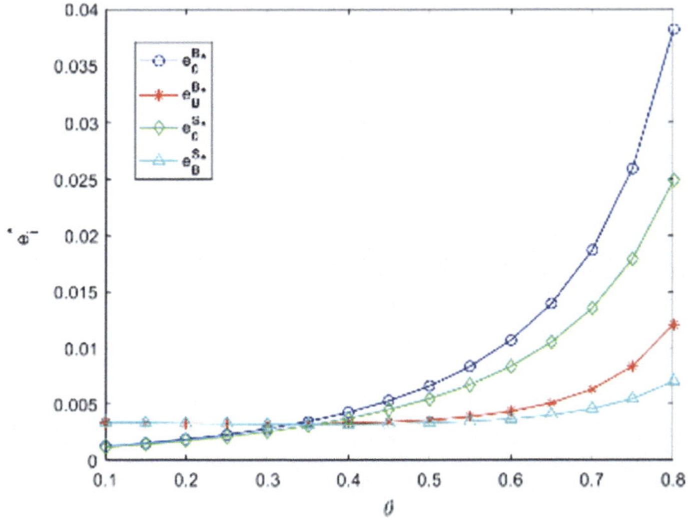


**Fig. 1** The impact of online product acceptance on service

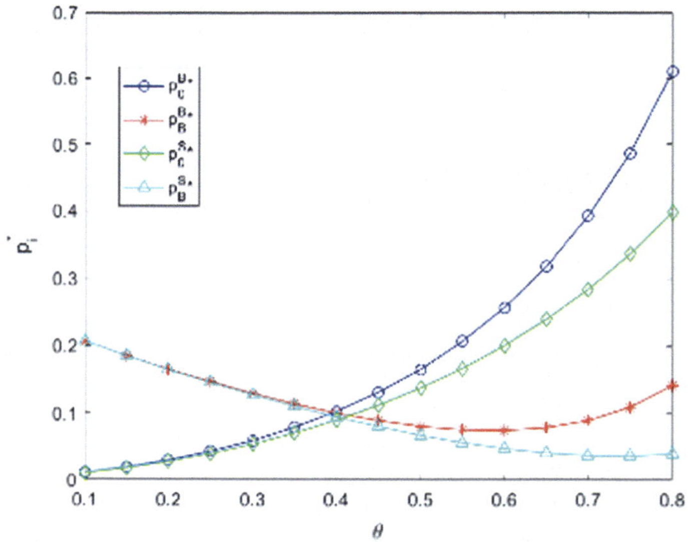


**Fig. 2** The impact of online product acceptance on pricing

while the dual-channel retailer keeps lowering his pricing either in Bertrand-Nash game or Stackelberg game. Similar with the above observation, both of the retailers' equilibrium pricing in Bertrand-Nash game is always higher than that in Stackelberg game. For the e-retailer, the dominant position in Stackelberg game allows her to set a higher price. While for the dual-channel retailer, a relative weak situation prompts him to improve his service level to compete with the e-retailer and thus sets a higher price as well.

(2) *The impact of r on the equilibrium service level and price:* From Fig. 3, we can intuitively get the observation that as $r$ increases within a reasonable range, the optimal service level of two retailers increase as well. This means that increasing cross-sales can not only make the dual-channel retailer improve his service, it can also stimulate the e-retailer to provide better service. Actually, the consistency between cross-sales effect and service level also reflects that the higher service level makes the cross-sales possible. We then study the impact of $r$ on the equilibrium price. Different from the above observation, the optimal price for the dual-channel retailer decreases as $r$ increases. While the e-retailer still improves her pricing as $r$ increases. As we have explained before, the more cross-sales happens, the more demand of the dual-channel generates. This means that the dual-channel retailer may adopt a quick buck strategy to earn stronger position in the competitive market. We also find that when the e-retailer dominates the market, her price will decrease at a larger scale, compared with that in the Bertrand-Nash game. However, her price is still higher than the dual-channel retailer. This is because her service level is also at a higher level compared to the dual-channel retailer, which means higher operational cost (Fig. 4).

(3) *The impact comparison of equilibrium profits for both retailers:* Figs. 5 and 6 show the equilibrium profits in Bertrand-Nash game and Stackelberg game, respectively. From which we can find that when the acceptance of online channel product is low, the profit of the dual-channel retailer is higher than the

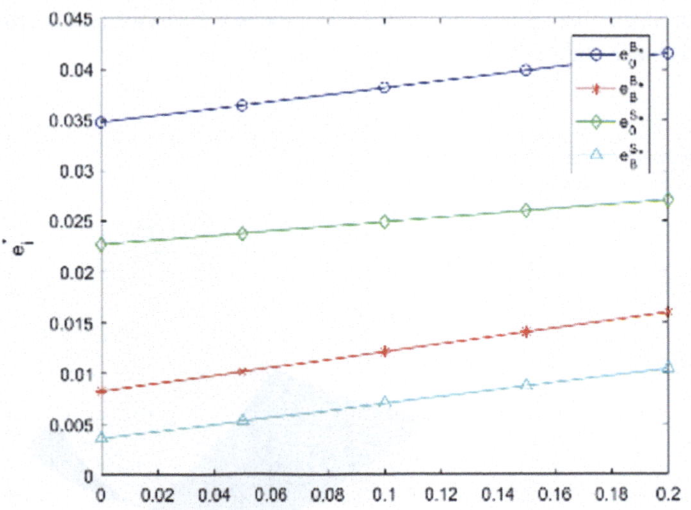


**Fig. 3** The impact of cross-sales effect on service

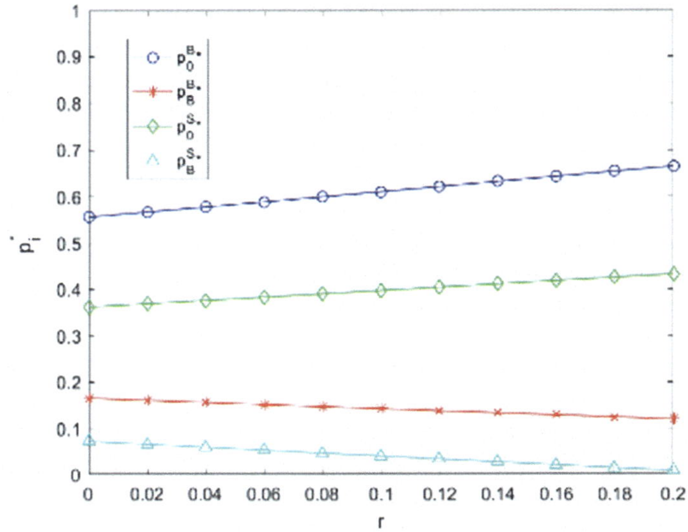


**Fig. 4** The impact of cross-sales effect on pricing

e-retailer. Yet once such acceptance increases, the profit of the e-retailer will soon surpass the dual-channel retailer and increase in a rapid growth. This is in accordance with the real practice that e-retailers absolutely have the superior position in a market that online shopping has overwhelming advantage. Figures 5 and 6 also imply that in Bertrand-Nash game, cross-sales has no obvious effect on the e-retailer's profit, while in Stackelberg game, cross-sales has positive influence on her profit. This is also in line with our above analysis that in Stackelberg game the e-retailer improves her service and keeps the price at reasonable range with first move advantage.

**Fig. 5** The profit comparison between the two retailers

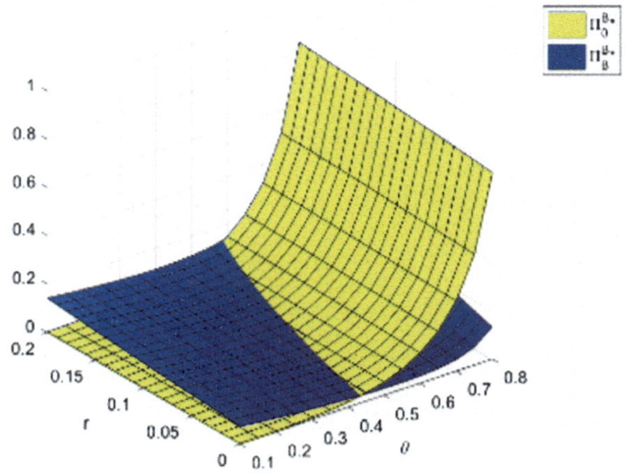

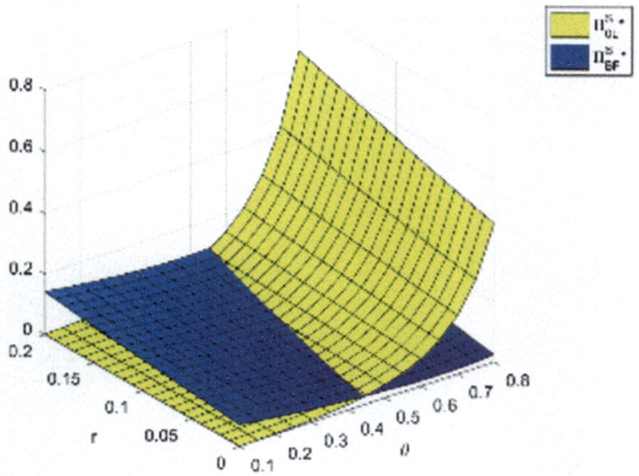


**Fig. 6** The profit comparison between the two game sequences

(4) *The comparison of equilibrium profits under different game sequences:* We investigate the two retailers' equilibrium profits under different game sequences. From Figs. 7 and 8, we can conclude that both of the retailers' profits in Bertrand-Nash game are higher than that in Stackelberg game. This seems anti-intuitive for the e-retailer because we used to accept that a dominant position in the market always means more profit. While the fact is that to maintain competitive advantage, the e-retailer must keep her price at relatively low stage to gain larger market share. Figure 8 also shows that cross-sales contribute profit margins to the dual-channel retailer, especially when the acceptance of online channel is high, such contribution can be obviously observed. Considering the above analysis of the impacts of $\theta$ and $r$ on the dual-channel retailer's pricing and service level, we can conclude when the acceptance of online channel is high enough, the dual-channel retailer can still gain larger profit even when the operation cost gets higher due to service

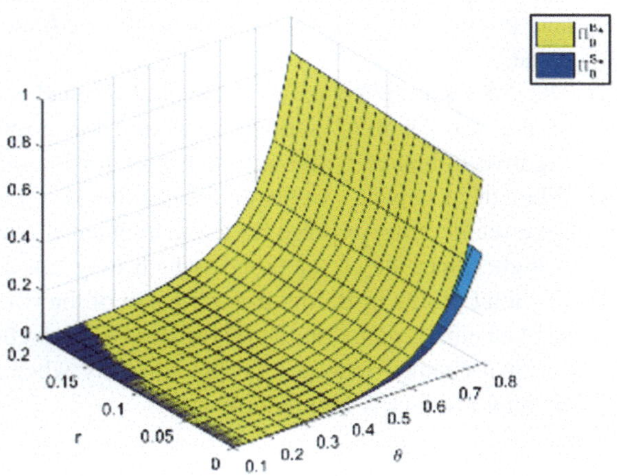


**Fig. 7** The e-retailer's profits under the two game sequences

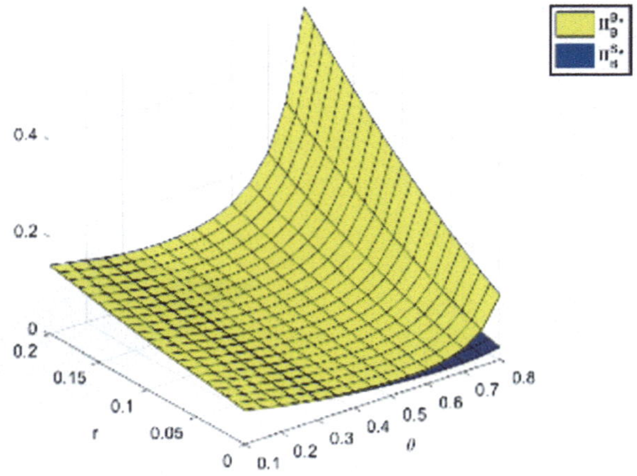


**Fig. 8** The omni-retailer's profits under the two game sequences

improvement, as we have shown before. This means that higher service level and lower price indeed brings enough consumer demand for the dual-channel retailer, which is also in accordance with the real practice.

# 5 Conclusion Remarks

Motivated by the competition lies between dual-channel retailer and e-retailer and cross-sales effect exists in dual-channel, we conduct an analytical study on equilibrium price and service strategy for both e-retailer and dual-channel retailer in different game sequences: Bertrand-Nash game and Stackelberg game. We obtain the optimal service level and pricing decisions under the two games. By numerical study we draw some conclusions as follows:

(1) The acceptance of online product positively influences the optimal service level and pricing for both retailers either in Bertrand-Nash game or Stackelberg game.
(2) The cross-sales effect exists in the dual-channel contributes to the improvement of the two retailers' service level and the e-retailer's pricing, while it has negative influence on the optimal price for the dual-channel retailer.
(3) When the acceptance of the online channel is at a higher level, the total profit of the e-retailer always precedes the dual-channel retailer's, even though her price is higher than the competitor's, which may cause a decrease in her demand.
(4) The total profits in Bertrand-Nash game of the two retailers are higher than that in Stackelberg game on the whole. Cross-sales effect has positive influence on the dual-channel retailer's profit especially when the acceptance of the online channel product is high enough.

For future research, we can extend to consider the case where competition also exists between the dual-channel and cross-sales can shift from the omni-channel retailer to the e-retailer, which means that when consumers pick-up items in brick-and-motor store and generate additional consumption, profit should contribute to omni-channel retailer may be lost. This is more interesting and closer to real situation. Besides, further discussion can be conducted when the leader of the Stackelberg game being the omni-channel retailer, and the assumed sequences about pricing and service decision can be also extended to different situations. Research opportunities abound in this emerging area.

**Acknowledgements** We would like to sincerely thank the three anonymous Reviewers and the Associate Editor for their constructive and helpful comments and suggestions that improve the paper.

## References

1. Wallace, W., Giese, L., & Johnson, L. (2004). Customer retailer loyalty in the context of multiple channel strategies. *Journal of Retailing, 80,* 249–263.
2. Zhang, P., He, Y., & Zhao, X. (2019). 'Preorder-online, pickup-in-store' strategy for a dual-channel retailer. *Transportation Research Part E: Logistics and Transportation Review, 122,* 27–47.
3. Hu, M., Xu, X., Xue, W., & Yang, Y. (2018). Demand pooling in omni-channel operations. SSRN Electric Journal.
4. Forrester (2014). Customer desires vs. retailer capabilities: minding the omnichannel commerce gap. Tech. Forrester Consulting.
5. Shi, X., Dong, C., & Cheng, T. C. E. (2018). Does the buy-online-and-pick-up-in-store strategy with pre-orders benefit a retailer with the consideration of returns? *International Journal of Production Economics, 206,* 134–145.
6. Gallino, S., & Moreno, A. (2014). Integration of online and offline channels in retail: The impact of sharing reliable inventory availability information. *Management Science, 60,* 1434–1451.
7. "Compete for high-end customers: Walmart launches private shopping guide", ebrun.com (2018). http://www.ebrun.com/20180612/281695.shtml. Accessed 12 Jun 2018.
8. Gao, F., & Su, X. (2016). Omnichannel retail operations with buy-online-and-pickup-in-store. *Management Science, 63,* 1–15.
9. Jin, M., Li, G., & Cheng, T. C. E. (2018). Buy online and pick up in-store: Design of the service area. *European Journal of Operational Research, 268,* 613–623.
10. Gao, F., & Su, X. (2016). Online and offline information for omnichannel retailing. *Manufacturing & Service Operations Management, 19,* 84–98.
11. Fan, C., Liu, Y., & Chen, X. (2018). Pricing and service cooperation in BOPS implementation: Considering channel competition and consumer behavior. *Journal of Systems Engineering, 33*(3), 388–397.
12. Guo, Y., Wang, K., & Chen, G. (2015). Research on traditional retailer's transformation and upgrading based on online and offline integration. *Chinese Journal of Management Science, 23,* 727–731.
13. Liu, Y., & Zhou, D. (2018). Is it always beneficial to implement BOPS: A comparative research with traditional dual-channel. *Operations Research and Management Science, 27*(2), 169–177.

14. Xun, X., & Jonathan, E. (2019). Examining customer channel selection intention in the omni-channel retail environment. *International Journal of Production Economics, 208,* 434–445.
15. Li, G., Li, L., & Sun, J. (2019). Pricing and service effort strategy in a dual-channel supply chain with showrooming effect. *Transportation Research Part E: Logistics and Transportation Review, 126,* 32–48.
16. Tsay, A., & Agrawal, N. (2004). Channel conflict and coordination in the E-commerce age. *Production and Operations Management Society, 13,* 93–110.
17. Chiang, K., Chhajed, D., & Hess, D. (2003). Direct marketing, indirect profits: A strategic analysis of dual-channel supply-chain design. *Management Science, 49,* 1–20.
18. Ali, S., Rahman, H., Tumpa, T., Rifat, A., & Pau, S. (2018). Examining price and service competition among retailers in a supply chain under potential demand disruption. *Journal of Retailing and Consumer Services, 40,* 40–47.
19. Kuksov, D., & Liao, C. (2018). When showrooming increases retailer profit. *Journal of Marketing Research, 55,* 459–473.

# Uncertain Random Programming Models for Chinese Postman Problem

**Qinqin Xu, Yuanguo Zhu, and Hongyan Yan**

**Abstract** Chinese postman problem (CPP) is a classic optimization problem which has been studied by researchers. Due to the indeterminate factors, such as the weather, traffic accidents, road conditions or sudden events, the indeterminacy should be taken into account. This paper takes advantage of chance theory to solve CPP in uncertain random situations, in which the weights are uncertain random variables. Considering various decision criteria, three uncertain random programming models are constructed and converted into corresponding deterministic forms based on chance theory. Subsequently, the models can be solved by classical algorithms or the ant colony optimization algorithm. Finally, a numerical example is given.

**Keywords** Chinese postman problem · Uncertain random variable · Uncertainty theory · Chance theory · Ant colony optimization algorithm

## 1 Introduction

Chinese postman problem (CPP) proposed by Kwan [1] is one of the typical combinatorial optimization problems. The postman is required to travel as short as possible, each street in the city is traveled at least once. Considering the delivery of practical problems, CPP has spawned many branches. Such as the postman problem on oriented graph, capacitated arc routing problem, windy postman problem and other derivative Chinese postman problems. Besides, CPP can also be applied to other optimization

This work is supported by the National Natural Science Foundation of China (No. 61673011) and Science Foundation of Jiangsu Province (China) for Young Scientists (No. BK20170916).

Q. Xu · Y. Zhu (✉)
School of Science, Nanjing University of Science and Technology, Nanjing, China
e-mail: ygzhu@njust.edu.cn

Q. Xu
e-mail: 18751969191@163.com

H. Yan
School of Science, Nanjing Forestry University, Nanjing, China
e-mail: melodymmyhy@163.com


J. Zhang et al. (eds.), *LISS2019*,
https://doi.org/10.1007/978-981-15-5682-1_47

problems so as to get the optimal solutions. The traditional CPP is an ideal optimization problem, that is, only certain factors are considered and some dynamic factors are ignored. Therefore, it provides a basis for further research and development.

In fact, the existence of indeterminacy is ineluctable. The travel time may be influenced by bad weather, reconstruction of roads, traffic congestion and other accidental factors. Owing to the fluctuation in fuel prices and parking fees, etc, the travel expenses are also not constant. There has been a lot of researches to describe such indeterminacy based on probability theory. Garlaschelli [2] introduced the weighted random graph (WRG) model. Gutin [3] proposed Chinese postman problem on edge-colored multi-graphs.

When there is no enough data to get a probability distribution, uncertainty theory was put forward by Liu to solve the optimization problems in uncertain situations [4]. Gao extended the shortest path problem to uncertain situations based on uncertainty theory [5]. With the development of uncertain theory, Sheng [7] proposed uncertain transportation problems in consideration of costs and demands. The research on such problem provides a reference for uncertain CPP. Then, Zhang [6] proposed three types of uncertain models by combining uncertainty theory with CPP. Various applications of uncertain programming problems are widely used in minimum cost flow problem [8], quadratic minimum spanning tree [9], minimum spanning tree problem [10], multi-objective portfolio selection problem [11] and multi-objective Chinese postman problem in uncertain network [12]. Considering above-mentioned studies in indeterministic environments, there does exist uncertain random optimization problems. To deal with such problems, Liu proposed chance theory, which is widely used to simulate complex situations of uncertainty and randomness [13]. Graph theory is a branch of mathematics. It takes graph as its object of study. Graphs in graph theory consist of a number of given vertices and edges connecting two vertices, which are usually used to describe a particular relationship between certain variables. Liu introduced chance theory into the network optimization with both uncertain and random factors, and put forward the concept of uncertain random network for the first time.

Compared with the traditional optimization problems, the optimization problems in uncertain random situations are more dynamic and optimal. The optimal solutions obtained by chance theory also accord with the practical problems to some extent. Uncertain CPP and random CPP have been studied in many literatures. However, the uncertainty is not taken into account in random CPP, and uncertain CPP also neglects the existence of randomness. There is no research on uncertain random CPP based on chance theory. Therefore, we will present uncertain random CPP and propose three uncertain random models. Furthermore, chance theory offers a tool to transform these models into their deterministic forms. In this paper, chance theory is mainly applied to deal with the expected value of objective function and constraints.

The sections are listed below. Section 2 introduces some necessary definitions and theorems of chance theory. Section 3, three types of uncertain random programming models are proposed. Section 4, the models and algorithms are verified by a numerical example. Section 5 summarizes this paper.

## 2 Preliminary

For uncertain random CPP, this section should introduce a lot of necessary properties and theorems in chance theory. These preliminary theorems can provide an effective approach to uncertain random CPP models. The relevant theorems and concepts needed for the model are shown below.

Let $(\Gamma, \mathcal{L}, \mathcal{M}) \times (\Omega, \mathcal{A}, \mathcal{P})$ be a chance space, and let $\Theta \in \mathcal{L} \times \mathcal{A}$ be an event. Then the chance measure of $\Theta$ is

$$\mathrm{Ch}\{\Theta\} = \int_0^1 \Pr\{\omega \in \Omega \mid \mathcal{M}\{\gamma \in \Gamma \mid (\gamma, \omega) \in \Theta\} \geq x\} dx.$$

In probability theory, the variable is defined as a random variable. In uncertainty theory, the variable is defined as an uncertain variable. In chance theory, Liu defined it as an uncertain random variable.

An uncertain random variable is a function $\xi$ from a chance space $(\Gamma, \mathcal{L}, \mathcal{M}) \times (\Omega, \mathcal{A}, \mathcal{P})$ to the set of real numbers such that $\xi \in \mathcal{B}$ is an event in $\mathcal{L} \times \mathcal{A}$ for any Borel set $\mathcal{B}$ of real numbers. For any $x \in \mathcal{R}$, an uncertain random variable $\xi$ can be characterized by its chance distribution, which is defined as:

$$\Phi(x) = \mathrm{Ch}\{\xi \leq x\}$$

To deal with the expected value and chance measure of an uncertain random variable, two essential theorems should be described.

**Theorem 1.** *Let $\eta_1, \eta_2, \cdots, \eta_m$ be independent random variables with probability distributions $\Psi_1, \Psi_2, \cdots, \Psi_m$, and let $\tau_1, \tau_2, \cdots, \tau_n$ be independent uncertain variables with regular uncertainty distributions $\Upsilon_1, \Upsilon_2, \cdots, \Upsilon_n$, respectively. if $f(x, \eta_1, \eta_2, \cdots, \eta_m, \tau_1, \tau_2, \cdots, \tau_n)$ is a strictly increasing function or decreasing function with respect to $\tau_1, \tau_2, \cdots, \tau_n$, then the expected function*

$$E[f(x, \eta_1, \eta_2, \cdots, \eta_m, \tau_1, \tau_2, \cdots, \tau_n)]$$

*is equal to*

$$\int_{R_m} \int_0^1 f(x, y_1, \cdots, y_m, \Upsilon_1^{-1}(\alpha), \cdots, \Upsilon_n^{-1}(\alpha)) d\alpha \, d\Psi_1(y_1) \cdots d\Psi_m(y_m).$$

**Theorem 2.** *Let $\eta_1, \eta_2, \cdots, \eta_m$ be independent random variables with probability distributions $\Psi_1, \Psi_2, \cdots, \Psi_n$, and let $\tau_1, \tau_2, \cdots, \tau_n$ be independent uncertain variables with regular uncertainty distributions $\Upsilon_1, \Upsilon_2, \cdots, \Upsilon_n$, respectively. if $g(x, \eta_1, \eta_2, \cdots, \eta_m, \Psi_1, \tau_1, \tau_2, \cdots, \tau_n)$ is a strictly increasing function with respect to $\tau_1, \tau_2, \cdots, \tau_n$, then the chance constraints*

$$Ch\{g(x, \eta_1, \eta_2, \cdots, \eta_m, \tau_1, \tau_2, \cdots, \tau_n) \leq 0\} \geq \alpha_j$$

*holds if and only if*

$$\int_{R_m} G(x, y_1, y_2, \cdots, y_m) d\Psi_1(y_1) \cdots d\Psi_m(y_m) \geq \alpha_j$$

*where $G(x, y_1, y_2, \cdots, y_m)$ is the root $\alpha$ of the equation*

$$g(x, y_1, y_2, \cdots, y_m, \Upsilon_1^{-1}(\alpha), \cdots, \Upsilon_n^{-1}(\alpha)) = 0.$$

## 3 Uncertain Random Programming Models for Chinese Postman Problem

Let $G = (V, E, \omega)$ be an undirected weighted network [6], in which $V = \{v_1, v_2 \cdots v_n\}$ is a finite set of vertices, $E$ is a finite set of edges $e_{ij}$. In the traditional CPP, vertices are used to represent specific service locations, and edges connecting two vertices are used to represent streets. For a weight $\omega_{ij}$ on $e_{ij}$ of $G$, it is nonnegative and finite.

Considering the coexistence of the randomness and uncertainty, this section presents uncertain random CPP and constructs three types of uncertain random models for CPP. Assume the weight $\omega_{ij}$ on $e_{ij}$ represents the travel time. In the traditional CPP, the travel time on edge $e_{ij}$ is constant. Depending on the size of the sample data, $\omega_{ij}$ is uncertain or random. Let $G = (V, U, R, \omega)$, where $V = \{v_1, v_2 \cdots v_n\}$ is a finite set of vertices, $U$ is a finite set of all edges with uncertain travel times, $R$ is a finite set of edges with random travel times, $\xi_{ij}$ is the travel time on $e_{ij}$ which is an uncertain random variable based on chance theory.

### 3.1 Uncertain Random Programming Model 1

Uncertain random CPP is to let a postman from the post office deliver letters in the shortest route, and each street should be traveled at least once. The objective function of model 1 is to minimize the total travel time. Here, $f(x, \xi)$ represents the total time, $f(x, \xi)$ is defined as

$$f(x, \xi) = \sum_{k=1}^{K} \sum_{(e_{ij}) \in U \cup R} x_{ij}^{(k)} \xi_{ij} \tag{1}$$

where $K$ and $x_{ij}^{(k)}$ are defined as

$$K = \max_{(e_{ij})\in U\cup R} \sum_k x_{ij}^{(k)}$$

For the decision variable, it is defined as

$$x_{ij}^{(k)} = \begin{cases} 1, & \text{if a postman travels } e_{ij} \text{ for the } k\text{th order} \\ 0, & \text{else.} \end{cases}$$

The model 1 is described below:

$$\begin{cases} \min E\left[\sum_{k=1}^{K} \sum_{(e_{ij})\in U\cup R} x_{ij}^{(k)} \xi_{ij}\right] \\ \text{subject to} \\ \sum_{k=1}^{K} \sum_{(e_{ij})\in U\cup R} (x_{ij}^{(k)} - x_{ji}^{(k)}) = 0 \\ \sum_{k=1}^{K} (x_{ij}^{(k)} + x_{ji}^{(k)}) \geq 1. \end{cases} \tag{2}$$

In model 1, the objective function represents the minimization of the total travel time. The first constraint ensures a closed loop, and the second constraint is another a requirement for CPP to be satisfied, that is, each edge must be traveled at least once.

Due to an uncertain random variable in model 1, it is necessary to transform model 1 into a deterministic model to get the optimal solution.

**Theorem 3.** *Let $\Psi_{ij}$ be the probability distribution of independent random travel time. Let $\Upsilon_{ij}$ be the uncertainty distribution of independent uncertain travel time. For any real number $y_{ij}$, model 1 can be converted into*

$$\begin{cases} \min \sum_{k=1}^{K} \int_{R_m} \int_0^1 \sum_{(e_{ij})\in R} x_{ij}^{(k)} y_{ij} \\ + \sum_{(e_{ij})\in U} x_{ij}^{(k)} \Upsilon_{ij}^{-1}(\alpha) d\alpha \prod_{(e_{ij})\in R} d\Psi_{ij}(y_{ij}) \\ \textit{subject to} \\ \sum_{k=1}^{K} \sum_{(e_{ij})\in U\cup R} (x_{ij}^{(k)} - x_{ji}^{(k)}) = 0 \\ \sum_{k=1}^{K} (x_{ij}^{(k)} + x_{ji}^{(k)}) \geq 1. \end{cases} \tag{3}$$

**Proof:** For (1), refer to chance theory, we have

$$f(x,\eta,\tau)=\sum_{k=1}^{K}\Big(\sum_{(e_{ij})\in R}x_{ij}^{(k)}\eta_{ij}+\sum_{(e_{ij})\in U}x_{ij}^{(k)}\tau_{ij}\Big)$$

Obviously, $f(x,\eta,\tau)$ is a strictly increasing function with respect to $\tau$. Refer to Theorem 1, we have

$$\begin{aligned}&E\left[\sum_{k=1}^{K}\Big(\sum_{(e_{ij})\in R}x_{ij}^{(k)}\eta_{ij}+\sum_{(e_{ij})\in U}x_{ij}^{(k)}\tau_{ij}\Big)\right]\\&=\sum_{k=1}^{K}\int_{R_m}\int_0^1\sum_{(e_{ij})\in R}x_{ij}^{(k)}y_{ij}\\&+\sum_{(e_{ij})\in U}x_{ij}^{(k)}\Upsilon_{ij}^{-1}(\alpha)d\alpha\prod_{e_{ij}\in R}d\Psi_{ij}(y_{ij})\end{aligned}$$

Hence, the theorem is verified.

Model 1 can be transformed into it's corresponding deterministic form by Theorem 3. To obtain the optimal solution, Dijkstra algorithm [14] and Fleury algorithm [15] served as typical traditional algorithms have been widely used. On the basis of Dijkstra algorithm, Fleury algorithm is applied to get the closed loop in an euler graph. Refer to such two algorithms, the classical algorithm for model 1 is presented below.

**Algorithm 1.** The classical algorithm for model 1.

1: Input the set of odd-points, $V_0=\{v|v\in V, d(v)=1(mod\ 2)\}$.
2: Calculate each pair of odd-points' distance $d(u,v)$ by Dijkstra algorithm.
3: Construct complete graph $K_{|v_0|}$, $V(K_{|v_0|})=V_0$ with the weight $\omega(u,v)=d(u,v)$.
4: Select the perfect match $M$ of $K_{|v_0|}$ with a minimum weight.
5: For $G$, choose the shortest path between each pair of vertices of $M$.
6: Repeat the shortest path.
7: Get a closed loop in $G'$ by Fleury algorithm. The algorithm ends.

## 3.2 Uncertain Random Programming Model 2

Extended to other optimization problems related to CPP, such as a logistics company transport goods by aircraft. If they arrive the destination early, there may not be a runway suitable for landing. Therefore, the goal is not to minimize the total time. Take it into account in this section, the decision criterion is that the postman needs to finish the delivery at time $T^*$. Inevitably, the task may be completed ahead of schedule or later. There should be a deviation between the actual travel time and the target travel time. Consequently, the objective function of model 2 is to minimize the deviation. The positive deviation and negative deviation are introduced below.

$$(f(x,\xi)-T^*)^+ = \begin{cases} f(x,\xi)-T^*, & \text{if } f(x,\xi) > T^* \\ 0, & \text{if } f(x,\xi) < T^* \end{cases}$$

$$(f(x,\xi)-T^*)^- = \begin{cases} T^*-f(x,\xi), & \text{if } f(x,\xi) < T^* \\ 0, & \text{if } f(x,\xi) > T^* \end{cases}$$

Here, we define the objective function as

$$E[(f(x,\xi)-T^*)^+] + E[(f(x,\xi)-T^*)^-]$$

The model 2 can be constructed as

$$\begin{cases} \min\{E[(f(x,\xi)-T^*)^+] + E[f((x,\xi)-T^*)^-]\} \\ \text{subject to} \\ \sum\limits_{k=1}^{K} \sum\limits_{(e_{ij})\in U\cup R} (x_{ij}^{(k)} - x_{ji}^{(k)}) = 0 \\ \sum\limits_{k=1}^{K} (x_{ij}^{(k)} + x_{ji}^{(k)}) \geq 1 \end{cases} \tag{4}$$

In model 2, the objective function is to minimize the deviation, and the constraints of model 2 are the same as model 1. Model 2 could also be converted equivalently in the following way.

**Theorem 4.** *Let $\Psi_{ij}$ be the probability distribution of independent random travel time, and $\Upsilon_{ij}$ be the uncertainty distribution of independent travel time. For real number $y_{ij}$, model 2 equivalently becomes*

$$\begin{cases} \min(\sum_{k=1}^{K} \int_{R_m} \int_0^1 \sum_{(e_{ij})\in R} x_{ij}^{(k)} y_{ij} \\ + \sum_{(e_{ij})\in U} x_{ij}^{(k)} \Upsilon_{ij}^{-1}(\alpha) d\alpha \prod_{(e_{ij})\in R} d\Psi_{ij}(y_{ij}) - T^*)^+ \\ +(\sum_{k=1}^{K} \int_{R_m} \int_0^1 \sum_{(e_{ij})\in R} x_{ij}^{(k)} y_{ij} + \\ \sum_{(e_{ij})\in U} x_{ij}^{(k)} \Upsilon_{ij}^{-1}(\alpha) d\alpha \prod_{(e_{ij})\in R} d\Psi_{ij}(y_{ij}) - T^*)^- \\ \text{subject to} \\ \sum_{k=1}^{K} \sum_{(e_{ij})\in U\cup R} (x_{ij}^{(k)} - x_{ji}^{(k)}) = 0 \\ \sum_{k=1}^{K} (x_{ij}^{(k)} + x_{ji}^{(k)}) \geq 1. \end{cases} \tag{5}$$

**Proof:** Since $f(x, \eta, \tau)$ is a strictly increasing function with respect to $\tau$, then $f(x, \eta, \tau) - T^*)^+$ is also a strictly increasing function with respect to $\tau$. Refer to Theorem 1

$$E[(f(x, \xi) - T^*)^+]$$
$$= \left[ \sum_{k=1}^{K} \int_{R_m} \int_0^1 \sum_{(e_{ij})\in R} x_{ij}^{(k)} y_{ij} \right.$$
$$\left. + \sum_{(e_{ij})\in U} x_{ij}^{(k)} \Upsilon_{ij}^{-1}(\alpha) d\alpha \prod_{(e_{ij})\in R} d\Psi_{ij}(y_{ij}) - T^* \right]^+$$

Traditional algorithms are to obtain the shortest route in deterministic network. This section presents the ant colony optimization algorithm to get the optimal solution for model 2. The ant colony optimization algorithm can obtain the optimal solution directly by setting the constraints, avoiding two-step process in the common method.

The post office is located in $v_1$, $m$ postmen serve the same area. Each postman selects the travel direction at the initial time randomly. It may be assumed that the selection probability in each direction is the same. Some variables are defined below

$$p_{ij}(k) = \begin{cases} \frac{\tau_{ij}^{\alpha}(t) \cdot \eta_{ij}^{\beta}(t)}{\sum_{k\in adj(t)} \tau_{ij}^{\alpha}(t) \cdot \eta_{ik}^{\beta}(t)}, & j \in adj(i) \\ 0, & \text{else.} \end{cases}$$

$p_{ij}(k)$ represents the probability of selection when the $k$th postman move from $v_i$ to $v_j$. $t$ represents the time when postman is in $v_1$. $\Delta\tau_{ij}(t)$ represents the increment of the pheromone on edge $e_{ij}$. $\alpha$ is the importance of accumulated pheromone. $\beta$ represents heuristic factor in the selection. $\eta_{ij}$ is the expected value of $e_{ij}$. $adj(i)$ represents the set of $v_j$ where $v_i$ and $v_j$ are connected. $(1-\rho)$ indicates the depletion of pheromone.

At initial moment, $\tau_{ij}^k = C$, $\Delta\tau_{ij}^k = 0$, for $\rho \in (0, 1)$. The pheromone on each path adjusts to

$$\tau_{ij}(t+n) = \rho \cdot \tau_{ij}(t) + \Delta\tau_{ij}$$

$$\Delta\,\tau_{ij} = \sum_{k=1}^{m} \Delta\tau_{ij}^k$$

$$\Delta\tau_{ij}^k = \begin{cases} \frac{Q}{L_k}, & \text{if the } k\text{th ant travels the edge } e_{ij} \\ 0, & \text{else.} \end{cases}$$

Then, the ant colony optimization algorithm for model 2 can be summarized as follows.

**Algorithm 2.** The ant colony optimization algorithm for model 2.

1: Input the upper triangular adjacency matrix of graph $G$, if there is an odd-degree node, turn to step 2, otherwise the algorithm ends.
2: From the starting point, the ant chooses next vertex until every street is traveled at least once and keep the current optimal route.
3: Update the pheromone and modify the probability of selection.
4: Repeat steps 2-3 (set to 100 times).
5: Output the optimal route.

### *3.3 Uncertain Random Programming Model 3*

Owing to the fluctuation in fuel prices and parking fees, etc., the travel expense $\rho_{ij}$ is also not constant, which can be considered as a random variable. The probability distribution of travel cost can be obtained from history data. Besides, the travel allowance $\zeta_{ij}$ can be considered as an uncertain variable due to the lack of historical data. There is no traffic allowance when the postman travels through $e_{ij}$ twice.

In this model, the postman needs to maximize the profit as much as possible, the target profit cannot be less than $M$. The postman not only makes sure that the profit meets the constraint, but also spends as little time as possible. Model 3 can be presented below

$$
\begin{cases}
\min E\left[\sum_{k=1}^{K} \sum_{(e_{ij})\in U\cup R} x_{ij}^{(k)}\xi_{ij}\right] \\
\text{subject to} \\
\sum_{k=1}^{K} \sum_{(e_{ij})\in U\cup R} (x_{ij}^{(k)} - x_{ji}^{(k)}) = 0 \\
\sum_{k=1}^{K} (x_{ij}^{(k)} + x_{ji}^{(k)}) \geq 1 \\
\mathrm{Ch}\{M - \sum_{k=1}^{K} \sum_{(e_{ij})\in U\cup R} x_{ij}^{(k)}(\zeta_{ij} - \rho_{ij}) \leq 0\} \geq \alpha.
\end{cases}
\tag{6}
$$

In model 3, the objective and the first two constraints are the same as model 1, the third constraint is that the profit cannot be less than $M$. Using preliminary theorems, model 3 could also be converted equivalently.

**Theorem 5.** *Let $\Omega_{ij}$ be the probability distribution of independent random travel cost, let $\Phi_{ij}$ be the uncertainty distribution of independent travel allowance. For real number $y_{ij}$, model 3 can be transported to*

$$
\begin{cases}
\min \sum_{k=1}^{K} \int_{R_m} \int_0^1 \sum_{(e_{ij})\in R} x_{ij}^{(k)} y_{ij} \\
+ \sum_{(e_{ij})\in U} x_{ij}^{(k)} \Upsilon_{ij}^{-1}(\alpha) d\alpha \prod_{(e_{ij})\in R} d\Psi_{ij}(y_{ij}) \\
\textit{subject to} \\
\sum_{k=1}^{K} \sum_{(e_{ij})\in U\cup R} (x_{ij}^{(k)} - x_{ji}^{(k)}) = 0 \\
\sum_{k=1}^{K} (x_{ij}^{(k)} + x_{ji}^{(k)}) \geq 1 \\
\int_{R_m} G(x_{ij}^{(k)}, y_{ij}) \prod_{(e_{ij})\in R} d\Omega_{ij}(y_{ij}) \geq \alpha
\end{cases}
\tag{7}
$$

*where $G(x_{ij}^{(k)}, y_{ij})$ is the root $\alpha$ of the equation*

$$
\sum_{k=1}^{K} [\sum_{(e_{ij})\in U} x_{ij}^{(k)} \Upsilon_{ij}^{-1}(1-\alpha) - \sum_{(e_{ij})\in R} x_{ij}^{(k)} y_{ij}] = M
$$

**Proof:** The objective and the first two constraints are the same as model 1, the expected value can be obtained by Theorem 3. For the third constraint, it is a strictly decreasing function with respect to $\zeta_{ij}$. According to Theorem 2, the uncertain random constraint holds with confidence level $\alpha$, where $\alpha$ is a constant. Then the chance constraint can be defined by

$$\text{Ch}\{M - \sum_{k=1}^{K} \sum_{(e_{ij}) \in U \cup R} x_{ij}^{(k)} (\zeta_{ij} - \rho_{ij}) \le 0\} \ge \alpha$$

holds if and only if

$$\int_{R_m} G(x_{ij}^{(k)}, y_{ij}) \prod_{(e_{ij}) \in R} d\Omega_{ij}(y_{ij}) \ge \alpha$$

Hence, the theorem is verified.

Obviously, the objective function and first two constraints of model 3 are the same as of model 1. Consequently, we can refer to Algorithm 1 to solve model 3.

**Algorithm 3.** The improved traditional algorithm for model 3.

1: Obtain the optimal solution of model 1 by Algorithm 1.
2: If the solution satisfies the third constraint of model 3, then the solution is the final optimal solution, and the algorithm ends. If not, turn to step 3.
3: Remove the obtained solution from the feasible solutions of model 1 and re-screen the optimal solution.
4: Turn to step 2.
5: Output the optimal solution.

## 4 Numerical Experiment

In order to verify the algorithms, an example should be proposed. The relevant uncertain parameters of models are considered as a linear variable $\mathcal{L}(a, b)$, a zigzag variable $\mathcal{Z}(a, b, c)$ and a normal uncertain variable $\mathcal{N}(e, \sigma)$. Some relevant random parameters are also expressed in a uniform variable $\mathcal{U}(a, b)$ and a normal random variable $\mathcal{N}(u, \sigma^2)$.

**Example 1.** *The postman office is located in $v_1$, a postman to deliver letters from $v_1$. When he returns to $v_1$, each street should be traveled at least once. To choose a reasonable route for the postman based on three models proposed above, an undirected connected network G is shown in Fig. 1. The relevant uncertain parameters of $e_{ij}$ are shown in Tables 1 and 2. Set some parameters: $T^* = 650$, $M = 500$ and $a = 0.95$.*

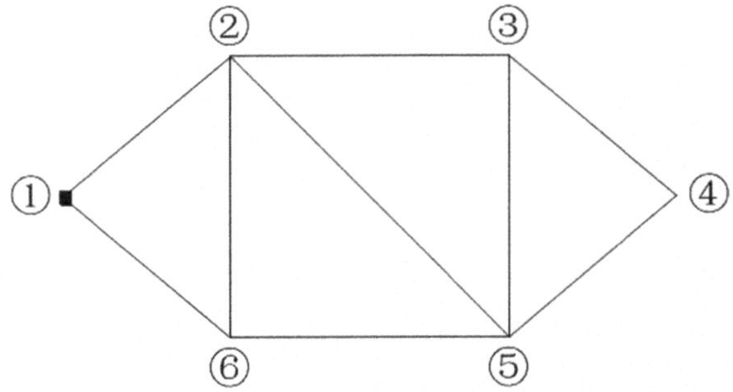


**Fig. 1** Undirected network $G$.

**Table 1** List of travel time

| Edge | Uncertain travel time | Random travel time |
|---|---|---|
| $e_{12}$ | $\mathcal{Z}(48, 50, 52.7)$ | - |
| $e_{23}$ | $\mathcal{Z}(48.9, 52.4, 53.2)$ | - |
| $e_{25}$ | $\mathcal{N}(71, 8.2)$ | - |
| $e_{35}$ | $\mathcal{N}(42.6, 7.2)$ | - |
| $e_{56}$ | $\mathcal{L}(55, 58)$ | - |
| $e_{16}$ | - | $\mathcal{U}(52, 54)$ |
| $e_{26}$ | - | $\mathcal{N}(68.2, 8.2)$ |
| $e_{34}$ | - | $\mathcal{N}(70.7, 10.8)$ |
| $e_{45}$ | - | $\mathcal{U}(67, 70)$ |

**Table 2** List of travel cost and travel allowance

| Edge | Random travel expense | Uncertain travel allowance |
|---|---|---|
| $e_{12}$ | $\mathcal{U}(44, 52)$ | $\mathcal{L}(89, 105.8)$ |
| $e_{23}$ | $\mathcal{U}(38, 40)$ | $\mathcal{Z}(41.5, 46, 70.7)$ |
| $e_{25}$ | $\mathcal{N}(40, 4.2)$ | $\mathcal{L}(78.6, 84.8)$ |
| $e_{35}$ | $\mathcal{N}(40, 4.3)$ | $\mathcal{N}(104.1, 8.6)$ |
| $e_{56}$ | $\mathcal{U}(34, 40)$ | $\mathcal{L}(68.2, 106)$ |
| $e_{16}$ | $\mathcal{N}(38, 6.2)$ | $\mathcal{Z}(96, 108.6, 121.2)$ |
| $e_{26}$ | $\mathcal{N}(36, 4.2)$ | $\mathcal{N}(72.1, 10.4)$ |
| $e_{34}$ | $\mathcal{U}(42.4, 48.6)$ | $\mathcal{Z}(67, 80.4, 96.6)$ |
| $e_{45}$ | $\mathcal{N}(32.4, 4.2)$ | $\mathcal{L}(68.2, 77.9)$ |

By Algorithm 1, we can get the optimal route of model 1, $v_1 \to v_2 \to v_6 \to v_5 \to v_2 \to v_3 \to v_5 \to v_3 \to v_4 \to v_5 \to v_6 \to v_1$. The $e_{56}$ and $e_{35}$ are traveled twice, the rest of the edges are traveled once, and the travel time of the route is 631.5.

By Algorithm 2, we can get the optimal route of model 2, $v_1 \to v_2 \to v_6 \to v_2 \to v_3 \to v_2 \to v_5 \to v_3 \to v_4 \to v_5 \to v_6 \to v_1$. The $e_{23}$ and $e_{26}$ are traveled twice, the rest of the edges are traveled once, and the travel time of the route is 652.325.

By Algorithm 3, we can get the optimal route of model 3, $v_1 \to v_6 \to v_5 \to v_2 \to v_5 \to v_6 \to v_2 \to v_3 \to v_5 \to v_4 \to v_3 \to v_2 \to v_1$. The $e_{23}$, $e_{56}$ and $e_{52}$ are traveled twice, the rest of the edges are traveled once, the travel time of the route is 711.625, and the travel profit of the route is 505.15.

## 5 Conclusion

Considering various decision criteria under the uncertain random situations, three uncertain random CPP models are constructed in order to assign routes. Model 1 is to obtain a route by minimizing travel time, model 2 is to obtain a route by minimizing the deviation between actual travel time and target time, model 3 is to get a route by minimizing travel time with consideration of cost. This paper serves a powerful tool for similar problems such as garbage collection, street cleaning, milk delivery, school bus scheduling, etc.

Lastly, some suggestions are put forward for the future research. First, uncertain random CPP can be studied in directed network. Second, different branches of CPP, (Rural CPP, Windy CPP, *k* person CPP, etc.) can also be extended to uncertain random domains. Third, extensions of various applications of uncertain random CPP, including the route of logistics vehicle, watering cart and garbage collector, deserves to be considered as our future research.

## References

1. Kwan, M. K. (1962). Graphic programming using odd or even points. *Chinese Mathematics*, *1*, 273–277.
2. Garlaschelli, D. (2009). The weighted random graph model. *New Journal of Physics*, *11*(7), 1–10.
3. Gutin, G., Jones, M., & Sheng, B. (2017). Chinese postman problem on edge-colored multigraphs. *Discrete Applied Mathematics*, *217*(2), 196–202.
4. Liu, B. (2007). *Uncertainty theory* (2nd ed.). Berlin: Springer-Verlag.
5. Gao, Y. (2011). Shortest path problem with uncertain arc lengths. *Computers Mathematics with Applications*, *62*(6), 2591–2600.
6. Zhang, B., & Peng, J. (2012). Uncertain programming model for Chinese postman problem with uncertain weights. *Industrial Engineeering & Management Systems*, *11*(1), 18–25.
7. Sheng, Y., & Yao, K. (2012). A transportation model with uncertain sots and demands. *Information*, *15*(8), 3179–3186.
8. Ding, S. (2014). Uncertain minimum cost flow problem. *Soft Computing*, *18*(11), 2201–2207.

9. Zhou, J., He, X., & Wang, K. (2014). Uncertain quadratic minimum spanning tree problem. *Journal of Communications*, *9*(5), 385–390.
10. Zhou, J., Chen, L., & Wang, K. (2015). Path optimality conditions for minimum spanning tree problem with uncertain edge weights. *International Journal of Uncertainty Fuzziness and Knowledge-Based Systems*, *23*(1), 49–71.
11. Kar, M. B., Majumder, S., Kar, S., & Pal, T. (2017). Cross-entropy based multi-objective uncertain portfolio selection problem. *Journal of Intelligent & Fuzzy System*, *32*(6), 4467–4483.
12. Majumder, S., Kar, S., & Pal, T. (2018). Uncertain multi-objective Chinese postman problem. *Soft Computing*, *4*(12), 1–16.
13. Liu, Y. (2013). Uncertain random variables: a mixture of uncertainty and randomness. *Soft Computing*, *17*(4), 625–634.
14. Dijkstra, E. W. (1959). A note on two problems in connexion with graphs. *Numerische Mathematik*, *1*(1), 269–271.
15. Floyd, R. W. (1969). Algorithm 97: Shortest path. *Communication of ACM*, *5*(6), 345.

# "Choose for No Choose"—Random-Selecting Option for the Trolley Problem in Autonomous Driving

**Liang Zhao and Wenlong Li**

**Abstract** Trolley Problem is a famous thought experiment asking if it is ethically right to divert a runaway trolley to kill one person in order to save another five people. It was first formally stated by P. Foot in 1967, then extensively studied by many researchers with many variants. These days it is often discussed in the context for ethics of Artificial Intelligence (AI), in particular, for autonomous driving (see e.g., "The Moral Machine Experiment," nature 563, 59–64 (2018)). For this dilemma situation, previous studies focus on two options: saving the larger crowd (the Utilitarianism) or doing nothing (the Deontology or Kantianism). This paper proposes a third option: random selecting, and argues it fits the dilemma situation for autonomous driving better than the previous options.

**Keywords** Random selecting · Trolley problem · Autonomous vehicle · Utilitarianism · Deontology · Kantianism

## 1 Introduction

Recent achievement of Artificial Intelligence (AI) is making a profound impact on our daily life. It is widely considered that in the future, AI (i.e., algorithms) may deal with even ethical decisions that are currently considered only by our human beings [2, 4–6, 11–17]. For instance, the trolley problem [3, 18] must be solved before massive autonomous vehicles can drive on the roads (see, e.g., [1] and Max Tegmarks "Life 3.0"), because traffic accidents cannot be completely eliminated by new technology [16].

L. Zhao (✉) · W. Li
Graduate School of Advanced Integrated Studies in Human Survivability (Shishu-Kan), Kyoto University, Kyoto, Japan
e-mail: liangzhao@acm.org

W. Li
e-mail: li.wenlong.77n@st.kyoto-u.ac.jp



J. Zhang et al. (eds.), *LISS2019*,
https://doi.org/10.1007/978-981-15-5682-1_48

The trolley problem was first formally stated by P. Foot in 1967 [3], which is now referred as the "Switch" version of the problem. In this original version, it is asked if one should switch a lever to kill one person in order to save another five people, supposing the runaway tram cannot be safely stopped. See the illustration in Fig. 1. The trolley problem was then extensively studied by many researchers for many variants (see an introduction from [1, 18]).

An interesting fact about the trolley problem is that in different scenarios, people's selection can be different. Kagan argued that, in order to save the five, there are differences on how to think on harming the one [10]. For example, in the original Switch version, harming the one is a side effect, whereas in other cases it may be a part of the saving action. Unger argued that different responses are based more on psychology of ethics, but he also agreed that this kind of difference in the scenario cannot make a moral difference [19]. In this paper, for simplicity, we only consider the original Switch version.

Nowadays, due to the development of autonomous vehicles, the trolley problem gets more attention by the public [1, 4, 5, 9, 11–17]. While this dilemma situation may be rare in the future [15], obviously it is not impossible to happen, and ethical guidelines for autonomous vehicles do not depend on the frequency. Therefore autonomous vehicles (or rather the programmer of their operating systems) must consider these ethical decisions [1, 14], even if they do not need to know good options [15].

Unfortunately, like most other ethical problems, the trolley problem has no simple solution. Before the rising of AI, ethicists discussed on what kind of ethical principle that *human* should follow, where the central question asks if it is morally right to harm an individual if doing so produces a greater utility for others. For that purpose, the literature has discussed two options based on the Utilitarianism and the Deontology (Kantianism) respectively.

The first option, Utilitarianism-based option, is to maximize the total utility by taking the result into consideration, therefore, in the case of trolley problem (the Switch version), it is to switch the lever to kill one since doing so can save five. On the other hand, the second option, Deontology-based option claims it is right only when one makes decision without considering the result, therefore, if killing is wrong then one should not do that. In the case of the trolley problem, this is considered to "do nothing", hence letting the runaway tram to kill the five.

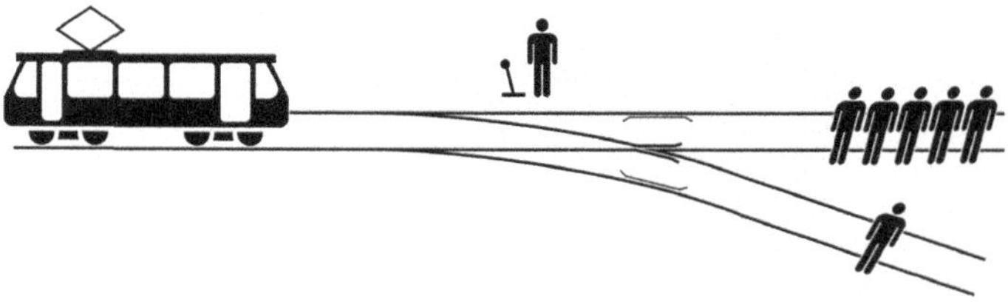

**Fig. 1** An illustration of the original "Switch" version of the trolley problem [3], created by McGeddon—Own work, CC BY-SA 4.0, https://commons.wikimedia.org/w/index.php?curid=52237245

We consider that, in the case of autonomous vehicles, both of the above options are not wise. In particular, the "do nothing" option is strange, because it actually chooses to harm the five (notice that we assumed that the possible results are known before making any decision). It may be morally right for a human but definitely not for an autonomous vehicle, which must decide which side should be sacrificed.

To solve this dilemma situation, this paper proposes a third option: *random selecting*, i.e., to randomly select a side to be sacrificed. We argue this fits the dilemma case for autonomous vehicles better than the above two options. In the following, we will first compare these three options in Sect. 2, then discuss the practical issues in autonomous vehicles in Sect. 3 and finally provide a conclusion and future work in Sect. 4.

## 2 Comparing Options for the Trolley Problem

In this section we compare the three options for the trolley problem: the utilitarian approach of maximizing the utility, the so-called deontological approach of "do nothing" and our proposal of random selecting.

### *2.1 Utilitarian Approach*

Based on the philosophy of Jeremy Bentham, the fundamental idea of the utilitarian approach is that an action or behavior is morally correct if the greatest benefit or the greatest number of people could be achieved. When applied to a possible crash of an autonomous vehicle, this means that a utilitarian algorithm should be designed in such a way that the cost of the accident should be as low as possible. This means a reaction to harm the smallest number of people or minimize the cost measured in money. In particular, the control algorithm may decide that the occupant of the vehicle should be sacrificed if it is only one human life instead of, for example, two or more pedestrians. Therefore, if utilitarianism is adopted for autonomous vehicles, it is quite possible that many people would incline not buying it, even if they know that an ordinary vehicle has higher risk to kill more people.

Another major problem of utilitarianism is that usually it is difficult to quantify societal expectations on ethical principles to guide machine behavior. For example, Awad et al. conducted a Moral Machine Experiment trying to discover common opinions of people all over the world about the ethics of autonomous vehicles against the trolley problem [1], and from [1] it is reported that "strongest preferences are observed for sparing humans over animals, sparing more lives, and sparing young lives." Other preferences are also observed including big difference between East Asia and South America to spare younger people, the preference for sparing higher status people, and so on [8].

All the patterns of these similarities and differences suggest that lots of efforts are needed to reach a consensus on algorithm standards for the public. In fact, we cannot expect the existence of a (deterministic) algorithm because the real situation is always more complicated than those simple circumstance. That approach, in general, can be described as a ranking system: with the development of technology we may have the ability to develop a kind of ranking system based on the utilitarian principle, and then decide who to sacrifice by comparing their scores. However, this approach leads to challenges to other more fundamental principles such as "Everyone should be equal."

Even a regulation or law could be created based on the consideration of utilitarian principles, some people may refuse to buy a vehicle which may sacrifice the passenger inside the vehicle as stated before (especially in the States). Those people may continue to drive ordinary vehicles which have higher risks to cause more accidents. Therefore the utilitarian approach is not easy to implement for autonomous vehicles.

And, as known by the tragedy of the commons traditionally raised in 1833 by the British economist Lloyd [7], "by making the individually rational choice of prioritizing their own safety, they may collectively be diminishing the common good" [17]. In the case of autonomous vehicles, this means that the vehicle manufacturers may design vehicles to maximize the safety of people inside the vehicles, and by the usage of modern machine learning technology, the vehicles may be able to learn automatically to do so (i.e., maximizing the safety of people inside which may have to increase the risk of pedestrians). This is called the "tragedy of the algorithmic commons" in [17].

## 2.2 *Deontological Approach*

In a deontological approach, autonomous vehicles should follow duty-bound principles, like "Thou shalt not kill" (Kant). A naive conclusion is that an autonomous vehicle should not take any special action that harms a human, hence in the case of the trolley problem, it is considered that an algorithm should let the vehicle take its current course (i.e., do nothing), even if it is going to harm more people [18].

Unfortunately, however, this "do nothing" option is strange. Firstly, it actually does something, i.e., it chooses the action to let the vehicle take it current course, which means it chooses the action to harm five people. Therefore this "do nothing" option, while it looks deontological, it is actually not a true deontology. Moreover, supposing that in a deontological approach, an autonomous vehicle follows certain rules, e.g., Asimov's Three Laws of Robotics [4], since in the trolley problem the machine has to choose one of two evils [15], it is not possible to follow all rules. For example, the first law of Asimov's laws says "a robot may not injure a human being or, through inaction, allow a human being to come to harm" cannot be abided.

Finally, while deontological ethics can provide guidance for many situations, it is rather unsuitable for programming crash algorithms since no rule can ever cover all aspects of road traffic, and it is difficult to "transform" complex human ethics in a set of rules [5]. For example, in the traditional case of trolley problem, one person may decide not to interfere the route based on what Kant said: one shall not kill. But as stated before, if the autonomous vehicle decides to do nothing and continue the journey, this action already made a choice: kill five and save one. In other words, in the case of autonomous vehicles, all choices we are discussing are getting involved. Based on this observation, one may say that if we must to select one, it is acceptable to choose by numbers or other standards (utilitarian). This looks to have no better solution. But actually, they missed a third option: random selecting.

## 2.3 *Random Selecting*

We propose a third option to the trolley problem: random selecting, i.e., randomly select a side to be sacrificed. At the first glance, it may look silly and irresponsible. But this idea is exactly based on the principle of deontological ethics. Because we value all the lives of human beings, there is no standard which we can use to take the right to live from anyone. But in the dilemma situation, there should be a way out, or in another words, there must be an evil to be done (because the autonomous vehicle knows the possible result). Random selecting assumes all stakeholders have the same risk to be harmed. Walkers and passengers are all not involved in the driving decision making process, but when they decide to use public resources such as road infrastructure or autonomous vehicle system, it is by default assumed that they admit the existence of risk, no matter how high (or low) the risk may be. In other words, they are involved as well. Just like the daily life of everyone today, when one walks on the street, he or she is innocent to take any harm, but there is still a (low) possibility that some vehicle (driven by a human) hits him or her by accident.

Therefore, we conclude, based on the deontological consideration, random selecting is an action for no choose, which we call "*Choose for no choose*." Because of the respect for life, any standards or bias will be immoral to take. But a random selecting algorithm has the ability to guarantee an equal possibility for everyone without any bias. Therefore, we argue random selecting is the true choice from deontological ethics. Meanwhile, it can avoid the criticism from utilitarian that deontological ethics has no consideration of the total public benefit, since the expected cost (which is $(5 + 1)/2 = 3$ people in the Switch version of the trolley problem) is lower than the cost (5 people) of doing nothing. In the next section we will discuss from practical aspect of random selecting.

# 3 Practical Aspect of Random Selecting

So far, we discussed that random selecting also follows the deontological principle but has less disadvantage than the conventional "do nothing" option. In this section, we focus on practical aspect and issues on implementation.

## *3.1 Ethical Rules for Autonomous Vehicles*

In 2017, German Ethics Commission on Automated and Connected Driving has proposed ethical rules on autonomous vehicle (BMVI 2017b, see [15]), which try for the first time ever to provide official ethical guidelines for autonomous vehicles. These rules says that, in dilemma situations,

- The protection of human life should have top priority over animals' life or property (Ethical Guideline 7, [15]).
- "... any distinction based on personal features (age, gender, physical, or mental constitution) is strictly prohibited." (Ethical Guideline 9, [15]).

On the other hand, it is interesting that there is no clear stance on whether autonomous vehicles should be programmed to sacrifice the few to spare the many (Ethical Guideline 9, [15]). In fact, Ethical Guideline 8 [15] clearly stated that "a decision of deliberately sacrificing specific lives should not be taken by a programmer." Therefore obviously the only option following this guideline is our "random selecting" for unavoidable, dilemma situations like the trolley problem.

## *3.2 No Solution, no Problem*

We all know that it is unrealistic to design a perfect driving system, but we can have high reliable designs. For example, based on calculation, a well-functioning engine could be guaranteed on the possibility as high as 99.9%. Thus, someone may say the dilemma case for autonomous vehicles is rare, hence there is no real need to consider a solution for that kind of dilemma [9].

But what we should know is that it is still possible for that kind of situation to happen. When someone encounters this rare case, if there is no consensus in society, the lawsuit for the damage may be hard to settle, whose damage will not be easy to be covered. There should be algorithm settings for autonomous vehicles to implement when the dilemma happens. Any option other than our random-selecting must face the possible lawsuit says it intends to kill a special type of people (i.e., biased). In those cases, random selecting can be the only un-biased algorithm.

### 3.3 *Random Selecting*

Even random selecting may not be the best choice in common sense, it is still a practical solution following the deontological principle given the controversy of the utilitarian approach. If the "random selecting" option is adopted in the future, people may learn from the real experience of implementation. One possible result is that the situation is rare to happen and people are willing to accept the principle of random selecting. And the challenge to the universal value of humanity will not happen.

Comparing to the utilitarian approach, it is hard for the families of the victim to accept that their family member was killed because of some others preference. It seems impossible to decide a preference to follow and also impossible for the harmed one to accept the results. By contrast, we can find no one to blame if the outcome was made by random selecting. In this content, outcome of random choice is similar to the case of negligent homicide, whose definition is the killing of another person through gross negligence or without malice. This illustrates the necessity for autonomous vehicles to find a way out in reality. As we all know, a random choice may be seen as negligent homicide, which is a criminal, but it is quite different to kill someone because of bias, which may equal to murder.

## 4 Conclusion

This paper considered the trolley problem and proposed an option—random selecting—and compared it with two well-known ethical principles: utilitarian approach (i.e., saving the larger crowd) and the deontological approach (i.e., doing nothing). We conclude that random selecting is a better and practical deontological approach in the case of autonomous vehicles. As future works, we consider the final settlement of this problem requires more public consensus to implement and more feedback from the experience. Finally, letting machine (or AI) to do random selecting in other situation may generate serious results, thus how to develop a safe AI (or Beneficial AI) is an urgent task.

## References

1. Awad, E., Dsouza, S., Kim, S., Schulz, J., Henrich, J., Shariff, A., et al. (2018). The moral machine experiment. *Nature, 563,* 59–64.
2. Etzioni, A., & Etzioni, O. (2017). Incorporating ethics into artificial intelligence. *Journal of Ethics, 21*(4), 403–418.
3. Foot, P. (1978). The problem of abortion and the doctrine of the double effect. In *Virtues and Vices*, Basil Blackwell, Oxford. originally appeared in the Oxford Review, No. 5 1967.

4. Gerdes, J.C., & Thornton, S.M. (2015) Implementable ethics for autonomous vehicles. In M. Maurer, C. Gerdes, B. Lenz, H. Winner (Eds.), *Autonomous Driving. Technical, Legal and Social Aspects* (pp. 687–706). Springer, Heidelberg.
5. Goodall, N. (2014). Ethical decision making during automated vehicle crashes. *Transportation Research Record, 2424*(1), 58–65.
6. Harari, Y. N. (2017). *Homo deus: A brief history of tomorrow*. London: Vintage.
7. Hardin, G. (1968). The tragedy of the commons. *Science, 162*(3859), 1243–1248.
8. Hofstede, G. (2003). *Cultures consequences: Comparing values, behaviors, institutions and organizations across nations*. Thousand Oaks: Sage.
9. Johansson, R., & Nilsson, J. (2016) Disarming the trolley problem—why self-driving vehicles do not need to choose whom to kill. In *Proceedings VEHICLES 2016—Critical Automotive Applications: Robustness and Safety*, Gothenburg, Sweden.
10. Kagan, S. (1989). *The Limits of Morality*. Oxford: Oxford U. Press.
11. Keeling, G. (2019). Why trolley problems matter for the ethics of automated vehicles. *Sci. Eng. Ethics*. https://doi.org/10.1007/s11948-019-00096-1.
12. Kumfer, W., & Burgess, R. (2015). Investigation into the role of rational ethics in crashes of automated vehicles. *Transportation Research Record, 2489*(1), 130–136.
13. Leben, D. (2017). A rawlsian algorithm for autonomous vehicles. *Ethics and Information Technology, 19*(2), 107–115.
14. Lin, P. (2015). Why ethics matters for autonomous vehicles. In M. Maurer, C. Gerdes. B. Lenz, H. Winner (Eds.), *Autonomous Driving. Technical, Legal and Social Aspects* (pp. 69–85). Springer, Deutschland.
15. Luetge, C. (2017). The German Ethics Code for automated and connected driving. *Philosophy & Technology, 30,* 547–558.
16. Nyholm, S., & Smids, J. (2016). The ethics of accident-algorithms for self- driving vehicles: an applied trolley problem? *Ethical Theory and Moral Practice, 19*(5), 1275–1289.
17. Rahwan, I. (2016). "What Moral Decisions Should Autonomous Vehicle Made? - Ted", *Ted*, 2016. https://www.ted.com/talks/iyad_rahwan_what_moral_decisions_should_autonomous_vehicles_make.
18. "Trolley Problem - Wikipedia", *Wikipedia*, 2019. Retrieved in May 28, 2019, https://en.wikipedia.org/wiki/Trolley_Problem.
19. Unger, P. (1996). *Living High and Letting Die*. Oxford: Oxford U. Press.

# Effects of Upward and Downward Social Comparison on Productivity in Cell Production System

**Yuren Cao, Yanwen Dong, and Qin Zhu**

**Abstract** A lot of companies have introduced cell production systems, and realized empirically that it is very effective to prompt workers to compare themselves with or compete against others for improving cells' performance. As there is still insufficient researches to deal with human-related factors in cell production systems, fewer academic studies have paid attention to address the effect of social comparison in cell production systems. This study intends to explore effective social comparisons for improvement of cell production systems, and conduct an academic examination on the effects of upward and downward social comparison on the function of production cells. We design and conduct a cell production experiment. Based on the experiment result of 70 workers, we conduct a three-way ANOVA and two Bonferroni's multiple comparison tests. Through our experiment and analysis, it is clear that downward comparison has a stronger effect to enhance the productivity of production cells comparing to the upward comparison.

**Keywords** Cell production · Cellular manufacturing · Human factor · Social comparison · Upward and downward

---

This work was supported by JSPS KAKENHI Grant Number JP16K03848 and the National Natural Science Foundation of China (No. 71563028). It is also partly supported by Fukushima University Fund for the promotion of academic activities No. 19FE012.

Y. Cao (✉)
Graduate School of Symbiotic Systems Sciences, Fukushima University, Fukushima, Japan
e-mail: s1870037@ipc.fukushima-u.ac.jp

Y. Dong
Cluster of Science and Technology, Fukushima University, Fukushima, Japan
e-mail: dong@sss.fukushima-u.ac.jp

Q. Zhu
School of Economics and Management, Nanchang University, Nanchang, China
e-mail: zhuqin@ncu.edu.cn



J. Zhang et al. (eds.), *LISS2019*,
https://doi.org/10.1007/978-981-15-5682-1_49

## 1 Introduction

Cell production or cell production system is an important lean manufacturing system, in which the workers are divided into self-contained teams to complete a specific manufacturing process or product. Cell production has enabled greater flexibility to produce a high variety of low demand products while maintaining higher productivity in mass production. Numerous organizations have applied the concept of cell production in manufacturing and service processes. The successful implementation of cell production will bring improved benefits such as reducing delivery lead times and inventory of products, as well as remarkable improvements in product quality, scheduling, space utilization, operational control and employee morale [1, 2].

Industry and academia have widely accepted that both technical and human factors have a major role to play in the successful implementation of cell production systems. A lot of researchers have argued that people who will eventually manipulate, control, support and maintain the manufacturing cells should actively participate in their design and development in order to successfully implement cell production [3, 4]. Wemmerlov, et al. [5] surveyed 46 user plants with 126 cells and clarified that implementation of cell production is not simply a rearrangement of the factory layout; instead, it is a complex reorganization that involves organizational and human aspects. They also concluded that most of the problems faced by companies implementing cells were related to people rather than technical issues.

We conducted a series of experiments to study the impact of various human factors on the performance of production cells [6–8]. One of our research results indicated that two-thirds of the variance in productivity of production cells were decided by workers' individual differences [6], and there is three to five times of difference in productivity between the fastest workers and the slowest ones. Since the performance of production cells depend largely on workers, it is very important to introduce an effective mechanism to maintain the motivation of workers and encourage their production behavior to successfully implement cell production. In practice, Japanese companies that introduced cell production systems have made a lot of efforts to motivate workers for improving productivity. They have realized empirically that it is very effective to prompt workers to compare themselves with or compete against others for improving the productivity of production cells.

At the same time, many researchers have paid a lot of attentions to the effect of social comparison in the workplace. The social comparison seems to be embedded deeply into the fabric of organizational life and plays a critical role in influencing our behavior at work. As Kramer et al. [9] concluded in their report, social incentives in the workplace matter and symbolic rewards that trigger pride and shame can motivate individuals to work harder. Gino and Staats [10] explored how managers can use performance feedback to sustain employees' motivation and performance in organizations, and put forward several opinions on the role of performance feedback in motivating productivity in repetitive tasks. Cohn et al. [11] examined how workers respond to wage cuts, and whether their demands depends on the wages paid to their colleagues. They argued that social comparison has an asymmetric effects on efforts.

We have also investigated the impact of social comparison on the performance of production cells [12], and it is clear that through announcing relative performance information, the rank in assembly times for each worker, the competition among workers could be prompted effectively and as the result, the productivity could be increased by averagely 20%. In this paper, we improve our previous research and make an experimental study to investigate further the effects of upward and downward social comparison in cell production systems. We put our emphasis on clarifying how differently the comparison direction affects the performance of production cells.

This organizational structure of this paper is as follows. At the beginning of, we give a brief introduction of social comparison, the purpose and contribution of this study are elaborated and then outline the cell production experiment design. Next, we describe the cell experiment result briefly and then compare the effects of upward and downward social comparison on the productivity of cells. At last, we give some concluding remarks.

## 2 Social Comparison Theory and Comparison Direction

Social comparison is a central feature of human social life. In 1954, social psychologist Festinger [13] initially proposed the term social comparison and outlined a detailed theory of social comparison. He argued that people are inherently motivated to compare themselves with others and averse to "looking bad" in these comparisons, even when there are no consequences beyond the individuals' private awareness of the comparisons. Over the past five decades, the research on social comparison has undergone numerous transitions and reformulations. Many different methods, including interviewing people about their comparison habits and preferences and confronting individuals with vivid social comparison information, have been used in various researches on social comparison. A series of major theoretical developments such as classic social comparison theory, fear-affiliation theory, downward comparison theory, social comparison as social cognition, and individual differences in social comparison has been conducted [14].

The concept of downward social comparison was introduced by Wills [15]. When people look to another individual or group that they consider to be worse off than themselves in order to feel better about their self or their personal situation, they are making a downward social comparison. On the contrary, upward social comparison is a comparison with others who are better off or superior. Downward comparison theory emphasizes the positive effects of comparisons in increasing one's subjective well-being. Upward social comparisons are made to self-evaluate and self-improve in the hopes that self-enhancement will also occur. In an upward social comparison, people want to believe themselves to be part of the elite or superior and make comparisons highlighting the similarities between themselves and the comparison group.

Moreover, the social comparison theory has been applied to explain or solve some human-related issues such as getting cancer, eating disorders, the task division at home, or job satisfaction. Intending to encourage productive behavior, many organizations have prompted and guided social comparisons via the dissemination of relative performance information (RPI) or introducing competitive monetary incentives [16]. A lot of Japanese companies that introduced cell production systems have made various efforts to prompt workers to compare or compete against others for the improvement of the performance of production cells. However, as there are still insufficient researches to deal with human-related factors in cell production systems, fewer academic studies have been reported so far to examine the impact of social comparison on the performance of production cells.

## 3 Aim and Contribution of This Study

This study intends to improve our previous research [9] and conduct an academic examination on the effects of upward and downward social comparison on the performance of production cells. The main aim is to make the following contributions to the literature:

- To the best of our knowledge, there is no research reported so far to compare the effects of upward and downward social comparison on the performance of production cells, this study is the first attempt.
- As human-related problems are typically difficult to quantify, previous studies have mostly used questionnaire survey or case study methods. Because of this, the evidence may change with who answered the questionnaire. In order to measure the effects of comparison direction precisely, this paper applies the experimental study method. We design and conduct a cell production experiment, and use the assembly time to measure the performance of production cells. Thus, we can not only examine how differently the comparison direction affects the performance of production cells, but we can also measure to what extent the performance of production cells could be improved through introducing upward and downward social comparison.
- In the actual cell production systems, there are many factors affecting the performance of production cells, and it is not easy to measure the effect coming from just one factor. This is likely the main reason that there is no any report published so far that address the effects of social comparison although many Japanese firms have practiced internal competition as a motivating device for workers. In this study, we can identify and measure only the effect of comparison direction through designing the cell production experiment. Thus we can give a reasonable insight into how upward and downward social comparison can motivate workers and enhance the performance of production cells.

## 4 Cell Production Experiment Design

We design a laboratory experiment and use a toy robot consisting of LEGO Mindstorms as a virtual product (see Fig. 1). The toy robot consists of 106 pieces of parts and its assembly process is divided into 17 tasks. The workers are required to complete the assembling tasks according to the following three steps:

[Step 1] We give the workers an assembly manual, which consists of 17 figures, one figure for one assembling task. Then, an instructor gives the workers some instructions or demonstrations on the assembling tasks of the toy robot.
[Step 2] The workers assemble the robots in a one-person cell mode. When assembling, workers measure the operation time required to complete each task.
[Step 3] The assembly time of toy robot is calculated according to the sum of the operation time of all tasks. In order to study the learning effect, the assembling and time measurement are repeated five times.

We prompt upward and downward social comparison through disseminating the relative performance information according to the following two procedures:

(1) *Upward comparison:* we check the assembly times for a group of the workers every 15 min, and determine the rank of each worker in the group according to the shortest assembly time. Then we show the names of the workers who are in the upper half of the ranking on a big-screen video projector. But we don't provide any information related to the workers who are in the lower half of the ranking.
(2) *Downward comparison:* we check the assembly times for a group of the workers every 15 min, and determine the rank of each worker in the group according to the longest assembly time. Then we show the names of the workers who are in the upper half of the ranking on a big-screen video projector. But we don't provide any information related to the workers who are in the lower half of the ranking.

**Fig. 1** The toy robot as the virtual product

As the workers leave the laboratory at once they complete five times of assembling, we determine the ranking only for the workers who are still doing the assembling in the laboratory.

## 5 Basic Statistics of Assembly Times

We conducted the experiments from October 2018 to January 2019. As the workers, 70 students of Fukushima university participated in the experiment. For the limited space of the laboratory, the workers were divided into four groups, two for upward comparison and another two for downward comparison. Moreover, as there were two instructors to organize the experiment, we attempted the following two instructing methods together with the two instructors.

(1) *Simple instruction:* The instructor outlines some point of the assembly tasks, and the workers complete the assembling tasks mainly according to the assembly manual.
(2) *Demonstration:* The instructor demonstrates the order and essentials of the assembling process through assembling one toy robot. Following the instructor's demonstration, the workers assemble one toy robot too. After the instruction, the workers complete the assembling tasks according to the assembly manual.

Table 1 shows the basic statistics on the assembly time for the 70 one-person cell, where *n* is the number of workers in each group, "Downward" corresponds to the downward comparison and "Upward" to the upward comparison; "Simple" corresponds to the simple instruction method and "Demo" to the demonstration method of instructing. And, experience represents the order of the assembling and time measurement.

From Table 1, it is clear that:

- The average number of toy robots assembled for the first time was 10.99, 9.57, 11.61 and 11.71 (min) for the four groups respectively; these assembly times got shorter into 7.49, 6.77, 7.26 and 7.50 (min) on the fifth experience. With the reduction of assembly time and the increase of workers experience, a significant learning effect could be confirmed. This trend can be more clearly observed from Fig. 2. It shows the average assembly time that decreases along with experience increasing, where the average was calculated for each two groups of downward and upward comparison respectively.
- Comparing the average of minimum and maximum of assembly times for each group, there are 2.58, 2.27, 1.77 and 1.93 time of difference for the four groups respectively. This indicates that there is a big difference between the slowest and the fastest workers, and therefore workers' aptitude gives a strong influence on the production cells performance.

- As most of the means of assembly times for the two groups of upward comparison are larger than that for the two groups of downward comparison, this suggests that downward comparison is more effective to shorten assembly times than upward comparison.

**Table 1** Basic statistics of the assembly times (min)

| Statistics | Group (*n*) | Experience | | | | | *Average* |
|---|---|---|---|---|---|---|---|
| | | *1st* | *2nd* | *3rd* | *4th* | *5th* | |
| Mean | Downward + Simple (18) | 10.99 | 9.14 | 8.38 | 7.82 | 7.49 | 8.77 |
| | Downward + Demo (17) | 9.57 | 8.26 | 7.38 | 7.00 | 6.77 | 7.79 |
| | Upward + Simple (18) | 11.61 | 9.85 | 8.20 | 7.85 | 7.26 | 8.95 |
| | Upward + Demo (17) | 11.71 | 9.56 | 8.26 | 7.85 | 7.50 | 8.98 |
| Std. deviation | Downward + Simple (18) | 2.68 | 2.32 | 2.44 | 2.03 | 2.03 | 2.17 |
| | Downward + Demo (17) | 2.15 | 1.81 | 1.69 | 1.53 | 1.30 | 1.59 |
| | Upward + Simple (18) | 2.80 | 1.88 | 1.36 | 1.55 | 1.22 | 1.43 |
| | Upward + Demo (17) | 3.10 | 2.50 | 1.87 | 1.47 | 1.53 | 1.84 |
| Minimum | Downward + Simple (18) | 6.98 | 5.28 | 4.47 | 4.47 | 4.43 | 5.23 |
| | Downward + Demo (17) | 5.80 | 5.55 | 4.82 | 4.32 | 4.32 | 4.96 |
| | Upward + Simple (18) | 8.13 | 7.85 | 6.18 | 5.98 | 5.53 | 7.04 |
| | Upward + Demo (17) | 7.83 | 6.02 | 5.90 | 5.82 | 5.70 | 6.41 |
| Maximum | Downward + Simple (18) | 15.50 | 15.12 | 14.28 | 12.63 | 10.35 | 13.50 |
| | Downward + Demo (17) | 14.75 | 12.18 | 10.73 | 10.55 | 9.52 | 11.28 |
| | Upward + Simple (18) | 19.00 | 14.05 | 11.57 | 11.82 | 10.18 | 12.43 |
| | Upward + Demo (17) | 18.17 | 14.35 | 11.52 | 10.28 | 10.43 | 12.40 |

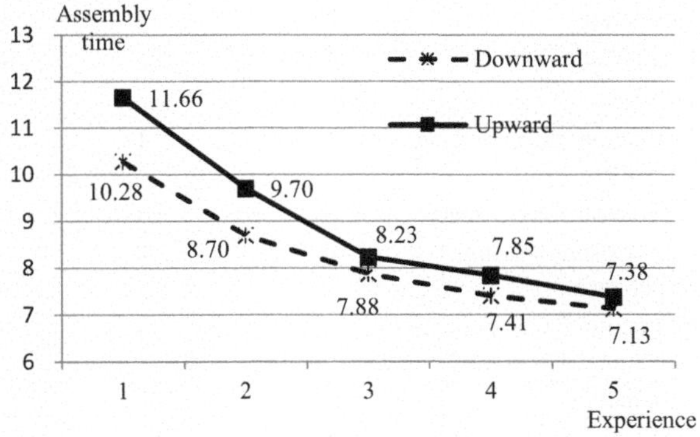


**Fig. 2** Average assembly time and experience

# 6 Verification of Main Effect and Interaction

In order to verify quantitatively the effect of comparison direction and other factors, we conducted a three-way ANOVA analysis, in which the dependent variable is assembly time, the three factors are comparison direction (upward or downward), instruction method (simple instruction or demonstration) and experience (learning effect). The result is shown in Table 2, three main effects and one interaction are statistically significant.

Concretely, we could observe that:

- Instruction method has a significant main effect on assembly times with *p*-value of 3.0%, it accounts for 0.94% (=19.69/2097.61) of the total variance. In addition to that interaction of instruction and comparison direction is also statistically significant with *p*-value of 2.3%, it is obvious that instruction method has a very weak impact on the assembly times and the effect changes with comparison direction.
- Comparison direction is also a significant main factor to affect assembly times with *p*-value of 0.2%, it accounts for 1.95% (=40.99/2097.61) of the total variance. As described above, because of the interaction effect of instruction and comparison direction, the effect of comparison direction depends also on the instruction method.
- Experience (learning effect) has a comparatively strong impact on assembly times with *p*-value of 0.0%. it accounts for 30.08% (=630.95/2097.61) of the total variance, and there is a marked rising from our previous research (20.0%) [6].

**Table 2** Three-way anova

| Factors | Sum of squares (Type III) | Degree of freedom | Mean square | *F* statistic | *p*-value |
|---|---|---|---|---|---|
| Instruction | 19.69 | 1 | 19.69 | 4.77 | 3.0%* |
| Comparison direction | 40.99 | 1 | 40.99 | 9.92 | 0.2%** |
| Experience | 630.95 | 4 | 157.74 | 38.18 | 0.0%** |
| Instruction * Comparison direction | 21.56 | 1 | 21.56 | 5.22 | 2.3%* |
| Instruction * Experience | 1.87 | 4 | 0.47 | 0.11 | 97.8% |
| Comparison direction * Experience | 16.57 | 4 | 4.14 | 1.00 | 40.6% |
| Instruction * Comparison direction * Experience | 2.08 | 4 | 0.52 | 0.13 | 97.3% |
| Residual | 1363.36 | 330 | 4.13 | | |
| Total | 2097.61 | 349 | | | |

*p<0.05, **p<0.01

- Residual (error) accounts for 65.0% (=40.99/2097.61) of total variance. According to our previous research, it mainly resulted from the workers' individual difference or their aptitude.

## 7 Effect of Comparison Direction

In order to clarify the effect of comparison direction, we conducted two Bonferroni's multiple comparison tests. Table 3 is to compare the effect of downward and upward comparison according to each instruction method, and Table 4 is to do the same comparison according to every experience. The column of "Difference in means (%)" in Tables 3 and 4 is the difference in average assembly time for the workers conducting upward comparison from that for the workers conducting downward comparison, and the rate (%) expressed in parentheses is the rate of the difference to the mean of assembly time for workers conducting upward comparison.

From Table 3, it is clear that if the instruction method is a demonstration, there is a significant difference in the effects of downward and upward comparison. Comparing to upward comparison, conducting downward comparison could shorten assembly time by 13.2%. However, if the instruction method is simply to give workers an assembly manual and outline some points, there is no significant difference between downward and upward comparison. We think the reason for this result is the method of conducting a social comparison. To prompt social comparison, we just showed the names of the workers who are in the upper half of the ranking on a big-screen video projector. If the workers didn't look at the projector screen, they wouldn't access to the ranking information. In the case of simple instruction, the workers completed the assembling tasks mainly according to the manual. The most likely scenario was that the workers concentrated on the manual and didn't have time to look at the projector screen. As the result, neither downward comparison nor upward comparison affects the performance of production cells. Therefore, there is no significant difference in average assembly time between the two comparison directions.

As shown in Table 4, the means of assembly times for the groups of workers conducting downward and upward comparison respectively have significant difference only on the first experience and the second experience. The effects of

**Table 3** Bonferroni's multiple comparison test by instruction

| Instruction | Mean for downward | Mean for upward | Difference in means (%) | Test statistic | *p*-value |
|---|---|---|---|---|---|
| Simple | 8.77 | 8.95 | 0.19 (2.1%) | 0.62 | 53.5% |
| Demonstration | 7.79 | 8.98 | 1.18 (13.2%) | 3.79 | 0.0%** |

**$p<0.01$

**Table 4** Bonferroni's multiple comparison test by experience

| Experience | Mean for downward | Mean for upward | Difference in means (%) | Test statistic | *p*-value |
|---|---|---|---|---|---|
| 1 | 10.28 | 11.66 | 1.38 (11.8%) | 2.84 | 0.5%** |
| 2 | 8.70 | 9.70 | 1.00 (10.3%) | 2.06 | 4.0%* |
| 3 | 7.88 | 8.23 | 0.35 (4.3%) | 0.72 | 47.0% |
| 4 | 7.41 | 7.85 | 0.44 (5.6%) | 0.90 | 36.8% |
| 5 | 7.13 | 7.38 | 0.25 (3.4%) | 0.51 | 60.8% |

** $p<0.01$, *$p<0.05$

downward and upward comparison are not significant from the third experience. This result was caused by the fact that because the workers left the laboratory at once they completed five times of assembling, the number of workers was decreasing as the experiment proceeded. On the first experience and the second experience, all of the workers were ranked by their performance. On the third experience or the later, the workers with higher performance have left the laboratory and fewer workers were ranked. The smaller the number of workers, the smaller the difference in their performance and the weaker the impact. Therefore, the effect of social comparison was weaken as the experience increased. As the result, there is less difference observed in average assembly time for different comparison direction. This implies that social comparison method in our experiment design should be refined.

## 8 Concluding Remarks

This study designed a laboratory experiment for cell production to examine the effects of downward and upward comparison. Based on the experiment result of 70 workers, we conducted a three-way ANOVA and two Bonferroni's multiple comparison tests. Through our experiment and analysis, it is obvious that:

- The main influences of three factors are comparison direction (upward or downward), instruction method (simple instruction or demonstration) and experience (learning effect), and one interaction (instruction method and comparison direction) are statistically significant.
- If the instruction method is a demonstration, there is a significant difference in the effects of downward and upward comparison. However, if the instruction method is simply to give workers an assembly manual and outline some points, there is no significant difference between downward comparison and upward comparison.
- Because of the method to show the relative performance information, the effects of downward and upward comparison are significantly different only on the first experience and the second experience.

Our further researches include that: (1) improving the method to show the relative performance information more effectively; (2) designing and conducting more effective downward comparison; (3) investigating the relationship between workers' personality and the effects of social comparison.

**Acknowledgements** We are grateful to Professor Munenori Kakehi for his valuable supporting and organizing of a part of our experiment.

## References

1. Yin, Y., Stecke, K. E., Swink, M., & Kaku, I. (2017). Lessons from seru production on manufacturing competitively in a high cost environment. *Journal of Operations Management, 49*(1), 67–76.
2. Yu, Y., & Tang, J. (2019). Review of seru production. *Frontiers of Engineering Management, 1*(1), 1–10.
3. Bopaya, B., Poonsiri, A., Kim, L. N., Bryan, A. N., & Wipawee, T. (2005). Human related issues in manufacturing cell design, implementation, and operation: a review and survey. *Computers & Industrial Engineering, 48*(3), 507–523.
4. Charlene, A. Y., & Harold, J. S. (2002). Cellular manufacturing for small businesses: key cultural factors that impact the conversion process. *Journal of Operations Management, 20* (5), 593–617.
5. Wemmerlov, U., & Johnson, D. J. (1997). Cellular manufacturing at 46 user plants: implementation experiences and performance improvements. *International Journal of Production Research, 35*(1), 29–49.
6. Hao, X., Haraguchi, H., & Dong, Y. (2013). An experimental study of human factors' impact in cellular manufacturing and production line system. *Information, 16*(7), 4509–4526.
7. Dong, Y., & Hao, X. (2016). Experimental study and statistical analysis of human factors' impact in cell production system. In R. J. Fonseca, G.-W. Weber, & J. Telhada (Eds.), *Computational management science, state of the art 2014. Lecture notes in economics and mathematical systems* (Vol. 682, pp. 107–113). Cham: Springer.
8. Dong, Y., Sato, S., Kumar, V., & Hoshino, K. (2016). Definition and verification of workers' aptitude toward assembly tasks in production cells. *Innovation and Supply Chain Management, 10*(1), 43–52.
9. Kramer, S., Maas, V. S., & van Rinsum, M. (2016). Relative performance information, rank ordering and employee performance: a research note. *Management Accounting Research, 33,* 16–24.
10. Gino, F., Staats, B.R. (2011). *Driven by social comparisons: how feedback about coworkers' effort influences individual productivity* (Harvard Business School NOM Unit Working Paper 11-078).
11. Cohn, A., Fehr, E., Herrmann, B., Schneider, F. (2012). *Social comparison and effort provision: evidence from a field experiment* (IZA Discussion Paper no. 5550). Bonn: Institute for the Study of Labor.
12. Dong, Y., Kakehi, M. (2018). An investigation on influence of competition among workers on productivity in cell production system. In *The Fourteenth International Conference on Industrial Management*, 12–14 September, Hangzhou, China.

13. Festinger, L. (1954). A theory of social comparison processes. *Human Relations, 7*(2), 117–140.
14. Buunk, A. P., & Gibbons, F. X. (2007). Social comparison: the end of a theory and the emergence of a field. *Organizational Behavior and Human Decision Processes, 102*(1), 3–21.
15. Wills, T. A. (1981). Downward comparison principles in social psychology. *Psychological Bulletin, 90*(2), 245–271.
16. Joan, L. (2016). Cooperation and competition among employees: experimental evidence on the role of management control systems. *Management Accounting Research, 31,* 75–85.

# Content-Based Weibo User Interest Recognition

**Wei Wang, Sen Wu, and Qingyao Zhang**

**Abstract** In recent years, more and more people use Weibo as an important tool to get new information. It is very important to identify the interest of Weibo user. In this paper we propose a content-based interest category recognition model for Weibo user. Firstly, the user data of Weibo was collected and the characteristics of text content were extracted by word2vec. Then, K-means clustering algorithm was used to cluster the user's characteristic data of the Weibo, and the user interest label was marked according to the clustering results. Finally, the random forest classification algorithm was used to train and classify the Weibo data to identify the interest of the user. The application results show that the performance of the content-based user interest identification method proposed in this paper, measured by the mean of F-measure, recall rate, and precision, is improved compared with the traditional method, which indicates that the proposed method is effective for the interest identification of Weibo user.

**Keywords** Data mining · Interest recognition · Word2vec · K-means · Classification

Supported by National Natural Science Foundation of China (NSFC) under Grant No.71271027.

W. Wang · S. Wu (✉) · Q. Zhang
Donlinks School of Economics and Management,
University of Science and Technology Beijing, Beijing, China
e-mail: wusen@mange.ustb.edu.cn

W. Wang
e-mail: wangweiustb@163.com

Q. Zhang
e-mail: zhangqingyao@imdada.cn



J. Zhang et al. (eds.), *LISS2019*,
https://doi.org/10.1007/978-981-15-5682-1_50

# 1 Introduction

In the past ten years, China's mobile Internet industry has developed rapidly. As shown by relevant data, the number of Internet users in China has reached 802 million by 2018, and the proportion of mobile Internet access accounts for 98.3% [1]. By December 2018, the monthly active users of Weibo have increased to 431 million, and the mobile terminal access accounts for 93%. The most important thing is that Weibo has become one of the top ten social platforms for users [2]. There is no doubt that the reach of the mobile Internet has extended to all aspects of people's daily lives. Weibo is one of the most representative platforms in the social networking platform. It is composed of thousands of Weibos, each of which has a length of no more than 140 words. Because Weibo is simple and easy to understand in the social network platform, user can publish, obtain and share information in a short time, making the update speed of information fast. In a word, Weibo has become an important channel for users to get the latest information and social trends.

The content posted by users on Weibo involves a variety of topics, which creates an information explosion problem and can't meet all users' interest. Due to the limited time for each user to get information every day, it is difficult for users to have enough time to choose their favorite information when facing thousands of Weibos every day. Only by accurately identifying user interest, can the platform better recommend relevant topics and accurately advertise user, and improve user's viscosity and goodwill. Therefore, it is of great practical value to classify and identify Weibo user interest.

To identify the interest of Weibo user, the characteristics of Weibo should be accurately extracted and vectorized. The text representation model is the basis of text recognition. Since the computer is not able to process unstructured data (such as text) well, we need to restructure the unstructured data according to some specific rules before performing feature extraction. The text representation model based on Boolean, vector space and topic has been widely used in natural language processing (NLP). Word2vec is a tool to transform words into vector form, and it can also introduce semantic features, which is good at short text classification [3, 4]. Rong et al. introduce the principle of word2vec and give a detailed derivation and explanation of the model parameter updating equation [5, 6]. At the same time, they created an interactive presentation to make it easy and intuitive to understand the model. Word2vec has been used in many ways since it was introduced in 2013. Word2vec can be used to calculate the word vector of microblogs [7]. Bai et al. calculated semantic similarity through word2vec, established an emotion dictionary, and used the constructed emotion dictionary to classify the test text [8]. Zhang et al. studied the classification of short microblog texts based on the word2vec model [9]. The word2vec model can also implement the vectorization of emotional words and their microblog sentences [10]. It can be seen that word2vec has great potential in the field of text recognition. The difference between the content-based study on the classification and recognition of Weibo user interest proposed in this paper and the methods in the literature mentioned above is that a new framework of the classification system of Weibo user interest is used in this study.

This paper tries to recognize Weibo user interest. First of all, we quantize the text words in order to extract the vector features of each Weibo. Then, we use K-means clustering algorithm [11] to cluster the users' Weibo characteristic data and mark the users' interest label according to the clustering results. Finally, we use data mining classification algorithm of the random forest algorithm [12–14] to conduct training and classification prediction on Weibo data and complete accurate classification and recognition of Weibo user interest.

# 2 Related Work

## *2.1 User Interest Classification Model Framework*

The interest classification and recognition model of Weibo user is mainly to classify and recognize the interest of Weibo user through relevant methods of data mining. The content-based interest classification and recognition model of Weibo user in this paper mainly includes the following three parts.

Python was used to crawl the information of Weibo users and obtain the information documents of Weibo users. The information of Weibo users is preliminarily screened out, and the interest labels are manually marked on the Weibo data according to the user information and Weibo content.

Data preprocessing is carried out to extract the Weibo content information in the Weibo information document. Then, the operation of word segmentation and removing stop words is carried out on the Weibo, which includes meaningless stop words, such as punctuation marks, pronouns, modal words, adverbs, and conjunctions. We remove these stop words according to the rules of part of speech. Furthermore, word2vec is used for feature extraction to transform the text into a vector that can be recognized by the computer, and the eigenvector matrix is obtained. Finally, K-means clustering algorithm is selected to cluster the characteristic data of Weibo by users and mark the category of user interest label according to the clustering results.

The random forest classifier was used for training and classification prediction of Weibo users data.

The specific model framework of this paper is shown in Fig. 1.

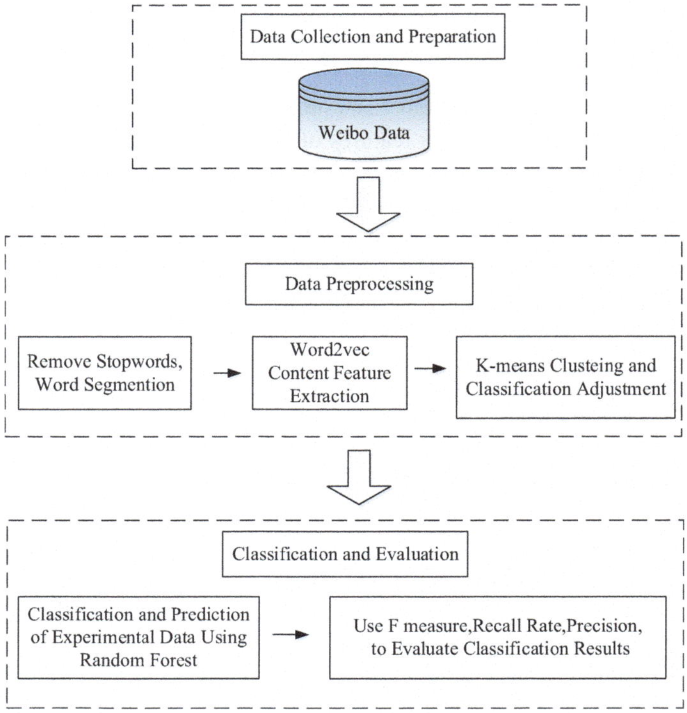


**Fig. 1** Schematic diagram of the framework of Weibo user interest recognition model

## 2.2 *Content Feature Generation*

Considering that word2vec is widely used in microblog, this paper uses word2vec method to quantize the words, which can map the content features of Weibos into a specified number of vectors. The word2vec model extracts word vectors based on contextual information of words in the text, and the generated word vectors carry contextual semantic information. The model mainly has two training models: CBOW and Skip-Gram. The CBOW model inputs the word vector of the surrounding words and outputs the word vector of the current word, which is to predict the current word by surrounding words. However, the Skip-Gram model is opposite to the CBOW model, its input is the word vector of the current word and the output is the word vector of the surrounding words, which are to predict surrounding words by the current word.. The Skip-Gram model is more accurate than the CBOW model, so this paper uses Skip-Gram as the experimental method. The schematic diagram of CBOW and Skip-Gram model in this paper is shown in Fig. 2.

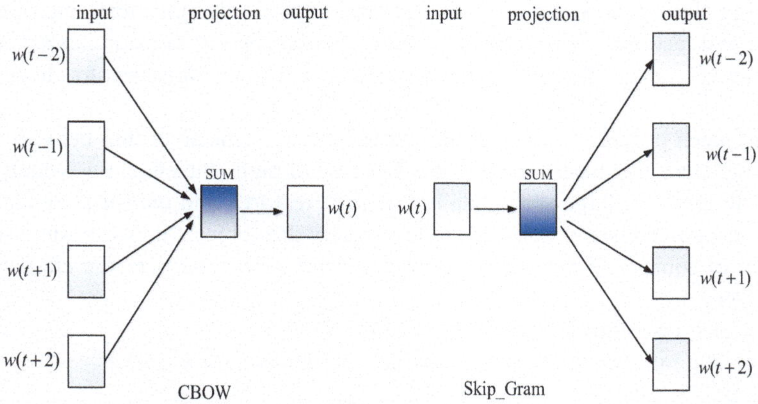


**Fig. 2** Schematic diagram of CBOW and Skip-Gram model

In the specific process of word vectorization, word2vec is used to first convert each word in the text after word segmentation into a word vector. Then the average value of each word vector in Weibo is taken as the vector feature of each Weibo. All the Weibo features of a user constitute the characteristics of a user.

The vector feature of each user's Weibo is to sum and then average the corresponding word vector of each Weibo, which is used to represent the vector feature of each Weibo. In a given Weibo corpus, any Weibo $S_j = \{w_1, w_2, \cdots, w_m\}$ is composed of $m$ characteristic words. The word vector corresponding to the characteristic word in this Weibo is $V_{w_i} = (v_{i1}, v_{i2}, \cdots, v_{in})$, then the characteristic vector of this Weibo can be expressed as.

$$\bar{S}_j = \frac{1}{m}\sum_{i=1}^{m} V_{w_i} \tag{1}$$

where $w_i$ represents the $i$ characteristic word in the Weibo, $m$ represents the number of the Weibo words, $V_{w_i}$ represents the word vector of the characteristic word, and $n$ is the word vector dimension.

## 2.3 *User Interest Category Optimization*

Traditionally, the label is manually labeled according to the user information and the content of Weibo, so the label obtained normally has some errors. Firstly, the user may have an unclear understanding of the category to which he belongs to. Secondly, the user interest is transferred. Finally, the user interest is ambiguous in the above categories. The solution is to cluster user characteristics and adjust user labels according to the clustering results. K-means clustering algorithm is a hard clustering algorithm, which is a classical algorithm to solve the clustering problem, simple, and fast. For processing large data sets, this algorithm maintains scalability and high efficiency. Therefore, this paper uses K-means algorithm to cluster user data.

MacQueen proposed K-means algorithm in 1967. This algorithm belongs to the clustering algorithm based on partition. Because of high efficiency, it is widely used in the clustering of large-scale data in scientific research and industrial application.

The idea of K-means algorithm is to divide $n$ data objects into $k$ clusters so that the sum of squares of data points in each cluster to the center of the class can be minimized.

K-means algorithm procedure is as follows.

Input: the number of clusters $k$ and data set containing $n$ objects.

Output: $k$ clusters.

Step1: The $k$ initial cluster centers are randomly selected from $n$ objects. For the convenience of description, the notation is introduced:

$$X = \{x_i | x_i \in R^m, i = 1, 2, \cdots, n\} \quad (2)$$

where X represents the data set to be clustered.

Step2: Calculate the similarity (distance) between each object and each cluster center, and assign all objects to the category with the highest similarity matching (closest distance). The distance measurement is often calculated by the European formula:

$$d(x_i, x_j) = \sqrt{(x_{i1} - x_{j1})^2 + (x_{i2} - x_{j2})^2 + \cdots + (x_{im} - x_{jm})^2} \quad (3)$$

where $x_i = (x_{i1}, x_{i2}, \cdots, x_{im})$ and $x_j = (x_{j1}, x_{j2}, \cdots, x_{jm})$ represents two $m$-dimensional objects.

Step3: The $k$ clustering centers are recalculated. The calculation formula is as follows.

$$z_j = \frac{1}{n_j} \sum_{x \in X_j} x \quad (4)$$

where $n_j$ represents the number of objects in the same class.

Step4: Compared with the $k$ cluster centers obtained in the previous time, if the cluster center changes (the clustering criterion function does not converge), repeat Step2 until no cluster center changes. Generally, the error squared criterion function is used as the clustering criterion function as follows.

$$J = \sum_{i}^{k} \sum_{j}^{n_i} d(x_j, z_i) \quad (5)$$

where $J$ is the sum of the mean square deviations of all objects in the data set. The clustering criteria shown are intended to make the same class obtained as compact as possible without separating the classes as far as possible.

Step5: Output $k$ clusters.

The specific classification optimization flow diagram proposed in this paper is as shown in Fig. 3.

Step1: According to the content and information of Weibo, we manually label the user interest categories.
Step2: Carry out data preprocessing for the Weibo text, and extract the feature vector of user Weibo content with word2vec.
Step3: Take the number of categories of manual classification as the $k$ value in K-means algorithm and conduct clustering.
Step4: Merge similar categories according to the clustering distribution, determine the new $k$ value after the combination, and return to Step3.
Step5: Until the clustering results are stable and output the clustering results.
Step6: Visualize the categories of clustering, we analyze the visualized results of clustering and adjust the categories output.

**Fig. 3** Classification system optimization flow diagram

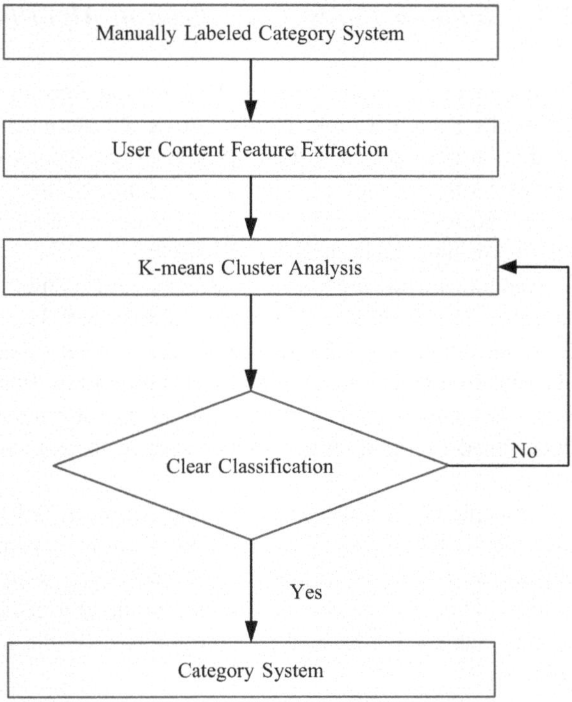

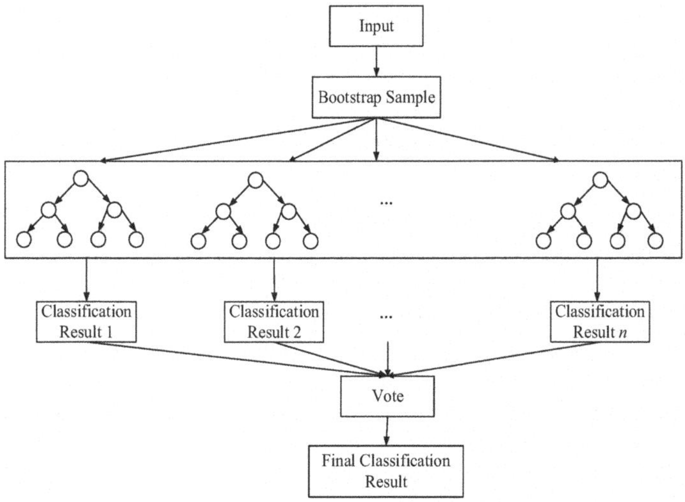


**Fig. 4** Random forest algorithm diagram

## 2.4 Random Forest Classification Algorithm

In this paper, the random forest classification algorithm is used to train and classify Weibo user interest. Several advantages of random forests are as follows. (1) high prediction accuracy, and it can produce high accuracy classifiers; (2) fast learning process; (3) insensitive to noise, and difficult to overfit. Therefore, the Weibo feature vectors obtained by classification optimization are classified into Weibo user interest using a random forest classifier.

Random forest refers to a classifier that uses multiple trees to train and predict samples. The classifier was firstly combined with Leo Breiman, and Adele Cutler's Bagging integrated learning theory and random subspace method to combine the classification trees into a random forest algorithm. In machine learning, the random forest is a classifier containing multiple decision trees. And the output category is determined by the number of categories of the categories output by the individual trees.

Principle of the random forest: random forest (RF) is a combined classifier that uses the bootstrap resampling method to extract $n$ samples from the original sample and model the decision tree for each bootstrap sample to obtain $n$ decision, tree models. Then, these decision trees are grouped together, and the final classification of each sample is obtained by voting based on the results of the $n$ classifications (Fig. 4).

**Table 1** Sample interest classification of table

| Classification result | True result | |
|---|---|---|
| | X true | X false |
| For X | *a* | *b* |
| Non-X | *c* | *d* |

## 2.5 Classification Result Evaluation Method

The evaluation of the Weibo text classification effect is measured and characterized by precision, recall rate, and F measure. Suppose that you want to see the classification results of the interest *X* class.

Precision refers to the ratio of the identified users interest in the proportion of all users, viz. the proportion of users whose interests are correctly classified by the classifier. The specific calculation formula is shown as follows.

$$\text{Precision} = \frac{a}{a+b} \tag{6}$$

where *a* refers to the number of users correctly classified as *X*, *b* refers to the number of users who are incorrectly classified into *a* category, *c* for users who are misclassified as other interest classes (Table 1).

The recall rate, also known as the full rate of recall, is used to measure the proportion of user interest that is correctly classified by the classifier.

$$\text{Recall} = \frac{a}{a+c} \tag{7}$$

F measure is the harmonic mean of Precision and Recall and is a comprehensive evaluation indicator.

$$\text{F} = \frac{2\,\text{Pr ecision} \times \text{Recall}}{\text{Precision} + \text{Recall}} \tag{8}$$

Finally, the average metric can be used to measure the average classification performance of the classifier across all categories.

# 3 Result Analysis

## 3.1 Data Preparation

There are too many types of interests covered on Weibo, results in many types of interest corresponding to users. Simply selecting the Weibo data of ordinary users will make the experimental data contain more noise, compared with the authenticated users. It is easier to distinguish in categories. According to the obvious principle of classification, this paper selects 1033 Weibos of 251 authenticated users. The collected data includes the basic information of the user and the content published by the user, the time of publication, and the tools published. The obvious error information such as repeated garbled data in the data of the selected user is initially removed, and then the initial classification according to the manual method is determined into 8 categories, and then each user is manually assigned to the closest category. The data volume and label of each category are as follows: current affairs, star, constellation, game, finance, science and technology, sports, live broadcast, as shown in Tables 2 and 3.

The collected Weibo corpus needs to be converted into a form that word2vec can handle. Firstly, the text should be preprocessed. The preprocessing mainly includes: word segmentation and removing punctuation stop words. Jieba, word segmentation toolkit in Python, is used for word segmentation. According to the stop word list to remove the stop words, considering that the string does not contribute much to the classification, all the strings are filtered and the words that have appeared in all classes are filtered. Weibo user interest classification belongs to short text classification and belongs to supervised learning. For the feature extraction of Weibo content, the word feature is extracted from the Weibo text using the word2vec model in the Gensim package. Considering the vectorization value of the microblog text, the data attribute is multidimensional and cannot be directly visualized. Through the current t-SNE method in manifold learning [15], the dimensionality reduction is visualized. The preliminary Weibo user interest classification data is manually classified into 8 types and visualization results are shown as in Fig. 5.

**Table 2** Labels of classification

| Label | Number | Label | Number |
|---|---|---|---|
| Current events | 1 | Finance | 5 |
| Star | 2 | Science and technology | 6 |
| Constellation | 3 | Sports | 7 |
| Game | 4 | Live broadcast | 8 |

**Table 3** Classification of manual Weibo data

| Label | Number of Weibos | Label | Number of Weibos |
|---|---|---|---|
| Current events | 440 | Finance | 75 |
| Star | 64 | Science and technology | 70 |
| Constellation | 112 | Sports | 120 |
| Game | 81 | Broadcast | 71 |

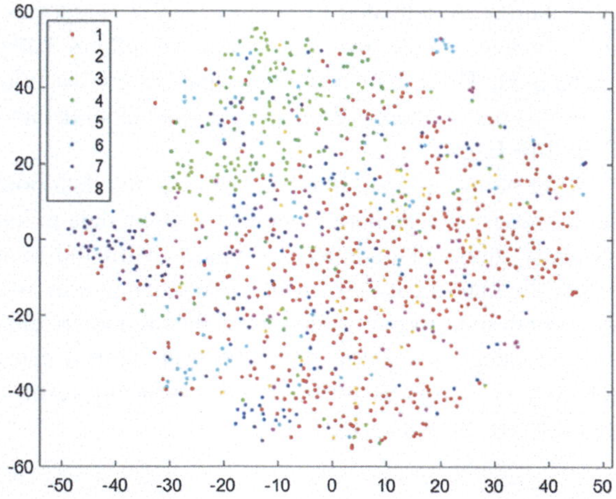


**Fig. 5** Manual 8 class initial feature visualization

In an ideal state, the points of each color (the points of the initial class) should be together, and the classification of points of other classes is obvious. In combination with the figure, we can know that the labels of manual labels are not accurate, and some data points are scattered. Some data points are relatively close in distribution but are divided into different classes, so it is necessary to be adjusted.

## *3.2 User Interest Category Adjustment*

Before training the interest recognition model, to determine the correctness of the category labeling, K-means algorithm is used to cluster the vector features of the users' Weibo. The clustering results do not fully represent the correct classification results but can be compared with the artificially labeled categories to determine the correct category.

The specific method of Weibo user interest category optimization is to use K-means algorithm to cluster Weibo user interest. The initial value of $k$ is set to the initial user interest class number 8, which is reduced by one each time. The iterative to clustering effect is relatively stable, and finally, the clustering effect is stable at $k = 5$, so the final number of categories is determined to be 5. During the clustering process, it is found that there are several clusters with similar distributions. At this time, the two subclasses will be adjusted into one parent class.

Comparing the visual distribution results of clusters with $k = 5$ with the results of visual distributions marked with 8 categories, analyzing the similarities and differences of the categories of users in these two results, and adjusting them to new ones based on the category labels of the original 8 categories. Most of the users in Category 3 and Category 4 are grouped into one category, and the original Category 3 and 4 label names are adjusted to entertainment; the Category 5 and Category 6 label names are adjusted to Finance, and 7 categories and The 8 categories of tags are adjusted to the event, the current affairs category and the star class are unchanged. The list of new tags obtained are as follows (Table 4).

In this paper, the $k = 5$ clustering table label list and visualization are obtained as shown in Fig. 6.

After being grouped into 5 categories, the data points of the users with the same color are visually gathered together, and the data points with similar distribution are basically gathered together. Compared with the manual 8 types of visualization maps, the classification of user interest categories is clearer and more in line with the actual situation. It is indicated that K-means clustering can optimize the user interest category system. After the user interest category label is determined, the category system is established. The clustering result category Weibo quantity histograms are shown in Fig. 7.

**Table 4** Adjusted classification label

| Label | Number |
|---|---|
| Science and finance | 1 |
| Entertainment | 2 |
| Competition | 3 |
| Star | 4 |
| Current events | 5 |

**Fig. 6** Feature visualization of $k = 5$ clustering

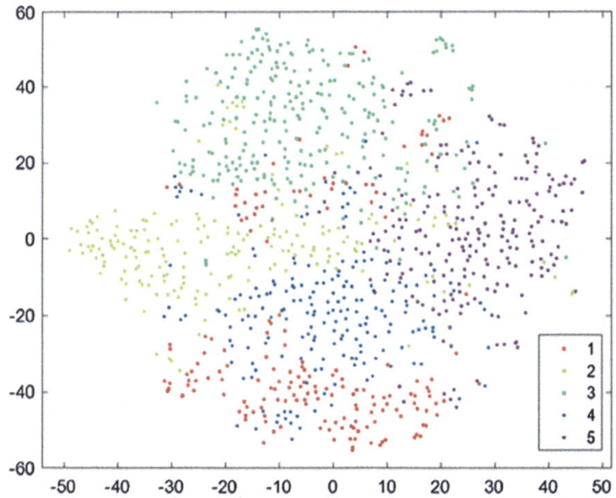

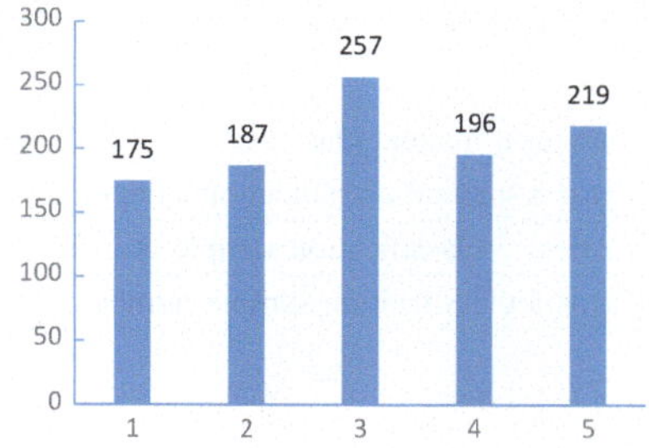


**Fig. 7** $k = 5$ clustering result category Weibo quantity histograms

## *3.3 Training and Testing of Weibo User Recognition Model*

According to the application of random forests in multivariate classification, this paper uses the random forest method in data mining as the classification algorithm. In this paper, 1033 pieces of data have been marked, and 10 fold cross-validation method is exploited. 90% of the data is used as the training set, and 10% of the data is used as the test set. 930 training data and 103 test data are obtained.

The application results show that the accuracy rate of random forest manual 8 classification is 55.37%, the error rate is 44.63%, the classification correct sample is 572, the error sample is 461, the random forest 5 classification accuracy can reach 88.27%, and the correct classification sample number is 912, the error rate is 11.72%, and the number of samples for the error is 121.

Weibo text used in this paper is not a formal written text. The content published by the user will not be as clear and focused as the written text, and it may be that the theme is cluttered and fuzzy. These factors all make the classification difficult. To further study the classification effect of each class, the F-measure, recall rate and precision are used as the evaluation indicators of the random forest classifier effect. The classification effect is as follows.

It can be seen from the Tables 5, 6 and 7 that the accuracy index of the manual identification of 8 categories of interest is only 0.554, and the average of the accuracy, recall rate and F-measure in the 8 categories application results are 0.554, 0.267 and 0.532. It is indicated that the classification effect of the 8 categories is not good; after K-means clustering algorithm is used to optimize the user interest category to 5 categories, the overall accuracy of the classification reaches 0.883. The average precision, recall, and F-measure in the 5 categories application results are 0.883, 0.883, and 0.882. The highest category precision is competition category, and the precision is 0.899; the lowest is entertainment, and the precision is 0.863. The highest recall rate is science and finance category, and the recall rate is 0.956; the lowest is the competition and the recall rate is 0.801; the highest of F-measure is the class of science and finance, and the F-measure is 0.919; the lowest is the competition class, and the F-measure is 0.847.

Table 5 8 manual classification and 5 classification results

| Item | Number | Percentage (%) |
|---|---|---|
| Correct 8 manual classification sample number | 572 | 55.37 |
| Error 8 manual classification sample number | 461 | 44.63 |
| Correct 5 classification sample number | 912 | 88.27 |
| Error 5 classification sample number | 121 | 11.72 |

Table 6 Random forest 8 manual classification results

| Categories label | Evaluation index | | |
|---|---|---|---|
| | Precision | Recall | F |
| Current events | 0.535 | 0.930 | 0.679 |
| Star | 0.001 | 0.001 | 0.001 |
| Constellation | 0.581 | 0.637 | 0.608 |
| Game | 0.517 | 0.188 | 0.275 |
| Finance | 0.909 | 0.267 | 0.412 |
| Science and technology | 0.579 | 0.157 | 0.247 |
| Sports | 0.620 | 0.367 | 0.461 |
| Live broadcast | 0.333 | 0.014 | 0.027 |
| Average | 0.554 | 0.267 | 0.532 |

Table 7 Random forest 5 classification results

| Categories label | Evaluation index | | |
|---|---|---|---|
| | Precision | Recall | F |
| Science and finance | 0.885 | 0.956 | 0.919 |
| Entertainment | 0.863 | 0.919 | 0.890 |
| Competition | 0.899 | 0.801 | 0.847 |
| Star | 0.893 | 0.863 | 0.878 |
| Current events | 0.882 | 0.831 | 0.856 |
| Average | 0.883 | 0.883 | 0.882 |

It can be obtained from the results that the classification precision, recall rate, and F-measure of the 5 category results are all greater than the 8 category results. In the 5-category results, the values of the three evaluation indicators increased, the precision average increased from 0.554 to 0.883, the recall average increased from 0.267 to 0.883, and the F-measure increased from 0.532 to 0.882. In summary, compared with before category optimization, the optimization of interest categories of Weibo user using K-means clustering algorithm and t-SNE visualization method can improve the classification accuracy of user interest category, indicating that the proposed method is effective.

# 4 Conclusion

Aiming at the problem of Weibo user interest classification and recognition, this paper proposes a classification model based on content-based Weibo user interest recognition. Firstly, we use the word segmentation and the method of removing the stop words to preprocess the Weibo published by Weibo users, and then we use the word2vec method to vectorize the Weibo user text. Furthermore, K-means clustering algorithm and t-SNE were used to cluster the interest category into 5 categories, and we label the user interest by clustering results. Finally, the random forest method was used to train and classify the Weibo data. The mean value of the evaluation indexes of our proposed method is higher than that of the traditional manual classification method, which validates that the content-based Weibo interest recognition model proposed in this paper is effective. However, the Weibo corpus collected in this paper is offline text information, and the dynamic user interest recognition and classification of real-time Weibo will be further studied.

# References


1. China's 802 million Internet users surf the Internet on mobile phones as high as 98.3% - Sohu News, Sohu News, 2018. https://www.sohu.com/a/249195943_100045145. Accessed 21 Aug 2018.
2. Weibo has 430 million active users - Sohu News, Sohu News, 2019. https://www.sohu.com/a/292443492_465976. Accessed 30 Jan 2019.
3. Mikolov, T., Chen, K., & Corrado, G. (2013). Efficient estimation of word representations in vector space. In *Proceedings of the International Conference on Learning Representations*, Arizona (pp. 1301–3781).
4. Mikolov, T., Sutskever, I., & Chen, K. (2013). Distributed representations of words and phrases and their compositionality. In *Proceedings of the Neural Information Processing Systems, NLPS*, Nevada (pp. 3111–3119).
5. Rong, X. (2014). Word2vec parameter learning explained. *Computer Science*, 1411–2738.
6. Lilleberg, J., Zhu, Y., & Zhang, Y. (2015). Support vector machines and Word2vec for text classification with semantic feature. In *2015 IEEE 14th International Conference on Cognitive Informatics & Cognitive Computing*, Beijing, New York (pp. 136–140).
7. Li, R., Zhang, Q., & Li, J. (2017). Microblog sentiment analysis based on weighted Word2vec. *Communications Technology, 50*(3), 502–506.
8. Zhang, Q., Li, R., & Li, J. (2017). Research of microblog short text classification based on Word2vec. *Netinfo Security, 1,* 57–62.
9. Bai, X., Fu, C., & Zhang, S. (2014) A study on sentiment computing and classification of sina microblog with Word2vec. In *IEEE International Congress on Big Data* (pp. 358–363). IEEE, Anchorage.
10. MacQueen, J. (1967). Some methods for classification and analysis of multivariate observations. In *Proceedings of the fifth Berkeley symposium on mathematical statistics and probability* (Vol. 1, pp. 281–297).
11. Wang, J., & Luo, L. (2018). Research on chinese short text classification based on Word2Vec. *Computer Systems & Applications, 27*(5), 209–215.
12. Kwok, S. W., & Carter, C. (1990). Multiple decision trees. *Machine Intelligence and Pattern Recognition, 9,* 327–335.

13. Ho, T. K. (1998). The random subspace method for constructing decision forests. In *IEEE Transactions on Pattern Analysis and Machine Intelligence*, (Vol. 20, pp. 832–844).
14. Breiman, L. (2001). Random forests. *Machine Learning, 45*(1), 5–32.
15. Dimitriadis, G., Neto, J. P., & Kampff, A. R. (2018). T-SNE visualization of large-scale neural recordings. *Neural Computation, 30*(7), 1750–1774.

# Distribution of the Profit in Used Mobile Phone Recovery Based on EPR

Yufeng Zhuang, Rong Huang, and Xudong Wang

**Abstract** In 2018, more than 300 million used mobile phones were produced, but the recovery rate of regular channels was less than 5%. Most of the recyclers were unwilling to recycle low-value used mobile phones, which led to waste of resources and potential environmental hazards. In recent years, the extended producer responsibility proposed by environmentalists can just solve this problem. At the same time, the research shows that China's express delivery efficiency is low, far from that of developed countries, and the return empty-load rate is high. Therefore, this paper assumes that manufacturers and logistics enterprises cooperate. The recovery is led by manufacturer, and meanwhile logistics enterprises are responsible for reverse supply chain, to solve the problem of recycling used mobile phones. At the same time, we should solve the problem of benefit distribution under this kind of cooperation. Nash negotiation model is chosen as the most suitable profit distribution model, and the model is improved. Finally, an example is given to verify the rationality of the algorithm.

**Keywords** Reverse logistics · EPR · Used mobile phone · Profit distribution

## 1 Introduction

Data from the Ministry of Industry and Information Technology show that by the end of 2018, 414 million mobile phones had been shipped in the domestic market, and the number of used mobile phones had exceeded 300 million. Waste mobile phones have high recycling value, which can reduce the waste of resources. At the

Y. Zhuang · R. Huang · X. Wang (✉)
School of Automation, Beijing University of Posts and Telecommunication, Beijing, China
e-mail: xudong_king@163.com

Y. Zhuang
e-mail: zhuangyf@bupt.edu.cn

R. Huang
e-mail: rongerhuang@126.com



J. Zhang et al. (eds.), *LISS2019*,
https://doi.org/10.1007/978-981-15-5682-1_51

same time, the waste mobile phones contain dangerous substances such as batteries, which may cause hidden dangers to the environment if they are discarded at will.

Although the recycling of waste mobile phones has experienced more than ten years of development, there are still many problems to be solved urgently, such as imperfect laws and regulations, low participation of consumers and low recycling efficiency. In fact, research shows that the recovery rate of the normal channel of used mobile phones is less than 5% [1].

In recent years, the EPR (extended producer responsibility) proposed by environmental economists can solve the problems mentioned above. It requires producers to be responsible for the environmental impact of the product throughout its life cycle, and in particular to require producers to take responsibility for waste disposal. Western countries are relatively mature in this respect. Most of them have formulated relevant laws and regulations on waste product recycling, stipulating extended producer responsibility. In China, until 2017, the Development and Reform Commission launched the extended Producer Responsibility Scheme, pointing out that by 2025, more than 50% of the waste products will be effectively recycled and utilized, and at the same time, the extended producer responsibility of key categories will be further improved.

At the same time, third-party logistics has played an important role in the recycling of electronic waste, and has formed a relatively mature recycling model, which has become one of the effective ways of sustainable development.

Based on the above background, this paper use EPR, and the recovery of used mobile phones will be done through the cooperation between manufacturers and logistics enterprises, The course is led by manufacturers, and meanwhile logistics enterprises will be responsible for reverse supply chain, to solve the problem of recycling used mobile phones. At the same time, to solve the problem of interest distribution under this kind of cooperation, Nash negotiation model is chosen as the most suitable one, and the model is improved. Finally, an example is given to verify the rationality of the algorithm.

## 2 Literature Review

### *2.1 About Reverse Logistcs, Recovery Logistics, Return logistics*

Compared with forward logistics, reverse logistics has existed for a long time, but the definition of reverse logistics is less than 30 years. In developed countries such as Europe and the United States, the understanding and research of reverse logistics is earlier than in China. After years of development, the definition and understanding of reverse logistics is also developing. In 1992, the American Logistics Management Association (CLM) formally put forward the concept of reverse logistics. It introduced that reverse logistics is a series of logistics activities involving the return or maintenance of products, the reuse of waste products and the disposal of waste. Reverse logistics includes return logistics and recovery logistics.

Return logistics is related to the flow of normal products. It is the logistics generated after the return of the purchased products due to unsatisfactory or unqualified products. For example, the goods purchased online are returned due to quality problems or personal reasons, and the returned goods flow due to wrong delivery of orders.

Recovery logistics refers to the logistics that enterprises or other institutions recycle used waste goods to consumers.

## 2.2 About EPR

The theory system of extended producer responsibility abroad has been studied earlier. Wilmshurst et al. [2], focused on the field of packaging recycling based on the extended producer responsibility. Spicer et al. [3] discussed three recycling modes under EPR: manufacturer recycling, joint recycling and third-party recycling, and compared them qualitatively. Forslind [4] applies manufacturer responsibility to the automotive industry, pointing out that besides the automotive manufacturer's responsibility to dispose of abandoned vehicles on its own initiative, automotive consumers also have the responsibility to send abandoned vehicles to the automotive manufacturer on their own initiative.

Most domestic scholars have translated the achievements of the extended producer responsibility system abroad, while some scholars have combined the extended producer responsibility system with the domestic recycling of renewable resources and put forward constructive suggestions. Tong [5] introduced the concept of the extended producer responsibility system, and the practical experience of developed countries in recent years. Combining with the current situation of China's recycling industry, Tong put forward the great significance of this policy for China's recycling industry. Jiang [6] analyzed the problems existing in the recycling of abandoned automobiles in China and the reasons for the problems. At the same time, he established a reverse logistics network of abandoned automobiles based on the extended producer responsibility system.

## 2.3 About Recycling Mode of Used Mobile Phone

Morrell [7], Meade [8], Spicer [9] analyzed the mode of third-party reverse logistics in this paper. The third party reverse logistics company takes over the reverse logistics business of the product enterprise, and deals with discarded goods. Manufacturers are required to pay a fee to third-party reverse logistics providers to ensure that their products are handled in accordance with legislation and environmental protection. Sun [10] analyzed three operating modes of reverse logistics, and introduced self-built reverse logistics, joint operation reverse logistics and third-party reverse logistics respectively. Xu [11] analyzed in detail three commonly used reverse logistics modes: self-run, joint operation and outsourcing, studied the selection and evaluation problems of the three modes, and analyzed the influencing factors and selection methods. Chang [12] studied the government responsibility system, production responsibility system and third-party responsibility system under the extended producer responsibility system, and introduced the characteristics and applicability of each recovery mode. Xia and Wang [13] put forward a joint recycling model that is conducive to giving full play to the main responsibility by studying the successful recycling model of mobile phones abroad, namely the extended producer responsibility system.

## 2.4 About Profit Distribution

In the supply chain, the distribution of interests among partners is particularly important, because scientific and rational distribution of interests can effectively avoid conflicts among different stakeholders, increase the stability of cooperation, and achieve win-win result among members. At present, most of the research results are to solve the problem of profit distribution among the partners in the forward supply chain, while the research on the problem of profit distribution in the reverse supply chain is less.

Wang [14] and others first elaborated the important position of manufacturers, distributors and customers in reverse supply chain, then studied the profit distribution under the participation of manufacturers, the profit distribution without the participation of manufacturers, and the profit distribution under centralized decision-making. Wang et al. [15] used Lagrange mean value theorem to solve the profit distribution problem of both parties involved in reverse logistics. Gong et al. [16] used Nash negotiation model to distribute profits among participants in reverse supply chain, and improved the model according to the capital and status of participants.

# 3 Recycling Mode

## 3.1 Three Kinds of Recycling Mode

The extended producer responsibility clearly stipulates that the manufacturer is responsible for the recycling of waste mobile phones, which establishes the main position of mobile phone manufacturers in the reverse logistics of mobile phones. According to the ways in which mobile phone manufacturers participate in recycling, mobile phone recycling modes can be divided into the following three types:

(1) *Producer self-operated recycling model:* The producer's self-operated recycling mode refers to the establishment of a reverse logistics system by the manufacturer of electronic products, handling the return of the products and recycling of used and discarded goods, and is solely responsible for the reverse logistics business of the enterprise.
(2) *Producer Joint Recycling Model:* The joint recovery mode of producers in reverse logistics of electronic products refers to the joint venture of electronic products with the same industry to establish reverse logistics system. The established reverse logistics system is jointly managed by the cooperative enterprises, and a joint organization is formed among the cooperative enterprises, which is responsible for the recovery task.
(3) *Third-party logistics Recycling Model:* The third party recycling mode of electronic products is also known as the outsourcing mode of reverse logistics, which means that enterprises do not directly participate in the reverse logistics recycling process, but pay a fee to the specialized reverse logistics company for product recycling and processing. The third party reverse logistics enterprise is fully responsible for the recovery and treatment of products, which can be handed over to the original manufacturer or transferred to a specialized third party manufacturer.

By comparing the three main recycling modes, the advantages of the third-party recycling mode are relatively prominent. The third-party recycling mode has lower cost, lower risk, faster information feedback, less difficulty in logistics management and higher recycling efficiency, which is suitable for the recycling of most products. Therefore, this paper adopts the recycling mode of cooperation between producers and logistics enterprises.

### 3.2 Cooperation Cases Between Manufacturers and Logistics Enterprises

Faced with the huge waste mobile phone market, some manufacturers recycle waste mobile phones through direct stores and network channels. The model of mobile phone manufacturers establishing network recycling platform to recycle used mobile phones only began to appear in China in 2015. At the end of July 2015, meizu released a product "meizu blue 2", and announced meizu's "mCycle for home" mobile phone recycling and replacement plan, through cooperation with "repurchase network" to recycle used mobile phones and exchange old ones for new ones. Consumers submit the basic information of used mobile phones on the website, and "repurchase network" is responsible for the pricing of used mobile phones. Meizu feedbacks the quotation to consumers, and if the consumers confirm the transaction, meizu staff will inform the express company to collect it at the door through the sf-express appointment. Consumers can receive "meizu repurchase coupons" at the corresponding price, which are used to purchase equivalent goods on meizu's official online platform.

## 3.3 *The Cooperation Mode Between Manufacturers and Logistics Enterprises in this Paper*

This paper adopts the recycling mode of cooperation between manufacturers and logistics enterprises, which is dominated by manufacturers and logistics enterprises are responsible for the logistics of recycling process. Regulated by government departments and the public, manufacturers are required to recycle used mobile phones with the same proportion of market share. At the same time, two recycling schemes are designed. For high value used mobile phones, cash settlement method can be adopted for low value mobile phones, old ones can be exchanged for new ones for subsidies. To some extent, this can increase the sales volume of manufacturers, which is a win-win result. Recycling channels can be covered both online and offline. Recycling can be carried out online through mobile official website or shopping platform, while recycling can be carried out offline through mobile retail stores.

# 4 Research Hypotheses

The common profit distribution patterns among partners are as follows:

A. *Fixed Payment Mode*

The main approach of this model is that the main members of the alliance pay the remaining fixed profits of the enterprise according to the pre-signed contracts or agreements, and the rest are entirely enjoyed by the main members.

B. *Output Sharing Model*

This model means that all partners can share their profits in accordance with a clear proportion. The proportion is related to cost and risk.

C. *Mixed Model*

This model is a combination of the two models. In addition to paying certain remuneration to other members, the main members will also pay some remuneration to other members proportionally from the proceeds.

In the actual situation of this paper, the output sharing model is more suitable. Firstly, from the perspective of manufacturers, we do not have to bear the cost risk of recycling low-value used mobile phones; for logistics enterprises, this can share more profits and improve the enthusiasm of participation.

From the current research point of view, there are many achievements in the cooperative benefit allocation of the forward supply chain, which can be used as the basis for the study of the cooperative benefit allocation of the reverse supply chain, such as Shapley value method and Nash negotiation model, as well as the establishment of group focus model, Raiffa's ruling and so on.

The Shapley Value Method and Nash Negotiation Model mentioned above are game theory. Game theory refers to the discipline of implementing strategies by using relevant methods among teams under certain conditions. There are many game situations in our daily life, such as playing chess.

Game theory is divided into cooperative game and non-cooperative game. If the participants can reach a binding agreement in the course of the game, it is called cooperative game, otherwise it is called non-cooperative game. For example, two enterprises cooperate to build a production line. Party A provides production technology and Party B provides plant and equipment. The non-cooperative game is formed for the value evaluation of technology, plant and equipment. If Party B can know Party A's true evaluation of technology, it can maximize its own evaluation value, and so can Party A. Participants give priority to their own interests and make choices, which is non-cooperative game.

(1) *Shapley value method*

Shapley value method, proposed by Shapley in 1953, is one of the commonly used methods to solve the profit distribution problem between two or more cooperators. When n subjects cooperate to accomplish certain economic activities, each combination of cooperation methods can achieve benefits, and the number of cooperation increases, will not reduce benefits. So all n people cooperate to get the best benefit.

The traditional Shapley value method is based on the assumption of the same risk, investment and other factors for each subject, without taking into account the differences between each subject. Usually, analytic hierarchy process, network analysis and fuzzy mathematics can be used to estimate the possible benefits of various alliance combinations, and then proceed to benefit distribution.

At the same time, one of the preconditions of using Shapley value method is that it needs to calculate the profit of each different cooperation combination, but in practice, it is difficult to do so.

In the case of this paper, logistics enterprises and manufacturers are two different entities. It is impossible to describe the risk-sharing factors and costs of alliance members in a quantitative way.

At the same time, logistics enterprises and manufacturers cannot reach a binding agreement in the game process. Throughout the process, logistics enterprises provide front logistics for the recycling of used mobile phones. Manufacturers are responsible for the disposal and resale of used mobile phones, whether sold directly or downstream. During the whole process, the profit of the used mobile phone is relatively stable, when the participants choose their own actions, they will give priority to how to safeguard their own interests, rather than maximizing the collective interests.

(2) *Nash negotiation model*

Nash negotiation model regards all parties involved in the distribution as a cooperative game alliance. By reaching an implementable agreement, in order to improve the overall profits, the two parties start negotiations and finally reach a satisfactory benefit distribution plan.

(3) *Raiffa's ruling*

Raiffa's ruling was proposed by Elliot and Archibald. Because other profit distribution models only consider the fairness of income distribution, but Raiffa's ruling takes into account the limitations of the minimum and maximum profits. This method not only absorbs the essence of Shapley value method, but also guarantees the interests of the weak.

There are many achievements in cooperative benefit allocation of forward supply chain, which can be used as the basis for the study of cooperative benefit allocation of reverse supply chain, such as Shapley value method and Nash negotiation model, as well as Raiffa's ruling and so on. At the same time, one of the preconditions of using Shapley value method is that it needs to calculate the profit of each different cooperation combination, but in practice, it is difficult to do so.

In the case of this paper, logistics enterprises and manufacturers are two different entities. It is impossible to describe the risk-sharing factors and costs of alliance members in a quantitative way. At the same time, logistics enterprises and manufacturers cannot reach a binding agreement in the game process. It is a non-cooperative game. Throughout the process, logistics enterprises provide front logistics for the recycling of used mobile phones. Manufacturers are responsible for the disposal and resale of used mobile phones, whether sold directly or downstream. During the whole process, the profit of the used mobile phone is relatively stable, when the participants choose their own actions, they will give priority to how to safeguard their own interests, rather than maximizing the collective interests. Based on the above analysis, this paper will adopt Nash negotiation model to allocate the cooperative benefits of reverse supply chain, and take into account the status weight of both partners participating in reverse supply chain, and use the improved Nash negotiation model to allocate profits.

# 5 Model

## *5.1 The Nash Model*

The extended producer responsibility clearly stipulates that the manufacturer is responsible for the recycling of waste mobile phones, which establishes the main position of mobile phone manufacturers in the reverse logistics of mobile phones. According to the ways in which mobile phone manufacturers participate in recycling, mobile phone recycling modes.

The problem is described as follows:

The manufacturer evaluates the actual situation of the used mobile phone, sets the quotation of the discarder, and obtains the net profit in the whole logistics process. The participants in the whole process are manufacturers and logistics enterprises. Assuming that the total profit available for distribution is 1, $v_1$ and $v_2$ represent the profit allocation quota for each manufacturer and logistics enterprise respectively, $v_1 + v_2 = 1$.

In the negotiations, both manufacturers and logistics enterprises have the lowest acceptable bottom line, i.e. the minimum acceptable profits, which are $v_1^*$ and $v_2^*$ respectively, and meet the requirements of $v_1^* + v_2^* \leq 1$. At the same time, manufacturers and logistics enterprises have their own best plan, that is, the distribution plan they advocate. The distribution scheme proposed by the manufacturer is represented by $(v_{11}, v_{12})$, $(v_{21}, v_{22})$ is represented by the distribution scheme proposed by the logistics enterprises, and there is $v_{11} + v_{22} \geq 1$.

Assuming that the two parties are in a complete information game, the following basic assumptions are made:

- Assuming that there are two subjects in the whole negotiation process, the set of participants $N = \{1, 2\}$. All the optional policy sets of each participant are expressed as policy spaces, which are expressed as $T_t$. Then the policy spaces of the two participants are Cartesian product $T = T_1 * T_2$.
- There is no information concealment between the two sides, and the negotiation status of the two sides is equal. Both sides have infinite patience, while the gains are not reduced by the prolongation of negotiation time.
- Under any strategy combination $t(t \subseteq T)$, the benefit $v_t$ of each participant can be quantified. And under a certain strategy combination, the benefit $v_t$ of both participants satisfies the condition: $v_1 + v_2 = V$.
- If the negotiation between two participants breaks down and no cooperation agreement is reached, the benefit obtained by the two participants when they do not cooperate is expressed in terms of $(v_1^*, v_2^*)$. $(\overline{v_1}, \overline{v_2})$ express the final solution of cooperative negotiation, it represents the final profit of participant 1 and participant 2 respectively.

On the basis of these assumptions, Nash proves and gives the following conclusions:

If $(v_1, v_2) \in T$ satisfies $v_1 \geq v_1^*$, $v_2 \geq v_2^*$, then the product maximization problem.

$$max\, g(v_1, v_2) \triangleq (v_1 - v_1^*)(v_2 - v_2^*) \tag{1}$$

There exists a unique optimal solution.

By establishing Lagrange multiplier function, the equilibrium solution of the Nash negotiation model can be obtained as follows:

$$\begin{cases} \overline{v_1} = v_1^* + \frac{1}{2}\left(V - v_1^* - v_2^*\right) \\ \overline{v_2} = v_2^* + \frac{1}{2}\left(V - v_1^* - v_2^*\right) \end{cases} \quad (2)$$

The above Nash model will seek the equalitarian equilibrium solution for all members of the alliance, but often in the alliance, the participants investment costs, negotiating positions and risks are different. Therefore, considering the actual situation, this paper will consider the status differences of the two participants and their respective satisfaction.

## 5.2 The Improved Nash Model

(1) *Introduction of negotiation status factor*

In the process of cooperation between manufacturers and logistics enterprises, the status is different. The final result of profit distribution will be affected by status. Therefore, the negotiation status factor w is introduced here. The negotiation status of manufacturers and logistics enterprises is expressed by $w_1$ and $w_2$ respectively, and $w_1 + w_2 = 1$. There are many factors affecting the size of w, including the investment cost and risk of manufacturers and logistics enterprises. W is proportional to investment cost and risk.

(2) *Introduction of Satisfaction*

If the maximum expected return value of the two participants is $v_{max}$, and when the return value is $v_i$, the satisfaction degree is $f_i = v_i/v_{max}$. If the minimum received return value is $v_{min}$, the minimum received satisfaction degree is $f_{min} = v_{min}/v_{max}$.

So the product maximization problem:

$$max\, g(v_1, v_2) \triangleq \left(\frac{v_1}{v_{11}} - \frac{v_1^*}{v_{11}}\right)^{w_1}\left(\frac{v_2}{v_{22}} - \frac{v_2^*}{v_{22}}\right)^{w_2} \quad (3)$$

There exists a unique optimal solution:

$$\begin{cases} \overline{v_1} = v_1^* + \left(1 - v_1^* - v_2^*\right)\frac{w_1 v_{11}}{w_1 v_{11} + w_2 v_{22}} \\ \overline{v_2} = v_2^* + \left(1 - v_1^* - v_2^*\right)\frac{w_2 v_{22}}{w_1 v_{11} + w_2 v_{22}} \end{cases} \quad (4)$$

The above formula shows the specific distribution of the profit of each participant in the improved Nash model. The result consists of two parts. The first half of the right side of the equation is the lowest bottom line received by the negotiators. The second part takes into account the negotiating position of the negotiators. Considering the satisfaction of both parties, the model is closer to the actual and more operable after the adjustment of the distribution.

# 6 Empirical Results

## 6.1 *The Calculating Method of the Profit of Used Mobile Phone*

First of all, before the profit distribution, we need to know the specific calculation method of the profit of each phone. Suppose the main parameters of used mobile phones are as Table 1 shows:

Based on the following assumptions, the calculation method is different for each case:

(1) *Used mobile phones that can be resale directly:*

$$P_1 = T_a - T_b - L - A_2 - C_1 \tag{5}$$

(2) *Used mobile phones that can be resale after renovation:*

$$P_2 = T_a - T_b - L - A_1 - A_2 - C_1 \tag{6}$$

(3) *Used mobile phones that must be remanufactured:*

$$P_3 = C_2 - C_3 - C_5 - C_1 - A_1 - A_2 - L \tag{7}$$

(4) *Used mobile phones that must be recycled for environmental protection:*

$$P_4 = C_4 - C_5 - C_1 - A_1 - A_2 - L \tag{8}$$

**Table 1** The main parameters of used phones' profit calculation

| Variables | Meanings |
|---|---|
| $T_b$ | Recycling price of used mobile phones from customers |
| $T_a$ | The selling price of second-hand mobile phones |
| L | The fixed cost invested by the manufacturer, including the manpower and material resources of the second sale and the transportation expense, etc. |
| $A_1$ | Variable recovery costs paid by the manufacturer to the processor |
| $A_2$ | Testing costs paid by the manufacturer |
| $C_1$ | transportation costs of products paid by logistics enterprises |
| $C_2$ | The cost of making a phone from new material |
| $C_3$ | The cost of producing a mobile phone by the manufacturer using parts of the recycled used mobile phone |
| $C_4$ | The value of precious metals obtained after the complete end-of-life mobile phone has been processed (generally recyclable) |
| $C_5$ | The amount of a subsidy given to customers |

## 6.2 Example of Profit Distribution Model

An example is introduced and Nash negotiation model is used for calculation. Assuming that the profit is 180 yuan, the distribution scheme proposed by the manufacturer is (170,10), the minimum profit accepted is 160 yuan, the distribution scheme proposed by the logistics enterprise is (150,30), and the minimum profit accepted is 8 yuan.

(1) *According to the above data, using formula* (3)

$$\left(v_1^*, v_2^*\right) = \left(\frac{8}{9}, \frac{2}{45}\right), \left(v_{11}, v_{22}\right) = \left(\frac{17}{18}, \frac{3}{18}\right)$$

(2) *Firstly, without considering the factors of negotiation status and satisfaction, the allocation ratio should be as follows:*

$$\begin{cases} \overline{v_1} = v_1^* + \frac{1}{2}\left(V - v_1^* - v_2^*\right) = 0.922 \\ \overline{v_2} = v_2^* + \frac{1}{2}\left(V - v_1^* - v_2^*\right) = 0.078 \end{cases} \tag{9}$$

(3) *Considering negotiation status and satisfaction, it can be concluded that the distribution ratio of the two is as follows:*

$$\begin{cases} \overline{v_1} = \frac{8}{9} + \left(1 - \frac{8}{9} - \frac{2}{45}\right) \frac{w_1 v_{11}}{w_1 v_{11} + w_2 v_{22}} \\ \overline{v_2} = \frac{2}{45} + \left(1 - \frac{8}{9} - \frac{2}{45}\right) \frac{w_2 v_{22}}{w_1 v_{11} + w_2 v_{22}} \end{cases} \tag{10}$$

For different values, the following table can be obtained (Table 2):

As a result, the manufacturer invests more manpower, material resources and risk costs in the whole process, so the manufacturer has a higher negotiating position, and the proportion of profits is gradually increasing. This shows that the profit in the model is proportional to the negotiating position. That is, the profit distribution of the improved model is positively correlated with the risk cost, and the manufacturer gains more profits in the process. This is also in line with the actual situation, which shows that the improved algorithm has a certain practical application.

This paper mainly considers the algorithmic innovation in the recovery mode. Generally, there are many studies on the profit distribution among the partners in

**Table 2** Profit distribution in different negotiating positions

| Parameter | $(v_{11}, v_{12})$ | $(\overline{v_1}, \overline{v_2})$ |
|---|---|---|
| Distribution ratio | (0.2, 0.8) | (0.908, 0.092) |
| | (0.4, 0.6) | (0.915, 0.085) |
| | (0.6, 0.4) | (0.919, 0.081) |
| | (0.8, 0.2) | (0.921, 0.079) |

the forward supply chain, but few on the algorithm of the profit distribution among the partners in the reverse supply chain. Therefore, this paper applies the profit allocation algorithm between the partners of the forward supply chain to the profit allocation problem of the reverse supply chain.

In the case of this paper, logistics enterprises and manufacturers are two different entities. It is impossible to describe the risk sharing factors and costs of alliance members by a comparative quantitative method.

At the same time, logistics enterprises and manufacturers cannot reach a binding agreement in the game process. Non-cooperative game. Throughout the process, logistics enterprises provide the front logistics for the recycling of used mobile phones. Manufacturers are responsible for the disposal and resale of used mobile phones, whether sold directly or downstream. During the whole process, the profit of the used mobile phone is relatively stable, when the participants choose their own actions, they will give priority to how to safeguard their own interests, rather than maximizing the collective interests. So Nash model has some applicability.

**Acknowledgements** I would like to thank my teacher, Professor Yufeng Zhuang. During the completion of the thesis, Mr. Zhuang put forward a lot of constructive opinions to help us improve our thesis. The rigorous attitude of the teacher and the broad thinking have given me great encouragement and encouragement, so that I can complete the graduation design more seriously.

## References

1. Silent Giant Mine: Research on Mobile Phone Recycling Market in 2017 (2018). https://weibo.com/ttarticle/p/show?id=2309404214928201399399. Accessed 10 Apr 2019.
2. Wilmshurst, N. R., & Newson, P. L. (1996). Packing tomorrow's challenge. *Journal of Logistics Focus, 4*(1), 13–15.
3. Spicer, A. J., & Johnson, M. R. (2004). Third-party demanufacturing as a solution for extended product responsibility. *Journal of Cleaner Production, 12*(1), 37–45.
4. Forslind, K. H. (2005). Implementing extended producer responsibility: The case of Swedens car scrapping scheme. *Journal of Cleaner Production, 1*(136), 19–29.
5. Tong, X. (2003). Extended producer responsibility principle in electronic waste management. *Journal of Environmental Management in China, 1*(4), 1–4.
6. Jiang, M. Y. (2011). *Research on reverse logistics network of scrapped vehicles based on EPR*. Chongqing University.
7. Merrel, A. L. (2001). The forgotten child of the supply chain. *Journal of Modern Materials Handing, 5*(1), 33–36.
8. Mecade, L., & Sarkis, J. (2002). A conceptual model for selecting and evaluating third-PArty reverse logistics providers. *Journal of Supply Chain Management, 7*(5), 283–295.
9. Spicer, A. J., & Johnson, M. R. (2004). Third Party demanufacturing as a solution for extended producer responsibility. *Journal of Cleaner Production, 12*(4), 37–45.
10. Sun, S. L. (2008). Research on reverse logistics operation mode of electronic products. *Journal of Logistics Engineering, 1*(4), 88–90.
11. Xu, J., & Zhang, Y. L. (2006). Research on the mode decision of reverse logistics of waste electronic products. *Journal of Modern Logistics, 29*(128), 14–16.
12. Chang, X. Y., Fan, T. J., & Huang, J. Y. (2006). Reverse logistics management mode based on extended producer responsibility. *Journal of Modern Management Science, 5*(12), 35–37.

13. Xia, X. Q., & Wang, D. P. (2011). Research on reverse logistics mode of mobile phone recycling in China. *Journal of Beijing Jiaotong University, 4*(8), 285–294.
14. Wang, W. B. (2006). *Study on evaluation and optimization of agricultural logistics system*. Shandong Normal University (2006).
15. Wang, X., Dai, Y., Yang, J. Z., & Lin, Y. (2009). Research on benefit distribution mechanism of reverse supply chain based on circular economy. *Journal of Chongqing University, 15*(6), 52–56.
16. Gong, W. W., Ge, C. C., Chen, J. X., & Pan, J. G. (2011). Benefit distribution model of three-level reverse supply chain cooperation based on nash negotiation. *Journal of Industrial Engineering and Management, 16*(3), 16–21.

# Operating High-Powered Automated Vehicle Storage and Retrieval Systems in Multi-deep Storage

**Andreas Habl, Valentin Plapp, and Johannes Fottner**

**Abstract** Automated vehicle storage and retrieval systems (AVSRS) allow for a flexible and powerful way to supply picking or manufacturing areas based on the goods-to-person principle. AVSRS with tier- and aisle-captive vehicles achieve the highest throughput capacity which can be further increased by deploying more than one vehicle on each tier of an aisle. With these high-powered AVSRS, it is possible to compensate for the reduction of throughput capacity in multi-deep storage systems where relocation operations occur frequently and result in more transportation tasks. In this work, we present and compare different algorithms for efficiently accomplishing relocation operations in high-powered AVSRS. We conduct a series of simulation experiments to analyze the performance of these algorithms and evaluate the application of high-powered AVSRS in multi-deep storage.

**Keywords** Automated vehicle storage and retrieval systems · Shuttle systems · Discrete-event simulation · Scheduling

## 1 Introduction

Automated vehicle storage and retrieval systems (AVSRS) are used for storing transportation units (TUs) such as containers, boxes or trays in order to supply picking or manufacturing areas based on the goods-to-person principle [1]. As a key characteristic of this technology, horizontal and vertical transport are largely separated from each other and are thus executed by shuttle and lift vehicles, respectively [2].

A. Habl (✉) · V. Plapp · J. Fottner
Chair of Materials Handling, Material Flow, Logistics, Technical University of Munich, Munich, Germany
e-mail: andreas.habl@tum.de

V. Plapp
e-mail: valentin.plapp@tum.de

J. Fottner
e-mail: j.fottner@tum.de


J. Zhang et al. (eds.), *LISS2019*,
https://doi.org/10.1007/978-981-15-5682-1_52

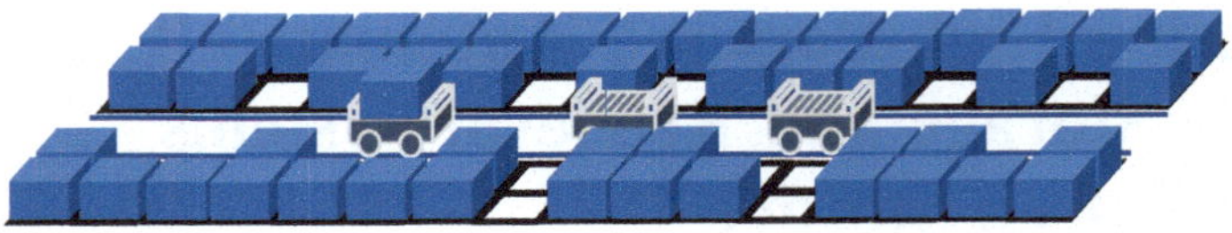

**Fig. 1** Several shuttle vehicles operating on a tier of an aisle

Due to the decoupling of horizontal and vertical transports, AVSRS can achieve a significantly higher throughput capacity compared to conventional automated small parts storage systems. As derived already in [3], AVSRS with tier- and aisle-captive vehicles can achieve the greatest throughput. However, the throughput capacity of AVSRS could be further increased by deploying multiple vehicles on every tier of an aisle (see Fig. 1). The resulting version of high-powered AVSRS provides a powerful solution in rapidly changing environments. Since additional vehicles could also be added in an already existing system, high-powered AVSRS are able to flexibly adapt to vehicle fleet size and throughput capacity in the horizontal direction, respectively. As a consequence, the throughput capacity of existing AVSRS can be subsequently increased and transformed into high-powered AVSRS.

In order to reduce storage area by maximizing storage density, multi-deep storage is applied to AVSRS in many cases. However, multi-deep storage systems usually face a loss of throughput capacity since relocation operations may cause additional transportation tasks. With the help of high-powered AVSRS, the reduction of throughput capacity could be compensated for.

The remainder of the paper is organized as follows: In the following section, we reveal related work in the respective areas. In Sect. 3, we present different algorithms for efficiently coordinating vehicles and executing relocation operations. In Sect. 4, we compare the performance of the different relocation strategies in the context of the conducted simulation study. Finally, a conclusion and outlook follow in Sect. 5.

## 2 Related Work

The following section initially contains a review of literature in the field of the controlling of AVSRS. Since scheduling problems with non-crossing constraints also occur in other applications, existing literature in the respective research areas is presented. Finally, a brief overview about relocation strategies in automated warehouses is provided.

## 2.1 Control of Automated Vehicle Storage and Retrieval Systems

Especially when deploying more than one vehicle on each tier of an aisle, the applied control strategies to AVSRS are crucial for the efficiency of the system. However, the scheduling of vehicles in AVSRS is mostly realized by priority rules.

The most related paper to this issue is presented in [4] by scheduling two non-passing lift vehicles that share a common mast. This paper has been improved in [5] by considering the acceleration and deceleration of the lift vehicles. Since both studies focused on the lifting system, no relocation operations had to be considered. In addition, the number of lift vehicles has been limited to two vehicles, whereas in high-powered AVSRS, there can be more than two vehicles operating on a tier of an aisle.

## 2.2 Control Algorithms in Other Applications

In high-powered AVSRS, several vehicles move along the same rail. In order to prevent collisions and blockings between the vehicles, scheduling algorithms need to be applied in such systems. However, we can find similar scheduling problems with non-crossing constraints in other applications like port and factory cranes (crane scheduling problem), elevators, and industrial robots.

Reference [6] present a dynamic programming algorithm and a related beam search heuristic to schedule two gantry cranes that move along one rail while avoiding interferences. Reference [7] present a heuristic algorithm to schedule several factory cranes on a common track by using a block sequencing method. Reference [8] present an optimized zoning strategy for controlling two elevator cars in one shaft by using a genetic algorithm to determine the optimal zoning. Reference [9] present exact and heuristic algorithms to schedule two robots operating on a common line.

## 2.3 Relocation Strategies in Automated Warehouses

In multi-deep storage systems relocation operations may become necessary if TUs block a retrieving TU in the rack. To efficiently accomplish the additional transportation tasks, relocation strategies need to be applied. Especially in the field of conventional automated storage and retrieval systems, relocation strategies have already been addressed in [10–12]. However, relocation strategies in AVSRS have been addressed only in double-deep AVSRS [13] or have not been necessary because multi-deep AVSRS are limited to homogeneous storage channels [14].

# 3 Control Algorithms for Operating High-Powered AVSRS in Multi-deep Storage

In this section, we present the control algorithm that accomplishes the scheduling of vehicles and we focus on emerging challenges when deploying several vehicles on one tier of an aisle as well as on the adaption to multi-deep AVSRS. Subsequently, we describe different relocation strategies that can be applied in multi-deep AVSRS.

## 3.1 Scheduling Algorithm

In high-powered AVSRS, multiple vehicles move along a common rail. The vehicles need to be coordinated in such a way that blockings and collisions among themselves can be avoided as well as empty driving times and waiting times minimized. Similar to the aforementioned crane scheduling problem, the vehicle scheduling problem in AVSRS can be regarded as a special type of general machine scheduling problem [15], which is characterized as NP-hard [16]. The vehicle scheduling problem may be stated as: $n$ jobs $\{j_1, j_2, \ldots, j_n\}$ have to be processed by $m$ vehicles $\{s_1, s_2, \ldots, s_m\}$.

In order to find a schedule that optimizes the processing with regards to the minimization of completion time, a scheduling algorithm needs to be applied. In this work, the scheduling algorithm relies on the block sequencing algorithm for scheduling gantry cranes that is presented in [7] and it is adapted to AVSRS in [17].

Based on the block sequencing approach and according to the selected block size, a number of jobs (block) is extracted from the job list for processing. The algorithm creates a decision tree of possible states (partial and complete schedules) to assign the extracted jobs to the vehicles and it selects the schedule with the lowest completion time. The final schedule is then forwarded to execution where vehicles complete the jobs. After the last job has been finished, the next iteration starts by extracting the next block of jobs from the job list.

Since the number of possible states grows exponentially, the following heuristics is developed in [7] in order to improve the algorithm's efficiency:

- Pruning inferior partial schedules (dominance check)
- Chopping partial schedules when the heuristics threshold is passed

However, in order to apply the scheduling algorithm in multi-deep AVSRS, relocation has to be included by carrying out an additional process (see Fig. 2). When a block of jobs is extracted from the job list, all retrievals are copied into a retrieval list that is further processed by determining and sorting the number of necessary relocations for each retrieval. As long as the retrieval list contains retrievals requiring relocations, the respective relocations are added to job execution. Otherwise, the retrieval is added to job execution and removed from the retrieval list. After all retrievals are added to job execution, all storages can be

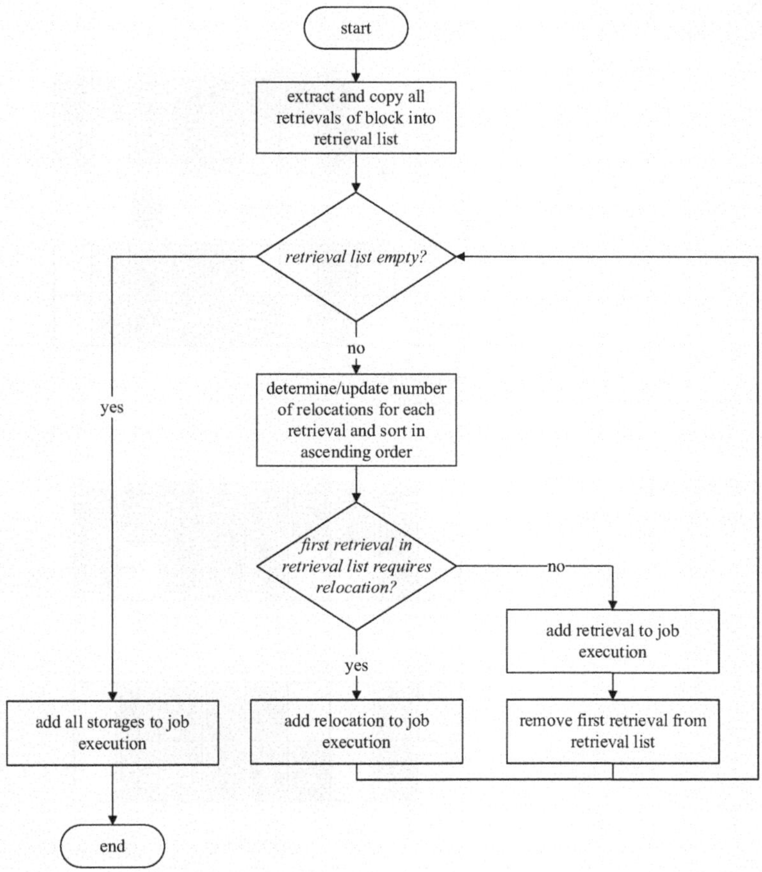


**Fig. 2** Process of including relocations in scheduling algorithm

processed. Since TUs are always stored in the last position, storage jobs don't require relocations and thus, they can be added to job execution without further processing.

## 3.2 *Relocation Strategies*

In multi-deep AVSRS, additional transportation tasks have to be executed by the vehicles, because there can be blocking TUs at retrievals. For an efficient execution of relocation tasks, relocation strategies need to be applied. In the following, we describe three strategies on the basis of Fig. 3, which represents a section of a tier in an aisle with lifting systems on the front side.

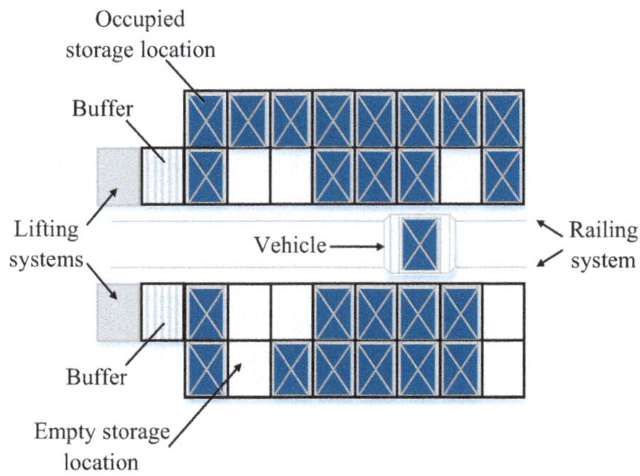


**Fig. 3** Section of a tier in an aisle of a AVSRS

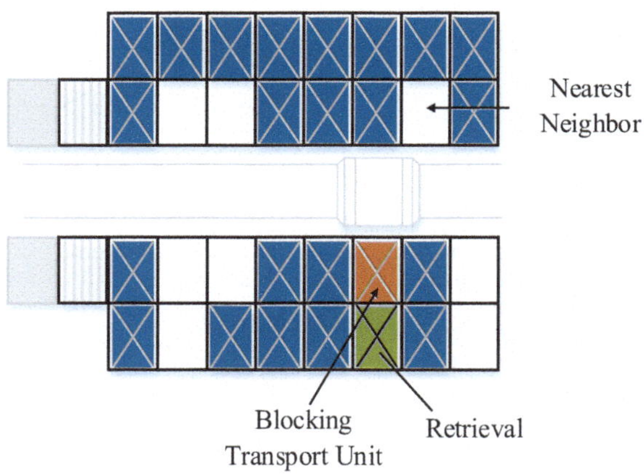


**Fig. 4** Relocation by applying the strategy nearest neighbor

(1) *Nearest neighbor*

The strategy Nearest Neighbor looks for an empty storage location that is closest to the blocking TU (see Fig. 4). This minimizes the distance that needs to be traveled by the vehicle in order to relocate the TU. Since this strategy doesn't require information about future jobs, it is a simple and robust solution for relocation operations. Only information about free storage locations and their positions have to be available.

(2) *Relocation close to retrieval*

This strategy tries to relocate the blocking TU to a storage location close to a future retrieval. As shown in Fig. 5, the blocking TU isn't relocated to the closest storage location (nearest neighbor) since there is a free storage location close to another retrieval. This saves time since one accelerating and decelerating process as well as

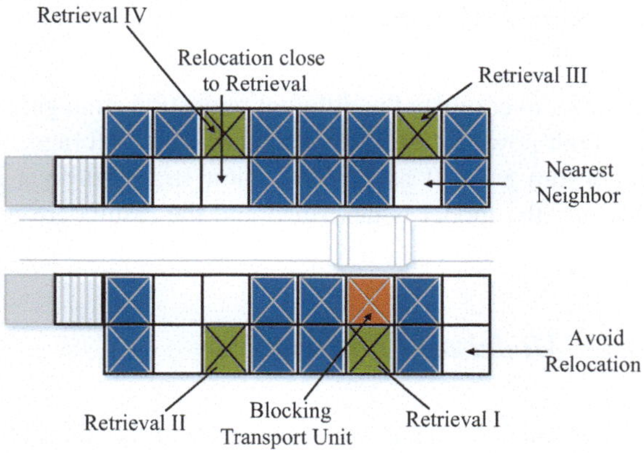


**Fig. 5** Relocation by applying the strategy relocation close to retrieval

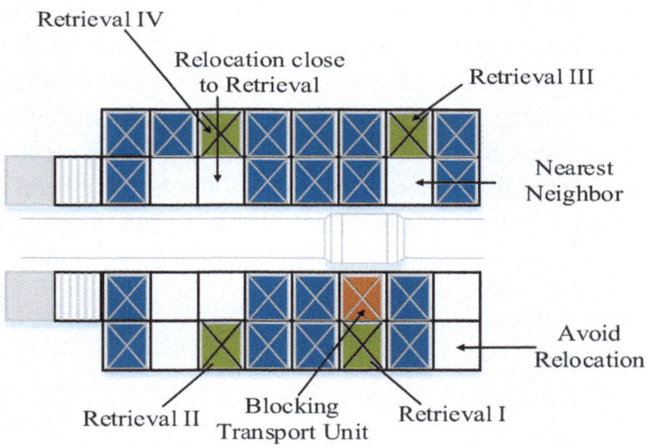


**Fig. 6** Relocation by applying the strategy avoid relocation

one alignment process can be omitted by the vehicle. However, a requirement for this strategy is the availability of information about free storage locations and their positions as well as future retrievals. In addition, the sequence of retrievals is not predefined.

(3) *Avoid relocation*

If information about future jobs is available, relocations can be avoided by storing blocking TUs only in storage locations that aren't affected by future retrievals. Figure 6 illustrates the difference from the other strategies. Both strategies Nearest Neighbor and Relocation close to Retrieval relocate the blocking TU to storage locations that interfere with upcoming retrievals. As a consequence, the strategy Avoid Relocation looks for a storage location without any known retrieval.

# 4 Simulation Study

In order to compare the different relocation strategies and to evaluate the application of high-powered AVSRS in multi-deep storage, we implemented the control algorithm as well as the relocation strategies in a simulation model. In the following, the model is described and the results are presented and evaluated.

## 4.1 Model and Parameters

The simulation model was created using the discrete-event simulation environment Tecnomatix Plant Simulation. It represents a segment of an AVSRS: One tier of an aisle with multi-deep storage.

Table 1 shows the parameter specifications of the individual items used in the simulation experiments. Kinematic parameters of the vehicles are based on [2].

In order to compare the different relocation strategies, a simulation experiment for every strategy is conducted. In each experiment, the number of vehicles is varied from 1 to 4. Since storage and retrieval jobs are created randomly, every experiment is repeated five times. As soon as 500 jobs have been completed by the vehicles, the experiment has finished and the time needed to carry out these jobs can be measured.

**Table 1** Simulation parameters

| | |
|---|---|
| *Vehicle* | |
| Velocity | 2 m/s |
| Acceleration | 2 m/s$^2$ |
| Deceleration | 2 m/s$^2$ |
| Pick-up time basis | 3 s |
| Pick-up time per depth | 1 s |
| *Rack* | |
| Number of aisles | 1 |
| Number of tiers | 1 |
| Number of bays | 200 |
| Distance between bays | 0.5 m |
| Initial/Min./Max. Storage ratio | 80/60/99% |
| Storage depth | 1–4 |
| *Lift* | |
| Number per aisle | 2 |
| Position in aisle | Centered |
| Number of transfer buffers per lift | 2 (storage and retrieval) |
| *Scheduling* | |
| Block size | 10 |
| Heuristics threshold | 400 states |
| Heuristics deletion | 40 states |

## *4.2 Simulation Results*

Figures 7, 8 and 9 show the obtained throughput by applying the relocation strategies Nearest Neighbor, Relocation close to Retrieval, and Avoid Relocation as well as a strategy based on random relocations with respect to the number of deployed vehicles. As can be seen, by deploying up to three vehicles, the throughput increases at every storage depth. Each of the presented relocation strategies achieves a higher throughput in comparison to a random relocation. The strategies Avoid Relocation as well as Nearest Neighbor allow for the highest throughput. With increasing storage depth, the number of relocations also increases and as a consequence, the difference between the strategies becomes greater.

**Fig. 7** Throughput depending on the relocation strategies and number of vehicles in double-deep storage

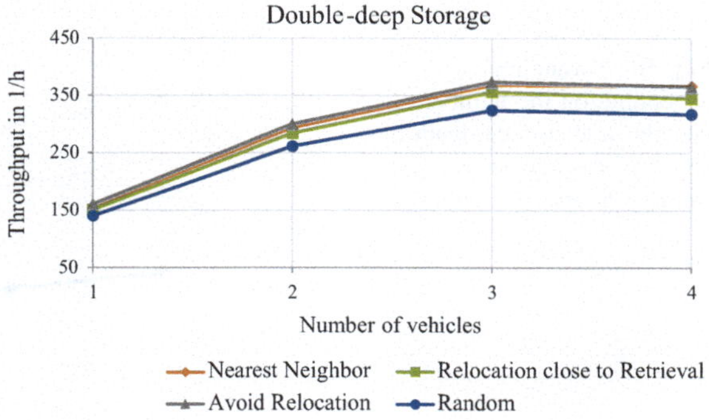


**Fig. 8** Throughput depending on the relocation strategies and number of vehicles in triple-deep storage

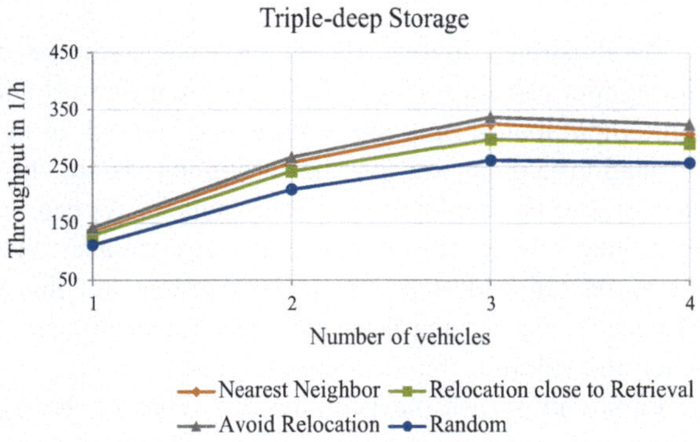

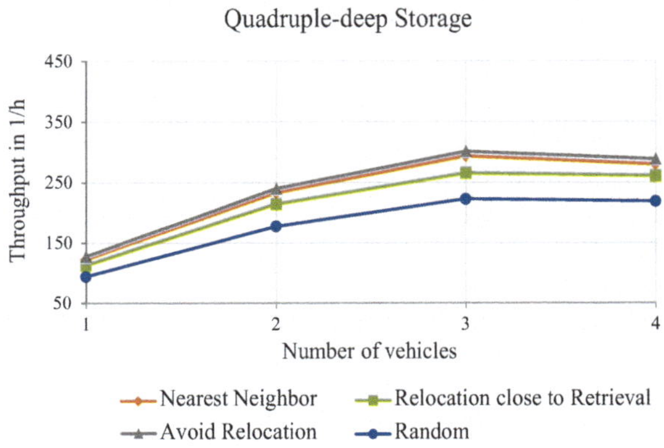


**Fig. 9** Throughput depending on the relocation strategies and number of vehicles in quadruple-deep storage

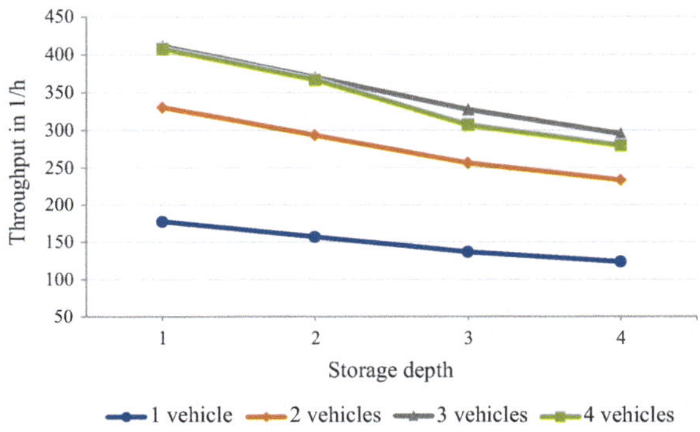


**Fig. 10** Throughput depending on the number of vehicles and storage depth by applying the relocation strategy Nearest Neighbor

As illustrated in Fig. 10, an increasing storage depth leads to a decrease in throughput capacity. Deploying more than one vehicle on a tier results in a higher throughput than if only one vehicle is deployed. In fact, by deploying two vehicles in quadruple-deep storage, the throughput is still higher than it is in single-deep storage and the deployment of one vehicle. As can be seen, when deploying more than three vehicles, throughput capacity decreases. Since more vehicles move along the same rail, sidesteps occur more often and this may affect the performance. Moreover, the applied heuristics may select not the optimum schedule, especially when the solution tree becomes large.

Figure 11 shows the percentage change of throughput by varying the storage depth and number of vehicles. The obtained throughput when deploying one vehicle in single-deep storage is selected as a reference. In single-deep storage, the largest percentage increase of throughput (85%) can be achieved by adding one more vehicle. Another vehicle raises the throughput by an additional 45%, which results in a total increase of 130% when deploying three vehicles. When storage depth increases, the loss of throughput can be prevented by adding one more vehicle.

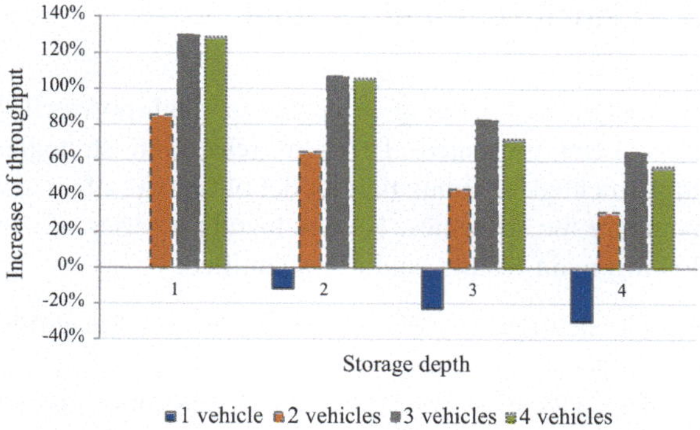


**Fig. 11** Increase of throughput depending on the number of vehicles and storage depth compared to obtained throughput with one vehicle in single-deep storage

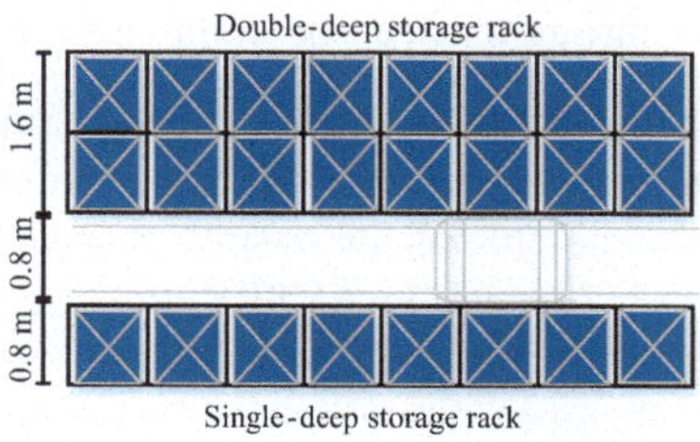


**Fig. 12** Dimensions of single-deep and double-deep storage racks

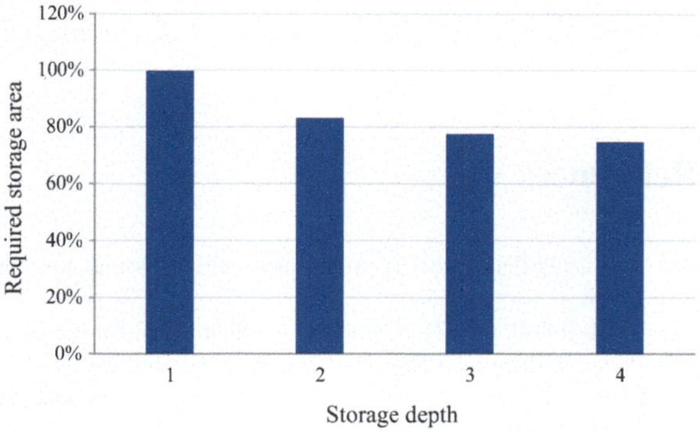


**Fig. 13** Required storage area depending on the storage depth when selecting a single-deep storage rack as a reference

Since multi-deep AVSRS can achieve a higher storage density, the increase of storage depth leads to a reduction of the required storage area. Several aisles can be consolidated and thus, the number of railing and lifting systems, as well as pre-storage area can be reduced. Assuming the various dimensions of multi-deep storage racks, which are exemplarily shown in Fig. 12, the required storage area can be reduced by 17% in double-deep storage, 22% in triple-deep storage, and 25% in quadruple-deep storage. Figure 13 shows the change of the required storage area based on the reduction of railing systems when selecting a single-deep storage rack as a reference of 100%.

## 5 Conclusion and Outlook

In this paper, operating strategies for high-powered AVSRS in multi-deep storage have been presented. Different relocation strategies have been developed and implemented. A simulation model of one tier of an aisle has been created to test and compare the strategies. By conducting a series of simulation experiments, the following main results could be identified:

- The deployment of additional vehicles can significantly increase the throughput capacity.
- The relocation strategies Avoid Relocation as well as Nearest Neighbor show the best performance.
- Multi-deep storage reduces throughput capacity. However, the loss of throughput can be compensated for by adding just one additional vehicle.

In future work, the created simulation model of one tier of an aisle will be integrated into an entire aisle with several tiers. At that point, lifting systems will become crucial for overall throughput capacity. Since vertical transport is usually the bottleneck in AVSRS, the effectivity and relevance of relocation strategies on each tier will then be revealed.

Another future work is the improvement of the control algorithm. Especially when deploying several vehicles in multi-deep storage, the state space becomes large and this results in long computation time. The improvement of the applied heuristics or the application of alternative control algorithms can alleviate this problem.

## References

1. VDI 2692 Part 1 (2015). Automated vehicle storage and retrieval systems for small unit loads, Beuth.
2. FEM 9.860 (2017). Cycle time calculation for automated vehicle storage and retrieval systems. European Materials Handling Federation.
3. Lienert, T., & Fottner, J. (2018). Routing-based sequencing applied to shuttle systems. In *21st International Conference 2018* (pp. 2949–2954).
4. Carlo, H. J., & Vis, I. F. A. (2012). Sequencing dynamic storage systems with multiple lifts and shuttles. *International Journal of Production Economics, 140,* 844–853.
5. Zhao, N., Luo, L., & Lodewijks, G. (2018). Scheduling two lifts on a common rail considering acceleration and deceleration in a shuttle based storage and retrieval system. *Computers & Industrial Engineering, 124,* 48–57.
6. Kress, D., Dornseifer, J., & Jaehn, F. (2019). An exact solution approach for scheduling cooperative gantry cranes. *European Journal of Operational Research, 273*(1), 82–101.
7. Peterson, B., Harjunkoski, I., Hoda, S., & Hooker, J. N. (2014). Scheduling multiple factory cranes on a common track. *Computers & Operations Research, 48,* 102–112.
8. Takahashi, S., Kita, H., Suzuki, H., Sudo, T., & Markon, S. (2003). Simulation-based optimization of a controller for multi-car elevators using a genetic algorithm for noisy fitness function. In *The 2003 Congress on Evolutionary Computation, CEC 2003*, Canberra, Australia (pp. 1582–1587).

9. Erdoğan, G., Battarra, M., & Laporte, G. (2014). Scheduling twin robots on a line. *Naval Research Logistics, 61*(2), 119–130.
10. Atz T. Eine algorithmenbasierte Methode zur ganzheitlichen Systemplanung automatischer Hochregallager. Dissertation, Technische Universität München. https://doi.org/10.2195/lj_proc_doerr_de_201610_01.
11. Dörr, K., & Furmans, K. Durchsatzbetrachtungen für doppeltiefe Lager unter dem Einsatz von zwei Lastaufnahmemitteln, 3-8316-0581-53-8316-0581-5.
12. Seemüller, S. (2006). Durchsatzberechnung automatischer Kleinteilelager im Umfeld des elektronischen Handels. München: Utz.
13. Lerher, T. (2016). Travel time model for double-deep shuttle-based storage and retrieval systems. *International Journal of Production Research, 54*(9), 2519–2540.
14. Tappia, E., Roy, D., de Koster, M.B.M., & Melacini, M. Modeling, analysis, and design insights for shuttle-based compact storage systems. Erasmus Research Institute of Management (ERIM), ERIM is the joint research institute of the Rotterdam School of Management, Erasmus University and the Erasmus School of Economics (ESE) at Erasmus University Rotterdam ERS-2015-010-LIS. https://ideas.repec.org/p/ems/eureri/78379.html.
15. Kemme, N. (2011). RMG crane scheduling and stacking. In J. W. Böse (Ed.), *Operations Research/Computer Science Interfaces Series, Handbook of Terminal Planning* (pp. 271–301). New York: Springer.
16. Maccarthy, B. L., & Liu, J. (1993). Addressing the gap in scheduling research: A review of optimization and heuristic methods in production scheduling. *International Journal of Production Research, 31*(1), 59–79.
17. Habl, A., Lienert, T., Pradines, G., & Fottner, J. (2019). Vehicle coordination and configuration in high-powered automated vehicle storage and retrieval systems. 18. ASIM-Fachtagung Simulation in Produktion und Logistik, Chemnitz (Submitted).

# Customer Perceived Value and Demand Preference of Cross-border E-Commerce Based on Platform Economy

**Li Xiong, Kun Wang, Xiongyi Li, and Mingming Liu**

**Abstract** Cross-border online shopping platform has become an important channel for consumers to choose imported products. This paper studies the four dimensional driving model of customer perceived value based on the customer's perceived value and Bayesian network. Python and SQL are used to capture product information and online reviews from November 2017 to April 2018. Through pre-cleaning, drying, desensitization, and integration, NLPIR is used to carry out keywords. The second-level driving factors are extracted and classified, and the Bayesian network model of the cross-border platform and non-cross-border platform is constructed. The structure and importance of the model are compared and analyzed, and user preferences of different platforms are observed. This paper finds that cross-border platform users are more sensitive to brand drivers, while non-cross-border platform users are more sensitive to value drivers. This conclusion can provide side support for the rapid introduction and effective implementation of cross-border policies in China.

**Keywords** Platform economy · Cross-border e-commerce · Customer perceived value · Online reviews · Bayesian network

This research is financially supported by Shanghai university think tank connotation construction project (N.62-0129-18-201).

L. Xiong (✉) · K. Wang · M. Liu
School of Management, Shanghai University, Shanghai, China
e-mail: xiongli8@shu.edu.cn

K. Wang
e-mail: wangkun1@shu.edu.cn

M. Liu
e-mail: 18817668909@163.com

X. Li
School of Finance, Shanghai University of Finance and Economics, Shanghai, China
e-mail: Lixyi1@163.com



J. Zhang et al. (eds.), *LISS2019*,
https://doi.org/10.1007/978-981-15-5682-1_53

# 1 Introduction

The platform economy is a new economic form. In recent years, the platform economy has become an important direction for the optimization and upgrading of commodity trading markets. It is an important measure for China to deepen supply-side structural reform and promote high-quality economic development. With the improvement of people's living standards, diversified consumption regions, and refined category demand, cross-border online shopping platforms have become an important channel for consumers to choose and buy imported products. According to statistics from the general administration of customs, the total volume of import and export commodities through the cross-border e-commerce management platform of customs in 2018 was 134.7 billion yuan, a increase of 50%, including 78.58 billion yuan of imports, a increase of 39.8%. With the rapid development of cross-border e-commerce, the quality of commodities has attracted much attention, and the quality of platforms has become the focus of customers' choice. What are the influencing factors that consumers choose from when buying overseas products? What are the problems with cross-border e-commerce import platforms? Currently, more and more research focuses on the consumption behavior of cross-border e-commerce retail import platforms. There are a few micro-level studies based on the fact data and consumer behavior research, and there is a lack of consumer behavior preference modeling based on the platform customer perceived value. Based on the theory of customer perceived value, we think about consumer behavior in the cross-border e-commerce retail studies. We import the workbench to build a four-dimensional driver model and then grab online reviews to analyze customer perceived value. We aim to expand the theory of customer perceived value further, adjust the past research perspectives and methods for business applications, and provides a new solution for the future study.

# 2 Literature Review

In terms of customer perceived value, Zeithaml divided customer perceived value into perceived profit and perceived profit loss through empirical research. Perceived profit includes product factor, service factor, technical factor, and other quality factors. Perceived profit and loss include the cost paid in the whole process of purchase, specifically price factor, maintenance factor, and waiting factor [1]. Similarly, Kotler interpreted and analyzed customer value from the perspective of the customer transfer value, which is divided into total customer and total customer value [2]. Sheth divided customer perceived value into functional value, cognitive value, emotional value, social value, and conditional value [3]. Kotler divided customer perceived value into product value, personnel value, service value, and image value [4]. Sweeney et al. divided customer perceived value into four aspects: quality value, emotional value, price value and, social value [5]. Rust constructed the customer perceived value model and proposes three drivers: general value, brand value, and relationship value [7]. Heinonen hold that customer perceived value refers to the perceived value of a customer in a specific context (at a specific time and place) to provide a specific product or service in a specific way, so the measurement of customer perceived value should be made from four aspects: time, space, function, and technology [6]. According to Jeffrey's research, customer perceived value in shopping is mainly composed of hedonic value and utilitarian value [8]. Chiu shared customers' perceived value with music value, utilitarian value, and perceived risk [9].

## 3 Four-Dimensional Driving Model Construction of Customer Perceived Value

Platform customer perceived value is an important predictor of consumer behavior, and theoretical and empirical studies have largely supported its influence on behavior orientation change. It can be reflected in two aspects:

Some researchers have studied that the perceived value of platform customers has a direct impact on the behavioral intention of consumers, and indirectly influences the behavioral intention of this and the future through user satisfaction [10, 11], and directly leads to the change of users' loyalty to the platform [12]. Ryu found that the fast-food store image has a significant impact on the perceived value of the platform and then ACTS on consumer behavior [13]. Other researchers focused on the influence or effect of a certain driving factor of platform customer perceived value, such as KIM focusing on foreign luxury fashion brands to test the impact of brand value on brand loyalty [14]. Similarly, Yoo's research in the field of luxury goods found that the dimensions of platform perceived value acquired by consumers are mainly reflected in pleasure, utility, creative achievement, and social value, and directly affect consumer satisfaction [15]. Jiang et al. further tested the effect of e-service quality dimensions on customer perceived value [16]. Hong, who based on the field of innovation, explored the hedonic value and utilitarian value in the context of smart watches, which is particularly important [17].

Above, for the cross-border e-commerce platform, the past theory model can't well cover the entire shopping process dimensions of customer perception. This paper combined the cross-border e-commerce platform to improve the above theory model We concluded that platform driving factors of customer perceived value driven by brand elements, the drive factors, value-driving elements and technical drivers of four aspects, become a platform 4 d drive model, as shown in Fig. 1.

Based on the perspective of consumers, this model takes the online platform reviews to recognize the comprehensive measurement of the behavioral needs and preferences of cross-border e-commerce platform import consumers. Value driving factors include product quality, price, etc. Brand drivers include brand recognition and reputation; Technical drivers include platform convenience and logistics feedback services.

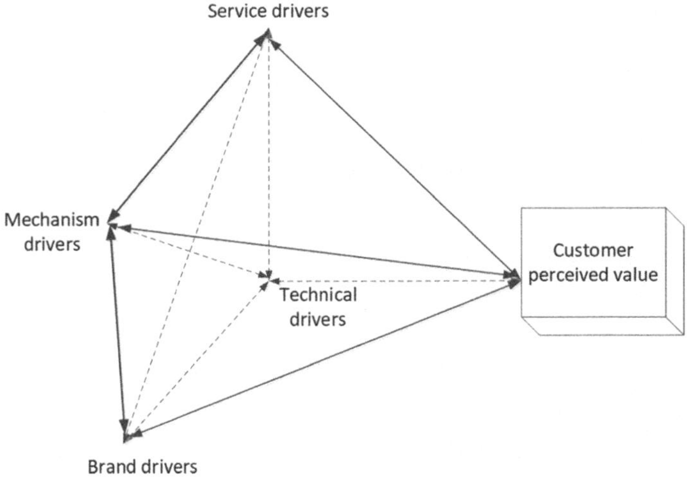


**Fig. 1** Four-dimensional driving model of customer perceived value.

# 4 Analysis on Customer Perceived Value Drivers of Cross-border E-commerce Import Platforms

## *4.1 The Data Source*

The main online reviews data captured in this paper include UserNick (user nickname), auctionSku (product property), cmsSource (source), rateDate (transaction date), and rateContent (reviews content) on the website of imported B2C cross-border e-commerce platform. Through consumer research, compare the main platform penetration DuDu and authenticity perception found that Tmall international, a comprehensive retail e-commerce platform, are easier to implement innovation of product variety and quality of analogy. Tmall is an independent domestic users shopping mall platform that has the same range of users, and the quality of the user and user-level consistent, so choose Tmall international, which is an import B2C cross-border e-commerce platform. According to import category of various beauty care in 2016 and 2017, more than 50% customers choose the hottest products under this category as the research object: choosing the first ten sales commodity on Tmall international platform as the research sample, capture the online reviews fetching period from November 2017 to April 2018. At the same time, analyzing the online reviews and find out customers' preference.

## 4.2 Data Collection

We use Python+Mysql to collect data, using Scrapy's URL parsing and recursive crawling method to achieve automatic data collection, grabbing pages are mainly product information data and online reviews data. The grabbed reviews include multiple dimensions, namely, text content and images of online reviews, product attributes, account nicknames of online reviews users, and online reviews publishing. According to the rules, the first ten sales of the liquid foundation/paste are F1: Revlon, F2: Thailand Mistine, F3: Dermacol damask, F4: Korea MISSHA, F5: Ai Jing age, F6: Revlon, F7: MaxFactor/honey Buddha, F8: Milani America, and F9: Ai Jing, Zhu: France, Paris. Tables are arranged in descending order of sales. Table 1 is the statistics of basic information on individual product sales platform pages, including product price (yuan), import tax (yuan), freight (yuan), monthly sales (pieces), cumulative online evaluation (bar), number of collections (popularity), etc. To some extent, it indirectly reflects the acceptance, popularity, and activity of platform products (Table 1). And we also conducted keyword selection and classification, as shown in Table 2.

Fetching the data on the international platform Tmall article 17820, article 1900 or so, each product Tmall is outside the same brand products, a total of nearly 176.94 million collection period from November to April 2018, 2017, as Tmall

**Table 1** Product information of foundation/cream on cross-border e-commerce platform.

| Serial number | Price (yuan) | Import duty | Freight | Monthly trading volume (Pieces) | Cumulative evaluation | Collection volume |
|---|---|---|---|---|---|---|
| F1 | 98.00 | Full tax-inclusive | 0.00 | 16736 | 94105 | 247048 |
| F2 | 65.00 | The highest selling price | Free postage | 15633 | 88985 | 291001 |
| F3 | 208.00 | Full tax package | 0.00 | 14923 | 18005 | 136488 |
| F4 | 65.00 | Businessmen assume | 0.00 | 10842 | 10975 | 3077 |
| F5 | 158.00 | 17.70-40.34 yuan | 0.00 | 9876 | 9065 | 532 |
| F6 | 89.00 | Full tax package | Full tax package | 5952 | 11569 | 77140 |
| F7 | 128.00 | Businessmen assume | 0.00 | 5718 | 3515 | 13931 |
| F8 | 118.00 | Businessmen assume | 0.00 | 5274 | 19221 | 93335 |
| F9 | 158.00 | Businessmen assume | 0.00 | 3743 | 4460 | 16251 |
| F10 | 99.00 | Businessmen assume | Full tax package | 2569 | 5898 | 69141 |

**Table 2** The second level driver index for liquid foundation/paste keywords classification on cross-border e-commerce platform.

| Driving index | Key words | Variable |
|---|---|---|
| Makeup effect | Natural, makeup feeling, naked makeup, makeup effect, fit, makeup, fit, smooth, glossy, false white, white, false face | $X_1$ |
| Concealer effect | Concealer, concealer, concealer strength | $X_2$ |
| Duration | Take off makeup, durable, hold makeup, durable | $X_3$ |
| Skin type | Mixed oil, oil skin, oily, dry | $X_4$ |
| Price | Price, par | $X_5$ |
| Product quality | Light, moist, waterproof, allergic, itchy face, fragrance, odor, powder tone | $X_6$ |
| Genuine security | Genuine, genuine, official | $X_7$ |
| Package | Packing, boxes, cases | $X_8$ |
| Merchant service | Customer service | $X_9$ |
| Logistics | Logistics, express delivery, delivery speed | $X_{10}$ |

international and Tmall display rules limit online reviews, and the data is fetched as shown on the under the conditions of partial data, therefore, this period of time the source data is far greater than visual data, this paper source data cleaning, the cleaning rules are as follows: (1) irrelevant evaluation: the reviews have nothing to do with the product, for example: "the clothes…". (2) Invalid evaluation: the reviews are made with invalid behaviors such as punctuation, emotion, filler words or superposition of modal words, such as: "following filler words …". (3) Advertising: advertising or commission publicity for the products sold in the reviews area (4) Repeated evaluation: if the same review content appears in the review area due to multiple orders placed at the same time or multiple products purchased at the same time, only one valid evaluation will be reserved. (5) Reduplication evaluation: the content of the reviews is copied and pasted to form a multi-word review, the reduplication evaluation is deleted, and only one effective evaluation is retained, which is similar to repeated evaluation.

## 4.3 Determination of Driving Factor Indicators

In this study, cosmetics on the import platform of cross-border e-commerce were selected to establish the second-level driving index of the model for empirical test. The second-level driving index is shown in Table 3.

In this table, it is found that the value-driving index is more detailed, and the merchants' focus is more complex, so it is refined and measured. The packaging is

**Table 3** Four-dimensional driving factors and second-level driving indicators of cross-border e-commerce platforms.

| The driving factor | Secondary driving index | Variable |
|---|---|---|
| Value drivers | Makeup effect | $X_1$ |
| | Concealer effect | $X_2$ |
| | Duration | $X_3$ |
| | Skin type | $X_4$ |
| | Price | $X_5$ |
| | Product quality | $X_6$ |
| Brand drivers | Genuine security | $X_7$ |
| | Package | $X_8$ |
| Merchant drivers | Merchant service | $X_9$ |
| Technical drivers | Logistics | $X_{10}$ |

the acceptance and recognition of consumers' product brands, so quality assurance and packaging are the secondary indicators of brand drivers. Service drivers are involved in online reviews primarily for merchants; Technology drivers logistics delivery, and feedback systems embody platforms and merchants. We counted the word frequency of keywords. Since some keywords have similar meanings, we grouped the keywords with similar meanings together and used them as indicators of driving factors. In order to determine influence import B2C cross-border e-commerce platform user preferences, the driving factors are shown in Table 2, as an index for later Bayesian network modeling source data.

We use the five-point calibration method of Likert scale based on the characteristic factors mentioned in the review content to convert the users of the cross-border e-commerce platform online to review the text data to digital data. We calculate the final score as the basis for judging consumer preferences, and at the same time it is also the basic data for modeling and analyzing customer perceived value.

## 5 Analysis of User Preference Model of Cross-border E-Commerce Platform

Bayesian network is an important branch of business intelligence that is used to predict uncertain problems. In this paper, after the data preprocessing, we clean and standardize the index data. Then, using Clementine software to apply the Bayesian network modeling process. We use the ten send level drive index as the input variables to build the training model, then calculate the overall score of each variable. In this work, we select the TAN method to construct the FUN Bayesian network, generate the network diagram, and calculate the conditional probability of each node. Finally, output the importance of each node driving factors. In this way, the consumer preference factors of imported B2C cross-border e-commerce users for products can be judged. Meanwhile, the data of non-cross-border platforms can be operated in this way to obtain the FUN1 Bayesian network. Fig. 2 for the Bayesian network of non-cross-border platform and Fig. 3 for the Bayesian network of cross-border platform.

Analysis of FUN Bayesian network model and FUN1 Bayesian network model found that the network structure is very consistent, various interactions between the two drive indicator further confirmed. We mainly focus on the output (the indicators importance) of FUN Bayesian model and FUN1 Bayesian model.

According to Fig. 4 and Fig. 5, the importance rank of the indicators in the FUN1 model is as follows: concealer effect, makeup effect, duration, quality assurance, merchant service, price, etc., and there is a big difference in the importance of each indicator. The importance rank of the indicators in the FUN model, in turn, is: concealer effect, quality assurance, logistics, packaging, makeup effect, duration, product quality, price, merchant service, and skin type. There is a ladder gap among different indicators, among which the most obvious are concealer effect, quality assurance and logistics. As shown in the model structure and diagram:

(1) The second-level index importance rankings of both non-cross-border platforms and cross-border platforms are ladder distributed, but there is a

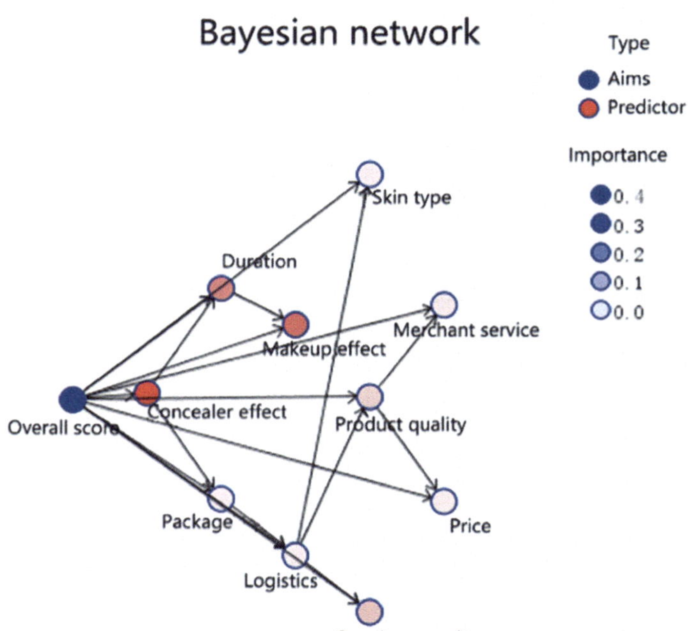


**Fig. 2** Non-cross-border e-commerce platform FUN1 Bayesian network.

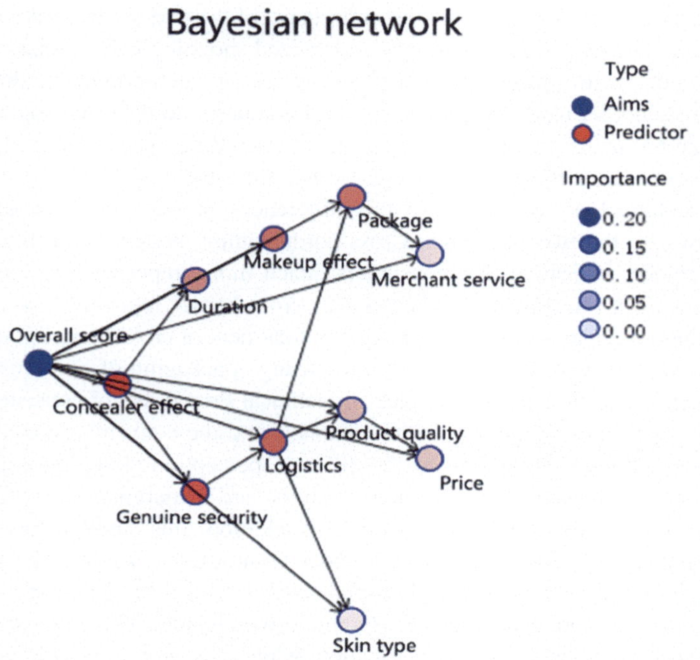


**Fig. 3** Cross-border e-commerce platform FUN1 Bayesian network.

significant difference in the value. The value of each index in cross-border platforms is more uniform, while the value of each index in non-cross-border platforms is more volatile, leading to the phenomenon of "supporting the moon", and some indicators can be almost omitted.

(2) The value drivers are focused by customers' online reviews of the two platforms. The importance of the indicators of the value drivers of the two platforms is ranked by concealer effect, makeup effect and duration. The difference lies in that the two-dimensional indicators of value drivers are different in different platforms. The two dimensional indicators of value drivers are interspersed with brand drivers and technology drivers in the online reviews of cross-border platforms, while there is no such phenomenon in cross-border platforms. In addition, the importance value of indicators in non-cross-border platforms is almost twice that of the same indicators in cross-border platforms.

(3) The brand drivers in both platforms online review importance ranking after the value driving factors. Still, the difference is authentic cross-border platform to ensure the index data is a platform for cross-border almost two times, and the index of packaging in cross-border platform importance ranking fourth, relatively, but is the lowest of the cross-border platform, almost negligible.

(4) The services driving factors in the importance of the two models have a similar ranking, this may and increasingly customized shopping habits, related service drivers in a platform for cross-border online reviews attention more, although in the cross-border platform more later but also superior to the price, logistics, and packaging such as secondary indexes, the obvious difference is that the latter low ranking but importance value is greater than the latter.
(5) The technical drivers have obvious differences in attention between online reviews on the two platforms. Cross-border online reviews rank the third in importance, indicating that consumers attach more importance to technology drivers. This phenomenon indicates that cross-border logistics costs are high and the time span is large. The packaging treatment of cross-border platforms is not consistent with the "secondary processing" packaging of non-cross-border platforms, and thus the importance of technical drivers is not consistent.
(6) Inconsistent knowledge acquisition channels and the convenience of customer service communication lead to the different perception shopping experience. The specific reasons need further analysis and experimental confirmation, which will be the next research goal. In addition, this study focuses on the importance of "genuine guarantee" under brand-driven factors. The purchase behavior on non-cross-border platforms is less sensitive to "genuine guarantee" and users are more concerned about value-driven factors. This also explains the relevance of the user's platform selection, while cross-border platform users are more sensitive to "genuine guarantee" as a secondary driving indicator. In terms of actual importance, the FUN model is close to the FUN1 model, but the FUN model pays more attention to quality.

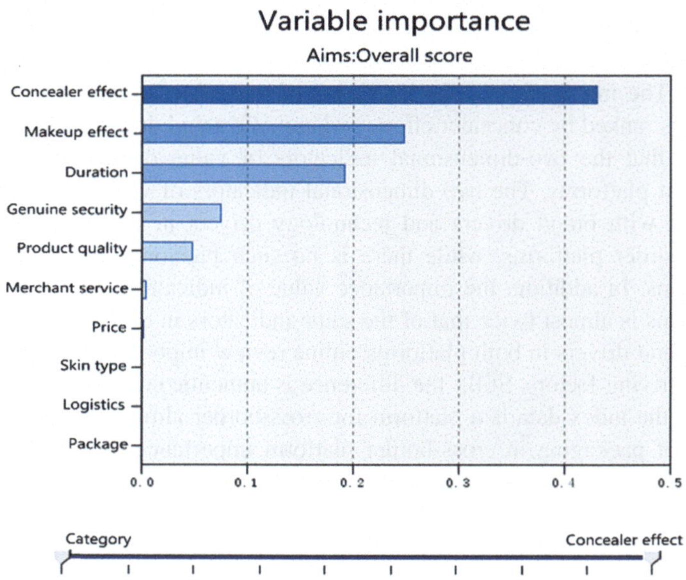


**Fig. 4** Importance of FUN1 Index for non-cross-border e-commerce platform.

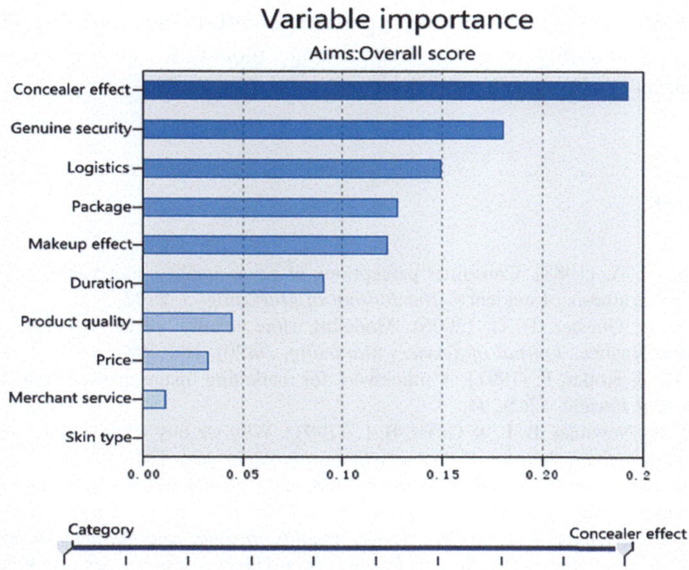


**Fig. 5** Importance of FUN Indicators for cross-border e-commerce platform.

## 6 Conclusion

In this paper, the text analysis is carried out by capturing the online reviews of the customers of the B2C website. We extract the commodity characteristics factors that customers are concerned about the online review, build the Bayesian network model of customer preferences through data training, and calculate the conditional probability distribution of the variable nodes. By adjusting the probability of parent nodes and sensitive factors, the consumption preference of customers under different evaluation results is analyzed. In the constructed network model, the parent nodes are overall evaluation, comfort degree, quality, feeling, and beautiful degree. Under the impact of the parent node, the customer evaluation of the sub nodes of each characteristic factor variable is larger in the probability of "moderate" and "good". Customer preferences of commodities mainly include four factors, comfort degree, quality, handle and beautiful degree.

Through the sensitivity analysis of the network model, we find that the sensitivity of the price is the largest, 165e+003, and the sensitivity of the appropriate degree, fabric workmanship, and handle are 640, 92.1, and 84, respectively. After increasing the probability of "good" to 1, the effect of these four sensitive factors on the probability of the remaining factors has a reverse effect. The probability of "good" has a certain increase, some factors probability of "bad" decreased, and the probability of "moderate" was risen. In the probability that all factors are "good", besides the sensitive factors, customers prefer to comfort degree, quality, color, and style. In the probability of every factor being "moderate", customers prefer the factors such as quality goods, match degree, logistics, style.

## References


1. Zeithaml, V. A. (1988). Consumer perceptions of price, quality, and value: A means-end model and synthesis of evidence. *The Journal of Marketing, 3,* 2–22.
2. Chen, S., & Quester, P. G. (2006). Modeling store loyalty: perceived value in market orientation practice. *Journal of Services Marketing, 20*(20), 188–198.
3. Barich, H., & Kotler, P. (1991). A framework for marketing image management. *MIT Sloan Management Review, 32*(2), 94.
4. Sheth, J. N., Newman, B. I., & Gross, B. L. (1991). Why we buy what we buy: A theory of consumption values. *Journal of Business Research, 22*(2), 159–170.
5. Sweeney, L. C. (1999). The role of perceived risk in the quality-value relationship: A study in a retail environment. *Journal of Retailing, 75*(1), 159–170.
6. Rust, R. T., & Oliver, R. L. (1994). *Service Quality: Insights and Managerial Implications from the Frontier* (pp. 1–19). Service Guality: New Directions in Theory and Practice.
7. Overby, J. W., & Lee, E. J. (2006). The effects of utilitarian and hedonic online shopping value on consumer preference and intentions. *Journal of Business Research, 59*(11), 1160–1166.
8. Heinonen, K. (2004). Reconceptualizing customer perceived value: the value of time and place. *Managing Service Quality: An International Journal, 14*(2), 205–215.
9. Chiu, C.M., Wang, E.T.G., & Fang, Y.H. (2014). Understanding customers' repeat purchase intentions in B2C e-commerce: The roles of utilitarian value, hedonic value and perceived risk. *Information Systems Journal, 24*(1), 85–114.
10. Liu, W. B., & Chen, R. Q. (2009). Managerial strategies of customer perceived value based on customer participation. *Journal of Wuhan University of Science and Technology (Social Science Edition), 11*(1), 49–54.
11. Grisaffe, D. (2004). A dozen problems with applied customer measurement. *Journal of Consumer Satisfaction, Dissatisfaction and Complaining Behavior, 17,* 1–15.
12. Ryu, K., Han, H., & Kim, T. H. (2008). Managerial strategies of customer perceived value based on customer participation. *The Relationships Among Overall Quick-Casual Restaurant Image, Perceived Value, Customer Satisfaction, and Behavioral Intentions, 27*(3), 459–569.
13. Kim, M., Kim, S., & Lee, Y. (2010). The effect of distribution channel diversification of foreign luxury fashion brands on consumers brand value and loyalty in the Korean market. *Journal of Retailing and Consumer Services, 17*(4), 286–293.
14. Yoo, J., & Park, M. (2016). The effects of e-mass customization on consumer perceived value, satisfaction, and loyalty toward luxury brands. *Journal of Business Research, 69*(12), 5775–5784.
15. Wang, C., Liu, J., & Wu, J. B. (2011). Building and research on consumer' s perceived utility model in network environment. *Chinese Journal of Management Science, 19*(03), 94–102.
16. Jiang, L., Jun, M., & Yang, Z. (2016). Customer-perceived value and loyalty: how do key service quality dimensions matter in the context of B2C e-commerce. *Service Business, 10* (02), 301–317.
17. Hong, J. C., Lin, P. H., & Hsieh, P. C. (2017). The effect of consumer innovativeness on perceived value and continuance intention to use smartwatch. *Computers in Human Behavior, 67,* 264–272.

# Decision-Making for RPA-Business Alignment

Bo Liu and Ning Zhang

**Abstract** Robotic process automation (RPA) is a software robotic technology that is set to transform business operations by automating repeatable process. RPA-business alignment can free up workers' time to focus on core competencies. How to do decision-making for RPA-business alignment? To answer this question, the aim of this research is to find the general decision-making path. With fsQCA method to review and analyze the decision-making factors during RPA deployment in real cases, there are two decision-making paths. Then, according to coupling evaluation, the two paths can reach RPA-business alignment that is in low coupling coordination phase.

**Keywords** Robotic process automation · Coupling mechanism · Fuzzy-set qualitative comparative analysis (fsQCA)

## 1 Introduction

Robotic process automation (RPA) is a software robot that can mimic human tasks [2, 3]. Although it is an emerging technology, it has been demonstrated through real practice to reduce operating costs, improve efficiency [4, 7], improve productivity and reduce human errors [1].

RPA-business alignment can prompt operation efficiency and decrease costs, and these two dimensions positively affect enterprise performance [5, 21]. Companies

This research was supported by the National Key R&D Program of China under grant 2017YFB1400701 and by the Discipline Construction Project of the Central University of Finance and Economics.

B. Liu (✉) · N. Zhang
School of Information, Central University of Finance and Economics, Beijing, China
e-mail: Neal.Liu@foxmail.com

N. Zhang
e-mail: Zhangning@cufe.edu.cn



J. Zhang et al. (eds.), *LISS2019*,
https://doi.org/10.1007/978-981-15-5682-1_54

have received good returns from RPA investment, as mainly reflected in decreased FTE and overtime work, and increased time for employees to carry out high value-added activities. It is the next coming wave of technology that enterprises have embraced RPA as a future technology that may replace present human labor [1]. RPA is a "lightweight" IT that can be integrated with businesses and provide benefits to the enterprise [24, 25].

This potential technology can support more professional knowledge fields [14], has been attracting increasing attention [15] and may become a core topic relevant to enterprise managers. But how to do making-decisions for RPA-business alignment? And how is the decision-making path? To answer these questions, we summarized the literature to address the key factors required to quantify and evaluate the decision-making path.

## 2 Reference Review

Researchers have conducted detailed case studies in many industries and have encountered various difficulties and obstacles during RPA deployment, thus generating valuable experiences [6, 11–13, 20–23]. RPA is a mature digital transformation and automatic upgrade plan for enterprises, and its simple configuration provides good flexibility and usability to automate daily tasks [2, 3]. In fact, most of fortune 500 companies worldwide have adopted RPA, thus producing a large amount of practical experience and demonstrating that RPA can improve efficiency and reduce operating costs [4–21].

A review of the existing literature indicates that researchers have performed primarily theoretical explorations or case studies, mainly based on their specific enterprise backgrounds. There is thus a lack of generality and the reviewed literature indicates that to increase RPA adoption, a primary research direction is required to realize RPA-business alignment [18]. This is also the advanced stage of RPA adoption.

Here, we further investigate various decision-making factors related to RPA-business alignment through the analysis of existing case study data and provide a comprehensive summary to obtain universal decision-making factors for further decision quantification.

In addition, in the existing research cases of RPA, enterprises have used RPA to build new enterprise capability without affecting the existing architecture, thus eventually forming a non-invasive import scheme [18, 19]; therefore. For example, in the case of Telefonica O2, RPA was initiated by departments outside IT, and IT started with a small scale deployment and continued with a large scale deployment [8].

RPA is also an innovation in the field of IT business, and RPA is playing its role and appears to be taking over the previous work of IT, thus posing a challenge to IT departments [6, 12, 13, 18, 19].

Therefore, research on decision-making factors can be referenced from existing information technology and IT-business alignment research results. Among them, the famous strategic maturity model (SAM) is an effective descriptive and normative tool for IT-business alignment [26, 27], which indicates the influence of the IT-business alignment for five possible determinants: environmental changes, decreased profits, external influences, management changes, and information ideas [28].

In addition, the latest research in this field adopts punctuated equilibrium theory to transition this alignment from static to dynamic and reveals that organizational inertia is another determinant [29].

Therefore, we used RPA for more detailed analysis and combined with the theory of existing IT-business alignment, exploring the key decision-making processes in automation and business alignment.

## 3 Research Method and Data Source

The qualitative comparative analysis (QCA) method, which is based on the logic of set theory with Boolean algebra, formalizes causal logic relationships for causality asymmetry and has good applicability in small-scale sample analysis, is regarded as a bridge between qualitative and quantitative methods [30, 31]. At present, the QCA method has good adaptability to identify critical impact paths and results. There are three QCA methods: crisp-set QCA [30, 31], fuzzy-set QCA (fsQCA) [33] and multi-value QCA [32]. For example, some researchers have conducted social movement researches to identify causality factors and results with the QCA method [34–39].

In this paper, some key decision conditions were difficult to be accurately quantify; therefore, among the three QCA methods, fuzzy-set QCA (fsQCA) was closest to the study objective. In an RPA literature review, the existing studies were all found to be qualitative regarding the basic concept and case study and showed a lack of generality.

To study RPA-business alignment to address the key factors and decision-making paths, we used the fsQCA method to analyze existing cases. Therefore, we fully reviewed the RPA cases and IT-business alignment references, extracted data and captured all key factors in RPA-business alignment.

RPA is still emerging and is only approximately 3 years old. Companies need more time to finish deployment. Existing data indicate that Chinese enterprises began RPA deployment in 2017 [41]; therefore, the number of cases is limited. By searching Google Scholar from 2015 to the present with the key words "RPA" and "Robotic Process Automation" with the language set to English or Chinese, we found a total of 33 cases, referring to 21 enterprises.

Among them, Europe accounted for 78% with 9 enterprises; 1 enterprise in America accounted for 11%; and 1 Chinese enterprise accounted for 11%, and the ratio was 7:1:1.

**Table 1** Case distribution

| Country | Continent | Qty. | Percent (%) | Percent by continent (%) |
|---|---|---|---|---|
| UK | Europe | 5 | 23.81 | 52.38 |
| Netherlands | | 1 | 4.76 | |
| Finland | | 1 | 4.76 | |
| Hungary | | 1 | 4.76 | |
| Sweden | | 1 | 4.76 | |
| One European country | | 1 | 4.76 | |
| Austria | | 1 | 4.76 | |
| US | America | 5 | 23.81 | 28.57 |
| Canada | | 1 | 4.76 | |
| Australia | Australia | 1 | 4.76 | 4.76 |
| China | Asia | 1 | 4.76 | 14.29 |
| India | | 1 | 4.76 | |
| Singapore | | 1 | 4.76 | |

Therefore, the small-scale sample support characteristics of QCA can be analyzed through qualitative comparison for RPA-business alignment (Table 1).

RPA, especially in Western Europe, was adopted earlier and lacked labor arbitrage because of high human costs in this region. Enterprise managers have aimed to promote efficiency cost savings. Owing to the large number of immigrants from Mexico and South America, the United States has some room for labor arbitrage and therefore has been slower than Europe to adopt the RPA requirement.

Because Chinese enterprises are increasingly involved in international competition, managers are encouraged to seek certain process automation (Table 2).

# 4 Fuzzy-Set Qualitative Comparative Analysis

## *4.1 Variable Construction*

We used the deductive method for RPA-business alignment research [52], with the existing literature conclusions regarding key factors for IT-business alignment, such as environmental changes, decreased profits, external influences, management changes, information ideas [28] and organizational inertia [29].

Therefore, for factors scores, 6 IT strategy consultants with more than 10 years experiences scored the casual factors of each target enterprise according to the Likert scale. The average is the final value of each factor. The value of information ideas takes the FTE saving ratio after RPA deployment, because perception transformation is commonly thought to improve operational efficiency.

**Table 2** Cases source of RPA adoption

| Study object | Author or data source |
|---|---|
| Xchanging | Willcocks L P, Lacity M, Craig A 2015 [11] |
| Royal DSM | Lacity M, Willcocks LP, Craig A 2016 [5] |
| SEB Bank | Lacity M, Willcocks LP, Craig A 2017 [18] |
| OpusCapita | Hellström A 2016 [7] |
| Radiant Law | Lacity MC, Willcocks L, 2016 [19] |
| Telefónica O2 | Lacity M, Willcocks LP, Craig A, 2015 [8] |
| Liberty Source | Lacity M, Khan S, Carmel E, 2016 [6] |
| European Energy | Lacity M, Willcocks LP, Craig A 2015 [21] |
| Vodafone | Salvatore R. 2016 [38] |
| Ascension Health | A.J. Hanna 2016 [39] |
| BNY Mellon | BNY Mellon 2017 [40] |
| SINOCHEM | SINOCHEM 2017 [41] |
| Davies Turner | Gould R 2018 [42] |
| MUFG Union Bank | KOFAX 2017 [43] |
| Leeds Building Society | Sumner S 2017 [44] |
| Coveris Holdings Sarl | Auxis website 2016 [45] |
| Wipro | Tarafdar M, Beath C. [46] |
| Circle K | Accenture 2017 [47] |
| Raiffeisen Bank | Makarchenko M, Nerkararian S, Shmeleva IA 2016 [48] |
| ANZ Bank | Paul Smith 2015 [49] |
| DBS Bank | DBS Website 2016 [50] |

For the result variable, RPA-business assignment, it is the number of RPA robots, because one of the main phenomena in RPA-business alignment is the appearance of robot employees (Table 3).

## 4.2 Data Calibration

fsQCA calculates the membership degree of each factor between 0 and 1. Because the data for the above variables are all positive, which have different dimension, the min-max normalization method is suitable for variable dimensionless, and each factor can be converted into a value from 0 to 1 for future fsQCA analysis. Then, it is determined whether each factor is necessary or mostly necessary for the outcome to occur. The expressions are:

**Table 3** Variable construction

| | Items | Abbr. | Description | Remark |
|---|---|---|---|---|
| Variables | Environmental changes | ENVSHI | Environment or business strategy change | Qualitative variable |
| | Decreased profits | LOWPER | Revenue decline | Qualitative variable |
| | External influences | INFOUT | Outsider impact | Qualitative variable |
| | Management changes | NWLDER | Leadership change | Qualitative variable |
| | Information ideas | PERTRA | Internal digital perception | Quantitative variable |
| | Organizational inertia | ORGINE | Organization barrier | Qualitative variable |
| | RPA-business alignment | RBALIG | IT-business alignment maturity | Quantitative variable |

$$Consistency(X_i \leq Y_i) = \sum[min(X_i, Y_i)]/\sum X_i \tag{1}$$

$$Coverge(X_i \leq Y_i) = \sum[\min(X_i, Y_i)]/\sum Y_i \tag{2}$$

According to fsQCA good practice recommendations for the necessary conditions, the *consist* score threshold should be above 0.9. For the necessary calculation, the fsQCA 2.5 software can be used [30, 31]. As shown in the table below, all consist scores are less than <0.9 (0.9 is the best threshold value), except that of environmental changes (ENVSHI) (0.9534 > 0.9). Therefore, the factor ENVSHI is a key and necessary decision-making factor for RPA-business alignment (Table 4).

**Table 4** Necessary analysis

| Conditions | Consistency | Coverage |
|---|---|---|
| ENVSHI | 0.9534 | 0.4902 |
| ~ENVSHI | 0.3947 | 0.4579 |
| LOWPER | 0.7178 | 0.5042 |
| ~LOWPER | 0.6971 | 0.5040 |
| INFOUT | 0.7298 | 0.5729 |
| ~INFOUT | 0.6752 | 0.4405 |
| NWLDER | 0.8924 | 0.4287 |
| ~NWLDER | 0.3964 | 0.5468 |
| PERTRA | 0.7390 | 0.4996 |
| ~PERTRA | 0.6365 | 0.4795 |
| ORGINE | 0.7220 | 0.5417 |
| ~ORGINE | 0.5699 | 0.3867 |

~means the opposite value of its own value. The result calculated by fsQCA 2.5 software

## 4.3 Sufficient Analysis

After exclusion of the necessary conditions, the remaining factors (decreased profits, external influences, management changes, information ideas and organizational inertia) were analyzed in different combinations in causal factor analysis with the Quine–McCluskey algorithm [51], and the factor configuration was sufficient to obtain results.

Thus, the fuzzy truth table generated (where the value of the result is 1, if the consistency was greater than 0.75; otherwise it is 0). Consequently, the decision-making paths of RPA-business alignment obtained. Then, according to the fuzzy truth table, fsQCA 2.5 software was used to calculate the decision path of possible decision factors of RPA-business alignment. Table 5 is fsQCA calculation results.

In the fsQCA analysis results above, through case analysis, we found two key decision-making paths. Because the decision-making paths affect the alignment process of RPA and business, there is causality between the alignment results' asymmetry; that is, two decision-making paths can enable internal implementation

**Table 5** Configurations for decision-making of RPA-business alignment

| Variables | Configurations (1) | Configurations (2) |
|---|---|---|
| LOWPER | ⊗ | ● |
| INFOUT | ⊗ | ● |
| NWLDER | ● | ● |
| PERTRA | ⊗ | ● |
| ORGINE | ● | ● |
| Consistency | 0.7960 | 0.8026 |
| Raw coverage | 0.3662 | 0.4711 |
| Unique coverage | 0.0682 | 0.1732 |
| Solution coverage | 0.5394 | |
| Solution consistency | 0.7735 | |
| Consistency cutoff | 0.75 | |
| Frequency cutoff | 0.7960 | |

⊗ indicates that the reason variable does not appear; ● indicates that the reason variable, in which the large ● represents the core conditions, •represents the non-core conditions; space indicates that the reason variable cannot appear.

of RPA and allow for smooth integration in the business. This result also reveals that the traditional quantitative regression method is unable to analyze this cause-and-effect asymmetry, thus briefly verifying the rationality and correctness of the fsQCA method used in this paper. Then, how can the pros and cons of the two key decision-making paths combined with multiple factors be evaluated? In addition, the two decision-making paths should have high practical and theoretical reference value for enterprise managers seeking to align RPA into enterprise business.

From the above analysis results, we identified two key decision path combinations of RPA-business alignment. The consistency of the first decision paths (consistency) is closed to the second (0.7960 < 0.8026), and the overall solution consistency (solution consistency) is 0.7735. The two paths have strong power to reach RPA-business alignment in enterprises.

The two key decision-making paths are as follows:

- **Decision-Making Path 1:**

  ENVSHI * ORGINE * LOWPER * INFOUT * NWLDER * PERTRA (Fig. 1).

- **Decision-Making Path 2:**

  ENVSHI * ORGINE * ~LOWPER * ~INFOUT * NWLDER * ~PERTRA (Fig. 2).

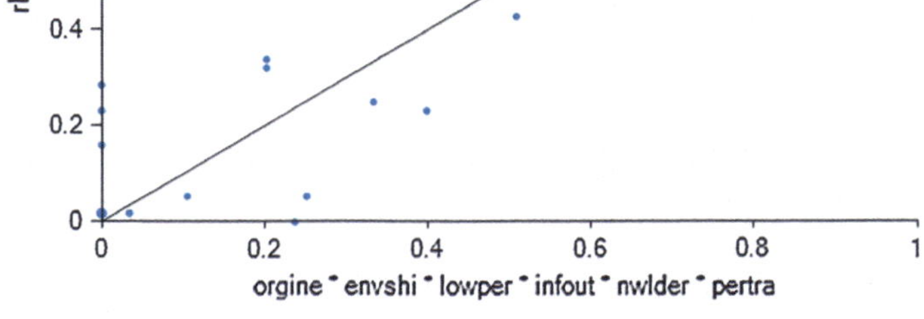


**Fig. 1** Decision-making path 1

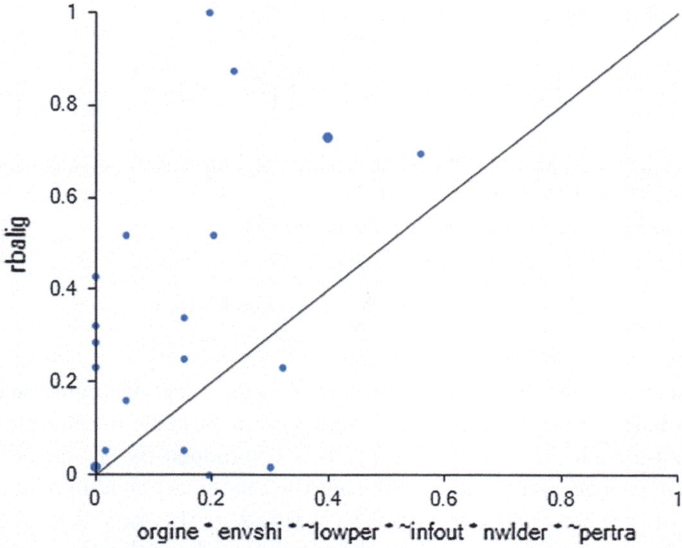


**Fig. 2** Decision-making path 2

# 5 Evaluation of Decision-Making Path

## *5.1 Evaluation Principles of Decision Influence Paths*

We found that the above two key decision-making paths can both achieve RPA-business alignment. Thus, all elements in each path must have a linkage relationship with the results. Similarly, some researchers have studied the coupling coordination relationship between innovation and business models [53]. Therefore, we sought to evaluate the relationship between each path and the alignment results through coupling theory. The evaluation function of decision-making factor expression is:

$$\mathrm{u}_i = \sum_{i=1}^{m} \lambda_{ij}\mathrm{u}_{ij}, \quad \sum_{j=1}^{gn} \lambda_{ij} = 1, \tag{3}$$

where $i = 1, 2, 3, \ldots 7$; $j = 1, 2, 3, \ldots 7$, $\mathrm{u}_{ij}$ represents the evaluation function of decision-making factors, and $\lambda_{ij}$ represents the weight of each indicator, and m and n represent the number of specific indicators. In this research, each decision-making path has six key decision factors and one outcome factor, so $m = n = 7$.

The coupling degree expression is:

$$C_n = \left\{(u_1, u_2, \ldots, u_m) / \prod (u_i + u_j)\right\}^{1/n} \tag{4}$$

where $u_m$ represents the comprehensive evaluation function of each key factor. $C$ is the coupling degree value between [0, 1].

The coupling coordination degree expression is:

$$D = \sqrt{C \times T}, T = \sum_{i=1}^{m} \alpha_i u_i, (i = 1, 2, 3, \ldots)_i, \tag{5}$$

where $D$ is the coupling coordination degree, $C$ is the coupling degree, and $T$ is the evaluation index of the corresponding factor system. $\alpha_i$ is the undetermined coefficient, because each decision-making patch is determined by six factors, and the environment change has been demonstrated to be a necessary condition for the result; therefore, $\alpha_1 = 0.3, \alpha_2 = 0.1, \alpha_3 = 0.2, \alpha_4 = 0.1, \alpha_5 = 0.1, \alpha_6 = 0.1, \alpha_7 = 0.1$.

According to the existing literature, the coupling and coordination degree can be divided into four intervals according to the value range (Table 6).

## 5.2 *Evaluation Results*

First, according to the information entropy method, the weight values of six decision factors of each critical path and the results of RPA-business alignment were calculated. The entropy weight method tested the rationality and effectiveness of the data. The results revealed a value of >0.01 for all entropy weight ρ, so all index data can be retained for the next analysis.

Second, according to the formula of coupling coefficient C (and then 3) and the coupling coordination degree D, formula (4) can be used to calculate each key decision factor of the path and the results of their comprehensive evaluation function, and can be used to further calculate the environment changes, decreased profits, external influences, management changes, information ideas and organizational inertia coupling between the RPA and business integration correlation between the C and D coupling coordination degree, thus laying a foundation for further comparative analysis (Tables 7 and 8).

**Table 6** Coupling coordination standard

| Classification | Low | Medium | High | Max |
|---|---|---|---|---|
| Interval | (0, 0.3] | (0.3, 0.5] | (0.5, 0.8] | (0.8, 1] |

**Table 7** Entropy weights of decision-making factor

| **Decision-Making Path 1:** ENVSHI * ORGINE * LOWPER * INFOUT * NWLDER * PERTRA | | | | | | |
|---|---|---|---|---|---|---|
| RBALIG | ORGINE | ENVSHI | LOWPER | INFOUT | NWLDER | PERTRA |
| 0.2836 | 0.2044 | 0.0632 | 0.0769 | 0.1458 | 0.0694 | 0.1566 |
| **Decision-Making Path 2:** ENVSHI * ORGINE * ~LOWPER * ~INFOUT * NWLDER * ~PERTRA | | | | | | |
| RBALIG | ORGINE | ENVSHI | ~LOWPER | ~INFOUT | NWLDER | ~PERTRA |
| 0.1886 | 0.136 | 0.0421 | 0.1745 | 0.1679 | 0.0462 | 0.2447 |

All entropy weight values had $\rho > 0.01$, entropy weight elimination test, all retention index

**Table 8** Coupling coordination degree for decision-making path

| Case | Decision-making path 1 | | Decision-making path 2 | |
|---|---|---|---|---|
| | Coupling degree (C) | Coupling coordination degree (D) | Coupling degree (C) | Coupling coordination degree (D) |
| Xchanging | 0.04 | 0.00 | 0.07 | 0.00 |
| Royal DSM | 0.08 | 0.09 | 0.06 | 0.00 |
| SEB bank. | 0.06 | 0.00 | 0.08 | 0.07 |
| OpusCapita | 0.03 | 0.00 | 0.06 | 0.00 |
| Radiant Law | 0.04 | 0.00 | 0.04 | 0.00 |
| Telefónica O2 | 0.08 | 0.09 | 0.05 | 0.08 |
| Liberty Source | 0.08 | 0.10 | 0.05 | 0.08 |
| European Energy | 0.10 | 0.11 | 0.06 | 0.00 |
| Vodafone (VSS) | 0.09 | 0.11 | 0.06 | 0.09 |
| Ascension Health | 0.07 | 0.00 | 0.06 | 0.00 |
| BNY Mellon | 0.04 | 0.07 | 0.04 | 0.06 |
| SINOCHEM | 0.12 | 0.12 | 0.08 | 0.10 |
| Davies Turner | 0.03 | 0.05 | 0.06 | 0.07 |
| MUFG Union Bank | 0.04 | 0.00 | 0.05 | 0.07 |
| Leeds Building Society | 0.07 | 0.10 | 0.06 | 0.09 |
| Coveris Holdings Sarl | 0.03 | 0.00 | 0.04 | 0.00 |
| Wipro | 0.12 | 0.12 | 0.09 | 0.11 |
| Circle K | 0.14 | 0.13 | 0.10 | 0.11 |
| Raiffeisen Bank | 0.05 | 0.07 | 0.04 | 0.07 |
| ANZ Bank | 0.10 | 0.10 | 0.10 | 0.11 |
| DBS Bank | 0.12 | 0.12 | 0.09 | 0.10 |

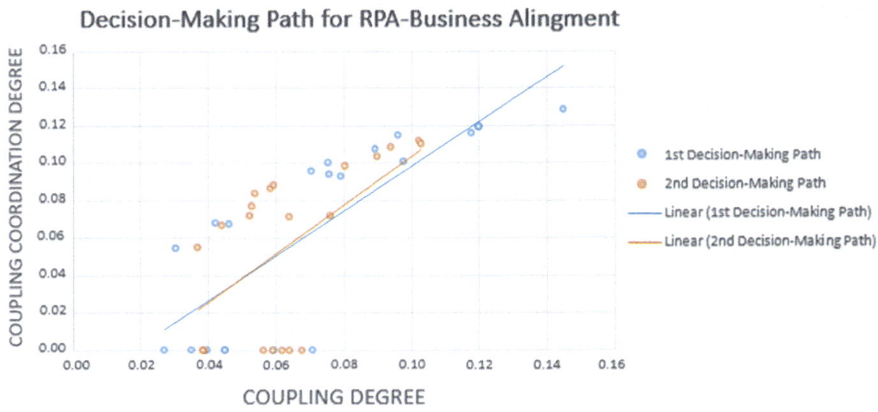


**Fig. 3** RPA-business alignment coupling coordination

According to coupling degree C and coupling degree D of the two key decision paths of RPA and business alignment calculated above, all values are less than 0.3; that is, all cases are in low coupling correlation degree or primary coupling correlation degree. This result is consistent with RPA's being only 3 years old; many enterprises are still in the initial deployment stage or the just completed stage. In addition, according to comparative analysis of the figure below, the development trend of the key decision path 2 (orange line) is better than that of path 1 (blue line), thus again verifying the results of the previous fsQCA; that is, the consistency of the decision path 2 is due to plan 1. Therefore, for the actual introduction of RPA into daily business operation, enterprise managers can consider the decision path 2 analysis (Figure 3).

## 6 Conclusion

Through the above fsQCA analysis and coupling evaluation, two decision-making paths for RPA-business alignment were found, with an overall solution consistency of 0.7735. The coverage of 0.5394 is low, because the RPA technology is newly emerging, and many enterprises have just completed the deployment that verified by coupling evaluation results.

**Decision-Making Path 1:** ENVSHI * ORGINE * LOWPER * INFOUT * NWLDER * PERTRA.

For achieving RPA-business alignment, enterprises should first adjust their business strategy and maintain foreign business direction by adopting broader science and technology management views, strengthening the internal efficiency consciousness to overcome any internal structural rigidity and decreases in performance. Finally, enterprise workers must realize that RPA-business alignment will ultimately improve the internal efficiency of the enterprise and eventually improve enterprise competitive advantage. In fsQCA, the consistency is 0.7960, the original coverage is 0.3662, and the case coverage effect is low.

**Decision-Making Path 2:** ENVSHI * ORGINE * ~LOWPER * ~INFOUT * NWLDER * ~PERTRA.

For achieving better RPA-business alignment, enterprises should first change their business strategy and simultaneously improve enterprise efficiency and maintain the stability of the global business strategy, using broader vision management to overcome the customer enterprise's internal lack of vision. Ultimately, RPA improves the internal efficiency of the enterprise and provides a competitive advantage. In fsQCA, the consistency and raw coverage are 0.8026 and 0.4711 respectively.

The consistency and case coverage of this decision-making path is closed to the first path. Moreover, evaluation of the above two decision-making paths through coupled coordination calculation indicated that the two paths have a low coupling coordination degree or initial coupling coordination degree. The fsQCA calculation results indicated that, because RPA technology is new, many enterprises have just finished deployment. According to the trend line in the coupling coordination, the two decision-making paths are closely each other. Therefore, for enterprises that plan to deploy RPA, this paper's conclusions may provide a reference. Similarly, two different methods of RPA-business alignment decision-making patch research indicate the same conclusion in this paper. Our research results also enrich the RPA-business alignment field.

## References


1. Lhuer, X. (2016). *The next acronym you need to know about: RPA (robotic process automation)*. McKinsey & Company.
2. Oliveira, J. (2016). Robotic Process Automation (RPA)-Part 2. Information Systems and Management Accounting, Chap. 21.
3. Berruti, F., Nixon, G., Taglioni, G., & Whiteman, R. (2017, March). *Intelligent process automation: the engine at the core of the next-generation operating model*. Digital McKinsey.
4. Willcocks, L., Lacity, M., & Craig, A. (2017). Robotic process automation: Strategic transformation lever for global business services? *Journal of Information Technology Teaching Cases, 7*(1), 17–28.
5. Lacity, M., Willcocks, L., & Craig, A. (2016). *Robotizing global financial shared services at royal DSM* (The Outsourcing Unit Working Research Paper Series, LSE Research Online Documents on Economics, Paper 02).
6. Lacity, M., Khan, S., & Carmel, E. (2016). Employing US military families to provide business process outsourcing services: a case study of impact sourcing and reshoring. In Communications of The AIS, CAIS2016 (Vol. 39, p. 9).
7. Aleksandre, A., & Esko, P. (2016). Turning robotic process automation into commercial success – Case OpusCapita. *Jouranl of Information Technology Teaching Cases, 6*(2), 67–74.
8. Willcocks, L., Lacity, M., & Craig, A. (2015). *Robotic process automation at Telefonica O2* (MIS Quarterlity Outsourcing Unit Working Research Paper Series, Paper 02).

9. Aguirre, S., & Rodriguez, A. (2017). Automation of a business process using robotic process automation (RPA): a case study. In *Proceedings of 4th Workshop on Engineering Applications, Applied Computer Sciences in Engineering, WEA 2017*, Cartagena Colombia (Vol. 1, pp. 65–71).
10. Zhang, N., & Liu, B. (2018). The key factors affecting RPA-business alignment. In *Proceedings of the 3rd International Conference on Crowd Science and Engineering, ICCSE 2018*, Singapore (Vol. 1, No. 10, pp. 1–6).
11. Willcocks, L., Lacity, M., & Craig, A. (2015). *Robotic process automation at Xchanging* (The Outsourcing Unit Working Research Paper Series, LSE Research Online Documents on Economics, Paper 03).
12. Herbert, I. P., Dhayalan, A., & Scott, A. (2016). The future of professional work: will you be replaced, or will you be sitting next to a robot? *Management Services Journal, 2134*(24457), 22–27.
13. Holder, C., Khurana, V., Harrison, F., & Jacobs, L. (2016). Robotics and law: Key legal and regulatory implications of the robotics age (Part I of II). *Computer Law & Security Review, 32* (3), 383–402.
14. Barnett, G. (2015). Robotic process automation: adding to the process transformation toolkit. White paper IT0022-0005, Ovum Consulting, Technical report.
15. Acemoglu, D., & Restrepo, P. (2017). Robots and Jobs: Evidence from US Labor Markets. National Bureau of Economic Research, No. 23285.
16. Willcocks, L., Craig, A., & Craig, A. (2015). *The IT function and robotic process automation* (London School of Economics and Political Science (LSE), Paper 6).
17. Herbert, I. P. (2016). How students can combine earning with learning through flexible business process sourcing: a proposition. Loughborough University, London, UK, Technical report 23729.
18. Lacity, M., Willcocks, L., & Craig, A. (2017). Service automation: cognitive virtual agents at SEB Bank. London Sch. Econ. Polit. Sci, pp. 1–29.
19. Lacity, M., & Willcocks, L. (2016). Rethinking legal services in the face of globalization and technology innovation: the case of radiant law. *Journal of Information Technology Teaching Cases, 6*(1), 15–22.
20. Ovum (2016). Robotic Process Automation offers scope for artificial intelligence. Wikipedia. Retrieved 01 June, 2016, from https://en.wikipedia.org/wiki/Robotic_process_automation, TMC.Intelligence.
21. Lacity, M., Willcocks, L., & Craig, A. (2015). *Robotic process automation: mature capabilities in the energy sector* (The Outsourcing Unit Working Research Paper Series, LSE Research Online Documents on Economics, Paper 03, Paper 06).
22. Evans, G. L. (2017). Disruptive technology and the board: the tip of the iceberg 1. *Economics and Business Review, 3*(1), 205–233.
23. Mesaglio, M., & Mingay, S. (2014). Bimodal IT: how to be digitally agile without making a mess. Gartner Website.
24. Luftman, J. (2000). Assessing business-IT alignment maturity. *Communications of the AIS, 4* (1), 99–128.
25. Luftman, J., & Kempaiah, R. (2007). An update on business-IT alignment:"A line" has been drawn. *MIS Quarterly Executive, 6*(3), 165–177.
26. Sabherwal, R., Hirschheim, R., & Goles, T. (2001). The dynamics of alignment: Insights from a punctuated equilibrium model. *Organization Science, 12*(2), 179–197.
27. Wang, N., Xue, Y., Liang, H., & Ge, S. (2011). The road to business-IT alignment: A case study of two Chinese companies. *Communications of the Association for Information Systems, 28*(26), 415–436.
28. Ragin, C. C. (2014). The comparative method: Moving beyond qualitative and quantitative strategies. *Social Forces, 67*(3), 827–829.
29. Ragin, C. C. (2009). Redesigning social inquiry: Fuzzy sets and beyond. *Social Forces, 88*(4), 1936–1938.

30. Cronqvist, L., & Berg-Schlosser, D. (2009). *Multi-value QCA (mvQCA), configurational comparative methods: qualitative comparative analysis (QCA) and related techniques* (Vol. 51, pp. 69–86). Thousand Oaks: SAGE Publications.
31. Fiss, P. C. (2007). A set-theoretic approach to organizational configurations. *Academy of Management Review, 32*(4), 1180–1198.
32. Ishida, A., Yonetani, M., & Kosaka, K. (2006). Determinants of linguistic human rights movements: An analysis of multiple causation of LHRs movements using a Boolean approach. *Social Forces, 84*(4), 1937–1955.
33. Watanabe, T. (2007). International comparison on the occurrence of social movements. *Journal of Business Research, 60*(7), 806–812.
34. Hagan, J., & Hansford-Bowles, S. (2005). From resistance to activism: The emergence and persistence of activism among American Vietnam war resisters in canada. *Social Movement Studies, 4*(3), 231–259.
35. Cress, D. M., & Snow, D. A. (2000). The outcomes of homeless mobilization: The influence of organization, disruption, political mediation, and framing. *American Journal of Sociology, 105*(4), 1063–1104.
36. Hicks, A. (2018). *Social democracy and welfare capitalism: A century of income security politics*. Ithaca: Cornell University Press.
37. Osa, M., Corduneanu-Huci, C., & Kousis, M. (2005). Linking economic and political opportunities in nondemocracies. *Contemporary Sociology, 35*(10), 171–201.
38. Salvatore, R. (2016). Vodafone Shared Services: Exploring RPA Opportunities. ssonetwork. Retrieved 01 August, 2016 from https://www.ssonetwork.com/technology-automation/articles/relieving-the-pressure-of-recruitment-through-rpa.
39. Hanna, A. J. (2016). RPA for Real: Ascension Health Takes on Leading Role. ssonetwork. Retrieved October 01, 2018 from https://www.ssonetwork.com/technology-automation/articles/rpa-for-real-ascension-health-takes-on-leading-rol.
40. The Rise of Robots (2017). BNY Mellon. Retrieved October 01, 2018 from https://www.bnymellon.com/us/en/who-we-are/people-report/innovate/the-rise-of-robots.jsp.
41. Sun, Y., & Dong, Z. Q. (2017). RPA: the only way to financial intelligence. *Xin Li Cai, 11*(28), 64–65.
42. Gould, R. (2017). How Robotics Technology is Modernizing Third-Party Logistics. Mhlnews. Retrieved October 01, 2018 from https://www.mhlnews.com/technology-automation/how-robotics-technology-modernizing-third-party-logistics.
43. Union Bank Accelerates Time to Revenue for Mortgage Loans (2017). Kofax. Retrieved October 01, 2018 from https://www.kofax.com/Learn/Case%20Studies/2015/Union%20Bank.
44. Sumner, S. (2017). Robots speed up the customer experience at Leeds Building Society. Computing. Retrieved October 01, 2018 from https://www.computing.co.uk/ctg/news/3018565/robots-speed-up-the-customer-experience-at-leeds-building-society.
45. RPA Enables 100% Process Efficiency and 45% Cost Reduction in AP for Global Plastic Manufacturer (2016). Auxis. Retrieved October 01, 2018 from https://www.auxis.com/case-studies/rpa-enables-100-process-efficiency-and-45-cost-reduction-in-ap-for-global-plastic-manufacturer.
46. Tarafdar, M., & Beath, C. (2018). Wipro limited: developing a cognitive DNA. In *Proceeding of 39th International Conference on Information System*, San Francisco (pp. 1–15).
47. Accelerate Business Results, Improve Employee and Customer Experience (2017). Accenture. Retrieved October 01, 2018 from https://newsroom.accenture.com/news/accenture-and-blue-prism-team-to-provide-robotic-process-automation-to-help-clients-accelerate-business-results-improve-employee-and-customer-experience.htm.
48. Makarchenko, M., Nerkararian, S., & Shmeleva, I. A. (2016). How traditional banks should work in smart city. In *1st Proceedings of International Conference on Digital Transformation and Global Society, DTGS 2016* (Vol. 1, pp. 123–134).
49. Smith, P. (2015). Rise of the machines as ANZ brings in robot workers to do the 'boring' jobs. AFR. Financial Review. Retrieved October 01, 2018 from https://www.afr.com/technology/rise-of-the-machines-as-anz-brings-in-robot-workers-to-do-the-boring-jobs-20150820-gj3fp6.

50. DBS Bank accelerates digitalisation transformation with robotics programme (2017). DBS bank. Retrieved October 01, 2018 from https://www.dbs.com/newsroom/DBS_Bank_accelerates_digitalisation_transformation_with_robotics_programme.
51. Jain, T.K., Kushwaha, D.S., & Misra, A.K. (2008). Optimization of the Quine-McCluskey method for the minimization of the boolean expressions. In *Proceeding of 4th International Conference on Autonomic and Autonomous Systems*, ICAS2008, Gosier, Guadeloupe ( pp. 165–168).
52. Ketchen, J., David, J., James Thomas, B., & Snow, C. C. (1993). Organization configurations and performance: A comparison of theoretical approaches. *Academy of management journal*, *36*(6), 1278–1313.
53. Yaoyuan, Q., Shufen, D., & Zehui, G. (2015). Research on the coupling relationship between technology innovation and business model innovation in enterprise. science & technology progress and policy.

# A New Performance Testing Scheme for Blockchain System

**Chongxuan Yuan and Jianming Zhu**

**Abstract** The development of blockchain technology is developing rapidly. From the appearance of Bitcoin, Ethereum and other blockchain systems to the Hyperledger and EOS platforms, the core concerns of the development of these systems is the performance. Bitcoin, Ethereum and other blockchain systems have been criticized a lot due to their lower TPS. Subsequent system platforms such as Hyperledger and EOS have upgraded TPS performance to thousands or even more. In addition, such a lot of blockchain projects claimed to have thousands to millions of throughput performance. It is difficult for market participants to verify whether the performance of such a system is in line with what they claimed, which raises concerns about the overall security of the blockchain system. At the same time, many projects ignore other indicators to pursuit only higher TPS performance, and cause the "Only Throughput" slogan. This paper proposes an application system performance test scheme for blockchain technology, gives an indicator system for performance testing and verification, explains the testing principles and methods of these indicators, and designs the framework and components of the test tools. This scheme can test whether the performance of the blockchain system meets the actual business needs, verify and evaluate the system from a technical point of view, then promote the real blockchain project.

**Keywords** Blockchain · Performance · Test indicator · System component

## 1 Introduction

With the development of blockchain technology rapidly, the industry has shifted from a wait-and-see attitude to an attempt to use blockchains.

---

C. Yuan (✉) · J. Zhu
School of Information, Central University of Finance and Economics, Beijing, China
e-mail: yuanchongxuan@foxmail.com

J. Zhu
e-mail: zjm@cufe.edu.cn



J. Zhang et al. (eds.), *LISS2019*,
https://doi.org/10.1007/978-981-15-5682-1_55

Government and website certificate agencies are good examples to explain, citizens trust government authority so that they trust the purchasing ability of the currency issued by the government. Although the performance indicators of the system are well, the information and network security issues are increasingly highlight, the traditional central structure poses higher risk, mainly in the system and moral aspects. Take the website certificate issuing institution as an example, once the central node is hacked and loses the root authentication key, it will rise system risk. Once the central organization has issued a certificate for selfish abusing, it will cause moral hazard, and all participants in this centralized structure will powerless. The risk of traditional trust structure mode can not be ignored.

Under the above background, the original intention of Haber et al. invented time stamps and package transactions with hash trees is to solve the system risk of traditional public trust forms [1], while Satoshi [2] tries using Bitcoin Blockchain to solve moral hazard due to statutory banknotes issued by the government. The new trust model provided by blockchain can be applied in many fields, such as the supply chain management [3]. Block information is voluntarily provided by each authenticated participant. Compared to the manufacturer be the only authority node providing supply chain information, the credibility of overall system using blockchain technology is higher than traditional models. While such distributed applications systems struggle to outperform the performance of traditional centralized systems, it should be recognized that losing performance to ensure network security is necessary.

This paper hopes to measure the performance of the overall blockchain system under the premise of using blockchain technology to ensure information and network security, and verify that the performance indicators can maintain the good operation of the whole system when it needs to complete specified services.

## 2 Related Work

### *2.1 Contradictions in Blockchain*

This section mainly describes the current research status of blockchain performance testing technology. There are two main contradictions in blockchain system:

- The main contradiction in blockchain technology is system poor performance. The poor performance of the blockchain system is mainly manifested in two aspects. First, the network congestion is slow, mainly focusing on high delay, excessive redundancy, long communication time, etc. Second, the implementation cost is high, mainly due to high hardware environment requirements.
- Another contradiction is fake bubble. At present, there are still many speculative ICO projects claimed to have thousands to even millions of throughputs. There are three problems—system with insufficient performance, good performance but no application, only ensuring throughput but other aspects are not guaranteed.

Every discussion about the scalability of blockchain cannot avoid the topic on the measurement method of performance indicators. Many scholars have proposed a scheme to extend the function of blockchain.

The performance test of the blockchain system was originally proposed by the Ethereum official when the Ethereum platform was released. They made a performance analysis and benchmark on the Ethereum network [4, 5]. Subsequent related scalability studies for existing blockchains (mainly focus on reinforce Bitcoin network and improved consensus algorithms) have emerged that have improved the performance of blockchain networks to some extent. Decker et al. foremost researched the Bitcoin network test report on how Bitcoin used multi-hop broadcast to spread transactions and prohibited updating a copy through the network with analyzing block size, block height, block generation time, etc. [6]. Croman and Decker et al. also analyzed the Bitcoin network and studied the performance of Bitcoin in terms of distributed blockchain scalability, including maximum throughput, latency, block generation time, and cost in per confirmed transaction [7]. In the PoW consensus blockchain, Gervais et al. measured the security and performance of PoW blockchains, analyzed the interactions between security and performance, captured some current PoW blockchains and compared how to trade off between security and performance by setting variables [8]. Kongrath and Dinh designed a mobile blockchain application whose performance analysis focused on verifying memory utilization rate of the process on the PoW process by setting up on-chain blocks containing different numbers of transactions to observe these transactions accordingly, finding that they have different execution time and resource consumption [9]. Gervais and Ghassan et al. introduced a new quantitative framework to analyze the security and performance impact of various consensus and network parameters in the PoW blockchain, and objectively compare the trade-off between performance and security of the blockchain [10]. In addition, Jason and Peter's main job is to compare the differences between blockchain-based and non-blockchain-based applications, focusing on response time, throughput, and network topology and so on overall performance rather than merely data-reading modules test [11].

However, the performance measurement of the blockchain system lacks a general testing method for multiple kind of blockchains, so that lacks a general performance test solution for various blockchains. This paper proposes a general blockchain test index and its implementation principles and methods.

## 2.2 Problems and Challenges

Whether to transform an existing network or designing a new blockchain, the research status of blockchain testing indicates that its scalability needs to be greatly improved. In addition, what are the problems with the existing improvement measures? Before eliciting the test architecture, we believe it is necessary to explain the difficulties and challenges in measuring the performance of blockchain systems.

The key point: various schemes proposed by scholars and industry have considered performance and security, but wondering on whether security, scalability and decentralization be an impossible triangle problem in the blockchain? Performance is often the most important consideration in terms of scalability. A system called Monoxide is composed by a 48,000 global nodes. In the test environment composed of these distributed nodes, the transaction volume on per second and the memory capacity is 1000 times higher than that of the Bitcoin network [12]. It may explains some difficulties and challenges on the performance test of blockchain. In the future, advanced techniques such as improved sharding and asynchronous consensus can be used to solve the impossible triangle problem. The enlightenment of the Monoxide environment deployment solution by designing a distributed node test structure like the open source project Apache Zookeeper solution that providing high-performance distributed application collaboration services [13].

Typical project: typical improvements in blockchain focus on these three aspects—scalability, security and decentralization. For the designated business areas, how to choose between these three aspects is particularly important.

- Stella. In September 2017, the European Central Bank and the Bank of Japan jointly issued a first phase project report on blockchain testing [14]. In May 2018, the second phase, a feasibility report on blockchain securities settlement was jointly issued [15]. The project research only analyzes the theoretical performance, lacks the plan of the performance verification scheme, and therefore has doubts about the overall performance and safety of the blockchain system.
- Hyperledger. Hyperledger Fabric provides the infrastructure and code, the official demonstrates its performance: Fabric achieves end-to-end throughput of more than 3,500 transactions per second, as well as other performance parameters provided to users [16, 17]. Li and Sforzin proposed a performance-improved blockchain solution based on the Fabric platform, using satellite chain technology running different consensus protocols in parallel to make the performance meet industry standards [18].
- Blockbench. The Blockbench project provides a private chain assessment methodology that tests private blockchain from both macro and micro indicators which is the first code tool to evaluate the performance of private chains in testing environments [19]. Tuan and Wang et al. provide test methods for TPS indicators in Blockbench, but only be limited to testing private blockchain.
- Bitcoin and Ethereum status quo. Weber and Gramoli et al. determined the availability limitations of two major blockchain systems, such as Ethereum and Bitcoin. The proposed suspension mechanism technology reduces the availability limit of the existing blockchain and it tests the effectiveness of the proposed technology through experiments [20].

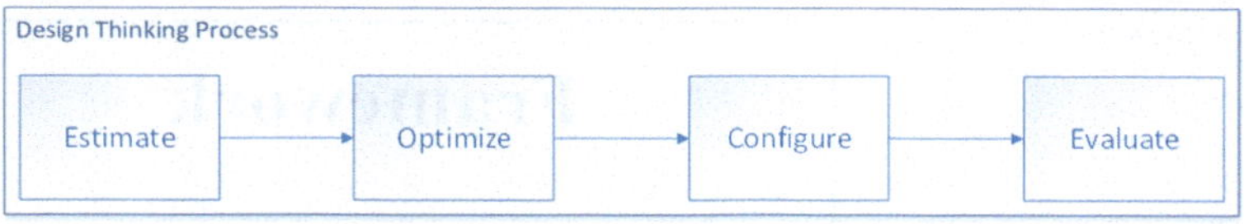


**Fig. 1** Design thinking process

# 3 Performance Test Scheme for Blockchain

In the context of Blockchain as a Service (BaaS), Blockchain Test-as-a-Service (BTaaS) becomes a viable solution. By launching the BTaaS service [21], the development team can quickly help users to verify performance and deploy distributed blockchain ledgers.

## *3.1 Design Thinking*

This section presents four design ideas for BTaaS general performance testing like Fig. 1.

- Estimation. Obtain the parameters of the existing blockchain system, including the number of nodes, net topology, block size, block generation rules, consensus algorithm, chain code, etc.
- Optimization. Verify and decompose the application layer before deployment, preserve the necessary applications, change the infrastructure, protocols, and consensus algorithms.
- Configuration. Quickly launch a synchronous test network with a specified number of configurable nodes in a test or production environment.
- Evaluation. Observe the performance of the blockchain system and its distributed applications in a controlled environment.

## *3.2 Propose the Test Framework*

The core of the above four design ideas is to use the adapter tool to simulate the real network configuration to generate the test blockchain network, embed the pre-programmed test cases and network optimization configuration into the test network, and follow the proposed performance metrics system and its testing principles and methods. Run the test program and finally get the test conclusion.

- Adapter Frame. According to Fig. 2, this paper proposes an adapter-based blockchain test framework.

Fig. 2 Framework

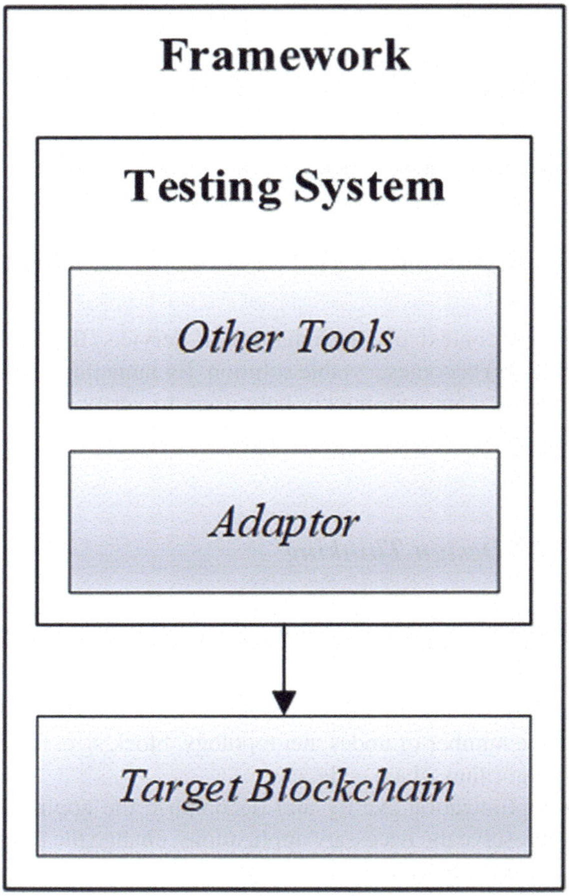


This framework designing provides theoretical support for blockchain testing, supporting tests of generic blockchain systems: firstly, giving a target blockchain, secondly adapting the corresponding configuration through this test system, then generating a simulated test network, and lastly targeting the blockchain based on pre-programed test cases. In the end, verify the performance indicators are up to standard.

- System Architecture. As shown in Fig. 3, the architecture of the blockchain test system is embodied as a target blockchain network, an adaptation layer, an interaction layer, and a testing layer from bottom to top. The respective functions are: the target blockchain network is the network to be tested; the adaptation layer is used to integrate the existing blockchain system into the framework; the interaction layer provides interfaces for data exchange for lower adaptation layer and upper testing layer; the testing layer contains test cases for multiple blockchain schemes and supports deploying custom case. Three experiments designed in this paper are examples of bank access services, ball games, digital copyrights. Each test provides a configuration file to define the back-end blockchain network and test parameters. The experimental results show relevant metrics and give experimental conclusions.

- Test Metrics. The test metrics provided by the system are shown in Fig. 4. They consist of four categories: TPS, delay, transaction execution success rate, and resource occupation. Among them, the delay metric composed of the maximum, minimum and average delay respectively, and the resource utilization is divided into three kinds of tests: CPU usage, memory usage, and I/O traffic.

The indicator system includes but not limited to the above metrics can be integrated into the system. State database monitoring multiple metrics such as the status data format. These monitoring methods are achieved through resource monitors in the test engine of the system components, which is based on the general implementation of Docker container monitoring technology [22].

**Fig. 3** System architecture

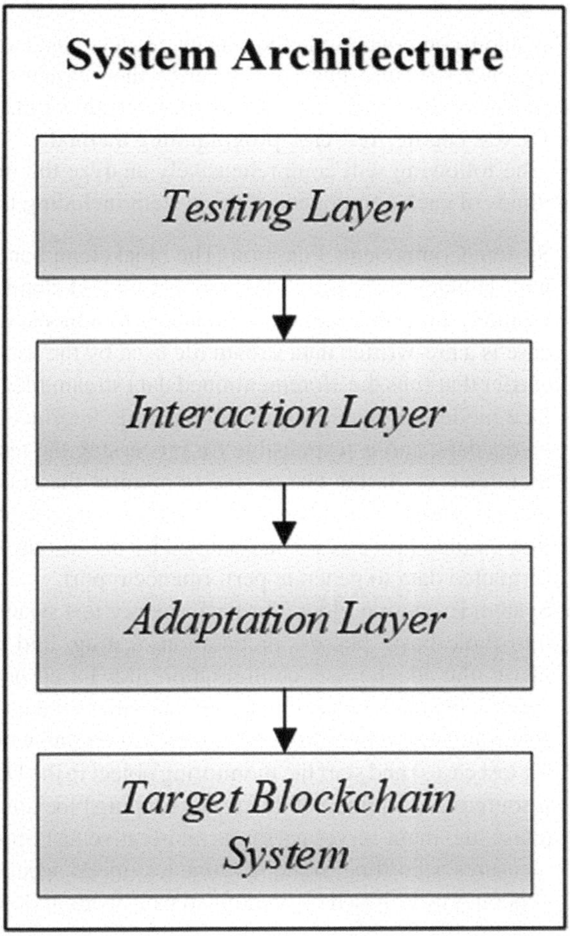

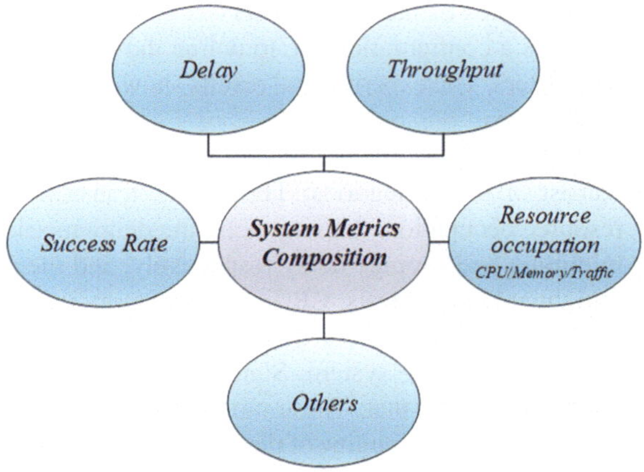


**Fig. 4** Test metrics composition

## 3.3 System Components Design

The main components include adapters, test cases and test engines. The system component structure diagram is given by the following Fig. 5. The core of the system needs to achieve three targets: Adapter design, Configuration files implementation in the test engine, Test case programming method.

The following will comprehensively analyze the implementation principles and methods of each component of the system including the above three targets.

- System Components Function. The blockchain benchmark performance test system includes test cases, a test engine, a blockchain interface module, a resource monitor, a report generation module, and a blockchain adapter module. The test case is a pre-written data stream file used by the user; the test engine is a program carrier that runs the aforementioned data stream file, wherein the blockchain interface module provides a set of compatible interfaces processing upper and lower layers data, and is responsible for processing the test case data files; the resource monitor is set in the test engine to monitor the computer resources occupied by the test engine which based on Docker Container technology; the report generation module processes data obtained by the resource monitor into a user-friendly formatted data to generate performance report.
- System Execution Flow. The performance test system execution steps are divided into three basic phases: preparation, testing, and reporting. In the preparation phase, the adapter sets configuration files based on the actual target blockchain system, the main server creates and initializes the internal blockchain test network according to configuration files, then test engine would deploy the smart contract (or test cases) and start the monitoring object in the Docker container to monitor the resource consumption of the device running blockchain system; During the testing phase, the main server performs an iterative test process, carries out performance measures according to the iterative test mode specified in the configuration files, generates tasks based on predefined workloads and assigns them to multiple clients

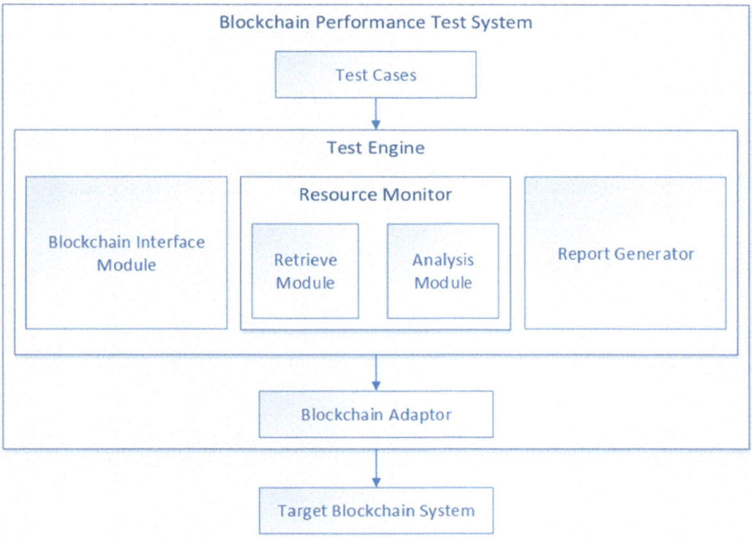


**Fig. 5** Blockchain performance test system components structure diagram

(or nodes) for execution, and stores performance statistics returned by clients (or nodes) for follow-up analysis; During the reporting phase, all statistics for each test iteration are analyzed and a formatted data report is automatically generated.

These three basic steps involve processing a set of data containing input and output data files. The detailed program flow is shown in Fig. 6.

The first step is to import predefined use cases. These use cases can ultimately be represented by a set of data files that can be read by the test engine. In the second step, the test engine uses several independent configuration files to process use cases, these configuration files specify the type and name of blockchain to be tested, the command scripts for starting and stopping, and detailed test methods (including number of iterations, iteration parameters, rate control). The rate control is performed by a rate controller that defines a set of interfaces for processing data on the input data stream at a specified rate, enabling the user to perform tests under a custom mechanism. In the third step, the test engine activates resource monitors while executing the test. The resource monitors include an retrieve module and an analysis module, and the retrieve module acquires resource consumption of the blockchain system to be tested in the test engine, including CPU usage, memory usage, and network I/O traffic, etc. The analysis module reads the performance statistics and calculates the throughput, transaction delay, and transaction success rate according to the specified data processing procedure. The specified blockchain adapter module records key metrics when the interface is invoked, such as the commit time and end time of the created transaction, and then provide these to the engine. In the fourth step, the report generation module obtains the data results

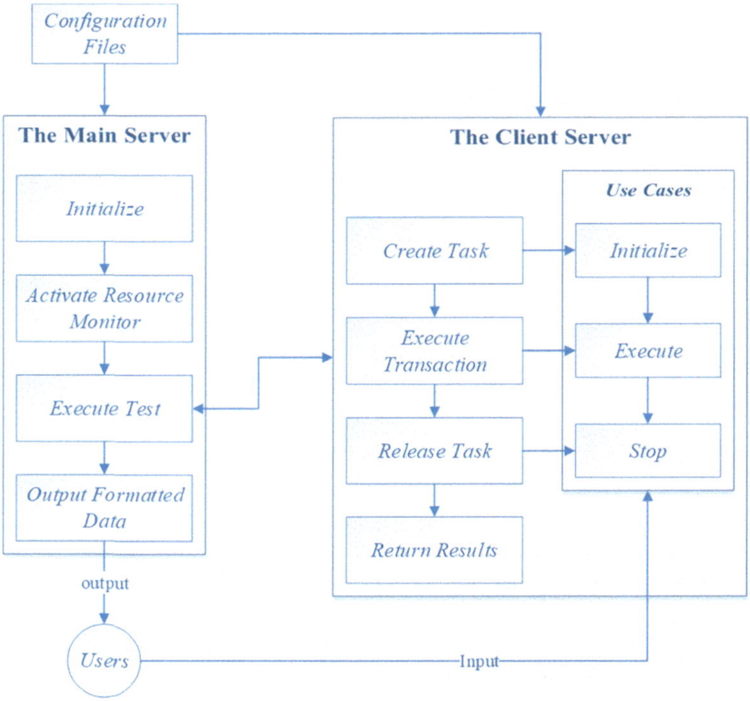


**Fig. 6** Performance test program execution process

from resource monitors and outputs the data to generate a formatted performance report.

- Test Engine Configuration File. The core designing of the test engine is to support reading and executing two kinds of configuration files (base profile and blockchain profile).

**Listing 1** Network configuration parameters

```
"blockchain": {
    "type": "fabric",
    "config": "./fabric.json"
  },
"command" : {
    "start": "docker-compose -f up",
    "end" : "docker-compose -f down,
             docker rm .."
  }
```

Listing 1 is a script of a configuration file can be used in a Fabric blockchain network. The main parameters are the specified blockchain type and detailed parameter configuration json file, using starting and stopping script commands based on Docker container technology.

**Listing 2** Test case parameters

```
"test": {
      "name": "test",
      "description : "customize",
      "clients": {
         "type": "local",
         "number": 16
      },
      "rounds": [{
            "label" : "test1",
            "txNumber" : [10000],
            "rateControl" : [{"type":
                  "fixed-rate", "opts":
                  {"tps" : 2000}}],
            "arguments": {  "test": 1000 },
            "callback" : "test1.js"
         }]
}
```

Listing 2 is to configure the parameters used to specify a target test case. The parameters include the use case name, description, number of client nodes and its type (divided into local and distributed environments).

**Listing 3** Resource monitor parameters

```
"monitor": {
      "type": ["docker", "process"],
      "docker":{
         "name": ["p0.o1 ",
            "192.168.1.2:5250/orderer"]
      },
      "process": [
         {
            "command": "node",
            "arguments": "local-client.js",
            "multiOutput": "avg"
         }
      ],
      "interval": 1
}
```

The configuration parameters represented in Listing 3 are used to specify the Docker resource monitors, including defining resource monitors. This example uses the Fabric as an instance to specify the container name, process information, and other parameters that are running in the Docker.

- System Deployment Mode. The deployment of blockchain performance testing systems is divided into sole and multi-machine modes. The multi-machine deployment idea of test blockchain performance proposed in this paper is reflected in the distributed deployment structure diagram called Fig. 7. Designing a distributed

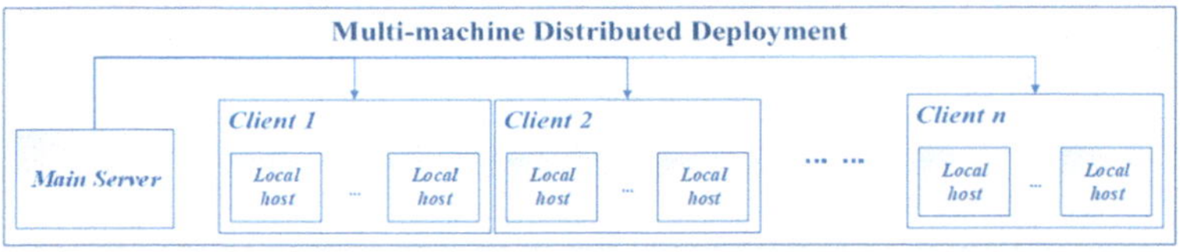


**Fig. 7** Multi-machine distributed deployment structure

node test structure is similar to the open source project Apache Zookeeper solution, which provides collaboration services for high-performance distributed applications [13].

- Use Case Designing. According to the blockchain performance test plan proposed in this paper, three test cases of bank business, a ball game and digital copyright are provided. Due to space limitations, this section only gives the design principles and methods of bank business test cases. The experimental analysis would also given later. In the traditional mode, the methods of benchmarking a specified system mainly with the help of pressure performance testing, the common content is to make capability verification, performance planning, performance tuning, pressure loading and drawing performance degradation curve. In the field of traditional relational database, Djellel et al. proposed a workload technology with standardized definitions, and used TPC-C to design a relational transaction database system [23]. Chaitanya et al. provided big data analysis and worked for database systems that store large-scale data. The workload provides a benchmark test [24]. In the emerging field of blockchain testing, some companies such as Huawei and Hyperchain have supported user-defined and rate-controlled concepts into Hyperledger, using Caliper tool to design benchmark test for Fabric, Iroha and Sawtooth [25]. Hyperledger deliver performance traffic machine concept is applied to the blockchain test, using the PTE SDK tool to send requests to one or more networks and receive responses from one or more networks, so that it can interact with the Fabric network to obtain status information.

The framework for use case designing in this paper has a similar advanced design to the traditional benchmarking framework, but its workload and main drivers are specifically designed for blockchain system. We write a dedicated smart contract for the blockchain system that needs to be tested as a test case, use the interface to specify the test flow.

Rate control [25] should include: a fixed rate controller, a fixed rate feedback controller, a PID rate controller, Composite rate controller, linear rate controller, recording controller, delay controller. The calculation theory of rate control is not complicated. Its calculation method is expressed as the following mathematical form:

$$rate = \frac{trans_{submit} - trans_{finish}}{unitTestNumber} \tag{1}$$

Bank business use cases are reflected in the transaction of opening accounts, deposit and withdrawal, and querying. The core of the use case is the implementation of the “read and write” program (Js code), the “write” program is used for account opening, deposit, etc. The “read” program is used for querying information. Listing 4 is an example of querying balance. Each use case implementation requires three functions, all of which return a Promise object: Init Function—The client will call it at the beginning of each test, by using the given blockchain object context and user-defined parameters that read from the base configuration file. Run Function—The transaction should be generated and committed here using system APIs. The client will call this function repeatedly based on the workload. Each call should only commit one transaction, and if multiple transactions are committed each time, the actual workload may be different from the configured workload; End Function—will be called at the end of each round of testing. Implement program cleaning work.

**Listing 4** A bank transaction use case instance

```
module.exports.info= "Query balance";
let bc, contx, accounts;
module.exports.init = function (
        blockchain, context, args) {
    let acc=require('./Operations.js');
    bc = blockchain;
    contx = context;
    accounts = acc.account_array;
    return Promise.resolve();
};
module.exports.run = function() {
    let acc_num = accounts[Math.floor(
        Math.random()*
        (accounts.length))];
    return bc.queryState(
        contx, 'bank', 'v1', acc_num);
};
module.exports.end = function() {
    return Promise.resolve();
}
```

## 4 Experiment

The experiment carried out four complete-experiments to eliminate accidental errors. Each complete experiment consisted of 10 groups of group-tests, totaling 40 groups of group-tests, including 23 groups of account opening and deposit transactions and 17 groups of query transactions. Each group-test consisted of 2000 unit tests that have 80,000 unit-tests in total, each unit-test performs an independent transaction execution, each unit-test separately calculates the delay metric, the maximum, the minimum and the average delay metric in each group-test statistics, and each

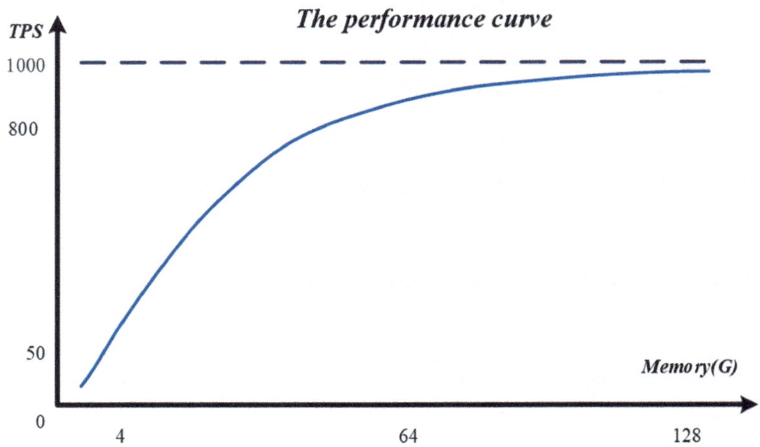


**Fig. 8** The performance curve varying with configuration

group-test respectively calculate memory consumption, CPU usage of rate, I/O traffic, disk read and write and other metrics. The experiment explores two questions:

- Relationship between business performance and hardware configuration. This paper finds that the system performance under the bank use case is basically linearly improved when the hardware configuration goes up. As the hardware configuration improves, the system performance will stay at a stable level finally. (By upgrading other configurations, it can break through 1000 TPS. The Fig. 8 only represent the relationship between memory configuration and system performance, it shows the curve of system performance from 4G memory to 128G memory:
- Overall performance of blockchain system for a certain business. The overall system under the bank business use case is maintained the performance at 800–1000 TPS basically, with 8000 transaction execution success rate on 99.725% and transaction minimum latency less than 0.01 s. The maximum delay time of querying is only 2.27 s, the average delay time of 34,000 querying unit tests is 0.0176 s, and the transaction success rate is 100%; the maximum delay time for more complicated transactions such as account opening and deposit are 99 s, the minimum delay reaches 0.5 second level, and the average delay time maintained at 35–50 s, and there are 220 errors in 46,000 complex transactions occurred. Table 1 is a complete experiment in a 128G memory environment, the performance results of the use case blockchain test.

  The rate of transaction input adopts the ladder control method. The transaction rate of account opening and deposit is increased by 50 TPS per time from 50 to 250 TPS. The querying transaction is controlled on 100 TPS in the first group-test, and the subsequent group-test would no longer limit the rate. The experiment finds that rate control on complex transaction is no effect due to the complexity and security of its business logic, it is close to the limit at around 18 TPS. The rate control of simple transaction such as querying can be controlled. By artificially

**Table 1** Performace test result

| Test | Name | Succ | Fail | Send rate (tps) | Max latency (s) | Min latency (s) | Avg latency (s) | Throughput (tps) |
|---|---|---|---|---|---|---|---|---|
| 1 | open | 2000 | 0 | 50.0 | 78.52 | 0.58 | 41.63 | 16.9 |
| 2 | deposit | 2000 | 0 | 100.2 | 95.68 | 0.66 | 49.78 | 17.3 |
| 3 | deposit | 1940 | 60 | 150.0 | 98.38 | 0.61 | 50.25 | 17.9 |
| 4 | deposit | 1980 | 20 | 200.1 | 99.87 | 0.56 | **51.45** | 17.7 |
| 5 | query | 2000 | 0 | 249.8 | 99.91 | 0.69 | 50.89 | 18.4 |
| 6 | query | 2000 | 0 | 100.2 | 0.08 | 0.01 | 0.01 | 100.1 |
| 7 | query | 2000 | 0 | 1002.5 | 0.38 | 0.01 | 0.06 | 995.5 |
| 8 | query | 2000 | 0 | 992.6 | 0.19 | 0.01 | 0.03 | 988.1 |
| 9 | query | 2000 | 0 | 999.0 | 0.23 | 0.01 | 0.04 | 990.6 |
| 10 | query | 2000 | 0 | 998.5 | 0.15 | 0.01 | 0.03 | 993.5 |

**Table 2** Account opening transaction resource occupation

| Name | Memory (max) (MB) | Memory (avg) (MB) | CPU (max) (%) | CPU (avg) (%) | Traffic in | Traffic out |
|---|---|---|---|---|---|---|
| peer0.org1 | 60.0 | 59.4 | 25.10 | 2.54 | 3.4 MB | 1.6 MB |
| peer0.org2 | 65.1 | 64.6 | 25.60 | 2.66 | 3.4 MB | 1.6 MB |
| peer0.org2 | 53.1 | 53.1 | 0.01 | 0.00 | 303 B | 262 B |
| peer0.org1 | 52.7 | 52.7 | 0.03 | 0.00 | 303 B | 262 B |
| orderer | 107.3 | 105.0 | 50.00 | 6.67 | 7.7 MB | 16.0 MB |
| ca.org1 | 12.3 | 12.3 | 0.00 | 0.00 | 0 B | 0 B |
| ca.org2 | 10.3 | 9.0 | 0.52 | 0.01 | 0 B | 0 B |

controlling the rate at 100 TPS, the system only processes about 100 querying transactions per second. Querying transaction with the rate control can observe that the highest throughput of the blockchain system reaches 995.5 TPS, which basically meets the business needs. The resource monitors of the Docker container return the results including memory usage, CPU usage, I/O traffic, etc. Tables 2 and 3 are the resource occupancy of each node in two groups of group-test (a group of account opening transactions and a group of querying transactions).

The resource occupancy of a group of group-test is counted. The simple querying transaction on the nodes does not use I/O traffic but consumes CPU. The more complex account opening transaction requires other nodes to endorse and reach a consensus through consensus algorithm so that produce a lot of I/O traffic. In addition, all nodes occupy a certain amount of memory because they are in active state.

**Table 3** Querying transaction resource occupation

| Name | Memory (max) (MB) | Memory (avg) | CPU (max) (%) | CPU (avg)(%) | Traffic in (B) | Traffic out (B) |
|---|---|---|---|---|---|---|
| peer0.org1 | 62.6 | 62.6 MB | 57.29 | 57.29 | 0 | 0 |
| peer0.org2 | 68.3 | 68.3 M | 60.24 | 60.24 | 0 | 0 |
| peer0.org2 | 53.2 | 53.2 MB | 0.00 | 0.00 | 0 | 0 |
| peer0.org1 | 52.7 | 52.7 MB | 0.00 | 0.00 | 0 | 0 |
| orderer | 131.2 | 131.2 MB | 0.00 | 0.00 | 0 | 0 |
| ca.org1 | 14.5 | 14.5 MB | 0.00 | 0.00 | 0 | 0 |
| ca.org2 | 10.3 | 10.3 MB | 0.00 | 0.00 | 0 | 0 |

## 5 Summary

Blockchain technology is improving, in the areas of ntech, e-commerce, supply chain management, e-government, they are considering using blockchain to solve security problems. In order to verify whether the performance of the blockchain system can meet the business needs and identify the fraudulent blockchain project in the market, in addition to strengthening supervision, this paper considers that it is especially important to test and evaluate the blockchain system from the technical aspect.

This paper proposes a new blockchain technology based application system performance test scheme. This scheme uses the adapter to build the target blockchain network, and builds a set of technically usable measurement systems through a four-layer basic architecture designing. It includes a set of metrics such as throughput, transaction delay, success rate, resource occupation, etc. From the specified transaction system service, a series of test cases that need to be written can be easily embedded into this test tool to provide data to the core test engine to analyze and produce user-friendly test results.

The contribution of this paper is that the proposed scheme is a new general test method for blockchain network. This paper verifies the effectiveness of this performance test scheme through experiments, and gives reasonable metrics for participating in performance testing. The program reflects four design principles: estimation, optimization, configuration, and evaluation. It can test whether the performance of the blockchain system meets actual business needs, verify and evaluate the blockchain system from a technical perspective, and promote valuable blockchain projects landed.

## References

1. Haber, S., & Stornetta, W. S. (1991). How to time-stamp a digital document. *Journal of Cryptology*, *3*(2), 99–111.
2. Satoshi, N. (2008). Bitcoin: a peer-to-peer electronic cash system

3. Chou, C. F., Lin, Y. P., Petway, J. R., & Ho, Y. F. (2017). Blockchain: the evolutionary next step for ICT e-agriculture. *Environments, 4*(3), 50.
4. Ethcore performance analysis. https://blog.ethcore.io/.
5. Ethereum benchmarks. https://github.com/ethereum/wiki/benchmarks.
6. Decker, C., & Wattenhofer, R. (2013, September 9–11). Information propagation in the bitcoin network. In *13th IEEE international conference on peer-to-peer computing (IEEE P2P 2013)*, Trento, Italy.
7. Croman, K., Decker, C., Eyal, I., Gencer, A. E., Juels, A., Kosba, A. E., Miller, A., Saxena, P., Shi, E., Sirer, E. G., Song, D., & Wattenhofer, R. (2016, February 26). On scaling decentralized blockchains - (a position paper). In *Financial cryptography and data security - FC 2016 international workshops, BITCOIN, VOTING, and WAHC*, Christ Church, Barbados.
8. Gervais, A., Karame, G. O., Wüst, K., Glykantzis, V., Ritzdorf, H. T., & Capkun, S. (2016, October 24). On the security and performance of proof of work blockchains. In *Proceedings of the 2016 ACM SIGSAC conference on computer and communications security*, Austria.
9. Suankaewmanee, K., Hoang, D. T., Niyato, D., Sawadsitang, S., Wang, P., & Han, Z. (2018, March 5–8). Performance analysis and application of mobile blockchain. In *2018 international conference on computing, networking and communications (ICNC 2018)*, Maui, HI, USA.
10. Gervais, A., Karame, G. O., Wüst, K., Glykantzis, V., & Ritzdorf, H. (2016, October 24–28). On the security and performance of proof of work blockchains. In *Proceedings of the 2016 ACM SIGSAC conference on computer and communications security*, Vienna, Austria.
11. Spasovski, J., Eklund, P. W. (2017, November 07–10). Proof of stake blockchain: performance and scalability for groupware communications. In *Proceedings of the 9th international conference on management of digital ecosystems (MEDES 2017)*, Bangkok, Thailand.
12. Wang, J., & Wang, H. (2019). Monoxide: scale out blockchains with asynchronous consensus zones. In *16th USENIX symposium on networked systems design and implementation*. NSDI.
13. Zookeeper. https://cwiki.apache.org/confluence/display/zookeeper.
14. Payment systems: liquidity saving mechanisms in a distributed ledger environment (2017).
15. BOJ/ECB joint research on distributed ledger technology (2018).
16. Oliveira, R., Felber, P., & Hu, Y. C. (2018). *Proceedings of the thirteenth EuroSys conference*, EuroSys, Porto, Portugal. ACM.
17. Hyperledger performance metrics (2018). https://hyperledger.org.
18. Li, W., Alessandro, S., Sergey, F., & Ghassan, K. (2017). Towards scalable and private industrial blockchains. In *Proceedings of the ACM workshop on blockchain system, cryptocurrencies and contracts*. ACM.
19. Dinh, T.T.A., Wang, J., Chen, G., Liu, R., Ooi, B. C., & Tan, K. (2017, May 14–19). BLOCKBENCH: a framework for analyzing private blockchains. In *Proceedings of the 2017 ACM international conference on management of data (SIGMOD 2017)*, Chicago, USA.
20. Ingo, W., Vincent, G., Alex, P., Mark, S., Ralph, H., Binh, T. A., & Paul, R. (2017). On availability for blockchain-based systems. In *IEEE 36th symposium on reliable distributed systems (SRDS)*. IEEE.
21. White block. https://www.whiteblock.io/.
22. Soltesz, S., Pötzl, H., Fiuczynski, M. E., Bavier, A. C., Peterson, L. (2007, March 21–23). Container-based operating system virtualization: a scalable, high-performance alternative to hypervisors. In *Proceedings of the 2007 EuroSys conference*, Lisbon, Porto, Portugal.
23. Difallah, D. E., Pavlo, A., Curino, C., & Cudré-Mauroux, P. (2013). OLTP-Bench: an extensible testbed for benchmarking relational databases. *Proceedings of the VLDB Endowment, PVLDB, 7*(4), 277–288.
24. Ghazal, A., Rabl, T., Hu, M., Raab, F., Poess, M., Crolotte, A., et al. (2013). BigBench: towards an industry standard benchmark for big data analytics (vol. 22).
25. Caliper. https://github.com/hyperledger/caliper.

# An Improved Two-Level Approach for the Collaborative Freight Delivery in Urban Areas

**Ahmed Karam, Sergey Tsiulin, Kristian Hegner Reinau, and Amr Eltawil**

**Abstract** Despite the negative consequences on society and environment, urban freight transport is critical for the prosperity of cities. Decision makers have considered the horizontal collaboration between carriers as a solution to reduce the total transportation cost and the related negative impacts. In literature, the collaborative freight delivery is modelled as a Multi-Depot Vehicle Routing Problem (MDVRP) that can be solved by being decomposed into two sub-problems, i.e. assignment problem and a set of vehicle routing problems. The assignment problem allocates customers to the nearest depots while vehicle routing problem determines the optimal routes to serve customers assigned to each depot. However, most of existing approaches did not consider the interrelation between these two sub-problems, which in turn impairs the solution quality of the overall problem. This paper presents an improved two-level mathematical modelling approach to evaluate and optimize the implementation of the collaborative freight distribution in urban areas. Unlike existing studies, the proposed approach considers the interrelation among the two-sub problems. A real-life case is used to illustrate the savings obtained from the

This research was partially funded by the Municipality of Aalborg through the research project No. 885086 "Collaborative Logistics in Aalborg—Opportunities, Challenges and the Road Ahead"

A. Karam (✉) · S. Tsiulin · K. H. Reinau
Freight Transport Research Group, Department of the Built Environment, Aalborg University, Aalborg 9220, Denmark
e-mail: akam@build.aau.dk

S. Tsiulin
e-mail: setsi@build.aau.dk

K. H. Reinau
e-mail: khr@build.aau.dk

A. Eltawil
Department of Industrial and Manufacturing Engineering,
Egypt-Japan University of Science and Technology, Alexandria, Egypt
e-mail: Eltawil@ejust.edu.eg

A. Karam
Department of Mechanical Engineering (Shoubra), Benha University, Benha 11672, Egypt



J. Zhang et al. (eds.), *LISS2019*,
https://doi.org/10.1007/978-981-15-5682-1_56

collaborative distribution. In addition, the significance of the proposed approach is demonstrated by a comparison against a similar approach found in literature.

**Keywords** City logistics · Carrier collaboration · Multi-vehicle routing problem · Order sharing · Freight transportation · Sustainability

## 1 Introduction

In the recent years, economic and industrial growth led to a dramatic increase in the volumes of transported goods in cities [1]. This in turn resulted in traffic congestion and environmental pollutions in urban areas [2]. Congestion and environmental issues appear primarily within city centers in Europe that are sensitive to any extra vehicular traffic due to infrastructure limitations. In addition, freight in-city transportations are responsible for 25% of overall $CO_2$ emissions and for 30–50% of other pollutants (Particulates, NOx) [3]. Moreover, such rates are expected to increase over time due to the continual growth of E-commerce activities and the world's population living in cities [4].

In response, many strategies have been proposed in order to increase the efficiency of freight distribution and reduce its related environmental and social impacts, particularly in the context of urban logistics. One of these strategies is the collaboration between carriers delivering goods in the same urban area. During the last decade, research on carriers' collaboration has gained an increasing interest of professional and scientific communities [5, 6].

This paper considers a collaborative distribution network in which goods (customers) of different carriers are pooled and shared between collaborating carriers. In addition, each carrier has a single depot where its vehicles start delivering goods to customers and return again to the depot. This situation can be modeled as a variant of the Capacitated Vehicle Routing Problem (CVRP), namely Multi-Depot Vehicle Routing Problem (MDVRP) in which more than one depot is to be considered [6]. The MDVRP is a well-known NP-hard problem [7]. To reduce its complexity, the problem can be decomposed into two sub-problems: firstly, customers must be assigned to depots (assignment problem); then routing of vehicles can be planned to serve customers assigned to the same depot (herein, CVRP is solved as many as the number of depots). Logically, it is more efficient to solve the two sub-problems holistically as a one problem. However, this holistic approach is not tractable computationally, especially in the collaborative environment where the size of problem increases dramatically. It should be noted that due to the interrelation of these two sub-problems, poor assignment solution leads to routing solutions of higher total distance [8]. However, most of existing decomposition-based approaches have been developed in a way that the interrelation of the sub-problems is not adequately considered [9, 10].

The main aim of this paper is to develop a two-level mathematical modelling approach for decomposing and solving the MDVRP in collaborative environment.

Compared to similar approaches in literature, the proposed approach considers the interrelation between the decomposed problems. Thus, this improves the solution quality of the overall problem as will be illustrated in this paper. In doing so, a new mathematical modeling approach is developed and validated using real-life data from a large city in a European country. The remainder of this paper is organized as follows: Sect. 2 provides a review of relevant studies. Section 3 discusses the framework of the proposed approach. Numerical experiments are introduced in Sect. 4. The conclusions and recommendations are in the last section.

## 2 Literature Review

The literature on collaborative logistics is very rich in several works. Most of existing works on collaborative logistics have focused on maritime and air transport, while research on collaborative logistics in road transportation is still rare and recent [10]. In general, there are two types of collaboration, i.e. vertical and horizontal collaboration. Vertical collaboration is made by actors in the same supply chain at different levels, an example can be found in [11] where the authors proposed a collaboration framework among carriers and the container port. The horizontal collaboration is made by actors in the supply network, working at the same level and providing similar services and products, an example can be found in [12] where a collaboration strategy was developed among adjacent container ports to reduce their operating costs by sharing container trucks among them. Herein, relevant literature on horizontal collaboration among freight carriers is breifly reviewed, specifically in the context of urban logistics. For a more boarder and detailed description of the relevant works, the reader can examine the survey in [13].

Two main types of horizontal collaboration can be distinguished in literature: order sharing and capacity sharing [13]. Order sharing means that customer orders from different carriers are pooled and reallocated amongst collaborating partners to decrease logistics costs and environmental impacts. Capacity sharing is realized by sharing vehicle capacities in order to reduce empty back hauls and to increase utilization rate of vehicles. In both types, carriers are motivated to share trucks and customers in order to enhance their individual turnovers. Similarly, a better service to customers can be achieved while reducing the environmental impacts of the transport activities. In spite of these benefits, allocating the savings among the collaborating partners still represents a major issue in collaborative freight transportation. Therefore, some studies employed methods of game theory to optimize allocating the joint cost savings between collaborating partners [14].

From modelling point of view, horizontal collaboration between carriers can be considered as an MDVRP [15]. In literature, several techniques were employed to solve the MDVRP. Mixed Integer Programming (MIP) models were used as in [16]. Meta-heuristic approaches such as genetic algorithm and a variable neighborhood search (VNS) algorithm were utilized in [17, 18]. As stated before, some studies solved the MDVRP through decomposing the overall problem into two sub-problems, i.e. assignment problem and routing problem [8–10]. The main advantage of the decomposition-based approach is reducing the complexity of the MDVRP and thus, more realistic aspects arising in the urban freight delivery can be considered. For example, reference [9] proposed a two-level heuristic approaches in which an MIP model is used to assign customers to depots at the first level while in the second level, an MIP model is used to find the optimal routes to serve the customers of each depot.

To sum up, the collaborative freight delivery has gained an increasing interest in recent literature. The MDVRP has been considered as a tool to perform such collaborative distributions in context of urban logistics. Even though the MDVRP has been solved using two-level hieratical approach as in [8–10], but the method of decomposing the overall problem, resulted in impairing the interrelation between the decomposed sub-problems. Therefore, this paper presents an improved two-level approach that could tackle the issues with existing approaches as will be elaborated in the remaining sections of this paper.

## 3 The Framework of the Proposed Approach

This section is divided into four parts. In the first part, the collaborative delivery problem is illustrated while the second part describes the proposed modelling approach. An illustrative example is introduced in the third part to show the difference between the proposed approach and a similar approach in literature. Finally, the solution methodology is illustrated in last part.

### *3.1 Problem Description*

This paper considers a number of carriers that delivers goods in the same urban region. In addition, these carriers are incentivized to collaborate to reduce their transportation costs and as a result, reducing the environmental externalities of their transportation activities. The current work addresses the collaborative freight distribution through order sharing, meaning that each carrier can exchange some of its delivery orders with other carriers. This collaborative strategy is modeled as an MDVRP in which each carrier has at least one depot which serves a number of customers with a set of vehicles that may not be identical. The locations of the depot and each customer are known in advance. Each vehicle must start from a depot and return to the same depot. Each customer is visited one and only one by a vehicle. The total demands of customers assigned to a vehicle must not exceed its capacity. It should be noted that collaboration may also require some logistics costs to transport the exchanged orders between depots of the collaborating carriers. However, these costs may not be applied or ignored when both carriers serve the same consolidation center or their depots are very adjacent to each other as considered in this paper.

## 3.2 The Proposed Approach

To overcome the NP-hardness of the MDVRP, we model it by a two-level mathematical modelling approach in which a mathematical model is used in each level. Figure 1 shows the solution procedures of the proposed approach. As a first step, data related to depots, customers, vehicles and other constraints are input to the proposed approach. In the second step, a new binary linear programming model is used to assign customers to depots. In the third step, the obtained assignment solution is input to solve the vehicle routing model which is proposed in [9]. Finally, the collaborative solutions for each company can be obtained in the fourth step. In the following, the two levels of the proposed approach are illustrated.

(1) *Assigning customers to depots*

A new binary linear programming model is formulated to assign customers to depots with considering two objectives. The first objective minimizes the total distance among customers and their assigned depots while the second objective minimizes the total distance among customers assigned to the same depot. Mathematical notations are as follows:

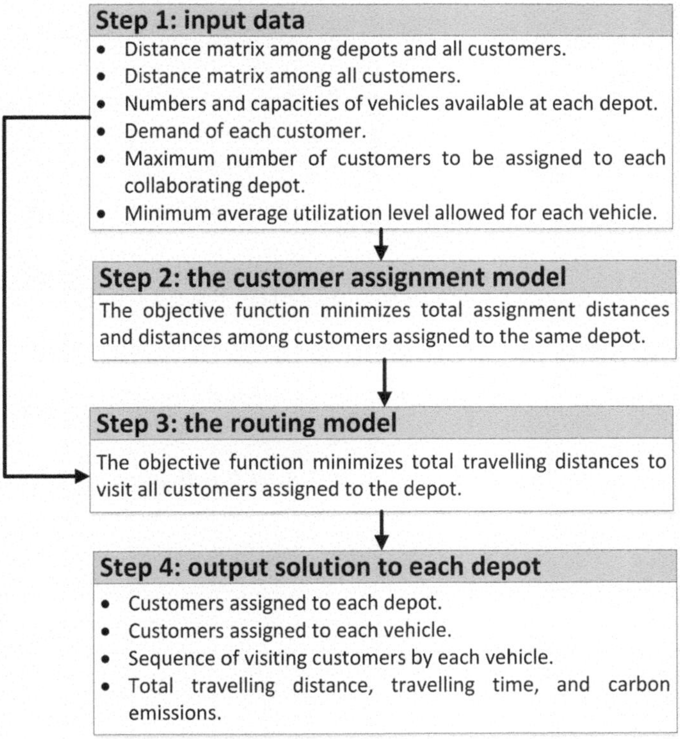


**Fig. 1** The description of the proposed approach

- Sets:
  A: set of all depots, A = {1,2,…,*a*}(indexed by *i*)
  B: set of all customers, B = {1,2,…,*b*}(indexed by *j* & *h*)

- Parameters:

$c_{ij}$: Average distance between depot *i* and customer *j*.
$c1_{jh}$: Distance between customer *j* and customer *h: $h \neq j$*.
$m_i$: Maximum number of customers that can be assigned to depot *i*, $\sum_{i=1}^{a} m_i = b$.
$TQ_i$: Total capacity of all vehicles at depot *i*, available to serve assigned customers.
$q_i$ : Demand of customer *i*.
Ut: Minimum limit of average vehicle utilization, %

- Decision variables:

$z_{ij}$: 1, if depot *i* serves customer *j*, 0 otherwise.
$y_{ijh}$: 1, if depot *i* serves customers *j* and *h*: $h \neq j$, 0 otherwise

The model:

$$f_1 : min \sum_{i=1}^{a} \sum_{j=1}^{b} c_{ij} \cdot z_{ij}$$

$$f_2 : min \sum_{i=1}^{a} \sum_{j=1}^{b} \sum_{h=1,h \neq j}^{b} c1_{ij}.y_{ijh}$$

Subject to

$$\sum_{i=1}^{a} z_{ij} = 1 \qquad \forall j \in B \tag{1}$$

$$\sum_{i=1}^{b} z_{ij} \leq m_i \qquad \forall i \in A \tag{2}$$

$$z_{ij} + z_{ih} - 1 \leq y_{ijh} \qquad \forall i \in A, \forall j, h \in B, j \neq h \tag{3}$$

$$\sum_{j=1}^{b} q_i \cdot z_{ij} \leq TQ_i \qquad \forall i \in A \tag{4}$$

$$\sum_{j=1}^{b} q_i \cdot z_{ij} \geq ut.TQ_i \qquad \forall i \in A \tag{5}$$

$$z_{ij}, y_{ijh} \in \{0, 1\} \qquad \forall i \in A, j, h \in B \tag{6}$$

The first objective $f_1$ aims at minimizing the total distance among depots and their assigned customer while the second objective $f_2$ minimizes the total distance among customers assigned to the same depot. Constraint (1) ensures that each customer is assigned to no more than one depot. Constraint (2) guarantees that the number of customers assigned to each depot does not exceed its maximum allowed number. Constraint (3) states that when customer $j$ and $h$ are assigned to the same depot $i$, $y_{ijh}$ equals to one. Constraint (4) ensures that the total demand of customers assigned to each depot does not exceed the total capacity of vehicles available at that depot. Constraint (5) forces average vehicle utilization of each depot to be not lower than a minimum specified limit. Constraint (6) defines domains for the decision variables.

(2) *Routing of vehicles for each depot*

In this level, the optimal routes to visit all customers assigned to each depot, are determined by using the MIP model introduced in [9]. This model is used to determine the optimal routes with objective of minimizing the total travelled distance. The travel time and $CO_2$ emissions are calculated based on the total travelled distance. The travel time (h) is calculated as a result of dividing average total distance (km) by travel speed (km/h) while the amount of $CO_2$ emissions (g) is calculated as a result of multiplying the average factor for the level of carbon emissions, 160 g/km by total distance (km).

## 3.3 An Illustrative Example

In this section, an example is provided to demonstrate the benefits of the proposed approach by a comparison against the approach introduced in [9]. The graphic illustration of the example is shown in Fig. 2. In this example, there are two depots

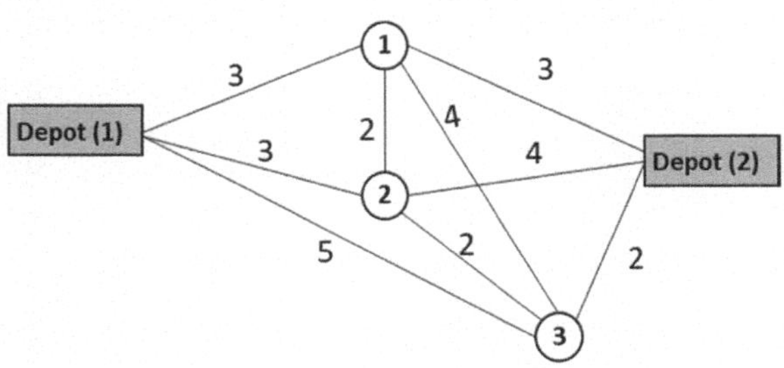


**Fig. 2** Graphic illustration of the example

**Table 1** Results of solving the example

| | | Results | | |
|---|---|---|---|---|
| | | *Assignment results* | *Routing results (distance)* | *Total distance* |
| The existing approach [9] | Depot (1) | Customer 2 | 6 | 15 |
| | Depot (2) | Customer 1<br>Customer 3 | 9 | |
| Our approach | Depot (1) | Customer 1<br>Customer 2 | 8 | 12 |
| | Depot (2) | Customer 3 | 4 | |

(carriers) and three customers to be served. To make the comparison simple, capacity of vehicles and demands of customers are not considered. Suppose also, that at most two customers can be served by any depot (constraint 2). The distances between customers and each other as well as depots are also shown in Fig. 2.

Table 1 shows the results of solving the example with the approach in [9] and our proposed approach. As can be noted, the proposed approach achieves a total distance of 12 distance unit which is less than the total distance obtained by the existing approach. The main reason for this improvement is due to the better assignment of customers to depots produced by the new MIP model of level (1). In other words, reference [9] assigns customers to depots with considering only objective $f_1$ while in the proposed approach; both $f_1$ and $f_2$ are considered when assigning customers to depots. The consideration of the objective $f_2$ results in improving the solution of the routing problem at the second level.

## 3.4 Solution Methodology

As stated before, two objective functions are considered when solving the assignment model. From preliminary experiments, it is noted that summing the two objectives into one objective function may be unsuitable as the value of $f_2$ is often higher than that of $f_1$, and therefore it is expected that $f_2$ will dominate $f_1$. This situation might lead to solutions focus only on minimizing $f_2$ at the expense of $f_1$. So, the two objectives are combined into one non-dimensional fitness function ($f_3$) with equal weights by using a function transformation method as follows:

$$f_3 = w\left(\frac{f_1}{f_1^*}\right) + (1 - w)\left(\frac{f_2}{f_2^*}\right) \tag{7}$$

Where $f_1^*$ and $f_2^*$ are minimum or approximate minimum values of $f_1$ and $f_2$ respectively while $w$ is the relative weight which we set to 0.5. After assigning costumers to each depot, the routing model will be used to solve the VRP for each depot separately, meaning that we solve the routing model as many as the number of depots.

## 4 Results and Analysis

In this section, the benefits of collaboration and the significance of the proposed approach are demonstrated by solving a real-life case. All the experiments are conducted on a PC with 2.3 GHz processor and 4 GB RAM working under windows 7 operating system. The MIP models are solved using CPLEX$^{\text{TM}}$12.2.

### 4.1 Description of the Case Study

This case study includes two carriers A and B. For confidentiality reasons, we cannot mention the names as well as locations of the depots or their customers. Therefore, all figures of customer locations and routing solutions are converted to schematic illustration by overlaying Google map with a white background. We consider 35 customers to be served by each carrier. The average driving speed in the delivery area is 25 km/h. A schematic illustration of customer locations is shown in Fig. 3. The case study is solved, assuming that each carrier can serve the customers with a single vehicle having a sufficient capacity to accommodate its assigned demands. It should be noted that the proposed approach can solve for multiple vehicles as well, we only consider a single vehicle in this case study as any

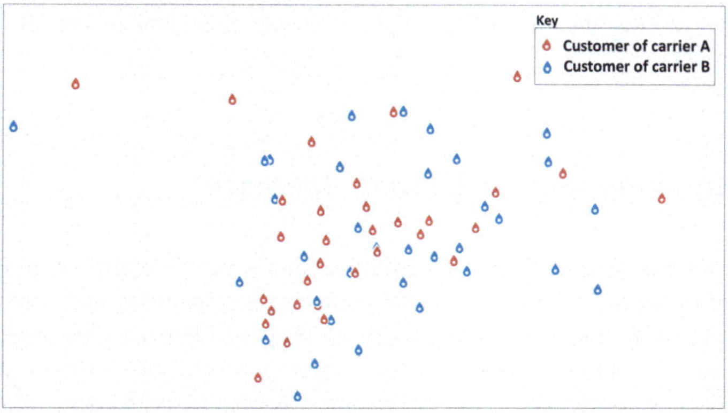


**Fig. 3** Schematic illustration of the customer locations

savings coming from one vehicle translate into similar savings for several vehicles. The depot–customer and customer-customer distance matrices are determined using real driving distances using Google Maps™ mapping service. To facilitate obtaining the distance matrices among customers and depots, we created a custom function in Google Apps Script and used it in Google spreadsheet. All data of this case study can be provided upon request to the corresponding author.

## *4.2 Non-collaborative vs. Collaborative Solutions*

In this section, we illustrate the benefits of collaboration in terms of total travelled distances, travel times and $CO_2$ emissions by solving the case study. In the non-collaborative scenario, each carrier plans the routing of its vehicles individually. In order to obtain the non-collaborative solutions, the MIP model of level 2 is solved for each carrier. In the collaborative scenario, we use the proposed approach, which firstly assigns the customers to one of the two carriers, and then the routing problem for each new assignment is solved. $m_i$ is set to 35 for all carriers, meaning that each carrier should keep the same amount of customers before and after collaboration. This setting may increase the motivation of carriers to participate in the collaborative distribution. The solution of the assignment model could be obtained in 20 minutes with an optimality gap of 5% while the optimal solutions of the routing model are obtained in few seconds. Figure 4 shows the delivery routes for each carrier in the non-collaborative and collaborative scenarios. Table 2 shows the numerical results for both scenarios. It can be noted that in the collaborative scenario, each carrier is reassigned to the customers whose locations are relatively closer to its depot with respecting all assignment constraints. The new assignments of customers to each carrier result in reducing the total travelled distance, and as a result, the travel times as well as amount of $CO_2$ emissions are also reduced. The improvements in travelled distances for carriers A and B are 22 and 27.5% respectively. It is worth noting that carrier B has slightly larger improvement in the total distance as the location of its depot is closer than that of carrier A to the delivery area.

## *4.3 Comparison with an Existing Approach*

The difference between the proposed approach and a similar approach in literature, is illustrated in Sect. 3. Herein, the performance of the proposed approach is tested using the case study. In doing so, the approach proposed in [9] is also applied to the case study and the obtained results are compared to ours. Table 3 shows the results of solving the case study by the two approaches. As can be noted from Table 3, our

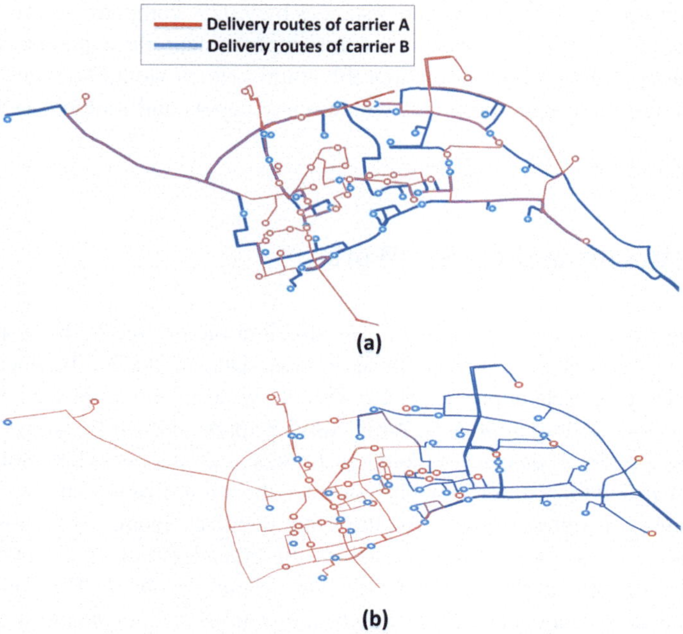


**Fig. 4** Schematic illustration of the solutions before collaboration **(a)** and after collaboration **(b)**

**Table 2** Results of non-collaborative and collaborative scenarios

| | Non-collaborative | | | Collaborative | | |
|---|---|---|---|---|---|---|
| | *Distance (Km)* | *Time (h)* | *$CO_2$ emissions (g)* | *Distance (Km)* | *Time (h)* | *$CO_2$ emissions (g)* |
| Carrier A | 40,967 | 1.64 | 6554.72 | 31,961 | 1.3 | 5113.76 |
| Carrier B | 32,738 | 1.31 | 5238.08 | 23,735 | 0.95 | 3797.6 |

**Table 3** Comparison of solutions obtained by our approach and the approach in [9]

| | Distance (Km) | | |
|---|---|---|---|
| | *Carrier A* | *Carrier B* | *Total* |
| An existing approach [9] | 35,330 | 25,890 | 61,220 |
| Our approach | 31,961 | 23,735 | 55,700 |
| Improvement (%) | 9.54% | 8.30% | 9.02% |

approach improves the total travelled distances by 9.02% compared to that obtained by the approach in [9]. The reasons for this improvement are explained before in part C of Sect. 3. It is worth noting that this improvement would be much larger in case of solving large-scale problems where many depots and multiple vehicles are included.

## 5 Conclusions and Future Work

This paper investigates the collaborative distribution of freight by solving the MDVRP in the context of urban transportation. Due to the NP-hardness of the problem, the proposed approach dealt with it by using an improved two-level mathematical modelling approach. While similar approaches in literature, consider minimizing the total distance among customers and depots when solving the assignment problem, we additionally consider the minimization of the total distances among customers assigned to the same depot by using a new assignment model. The new assignment model enables the consideration of the interrelation between the assignment problem and the vehicle routing problem. This interrelation was ignored in existing approaches, which in turn impaired, the solution quality of the overall problem.

The benefits of the proposed collaborative approach were illsutrated through solving a realistic case. The results showed that the collaborative distribution could reduce the total travelled distance of each carrier by an average of 24.75% compared to the non-collaborative distribution. Moreover, compared to the solution of this case by a similar approach found in literature, our approach improved the total travelled distances by 9.02%.

In this paper, the proposed solution approach depends on solving the mathematical models by using an exact solver. Therefore, solving the mathematical models by heuristic or Meta-heuristic algorithms, is recommended to be studied in the future research. In addition, future research may consider more realistic aspects such as the existence of contracted customers and estimating the $CO_2$ emissions based on speed as well as loads on the vehicles.

**Acknowledgements** This research was partially funded by the Municipality of Aalborg through the research project No. 885086 "Collaborative Logistics in Aalborg—Opportunities, Challenges and the Road Ahead". The authors would like to thank the two anayonoumus companies for providing their logistics data to test the developed approach.

## References


1. Tsiulin, S., Hilmola, O., & Goryaev, N. (2017). Barriers towards development of urban consolidation centres and their implementation: literature review. *World Review of Intermodal Transportation Research, 6,* 251–272.
2. Allen, J., et al. (2018). Enabling the freight traffic controller for collaborative multi-drop urban logistics: practical and theoretical challenges. *Transportation Research Record: Journal of the Transportation Research Board, 61,* 77–84.
3. ALICE (2013). Urban freight research roadmap. https://www.ertrac.org/uploads/documentsearch/id36/ERTRAC_Alice_Urban_Freight.pdf. Accessed 03 Nov 2018.
4. McGranahan, G., Satterthwaite, D. (2014). Urbanisation concepts and trends. Human Settlements working paper, International Institute for Environment and Development, London.
5. Amer, L.A., & Eltawil, A.B. (2015). Analysis of quantitative models of horizontal collaboration in supply chain network design strategies. In *International conference on industrial engineering and operations management* (pp. 1–10).
6. Gansterer, M., & Hartl, R. F. (2018). Collaborative vehicle routing: a survey. *European Journal of Operational Research, 268,* 1–12.
7. Toth, P., & Vigo, D. (2014). *Vehicle routing: problems, methods, and applications.* Philadelphia: Society for Industrial and Applied Mathematics.
8. Tansini, L., Urquhart, M.E., & Viera, O. (2001). Comparing assignment algorithms for the Multi-Depot VRP. *Reportes Técnicos* (pp. 01–08).
9. Montoya-Torres, J. R., Muñoz-Villamizar, A., & Vega-Mejía, C. A. (2016). On the impact of collaborative strategies for goods delivery in city logistics. *Production Planning & Control, 27,* 443–455.
10. Muñoz-Villamizar, A., Montoya-Torres, J. R., & Vega-Mejía, C. A. (2015). Non-collaborative versus collaborative last-mile delivery in urban systems with stochastic demands. *Procedia CIRP, 30,* 263–268.
11. Azab, A., Karam, A., & Eltawil, A. (2018). Impact of collaborative external truck scheduling on yard efficiency in container terminals. In G. Parlier, F. Liberatore, & M. Demange (Eds.), *Operations research and enterprise systems. ICORES 2017. Communications in computer and information science* (Vol. 884). Cham: Springer.
12. Karam, A., & Attia, E.-A. (2019). Integrating collaborative and outsourcing strategies for yard trucks assignment in ports with multiple container terminals. *International Journal of Logistics Systems and Management, 32,* 372–391.
13. Verdonck, L., Caris, A. N., Ramaekers, K., & Janssens, G. K. (2013). Collaborative logistics from the perspective of road transportation companies. *Transport Reviews, 33,* 700–719.
14. Lozano, S., Moreno, P., Adenso-Díaz, B., & Algaba, E. (2013). Cooperative game theory approach to allocating benefits of horizontal cooperation. *European Journal of Operational Research, 229*(2), 444–452.
15. Pérez-Bernabeu, E., Juan, A. A., Faulin, J., & Barrios, B. B. (2015). Horizontal cooperation in road transportation: a case illustrating savings in distance and greenhouse gas emissions. *International Transactions in Operational Research, 22,* 585–606.
16. Aras, N., Aksen, D., & Tekin, M. T. (2011). Selective multi-depot vehicle routing problem with pricing. *Transportation Research Part C: Emerging Technologies, 19*(5), 866–884.
17. Ho, W., Ho, G. T., Ji, P., & Lau, H. C. (2008). A hybrid genetic algorithm for the multi-depot vehicle routing problem. *Engineering Applications of Artificial Intelligence, 21*(4), 548–557.
18. Salhi, S., Imran, A., & Wassan, N. A. (2014). The multi-depot vehicle routing problem with heterogeneous vehicle fleet: formulation and a variable neighborhood search implementation. *Computers & Operations Research, 52,* 315–325.

# A Novel Course Recommendation Model Fusing Content-Based Recommendation and K-Means Clustering for Wisdom Education

**Penghui Cao and Dan Chang**

**Abstract** It has become a hot issue in digital research on how to recommend high-quality courses for users in the massive information of the wisdom education course recommendation platform. Traditional content-based recommendation algorithms usually have low accuracy. In this paper, we propose a novel recommendation model of fusing content-based recommendation and K-means clustering. The proposed model mainly contains three parts. Firstly, the users' data is converted into a high-dimensional space vector by Term frequency–inverse document frequency. Secondly, the model calculates the similarity between user course content model and course resources, and recommends the most frequent courses by clustering similar users. Finally, a comprehensive recommendation list is generated. The contribution of the proposed algorithm is that the performance of the content recommendation is improved by using the optimized K-means clustering method. In addition, the proposed algorithm is verified by online user data of the Chinese university MOOC platform, and the experimental results show that the recommendation model of fusing content-based recommendation and K-means clustering has high recommendation accuracy and certain practical value.

**Keywords** Wisdom education · Content-based recommendation · TF-IDF · K-means clustering

## 1 Introduction

In current society, information technology is widely used in various fields, so is the education industry. The goal of wisdom education, i.e. education informationization, is to realize the sharing, collaboration and interaction of educational modes.

P. Cao (✉) · D. Chang
School of Economics and Management, Beijing Jiaotong University, Beijing, China
e-mail: 18125477@bjtu.edu.cn

D. Chang
e-mail: dchang@bjtu.edu.cn


J. Zhang et al. (eds.), *LISS2019*,
https://doi.org/10.1007/978-981-15-5682-1_57

The rapid development of education informationization has brought about increasing content of course teaching resources and various data forms, which usually requires a lot of time and effort for users to find useful and targeted information [1].

Therefore, under the condition of increasing amount of educational resources and course data, the recommendation system applied in the field of education came into being and gradually becomes a research hotspot. The frequently used wisdom education platforms by users provide personalized resource recommendations for users based on the user's learning behavior and the characteristics of the course. The course recommendation system can analyze the user identity information, user course information, and certificate information, so as to find potential course resources and recommend some resources for the user. The content-based recommendation, with low dependence on the user's behavioral operation information, can solve the sparse matrix problem to a large extent. Users of the course platform often have their own areas of attention, and therefore the content-based recommendation can provide more appropriate course resources using the main recommendation algorithm.

However, traditional content-based recommendation algorithms have some limitations: First, the problem of validity. For example, the selection of feature items and the setting of related parameters will affect the effectiveness of the system. Secondly, the solely usage of content-based recommendation will make the system only recommend existing information, whereas it is unable to recommend courses that users might be interested in. In this paper, a recommendation model is proposed of fusion content and K-means clustering. Of all the platform courses, the user's records and identity information are extracted by feature segmentation and term frequency–inverse document frequency (TF-IDF), to form a space vector. By calculating the similarity between the user content model and the course resources, the recommendation is made; after the similar users are clustered by the K-means algorithm, the recommended resources of the user are found by using the most frequent items. The combination of two types of recommendations will form an end-user recommended course. It takes the advantages of content-based recommendation and K-means clustering algorithm. On the basis of user content information, it solves the problem that content-based recommendation algorithm cannot recommend novel resources and further improves the performance of recommendation system. Finally, the user and course data on the Chinese university MOOC platform can be used to verify and compare the fusion content and K-means clustering recommendation model, so as to improve the accuracy and efficiency of the recommendation.

## 2 Literature Review

### 2.1 *Wisdom Education*

Wisdom education is an evolved form of teaching and learning in the network environment. Existing research mainly include the following recommendations: content-based recommendation, collaborative filtering recommendation, social network-based recommendation, association rule-based recommendation, and so on [1]. With the advent of the era of big data, many scholars have studied the recommendation algorithm based on big data platform. Ding et al. used the fusion tensor model of "learner-resource" to construct the direct relationship between learners and resources by using high-order singular values to achieve a match [2]. Luo et al. used the item-based collaborative filtering algorithm with user behavior data, which achieved recommendation of related course resources to the user and therefore reduced the course resources existing in the "dark information", and improves the sharing effect of the course resources [3]. In order to solve the problem that the traditional collaborative filtering resource recommendation technology is ineffective in processing sparse data and consequently cannot accurately process the high-dimensional attributes of online learning users, Zhang et al. proposed a personalized recommendation system based on DBN in MOOC environment. This system utilizes DBN for high performance in function approximation, feature extraction, and prediction classification [4].

Albeit the above recommended strategies all have respective disadvantages, the recommendation for personalized resources in the field of wisdom education has been greatly improved to a certain extent. But most of the recommended algorithms utilized in other fields are not fully applicable to the field of wisdom education, where learners have a focused field of study, personalized interests, high latitude learning resources, and diverse data resources. This requires more professional and more instructive resources to these users by considering their historical learning content.

### 2.2 *Content-Based Recommendation*

Now the frequently-used collaborative filtering recommendation algorithm only considers user history score data, but does not include the user's own identity characteristics and course resource information. However, the content-based recommendation algorithm can make a more full use of the information, i.e., to recommend the course with the highest similarity to the course that the user was once interested in [5]. Applying the content-based algorithm to the field of wisdom education course resource recommendation can properly utilize the user's learning content and therefore identity more professional data to improve the performance of this algorithm. Nevertheless, this algorithm shows the problem of low accuracy and cold start. In order to solve this problem, many research scholars made research. In the field of e-commerce, Shan applied the k-means clustering algorithm to the traditional content-based recommendation. First, the products with similar characteristics are put together, and then the products closest to each cluster center to the target customers are found [6]. In the field of learning resources, Shu et al. proposed a content-based recommendation algorithm based on Convolutional Neural Network (CNN). In this algorithm, CNN and split Bregman iterative methods are used, therefore text information can be directly used for content-based recommendation without mark [7]. Therefore, it is meaningful to improve the traditional content-based recommendation algorithm.

## *2.3 K-Means Clustering*

The K-means algorithm is a typical unsupervised machine learning method. By calculating the Euclidean distance to determine the similarity degree, the optimal classification of the initial cluster center is obtained. This algorithm has the advantages of high efficiency, easy understanding and implementation, but it also has some limitations. For example, the reasonable k value in the algorithm is difficult to be determined, and the randomness of the algorithm initial cluster center will lead to unstable clustering results. In order to solve these problems, some scholars studied to improve the K-means algorithm. Tao and others improved the k-means clustering algorithm through the globalization idea, and got better clustering effect, and improved the stability of the algorithm [8]. In addition, the elbow point of the traditional elbow method is not clear. Wang et al. used the exponential function property and weight adjustment to propose an improved method based on the elbow method—ET-SSE algorithm. This algorithm can determine the k value more quickly and accurately than the elbow method [9].

In summary, most algorithms in the field of wisdom education course resource recommendation do not take into full account the detailed characteristics of the education course category and the characteristics of the learner's professional field. Therefore, this paper proposes a recommendation model of fusion content and K-means clustering. Firstly, based on the user content information, this model uses TF-IDF to extract the best feature items, which can improve the real-time performance of the algorithm and solve the problem of cold start. Then through the combined K-means clustering algorithm, this model recommends new courses for users and improves the accuracy of the algorithm.

## 3 The Construction of Recommendation Model of Fusion Content and K-Means Clustering

This paper proposes a fusion model based on user content information, which combines content and k clustering. Recommendations are made by similarity alignment and K-means clustering. The specific recommendation framework is shown in Fig. 1.

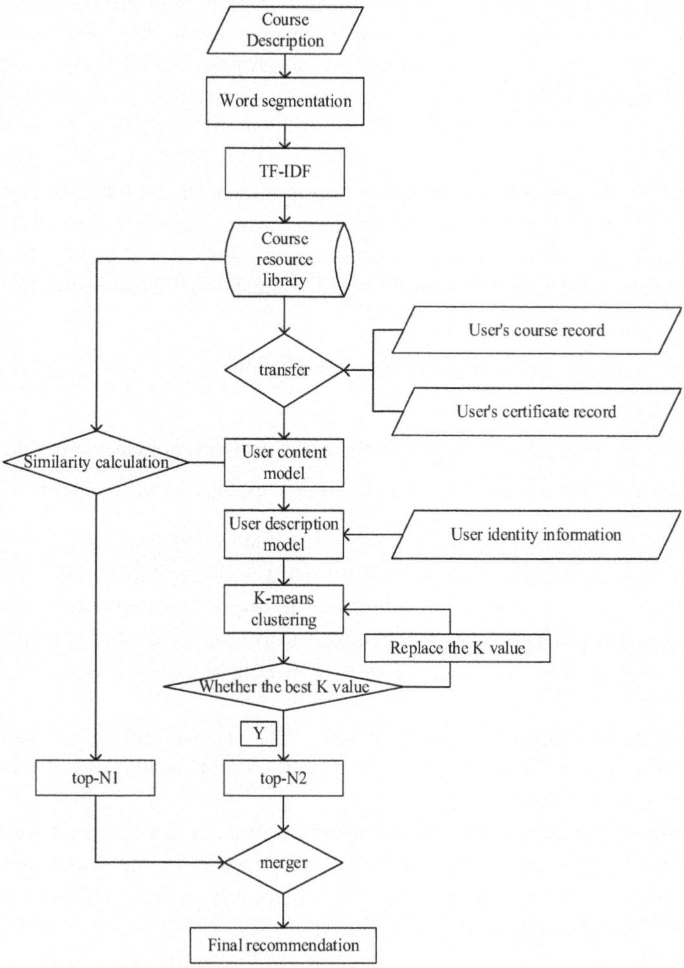


**Fig. 1** Recommended framework

## 3.1 Data Preprocessings

The user's identity information can determine its strong interest in a certain field. The higher the frequency of learning certain courses, the higher the user's interest in such courses. The user's obtaining of the course and certificate indicates that the user has great interest in the course. Therefore, the recommendation model of this paper uses the user's identity information, course selection information and the acquisition certificate record as the original data.

In this paper, the vector space model is used to describe the resource information, to remove the content-independent structure of the crawled information. Considering that the number of words in the resource description and the vector space dimension of the text are large, this paper extracts the feature items in the model, which can improve the efficiency of the program and therefore improve the accuracy of the algorithm.

This article uses the jieba library in Python for word segmentation, including the domain dictionary of the course and the stop words of the Chinese Academy of Sciences. *TF-IDF* is used to calculate the importance of the feature words in the text, and the first n% feature words are extracted, to ensure the objectivity of the description. The extracted feature words are used as vector components, and the frequency as component values constitutes a feature vector. The TF-IDF calculation formula is:

$$TFIDF_i(d) = tf_i(d) * \ln \frac{N}{n_i} \quad (1)$$

where $tf_i(d)$ indicates the number of times a feature word appears in this text, $\ln\left(\frac{N}{n_i}\right)$ indicates the number of times the feature word appears in all the text.

- Establish a course resource library D. The feature vectors of each course are divided into two groups, and the feature words extracted from TF-IDF are taken as vector components and the occurrence frequency as component values. The type, course name and school of course information are grouped into a group, and the weight is a; The description part of course information is a group, and the weight is b.
- Establish a user identity information vector $P_0$. The identity information words after word segmentation are taken as vector components and the occurrence frequency as component value.
- Establish a user course selection information vector $P_1$. Use (course name, school name) as a unique identifier, corresponding to the course information resource database, and take out the feature vector of the course, then combine all the course vectors for each user's course.
- Establish the user certificate course vector $P_2$. Using the course name to retrieve the unique identification in the course selection information, corresponding to the course information resource database, take out the course feature vector. Each course vector of all the certificate courses for each user is integrated.

### 3.2 *Recommendation Model of Fusion Content and K-Means Clustering*

(1) *Calculation based on content similarity*

The idea of content-based recommendation algorithm is to compare the similarity between the courses selected by users and all the courses, to recommend the best matched items by reasonably utilizing the content of the course resources. This algorithm was first applied to information retrieval technology, as it has little dependence on users' evaluation and comment on the course. Firstly, it is necessary to extract the user's course selection and obtain the content features of the certificate course, to establish the user's content vector. The content vector is obtained according to the characteristics of the course resource content formerly selected by the same user; finally, the similarity between the user's content vector and the item to be recommended is compared to recommend the most similar course resources.

(a) According to the data pre-processing, the user's course selection vector and the certificate course vector are obtained. And the user's content vector is obtained by combining the course selection vector and the certificate course vector with a certain weight.

$$pf = w * p_1 + v * p_2 \tag{2}$$

where $p_1$ is the user's course information vector, $p_2$ is the user's acquisition certificate vector, pf is the user's comprehensive content vector. By assigning different weights, the influence of the user course information and certificate information on the content vector can be adjusted.

(b) The recommendation algorithm calculates the performance evaluation index under this value for any recommended number by calculating the similarity between the content vector and the course resource.

(c) Through the similarity calculation, the top-$N_1$ courses with the highest similarity are recommended to the user, albeit it is very difficult to determine the $N_1$ value. In this algorithm, the initial value is predetermined and then adjusted by the accuracy of the test data recommendation.

(2) *K-means clustering based on content*

(a) A reasonable user description model collects user personal identity information to the user content vector. The degree of influence of each information can be adjusted by different weights.

$$pl = x * p_0 + y * p_f \tag{3}$$

where *pl* indicates the user description vector, $p_0$ indicates the user identity information vector, and *pf* indicates the user content vector.

(b) In the real problem, when the positive and negative samples in different scenes are offset, the conclusion that the artificially labeled samples are obtained as the training set after the condition is improved may be ineffective. Therefore, the algorithm uses K-means unsupervised learning method to cluster users, which avoids the impact of artificial labeling.
The main feature of the k-means algorithm is its computational efficiency and easy understandability, a as well as the wide application in daily life. However, this algorithm is flawed in selecting the k value, which is generally judged by the elbow method. The basic idea of the elbow method is to establish a correspondence between the k value and SSE, and then determine the optimal k value by observing the inflection point in the two-dimensional plane rectangular coordinate system. This method is computationally efficient and simple to understand than other k-value selection algorithms, albeit it has problems in practical applications: When applied to a specific data set, there will be an inconspicuous "elbow point". The actual value of k is a range *[a, b]* (*a, b* are positive integers, and $a < b$), so some clustering results will be greatly deviated, which will consequently affect the final clustering results. Since the choice of k value has randomness and uncertainty, this paper uses the improved k value selection algorithm *ET-SSE* to determine the k value, which can eliminate the artificial influence and therefore make the algorithm more objective.

(c) The data used in this algorithm is the user's course selection and acquisition information record. Therefore, the most frequent method is recommended. The course selection records of all users in the current user cluster are counted, and accordingly the $N_2$ courses with the highest frequency to this user are recommended.

(3) *Fusion recommendation model*
Now the frequently used collaborative filtering recommendation algorithm only considers the user history score data, but does not include the user's own characteristics and resource information. To address this problem, this paper uses content similarity calculation and unsupervised k-means cluster recommendation algorithm. The main steps are as follows:

(a) The similarity calculation is performed by the user course content vector composed of the user's course selection and certificate information as well as the course resource library, while the similarity is represented by the cosine value of the angle between the vectors. Recommend $N_1$ resources with the highest similarity to form a top-$N_1$ recommendation.

$$sim_u(pf, D_i) = \frac{\sum_{k=1}^{m} pf_k * D_{i_k}}{\sqrt{\sum_{K=1}^{m} pf_k^2 * \sum_{k=1}^{m} D_{i_k}^2}} \tag{4}$$

where $D_i$ indicates course vector in the course resource library, pf represents user content vector, m represents the dimension of the feature vector, $D_{i_k}$ expresses the weight of the kth word, $pf_k$ represents the weight of the kth word in the user content vector.

(b) Add user identity information to build user description content vector. The k-means algorithm clusters users by calculating the Euclidean distance between the users. The most frequent course will be recommended as a *top-N2* recommendation.

Calculate the Euclidean distance between users *u, v*

$$d = \sqrt{\sum_{i=1}^{n} (pl_{u_i} - pl_{u_i})} \quad (5)$$

where $pl_u$ represents the description vector of the user u, $pl_v$ represents the description vector of the user v, and n represents the number of components of the user description vector.

The K-means clustering algorithm flow is as follows:

Input: Databases and number of clusters k;

Output: k clusters.

step:

---

①Choose k objects arbitrarily as the initial cluster center

②Repeat.

③Calculate the Euclidean distance between a point and each cluster, and select the smallest Euclidean distance into the cluster.

④Update the average of the clusters based on the average of the clusters of the objects in the cluster.

⑤Until no change occurs.

---

Average of clusters

$$C_i = \frac{1}{L_i}\left\{\sum\nolimits_{J=1}^{L_1} t_1^{ij}, \sum\nolimits_{J=1}^{L_1} t_2^{ij}, \ldots, \sum\nolimits_{J=1}^{L_1} t_k^{ij}\right\} \quad (6)$$

where $L_i$ represents the number of users of the *i-th* cluster, and $t_k^{ij}$ represents the *k-t*h component of the *j* user in the *i* cluster.

Since the choice of *k* value has randomness and uncertainty, this paper uses the improved *K* value selection algorithm *ET-SSE* to determine the *K* value. Based on the total error tempering formula, the exponential function $e^x$ is introduced. The exponential function is very sensitive to the changes in the positive value of the exponent x. By amplifying the *SSE* and adjusting a certain weight, the accuracy of *K* can be further improved, which can eliminate any artificial influence and make the algorithm more objective.

Sum of squared errors *SSE* calculation formula:

$$sse = \sum\nolimits_{i=1}^{k} (p - c_i)^2 \quad (7)$$

where k is the number of clusters, p is the sample object in cluster $c_i$, and $c_i$ is the cluster center of the current cluster.

Improved *ET-SSE* formula:

$$ET\text{-}SSE = \sum\nolimits_{i=1}^{k} e^{\frac{(p-c_i)}{\theta}} \quad (8)$$

where k is the artificially set cluster number, $c_i$ is the center of the cluster $c_i$, θ is the weight responsible for adjusting the value, and p is the sample object in the cluster $c_i$.

(c) The top-N1 and top-N2 sequences are combined as the recommended resources for the fusion recommendation model.

$$top\text{-}N = top\text{-}N_1 \cup top\text{-}N_2 \quad (9)$$

(d) In order to make a scientific evaluation of the performance of the top-N recommendation system, some evaluation indicators are used in the text retrieval system for evaluation. The main evaluation indicators in the text retrieval system are:
Precision rate:

$$Precision = \frac{testing \cap top\text{-}N}{N} \quad (10)$$

Recall rate:

$$Recall = \frac{testing \cap top\text{-}N}{testing} \quad (11)$$

where testing is the user-selected course, and top-N is the recommended course resource.

The accuracy P is used to measure the probability that the recommended list predicts the user's favorite items; The recall rate R is used to measure the proportion of users' favorite items in the recommended list. The two indicators are combined to form a comprehensive index F-measure for evaluation, for balance consideration. The larger the F value, the stronger the ability of this algorithm to recommend.

Comprehensive evaluation index F-measure:

$$F\text{-}measure = \frac{2 * precision * recall}{precision + recall} \tag{12}$$

## 3.3 Model Optimization

The fusion content and K-means the cluster recommendation algorithm proposed in this paper comprehensively considers the influence of these two categories: Recommend the corresponding courses according to the similarity with the curriculum resources, and recommend the most frequent courses according to the user classification. The recommendation mechanism proposed in this paper can dynamically modify various parameters according to different platforms and courses, to improve the accuracy of the recommendation model.

Adjustable parameters in this model:

① The feature selection ratio after calculating TF-IDF: n%
② The type and description of course resources constitute the proportion: a:b
③ Select the combination ratio of the course and the certificate: w:v
④ User information and course portfolio ratio: x:y
⑤ Number of clusters K
⑥ Similarity recommendation and cluster recommendation ratio: $N_1$:$N_2$

(a) This paper selects the top 10% of the vocabulary as feature words; Yang and Pedersen pointed out that using the document frequency to reduce the feature space to 10%, the performance of the classification almost has no loss; after reducing the feature space to the original 1%, the classification performance has only suffered a small loss [10, 11].
(b) According to the characteristics of the course of the MOOC platform and the difficulty in obtaining the certificate, selecting appropriate weights for course selection information, certificate information, and user personal information.
(c) The sum of squared errors SSE for different cluster numbers K is shown in Fig. 2:
(d) *Choose the best* $N_1$ *and* $N_2$ *ratio.*

Through the improved ET-SSE amplification error squared effect, the best cluster number is selected: K = 5. The improved error squared ET-SSE for different cluster numbers K is shown in Fig. 3:

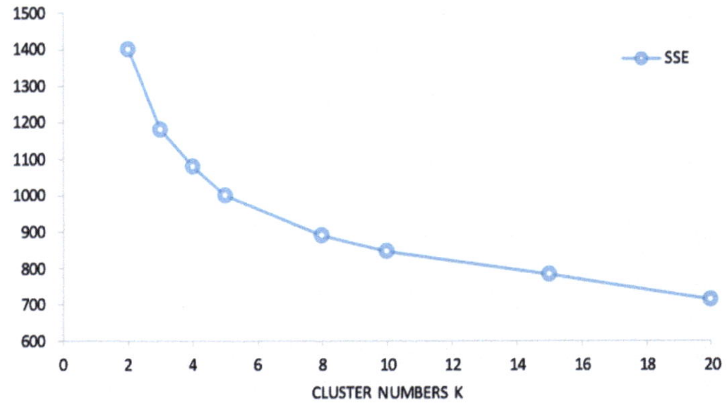


**Fig. 2** SSE for different cluster numbers K

**Fig. 3** ET-SSE for different cluster numbers K

In this paper, the total recommended resource number is N = 15, and the ratio of $N_1$ and $N_2$ is tested. The results are shown in Fig. 4.

As can be seen from Fig. 4 when the ratio of $N_1$:$N_2$ is 1:2, the recommendation system can obtain the best performance, so the ratio of $N_1$:$N_2$ is 1:2 in the subsequent experiments.

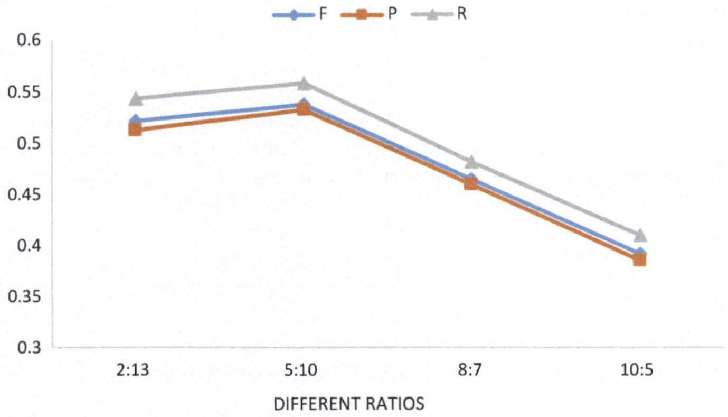


**Fig. 4** Recommended performance for different ratios

# 4 Model Application

This chapter uses the real user data on the Chinese university MOOC platform to test the fusion content and K-means clustering recommendation model, and then compares the performance indexes of each recommendation system.

## *4.1 Data Description*

The data in this paper uses crawler software to crawl user basic information and resource learning data on the Chinese university MOOC platform. The user set is the evaluation user of the front-end technology, program design and development, computer foundation and application, hardware and software system and principle courses on the platform. The data set has a total of 11,000 selective data and 4200 certificate data. The test user is the user whose study time in the data set is longer than 10 and the number of courses selected is greater than 10 and less than 20. Hardware configuration in the experiment: ASUS ZX50JX4720, 2.6 GHZ, 8 GB of memory.

The format of the captured information is as follows:

(a) *Course data is shown in Table* 1:
(b) *User data*

User identity information is shown in Table 2:
The user part selection information is shown in Table 3:
The user part certificate information is shown in Table 4:

**Table 1** Course data

| Type | Course name | Description | School |
|---|---|---|---|
| Economics | Futures and options | The volatility of global economic finance has created a favorable opportunity for the rapid development of the derivatives market. In order to achieve their respective functions of hedging, speculation and arbitrage, various market entities need derivatives as a powerful weapon. Our course will unveil its mystery, bring derivatives into thousands of households, help the workplace to compete, enrich your asset allocation options | Central University of Finance and Economics |

**Table 2** User identity information

| Username | User info | Duration |
|---|---|---|
| Eating is worse than learning | Student\|Nankai University-Finance Department | 174:25 |

**Table 3** User part selection information

| Username | Course name | School |
|---|---|---|
| Eating is worse than learning | Futures and options | Central University of Finance and Economics |
| Eating is worse than learning | Python web crawler and information extraction | Beijing Institute of Technology |
| Eating is worse than learning | Discrete mathematics | University of Electronic Science and Technology |
| Eating is worse than learning | High-level language programming (Python) CAP | Harbin Institute of Technology |

**Table 4** User part certificate information

| Username | Certificate course |
|---|---|
| Eating is worse than learning | Futures and options |
| Eating is worse than learning | Python language programming |
| Eating is worse than learning | Ten lectures on mathematics culture |
| Eating is worse than learning | Behavioral economics |

## 4.2 Application of Fusion Recommendation Model

The TF-IDF is used to obtain the importance of each word, and then the top 10% of the vocabulary is taken as the vector component in descending the order of importance. The number of occurrences of the feature words in the course description is used as component values. The course resource library consists of types and text descriptions. Take the "Futures and Options" course run by the Central University of Finance and Economics as an example: Table 5.

"Eating is worse than learning" user identity information vector $P_0$: Table 6.

"Eating is worse than learning" user course information vector $P_1$: Table 7.

"Eating is worse than learning" user certificate course vector $P_2$: Table 8.

The user's course and certificate information are integrated into a user content model by using (2). Next, the calculation of similarity is proceeded by (3) with the aid of user's content model and resources in the course resource library. Then, the best match is recommended. Take the user's "Eating is better than learning" and "Futures and Options" courses as an example:

"Eating is worse than learning" user content model pf: Table 9.

The user information model is formed by using the (4) to integrate the identity information on the basis of the user content model, and then (5) can be used to calculate the Euclidean distance among users.

$$\cos(pf, D_i) = \frac{1 * 1.4 + 0.8 * 0.96 + \ldots}{\sqrt{(1.4^2 + \ldots) * (1^2 + \ldots)}} \tag{13}$$

Take "learning is better than eating" and "shuo nan han" as examples: "Eating is worse than learning" user content model pl: Table 10.

$$d = \sqrt{(1-1)^2 + (1-0)^2 + (1-0)^2 + \ldots} \tag{14}$$

The recommended resources include five best matched resources and ten most frequent courses. The recommended results are shown in Table 11.

"Shuo nanhan" user content model pl: Table 12. The evaluation result of "Eating is worse than learning" is:

**Table 5** "Futures and options" vector

| Component | Value | Component | Value |
|---|---|---|---|
| Economics | 1 | Literature | 0 |
| Transaction | 0.8 | Risk | 0.65 |
| Mathematical | 0.5 | Philosophy | 0 |
| National boutique | 1 | General elective | 0 |
| Computer | 0 | Foreign language | 0 |

**Table 6** "Eating is worse than learning" identity vector

| Component | Value | Component | Value |
|---|---|---|---|
| Sudent | 1 | Risk | 0 |
| Worker | 0 | High duration | 1 |
| Nankai University | 1 | Medium duration | 0 |
| Financial | 1 | …… | …… |

**Table 7** "Eating is worse than learning" course vector

| Component | Value | Component | Value |
|---|---|---|---|
| Economics | 1 | Literature | 0 |
| Transaction | 0.8 | Risk | 1 |
| Mathematical | 1 | Philosophy | 0 |
| National boutique | 4 | General elective | 0 |
| Computer | 3 | Foreign language | 0 |

**Table 8** "Eating is worse than learning" certificate vector

| Component | Value | Component | Value |
|---|---|---|---|
| Economics | 2 | Literature | 0 |
| Transaction | 1.2 | Risk | 0.65 |
| Mathematical | 0.8 | Philosophy | 0 |
| National boutique | 4 | General elective | 0 |
| Computer | 2 | Foreign language | 0 |

**Table 9** "Eating is worse than learning" content model

| Component | Value | Component | Value |
|---|---|---|---|
| Economics | 1.4 | Literature | 0 |
| Transaction | 0.96 | Risk | 0.86 |
| Mathematical | 0.92 | Philosophy | 0 |
| National boutique | 4 | General elective | 0 |
| Computer | 2.6 | Foreign language | 0 |

**Table 10** "Eating is worse than learning" content model pl

| Component | Value | Component | Value |
|---|---|---|---|
| Student | 1 | Transaction | 0.96 |
| Nankai University | 1 | Mathematical | 0.92 |
| Financial | 1 | National boutique | 4 |
| High duration | 1 | Computer | 2.6 |
| Low duration | 0 | Risk | 0.86 |
| Economics | 1.4 | …… | …… |

**Table 11** The recommended results

| Username | Similarity recommendation | Cluster recommendation | |
|---|---|---|---|
| Eating is worse than learning | High-level language programming (Python) CAP, High-level language programming (Python), C language programming, C language programming, C programming methodology | Python web crawler and information extraction, discrete mathematics, High-level language programming (Python) CAP, Python language programming, Advanced Mathematics (2) | C language programming is advanced, data structure, Advanced Mathematics (1), Zero-based Java language, Programming and Algorithm (1) C language programming |

**Table 12** "Shuo Nanhan" user content model pl

| Component | Value | Component | Value |
|---|---|---|---|
| Student | 1 | Transaction | 0 |
| Nankai University | 0 | Mathematical | 0.32 |
| Financial | 0 | National boutique | 6 |
| High duration | 0 | Computer | 4.8 |
| Low duration | 1 | Risk | 0 |
| Economics | 0 | …… | …… |

$$P = \frac{10}{14} = 0.71 \tag{15}$$

$$R = \frac{10}{16} = 0.625 \tag{16}$$

$$F = 2 * P * \frac{R}{P + R} = 0.67 \tag{17}$$

## 4.3 Results and Analysis

(1) *Comparative analysis with the recommended model before optimization*

In order to verify that the model optimization has an improvement effect on the algorithm proposed in this paper, the same data is used to test the pre-optimized and optimized models, and compare the prediction accuracy P, recall rate R and comprehensive index F when different K values are selected.

Figure 5 show the prediction accuracy P of the three values for the three K-value models.

Figure 6 show the recall rate R of the three values for the three K-value models.

Figure 7 show the comprehensive index F of the three values for the three K-value models.

It can be seen that when K = 5, P, R, and F are the highest among the three sets of models. So the performance of the recommendation algorithm can be improved by model optimization.

It can be seen from the figures that when the number of recommended items is small, the prediction accuracy is high and the recall rate is low. As the number of recommendations increases, the prediction accuracy decreases and the recall rate increases; while the comprehensive index F is firstly increased and then decreased. When N = 15, the recommendation ability of the optimized model is obviously improved, and the best performance is obtained. Therefore, the number N of the best recommended items in the subsequent experiment is 15.

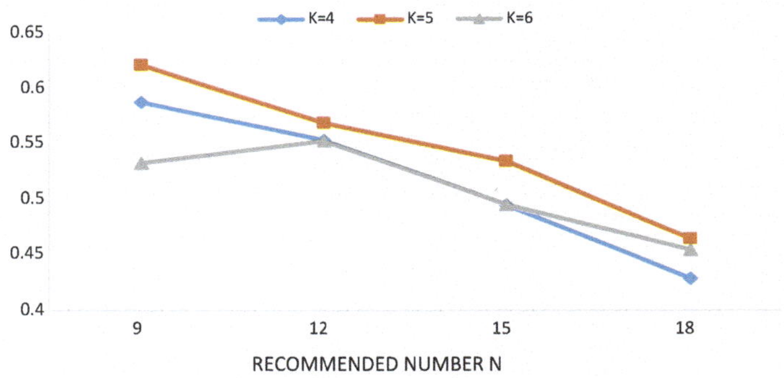


**Fig. 5** Prediction accuracy of different K values

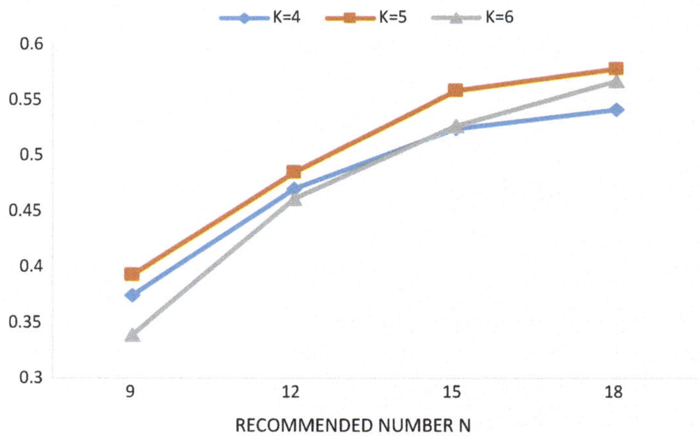


**Fig. 6** Recall rate for different K values

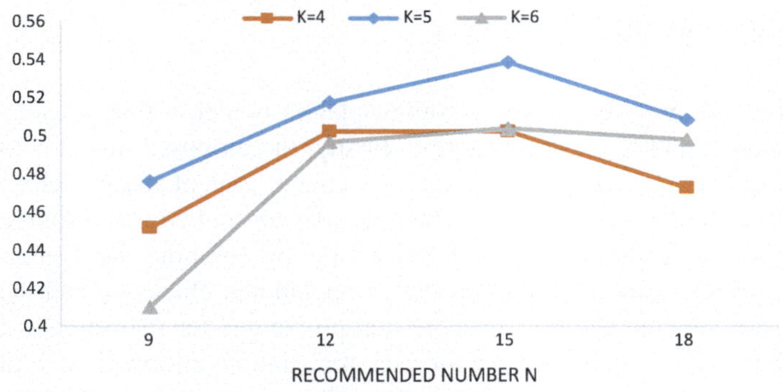


**Fig. 7** Comprehensive index F for different K values

(2) *Comparative analysis with traditional recommendation algorithm model*

In order to verify the fairness and objectivity of the recommendation experiments, this paper uses the improved content recommendation model, the traditional recommendation algorithm and the K-means clustering algorithm to recommend the same data crawled. The parameters K = 5, N = 15 are selected to compare the values of P, R and the F. The recommended results are compared as shown in Fig. 8.

It can be seen from the table that, compared with the traditional content recommendation algorithm and the K-means clustering algorithm, the recommendation algorithm proposed in this paper greatly improves the recommendation performance.

In order to fairly and objectively compare the recommendation efficiency of the two models, the traditional content-based recommendation model, and the recommendation model proposed in this paper, this paper compares the running time of the two models. The traditional model takes 473 s, while the model proposed here takes only 152 s, to calculate by extracting the best feature. It is obvious that the newly proposed recommendation model is better.

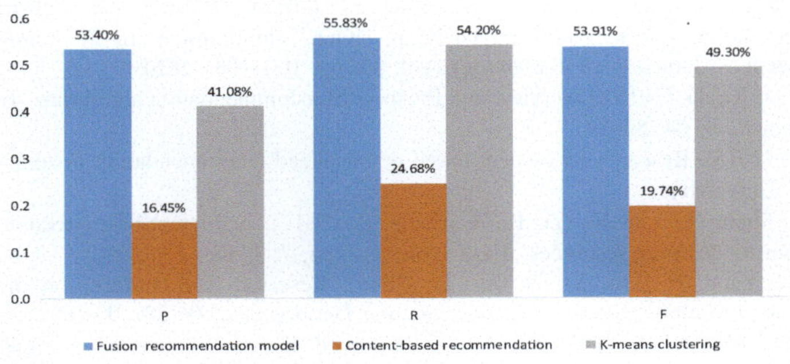


**Fig. 8** Comparison of recommended accuracy of different models

## 5 Conclusions and Prospects

In this paper, we propose a novel recommendation model of fusing content-based recommendation and K-means clustering. Firstly, the proposed model converts the user's course learning content into the user's course content space vector. Then it, calculates the similarity between content space vector and course resource library. Finally, it recommends the best matched course by combing the K-means algorithm. The proposed model improves the computational efficiency and the recommendation accuracy. In addition, the cold start problem of the mainstream algorithm can be alleviated via combining with user registration information. Finally, the experiments which use the online data of the Chinese University MOOC platform show that the proposed algorithm has good performance in terms of accuracy. Compared with the traditional recommendation methods, the accuracy of the proposed recommendation model increases from 19.74 to 53.91%, and the calculation time reduces from 473 to 153 s. Therefore, this proposed algorithm has certain practical values for the recommendation of wisdom education courses.

Although the proposed recommendation model in this paper shows good performance on the real dataset, it is still difficult to select the vector combination weights. In addition, there is no reasonable weighting method for other similar platforms. In subsequent studies, we will add more considerations to the setting of weights, which can make the model more applicable to other wisdom education platforms.

## References

1. Zheng, Q., Dong, B., Qian, B., et al. (2019). Status and development trend of wisdom education research. *Computer Research and Development, 01*(56), 213–228.
2. Ding, J., & Liu, H. (2018). Accurate recommendation of learning resources based on multidimensional association analysis in big data environment. *Journal of Electrotechnical Education, 02*(39), 53–59.
3. Luo, J., Zeng, D., Pan, Z., & Liu, B. (2018). Application research of curriculum teaching resource recommendation system based on big data platform. *Big Data Era, 03,* 44–48.
4. Zhang, H., Yang, H., Huang, T., & Zhan, G. (2017). DBNCF: personalized courses recommendation system based on DBN in MOOC environment. In *2017 International Symposium on Educational Technology* (vol. 33, pp. 106–108). IEEE.
5. Yu, W., & Xu, D. (2019). Overview and Prospect of recommendation algorithms. *Technology and Innovation, 04,* 50–52.
6. Shan, J. (2015). Research on content-based personalized recommendation system. Northeast Normal University.
7. Shu, J., Shen, X., Liu, H., Yi, B., & Zhang, Z. (2017). A content-based recommendation algorithm for learning resources. *Multimedia Systems, 1,* 1–11.
8. Tao, Y., Yang, F., Liu, Y., & Dai, B. (2018). Research and optimization of K-means clustering algorithm. *Computer Technology and Development, 06*(28), 90–92.
9. Wang, J., Ma, X., & Duan, G. (2019). Improved K-means clustering k-value selection algorithm. *Computer Engineering and Application, 55*(8), 27–33.

10. Yang, Y., Pedersen, J.O. (1997). A comparative study on feature selection in text categorization. In *Proceedings of ICML 1997, 14th International Conference on Machine Learning* (vol. 4, pp. 412–420). Nashville: Morgan Kaufmann Publishers Inc.
11. Sebastiani, F. (2002). Machine learning in automated text categorization. *ACM Computing Surveys, 34*(1), 1–47.

# Robot Task Allocation Model of Robot Picking System

**Teng Li and Shan Feng**

**Abstract** In the logistics robot picking system, how to assign tasks to each robot reasonably determines the running time and cost of the whole system. A robust optimization model for task allocation based on logistics robot picking system is presented in this paper. In the model, the task completion time is taken as the objective function, and the actual walking distance of the robot is used as the uncertainty factor. With the robust optimization theory, the robust peer-to-peer transformation was performed on the task allocation model with uncertain parameters. Then, the task allocation model is solved by the Hungary algorithm. Finally, the effectiveness of the model is verified by an example simulation. The optimized task completion time is not sensitive to the uncertainty distance and is relatively stable.

**Keywords** Logistics picking system · Robot task allocation · Robust optimization · Hungary algorithm

## 1 Introduction

With the development of the Internet economy, the e-commerce industry is paying more and more attention to efficient logistics systems in order to improve customers' requirements for delivery timeliness in the context of multi-category, small-volume, and high-frequency orders. The picking time accounts for most of the entire logistics operation time. Therefore, companies favor the improvement of picking efficiency. The traditional picking method is manual-oriented. This kind of "picker-to-rack" picking system is less efficient, so the new "rack-to-picker" picking system is increasingly favored [1]. The "rack-to-picker" picking system based on mobile robot represented by Kiva is studied in this paper. The robot task

T. Li · S. Feng (✉)
School of Management, Harbin University of Commerce, Harbin, China
e-mail: 439191623@qq.com

T. Li
e-mail: liteng_ha@126.com



J. Zhang et al. (eds.), *LISS2019*,
https://doi.org/10.1007/978-981-15-5682-1_58

allocation means that each task is allocated to the robot in the system for execution, and the robot transports the rack where the goods are located to the picking station for picking. That is, the goods are moving, people are not moving. Therefore, how to allocate tasks to the robot in multi-robot, multi-task, parallel operation system determines the operating efficiency and cost of the whole system.

In recent years, many scholars have studied the problem of multi-robot assignment in different application fields. D. H. Lee studied the task allocation of multi-robot system and proposed a resource-based task allocation algorithm [2]. During the task operation, the robot continuously consumed their resources. The robot calculated the tasks and accessed them according to their resource levels. The results showed that the unnecessary time and resource waste of the robot during the completion of the task has been reduced. Reference [3] proposed a clustering-based multi-robot task allocation method, which used genetic algorithm and imitation learning algorithm to solve the problem, the running time has been saved. Su studied the task allocation problem of multi-robot system, and proposed a distributed task allocation method. The model was established by using the theory of combined auction, which proved that this method can obtain reasonable result of task allocation [4].

Liu studied the dynamic task allocation problem and proposed a task allocation model based on contract network. It can be seen from the simulation results that the method can effectively reduce the communication cost of the robot to complete the task [5]. Cao studied the allocation of robot tasks in dynamic environment. They used the idea of utility function, an improved ant colony algorithm was proposed to solve the task allocation problem. It has verified by simulation that this method can obtain the task allocation result with the best cost [6]. Reference [7] and reference [8] aimed at load balancing, analyzed the task allocation problem, and realized the emerging task allocation. J. Guo used the task allocation of multi-objective set as the research problem, and realized the optimal allocation of the target point set by using the network graph-based allocation method [9]. Arif expressed the multi-robot task assignment problem as a generalization of multi-traveler problem and used Evolutionary Algorithm for optimal task allocation [10]. Li proposed a utility evaluation algorithm based on adaptive neuro-fuzzy inference system, and used Q learning to improve the learning efficiency of utility function and improve the quality of task allocation scheme [11].

## 2 Problem Description

The problem of "rack-to-picker" picking system ask allocation can be described as follows: after the system accepts the order, the order is processed in batches according to certain rules. The system needs to confirm the position of the rack where the goods are placed in each order, and then assign tasks to robots. During the robot's completion of the mission, the robot first moves from the initial position to the position of the rack, and transports it to the picking station. The picker picks the goods on the rack based to the cargo information and places them in the designated area. When all the picking stations in the system are occupied by other robots, the robot will wait in line until it is selected. After the picker completes the picking operation, the robot transports the rack to the original position of the rack and waits for the allocation of the next task at this position. The system requires that each robot can only carry one rack at a time, and each rack can only be transported by one robot at a time.

## 2.1 Assumptions

(1) At initial moment, all robots in the system are idle.
(2) During the completion of the mission, the power of all the robots in the system is always greater than 20%.
(3) The initial position of all robots in system is random.
(4) All tasks in the system are distributed in different shelves, and there is no shortage of goods.
(5) After the robot moves the shelf back, wait for the next task at the shelf position.
(6) Ignore waiting time of the robot at the picking station.
(7) All robots have the same speed and ignore the acceleration and deceleration of the robot.

## 2.2 Parameter Definition

(1) $R$ represents the robot set, and $m$ is the number of robots in the system. $R_i$ represents the $i$-th robot in the system, where $i \in [1, m]$; $T$ represents a task set, and $n$ is the number of tasks in the system. $T_j$ represents the $j$-th task in the system, where $j \in [1, n]$.
(2) $(x_i, y_i)$ is the initial position coordinates of the robot $R_i$; $(x_j, y_j)$ is the position coordinates of task $T_j$.
(3) $W$ represents the collection of road segments that robot $R_i$ completes $T_j$, $(i, j) \in W$.
(4) $D_{ij}$ indicates the nominal value of the distance traveled by the robot $R_i$ to complete the task $T_j$; the distance used here is the Manhattan distance, i.e.

$$D_{ij} = |x_i - y_i| + |x_j - y_j| \tag{1}$$

(5) $V_i$ indicates the walking speed of the robot $R_i$.

(6) $$X_{ij} = \begin{cases} 1, & \text{robot } Ri \text{ completes task } T_j \\ 0, & \text{otherwise} \end{cases}$$

(7) *Z indicates the total time to complete the task.*

# 3 Model of Logistics Robot Task Allocation

## 3.1 Ideal Robot Task Allocation Model of Picking System

The ultimate goal of the task allocation of the logistics robot picking system is to spend the least amount of time on the premise of completing the task, so the objective function obtained is:

$$\min Z = \sum_{i=1}^{m} \sum_{j=1}^{n} \frac{D_{ij}}{Vi} X_{ij} \tag{2}$$

In the logistics robot picking system, one robot is required to be completed at most, and one task is allowed to be completed by one robot. So, the constraints are as follows:

$$\sum_{i=1}^{m} X_{ij} = 1, i = 1, 2, 3.\ldots.m$$

$$\sum_{j=1}^{n} X_{ij} \leq 1, j = 1, 2, 3.\ldots.n$$

$$Xij \in [0, 1], i = 1, 2, 3, \ldots m, j = 1, 2, 3, \ldots n$$

## 3.2 Robot Task Allocation Model of Picking System Affected by Uncertain Factor

In the process of the robot completing the task, if multiple tasks are concentrated in a certain area at the same time, the walking path between the multiple robots will be conflicted. In order to reduce the collision between the robots or reduce the waiting time of the robot, the scheduling system will schedule the robot to walk other paths to complete the task. Regardless of whether the robot chooses to wait or change the path, will affect the completion time of the task, and thus affect the efficiency of the robot to complete the task. Therefore, the time and path change of the robot are converted into the increase of the walking distance, and the uncertainty of the walking distance caused by the robot in the process of completing the task due to scheduling, obstacle avoidance, path planning, etc. is considered. It can be seen that the actual walking distance of the robot to complete the task is not necessarily the Manhattan distance from which the robot completes the task, so the actual walking distance of the robot to complete the task is represented by the interval value [12]:

$\widehat{D}_{ij}$ indicates that the robot $R_i$ completes the deviation of the actual walking distance of task $T_j$ from its nominal value; $\widehat{D}_{ij} \geq 0$ indicates the uncertainty distance that the robot $R_i$ has completed the task $T_j$; $\widetilde{D}_{ij} \in \left[D_{ij}, D_{ij} + \widehat{D}_{ij}\right]$. Introducing parameters $J$, $J$ indicates the set of road segments affected by the uncertainty distance, introducing parameters $\Gamma$, $\Gamma \in [0, |J|]$, $\Gamma$ is used to control the conservative degree of uncertainty, let $\Gamma$ take an integer, When $\Gamma = 0$, regardless of the influence of distance deviation. At this time, the task allocation model is consistent with (2), the model is most sensitive to uncertain distances. As the value of $\Gamma$ increases, the robustness increases, and the sensitivity of the model to the uncertainty distance decreases. When $\Gamma = |J|$, taking into account the uncertainty distance deviation of all road segments, the model is least sensitive to the uncertainty distance, and the obtained task allocation result is the most conservative [13]. Therefore, the robust optimization model of task allocation:

$$\min Z = \sum_{i=1}^{m} \sum_{j=1}^{n} \frac{D_{ij}}{V_i} X_{ij} + \max_{\{J|J \in W, \Gamma = |J|\}} \sum_{i=1}^{m} \sum_{j=1}^{n} \frac{\widehat{D}_{ij}}{Vi} X_{ij} \tag{3}$$

$$s.t. \sum_{i=1}^{m} X_{ij} = 1, i = 1, 2, 3 \ldots\ldots m$$

$$\sum_{j=1}^{n} X_{ij} \leq 1, j = 1, 2, 3 \ldots\ldots n$$

$$X_{ij} \in [0, 1], i = 1, 2, 3, \ldots m, j = 1, 2, 3, \ldots n$$

Equation (3) considers the deviation of the uncertainty distance of the robot when completing the task, indicating that the task allocation is performed with the maximum deviation of the uncertainty, so that the robot has the shortest time to complete the task. The objective function contains the max term and cannot be solved directly. Therefore, this paper refers to the robust discrete transformation rule proposed [14], and transforms the objective function in the above model as follows:

$$Z^{IJ} = \min\{a + b\} + \Gamma \frac{\widehat{D}_{IJ}}{Vi} \tag{4}$$

where

$$a = \sum_{\{i=1:(i,j)\in W\}}^{\{I:(I,j)\in W\}} \sum_{\{j=1:(i,j)\in W\}}^{\{J:(i,J)\in W\}} \frac{\widehat{D}_{ij}}{Vi} X_{ij}$$

and

$$b = \sum_{\{i=1:(i,j)\in W\}}^{\{I:(I,j)\in W\}} \sum_{\{j=1:(i,j)\in W\}}^{\{J:(i,J)\in W\}} \frac{\widehat{D}_{ij} - \widehat{D}_{IJ}}{V_i} X_{ij},$$

$$s.t. \sum_{i=1}^{m} X_{ij} = 1, i = 1, 2, 3.\ldots..m$$

$$\sum_{j=1}^{n} X_{ij} \leq 1, j = 1, 2, 3.\ldots..n.$$

$$X_{ij} \in [0, 1], i = 1, 2, 3, \ldots m, j = 1, 2, 3, \ldots n.$$

The time optimal target value:

$$T = \min_{(I,J\in W)} Z^{ij} \tag{5}$$

$$s.t. \sum_{i=1}^{m} X_{ij} = 1, i = 1, 2, 3.\ldots..m$$

$$\sum_{j=1}^{n} X_{ij} \leq 1, j = 1, 2, 3.\ldots..n$$

$$X_{ij} \in [0, 1], i = 1, 2, 3, \ldots m, j = 1, 2, 3, \ldots n.$$

### 3.3 Solving Task Allocation Model Based on Hungarian Law

The Hungarian law is used to solve the assignment problem. Because the robot task assignment problem in this paper is consistent with the assignment problem, and a shelf can only be transported by one robot, one robot can only carry one shelf at the same time, so the problem is solved by the Hungarian algorithm [15]. Figure 1 shows the specific steps.

(1) Determine the coordinates of each robot and each task.
(2) Determine whether m and n are equal. If they are equal, construct the Manhattan distance between each robot and each task as matrix $D_{ij}$; if m is greater than n, assume that the virtual task constructs n-order matrix of $D_{ij}$ (filled with 0); if m less than n, assume that the virtual robot constructs an m-order matrix of $D_{ij}$ (filled with 0).
(3) In the distance matrix, find the smallest element of each row, and each element is different from the element.
(4) In the distance matrix, find the smallest element of each column, and each element of this column is different from the element.
(5) Overwrite all elements 0 in the matrix with a minimum number of lines. If the number of lines is less than the order of the matrix, find the smallest element value that is not covered in the matrix, and the value of each element that is not covered minus the minimum element. The value of each element at the intersection of the line plus the minimum element value, the values of other elements are unchanged, constructing a new distance matrix.
(6) Repeat step (5) until the number of lines is equal to the order of the matrix.

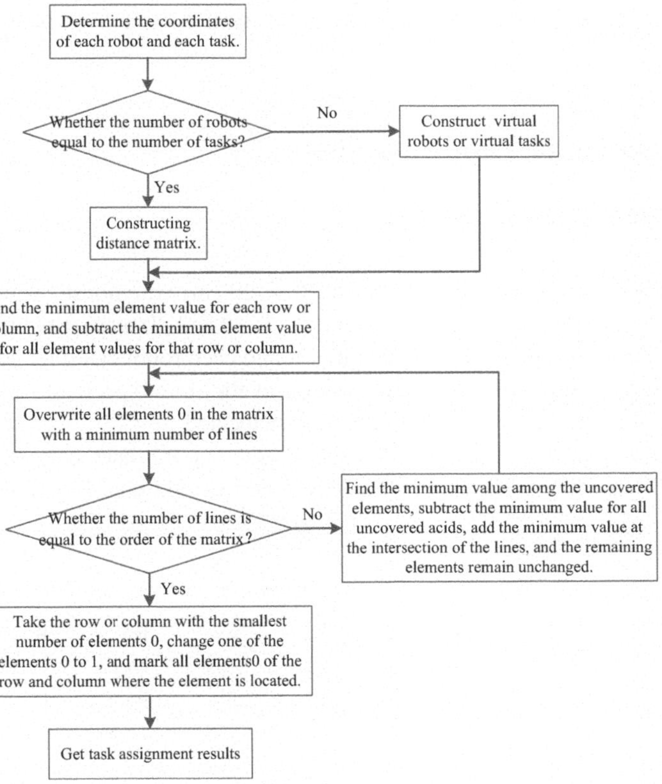


**Fig. 1** Flow of the Hungarian algorithm for solving the task assignment problem of robots

(7) Take the line with the least number of elements 0, mark the element 0 of the line, and mark the element 0 of the same column, repeat the above steps until the last line.
(8) Take the column with the least number of remaining elements 0, mark the element 0 of the column, and mark the remaining elements 0 of the same column, repeat the above steps until the last column.
(9) If m is equal to n, mark the number of m or the number of n elements 0; if m is greater than n, mark the number of n elements 0; if m is less than n, mark the number of m elements 0.
(10) Replace the value of the marked element 0 with 1, and replace the other element values in the matrix with 0, thus forming a matrix of robot task assignment. The value of 1 means that the robot completes the task, and the element value is 0 means this robot does not complete this task.
(11) Find the distance corresponding to each robot in the system to complete the task, and add the distance values to get the shortest distance for each robot in the system to complete the task.
(12) If m is equal to or greater than n, the task assignment ends; if m is less than n, continue to allocate the remaining tasks.
(13) Assign the completed task coordinate values to the corresponding robot and repeat steps (2)–(12) until all tasks in the system are assigned.

# 4 Simulation of Task Allocation

## *4.1 Instance Description*

The simulation environment is set in a 20 × 20 (m) warehouse, with 10 robots randomly distributed, and the number of tasks is 30 and 40 respectively. It is agreed that each task is completed by a single robot, and each robot can only complete one task at a time. Table 1 is the initial position of the robot specified randomly, and Table 2 is the shelf position where the task is located.

When the number of tasks is 30, Table 2 is the coordinates of each task.

When the number of tasks is 40, Table 3 is the additional coordinates of each task.

**Table 1** Robot position

| Robot Ri | R1 | R2 | R3 | R4 | R5 |
|---|---|---|---|---|---|
| Position | 4,2 | 16,8 | 14,10 | 19,12 | 2,11 |
| Robot Ri | R6 | R7 | R8 | R9 | R10 |
| Position | 7,19 | 15,4 | 18,13 | 16,3 | 5,18 |

**Table 2** Task position

| Task Tj | T1 | T2 | T3 | T4 | T5 | T6 |
|---|---|---|---|---|---|---|
| Position | 10,4 | 12,6 | 2,7 | 5,2 | 11,18 | 8,13 |
| Task Tj | T7 | T8 | T9 | T10 | T11 | T12 |
| Position | 8,10 | 4,7 | 9,2 | 3,17 | 4,18 | 7,6 |
| Task Tj | T13 | T14 | T15 | T16 | T17 | T18 |
| Position | 9,11 | 16,15 | 14,7 | 5,15 | 3,18 | 18,2 |
| Task Tj | T19 | T20 | T21 | T22 | T23 | T24 |
| Position | 14,6 | 13,3 | 2,19 | 12,7 | 14,14 | 5,19 |
| Task Tj | T25 | T26 | T27 | T28 | T29 | T30 |
| Position | 3,6 | 7,17 | 10,2 | 17,16 | 8,12 | 9,4 |

**Table 3** Task position

| Task Tj | **T31** | **T32** | **T33** | **T34** | **T35** | **T36** |
|---|---|---|---|---|---|---|
| Position | 20,20 | 8,14 | 19,7 | 15,8 | 4,17 | 10,10 |
| Task Tj | T37 | T38 | T39 | T40 | | |
| Position | 16,3 | 11,8 | 6,11 | 18,14 | | |

## *4.2 Analysis of Task Allocation Simulation Results Under Ideal Conditions*

When the number of robots is 10, the number of tasks is 30 and 40 respectively. The simulation results are as follows:

When the number of tasks is 30, Table 4 is the assignment result of the task.

When the number of tasks is 40, Table 5 is the assignment result of the task.

**Table 4** Task allocation result

| Robot Ri | R1 | R2 | R3 | R4 | R5 |
|---|---|---|---|---|---|
| Task Tj | T4 | T19 | T15 | T14 | T3 |
| Task Tj | T9 | T2 | T22 | T6 | T8 |
| Task Tj | T30 | T13 | T7 | T29 | T25 |
| Robot Ri | R6 | R7 | R8 | R9 | R10 |
| Task Tj | T24 | T23 | T28 | T18 | T11 |
| Task Tj | T21 | T20 | T5 | T27 | T17 |
| Task Tj | T16 | T1 | T26 | T12 | T10 |
| Total distance | 115 m | | | | |
| Speed | 2 m/s | | | | |
| Total time | 57.5 s | | | | |

**Table 5** Task allocation result

| Robot Ri | R1 | R2 | R3 | R4 | R5 |
|---|---|---|---|---|---|
| Task Tj | T4 | T34 | T15 | T33 | T39 |
| Task Tj | T9 | T38 | T22 | T19 | T7 |
| Task Tj | T27 | T36 | T12 | T2 | T29 |
| Task Tj | T25 | T13 | T3 | T8 | T6 |
| Robot Ri | R6 | R7 | R8 | R9 | R10 |
| Task Tj | T26 | T23 | T40 | T37 | T11 |
| Task Tj | T35 | T20 | T14 | T18 | T17 |
| Task Tj | T10 | T1 | T28 | T31 | T21 |
| Task Tj | T16 | T30 | T32 | T5 | T24 |
| Total distance | 142 m | | | | |
| Speed | 2 m/s | | | | |
| Total time | 71 s | | | | |

According to the simulation results, as the number of tasks increases, the total time for the robot to complete the task will also increase, and the task allocation result will be different each time.

## *4.3 Analysis of Simulation Results of Task Assignment Affected by Uncertain Factors*

When the number of robots is 10 and the number of tasks is 30, since the distance is uncertain, the deviation $\widehat{D}_{ij}\left(0 \leq \widehat{D}_{ij} \leq 0.5D_{ij}\right)$ of the nominal value is set to a randomly generated number, and the nominal value and the deviation constitute a robust model test example. The uncertainty distance matrix is taken as $\Gamma = 30$ and $\Gamma = 80$ respectively, to simulate the task allocation problem of logistics robots. The simulation results are as follows.

**Table 6** Task allocation result

| Robot Ri | R1 | R2 | R3 | R4 | R5 |
|---|---|---|---|---|---|
| Task Tj | T4 | T19 | T15 | T14 | T3 |
| Task Tj | T9 | T2 | T22 | T6 | T25 |
| Task Tj | T12 | T13 | T7 | T29 | T8 |
| Robot Ri | R6 | R7 | R8 | R9 | R10 |
| Task Tj | T3 | T26 | T23 | T28 | T18 |
| Task Tj | T10 | T20 | T5 | T27 | T17 |
| Task Tj | T21 | T30 | T16 | T1 | T24 |
| Total distance | 123.0 m | | | | |
| Speed | 2 m/s | | | | |
| Total time | 61.5 s | | | | |

**Table 7** Task allocation result

| Robot Ri | R1 | R2 | R3 | R4 | R5 |
|---|---|---|---|---|---|
| Task Tj | T4 | T15 | T22 | T14 | T3 |
| Task Tj | T9 | T19 | T2 | T5 | T25 |
| Task Tj | T27 | T7 | T13 | T11 | T7 |
| Robot Ri | R6 | R7 | R8 | R9 | R10 |
| Task Tj | T26 | T23 | T28 | T18 | T24 |
| Task Tj | T16 | T1 | T6 | T20 | T30 |
| Task Tj | T21 | T29 | T10 | T12 | T8 |
| Total distance | 155 m | | | | |
| Speed | 2 m/s | | | | |
| Total time | 77.5 s | | | | |

When the number of tasks is 30 and 30 distances change, Table 6 is the result of task allocation.

When the number of tasks is 30 and 80 distances change, Table 7 is the result of task allocation.

When taking $\Gamma = 1\text{–}300$, it means that the number of uncertain distances changes from 1 to 300, Fig. 2 is the simulation result. The abscissa represents $\Gamma$ and the ordinate represents the total time the robot has completed all tasks.

In order to compare with the total time of the robot in the ideal situation to complete the task, the distance is also randomly changed when the task assignment result is unchanged. Figure 3 is the comparison of the total time of the robot in both cases.

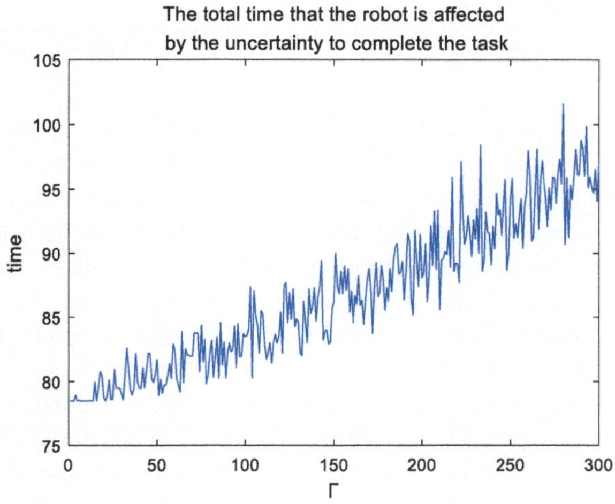


**Fig. 2** Total time of completion of tasks affected by uncertainties

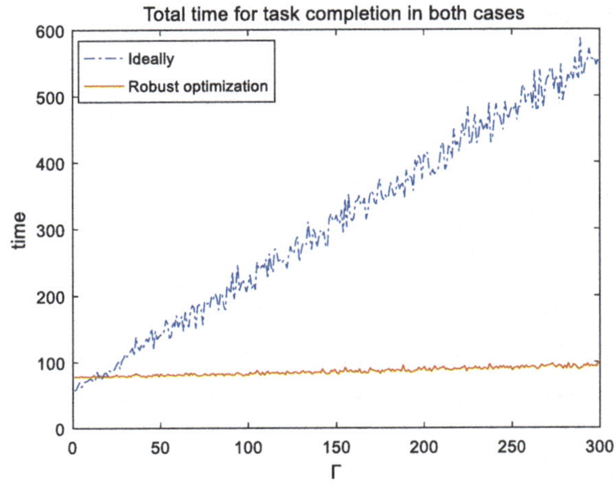


**Fig. 3** Total time for task completion in both cases

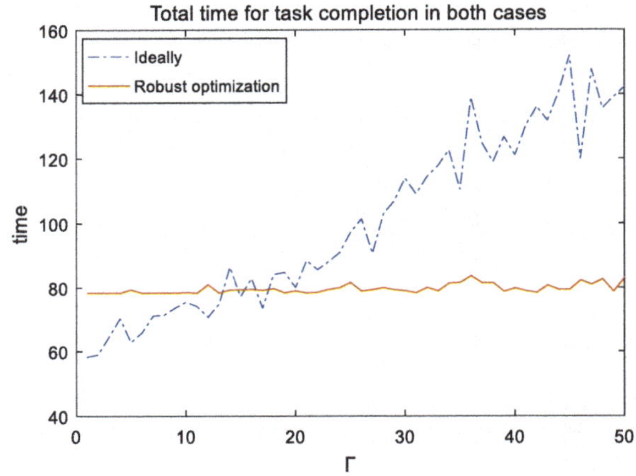


**Fig. 4** Partial enlargement of the total time of task completion in both cases

In Fig. 4, the abscissa represents t, the ordinate is the total time the robot completes the task, and the dashed line indicates the total time the robot completed the task in an ideal situation. When the distance changes randomly, the solid line indicates the total time that the robot completes the task after the robust optimization. It can be seen from the figure that when the distance changes under ideal conditions, the total time for the robot to complete the task is significantly more than the total time after the robust optimization and gradually increases. While the optimized curve is relatively stable and the fluctuation range is small. Figure 4 is a partial enlargement of Fig. 3. In the first part, the optimized curve is higher than the curve before optimization because the model is sensitive to the uncertainty distance. As the value increases, the sensitivity of the model to the uncertainty distance is reduced. The robustness of the understanding; when $\Gamma = 300$, all the sections will change randomly. At this time, the model is least sensitive to the uncertainty distance, and the obtained task assignment result makes the total time of completing the task the most stable.

## 5 Conclusion

In the "rack to picker" picking system, the key to improve the efficiency of the operation depends on how to assign tasks, which affects the operating efficiency and cost of the whole system. This paper considers the scheduling, obstacle avoidance and path planning of the robot in the process of completing the task. Based on the uncertain factors of walking distance, a robust optimization model based on walking distance uncertainty is established. The minimum total time of the robot is the objective function. The Hungarian algorithm is used to study the model and the optimization results are analyzed. By comparing the simulation results without considering the distance variation and the robust optimization, the result of the robust optimization of the task allocation results in the total time that the robot completes the task is not sensitive to the uncertainty distance factor. This means that it is the most conservative in the case of such task assignment results, and the model is least sensitive to uncertain distances. It improves the picking efficiency of the system and reduces the operating cost of the whole system. The result has a certain guiding effect on the task assignment of the robot, and at the same time realizes the picking sequence of the robot, which provides a reference for the design and application practice of the picking system.

## References

1. Ren, F. (2017). The rack to picker picking scheme and its innovative development. *Logistics Technology and Applications, 22,* 80–84.
2. Lee, D. H. (2018). Resource-based task allocation for multi-robot systems. *Robotics and Autonomous Systems, 103,* 151–161.
3. Janati, F., Abdollahi, F., & Ghidary, S.S. (2017). Multi-robot task allocation using clustering method. In *Robot Intelligence Technology and Applications 4* (pp. 233–247). Springer, Cham.
4. Su, L. Y., Mo, L. S., Li, X. P., & Du, F. (2013). Modeling and solving of task allocation problem for multi-robot system. *Journal of Central South University: Natural Science, 44,* 122–125.
5. Liu, Z. Q., Chen, S. Y., Shao, Z. Z., Zhang, Y., & Liu, Y. M. (2017). Study on dynamic task allocation algorithm with two-way screening in multi-robot hunting. *Mini-micro Systems, 38,* 1568–1572.
6. Cao, Z. H., Wu, B., Huang, Y. Q., & Deng, C. Y. (2013). Multi-robot task allocation based on improved ant colony algorithm. *Combined Machine Tool & Automatic Processing Technology, 02,* 34–37.
7. Zhou, J., & Mu, D. J. (2014). Research on task assignment of multi-robot system. *Journal of Northwest University (Natural Science), 44,* 403–410.

8. Zhou, J., Mu, D.J. (2014). Research on probability model of task assignment in multi-robot system. *Microprocessors, 35*, 77–83+88.
9. Guo, J., & Zhou, M. (2014). A multi-robot system task assignment planning algorithm. *Fire Power and Command System, 39,* 107–110.
10. Arif, M.U., Haider, S. (2017). An evolutionary traveling salesman approach for multi-robot task allocation. In *9th Conference Agents Artifical Intelligence, Porto* (pp. 567–574).
11. Li, P., Zhu, J. Y., & Yang, Y. M. (2013). Algorithm for task assignment utility evaluation in multi-robot systems. *Computer Engineering and Design, 34,* 3288–3292.
12. Xiong, R.Q., Ma, C.X. (2017). Robust optimization of dangerous goods distribution path in multi-distribution center. *Journal of Computer Applications, 37*, 1485–1490+1515.
13. Ma, C. R., & Ma, C. X. (2014). Robust optimization of hazardous materials transportation route in uncertain environment. *Chinese Journal of Safety Science, 24,* 91–96.
14. Bertsimas, D., & Sim, M. (2004). The price of robustness. *Operations Research, 52,* 35–53.
15. Li, Z.M. (2015). A note on the hungarian solution to assignment problem. *Journal of Xinjiang University (Natural Science Edition), 32*, 286–288+303.

# E-closed-loop Supply Chain Decision Model Considering Service Level Input

**Ying Yang and Yuan Tian**

**Abstract** This paper studied the decision-making of the E-closed-loop supply chain (E-CLSC) which is composed of the manufacturer, the retail e-commerce and consumers. Based on Stackelberg game and Nash game, we proposed centralized and decentralized decision-making models and considered three market structures in the decentralized system. First, this paper confirmed that the centralized decision model can maximize system profit, and in decentralized system, the market oriented by the manufacture is more beneficial for higher profits achieved for the system and this is also a market structure that consumer's favorite. In addition, based on the profit loss of supply chain members in decentralized decision model, we proposed a profit coordination mechanism. Second, we find that market leaders always obtain more profits from recycling and remanufacturing activities, and the highest waste product recycling rate can be achieved when retail e-commerce acts as a leader. Finally, based on the comparison of market demand in different market structures, we find that consumers no longer only pay attention to product prices in the E-CLSC, but also concerned about service levels, and are willing to pay higher prices in order to obtain better service.

**Keywords** E-closed-loop supply chain · Service level · Game · Decision

## 1 Introduction

As the potential of the e-commerce market continues to expand, more and more manufacturers are beginning to choose online sales channels, thus forming E-supply chain (E-SC). Different from traditional offline sales channel, consumers in the E-SC no longer only pay attention to product prices, but also focus on the service

Y. Yang (✉) · Y. Tian
School of Economics and Management, Beijing Jiaotong University, Beijing, China
e-mail: 18120597@bjtu.edu.cn

Y. Tian
e-mail: ytian@bjtu.edu.cn



J. Zhang et al. (eds.), *LISS2019*,
https://doi.org/10.1007/978-981-15-5682-1_59

level of e-commerce companies. And when the manufacturer recycles waste products through an e-commerce partner and combines it with the forward sales process, the E-CLSC are formed. It is widely accepted that closed-loop supply chain (CLSC) are more complicated than traditional manufacturing and selling process. on the one hand, the CLSC for remanufacturing, which involves the movement of products from upstream suppliers to downstream customers and the flow of used products back to the remanufacturers, combines the forward logistic with the reverse logistic, on the other hand, there also exist uncertainty of sources of used products [1, 2].

In short, unlike the traditional supply chain decision-making problem, E-CLSC members not only need to face the new market demand structure, but also be affected by the complexity of CLSC. Therefore, the study of E-CLSC decision-making problems can provide members with decision-making reference, in order to maximize their own profits or minimize losses when they face different market structures, and it can also clarify the new market positioning of the E-SC different from traditional ones and helps E-SC members accurately grasp changes in market demand.

The current research on the E-CLSC can be divided into two categories. The first category generally focuses on E-CLSC management. Werbach proposed a joint model based on the Internet to promote enterprises to focus on adjusting business strategies [3]. Siddiqui found that when transforming to E-SCs, enterprises should focus on innovation firstly, and later shift to focus on collaboration [4]. Battini provided a theoretical basis for the new research direction of CLSC management by combing literatures on the economics, modeling and management of CLSC [5].

Another group of E-CLSC research mainly focuses on decision-making problems. Savaskan found that when the manufacturer acts as the leader in CLSC, the model that retailers are responsible for recovery is optimal [6]. The study by Giovanni verified this conclusion, too [7]. Wang studied CLSC decision-making problem by considering government intervention [8]. Miao considered three models for CLSC with trade-ins and found that the decentralized model can achieve better environmental performance [9]. Yao only considered the impact of price on product demand when studying CLSC decision problems [10]. However, the service level factor is also taken into account when E-SCs decision-making problems are studied. Wang conducted a study on E-CLSC that consists of a manufacturer and a platform e-commerce [11]. Kong believed that in order to alleviate the two-channel conflict, offline retailers should also focus on improving service levels [12].

To sum up, although the existing supply chain research is very rich, service level factor was rarely considered in previous E-CLSC studies. In general, the contribution of this research is threefold. Firstly, different from the existing research, considering the fact that E-CLSC decisions are affected by the product price and service level, a series of decision-making models with different market structures were established. With the theoretical derivation, the characteristic of E-CLSC was exhaustively explored and optimal decisions of E-CLSC members were detailed illustrated. Secondly, indicating the centralized system is most beneficial to the system and proposing a profit coordination mechanism in the decentralized system.

Thirdly, the paper confirmed that consumer behavior in the E-SC is different from traditional one, which was significantly meaningful to market position in the E-CLSC.

## 2 The Model and Notations

The E-CLSC is consist of the manufacturer, the retail e-commerce and consumers. The flow of products in this supply chain is as follows: First, manufacturers sell products to the retail e-commerce at a wholesale price. Then, new products are sold to the market by retail e-commerce who also needs to provide transportation, warehousing and other services. In addition, the retail e-commerce is also responsible for recycling used products from the market. Finally, these used products are recycled by manufacturers from the retailer for remanufacturing and then remanufactured products together with new ones are put on the market. It is worth noting that the reason why the retail e-commerce is responsible for recycling is that its distribution and logistics networks are spread throughout the country generally and recycling activities are easy to implement. The E-CLSC system is shown in Fig. 1.

The notations and assumptions in this article are defined as follows: In the forward E-CLSC, manufacturers can both use raw materials and the waste to produce products, and consistent with the research hypothesis of Savaskan [6], there is no difference between the new product and the re-product, and they are both sold at the same price. The unit cost of producing new products is $c_m$, while the unit cost of producing re-products is $c_r$, $c = c_m - c_r$ is used to denote the manufacturer's unit cost savings from remanufacturing.

$f$ is the service level provided by retail e-commerce, and $\frac{k}{2}f^2$ is the service cost of the e-commerce, where $k$ is unit cost elasticity coefficient ($k > 0$). The use of quadratic functions to represent the cost of decision makers has been widely used in many economic and management literature [6, 13].

Supposing that the market demand is a linear function of price and service level, the market demand function is defined as $Q = D - ap + \pi f$, $D$ refers to potential maximum market demand, and $a > 0$ is the price sensitivity coefficient, and $\pi$ is the service level elastic coefficient, the retail price of products is denoted as $p$. Assuming

**Fig. 1** The E-CLSC system

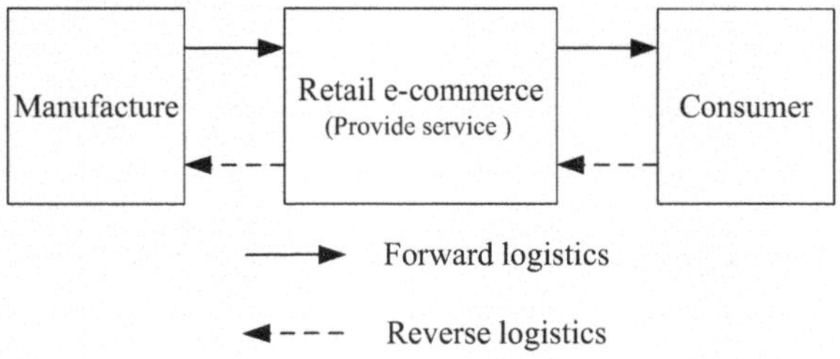

that the recovery amount is only related to the recovery price and is denoted as $Q_0 = kp_0$, where $p_0$ refers to the recycling price of the used product. In addition, the wholesale price and recovery rate are denoted as $w$ and $\delta = \frac{Q_0}{Q}$ respectively.

Based on the above assumptions and symbol descriptions, following profit functions can be obtained: The profit function of the manufacturer, retail e-commerce and supply chain are as follows:

$$\pi_{im} = (w - c_w)(Q - Q_0) + (w - c_r - b)Q_0 \tag{1}$$

$$\pi_{ie} = (p - w)Q + (b - p_0)Q_0 - \frac{k}{2}f^2 \tag{2}$$

$$\pi_i = (p - c_m)Q + (c - p_0)Q_0 - \frac{k}{2}f^2 \tag{3}$$

We use *m, e* to represent the manufacturer and e-commerce, respectively. $i = M, E, W$ are defined to represent the market dominated by manufacturers, e-commerce companies and no dominance respectively. To ensure the practical significance of the above problems, there is the definition of $ak > \pi^2$, $p_0 < b < b + c_r < c_m < w < p$.

## 3 Centralized Decision Model

In this paper, we only consider a supply chain consisting of one manufacturer and one e-commerce. In E-CLSC, the decision of each participant is as follows: manufacturers decide the wholesale price $w$ and the recycling subsidy $b$, and the retail e-commerce determines the service level $f$, the retail price $p$ and the recycling price $p_0$.

In the centralized system, the manufacturer and the retail e-commerce form an integrated firm, which means that they will make decisions with the goal of maximizing system profit and the wholesale price and recycling subsidy are internal factors. They determine the optimal $p$, $f$, $p_0$ *to* maximize the total value of the system:

According to the objective (3), we can obtain the optimal $p_C^*, f_C^*, p_{C0}^*$ as follows:

$$p_C^* = \frac{KD + c_m(ak - \pi^2)}{2ak - \pi^2} \tag{4}$$

$$f_C^* = \frac{\pi(D - ac_m)}{2ak - \pi^2} \tag{5}$$

$$p_{C0}^* = \frac{c}{2} \tag{6}$$

where subscript "C" denotes the centralized system, and the system profit is calculated as $\pi_C^* = \frac{k(D-ac_m)^2}{2(2ak-\pi^2)} + \frac{kc^2}{4}$.

## 4 Decentralized Decision Model

A. *Stackelberg Game*

(1) *Manufacturer oriented*

When the manufacturer acts as the Stackelberg leader, it will maximize its profits in prior, so the decision order is that manufacturers decide wholesale price $w$ and the recycling subsidy $b$, and then the retail e-commerce gives the service level $f$, the retail price $p$ and the recycling price $p_0$ according to manufacturers' pricing strategy, because of the assumption of complete information game. Manufacturers take into account the retail e-commerce's response when making decisions. As a result, reverse induction is used in two periods.

In the first period, we calculate the best-response function of the retail e-commerce for a given $p_E^{MR}$, $p_{E0}^{MR}$, and $f_E^{MR}$:

$$p_{M0} = \frac{b}{2} \tag{7}$$

$$f_M = \frac{\pi(D - aw)}{2ak - \pi^2} \tag{8}$$

$$p_M = \frac{(ak - \pi^2)w - kD}{2ak - \pi^2} \tag{9}$$

Then, substituting $p_M^{MR}, p_{M0}^{MR}, f_M^{MR}$ into (1), and we can obtain best decisions of the manufacturer for $w$ and $b$ and deduce the optimal value of $p_E^{MR}$, $p_{E0}^{MR}$, and $f_E^{MR}$ as follows:

$$w_M^* = \frac{D + ac_m}{2a} \tag{10}$$

$$b_M^* = \frac{c}{2} \tag{11}$$

$$p_{M0}^* = \frac{c}{4} \tag{12}$$

$$f_M^* = \frac{\pi(D - ac_m)}{2(2ak - \pi^2)} \tag{13}$$

$$p_M^* = \frac{2akD + (D + ac_m)(\pi^2 - ak)}{2a(2ak - \pi^2)} \tag{14}$$

In the second period, by substituting (10) (11) (12) (13) into the objective function, the optimal values $\pi_{Mm}^{MR*}, \pi_{Me}^{MR*}, \pi_M^{MR*}$ are calculated, which denotes the optimal profit of the manufacturer, retail e-commerce and supply chain respectively. Let $Q_M^{MR*}$ denotes optimal market demand, and to ensure that $Q_M^{MR*}$ is meaningful, $D > ac_m$ is established. The value of $\pi_{Mm}^{MR*}, \pi_{Me}^{MR*}$ and $Q_M^{MR*}$ are listed in Table 1.

(2) *Retail e-commerce oriented*

When the retail e-commerce acts as a leader, the decision order is that the retail e-commerce decides the service level *f*, the retail price *p* and the recycling price $p_0$ firstly, and then the manufacturer decides *w* and *b*. The backward induction method is also used. We suppose that the e-commerce earns *r* for selling one unit of product and *y* for recycling one unit of product, then $r = p - w$ and $y = b - p_0$ are obtained.

In the first period, we get the best-response function of *w* and *b* by bringing $p = r + w$ and $p_0 = b - y$ into (1):

$$w_E^{MR} = \frac{Q + ac_m}{a} \tag{15}$$

$$b_E = c - p_0 \tag{16}$$

Then, we substitute $w_E^{MR}$ and $b_E$ into (2), and optimal decisions of the retail e-commerce $P_E^{MR*}, f_E^{MR*}, P_{E0}^{MR*}$ and the value of $w_E^{MR*}$, $b_E^*$ will be obtained as follows:

$$p_{E0}^* = \frac{c}{4} \tag{17}$$

**Table 1** The optimal results of E-CLSC

| | Manufacturer oriented | Retail e-commerce oriented | No dominance |
|---|---|---|---|
| $Q^*$ | $\frac{ak(D-ac_m)}{2(2ak-\pi^2)}$ | $\frac{ak(D-ac_m)}{(4ak-\pi^2)}$ | $\frac{ak(D-ac_m)}{(3ak-\pi^2)}$ |
| $\pi_m^*$ | $\frac{k(D-ac_m)^2}{4(2ak-\pi^2)} + \frac{kc^2}{8}$ | $\frac{ak^2(D-ac_m)^2}{(4ak-\pi^2)^2} + \frac{kc^2}{16}$ | $\frac{ak^2(D-ac_m)^2}{(3ak-\pi^2)^2} + \frac{kc^2}{9}$ |
| $\pi_e^*$ | $\frac{k(D-ac_m)^2}{8(2ak-\pi^2)} + \frac{kc^2}{16}$ | $\frac{k(D-ac_m)^2}{2(4ak-\pi^2)} + \frac{kc^2}{8}$ | $\frac{k(2ak-\pi^2)(D-ac_m)^2}{2(3ak-\pi^2)^2} + \frac{kc^2}{9}$ |

$$f_E^* = \frac{\pi(D - ac_m)}{4ak - \pi^2} \tag{18}$$

$$p_E^* = \frac{3kD + c_m(ak - \pi^2)}{4ak - \pi^2} \tag{19}$$

$$w_E^* = \frac{kD + c_m(3ak - \pi^2)}{4ak - \pi^2} \tag{20}$$

$$b_E^* = \frac{3c}{4} \tag{21}$$

Similarly, further calculations can get $\pi_{Em}^*$, $\pi_{Ee}^*$ and $Q_E^*$, which are also listed in Table 1.

B. *Nash Game*

When there is no dominance, the manufacture and retail e-commerce take actions simultaneously, so it is easy to obtain the optimal retail price, recycling price, service level, wholesale price and recycling subsidy:

$$p_W^* = \frac{2kD + c_m(ak - \pi^2)}{3ak - \pi^2} \tag{22}$$

$$p_{WO}^* = \frac{c}{3} \tag{23}$$

$$f_W^* = \frac{\pi(D - ac_m)}{3ak - \pi^2} \tag{24}$$

$$w_W^* = \frac{kD + c_m(2ak - \pi^2)}{3ak - \pi^2} \tag{25}$$

$$b_W^* = \frac{2c}{3} \tag{26}$$

Similarly, we can also obtain $\pi_{Wm}^*$, $\pi_{We}^*$ and $Q_W^*$, which are also listed in Table 1.

## 5 Comparison and Numerical Stimulation

A. *Comparative Analysis of Models*

Based on the above analysis of optimal decision-making results, the following propositions can be drawn:

**Proposition 1.** The retail price satisfies: $p_E^* > p_W^* > p_C^* > p_M^*$.

**Proof.** Due to $D > ac_m, ak > \pi^2$, the inequality group $p_W^* - p_C^* = \frac{k(D-ac_m)(ak-\pi^2)}{(2ak-\pi^2)(3ak-\pi^2)} > 0$, $p_C^* - p_M^* = \frac{(D+3ac_m)(ak-\pi^2)}{2a(2ak-\pi^2)} > 0$ *and* $p_E^* - p_W^* = \frac{k(ak-\pi^2)(D-ac_m)}{(4ak-\pi^2)(3ak-\pi^2)} > 0$ could be obtained, and the Proposition 1 is established.

**Proposition 2.** The service level satisfies: $f_C^* > f_W^* > f_M^* > f_E^*$.

**Proof.** For $f_W^* - f_M^* = \frac{\pi(ak-\pi^2)(D-ac_m)}{2(3ak-\pi^2)(2ak-\pi^2)} > 0$ and $f_C^* - f_W^* = \frac{\pi ak(D-ac_m)}{(3ak-\pi^2)(2ak-\pi^2)} > 0$, $f_M^* - f_E^* = \frac{\pi^3(D-ac_m)}{2(4ak-\pi^2)(2ak-\pi^2)} > 0$, the Proposition 2 is established.

**Proposition 3.** The demand satisfies: $Q_C^* > Q_W^* > Q_M^* > Q_E^*$.

**Proof.** For $Q_W^* - Q_M^* = \frac{ak(ak-\pi^2)(D-ac_m)}{2(3ak-\pi^2)(2ak-\pi^2)} > 0$ and $Q_M^* - Q_E^* = \frac{ak\pi^2(D-ac_m)}{2(4ak-\pi^2)(2ak-\pi^2)} > 0$, $Q_C^* - Q_W^* = \frac{a^2k^2(D-ac_m)}{(3ak-\pi^2)(2ak-\pi^2)} > 0$ the Proposition 3 is established.

**Proposition 4.** The wholesale price satisfies: $w_M^* > w_W^* > w_E^*$.

The conclusion of above propositions can be summarized as follows. First, in the decentralized decision system, it is observed from Proposition 1 and 2 results that, compared with retail e-commerce as a market leader, when the manufacture is regarded as the leader, not only the retail price is the lowest, but the service level is also improved. Second, in the centralized decision system, not only the service level is the highest, but also the price is not high.

Proposition 3 shows that with the combined effect of prices and service levels, the market for joint decision making is most popular with consumers. In addition, according to the relationship of $p_W^* > p_C^* > p_M^*, f_C^* > f_W^* > f_M^*$ and $Q_C^* > Q_W^* > Q_M^*$, the conclusion can be observed as that consumers are willing to pay a higher price for better service. Therefore, different from traditional supply chains, consumers no longer only care about prices, but also focus on service levels in the E-CLSC.

Note that from Proposition 1 and 4, although the retailer can obtain the lowest wholesale price when itself dominates the market, it will not sell products at the lowest price, but take advantage of dominant strength to formulate a price for obtaining high profits. Conversely, when it is in a weak position, manufacturers will take use of dominant strength to increase wholesale prices, but this will not directly lead to a high retail price, however, retailers choose to sacrifice some of their profits to sell at low prices to increase sales, which is not only can increase the visibility of the e-commerce platform in the short term, but in the long run, it can also help to stabilize the cooperation relationship with manufacturers, thereby improving bargaining power and lowering the wholesale price level.

**Proposition 5.** The recycling price and recovery amount satisfy: $p^*_{M0} = p^*_{E0} < p^*_{W0} < p^*_{C0}$, $Q^*_{M0} = Q^*_{E0} < Q^*_{W0} < Q^*_{C0}$.

**Proposition 6.** The recovery rate satisfies: $\delta^*_E > \delta^*_W > \delta^*_M = \delta^*_C$.

**Proof.** For $\delta = \frac{Q_0}{Q}$, it is easy to prove Proposition 6.

**Proposition 7.** The recycling subsidy satisfies: $b^*_E > b^*_W > b^*_M$.

From above propositions, we have the following observations. Firstly, note that from Proposition 5, the recycling price and volume is highest in the centralized decision system, and in the decentralized one, compared to a market structure without leadership, the existence of a market leader will lead to the decline in the recycling price and total recycling volume of waste products, which in turn damage the profit of consumers. Secondly, Proposition 6 reveals that the highest recovery rate is achieved when retail e-commerce dominates the market. The reason is that when the retail e-commerce dominates the market by itself rather than being led by the manufacturer, it can obtain higher recycling subsidies from manufacturers due to higher bargaining power. Therefore, the increase in profit margin will stimulate the recycling enthusiasm of retail e-commerce, thereby a higher recycling rate.

**Proposition 8.** The profit of the manufacturer satisfies: $\pi^*_{Mm} > \pi^*_{Wm} > \pi^*_{Em}$.

**Proof.** For $\pi^*_{Mm} - \pi^*_{Wm} = \frac{a^2k^3 + 2ak^2\pi^2 + k\pi^4}{4(4ak - \pi^2)(2ak - \pi^2)} + \frac{kc^2}{72} > 0$, $\pi^*_{Wm} - \pi^*_{Em} = \frac{a^2k^3 + \left(7ak - 2\pi^2\right) + (D - ac_m)^2}{(4ak - \pi^2)^2(3ak - \pi^2)^2} + \frac{7kc^2}{144} > 0$, the Proposition 8 is established.

**Proposition 9.** The profit of the e-commerce satisfies: $\pi^*_{Ee} > \pi^*_{We} > \pi^*_{Me}$.

**Proof.** For $\pi^*_{We} - \pi^*_{Me} = \frac{k(D - ac_m)^2\left(7ak - \pi^2\right)\left(ak - \pi^2\right)}{8(3ak - \pi^2)^2(2ak - \pi^2)^2} + \frac{kc^2}{72} > 0$ and $\pi^*_{Ee} - \pi^*_{We} = \frac{a^2k^3(D - ac_m)^2}{2(4ak - \pi^2)(3ak - \pi^2)^2} + \frac{kc^2}{72} > 0$, the Proposition 9 is established.

**Proposition 10.** The profit of the supply chain satisfies: $\pi^*_C > \pi^*_W > \pi^*_M > \pi^*_E$.

**Proof.** For $\pi^*_M - \pi^*_E = \frac{k\pi^2\left(24ak - \pi^2\right)(D - ac_m)^2}{8(4ak - \pi^2)^2(2ak - \pi^2)} > 0$ *and* $\pi^*_W - \pi^*_M = \frac{k\pi^4 + 6\left(ak - \pi^2\right)(D - ac_m)^2}{8(2ak - \pi^2)(3ak - \pi^2)^2} + \frac{5kc^2}{144} > 0$, $\pi^*_C - \pi^*_W = \frac{a^2k^3(D - ac_m)^2}{2(2ak - \pi^2)(3ak - \pi^2)^2} > 0$, the Proposition 10 is established.

It follows from Proposition 8 and 9 that manufacturers and retail e-commerce can achieve the highest profit when themselves act as market leaders, however, if they are led by the other party, their profits will be greatly damaged. Observing the total profit of the supply chain, it can be found that no matter whether the market is dominated by either one, the overall profit cannot be optimized as the centralized decision, but the manufacturer oriented is better than the retailer in the decentralized system.

B. *Numerical Simulation*

In order to verify the rationality of above propositions and analyze the management implications deeply. This section simulates the example of waste product recycling and remanufacturing of an electronic product enterprise in China, and parameter settings are as follows $D = 100,\ c_m = 100,\ c_r = 5, a = 2,\ n = 1,\ \pi = 1, k = 1$ The simulation consists of three parts.

The first is to propose a profit coordination mechanism. The second performs parameter sensitivity analysis, which is to analyze the impact of consumer sensitivity on prices and service levels on decisions, market demand and profits, respectively. The third is to analyze the influence of manufacturer's unit cost savings from remanufacturing on recycling subsidy and market recycling price.

(1) *Profit coordination mechanism*

First, the results obtained by the example analysis are shown in Table 2. It can be seen that all the above propositions are verified. Second, through theoretical analysis and case verification, it can be seen that the decentralized decision-making system has been unable to achieve the optimal system profit regardless of who leads it, resulting in the whole system being passive. Therefore, an optimization mechanism should be designed to coordinate the profit distribution among supply chain members in order to achieve multi-party win-win.

Take the manufacturer-oriented market structure as an example to design a profit coordination mechanism. When the manufacturer acts as the leader, the increased system profit is:

$$U = \pi_C^* - \pi_M^* = \frac{k(D - ac_m)^2}{8(2ak - \pi^2)} + \frac{kc^2}{16} > 0 \tag{27}$$

Allocating profits under the coordination mechanism, assuming that the profit distributed by the manufacturer accounts for $t$ of $U$, and the retail e-commerce accounted for *(1 − t)*, the value of $t$ depends on the strength of both sides during negotiations.

Base on the profit coordination mechanism, the profit function of the manufacturer and retail e-commerce is:

$$\pi_m(t) = \pi_{Mm}^* + tU \tag{28}$$

**Table 2** Numerical simulation results

| Leader | $p^*$ | $f^*$ | $p_0^*$ | $w^*$ | $b^*$ | $Q^*$ | $\delta^*$ | $\pi_m^*$ | $\pi_e^*$ | $\pi^*$ |
|---|---|---|---|---|---|---|---|---|---|---|
| Manufacturer | 40 | 13 | 1.25 | 30 | 2.5 | 27 | 4.7% | 536 | 268 | 804 |
| Retail e-commerce | 44 | 11 | 1.25 | 21 | 3.7 | 23 | 5.5% | 262 | 460 | 722 |
| No dominance | 42 | 16 | 1.33 | 26 | 3.3 | 32 | 5.2% | 514 | 387 | 907 |

$$\pi_e(t) = \pi_{Me}^{*} + (1 - t)U \tag{29}$$

It can be seen that the party with strong bargaining power will always get a large proportion of the increased system profit, but regardless of any value that $t \in (0, 1)$, both companies can get better returns than decentralized decisions. Therefore, it is recommended that both parties should actively adopt a joint pricing strategy and distribute system profit rationally. Therefore, the profit of each member of the closed-loop supply chain will be increased.

(2) *Influence of a, π on equilibrium decisions and profits*

As shown by Fig. 2, we can summarize the following conclusions. First of all, profits of supply members and supply chain are monotonically increasing and decreasing with the increase of $\pi$ and $a$ respectively. Secondly, in terms of the respective profits of manufacturers and retail e-commerce, when the Nash game is played, the profits obtained by both parties are similar to those of the market dominated by themselves, but if the market is controlled by the other party, its own profits will be damaged greatly. Finally, observing the total profit of the supply chain, when the centralized decision is conducted by supply chain members and the system profit can be optimal.

Figure 3 shows that optimal wholesale price decision of the manufacturer is monotonically increasing with the increase of $a$ and decreasing with the increase of $\pi$ when the marker is oriented by the retail e-commerce or no dominance, however, it is independent of π when the manufacture acts as the market leader. Thus, the wholesale price in the market dominated by manufactures does not depend on the volatility of service sensitivity coefficient, which means that it is sufficient for the manufacture to make the optimal wholesale decision only by considering the price factor.

(3) *Influence of c on equilibrium decisions*

Note that from Fig. 4, it is observed that recycling subsidy and recycling price increase with unit remanufacturing cost savings $c$. In addition, the cost-saving benefits of remanufacturing will shift with the transfer of rights among supply chain members. First, when the manufacturer is a leader, retailers and consumers get very low profits, so the profit left to itself is relatively large. However, when the market leadership is transferred from the manufacturer to the retailer, $b$ will rise significantly, which means that retailers take use of their dominant rights to gain more profits, but at this time would not damage the consumer's income. Only when there is no dominance,$p_0$ will increase, thereby increasing consumer income.

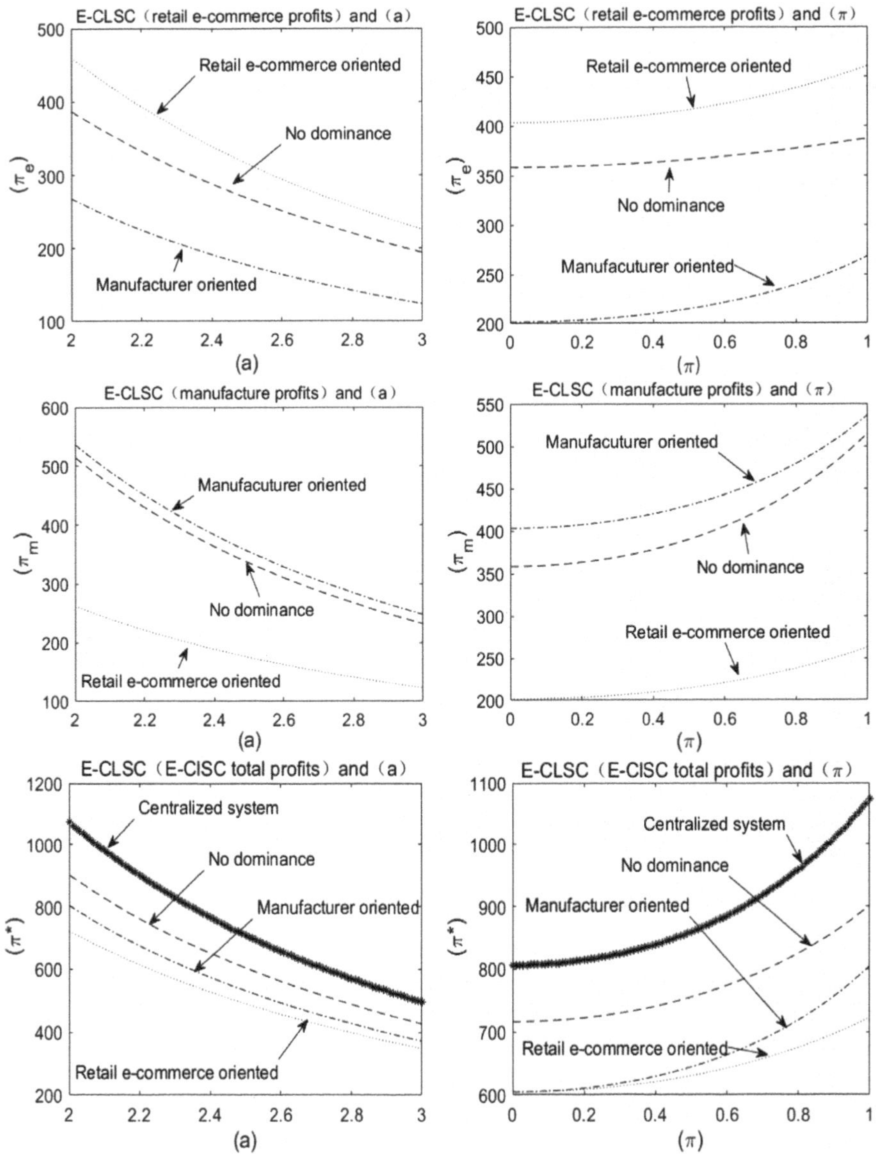


**Fig. 2** Relationship between $a$, $\pi$ and profits in E-CLSC

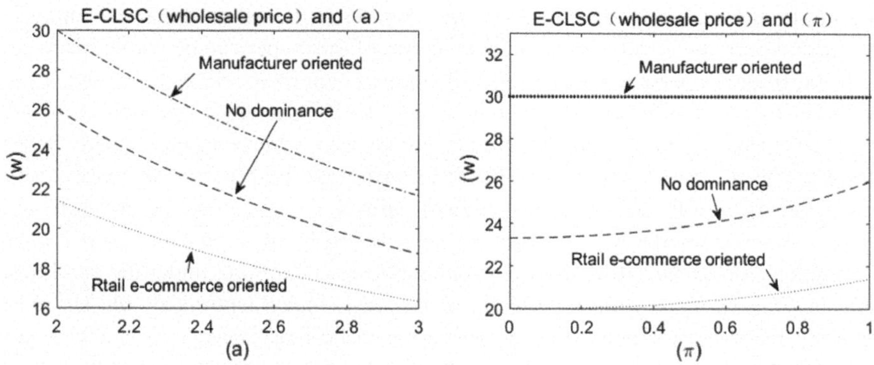


**Fig. 3** Relationship between $a$, $\pi$ and equilibrium decisions ($w$)

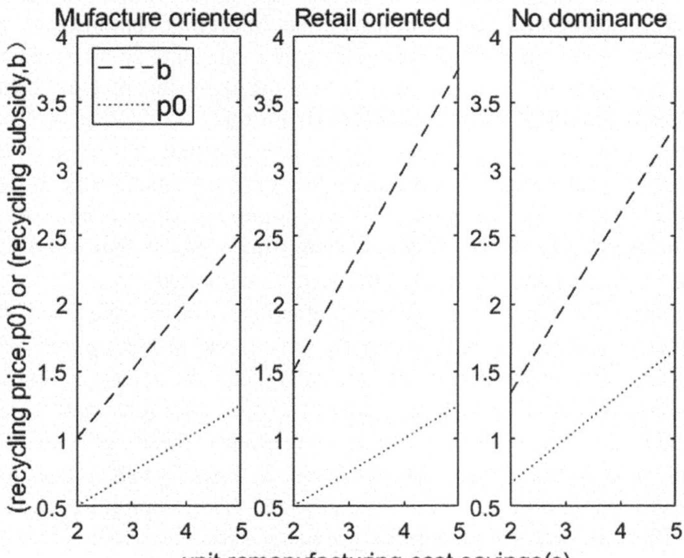


**Fig. 4** Relationship between $c$ and equilibrium decisions ($b$, $p_0$)

## 6 Conclusion

In this paper, we discuss the decision-making of the E-CLSC with recycling and remanufacturing plans based on the fact that the market demand is influenced by both product prices and service levels. In the case of real market competition, we consider centralized system and decentralized system respectively, and three kinds of market structure that is dominated by manufacturers, e-commerce enterprises and has no dominance in the decentralized are considered. Based on the above analysis and discussion, the conclusion and contribution of this paper can be summarized as:

First, from the perspective of profits of supply chain members and overall supply chain, this paper confirms that the centralized decision system is optimal, however, considering that joint decision making is not easy to achieve in reality, this paper designs a profit coordination mechanism to encourage both parties to make cooperative decisions. In addition, in the decentralized system, manufacturer-oriented market is more conducive to achieving optimal profit of the system, and manufacturers also maximize their profits simultaneously. Therefore, under the condition that the joint decision cannot be realized in reality, manufacturers should strive to improve their own competitiveness and strive to be market leaders, which is not only beneficial to themselves, but also to the overall profit of the supply chain. However, in order to stabilize the cooperative relationship, manufacturers must make up for the loss of profits of retail e-commerce to a certain extent. On the one hand, manufacturers can make up for the increase of recycling subsidies for retail e-commerce, which can also improve the recycling enthusiasm of retail e-commerce, thereby improving waste recycling rate and environmental performance, on the other hand, it can also be stabilized by the revenue cost sharing contract design to reduce the loss of retailers' profits.

Secondly, from the view of profit distribution, we find that the cost that manufacturers save from recycling and remanufacturing activities in the E-CLSC will be translates into the total benefits of manufacturers, retail e-commerce, and consumers. And the dominant party always obtains more profits, but when there is no dominance in the market, profits will be evenly distributed.

Finally, from the point of consumer preferences, on the one hand, the paper confirms that consumers are almost equally sensitive to product prices and service levels, on the other hand, consumers are very keen on a market dominated by manufacturers, because they will enjoy the lowest product price and better service level. In addition, the research work is significantly meaningful to the market orientation and recycling strategy development. In terms of market positioning, we find that consumers no longer focus solely on the price of products in E-SCs, but they are willing to pay a higher price for better services, which is an important instruction for E-SCs members about their decisions. As well as the important instruction for enterprises (especially manufactures) in recycling strategy development.

However, our research is restricted and there are three potential future extensions to improve model. One would be to analyze the competition among multiple manufacturers. The other potential direction is to consider the uncertainty associated with the market demand. The third is to study multi-period game.

## References


1. Agrawal, S., Singh, R. K., & Murtaza, Q. (2015). A literature review and perspectives in reverse logistics. *Resources, Conservation and Recycling, 97,* 76–92.
2. Rachih, H., Mhada, F. Z., & Chiheb, R. (2019). Meta-heuristics for reverse logistics: a literature review and perspectives. *Computers & Industrial Engineering, 127,* 45–62.
3. Werbach, K. (2000). Syndication–the emerging model for business in the Internet era. *Harvard Business Review, 78*(3), 84–93.
4. Siddiqui, A., & Raza, S. A. (2015). Electronic supply chains: status & perspective. *Computers & Industrial Engineering, 88,* 536–556.
5. Battini, D., Bogataj, M., & Choudhary, A. (2017). Closed loop supply chain (CLSC): Economics, modelling, management and control. *International Journal of Production Economics, 183,* 319–321.
6. Savaskan, R. C., & Wassenhove, L. V. (2006). Reverse channel design: The case of competing retailers. *Management Science, 52*(1), 1–14.
7. Giovanni, D. P., & Zaccour, G. (2014). A two-period game of a closed-loop supply chain. *European Journal of Operational Research, 232*(1), 22–40.
8. Wang, Y. Y. (2012). The closed-loop supply chain models analysis based on dual channel taking-backing government intervention. *Operations Research and Management Science, 21* (3), 250–255.
9. Miao, Z. W., et al. (2015). Models for closed-loop supply chain with trade ins. *Omega, 66,* 308–326.
10. Yao, F. M., & Teng, C. X. (2017). Decision models of closed-loop supply chain with dominant retailer considering fairness concern. *Control and Decision, 32,* 117–123.
11. Wang, Y. Y., & Li, J. (2018). Research on dominant models of E-CLSC based on network sale and recycle considering fairness concern. *Chinese Journal of Management Science, 26* (1), 139–151.
12. Kong, L. C., et al. (2017). Pricing and service decision of dual-channel operations in an O2O closed-loop supply chain. *Industrial Management & Data Systems, 117*(8), 1567–1588.
13. Ma, P., Shang, J., & Wang, H. Y. (2017). Enhancing corporate social responsibility: Contract design under information asymmetry. *Omega, 67,* 19–30.

# Intellectual Structure Detection Using Heterogeneous Data Clustering Method

**Yue Huang**

**Abstract** In recent years, with the widespread of the concept of big data, data mining has gradually been applied in different disciplines as well as bibliometrics. Bibliographic data analysis provides a way for researchers to know the intellectual structure of a specific area. However, most existing methods for bibliographic data analysis firstly compute the similarity matrix between authors or papers then conduct some kind of clustering method on the matrix, and they tend to ignore the external user-provided information. Enlightened by a newly proposed heterogeneous data clustering method, this paper proposed an algorithm called ISDHBD (Intellectual Structures Detection in Heterogeneous Bibliographic Data) which discovers intellectual structures from the perspective of heterogeneous networks. ISDHBD involves three main steps and takes papers from bibliographic data as the central objects while the other kinds of objects as attribute objects. Lastly, a running example extracted from CNKI (China National Knowledge Infrastructure) illustrated the principles and procedures of ISDHBD.

**Keywords** Data mining · Bibliographic data analysis · Clustering · Heterogeneous data · Intellectual structure

## 1 Introduction

In the era of "Internet+", big data has become the focus of attention. How to mine and use these massive information is the significance of scientists studying big data. Generally speaking, data mining refers to an engineered and systematic process of mining implicit and previously unknown but potentially useful information and patterns from large amounts of data. Through data mining, engineers can propose

Supported by BLCU Youth Talent Development Program.

Y. Huang (✉)
School of Information Science, Beijing Language and Culture University, Beijing, China
e-mail: huang.yuet@blcu.edu.cn



J. Zhang et al. (eds.), *LISS2019*,
https://doi.org/10.1007/978-981-15-5682-1_60

new algorithms and models, and medical institutions can develop new antibodies and drugs. We can say that data mining provides new ideas and methods for all disciplines as well as for bibliometrics.

Bibliometrics provides a way for researchers to know the intellectual structure or frontier of a certain area. Intellectual structure [1] is a network of clusters along with the relations, which is obtained by clustering through the building of a bibliographic coupling matrix based on the analysis of bibliographies. Each cluster corresponds to one research subfield (also known as topics). Whereas, the concept of the frontier of the discipline has been revised and enriched by other scholars since it was introduced by Price in 1965. Price believed that the research frontier was time-varying, that is, the research frontier changed with time [2]. For the discipline field, the process of changes in the frontier of research basically represents the development process of this discipline. There are many concepts related to the research frontier, such as emerging trends, emerging topics, and research hotspots. The concept of emerging trends was proposed by A. Kontostathis in 2003, which refers to the subject areas that have gradually attracted interests of people over time and are being discussed by more and more researchers [3]. The concept of emerging topics was proposed in 2002, which refers to a set of emerging subject areas represented by multiple keywords or phrases in a particular scientific research field. It represents the most promising research directions or trends in the field of discipline research [4]. In this paper, we believe that by detecting the intellectual structures of a specific area with different time periods, we can easily get the research hotspots.

The identification methods of the intellectual structure are roughly divided into two categories which are qualitative research method and quantitative research method. The qualitative research method is relatively mature and the quantitative research method is still developing and improving. The methods for quantitative bibliographic data analysis fall roughly into three categories, co-citation analysis [5, 6], coupling analysis [7, 8] and co-words analysis [9, 10]. And there have been used in various disciplines to help discover intellectual structure, such as information science, library science, knowledge management, supply chain management [11]. However, the conventional methods for bibliographic data analysis usually involves two steps, i.e., they first calculate the similarity matrix between authors or papers based on different quantitative bibliographic data analysis methods, then some kind of clustering method is conducted on this matrix to obtain the intellectual structure. Hence, we can see that, detecting the intellectual structure of a specific area is in essence the same as the process of clustering. By further thought, the bibliographic data is composed of heterogeneous kinds of objects, such as papers, authors, venues and keywords.

The aim of this paper is to apply heterogeneous data clustering methods directly to bibliographic data analysis, with a purpose to explore the possibility of using heterogeneous data clustering methods to detect intellectual structure.

## 2 Method

Heterogeneous information networks [12], which consist of multi-typed objects and links, are ubiquitous in the real world. Clustering on heterogeneous networks directly is of great importance [13], since it loses less information than heterogeneous-transformed homogeneous networks clustering.

Currently, most clustering methods for heterogeneous networks cluster objects based on internal information of the dataset. However, sometimes the object information provided by a dataset does not suffice to provide effective information for clustering in actual scenarios. This is quite similar to the thesaurus construction in bibliographic data analysis where it is common to combine two synonyms words before conduct bibliometric methods. Thus, in this paper, we use the newly proposed algorithm Heterogeneous Data Clustering Considering Multiple User-provided Constraints (UserHeteClus) [14] which consider the auxiliary external information to detect intellectual structure. In a star-structured heterogeneous networks, there has to be a key kind of objects. Apparently, in bibliographic data, the central objects are papers and all the other kinds of objects are attribute objects. Hence, the followings are some definitions.

**Definition 1: User Constraint on Object Relation in Heterogeneous Bibliographic Data.** For the given heterogeneous bibliographic data of DB = (P, A), assume that the user-provided constraint is on the objects of paper P = {Pi | i = 1, 2, …, n} and that the objects of paper eventually cluster as the set of Clus_p that contains N objects of paper clusters, described as P.Clus = {Clus1, Clus2, …, ClusN}. Suppose that there are two papers, i.e., Ps and Pt (s $\neq$ t). The must-link constraint and the cannot-link constraint of the user-provided constraint on object relation can be specifically described as follows:

(1) If Ps and Pt should be in the same cluster, then Must-link (Ps, Pt) = True, which can be specifically described as Ps $\in$ Clusi, Pt $\in$ Clusj, i = j;
(2) If Ps and Pt should not be in the same cluster, then Cannot-link (Ps, Pt) = True, which can be specifically described as Ps $\in$ Clusi, Pt $\in$ Clusj, i $\neq$ j.

Similarly, as illustrated in UserHeteClus [14], the must-link constraint and the cannot-link constraint are both Boolean functions and have the following properties. That is the must-link constraint and the cannot-link constraint of the user's two types of constraints have symmetry. And the must-link constraint and the cannot-link constraint of the user's two types of constraints have limited transitivity.

**Definition 2: User Constraint on Decisive Attributes in Heterogeneous Bibliographic Data.** For the given heterogeneous bibliographic data of $D_B = (P, A)$, assume that the user-provided constraint is for the objects of paper $P = \{P_i \mid i = 1, 2, \ldots, n\}$, that there are two papers, i.e., $P_s$ and $P_t$ ($s \neq t$), and that the objects of paper eventually cluster as the set of *Clus_p* that contains *N* central object clusters, described as $P.\ Clus = \{Clus_1, Clus_2, \ldots, Clus_N\}$. Suppose that there are two paper objects, i.e., $P_s$ and $P_t$ ($s \neq t$), and that the cluster attributions of $P_s$ and $P_t$ are represented by $P_s$. *Clus* and $P_t$. *Clus*, respectively. Then, a user-provided constraint on paper attributes can be described as $P_s.\ A_{kp} = P_t.\ A_{kp} \Leftrightarrow P_s.\ Clus = P_t.\ Clus$, in which $A_k$ is denoted as the decisive attribute.

**Definition 3: User Constraint on Attribute Value Equality in Heterogeneous Bibliographic Data.** For the given heterogeneous bibliographic data of $D_B = (P, A)$, assuming that the user-provided constraint is for the objects of paper $P = \{P_i \mid i = 1, 2, \ldots, n\}$ and that there are two papers, i.e., $P_s$ and $P_t$ ($s \neq t$), the user-provided constraint on attribute value equality can be described as $A_{kp} = A_{kq}$ ($p \neq q$).

**Definition 4: Similarity between Objects of Papers in terms of the *k*th Type of Attribute Object in Heterogeneous Bibliographic Data.** For the given heterogeneous bibliographic data of $D_B = (P, A)$, the calculation formula of the similarity, $S_k$ ($P_i$, $P_j$), between two papers, i.e., $P_i$ and $P_j$ in terms of the *k*th type of attribute object is as follows:

$$S_k(P_i, P_j) = \frac{2 \times |P_i.A_k.ID \cap P_j.A_k.ID|}{|P_i.A_k.ID| + |P_j.A_k.ID|} \quad (1)$$

where $P_i.\ A_k.\ ID$ represents the collection of IDs of the *k*th type of attribute object correlated to $P_i$ and $P_j.\ A_k.\ ID$ represents the collection of IDs of the *k*th type of attribute object correlated to $P_j$.

**Definition 5: Similarity between Objects of Paper in Heterogeneous Bibliographic Data.** For the given heterogeneous bibliographic data of $D_B = (P, A)$, the similarity, $S$ ($P_i$, $P_j$), between two papers $P_i$ and $P_j$, is the linear combinations of similarities between the two in terms of the $k$ th type of attribute object, with the following calculation formula:

$$S(P_i, P_j) = \sum\nolimits_{k=1}^{r} w_k \cdot S_k(P_i, P_j) \quad (2)$$

where $w_k$ is called the contribution coefficient, representing the contribution of the *k*th type of attribute object to the judgment of whether $P_i$ and $P_j$ are similar and satisfying: (1) $\sum\nolimits_{k=1}^{r} w_k = 1$ and (2) $w_k \geq 0$, $w_k \in R$, which jointly constitute the contributing coefficient vector $w = (w_1, w_2, w_k, \ldots, w_r)$.

**Definition 6: Similarity between Paper Clusters in Heterogeneous Bibliographic Data.** For the given heterogeneous bibliographic data of $D_B = (P, A)$ and paper clusters *Clus*, the similarity between paper object clusters $Clus_s$ and $Clus_t$, $S$ ($Clus_s$, $Clus_t$), the average of the similarities between papers contained in one cluster and those contained in another cluster, has the following formula:

$$S(Clus_s, Clus_t) = \frac{\sum_{i=1}^{|Clus_s|} \sum_{j=1}^{|Clus_t|} S(P_i, P_j)}{|Clus_s| \times |Clus_t|} \quad (3)$$

where S (Pi, Pj) represents the similarity between Pi and Pj.

Obviously, the value range of all these similarities is [0, 1], and it is easy to prove that it meets the three properties of similarity measurement.

Procedures for intellectual structures detection in heterogeneous bibliographic data (ISDHBD) are depicted as follows:

(1) Entity resolution of all kinds of objects in heterogeneous bibliographic data.
(2) Similarity measurement between papers.
(3) Paper clustering considering user-provided constraint.

## 3 Experimental Results

In this section, we use a running example to illustrate the principle and procedures of the proposed method ISDHBD.

### *3.1 Dataset*

In this section, we use an experimental dataset comes from CNKI (http://www.cnki.net/) to illustrate the principles and main procedures of ISDHBD as shown in Table 1. It contains 54 objects of four kinds, in which papers account for 8, authors account for 23, venues account for 8, keywords account for 33. Then, it is constructed as a star-structured heterogeneous network. In addition, the user-provided constraints are: the papers "Improved method of CABOSFV" and "High dimensional sparse data clustering based on sorting idea" belong to Must-link, while the keywords "CABOSFV algorithm". "CABOSFV" and "CABOSFV clustering" actually represent the same meaning.

### *3.2 Entity Resolution of All Kinds of Objects*

Firstly, papers, authors, venues and keywords are detected according to the names of all kinds of objects, which are shown in Tables 2, 3, 4, and 5.

**Table 1** Bibliographic data for ISDHBD

| ID | Title | Attributes | Attribute objects |
|---|---|---|---|
| 1 | Survey on feature dimension reduction for high-dimensional data | Author | Hu Jie |
| | | Venue | Application Research of Computers |
| | | Keywords | dimension reduction; machine learning; feature selection; feature abstraction; selection criteria |
| 2 | Research on methods of dimensionality reduction in high-dimensional data | Author | Yu Xiaosheng; Zhou Ning |
| | | Venue | Information Science |
| | | Keywords | high-dimensional data; dimension reduction; MDS; Isomap; LLE |
| 3 | The method of how to determine the threshold value of set dissimilarity in CABOSFV algorithm | Author | Song Yan; Xiao Qian |
| | | Venue | Ship Science and Technology |
| | | Keywords | clustering; CABOSFV algorithm; set dissimilarity; threshold |
| 4 | Improved method of CABOSFV | Author | Yin Jia; Wu Sen; Wang Shacheng |
| | | Venue | Information Research |
| | | Keywords | CABOSFV; sparse dissimilarity; clustering |
| 5 | Clustering algorithm based on set dissimilarity for high dimensional data of categorical attributes | Author | Wu Sen; Wei Guiying; Bai Chen; Zhang Guiqiong |
| | | Venue | Journal of University of Science and Technology Beijing |
| | | Keywords | clustering; high-dimensional; set; dissimilarity; data mining |
| 6 | High dimensional sparse data clustering based on sorting idea | Author | Zhu Qin; Gao Xuedong; Wu Sen; Chen Min; Chen Hua |
| | | Venue | Computer Engineering |
| | | Keywords | high dimensional sparse data; CABOSFV clustering; sorting |
| 7 | Application of interesting subspace mining algorithm in high-dimensional data clustering | Author | Yang Yin; Han Zhongming; Yang Lei |
| | | Venue | Computer Engineering |
| | | Keywords | interesting subspace; high-dimensional data; clustering; data mining |
| 8 | Improvement method of dimensionality reduction through mining density information | Author | Jia Hongzhe; Yan Deqin; Zhang Yan |
| | | Venue | Application Research of Computers |
| | | Keywords | manifold learning; dimension reduction; clustering; diffusion maps |

**Table 2** New ID for papers

| Title | ID |
|---|---|
| Survey on feature dimension reduction for high-dimensional data | P1 |
| Research on methods of dimensionality reduction in high-dimensional data | P2 |
| The method of how to determine the threshold value of set-square-difference in CABOSFV algorithm | P3 |
| Improved method of CABOSFV | P4 |
| Clustering algorithm based on set dissimilarity for high dimensional data of categorical attributes | P5 |
| High dimensional sparse data clustering based on sorting idea | P6 |
| Application of interesting subspace mining algorithm in high-dimensional data clustering | P7 |
| Improvement method of dimensionality reduction through mining density information | P8 |

**Table 3** New ID for authors

| Author name | ID | Author name | ID |
|---|---|---|---|
| Hu Jie | A1 | Zhu Qin | A12 |
| Yu Xiaosheng | A2 | Gao Xuedong | A13 |
| Zhou Ning | A3 | Chen Min | A14 |
| Song Yan | A4 | Chen Hua | A15 |
| Xiao Qian | A5 | Yang Ying | A16 |
| Yin Jia | A6 | Han Zhongming | A17 |
| Wu Sen | A7 | Yang Lei | A18 |
| Wang Shacheng | A8 | Jia Hongzhe | A19 |
| Wei Guiying | A9 | Yan Deqin | A20 |
| Bai Chen | A10 | Zhang Yan | A21 |
| Zhang Guiqiong | A11 | – | – |

**Table 4** New ID for venues

| Venue name | ID |
|---|---|
| Application Research of Computers | V1 |
| Information Science | V2 |
| Ship Science and Technology | V3 |
| Information Research | V4 |
| Journal of University of Science and Technology Beijing | V5 |
| Computer Engineering | V6 |

**Table 5** New ID for keywords

| Keywords | ID | Keywords | ID |
|---|---|---|---|
| Dimension reduction | T1 | CABOSFV | T14 |
| Machine learning | T2 | Sparse dissimilarity | T15 |
| Feature selection | T3 | High-dimensional | T16 |
| Feature abstraction | T4 | Set | T17 |
| Selection criteria | T5 | Dissimilarity | T18 |
| High-dimensional data | T6 | Data mining | T19 |
| MDS | T7 | High dimensional sparse data | T20 |
| Isomap | T8 | CABOSFV clustering | T21 |
| LLE | T9 | Sorting | T22 |
| Clustering | T10 | Interesting subspace | T23 |
| CABOSFV algorithm | T11 | Manifold learning | T24 |
| Sparse set dissimilarity | T12 | Diffusion maps | T25 |
| Threshold | T13 | – | – |

## 3.3 Similarity Measurement between Papers

Since papers are central objects in heterogeneous bibliographic data, in this section we compute the similarities between papers.

(1) *Papers represented by attribute objects*

Firstly, papers are represented by other kinds of objects, namely authors, venues, and keywords (Table 6).

(2) *Update the description of papers based on constraints*

According to the user-provided constraints where the keywords "CABOSFV algorithm". "CABOSFV" and "CABOSFV clustering" actually represent the same meaning, we update the papers representation as Table 7.

**Table 6** Papers represented by attribute objects

| ID | P_ID | A_ID | V_ID | K_ID |
|---|---|---|---|---|
| 1 | P1 | A1 | V1 | T1, T2, T3, T4, T5 |
| 2 | P2 | A2, A3 | V2 | T6, T1, T7, T8, T9 |
| 3 | P3 | A4, A5 | V3 | T10, T11, T12, T13 |
| 4 | P4 | A6, A7, A8 | V4 | T14, T15, T10 |
| 5 | P5 | A7, A9, A10, A11 | V5 | T10, T16, T17, T18, T19 |
| 6 | P6 | A12, A13, A7, A14, A15 | V6 | T20, T21, T22 |
| 7 | P7 | A16, A17, A18 | V6 | T23, T6, T10, T19 |
| 8 | P8 | A19, A20, A21 | V1 | T24, T1, T10, T25 |

**Table 7** Updated papers represented by attribute objects

| ID | P_ID | A_ID | V_ID | K_ID |
|---|---|---|---|---|
| 1 | P1 | A1 | V1 | T1, T2, T3, T4, T5 |
| 2 | P2 | A2, A3 | V2 | T6, T1, T7, T8, T9 |
| 3 | P3 | A4, A5 | V3 | T10, T11, T12, T13 |
| 4 | P4 | A6, A7, A8 | V4 | T11, T15, T10 |
| 5 | P5 | A7, A9, A10, A11 | V5 | T10, T16, T17, T18, T19 |
| 6 | P6 | A12, A13, A7, A14, A15 | V6 | T20, T11, T22 |
| 7 | P7 | A16, A17, A18 | V6 | T23, T6, T10, T19 |
| 8 | P8 | A19, A20, A21 | V1 | T24, T1, T10, T25 |

**Table 8** Similarity between papers

| $C_i$ | $C_j$ | $S_{\text{author}}$ | $S_{\text{venue}}$ | $S_{\text{keyword}}$ | $S$ |
|---|---|---|---|---|---|
| P1 | P2 | 0.000 | 0.000 | 0.200 | 0.140 |
| P1 | P3 | 0.000 | 0.000 | 0.000 | 0.000 |
| P1 | P4 | 0.000 | 0.000 | 0.000 | 0.000 |
| P1 | P5 | 0.000 | 0.000 | 0.000 | 0.000 |
| P1 | P6 | 0.000 | 0.000 | 0.000 | 0.000 |
| P1 | P7 | 0.000 | 0.000 | 0.000 | 0.000 |
| P1 | P8 | 0.000 | 1.000 | 0.222 | 0.256 |
| P2 | P3 | 0.000 | 0.000 | 0.000 | 0.000 |
| P2 | P4 | 0.000 | 0.000 | 0.000 | 0.000 |
| P2 | P5 | 0.000 | 0.000 | 0.000 | 0.000 |
| P2 | P6 | 0.000 | 0.000 | 0.000 | 0.000 |
| P2 | P7 | 0.000 | 0.000 | 0.222 | 0.156 |
| P2 | P8 | 0.000 | 0.000 | 0.222 | 0.156 |
| P3 | P4 | 0.000 | 0.000 | 0.571 | 0.400 |
| P3 | P5 | 0.000 | 0.000 | 0.222 | 0.156 |
| P3 | P6 | 0.000 | 0.000 | 0.286 | 0.200 |
| P3 | P7 | 0.000 | 0.000 | 0.250 | 0.175 |
| P3 | P8 | 0.000 | 0.000 | 0.250 | 0.175 |
| P4 | P5 | 0.286 | 0.000 | 0.250 | 0.232 |
| P4 | P6 | 0.250 | 0.000 | 0.333 | 0.283 |
| P4 | P7 | 0.000 | 0.000 | 0.286 | 0.200 |
| P4 | P8 | 0.000 | 0.000 | 0.286 | 0.200 |
| P5 | P6 | 0.222 | 0.000 | 0.000 | 0.044 |
| P5 | P7 | 0.000 | 0.000 | 0.444 | 0.311 |
| P5 | P8 | 0.000 | 0.000 | 0.222 | 0.156 |
| P6 | P7 | 0.000 | 1.000 | 0.000 | 0.100 |
| P6 | P8 | 0.000 | 0.000 | 0.000 | 0.000 |
| P7 | P8 | 0.000 | 0.000 | 0.250 | 0.175 |

(3) *Calculate S($P_i$, $P_j$) according to (1) and (2).*

Repeat calculating to get all similarities (Table 8). Take $S$ ($P_1$, $P_2$) as an example.

$$S_{Author}(P1,P2) = \frac{2 \times |\phi|}{|\{A1\}| + |\{A2,A3\}|} = 0$$

$$S_{Venue}(P1,P2) = \frac{2 \times |\phi|}{|\{V1\}| + |\{V2\}|} = 0$$

$$S_{Keyword}(P1,P2) = \frac{2 \times |\{T1\}|}{|\{T1,T2,T3,T4,T5\}| + |\{T6,T1,T7,T8,T9\}|} = \frac{2}{10}$$

If $w = (w_{Author}, w_{Venue}, w_{Keyword}) = (0.2, 0.1, 0.7)$, then

$$S(P1,P2) = w_{Author} \times S_{Author}(P1,P2) + w_{Venue} \times S_{Venue}(P1,P2) + w_{Keyword} \times S_{Keyword}(P1,P2) = 0.2 \times 0 + 0.1 \times 0 + 0.7 \times 2/10 = 0.140$$

## 3.4 Paper Clustering Considering User-provided Constraint

(1) *Sort the similarity between papers.*

As shown in Table 9, the similarity evaluation between papers is 0.125, thus the threshold could be 0.130.

**Table 9** Descended order of similarity between papers

| $P_i$ | $P_j$ | *Similarity* | $P_i$ | $P_j$ | *Similarity* |
|---|---|---|---|---|---|
| P3 | P4 | 0.400 | P5 | P8 | 0.156 |
| P5 | P7 | 0.311 | P1 | P2 | 0.140 |
| P4 | P6 | 0.283 | P6 | P7 | 0.100 |
| P1 | P8 | 0.256 | P5 | P6 | 0.044 |
| P4 | P5 | 0.232 | P1 | P3 | 0.000 |
| P3 | P6 | 0.200 | P1 | P4 | 0.000 |
| P4 | P7 | 0.200 | P1 | P5 | 0.000 |
| P4 | P8 | 0.200 | P1 | P6 | 0.000 |
| P3 | P7 | 0.175 | P1 | P7 | 0.000 |
| P3 | P8 | 0.175 | P2 | P3 | 0.000 |
| P7 | P8 | 0.175 | P2 | P4 | 0.000 |
| P2 | P7 | 0.156 | P2 | P5 | 0.000 |
| P2 | P8 | 0.156 | P2 | P6 | 0.000 |
| P3 | P5 | 0.156 | P6 | P8 | 0.000 |

(2) *Take each paper as a single cluster.*

Currently, {{P1}, {P2}, {P3}, {P4}, {P5}, {P6}, {P7}, {P8}}.

(3) *Combine the papers in Must-link.*

According to the user-provided constraints, the papers "Improved method of CABOSFV" and "High dimensional sparse data clustering based on sorting idea" belong to Must-link. Currently, {{P1}, {P2}, {P3}, {P4, P6}, {P5}, {P7}, {P8}}.

(4) *Compare the similarity between papers to cluster.*

P3 and P4 have the biggest similarity (0.400) and they are not in the same cluster, then calculate: (S (P3, P4) + S (P3, P6)) ÷ 2 = (0.400 + 0.200) ÷ 2 = 0.300 > 0.130, then they should be combined. Currently, {{P1}, {P2}, {P3, P4, P6}, {P5}, {P7}, {P8}}.

P5 and P7 have the biggest similarity (0.311) and they each is a single cluster, then they should be combined. Currently, {{P1}, {P2}, {P3, P4, P6}, {P5, P7}, {P8}}.

P4 and P6 have the biggest similarity (0.283), they are already in the same cluster.

P1 and P8 have the biggest similarity (0.256) and they each is a single cluster, then they should be combined. Currently, {{P1, P8}, {P2}, {P3, P4, P6}, {P5, P7}}.

P4 and P5 have the biggest similarity (0.232) and they are not in the same cluster, then calculate: (S(P3, P5) + S(P3, P7) + S(P4, P5) + S(P4, P7) + S(P5, P6) + S(P5, P7))/6 = (0.156 + 0.175 + 0.232 + 0.200 + 0.044 + 0.100)/6 = 0.151 > 0.130, then they should be combined. Currently, {{P1, P8}, {P2}, {P3, P4, P6, P5, P7}}.

P3 and P6 have the biggest similarity (0.200) and they are already in the same cluster.

P4 and P7 have the biggest similarity (0.200) and they are already in the same cluster.

P4 and P8 have the biggest similarity (0.200) and they are not in the same cluster, then calculate: (S(P1, P3) + S(P1, P4) + S(P1, P5) + S(P1, P6) + S(P1, P7) + S(P3, P8) + S(P4, P8) + S(P5, P8) + S(P6, P8) + S(P7, P8))/12 = (0 + 0 + 0 + 0 + 0 + 0 + 0.175 + 0.200 + 0.156 + 0 + 0.175)/12 < 0.130, thus they should not be combined.

Currently, {{P1, P8}, {P2}, {P3, P4, P6, P5, P7}}.

P2 and P8 have the biggest similarity (0.156) and they are not in the same cluster, then calculate: (S (P1, P2) + S (P2, P8))/2 = (0.140 + 0.156)/2 = 0.148 > 0.130, then they should be combined.

Thus, {{P1, P8, P2}, {P3, P4, P6, P5, P7}}.

Lastly, all papers are in one cluster.

It is the same as in hierarchical clustering methods, ISBHD also get a dendrogram tree. In this example, we take {{P1, P8, P2}, {P3, P4, P6, P5, P7}} as the intellectual structure detection result (shown in Fig. 1). We could see that it takes

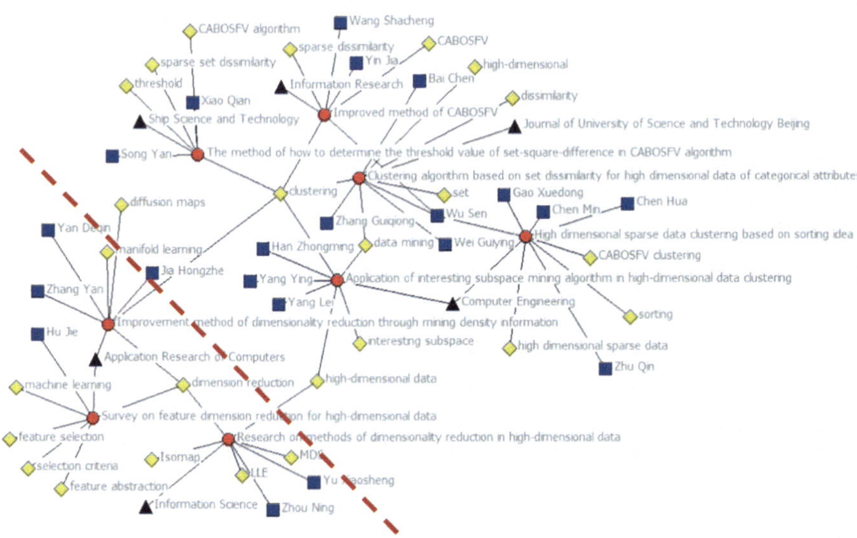


**Fig. 1** Intellectual structure detection result of the example

"Survey on feature dimension reduction for high-dimensional data", "Research on methods of dimensionality reduction in high-dimensional data" and "Improvement method of dimensionality reduction through mining density information" as a cluster and the others as another cluster, which is consistent with human judgement.

## 4 Conclusions

Bibliographic data analysis can detect intellectual structure of a specific area and they are heterogeneous. This paper proposed an algorithm called ISDHBD to directly cluster the multi-typed heterogeneous bibliographic data to obtain the hidden intellectual structure. ISDHBD takes "papers" in bibliographic data as the central objects and all the other kinds (such as authors, venues, keywords) as the attribute objects. Firstly, it needs the identification number of all kinds of objects in the heterogeneous bibliographic data. Then it calculates the similarities between the central object papers. Lastly, it clusters papers using external information provided by users. A running example extracted from CNKI has shown the working procedures and effectiveness of ISDHBD. For future study, large scale analysis of bibliographic data will further testify the effectiveness of ISDHBD.

## References

1. Huang, Y. (2018). Intellectual structure of research on data mining using bibliographic coupling analysis. In *Proceedings of Eighth IEEE International Conference on Logistics, Informatics and Service Science, Beijing* (pp. 205–209).
2. Price, D. J. D. S. (1965). Networks of scientific papers. *Science, 149*(3638), 510–515.
3. Kontostathis, A., Galitsky, L. M., Pottenger, W. M., Roy, S., & Phelps, D. J. (2004). *Survey of text mining* (pp. 185–224). New York: Springer.
4. Matsumura, N., Matsuo, Y., & Ohsawa, Y. (2002). Discovering emerging topics from WWW. *Journal of Contingencies and Crisis Management, 10*(2), 72–81.
5. Small, H. (1973). Co-citation in the scientific literature: A new measure of the relationship between two documents. *Journal of the Association for Information Science and Technology, 24*(4), 265–269.
6. White, H. D., & Griffith, B. C. (1981). Author cocitation: A literature measure of intellectual structure. *Journal of the American Society for Information Science, 32*(3), 163–171.
7. Kessler, M. M. (1963). Bibliographic coupling between scientific papers. *American Documentation, 14*(1), 10–25.
8. Weinberg, B. H. (1974). Bibliographic coupling: a review. *Information Storage & Retrieval, 10*(5), 189–196.
9. Morris, S. A., & Yen, G. G. (2004). Crossmaps: Visualization of overlapping relationships in collections of journal papers. In *Proceedings of National Academy of Sciences of the USA* (vol. 101, pp. 5291–5296).
10. Bhattacharya, S., & Basu, P. K. (1998). Mapping a research area at the micro level using co-word analysis. *Scientometrics, 43*(3), 359–372.
11. Fahimnia, B., Sarkis, J., & Davarzani, H. (2015). Green supply chain management: A review and bibliometric analysis. *International Journal of Production Economics, 162,* 101–114.
12. Sun, Y., Yu, Y., & Han, J. (2009). Ranking-based clustering of heterogeneous information networks with star network schema. In *Proceedings 15th ACM SIGKDD International Conference on Knowledge Discovery and Data Mining, Paris* (pp. 797–806).
13. Huang, Y. (2014). Clustering on heterogeneous networks. *Wiley Interdisciplinary Reviews: Data Mining and Knowledge Discovery, 4*(3), 213–233.
14. Huang, Y. (2019). Heterogeneous data clustering considering multiple user-provided constraints. *International Journal of Computers Communications & Control, 14*(2), 170–182.

# Improving the Utilization Rate of Intelligent Express Cabinet Based on Policy Discussion and Revenue Analysis

**Erkang Zhou**

**Abstract** In this paper, we have studied how to reduce the service time of the intelligent express cabinet service system and improve the utilization rate of intelligent express cabinet. By considering the intelligent express cabinet service system as a queuing system, the current status of the use of intelligent express cabinet is analyzed. Then we have conducted policy discussions, collected user-fetched pick-up time data, and finally performed revenue analysis. It is concluded that under the dual policy of reward and punishment, the utilization rate of the intelligent express cabinet can be improved without increasing the cost, and suggestions are given to the operator of the express cabinet.

**Keywords** Queueing theory · Intelligent express cabinet · Policy discussion · Revenue analysis

## 1 Introduction

The emergence of intelligent express cabinets can help the recipients keep the express in a short time, which is safe and reliable. It also solves the troubles of many parties and brings convenience to everyone hence more and more customers tend to use it. As for the recipients, they can enjoy a simple self-service experience, while from the perspective of the courier, it saves the time for the courier to wait for the recipient to take the express and improve the work efficiency. As for the courier company, the cost of communication between the courier and the recipient is reduced and the difficulty of recruiting the courier is solved. Finally, centralized delivery is realized, which is convenient, fast and effective.

With the development of intelligent express cabinets, there are naturally some problems. After the express is put into the intelligent express cabinet, some customers can take it away as soon as possible, but sometimes it takes a long time for

---

E. Zhou (✉)
School of Economics and Management, Beijing Jiaotong University, Beijing, China
e-mail: 2601265199@qq.com



J. Zhang et al. (eds.), *LISS2019*,
https://doi.org/10.1007/978-981-15-5682-1_61

them to take it away, others are even too lazy to take it. As a result, there are not enough cabinets available after the arrival of the next batch of express, while manual delivery will increase the delivery cost. Increasing the number of express cabinets to meet the demand of customers is practicable, but the cost of facilities is relatively large. The cost of one intelligent express cabinet is about 30,000 yuan, and the operational expenses of each express cabinet are about 10,000 yuan per year including electricity, network, maintenance, property charges.

In addition, as the number of express cabinets increases, the relative utilization rate will decrease. It is also feasible to reduce the time of the express being in the express cabinet by promoting the customer to pick it up, thereby improving the utilization rate of the intelligent cabinet.

Correspondingly, it also brings certain costs, such as the cost of providing incentives to customers.

Therefore, the main problem at present is to weigh the increased utilization rate against the cost paid after providing a certain policy. In theory, it is a trade-off problem.

This paper aims to study how to improve the utilization rate of intelligent express cabinet by facilitating customers' pick-up. The process of accessing express through the intelligent express cabinet is regarded as a queuing system. It is studied that how to improve the efficiency of the intelligent express cabinet by policy discussion and revenue analysis, based on the queuing theory model.

## 2 Previous Work

With the rise of intelligent express cabinets, there are many studies focusing on intelligent express cabinets. In the research on the concept and development status of intelligent express cabinets, Shang introduced the definition and operation flow of intelligent express cabinets, and analyzed intelligent express cabinet operation mode and existing problems in China [1]. Reference [2] introduced the current research and application status of intelligent express cabinets, put forward the application and advantages of intelligent express cabinet self-service and the existing problems of intelligent express self-service at the present stage, and suggestions for its development. Reference [3] proposed to develop standards for intelligent express cabinets and charge a certain management fee for overdue items, and guide consumers to make rational use of resources such as public facilities through the analysis of the status quo of intelligent express cabinets.

In the research of intelligent express cabinet layout planning, Morganti, Dablanc and Fortin introduced the network planning of the French pickup point, analyzed the influence of transportation nodes and population density on the delivery of pickup points, and pointed out that France is a country with the highest usage of pickup point in Europe [4]. Reference [5] studied the layout planning of intelligent express cabinets, using fuzzy data analysis, operations research, transportation planning theory and other methods, combined with logistics, traffic engineering and other subjects. Reference [6] proposed a location model for pickup point based on cost, and demonstrated the feasibility of the model by cost estimation.

In the research on the size of the intelligent express cabinet, Shang took the size ratio of the grid of the express cabinet as the research direction, and qualitatively analyzed the differences in geographical location and population structure of different regions according to the market segmentation theory. And then it conducted surveys on couriers, collected data on the size of express and conducted cluster analysis to study the size ratio of express cabinets in different regions [7].

In the research on system designing of intelligent express cabinets, Wang and Huang designed a new intelligent express terminal based on Android and STM32, which can realize 24-h self-service mailing, self-service dispatch and self-service pickup. The system realizes automatic access of the express through the fully automatic mechanical system, optimizes the storage of the express through the optimized space allocation algorithm, which can improve the utilization rate of intelligent express cabinet [8]. Reference [9] focused on the analysis of the overall system structure of intelligent express cabinet system and decomposed the functions of intelligent express cabinet terminals and servers, designed the terminal system of intelligent express cabinet by selecting Smart210 board as the target hardware platform, which is based on the ARM Cortex-A8 core. Reference [10] designed an intelligent express cabinet that can provide diversified services based on the "Internet+" information service platform according to the analysis of the current market demand and usage of intelligent express cabinet.

In the "Attended" and "Unattended" studies of express delivery, Punakivi, Yrjola, and Holmstrom proposed a "centralized point of delivery" (collection and delivery point, CDP) model to solve the last mile issue [11]. Reference [12] comparatively analyzed the cost of two "centralized delivery point" CDP models and found that unattended delivery points could save nearly one-third of the logistics costs compared to attended, and small packages using "centralized delivery point" CDP are more efficient in delivering, while the implementation of unattended delivery points (unattended CDP) is difficult.

In related research on identified service systems, zhao studied the frame and the message delivery of the system based on the Queueing Theory, Second, the modified $M/E_k/1$ model has been built to analyze the performance of the email system when the email message delivery system is conforming to the Queueing Theory [13]. Reference [14] presented an analysis for an $M/E_k/1$ queuing system with balking and state-dependent service, customers are served with two different rates depending on the number of customers in the system, entering the queue or balking with a constant probability, then obtained the equilibrium condition of the system by formulating the queuing model as a quasi-birth and death (QBD) process. Reference [15] considered the single-channel queueing equations characterized by Poisson-distributed arrivals and Erlang service times, and obtained the transient phase probabilities in terms of a new generalization of the modified Bessel function, and then evaluated the mean waiting time in the queue. Reference [16] considered a control policy for $M/E_k/1$ queue in which the server turns up after a random period and service initiation does not depend on the arrival pattern, but service rate provided by the server is proportional to the customers in the queue.

In order to improve the efficiency of the use of intelligent express cabinets, domestic research can effectively achieve their goals through the layout planning of intelligent express cabinets, the optimization of their size ratio and the design of more efficient intelligent express cabinet system. However, they do not involve customers and there is no participation of customers, in fact, in the intelligent express cabinet service system, the most important thing is the process of customers' pickup.

For a more comprehensive study of how to improve the efficiency of intelligent express cabinets, this paper considers the time factor and uses the model of queuing theory in operations research to improve the efficiency of intelligent express cabinet by improving its service rate.

# 3 Queueing Theory Model of Intelligent Express Cabinet

## *3.1 Intelligent Express Cabinet Service System*

In the intelligent express cabinet service system, the whole process is as follows: First, the parcels arrive, and then the staff sorts and accepts the parcels. After the acceptance, the parcels are scanned and put into the intelligent express cabinet. If there is no empty cabinet, then wait for the customer to pick up the parcels, and then put the rest of them into the express cabinet with the next batch of parcels. When the customer receives the pick-up message, he/she arrives at the express cabinet within a certain time and takes away the express, which will be the end of the whole process. It can be simplified as follows: Express arrives → Staff acceptance → The staff puts the accepted parcels into the express cabinet → Notify the customer to pick up the item → The customer picks up the item (Fig. 1).

According to the foregoing, the whole process of accessing express delivery through the intelligent express cabinet can be regarded as a queuing system, the intelligent express cabinet is used as a service organization in the queuing system, and the parcel to be stored is used as a customer. In addition, the process of the parcel being delivered according to the order of delivering time can be regarded as the parcel waiting for the service organization to perform the service, and the model for accessing the parcel through the intelligent express cabinet based on the formula can be expressed by the equation:

$$X/Y/C/\infty/\infty/FCFS \tag{1}$$

Among them, the time interval for the staff to put the parcel into the express cabinet is the distribution of the customer arrival time interval, denoted as *X*; The distribution of the service time is from the delivery to the removal of the parcel, denoted as *Y*; The number of cabinets in the intelligent express cabinet system

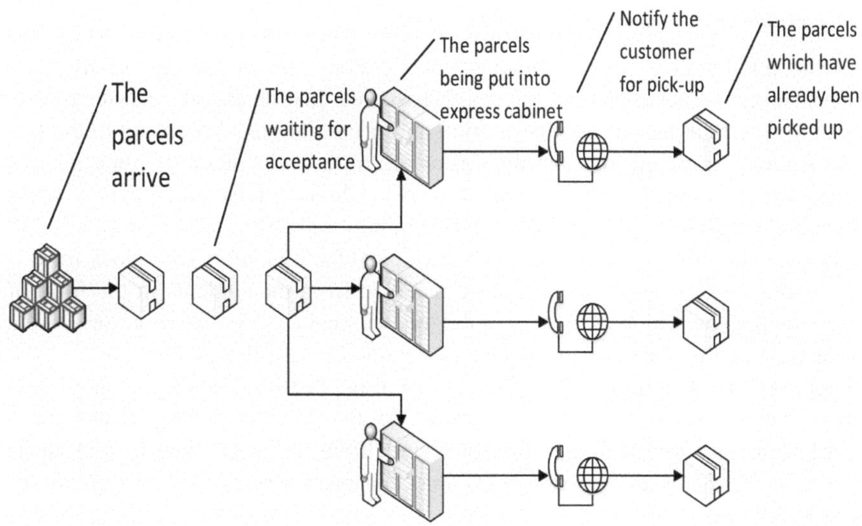


**Fig. 1** Intelligent express cabinet service system flow chart

selected in this paper is constant, that is, the number of service desks (denoted as $Z$) is $C$ ($C > 1$); the customer source (denoted as $B$) is infinite; Assume that the system capacity (denoted as $A$) is infinite; the delivery process is generally delivered in the order of arrival, so the service rule (denoted as $F$) is a first-come, first-served service.

This paper assumes that the process of putting the parcel into the express cabinet and the customer's taking it are continuous, regardless of the idle time at night. After the parcel is accepted, it will be placed in the free express cabinet. If all the express cabinets are occupied, it is considered that the parcel is waiting in line.

## 3.2 *Intelligent Express Cabinet System Analysis*

(1) *Determination of the distribution of arrival time intervals*

This paper mainly analyzes the law of the process of the staff's putting the parcel into the intelligent express cabinet service system. Firstly, in the non-overlapping time, the parcels that need to be placed are independent; Secondly, for a sufficiently small time interval $[t, t + \Delta t]$, the probability that a parcel needs to be serviced is independent of t, which is only proportional to the time interval, furthermore, the probability that two or more parcels need to be serviced is negligible. The random event, the staff's putting the parcels into the express cabinet, appears randomly and independently at a fixed mean instantaneous rate, that is, the delivery law of parcels satisfies the inefficiency, the stability and the generality. It can be seen from the above analysis that the delivery law of express delivery is Poisson distribution.

(2) *Determination of service time distribution*

In the intelligent express cabinet service system, the probability of the customer's taking the parcel is random, so it is assumed that the parcel's service time is subject to a certain probability distribution and is continuous. According to the law of the students' pick-up, the distribution of the time of the students' pick-up was investigated. Combined with the staff's working hours of the intelligent express cabinet called "Jinlinbao", the time is divided into 4 segments, 9 to 12 and 14 to 18 for normal class time, 12 to 14 for lunch break, and the time after 18 o'clock for rest time. A total of 127 questionnaires were received from the students in the form of questionnaires, of which 127 were valid questionnaires. The distribution of the service time of the current service system obtained is as follows (Fig. 2):

Therefore, it is assumed that the service time distribution of the intelligent express cabinet service system is subject to the k-order Erlang distribution. According to the current distribution of service time, the peak time of pick-up is between 12:00–14:00 and 18:00–21:00, and the mean service time of current situation is 5.85 h.

(3) *Indicators of the intelligent express cabinet service system*

We define

$\lambda$ = the arrival rate of the parcels,
$C$ = number of service desks,
$\mu$ = the mean service rate of the entire system,
$\rho$ = service intensity, which is the ratio of the mean arrival rate to the mean service rate.

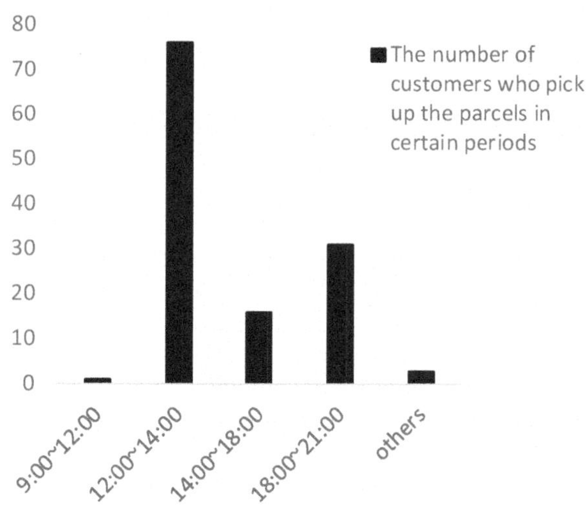


**Fig. 2** The service time distribution of the service system

$$\sum_{j=C}^{\infty} p_j = \frac{(\lambda/\mu)^C}{C!(1-\rho)} p_0 \tag{9}$$

The approximate solution to the mean waiting queue length is

$$Lq = \frac{\rho(\lambda/\mu)^C}{C!(1-\rho)^2} p_0 \left\{ (1-\rho)C\mu \int_0^{\infty} e^{-K\mu Ct} \left[ 1 + \frac{K-1}{K} \frac{K\mu t 1!}{+} \frac{K-2}{K} \frac{(K\mu t)^2}{2!} + \ldots + \frac{1}{K} \frac{(K\mu t)^{K-1}}{(K-1)!} \right]^C dt + \frac{\rho}{2}\left(1 + \frac{1}{K}\right) \right\} \tag{10}$$

The approximate solution to the mean queue length is $L_s = L_s + \lambda/\mu$.

The waiting time is the time from the arrival of the parcel to the time when it is put into the cabinet. The time of stay is the time from the arrival of the parcel to the time it is taken.

The mean waiting queue length $(L_q)$ and the mean waiting time $(W_q)$ of the $M/E_k/C$ queuing model satisfy the Little formula $L_q = \lambda\, W_q$. Hence the approximate solution to the mean waiting time of the express cabinet system is $W_q = L_q/\lambda$. In turn, the approximate solution to the mean stay time of the system is $W_s = W_q + 1/\mu$.

In summary, for the $M/E_k/C$ queuing model, if $\rho < 1$, then under the approximation hypothesis (AH), the approximation to the mean waiting time of the parcel in the system is

$$W_q = \frac{(\lambda/\mu)^C}{C!C\mu(1-\rho)^2} p_0 \left\{ (1-\rho)C\mu \int_0^{\infty} e^{-K\mu Ct} \left[ 1 + \frac{K-1}{K} \frac{K\mu t}{1!} + \frac{K-2}{K} \frac{(Kmut)^2}{2!} + \ldots + \frac{1}{K} \frac{(K\mu t)^{K-1}}{(K-1)!} \right]^C dt + \frac{\rho}{2}\left(1 + \frac{1}{K}\right) \right\} \tag{11}$$

The approximate solution to the mean stay time is

$$W_s = W_q + 1/\mu. \tag{12}$$

# 4 Policy Discussion and Revenue Analysis

## 4.1 *Revenue Model*

In the intelligent express cabinet service system, unlike the general random service system, reducing the service time of the parcel in the cabinet depends on the customer's pickup, rather than depending on the service efficiency of the service organization. Therefore, this paper aims to improve the enthusiasm of customers to take their parcels by policy discussion, and thus improve the utilization rate of intelligent express cabinet.

The current service time distribution of the system is as shown above. According to the time law of customers' pickup, the changes in the distribution of service time are observed by giving customers different policies. Nowadays, the sources of revenue for intelligent express cabinets mainly depends on the advertising revenue, the income from parcels, the usage fee charged to the courier, etc., but these sources of funds are quite limited, and they cannot be balanced compared with the huge cost. Therefore, the reward policy of this paper is to provide customers with points. When customers need to send parcels, they can use a certain amount of points to realize the reward, one point equals to one cent. As for customers, they get discounts when they need to send parcels. As for operators of intelligent express cabinets, this policy can promote customer consumption, and on the other hand, it will bring certain profits to the company. The penalty policy is to impose a certain amount of fine on customers, prompting them to take their parcels earlier.

In this section, the profit model of the intelligent express cabinet is constructed, and the related parameters, including the income of the intelligent express cabinet and the cost of the relevant policy, are discussed and analyzed.

Assume that the revenue per cabinet is r, there are a total of C express cabinets, regardless of the impact of the size of the express cabinet on the income, the service intensity of the intelligent express cabinet is ρ ($0 < \rho < 1$); According to the four periods divided, the probability of customers' pickup from morning to night is $p_1, p_2, p_3, p_4$, the corresponding reward points are $a_1$, $a_2, a_3, a_4$, and the corresponding fines are $b_1, b_2, b_3, b_4$. The revenue of the intelligent express cabinet is $(\mathrm{r} + p_1b_1 + p_2b_2 + p_3b_3 + p_4b_4)^* C\rho$, and the cost of the intelligent express cabinet after the relevant policy is formulated is $(p_1a_1 + p_2a_2 + p_3a_3 + p_4a_4)^* C\rho$. This section only considers the costs and income associated with customers' pickup, and other factors such as utilities and advertising revenue are not considered.

In summary, the revenue model of the intelligent express cabinet is:

$$
\begin{aligned}
Z = (r + p_1b_1 + p_2b_2 + p_3b_3 + p_4b_4) * C\rho - p_1a_1 + \\
+ p_2a_2 + p_3a_3 + p_4a_4) * C\rho, (0 < \rho < 1)
\end{aligned}
\tag{13}
$$

The existing income of the intelligent express cabinet reaching steady state is $Z_0 = \mathrm{r}C\rho_0$. Assume that the initial arrival rate of the parcels in the system is $\lambda_0 = 10$, the number of cabinets is C = 100, and from 3.2.2, $\mu_0 = 0.17$, $\rho_0 = \lambda_0 / C\mu_0 = 0.588$ is obtained, so the current revenue of the intelligent express cabinet is $Z_0 = 58.8r$.

## 4.2 Policy Discussion

The specific policy discussions are as follows:

(1) Provide customers with a reward policy while the original rules are unchanged. Customers will get 15 points for rewards if they pick up parcels before 12 o'clock, 10 points for rewards between 12 and 2 o'clock, and 5 points for rewards between 14 and 18 o'clock and points reward after 18 o'clock.

According to the data obtained by the survey, the service time distribution of the system obtained after providing the reward policy is as follows (Fig. 3):

According to the new service time distribution of the system, customers are more inclined to take the parcels from 9:00 to 12:00, and the mean service time μ of the system is reduced. According to (12), the mean stay time of the system decreases with the decrease of the mean service time μ, and the new mean service time of the system becomes 4.73 h, with an average decrease of 1.12 h. And as the parcels are taken earlier, the staff can also put more parcels into the cabinets, assuming that the staff will put them into the cabinets at the rate of $\lambda_1 = 13$, then the service intensity of the intelligent express cabinet is $\rho_1 = \lambda_1 / C\mu_1 = 0.615$.

The revenue of the intelligent express cabinet under this policy is $Z_1 = rC\rho_1 - (p_1a_1 + p_2a_2 + p_3a_3 + p_4a_4) * C\rho_1 = 61.5r - 5.617$. When $Z_1 - Z_0 > 0$, that is, $r > 2.08$, this policy can increase the utilization rate of the intelligent express cabinet without increasing the cost.

(2) Provide customers with a penalty policy while the original rules are unchanged. Customers who take the parcels before 12 o'clock will not be punished, a penalty of 0.05 yuan is required to take the parcels between 12 o'clock and 2 o'clock, a penalty of 0.1 yuan is required to take the parcels between 14 o'clock and 18 o'clock, and a penalty of 0.15 yuan is required after 18 o'clock. According to the data obtained by the survey, the service time distribution of the system obtained after providing the penalty policy is as follows (Fig. 4):

According to the new service time distribution, the customers who prefer to pick up the parcels from 9:00 to 12:00 are greatly increased, and the mean service time of the system is reduced. According to (12), the mean stay time of the system decreases with the decrease of the mean service time μ, and the new mean service time becomes 3.96 h, with an average decrease of 1.89 h. Assuming that the staff will put parcels into the cabinets at the rate of $\lambda_2 = 16$, then the service intensity of the intelligent express cabinet is $\rho_2 = \lambda_2 / C\mu_2 = 0.634$.

**Fig. 3** The new service time distribution of the service system with a reward policy

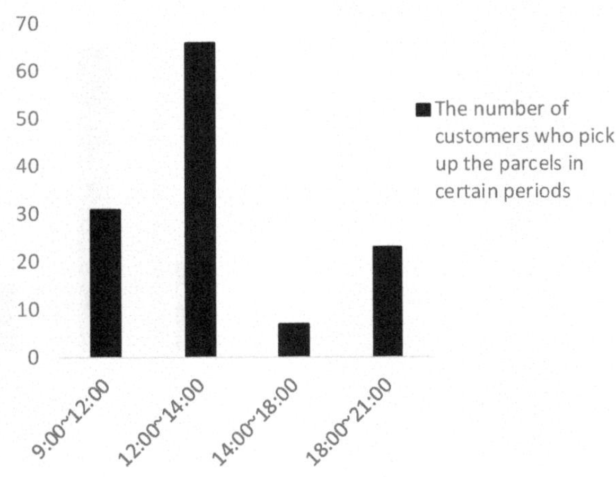

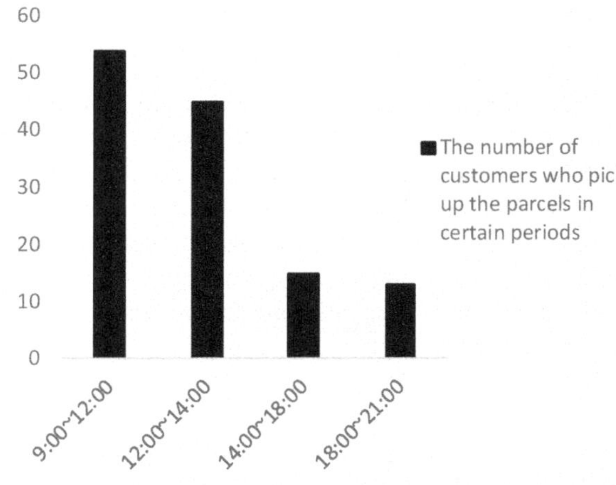


**Fig. 4** The new service time distribution of the service system with a penalty policy

The revenue of the intelligent express cabinet under this policy is $Z_2 = (r + p_1b_1 + p_2b_2 + p_3b_3 + p_4b_4)^* C\rho_2 = 63.4r + 2.846$. It can be seen that the penalty policy can increase the utilization rate of the express cabinet without increasing the cost.

However, in this case, there may be customers who choose not to use the intelligent express cabinet, so the trade-offs must be weighed.

(3) Now combine the reward policy with the penalty policy: customers will get 10 points for rewards if they pick up parcels before 12 o'clock, 5 points for rewards between 12 and 2 o'clock, while a penalty of 0.05 yuan is required to take the parcels between 14 o'clock and 18 o'clock, and a penalty of 0.10 yuan is required after 18 o'clock. According to the data obtained by the survey, the service time distribution of the system obtained after providing a policy combining reward with penalty is as follows (Fig. 5):

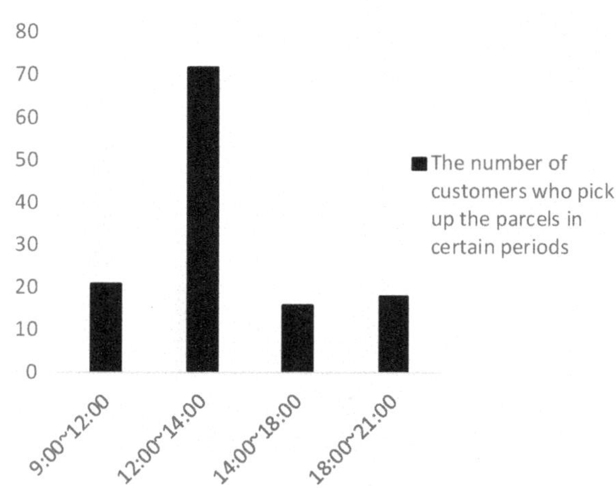


**Fig. 5** The new service time distribution of the service system with a policy combining reward with penalty

According to the new service time distribution, there are fewer customers who tend to pick up the parcels between 18:00 and 21:00, and there are more customers who tend to pick up the parcels between 9:00 and 12:00. The mean service time μ of the system is reduced. Similarly, the mean stay time of the system decreases with the decrease of the mean service time μ according to (12). The new mean service time is 4.89 h, with an average decrease of 0.96 h. Assuming that the staff will put parcels into the cabinets at the rate of $\lambda_3 = 13$, then the service intensity of the intelligent express cabinet is $\rho_3 = \lambda_3 / C\mu_3 = 0.636$.

The revenue of the intelligent express cabinet under this policy is $Z_3 = (r + p_3 b_3 + p_4 b_4) * C\rho_2 - (p_1 a_1 + p_2 a_2) * C\rho_3 = 63.6r - 1.553$. When $Z_3 - Z_0 > 0$, that is, $r > 0.324$, the policy can increase the utilization rate of the intelligent express cabinet without increasing the cost.

According to the data obtained in the surveys above, comparing the three policies given, it is found that the customer is more sensitive to the penalty policy. As for the intelligent express cabinet service system, it can significantly reduce the mean service time of the system and improve the utilization rate of the intelligent express cabinet, but the gains and losses must be weighed.

For the policy combining reward with penalty, when r > 0.324, the utilization rate of the intelligent express cabinet can be improved without increasing the cost. According to research on the current market, it is found that the income of one single cabinet is generally r = 0.5 yuan. Therefore, this paper believes that the policy combining reward with penalty can improve the utilization rate of the intelligent express cabinet without losing customers and increasing costs, which is preferable.

As a result, this paper suggests that operators of intelligent express cabinets can increase the enthusiasm of customers' pick-up by appropriately applying the policy combining reward with penalty, thereby improving the operational efficiency of intelligent express cabinets. The specific policy is that customers will receive a certain amount of points for rewards if they pick up the parcels within a certain period of time, while a certain amount of fines are required to take the parcels if the parcels are not taken at the specified time, and different fines will be charged to the customers according to different time periods.

## 5 Conclusion and Outlook

As a logistics service facility, the intelligent express cabinet is a symbol of the development of the logistics express industry to the electronic industry. It is also the specific practice of developing a smart city. It needs to use advanced science and technology and means to realize smart management and promote social progress. This paper has referred to studies about the development trend of intelligent express cabinet, system design and the model of the determined stochastic service system and so on, investigated the user's behavior of pick-up, and selected the best policy that can improve the utilization rate of intelligent express cabinet through policy discussion and benefit analysis, and finally made recommendations to the operators of intelligent express cabinets. By appropriately applying the policy combining reward with penalty, to improve the enthusiasm of customers' pickup, thereby improving the utilization rate of intelligent express cabinets.

Although this paper has achieved certain research results, due to limited time and energy, it has not been able to continue to study in depth. There can be some improvements, such as when collecting data, the research object is only the students of the school, the pick-up laws of different people are different, and the data is limited; This paper can also continue to conduct in-depth research, for instance, the amount of rewards and fines to maximize the benefits of the intelligent express cabinet can be studied after giving the optimal policy.

## References

1. Shang, Y. (2016). Analysis of the status quo of the development of intelligent express cabinets in China. *Management and Administration, 8,* 36–38.
2. Liu, L. (2015). Application and development direction of intelligent self-service express cabinet. *Logistics Engineering and Management, 38,* 54–55.
3. Wang, J., & Zou, E. (2015). Analysis of application of smart express parcel cabinets in online shopping end point logistics. *Logistics Technology, 34,* 58–60.
4. Morganti, E., Dablanc, L., & Fortin, F. (2014). Final deliveries for online shopping: The deployment of pickup point networks in urban and suburban areas. *Research in Transportation Business & Management, 11,* 23–31.
5. Feng, B. (2015). *Research on layout and operation mode of intelligent express mail box*. M.S. thesis, Beijing Jiaotong University, Beijing, China.
6. Li, M. (2014). Research on pick-up mode in B2C logistics distribution. *Logistics Engineering and Management, 36,* 68–71.
7. Shang, Y. (2017). *Research on size matching optimization and income distribution of intelligent express cabinet based on customer satisfaction*. M.S. thesis, Zhejiang Sci-Tech University, Zhejiang, China.
8. Wang, P., & Huang, Z. (2014). Research and design of intelligent express terminal based on Android and STM32. *Science Mosaic, 8,* 237–242.
9. Zou, H. (2016). *Design and implementation of intelligent express terminal based on Smart210*. M.S. thesis, Anhui University, Anhui, China.
10. Song, Y., Yin, H., & Fang, J. (2018). The application and function design of intelligent express boxes based on Internet +. *Journal of Huizhou University, 38,* 72–77.
11. Punakivi, M., Yrjola, H., & Holmstrom, J. (2001). Solving the last mile issue: reception box or delivery box? *International Journal of Physical Distribution & Logistics Management, 31,* 427–439.
12. Vesa, K., & Punakivi, M. (2002). Developing cost-effective operations for the e-grocery supply chain. *International Journal of Logistics Research and Applications, 5,* 285–298.
13. Zhao, D. (2018). Analysis of E-mail system based on queuing theory model. *Computer Engineering, 44,* 309–313.
14. Yue, D., Li, C., & Yue, W. (2006). The matrix-geometric solution of the M/Ek/1 queue with balking and state-dependent service. *Nonlinear Dynamics and Systems Theory, 3,* 295–308.
15. Griffiths, J. D., Leonenko, G. M., & Williams, J. E. (2006). The transient solution to M/Ek/1M/Ek/1 queue. *Operations Research Letters, 34,* 349–354.
16. Dhyani, I., & Jain, M. (2001). Control policy for M/Ek/1 queueing system. *Journal of Statistics and Management Systems, 4,* 10.

# Coordination Between Regional Logistics and Regional Economy in Guangdong-Hong Kong-Macao Greater Bay Area

**Jiajie Ye and Jiangxue Di**

**Abstract** Regional logistics is a sub-system of regional economy. Studying on the coordination between regional logistics and regional economy will help to learn the development of regional economy. Taking Guangdong-Hong Kong-Macao Greater Bay Area as an example, this paper selects the data of 10 years from 2008 to 2017, established a multiple linear regression model, analyzed the coordination between regional logistics and regional economy to carried out prediction on this basis. The results show that Guangdong-Hong Kong-Macao Greater Bay Area has a good coordination between regional logistics and economy, and there is a significant positive influence between them by forming a mutually reinforcing relationship.

**Keywords** Regional logistics · Regional economy · Coordination · Multiple linear regression · Guangdong-Hong Kong-Macao Greater Bay Area

## 1 Introduction

Regional economy combines all kinds of economic activities (including logistics activity) in a region, and the main purpose of developing regional logistics is to maximize the advantages of logistics facilities in the region, realizing the reasonable connection among spatial benefits, time benefits, and various logistics links, and to promote the economic development of the region [1]. The study of the coordination of the two is helpful to deeply analyze the economic development of a region, so as to improve the quality and efficiency of economic development. The coordination between regional logistics and regional economy provides a new way to realize high-quality development of Guangdong-Hong Kong-Macao Greater Bay Area's economy.

---

J. Ye · J. Di (✉)
Hubei University of Chinese Medicine, Wuhan, China
e-mail: 791424893@qq.com

J. Ye
e-mail: 1018861997@qq.com


J. Zhang et al. (eds.), *LISS2019*,
https://doi.org/10.1007/978-981-15-5682-1_62

Many scholars have done extensive research on the relationship between regional logistics and regional economy. Among them, Yan and Yuan [2] established a multivariate linear regression model by taking data of freight volume, freight turnover, passenger transport turnover, traffic fixed assets investment, etc., as the representative indicators of regional logistics and taking annual GDP as the representative index of regional economic development on research of interaction between regional logistics and regional economy. We can conclude that the regional logistic has positive impact on regional economy, and they put forward some suggestions on how to use regional logistics to promote the development of regional economy. Zeng et al. [3] established regression equation by selecting passenger volume, total volume of post and telecommunications business, turnover of goods, added value of tertiary industry, regional gross product, and industrial value added as research indicators, and using the least square method. The conclusion is that the regional economy has an obvious driving effect on the development of regional logistics, the development of regional logistics industry has no significant driving effect on economic growth, and the development level of logistics industry lags behind. Tan et al. [4] made a statistical analysis for the gross value of production, logistics capacity, fixed asset investment, working labor force, and intangible scientific and technological progress in the region. The result shows that it is an important task for the research area to adjust the structure of logistics system reasonably, strengthen the information construction of logistics system, and build a modern logistics system as soon as possible while strengthening the construction of logistics infrastructure.

It can be seen that GDP is usually used as an index to measure the level of regional economy development, but the academic community has not yet had a unified definition standard in studying the logistics level of a certain region. At the same time, in the macro statistics of the domestic economy and industry, there is no independent accounting for logistics, it needs to combine the actual situation of the study region and the researchers' understanding about regional logistics to select the indicators. The indicators selected in this paper combine the actual situation of Guangdong-Hong Kong-Macao Greater Bay Area logistics industry in a certain extent.

## 2 General Situation of Guangdong-Hong Kong-Macao Greater Bay Area

By the end of 2017, Guangdong-Hong Kong-Macao Greater Bay Area's total GDP had exceeded 10 trillion (RMB, the same below), accounting for about 12.87% of the country's total GDP. The average density of road network within the region is 9.5 km/100 $m^2$, and the mileage of expressway is more than 4300 km; The mileage of inland waterway exceeds 6000 km, and the container throughput of coastal port is more than 80 thousand TEU. The passenger throughput of civil aviation exceeds

200 million, and the annual cargo and mail throughput capacity is about 8.07 million tons. The mileage of the railway is 2179 km, of which high-speed rail accounts for 64.8%. In terms of policy, in the outline of the development plan for Guangdong-Hong Kong-Macao Greater Bay Area issued by the National Development and Reform Commission in 2019, it focused on "building a modern integrated transportation system," "constructing a modern freight logistics system", and speeding up the development of railway-water carriage, highway-railways, air-railway freight, combined maritime-riverine shipment, and 'one sheet system' inter-modal service [5].

However, the freight transportation in the Greater Bay Area mainly depends on road transportation, seamless docking, or efficient docking of multi-modal transportation, such as connecting the sea and air transportation, etc., which cannot be effectively realized. On the other hand, the logistics industry in Greater Bay Area is still insufficient in management, modern logistics technology, information technology system application and so on.

The above situation shows that although Guangdong-Hong Kong-Macao Greater Bay Area has a developed economy and excellent regional logistics base, there are still some shortcomings in the development of regional logistics, and the coordination between regional logistics and regional economy is worthy of further exploration.

# 3 Research Method

## 3.1 Model Selection

Referring to [2], Analysis of the Interactive Relationship between Regional Logistics and Regional Economy of Hubei Province and Zeng et al. [3] Empirical analysis of Relationship Between the Regional Economy and Regional Logistics Development in Gansu Province and other literature on this filed, all of which use linear regression model to analyze the relationship between regional logistics and regional economy. As the logistics level is a synthesis of many factors, the correlation degree between the independent variable and the dependent variable can be displayed intuitively by using regression analysis method, so this paper established the multivariate linear regression model by selecting Guangdong-Hong Kong-Macao Greater Bay Area's relevant logistics index and economic indexes used to represent the regional logistics development level and the regional economic development level from 2008 to 2017, and then carrying on the linear regression analysis. The model is shown as follows:

$$Y = \beta_0 + \beta_1 X_1 + \beta_2 X_2 + \beta_3 X_3 + \beta_4 X_4 + \ldots \beta_n X_n + \varepsilon \quad (1)$$

From (1), $Y$ represents the dependent variable, $X_n$ represent the independent variables, $\beta_0$ represents the constant term, $\beta_n$ represent the coefficients, respectively, of $X_n$, and $\varepsilon$ represents the random error term.

## 3.2 Variable Selection

First of all, the total retail sales of social consumer goods represent the sum of the amount of goods used by urban and rural residents for living consumption and social groups for public consumption, which is positively correlated with economic development in a certain extent. At the same time, the resources flow between regions, when the specific performance of the circulation of retail goods, the greater the demand for retail goods, the faster the speed of circulation, the more the development of logistics. Therefore, the index of total retail sales of social consumer goods is related to the level of economy and logistics. Secondly, due to the large number of ports in Guangdong-Hong Kong-Macao Greater Bay Area, port logistics has an important impact on the regional economy, so the index of Freight Throughput of Ports is selected. Thirdly, road traffic infrastructure is the foundation of logistics, especially in solving the problem of "last kilometer" of logistics, the highway with traffic ability is the next key point that cannot be bypassed. Finally, the increase in the scale of logistics will certainly cause an increase in the number of people employed in the logistics industry, so this indicator of the number of employed persons in the transport, storage, and postal services can effectively reflect the change of logistics scale of Guangdong-Hong Kong-Macao Greater Bay Area from 2008 to 2017.

To sum up, combined with the author's understanding of the concepts related to regional economy and regional logistics, after summing up the relevant data of Guangdong-Hong Kong-Macao Greater Bay Area from 2008 to 2017, the following selected indicators in this paper as below: GDP, total retail sales of social consumer goods, freight throughput of ports, length of highways, and the number of employed persons in the transport, storage, and postal services. Among them, GDP, as a dependent variable, is an index to measure Guangdong-Hong Kong-Macao Greater Bay Area's regional economic development. The total retail sales of social consumer goods, freight throughput of ports, length of high ways and the number of employed persons in the transport, storage, and postal services are four independent variables, which represent Guangdong-Hong Kong-Macao Greater Bay Area's logistics level.

# 4 Empirical Test and Results

## 4.1 Data Processing

Table 1 shows the GDP, the total retail sales of social consumer goods, freight throughput of ports, length of highways and the number of employees about transport, storage and postal services in Guangdong-Hong Kong-Macao Greater Bay Area. The total retail sales of social consumer goods, freight throughput of ports, length of highways, and the number of employees about transport, storage and postal services are recorded as $X_1$, $X_{2,}$ $X_3$, and $X_4$, respectively.

In order to avoid the influence of excessive abnormal fluctuation of time series data and make the abnormal data shrink to the expected range, this paper takes logarithmic processing for GDP, which is recorded as *lnGDP*. At the same time, the unit root test is used to test the stationarity of the time series of each independent variable before the study, so as to avoid the occurrence of invalid results due to the existence of pseudo-regression. In this paper adopts ADF unit root test method to test the stationarity of time series. Carried out the root test of *lnGDP*, $X_1$, $X_2$, $X_3$, and $X_4$ by using Eviews 8.0. software.

The test results are shown in Table 2 below (where the critical value is taken as 5% confidence level, the test criterion is whether or not p value is greater than 0.05, If $p > 0.05$, the test fails and the sequence is unstable). The results show that the *lnGDP*, the total retail sales of social consumer goods, freight throughput of ports, length of highways and the number of employed persons in the transport, storage, and postal services are all non-stable, while some independent variables still have

**Table 1** Annual data of Guangdong-Hong Kong-Macau Greater Bay Area's main indicators for measuring economic level and logistics level

| Year | GDP (RMB trillion) | X1 (RMB trillion) | X2 (trillion tons) | X3 (ten thousand kilometers) | X4 (thousand persons) |
|---|---|---|---|---|---|
| 2008 | 4.664708 | 1.25 | 10.68 | 5.59 | 547.02 |
| 2009 | 4.861988 | 1.39 | 10.80 | 5.67 | 595.46 |
| 2010 | 5.557599 | 1.61 | 12.58 | 5.83 | 629.14 |
| 2011 | 6.26147 | 1.94 | 13.46 | 6.02 | 697.02 |
| 2012 | 6.765227 | 2.08 | 13.68 | 6.24 | 696.12 |
| 2013 | 7.395885 | 2.31 | 15.05 | 6.39 | 892.54 |
| 2014 | 7.970996 | 2.51 | 15.64 | 6.41 | 910.63 |
| 2015 | 8.490038 | 2.70 | 15.48 | 6.56 | 887.27 |
| 2016 | 9.272479 | 2.93 | 16.01 | 6.62 | 880.48 |
| 2017 | 10.21713 | 3.17 | 17.61 | 6.67 | 891.25 |

By the end of 2017, date source: Guangdong-Hong Kong-Macau Greater Bay Area Municipal Bureau of Statistics, Census and Statistics Department of Hong Kong Special Administrative region and DSEC of Macau Special Administrative Region

**Table 2** ADF unit root test results

| Variable | P value | Conclusion |
|---|---|---|
| *lnGDP* | 1.00000 | Unstable |
| d-1*lnGDP* | 0.58650 | Unstable |
| d-2*lnGDP* | 0.00390 | Stable |
| $X_1$ | 1.00000 | Unstable |
| d-1$X_1$ | 0.57120 | Unstable |
| d-2$X_1$ | 0.00010 | Stable |
| $X_2$ | 0.99700 | Unstable |
| d-1$X_2$ | 0.22470 | Unstable |
| d-2$X_2$ | 0.00120 | Stable |
| $X_3$ | 0.99980 | Unstable |
| d-1$X_3$ | 0.31300 | Unstable |
| d-2$X_3$ | 0.00330 | Stable |
| $X_4$ | 0.94720 | Unstable |
| d-1$X_4$ | 0.02660 | Stable |
| d-2$X_4$ | 0.00080 | Stable |

**Table 3** Correlations

| | | lnGDP |
|---|---|---|
| Pearson correlation | lnGDP | 1.000 |
| | $X_1$ | 0.998 |
| | $X_2$ | 0.989 |
| | $X_3$ | 0.988 |
| | $X_4$ | 0.929 |

unit roots after the first order difference, and is non-stable; After the second order difference of the original sequence, the sequence is stable and the five variables belong to the second order integrated sequence.

## 4.2 Model Building

Table 3 shows that the correlation coefficient between *lnGDP* and the total retail sales of social consumer goods is the highest (0.998). The second shows the correlation coefficient between *lnGDP* and Freight throughput of port (0.989), the third indicates the correlation coefficient between *lnGDP* and length of highways (0.988), and finally the correlation coefficient between *lnGDP* and the number of employed persons in the transport, storage, and postal services (0.929). In addition, the p values corresponding to the four correlation coefficients are 2.2791E−11, 3.7077E−8, 3.9042E−8, 0.000050, respectively, all less than 0.05. The passed one-tailed test shows that there is a significant correlation between *lnGDP* and the four indexes. The simple correlation coefficients between *lnGDP* and $X_1$, $X_2$, $X_3$, $X_4$ are all above 0.9. This shows that the total retail sales of social consumer goods, freight throughput of ports, length of highways and the number of employed persons in the transport, storage, and postal services is highly positively correlated with Guangdong-Hong Kong-Macao Greater Bay Area's GDP. Based on this, a multivariate linear regression model is established, which is as follows:

$$lnGDPi = \beta_0 + \beta_1 X_{1i} + \beta_2 X_{2i} + \beta_3 X_{3i} + \beta_4 X_{4i} + \varepsilon_i \quad (2)$$

s.t.

$$i = 2008, 2009, \ldots, 2017$$

From (2), *lnGDP* represents the dependent variable, $X_1$, $X_2$, $X_3$, and $X_4$ represent the independent variables, $\beta_0$ represents the constant term, $\beta_1$, $\beta_2$, $\beta_3$, and $\beta_4$ represent the coefficients, respectively, of $X_1$, $X_2$, $X_3$, $X_4$, and $\varepsilon i$ represents the random error term.

Since the multiple collinearity problem between the independent variables, the stepwise regression method is then used. The SPSS 22.0 system uses the stepwise regression method to make the independent variable enter the equation. The probability of judging the F value is $p \leq 0.050$ when the independent variable enters into, and the independent variable is eliminated when $p \geq 0.100$.

The first stepped into the equation is the total retail sales of social consumer goods ($X_1$), the second stepped is freight throughput of port ($X_2$), and the third stepped is length of highways ($X_3$). Because the F value of the number of employed persons in the transport, storage, and postal services ($X_4$) is always $p \geq 0.050$, this variable is excluded from the range of independent variables of linear regression equation. Therefore, the multiple linear regression model is composed of three independent variables, namely, the total retail sales of social consumer goods ($X_1$), freight throughput of port ($X_2$) and the third stepped is length of highways ($X_3$).

Table 4 gives the relevant data of the fitting degree of the regression model. According to the notes of the table, the first model is the unitary linear regression model between *lnGDP* and the total retail sales of social consumer goods, and the second model is the binary linear regression model between *lnGDP* and the total retail sales of social consumer goods and freight throughput of port. The third model is a ternary linear regression model between *lnGDP* and total retail sales of social consumer goods, fright throughput port, and length of highways.

**Table 4** Model fitting results

| Model | R | R square | Adjusted R square | Std. error of the estimate |
|---|---|---|---|---|
| 1 | 0.998 | 0.996 | 0.996 | 0.0169832 |
| 2 | 0.999 | 0.998 | 0.998 | 0.0123975 |
| 3 | 1.000 | 0.999 | 0.999 | 0.0077325 |

The determination coefficient $R$ of the three models is 0.996, 0.998, and 0.999. The goodness of fit is getting better and better. The third model can already explain 99.9% of the variation, and the error of standard estimation declines gradually, from 0.01698 down to 0.00773.

According to the variance analysis table for F test of the three equations obtained by the stepwise regression method, which is used to test the significance of the linear relationship between independent variables and dependent variables, the F test values of the three equations are 2219.220, 2086.295, and 3579.319, the p values corresponding to three probability are 4.5582E−11, 1.9225E−10 and 3.8091E−10. Take the significant level $\alpha = 0.05$, that is, 5% confidence level, all p values are less than 0.05, indicating that all of them have passed the F test. The linear relationship between the independent variable and the dependent variable is significant, and the model can be designed as a linear model. So there is, (2) turns to (3):

$$lnGDP_i = \beta_0 + \beta_{1i} + \beta_2 X_{2i} + \beta_3 X_{3i} + \varepsilon_i \tag{3}$$

$$i = 2008, 2009, \ldots, 2017$$

From (3), *lnGDP* represents the dependent variable, $X_1$, $X_2$, and $X_3$ represent the independent variables, $\beta_0$ represents the constant term, $\beta_1$, $\beta_2$, and $\beta_3$ represent the coefficients, respectively, of $X_1$, $X_2$, $X_3$, and $\varepsilon i$ represents the random error term.

From the above data, it can be seen that taking *lnGDP* as the dependent variable, the total retail sales of social consumer goods ($X_1$), freight throughput of port ($X_2$), and the third stepped is length of highways ($X_3$) as independent variables, a multiple linear regression equation can be established to explain the relationship between relevant logistics indicators and economic development. Table 7 shows the specific values of parameters in (3) (Table 5):

The empirical regression equation obtained by the stepwise regression method is as follows:

$$lnGDP = 9.502 + 0.26X_1 + 0.022X_2 + 0.123X_3 \tag{4}$$

From the above, it can be seen that the empirical regression equations of *lnGDP* and $X_1$, $X_2$, $X_3$, and $X_4$ have passed the significance test, and the goodness of fit is getting higher and higher, but the equation still can't be used for analysis and

**Table 5** Regression coefficient

| Model | Unstandardized | Coefficient | Standardized coefficient |
|---|---|---|---|
| | B | Std. error | Beta |
| (Constant) | 9.502 | 0.169 | – |
| $X_1$ | 0.260 | 0.027 | 0.633 |
| $X_2$ | 0.022 | 0.006 | 0.189 |
| $X_3$ | 0.123 | 0.035 | 0.183 |

prediction. In the process of establishing the empirical regression equation, a series of theoretical assumptions are made on the random error term ε, including taking the residual error as the estimated value of the random error, etc. If the hypothesis is not satisfied, the equation is lack of basis and meaning. Therefore, it is necessary to analyze the residual error of the model and investigate the adaptability of the empirical regression equation to the theoretical hypothesis.

According to the residual statistical scale, the results show that the average residual is −1.7764E−16, which is approximately regarded as 0, and there is no systematic error. The absolute value of culling residual is less than 3, which can be considered as no abnormal value.

The histogram of standardized residual error is in good agreement with the normal curve, and the points basically distributed near the diagonal line of the p-p diagram. It can be concluded that the residual error approximately obeys the normal distribution. It can be seen from the residual diagram, with standardized residuals as the *X* axis and takes dependent variables as the *Y* axis, that the standardized residual is basically randomly distributed and is between ±2 standard deviation, the equation satisfies the assumption that the residual error is homogeneous.

After taking residual analysis, the empirical regression equation basically satisfies the assumption that the residual error is regarded as the estimated value of random error term ε in the process of establishing the equation, and the equation can be used for analysis and prediction.

## 5 Result Analysis and Prediction

### 5.1 Coordination Analysis

In this paper, the model is established by the selection of representative indexes. In the process of the modeling, because there is a certain problem of multiple collinearity between independent variables, in order to eliminate the influence of multiple collinear equations, stepwise regression method is used to eliminate the independent variable X4, that is the number of employed persons in the transport, storage, and postal services Although this independent variable is excluded, the Pearson correlation coefficient between this index and the dependent variable, *lnGDP*, is also reached by 0.929, and it can be considered that the variable has a highly positive correlation with the level of economic development.

Through a series of processes, such as ADF unit root test, multiple collinearity test and correction, residual analysis and so on, the results show that: The coefficient of the total retail sales of social consumer goods is 0.26, that is, for every 1 percentage point increase (or decrease) of the total retail sales of social consumer goods, the level of social and economic development of Guangdong-Hong Kong-Macao Greater Bay Area will increase (or decrease) by 0.26 percentage points. Similarly, the impact of freight throughput of port and length of highways on economic development is 0.022 percentage points and 0.123 percentage points, respectively.

In general, the total retail sales of social consumer goods have the greatest influence on Guangdong-Hong Kong-Macao Greater Bay Area's economic development level, reaching by 0.26, which plays the most important role in promoting economic development among the three independent variables. The influence coefficient of freight throughput of port on economy is 0.022. The length of highways in a region has a direct impact on the economic development of the region, which reached 1:0.123 in Guangdong-Hong Kong-Macao Greater Bay Area.

Generally speaking, the coordination between Guangdong-Hong Kong-Macao Greater Bay Area's regional logistics and regional economy is still strong, and there is a positive relationship between them. Regional logistics plays a strong role in promoting the regional economy. And economic development feeds back on the development of logistics, the two promote and improve with each other in a virtuous circle.

## 5.2 Prediction

According to the previous model (4), the following will predict the impact of logistics level on economic development in the next stage.

The first is the changes of independent variable. The formula for calculating the average growth rate is:

$$\bar{G} = \sqrt[n]{\frac{Y1}{Y0} \times \frac{Y2}{Y1} \times \ldots \times \frac{Yn}{Yn-1}} - 1 = \sqrt[n]{\frac{Yn}{Y0}} - 1 \quad (5)$$

According to the equation, *Yn* is the final data, that is, the data of each independent variable in year 2017. *Y*0 is the base period data, the data of each independent variable in year 2008. According to the data in Table 1, the average annual growth rate of *lnGDP* is 4.2%, while the average annual growth rate of the total retail sales of social consumer goods is 9.75%, and the average annual growth rate of the freight throughput of ports is 5.13%. The average annual growth rate of length of highways is 1.78%. Therefore, when Guangdong-Hong Kong-Macao Greater Bay Area's total retail sales of social consumer goods was RMB 3.17 trillion, the freight throughput of port was 17.61 trillion tons, and the length of highways was 66.7 thousand kilometers, according to the above data, it can be predicted that Guangdong-Hong Kong-Macao Greater Bay Area's total retail sales of social consumer goods will be RMB 3.48 trillion, the freight throughput of port was will be 18.51 trillion tons, and the length of highways will be 67.9 thousand kilometers (shown in Table 6).

When the above data are fitted into (4), it is calculated that the value of *lnGDP* in the next phase is about 11.649, and the scale of the economy was forecast to grow by 4.95% over the previous year. Table 7 is a comparison between the annual

**Table 6** Predictive values for each independent variable

| | $X_1$ | $X_2$ | $X_3$ |
|---|---|---|---|
| Predictive values | 3.48 | 18.51 | 6.79 |
| Average annual growth rate | 9.75% | 5.13% | 1.78% |

The predicted values are based on 2017

**Table 7** Comparison of two predictions

| | *lnGDP* |
|---|---|
| Average annual growth rate | 12.476 |
| Linear regression model | 11.649 |

average growth rate and the linear regression model when the value of *lnGDP* and the growth rate are annual average growth rate and linear regression model.

According to the predicted results, Guangdong-Hong Kong-Macao Greater Bay Area's economy will continue growing in the next stage. Under the macro-control of the government, the probability of seriously deviating from the operational range is relatively small, and it is expected to maintain the momentum of sustained, healthy, and stable growth. As the development of Guangdong-Hong Kong-Macao Greater Bay Area is not yet mature, much cooperation has not yet been carried out, including the centralization of statistical data and the standardization of statistical indicators. Therefore, in this study, due to the lack of statistical data and differences in statistical indicators in some areas, this present paper only selects five indexes of GDP, the total retail sales f social consumer goods, freight throughput of ports, length of highways, and the number of employed persons in the transport, storage, and postal services to establish the model. The results of the model only show the effect of the regional logistics level represented by these indicators on the regional economic development, and the above prediction results are only used as a reference.

## References

1. Liu, M., & Li, L. (2007). Analysis of interactive functions between regional logistics and regional economy. *Industrial Technology Economic, 26*(3), 40–42.
2. Yan, C., & Yuan, L. (2018). Analysis of interactive relationship between regional logistics and regional economy of Hubei Province. *Value Engineering, 7*(15), 7–10.
3. Zeng, J., Zhang, J., & Qian, Y. (2015). Empirical analysis of relationship between the regional economy and regional logistics development in Gansu Province. *Logistics Management, 38*(2), 1–5.
4. Tan, M., Feng, L., & Ge, Y. (2003). Research on the contribution of logistics ability to regional economy. *Modern Economic Research, 8*, 22–24.
5. The Central Committee of the Communist Party of China and the State Council of the People's Republic of China. (2019). The CPC Central Committee and the State Council Print and Issue the Outline Development Plan for Guangdong-Hong Kong-Macao Greater Bay Area. Gazette of the State Council of the People's Republic of China (no. 7, pp. 4–25).

# Location Selection of the Terminal Common Distribution Center of Express Delivery Enterprise

**Dongyu Cui, Yuan Tian, and Lang Xiong**

**Abstract** Mastering the structural characteristics of the express terminal delivery network has an important role in reasonable layout of the terminal common distribution center of express delivery enterprise. This paper takes the layout of the terminal common distribution center of express delivery enterprise as the research object. Under the premise of quantitative analysis of the terminal delivery network structure, and considered the importance of each node in its network. Then, combined with the network operation economy and network invulnerability, a multi-target location model of the terminal common distribution center of the express delivery enterprise is established and solved by genetic algorithm. Finally, taking the express terminal delivery network in Haidian District of Beijing as an example, the quantity, location and corresponding distribution range of the terminal common distribution center of express delivery enterprise are determined. This research introduces the joint distribution concept into the express terminal distribution, and integrates the existing express terminal delivery resources to reduce the express delivery cost and improve the service level of the express delivery industry. It has certain theoretical value and practical significance.

**Keywords** Express delivery enterprise · Common distribution · Social network analysis · Network invulnerability

D. Cui (✉) · Y. Tian
School of Economics and Management, Beijing Jiaotong University, Beijing, China
e-mail: cdydongdong@163.com

Y. Tian
e-mail: ytian@bjtu.edu.cn

L. Xiong
School of Economics and Management, Harbin Institute of Technology, Shenzhen, China
e-mail: 13121195155@163.com



J. Zhang et al. (eds.), *LISS2019*,
https://doi.org/10.1007/978-981-15-5682-1_63

## 1 Introduction

The rapid development of e-commerce has promoted a large number of private express delivery companies. At the same time, it has gradually exposed many problems such as disorderly competition, serious homogenization, high distribution costs and unreasonable layout of nodes in the express delivery industry. In particular, the problem of terminal distribution is more prominent. In order to seize the consumer market, many express delivery companies have set up a large number of end distribution nodes in the same service area, resulting in serious waste of resources and external uneconomic phenomena. Coupled with the impact of some special e-commerce festivals and unpredictable emergencies on the express terminal delivery network, it directly affects the quality evaluation of express delivery services. Therefore, many domestic express delivery companies have tried to introduce the common distribution concept into the end of the express delivery process, through the establishment of a common distribution alliance to integrate the existing express terminal delivery resources, and to rationally layout the terminal co-distribution center of express delivery. The aim is to solve a series of problems such as high delivery cost, low distribution efficiency and uneconomical outside the city at the end of express delivery from the root cause, and provide guarantee for the healthy development of the express delivery industry.

The terminal co-distribution of the express delivery belongs to the category of urban distribution. According to the different methods of customer pickup, the co-distribution mode of the express delivery is divided into two types: home delivery and customer self-acquisition. For the home delivery mode, in addition to the rise of the C2C crowdsourcing delivery model in the past two years [1]. The terminal co-distribution of the express delivery is mainly based on the optimization of the vehicle path, and providing consumers with "door-to-door" delivery services. Kovacs [2], Spliet [3], Ni [4] and Cao [5] combined with factors such as large terminal distribution, vehicle capacity limitation, and time uncertainty, etc., with the total cost or customer satisfaction as the goal, the end delivery path and vehicle scheduling of the express delivery enterprise has been optimized. At the same time, with the enhancement of environmental awareness and the development of electric vehicle technology, Xiao also studied the route optimization and scheduling of green vehicles [6]. For the customer self-pickup mode, the layout of the express terminal self-service site are mainly studied. Zhang [7], Feng [8] and Sun [9] analyzed the demand characteristics, implementation difficulty, cost input and other factors of the express terminal delivery, proposed the application of express delivery enterprises to cooperate with communities and schools to establish the express terminal self-service site. Sun proposed that the terminal co-delivery operation mode of express delivery enterprises should be based on the construction of self-pickup site, supplemented by the express self-container [10]. On the other hand, in the research of site selection for distribution centers. Chen analyzed the location problem of the same city express delivery center based on degree centrality, intermediate center degree and near center degree [11]. Zhang used the complex network theory to comprehensively analyze the structural characteristics of the express delivery network, and determined the important nodes in the distribution network [12]. Sheng used quantum particle swarm optimization to solve the problem of optimal solution for logistics distribution center location [13].

The existing literature on the terminal co-distribution of express delivery companies is more limited. Among them, the problem of site selection of the terminal common distribution center, or combined with path planning, or the construction of mathematical models with the goal of cost and consumer satisfaction, or focus on the improvement of the algorithm. However, these site selection studies rarely consider the interrelationship between nodes and the impact of node layout on the overall network. At the same time, it also neglects the impact and destructiveness of uncertain factors such as e-commerce special festivals and traffic conditions on the express delivery network. Based on the existing research, the innovation of this paper is mainly reflected in the following two points. First, the social network analysis method is applied to the express terminal delivery network. The structure of the express terminal delivery network is quantitatively analyzed from the perspective of relationship, and the importance degree of the nodes in the existing network are considered. Provide a new perspective for site selection research of the express terminal delivery center. Secondly, based on the site selection of express delivery nodes focusing on operational costs and customer time satisfaction, the new goal of introducing network invulnerability is more practical.

## 2 Problem and Model

### *2.1 Problem Description and Basic Assumptions*

The layout problem of the terminal co-distribution center of the express delivery enterprise can be described as: within a certain service area where the number of customer groups and demand are known, through the quantitative analysis of the existing express terminal delivery network structure, considering the importance degree of the nodes in its network, select the alternative points for establishing the terminal co-distribution center, then, combined with the economic and stability requirements of the express terminal delivery, determined the number, location, and corresponding delivery range of the terminal co-distribution center from the candidate points.

Based on the characteristics of the problem, the relevant basic assumptions made are as follows: in the site selection area, the geographical location, distribution range and express demand of the end delivery nodes of each express company are known; only consider the delivery process from the terminal co-distribution center to the demand point; the identified terminal co-distribution center can meet the demand of all demand points within its coverage, and one demand point can only be provided by at most one terminal co-distribution center; the number of vehicles required to deliver the goods during the delivery task shall not exceed the total number of vehicles configured by the terminal co-distribution center; a van or electric tricycle can be selected for the delivery task, but only one of them can be selected; the traffic haul coefficients of all distribution lines are the same.

## 2.2 Symbol Definition

The symbols used in the model are shown in Table 1.

## 2.3 Models

Based on the axis-spoke network structure of express delivery, this paper establishes the alternative point location model and multi-objective location model of the terminal co-distribution center.

**Table 1** The symbols used in the model

| Symbols | Definition |
|---|---|
| $N$ | The sum of the number of nodes in the terminal delivery network of the express delivery |
| $n$ | Number of the terminal co-distribution centers of express delivery $w_i$, $i = 1, 2, \ldots, n$ |
| $m$ | Number of demand points $w_j$, $j = 1, 2, \ldots, m$ |
| $q_{ij}$ | The amount of goods delivered from the terminal co-distribution center $w_i$ to the demand point $w_j$ |
| $d_{ij}$ | Delivery distance from the terminal co-distribution center $w_i$ to the demand point $w_j$ |
| $d_{ij}^*$ | The ideal delivery radius for the terminal co-distribution center |
| $u_{ij}$ | Penalty factor when the delivery distance exceeds the service radius of the terminal co-distribution center |
| $p_{ij}$ | Fuel consumption cost per unit distance of small vans |
| $p_{ij}^*$ | Electric tricycle cost per unit distance |
| $l(w_i)$ | The total number of nodes in the network that have direct business with node $w_i$ |
| $g_{kj}(w_i)$ | Number of shortest delivery paths between two nodes $w_k$ and $w_j$ containing node $w_i$, $k \neq i \neq j, k < j$ |
| $g_{kj}$ | Number of shortest delivery paths between nodes $w_k$ and $w_j$ |
| $h_i(w_i, w_j)$ | Shortest distance between node $w_i$ and node $w_j$ |
| $d(w_i, w_j)$ | The number of the fewest edges passed from node $w_i$ to node $w_j$ |

*(1) Alternative point location model of the terminal co-distribution center of express delivery*

The contact mode and connection strength between the nodes in the terminal delivery network of the express delivery enterprise determine the degree of control and utilization of the existing network resources, and then determine the distribution efficiency of the entire distribution network. However, on the location of the express terminal delivery nodes, few scholars use the central analysis theory to study it. That is to say, the central analysis theory can provide new ideas for the site selection of nodes in the express delivery network. The centrality analysis measures the importance of a node in its network from three aspects, includes the degree to which a node is in direct contact with other nodes in its network, the degree of mediation as a "bridge", and the importance of geographic location. They are called the point center degree, the intermediate center degree and the close center degree.

*(a) The degree of point center*

Combined with the characteristics of the terminal delivery network of the express delivery enterprise, the degree of point center indicates the distribution relationship among the nodes in the network. That is, the more nodes in the network that have a business relationship with the terminal distribution center, the higher the importance of the distribution center in the network. Its calculation formula is,

$$C_P(w_i) = l(w_i)/(N-1)$$

*(b) The degree of intermediate center*

If a node in the express terminal delivery network is at the shortest path distance from the terminal distribution center to other nodes, the node has a higher intermediate center. That is, it has the role of a bridge to communicate with other nodes. The shortest path distance here refers to the actual shortest delivery distance between nodes. Its calculation formula is,

$$C_I(w_i) = \sum_{k=1}^{N}\sum_{j=1}^{N}\left[\sum_{k<j} g_{kj}(w_i)/g_{kj}\right]/[(N-1)(N-2)]$$

*(c) The degree of close center*

In the express terminal delivery network, the demand point is limited by the terminal distribution center, and relies on the terminal distribution center to obtain information and complete the service. But the terminal distribution center has direct business contact with most demand points, without the need of other nodes. When calculating close-centrality, the focus is on shortcuts rather than direct relationships. That is, if the sum of the distances from the terminal distribution center to many other demand points in the network is smaller, the more it does not need to rely on other nodes in the terminal delivery process, the more important the role of the distribution center in the network is. Its calculation formula is

$$C_C(w_i) = \sum_{j=1}^{N} h_i(w_i, w_j)/(N-1)$$

In order to eliminate the influence of data processing and comprehensive analysis due to different units of centrality, the calculated data needs to be dimensionless. In addition, because the different networks need to evaluate according to the characteristics of the network itself when evaluating the centrality. Therefore, when the comprehensive evaluation, it is necessary to set the corresponding weight. Finally, the centrality of the nodes is comprehensively scored, and the nodes with higher scores are selected as the alternative terminal co-distribution center. The location model of the alternate point of the terminal co-distribution center of the express delivery is as follows:

$$\Pi_i = \omega_1 C_P + \omega_2 C_I + \omega_3 C_C \tag{1}$$

*(2) Multi-objective location model of the terminal co-distribution center of express delivery*

The task of the terminal co-distribution center of the express delivery is to realize the systemization and scale of the express terminal delivery, so as to further reduce the express delivery cost and improve the service efficiency of the express delivery industry. That is, economics and timeliness are the main objectives of the location modeling of the express terminal co-distribution center. In addition, with the development of e-commerce platform and people's recognition of the express delivery industry, the impact of the rapid increase in demand has become more and more serious to the express terminal delivery network. Therefore, in the site selection process of the terminal co-distribution center of the express delivery, in addition to considering the operating cost of the network, the invulnerability of the entire network should be considered. And how to balance the contradiction between the operating cost of the terminal delivery network and the network's invulnerability has become the key to the location of the express terminal co-distribution center. This section analyzes the location of the express terminal co-distribution center from the perspective of network operation economy and network invulnerability, and establishes a multi-objective location model as follows,

$$
\begin{aligned}
\min Z(L) &= A_1 + A_2 + A_3 + A_4 + A_5 \\
&= \sum_{i=1}^{n} x_i \cdot \Gamma + \sum_{i=1}^{n} \sum_{j=1}^{m} q_{ij} \cdot v \\
&+ \alpha \sum_{i=1}^{n} \sum_{j=1}^{m} c_{ij} q_{ij} d_{ij} + \alpha \left[ \beta \sum_{i=1}^{n} \sum_{j=1}^{m} d_{ij} p_{ij} + (1-\beta) \sum_{i=1}^{n} \sum_{j=1}^{m} d_{ij} p_{ij}^{*} \right] \\
&+ s \sum_{i=1}^{n} \lambda_i B_i (1+e_i)^{r_i+1} \left[ 1 - (1+e_i)^{T_i} (1+f)^{-T_i} \right] / (f - e_i) \\
&+ \theta \sum_{i=1}^{n} \sum_{j=1}^{m} t_{ij} q_{ij} \\
&+ \sum_{i=1}^{n} \sum_{j=1}^{m} \max\left\{ d_{ij} - d_{ij}^{*}, 0 \right\} u_{ij} q_{ij}
\end{aligned} \tag{2}
$$

$$
\max E(L) = \frac{1}{N(N-1)} \sum_{i,j \in N, i \neq j} \frac{1}{d(w_i, w_j)} \tag{3}
$$

$s.t.$

$$
\sum_{i=1}^{n} \lambda_i \leq n \tag{4}
$$

$$
\sum_{i=1}^{n} \lambda_i \varphi_{ij} = 1 \tag{5}
$$

$$
\lambda_i = 0, \sum_{j=1}^{m} q_{ij} = 0 \tag{6}
$$

$$
\sum_{i=1}^{n} q_{ij} \geq Q_j \tag{7}
$$

$$
\varphi_{ij} = \begin{cases} 1, d_{ij} \leq 5 \text{ Demand point is covered} \\ 0, d_{ij} > 5 \text{ Demand point is not covered} \end{cases} \tag{8}
$$

$$
\lambda_i = \begin{cases} 1, \text{The terminal co} - \text{distribution center is selected} \\ 0, \text{The terminal co} - \text{distribution center is not selected} \end{cases} \tag{9}
$$

$$
\beta = \begin{cases} 1, \text{Use small box truck} \\ 0, \text{Use electric tricycle} \end{cases} \tag{10}
$$

$$\sum_{j=1}^{m} \varphi_{ij} q_{ij} \leq E_i \tag{11}$$

$$0 \leq t_{ij} \leq 24 \tag{12}$$

$$q_{ij} \geq 0 \tag{13}$$

Among them, the Eq. (2) indicates that the total operating cost of the terminal delivery network of express delivery enterprise is the lowest, including labor costs, distribution costs, opportunity costs, time costs, and penalty costs. Equation (3) indicates that the network invulnerability of the terminal delivery network of the express delivery enterprise is the best. Equation (4) indicates that the final number of the terminal co-distribution centers cannot exceed the number of candidate points. Equation (5) means that each demand point can only be serviced by a terminal co-distribution center. Equation (6) indicates that if a terminal co-distribution center $w_i$ is not selected, the traffic volume is zero. Equation (7) indicates that the total amount of goods delivered to the demand point by all the terminal co-distribution centers in the end delivery network meets the total demand $Q_j$ of all demand points. Equations (8), (9) and (10) are 0–1 variables. Equation (11) indicates that the total amount of goods delivered by the terminal co-distribution center $w_i$ to the demand point $w_j$ is not greater than its service capability $E_i$. Equation (12) means that the time when the goods arrive at the terminal co-distribution center $w_i$ and then to the demand point $w_j$ must not exceed 24 h. Equation (13) is the value constraint of the variable.

## 3 Model Algorithm

When selecting the candidate points of the terminal co-distribution center of express delivery, it is necessary to determine the weight of the node's centrality. In order to objectively reflect the importance of nodes in the network, this paper uses the entropy method to determine the relevant weight values. In addition, because the multi-target location model of the express terminal co-distribution center is relatively complicated, multiple variables and constraints are involved, and the requirements for the output solution are high. Therefore, this paper uses the genetic algorithm to search for the optimal solution under the condition of satisfying the constraint. The specific design steps of the genetic algorithm are,

Step1: Determine the coding scheme. The coding scheme of the genetic algorithm is designed as follows: if there are n terminal co-distribution centers of express delivery, and m demand points, the coding length is $L = n + n * m$, and the gene is a real number in the interval [−1, 1].

For example: assuming n = 4, m = 6, the length of the code is $L = 4 + 4 * 6 = 28$, then a legal chromosome can be expressed as: [*−0.45*, *0.09*,

*0.26*,*−0.93*, 0.98, −0.84, 0.61, −0.07, 0.21, 0.52, −0.15, 0.36, 0.44, −0.31, 0.78, 0.11, 0.27, −0.13, 0.56, −0.69, 0.22, 0.81, 0.33, 0.71, 0.46, −0.25, 0.64, 0.37].

Coding meaning: convert the italic parts −0.45, 0.09, 0.26, −0.93 into a matrix $K_i$,

$$K_i = \begin{bmatrix} -0.45 & 0.09 \\ 0.26 & -0.93 \end{bmatrix}$$

When $K_i < 0$, it means that the terminal co-distribution center $w_i$ is not selected; when $K_{ij} \geqslant 0$, it means that the terminal co-distribution center $w_i$ is selected.

Convert the non-italic part 0.98, −0.84, 0.61, −0.07, 0.21, 0.52, −0.15, 0.36, 0.44, −0.31, 0.78, 0.11, 0.27, −0.13, 0.56, −0.69, 0.22, 0.81, 0.33, 0.71, 0.46, −0.25, 0.64, 0.37 into a matrix $K_{ij}$,

$$K_{ij} = \begin{bmatrix} 0.98 & -0.84 & 0.61 & -0.07 \\ 0.21 & 0.52 & -0.15 & 0.36 \\ 0.44 & -0.31 & 0.78 & 0.11 \\ 0.27 & -0.13 & 0.56 & -0.69 \\ 0.22 & 0.81 & 0.33 & 0.71 \\ 0.46 & -0.25 & 0.64 & 0.37 \end{bmatrix}$$

Then the code corresponding to the first demand point is 0.98, −0.84, 0.61, −0.07, and so on. When $K_{ij} < 0$, it means that the terminal co-distribution center $w_i$ does not provide service for the demand point $w_j$; when $K_{ij} \geqslant 0$, it means that the terminal co-distribution center $w_i$ provides the service for the demand point $w_j$.

Step 2: Initialize the population. The solution that satisfies the condition is reflected by the initialized population according to the existing known conditions to improve the computational efficiency of the optimal solution.

Step 3: Determine the fitness function. In the solution process, the lowest network operation cost is the main goal, and the network invulnerability is the secondary target. According to the genetic algorithm's requirement for the non-negative of the fitness function value, combined with the main objective function of the model. The fitness function of this study is 1/total cost.

Step 4: Choose. The probability that $i$ is selected is:

$$P_i = f_i / \sum_{n=1}^{N} f_i$$

Step 5: Cross. This paper sets the crossover rate of the genetic algorithm to 0.85.

Step 6: Variation. This paper sets the mutation rate of the genetic algorithm to 0.05.

Step 7: Terminate the condition judgment. In this paper, the termination criterion of the algorithm is that the number of iterations is 1000. If this condition is reached, the algorithm terminates.

## 4 Case Analysis

This paper takes the four private express delivery companies in Beijing Haidian District, such as Yunda Express, Shentong Express, ShunFeng Express and Zhongtong Express as examples to establish a joint distribution alliance for express delivery enterprises. Integrate the existing terminal distribution resources in Haidian District, and rationally layout the terminal co-distribution center to be responsible for the delivery of express mail in Haidian District. Through field research, we have mastered the relevant situation of 76 express delivery branches (function equivalent to the terminal distribution center) in Haidian District. Through the sorting, the 76 express delivery branches currently have 3,590 distribution routes, responsible for the delivery of 1,578 demand points. Among them, 19 Yunda Express Delivery Branches are responsible for delivering 1,279 demand points, 11 Shentong Express Distribution Branches are responsible for delivering 884 demand points, 20 ShunFeng Express Distribution Branches are responsible for delivering 605 demand points, 26 Zhongtong Express Distribution Branches are responsible for delivering 822 demand points, as shown in Fig. 1, which is used as the empirical basis data for the verification of the relevant model. The relevant constant parameters required for the rest of the calculations are set according to the market conditions of the current year.

### *4.1 Analysis of the Terminal Delivery Network Structure of Express Enterprises in Haidian District*

In order to study the structure of the existing express delivery network and the connection between nodes in its network in Haidian District. This paper uses the social network analysis method to process the collected data, and quantitatively analyzes the structure of the terminal delivery network of the express delivery enterprise.

With the help of the network analysis function in the social network analysis software UCINET, the network density of the terminal delivery network of the express delivery enterprise in Haidian District is 0.0026, and the E-I index is −0.889, as shown in Fig. 2. It shows that the links between the nodes in the terminal delivery network of the existing express delivery enterprises in Haidian District are very limited. At the same time, there are many independent small groups in the terminal delivery network, and there is a lack of communication between the groups.

Further, through the visual data analysis function in UCINET, the centrality of each node in the network is studied, as shown in Figs. 3, 4 and 5. According to the analysis, the distribution center of each express delivery company in the existing terminal delivery network in Haidian District has a higher degree of point (Fig. 3). In combination with the actual situation, each express delivery branch is responsible

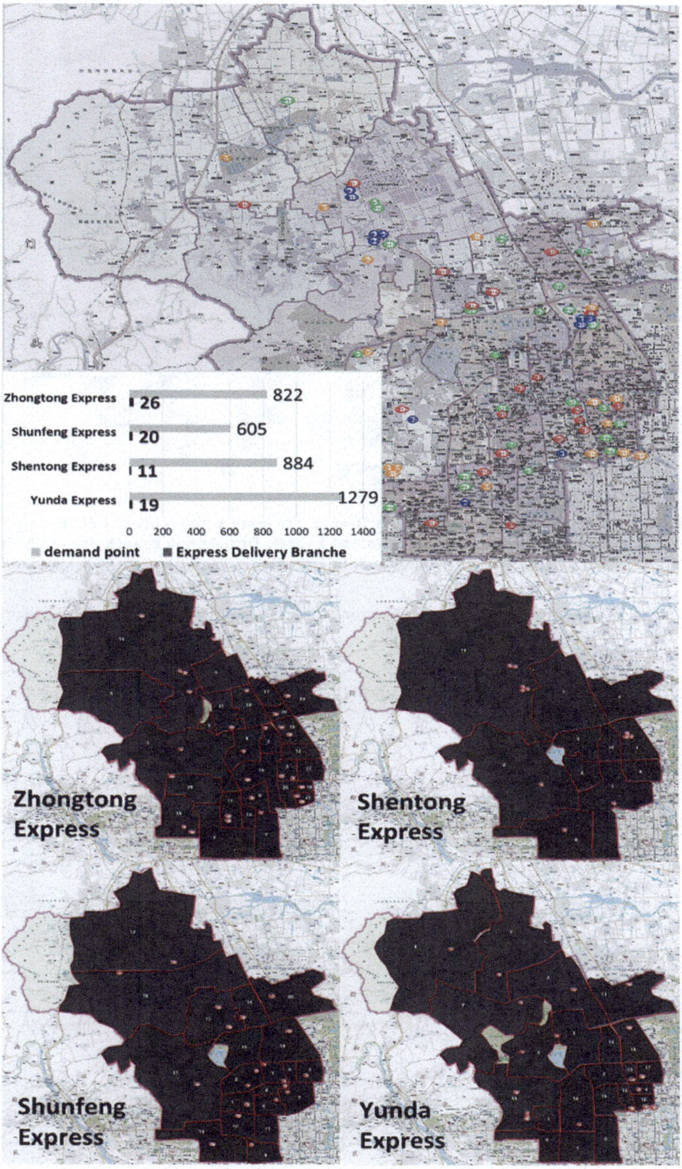


**Fig. 1** The map of distribution areas of express delivery branches in Haidian District

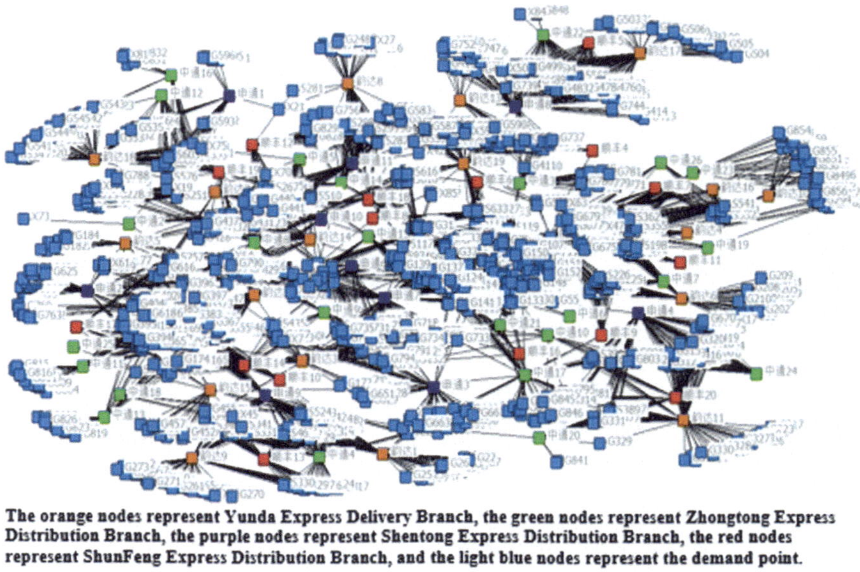


**Fig. 2** Structure of the terminal delivery network of express delivery enterprises in Haidian District

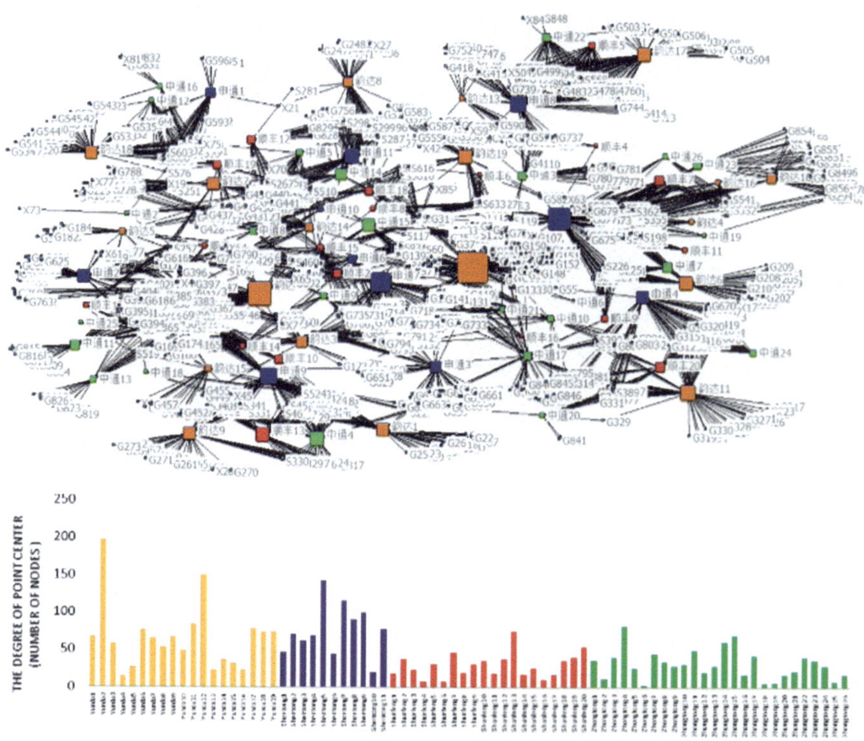


**Fig. 3** Analysis results of the point center

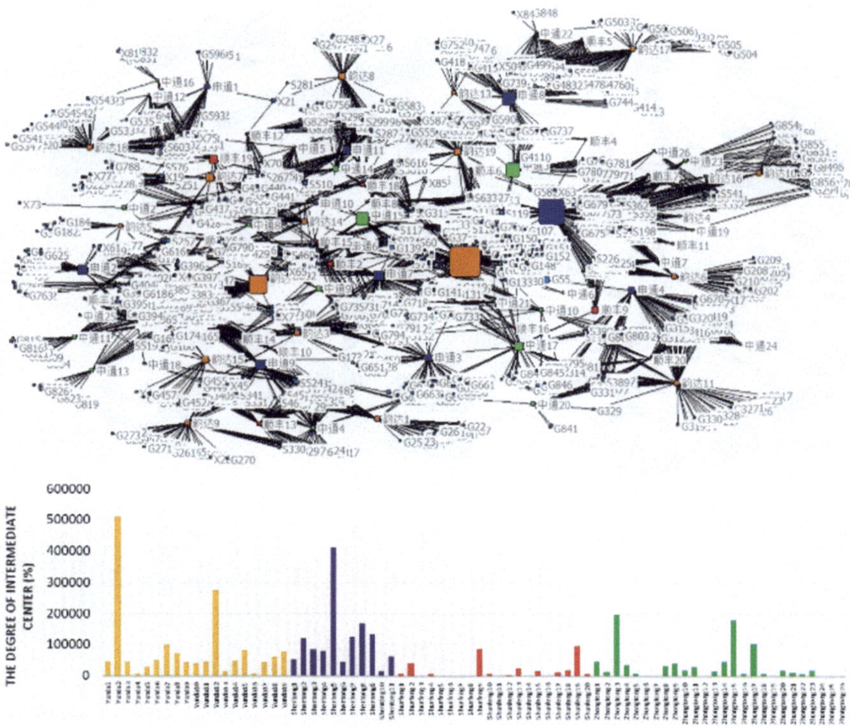

Fig. 4 Analysis results of the intermediate center

for the distribution of goods in different areas in its network, and is directly in contact with many demand points. It is at the center of the terminal delivery network of the respective express company, which determines the distribution efficiency of the goods and the company's service level. Among them, the point center of Yunda Express and Shentong Express Delivery Branches are generally higher than that of ShunFeng Express and Zhongtong Express, indicating that Yunda Express and Shentong Express have relatively better development and market share in Haidian District. In addition, from the analysis of the intermediate center degree (Fig. 4) and the close center degree (Fig. 5), it can be seen that in the current delivery network of the express delivery enterprise in Haidian District, each express delivery branch has relatively high independence and self-contained groups. And there are very few nodes that can serve as a bridge for communication.

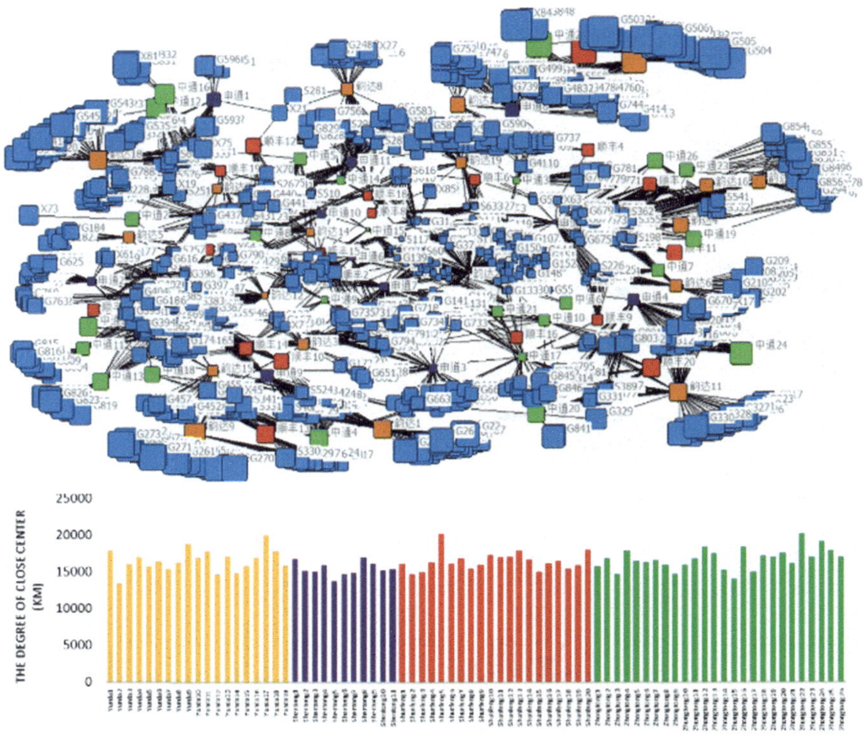


**Fig. 5** Analysis results of the close center

In summary, in the end of the express delivery process in Haidian District, each express company has a waste of resources that are repeatedly constructed at the terminal distribution center, isolated from each other, and repeatedly delivered to the same demand point, which is not conducive to the overall network and the development of individuals in the network. Therefore, it is possible to consider establishing the terminal co-distribution center in Haidian District to re-integrate the existing end-delivery resources of express delivery enterprises. Realize the systemization and scale of express delivery in Haidian District, reduce distribution costs and improve the service level of the express delivery industry.

## *4.2 Determination of the Terminal Co-distribution Center of the Express Delivery Enterprise in Haidian District*

Using the entropy method, the data matrix calculation can be used to obtain the weight values of the point center degree, the intermediate center degree and the close center degree of the nodes in the express delivery network in the current

**Table 2** Central comprehensive scores of the express delivery branches in haidian district

| Distribution Branch | Total score | Ranking | Distribution branch | Totals core | Ranking | Distribution branch | Total s core | Ranking | Distribution branch | Totals core | Ranking |
|---|---|---|---|---|---|---|---|---|---|---|---|
| Yunda2 | 31.442 | 1 | Yunda18 | 4.532 | 20 | Yunda5 | 2.128 | 39 | Zhongtong16 | 0.757 | 58 |
| Shentong5 | 25.044 | 2 | Yunda6 | 4.014 | 21 | Zhongtong23 | 1.580 | 40 | Zhongtong7 | 0.727 | 59 |
| Yunda12 | 17.654 | 3 | Yunda11 | 3.866 | 22 | Shunfeng18 | 1.506 | 41 | Yunda4 | 0.655 | 60 |
| Zhongtong3 | 11.621 | 4 | Shentong1 | 3.657 | 23 | Zhongtong10 | 1.408 | 42 | Shunfeng1 | 0.556 | 61 |
| Zhongtong15 | 11.087 | 5 | Yunda1 | 3.621 | 24 | Zhongtong13 | 1.347 | 43 | Zhongtong18 | 0.497 | 62 |
| Shentong8 | 10.803 | 6 | Yunda17 | 3.596 | 25 | Zhongtong20 | 1.243 | 44 | Shunfeng5 | 0.460 | 63 |
| Shentong9 | 8.864 | 7 | Zhongtong14 | 3.477 | 26 | Shunfeng15 | 1.214 | 45 | Shunfeng4 | 0.460 | 64 |
| Shentong7 | 8.719 | 8 | Yunda3 | 3.466 | 27 | Shentong10 | 1.176 | 46 | Shunfeng3 | 0.436 | 65 |
| Shentong2 | 7.841 | 9 | Yunda9 | 3.414 | 28 | Yunda13 | 1.159 | 47 | Zhongtong24 | 0.416 | 66 |
| Yunda7 | 6.643 | 10 | Yunda14 | 3.298 | 29 | Shunfeng20 | 1.081 | 48 | Shunfeng8 | 0.384 | 67 |
| Zhongtong17 | 6.427 | 11 | Shentong6 | 3.160 | 30 | Zhongtong2 | 0.959 | 49 | Zhongtong12 | 0.357 | 68 |
| Shunfeng19 | 5.953 | 12 | Zhongtong1 | 3.149 | 31 | Zhongtong21 | 0.957 | 50 | Shunfeng14 | 0.327 | 69 |
| Shentong3 | 5.643 | 13 | Zhongtong4 | 3.118 | 32 | Zhongtong22 | 0.904 | 51 | Shunfeng11 | 0.271 | 70 |
| Shentong4 | 5.525 | 14 | Yunda10 | 3.041 | 33 | Zhongtong5 | 0.868 | 52 | Zhongtong26 | 0.259 | 71 |
| Yunda19 | 5.402 | 15 | Shunfeng2 | 2.926 | 34 | Shunfeng17 | 0.836 | 53 | Shunfeng16 | 0.190 | 72 |
| Shunfeng9 | 5.287 | 16 | Zhongtong9 | 2.702 | 35 | Shunfeng10 | 0.833 | 54 | Shunfeng6 | 0.154 | 73 |
| Yunda15 | 5.145 | 17 | Shunfeng13 | 2.545 | 36 | Yunda16 | 0.804 | 55 | Zhongtong19 | 0.133 | 74 |
| Yunda8 | 4.910 | 18 | Zhongtong11 | 2.398 | 37 | Shunfeng12 | 0.796 | 56 | Zhongtong25 | 0.117 | 75 |
| Shentong11 | 4.619 | 19 | Zhongtong8 | 2.305 | 38 | Shunfeng7 | 0.769 | 57 | Zhongtong6 | 0.043 | 76 |

**Table 3** Simulation results of the terminal co-distribution center

| Objective function | Cost/10,000 yuan | Network invulnerability |
|---|---|---|
| Lowest cost | 6131.69 | 0.612 |
| Highest network invulnerability | 7047.65 | 0.898 |
| Optimal cost and network invulnerability | 6557.61 | 0.813 |

Haidian area, which are 0.233, 0.763, and 0.004. Bringing the weight value into (1), we can get the central comprehensive score of each express delivery branch in its express delivery network, see Table 2. The higher the central comprehensive score indicates the higher the importance of the express delivery branch in its terminal delivery network.

Through Table 2, a total of 39 express delivery branches with a score of more than 2 in the central comprehensive evaluation are initially selected as an alternative terminal co-distribution center. The genetic algorithm can be used to calculate the simulation results of the terminal co-distribution center location in the express delivery network under different objective functions, as shown in Table 3.

It can be seen from the location selection results in Table 3 that the network invulnerability is only 0.612 when the total operating cost of the terminal co-delivery network of express delivery is the lowest. It indicates that the terminal co-delivery network will cause large-scale network function failure when attacked. At this time, the express delivery network cannot maintain the original distribution capability, and the express service level of Haidian District is not guaranteed. When considering the network operation cost and network invulnerability at the same time, although the total operating cost of the terminal co-delivery network increased to 65.576 million yuan, the network invulnerability reached 0.813. It indicates that the terminal co-delivery network of express delivery enterprise has strong invulnerability. When the network is attacked, the network can still maintain a relatively stable state, and the network operation cost can be relatively low under the premise of ensuring the express delivery service level.

In summary, when the network operation cost and network invulnerability reach a relative balance, there are 15 express delivery branches in Haidian District that have chosen to establish the terminal co-distribution center of express delivery enterprise. And the feasibility distribution scheme corresponding to these 15 express terminal co-distribution centers in Haidian District is shown in Fig. 6.

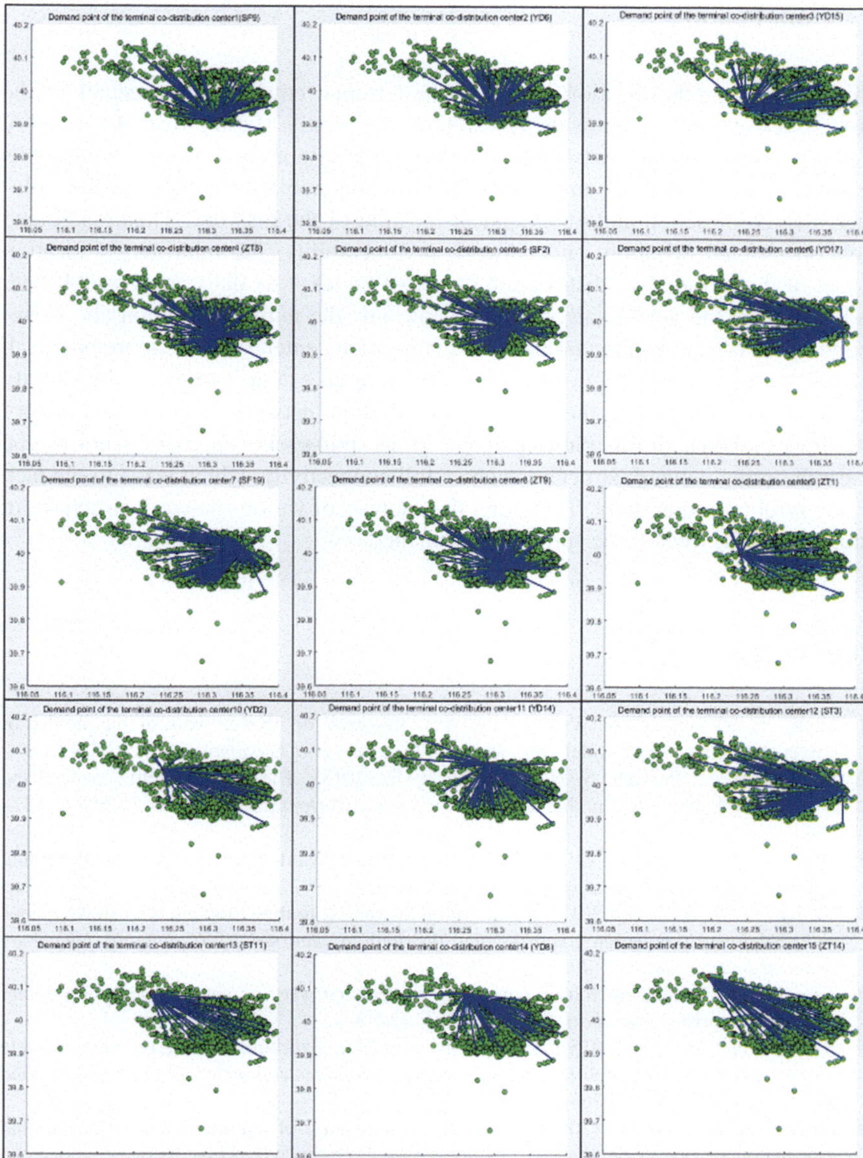

**Fig. 6** Feasible distribution scheme for each terminal co-distribution center of express delivery in Haidian District Conclusions

## 5 Conclusion

This paper takes the location layout of the terminal co-distribution center of the express delivery enterprise as the research object. Firstly, starting from the integrity of the terminal delivery network, considering the importance of each node in its network, it is found that some nodes have incomparable advantages in the influence, degree of action and geographical location of the network. Then, combined with operational economy and network invulnerability, the multi-target location model of the terminal co-distribution center of the express delivery is established, and then used the genetic algorithm to determine the number, location and corresponding service area of the terminal co-distribution center, which is more practical. Finally, the Haidian District of Beijing is taken as an example to prove the validity of the model. However, the terminal distribution network of express delivery enterprises studied in this paper belongs to the axis-spoke structure, ignoring the research on the network structure such as full connectivity, which leads to the lack of universality in this study. Therefore, the analysis of various network structures in future research needs to be further strengthened.

## References


1. Tang, X., Zhang, D., & Wang, Y. (2018). Research on the crowd-sourcing model of university campus express delivery. *Railway Transport and Economy 40*(1), 51–55+69.
2. Kovacs, A. A., Parragh, S. N., & Hartl, R. F. (2015). The multi-objective generalized consistent vehicle routing problem. *European Journal of Operational Research, 247*(2), 441–458.
3. Spliet, R., & Desaulniers, G. (2015). The discrete time window assignment vehicle routing problem. *European Journal of Operational Research, 244*(2), 379–391.
4. Ni, L., Liu, K., & Tu, Z. (2017). Optimization on vehicle routing problem with simultaneous pickup-delivery for urban express joint distribution. *Journal of Chongqing University (Natural Science Edition), 40*(10), 30–39.
5. Cao, X. (2017). Vehicle robust scheduling of joint delivery of single distribution center express enterprises. *Transport Research, 3*(4), 42–48.
6. Xiao, Y., & Konak, A. (2015). A Simulating annealing algorithm to solve the green vehicle routing and scheduling problem with hierarchical objectives and weighted tardiness. *Applied Soft Computing, 34,* 372–388.
7. Zhang, D., & Zhang, Y. (2016). Research on innovation of community logistics terminal distribution service mode. *Journal of Commercial Economics, 1,* 114–116.
8. Feng, T., & Chen, W. (2017). The status quo and countermeasures of the "last kilometer" distribution of university campus express delivery——Taking the university town of economic development zone in Hefei as an example. *China Management Informationization, 20*(10), 87–88.
9. Sun, H., & Yan, C. (2019). Research on the terminal network layout of urban express delivery under joint distribution. *Journal of Wuhan University of Technology (Information & Management Engineering) 41*(2), 186–190+196.
10. Sun, Z., & Zheng, Z. (2017). Research on the application of joint distribution theory to urban end logistics. *Technology & Economy in Areas of Communications 19*(5), 56–59+64.

11. Chen, Q., Hu, J., Wu, J., et al. (2016). Study on location selection problem of city-wide express distribution center based on centrality. *Journal of Zhejiang Sci-Tech University (Social Sciences Edition), 36*(1), 42–48.
12. Zhang, J., Qin, D., & Tang, H. (2018). Empirical research on complex structural properties of distribution network of express enterprise. *Chinese Journal of Systems Science, 26*(3), 86–91.
13. Sheng, L. (2018). Location of logistics distribution center based on quantum particle swarm optimization. *Science Technology and Engineering, 11,* 183–187.

# Radiation Range and Carrying Capacity of Logistics Core City: The Case of Xi'an, China

**Yaqi Zhang, Zhaolei Li, Dan Wei, and Yeye Yin**

**Abstract** The investment size and construction scale for logistics infrastructure of a city depend on its logistics service scope and logistics carrying capacity primarily, especially for the core city on the Belt and Road Initiative of China at present. We propose a comprehensive field intensity and gravity model in order to demarcate the logistics service scope of the core city according to the field intensity factor, medium factor and interaction factor based on radiation theory. In addition, the logistics service scope is further demarcated into direct radiation scope, indirect radiation scope and extended radiation scope of the importance degree. Then, depending on the demarcation of logistics radiation range, the influencing factors of logistics carrying capacity in Xi'an were analyzed using the gray incidence matrix method. The measurement model of the logistics carrying capacity based on the factor correlation was constructed. The rationality of the model is verified. Xi'an is provided to illustrate the practicability and effectiveness of this method as an example. Furthermore, some suggestions about logistics are provided for core cities on the Belt and Road Initiative.

**Keywords** Logistics core city · The belt and road · Logistics carrying capacity · Logistics radiation range

Y. Zhang
School of Economics and Management, Xi'an University of Technology, Xi'an, China
e-mail: yaqi0096@163.com

Z. Li (✉) · D. Wei · Y. Yin
School of Economics and Management, Chang'an University, Xi'an, China
e-mail: lizhaolei@chd.edu.cn

D. Wei
e-mail: wsweidan@163.com

Y. Yin
e-mail: 13294164984@163.com



J. Zhang et al. (eds.), *LISS2019*,
https://doi.org/10.1007/978-981-15-5682-1_64

# 1 Introduction

Over the past two decades, the Yangtze River Delta, Pearl River Delta and Circum-Bohai-Sea Region in China have acquired outstanding achievements due to their location advantage, resource advantage and policy advantage. Recently, central government of China supports relatively backward area by regional rejuvenation plans constantly, such as the Western Development, the Rise of Central Plains, and the Northeast Revitalization. So far, the effect is not ideal and even has a polarizing trend. The GDP growth rate of Liaoning province was-2.5% in 2016, which was the only negative growth one. The initial purpose of the Belt and Road Initiative is to connect all the separate economic regions together and is to promote export-oriented economic development. Although it probably needs long run efforts for every country to realize the prospects of the Belt and Road Initiative, central government of China has promoted it desperately. That was in order to reinforce the connection and exchange between each separate economic region, and to attain the whole prosperity and sustainable development of Chinese economy.

Therefore, logistics infrastructures, transport corridors and information communication networks are constructed in the core cities on the Belt and Road Initiative vigorously. These investments will help these cities get the initiative because they could gain political and financial support from central government of China. And further enhance the city's logistics carrying capacity. For example, Xi'an has promoted the construction of the inland port and free trade zones. Xinjiang has strengthened the construction of the logistics park and frontier port. In addition, some cities have opened special train to Central Europe for gaining the attention of central government, as shown in Fig. 1. Consequently, logistics competition has become the determinant factor of getting policy and fund support from central government. Each the core city on the Belt and Road Initiative is trying every means to improve logistics competitiveness and broaden logistics hinterland. How to demarcate the logistics service scope and measure the logistics carrying capacity of a core city are problems that must be solved first. Because the logistics services scope is directly related to how much logistics quantity will converge to the core city. And the logistics carrying capacity is directly related to whether the core city could meet the logistics demands. According to the logistics services scope, logistics demand and the existing logistics carrying capacity can scientifically determine the additional logistics investment. If not, it will be a blind logistics investment. In order to avoid blind and excessive logistics investment in core cities on the Belt and Road Initiative, the objective of this study is to develop a measure method of radiation range (service scope) and carrying capacity for logistics core cities, and give some useful logistics advices for the core cities on the Belt and Road Initiative.

The efficiency of logistics network is extremely important for promoting the city's or country's competitiveness [1]. The span, depth, and density of logistics service network has significant effects for gathering logistics resources and promoting regional economy [2].

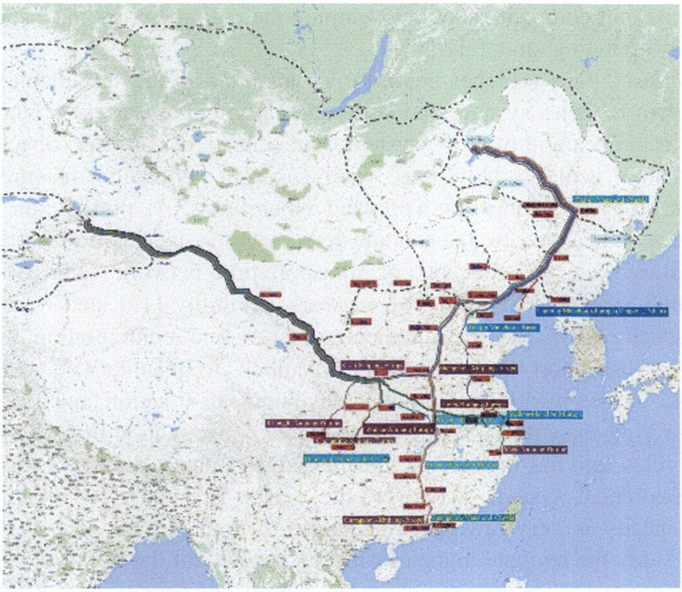

**Fig. 1** China-Europe special freight railway lines

BRI (the Belt and Road Initiative) as one of the "major new international initiatives address logistics issues", it is essential to investigate how the BRI affects international logistics network and further explore the implication for decision-makers. Sheu et al. presents a spatial-temporal logistics interaction model integrated with Markov chain to address the dynamic and stochastic challenges that underlie the problem of international logistic network reconfiguration induced by the One Belt-One Road initiative [3]. It is clear that extending and exerting the function of logistics radiation in the core city and forming a band economic zone with dense point, linear extension and surface radiation can help the economic development of the integration of production and trade along the road.

Research on the logistics radiation range. Liu He established the index system of hub node selection, used principal component analysis to determine the hub city and hub city in Haixi District, measured the logistics gravity intensity of hub city to hub city by gravity model, analyzed the radiation scope of each hub city and the construction of regional logistics corridor, and constructed the hub-and-spoke logistics network in Haixi District [4]. It is insufficient for the logistics hinterland demarcation study at present. Most of the researches focus on the application of methods, such as Gravity Model [5], Voronoi Diagram [6-7], Breakpoint model and Thiessen Polygon [8].

In addition, the researches on port-hinterland have gained wide attention as well as inland city's hinterland, especially after the emerging concept of dry port [9-10]. The dry port is conducive to expanding the scope from seaport to hinterland, improving overall competitiveness, developing the regional economics and making the logistics operating smoothly [11].

On this basis, the researches begin to pay attention to the sustainable development of urban logistics system, and study the carrying capacity of urban logistics system from the bearability of the system. The concept of carrying capacity comes from population ecology. Now it is widely used in population, resources, environment, ecology, economy, network and other fields. At present, the carrying capacity is mainly concentrated in industries, sectors such as transportation and logistics. Based on the concepts of transportation network capacity [12–15], transportation system flexibility [16], network flexibility [17], and transportation reserve capacity [18], the measurement methods and configuration models of transportation (transport, logistics) system facilities are quantitatively studied.

From the above-mentioned research development situation and trend, urban logistics carrying capacity has attracted more and more attention from academia and business circles, and has become an indispensable research content in the complex logistics theoretical system. However, the existing research on urban logistics carrying capacity is very few, the calculation method is very few, and it is not suitable to study the regional logistics carrying capacity of the Silk Road Economic Belt with large span. The logistics carrying capacity of core city on the Silk Road Economic Belt needs to be further studied.

As the starting point of the Silk Road Economic Belt, the connecting point of the Silk Road Economic Belt and the Maritime Silk Road Economic Belt, Xi'an has obvious geographical advantages and should play an important role as a "bridgehead". Therefore, we take Xi'an as a research case. In order to optimize the accurate demarcation of logistics hinterland, and to measure the logistics carrying capacity of Xi'an on the background of the Belt and Road Initiative. We compare the logistics basic data of the core city on the Silk Road from the different transportation modes and identifies the bottlenecks that restrict the Shaanxi regional logistics carrying capacity. We propose a comprehensive field intensity and gravity model to demarcate the logistics hinterland scope of the core city combined with influence factors based on radiation theory, and we measure the logistics carrying capacity by grey incidence matrix method with the case of Xi'an, which is an important dry port and the origin city of the Silk Road Economic Belt.

## 2 Radiation Model of Logistics Core City

### *2.1 Influence Factors of Logistics Radiation Range*

The concept of radiation originates from physics, which means the process that high energy transfer to low energy through a certain medium. The use of the term, radiation, in this study means transferring the energy (logistics technology, information, management and service from all those core cities) to the outside through radiation medium (logistics nodes, transport lines, information networks and so on). Meanwhile, the logistics radiation range of core cities means the origin-destination scope where logistics services originate and disappear. The intensity factor, medium factor and interaction factor influence the radiation scope significantly. The radiation mechanism of logistics radiation range is shown in Fig. 2.

*(1) The field intensity factor*

The field intensity means the size of acting force from the radiation source to some point in the radiation range, which reflects the logistics energy size of logistics core city. The field intensity is the most important influence factor of radiating field, which is showed as the overall scale, service level, information degree, labor quality, logistics facilities and equipment. The comprehensive level of these sub-factors reflects the degree of logistics energy. Equation (1) is used to calculate the field intensity of the core city.

$$S_i = \frac{Z_i}{d_i^{\alpha}} \tag{1}$$

Where $S_i$ is the logistics field intensity of the core city on point $i$; $Z_i$ is the core city's logistics energy; $d_i$ is the distance between the core city and the demanding point $i$; $\alpha$ is the friction coefficient of distance.

*(2) The transferring medium factor*

The transferring medium refers to the transportation systems and information networks. The space movement of cargo is the final form of logistics, so transportation system is a significant factor. Different grade transportation system and different transportation mode influences the logistics radiation ranges directly. That means the distance will be different under different transportation system with the same amount of time and the same origin. In other words, time can be used to

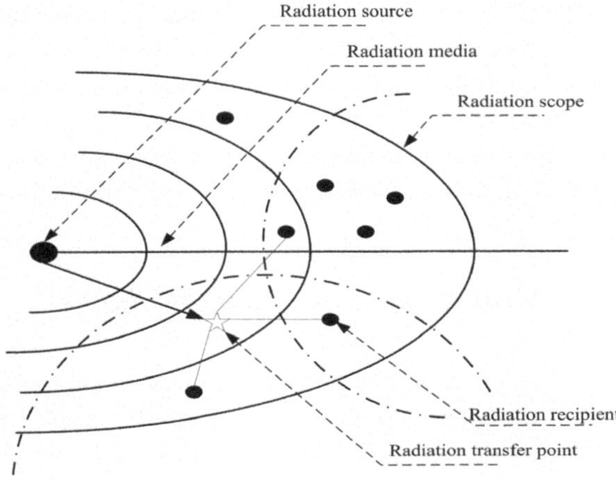


**Fig. 2** The radiation mechanism of logistics radiation range

measure the reachable of traffic facilities. Firstly, in order to determine the speed by different transportation mode and traffic facilities, the regional transportation network is analyzed and concluded systematically. Then, the reachable distance network can be calculated based on the transportation network and different time sets. In practical application, traffic reachable can be specified accord with different logistics service, such as 6-H Circle of Transportation, 8-H Circle of Transportation and so on. It should be noted that the distribution of logistics radiation range through transport corridor doesn't extend as layer but as zonal.

*(3) The interaction factors*

The interactive effect means the connecting and communicating process among the core city and other demanding nodes. This interactive effect reflects in capital flow, information flow, talent flow and material flow specifically. With the analysis of the direction and frequency of the four flows, the interaction degree among the core city and other nodes can be clearly investigated. Therefore, the logistics radiation range, which is divided by interaction degree, can reflect the interaction degree among the core city and other demanding nodes objectively. Since the competition is more than interaction among core cities, and the layouts of those cities are exclusive, the interactive effect is an inherent economic connection. Thus, the regional economic scale or supporting industry scale should be chosen as the main index. Based on the gravity model, the interaction model is established as (2).

$$F_i = \frac{K_i Q q_i}{f(d_i) t_i^{\alpha}} \tag{2}$$

Where $F_i$ is the interaction degree between the core city and the logistics demanding point $i$; $Q$ is the logistics quality of the core city; $q_i$ is the logistics quality of logistics demanding point; $K_i = \frac{Q}{(Q+q_i)}$; $f(d_i)$ is the traffic resistance function between the core city and node $i$; $t_i$ is the transit time between the core city and radiate node $i$; $\alpha$ is the empirical coefficient and $\alpha = 2.0$.

In addition, two aspects should be emphasized. Firstly, if the interactive effect is not from the core city (radiation source) directly but from the radiation transfer point, then the interactive effect needs to be calculated piecewise. Secondly, according to the gratify model, there are all existing interactive effect between the core city and each node demanding point (the values greater than zero). Hence, an interactive degree standard should be set, and if any node city's logistics demand lower than the standard, it was excluded from the logistics hinterland.

## 2.2 Measurement of Radiation Range

Based on the degree of connection between the logistics core city and demanding nodes, the logistics radiation range can be divided into direct logistics radiation scope, indirect logistics radiation scope and extensive logistics radiation scope.

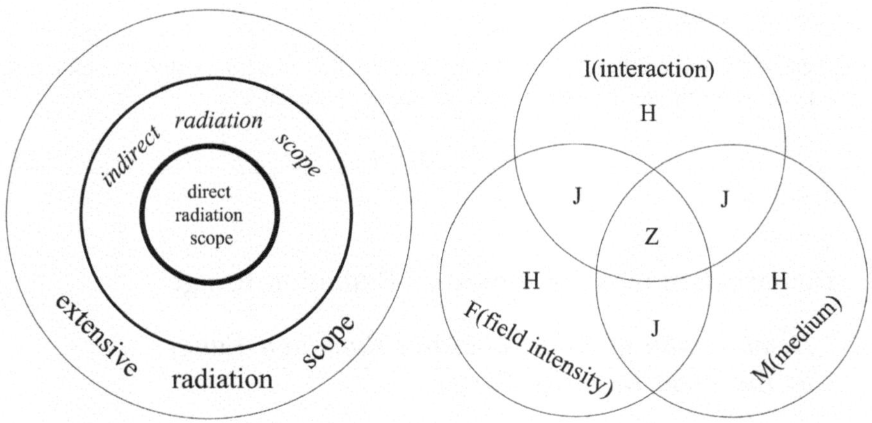


**Fig. 3** Logistics radiation range of the core city

With the analysis of influential factors, we can find that different factors have their significant influence. Different factors have different focus. The field intensity factor demarcates the radiation scope via the logistics energy size of the core city. The medium factor demarcates the radiation scope by the reachability and convenience of the transportation network. The interaction factor demarcates the radiation scope through the degree of economic connection between the core city and its hinterland. We can get different results by different dividing methods. Therefore, we calculate three radiation scope separately and then integrated three factors to obtain optimal logistics hinterland scope. Take the intersection set as the direct radiation scope, the complementary set as the indirect radiation scope and the union set as the extensive radiation scope. The logistics radiation range of the core city is showed as Fig. 3.

*(1) Direct radiation scope*

Based on the three different zones calculated by three factors, direct radiation scope is overlaid three of them and takes intersection, shows as (3).

$$Z = I \cap M \cap F \tag{3}$$

*(2) Indirect radiation scope*

Based on the three different zones calculated by three factors, indirect radiation scope is overlaid three of them and takes complement, shows as (4).

$$J = \overline{1 - [(I \cap M) \cup (M \cap F) \cup (F \cap I)]} \tag{4}$$

*(3) Extensive radiation scope*

Based on the three different zones calculated by three factors, extensive radiation scope is overlaid three of them and takes union, shows as (5).

$$H = I \cup M \cup F \tag{5}$$

# 3 Demarcation of Xi'an Logistics Radiation Range

## *3.1 Demarcation of Xi'an Logistics Radiation Range by the Field Intensity Factor*

Using logistics energy data of the core city and (1), we can get (the field intensity radius of the core city). Setting the core city as the central point, we can get logistics radiation range of the core city. However, considering the availability of data and the characteristics of equation, there are many factors influence logistics energy of the core city. Therefore, the freight volume can be used as the logistics energy index in practical application. The (1) can be transformed to (6).

$$s_i = \frac{Q}{d_i^{\alpha}} \tag{6}$$

Where $s_i$ (ton/km$^2$) is the logistics field intensity of the core city $i$; $Q$ (ton) is the freight volume; $d_i$ (km) is the distance between the core city and the destination $i$; $\alpha$ is the friction coefficient of distance, and $\alpha = 2.0$. Due to the function curve, the distance data for obviously attenuation is the field intensity radius of the core city.

According to the data from Xi'an Bureau of Statistics, the number of freight volumes is 462.69 million tons in 2015, which includes 8.47 million tons freight volumes by railway, 454.01 million tons freight volume by truck and 0.21 million ton by other ways. The attenuation data of Xi'an logistics field intensity with distance is drawn by (6), as shown in Fig. 4. The logistics field intensity attenuates obviously beyond 300 km, so the logistics radiation range by the field intensity factor can be demarcated which is the scope of a radius of 300 km, Xi'an as the center of the circle. This scope could cover Guanzhong-Tianshui economic zone basically.

## *3.2 Demarcation of Xi'an Logistics Radiation Range by the Medium Factor*

Xi'an is located in the center of Chinese land territory and in the intersecting part of the central plains economic zone and the northwest economic zone in China. Besides, Xi'an is in the main path which is from the northwest to the central plains,

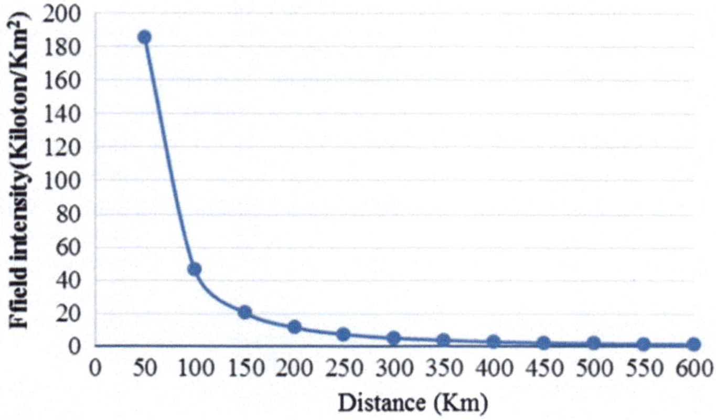


**Fig. 4** The graphs represent the attenuation of Xi'an logistics field intensity with different distance for Q = 462.69 and $\alpha = 2.0$

north China and east China. As the original city of the Silk Road Economic Belt, Xi'an occupies the important strategic position with linking the west and the east, connecting the north and the south. The radiation scope of 8-hour cargo transportation circle of Xi'an is shown as Table 1.

According to the 8-h cargo transportation circle data of Xi'an, its logistics radiation range is vast, which is north to Yinchuan, Ningxia autonomous region and Ordos, Inner Mongolia autonomous region; east to Shangqiu, Henan province and Shijiazhuang, Hebei province; south to Chongqing municipality and Wuhan, Hubei province; west to Chengdu, Sichuan province.

## 3.3 *Demarcation of Xi'an Logistics Radiation Range by the Interaction Factor*

In order to describe the interactive action between Xi'an and other cities, the road freight volume, railway freight volume, GDP, population density, gross value of industrial output, total sales of wholesale and retail trade, retail sales of consumer goods and total export-import volume are selected to be the indexes of interaction factor. The calculation results and ranks of interactive action intensity between Xi'an and other cities in (2) are shown in Table 2. The top 15 cities in each index are included in the logistics radiation range by the interaction factor.

**Table 1** The Hinterland scope of 8-H transportation circle of Xi'an

| Transportation mode | Route | Transportation speed (km/h) | 8-h distance (km) | Reach city |
|---|---|---|---|---|
| Expressway | Shanghai–Xi'an Expressway(G4) | 95 | 760 | Hefei & Lu'an, Anhui province |
| | Fuzhou–Yinchuan Expressway(G70) | 90 | 720 | South to Wuhan, Hubei province; north to Yinchuan, Ningxia |
| | Baotou–Maoming Expressway(G65) | 95 | 760 | South to Chongqing; north to Ordos, Inner Mongolia |
| | Beijing–Kunming Expressway(G5) | 95 | 760 | South to Chengdu; north to Shijiazhuang, Hebei province |
| | Lianyungang–Khorgas Expressway(G30) | 90 | 720 | South to Shangqiu, Henan province; north to Yongchang, Gansu province |
| Railway | Lanzhou-Lianyungang Railway | 80 | 640 | South to Shangqiu, Henan province; north to Lanzhou, Gansu province |
| | Zhengzhou-Xi'an railway | 80 | 640 | Zhengzhou, Henan province |
| | Xi'an-Pingliang railway | 80 | 640 | Pinliang, Gansu province |
| | Houma–Xi'an Railway | 80 | 640 | Houma, Shanxi province |
| | Baotou–Xi'an Railway | 80 | 640 | Baotou, Inner Mongolia |

## *3.4 Demarcation Integrated Logistics Radiation Range of Xi'an*

Based on the analysis above, we can take intersection, complement and union of the three different zones calculated by three factors separately to gain direct radiation scope, indirect radiation scope and extensive radiation scope of Xi'an, as shown in Fig. 5.

The direct radiation scope of Xi'an is determined by the field intensity factors. It can be seen that the location is important for regional logistics hinterland. The medium factor further extends the radiation scope of Xi'an. 11 provinces, municipalities directly under the central Government of China and municipality burgs make up the indirect radiation scope. Relative to direct radiation scope, indirect radiation scope has a vast territory but weaker intensity. The interaction factor

**Table 2** The gravity and rank of interactive effect index between Xi'an and other cities

| Area | City | Road volume of freight | | Railway volume of freight | | GDP | | Population density | | Gross value of industrial output | | Total sales of wholesale and retail trade | | Retail sales of consumer goods | | Total export-import volume | |
|---|---|---|---|---|---|---|---|---|---|---|---|---|---|---|---|---|---|
| | | Gravity | Rank | Gravity | Rank | Gravity | Rank | Gravity | Rank | Gravity | Rank | Gravity | Rank | Gravity | Rank | Gravity | Rank |
| The silk road economic belt | Baoji | 171.35 | 2 | 29.03 | 2 | 153.55 | 2 | 413552.1 | 2 | 105.59 | 2 | 24.19 | 5 | 24.58 | 2 | 265.3 | 3 |
| | Xianyang | 5298.33 | 1 | 1126.8 | 1 | 5144.92 | 1 | 16276094.4 | 1 | 4840.92 | 1 | 455.67 | 1 | 851.26 | 1 | 4559.6 | 1 |
| | Tianshui | 31.39 | 3 | 5.35 | 5 | 21.74 | 6 | 29112.93 | 5 | 10.06 | 6 | 4.4 | 24 | 4.87 | 6 | 31.81 | 34 |
| | Lanzhou | 21.5 | 6 | 3.84 | 8 | 30.04 | 5 | 34192.21 | 3 | 20.36 | 3 | 28.66 | 2 | 8.05 | 4 | 107.78 | 11 |
| | Wuwei | – | – | – | – | 2.09 | 41 | 3078 | 33 | 0.5 | 45 | 0.41 | 43 | 0.23 | 43 | 0.94 | 45 |
| | Jiayuguan | 2.04 | 40 | 0.52 | 27 | 0.91 | 45 | 780.52 | 44 | 1.34 | 40 | 0.35 | 44 | 0.14 | 44 | 1.47 | 44 |
| | Xining | 5.93 | 22 | 0.47 | 28 | 4.11 | 30 | 24228.84 | 8 | 1.6 | 37 | 4.45 | 23 | 1.4 | 21 | 19.22 | 39 |
| | Yinchuan | 7.62 | 17 | 2.12 | 15 | 13.27 | 11 | 23540.56 | 9 | 8.49 | 12 | 7.73 | 17 | 3.37 | 9 | 95.99 | 14 |
| | Urumchi | 1.4 | 43 | 0.77 | 23 | 1.56 | 42 | 329.56 | 45 | 0.85 | 43 | 1.22 | 42 | 0.35 | 41 | 6.55 | 42 |
| | Chongqing | 28.39 | 4 | 8.48 | 4 | 35.21 | 3 | 14321.95 | 12 | 20.01 | 4 | 27.93 | 3 | 8.59 | 3 | 337.78 | 2 |
| | Chengdu | 24.08 | 5 | 9.18 | 3 | 30.08 | 4 | 30471.43 | 4 | 13.09 | 5 | 24.28 | 4 | 7.84 | 5 | 250.57 | 4 |
| | Kunming | 4.86 | 27 | 1.03 | 19 | 4.73 | 27 | 2372.36 | 37 | 2.11 | 34 | 0 | 45 | 1.12 | 27 | 25.42 | 37 |
| | Qujing | 3.34 | 34 | 0.4 | 31 | 1.43 | 44 | 2121.97 | 39 | 0.87 | 42 | 1.6 | 37 | 0.13 | 45 | 1.98 | 43 |
| | Nanning | 2.53 | 39 | 0.13 | 39 | 2.28 | 39 | 2293 | 38 | 0.81 | 44 | 2.05 | 34 | 0.65 | 35 | 10.91 | 41 |
| 21st-Century maritime silk road | Xuzhou | 12.34 | 10 | 4.2 | 7 | 14.76 | 9 | 19751.39 | 10 | 8.77 | 9 | 8.58 | 15 | 2.87 | 17 | 57.03 | 20 |
| | Lianyungang | 6.2 | 21 | 1.8 | 16 | 4.31 | 29 | 7697.89 | 15 | 2.85 | 27 | 4.29 | 26 | 0.96 | 29 | 50.62 | 23 |
| | Hangzhou | 6.97 | 19 | 0.63 | 25 | 11.86 | 17 | 7061.56 | 17 | 7.11 | 14 | 10.7 | 10 | 2.52 | 18 | 98.03 | 13 |
| | Ningbo | 4.97 | 26 | 0.65 | 24 | 5.98 | 22 | 4829.33 | 27 | 3.92 | 22 | 4.71 | 21 | 1.11 | 28 | 76.07 | 17 |
| | Wenzhou | 3.4 | 33 | 0.05 | 41 | 3.2 | 35 | 3058.61 | 34 | 1.94 | 36 | 2.14 | 33 | 0.92 | 30 | 27.32 | 36 |
| | Xiamen | 1.24 | 44 | 0.15 | 38 | 3.73 | 33 | 2431.15 | 36 | 2.56 | 29 | 3.24 | 29 | 0.76 | 33 | 46.37 | 26 |
| | Quanzhou | 2.82 | 36 | 0.17 | 36 | 2.24 | 40 | 4106.03 | 28 | 1.26 | 41 | 1.28 | 41 | 0.43 | 40 | 33.01 | 33 |
| | Fuzhou | 3.23 | 35 | 0.19 | 35 | 4.01 | 31 | 3900.43 | 29 | 2.07 | 35 | 3.65 | 28 | 0.92 | 30 | 37.51 | 30 |
| | Guangzhou | 4.6 | 28 | 0.83 | 21 | 5.92 | 23 | 3316.56 | 30 | 3.15 | 24 | 4.91 | 20 | 1.51 | 20 | 47.36 | 25 |
| | Jinan | 12.68 | 9 | 4.43 | 6 | 13.24 | 12 | 7488.87 | 16 | 6.8 | 15 | 8.23 | 16 | 3.05 | 13 | 45.82 | 27 |
| | Qingdao | 9.58 | 14 | 2.29 | 14 | 6.93 | 21 | 6306.18 | 22 | 4.82 | 21 | 5.28 | 19 | 1.24 | 25 | 72.47 | 18 |

(continued)

**Table 2** (continued)

| Area | City | Road volume of freight | | Railway volume of freight | | GDP | | Population density | | Gross value of industrial output | | Total sales of wholesale and retail trade | | Retail sales of consumer goods | | Total export-import volume | |
|---|---|---|---|---|---|---|---|---|---|---|---|---|---|---|---|---|---|
| | | Gravity | Rank | Gravity | Rank | Gravity | Rank | Gravity | Rank | Gravity | Rank | Gravity | Rank | Gravity | Rank | Gravity | Rank |
| Yangtze river delta | Shanghai | 7.51 | 18 | 1.01 | 20 | 12.58 | 13 | 7882.91 | 14 | 6.68 | 16 | 9.62 | 12 | 3.04 | 14 | 105.74 | 12 |
| | Nanjing | 9.44 | 15 | 1.3 | 17 | 13.59 | 10 | 9503.72 | 13 | 8.72 | 10 | 11.97 | 8 | 3.21 | 10 | 114.55 | 8 |
| | Hefei | 6.92 | 20 | 0.58 | 26 | 12.11 | 16 | 24974.72 | 6 | 8.62 | 11 | 11.66 | 9 | 3.19 | 11 | 134.97 | 7 |
| | Suzhou | 5.14 | 25 | 0.23 | 34 | 7.94 | 20 | 6353.25 | 21 | 5.62 | 19 | 4.48 | 22 | 1.33 | 22 | 113.82 | 9 |
| | Wuxi | 5.92 | 23 | 0.27 | 32 | 9.52 | 19 | 6864.83 | 18 | 6.09 | 17 | 6.87 | 18 | 1.92 | 19 | 94.42 | 15 |
| | Wuhan | 10.31 | 13 | 3.07 | 10 | 11.56 | 18 | 5332.42 | 25 | 5.91 | 18 | 9.49 | 13 | 3 | 15 | 66.89 | 19 |
| | Changsha | 18.11 | 7 | 0.46 | 30 | 12.58 | 13 | 18741.92 | 11 | 5.19 | 20 | 9.13 | 14 | 3.5 | 8 | 171.06 | 5 |
| | Shengzhen | 2.64 | 37 | 0.16 | 37 | 5.28 | 24 | 1796.96 | 41 | 3.06 | 25 | 3.19 | 30 | 1.19 | 26 | 50.65 | 22 |
| | Zhuhai | 1.57 | 42 | 0 | 42 | 2.66 | 37 | 1391.14 | 43 | 2.17 | 33 | 1.65 | 36 | 0.54 | 37 | 35.11 | 32 |
| | Foshan | 3.83 | 31 | 0.07 | 40 | 1.44 | 43 | 5323.89 | 26 | 1.38 | 38 | 1.53 | 38 | 0.32 | 42 | 36.67 | 31 |
| | Dongguan | 2.6 | 38 | 0 | 42 | 3.53 | 34 | 1788.88 | 42 | 2.59 | 28 | 1.97 | 35 | 0.73 | 34 | 49.19 | 24 |
| | Zhongshan | 2.02 | 41 | 0 | 42 | 2.62 | 38 | 1995.38 | 40 | 2.18 | 32 | 1.42 | 40 | 0.5 | 38 | 30.9 | 35 |
| Bo Hai coastal region | Beijing | 10.75 | 12 | 2.85 | 11 | 18.36 | 7 | 6498.55 | 20 | 9.75 | 7 | 14.82 | 7 | 4.84 | 7 | 161.08 | 6 |
| | Tianjin | 11 | 11 | 2.71 | 12 | 12.56 | 15 | 5927.88 | 24 | 7.45 | 13 | 9.74 | 11 | 3.19 | 11 | 108.49 | 10 |
| | Shijiazhuang | 15.27 | 8 | 3.72 | 9 | 14.88 | 8 | 24599.08 | 7 | 9 | 8 | 15.76 | 6 | 2.98 | 16 | 84.58 | 16 |
| | Tangshan | 7.87 | 16 | 2.45 | 13 | 5.04 | 26 | 6592.35 | 19 | 3.19 | 23 | 3.02 | 31 | 0.88 | 32 | 38.96 | 29 |
| | Qinhuangdao | 4.49 | 29 | 1.11 | 18 | 2.67 | 36 | 6032.14 | 23 | 1.38 | 38 | 1.47 | 39 | 0.49 | 39 | 13.58 | 40 |
| | Shenyang | 4.12 | 30 | 0.27 | 32 | 4.65 | 28 | 3112.62 | 32 | 2.3 | 30 | 4.36 | 25 | 1.25 | 24 | 19.7 | 38 |
| | Dalian | 3.53 | 32 | 0.82 | 22 | 5.09 | 25 | 3127.92 | 31 | 2.96 | 26 | 3.75 | 27 | 1.27 | 23 | 43.75 | 28 |
| | Yantai | 5.46 | 24 | 0.47 | 28 | 3.94 | 32 | 2694.15 | 35 | 2.25 | 31 | 2.52 | 32 | 0.65 | 35 | 56.16 | 21 |

*Note* Road volume of freight gravity is calculated by road transportation time; Railway volume of freight gravity is calculated by railway transportation time; other gravities are calculated by road transportation time which is less than 7 h, or by railway transportation time which is more than 7 h

**Table 3** Logistics demand analysis of Xi'an district

| District name<br>Sequence item | Xincheng District | Beilin District | Lianhu District | Weiyang District | Yanta District | Baqiao District | Yanliang District | Lintong District | Changan District |
|---|---|---|---|---|---|---|---|---|---|
| The population density [million/km2], m | 2.06 | 3.77 | 1.68 | 0.23 | 0.74 | 0.18 | 0.10 | 0.08 | 0.07 |
| Population [10000 people], r | 64 | 83 | 64 | 61 | 112 | 55 | 25 | 67 | 103 |
| Per capita income of urban residents [yuan], c | 3334.50 | 3400.33 | 3400.16 | 3283.08 | 3452.33 | 3167.58 | 3377.75 | 2683.08 | 2920.67 |
| GDP [100 million yuan/month], g | 42.01 | 56.39 | 48.03 | 60.43 | 96.52 | 27.69 | 15.99 | 16.03 | 42.73 |
| Logistics flow [package/day], d | 388.08 | 154.00 | 126.31 | 713.08 | 617.85 | 85.92 | 30.23 | 34.154 | 159.08 |
| Rotation volume of freight transport[kg], h | 3498.88 | 1146.06 | 1037.08 | 5054.66 | 5352.22 | 607.58 | 125.21 | 246.55 | 1207.93 |
| Daily return [yuan], s | 13980.60 | 5454.00 | 2738.83 | 19053.00 | 16112.50 | 2961.17 | 891.33 | 1031.00 | 4508.83 |
| Logistics service node, j | 5 | 1 | 2 | 6 | 6 | 2 | 1 | 2 | 3 |
| Logistics flow/node, L | 77.62 | 154.00 | 63.15 | 118.85 | 102.97 | 42.96 | 30.23 | 17.08 | 53.03 |
| Daily return/node, W | 2796.13 | 5454.00 | 1369.41 | 3175.50 | 2685.42 | 1480.59 | 891.33 | 515.50 | 1502.94 |

tremendously extends the logistics hinterland of Xi'an to some developed province of southeast coast and the northeast. The extensive radiation scope of Xi'an connects economic belt surrounding the Circum-Bohai-Sea Region and Yangtze River Delta. Including many vital node cities in the Belt and Road Initiative, the extensive radiation scope is the widest but its intensity is the weakest.

## 4 Logistics Carrying Capacity Model of Xi'an

The model of carrying capacity is constructed to measure the current level of carrying capacity of the regional logistics based on the above study. Combining with the demand forecasting, the reasonably logistics construction resources could be allocated in order to enhance the carrying capacity of regional logistics. Taking Xi'an as an example, the carrying capacity of the logistics core city is studied.

The objective condition of logistics carrying capacity is mainly determined by the geographical location of logistics nodes. Its capacity depends on the number of the local population and the level of logistics demand. The problems of logistics nodes in the construction process are often based on experience to solve, due to the lack of macro planning, leading to waste of resources and lack of resources and so on. Therefore, firstly we should measure the carrying capacity of the regional logistics network, according to the carrying capacity of the logistics system, combined with the regional logistics demand to carry out the logistics construction, and enhance the carrying capacity of regional logistics.

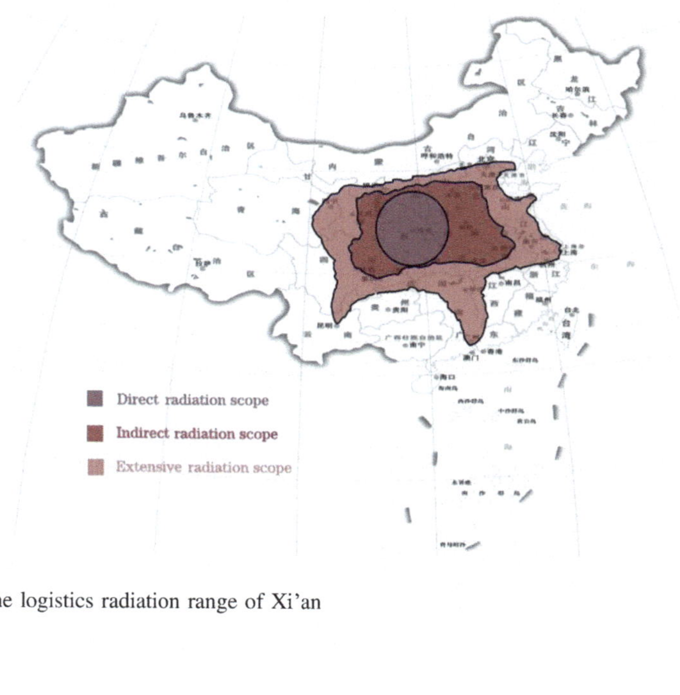


**Fig. 5** The logistics radiation range of Xi'an

### 4.1 Influence of Various Factors on Carrying Capacity by Grey Incidence Matrix Method

The processing capacity of logistics nodes is generally adapted to the factors such as the density of the local population, the total population, the per capita income of the urban residents, and the regional production value. And how these factors affect logistics flow, revenue and rotation volume of freight transport is not obvious, so the method of grey incidence matrix is used to analyze. The grey relational analysis method regards the factor value of the research object and the influence factors as the point on a curve, compares the closeness between them, and quantifies them separately; calculate the correlation degree of the degree of closeness between the object of study and the influencing factors of the object to be identified. At the last, by comparing the degree of correlation, the influence of the object to be identified with the object of study is judged.

*(1) Algorithm steps of grey incidence system method*

The grey relational system theory puts forward the concept of grey correlation analysis for each subsystem, and intends to seek the numerical relationship among subsystems by means of certain methods. Grey correlation analysis provides a quantitative index of the development of a system, which is suitable for dynamic process analysis.

(a) *Determine the reference sequence reflecting the system behavior characteristics and the comparative sequence affecting the system behavior:* The reference sequence is the local population density, the total population, the urban residents' per capita income and the local GDP; the comparison sequence is the logistics flow, rotation volume of freight transport and daily income.
*(b) Dimensionless treatment of reference sequence and comparison sequence.*
*(c) Seeking grey correlation coefficient of the reference sequence and comparative sequence* $\xi(X_i)$: The degree of association is the degree of difference between the geometric shapes of the curves. A reference sequence $x_0$ has a number of comparative sequences $x_1, x_2 . \;\; . \;\; . x_n$. In the formula for the degree of difference between the comparative sequence and the reference sequence at each moment (that is, the points on the curve), ρ is the resolution coefficient, generally between $0 \sim 1$, usually take 0.5. $\Delta(\min)$ means the minimum difference of class two, $\Delta(\max)$ means the maximum difference of class two, $\Delta_{0i}(K)$ is the absolute difference between each point on the $x_i$ curve of comparison sequence and each point on the $x_0$ curve of the reference sequence. Therefore, the calculation formula of the association degree is simplified as follows:

$$\xi_{0i} = \frac{\Delta(\min) + \rho\Delta(\max)}{\Delta_{0i}(K) + \rho\Delta(\max)} \tag{7}$$

*(d) Seeking relevancy $r_i$*: Because the correlation coefficient is the correlation degree between the comparison sequence and the reference sequence at each moment, so it's numerical more than one. Therefore, it is necessary to concentrate the correlation coefficients of each moment into a numerical value to make a holistic comparison, That is, the average value of the correlation coefficient is used as the quantitative representation of the correlation degree between the comparison sequence and the reference sequence, the formula of correlation degree $r_i$ is as follows:

$$r_i = \frac{1}{N} \sum_{K=1}^{N} \xi_i(K) \tag{8}$$

The closer the value of $r_i$ to 1, the better the correlation

*(e) The ranking of correlation degree:* The degree of association between factors is mainly described by the ranking of the degree of association. The m subsequence is arranged in order according to the correlation degree of the same parent sequence, and then the association order is formed, denoted by $\{x\}$, which reflects the "good and bad" relationship of the subsequence for the parent sequence. If $r_{0i} > r_{0j}$, then $\{x_i\}$ is superior to $\{x_0\}$ for the same parent sequence $\{x_j\}$; $r_{0i}$ is representing the eigenvalues of the i subsequence to the parent sequence.

*(2)* Matlab realization of grey relational analysis and construction of logistics carrying capacity model.

Based on the above algorithm, the local population density, the total population, the per capita income of urban residents and the regional GDP are taken as the reference sequence. The logistics flow, rotation volume of freight transport and daily income are compared. Taking Xi'an city as an example, the corresponding data of 9 districts were collected, and the data of logistics flow, freight turnover and daily income of a logistics company in Xi'an were forwarded to all parts of the country in March, 2017.

Programming with Matlab the grey incidence matrix is as follows: (Table 4)

It can be found that the logistics flow, rotation volume of freight transport and daily income are most closely related to population density and GDP in this region from the Fig. 6.

**Table 4** The incidence matrix

| $r_{ij}$ | $d$ | $h$ | $s$ |
|---|---|---|---|
| $m$ | 0.7097 | 0.7270 | 0.7004 |
| $r$ | 0.6193 | 0.5925 | 0.5615 |
| $c$ | 0.5385 | 0.5476 | 0.5810 |
| $g$ | 0.7126 | 0.7275 | 0.6729 |

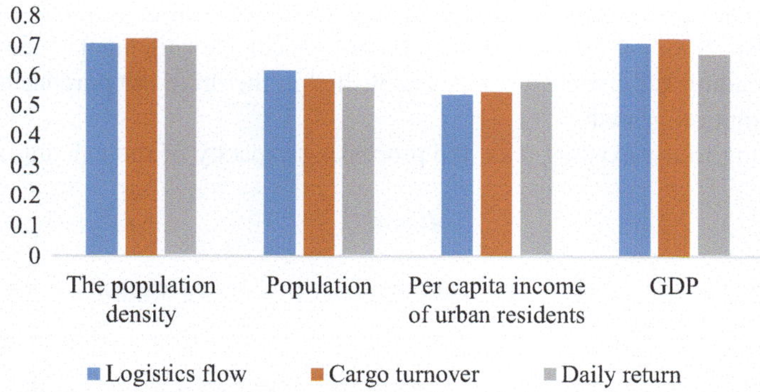


**Fig. 6** The effect diagram of each factor on logistics quantity

According to the grey incidence matrix, the ranking of the processing requirements of the nodes (i.e. the order of the odd number) can be described by the equation:

$$d = 0.7097m + 0.6193r + 0.5385c + 0.7126g + \varepsilon \tag{9}$$

ε is the correction factor.

The ranking of the logistics carrying capacity should be consistent with the logistics demand, but only a ratio coefficient K. K is determined by the level of regional economic development and the speed of economic development. Therefore, the measurement model of logistics carrying capacity is as follows:

$$F = k \cdot d \tag{10}$$

The number of nodes to be established here is:

$$j \propto d \tag{11}$$

With the construction of "The Belt and Road" and the growth of logistics demand we can determine the number and location of the construction of regional logistics nodes by using this method, in order to ensure the node warehouse suitable, not a waste of resources.

### 4.2 *Empirical Analysis of Logistics Carrying Capacity in Xi'an*

Table 3 shows that the income of urban residents is not very different, and the GDP has an obvious difference in the district of Xincheng, Beilin, Lianhu, Baqiao,

Weiyang, Yanta, Yanliang. The reason is that Weiyang District as a new city center which Xi'an National Economic and Technological Development Zone and the Xi'an new administrative center was established in the drive the purchasing power and consumption growth.

According to the existing data, the processing capacity of the existing nodes is:

$$L = d/j \tag{12}$$

According to the equation and data, the daily processing capacity of the node is calculated as follows:

$$L(Beilin) > L(Weiyang) > L(Yanta) > L(Xincheng) > L(Lianhu) \\ > L(Changan) > L(Baqiao) > L(Yanliang) > L(Lintong)$$

The revenue value of each node is approximately calculated by the following equation:

$$W = r/j \tag{13}$$

According to the equation and data, the node revenue is sorted as follows:

$$W(Beilin) > W(Weiyang) > W(Xincheng) > W(Yanta) \\ > W(Changan) > W(Baqiao) > W(Lianhu) \\ > W(Yanliang) > W(Lintong)$$

The two sets of sorting are approximately the same, so the point of a big deal of processing is also large, and the data fitting is done by Matlab, the fitting result is shown as Fig. 7.

As showed in the Fig. 6, processing capacity and the revenue value of each node are approximately linear. The processing capacity of Beilin, Weiyang, and Yanta District is large, almost > 100. They should add more nodes or increase the staff, to ensure no blasting warehouse. However, the processing capacity of each node in Lintong district is low, which may result in waste of resources, so a node can be revoked.

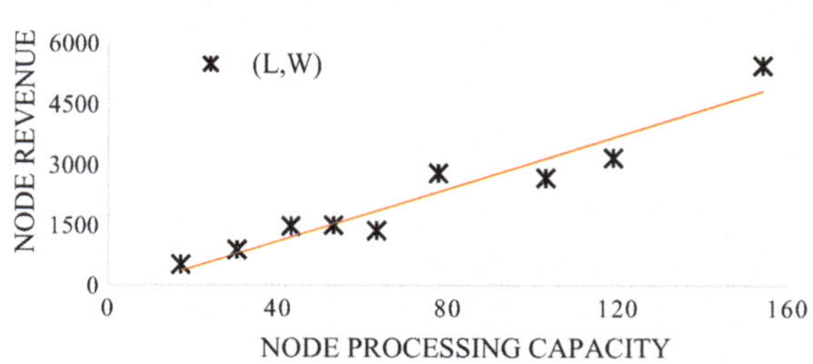


**Fig. 7** Correlation analysis between node processing capacity and revenue

## 5 Conclusion

The core city diffuse capital flow, information flow and material flow to its relevant node cities through transferring medium, which reflects strong attraction and radiate effects. The study finds that demarcation the logistics hinterland of core cities and its logistics carrying capacity have significant influence on the planning and construction of the Belt and Road Initiative. We can draw the conclusion from the analysis above.

The field intensity factor is a decisive factor for radiation range. In order to improve the logistics energy of the core city, there are two effective measures. On the one hand, we can promote the development of manufacturing and commercial trade, so as to facilitate the logistics business. On the other hand, we can reinforce the development of logistics industry as well as infrastructure, thus the logistics service will be improved and the logistics hub will be established.

The medium factor is important for the span of logistics hinterland. The outline of logistics radiation range is zonal which is along the transport corridors. In the Belt and Road Initiative, transport corridor is the guide line for the core cities. Therefore, based on the main transport corridors, more branch networks are needed, which will improve the convenience and reachability of transportation network. In addition, apart from transport system, we find that information network can improve logistics efficiency, reduce logistics costs and widen logistics hinterland. The discussion has been limited to gain data for quantitative analyzing the impact of the information medium. However, the logistics information platform has become one of the necessary requisites for core cities. Such as Guiyang city, it has become a regional central city for taking "big data in logistics" as a resort.

The interaction factor is related to the sustainability of logistics hinterland closely. The demand of logistics is random and fluctuant, then the logistics radiation range will vary with the demand. In order to make sure the sustainability of logistics hinterland, the stable connection among node cities is needed, especially a steady long-term trade operation with countries along the Belt and Road Initiative.

The measurement model of logistics core city is constructed to measure the logistics carrying capacity at the current situation. Combining the demand forecasting, reasonably allocate logistics construction resources, and enhance the carrying capacity of regional logistics. The study found that the logistics carrying capacity of various urban areas in Xi'an is inconsistent with the demand. The logistics carrying capacity can be regulated by adjusting the number of nodes, the number of employees, facilities and equipment.

**Acknowledgment** This work was supported by Natural Science Basic Research Plan in Shaanxi Province of China (Program No. 2020JM-257) and China Postdoctoral Science Foundation (Program No. 2017M623183).

## References


1. Mesa-Arango, R., & Ukkusuri, S. V. (2015). Demand clustering in freight logistics networks. *Transportation Research Part E: Logistics & Transportation Review, 81,* 36–51.
2. Bensassi, S., Márquez-Ramos, L., Martínez-Zarzoso, I., et al. (2015). Relationship between logistics infrastructure and trade: Evidence from Spanish regional export. *Transportation Research Part A: Policy & Practice, 72,* 47–61.
3. Sheu, J. B., & Kundu, T. (2018). Forecasting time-varying logistics distribution flows in the One Belt-One Road strategic context. *Transportation Research Part E: Logistics & Transportation Review, 117,* 5–22.
4. He, X. F., & Jing, Y. W. (2013). Empirical Study on influencing factors of Sichuan logistics development based on grey relation degree model. *Applied Mechanics and Materials, 253–255,* 1492–1495.
5. Heuvel, F. P. V. D., Rivera, L., Donselaar, K. H. V., et al. (2014). Relationship between freight accessibility and logistics employment in US counties. *Transportation Research Part A: Policy & Practice, 59,* 91–105.
6. Xie, W. J., & Ouyang, Y. F. (2015). Optimal layout of transshipment facility locations on an infinite homogeneous plane. *Transportation Research Part B: Methodological, 75,* 74–88.
7. Long, X. Q., Zhang, Y. B., & Chen, Y. R. (2011). Using voronoi diagram in construction the scope of logistics park hinterland: An engineering application. *Systems Engineering Procedia, 2,* 69–76.
8. Bowen, J. T. (2012). A spatial analysis of FedEx and UPS: hubs, spokes, and network structure. *Journal of Transport Geography, 24,* 419–431.
9. Shi, X., & Li, H. (2016). Developing the port hinterland: Different perspectives and their application to Shenzhen Port, China. *Research in Transportation Business & Management, 19,* 42–50.
10. Awad-Núñez, S., González-Cancelas, N., Soler-Flores, F., et al. (2016). How should the sustainability of the location of dry ports be measured? A proposed methodology using Bayesian networks and multi-criteria decision analysis. *Transportation Research Procedia, 14,* 936–944.
11. Wang, X. C., Meng, Q., & Miao, L. X. (2016). Delimiting port hinterlands based on intermodal network flows: Model and algorithm. *Transportation Research Part E: Logistics & Transportation Review, 88,* 32–51.
12. Chung, K., Rudjanakanoknad, J., & Cassidy, M. J. (2007). Relation between traffic density and capacity drop at three freeway bottlenecks. *Transportation Research Part B: Methodological, 41*(1), 82–95.
13. Nicol, D. J. (1980). Determining the capacity of a transportation system-Passenger airlines. *Omega, 8*(5), 545–552.
14. Yang, H., & Bell, M. G. H. (1998). A capacity paradox in network design and how to avoid it. *Transportation Research Part A: Policy & Practice, 32*(7), 539–545.
15. Yang, H., Bell, M. G. H., & Meng, Q. (2000). Modeling the capacity and level of service of urban transportation networks. *Transportation Research Part B: Methodological, 34*(4), 255–275.
16. Morlok, E. K., & Chang, D. J. (2004). Measuring capacity flexibility of a transportation system. *Transportation Research Part A: Policy & Practice, 38*(6), 405–420.
17. Bassamboo, A., Randhawa, R. S., & Van Mieghem, J. A. (2010). Optimal flexibility configurations in newsvendor networks: going beyond chaining and pairing. *Management Science, 56*(8), 1285–1303.
18. Xiao, L. L., Liu, R. H., & Huang, H. J. (2014). Stochastic bottleneck capacity, merging traffic and morning commute. *Transportation Research Part E: Logistics & Transportation Review, 64,* 48–70.

# Thoughts and Construction on Improving the Quality of Training Postgraduates in Different Places

**Bin Dai and Ye Chen**

**Abstract** The construction of research institutes in different places is an important way for universities to realize the industry-education integration, and it is also in line with the requirements of the construction of "double first-class". The research institutes have positive significance for the development of the subjects, the incremental development and aggregate resources. But it is also faces some problems such as insufficient input of the adviser, insufficient academic atmosphere, insufficient support of the courses and the influence of the quality of the enrollment. This paper puts forward the concrete ways and methods to improve the quality of postgraduates in different places from four aspects: industry-education integration, employment-oriented enrollment, training process and mode diversification, and teaching staff motivation system.

**Keywords** Double first-class · Industry-education integration · Research institutes in different places · Training of postgraduates in different places

## 1 Introduction

At the National Education Conference on Teachers' Day in 2018, General Secretary Xi Jinping delivered an important speech on whom to train, how to train and for whom to train. It points out the direction for speeding up the modernization of education, building a powerful country in education, and running a satisfactory

This work was supported by 2018 Special Fund for Postgraduate Education and Development Research of Beihang University. Fund number: BUAA2018015.

B. Dai (✉)
Development and Planning Department, Beihang University, Beijing, China
e-mail: daibin@buaa.edu.cn

Y. Chen
School of Public Administration, Beihang University, Beijing, China
e-mail: qq5101575@126.com



J. Zhang et al. (eds.), *LISS2019*,
https://doi.org/10.1007/978-981-15-5682-1_65

education for the people [1]. At present, in order to deeply study and implement Xi Jinping's thought of socialism with Chinese characteristics in the new era and the spirit of the 19th CPC National Congress, and to promote the close connection between education, science, technology, and industry, universities promote the integration of industry and education actively [2].

Therefore, some universities have explored the construction of research institutes in different places to carry out the work of cultivating postgraduates [3]. This way can narrow the distance between talent training and industrial development, and at the same time, it can also enable universities to make use of the support given by the local government and enterprises to promote the development of scientific researches, disciplines and schools [4]. It is of great significance to understand the situation of postgraduates training in different places, to clarify the existing problems, and to find the corresponding countermeasures for perfecting the training work of postgraduates in different places, better realizing the integration of production and education, and promoting the construction of double first-class students [5].

## 2 Literature Review

The integration of production and education has become an important research topic [6]. China's economy has entered a new normal of steady development, enterprises have reduced the use of personnel, and the difficult of employment for college graduates has become prominent problem [7]. In this context, universities began to attach importance to the integration of production and education in the training process [8]. At the same time, due to the problems in the development of higher education and vocational education, China has entered a new stage of innovation-driven development, the acceleration of technological revolution and the cluster breakthrough of new technology. The integration of production and education has become a national strategy [9].

Although the integration of industry and education is becoming a strategic choice to solve the main contradiction between industry with higher education, the imbalance of economic and social development [10]. However, in the process of promotion, we still face some problems, such as the restriction of traditional thinking inertia and institutional mechanism [11].

The main reasons for these problems were influenced by three kinds of institutional logic: the national logic with the conflict of power, the market logic with property rights and benefits, and the university logic with discipline and discipline [12]. These three kinds of logic contradict the development goal of "integration of production and education", which leads to the institutional dilemma of "integration of production and education" [13].

In order to deepen the integration of production and education, (1) finding a systematic breakthrough, legislation, the construction of teachers and give full play to the role of private education in the leading position to deepen the implementation of the integration of production and education [14]. (2) the government to "set up the stage" and for the market to "sing the show". Under the "four-in-one" integration mode of production and education, the cooperative dynamic mechanism of the integration of production and education was solved [15]. Public universities and private universities have different development basis and policy background [16]. The state should not only actively promote the market-oriented "introduction" mechanism of public schools [17], but also pay attention to the inherent advantages and incentive policies of the integration of production and education in private universities [18]. (3) The government, industry, university, enterprise and intermediary were the five cooperative subjects to deepen the integration of production and education [19]. The key factors affecting the deepening of production and education integration were talent, technology, concepts, systems, projects, funds, vision, institutions, platforms and information [20]. The integration of production and education was not only the integration of various subjects, but also the integration of key factors [21].

The construction of research institutes in different places was a specific form of the integration of production and education, and its core purpose was to improve the quality of graduate training [22]. At present, some studies have discussed the quality of graduate training. Graduate training was a systematic work, therefore, the evaluation of training quality also needs to be systematically promoted [23]. The quality of students, training, degree and development were the main factors that affected the quality of graduate training (Fig. 1). Among them, the quality of development involves the employment feedback mechanism, which was related to the integration of production and education [24].

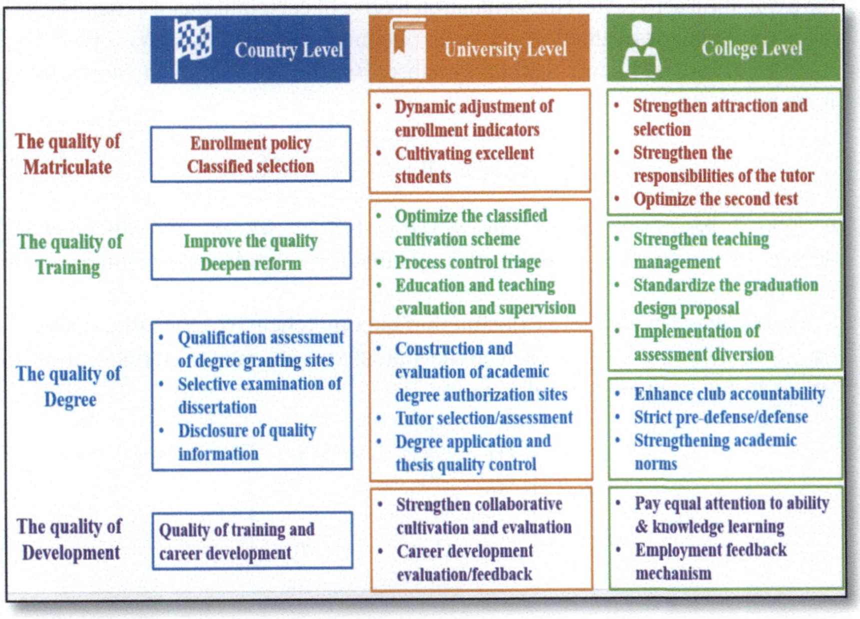


**Fig. 1** Evaluation and Assurance system of Educational quality

At the same time, some studies began to pay attention to the evaluation of graduate training quality from the quantitative level. Some researchers have evaluated the quality of management graduate education with Fuzzy Analytic Hierarchy Process (FAHP) [25]. The specific steps for conducting a comprehensive evaluation as follows:

(1) Determine the set of factors. According to the established hierarchical structure of the ladder, the factor set U = $\{u_1, u_2, \ldots\ldots, u_m\}$, U = $\{u_{i1}, u_{i2}, \ldots\ldots, u_{is}\}$, $m$ is the number of the first index, and $s$ is the number of the second index under the first index $i$.
(2) Establish the evaluation set. The evaluation model V = $\{v_1, v_2, \ldots\ldots, v_m\}$, $v_1$, $v_2$, $\ldots v_m$, $n$ represented the corresponding evaluation grades, such as excellent, good, medium, poor and so on.
(3) Introduce the weight set respectively. According to the weight in the evaluation index system, the weight set of the first level can be obtained, and the weight set of the second level W = $\{w_{i1}, w_{i2}, \ldots, w_{is}\}$.
(4) Single factor evaluation. Based on the evaluation of each evaluator, the single factor evaluation matrix $R_i$ is established, as in (1).

$$R_i = \begin{bmatrix} r_{i11} & r_{i12} & \cdots & r_{i1n} \\ r_{i21} & r_{i22} & \cdots & r_{i2n} \\ \cdots & \cdots & \cdots & \cdots \\ r_{is1} & r_{is2} & \cdots & r_{isn} \end{bmatrix} \quad (1)$$

$0 \leqslant r_{isy} \leqslant 1 (x = 1, 2, \cdots, s; y = 1, 2, \cdots, s)$, $s$ is the number of secondary indexes that corresponding to each primary index, $n$ is the number of elements in the evaluation set $V$. After the evaluation matrix of each unit factor is determined, each single factor evaluation vector is obtained based on $B_i = W_i \times R_i = \{b_{i1}, b_{i2} \ldots, b_{in}\}$, and the hierarchical fuzzy subset Bi* can be obtained by normalizing it, as in (2).

$$B_i^* = \{b_{i1}^*, b_{i2}^*, \ldots\ldots, b_{in}^*\} \quad (2)$$

$t = b_{i1} + b_{i2} + \ldots + b_{in}, 0 \leqslant b_{ij}^* \leqslant 1 (j = 1, 2, \ldots, n)$, is the membership degree of grade fuzzy subset $B_i$* obtained by grade $v_j$ to single factor evaluation.
(5) Comprehensive evaluation. Construction of comprehensive evaluation matrix R based on fuzzy subset of each grade obtained from vector normalization of single factor evaluation results, as in (3).

$$R = \begin{bmatrix} r_{11} & r_{12} & \cdots & r_{1n} \\ r_{21} & r_{22} & \cdots & r_{2n} \\ \cdots & \cdots & \cdots & \cdots \\ r_{m1} & r_{m2} & \cdots & r_{mn} \end{bmatrix} \quad (3)$$

Then, through W × R={$b_1$,$b_2$…,$b_n$} gets the comprehensive evaluation vector B, and the grade confusion set B* can be obtained by normalization, as in (4).

$$B^* = \{b_1{}^*, b_2{}^*, \ldots\ldots, b_n{}^*\} \quad (4)$$

$t = b_1 + b_2 + \ldots + bn, 0 \leqslant b_j{}^* \leqslant 1(j = 1, 2, \ldots, n)$, is the membership degree of the grade fuzzy set B* obtained by grade $v_j$ to the comprehensive evaluation.

In addition, a research has compared and analyzed the quality evaluation of graduate education in Britain, the United States and Australia, obtained the corresponding data through the graduate experience survey. The dimensional indicators of the three countries were similar to each other (Table 1). Among them, the

**Table 1** Common indicators of investigation on postgraduate education quality in three countries

| Main dimension | Main indicators |
|---|---|
| Mentor guidance | Mentors provide useful advice and guidance in all aspects |
| | Mentors provide both useful progress feedback to guide graduate research |
| | The tutor has enough professional and technical level to guide the students |
| | Mentors provide guidance and advice on career choices |
| Courses | High quality teaching (teaching methods, teaching skills, teaching incentives, etc.) |
| | Effective teaching feedback (teacher-student interaction, teaching evaluation is feedback, etc.) |
| | Students' participation and input in and out of class |
| Environment and resources | Cultural atmosphere (respect for students, encourage communication, etc.) |
| | Facilities and resources (learning venues, libraries, equipment, etc.) |
| Organization and management | Students' evaluation on the arrangement and management of training links |
| | Understanding and evaluation of degree thesis defense procedure and Standard |
| Skills and professional development | General skill development (communication, writing, teamwork, etc.) |
| | Research capacity improvement (project management, innovation, analysis, etc.) |
| | Professional development (professional and professional confidence) |
| Comprehensive satisfaction degree | Overall study satisfaction and evaluation of satisfaction with each link and element |

dimension of "skill and professional development" is also related to the integration of production and education [26].

The research findings have important reference significance for this study. Of course, this study is not focus on the overall quality of graduate training, but a specific analysis of the integration of production and education.

# 3 The Significance of the Construction of Research Institutes in Different Places Under the Background of the Integration of Industry and Education

As an important part of the integration of industry and education, the construction of research institutes in different places is of great significance to the cultivation of postgraduates.

## *3.1 Improve the Quality of Subjects*

The improvement of discipline quality was reflected in two aspects. On the one hand, improve the quality of talent training. The quality of graduates is an important index to measure the quality of talent training, and the judgment of the quality of graduates is based on the evaluation of employers. The construction of research institutes in different places makes some of the superior disciplines and teachers of the school transfer to the outside world, and close relations have been established with the development of local industries, so that scientific researchers can better serve the local social and economic development [27]. In this process, the cultivation of postgraduates will gradually form a strong interaction with the development of the industry, thus promoting the formation of customized employment intentions between local enterprises and graduates.

In this way, enterprises increase the opportunity and probability of obtaining "ready-to-use" talents, and then make a satisfactory evaluation of the quality of graduates, and affirm the quality of talent training in the subjects in which graduates are located and in schools as a whole. Furthermore it lay an important foundation for further improvement in the future. On the other hand, the construction of research institutes in different places improves the reputation of social service and discipline. The positive evaluation of talent quality by enterprises and the government will urge schools to pay more attention to the construction of relevant disciplines, improve the process of student training, strengthen the construction of teaching staff, and provide more and stronger support for education and scientific researches. At the same time, it will also attract more students to apply, which will provide a sustainable driving force for the development of the subject.

## 3.2 Incremental Development

Incremental development is reflected in two aspects: scientific research projects and scientific research teams. On the one hand, the increase of new projects and the transformation of mature scientific research achievements. Some dominant subject and teachers of the school have been transferred to the research institutes in different places, and made the combination of scientific research with local industries development. Guided by actual demands, scientific research can solve the problems in the process of industrial development more specifically and pertinently, which will become the source of crosswise tasks and promote the growth of scientific research funds. This growth effect is difficult to achieve in Alma Mater. At the same time, the mature scientific research achievements of Alma Mater teachers can come to the research institutes in different places with the postgraduates, and make full use of support to carry out the transformation of the results, and obtain more economic and social benefits [28].

On the other hand, promote the professional construction of professional scientific research team. The scientific research work of the research institute in different places has clear goal orientation, that is, to serve the local industries and enterprises. The scientific research team of Alma Mater was extensiveness and dispersion, The scientific research team of the research institutes in different places can better concentrate its time, energy and resources into a number of specific fields, which improves the professionalism of the scientific research team in some fields. At the same time, the scientific research team undertakes the task of training postgraduates, so it can also teach more professional research ability and research methods to postgraduates, and improve the professional ability of them.

## 3.3 Aggregate Resources

Aggregate resources not only refers to the support from local enterprises and governments, but also the external absorption of other resources.

On the one hand, many domestic universities gathered in central cities, but the resources of the central cities were limited, and the support provided to universities was weak. Research institutes in different places have sufficient resources from local governments and enterprises to provide policies, funds, equipment, infrastructure, and so on, which provide rich resources for the construction of relevant disciplines and the training of talents [29]. It also lays a sufficient resource foundation for the development of the institutes and the universities. For example, some scientific research projects of Alma Mater can be carried out by using the advanced and perfect experimental equipment of the research institutes in different places.

On the other hand, On the other hand, the construction of research institutes in different places have also strengthened the attraction of foreign resources. With the formation of the cooperative relationship, the superior resources of universities and localities were combined. Some universities, enterprises, scientific research projects, etc. will take a look at the platform of the research institutes and actively cooperate. At the same time, more cooperation subjects and cooperation projects will bring more resources to the research institutes and provide more powerful support for postgraduate education of research institutes in different places.

# 4 Challenges in the Training of Research Institutes in Different Places

At present, Tsinghua University, Peking University and Beihang University are promoting the construction of the research institutes in different places and carrying out postgraduates training in different places. However there are some problems in the process.

## 4.1 *Insufficient Adviser's Invest*

Due to the fact that there is a certain distance between the city of the research institutes in different places and the city where the university is located, the face-to-face communication between the postgraduates and the advisers is difficult. Although advisers and postgraduates can communicate with each other through mail, telephone or other means to discuss academic issues, but the effect is difficult to compare with that of face-to-face communication. More importantly, it is difficult to achieve regular face-to-face communication with advisers, which also means that postgraduates are unable to participate in adviser's research and lose the opportunity to learn and improve themselves. Thus it can be seen that the above objective situation is difficult to establish a close relationship between postgraduates and advisers in different research institutes. The instruct of advisers to postgraduates is more like a part-time behavior, which will undoubtedly affect the quality of postgraduates' training.

## 4.2 *Insufficient Academic Atmosphere*

Although training postgraduates in different places close the relationship between training work and industrial development and provides a good learning and development platform for them. However, due to the excessive industrial atmosphere, the academic atmosphere has been affected. For example, some postgraduates trained in different places are required to enter the research institutes to carry out specific research work after the completion of their courses [5]. In this way, the communication between these postgraduates and senior will be separated.

At present, the main scientific research energy of most advisers are focused on the headquarters of the universities, and the postgraduates instructed by them are also mainly in the main area. Students from different research institutes can get the help of adviser or senior sisters and brothers apprentice in time after admission. The ability and experience of research can be accumulated quickly through continuous team learning. However, after entering research institutes in different places, postgraduates have to adapt to a completely different research state, unable to participate in team learning, which will inevitably weaken the improvement of personal research ability and the development of research work.

### 4.3 Insufficient Curriculum Support

At present, different universities have different programs for postgraduates' courses in different places. In some universities the postgraduates have to complete the theoretical courses in the headquarters of the universities and enter the research institutes in different places to carry out researches. Some other universities complete the teaching and scientific researches in research institutes. However, due to the limitation of teaching resources and the focus of teaching management, any way will face the problem of insufficient curriculum support. Because most of the courses are arranged in the headquarters of the universities, and the time span of the courses is basically one semester, these students need to stay in different places for most of the time, once they are unable to restudy in time, it will inevitably affect graduation and will not be conducive to the smooth development of the overall training work.

### 4.4 Affect the Quality of Enrollment

The quality of enrollment has an important impact on the quality of training. Compared with the headquarters of the university, the quality of postgraduate enrollment in research institutes in different places cannot be guaranteed effectively. Because the headquarters of the universities are concentrated in the central big city and have the advantage of regional support, it has most attraction [6]. The increase of the number of examinees will inevitably aggravate the competition, some of the non-advantage of the universities, the enrollment scores of non-top disciplines will also make the competition more intense because of the regional advantage. Of course, the fierce competition also ensures the quality of the enrollment, so that the future training work can be maintained at a high level. However, due to the fact that the scientific research focus of postgraduates in research institutes in different places is not in the headquarters of the universities, even the whole study and scientific research process, so that the candidates will consider more factors. The research institutes in different places no longer has the advantage of the location of the headquarters of universities, and the intensity of applying for the examination cannot be compared with the enrollment of the headquarters of universities. With the weakening of the competition and the reduction of the number of candidates, the quality of enrollment cannot be guaranteed naturally.

# 5 Thoughts and Suggestions on Improving the Quality of Postgraduate Education in Different Places

In view of the above problems, in order to improve the quality of postgraduates in different places, it is necessary to take improvement measures from the following four aspects.

## *5.1 Industry-Education Integration*

The landing of research institutes in different places is bound to provide full support for the development of local industries, enterprises and economic society, also, deepen the degree of integration of production and education.

There are three specific ways to realize the integration of production and education (Fig. 2), that is, cooperation with enterprises, transformation of achievements, multi-level cooperation. Each way can strengthen the theoretical level and practical ability of postgraduates from different dimensions, which is of positive to strengthen the quality of training.

First of all, cooperate with enterprises actively. In order to serve the local industry, it is necessary to strengthen the cooperation with local enterprise in the process of cultivation of postgraduates in research institutes in different places. The specific ways include the double tutor system, the construction of joint laboratory

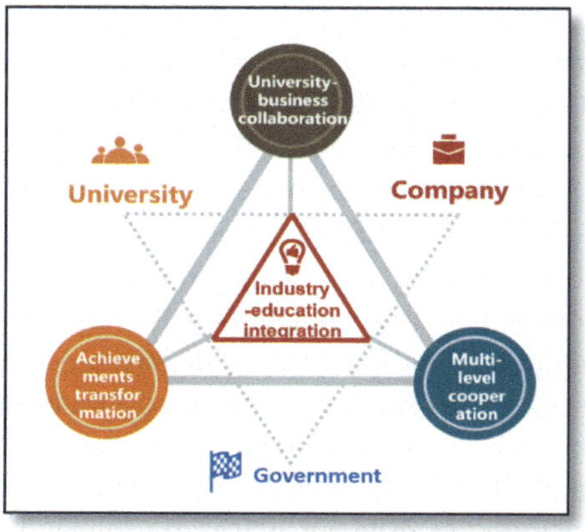


**Fig. 2** Ways of integration of production and education

and the implementation of joint training. The research institutes in different places and the advisers should provide sufficient theoretical guidance to the postgraduates to improve their scientific research ability. Meanwhile the enterprises should provide sufficient practical opportunities and practical guidance to the postgraduates. Different ways of cooperation make postgraduates not only extract theoretical problems, think theoretically and stimulate the motivation of theoretical learning, but also apply the knowledge mastered in theoretical learning to practice, and constantly improve their scientific research and practical abilities.

Secondly, we should promote the transformation of achievements actively. At present, many universities faculty have accumulated much research results, but affected by many factors, the transformation of results is not ideal. It is suggested that the postgraduates could bring the scientific research results to the research institutes in different places, make full use of local funds, technology and policy support to promote the transformation of the achievements. At the same time, the process of achievement transformation is also a scientific research process. Students can carry out scientific research training through the corresponding practice to improve the ability to find and solve problems.

Finally, carry out multilevel cooperation and accumulate resources. In order to improve the construction and development of research institutes in different places, the research institutes in different places could establish cooperative relations with universities, research institutes and enterprises with the help of local government. So that we can better carry out academic exchanges, teacher exchanges, joint training, scientific research practice base, scientific research achievements transformation and other work.

## 5.2 Employment-Oriented Enrollment

In order to clarify this orientation, must emphasis on the ideas transfer at enrollment stage and opportunities provision in the training process.

In order to better serve local industries and enterprises and improve the matching degree between talent training and demand, enrollment work needs to pay attention to employment-oriented. On the one hand, publicize the concept of enrollment accurately. In the stage of enrollment, we must convey the characteristics of research institutes in different places, the relationship with enterprises, employment-oriented training objectives and other information effectively, so that the candidates can fully understand and lay the foundation for the smooth development of the follow-up training work. Of course, in the specific operation, we can try to take the local and surrounding areas as the main source of candidates, at the same time radiate the whole country, improve the pertinence of enrollment work. On the other hand, in the process of training, research institutes in different places need to provide postgraduates with rich channels of practice and employment through various ways. Only when the stůdents are considered in many ways can we absorb and train more and better postgraduates, and then improve the matching degree between the postgraduates' training and the needs of the local industry.

## 5.3 *Diversified Training Process and Ways*

Exploring diversified training processes and methods play an important role in ensuring the training quality of postgraduates in different places. The main breakthrough points are advisers guarantee and the construction of scientific research team.

In terms of adviser security (Fig. 3), it is important to establish an effective adviser assessment mechanism at first. The training in different places makes the face-to-face communication between postgraduates and advisers is difficult and the lack of such communication will inevitably affect the guidance of advisers to postgraduates and the progress of students' studies.

Therefore, the indicators of face-to-face guidance can be included in the work assessment of advisers, so that the importance of face-to-face guidance between advisers and postgraduates trained in different places can be clearly defined. Once they do not meet the requirements, the performance appraisal of advisers will be affected. This will give some supervision to the advisers. To achieve the goal of face-to-face communication with postgraduates, advisers can refer to the following ways. Help postgraduates to take advantage of winter and summer vacation to study in the headquarters of the universities by participating in academic activities, so as to facilitate the communication between postgraduates and advisers and better receive the guidance of advisers. Guided by the development of scientific research projects, some of the work can be completed by advisers and postgraduates in research institutes in different places, which not only promotes the process of the project, but also strengthens the communication between advisers and postgraduates. It also makes full use of the advantages of scientific research facilities of

**Fig. 3** Tutor guarantee mechanism

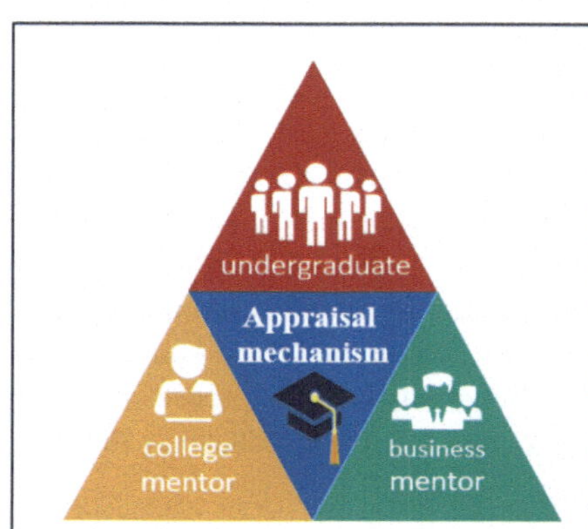

research institutes in different places. It is also necessary to implement the system of deputy tutor in enterprises. Inviting the enterprise managers with high professional level and rich experience as the second adviser of the postgraduates to provide more opportunities for postgraduates to receive guidance face to face, which is also conducive to the postgraduates to broaden their horizons. Promoting postgraduates to better combine theory with practice, and improve postgraduates' ability to solve practical problems.

In the terms of construction of scientific research team. The scale and specialization of scientific research team will not only affect the quality of postgraduates training, but also affect the effect of cooperation with local enterprises. Therefore, it is necessary to continue to strengthen the professional construction of existing scientific research teams and enhance the ability to provide knowledge and technical support to local industries and enterprises. In addition, research institutes in different places can set up their own or jointly set up postdoctoral workstations with enterprises, absorb more talents, and strengthen the scientific research team. The improvement of the construction of scientific research team can not only provide postgraduates with good theoretical guidance, make up for the lack of communication between advisers and postgraduates, but also provide better services for local industries and enterprises.

## *5.4 An Effective Incentive System for Teaching Staff*

The construction of teaching staff is also an important factor affecting the training of postgraduates in research institutes in different places. Only with a strong targeted incentive system can the research institutes in different places attract more needed talents and better retain the talents.

Taking a survey on the expatriate incentive mechanism of teaching staff in BUAA as an example. As shown in Table 2, 18 persons were "very much agreed" that they would be sent to participate in school-land cooperation, it takes up 7%; 97 persons were "more agreed, and specific discussions can be held", it takes up take up 39%; the cumulative proportion of people with a tendency to "agree" was 46%. It can be seen that at present, the willingness of teaching staff to work in remote research institutes is not strong.

According to the survey, there are four kinds of factors that affect the willingness of BUAA teaching staff.

**Personal Development**

There are 106 peoples believe that the assignment is very or relatively helpful to personal development among the 115 teaching staff who have intention to "agree to assignment", it takes up 92%. To this end, if the individual can obtain development space will tend to agree to assignment.

**Table 2** Survey results

| Survey results | Persons | Proportion |
|---|---|---|
| Agree | 18 | 7% |
| Comparative agree | 97 | 39% |
| Comparative disagree | 104 | 42% |
| Disagree | 27 | 12% |
| Total | 246 | 100% |

### Salary Level

The average expected annual salary of teaching is 328, 000 yuan. Among them, the lowest expected annual salary is 80, 000 yuan, the highest is 1 million yuan, the choice of most people are between 300,000 yuan and 500,000 yuan. The expected annual salary is also an important factor.

### Living Security

The living security mainly includes two aspects: the accompanying family members and the housing security of the assignment location. In terms of accompanying family members, the survey shows (Fig. 4), whether for all teaching staff or those who tend to "agree" to the assignment, the proportion of going with their children is the highest; The lowest proportion is going with elderly. Another key living security factor is the housing security of the assignment location. According to the survey, the overall demand for housing in the school headquarters is as high as 95%.

**Fig. 4** Distribution situation accompanying family members

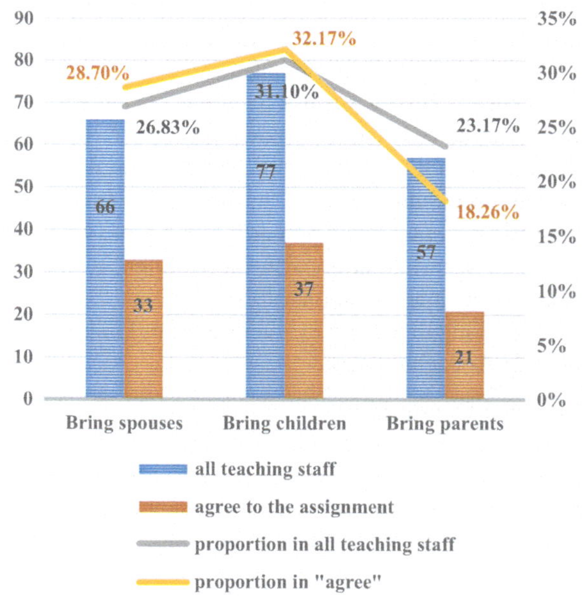

**Table 3** Evaluation form of assignment location

| Consent degree | Persons | Proportion |
|---|---|---|
| Comprehensive influence of Universities in the local area | 136 | 55.28% |
| Degree of local economic and social development | 132 | 53.66% |
| Local regional ecological and human environment | 176 | 71.54% |
| One's place of origin or social relationship | 143 | 58.13% |
| Other factors | 62 | 25.2% |

**Assigning Location**

The survey found (Table 3) that 176 peoples think the most important factor is "the degree of local economic and social development," which determines the material basis for the development of school-land cooperation, and also determines the local living standards of expatriate teachers, staff and their accompanying families, it takes up 71%; Secondly, the proportion of factors such as "local regional ecological and human environment", "Subject layout of School-Land Cooperation Base", "comprehensive influence of universities in the local area" are similar that around 55%; Thirdly", "One's place of origin or social relationship" is relatively low, indicating that the impact of this factor is relatively small.

When strengthening the construction of incentive system, we need to pay attention to the above four factors.

Therefore, on the one hand, each research institute in different places should further refine the existing incentive system. At the same time, we should also design an incentive mechanism in accordance with its own characteristics actively, such as salary system, insurance of family members and vocational development. In addition, in order to achieve sustainable development, research institutes in different places must constantly supplement the teaching staff to meet the needs of the construction of the teaching staff.

## 6 Conclusion

The research focus on the problems such as insufficient input of the adviser, insufficient academic atmosphere, insufficient support of the courses and the influence of the quality of the enrollment in the "Industry-education Integration". This paper puts forward the concrete ways and methods to improve the quality of postgraduates in different places from four aspects: industry-education integration, employment-oriented enrollment, training process and mode diversification, and teaching staff motivation system.

## References


1. Cai, W. H., Tao, W., & Dai, R. D., et al. (2013). Special zone of higher education: a case study of Shenzhen University City and Suzhou higher education park. *Education Teaching Forum, 20,* 4–7.
2. Gong, M. L., & Yang, J. (2017). The difficulties and their corresponding measures on the different area campus scientific research in the multi-campus university mode—Taking Zhuhai Campus of some universities as an example. *Science and Technology Management Research, 37*(12), 91–96.
3. Qin, W. Q., Peng, X. F., & Li, J. J. (2012). How to maintain harmonious relations between postgraduates and their tutor who lives and works in other part of the country—take Jishou University for example. *Academic Journal of Shanxi Provincial Committee Party School of C. P.C., 6,* 114–117.
4. Shi, Q. H., & Kang, M. (2017). Research and comments on universities with multi-campuses in different cities. *Journal of National Academy of Education Administration, 7,* 21–27.
5. Sang, F., Ma, H. Y., & Yang, J. M., et al. (2014). A preliminary study on the employment of postgraduate students under the mode of running a school in different places. *China Postgraduates, 10,* 51–53.
6. Zhang, Y., Zhang, J., & Qin, L., et al. (2013). Discussion on the management of postgraduate education based on running a school in different places. *Science and Technology Innovation Herald, 20,* 170.
7. Liu, J. P., Song, X., & Yang, Z., et al. (2019). Construction of university practical teaching system using principle of "integration of production and education, cooperation by school and enterprises". *Research and Exploration in laboratory, 38*(4), 230–245.
8. Sheng, Z. F. (2018). Newly-established industry-education integration of the undergraduate colleges and universities in new era. *Forum on Contemporary Education, 3,* 25–31.
9. Chen, F. (2018). The integration of production and education: the path of deepening and evolution. *China Higher Education, 1*(13), 13–16.
10. Yuan, J. Y. (2018). Industry-university integration in china: historical evolution and ultima strategic choice. *China Higher Education Research, 4,* 55–57.
11. Huang, L., & Sui, G. H. (2019). Research on cracking mechanism of the integration of production and education in colleges under background of applied transformation. *Heilongjiang Researches on Higher Education, 37*(2), 89–93.
12. Chen, Y. G., Zhuo, Z., & Liu, S. J., et al. (2015). Analysis of the framework of postgraduate education quality evaluation in industry-characteristic universities from the perspective of financial institutions. *Journal of Graduate Education, 3,* 65–71.
13. Cao, B. Z. (2018). Practical exploration of innovation model of industry-education integration and school-enterprise cooperation in the new era. *Vocational and Technical Education, 39* (14), 16–18.
14. Chen, S. (2019). Further reforms of the integration of production and education. *Journal of HeBei Normal University (Educational Science Edition), 21*(3), 19–21.
15. Fan, T., Liang, C. J., & Zeng, Q. D. (2016). Construction and application of comprehensive quality evaluation system of graduate students based on fuzzy comprehensive evaluation method. *Heilongjiang Researches on Higher Education, 11,* 85–90.
16. Li, Y., & Liu, L. (2015). The graduate quality assurance system of the university of Ottawa and its inspiration. *Journal of Graduate Education, 5,* 84–89.
17. Liao, W. W., Chen, S. T., & Liao, B. H. (2016). A probe into the construction of self-evaluation system of postgraduate education quality from the perspective of educational ecology. *Academic and Postgraduate Education, 11,* 1–6.
18. Yang, K. R. (2018). integration of production and education: problems, policies and strategic paths. *Heilongjiang Researches on Higher Education, 36*(5), 35–37.
19. Wang, C. Y., & Qiao, G. (2017). On construction of quality evaluation indicator system for provincial postgraduate education. *Journal of Graduate Education, 1,* 58–65.

20. Qiao, G., & Fu, H. F. (2016). On China's academic postgraduate education quality evaluation from the perspective of current postgraduate students. *Journal of Graduate Education, 5,* 60–65.
21. Wang, B. Y. (2018). Deepening industry-education integration: collaborative subjects and impacting factors. *Vocational and Technical Education, 39*(18), 28–33.
22. Rong, L., & Ying, D. F. (2018). Empirical analysis on graduate education quality assurance and innovation ability development—based on the graduate education satisfaction survey 2017. *Educational Research, 39*(9), 95–102.
23. Li, S., Li, Y., & Wang, H. Y. (2016). study on the construction of a model for postgraduate education quality evaluation index system—application and case demonstration of PSO-AHP methods. *Journal of Graduate Education, 5,* 53–59.
24. Li, X. K., & Li, P. (2018). exploration on the innovation and development of local colleges and universities under the view of integration of industry and education—based on the practice of Wenzhou University. *Journal of National Academy of Education Administration, 4,* 53–57.
25. Xiong, Z. D., Li, J., & Liu, X. L. (2016). Research on the quality evaluation system of top and innovative graduate education in management. *Hunan Social Sciences, 3,* 136–139.
26. Jin, W. (2017). Quality evaluation of postgraduate education centering on "learning"-based on the comparative analysis of Britain, the United States and Australia. *Higher Education Exploration, 3,* 86–90.
27. Bai, Y. X. (2019). The development dilemma of "integration of production and education" organization in high-level industry-based Universities: Analysis based on multiple institutional logic. *China Higher Education Research, 4,* 86–91.
28. Yang, S. R., Yang, Y., & Bao, C. X. (2019). Research on cultivating innovative talents of industry-education integration in Canada. *Modern Educational Technology, 29*(1), 120–126.
29. Li, Y. Z. (2018). Research on the institutional environment and optimization of the integration of production and education in China. *Vocational and Technical Education Forum, 8,* 33–38.

# Pricing Strategy of Manufacturer Supply Chain Based on the Product Green Degree

**Lili Du and Yisong Li**

**Abstract** Based on the influence of product green degree and green product price on the market demand of products, this paper develops a pricing strategy for Cournot competition between manufacturers in a supply chain system consisting of two manufacturers and one retailer. Since the manufacturer is the leader, the Cournot game model and the manufacturer Stackelberg game model are constructed. This paper focuses on the results obtained through numerical analysis. The study has shown that the three factors of the market scale of green products, consumers' green preferences and manufacturers' competition degree have positive effects on the wholesale prices, retail prices, product green degree and manufacturer and retailers' profits. The market scale of green products is the most important factor that can enhance the effects of other factors.

**Keywords** Product green degree · Manufacturer competition · Pricing strategy · Game theory

## 1 Introduction

With the rapid development of the global economy and the progress of society, the shortage of resources and the deterioration of the ecological environment have become serious challenges that people have to face. Countries are actively exploring the path of sustainable development. At the same time, with the increasing income level and quality of life, as well as the active publicity of the

---

This paper was financially supported by China Railway (Grant No. B19D00010: Optimization of railway material's management mode).

---

L. Du (✉) · Y. Li
School of Economics and Management, Beijing Jiaotong University, Beijing, China
e-mail: 18125505@bjtu.edu.cn

Y. Li
e-mail: ysli@bjtu.edu.cn



J. Zhang et al. (eds.), *LISS2019*,
https://doi.org/10.1007/978-981-15-5682-1_66

green consumption concept, the consumers began to accept the concept of green consumption and more favored the purchase of green products. Therefore, enterprises should not only consider product pricing but also consider the rational formulation of product green degree when producing and selling green products. How to maximize the green degree and the price of green products under the competition between supply chain member companies within the acceptable range of consumers, so as to maximize the economic benefits of enterprises and supply chains is necessary to solve the problem.

At present, domestic and foreign scholars have many new researches on the supply chain considering the product green degree. But the research object is mainly a secondary supply chain of a manufacturer and a retailer. For example, Reference [1] mainly consider the green supply chain game model of four different game relationships that introduce product green degree, focusing on the comparative analysis of models in terms of product green degree, product price and wholesale price. In addition, the researchers also introduced other influencing factors to study supply chain pricing, coordination, and income distribution. Reference [2] for manufacturers to avoid loss aversion, retailers are risk-neutral secondary green supply chain to study loss aversion and green efficiency coefficient on retail price, wholesale price, green degree, member profit And the impact of green supply chain profits. Reference [3] introduced the service level factor and proposed a supply chain coordination strategy based on two pricing contracts to explore the impact of green product R&D costs and service costs on pricing and profit. Reference [4] studied the impact of fair concerns on product green degree, profitability of the supply chain, and overall profit. Reference [5] consider the impact of targeted advertising investment and product green degree on market demand in the big data environment, and develop a green supply chain pricing strategy. Reference [6] studied the green product pricing strategy of government intervention under the fuzzy uncertainty of manufacturing cost and consumer demand. Reference [7] study the green degree and pricing of products when non-green products compete with green products based on an online and offline dual-channel supply chain. Reference [8] introduced e-commerce into green supply chain management, and studied the pricing and green strategies of supply chain members under the concentration and dispersion of dual-channel green supply chains. Reference [9] studied the pricing and green issues of a dual-channel green supply chain when market demand was interrupted. Studies have shown that the market scale caused by interference, the green cost is reduced, and the loyalty of customers to the retail channel is reduced, which is beneficial to the supply chain and can better improve the product green degree.

There are also some studies on more complex supply chain systems. Reference [10] Consider the influence of the choice of manufacturer's competitive relationship and different R&D models in the second-order green supply chain composed of two manufacturers and individual retailer on the different decisions of supply chain members. Reference [11] studied the impact of product green degree and sales efforts on firm decision-making and profit acquisition in a retailer-led closed-loop supply chain. Reference [12] began to consider three-level supply chain coordination based on product green degree.

## 2 Problem Definition and Model

### 2.1 Problem Description

This paper examines a supply chain system consisting of two manufacturers and one retailer, each of which produces a green product, with retail prices and product green degree for their respective green products. Providing alternative green products, retailers set retail prices for green products and sell two green products to the green consumer market. Pursue the maximization of their respective interests. There is a Cournot competition between the two manufacturers, and there is a Stackelberg game between the manufacturer and the retailer. Manufacturers are dominant throughout the supply chain, with manufacturers as leaders and retailers as followers. The supply chain structure as shown in Fig. 1.

This article will focus on the size of the green product market, the competition degree between manufacturers, the impact of consumer green preferences on the profitability of manufacturers and retailers, and the choice of strategies (Fig. 1).

### 2.2 Parameters

Set the manufacturer to $M_i$, the green product to $G_i$, and the retailer to $R$, where $i = 1, 2$.

$c_i$ : the manufacturing cost of the green product;
$w_i$ : the wholesale price of green product;
$w_i^*$: the optimal wholesale price for green product;
$p_i$ : the retail prices for green product;
$p_i^*$ : the optimal retail prices for green product;

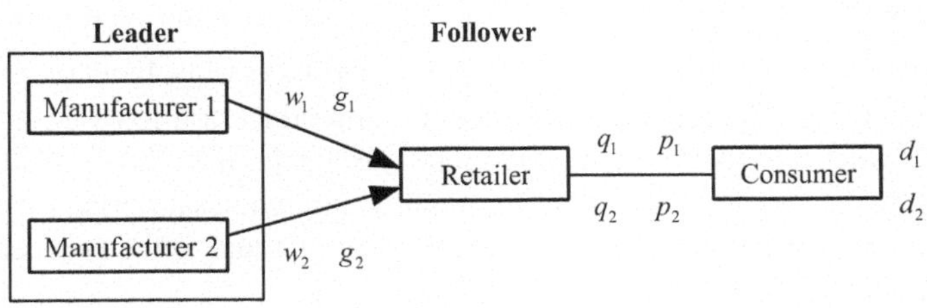


**Fig. 1** Supply chain structure with Cournot competition between two manufacturers

$d_i$ : the market scale of green product;
$q_i$ : the market demand for green product;
$g_i$ : product green degree;
$g_i^*$ : the optimal product green degree;
$e$ : The sensitivity coefficient of consumers to the green degree of products, that is, the green preference of consumers and $e > 0$;
$\theta$ : The degree of competition between manufacturers, $0 < \theta < 1$;
$\pi_{Mi}$ : the green manufacturer's profit;
$\pi_R$ : the retailer's profit

## 2.3 Assumption

In order to comply with the research questions in this paper, without violating the objective laws and not affecting the actual meaning, the following assumptions are made to turn complex problems into more feasible problems.

- The information is completely symmetrical and the information can be shared among the members in the supply chain.
- Two green products have their own market scale. and regardless of special circumstances such as shortage, the order quantity of a green product retailer is equal to the supply quantity of its manufacturer. In this paper, a linear demand function is used to consider that the product demand is determined by the retail price of the product and the green degree of the product, as well as the retail price of the alternative product and the green degree of the product.

  The market demand function of green product,

$$q_i = d_i - p_i + eg_i + \theta(p_j - eg_j)\ (i \in \{1, 2\}, j = 3 - i) \tag{1}$$

- The competitive relationship between the two manufacturers belongs to Cournot competition, in which case the decision is made at the same time. Manufacturers have always been leaders in the supply chain system, while retailers are followers.
- According to Liu [10]'s related research, it is set that the total cost of R&D investment of green products is $\frac{g_i^2}{2}$, which is the increasing function of product green degree $g$. The manufacturer produces a product at a fixed cost.
- In order to ensure that the model is more reasonable, the market retail price of the green product is greater than or equal to the wholesale price of the green product, and the wholesale price is greater than or equal to the production cost, that is, $p_i \geq w_i \geq c_i$.

## 2.4 Modeling

The two oligarch manufacturers compete with each other and there is no collusion, but they all know what action the other party will take. Thus, according to the Cournot model, the two manufacturers simultaneously determine the optimal wholesale price $w_1^*$, $w_2^*$ and product green degree $g_1^*$, $g_2^*$, and finally determine the result of the manufacturer Stackelberg game equilibrium between the manufacturer and the retailer. The optimal retail price is $p_1^*$, $p_2^*$, to maximize its profit and find the corresponding profit result.

First, establish the manufacturer profit function and the retailer profit function according to (1).

$$\max_{w_1,g_1} \pi_{M1} = (w_1 - c_1)[d_1 - p_1 + eg_1 + \theta(p_2 - eg_2)] - \frac{g_1^2}{2} \tag{2}$$

$$\max_{w_2,g_2} \pi_{M2} = (w_2 - c_2)[d_2 - p_2 + eg_2 + \theta(p_1 - eg_1)] - \frac{g_2^2}{2} \tag{3}$$

$$\begin{aligned} \max_{p_1,p_2} \pi_R = {} & (p_1 - w_1)[d_1 - p_1 + eg_1 + \theta(p_2 - eg_2)] \\ & + (p_2 - w_2)[d_2 - p_2 + eg_2 + \theta(p_1 - eg_1)] \end{aligned} \tag{4}$$

According to the game theory, the inverse induction method is used to solve the first-order partial derivatives $\frac{\partial \pi_R}{\partial p_1}$ and $\frac{\partial \pi_R}{\partial p_2}$ by using the retailer profit function to determine the retail price $p_1$ and $p_2$ of the market. Find that the Hessian matrix of $\pi_R$ with respect to $p_1$ and $p_2$, the first order main sub-form $|H_1| = -1 < 0$ and the second order main sub-form $|H_2| = 1 - \theta^2 > 0$ are obtained. So $H(\pi_R)$ is a negative fixed matrix and $\pi_R$ has a maximum value. When the equations $\frac{\partial \pi_R}{\partial p_1} = 0$ and $\frac{\partial \pi_R}{\partial p_2} = 0$ are respectively obtained,

$$p_1 = \frac{d_1 + e(g_1 - \theta g_2) + \theta(2p_2 - w_2) + w_1}{2} \tag{5}$$

$$p_2 = \frac{d_2 + e(g_2 - \theta g_1) + \theta(2p_1 - w_1) + w_2}{2} \tag{6}$$

In the second step, (5) and (6) are substituted into (2) and (3) can be drawn corresponding to each manufacturer's profit function. Find the first-order partial derivatives $\frac{\partial \pi_{M1}}{\partial w_1}$ and $\frac{\partial \pi_{M1}}{\partial g_1}$ for $w_1$ and $g_1$ according to the profit function of the manufacturer $M_1$. From the Hessian matrix of $\pi_{M1}$ on $w_1$ and $g_1$, the first-order order main sub-form $|H_1| = -1 < 0$ and the second-order order main sub-form $|H_2| = 1 - \frac{e^2}{4} > 0$ are obtained, so $H(\pi_{M1})$ is a negative fixed matrix, and $\pi_{M1}$ has a maximum value. Let $\frac{\partial \pi_{M1}}{\partial w_1} = 0$ and $\frac{\partial \pi_{M1}}{\partial g_1} = 0$ respectively, and get the expressions of

$w_1$ and $g_1$. According to the above derivation process, the same expression of $w_2$ and $g_2$ can be obtained. The four wholesale expressions can get the best wholesale price $w_1^*$ and $w_2^*$, the best product green degree $g_1^*$ and $g_2^*$, where $A = 2 - e^2$, $B = 4 - e^2$,

$$w_1^* = \frac{2\theta A d_2 + 2Bd_1 + 4\theta c_2 + \left(B + \theta^2 e^2\right)Ac_1}{B^2 - \theta^2 A^2} \tag{7}$$

$$w_2^* = \frac{2\theta A d_1 + 2Bd_2 + 4\theta c_1 + \left(B + \theta^2 e^2\right)Ac_2}{B^2 - \theta^2 A^2} \tag{8}$$

$$g_1^* = \frac{\theta eAd_2 + eBd_1 + 2\theta ec_2 + \left(\theta^2 A - B\right)ec_1}{B^2 - \theta^2 A^2} \tag{9}$$

$$g_2^* = \frac{\theta eAd_1 + eBd_2 + 2\theta ec_1 + \left(\theta^2 A - B\right)ec_2}{B^2 - \theta^2 A^2} \tag{10}$$

The best retail price $p_1^*$ and $p_2^*$ is available for simultaneous Eq. (7)–(10).

$$p_1^* = \frac{1}{2}\left[\frac{\theta d_2 + d_1}{1 - \theta^2} + \frac{(2 + e^2)(\theta Ad_2 + Bd_1 + 2\theta c_2) + \left(2\theta^2 e^2 A - e^2 B + AB\right)c_1}{B^2 - \theta^2 A^2}\right] \tag{11}$$

$$p_2^* = \frac{1}{2}\left[\frac{\theta d_1 + d_2}{1 - \theta^2} + \frac{(2 + e^2)(\theta Ad_1 + Bd_2 + 2\theta c_1)}{B^2 - \theta^2 A^2} + \frac{\left(2\theta^2 e^2 A - e^2 B + AB\right)c_2}{B^2 - \theta^2 A^2}\right] \tag{12}$$

Substituting the optimal wholesale price $w_1^*$, $w_2^*$ and the optimal retail price $p_1^*$, $p_2^*$ into the Eq. (2)–(4) can obtain the maximum profit of the manufacturer and the retailer respectively.

$$\pi_{M1} = \frac{B}{2}\left[\frac{\theta Ad_2 + Bd_1 + 2\theta c_2 + \left(\theta^2 A - B\right)c_1}{B^2 - \theta^2 A^2}\right]^2 \tag{13}$$

$$\pi_{M2} = \frac{B}{2}\left[\frac{\theta Ad_1 + Bd_2 + 2\theta c_1 + \left(\theta^2 A - B\right)c_2}{B^2 - \theta^2 A^2}\right]^2 \tag{14}$$

$$
\begin{aligned}
\pi_R = & \left[\frac{2\theta(A+B)d_2 + \left(2B + 2\theta^2 A\right)d_1}{\left(1-\theta^2\right)\left(B^2 - \theta^2 A^2\right)} - \frac{2Bc_1 + 2\theta Ac_2}{B^2 - \theta^2 A^2}\right] \\
& \times \frac{\theta Ad_2 + Bd_1 + 2\theta c_2 + \left(\theta^2 A - B\right)c_1}{B^2 - \theta^2 A^2} \\
& + \left[\frac{2\theta(A+B)d_1 + \left(2B + 2\theta^2 A\right)d_2}{\left(1-\theta^2\right)\left(B^2 - \theta^2 A^2\right)} - \frac{2Bc_2 + 2\theta Ac_1}{B^2 - \theta^2 A^2}\right] \\
& \times \frac{\theta Ad_1 + Bd_2 + 2\theta c_1 + \left(\theta^2 A - B\right)c_2}{B^2 - \theta^2 A^2}
\end{aligned} \tag{15}
$$

# 3 Numerical Analysis

In order to more intuitively study the impact of various parameter changes on the profit and decision of manufacturers and retailer, complex models are transformed here, the relevant parameters are assigned, and more accurate analysis is performed through simulation. Therefore, in the study, the market scale of green products $d_i(i = 1, 2)$, the green preference of consumers $e$, the competition degree of manufacturers $\theta$, the impact of green product pricing, product green degree and profit were considered. However, in the discussion process of $e$ and $\theta$, it is necessary to set the market scale of different green products reasonably, and the comparative analysis from $d_1 = d_2$, $d_1 < d_2$ and $d_1 > d_2$ is more comprehensive. The related parameters are assumed to be $\theta = 0.5$, $e = 0.3$, $c_1 = 5$, $c_2 = 3$ and $d_i = \{20, 30\}$.

## *3.1 The Influence of Factors on Prices*

From Figs. 2, 3 and 4, the market scale of green products, consumers' green preference and competition degree of manufacturers all have a positive impact on the price of green products. Among them, manufacturers' competition degree has the greatest positive impact on the wholesale price and retail price of the two kinds of green products, that is, the price increase is the biggest. The price of green products has a relatively slow growth under the influence of consumers' green preference. At the same time, the increase of the market scale of one green product directly increases the price of the green product, and indirectly promotes the price of another green product. The influence of the two factors, consumers' green preference and manufacturers' competition degree, is more obvious under the superposition of the difference of market scale.

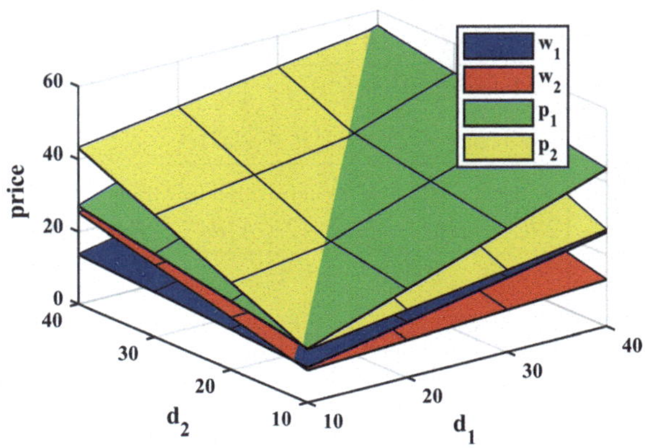


**Fig. 2** The Influence of market scale on prices

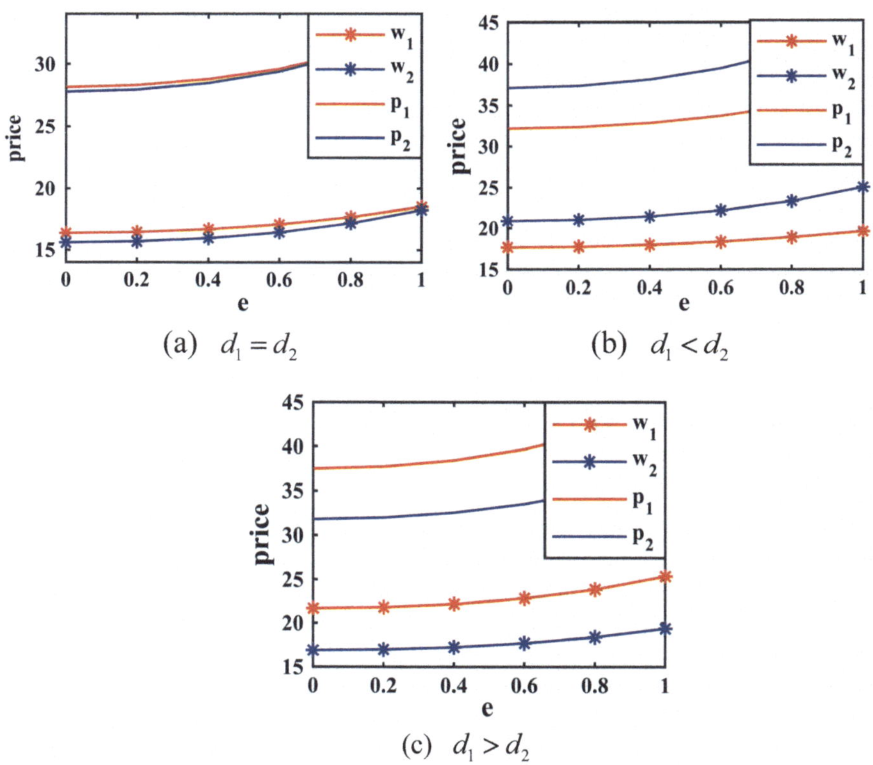


**Fig. 3** The influence of consumers' green preference on prices in different market scale

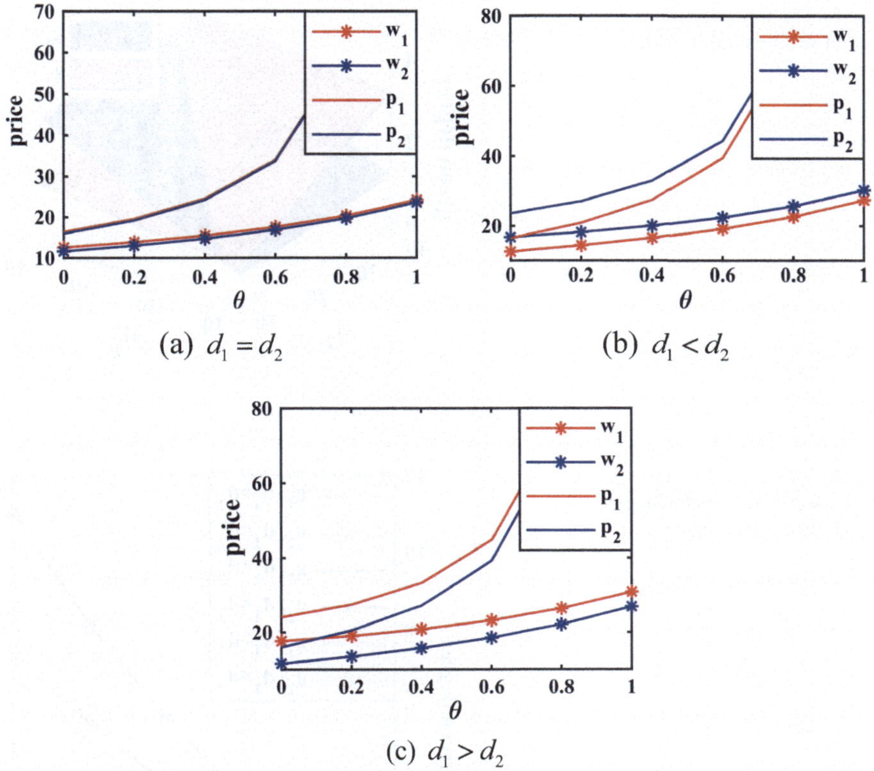


**Fig. 4** The influence of manufacturers' competition degree on prices in different market scale

## 3.2 The Influence of Factors on Product Green Degree

From Figs. 5, 6 and 7, the market scale of green products, consumers' green preference and manufacturers' competition degree all have a positive impact on the green degree of products. Among them, with the increase of consumers' green preference, product green degree increases the most, which can produce the green gap between products. The higher manufacturers' competition degree, product green degree will be steadily improved, and the green gap between the two green products is almost always the same. In the absence of competition between manufacturers, there will also be products that contain the green degree of the product, as shown in Fig. 7. At this time, green products with a higher degree of green products will always maintain a dominant position. From the point of market scale, the increase of one green product's market scale makes this product green degree increase greatly, and it also plays a certain role in promoting the green degree of another green product. Especially for the green products in the larger market scale, the increase of product green degree has been greatly improved under the action of consumers' green preference or manufacturers' competition degree.

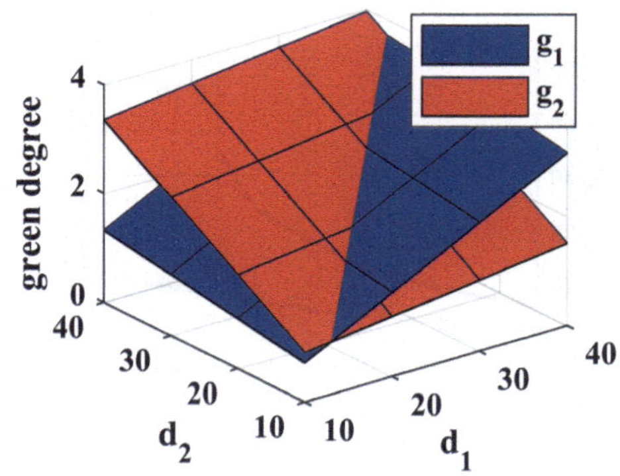


**Fig. 5** The influence of market scale on product green degree

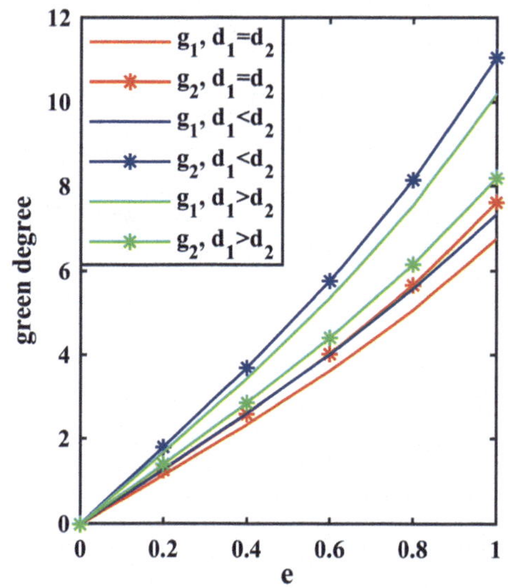


**Fig. 6** The influence of consumers' green preference on product green degree

## 3.3 *The Influence of Factors on Profits*

From Figs. 8, 9 and 10, the market scale of green products, consumers' green preference and manufacturers' competition degree all have a positive impact on the profits of manufacturers and retailer. Because the retailer' profit comes from two green products, under the influence of factors, the increase of retailer' profit is much higher than the sum of the two manufacturers' profit growth. Among them, in the case of increased competition between manufacturers, manufacturers and retailer

**Fig. 7** The influence of manufacturers' competition degree on product green degree

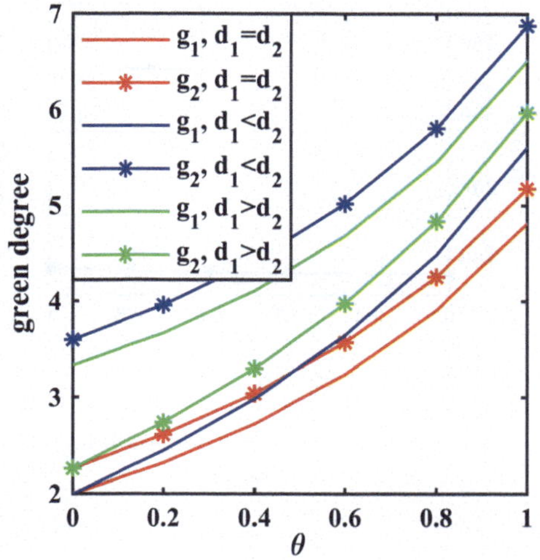


**Fig. 8** The influence of market scale on profits

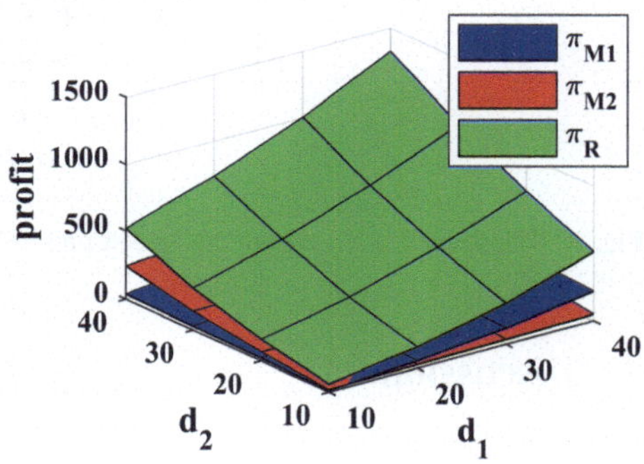


have the largest increase in profits. However, consumers' green preference has a greater positive impact on retailer's profits and less positive impact on manufacturers' profits. The increase of market scale will also lead to an increase in the profits of manufacturers and retailer. Once the market scale of a green product increases, it will cause manufacturers and retailer to make more profits from it, and the manufacturers on this side of the market will make more profits. Under the action of consumers' green preference and manufacturers' competition degree, if the market scale of a certain green product increases, the overall profit of the manufacturers and retailer will be greatly higher than that of each member when the market scale is the same.

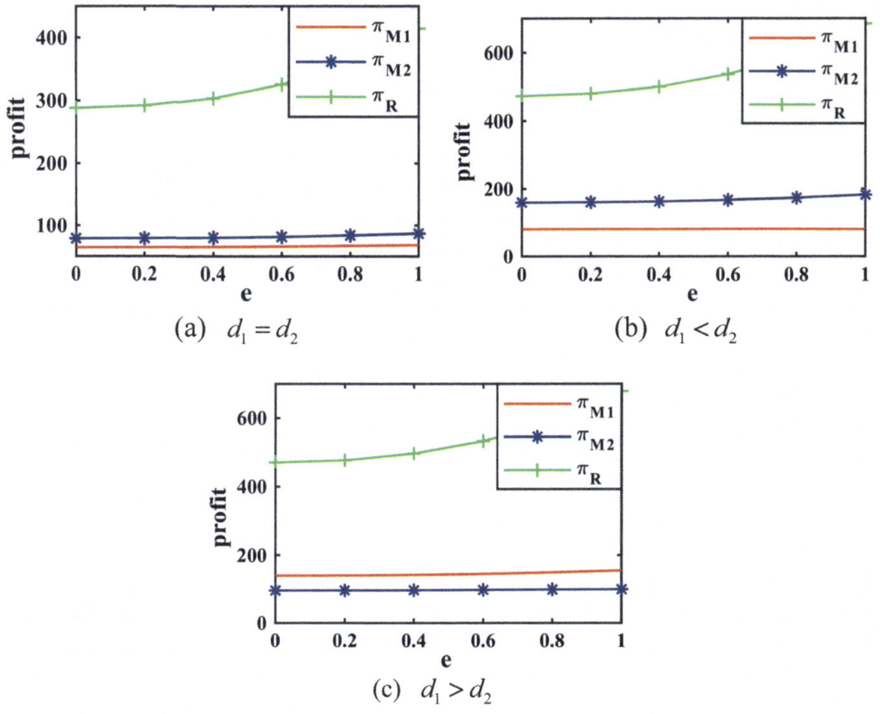


**Fig. 9** The influence of consumers' green preference on prices in different market scale

## 4 Discussion

The results of the above numerical analysis have certain guiding effect on policy making and enterprise decision-making, so some Suggestions will be put forward from these two perspectives:

From a policy perspective, the current government is applied on the main three types of green tax subsidies, subsidies for consumers and manufacturers and subsidies for green products in the early stages of the green product development, this kind of subsidy policy to a certain extent to businesses and consumers are very good, can make the green product prices remain within the scope of the consumer can accept reasonable growth, make the enterprise have more money and technology into production research and development for product green degree is high, to promote consumers to buy green products, thereby rapidly expanding green consumer goods market, to all the members of the supply chain enterprise's profit maximization. But late development of green product market mature, green product emerge in endlessly in consumer preference is not outstanding, green subsidy policy at this time will no longer apply to second, the government's response to each kind of green products to develop more specific green product evaluation

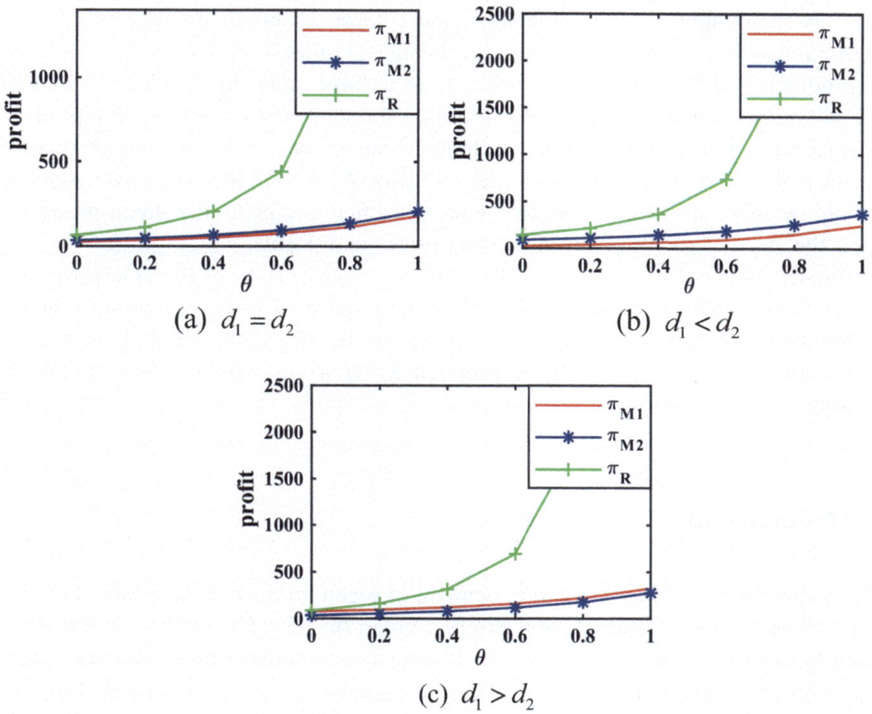


**Fig. 10** The influence of manufacturers' competition degree on prices in different market scale

standard, to ensure consumers to buy the real high quality green products, increase the consumer's trust in the green products at the same time, the government should develop green products market norms, promote the benign competition between enterprises, to maintain the market vigor, higher product prices and profits is the most important is that the government should formulate a lasting publicity policy, through the network media, TV and other channels vigorously publicize the concept of green consumption and the positive role of green products in environmental protection, strengthen consumers' preference for high green products, make them the driving force for manufacturers to produce and develop green products, and expand the market scale of green products.

From the point of view of the enterprise, the manufacturer should timely feedback, according to the opponents change between each other, make the product green degree of green product pricing again and again to reach equilibrium, manufacturers and retailer can gain maximum profit maker should pay attention to own the product quality, completes the production and research and development in catering to the market demand and respond to a competitor's situation changes, the manufacturers can be appropriately raise product green degree, enlarge the diversity of similar green products, increase the wealth of green product market, continuously meet the needs of the consumers' psychology and behavior, prompting the price is reasonable. At the same time, the enterprise should do well in product marketing strategy, improve the brand influence, in order to rapidly expand its scale of production of green products market, so as to have more say in pricing, and also can get more economic benefits of enterprises should pay attention to green product advertising, highlighting product highlights and green environmental protection performance of green consumption idea and way of life with memory advertisement should be innovative and to attract more consumers prefer to buy green products, keep the steady growth of green product prices and profits.

Finally, under the policy support and the correct decision of the enterprise, the enterprise can obtain the reasonable promotion space of price and product green degree by expanding the market scale, increasing the degree of benign competition and improving consumers' green preference, so as to maximize its economic benefits.

## 5 Conclusion

This paper proposes that the market demand of green products is determined by the retail price of green products and product green degree. The study focused on a manufacturer-led supply chain in which a retailer simultaneously sells two alternative green products that are produced by manufacturers with Cournot competition. Through the numerical analysis, the influence of factors on the price, product green degree, profit and decision-making of the manufacturer and the retailer are obtained. Research shows that:

The market scale of green products, consumers' green preference and manufacturers' competition degree have a positive impact on prices, product green degree and the profits of members. The manufacturers' competition degree is the main factors, which has the greatest impact on the product green degree and each member's profit. Consumers' green preference is a secondary factor, with the greatest influence on product green degree. In the case of other factors, the difference in the change of market scale leads to an enhanced effect of them. Therefore, the market scale is the most important factor.

The original green tax subsidy policy is only at applicable in the early stage of the development of green products. The government should also formulate green product evaluation standards and green product market norms to ensure the interests of consumers, promote the benign competition among enterprises. The continuous publicity policy of green consumption concept is further increase consumers' green preference, expand the market scale is an important means of as a result, businesses can and product green degree of green product price is reasonable, in order to gain more profit.

Take the initiative to form the benign competition between manufacturers, to do the research and development and production of the green products, the unique product marketing strategy and the outstanding product of green advertising can promote green products reasonable price and the product green degree, enterprise's profit increase.

There are shortcomings in this paper. The model does not take into account the incomplete information among supply chain members so that it does not fully reflect the complexity of the actual situation. And the market demand function is not entirely linear in reality. The adjustment of the model can be strengthened in the next step of the study.

## References

1. Jiang, S. Y., & Li, S. C. (2015). Green supply chain game model and revenue sharing contract considering product green degree. *Chinese Journal of Management Science, 23*(6), 169–176.
2. Feng, Z. W., & Tan, C. Q. (2019). Pricing, green degree and coordination decisions in a green supply chain with loss aversion. *Mathematics, 7*(3), 1–25.
3. Zhang, D. D., & Wang, H. D. (2017). Supply chain decision-making and coordination of demand-dependent product green degree and service level. *Frontiers of Social Science, 6*(2), 185–194.
4. Ping, S., Bo, Y., & Song, S. (2016). Green supply chain pricing and product green degree decision based on fairness. *Systems Engineering-Theory & Practice, 36*(8), 1937–1950.
5. Liu, P., & Yi, S. P. (2017). Pricing policies of green supply chain considering targeted advertising and product green degree in the big data environment. *Journal of Cleaner Production, 164,* 1614–1622.
6. Yang, D. Y., & Xiao, T. J. (2017). Pricing and green level decisions of a green supply chain with governmental interventions under fuzzy uncertainties. *Journal of Cleaner Production, 149,* 1174–1187.
7. Jamali, M. B., & Rasti-Barzoki, M. (2018). A game theoretic approach for green and non-green product pricing in chain-to-chain competitive sustainable and regular dual-channel supply chains. *Journal of Cleaner Production, 170,* 1029–1043.
8. Li, B., Zhu, M. Y., Jiang, Y. S., & Li, Z. H. (2016). Pricing policies of a competitive dual-channel green supply chain. *Journal of Cleaner Production, 112,* 2029–2042.
9. Rahmani, K., & Yavari, M. (2019). Pricing policies for a dual-channel green supply chain under demand disruptions. *Computers & Industrial Engineering, 127,* 493–510.
10. Liu, H. Y., & Yan, S. F. (2017). Horizontal competition game and pricing strategy of supply chain considering product green degree. *Industrial Engineering and Management, 22*(4), 91–99+114.
11. Gao, J. H., Han, H. S., Hou, L. T., & Wang, H. Y. (2015). Retailer-led closed-loop supply chain decision-making considering product green degree and sales effort. *Management Review, 27*(4), 187–196.
12. Zhang, C. T., & Liu, L. P. (2013). Research on coordination mechanism in three-level green supply chain under non-cooperative game. *Applied Mathematical Modelling, 37,* 3369–3379.

# The Bankability of Bike-Sharing

**Tanna Lai, Shiying Shi, and Shengyue Hao**

**Abstract** Bankability is the key basis for project investment and financing decisions. This study uses investment science, engineering economics and public product theory to analyze the basic attributes, financing connotation and evaluation indicators of bike-sharing projects. With the help of engineering financial management theory, the model of bike-sharing project financing system is constructed. This paper uses a city's bike-sharing project as a case and uses system dynamics simulation to simulate the structure of funds, the user volume, the cost of deposits, the path of impact on the financing of the project. The research results show that: (1) profitability is an important dimension reflecting the bankability of the project, which indicates the direction of the project's bankability evaluation; (2) the user volume and the deposit fee are factors that affect the bankability of the bike-sharing project. This is also an important object of bike-sharing project management. Based on these, the article provides the key recommendations for the bike-sharing project's investor and the management of the operating enterprises.

**Keywords** The project of bike-sharing · System dynamics · ROE · Engineering economics

T. Lai (✉) · S. Hao
School of Economics and Management, Beijing Jiaotong University, Beijing, China
e-mail: 872682045@qq.com

S. Hao
e-mail: haoshyue@bjtu.edu.cn

S. Shi
School of Civil Engineering, Henan Polytechnic University, Jiaozuo, China
e-mail: shishiying@hpu.edu.cn



J. Zhang et al. (eds.), *LISS2019*,
https://doi.org/10.1007/978-981-15-5682-1_67

## 1 Introduction

Chinese sharing economy has been developing rapidly since 2015. In the government report, Premier Li pointed out that it is necessary to promote the rapid development of the "new economy" including the sharing economy vigorously. The sharing economy [1] uses the Internet as the platform to connect buyers and sellers indirectly that were isolated from each other, making full use of social idle resources. As a new economic model, the sharing economy is not only supported by the government, but also receives the attention of the society.

Bike-sharing is the typical product of the sharing economy, with green and economic characteristics. At this stage, the city is polluted, traffic is crowded, therefore the emergence of bike-sharing alleviates the pressure on the urban environment, conforms to the concept of green environmental protection, and solves the problem of the "last mile" of urban travel.

Bike-sharing use the Internet platform to provide the service of bicycle for the urban residents. It is a new model for the integration of mobile network and the rental for bicycles [2]. Bike-sharing can effectively solve problem about short-distance travel in society and is considered as "demand-driven public service innovation" [3]. The deposit of bike-sharing is charged for every user. The operation mole of "one asset, multiple deposits" is the most important values of the bike-sharing. A bicycle can be used by different users. The number of users can be much larger than the amount of vehicles. The total amount of deposits will be much greater than the value of the bicycle itself. This mode of operation is the basis for the profitability of bike-sharing companies [4]. On the other hand, the data can reflect the behavior of consumers. It is valuable for many business models. For example, bike-sharing with positioning systems can accurately reflect consumer travel behaviors. Bike-sharing companies can make huge profits by working together with mobile payment companies [5].

Therefore, bike-sharing companies get the favor of many investors As of March 2018, ofo has completed two rounds of financing with Alibaba and obtained a total of 1.77 billion financing from Alibaba. In 2018, the US Mission announced the acquisition of Mobai for 2.7 billion CNY; Yongan and Harrow After the merger into Yonganxing Low Carbon Technology Company, the first round of financing valuation reached 1.468 billion. This reflects the fact that the shared bicycle financing boom is still high.

However, in December 2018, These small yellow cars were caught in the financial chain disruption, and a large number of users returned the deposit. This has brought warnings to the development of financing for shared bicycle projects.

# 2 The Theoretical Study

## 2.1 *The Research of Bike-Sharing*

There are some academics studying different areas about bike-sharing. He analyzed the relationship between bike-sharing and the sharing economy. Huang analyzed the willingness of the users of bike-sharing by setting mathematical methods [6]. Angelopoulos et al. used a city in Greece as an example to study the operation mode of bike-sharing in this city, and mentioned the cost of the system [7]. Leonardo et al. studied how to build a management model for bike-sharing [8].

## 2.2 *The Research of Bankability*

Bankability can be understood as "whether the project can be financed at all, whether it has a positive or obvious cash flow". From the theory of technical economics, the continuous cash flow of a construction project itself is the basic guarantee for the project to obtain loans; from the perspective of bank loans, when a project can persuade lending to provide loans to the project company, the project is "Bankability project" [9, 10].

## 2.3 *The Research of the Bankability of Bike-Sharing*

At present, there are few studies on the combination of bike-sharing and bankability at home and abroad. Duan analyzed the behavior of investors according to the bankability scheme of bike-sharing [11]. Yang et al. from the perspective of consumer economics to study the approach to solve the bike-sharing capital problems [12]. Zhou et al. studied the financial problems under the sharing economy mode and analyzed the capital mode of bike-sharing [1]. Fan guiling et al. studied the Internet finance model under the sharing economy [13].

At present, there are few studies on the bankability of bike-sharing, many research topics focus on the management of bike-sharing. Therefore, this paper will focus on the study of the bankability of bike-sharing, and provide a reference for the related projects.

## 2.4 *Bankability Research on Bike-Sharing Projects*

The judgment of the bankability of the project mainly includes the way of raising funds, the source of funds and the feasibility of the funding plan. Project's bankability is to evaluate the project from the standpoint of investors or creditors and analyze whether it is necessary to invest in the project. To evaluate the financing of the project, it is necessary to adopt the scientific evaluation theory and method of financial management for the characteristics of each stage of the project life cycle, adopt standardized evaluation indicators, follow the evaluation procedure, and conduct multiple indicators and multiple methods for investors.

To study the bankability of bike-sharing projects, it is necessary to evaluate the economic feasibility of shared bicycle projects, investor profitability, project fund repayment ability, etc., involving project production cost, economic income, two dependent variables, and total project Investment, operating costs, maintenance costs, bicycle riding income, deposit income and other independent variables.

### 2.5 System Dynamics

The concept of System Dynamics (SD) was first proposed by Professor Forrester of the Massachusetts Institute of Technology (MIT) in 1956. In the early days of its establishment, SD was mainly used in industrial enterprise management, and later gradually developed into an important method for studying complex system engineering. In the system dynamics, system behavior patterns and characteristics mainly depend on its internal dynamic structure and feedback mechanism [14]. The main object of system dynamics is the open giant system, which can solve the complex system with many factors. When building the system dynamics model, the process mainly includes the establishment of the system equation and the drawing of the stock and flow diagram. After that, the system stability is ensured through sensitivity analysis and validity test. Finally, scientific, targeted and representative decisions and Suggestions are put forward for the research object [15]. Due to nonlinear factors, high-order, more complex systems often exhibit a variety of dynamic characteristics, so system dynamics can be used as practical systems to study more complex problems in the field of social, economic and management. Both get important applications.

In the bankability system of bike-sharing, there is a complex, nonlinear and dynamic relationship between dependent variable and independent variable. Simple mathematical method is difficult to achieve more complex mathematical relationship calculation. Therefore, this paper chooses the system dynamics to study the bankability system of bike-sharing.

## 3 Model Establishment

### 3.1 System Analysis

Based on the system dynamics, establish a bankability model of the bike-sharing project to solve the following problems.

Showing the relationship between production costs, economic income and bankability of bike-sharing projects.

By changing the key variables of the system, analyzing the impact of changes in different variables on the indicators of bankability.

Through the data simulation results, the model will give the advice about the bankability of the bike-sharing project.

## 3.2 System Module Classification

In order to achieving the research goal, the bankability model of the bike-sharing project constructed in this paper mainly includes two subsystems: economic income and production cost. It's necessary to confirm the relationship between the two variables in the two subsystems and analyze the effect between with each factor, calculating the bankability evaluation indicator by the end.

(1) *Economic income system module:* Economic income mainly represents the cash inflow portion of the research, and sustainable cash flow is a sign of project profitability. According to the income from operating business, it is divided into two parts: main business income and other business income. The main business income of bike-sharing is from the income of the users mostly, which consists of deposit income, bicycle payment income and recharge income. Other business income includes bicycle residual value recovery income, government subsidy income, and advertising revenue.

(2) *Production cost system module:* The production cost mainly represents the part of the cash outflow in the financing research, which is divided into two parts: operating cost and construction cost according to the expenses incurred during the operation of the bike-sharing. The construction cost is the cost of bank loans, and the operating costs mainly include financial expenses, bicycle placement costs, and management fees.

## 3.3 Construct a System Causal Map

There are some influence variables in the bike-sharing project system, and the system causal map (Fig. 1) is constructed by analyzing the influence between each variable. The causal plots are plotted using Vensim software, where a positive arrow indicates a positive feedback relationship between variables, and a negative arrow indicates a negative feedback relationship between variables.

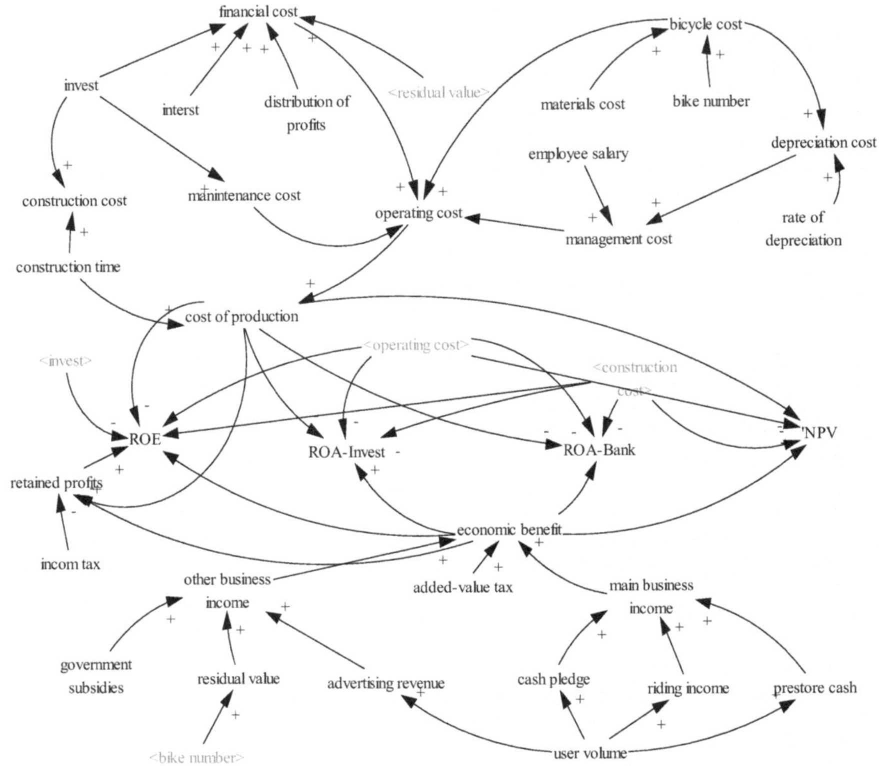


**Fig. 1** The causality of the bankability system of bike-sharing project

## 3.4 Building a System Cash Flow Model

After the relationship between each variable in the project and the model structure are determined and according to the causal flow map, the project cash flow model is constructed to visually represent the individual cash flows in the project (Fig. 2). Production cost, economic income are state variables, the others are auxiliary variables.

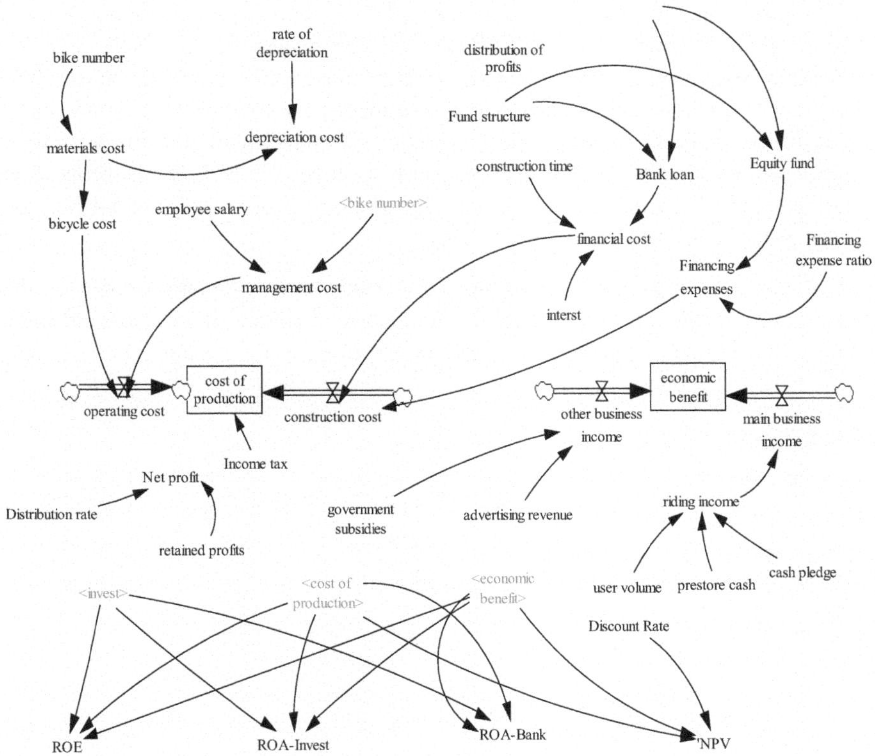


**Fig. 2** System flow diagram

# 4 Case Study

## *4.1 Case Overview*

Changzhou Yonga1nxing Public Bicycle Company obtained the operation right to bike-sharing in a certain city. The bid-winning unit is required to complete the delivery and construction of 5,000 bike-sharing and their auxiliary equipment within 30 days from the date of signing the cooper.

## *4.2 Data Collection and Discussion*

(1) *Questionnaire data:* According to the factors affecting the bankability of bike-sharing, the questionnaires for the decision factors of bike-sharing were designed, and the relevant data of shared bicycle riding income was obtained. 500 questionnaires were distributed, and 110 valid copies were returned, and

the questionnaire collection efficiency was 22%. The age of the respondents is concentrated between 18 and 45 years old; the male ratio is 44.33%, and the female proportion is 55.67%; the occupation constitutes a professional (teacher/doctor, etc.), corporate employees, government personnel (civil servants, etc.), Students, workers (waiters, etc.), freelance. Among them, the proportion of professionals is 12.72%, the proportion of employees is 27.27%, the proportion of government personnel is 22.73%, the proportion of workers is 5.45%, and the proportion of freelance is 17.27%.

The data related to the calculation of the bankability evaluation index in the questionnaire survey are: time of riding, frequency of riding, the way of rent bike, the way of payment. Data as in the table below (Tables 1, 2, 3 and 4):

**Table 1** Time of riding

| Single ride time (h) | Sample size | Total time (h) | Average time (h) |
|---|---|---|---|
| 0,0.5 | 49 | 24.5 | 0.8 |
| 0.5,1 | 57 | 57 | |
| 1,2 | 3 | 4.5 | |
| 2,+∞ | 1 | 2 | |
| Sum | 110 | 88 | |

**Table 2** Frequency of riding

| Frequency | Sample size | Total | Average |
|---|---|---|---|
| Once a year or less | 39 | 19.5 | 2.85 |
| At least once a day | 3 | 90 | |
| 1–3 times a week | 17 | 153 | |
| 1 time a month | 51 | 51 | |
| Sum | 110 | 313.5 | |

**Table 3** The way of rent bike

| The way of rent bike | Sample size |
|---|---|
| Deposit | 34.55% |
| Transportation card | 0.91% |
| Free | 64.54% |

**Table 4** The way of payment

| The way of payment | Sample size |
|---|---|
| Payment after the end of riding | 86.36% |
| Buy the card of payment | 10.91% |
| Save cash in the bike-sharing APP | 2.73% |

(2) *Relationship expression:* There are many factors affecting project bankability, such as operation mode, investment plan, policy regulations, etc. In the analysis of project bankability, the profitability of project investment should be considered according to relevant influencing factors. Project income [5] can be judged according to financial evaluation indexes. The indicators to examine project profitability include: net present value (NPV), return on assets (ROA), return on equity (ROE), etc. The evaluation index of findability is an important symbol of the findability of the project. In this system, the financial evaluation indexes are NPV, ROA-bank, ROA-invest and ROE. According to the mathematical formula of the four evaluation indexes, the value is calculated and used as the final output variable of the financing system of Shared bicycle project.

NPV = economic income − cost of production
ROA-Bank = pre-tax profit/total Bank loans
ROA-Invest = pre-tax profit/total project investment
ROE = net profit/total equity
Economic income = INTEG [(main business income + other business income − VAT)/(1 + discount rate, 0]time
Product cost = INTEG [(construction cost operation cost)/(1 + discount rate), 0]time
Main business = deposit income + cycling payment income + recharge deposit
Other business income = government subsidy income + advertising income + salvage value income
Pre-tax profit = main business income + other business income − construction cost − operation cost
Net profit = main business income + other business income − income tax

## 4.3 Operating Model

Using Vensim to analyze the financing of the Shared bike project in the case, the results are as follows:

(1) *NPV:* As shown in Fig. 3, the net present value of the bike-sharing project decreased in the first year, and increased from the second year. The final net present value of the project was RMB 6,962,900. In the 7.78, it shows the net present value is 0 and the project's dynamic payback period is 7.78 years. After 7.78, the net present value is greater than 0, it shows that the project profitability is good.

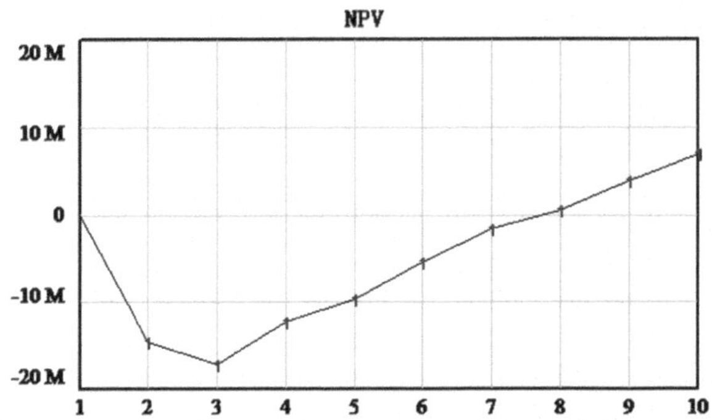


**Fig. 3** The trend of NPV

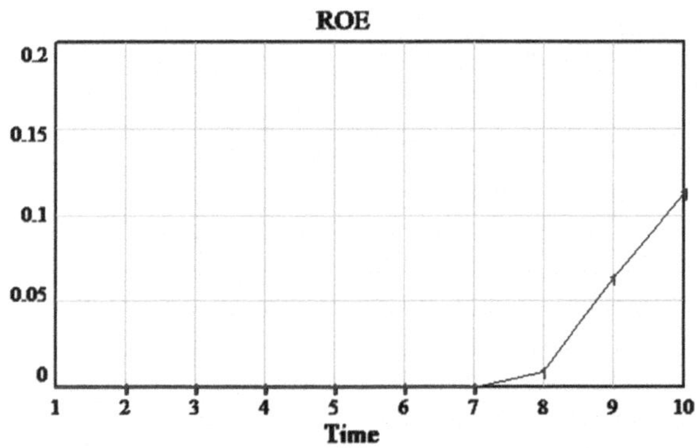


**Fig. 4** The trend chart of ROE

(2) *ROE:* As shown in Fig. 4, ROE was 0 in the first six years, and has been increasing since the sixth year. The final result was 11.29%. The ROE during the time shows a good growth trend, indicating that the project has a good profitability.

(3) *ROA:* As shown in Fig. 5, ROA-invest, ROA-bank are fluctuated a lot during the whole operation, but ROA-invest generally maintained around 11%, and the return on total assets in the last year of the project was 12.84%. The final ROA-bank value was 42.8%. ROA-bank is higher because the small proportion of Bank loans in the investment structure.

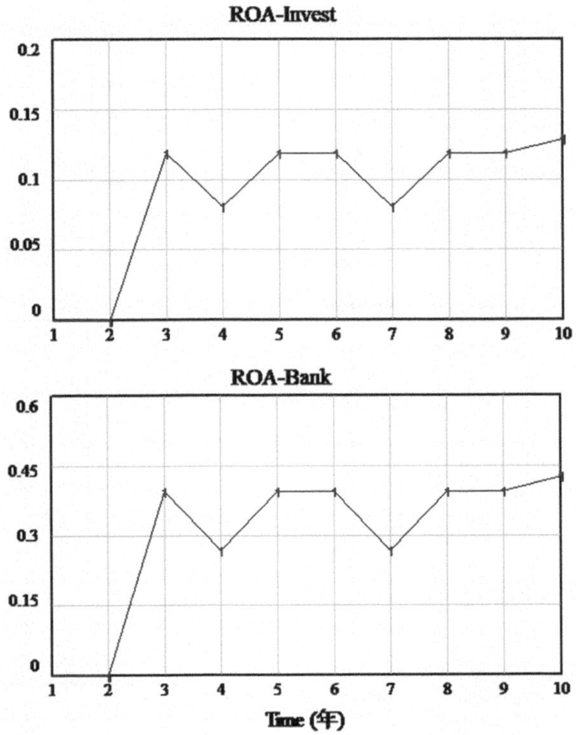


**Fig. 5** The trend chart of ROA

## 4.4 Model Checking

As shown in Fig. 6, the model is tested by using the method of extreme value test. The paper reduced the number of users in the model from 65,360 to 0, and run the model:

With the user volume reducing to 0, the net present value is also decreasing from 6,966,800 CNY to −47,722,400 CNY. The downward trending is obvious. ROE, ROA-Invest, and ROA-Bank are also decreasing to 0 as the number of users decreasing. The bankability evaluation index is the same as the changing trend of variable. Therefore the model is reliable.

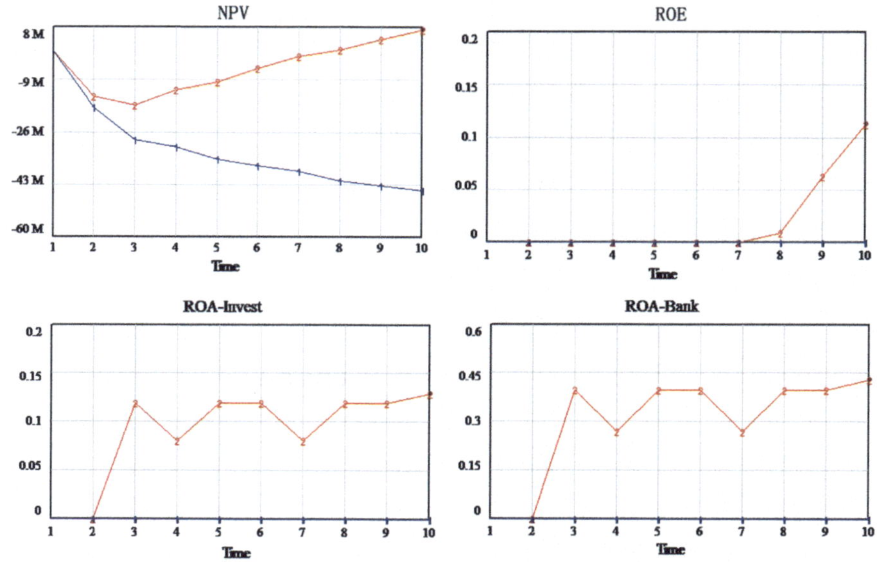


**Fig. 6** Diagram of model test

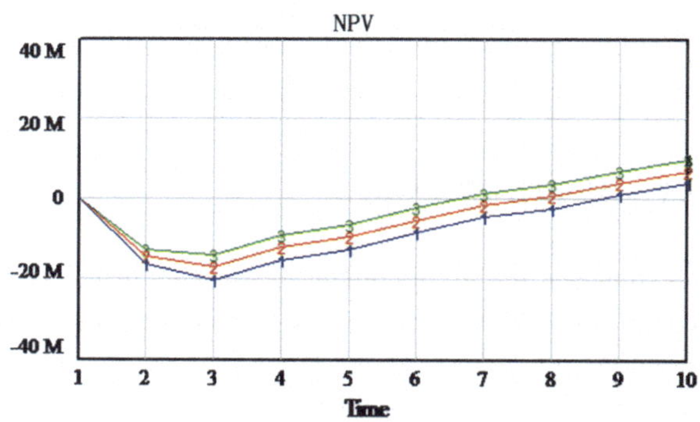


**Fig. 7** The trend chart of the change of NPV by adjusting fund structure

## 4.5 Strategic Analysis

(1) *Adjusting the fund structure:* The proportion of the capital structure in the bike-sharing project was adjusted to 0.35 (1) and 0.25 (3). The original capital structure is 0.3 (2).

(a) *NPV:* As shown in Fig. 7, the adjustment of the capital structure and the change in the net present value are consistent with the original capital structure of 0.3. When the ratio is 0.25, the net present value increases to 999.80 million CNY; when the ratio of found structure become to 0.35, the net present value drops to 3,927,700 CNY. The results show that he lower the project liabilities, the higher the net present value, the lower the liabilities, the lower the production cost of the project, and the higher the net present value.

(b) *ROE:* As shown in Fig. 8, when the funding structure fall to 0.25. The ROE increased to a final value of 15.0%; when it increase to 0.35, the ROE decreased to 6.9%. According to the formula of ROE = net profit/equity funds, if the proportion of bank loans decreases and the equity funds increase, the ROE may decline. However, in the bike-sharing project, the construction cost of bank loans is relatively large. When the construction cost reduced, the net profit increased, and the increasing for the net profit is greater than the increasing in equity funds, which cause the increasing of ROE. According to Yonganxing's bike-sharing financial report, the company's assets and liabilities account for a small proportion, indicating that investors should reasonably choose the proportion of bank loans according to their own financial capabilities, and a small amount of liabilities can obtain higher profits (Fig. 8).

(c) *ROA:* As shown in Fig. 9, ROA-Invest has little change trend, and the change of found structure has little effect on ROA-Invest. After three years, with the increase of the proportion of capital structure, ROA-Bank is decreasing.

(2) *User volum:* The paper adjusts the user volume amount to 70,000 people (1), 60,000 people (3), and 50,000 people (4).

(a) *NPV:* As shown in Fig. 10, the net present value increased with the number of users' increasing. When the number of users increased to 70,000, the net present value increased to 10,632,100 CNY, and when the number of users decreased to 60,000, the net present value decreased to 2,724,300 CNY, and when the number of users decreased to 50,000, the net present value remained negative and finally decreased to −5,183,500 CNY. The number of users has a great impact on the economic income of the project. If the number of users increases, the economic income of the project will increase.

(b) *ROE:* As shown in Fig. 11, ROE increases with the number of users. When the number of users increased to 70,000 and the ROE increased to 17.1%.

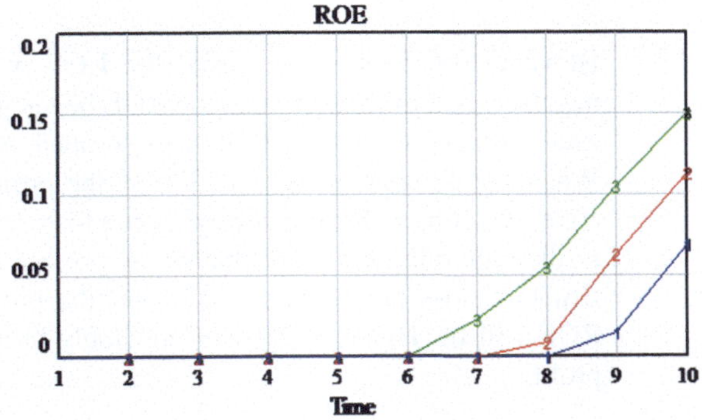


**Fig. 8** The trend chart of the change of ROE by adjusting fund structure

Fig. 9 The trend chart of the change of ROA by adjusting fund structure

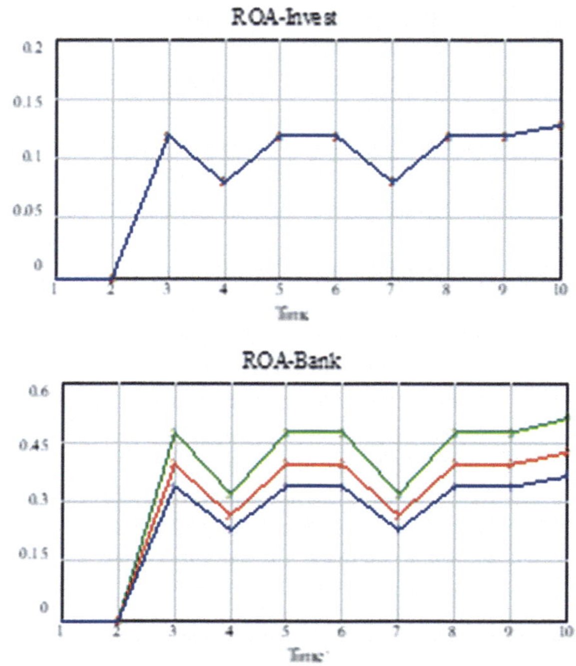


Fig. 10 The trend chart of the change of NPV by adjusting user volume

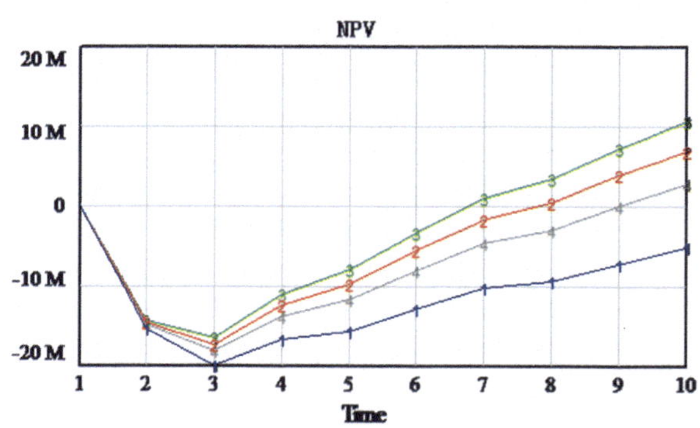


However, in the first six years, the ROE was 0 and the net profit was negative. After the sixth year, the ROE began to grow. When the number of users dropped to 60,000, ROE also dropped, and the final result was 4.5%. When the number of users is 50000, the project net profit is always negative, resulting in ROE is always going to be 0. The changing trend of ROE is directly related to the change of project net profit. When the found structure does not change, the higher the project net profit is, the higher ROE will be. Higher users will inevitably bring higher income and higher profit.

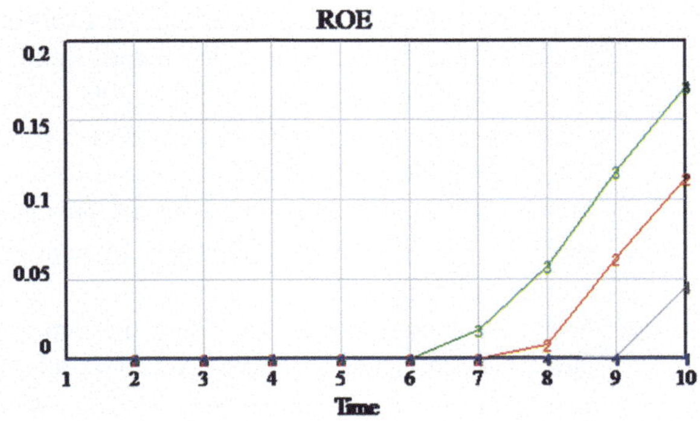


**Fig. 11** The trend chart of the change of ROE by adjusting user volume

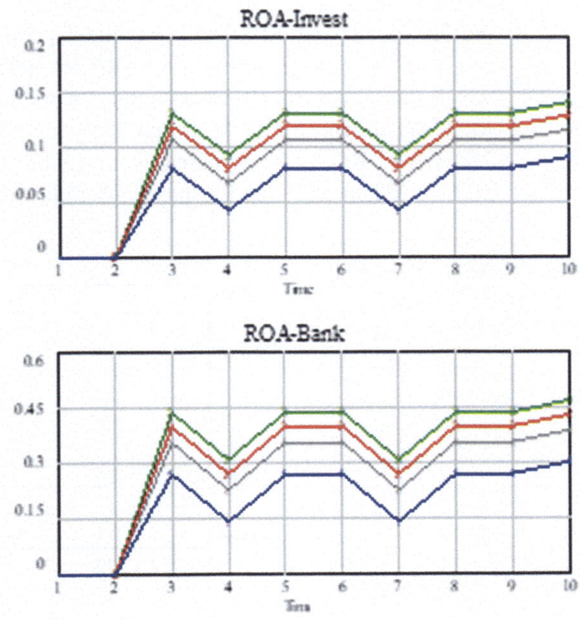


**Fig. 12** The trend chart of the change of ROA by adjusting user volume

(c) *ROA:* As shown in Fig. 12, compared with the net present value and the ROE, the ROA also increases when the number of user is increasing, and the ROA decreases. When the number of users increased to 70,000, ROA-Invest increased to 13.9%, ROA-Bank increased to 46.6%; when the number of users dropped to 60,000, ROA-Invest fell to 11.5%, ROA-Bank decreased to 38.3%, and the number of users was 50,000. At the time of the ROA-Invest, the ROA-Bank fell to 9.0% and the ROA-Bank fell to 30.1%.

(3) *Deposit:* The deposit fee for each person who pays for the shared bicycle is adjusted, and the deposit fee for each person is adjusted to 0 (1), 150 (2), and 250 (3).

(a) *NPV:* As shown in Fig. 13, when the everyone's deposit fee is 0, the net present value has been on a downward trend. In the last year of operation, the net present value is – 20,688,200 CNY. When the deposit fee per person is reduced to 150 CNY, the net present value finally drops to 50,104 CNY. At $300, the net present value increases to $13,875,600. Deposit income has a great impact on the net present value. If there is no deposit income, it is difficult for the project to achieve the expected income during the operation period.
(b) *ROE:* As shown in Fig. 14, when the deposit fee per person increases to 300 CNY, the ROE increases to 22.1%; when the deposit fee decreases to 150 CNY, the ROE decreases to 0.001%. When no deposit fee is charged, ROE is 0. ROE increases with the increase of deposit income, and deposit income has a great influence on the change of ROE. Although the proportion of users who choose to pay the deposit is less than 35%, the deposit accounts for a large proportion of the revenue of the Shared bike project, which can make the return on equity of the project higher.
(c) *ROA:* As shown in Fig. 15, when everyone's deposit payment increases to 250 CNY, ROA-invest increases to 15.1% and ROA-bank increases to 50.6%. When the deposit payment reduced to 150 CNY, the deposit

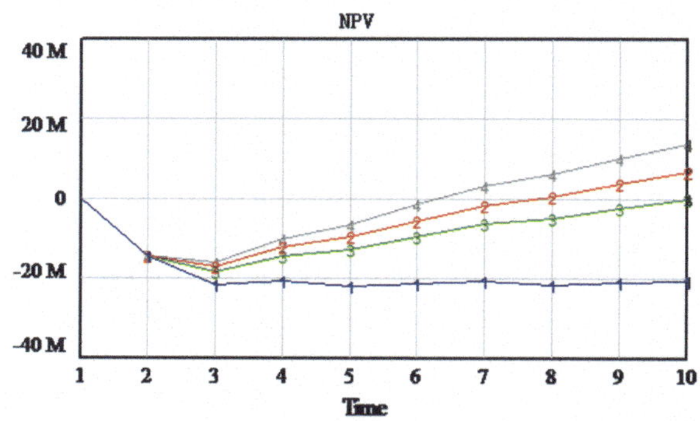


**Fig. 13** The trend chart of the change of NPV by adjusting deposit

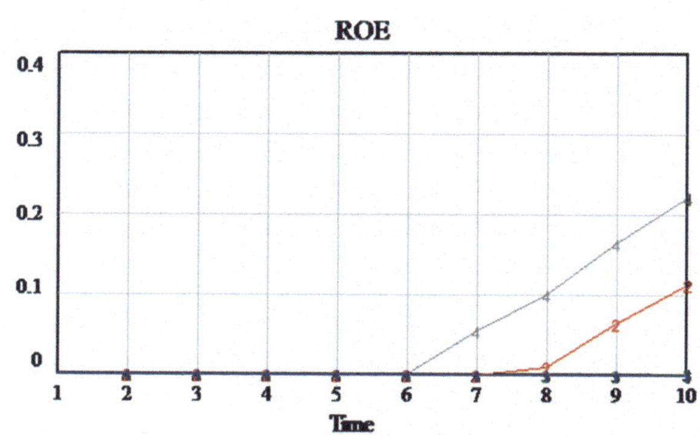


**Fig. 14** The trend chart of the change of ROE by adjusting deposit

**Fig. 15** The trend chart of the change of ROA by adjusting deposit

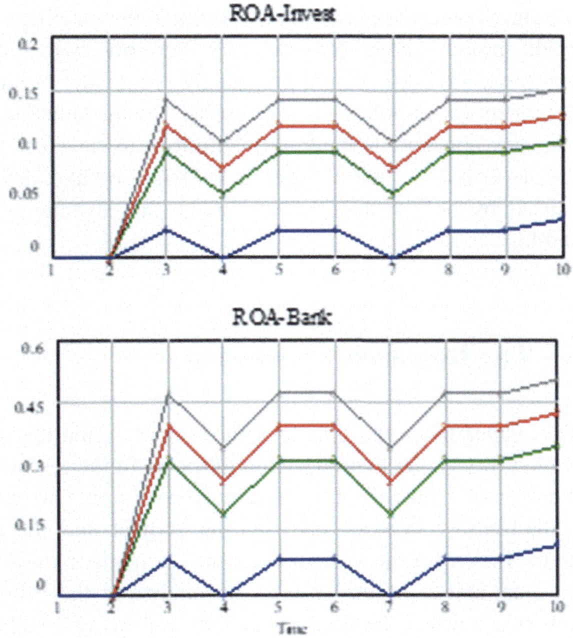


payment in ROA-invest reduced to 10.5% and the deposit payment in ROA-invest reduced to 35.0%. When the project does not charge deposit, ROA-invest drops to 3.4% and ROA-bank to 11.6%. SO if there is no deposit income, the project only relies on other income, and the return on total assets will be lower.

(4) *The results discussed:* Through the adjustment of key variables, it can be seen that the changes of different variables lead to different changes in the evaluation index of bankability, especially when the number of users and each deposit fee are adjusted, the evaluation index of the bankability of bike-sharing project has the most significant change. Therefore, according to the research results, the following suggestions are put forward for the bike-sharing project:

The number of users is the most basic element for a project to gain profits. When considering investment, investors should make full research on the number of local users, predict the development trend of the number of users in the future, and make a relatively accurate judgment on the project's bankability. For shared-bike enterprises, attractive and targeted plans should be made when promoting shared-bike projects, so as to attract more users to use Shared bikes and obtain continuous cash flow.

At present in the bike-sharing market, deposit income is unstable. However, as the shared-bike industry becomes more and more stable, the management system of deposit is becoming more and more standardized, and the shared-bike market is developing towards the direction of no-deposit payment. For example, when users use the Shared bike invested by Alibaba, they can ride without deposit with sesame credit points above 600 points. Therefore, when considering the bankability, enterprise investors can appropriately reduce the proportion of deposit income and obtain more reliable analysis results. On the other hand, bike-sharing enterprises can seek appropriate banking institutions to carry out payment cooperation, reduce the deposit fees, reduce the dependence on the deposit income, avoid the risk caused by policy changes, and make the revenue of bike-sharing projects more stable.

## 5 The Research Conclusion

This paper takes the bike-sharing project as the research object and the project's bankability as the main research purpose. In the research process, it mainly uses the theories of bankability and financial evaluation to construct each element of the bankability system of bike-sharing project and the causal relationship between them. Based on the system dynamics, a mathematical model was built to simulate the bankability of the case. After analyzing the different key variable's changes which can affect the final result. The following conclusions can be drawn:

This paper establishes the concept model of the bankability system for bike-sharing project through data analysis and data collection. Two modules which are the production cost and economic income are established, and the relationship between each variable is analyzed and established. The evaluation index of the bankability for bike-sharing project is obtained through model operation, which provides a reference for the bankability analysis for bike-sharing project.

The user volume and deposit income are the key variables for the bike-sharing project. The capital structure, the number of users and the deposit fee paid by each person in the bike-sharing project are adjusted respectively. The operation results show that the user volume has the most obvious influence on the evaluation index of project's bankability, followed by the deposit income, which are the most significant variables for the evaluation index of bankability.

In this paper, the study about bankability of bike-sharing project is mainly through the methods of literature review and model establishment. However, there are still some deficiencies in the research process, which can be researched in the future: the composition of project cost and economic income variables can be enriched. When this paper considers the cost of bike-sharing and economic income, it is mainly based on the case study in this paper to analyze and construct the mathematical relationship. However, the actual cost and economic income may be more than those listed in the system of this paper. In the future, corresponding cost and economic income variables can be added according to the actual situation to build a more comprehensive mathematical relationship.

## References


1. Zhou, G., & Zhang, D. (2018). Financial problems under the sharing economy model – a case study of sharing bicycle industry. *Financial Innovation Research Series of Private Enterprises, 1,* 153–155.
2. Sun, Z., Li, C., & Yan, M. (2018). Financial support for shared bicycle development. *Commercial Bank, 4,* 45–46.
3. Yang, L., & Zhu, D. (2018). Analysis of factors affecting the intention of parking behavior of shared bicycles under the framework of extended plan behavior. *Population, Resources and Environment of China, 28*(4), 125–133.
4. Xu, H. (2017). The criminal jurisprudence evaluation of the sharing of the 'deposit pool' of bicycles. *Law* (12), 124–132.
5. Gan, H., & Ruan, C. (2018). Research on reverse public-private cooperation under the supply of public service private sector. *Inner Mongolia Social Sciences (Chinese Edition), 39*(2), 29–35.
6. He, T. (2018). Discussion on sharing bicycle phenomenon and sharing economy development. *Technology Economics and Management Research, 8,* 99–104.
7. Angelopoulos, A., Gavalas, D., Konstantopoulos, C., Kypriadis, D., & Pantziou, G. (2016). An optimization model for the strategic design of a bicycle sharing system. In *Pan-Hellenic Conference on Informatics* (no. 5, pp. 25–30).
8. Leonardo, C., Rosalia, C., Michele, O., & Yuen, S. W. (2018). A modeling framework for the dynamic management of free-floating bike-sharing systems. *Transportation Research Part C-Emerging Technologies, 87,* 159–182.
9. Gatti, S. (2013). *Project finance in theory and practice: designing, structuring, and financing private and public projects* (24th ed., Vol. 7, pp. 25–26). Cambridge: Academic Press.
10. Ye, S., Shi, S., Liu, H., & Tang, T. (2017). Considering the transaction cost of public projects can be financing study. *Journal of Accounting Monthly, 5,* 3–8.
11. Duan, X., & Lin, D. (2017). The 'irrational' investment of venture capitalists from the perspective of shared bicycle financing. *Friends of Accounting* (24), 7–12.
12. Yang, J., & Huang, Y. (2017). Consumption economics and countermeasures of 'net rental bicycles'. *Consumer Economy, 33*(6), 35–40.
13. Fan, G., & Yan, F. (2017). Analysis of the new business model of internet finance in the age of shared economy. *Journal of Tianjin Administrative College, 19*(3), 29–35.
14. Wang, F. (2009). *System dynamics (revised edition)* (17th ed.). Shanghai: Shanghai University of Finance and Economics Press.
15. Liu, L. (2011). *Engineering economics (second edition)* (pp. 53–56). Beijing: China Building Industry Press.

# Technical Efficiency Analysis of Bus Companies Based on Stochastic Frontier Analysis and Data Envelopment Analysis

**Feras Tayeh and Shujun Ye**

**Abstract** An efficiency evaluation is a leading tool of assessing transportation performance, seeking to examine the valued outcomes of transportation system in relation to the resources, thus having an extreme importance for policymakers. Technical efficiency evaluation of Istanbul bus companies is the ultimate objective for the sake of identification the passenger performance, and to identify how concern factors influence the technical efficiency. Additionally, to find out the influence of the competitive transportation modes and distance factors that each bus line connects. The present study employs Stochastic Frontier Analysis with production function; the results of the current study are compared to those obtained by Data Envelopment Analysis.

**Keywords** Technical efficiency · Stochastic frontier analysis · Production function · Data envelopment analysis · Trolley buses passenger

## 1 Introduction

Increasing concerns about fiscal sustainability transportation have brought the issue of transportation system efficiency to the forefront of policy discussions at both national and international level. Numerous highways have significantly lost business to other transportation modes efficiency, such as railways and air over the past decades. Development of passenger transport is dependent on the economic play a critical role in transport planning authority's decision in policy development and resource allocation. In Turkey, road transportation is the main mode of passenger transportation and has one of the most developed road networks in its region, passengers transport increased by 4.36% which is reflected in the increase in the

F. Tayeh (✉) · S. Ye
School of Economics and Management, Beijing Jiao Tong University, Beijing, China
e-mail: elkbeeer11@gmail.com

S. Ye
e-mail: shjye@bjtu.edu.cn


J. Zhang et al. (eds.), *LISS2019*,
https://doi.org/10.1007/978-981-15-5682-1_68

number of driver's licenses of over 24 million in 2018. Istanbul is a big city with more than 15 million inhabitants. Undoubtedly, in Istanbul of these extents, the travel is one of the major troubles. Most of the transportation networks modes on land in Istanbul consist of trolley buses, private cars and taxis, which generate a big stress on the passenger's movement intercity and traveling to other cities. According to Turkish State Railways Annual Statistics in 2018 Istanbul had 20 trolley bus lines to connect main cities in turkey, which covered more than 6800 km, the fleet consisted of 3059 trolley buses, more than 28 million travelers use trolley buses and approximately 2800 employees yearly. For that, evaluating gaps and efficiency trolley buses is critical issue to the decision makers in transport sector.

Technical efficiency evaluation of Istanbul bus companies is the ultimate objective for the sake of identification the passenger performance in 2018, and to identify how concern factors influence the technical efficiency. The present study employs Stochastic Frontier Analysis with the production function. Monthly survey of twenty lines with 240 surveys, one output is passenger-kilometers as dependent variable, and three inputs are labor, vehicle-kilometers, and vehicles trolley and coaches number, of each line as independent variables. Additionally, the power of other factors on the competitive transport modes and distance that each line connects is conducted. The results of the current study are compared to those obtained (DEA-CCR).

## 2 Literature Review

The significance of evaluating transport modes concurrently stems from the fact that transport networks are considered as the pillar of sustainable metropolitan improvement, Stamos report on south east Europe explores its contribution to the rail transport enhancement [1]. Using the populated areas to study the urban rail networks system with currently travel positions and the circumstances such as London underground and British rail, an additional model to integrate and identify safety level was carried out by [2]. Although, there is a number of publications on DEA to evaluate efficiency within the various fields of transport [3]. Loizides presented the cost structure of 10 European countries based on the general index of technical change Between 1970 and 1992, and made the following rankings of the most productive countries; Germany, United Kingdom, Belgium, France, Italy, Netherlands, Greece, Denmark, Luxembourg, and Portugal [4]. Suarez quantified the efficiency level of European railway companies, examined key indicators (passengers, freight, kilometers of lines, the percentage of electrification and the percentage of kilometers of double line) and indicates that the most highly developed railway companies are found in Austria, Italy and Germany [5].

Pedro in some European countries such as Sweden, Britain, Italy, France and Germany, they observed the revenue from rail passenger market and public subsidies; and improvements in rail technology are a key driver of productivity growth in the railway transport sector [6]. Other studies examined a detailed analysis of previous empirical researches on data envelopment analysis (DEA) which uses as a tool to measure and a technique of linear programming to analysis the comparative efficiency of other decision-making units, with efficiency levels lie between (0–1) for the most efficient alternative [7]. This method of Data Envelopment Analysis and its extension are skilled to deal different variables of inputs and outputs; and discover other dealings that may be concluded with other approaches but assuming that all variables of input and output data are precisely recognized [8]. As a useful method for analyzing the relative performance within a group of organizations, Data Envelopment Analysis method discovered practical association between competition framework and productive effectiveness of decision-making units by designing construct efficiency scores and regressing these scores against variables. Bojović affirmed that DEA analysis provide indications for measuring and monitoring efficiency while improving the performance and ability to reach the efficiency frontier [9]. To determine efficiency using data envelopment analysis method and Tobit regression [10], examined 31 transport firms between 2000 to 2009 that offers passenger and other services in world, the results show that some transport companies in France, Japan, Luxemburg, and Spain in Western Europe operate effectively using six input variables. Bråthen identified that tool of measurement is used to evaluate and compare the relationships different road networks to discover the performance and comparing with other positions for technical support and help the policymakers to enable them take essential actions to make highway system efficient [11].

## 3 Research Methodology

Several methodologies have been used to estimate and analysis efficiency within the various fields of transport: The method needs a practical shape to determine the production function with contaminated data and evaluation errors and other noise [12, 13].

### 3.1 Stochastic Frontier Analysis (SFA)

Farrell recognized the explanation of technical efficiency, the empirical function was comparatively limited [13]. Aigner, Lovell, and Schmidt introduced the stochastic frontier production function [14], Meeusen and van Den Broeck offered the Cobb-Douglas production function with a composed multiplicative disturbance term [15], they presented as:

$$y = f(x, \beta)\, exp\,(\varepsilon), (\varepsilon) = (\nu - \mu), \mu > 0 \tag{1}$$

$y$: the observed output quantity.
$f$: the deterministic part of the frontier production.
$\beta$: vector of the input quantities.
$x$: vector of parameters to be estimated.
$v$: symmetrical random error.
$\mu$: One-sided non-negative random error

Technical efficiency is calculated:

$$TE = y/[f(x)\,exp\,(v)] = exp(-u) \tag{2}$$

Where $TE$ has a value between 0 and 1, with 1 defining a technically efficient firm. More specifically, from Eq. (1) is written as:

$$ln\,(y) = ln\,[f(x)] + v - u \tag{2a}$$

$$ln\,(y) = -u + ln\,[f(x)] + (v - u + \mu) \tag{2b}$$

$$\mu = E(u) > 0$$

Estimation of (2) by OLS gives the residuals $e_i, i = 1, 2\ldots n$.

The second and third central moments of the residuals, $m_2(e)$, $m_3(e)$ respectively, are considered, as follows:

$$m_2(e) = \left[\frac{1}{N-K}\right] * \sum e_i^2 \tag{3a}$$

$$m_3(e) = \left[\frac{1}{N-K}\right] * \sum e_i^3 \tag{3b}$$

$N$: number of observations. $k$: number of regressors. Then, we estimate $\sigma_u^2$ and $\sigma_v^2$ by using the formula.

$$\sigma_u^2 = [(\pi/2)[\pi/(\pi - 4)]\,m_2(e)]^{2/3} \tag{4}$$

$$\sigma_v^2 = m_2(e) - [(\pi - 2)/(\pi)]\sigma_u^2 \tag{5}$$

The point measure of technical efficiency is

$$\begin{aligned} TE_i &= E(\exp\{-u_i\}/\varepsilon_i) \\ &= \left[1 - F\left[\sigma - \left(M_i^*/\sigma\right)\right]/\left[1 - F\left(-M_i^*/\sigma\right)\right] \exp\left[-M_i^* + (\sigma 2/2)\right]\right. \end{aligned} \tag{6}$$

$$M_i^* = \left(-\sigma_u^2 \varepsilon_i\right)\left(\sigma_u^2 + \sigma_v^2\right) - 1 \quad (6a)$$

$$\sigma 2 = \sigma_u^2 \sigma_v^2 (\sigma_u^2 + \sigma_v^2) - 1 \quad (6b)$$

## 3.2 The DEA Input Oriented Model

With the most fundamental DEA method is DEA-CCR [16]. This method is very important because it process with constant returns to scale (CRS) to find that the observed DMUs work at the large amount efficient scale size [17]. The mathematical function of DEA-CCR method is offered with number of Decision Making Units can be measured. Each DMU has inputs, and different outputs [18]: $v_i$ ($i$ = 1, 2,..., $n$) as input and $u_r$ ($r$ = 1, 2,..., $q$) as output

$$max \sum_{r=1}^{q} u_r y_{rk} \quad (7)$$

$$s.t. \sum_{r=1}^{q} u_r y_{rk} - \sum_{i=1}^{m} v_i x_{ij} \leq 0 \quad (7a)$$

$$\sum_{i=1}^{m} v_i x_{ik} = 1 \quad (7b)$$

$$i = 1, 2 \ldots m; r = 1, 2 \ldots, q; j = 1, 2 \ldots, n$$

To examine the weak efficiency, dual model of (7) is formulated as:

$$min\ \theta \quad (8)$$

$$s.t. \sum_{j=1}^{n} \lambda_j x_{ij} \leq \theta\, x_{ik} \quad (8a)$$

$$\sum_{j=1}^{n} \lambda_j y_{rj} \geq \lambda_{rk} \quad (8b)$$

$$\lambda \geq 0 \quad (8c)$$

the panel data set consists of 240 monthly buses movement observations of the 20 buses network of Istanbul bus companies in 2018, which cover a large surface from Istanbul to the other cities in Turkey Table 1.

Table 2 shows inputs and outputs variables used in the study, dataset consists of following variables. One output is passenger-kilometers as dependent variable. The inputs are labor vehicle-kilometers and vehicles trolley buses number of each line as independent variables. Every one of these inputs expresses the practical conditions of each line of Istanbul. Moreover, to evaluate the probability impact of some other factors, two dummy variables were presented in Table 3.

**Table 1** Trolley bus lines

| Line | DMU |
|---|---|
| Line 1 | Istanbul - Adapazari |
| Line 2 | Istanbul - Afyonkarahisar |
| Line 3 | Istanbul - Alanya |
| Line 4 | Istanbul - Antalya |
| Line 5 | Istanbul - Aydin |
| Line 6 | Istanbul - Balikesir |
| Line 7 | Istanbul - Bodrum |
| Line 8 | Istanbul - Bolu |
| Line 9 | Istanbul - Burdur |
| Line 10 | Istanbul - Düzce |
| Line 11 | Istanbul - Eskişehir |
| Line 12 | Istanbul - Gebze |
| Line 13 | Istanbul - Izmir |
| Line 14 | Istanbul - Izmit |
| Line 15 | Istanbul - Kütahya |
| Line 16 | Istanbul - Manavgat |
| Line 17 | Istanbul - Manisa |
| Line 18 | Istanbul - Ordu |
| Line 19 | Istanbul - Samsun |
| Line 20 | Istanbul - Serik |

**Table 2** Input - outputs used to evaluate technical efficiency

| Inputs | Outputs |
|---|---|
| Labor | Total passenger-kilometers |
| Vehicle-kilometers | – |
| Vehicles trolley buses number | – |

The influence of the competitive transportation modes and distance factors that each bus line connects such as railway influence in passengers movements decision provides other options to be chosen, these options are conducted in this study as dummy variables, the influence of Istanbul railway, passengers prefer to use Istanbul railway, in this framework, we make the supposition that the lines that serve areas directly connected with the railway are negatively affected, where the dummy variables take two values (0, 1), other dummy variable is the impact of the distance.

**Table 3** Dummy variables used in the analysis

| DMU | Route | Dummy 1 | Dummy 2 |
|---|---|---|---|
| Line 1 | Istanbul - Adapazari | 1 | 1 |
| Line 2 | Istanbul - Afyonkarahisar | 1 | 1 |
| Line 3 | Istanbul - Alanya | 1 | 1 |
| Line 4 | Istanbul - Antalya | 1 | 1 |
| Line 5 | Istanbul - Aydin | 1 | 0 |
| Line 6 | Istanbul - Balikesir | 1 | 0 |
| Line 7 | Istanbul - Bodrum | 1 | 0 |
| Line 8 | Istanbul - Bolu | 1 | 1 |
| Line 9 | Istanbul - Burdur | 1 | 1 |
| Line 10 | Istanbul - Düzce | 1 | 1 |
| Line 11 | Istanbul - Eskişehir | 1 | 1 |
| Line 12 | Istanbul - Gebze | 0 | 0 |
| Line 13 | Istanbul - Izmir | 1 | 0 |
| Line 14 | Istanbul - Izmit | 0 | 0 |
| Line 15 | Istanbul - Kütahya | 1 | 1 |
| Line 16 | Istanbul - Manavgat | 1 | 1 |
| Line 17 | Istanbul - Manisa | 1 | 0 |
| Line 18 | Istanbul - Ordu | 1 | 1 |
| Line 19 | Istanbul - Samsun | 1 | 0 |
| Line 20 | Istanbul - Serik | 1 | 1 |

# 4 Results and Discussion

The production function to convert the inputs into outputs can be showed by log-linear Cobb-Douglas measurement, the logarithmic stochastic functions presented as

$$lny_{it} = \beta_0 + \sum \beta_n lnx_{nit} + v_{it} - u_i$$
$$i = 1\ldots N, t = 1\ldots T$$

$y_{it}$ and $x_{it}$ are observed output and inputs of the $i^{th}$ unit in year $t$.
$u_i$: Non-negative time-invariant random variables.

$$lny_{it} = \beta_0 + \beta_1 lnx_{1it} + \beta_3 lnx_{3it} + v - u$$

$v_{it}$ Random variables of $i^{th}$ unit in year t.

From (1), the fitted functional form is:

$$lny_{it} = \beta_0 + \beta_1 lnx_{1it} + \beta_2 lnx_{2it} + \beta_3 lnx_{3it} + \beta_4 d1 + \beta_5 d2 + v - u$$

$y$: dependent variable and donate to output
$x_1$: inputs of labor
$x_2$: vehicle-kilometers
$x_3$: trolley bus and Coaches number
$d_1$: dummy variable donates to the influence of other mode in transportation such as Istanbul railway, and $d_2$: the influence of the distance and covered areas.

## 4.1 Descriptive Statistics and Correlation in Dataset

Table 4 presents the descriptive statistics of variables used in the study and contains the descriptive statistics of the variables used in the study. They include the sample mean, median, standard deviation, minimum and maximum value for each of the variables.

Table 5 shows the variables which have influence on Passengers movement are vehicle-kilometers, total labor, available vehicles, and influence of dummy variables, that each line has impact significant coefficients. Vehicle-kilometers variable is statistically significant. With an increase by 1% of Vehicle-kilometers value, passengers will decrease by 1.02% on average. The total labor is highly statistically

**Table 4** The descriptive statistics

| Characteristics | $y$ | $x_1$ | $x_2$ | $x_3$ | $d_1$ | $d_2$ |
|---|---|---|---|---|---|---|
| Mean | 112990.0 | 340.1 | 141.1 | 2824.7 | 0.7 | 0.75 |
| Median | 77900.0 | 291.1 | 97.12 | 1947.5 | 1 | 1 |
| Maximum | 456000.0 | 753.1 | 570 | 11400 | 1 | 1 |
| Minimum | 14400.0 | 45.4 | 18 | 360 | 0 | 0 |
| Std. Dev. | 98792.8 | 199.9 | 123.5 | 2469.8 | 0.4 | 0.44 |
| Skewness | 2.14925 | 0.47 | 2.40 | 2.14 | −0.8 | −1.1 |
| Kutosis | 7.40526 | 2.1 | 7.40 | 7.40 | 1.7 | 2.34 |
| Jarque-Bera | 378.834 | 17.57 | 379.30 | 378.8 | 46 | 57.8 |
| Probability | 0 | 0 | 0 | 0 | 0 | 0 |
| Sum | 271176 | 8162 | 3388 | 6779 | 168 | 180 |
| Sum Sq. Dev. | 2.33E+9 | 9557885 | 364476 | 1.46E+9 | 50 | 45 |
| Observations | 240 | 240 | 240 | 240 | 240 | 240 |

**Table 5** Analysis of variance

| Variable | Coefficient | Std. error | t-Statistic | Prob |
|---|---|---|---|---|
| C | 12118072 | 4428893 | 2.736140 | 0.0067 |
| $ln\_x_1$ | −27906.38 | 3137.751 | −8.893752 | 0.0000 |
| $ln\_x_2$ | 4255228 | 1477585 | 2.879853 | 0.0043 |
| $ln\_x_3$ | −4136840 | 1477597 | −2.799709 | 0.0055 |
| $d_1$ | 15962.28 | 5633.912 | 2.833250 | 0.0050 |
| $d_2$ | −24473.16 | 6258.46 | −3.910409 | 0.0001 |
| R-squared | 0.886364 | | | |
| Adjusted R-squared | 0.883936 | | | |
| F-statistic | 365.0414 | | | |
| Prob (F-statistic) | 0.000000 | | | |

**Table 6** Correlation in the dataset

| Independent variable | Parameters | Estimate | t-Statistic | P-Value |
|---|---|---|---|---|
| C | $\beta_0$ | 1.96E+13 | 2.736140 | 0.0067 |
| $x_1$ | $\beta_1$ | −1.02E+09 | −8.893752 | 0.0000 |
| $x_2$ | $\beta_2$ | 6.54E+12 | 2.879853 | 0.0043 |
| $x_3$ | $\beta_3$ | −6.54E+12 | −2.2799709 | 0.0055 |
| $d_1$ | $\beta_4$ | −2.11E+09 | 2.736140 | 0.0050 |
| $d_2$ | $\beta_5$ | 3.45E+08 | −3910409 | 0.0001 |

significant with an increase by 1% of the total labor; passengers will increase by 6.54% on average. The available vehicles variable is also highly statistically significant Table 6.

## 4.2 *Technical Efficiency Scores and Lines Ranking*

According to The Stochastic Frontier Analysis (SFA) Table 7 presents technical efficiency scores (TE). The results lie between 0.83 and 0.94, with an average equal to 0.9055, lines of Istanbul - Aydin, Istanbul - Manavgat, Istanbul - Alanya, Istanbul - Burdur, and Istanbul - Ordu are the highest technical efficiency scores, while Istanbul - Afyonkarahisar line is the lowest technical efficiency scores one, lines of Istanbul and Istanbul - Izmit, which are subjective by the operation of the Istanbul railway, are not found to be between the highest technical efficiency scores.

**Table 7** Technical efficiency scores and lines ranking

| Line | Line | Technical efficiency | Ranking | Line |
|---|---|---|---|---|
| Istanbul - Adapazari | Line 1 | 0.91 | 1 | Istanbul - Aydin |
| Istanbul - Afyonkarahisar | Line 2 | 0.83 | 1 | Istanbul - Manavgat |
| Istanbul - Alanya | Line 3 | 0.93 | 2 | Istanbul - Alanya |
| Istanbul - Antalya | Line 4 | 0.92 | 2 | Istanbul - Burdur |
| Istanbul - Aydin | Line 5 | 0.94 | 2 | Istanbul - Ordu |
| Istanbul - Balikesir | Line 6 | 0.91 | 3 | Istanbul - Antalya |
| Istanbul - Bodrum | Line 7 | 0.92 | 3 | Istanbul - Bodrum |
| Istanbul - Bolu | Line 8 | 0.87 | 3 | Istanbul - Izmir |
| Istanbul - Burdur | Line 9 | 0.93 | 3 | Istanbul - Izmit |
| Istanbul - Düzce | Line 10 | 0.90 | 3 | Istanbul - Samsun |
| Istanbul - Eskişehir | Line 11 | 0.88 | 4 | Istanbul - Adapazari |
| Istanbul - Gebze | Line 12 | 0.84 | 4 | Istanbul - Balikesir |
| Istanbul - Izmir | Line 13 | 0.92 | 5 | Istanbul - Düzce |
| Istanbul - Izmit | Line 14 | 0.92 | 5 | Istanbul - Kütahya |
| Istanbul - Kütahya | Line 15 | 0.90 | 5 | Istanbul - Manisa |
| Istanbul - Manavgat | Line 16 | 0.94 | 5 | Istanbul - Serik |
| Istanbul - Manisa | Line 17 | 0.90 | 6 | Istanbul - Eskişehir |
| Istanbul - Ordu | Line 18 | 0.93 | 7 | Istanbul - Bolu |
| Istanbul - Samsun | Line 19 | 0.92 | 8 | Istanbul - Gebze |
| Istanbul - Serik | Line 20 | 0.90 | 9 | Istanbul - Afyonkarahisar |

## *4.3 Comparison with DEA*

With an average equal to 0.879. The most inefficient line regarding the CRATE-DEA scores were line Istanbul - Serik and Istanbul - Afyonkarahisar lines Table 8.

To compare the results from the two methods (SFA and DEA), we rank the buses lines in both methods. The results are consistent; lines of Istanbul - Alanya, Istanbul - Antalya, Istanbul - Aydin, Istanbul - Burdur, Istanbul - Manavgat, and Istanbul-Ordu are among the most efficient lines. Furthermore, lines of Istanbul - Afyonkarahisar, Istanbul - Bolu, Istanbul - Gebze, and Istanbul are among the least efficient lines in both methods Table 9.

**Table 8** Line ranking of DEA and SFA

| Ranking | SFA | CCR (TE) |
|---|---|---|
| 1 | Istanbul - Aydin | Istanbul - Burdur |
| 2 | Istanbul - Manavgat | Istanbul - Manavgat |
| 3 | Istanbul - Alanya | Istanbul - Ordu |
| 4 | Istanbul - Burdur | Istanbul - Alanya |
| 5 | Istanbul - Ordu | Istanbul - Aydin |
| 6 | Istanbul - Antalya | Istanbul - Antalya |
| 7 | Istanbul - Bodrum | Istanbul - Izmit |
| 8 | Istanbul - Izmir | Istanbul - Samsun |
| 9 | Istanbul - Izmit | Istanbul - Kütahya |
| 10 | Istanbul - Samsun | Istanbul - Manisa |
| 11 | Istanbul - Adapazari | Istanbul - Adapazari |
| 12 | Istanbul - Balikesir | Istanbul - Eskişehir |
| 13 | Istanbul - Düzce | Istanbul - Balikesir |
| 14 | Istanbul - Kütahya | Istanbul - Düzce |
| 15 | Istanbul - Manisa | Istanbul - Bolu |
| 16 | Istanbul - Serik | Istanbul - Bodrum |
| 17 | Istanbul - Eskişehir | Istanbul - Izmir |
| 18 | Istanbul - Bolu | Istanbul - Gebze |
| 19 | Istanbul - Gebze | Istanbul - Afyonkarahisar |
| 20 | Istanbul - Afyonkarahisar | Istanbul - Serik |

**Table 9** Technical efficiency measures and line rankings

| Line | Technical efficiency | Line | Ranking |
|---|---|---|---|
| Istanbul - Adapazari | 0.90 | Istanbul - Burdur | 1 |
| Istanbul - Afyonkarahisar | 0.71 | Istanbul - Manavgat | 1 |
| Istanbul - Alanya | 0.98 | Istanbul - Ordu | 1 |
| Istanbul - Antalya | 0.96 | Istanbul - Alanya | 2 |
| Istanbul - Aydin | 0.98 | Istanbul - Aydin | 2 |
| Istanbu - Balikesir | 0.87 | Istanbul - Antalya | 3 |
| Istanbul - Bodrum | 0.79 | Istanbul - Izmit | 4 |
| Istanbul - Bolu | 0.82 | Istanbul - Samsun | 5 |
| Istanbul - Burdur | 1 | Istanbul - Kütahya | 6 |
| Istanbul - Düzce | 0.84 | Istanbul - Manisa | 6 |
| Istanbul - Eskişehir | 0.90 | Istanbul - Adapazari | 7 |
| Istanbul - Gebze | 0.71 | Istanbul - Eskişehir | 7 |
| Istanbul - Izmir | 0.75 | Istanbul - Balikesir | 8 |
| Istanbul - Izmit | 0.95 | Istanbul - Düzce | 9 |
| Istanbul - Kütahya | 0.91 | Istanbul - Bolu | 10 |

(continued)

**Table 9** (continued)

| Line | Technical efficiency | Line | Ranking |
|---|---|---|---|
| Istanbul - Manavgat | 1 | Istanbul - Bodrum | 11 |
| Istanbul - Manisa | 0.91 | Istanbul - Izmir | 12 |
| Istanbul - Ordu | 1 | Istanbul - Gebze | 13 |
| Istanbul - Samsun | 0.93 | Istanbul - Afyonkarahisar | 13 |
| Istanbul - Serik | 0.67 | Istanbul - Serik | 14 |
| Mean | 0.879 | | |

## 5 Conclusion and Recommendations

Technical efficiency evaluation of Istanbul bus companies is the ultimate objective for the sake of identification the passenger performance, for the sake of identification the passenger performance of Istanbul bus companies, and to identify how concern factors influence the technical efficiency. The present study employs Stochastic Frontier Analysis with the production function Twenty lines with 240 survey, one output is passenger-kilometers as dependent variable, and three inputs are labor, vehicle-kilometers, and vehicles trolley and Coaches number, of each line as independent variables. Additionally, the power of other factors on the competitive transport modes and distance that each line connects is conducted. The results of the current study are compared to those obtained (DEA-CCR). Panel data showed well fitted According to production function of Cobb-Douglas and the probability value of the inputs and output variables were significantly efficient. Moreover, the explanatory control of certain dummy factors of the competitive transport modes and distance also were confirmed with significant probability value. The scores of technical efficiencies were ranked and provided variety scale efficiency in high levels, with average technical efficiency equal to 0.9055 and 0.879 for the Stochastic Frontier Analysis and data envelopment analysis methods, respectively.

The differences between the technical efficiency of each line could be explained by a number of reasons. The first reason is the total vehicle-kilometers and the coverage level that each line connects. The Stochastic Frontier Analysis outcomes pointed that the total of vehicle-kilometers of each line has positively impacts on the technical efficiency. Lines of Istanbul - Alanya, Istanbul - Antalya, Istanbul - Aydin, Istanbul - Burdur, Istanbul - Manavgat, and Istanbul - Orduare are among the most efficient lines, Furthermore, lines Istanbul - Afyonkarahisar, Istanbul - Bolu, Istanbul - Gebze, and Istanbul are among the least efficient lines in both methods. The second reason is the competitive transport modes that the areas are covered by buses network are also covered by other modes, like Istanbul railway. The results pointed that Istanbul - Gebze line which is influenced by the operation of the Istanbul railway, and serves areas near Istanbul city is not found to be among to the most efficient ones. This study confirmed the significant of these factors for the future of passenger's performance. Since the wide development and rapidly expand of Istanbul railway a future strategic planning of the buses network would be appropriate, particularly minibuses, taxis and private automobiles also are available in Istanbul.

## References

1. Stamos, I., Myrovali, G., & Aifadopoulou, G. (2016). Formulation of a roadmap towards the enhancement of international rail passenger transport – The South East Europe example. *Journal of Rail Transport Planning and Management, 6*(2), 89–98.
2. Evans, A. W., & Morrison, A. D. (1997). Incorporating accident risk and disruption in economic models of public transport. *Journal of Transport, Economics and Policy, 2541,* 46–55.
3. Lan, L. W., & Lin, E. T. J. (2006). Performance measurement for railway transport: Stochastic distance functions with inefficiency and ineffectiveness effects. *Journal of Transport, Economics and Policy, 40,* 383–408.
4. Munda, G. (2004). Social multi-criteria evaluation: Methodological foundations and operational consequences. *European Journal of Operational Research, 158,* 662–677.
5. Jorge, J., & Suarez, C. (2003). Has the efficiency of European railway companies been improved? *European Business Review, 15,* 213–220.
6. Cantos, P., & Maudos, J. (2001). Regulation and efficiency: The case of European railways. *Transportation Research Part A: Policy and Practice, 35,* 459–472.
7. Thanassoulis, E., Kortelainen, M., & Allen, R. (2012). Improving envelopment in Data Envelopment Analysis under variable returns to scale. *European Journal of Operational Research, 218*(1), 175–185.
8. O'Donnell, C. J. (2018). Stochastic frontier analysis. In *Productivity and Efficiency Analysis* (Vol. 63, pp. 991–1011).
9. Bogetoft P., & Otto L. (2011). Stochastic frontier analysis SFA. In International Series in Operations Research and Management Science (Vol. 5, p. 36).
10. Bauer, P. W. (1990). Recent developments in the econometric estimation of frontiers. *Journal Economic, 46,* 39–56.
11. Farrell, M. J. (1957). The measurement of productive efficiency. *Journal of the Royal Statistical Society, 120,* 253–290.
12. Aigner, D., Lovell, C. A. K., & Schmidt, P. (1977). Formulation and estimation of stochastic frontier production function models. *Journal Economic, 6,* 21–37.
13. Meeusen, W., & van Den Broeck, J. (1977). Efficiency estimation from Cobb-Douglas production functions with composed error. *International Economic Review (Philadelphia), 18,* 435–444.
14. Battese, G. E., & Coelli, T. J. (1988). Prediction of firm-level technical efficiencies with a generalized frontier production function and panel data. *Journal Economic, 38,* 387–399.
15. Coelli, T. J., Prasada Rao, D. S., O'Donnell, C. J., & Battese, G. E. (2005). *An introduction to efficiency and productivity analysis* (Vol. 7, pp. 11–37). Boston: Springer.
16. Charnes, A., Cooper, W. W., & Rhodes, E. (1978). Measuring the efficiency of decision making units. *European Journal of Operational Research, 2,* 429–444.
17. Markovits-Somogyi, R. (2011). Data envelopment analysis and its key variants utilized in the transport sector. *Periodica Polytechnica Transportation Engineering, 39,* 1366–1371.
18. Banker, R. D., Cooper, W. W., Seiford, L. M., & Zhu, J. (2004). Return to scale in DEA. In *Handbook on data envelopment analysis* (Vol. 154, pp. 345–362).

The manufacturer's authorised representative in the EU is Springer Nature Customer Service Centre GmbH, Europaplatz 3, 69115 Heidelberg, Germany. If you have any concerns regarding our products, please contact ProductSafety@springernature.com

Printed and bound by CPI Group (UK) Ltd, Croydon, CR0 4YY
30/09/2026
02225511-0001